LONGMAN
DICTIONARY
OF
GEOGRAPHY
human and physical

LONGMAN DICTIONARY OF GEOGRAPHY
human and physical

Audrey N. Clark

Longman

LONGMAN GROUP UK LIMITED
Longman House, Burnt Mill, Harlow, Essex CM20 2JE, England
and Associated Companies throughout the World.

First published 1985
Paperback edition published 1987
Second impression 1988
Cased ISBN 0 582 35261 4
Paperback ISBN 0 582 01779 3

Set in 9/10½pt Times Roman, Monophoto Lasercomp

Printed in Great Britain
by Butler & Tanner Ltd, Frome and London

British Library Cataloguing in Publication Data

Clark, Audrey N.
 Longman dictionary of geography.
 1. Geography—Dictionaries
 I. Title
 910'.3'21   G63
 ISBN 0-582-35261-4

British Library Cataloguing in Publication Data

Clark, Audrey N.
 Longman dictionary of geography: human and physical.
 1. Geography—Dictionaries
 I. Title
 910'.3'21   G63
 ISBN 0-582-01779-3

Library of Congress Cataloguing in Publication Data

Clark, Audrey N.
 Longman dictionary of geography.
 1. Geography—Dictionaries.   I. Title.
 G63.C56 1985   910'.3'21   85-4290
 ISBN 0-582-35261-4

Library of Congress Cataloguing in Publication Data

Clark, Audrey N.
 Longman dictionary of geography.
 1. Geography—Dictionaries.   I. Title.
 [G63.C56 1987]   910'.3'21   87-22667
 ISBN 0-582-01779-3 (pbk.)

Contents

Preface

This Dictionary, the first to deal with the major aspects of geography in one volume, covers the terms commonly used in geographical writing over the past 100 years. Over that time geography has progressed from being an uncomplicated branch of knowledge concerned mainly with exploration, the description of the physical characteristics of the earth and its people, and with products and trade, to become what it is today: a complex academic discipline and applied science, with many specialist branches, straddling the humanities, the natural sciences and the social sciences. The entry on geography covers the scope of the subject today and lists the other fields of knowledge with which it is now associated.

The Dictionary aims to be useful not only to students at school and university but also to professional geographers, especially to those who, now engaged in specialist studies, may need a reminder of once-familiar general terms and an introduction to terms used in specialist fields other than their own. It should also serve readers who have a general interest in environmental problems and in the world's resources and their use.

It includes terms that are purely geographical (British, and American terms differing markedly from the British), terms adopted from related fields of knowledge, and others absorbed from foreign languages. It also covers the plant products and other commodities that (appearing today in international trade statistics) figure so prominently in late–nineteenth, early–twentieth century works. Cross–references (conceptual and factual) enhance the definitions and unify the text; and to make the Dictionary self–contained, useful especially to those whose first language is not English, terms used in the definitions but not in popular use are explained. The vocabulary used is appropriate to the scholastic level of each term.

It must be stressed that the Dictionary does not contain terms that are strictly specific to other disciplines or specialist fields (such as cartography, geology, geomorphology) because such terms are unlikely generally to appear in geographical works; nor does it cover material proper to an atlas or gazetteer. It does, however, include an Appendix giving the Greek and Latin roots commonly used in the construction of English terms. It is hoped this will not only ease the interpretation and understanding of terms not appearing in the Dictionary, but also serve those who are driven to invent new ones. The origin and use of many geographical terms is recorded in L. Dudley Stamp and Audrey N. Clark (eds). *A Glossary of Geographical Terms*, 3rd edn, Longman, 1979, in which authors are cited. It is unfortunately impossible similarly to acknowledge the sources of the great number of works studied in the many years of preparation of the Dictionary. The definitions given are based on consensus, on general agreement on the use of the term by many authors in many different contexts, as viewed at the present time. I have tried not to quote any individual author verbatim; but if I have anywhere accidentally not succeeded I apologise most sincerely for the lack of personal acknowledgement, and ask forgiveness.

The lot of the lexicographer is not a happy one: while 'every other author may aspire to praise, the lexicographer can only hope to escape reproach'. Constructive criticism and suggestions for additional entries will of course be most warmly welcomed.

Audrey N. Clark

Abbreviations
used in the text

G   A reference to the term appears in L. Dudley Stamp and Audrey N. Clark (eds). *A Glossary of Geographical Terms*, 3rd edn, Longman, 1979. The Glossary, first published in 1961, is complementary to the Dictionary. It is concerned mainly with tracing the origin and changing use of selected geographical terms, the entries being supported by quotations and full references to sources.

C   For additional information, see L. Dudley Stamp. *Chisholm's Handbook of Commercial Geography*, 20th edn revised by G. Noel Blake and Audrey N. Clark, Longman, 1980.

cc	cubic centimetre
cm	centimetre
cu	cubic
ft	foot
gm	gram
ha	hectare
in	inch
km	kilometre
lb	pound (avoirdupois or troy)
m	metre
mb	millibar
mi	mile
mm	millimetre
mn	million
oz	ounce
sq	square
yd	yard

Cross references: (*see* Preface) are printed in SMALL CAPITAL LETTERS, in some cases followed by numbers. These indicate relevant items in the main reference cited.

A

aa (Hawaiian) a rough scoriaceous lava flow (SCORIA). In many cases its chemical composition is identical to that of PAHOEHOE. PILLOW LAVA, ROPY (CORDED) LAVA. G.

abaca Manila hemp, a tropical hard fibre so called from Manila, its chief place of export. It is obtained from the long leaves of *Musa textilis*, a perennial herb native to the Moluccas and Philippine Islands. It makes strong, durable rope, the finer fibres being used in the making of textiles and speciality papers.

Abbevillian epoch sometimes termed Chellean epoch, the earliest cultural period of PALAEOLITHIC times, when stone axes with two faces came into use.

abi (Indian subcontinent: Urdu) land irrigated by basket or SAQIA from tanks, pools or streams. Abi-sailābā is land irrigated by the flooding of a river. G.

abîme (French) a wide, deep shaft in LIMESTONE, with vertical or overhanging walls, opening underground to a channel. AVEN. G.

abiocoen in ecology, all the non-living parts of the environment. BIOCOEN, HOLOCOEN.

abiogenesis the hypothetical production of living matter from non-living matter. BIOGENESIS.

abioseston tripton. SESTON.

ablation the action or process of carrying away or removal, applied to **1.** the loss which a GLACIER undergoes through melting (the result of SOLAR RADIATION conducted by solid debris on the surface or flanking walls, or of rain falling on the surface, or of relatively warm water from melting streams), or through SUBLIMATION under favourable temperature, humidity and wind conditions, or through ABRASION (by sharp ice particles blown along the surface), or through CALVING of ICEBERGS on the edge of tidal water. The character of the ice and the quantity of debris lying on the surface affect the speed of melting, and the term differential ablation is applied to this phenomenon. ALIMENTATION **2.** the work of wind in removing fine rock debris in sandy areas (but for this process the term DEFLATION is usually preferred). G.

ablation moraine the pile of rock waste (superglacial TILL) left on the surface of a GLACIER after ABLATION, sometimes so thick and continuous as to hide the glacier ice completely. G.

ablation till rock debris lying within or on stagnant ice (ABLATION MORAINE) which is deposited in situ as the ice melts, the melt-water washing away the finer rock particles so that the remaining particles are coarser than those in LODGEMENT TILL. It has been proposed that the use of the term ablation till should be abandoned. TILL.

Abney level a convenient, reasonably accurate, surveying instrument, a type of CLINOMETER in which the bubble in a spirit level mounted over a sighting tube is reflected in the eyepiece. The INCLINATION-1 of a line between the observer and a distant point can be calculated by measuring the linear distance between the two points, their

1

difference in height being calculated by using the tangent ratio.

aboriginal an ABORIGINE. G.

aboriginal *adj.* of or pertaining to an aboriginal, ABORIGINE or aborigines, or to a physical feature judged, on the available evidence, to be the first, the earliest. G.

aborigine (Latin ab origine, from the beginning; pl. aborigines), **1.** a person who is believed, on contemporary evidence, to have been one of the original inhabitants of an area **2.** a descendant of such an individual **3.** one of the inhabitants occupying an area at the time of the arrival of European colonizers **4.** occasionally applied to a plant or animal believed on available evidence to have been one of the first or earliest in an area. ABORIGINAL. G.

abrasion the act or process of wearing down or wearing away by friction, brought about for example by a river dragging stones along its bed, by the sea carrying sand and shingle along the coast, by glaciers with rocks frozen into their mass, by hard ice particles blown by wind over an ice surface. The process involved in the gradual reduction in the size of pebbles as one rubs against another is perhaps better termed ATTRITION. Abrasion results from the CORRASION of rock. TRANSPORTATION-2. G.

abrasion platform, abrasion plane a wave-cut platform, a flat, nearly smooth, rock platform extending from the foot of a sea-cliff and formed by the erosive action of BREAKERS in carrying shingle along the coast, the seaward extension of a WAVE-CUT BENCH. The surface is usually bare, but in some cases may be covered with a thin layer of fine rock particles. Some authors prefer the term wave-cut platform, which indicates the agent of ABRASION. CONTINENTAL TERRACE, SHOREFACE TERRACE. G.

abri (French) a shallow cave or rock shelter, especially one in a cliff face which afforded shelter to people in PREHISTORIC times.

abscissa in mathematics, the horizontal or x coordinate in a plane COORDINATE system. ORDINATE.

absolute age the age of a rock, mineral or fossil as revealed by radiometric dating (RADIOMETRIC AGE) or by the counting of VARVES. RELATIVE AGE. G.

absolute difference in statistics, the difference between an observation and the mean value of a set of observations (or between the values of two variables). The direction is disregarded: if there is a negative sign it is eliminated or ignored. The sum of the absolute differences divided by the number of differences gives a measure of dispersion termed the MEAN DEVIATION.

absolute drought in UK, a period of at least fifteen consecutive days on none of which more than 0.25 mm (0.01 in) of rain falls; in USA such a period is termed a DRY SPELL. DROUGHT.

absolute humidity the amount of WATER VAPOUR in a unit mass of air, usually measured by the grams of water vapour present in one cubic metre of the air (or in grains per cubic foot: 1 grain = 0.0648 gram). The amount of water vapour held by a body of air is governed by temperature and pressure: it becomes saturated at DEW-POINT. The lower the temperature the lower the absolute humidity; thus over the land absolute humidity is highest near the equator and lowest in central Asia in winter. HUMIDITY, MIXING RATIO, RELATIVE HUMIDITY, SPECIFIC HUMIDITY. G.

absolute instability the state of an air mass, holding any amount of WATER VAPOUR (CONDITIONAL INSTABILITY), in which the ENVIRONMENTAL LAPSE RATE is higher than the DRY ADIABATIC LAPSE RATE of the atmosphere surrounding it, with the result that it is unstable.

absolute location the LOCATION-1 of a

point which can be expressed exactly by a GRID-1 reference. RELATIVE LOCA-TION.

absolute range the difference between the extremes of temperature, rainfall, etc., the highest and lowest values ever experienced at a place. RANGE-4.

absolute relic RELIC DISTRIBUTION.

absolute stability the state of an air mass in which the ENVIRONMENTAL LAPSE RATE is lower than the SATU-RATED ADIABATIC LAPSE RATE of the atmosphere surrounding it, with the result that it is STABLE, *adj.*

absolute scale a thermodynamic temperature scale in which the lower fixed temperature and the interval corresponds with that on the CELSIUS SCALE. The absolute scale is sometimes termed the KELVIN SCALE, temperature being measured in degrees A (Absolute) or KELVINS.

absolute zero the lowest temperature possible, the point at which all thermal molecular movement ceases, the zero point being 273.15 Celsius degrees below the ice point. This puts the ICE POINT (the melting point of pure ice at standard pressure) at 273.15 K. KELVIN SCALE.

absondering the jointing caused by cooling and contraction in IGNEOUS ROCKS which divides them into approximately regular posts. G.

absorption the physical process in which a material or system (the absorbent) takes into itself and holds another (the absorbate). ADSORPTION.

abstraction 1. the act of taking away, of reducing to a minimum, of extracting the essence of an idea. MODEL-4 2. the product of such acts 3. the formation of an idea, separate from material things etc. REIFICATION 4. a term regarded by some authors as synonymous with RIVER CAPTURE, which involves HEADWARD EROSION; but others apply it to the simplest type of river piracy, in which one stream, by its more vigorous action, widens its valley and

captures another. It is suggested that it may be better to avoid this specialized application of a word commonly and widely used in a general sense. G.

abyssal (abyss or abysm, an apparently bottomless chasm) *adj.* applied to 1. the deepest regions (and associated phenomena) of the ocean, usually referring to ocean depths between 2200 and 5500 m (1200 and 3000 fathoms: 7220 and 18 045 ft) but also, more loosely, the depths of 1800 m (1000 fathoms: 5900 ft) or 900 m (500 fathoms: 2950 ft) or, very loosely indeed, the ocean floor in general. DEEP 2. less frequently, the deepest region of a lake. ABYSSAL ZONE. G.

abyssal benthos, abyssobenthos the plants and animals living on the ocean floor at very great depths. BENTHIC DIVISION, BENTHOS.

abyssal deposits deposits on the very deep parts of the ocean floor. ABYSSAL PLAIN, OOZE. G.

abyssal plain a DEEP-SEA PLAIN, a very large, relatively level area of the deep ocean floor, covered with a thin layer of sediment. OOZE.

abyssal rock term now usually discarded in favour of PLUTONIC ROCK. HYPABYSSAL ROCK. G.

abyssal zone 1. the deepest regions of the ocean, depth not precisely defined, but generally considered to be below 1000 m (550 fathoms: 3300 ft) 2. the zone in a lake not effectively penetrated by light. G.

Acacia a GENUS of leguminous (LEG-UMINOSAE) woody plants with over 500 SPECIES many of which are trees. Some form the dominant vegetation in arid and semi-arid areas; some yield useful products, e.g. GUM, bark for tanning (WATTLE-3), TIMBER-1.

accelerated erosion SOIL EROSION occurring at a rate exceeding the rate at which a SOIL HORIZON can be formed from the parent material. This is commonly due to human activity.

access means of approach, sometimes

restricted to the physical means of approach. ACCESSIBILITY.

accessibility the state or quality of ease of approach, the relative opportunity for interaction and contact, used in a physical sense (the ease of getting to a place, a facility or a service, measurable in terms of the distance to be travelled, the cost involved, the time taken) or a social sense (the extent to which different social groups are able to obtain goods, facilities, services, regardless of geographical location). COST SPACE, EFFECTIVE ACCESSIBILITY, GRAPH THEORY, INTERVENING LOCATION EFFECT, LOCATIONAL ACCESSIBILITY, SHADOW EFFECT, SPACE-TIME CONSTRAINTS, TIME DISTANCE. G.

accessibility isopleth a line drawn on a map linking points of equal distance and/or of equal time taken to travel from a central point to a point in the surrounding area.

accessory mineral any of the many varied minerals occurring widespread and in very small quantity in a rock, which may reveal the origin of the rock but which do not affect its essential character. They are disregarded in the classification of the rock. ESSENTIAL MINERAL.

access point GRAPH-2, ROUTE-3.

accident a term given particular meaning by W. M. Davis, who applied it to irregular or unexpected happenings, e.g. climatic changes, volcanic eruptions, which upset the development of his 'normal' or 'ideal' CYCLE OF EROSION. G.

accidented relief rugged and irregular relief. G.

acclimatization the process by which plants and animals (including human beings) become, or are made to become, accustomed to climatic conditions unnatural to them (i.e. which are different from those of their native environment) so that they are able to live and reproduce. The term is sometimes extended to cover other environmental conditions, e.g. of soil. G.

accommodation a style of, or the stage reached in, INTEGRATION in society, in which conflict is reduced or avoided, the dominant group tolerating the differences of sub-groups such as ETHNIC MINORITIES. ASSIMILATION.

accommodation land a small enclosed field on the outskirts of a town where animals can be kept or 'accommodated' for short periods after being purchased by butchers for slaughter. G.

accommodation unit in physical planning, a housing unit occupied by one household (defined as consisting of an individual, or individuals sharing, as a family or otherwise) irrespective of whether that unit is a separate house, maisonette, flat or apartment and whether permanent or temporary. G.

accordance of summit levels accordant or concordant summit levels, the general conformity of mountain peaks or hill tops to approximately the same level. This may indicate that they are the remnants of a PENEPLAIN dissected by erosion, or that stream erosion combined with WEATHERING in an area of regularly spaced valleys has uniformly levelled off the summits. GIPFELFLUR, SUMMIT PLAIN, SUPERIMPOSED PROFILE. G.

accordant drainage conditions when the surface drainage is directly related to the DIP of the underlying STRATA. G.

accordant junction a river junction in which a tributary joins the main stream at exactly the appropriate level, i.e. the tributary grades its course to the level of its outfall. It was pointed out by Hutton and repeated by Playfair in his Law of Accordant Junctions 1802 that where tributaries join main rivers there is usually no sudden drop but an accordance of levels at the junction (PLAYFAIR'S LAW). DISCORDANT JUNCTION. G.

accretion the accumulation of particles, as in SEDIMENTATION-1, or in the process of growth by external addition, e.g. in the formation of crystals, or in the growth of ice particles by the addition of very small water particles.

acculturation 1. a process of culture change in which more or less continuous contact between two or more culturally distinct groups or societies produces the results outlined under CULTURE CONTACT **2.** the state which results from such a process. ENCULTURATION, NATIVISM, PLURALISM, PLURALISTIC INTEGRATION, PLURAL SOCIETY.

accumulated temperature the sum of the temperatures above or below a selected temperature measured over a period of time (DAY DEGREE). For example, 6°C (43°F) is the CRITICAL TEMPERATURE at which the growth of vegetation is commonly stimulated in midlatitude conditions: at lower temperatures growth is dormant. A temperature of 6°C (43°F) is therefore the basal crucial norm. If the mean daily temperature rises above this basal temperature each degree of the excess is known as the DAY DEGREE. The number of day degrees can be added together for periods of a week, a month or a growing season and the figures so obtained are referred to as the accumulated temperature. G.

accumulation, mountains of a term used in a once-popular classification of mountains and hills, mountains of accumulation resulting fom the piling up of material (e.g.volcanic material) on the earth's surface, mountains of circumdenudation resulting from the denudation or wearing away of a former large mass; and tectonic mountains resulting from the riding up of parts of the earth's surface. G.

acetic acid a colourless, liquid ACID, the chief acid of vinegar, used in textile, pigment, and paint manufacture, the derivative cellulose acetate being used in the plastics industry.

Acheulian epoch a cultural epoch of PALAEOLITHIC times when, in addition to the two-faced axes of the ABBEVILLIAN cultural epoch, flake implements came into use.

acid in chemistry, a compound containing HYDROGEN which, dissolved in water, provides hydrogen IONS (PROTONS); or a MOLECULE or ion which can give up protons to a BASE-2.

acid *adj.* **1.** relating to, or having the characteristics of, an ACID **2.** acid producing **3.** having an astringent, sharp, sour taste.

acid brown forest soil BROWN FOREST SOIL.

acidic *adj.* **1.** acid-forming **2.** ACID, *adj.*

acidic rock ACID ROCK.

acidify to increase the concentration of hydrogen IONS in a solution.

acidity the state or quality of being acid. ACID, pH.

acid lava a viscous, molten IGNEOUS material, with a high content of SILICA and a high melting point, which emerges slowly from a volcanic vent, cools and solidifies quickly, tending to form a plug or solid cone with steep sides (ACID ROCK, LAVA). Many acid lavas solidify into a glassy rather than a crystalline form and are termed OBSIDIAN; but some glassy lavas are less acid in composition. The term is obsolescent, but still widely used.

acidophile a CALCIFUGE, a plant which flourishes on an ACID SOIL.

acid peat BOG PEAT, PEAT.

acid rain precipitation, with a pH value of 5.6 or lower, charged with an excessive amount of acid droplets, formed particularly when oxides of SULPHUR and of NITROGEN released by combustion, especially by the burning of HYDROCARBONS, are converted to ACIDS in the atmosphere. Many animals and plants cannot tolerate the excessive acidity of such precipitation which may make over-acidic soils that are already acid (ACID SOIL), wash ALUMINIUM and other metals out of the ground, thereby polluting rivers and lakes, and greatly damage by chemical process the exterior of buildings. POLLUTION, SULPHUR DIOXIDE, SULPHURIC ACID, SULPHUR TRIOXIDE.

acid rock, acidic rock an IGNEOUS ROCK now defined as containing more than 10 per cent free QUARTZ. Acid rock used to be classified as an igneous rock with more than 66 per cent free or combined SILICA in its chemical composition, or consisting of silicate minerals such as FELDSPAR, MICA. The excess silica commonly appears as 10 per cent or more of free QUARTZ. As ORTHOCLASE FELDSPAR has 66 per cent of silica, the total in an acid magma is in excess of this amount, hence the terms oversaturated or persilicic rock preferred by some authors as alternatives to acid or acidic rock. The typical acid PLUTONIC ROCK is GRANITE, the HYPABYSSAL is QUARTZ PORPHYRY, the VOLCANIC is RHYOLITE. ACID LAVA, BASIC ROCK, INTERMEDIATE ROCK, ULTRABASIC ROCK. G.

acid soil a soil in which the pH or HYDROGEN ION concentration is below 7.2 (neutrality). ACID RAIN, ALKALI SOIL, ALKALINE SOIL, pH.

aclinic *adj.* without dipping or INCLINATION-2, thus in a horizontal position, usually restricted in use to geological and magnetic phenomena. G.

aclinic line an imaginary line close to the terrestrial equator and similarly encircling the earth, on which the needle of a magnetic COMPASS has no dip, i.e. it lies horizontally.

aclinic structure in geology, beds without DIP, lying in a horizontal position. G.

acre a unit of area in British measures, formerly the area of land (customary acre) which could be ploughed in a day by a team of oxen, later defined by statute as 4840 sq yds (0.4 hectare). An acre is sometimes divided into 4 roods, each rood into 40 (square) poles or perches; 640 acres equal one sq mi. CONVERSION TABLES, YARDLAND. G.

acre-foot the volume of any substance (e.g.water) needed to cover an ACRE of land to a depth of one FOOT (43 560 cu ft = 1232.75 cu m).

acronym a name made up of the initial letters of an official title, e.g.United Nations Educational, Scientific and Cultural Organization, acronym UNESCO.

acrylic fibre any of the synthetic fibres commonly used in making yarn for knitted goods and textiles, obtained by polymerization of acrylonitrile (the nitrile of acrylic acid), a poisonous, colourless and flammable liquid which is also used in the plastics industry. ACRYLIC RESIN.

acrylic resin any of the thermoplastic resins obtained by polymerization of single molecule compounds of acrylic acid (ACRYLIC FIBRE), widely used in industry and in the manufacture of consumer goods, being resistant to mild acid, alcohol, alkali and light, but not to oxidizing ACIDS.

actinic rays ELECTROMAGNETIC RADIATIONS which can produce photochemical changes. They include VISIBLE LIGHT and INFRA-RED, ultraviolet (ULTRAVIOLET RADIATION) and X RAYS.

actinology the scientific study of the chemical effects of LIGHT and RADIATIONS.

actinometer an instrument for measuring RADIATION, especially SOLAR RADIATION.

actinomycetes a group of BACTERIA with cells forming fine filaments, flourishing in well-aerated soils, where they assist in the decomposition of organic matter. HUMIDIFICATION.

action 1. the process of doing, of acting 2. the effect of a force 3. the way something (e.g. a muscle) acts 4. the working of one thing on another (e.g.of an acid on limestone, as in CARBONATION-SOLUTION) 4. in sociology, all human behaviour in so far as the ACTOR attaches a subjective meaning to it.

action space the AREA-1 known to an

individual, within which an individual acts, deciding where to live, to play, to shop, the LOCATION of each activity being perceived in relation to that of the others and assessed for PLACE UTILITY. ACTIVITY SPACE, SEARCH SPACE.

active *adj.* working, busy, energetic or effective, as opposed to passive, quiet, dormant or dead.

active cave in limestone country, a CAVE-1 which still has water flowing through it. G.

active glacier a GLACIER subject to a high rate of ALIMENTATION due to heavy snowfall in winter and a warm summer, resulting in rapid ABLATION and the transport of large amounts of debris. PASSIVE GLACIER.

active layer in soil, mollisol, the annually thawed layer in PERMAFROST regions, i.e. the layer of the soil,from the surface down, which thaws out in summer and freezes in winter. Below it lies the permanently frozen layer. G.

active permafrost the layer of PERMA-FROST which, having been thawed by artificial or natural causes, is able to return to permafrost under existing climatic conditions. G.

active volcano a VOLCANO which is liable to erupt, as opposed to a DOR-MANT VOLCANO. G.

activity the state of being, or the capacity for being, ACTIVE.

activity space the area of frequent contact, the area within which most of a person's activities are carried out, within which the individual comes most frequently into contact with others and with the natural and other features of the ENVIRONMENT, comprising all the places visited in RECIPROCAL MOVE-MENT. ACTION SPACE, AWARENESS SPACE. G.

actor one who acts, who performs an ACTION-1,4.

actual *adj.* existing in act or as a fact, as opposed to potential or theoretical.

actual isotherm an isotherm (ISO-)

based on actual temperatures, not on temperatures adjusted to sea-level by taking into account the altitude of the recording station.

adaptation 1. in biology, the act or process by which an organism becomes fitted to its environment, thereby improving its chances of survival and of leaving descendants in the environment it is inhabiting **2.** a characteristic (an inherited or acquired structure or function) which makes an organism so fitted **3.** the condition resulting from the modifying act, process or characteristic **4.** in sensory physiology, the process by which there is a reduction in the excitability, the response, of a sense organ to a uniformly continuously applied stimulus.

adaptive behaviour BEHAVIOUR-1 that involves deliberate, rational choice to bring about satisfactory or OPTIMUM conditions, e.g. the behaviour of a firm in deliberately and rationally choosing an economically satisfactory or OPTI-MUM LOCATION for an operation. ADOPTIVE BEHAVIOUR.

adaptive radiation the fanning out of evolutionary lines, the evolution from primitive stock to divergent forms as organisms are adapted by changes in structure to exploit a wide variety of different, new HABITATS or habits of life. A new species may result if FER-TILE-2 individuals able to breed with each other become isolated in their new habitats. EVOLUTION.

additive a substance added in small quantity to something else so as to improve it in some way, e.g. a preservative added to foods to delay change or decomposition; or a substance designed to improve flavour, colour, etc.

additive *adj.* characterized by joining, or being added to something else in order to increase the number, size, etc.

additive model REGRESSION MODEL.

adiabatic *adj.* of the physical change during which no heat leaves or enters the SYSTEM-1. This commonly occurs in an ascending or a descending parcel of air in the ATMOSPHERE-1. An ascending body of air expands and cools, the constituent MOLECULES being dispersed; a descending body of air contracts and becomes warmer, the molecules being compressed; but there is no net gain or loss of heat. ADIABATIC GRADIENT, DIABATIC, POTENTIAL TEMPERATURE. G.

adiabatic gradient the rate at which the temperature of an AIR MASS, ascending or descending, changes in response to expansion or compression of that air mass. ADIABATIC, DRY ADIABATIC LAPSE RATE, DYNAMIC LAPSE RATE, ENVIRONMENTAL LAPSE RATE, LAPSE RATE, SATURATED ADIABATIC LAPSE RATE. G.

adit a horizontal or nearly horizontal passage or working dug into a hillside from the surface for the purpose of extracting minerals; hence adit mine, adit mining. G.

administrative principle one of the principles used by W. Christaller to account for the varying levels and distribution of CENTRAL PLACES in a CENTRAL PLACE SYSTEM. Assuming the size of the COMPLEMENTARY REGION of a central place to be governed by consideration of the most effective political or administrative control, there will be a clearcut separation of the higher order central place from the neighbouring lower order central places, i.e. each lower order central place will come within the complementary region of the single high order central place, and will thus be served entirely by this one, dominant higher order centre, which is the seat of administration. This results in a K-7 hierarchy. K-VALUE, MARKETING PRINCIPLE, TRAFFIC PRINCIPLE.

adobe (Spanish) a term adopted in USA and Mexico and applied to 1. a LOESS-like, in some cases CALCAREOUS, clay, brought by the wind from the rock waste of deserts and glaciated

areas, deposited on the plains and in the basins of western USA and in arid parts of Mexico, and used for making unburnt, sun-dried bricks 2. the sun-dried bricks made from such deposits 3. a structure, e.g. a dwelling, made from such bricks or deposits 4. soils derived from such deposits. G.

adolescence 1. in general, the period in life when a child changes to an adult 2. the stage in W. M. Davis's CYCLE OF EROSION between youth and maturity, when the characteristic features of maturity are beginning to to develop.

adolescent *adj.* used by Griffith Taylor in his classification of towns. URBAN HIERARCHY. G.

adoptive behaviour BEHAVIOUR-1 that displays a lack of conscious choice, e.g. in the location of an activity, the behaviour of a firm in not making an economically sound choice for the location of an activity (ADAPTIVE BEHAVIOUR) but in allowing its location to happen by chance or subjective choice, e.g. because the founder happened to live in the area.

adret (originally French dialect, now adopted as a technical term; Italian adretto; German Sonnenseite) the sunny, warm slope of a hill or valleyside, as opposed to the UBAC or shady side. ASPECT, EXPOSURE. G.

adsorption a physical or chemical process by which MOLECULES of a gas, liquid or dissolved substance adhere to the surface of another. ABSORPTION.

advanced dune a small SAND DUNE formed round an obstacle and lying in front, i.e. to windward, of a larger dune from which it is separated by an eddying wind.

advanced industrial country AIC.

advection the horizontal movement of air, water or other fluid, applied especially to masses of air and water, in contrast to CONVECTION, applied to vertical movement. G.

advection fog fog formed when a warm, moisture-laden air stream moves horizontally over a cooler surface, so that its lower layers are chilled below DEW-POINT. G.

adventitious *adj.* appearing casually or in unusual places, something added from without, applied since 1949 in the study of rural population, i.e. primary population (farmers and farm workers), secondary (people serving the primary group) and adventitious (the population living in rural areas solely from choice, e.g. retired people). G.

adventive cone, adventitious cone sometimes termed a parasitic cone, a subsidiary CONE appearing on the slopes of a large volcanic cone. G.

adventive crater lateral crater or parasitic crater, a CRATER which opens on the flanks of a great CONE. G.

aeolian, eolian *adj.* associated with Aeolus, the Greek god of the winds, hence related to wind action, i.e. borne, deposited, produced or eroded by wind. G.

aeon, eon an immeasurably long period of time. CRYPTOZOIC, PHANEROZOIC.

aeration any process by which a substance becomes impregnated with air or another gas.

aeration zone the zone with PERMEABLE rocks, including the soil, which overlies the permanent WATER TABLE or zone of saturation (SATURATED-2) and in which interstices are partly filled with ground air. It is termed the vadose zone by some authors because it is associated with VADOSE WATER. G.

aerial *adj.* of or related to, existing in, moving in, the air.

aerial photograph a photograph taken from the air at an oblique or vertical angle, particularly useful in map making. ARCHAEOLOGY, GEOMORPHOLOGY, PHOTOGRAMMETRY, REMOTE SENSING, STEREOSCOPE.

aerobe an organism able to live only in the presence of FREE OXYGEN. AEROBIC, ANAEROBE, ANAEROBIC.

aerobic *adj.* active, living or respiring only in the presence of FREE OXYGEN. AEROBE, ANAEROBE, ANAEROBIC.

aerobic respiration the process in which organisms use gaseous or dissolved oxygen (FREE OXYGEN) to release energy by the chemical breakdown of food substances. ANAEROBIC RESPIRATION, RESPIRATION.

aerology the study of the atmosphere through its complete extent, including the upper layers, and also including the more general studies covered by METEOROLOGY; but sometimes confined to the study of the upper layers. AERONOMY, CLIMATOLOGY. G.

aeronomy the scientific study of the upper ATMOSPHERE-1, i.e. above the 50 km (30 mi) level.

aeroplankton minute organisms (spores, bacteria and other micro-organisms) floating freely in the atmosphere. PLANKTON.

aerosols 1. a system of ultra-microscopic solid or liquid colloidal particles suspended in a gas or in the ATMOSPHERE-1, appearing as mist, fog etc. In the atmosphere aerosols reduce the amount of SOLAR RADIATION reaching the earth's surface, and may thus have a cooling effect. POLLUTION 2. a substance held under pressure in a container and released by a spraying device; or such a container itself.

aerosphere the entire ATMOSPHERE, with all its layers, enveloping the earth. The term was introduced because some authors restricted the term atmosphere to the lowest layer; but it has not been generally adopted.

aerotropism the growth response of plants towards a source of air. TROPISM.

aesthenosphere erroneous spelling of ASTHENOSPHERE.

aestivation, estivation 1. in biology, the state of torpidity of some organisms during the hottest period, the summer. HIBERNATION 2. in botany, the arrangement of petals in the bud of a flower.

aetiology ETIOLOGY.

affinity 1. a close link or connexion 2. relationship by marriage 3. a strong liking or attraction between persons 4. in biology, the relationship between SPECIES that indicates a common origin.

affluent a stream flowing into another stream, an obsolescent term for TRIBUTARY. CONFLUENT. G.

afforestation 1. historically, the result of the declaration of English medieval kings that a tract of land should be subject to forest laws (FOREST-3), or their action in making this declaration 2. the clearing of land of sheep and cattle in Scotland in the mid-nineteenth century so that deer forests (FOREST-4) could be established 3. the planting of land, not formerly so covered, with trees to make a forest for commercial or other purposes. FOREST, REFORESTATION. G.

after-glow, afterglow the soft diffused light remaining after the light source has gone, e.g. as in the evening sky after the sun has set and sunk some 3° below the horizon, apparently due to the scattering of light by dust particles in the atmosphere. ALPINE GLOW. G.

aftermath the grass crop cut from the second, lesser growth of grass which follows the first cutting.

aftershock one of a series of small shocks following a main EARTHQUAKE and originating at or close to its SEISMIC FOCUS.

agar-agar a gelatinous substance obtained from SEAWEEDS native to the waters around Sri Lanka, Malaysia, Java and Japan (especially from the seaweed *Gracilaria lichenoides*). Having been mixed with hot water and cooled, it sets to a firm jelly, suitable for use in cooking, the manufacture of fine silk and paper, in pharmaceutical products and, in bacteriology, as a culture medium.

agate a very hard CRYPTO-CRYSTALLINE form of SILICA, a variety of CHALCEDONY, composed of layers of different colours resulting from the infiltration of other minerals. Agate is used for ornaments, for mortars in which hard substances are ground, and as knife-edges of chemical balances.

agave a tropical plant, member of *Agave*, family Amaryllidaceae, cultivated in tropical regions for its fibre (particularly the fibre of *Agave rigida* and *A. sisalana*) known as SISAL and HENEQUEN, for its sap (the source of intoxicants such as pulque, mescal, tequila in Mexico), and for ornamental purposes.

age 1. in general, the length of time for which an organism has lived or an object existed, exceptions being a life of markedly short duration (e.g. a butterfly living for a day or two is said to have a LIFESPAN, not an age) and the life of subatomic particles and radioactive species, to which the term HALF-LIFE is applied 2. in geology, a very long time in the earth's history during which FORMATIONS-2 were laid down, a subdivision of an epoch (GEOLOGICAL TIMESCALE) 3. in ethnology, a period of time based on cultural criteria, e.g. NEOLITHIC AGE.

age and area hypothesis in anthropology, a theoretical, empirical inference that the age of culture traits can be determined by studying their distribution over a large area, based on the assumption that culture traits undergo DIFFUSION-1 from a single centre or from multiple centres, and that cultures on the periphery may thus show traits characteristic of the centre at an earlier period of time. Anthropologists tend to disagree on the extent to which this hypothesis can be successfully used in the reconstruction of cultural history.

agglomerate in geology, a rock consisting of angular fragments mainly exceeding 2 cm (0.75 in) in diameter thrown out of an erupting VOLCANO and more or less solidified in a matrix of ASH or TUFF. BRECCIA, CONGLOMERATE, PYROCLAST. G.

agglomeration 1. in general, a gathering together into a mass 2. in soil science, a loose gathering together of soil particles which, when more closely united, form an AGGREGATE-1 3. in urban studies, an urban area, usually unplanned and formless, composed of formerly separate suburbs, villages or small towns which have expanded and coalesced 4. a concentration of productive enterprises at a certain location, e.g. in a large town or industrial region, where each enterprise may enjoy the benefits of the readily available labour, good transport and service facilities, the large market, the proximity of allied and other enterprises, the concentration itself producing ECONOMIES OF SCALE. AGGLOMERATION ECONOMIES, COMPLEMENTARITY, DEGLOMERATION, DISECONOMIES OF SCALE, EXTERNAL ECONOMIES. G.

agglomeration economies the savings to the individual productive enterprise that come from operating in the same location as others (thereby sharing specialist servicing industries, specialist financial services and public utilities) or from serving a growing, large market occupying a small, compact geographical area. AGGLOMERATION-4, CENTRIPETAL AND CENTRIFUGAL FORCES, DEGLOMERATION.

aggradation 1. a process in which a land surface is built up by deposition and accumulation of detritus, rock waste, sand, alluvium etc. derived from DENUDATION, thus the opposite of DEGRADATION, e.g. an alluvial river plain is an aggraded surface (PROGRADATION) 2. a process in which PERMAFROST grows, the upper surface of the permafrost rising and accumulating ice on the under surface of the ACTIVE LAYER; termed aggradational ice, it appears as white horizontal bands interspersed with horizontal dirt bands. G.

aggradational ice AGGRADATION.

aggregate 1. a total derived by addition, e.g. in statistics, POPULATION-4 2. in geology, a cluster of mineral particles, i.e. a rock 3. in soil science, a single mass or cluster of soil particles which acts as a unit. CRUMB STRUCTURE 4. a construction material made from crushed stone or gravel and sand, used in making CONCRETE, in road surfacing etc. PAVED SURFACE. G.

aggregate demand in Keynesian economic analysis, the value of the total planned expenditure in an economy. KEYNESIANISM.

aggregate information field all the information fields of an aggregate of population, indicating the frequency of contacts over different distances. MEAN INFORMATION FIELD, PRIVATE INFORMATION FIELD.

aggressive water water capable of dissolving rock. In KARST terminology the term usually implies that the water has in it dissolved CARBON DIOXIDE (CARBONIC ACID) or, rarely, other acids.

agistment 1. the pasturing of the animals of one person on the land of another on payment of a due 2. on COMMONS, the letting of a commoner's right to graze stock to one who does not hold such a right 3. the profit made from these practices, or the amount of the due 4. the natural herbage of a FOREST or the right to graze it. G.

agland the edge of an area of cultivated land.

agonic line the irregular line passing through the earth's north and south MAGNETIC POLES, along which the COMPASS needle points to TRUE NORTH, i.e. the line of no MAGNETIC DECLINATION. G.

agrarian 1. one who favours the redistribution of landed property. AGRARIANISM, LAND REFORM 2. an agrarian law.

agrarian *adj.* 1. of or relating to land, especially to its cultivation, management and distribution in a system of LAND TENURE 2. of, relating to, or connected with, landed property **3.** in botany, growing wild. G.

agrarianism the theory or practice of redistribution of landed property on an equitable basis, or of facilitating the more productive cultivation of land. AGRARIAN REFORM.

agrarian reform LAND REFORM combined with changes designed to improve rural life, e.g. by the provision of facilities for better education and social life as well as for more productive cultivation. AGRARIANISM.

agribusiness 1. all the operations and processes involved in running a FARM as a commercial enterprise, i.e. the making and transport of commodities used in production on the farm; the production on the farm; the handling, processing, storage of the items to be consumed on and off the farm; the distribution from the farm of the items offered for sale; and the buying and selling, the keeping of records and accounts etc. relating to all those operations and processes **2.** specifically, a highly commercial, efficiently organized, business-like CAPITAL-INTENSIVE farming enterprise using up-to-date technology in equipment and production methods to achieve the highest possible output of produce of a consistently high standard which, in some cases, is sold under contract to large-scale customers.

agric horizon (American) an ILLUVIAL SOIL HORIZON formed by long-continued cultivation, consisting of silt, clay and humus, and lying usually immediately below the ploughed surface.

agricultural *adj.* related to, characteristic of, AGRICULTURE.

agricultural area 1. an area of land used for farming, including arable land, improved or unimproved grassland and other pasture **2.** in FAO statistics, arable land and land under tree crops, plus permanent meadows and pastures. Unimproved grassland presents a difficulty: where it corresponds to 'range land' (RANGE-3), or the open natural grassland of the tropical regions, it is normally excluded. G.

agricultural climatology, agroclimatology the scientific study of CLIMATE specifically in relation to AGRICULTURE.

agricultural region a large area of land in which agricultural conditions and practices are similar enough to give it a special character, making it distinct from adjacent areas. NATURAL REGION, REGION. G.

agricultural revolution a period in agricultural development characterized by change and INNOVATION in techniques, plant breeding, animal husbandry, etc. GREEN REVOLUTION, HIGH FARMING.

agriculture the science and art of cultivating the soil and the rearing of livestock, equivalent to FARMING. Some authors include FORESTRY, others exclude RANCHING; others restrict the term to the cultivation of the land. AGRIBUSINESS, AGRICULTURAL AREA, AGRICULTURAL REGION, CROP FARMING, EXTENSIVE AGRICULTURE, EXTRACTIVE AGRICULTURE, INTENSIVE AGRICULTURE, LIVESTOCK FARMING, MIXED FARMING. G.

agro-climatology the scientific study of climatic elements in relation to AGRICULTURE.

agro-forestal, agroforestal *adj.* pertaining to AGRICULTURE and FORESTRY. Some authors prefer the even more precise term AGRO-SILVO-PASTORAL. Both terms avoid the ambiguity of 'agricultural', which may or may not include forestry. G.

agronomy the scientific study of the processes of agricultural production, including the study of soils, soil management, and the theory and practice of crop production.

agro-silvo-pastoral *adj.* pertaining to agriculture, forestry and grazing. AGRO-FORESTAL. G.

agrostology the scientific study of grasses.

agro-town a rural settlement in which most of the employed population is engaged in AGRICULTURE.

A horizon the surface layer in the SOIL containing HUMUS, an ELUVIAL LAYER from which minerals etc. are leached. The A horizon is subdivided by some soil scientists into A_{00}, a layer of litter; A_0, a layer with new organic matter; A_1, an organically rich layer; A_2, a leached layer; and A_3, the layer grading into the B HORIZON. The entry on SOIL HORIZON shows further refinements in the classification of the A horizon. L LAYER, MINERAL HORIZON, O HORIZONS, SOIL PROFILE.

ahqāf (Arabia: Arabic) hot desert country with very soft DUNES which make travelling impossible except along defined, narrow belts. G.

aiguille (French, a needle) a sharp, slender peak of rock resembling a needle, in mountainous country. GENDARME. G.

AIC advanced industrial country. INDUSTRIALIZATION-1, LDC, NIC, UNDERDEVELOPMENT.

air 1. the mixture of gases enveloping the earth and forming the ATMOSPHERE **2.** that air considered as a medium for the transmission of radio waves or for the operation of aircraft.

air drainage the downward movement of cold AIR-1 from higher to lower areas. FROST POCKET, KATABATIC WIND. G.

airfield a tract of land where aircraft can take off and land, be accommodated, serviced and maintained. AIRSTRIP, AIRPORT.

air frost air with a temperature at or below 0°C (32°F) as recorded at the level of a METEOROLOGICAL SCREEN, i.e. at about 1 m (4 ft) above the ground. GROUND FROST.

air-gap WIND-GAP. G.

airglow the very dim, weak light of the night sky resulting from the radiation of light from high levels of the atmosphere. G.

airline 1. a commercial organization providing regular air service for the carriage of goods and passengers along regular routes **2.** the route covered by such a service.

airliner a large passenger aircraft.

air mass, air-mass, airmass a mobile HOMOGENEOUS mass of air in the atmosphere, bounded by FRONTS. It may be very large, spreading over hundreds of sq km, or quite small and local; but it has distinct characteristics of LAPSE RATE and HUMIDITY (dry or moist) and temperature (hot or cold) derived from the region from which it originates, but which may be modified by long travel. There are many detailed classifications: a very broad general division identifies polar or tropical (based on temperature), maritime (on humidity, having taken up moisture from the ocean), continental (dry, of land, i.e. continental, origin). Such air masses, large and small, meet one another along fronts, the frontal surfaces which form the boundaries but which are often smoothed out into transition zones. Local weather is explained in terms of the movement of air masses. G.

air pollution POLLUTION.

airport an AIRFIELD designed for use by civil or commercial aircraft and equipped with facilities for the handling of passengers and goods. There is an official international classification of airports, revised frequently, according to the facilities available. INTERNATIONAL AIRPORT, AIRSTRIP. G.

air-stream a wind, a current of air flowing from an identifiable source.

airstrip a narrow tract of land cleared of obstructions so that aircraft may take off and land. AIRFIELD, AIRPORT. G.

airway 1. an air lane **2.** an AIRLINE.

ait, eyot a small island, especially one in a river. G.

aitogenic *adj.* PARATONIC.

akee *Blighia sapida*, a medium-sized tree native to West Africa, introduced in Jamaica in the late eighteenth century. It bears fruit containing three seeds surrounded by a fleshy, cream-coloured coating (the aril), that can be eaten raw but which is usually boiled or fried. Over-ripe and under-ripe fruits and the pink tissue joining the aril to the seed are very poisonous.

alabaster a fine-grained GYPSUM, translucent, soft, usually white, polishes well, used particularly for ornaments.

albedo the ratio of total solar electromagnetic radiation (SOLAR RADIATION) falling on a surface to the amount reflected from it, expressed as a percentage or a decimal. The average albedo of the earth is about 0.4 (40 per cent), i.e. four-tenths of solar radiation is reflected from the earth into space. G.

albic horizon a light-coloured SOIL HORIZON, in many cases sandy, the paleness resulting from the removal of IRON oxides and CLAY.

albite a white, sodic PLAGIOCLASE FELDSPAR, a sodium aluminium silicate, $NaAlSi_3O_8$, occurring mainly in rocks rich in ALKALI, e.g. GRANITE. The opalescent variety is MOONSTONE.

alcohol C_2H_5OH, ethyl alcohol, a colourless, volatile, inflammable and intoxicating liquid, obtained commercially by distilling fermented liquor and by the hydration of ETHYLENE, used as a fuel or solvent and in the making of organic chemicals.

alcove a crescent-shaped, steep-sided recess in rock occurring particularly in the Snake River Valley, USA, as a result of HEADWARD EROSION and SPRING SAPPING of the BASALT. G.

alcrete DURICRUST in which ALUMINIUM hydroxide predominates. It is akin to BAUXITE.

Aleutian low a sub-polar atmospheric low pressure area (LOW, ATMO-SPHERIC) lying over the North Pacific, particularly in winter, characterized by a series of swiftly moving separate lows interspersed by occasional HIGH pressure systems. ICELANDIC LOW.

alfalfa *Medicago sativa*, lucerne, a valuable perennial fodder plant of the LEGUMINOSAE family, with long roots enabling it to thrive in dry conditions in USA, temperate areas of South America and Asia, as well as in the Mediterrean region. It may be cut four or five times in a season.

alfisols in SOIL CLASSIFICATION, USA, an order of soils which are relatively young and acid with a CLAY B HORIZON. Alfisols commonly occur under DECIDUOUS forest and are associated with HUMID, sub-humid TEMPERATE and SUBTROPICAL climates.

algae pl. **alga** sing. a large group of simple PHOTOSYNTHETIC, non-vascular plants with unicellular organs of reproduction. The members vary widely in size and form, from the unicellular microscopic to the multicellular filamentous or ribbon-like with a more complex internal structure, most containing pigments in addition to the CHLOROPHYLL, which is often masked. The aquatic, e.g. DIATOMS, SEAWEEDS, live in fresh or salt water, the planktonic algae (PLANKTON) forming the bases of the aquatic food chains, and the seaweed providing a source of human and livestock food as well as a source of substances used industrially (AGAR-AGAR, ALGIN). The terrestrial algae live in damp places, e.g. damp parts of walls or tree trunks, and in soil. LICHEN, THALLOPHYTA.

algal bloom, phytoplankton bloom the sudden increase of microscopic ALGAE in bodies of water due to an increased supply of nutrients. EUTROPHICATION.

algin any of the soluble colloidal salts of ALGINIC ACID, especially the sodium salt, extracted from various SEAWEEDS (especially from the giant

KELP and KNOTTED WRACK) on account of its emulsifying, thickening, stabilizing and moisture-holding properties. It is used in the pharmaceutical, cosmetic, food, paper and textile industries. ALGAE.

alginate a sort of ALGINIC ACID.

alginic acid an insoluble, non-toxic acid occurring in the cell walls of some brown marine algae (SEAWEEDS), widely used in industry. ALGIN.

algorithm a procedure for performing a complex operation, for achieving a desired solution, by carrying out a carefully determined sequence of simpler operations that lead to the desired result.

alidade 1. a surveying instrument used (with a PLANE-TABLE) to determine direction by viewing distant objects and noting angular measurements. It consists of a rule with sights at each end or with a telescope mounted parallel to it 2. the index of any graduated surveying or measuring instrument, e.g. a QUADRANT-4 or a SEXTANT. G.

alien a non-naturalized foreigner, i.e. a foreigner who is not a citizen of the country in which he/she is living.

alien *adj.* 1. foreign, belonging to another country 2. alien from, differing in nature or character 3. alien to, so different as to be contrary or opposed in nature or character.

alienation 1. in social sciences, in general, an estrangement or separation between parts or the whole of the personality and those aspects of experience which are significant, e.g. a sense of estrangement or separation from society or of a lack of power to bring about social change; or a depersonalization of the individual in a large bureaucratic society. ANOMIE, MASS SOCIETY 2. in Marxism, the estrangement of workers from their labour on account of the relations of economic production and the class system (MODE OF PRODUCTION), a condition in which the workers, losing control over

the processes of work and the products of their labour, become dehumanized things, estranged from nature and from themselves.

alimentation in glaciology, the building-up of snow on an area of FIRN by snowfall and by the effects of avalanches and the refreezing of meltwater, which may feed an outward moving GLACIER. A glacier advances if alimentation near its source exceeds ABLATION at the other end; it remains stationary if the two processes are in balance; and it retreats or recedes if alimentation is less than ablation. ACTIVE GLACIER.

alio (French) IMPERVIOUS, FERRUGINOUS crust formed at a fairly constant depth of some 1 to 2 m (3 to 6 ft) below the soil surface by the deposition of salts of IRON carried by subterranean water. HARD PAN. G.

alkali 1. in chemistry, a usually soluble, strongly basic (BASE-2) hydroxide or carbonate of the alkali metals (i.e. of CAESIUM, FRANCIUM, LITHIUM, POTASSIUM, RUBIDIUM, SODIUM) or of the ALKALINE EARTH METALS (i.e. of CALCIUM or BARIUM) 2. loosely applied to a substance with properties similar to those of a true alkali.

alkalifiable *adj.* capable of being converted to an ALKALI-1, or of being made ALKALINE.

alkali feldspar FELDSPAR.

alkali flat a level area in an arid region with an incrustation of ALKALI salts formed as a result of the evaporation of a former lake. PLAYA. G.

alkaline *adj.* having the properties of an ALKALI-1.

alkaline earth metals a group of metals consisting of BARIUM, BERYLLIUM, CALCIUM, MAGNESIUM, RADIUM and STRONTIUM, the hydroxides of which are weaker BASES-2 than those of the alkali metals (ALKALI-1).

alkaline rock, alkali rock in petrology, an IGNEOUS ROCK low in CALCIUM but rich in alkali metals (ALKALI-1), indicated by the presence of sodium-rich PYROXENES and AMPHIBOLES and SODIUM and/or FELDSPARS rich in POTASSIUM. The opposite of alkali in petrology is not ACIDIC but calcalkali or calcalkaline, applied to an igneous rock in which the feldspar is rich in calcium.

alkaline soil any soil which is alkaline in reaction, precisely above pH 7.0, in practice above pH 7.3, neutrality being 7.2. ACID SOIL, HYDROGEN ION, pH. G.

alkalinity the state of being alkaline, i.e. of having a pH above 7.2. BASE-2, pH.

alkali soil soil containing ALKALI-1 salts, usually SODIUM carbonate, with a pH value of 8.5 and higher, likely to show surface incrustations of ALKALI-1, pH. G.

alkaloid any of a group of organic BASES-2 containing nitrogen in a cyclic structure and occurring in some flowering plants valued for medicinal or poisonous qualities, e.g. ATROPINE from *Atropa Belladonna*, morphine from the OPIUM POPPY, NICOTINE from the TOBACCO plant, QUININE from the bark of *Cinchona* species.

Allies, The 1. the 23 countries (including the British Empire, France, Italy, Russia and USA) allied against the CENTRAL POWERS in the First World War 2. the 49 countries (including Great Britain and Commonwealth countries, Belgium, China, Denmark, France, Greece, Netherlands, Norway, USA, USSR and Yugoslavia) allied against the AXIS POWERS in the Second World War.

allochthon something (especially a rock) transported, not in its original place, the opposite of an AUTO-CHTHON.

allochthonous *adj.* transported, not in place of origin, the opposite of AUTOCHTHONOUS, applied to 1. transported fossil plants or to organic deposits (e.g. some COALS) formed from them 2. rocks which have travelled far from

their place of origin, especially if moved by a tectonic process, e.g. overthrusting, recumbent folding, or sliding by gravity. G.

allogenic *adj.* applied 1. in geology, as an alternative term to ALLOTHIGENIC, the opposite of AUTHIGENIC, AUTHIGENOUS 2. in biology, having different genes 3. in ecology, produced by external factors, the opposite of AUTOGENIC.

allogenic succession a plant SUCCESSION produced by changes in the environment, by external factors, not by changes produced by the plants themselves, as in AUTOGENIC SUCCESSION.

allometric growth the systematic differential growth of parts within a complex growth structure, so that as the SYSTEM-1,2,3 grows as a whole, the ratios between each part and the whole stay constant. Thus the growth rate of a part is proportional to the growth rate of the system as a whole, e.g. in urban studies, population growth increases up to a certain distance from a CENTRAL PLACE; in STREAM ORDER, as the number of stream segments grows the proportion falling into each stream order stays constant. GENERAL SYSTEMS THEORY. G.

allopatric *adj.* applied to different SPECIES-1 or sub-species of which the areas of distribution do not overlap or coincide. SYMPATRIC.

allopatric speciation PUNCTUATED EQUILIBRIUM.

allothigenic, allothigenous *adj.* originating at a distance, not AUTHIGENIC, AUTHIGENOUS, applied to 1. constituents in a SEDIMENTARY ROCK which were formed outside and before the rock of which they are now a part, e.g. the pebbles in a CONGLOMERATE. ALLOGENIC 2. rounded crystals in an IGNEOUS ROCK which have come from some previously consolidated rock 3. streams deriving much of their water from afar. G.

allotment 1. a share ⸬f something allocated to a person (FIELD GARDEN ALLOTMENT) **2.** specifically today in Britain, the term is usually applied to small parcels of land as defined by the Allotments Act 1922, which identifies an 'allotment garden' under Section 22 (1) as … 'not exceeding 40 poles in extent which is wholly or mainly cultivated by the occupier for the production of vegetables and fruit crops for consumption by himself or his family', i.e. a plot under 0.1 ha (0.25 acre); and 'allotment' under Section 3 (7) … 'any parcel of land whether attached to a cottage or not, of not more than 2 acres in extent held by a tenant under a landlord and cultivated as a farm or a garden, or partly as a garden and partly as a farm'. If the parcel of land is rented from a local authority the limit of 0.81 ha (2 acres) is increased to 2.02 ha (5 acres). Such allotments are characteristic of rural areas and are often used to supplement the income of the tenant **3.** formerly 'allotment' was used especially in connexion with land enclosure. Under the enclosure awards in England and Wales certain portions of land were set aside for the use of villagers who would otherwise have become landless. They were given certain COMMON RIGHTS, including the right of pasturage for certain animals; but in addition 'allotments' were set aside for specific purposes, such as peat lands as fuel allotments (where peat could be dug for fuel), or simply lands which could be cultivated by the poor (poor's allotments). G.

allotriomorphic *adj.* applied to the minerals in IGNEOUS ROCK which do not have the characteristic CRYSTALLINE form. IDIOMORPHIC.

alloy a metallic substance of desired character made of two or more metallic elements, the physical properties of the alloy differing from those of the constituents, e.g. a small quantity of MANGANESE alloyed with STEEL renders

steel especially hard. Brass is an alloy of COPPER and ZINC. BRONZE, an alloy of copper and TIN, was the earliest alloy. BRONZE AGE.

allspice *Pimenta dioica*, a small tropical tree native to the West Indies and Central America. It is cultivated especially in Jamaica where wild trees also give a useful crop of the hard, unripe, dried berries that provide the spice known as allspice, so called because the pungent aromatic flavour resembles that of a blend of cinnamon, cloves, mace and nutmeg (hence the French name, quatre épices). Allspice is used in chutneys, sauces, etc.

alluvial *adj.* of, pertaining to, or consisting of, ALLUVIUM.

alluvial cone an ALLUVIAL FAN with a steep slope, particularly likely to be built up if most of the water of the stream sinks into a porous deposit, so that nearly all the load is dropped and the structure gains height, e.g. in arid or semi-arid conditions. G.

alluvial fan a fan-shaped deposit of coarse alluvium (sand and gravel), the apex pointing upstream, laid down by a stream where it issues from a constricted course, e.g. from a gorge, on to a more open valley or to a plain. ALLUVIAL CONE. G.

alluvial flat a level, nearly horizontal, tract of land, bordering a river, which receives the ALLUVIUM deposited by the river in flood. G.

alluvial soil an AZONAL SOIL, a FLU-VISOL forming on newly deposited ALLUVIUM.

alluvial terrace part of an ALLUVIAL FLAT, in some cases paired by another on the opposite side of the river, left standing as the river cuts down its bed following REJUVENATION. Some authors use the term as a synonym for RIVER TERRACE, but this can be justified only if 'alluvial' is applied very loosely to gravel and coarse sand as well as to the fine-grained deposits of alluvium. Many river terraces are not strictly alluvial terraces because they consist essentially of gravel. G.

alluvion 1. a term now rarely used, formerly applied to the wash of water against a bank or shore, to floodwater, or to material deposited by floodwater **2.** in law, the formation of new land by the action of flowing water, by the deposit of ALLUVIUM. DILUVION. G.

alluvium (Latin alluvius, washed against) **1.** broadly applied to the unconsolidated, loose material (not only silt but also the gravel and sand) brought down by a river and deposited in its bed, floodplain, delta or estuary, or in a lake, or laid down as a cone or fan **2.** more specifically restricted to the fine-grained deposits, in texture the silt or silty-clay (GRADED SEDIMENTS), so laid down. These are rich in mineral content and so form some of the most fertile soils, of the highest agricultural value, in the world. ALLUVIAL, DILUVIUM. G.

almanac, almanack 1. a book or table giving information on the calendar (days, weeks, months) with details of astronomical data (rising and setting of the sun, phases of the moon, eclipses, etc.) and tides **2.** *Ephemeris and Nautical Almanac*, issued by chief governments two or three years in advance, containing detailed tables of accurately predicted positions of the heavenly bodies, with times of celestial phenomena and other data, used by astronomers and navigators **3.** popularly and commonly applied to an annual work of reference giving general information and statistics, and recording the major events of the year prior to publication, e.g. *Whitaker's Almanack* in the UK. G.

almond *Prunus amygdalus* or *communis*, synonymous, a small tree, native to eastern Mediterranean lands, naturalized and cultivated in southern Europe and western Asia, and prized for its NUTS. The two main varieties are the bitter almond (*Prunus communis*, var-

iety amara), the main source of almond oil, used in flavouring and emollients for the skin; and the sweet almond (*Prunus communis*, variety dulcis), many cultivars, producing edible nuts used especially by bakers and confectioners.

almwind a strong and sometimes blustery FOHN-type wind, blowing from the south in the Tatra mountains down into the foreland of southern Poland, causing avalanches towards the end of winter and in spring.

Alonso model a MODEL-4 suggested by William Alonso in the 1960s to explain the variations in land values, land use and land use density in different districts within an urban area. Accepting VON THUNEN'S MODEL in explanation of the pattern of agricultural land use, he applied the principles of that model to the urban land use pattern. He assumed industry to be concentrated at the city centre, and physical ACCESSIBILITY (and thus transport costs) to be the prime consideration of those setting up industry as well as those working and living in the urban area. From this he suggested that a simultaneous resolution of BID PRICE CURVES, each related to a different category of land user, explains the variations in land values and land use, etc., and thus the urban land use pattern.

alp 1. high mountain pasture, snow-covered in winter, usually above the TREE LINE, on a gently sloping BENCH or SHOULDER-2, commonly on the side of a U-SHAPED VALLEY where there is an abrupt change of slope. In summer it provides rich pasture, and in some cases a site for a seasonal dwelling for herdsmen. It may also afford a site for permanent housing or a winter-sports centre **2.** pl. alps, high, especially snow-capped, mountains broadly similar to the Alps of Switzerland and adjacent European countries. G.

alpaca 1. a small South American RUMINANT MAMMAL, a species of

LLAMA, living both wild and domesticated on the high plateaus of the Andes and yielding a fine, soft wool **2.** the wool itself or the cloth made from it.

Alpides, Alpids a general term applied to the Alpine fold mountains. They have the same general trend as the ALTAIDES. -IDES. G.

alpine *adj.* **1.** of, pertaining to, or characteristic of, the Alps or any similar high mountains **2.** the parts of a mountain above the TREE LINE and below permanent snow, or the plants and animals living in that zone **3.** any processes connected with or a product of ALPINE OROGENY. G.

alpine glacier a VALLEY GLACIER, a type of GLACIER, an ice river, formed in an amphitheatre among mountain summits, descending a mountain valley and ending by melting or spreading out into a PIEDMONT GLACIER. GLACIAL TROUGH. G.

alpine glow the rosy flush on mountain peaks, especially the snow-covered, immediately following the sunset (displaying shades of orange) or before sunrise (with shades of purple) when the sun is 2° above the horizon. AFTER-GLOW.

Alpine orogeny, Alpine earth movements, Alpine revolution the great mountain-building movements which took place mainly in the Tertiary period, culminating in the Miocene (GEOLOGICAL TIME-SCALE) and resulted in the creation of the Alpides or Alpine systems of the Alps, Carpathians, Balkan Mountains, Pyrenees, Atlas and other great chains. OROGENESIS. G.

Altaides, Altaids, Altaid orogeny the great mountain chains stretching through central Europe and Asia, the result of the Altaid OROGENY which took place in late Carboniferous into Permian times (GEOLOGICAL TIME-SCALE). The name is derived from the Altai mountains of central Asia, which are typical of the system. ALPIDES, ARMORICAN OROGENY, HERCYNIAN, VARISCAN. G.

alternative hypothesis any admissible HYPOTHESIS alternative to the one being tested. NULL HYPOTHESIS, STATISTICAL TEST.

alternative technology the TECHNOLOGY-3 concerned with the technical processes that make use of renewable rather than non-renewable NATURAL RESOURCES and which are therefore regarded as ecologically sustainable. APPROPRIATE TECHNOLOGY.

altigraph a self-recording ALTIMETER.

altimeter an instrument for measuring altitude, used in aircraft and by surveyors on land, which employs the fall in atmospheric pressure with height above sea-level (averaging 34 mb: 1 in of mercury for each 300 m: 1000 ft) as an index, e.g. an ANEROID BAROMETER adapted for height. Electronic techniques are used in a radio altimeter. G.

altimetric frequency curve, altimetric frequency graph a curve constructed to show generalized altitudes in an area. It may be made by dividing a map into small squares and taking the highest point, or the average of the highest and lowest points, in each; or by using specific heights above sea-level; but various other methods are employed. If the measurements are shown on a graph, heights above sea-level appear on the horizontal scale, the percentage frequency on the vertical. G.

altiplanation a process of soil or earth movement in cold regions which tends to produce terraces or flattened areas consisting of an accumulation of loose rock, the terraces sometimes being termed EQUIPLANATION terraces. G.

altiplano (Spanish: South America) a high INTERMONTANE plateau in the Andes, specifically in western Bolivia. G.

altitude 1. the angular height of a star or other heavenly body measured in degrees from the plane of the observer's horizon **2.** the height of a point above mean sea-level, measured vertically **3.** the height of a mountain or hill above its base or surrounding plain, measured vertically from base to summit. G.

altitudinal zone VERTICAL ZONE.

altocumulus, altocumulus castellatus a cloud formation of middle altitudes (2400 to 6000 m: 8000 to 20 000 ft), consisting of small, fleecy, globular, relatively thin patches of CLOUD, the edges sometimes merging, and sometimes appearing as a MACKEREL SKY, usually indicating fine weather. Sometimes the tops of the small clouds become turrets, a formation termed altocumulus castellatus, indicating instability in the upper air, and thunderstorms.

altostratus a cloud formation of middle altitudes (2400 to 6000 m: 8000 to 20 000 ft), a wide expanse of continuous, uniformly grey, thick or thin, flat CLOUD, through which the sun or moon may usually dimly be seen, associated with an oncoming WARM FRONT and indicating rain. CIRROSTRATUS.

alumina ALUMINIUM oxide. SIAL.

aluminium (American aluminum) a very light, white, ductile, malleable, metallic element, durable and resistant to corrosion, occurring widely in nature but difficult to separate from its ores. Formerly obtained from the mineral CRYOLITE, it is now extracted mainly from the ore BAUXITE. The ore is subjected to very high temperatures in an electric furnace to produce alumina (aluminium trioxide, Al_2O_3), smelted to produce aluminium (some 4 tonnes of bauxite producing 2 tonnes of alumina and 1 tonne of aluminium). Aluminium is extensively used pure or in alloys in manufacturing motor vehicles, lightweight containers, aircraft, electrical apparatus, cooking utensils, thin foil for wrapping purposes, etc. C.

alvar (Swedish) **1.** a treeless limestone surface or area, bare or covered with only a thin layer of soil **2.** synonymous with alvarmark, the RENDZINA of the Baltic Sea islands and of northwestern Estonia. G.

a.m. ante meridiem, the time period after midnight (2400 hours) and before noon (1200 hours). P.M.

āmān (Indian subcontinent: Bengali) winter rice or rice-crop, harvested at the beginning of the āmān season, the cool dry season, June to November. G.

ambari *Hibiscus cannabinus*, an annual or perennial plant native to southeast Asia, providing fibre akin to JUTE.

amber the yellowish, translucent, somewhat brittle FOSSIL resin of certain extinct coniferous trees, found especially along the Baltic coasts. It is soluble in SULPHURIC ACID, slightly magnetic when subjected to friction; and is used in jewellery and ornaments.

ambergris a waxy substance formed as a secretion in the intestines of sperm whales; sometimes used in the making of perfume.

ambient *adj.* surrounding, lying around, encompassing, or on all sides of, a phenomenon, organism, or object.

ambient temperature the TEMPERATURE of the atmosphere immediately surrounding something specific, e.g. a CLOUD or a THERMAL.

ambilineal UNILINEAL.

amelioration an improvement, the condition of being made better, e.g. climatic amelioration, an improvement or a modification of climatic conditions over a period of time. The result may be that certain climatic factors hostile to various life-forms become less stringent, so that those life-forms spread into hitherto prohibitive areas. But those same factors may have favoured other life-forms, and their modification may result in the retreat of such life-forms from areas previously favourable to them.

amenity something in, or some quality of, the environment which is perceived as pleasant and attractive, which makes life agreeable, satisfying, for people.

American wild rice Indian rice, tuscarora rice, *Zizania aquatica*, a robust, tall (up to 4 m: 12 ft) and ornamental annual aquatic grass, native to eastern North America, providing food especially for water fowl. It is said to be richer in protein and vitamins than most other cereals, but is difficult and thus expensive to harvest. For human consumption it is steamed or boiled for savoury dishes and puddings, or 'popped' in hot, deep oil.

amethyst a hard, crystalline, transparent QUARTZ, tinted reddish-purple or bluish-violet probably by oxide of MANGANESE. As a SEMI-PRECIOUS STONE it is much used in jewellery.

ametoecious (monoxenic) parasite a PARASITE which needs a specific host in order to live. METOECIOUS.

ammonia NH_3, a gaseous compound of NITROGEN and HYDROGEN, soluble in water, producing an ALKALINE solution, widely used in industry, and as a fertilizer directly or as the basis of fertilizer compounds. AMMONIUM NITRATE.

ammonium chloride NH_4Cl, a volatile, white crystalline salt, used in dry cells.

ammonium nitrate NH_4NO_3, a crystalline salt used in herbicides, insecticides, explosives; and as a fertilizer (AMMONIA). Excessive use as a fertilizer may lead to so much nitrate in the RUN OFF from the land that POLLUTION and EUTROPHICATION occur.

ammonium sulphate $(NH_4)_2SO_4$, a white, crystalline salt, a by-product of the coal gas industry, produced when the AMMONIA obtained from the distillation of coal is put into sulphuric acid. It is valued as a nitrogenous fertilizer.

amorphous *adj.* formless **1.** in geology, unstratified **2.** of a mineral, noncrystalline.

ampere A the basic SI unit of ELECTRICITY, defined from a NEWTON and a METRE.

amphibian a member of Amphibia, a class of VERTEBRATE animals which spend the early stages of their life cycle in water and the adult stage in aquatic or damp habitats (e.g. frogs, toads, salamanders).

amphibole a mineral in the group of FERROMAGNESIAN silicates, occurring in many IGNEOUS and METAMORPHIC rocks, HORNBLENDE being the most common. JADE.

amphidromic point a NODE-1 at the centre of an AMPHIDROMIC SYSTEM. It is theoretically possible for an amphidromic point to be located on land, where it is termed a degenerate amphidromic point. BINODAL TIDAL UNIT.

amphidromic system a unit area in the ocean in the OSCILLATORY WAVE THEORY OF TIDES. AMPHIDROMIC POINT.

amygdale a VESICLE-1 which has been filled with a SECONDARY MINERAL.

anabatic *adj.* applied to air moving upward, due to CONVECTION, as opposed to katabatic, moving downward (KATABATIC WIND), e.g. applied to a local breeze which by day blows up a warm valley slope in mountain regions, the air on the slope being heated more rapidly and to a greater extent than air lying above the valley floor, thus creating a CONVECTION CURRENT as the cooler valley air moves up to replace the rising warmer, lighter air. If one side of the valley is warmer than the other the circulation is asymmetric. G.

anabranch a branch stream which leaves a river and re-enters it lower down. A term now rarely used. G.

anaclinal *adj.* contrary to the DIP of the surface rocks, applied particularly to rivers or their valleys which run in a direction opposite to that of the dip of the rocks which they cross. The opposite is CATACLINAL. The terms anaclinal and cataclinal are now rarely used. G.

anaerobe an organism able to live in an absence of air or FREE OXYGEN. AEROBE, ANAEROBIC.

anaerobic *adj.* living or active only in the absence of FREE OXYGEN. AEROBE, AEROBIC, ANAEROBE.

anaerobic decomposition organic disintegration in the absence of air.

anaerobic respiration the process by

which organisms, without taking in gaseous or dissolved oxygen (FREE OXYGEN), release energy by the chemical breakdown of food substances, e.g. the breakdown of sugar to ethyl alcohol and carbon dioxide by yeast (FERMEN-TATION). Many organisms are able to switch to anaerobic respiration when lack of oxygen inhibits AEROBIC RESPI-RATION; but others never take in FREE OXYGEN. RESPIRATION.

anafront a cold FRONT in which warm air is in the main rising over a wedge of cold air.

anaglyph a device to give a three-dimensional effect to a two-dimensional image. The image is printed twice, slightly offset, in complementary colours (normally red and green, as these are visually of equal density). When viewed through binocular filters of the same colours an illusion of a third dimension is created. The device is used particularly for maps and photographs to enhance the illustration of land forms and relief. G.

analemma a graph showing the daily DECLINATION of the sun on the vertical scale and the EQUATION OF TIME on the horizontal. It has a central axis, to the left of which the sun time fast is plotted, to the right the sun slow. G.

analogue 1. a thing that is similar to, corresponds in some respect to, some other thing, the implication being that the likeness is systematic or structured; e.g. in long-range WEATHER FORECAST-ING predictions of future repeating patterns of weather can be made by analysing and comparing synoptic patterns (SYNOPSIS) and sequences of weather already experienced 2. in biology, an organ of an animal or plant which is of different origin but which functions in a way similar to that of a corresponding organ in another animal or plant.

analogue model (American analog model) a MODEL-4 in which selected phenomena are represented by similar but different phenomena, e.g. clusters of people are represented by clusters of points. ANALOGUE THEORY, ICONIC MODEL, SYMBOLIC MODEL. G.

analogue theory a theory of MODEL-4 building which covers the selection of certain ELEMENTS-2 from actual observation in the real world and the transformation of those elements into a simplified, structured representation of a particular system. ANALOGUE, AN-ALOGUE MODEL.

analysis the breaking-down of a complex whole into its various parts. SYN-THESIS.

analysis of variance VARIANCE ANALY-SIS.

anamorphism KATAMORPHISM.

anarchism (Greek anarchos, without government) a doctrine, philosophy or political theory which maintains that the freedom of the individual should be absolute, that political authority (government and law), being unnecessary and undesirable, should be swept away and replaced by voluntary co-operation. The attitude to the possession of property varies. Individualistic anarchists consider property in material things should be vested in each individual; communist anarchists consider, generally, that property should be administered by voluntary groups.

anastomosis 1. in biology, the inter-communication between vessels or channels in plants and animals by means of connecting branches 2. in geomorphology, applied to rivers which divide and reunite continuously, producing a net-like mass of branches (BRAIDED RIVER COURSE), usually caused by excessive deposition of alluvial material in the stream. G.

anataxis a form of METAMORPHISM in which the structure of the original rock is totally destroyed and there is partial melting.

anchorage 1. a place where ships may ride at anchor 2. the fee payable for the use of such a facility.

anatomy 1. the scientific study of the structure of plants and animals, especially of the bodies and body parts of animals **2.** the structure of an organ **3.** the science of dissection.

anchor ice sometimes termed bottom ice, submerged ice which is attached to the underlying bed of the ocean, lake or stream when the water above and/or around it is not frozen. G.

andesite a pale grey, fine-grained IG-NEOUS ROCK of INTERMEDIATE composition, 52 to 65 per cent SILICA, composed essentially of a PLAGIO-CLASE FELDSPAR, with a PYROXENE, HORNBLENDE or BIOTITE, and some glassy base. It is usually EXTRUSIVE, but also occurs in small INTRUSIONS; and it forms new continental crust at SUBDUCTION ZONES. ANDESITE LINE, PLATE TECTONICS.

andesite line a boundary passing through the Pacific Ocean from Alaska, Japan, Marianas, Bismarck Archipelago to Fiji, Tonga and New Zealand, separating two PETROGRAPHIC PROVINCES, the one to the west of the line being predominantly andesitic (AN-DESITE) and intermediate, with rocks 52 to 65 per cent SILICA; the one to the east being basaltic (BASALT) and basic, with rocks less than 52 per cent silica. ATLANTIC SUITE, PACIFIC SUITE. G.

andosol, ando-soil in soil science, a dark coloured volcanic soil occurring in Japan. G.

anecumene, anekumene, anoecumene, anoekumene, anokumene the part of the earth's surface which is uninhabited, or only temporarily inhabited, by people. ECUMENE, NEGATIVE AREA. G.

anemometer an instrument for measuring and recording the strength and direction of wind. There are two types in general use. The simpler is known as a pressure plate anemometer, in which a thin, small sheet of metal is suspended from a metal arm in such a way that if it is placed face (at right angles) to the wind the bottom edge is blown upward. The angle of its position is noted, and the speed (velocity) of the wind, in kph or mph, read off from tables. In the cup anemometer three or four cups are set at the ends of horizontal arms that cross and pivot at the top of a vertical rotating shaft. The wind rotates the arms, and their movement turns the shaft, to which a meter is attached to record the wind velocity in metres per second, or mph, or kph. A more elaborate, self-recording anemometer, known as an anemograph, continuously traces and records. The most common type of this uses the Dines tube, the opening of which is made as a vane, always faced to the wind. The ensuing pressure passes down the tube to a float that carries a pen which traces wind velocity on paper surrounding a rotating drum.

aneroid barometer BAROMETER.

Angerdorf (German, a green village) a village built around a central, usually oblong, village green. GREEN VILLAGE. G.

angiosperm a seed-bearing plant of the class Angiospermae, producing seeds enclosed in an ovary (unlike a GYM-NOSPERM). The angiosperms succeeded the gymnosperms as dominant land plants, having gained ascendancy by the end of the CRETACEOUS period. Most seed-bearing plants today are angiosperms.

angle of repose, angle of rest the steepest slope at which a mass of downward-moving unconsolidated rock debris remains stationary. REPOSE SLOPE.

Angora goat a breed of domestic goat, native to Turkey (taking its name from the town, Angora, the ancient name for Ankara), introduced to South Africa. Its long, fine, silky and soft coat (mohair) is used in the manufacture of YARN and fine textiles. The soft, fluffy fabric or yarn known as angora is made from the silky hair of the Angora goat or of the Angora rabbit, now com-

monly combined with SYNTHETIC
FIBRES. C.

angular distance the distance between
two points in terms of the angle they
subtend at a third point.

angular unconformity in geology, an
unconformity (UNCOMFORMABLE) in
which the older set of bedded rocks dip
at a different angle (usually a steeper
angle) from that of the younger, overlying strata.

angulate drainage a modified TRELLIS
DRAINAGE pattern, the tributaries meeting the main streams at acute or obtuse
angles as a result of jointing, sometimes
of faulting, of the rocks. FAULT,
JOINT. G.

anhydrite (Greek anhydros, without
water) a naturally occurring mineral,
anhydrous calcium sulphate, $CaSO_4$
(calcium sulphate lacking its water of
crystallization), found in SEDIMEN-
TARY ROCKS and used in the manufacture of AMMONIUM SULPHATE, SUL-
PHURIC ACID and plaster. GYPSUM.

anhydrous adj. having no water, specifically applied to oxides, salts and other
compounds which lack water of crystallization or combined water. IONIC
compounds with water of crystallization in their strcture can lose it, and
become anhydrous, e.g. sodium sulphate decahydrate can lose water and
become anhydrous sodium sulphate.
HYDRATE.

anicut, annicut, anaicut (Indian subcontinent: Anglo-Indian corruption
from Tamil) a weir or dam built to
regulate the flow of a river. G.

aniline an aromatic, organic chemical,
obtained from COAL TAR or (more
usually) synthesized by reduction of
nitrobenzene, forming the basis of
dyestuffs, plastics, drugs, etc.

animal **1.** any member of the kingdom
Animalia (CLASSIFICATION OF OR-
GANISMS), the typical animal being
defined as a living organism with cell
walls not composed of cellulose, with
means of independent locomotion, with

a nervous system which may or may
not be centralized, that is incapable of
PHOTOSYNTHESIS and usually needs
complex organic food substances; but
some animals (e.g.sponges) do not
conform to those criteria, and exhibit
plant-like characteristics. CARNIVORE,
DIGITIGRADE, DIVERSIVORE, FOOD
CHAIN, HERBIVORE, HOMIOTHERMY,
MACROPHAGOUS, MICROPHAGOUS,
MONOPHAGOUS, OLIGOPHAGOUS,
OMNIVORE, PARASITE, PLANKTON,
POIKILOMOTIC, POIKILOTHERMIC,
POLYPHAGOUS, PLANT, PLANTI-
GRADE, UNGULIGRADE, XYLOPHA-
GOUS, ZOOPHAGOUS **2.** any member
of Animalia other than a human being.

anion an ION which, having gained one
or more ELECTRONS, has a negative
charge. CATION.

anise Pimpinella anisum, an annual
herb of the Hemlock family, native to
lands bordering the eastern Mediterranean, but long cultivated in other
parts of Asia, Europe and North
America for the sake of its seed
(aniseed), which is used in flavouring
confectionery and cakes, etc. The oil
(mainly anethol) obtained by distilling
the seeds gives the distinctive flavour to
anisette (a liqueur), to some spirits of
eastern countries, and to pernod. It has
some medicinal properties and is used
in medical remedies. Oil of anise is also
present in star anise, Illicium verum, a
tree of the Magnolia family, native to
southwest China. G.

anisotropic ISOTROPIC.

annual adj.as applied to plants, existing
for only one year, i.e. a plant that
completes its life-cycle, from seed germination to seed production, within
one season. BIENNIAL, EPHEMERAL,
PERENNIAL.

annual growth rings the annual increase
of secondary wood seen as concentric
rings in a cross-section of the stem or
trunk of a woody plant. The small-
celled rings formed towards the end of
one growing season, e.g. late summer,

contrast with the succeeding wide-celled rings formed early in the next, e.g. spring. By counting these growth rings the age of the plant can be estimated; and by studying their width it is possible to infer the environmental (particularly the climatic) conditions prevailing at the time when the rings were being formed. DENDROCHRON-OLOGY.

annular *adj.* shaped like a ring.

annular drainage a drainage pattern in which the CONSEQUENT STREAMS radiate, as the spokes of a wheel, and the SUBSEQUENT STREAMS, eroding their valleys in the weaker strata, seem to make a series of disrupted, concentric circles as they flow into the consequent streams. Annular drainage is especially likely to occur around dissected DOMES. G.

annular eclipse an ECLIPSE in which the MOON, moving between the earth and the sun, has a ring of light around it.

anomalistic cycle of tides the tide cycle caused by the changing distance of the moon from the earth in its passage from one PERIGEE to the next (approximately 27.55 days). ANOMALISTIC MONTH.

anomalistic month the average time taken by the passing of the moon from one PERIGEE to the next (approximately 27.55 days). ANOMALISTIC CYCLE OF TIDES.

anomalistic year the period of time between two successive PERIHELIONS, i.e. 365.25964 SOLAR DAYS.

anomalous watershed a WATERSHED in a mountain region which does not run along the crest of the highest range of a mountain chain. NORMAL WATERSHED. G.

anomaly a departure from the normal, predicted, or uniform state, value, etc., applied particularly in meteorology to temperature (TEMPERATURE ANOMALY). ANTIPLEION, GRAVITY ANOMALY, ISANOMAL, MAGNETIC ANOMALY, NEGATIVE ANOMALY, PLEION, POSITIVE ANOMALY, SALINITY ANOMALY, THERMOMEION. G.

anomie, anomy (Greek, without law) **1.** a state or condition (considered by some authors in the 1950s and 1960s to bear some similarity to ALIENATION) in which an individual, having lost traditional 'moorings', is prone to disorientation or psychic disorder **2.** a state or condition of a society in which commonly accepted SOCIAL NORMS are either absent or unclear, conflicting or not integrated, leading to a lack of order and control in social life. This condition is sometimes assumed to occur in societies which are rapidly becoming urbanized and industralized. URBANIZATION, INDUSTRIALIZATION.

Antarctic, antarctic strictly an adj. applied to the south polar regions, but used as a noun (the Antarctic) to denote the region lying within the ANTARCTIC CIRCLE (66°32′S). ANTARCTICA. G.

Antarctic *adj.* of, pertaining to, or characteristic of, the landscape, climatic conditions, animal and plant life etc. occurring roughly within the ANTARCTIC. G

Antarctica the continental land area of approximately 11.5 mn sq km (4.4 mn sq mi) surrounding the SOUTH POLE. The overlying ICE CAP is so thick in some places that it is believed its base may be below sea level; and it is possible that the land itself may be in two parts (East and West Antarctica). But in the prevailing conditions the ice may be considered as a rock, part of the land area. G.

Antarctic air mass an exceedingly cold air mass, symbol AA, originating over the ANTARCTIC ocean. It should not be confused with the POLAR AIR MASS.

Antarctic circle latitude 66°32′S, in use commonly 66°30′S, where, due to the inclination of the axis of the earth, the sun does not set about 22 December

(SUMMER SOLSTICE in the southern hemisphere) and does not rise above the horizon about 21 June (WINTER SOLSTICE in the southern hemisphere). The number of days such as this, without sun in winter, increases southwards until at the SOUTH POLE six months of darkness follow six months of daylight. At any given date the conditions are the reverse of those in the ARCTIC. ARCTIC CIRCLE. G.

Antarctic convergence a well marked natural boundary in the oceans round ANTARCTICA (the position shifting a little with the seasons, being roughly parallel to the isotherm of 10°C (50°F) in the warmest month) where cold heavy waters from the south sink below warmer waters, with a consequent change of sea and air temperatures, and of animal and plant life. G.

antecedent *adj.* existing, happening, or going before in time or other sequence.

antecedent boundary a political BOUNDARY established in a lightly peopled (or unpopulated) area, i.e. before the development of settlements, language, culture, created a CULTURAL LANDSCAPE. Totally antecedent (termed pioneer by some authors) boundaries are found where the line was drawn before any settlement took place. RELICT BOUNDARY, SUBSEQUENT BOUNDARY, SUPERIMPOSED BOUNDARY. G.

antecedent drainage, antecedent river a drainage system that was in existence before the present form of the land surface was established, and which maintains its original direction despite slow, localized uplift in its path which might turn it. It succeeds in the contest because it is able to cut its way through the barrier as rapidly as the barrier is formed. INSEQUENT DRAINAGE. G.

antelope 1. a RUMINANT animal resembling a DEER, especially a member of the family Bovidae, having cylindrical, unbranched, hollow horns, native to Africa, Asia, and North America **2.**

the hide of an antelope used as leather. LIVESTOCK.

ante meridiem A.M..

anthracite a hard, lustrous variety of COAL with a high proportion (about 85 to 93 per cent) of CARBON and a low proportion of volatile matter. It does not easily ignite, but when alight it burns with an almost clear, smokeless flame, giving out great heat. G.

anthropocentric *adj.* centred on human beings, regarding human beings as the reason for the existence of the world and/or as a measure of all value. EGOCENTRISM, ETHNOCENTRISM.

anthropogenic *adj.* **1.** of or relating to ANTHROPOGENY **2.** having an origin in human activity.

anthropogenic relic RELIC DISTRIBUTION.

anthropogeny the study of the origin of human beings. ANTHROPOGONY.

anthropogony 1. the origin of human beings **2.** an enquiry into, or an explanation concerned with, the origin of human beings. ANTHROPOGENY.

anthropogeography literally the geography of human beings, a concept of human geography developed by Friederich Ratzel in two volumes, 1882 and 1891. Each volume dealt with the relationship between the natural conditions of the earth and human culture, but the first put the emphasis on the natural conditions studied in relation to human culture, the second with the geographical distribution of human communities in relation to their environment. The term is now rarely used, but if it is, it is usually confined to what may be termed the anthropological aspects of human geography. ANTHROPOLOGY. G.

anthropogeomorphology the study of landforms and processes resulting from human activities.

anthropology the scientific study of human beings as animals and as members of SOCIETY-2 (thus including ARCHAELOGY, ETHNOLOGY, ETHNOGRAPHY) and no longer restricted, as formerly, to what is now termed anthropometry (the study concerned with the physical measurements of people, especially of the skull and other bones, and the evolution of the species from anthropoid ancestors). Anthropology today thus includes the study of the origins of human beings, the distribution of various ethnic groups, the development of technical skills, religion and folk lore and other aspects of culture, as well as languages and social systems. SOCIAL SCIENCES. G.

anthropometry ANTHROPOLOGY.

anthropomorphic soil an INTRAZONAL SOIL created by human activities.

anticentre EPICENTRE.

anticlinal *adj.* of, or pertaining to, an ANTICLINE. Formerly, but no longer, used as a noun, having been superseded by ANTICLINE.

anticlinal ridge a ridge or uplifted tract of country which corresponds with an ANTICLINE in the underlying strata. G.

anticlinal valley a valley which follows an anticlinal axis. ANTICLINE, BREACHED ANTICLINE. G.

anticline an arch-shaped upfold caused by COMPRESSION in the rocks of the earth's crust, the beds dipping down and away from the central line (termed the anticlinal axis) but not always symmetrically. The term OVERFOLD is applied to an anticline pushed over on its side. ANTICLINORIUM, AXIS OF FOLD, BREACHED ANTICLINE, PITCH-3, SYNCLINE. G.

anticlinorium a great arch of STRATA composed of numerous subordinate wrinkles or ANTICLINES and SYNCLINES, the opposite of SYNCLINORIUM. G.

anti-collectivist COLLECTIVIST.

anticyclone a high pressure system (high in relation to the pressure of the surrounding air), commonly termed a high, a slow-moving atmospheric condition in which the barometric pressure is high towards the centre of the system

and progressively declines outwards. It generally appears on a weather chart as a series of concentric, closed isobars (ISO-), fairly widely spaced. Winds are light and variable, blowing round such a system in a clockwise direction in the northern hemisphere, the reverse (counterclockwise or anticlockwise) in the southern (BUYS BALLOT'S LAW). There may be a calm area at the centre, where pressure is highest. The temporary anticyclones of temperate latitudes (not to be confused with the permanent, subtropical belts of high atmospheric pressure of HORSE LATITUDES) bring generally settled weather with light winds, warm and sunny in summer, in winter cold and frosty or foggy (RADIATION FOG). DEPRESSION-3. G.

anti-dip stream a stream flowing in a direction approximately opposite to that of the DIP of the surface rocks of an area. It is frequently, but not necessarily, an obsequent stream (OBSEQUENT DRAINAGE). ANACLINAL. G.

antidune a sand wave, a transient form of ripple in the sand of the bed of a highly-loaded, swiftly-flowing stream (analogous to a SAND DUNE) which moves progressively upstream because the steeper slope, facing upstream, is constantly being built-up by the addition of sediments carried by the flow of water which washes away sediments from the gentler, downstream slope. G.

antimeridian MERIDIAN.

antimony a metal obtained mainly from the ore stibnite, Sb_2S_3, used in various ALLOYS to give hardness to softer metals, in making antimonial lead for batteries, in BABBITT METAL, in white metal, in PEWTER, as a pigment in paint, and for fire-proofing. Some compounds are used medicinally.

antinode NODE-6.

antipleion an area with a NEGATIVE ANOMALY in relation to a climatic element, especially in relation to TEMPERATURE. ISANOMAL, MEION,

PLEION, THERMOMEION, THERMO-
PLEION.

antipodal *adj.* of, or relating to, the
ANTIPODES.

antipodes (Latin from Greek, having
the feet opposite) places on the earth's
surface diametrically opposite each
other, or that part of the earth's surface
which is diametrically opposite the
observer. The British apply the term to
those lands on the earth's surface which
are approximately diametrically opposite the British Isles, i.e. Australia and
New Zealand.

anti-trade wind, anti-trade 1. a term
now applied only to the westerly winds
of the upper atmosphere (at a height
exceeding 2000 m: 6500 ft), blowing
above and in a direction (i.e. westerly)
opposite to that of the surface winds of
middle latitudes known as the TRADE
WINDS **2.** formerly, but no longer,
applied to the surface westerly winds
(now termed WESTERLIES) of middle
latitudes, also termed counter-trades
(also an obsolete term). G.

anti-urbanism a very strong dislike of,
an aversion to, large towns and cities.
URBANISM.

anvil cloud, incus cloud the flattened
head of a large convective cloud (of a
CUMULONIMBUS cloud), levelled by
winds at a height of some 6000 m
(20 000 ft), the lowest level of the
STRATOSPHERE. ICE ANVIL.

AONB Area of Outstanding Natural
Beauty, in British planning, an area
designated under the National Parks
and Access to the Countryside Act
1949 in order to 'preserve and enhance
its natural beauty'. An AONB is usually smaller than a NATIONAL PARK,
and is not managed in the same way;
but the designation gives local planning
authorities the power to operate some
development control.

apartheid (Afrikaans, separation, sep-
arateness) racial segregation, the policy
of the White government of the
Republic of South Africa, under which

White, African, Asian and Coloured
(people of mixed ethnic groups) communities are segregated on the principle
that each group should develop its own
culture and society. Apartheid led to
the withdrawal of the Republic from
the Commonwealth when other members expressed their disapproval of the
doctrine. It also led to the Republic's
setting up of nominally independent
Bantustans, i.e. African homelands
within the Republic, which are not in
fact economically viable, and are not
officially recognized internationally. G.

apatite a group of minerals occurring
in IGNEOUS ROCKS, comprising metamorphosed LIMESTONES, a naturally
occurring CALCIUM PHOSPHATE containing fluoride or chloride, forming
one of the main sources of PHOSPHO-
RUS and used particularly in making
fertilizers. PHOSPHATE, PHOSPHATE
ROCK.

aphanite, aphinite any fine-grained IG-
NEOUS ROCK or GROUND-MASS in
which the constituent grains are so
small they cannot be distinguished by
the naked eye. G.

aphanitic, aphinitic *adj.* of, relating to,
or characteristic of, APHANITE.

aphelion (Greek apo, from; helios, the
sun) that point farthest from the sun in
the ORBIT of a comet or planet. The
earth arrives at aphelion on 4 July,
when it is some 152 mn km (94.5 mn
mi) distant from the SUN. APSIS,
PERIHELION. G.

apheliotropism growth response of
plants away from the sun, an example
of negative TROPISM.

aphotic *adj.* without light.

aphotic zone the zone of the ocean
below 800 m (437 fathoms: 2625 ft)
where the intensity of sunlight is insufficient for PHOTOSYNTHESIS to take
place. DISPHOTIC, EUPHOTIC, PE-
LAGIC, PHOTIC. G.

apiary a bee farm. APICULTURE.

apiculture the rearing and keeping of
bees.

apogean tide the tidal condition characteristic of a period when the moon is at its APOGEE, so that its gravitational pull is reduced, resulting in lower high tides and higher low tides, with a tidal range smaller than is usual. PERIGEAN TIDE, TIDE.

apogee 1. the point farthest from the earth in the orbit of a planet (especially of the moon) or satellite **2.** the greatest distance of the sun from the earth when the earth is in APHELION. APSIS, PERIGEE.

apophysis in geology, a direct offshoot (e.g. a vein, dyke, tongue) from a larger INTRUSION of IGNEOUS ROCK, the connexion between the two being traceable.

apothecaries' measures and weights British measures and weights used in dispensing drugs, etc.

Appalachian orogeny, revolution (from the Appalachian Mountains, USA) a period of intense earth movements which occurred between the Permian and Triassic periods (GEOLOGICAL TIMESCALE) in North America, equivalent to but slightly later than the ARMORICAN and HERCYNIAN movements in Europe. G.

apparent dip the DIP-2 as seen by the observer (which is not necessarily the TRUE DIP, the angle of maximum slope). Thus apparent dip may be defined as the amount and direction of DIP-2. G.

apparent time local solar time, or LOCAL TIME, as shown by the apparent diurnal movement of the real sun, e.g. as by a sundial. Apparent noon is the instant when the sun's centre reaches the highest point in its apparent daily course over any place on the earth's surface (i.e. when the sun crosses the MERIDIAN of that place) and the shadows of vertical objects are at their shortest. Because the sun's orbit is elliptical and inclined towards the equator, the period between two successive diurnal crossings is not con-

stant; thus MEAN SOLAR TIME is a more useful measure. ANALEMMA, EQUATION OF TIME. G.

apple *Pyrus malus*, a small DECIDUOUS tree of temperate areas bearing firm, edible, juicy fruit, widely grown and known to have been used for human food for at least 3000 years. The fruit (a POME) is generally round and ranges in flavour from acid and sour (best for cooking) to sweet. There are many varieties in cultivation. The fruit travels well, can be dried or frozen; and is eaten raw or cooked. It is used in preserves, chutney, pickles, etc. CIDER is made from the smaller, acid varieties.

applejack in North America, an alcoholic liquor distilled from CIDER. CALVADOS.

Appleton layer F2 layer, the upper stratum of the IONOSPHERE (i.e. above the HEAVISIDE layer, the E layer) about 300 km (190 mi) above the surface of the earth, the layer in which ionization by solar radiation results in the refraction and reflection of short RADIO-WAVES back to earth. It was identified by Sir Edward Victor Appleton, 1892-1965, an English physicist and 1947 Nobel prizewinner.

applied climatology the application of the scientific study of weather conditions to the physical, economic and cultural problems of human beings in relation to their environment.

applied geography the application of geographical knowledge and techniques to the solution of economic and social problems on a local to a world scale, in such fields as town and country planning, land use, location policy, under-development, and population studies, etc. G.

applied geomorphology the application of geomorphological methods of survey and analysis to the solving of problems in the physical environment, e.g. problems arising from flood control, land reclamation, etc.

apposed glacier a GLACIER resulting

from the merging of two separate glaciers.

appropriate technology 1. the TECH-NOLOGY-3 concerned with technical processes which are suited to the degree of INDUSTRIALIZATION in an area or country **2.** used by some authors as synonymous with ALTERNATIVE TECHNOLOGY. G.

apricot *Prunus armeniaca*, a DECIDU-OUS tree with edible, pale yellow to deep orange coloured fruit, native to China, but widely cultivated through-out the warm temperate zone, especi-ally in China, Japan, the eastern Medi-terranean area, north Africa and Cali-fornia. It will survive in areas with lower temperatures (e.g. in Britain) if protected from severe frost. The fruit is eaten raw or cooked, canned, dried, made into jam, or crystallized.

apron OUTWASH APRON, FAN, or PLAIN, a spread of ALLUVIUM depos-ited by streams, especially by those from a melting glacier. G.

apsis pl. **apsides** a critical point on an ORBIT in relation to the centre of attraction, the higher apsis being the farthest point from that centre, the lower apsis the nearest to it, a straight line joining the two points being termed the line of apsides. Thus for a planet in the SOLAR SYSTEM the apsides are the APHELION and PERIHELION; and for the moon (in relation to the earth) the APOGEE and PERIGEE.

aquaculture the management of aqua-tic environments for the production of organic materials, mainly concentrated at present on PISCICULTURE, fish farming, the controlled breeding and rearing of fish (freshwater or marine) for commercial purposes. FISH FARM.

aquamarine BERYL.

aqua regia a corrosive mixture of con-centrated hydrochloric and nitric acids in the ratio of 3 or 4 to 1, which is capable of attacking metals such as GOLD and PLATINUM (NOBLE METALS), forming soluble chlorides.

aquatic *adj.* living in, growing in, or frequenting water. ESTUARINE, MARINE, RIPARIAN, TERRESTRIAL.

aqua vitae, aquavit an alcoholic drink, usually colourless, a liqueur fortified with alcohol derived from wine or cider.

aqueduct an artificial channel built to carry water from one place to another, especially such a channel supported on a series of arches and spanning a valley or river. VIADUCT.

aquiclude a porous rock which, al-though usually PERMEABLE, becomes impermeable because of the saturation of its pores by water, e.g. a SATU-RATED CLAY or SHALE. AQUIFER, AQUIFUGE, POROSITY. G.

aquifer, aquafer a water-bearing stra-tum of rock, sufficiently porous to carry the water and sufficiently coarse to release the water and permit its use. Hence an aquiferous rock, one convey-ing or yielding water. AQUICLUDE, AQUIFUGE, ARTESIAN BASIN, ARTI-FICIAL RECHARGE. G.

aquifuge an IMPERMEABLE rock stra-tum which not only obstructs the passage of water but cannot absorb it, e.g. GRANITE; an American term not in general use. G.

Arab Common Market an organization established in January 1965 by Iraq, Jordan, Syria and Egypt, with the object of achieving economic coopera-tion among all members of the ARAB LEAGUE.

Arab League a confederation of sover-eign Arab states, established in March 1945, the founder members being Egypt, Iraq, Saudi Arabia, Syria, Leba-non, Jordan and Yemen. They were joined in 1976 by Algeria, Bahrain, Kuwait, Libya, Mauritania, Morocco, Oman, Qatar, Somalia, Sudan, Tunisia, United Arab Emirates and the Yemen Peoples Democratic Republic. ARAB COMMON MARKET.

arable farming ARABLE LAND, CROP FARMING, FARMING.

arable land land capable of being ploughed. The term is sometimes used in this sense but more usually applied to agricultural land which is tilled for crops, though not necessarily each year, e.g. ploughland, market gardens, vineyards, temporary FALLOW land, and ROTATION GRASS; orchards are sometimes included, gardens attached to houses are usually excluded. In British agricultural statistics land is distinguished as arable, permanent grass, rough grazing and woodland. AGRICULTURE, CROP FARMING, FAO, FARMING. G.

aragonite (from Aragon, Spain) a mineral, one of the crystalline forms of CALCIUM CARBONATE, occurring in SEDIMENTARY ROCKS and in many shells. The crystals are ORTHORHOMBIC (i.e. they have three unequal axes of symmetry at right angles to each other). CALCITE.

arboreal *adj.* of, relating to, or living in, TREES.

arboretum a place set aside for the cultivation, display or study of trees, in some cases including rare SPECIES.

arboriculture the cultivation of TREES and SHRUBS.

arc 1. part of the circumference of a circle, or of any other curved line 2. in astronomy, part of the circular path that the sun or other celestial body appears to follow when seen from the earth, termed diurnal arc when above the horizon, nocturnal arc when below the horizon 3. in TOPOLOGY, a line, not necessarily curved, joining two NODES- 5. ARC FURNACE, ELECTRIC ARC, ISLAND ARC.

arc furnace any furnace in which heat is supplied by an ELECTRIC ARC, as distinct from one heated externally or by a fuel, as in a BLAST FURNACE.

arch a curved opening formed in a rock mass by any of several processes, e.g. by the collapse of the roof in a limestone CAVERN, by marine erosion of projecting rocks (CAVE-1, STACK), or by weathering of a relatively soft layer in a rock mass.

Archaean, Archean *adj.* 1. in common use, of or relating to the rocks of PRECAMBRIAN time, i.e. the oldest known rocks in the earth's crust 2. sometimes restricted to the rocks of the ARCHAEOZOIC era 3. specifically, by North American geologists, the earlier of two eras of Precambrian time. GEOLOGICAL TIMESCALE.

archaeo-, archeo- derived from the Greek, ancient or primitive, hence used in such terms as ARCHAEOLOGY, ARCHAEAN rocks, archaeomagnetism (PALAEOMAGNETISM). G.

archaeology, archeology the study of prehistory and of ancient history generally, based primarily on the examination of physical remains, but also (particularly in the study of prehistory) taking into account the effects of human beings on the ECOSYSTEM, possible even if material remains (e.g. ARTIFACTS) are absent or insufficient. INDUSTRIAL ARCHAEOLOGY. G.

Archaeozoic *adj.* 1. used by those geologists who divide Precambrian geological time and associated rocks into three eras, the earliest EOZOIC, followed by ARCHAEOZOIC, then by PROTEROZOIC 2. by other geologists applied to all time and rocks before the beginning of the PALAEOZOIC. GEOLOGICAL TIMESCALE.

archetype a prototype, the original MODEL-1 serving as a base from which other models are developed or to which other models conform.

Archimedes' screw a simple device used in PERENNIAL IRRIGATION. A spiral screw revolves inside a close-fitting sleeve, the lower end of which is put in water at a low angle, the screw being turned at the higher end so that the water travels up the screw to flow out at that, higher, end.

archipelago (Italian, the original sea) applied initially to the Aegean Sea which is studded with islands, then to any sea studded with islands, now applied solely to a group of islands.

Arctic, arctic (belonging to Arctos, the constellation of the Bear) strictly an adj. applied to the north polar regions but used as a noun (the Arctic) to denote the region lying within the ARCTIC CIRCLE, or to the landscape, climatic conditions, animal and plant life found roughly within that area; and often used loosely just to mean very cold. G.

Arctic air mass an exceedingly cold air mass, symbol A, originating over the Arctic Ocean. It should not be confused with the POLAR AIR MASS. ARCTIC FRONT.

Arctic circle latitude 66°32′N, in use commonly 66°30′N, where, due to the inclination of the axis of the earth, the sun in the northern hemisphere does not set about 21 June (summer SOLSTICE) and does not rise above the horizon about 22 December (winter solstice). The number of days such as this, without sun in winter, increases northwards until at the NORTH POLE six months of darkness follow six months of daylight. At any given date the conditions are the reverse of those in the ANTARCTIC. G.

Arctic front a not very active, almost permanent frontal zone (FRONT), lying to the north of the POLAR FRONT, in which cold air from the ARCTIC in the northern hemisphere meets less cold air in latitudes 50°N to 60°N. Its inactivity is due to the fact that the temperature difference between the ARCTIC AIR MASS and the POLAR AIR MASS is small.

Arctic prairies TUNDRA.

Arctic smoke fog occurring when icy-cold air from the land passes over warmer air lying over water and causes the water in the air to condense.

Arctogea ZOOGEOGRAPHICAL REGION.

arcuate delta a fan-shaped DELTA with the rounded outer margin, the arc of the fan, spreading into the sea, e.g. the Nile delta. G.

arcus cloud a dense, dark horizontal CLOUD forming an arch ahead of a CUMULONIMBUS.

area **1.** a part, space, tract or region of the earth's surface, of any size, a term used loosely or sometimes precisely (e.g. in British planning law, an Area of Outstanding Natural Beauty: AONB) **2.** a sunken, small piece of enclosed land, a yard or court adjoining and giving access to the basement of a dwelling **3.** in measures, the extent in two dimensional space, i.e. the extent of a surface contained within given limits, calculated by the use of any of various formulae, e.g. for a square, circle, ellipse, etc. **4.** a sphere of operation, e.g. sterling area **5.** a mental image of extent, possibilities, or range.

areal differentiation the varied nature of the earth's surface, apparent in the character, pattern, and interrelationship of relief, climate, soil, vegetation, land use, population distribution, and so on, which together produce a mosaic of dissimilar units. Areal differentiation is closely connected to the difficulty of defining REGIONS. DEVIATION-2, ECONOMIC GEOGRAPHY, REGIONAL GEOGRAPHY. G.

Area of Outstanding Natural Beauty AONB.

area strip mining (American) a form of STRIP MINING common in regions where a coal seam lies nearly horizontally under the land surface.

areg (Arabic, pl. of ERG). G.

areic, aretic *adj.* without flow, applied to desert regions where the rainfall (if any) is so slight that it sinks into the ground, or evaporates, so that there are no flowing streams. ENDOREIC, EXOREIC. G.

arena a term applied to a shallow basin in east Africa, the resistant rocks of the rim enclosing the central area where less resistant rocks have been eroded. G.

arenaceous (Latin arena, open central area of amphitheatre, strewn with sand) *adj.* applied to **1.** a rock of sandy texture, consisting mainly of grains of SAND (PSAMMITIC) which may be loose or cemented; or a rock basin with a sandy floor. ARGILLACEOUS **2.** a plant growing in sand or on sandy soil. G.

arenization a WEATHERING process in solid rock whereby a deep, sandy REGOLITH is formed. It is common in humid tropical regions where high temperatures and the presence of ground water provide conditions ideal for CORROSION.

arête (French) a sharp, narrow, steep-sided mountain ridge, especially one formed when two CIRQUES have been developed back to back. An alternative term, used especially in the USA, is comb-ridge. ARRIS. G.

aretic drainage AREIC. G.

argillaceous (Latin argilla, clay) *adj.* clayey, containing clay, or clay-like in composition or texture, applied especially to a SEDIMENTARY ROCK in which clay minerals (CLAY-2) predominate. ARENACEOUS, PELITE, PELITIC. G.

argillic brown earths BROWN SOILS.

argillic horizon a SOIL HORIZON (commonly the B horizon) in which CLAY minerals have accumulated by ILLUVIATION.

argillite a fine-grained SEDIMENTARY ROCK composed mainly of clay minerals (CLAY-2), e.g. MUDSTONE.

argol, argal, argul (China: Mongolia) dried cattle dung used as fuel, especially in the STEPPE of central Asia. G.

arid *adj.* dry, parched, lacking moisture, applied especially to climate or land (ARID LAND), the main factors being insufficient rainfall and a rate of EVAPORATION exceeding that of the PRECIPITATION. G.

aridisols, aridosols in SOIL CLASSIFICATION, USA, an order of soils with generally mineral profiles, with or without ARGILLIC HORIZONS, in which an accumulation of soluble salts or carbonates is common. Such soils are characteristic of DESERT and ARID regions.

aridity the state or quality of being ARID. In attempts to give arid and aridity a precise definition in relation to climate or land, several authors have suggested an index of aridity, based mainly on values of TEMPERATURE-2 and PRECIPITATION, including its REGIME. RAIN FACTOR. G.

arid land, arid zone land where, due to insufficient PRECIPITATION and a rate of EVAPORATION exceeding that of the precipitation, there is little or no natural vegetation, and agriculture is possible only with the aid of IRRIGATION.

Aries, First Point of RIGHT ASCENSION, ZODIAC.

aristocracy (Greek aristos, best; kratia, rule) **1.** government by a small, privileged, generally hereditary, class, considered to be the ELITE and designated as best at ruling **2.** a state so governed **3.** the members of such a ruling class. AUTOCRACY, AUTARCHY, DEMOCRACY, GERONTOCRACY, HAGIOCRACY, MONARCHY, OLIGARCHY, PLUTOCRACY, REPUBLIC, THEOCRACY.

arithmetical scale FLOW-LINE MAP.

arithmetic equality an equal share for each one. PROPORTIONAL EQUALITY.

arithmetic mean commonly termed the MEAN, an AVERAGE, calculated by adding together several quantities and dividing by the number of those quantities, e.g. the total of the values of a variable for all the observations in a data set divided by the total number of observations. CENTRAL TENDENCY, COEFFICIENT OF VARIATION, MEDIAN, MODE.

arithmetic progression a series of numbers each of which is greater or smaller than the one before it by the same amount, e.g. 2, 4, 6, 8 or 8, 6, 4, 2. GEOMETRIC PROGRESSION, LINEAR GROWTH.

arkose a coarse-grained SANDSTONE or GRIT derived from the swift breaking down of GRANITE or GNEISS and characterized by a considerable proportion of fragments of FELDSPAR, which have been altered little, if at all, by WEATHERING. G.

Armorican (from Armorica, the old name for Brittany) *adj.* applied to the great earth-building movements and mountain-building period (corresponding to the VARISCAN of central Europe), starting in Carboniferous times and continuing into Permian, which created mountain chains (sometimes considered as part of the ALTAIDES) in Brittany, southwest England and Ireland and throughout northern Europe. The remnants of the mountain masses form Armorican massifs. Through large areas the folds were east to west, which is accordingly termed the Armorican trend; where later folding took place along the same lines, the term Armoricanoid is used. HERCYNIAN, from the Harz Mountains of Germany, is used by some authors as synonymous with Armorican; but others consider that the Armorican mountains constitute the western sector of the Hercynian system, the Variscan the eastern. G.

arris, arridge (Lake District, England) a local term for ARETE.

arrack (Arabic araq, juice) a spirit distilled from toddy, the fermented sap of the COCONUT palm or PALMYRA palm, or from fermented RICE etc.

array in statistics, **1.** an explicit display of a set of observations, all the values in a set of data displayed together (MATRIX). A simple form is a table in which the VARIABLE is divided into a number of CATEGORIES-3, the number of observations in each category being counted. The number of observations or the number of times a variable occurs is termed the FREQUENCY. Thus a frequency array is an array of frequencies according to variate values, i.e. a FREQUENCY DISTRIBUTION **2.** the individual frequency distributions

which form the separate rows and columns of a bivariate frequency table (FREQUENCY DISTRIBUTION).

arrowroot a fine-grained, easily digested starch flour made from some starchy edible root. True arrowroot is made from the swollen rhizomes of *Maranta arundinacea*, a tropical PERENNIAL, HERBACEOUS plant, cultivated particularly in St Vincent, West Indies, which provides most of the world's supply.

arroyo (Spanish) a periodically dry watercourse. NALA, WADI. G.

arsenic a semi-metallic element of which various compounds are used in medicine and in the manufacture of herbicides and insecticides. Arsenopyrite, FeAsS, occurring in HYDROTHERMAL deposits, is the principal ore of arsenic. CHALCOPHILE.

artesian basin a synclinal (SYNCLINE) basin in the earth's crust, in some places very large (e.g. in Australia), in which one or more PERMEABLE, POROUS water-bearing strata or AQUIFERS lie between IMPERMEABLE strata, the whole being folded to form the synclinal basin. ARTESIAN WELL, RECHARGE. G.

artesian well 1. a perpendicular boring sunk through the upper IMPERMEABLE layer in an ARTESIAN BASIN to reach the AQUIFER. If the outlet of the well lies at a level lower than the WATER TABLE of the aquifer at the margins of the basin, water will rise in the well under HYDROSTATIC PRESSURE (PIEZOMETRIC LEVEL). The name is derived from Artois, France, where such wells were sunk in the twelfth century **2.** a term loosely applied to any deep well in which water rises under pressure, but not necessarily to the surface. Some authors distinguish such wells as sub-artesian, the water having to be pumped to the surface, as distinct from the true artesian, naturally flowing, wells. G.

artichoke 1. globe artichoke, *Cynara*

scolymus, a PERENNIAL plant related to the thistle, with edible flower heads, native to the Mediterranean region **2.** Jerusalem artichoke, *Helianthus tuberosus*, native to North America, related to the sunflower. It produces edible, sweet, underground stem-tubers, irregularly rounded and knobbly. Both types are cooked before being eaten.

articulacy the quality or state of being able to speak fluently and intelligibly in well-arranged words pronounced clearly and distinctly. GRAPHICACY, LITERACY, NUMERACY.

artifact, artefact a simple object made by human workmanship, especially one related to prehistoric time, e.g. a stone tool. DATING, HUXLEY'S MODEL. G.

artificial *adj.* **1.** made by a person, by human activity, as opposed to NATURAL **2.** synthetic **3.** something made to imitate a natural product.

artificial fibre FIBRE-2, SYNTHETIC FIBRE.

artificial radioactivity, induced radioactivity a process in which radioactive ISOTOPES are prepared by bombarding stable isotopes with particles of high energy. RADIOACTIVITY.

artificial recharge the introduction of surface water into an AQUIFER by means of recharge wells. RECHARGE.

ås, ose (Swedish, pl. åsar) a term superseded by ESKER in English works. G.

asbestos certain fibrous silicate minerals, mainly calcium magnesium silicate, which split into fine silky hairs and occur usually in veins in the earth's crust. The hairs are incombustible, resistant to electricity, have a low conductivity, will not corrode or decay, and can be woven (chrysotile, a SERPENTINE mineral, is commercially the most important). Used in manufacturing when these qualities are needed, e.g. in flameproof fabrics, building materials, heat insulation in machinery, asbestos is falling from favour because some types are deemed to be a danger to health.

ascension of a celestial body, RIGHT ASCENSION.

ASEAN Association of South East Asian Nations, formed August 1967 with the aim of developing political and economic cooperation among five southeast Asian states: Indonesia, Malaysia, Philippines, Singapore, Thailand.

ash **1.** the powdery residue left after a substance has been burnt **2.** the material known as VOLCANIC ASH ejected from the crater of a VOLCANO in eruption. This is not true ash: it is not produced by burning, but consists of fine particles (the majority of which measure less than 4 mm in their long dimension) of pulverized lava.

ash cone the volcanic cone formed by ASH-2 ejected during eruption. CINDER CONE. G.

ash flow NUEE ARDENTE, PELEAN ERUPTION.

ashlar **1.** a square block of building stone, hewn and squared by a mason **2.** masonry consisting of such stones **3.** a thin stone slab cut from such a stone and used to face (i.e. to form a facing on) a wall of inferior material, e.g. rubble, or inferior bricks, etc.

asparagus *Asparagus officinalis*, a PERENNIAL plant, native to the eastern Mediterranean region, cultivated there and in cooler areas for its succulent young shoots, eaten when cooked.

aspect the direction in which a thing faces, particularly applied to slopes in relation to the sun on account of its affect on settlement and plant growth. ADRET, UBAC. G.

asphalt a viscous, brownish-black bituminous substance, a mixture of HYDROCARBONS, occurring naturally (e.g. in Trinidad Pitch Lake, or the Athabaska tar sands, Canada) or as an industrial residue in the refining of some varieties of petroleum. It is used mainly as a surfacing material for roads, a waterproofing material for flat roofs, and in some fungicides and paints.

aspiration **1.** ambition, or an ambition, a strong desire for **2.** the object of such a desire.

aspiration region the area defined by the minimum requirements an individual is willing to accept and the maximum aspired to in seeking to meet a specific need or specific needs in a search, e.g. for new housing. SEARCH BEHAVIOUR, SEARCH SPACE.

ass, donkey a mammalian quadruped of the genus *Equus*, especially *E. asinus*, the donkey, domesticated since ancient times, sure-footed, used as a draught and transport animal in rough country. The ass needs less food, of coarser quality, than that required by the horse. A she-ass mated with a male horse (stallion) produces a hybrid termed a hinny. A male ass (jackass) mated with a mare produces a (usually sterile) hybrid termed a mule. A hinny is smaller and more docile than a mule, but not so strong, and thus is not such a good pack animal.

assart a piece of land cleared of trees or bushes so as to provide land for cultivation, a term applied especially to early and medieval clearing in England. G.

assemblage **1.** a number of persons or things gathered together **2.** in archaeology, a group of ARTIFACTS found close together in such a way as to suggest that they were likely to have been used at the same time.

assembly the smallest unit in the hierarchy of plant and animal communities.

asset stripping a process in which companies or firms with assets (especially land and buildings) which are undervalued (a fact reflected in the share value or in the purchase price of the enterprise) are bought for the sole purpose of giving the buyer the highest possible profit on the disposal of those assets. The buyer achieves this by closure and sale or redevelopment of the most valuable assets. The remainder of the company or firm, being of little interest to the buyer (the asset stripper), is disposed of as conveniently as possible.

assimilation in society, the process by which various groups in society merge, lose their distinguishing characteristics, their separate identities, are absorbed one with another in a 'melting pot', and become culturally HOMOGENEOUS-1. ACCULTURATION, ACCOMMODATION, CULTURE CONTACT, INTEGRATION, PLURAL SOCIETY.

association **1.** (with) in general, the act of becoming allied to, joining, or combining; or the condition of being allied to, combined, or joined with someone or something **2.** of plants, a stable, climax plant community, of definite floristic composition, usually with one or more DOMINANT SPECIES living together in close interdependence and related to certain physical conditions, named according to the dominant species, e.g. heath association. This original definition, applied to the largest natural group recognized, is not now everywhere accepted, being applied also to very small natural vegetation units in Europe, especially in Sweden. COMMUNITY **3.** of people, an organized group of individuals united by a common interest or objective. G.

Association of South East Asian Nations ASEAN.

asteroid a planetoid, one of some 1500 or more small PLANETS of unknown origin in orbit in the SOLAR SYSTEM, mainly between Mars and Jupiter. Ceres is the largest asteroid, with a diameter of some 770 km (480 mi).

asthenosphere (erroneously spelled aesthenosphere) a weak sphere, the zone of hot rock, believed to be in a plastic condition, underlying the solid LITHOSPHERE (the earth's CRUST), the top of the zone lying some 70 to 150 km (45 to 95 mi), the bottom some 200 to 360 km (125 to 225 mi), below the earth's surface. It is sometimes termed the soft layer of mantle, or the low velocity zone (LVZ), the latter because earthquake shock waves travel in it at reduced speed. Horizontal currents in this zone may be associated with plate movements. PLATE TECTONICS, GUTENBERG CHANNEL, GUTENBERG DISCONTINUITY. G.

astrakhan, astrachan the skin of the newborn lambs of KARAKUL or astrakhan sheep. PERSIAN LAMB.

astrolabe an historic instrument known to the ancient Greeks, formerly used to measure the altitude of any heavenly body or to fix LATITUDE by observing the apparent passage of the sun across the MERIDIAN at midday. The forerunner of the SEXTANT, it consisted of a metal circular ring or flat plate on which the CELESTIAL SPHERE was projected, fitted with a sight, and on the edge marked with degrees.

astronaut one who travels outside the earth's atmosphere, a space traveller. The alternative term, cosmonaut, is preferred in the USSR.

astronomical twilight TWILIGHT.

astronomical year SOLAR YEAR.

astronomic unit the unit of distance used in measuring distance within the solar system, being equal to 149 685 270 km (93 012 657 mi), the mean radius of the earth's orbit around the sun.

astronomy the scientific study of the heavenly bodies, the oldest exact science.

astrophysics the branch of astronomy concerned with the scientific study of the physical and chemical nature of heavenly bodies and the gases and particles between them, their origin and evolution, by applying the laws of physics established on earth to the information received from space by COSMIC RAYS and electromagnetic radiation. ELECTROMAGNETIC WAVE.

asymmetric, asymmetrical *adj.* showing lack of symmetry. ASSYMETRY.

asymmetrical fold a FOLD with one limb dipping away from the axis (AXIS OF FOLD) more steeply than the other. OVERFOLD.

asymmetrical valley a valley with one side sloping more steeply than the other.

asymmetry without SYMMETRY, the state of lacking the capability of being divided into two or more exactly similar and equal parts.

asymptote in mathematics, a line which approaches but never touches a plane curve within any finite distance. At infinity it would touch but not intersect the curve.

asymptotic, asymptotical *adj.* of, pertaining to, or having the characteristics of, an ASYMPTOTE.

atavism (Latin *atavus*, an ancestor) **1.** in biology, the reappearance of, or the tendency to reproduce, ancestral features (including disease) missing from intermediate generations in plants or animals; or **2.** a resemblance in appearance to remote ancestors rather than to parents **3.** in human beings, a reversion to primitive instincts, or to a primitive way of life. G.

Atlantic polar front the POLAR FRONT between the polar maritime and tropical maritime (TROPICAL AIR MASS) air masses over the North Atlantic ocean.

Atlantic stage, of climate the sudden development of a mild moist phase following the dry cold Boreal phase during the retreat of the Great ICE AGE, lasting from about 5500 to 3000 BC, sometimes termed the Megathermal Period. In Britain a general rise of about 3 m (10 ft) in sea level resulting from the meeting of the ice sheets, led to the formation of the Strait of Dover c.5000 BC, to the growth of mixed oak forest, and to the formation of peat. FLANDRIAN, PRE-BOREAL, SUB-BOREAL.

Atlantic suite a PETROGRAPHIC PROVINCE distinguished by the ATLANTIC TYPE OF COASTLINE, areas of BLOCK FAULTING and rocks rich in ALKALIS (sodium and potassium). ANDESITE

LINE, PACIFIC SUITE, SPILITIC SUITE. G.

Atlantic type of coastline a type of coastline developed where the main folds and trend lines, often faulted, run at right angles or obliquely to the coastline, in contrast to the Pacific type (CONCORDANT COAST, LONGITUDINAL COAST) where they run parallel to the coastline. G.

atlas a uniform collection of maps bound in a volume or (rare use) a bound volume of pictures, engravings, etc., apparently so named because some early collections (notably one by Mercator in the sixteenth century) showed on the title page a representation of a member of the older family of Greek gods, named Atlas in Latin, supporting the heavens on his shoulders. The term atlas first appeared on the general title page of Mercator's *Atlas* 1595. G.

atmosphere 1. the air or mixture of gases, roughly (by volume) 20 per cent OXYGEN, 79 per cent NITROGEN, with 0.03 per cent CARBON DIOXIDE and traces of argon, krypton, xenon, neon, helium, as well as water vapour, ammonia, ozone, organic matter, some salts and solid particles in suspension, which envelops the earth (ASTHENOSPHERE, HYDROSPHERE, LITHOSPHERE). Various concentric layers are identified, based on such criteria as rate of temperature change, composition, electrical nature. The lower atmosphere is the TROPOSPHERE, then comes the TROPOPAUSE, above which is the upper atmosphere or STRATOSPHERE, passing into the STRATOPAUSE, MESOSPHERE, MESOPAUSE, IONOSPHERE (thermosphere), with the upper zone, the EXOSPHERE and the MAGNETOSPHERE **2.** a unit of pressure, a measure of air pressure, one atmosphere being equated with the pressure exerted by the weight of a column of 760 mm (29.92 in) of MERCURY at 0°C (32°F) under standard gravity, at sea-level, or to a weight of air of 1033.3 gm per sq

cm (14.66 lb per sq in), the average pressure over the earth's surface under those conditions. ATMOSPHERIC PRESSURE. G.

atmospheric cell 1. a large, three dimensional air mass of HIGH or LOW atmospheric pressure created by the disturbance of the planetary pressure system which is due mainly to the unequal solar heating of the irregularly distributed continents and oceans **2.** a vertical circulation cell in the TROPOPAUSE associated with meridional circulation (MERIDIONAL FLOW) of the atmosphere, e.g. the HADLEY CELL.

atmospheric circulation the general circulation of the ATMOSPHERE-1, covering all the movement of the air enveloping the earth in the lower and upper atmosphere resulting from the unequal heating of the earth and its atmosphere brought about mainly by the imbalance of SOLAR RADIATION between lower and higher latitudes, the differences in energy distribution in the atmosphere and the tendency for these differences to be smoothed out, and the angular momentum of the earth and its atmosphere.

atmospheric instability a state of the ATMOSPHERE-1 occurring when a body of air with a LAPSE RATE higher than that of the DRY ADIABATIC LAPSE RATE and warmer than overlying air, rises and expands. The adiabatic cooling of such moist air as it ascends disturbs the atmosphere, frequently gives rise to deep clouds, and leads to PRECIPITATION. ADIABATIC, SATURATED ADIABATIC LAPSE RATE.

atmospheric pressure the pressure exerted by the weight of the ATMOSPHERE-1 on the earth's surface, decreasing with height above sea-level and varying with weather conditions. The unit of measurement used is the ATMOSPHERE-2, expressed in millibars (mb): 1 atmosphere = 1013.25 mb (1000 mb = 1 BAR = 1 mn DYNES per sq cm), indicated by a BAROMETER.

atmospheric weathering WEATHERING.

atmospheric window the parts of long-wave INFRA-RED radiation within which TERRESTRIAL RADIATION escapes to outer space because, cloud cover being thin or non-existent, it is not absorbed by WATER VAPOUR and CARBON DIOXIDE present in clouds in the atmosphere. WINDOW-2.

atoll a circular or almost circular CORAL REEF, the crest lying at a low height above sea-level, sometimes interrupted, enclosing a central LAGOON. Atolls are most common in the central and west Pacific ocean. G.

atollon a small ATOLL on the margin of a larger one. G.

atom the smallest particle of an ELEMENT-6 which exhibits the properties of that element, consisting of a NUCLEUS-3 and extra-nuclear ELECTRONS. FREE RADICAL, MOLECULE, NUCLEAR ENERGY, PROTON.

atomic energy NUCLEAR ENERGY.

atropine a crystalline ALKALOID present in the berries of *Atropa Belladona*, a plant of CALCAREOUS soil. It is used medicinally to dilate the pupil of the eye and to relieve muscular spasm. PILOCARPINE.

attached dune a SAND DUNE formed around an obstacle lying in the path of blown sand in ARID LANDS, the obstacle forming the nucleus that 'fixes' the dune.

attar of roses (Persian itr, perfume, from Arabic) an ESSENTIAL OIL with a sweet fragrance, distilled from the petals of the rose *Rosa damascena*.

attitude 1. a state of mind, a relatively enduring tendency to perceive, feel or behave towards certain people or events in a particular manner. Attitudes play an important part in PERCEPTION, and they affect PREFERENCES and choice of GOALS 2. in geology, the disposition of a rock STRATUM, horizontal or tilted, e.g. the relationship of a BEDDING PLANE to the horizontal plane in terms of its DIP and STRIKE.

attribute 1. a quality proper to, or characteristic of, a person or thing 2. in statistics, a QUALITATIVE VARIABLE which can take only certain fixed values, sometimes termed a DISCRETE VARIABLE. If the CATEGORIES-3 cannot be ordered (e.g. in the case of religious affiliation), the variables are measured at nominal value; if they can be ordered (e.g. in the case of age groups), the variables are measured at ordinal level, and in this case they are sometimes termed ORDERED ATTRIBUTES. Attributes which take only two values (e.g. male or female) are sometimes termed dichotomous variables. MEASUREMENT IN STATISTICS, NOMINAL SCALE, ORDINAL SCALE.

attrition the act or process of wearing away by rubbing or friction of one thing against another, each being affected, e.g. occurring when pebbles carried along by running water, by sea waves, or by wind, rub against one another. ABRASION, CORRASION, TRANSPORTATION. G.

aubergine *Solanum melongena*, eggplant, a PERENNIAL plant, usually grown as an ANNUAL, native to tropical Asia but widely grown in tropical, subtropical and warm temperate regions for its fruit, usually eaten cooked.

Aufeis (German; Russian naled, pl. naledee) surface icing, a term applied in periglacial studies to 1. ice incrustation occurring when the stream flowing underneath a frozen river in winter is too great for its channel so that HYDROSTATIC PRESSURE forces the water to permeate the ALLUVIUM on each bank, and ultimately to rise to the surface through the river ice or through the ground. This water freezes in extensive, successive sheets of ice, sometimes even 4 m (13 ft) thick, termed river icing if occurring over the river ice, ground icing if over the ground 2. ice formed when water from overflowing streams freezes on the surface of the adjoining land 3. spring water when it freezes on the land surface. G.

auger a tool for boring or drilling into the soil or rocks of the earth's surface in order to collect a core of either or both for analysis.

augite a complex calcium magnesium aluminous SILICATE, a mineral of the PYROXENE group, occurring in many IGNEOUS ROCKS. GABBRO.

aureole METAMORPHIC AUREOLE.

Aurignacian epoch (from Aurignac, a place in the Pyrenees) a PALAEO-LITHIC cultural period, characterized by the use of bone tools and well finished, finely pointed flint tools, and by cave paintings, the work of the tall, long-skulled Cro-Magnon people.

aurora australis, aurora borealis (the former in the southern, the latter in the northern hemisphere) the spectacular coloured lights, probably electro-magnetic in origin in the IONOSPHERE, seen near the horizon in the night sky in high latitudes. They are red, green and white, and they ascend in streaks and sheets, roughly in the shape of a fan. The phenomenon is known as Northern Lights in the northern hemisphere.

aūs (Indian subcontinent: Bengali) in Bangladeshj, autumn rice or rice crop, grown on the generally coarser soils of the ground higher than that producing the AMAN and depending for at least half of its growing period on pre-monsoon (March to May) rainfall. G.

Ausland (German, foreign country) specifically applied in English to that part of the UMLAND of a city lying in a foreign country. G.

austral *adj.* southern, belonging to the south.

Australite TEXTITE. G.

Autan VENT D'AUTAN.

autarchy (Greek autarchia: autos, self; archos, ruler) **1.** absolute sovereignty, despotism, the rule of an autocrat, a despot, an absolute ruler **2.** a state under such rule. AUTARKY, AUTO-CRACY. G.

autarky (Greek autarkeia: autos, self; arkeein, to suffice) economic self-suffici-ency, particularly national economic self-sufficiency as the object of national policy. Autarky should be distin-guished from AUTARCHY, but unfortu-nately autarky and autarchy are com-monly used as alternative spellings of the same word. G.

autecology the scientific study of the relationship between a single species and its environment, as opposed to SYNECOLOGY. ECOLOGY. G.

authigenesis the formation of a MIN-ERAL in situ during or after the laying down of the sediment in which it occurs.

authigenic, authigenous *adj.* formed in place, applied especially to a mineral formed in situ during or after the formation of the rock of which it forms a part, e.g. the crystals in an IGNEOUS ROCK, in contrast to ALLOGENIC, AL-LOTHIGENIC. AUTHIGENESIS, SEC-ONDARY MINERAL. G.

autobahn a German motorway, con-structed specifically for and restricted to motor traffic, usually with at least two up and two down lanes separated by a central strip, without surface crossroads, and with controlled access. The term was used in Britain until the opening of the first UK MOTORWAY in 1959 led to its gradual falling into disuse. AUTOSTRADA. G.

autochthon (Greek autochthonic, sprung from the land itself) **1.** the earliest known inhabitant of a place **2.** an animal or plant native to a place, i.e. not introduced **3.** a rock in its original place of deposition. ALLOCHTHON, ALLOCHTHONOUS, AUTOCHTHON-OUS. G.

autochthonous, autochthonal, autochthon-ic *adj.* applied **1.** in biology, to organ-isms in the soil which are not affected by the addition of organic matter to the soil (ZYMOGENOUS) **2.** in geology, to rocks (or organic material such as COAL) and their constituents formed in situ, in the place where they are now found. The strata may have been subjected to earth movements resulting in FAULTING and FOLDING but this has not moved such rocks, etc., far from their place of origin. G.

autoclastic *adj.* applied to CLASTIC or broken rocks originating in situ, especially to the CRUSH BRECCIAS found in FAULTS. G.

autocorrelation in statistics, the internal correlation between members of a series of OBSERVATIONS-3 ordered in space or time. If, in a mapped variable, a spatial pattern caused by geographical proximity is displayed (e.g. if the values for some variable form clusters in neighbouring observations) the autocorrelation is said to be spatial.

autocracy 1. government by a single, absolute ruler, i.e. by an autocrat, a despot 2. a state governed by an autocrat. ARISTOCRACY, AUTARCHY, DEMOCRACY, GERONTOCRACY, MONARCHY, OLIGARCHY, PLUTO-CRACY, REPUBLIC, THEOCRACY.

autoecious parasite a PARASITE which lives on an individual host for the whole duration of its life cycle.

autogenic, autogenous (from autogenesis, spontaneous generation) *adj.* produced by an organism itself, not by an external influence.

autogenic change a self-induced change produced by natural processes. G.

autogenic factor the direct effect of the members of a plant community on each other (e.g. by competition), or of the community on its own habitat (e.g. by deposition of humus).

autogenic succession a plant SUCCES-SION-2 produced by changes brought about by the plants themselves. ALLO-GENIC SUCCESSION.

autolysis the disintegration of dead cell tissue in plants and animals brought about by ENZYMES in the cells themselves.

autometamorphism the METAMOR-PHISM taking place in IGNEOUS mater-

ial and the surrounding COUNTRY ROCK during cooling in PNEUMATO-LYSIS and in HYDROTHERMAL PRO-CESSES. METASOMATISM.

autonomic *adj.* originating within an organism, without external stimuli, e.g. applied to plant movement resulting from internal stimuli, the opposite of PARATONIC.

autonomous *adj.* self-governing, usually applied to a state which is not a sovereign state or absolutely independent, but one which is self-governing in home affairs while being under the control of a larger (and sovereign) state. AUTONOMY.

autonomy self-government, the right of self-government. AUTONOMOUS. G.

autopiracy, auto-piracy the simplification of a river course occurring when the upper part of the river is captured by the lower (RIVER CAPTURE), e.g. as in a cut-off MEANDER. G.

autostrada (Italian, a motorway) AUTOBAHN, MOTORWAY. G.

autotroph an organism which is AUTO-TROPHIC.

autotrophic *adj.* applied to an organism capable of producing its own organic substances (i.e. without recourse to outside organic substances) from inorganic compounds, using energy from the sun (PHOTOTROPHIC) or from chemical reaction (CHEMOTROPHIC). Most CHLOROPHYLL-containing plants and some BACTERIA are autotrophic; all other organisms are HET-EROTROPHIC. FOOD CHAIN, HOLO-PHYTIC, MIXOTROPHIC, PHOTOSYN-THESIS.

autumn (American Fall) the third SEA-SON of the year in MIDLATITUDES, variously defined. Properly from the autumnal EQUINOX to the winter SOL-STICE, i.e. from 21-22 September to 21-22 December in the northern hemisphere, from 21-22 March to 21-22 June in the southern. But in the northern hemisphere popularly regarded as September, October and

November; in the southern hemisphere, February, March and April. G.

available relief local relief, the vertical distance from the original, fairly flat upland surface (i.e. from the height of a surface which is being dissected) to the local base level (the valley floors) of the dissecting streams. RELIEF. G.

avalanche a French dialect term originally applied to a large mass of snow mixed with earth, stones and ice loosened from a mountainside and falling swiftly by gravity to the valley below (SNOW-AVALANCHE, STAUB-LAWINE). The term avalanche is now usually restricted and applied to a fall of a mass of snow, ice and FIRN, being qualified if used to cover similar movements of other materials, e.g. ROCK AVALANCHE (better termed a LANDSLIDE), sand avalanche (PLINTH). G.

avalanche cone a mass of material deposited by an AVALANCHE, including not only the SNOW, ICE, or FIRN but also everything torn away and carried along by the avalanche. G.

avalanche wind the rush of air produced by and preceding an AVALANCHE in its descent.

aven (French) **1.** a vertical or inclined shaft (smaller than an ABIME) in limestone leading down, generally from the land surface, to a cave passage (GOUFFRE) **2.** in England, an enlarged vertical joint in the roof of a cave passage, narrowing upwards. G.

average 1. ARITHMETIC MEAN **2.** an undefined measure of CENTRAL TENDENCY of the values in a DATA SET or FREQUENCY DISTRIBUTION.

avian *adj.* of, or relating to, BIRDS.

avocado pear, alligator pear *Persea americana*, a tropical, small or medium-sized tree, of which there are some 400 species, native to central America, now widely cultivated in tropical and subtropical regions for its edible fruit, a DRUPE, the size and shape of a PEAR, which contains more protein than any

other fruit, and up to 30 per cent digestible fat. It is usually eaten raw.

avoirdupois a system of measuring weight in English-speaking countries based on a POUND of 16 ounces (OZ). 1 lb avoirdupois = approximately 1.215 lb troy.

avulsion a legal term applied in Britain to the sudden separation of land from the estate of one person and its addition to that of another by a change in the course of a river or change in a seashore. The land so separated continues in the original ownership.

awareness space all the locations about which a person has knowledge above a minimum level even without visiting some of them (INDIRECT CONTACT SPACE). Awareness space includes ACTIVITY SPACE, and its area enlarges as new locations are discovered and/or new information is gathered. SEARCH SPACE.

axial plane in geology, the imaginary surface dividing the limbs of a FOLD as symmetrically as possible, and passing through the axis of the fold. Different types of fold are identified by the INCLINATION-2 from the vertical of the axial plane.

axis 1. a real or imaginary line around which a thing rotates **2.** one of the reference lines in a COORDINATE system **3.** an alliance between countries which aims to ensure a common policy, e.g. AXIS POWERS.

axis of the earth the diameter between the NORTH POLE and the SOUTH POLE, tilted at an angle of about 66°30′ to the plane of the earth's ORBIT, around which the earth rotates (anticlockwise) once in every 24 hours. ORBIT OF THE EARTH, PRECESSION OF THE EQUINOXES.

axis of a fold the central line of a FOLD, the crest from which STRATA dip downwards and away in an ANTICLINE, or the central line of the lowest depth of the trough from which strata rise in opposite directions in a SYN-CLINE. AXIAL PLANE.

Axis Powers, The Axis the alliance of Germany, Italy, Hungary, Bulgaria, Romania, Finland and Japan in the Second World War. ALLIES.

ayacut (Indian subcontinent: Urdu)a term from southern India applied to the area irrigated from a weir or ANICUT. G.

ayala a local, very strong, moist wind, sometimes hot, blowing in the Central Massif, France, resembling the MARIN.

'azbeh (Arabic: Israel) a summer dwelling for those in charge of the herds and flocks sent down to graze on the plain. G.

azimuth **1.** in surveying, a BEARING read clockwise from TRUE NORTH in degrees (0° to 360°) **2.** in astronomy, the azimuth of a heavenly body, the horizontal arc expressed as the angle between the MERIDIAN plane of the observer and the vertical plane (cutting the vertical circle) passing through the heavenly body (all GREAT CIRCLES passing through the observer's ZENITH are necessarily perpendicular to the celestial horizon and are termed vertical circles), measured the nearest way from the meridian plane, i.e. in degrees (0° to 180°) eastward or westward from the Pole.

azimuth circle a GREAT CIRCLE in the heavens which passes through the ZENITH and the NADIR.

azimuth compass a magnetic COMPASS fitted with sights for measuring the azimuth angle, i.e. the BEARING.

azimuthal map projection a MAP PROJECTION in which all bearings are laid off correctly from the central point of the map, so that all points on the map are true in distance and direction from the centre. G.

azoic without life, hence applied to PRECAMBRIAN rocks which were once thought to lack organic remains. G.

azonal soil a young soil that lacks marked horizons, commonly because insufficient time has elapsed for climate and vegetation to effect a differentiation of such horizons, e.g. a recent alluvial soil. In contrast, mature soils (ZONAL SOIL) group themselves according to the great climatic zones of the earth's surface. IMMATURE SOIL, INTRAZONAL SOIL, SOIL CLASSIFICATION, SOIL PROFILE. G.

Azores high a subtropical ANTICYCLONE stationed generally over the eastern sector of the North Atlantic ocean, more persistent in the summer than in the winter of the northern hemisphere, and in summer usually extending so far northeast as to affect western Europe, including the British Isles.

B

Babbit metal an ALLOY of TIN, COP-PER, ANTIMONY, sometimes with LEAD, formerly used in low-friction bearings.

bach (New Zealand) a beach or lake-side holiday home in North Island, termed a crib in South Island. G.

backing of wind a change of direction of the wind in a cyclonic (anti-clockwise) direction, i.e. from north through west to south in the northern hemisphere (the opposite of VEERING), where the wind is said to 'back' at a place north of the centre of a DEPRES-SION-3 travelling eastward. In the southern hemisphere the direction of the wind in relation to pressure systems of closed isobars is reversed, so that a backing wind in the southern hemi-sphere is equivalent to a veering wind in the northern hemisphere. G.

backset bed a term applied mainly in the USA to a deposit of sand on the windward slope of a SAND DUNE, com-monly fixed by vegetation.

backshore the land lying inland from the average HIGH-WATER line to the COASTLINE, bordered seawards by the FORESHORE (the land between the low-est LOW-WATER mark and the average high-water mark), and to seaward of which lies the offshore zone (below low-water mark to the depth at which sub-stantial movement of beach material ceases). OFFSHORE-3. G.

back slope, back-slope the gentler slope of a CUESTA. It is termed DIP SLOPE by many authors, but the term back slope is more accurate because the slope rarely coincides exactly with the DIP of the rocks. FEATHER-EDGE. G.

backswamp, backswamp deposits the tract of low, swampy land lying on the FLOODPLAIN of an alluvial river be-tween the natural LEVEES or banks of the river and the BLUFFS; and the layers of silt and clay deposited there. In the USA the term is used particu-larly in relation to the floodplain of the Mississippi. G.

back wall, back-wall the steep wall at the back of a CIRQUE.

backwash, back-wash 1. the seaward flow of a body of water down the slope after a WAVE-3 has broken on the beach, in contrast to SWASH **2.** the drag of a receding wave. LONGSHORE DRIFT. G.

backwash and spread effects CIRCU-LAR AND CUMULATIVE GROWTH.

backwasting the recession of an ICE FRONT, due, e.g., to ABLATION.

backwater loosely applied to any stretch of quiet water connected with, but little affected by, the main stream of a river. Hence, figuratively, an isolated place, especially one cut off from new ideas, new ways.

backwoods wild uncleared land at the back of an established settlement, ap-plied especially in the pioneer days of the USA and Canada, extended to any sparsely inhabited land far from urban areas.

bacon the flesh from the back and sides of a PIG, pickled and smoked, or dry cured.

bacteria BACTERIUM.

bacterium pl. bacteria, a member of a very large group of microscopic unicel-lular or multicellular organisms classified as PLANTS or PROTISTA. Bacteria are AUTOTROPHIC, SAPROPHYTIC or PARASITIC in nutrition (some being SYMBIOTIC), and they live in very large numbers in favourable habitats. Their activities are of major importance to human beings, positively in soil in the breakdown of organic matter (ACTINO-MYCETES, NITROGEN CYCLE, NITRO-GEN FIXATION), in sewage disposal, and as source of antibiotics; and negatively, e.g. as agents of plant disease and as the cause of serious diseases in animals, including human beings, e.g. tuberculosis, typhoid, pneumonia. THAL-LOPHYTA.

bacteriology the scientific study of BACTERIA.

badlands originally and specifically applied to a large, dry region in South Dakota, USA, where erosion of the nearly horizontal, unconsolidated sedimentary beds resulted in a land of narrow ravines, sharp crests and pinnacles, devoid or almost devoid of vegetation. This highly dissected landscape was well described by early French travellers as 'mauvaises terres à traverser' (bad lands to cross). The term is now applied to similar lands elsewhere, e.g. in Algeria and Morocco. G.

bādōb (Sudan: Arabic) stoneless, clay soils east of the White Nile in Sudan, characterized by very wide cracks in the dry season. G.

bagasse the fibrous residue left when SUGAR CANE is crushed for the extraction of the juice from which sugar is made. It has low economic value, but it can be used as fuel and in paper making, in making types of fibreboard and, suitably treated, as a fertilizer.

baguio a local name for a tropical CY-CLONE in the area of the Philippine Islands, occurring usually from July to November. G.

bahada BAJADA.

bahru (Malay) a town. G.

baile (Highland Scots) FERMTOUN.

baite 1. a shelter in the upper part of high ALPINE-2 pastures, used as temporary shelter for hay, the conversion of wood to charcoal, etc. It is often merely a natural hollow covered with branches 2. the zone where these shelters may be found, characterized by thin pasture, up to some 2000 to 2400 m (6500 to 8000 ft). G.

bai-u, bai-u rains (Japanese) the season of the heaviest rainfall in Japan and parts of China, occurring when the plums are ripening, mid-June to mid-July, i.e. late spring and early summer, hence termed plum rains. MAI-YU. G.

bajada, bahada (Spanish) a continuous apron of gently sloping sediments, e.g. gravel and coarse sand, formed by the merging of ALLUVIAL FANS laid down by swollen streams from a series of mountain streams where they debouch on a plain at the base of a mountain range in an arid or semi-arid area. G.

bajir (bayir in Lop basin) a lake among sandhills in central Asia. G.

bakkeöer (Danish) hill-island, a mature hill of varying size, composed of glacial MORAINE (mainly sand), rising from an OUTWASH APRON of a later glacial period.

balance of nature the relationship between all the component parts of the BIOSPHERE-3 which, by interaction one with another, ensure that it is in a state of equilibrium. The balance is delicate and can be upset by human activity.

balance of payments 1. the relationship between the credits of one nation or group of nations against all other trading partners and the debits of that nation or group of nations to all other trading partners over a specified period of time 2. a systematic record of all economic transactions between one nation or group of nations and all other nations with which it has contacts. Theoretically total debits and total credits must balance. Included are all goods (visible, such as manufactured goods, raw materials bullion, etc.) and services (invisible, covering transport, banking, insurance, interest payments, tourism etc. as well as the flow of capital). BALANCE OF TRADE, TERMS OF TRADE. G.

balance of trade the relationship between a nation's total visible imports and exports, i.e. of goods. BALANCE OF PAYMENTS, TERMS OF TRADE. G.

balata (Spanish) the rubber-like exudation from *Mimusops globosa*, a tree native to South America (especially to Guyana). Balata was used in making golf-balls, insulation, etc., but its importance declined when the cultivation of the RUBBER tree, *Hevea brasiliensis*, increased.

balk, baulk a piece of unploughed land used for grazing and giving access to the ploughed parts of an open field under the medieval system, the minor ones being used as boundaries to separate one man's strip from that of another, the major to separate groups of strips. The larger balks, primarily grass-covered 'occupation roads', were usually termed town balks or common balks to distinguish them from the minor balks between strips. G.

balkanization the division of an area into small units, as in the BALKANS, sometimes implying mutual hostility among such units, or between those units and others outside the area.

Balkans the countries of the mountainous Balkan peninsula, now comprising Albania, Bulgaria, Greece, Romania, Turkey-in-Europe, Yugoslavia, or the peninsula itself, which lies in southeast Europe, bounded by the Adriatic, the Aegean, the Mediterranean, the Sea of Marmara and the Black Sea, and in the north by the rivers Danube and Sava. In the late nineteenth and early twentieth centuries the area was notorious for the hostile relations among what were then the constituent states, and between those states and other powers, hence the implication of the term BALKANIZATION.

ball lightning an unusual form of LIGHTNING, taking the shape of a mobile or stationary, incandescent ball, the cause of which is uncertain.

balloon-sonde, ballon-sonde (French) a hydrogen-filled balloon carrying self-recording meteorological instruments, sent up to register conditions in the atmosphere. The balloon itself eventually bursts, but the instruments descend to earth by parachute. RADIOSONDE, RAWINSONDE, ROCKETSONDE, SONDE.

ballstone a term applied in Shropshire, England, to an irregular lenticular mass of unstratified LIMESTONE consisting of corals, etc., in a matrix of CALCAREOUS mud, and occurring in Wenlock and other PALAEOZOIC limestones. G.

balm *Melissa officinalis*, a PERENNIAL fragrant herb, native to the Mediterranean region, widely grown in temperate lands for its aromatic leaves. It was once of medicinal importance, but is now more valued as a flavouring for beverages, including wine cups.

balma (Spanish) a concave cliff forming a shelter under the overhanging rock. G.

balm (balsam) of Gilead a gum (balsam) valued from ancient times for its aromatic and medicinal properties, obtained from the exuding sap of the wood and the juice from the berries of *Commiphora opobalsamum*, a small evergreen tree, native to northeastern Africa and southwest Asia.

balneology the science of bathing, the study of mineral springs and the healing effects of their waters. G.

balsa *Ochroma lagopus*, a South American tree with extremely light wood, used for rafts, floats, etc. especially in Ecuador and Bolivia, and, more widely, in model-making.

balsam 1. a mixture of gums in volatile oils, used medicinally or in perfumery 2. any plant yielding the gum or resin so used, e.g. BALM (BALSAM) OF GILEAD.

Baltic shield the ARCHAEAN-1 platform of Finland and eastern Scandinavia. SHIELD. G.

bamboo an arborescent grass of the genus *Bambusa*, some 30 species, family Gramineae, native to tropical and subtropical regions. The economically important varieties have edible shoots and grain and/or hollow, hard, durable stems used, especially in Asia, for tools, furniture, mats and papermaking, etc., and in buildings and their construction.

bañada, bañado (South America: Spanish) 1. marshes in the porous LOESS of the PAMPAS 2. in Argentina, cropland irrigated by annual floods on the southern border of the CHACO. G.

banana any of the genus *Musa*, a giant herb native to tropical areas, now widely cultivated there for its nutritious, sweet fruit (a staple food in some tropical countries) which grows in bunches or stems about 10 to 12 half-spirals, known as 'hands' and markedly separated from each other, each hand consisting of 12 to 16 'fingers'. The export trade is important internationally and vital to some banana-growing countries. It is large-scale and highly organized. The bunches are cut when the fruit is green, before ripening, transported in chambers cooled at a constant temperature, as near as possible to 10° to 11°C (51°F) and carefully ripened on arrival at the importing country, the skin of the ripe fruit being yellow. Dessert bananas, with a high sugar content, are usually eaten raw in growing and importing countries; PLAINTAINS are eaten cooked in growing countries, usually (but not always) raw in importing.

band ELECTROMAGNETIC SPECTRUM.

banded ironstone a PRECAMBRIAN deposit consisting of alternating, very thin layers of CHERT (rich in fossils) and HAEMATITE. If not leached, it is termed TACONITE, an ore which, after treatment (BENEFICIATION) is suitable for smelting; if leached it produces an ore of 55 per cent or more iron.

bāngar, bhāngar (Indian subcontinent: Urdu-Hindi; Panjabi) 1. old alluvium of the northern area of the Indian subcontinent, usually with KANKAR 2. a high area in a river valley built up of old alluvium and beyond the reach of river floods. In Panjabi it is also termed manjha. KHADAR. G.

banjar (Indian subcontinent: Urdu) waste land in India, termed banjar jadid (new fallow) if not cultivated for four harvests, banjar kadim (old fallow) if not cultivated for eight; but the latter also includes all 'cultivable' waste whether it has been ploughed or not. CHACHAR, PARANTI, POLAJ. G.

banjir (Java) a mass of water, laden with boulders and mud, sweeping suddenly down a hillside into the valley. G.

bank 1. sloping ground bordering a river, stream or lake 2. in the north of England, a hill or hillside 3. an elevation in the floor of a river or a shallow sea, usually of sand, mud, gravel (not of solid rock or CORAL), in some cases connected with the shore, but not suficiently near the surface as to be dangerous to shipping. G.

bank caving the undercutting and erosion of the BANK-1 by water flowing on the outside curve of a river, resulting in the washing downstream of the material so dislodged. LATERAL EROSION.

banket (South Africa) an Afrikaans term applied to the gold-bearing pebble conglomerate of Witwatersrand, derived from a Dutch term applied to a sweetmeat, almond hardbake, with which the rock has a fancied resemblance. G.

bankfull stage in river flow, the stage when the channel is completely filled with water, from bank to bank, the stage before the river overflows. FLOOD STAGE, OVERBANK STAGE, STREAM STAGE.

banner cloud a CLOUD touching and flowing out like a banner on the LEE side of a mountain peak in clear sky. It occurs when water vapour in a forced up-draught of warm air from the mountainside condenses and the cooled air descends downwind, to be warmed, so that it rises again. Descent and ascent continue downwind until the ultimate complete evaporation of the water droplets. The best known example is the cloud of the Matterhorn. HELM WIND, LEE WAVE. G.

banto faro (West Africa: Manding, beyond swamps) grasslands inundated in the flood season but above water in the dry season, when their coarse grasses wither. G

bar 1. loosely applied to a marine deposit of mud, sand, shingle, covered by water at least at high tide (BARRIER), e.g. across an estuary, parallel to the shore (longshore or offshore bar), across a bay (BAY BAR), between an island and the mainland (TOMBOLO), across the access to a harbour (harbour bar) 2. in USA, the deposits of ALLUVIUM etc. in streams, river mouths and some lakes 3. in meteorology, the unit of ATMOSPHERIC PRESSURE equivalent to 1 mn DYNES per sq cm: 29.5306 in or 750.076 mm of mercury at 0°C in latitude 45°N. The unit commonly used is the millibar (mb), a thousandth part of a bar. In SI 1 bar = 10 NEWTONS. ATMOSPHERE, BAROMETER.

bārāni (Indian subcontinent: Urdu-Hindi; Panjabi) cultivated or crop land entirely dependent on irrigation. In a land use survey of the Indus basin, 1954-5, cropland was classified under irrigation systems: perrenial, seasonal (from inundation canals), SAILABA (land watered by surface run-off trapped in field BUNDS) and bārāni (land dependent for water supply on normal rainfall in all seasons). G.

barbed drainage a pattern of drainage in which tributaries meet the main stream at obtuse angles, i.e. at such angles that their flow appears to be directed to the source of the main stream. It is caused by RIVER CAP-

TURE which has reversed the direction of flow of the main stream. G.

barchan, barchane, barkan, barkhan, balqan (Russian; originally a Turki term from central Asia, applied to a sandhill in the Kirgiz steppe) a crescent-shaped dune of shifting sand formed when the direction of the wind varies only very slightly or not at all. The windward side is convex, with a gentle slope, the steeper leeside is concave, and the 'horns' point downwind. It travels if an adequate supply of sand is maintained; it occurs singly, or in groups; and the height ranges from quite low to over 30 m (100 ft). PARABOLIC DUNE. G.

bare fallow fallow land left without a crop for a whole season. FALLOW, GREEN FALLOW. G.

bar graph, bar chart a diagram drawn to display data. It consists of a series of bars or columns, representing categories, the length of each bar or column being proportional to the quantity represented. The bars are set horizontally (in some cases, e.g. POPULATION PYRAMID, taking the form of a pyramid) or vertically (as columns), the relative importance of each category becoming immediately clear. All the bars have the same width, and may be separated from each other by small spaces, to emphasize that each category is distinct. The bars may show total values, or they may be divided (as in an ENERGY PYRAMID) to show the constituent parts of the total values.

barium a silver-white, toxic metallic element of the ALKALINE EARTH METAL group, occurring in combination with other elements, used in making insecticides, paints, special steel. BARYTES, FELDSPAR.

barkhan BARCHAN.

barley a cultivated grass of the genus *Hordeum*. It is one of the food grains, but does not make good bread. Flat barley cakes (the barley loaves of the Bible) are made and eaten in north

Africa and some eastern countries, but a major use of barley is, as formerly, in the brewing of beer and the making of whisky (a fact recognized by the old English terms 'bread corn', applied to wheat and rye; and 'drink corn', applied to barley). But today barley is principally grown for animal feed. Barley flourishes generally under conditions similar to those which suit WHEAT, but it tolerates poorer and lighter soils. It can take advantage of the long hours of daylight in northern latitudes, so 'arctic' barley grows farther north than any other grain. It will grow higher up mountain slopes than wheat, but it is even less tolerant than wheat of damp conditions. The cultivars commonly grown in the British Isles come from *Hordeum distichon* (two-rowed barley); in colder regions, where hardiness is needed, *Hordeum vulgare* (six-rowed barley) is favoured. Carbonized remains of the latter date back to NEOLITHIC and BRONZE ages; and this species is depicted on Greek and Roman coins of c.500 BC.

barn   a farm building used for the storage of grain, hay, farm machinery and implements, sometimes for housing livestock.

barogram   a continuous record of changes of ATMOSPHERIC PRESSURE as measured by an aneroid BAROMETER, indicated by a curve drawn on scaled paper.

barograph   a self-recording BAROMETER, in which a pen linked to an aneroid barometer records changes in ATMOSPHERIC PRESSURE on a revolving cylinder which makes one complete revolution each week.

barometer   an instrument, of which there are several forms, for measuring ATMOSPHERIC PRESSURE, used in estimating height above sea level and in weather forecasting. In the mercury barometer the weight of a column of mercury is balanced against that of a column of the atmosphere, with adjustments made for latitude (standardized to 46°N), temperature (to 12°C: 53°F), altitude (average decrease of 33.9 mb: 1 in mercury for each 275 m: 900 ft for the first 1000 m: 3300 ft, progressively decreasing thereafter with height) and for any peculiarity of the (usually sensitive) individual instrument. A vertical tube is sealed at the top end, with the bottom, open end, standing in a container holding MERCURY; atmospheric pressure is measured by the vertical height of the column of mercury which the atmosphere will support. In the aneroid (without liquid) barometer, used in ALTIMETERS and BAROGRAPHS, a shallow metal box or cylinder is nearly exhausted of air. In one type, this metal box has flexible sides which expand and contract with changing air pressure; in another the thin metal upper face (the top), which is corrugated, is held up by an external clip spring; as the air pressure changes it upsets the balance between the spring and the pressure in the box/cylinder, so that the top moves. In either type, the movements are magnified and transmitted by a system of levers to a pointer moving over a calibrated scale. The aneroid barometer generally available is not capable of absolute accuracy, and is not to be compared with the precision models used in aviation and meteorology.

barometric gradient   pressure gradient, the amount of change in ATMOSPHERIC PRESSURE between two points, indicated by the distance apart on a level surface of the isobars (ISO-) on a SYNOPTIC CHART. Closely spaced isobars indicate great differences in pressure and therefore a steep gradient, associated with strong winds, in some cases with TROPICAL REVOLVING STORMS; isobars with wide intervening spaces indicate only small differences, and thus a gentle gradient.

barometric tendency   the character (increasing or decreasing) and amount of change in ATMOSPHERIC PRESSURE during a specified period, usually of three hours.

barrage **1.** a large structure, usually of masonry or concrete, occasionally of earth, built to hold up a large quantity of water, especially for irrigation. A DAM similarly impounds water, but some authors use the term dam if the generation of power is involved, barrage if it is not. Others distinguish the two by the duration of water storage, i.e. a barrage serves annual storage of floodwater only, a dam has perennial use **2.** part of a tidal power station (TIDAL BARRAGE). G.

barranca, barranco (Spanish) in USA, a deep ravine with precipitous sides, caused by water erosion, occurring especially on the slopes of volcanic CONES where many may lie close together, radiating outwards and becoming deeper towards the base of the cone. G.

barrel as a unit of measure used in the petroleum industry. CONVERSION TABLES.

barren lands a general term applied to a relatively open, infertile tract of land, supporting a sparse vegetation. G.

Barrens, The a term formerly applied to the TUNDRA of northern Canada, evoking protests from Stefansson (among others) who considered it to be derogatory, and who introduced the now preferred term, arctic prairies. G.

barrier something that hinders or prevents access or advance. Referring to coastal features, some authors urge that a bank of mud, sand, shingle, etc. should be termed a BAR-1 if it is submerged at least at high tide, a barrier if it lies above high tide level. BARRIER BEACH, BARRIER CHAIN, BARRIER ISLAND, BARRIERS AND DIFFUSION WAVES, ICE BARRIER. G.

barrier beach a long, narrow, sandy ridge, lying above high tide level and parallel to the coast from which it is separated by a LAGOON. The term is used by some authors, especially in the USA, as synonymous with SPIT. BARRIER CHAIN, BARRIER ISLAND. G.

barrier chain a series of BARRIER ISLANDS and BARRIER BEACHES extending a considerable distance along a coast. G.

barrier island a feature similar to a BARRIER BEACH but consisting of several ridges, commonly with DUNES, vegetation and swampy areas on the LAGOON side. Some American authors use the term barrier spit if the barrier island is at one end joined to the mainland. SPIT. G.

barrier lake a LAKE formed by a natural obstruction across a valley, e.g. by an AVALANCHE, ROCK FALL, ALLUVIAL deposits, TERMINAL MORAINE, or by a DAM formed by a build-up of VEGETATION, ICE, LAVA, etc.

barrier reef a CORAL REEF skirting a shore and some distance from it, so that it acts as a barrier between the open ocean and the sheltered LAGOON lying between the reef and the coast. On a large scale the Great Barrier Reef, stretching for over 1600 km (1000 mi) off the coast of Queensland, Australia, functions in that way. G.

barriers and diffusion waves barriers that act as a drag on the process of DIFFUSION-1 are commonly classified according to the decreasing amount of drag exhibited, i.e. the superabsorbing BARRIER (absorbing the message but destroying the transmitter), the absorbing (absorbing the message but not affecting the transmitter), the reflecting (not absorbing the message, but allowing the transmitter to transmit a new message in the same time period) or the direct reflecting (not absorbing the message but deflecting it to the available cell nearest to the transmitter). RECOVERY RATE.

barrio SHANTY TOWN.

barrow, tumulus a prehistoric mound of earth, piled over a burial ground, common in the British Isles and other parts of western Europe. NEOLITHIC.

barter 1. trade by exchange, without the use of money **2.** the thing so exchanged.

barysphere loosely applied, sometimes to the dense mass (possibly of nickel-iron) believed to occupy the CORE of the earth below the MANTLE, sometimes to the mantle only, sometimes (and this is preferable) to the core and mantle together, i.e. all of the earth's interior beneath the LITHOSPHERE. CENTROSPHERE. G.

barytes native BARIUM sulphate, used in pharmaceuticl, paint and glass manufacture.

basal, basic *adj.* pertaining to, situated at, forming the BASE-1. BASIC.

basal complex BASEMENT COMPLEX.

basal conglomerate the CONGLOMERATE commonly formed at the beginning of a cycle of sedimentation, hence at the base of a series of strata which lie unconformably on older rocks. G.

basal platform BASAL SURFACE.

basal sapping 1. the breakdown of rocks forming the back wall of a CIRQUE, probably caused by melt-water which, trickling down the BERGSCHRUND, freezes by night, thaws by day, thus shattering the rock wall **2.** the wearing down of the lower parts of slopes by chemical weathering, water seepage, erosion by swiftly flowing water, especially in tropical regions, resulting in sudden changes in the slope profile. CAVERNOUS WEATHERING, TAFONI.

basal slip slipping over the rock floor by the ice at the base of a GLACIER, due to the force of gravity and the weight of the ice in the upper part of the glacier.

basal surface the boundary between actively weathered rock and the sound, unweathered rock on which it lies. D. L. Linton termed it the basal platform; and some authors prefer the alternative term, weathering front. SHIELD INSELBERG, WEATHERING. G.

basalt 1. a fine-grained black or dark grey IGNEOUS ROCK belonging to the basic group (BASIC ROCK), with 45 to 52 per cent of SILICA, mainly sodium or potassium alumino-silicates with some iron (FELDSPAR). When erupted from volcanic vents or fissures it tends to be very fluid and to flood evenly over large areas before consolidating, hence many LAVA plains or plateaus are of basalt. It may solidify into perfect hexagonal columns (e.g. Giant's Causeway). SPHEROIDAL WEATHERING **2.** a type of black stoneware invented by Wedgwood, the English potter. G.

basaltic *adj.* of, pertaining to, or consisting of BASALT-1. HAWAIIAN VOLCANIC ERUPTION, PILLOW LAVA.

basaltic lava the most common LAVA, forming 90 per cent of all lavas, consisting mainly of BASALT-1. HAWAIIAN VOLCANIC ERUPTION, PILLOW LAVA.

basal till TILL, in many cases with a high CLAY content, carried underneath or deposited by a moving GLACIER.

base 1. in general, the bottom, the lowest part or that on which something stands or rests **2.** in chemistry, a substance which reacts with an acid to form a salt and water only; or a substance which, dissolved in water, provides hydroxyl IONS from its own MOLECULES; or a molecule or ion which accepts PROTONS. BASE CATION **3.** in statistics, a number or magnitude used as a standard reference. G.

base cation (base) any one of the CATIONS in a soil solution that is also a PLANT NUTRIENT, the most important being the cations of CALCIUM, MAGNESIUM, POTASSIUM and SODIUM. BASE STATUS OF SOILS.

base data a set of data serving as a base line against which changes may be measured.

base exchange a reaction which some insoluble elements in the soil undergo when in contact with a salt solution, the

CATIONS (positively charged ions) of the salt replacing the bases (BASE-2) from the soil. CATION EXCHANGE. G.

base flow a term applied particularly in North America to that part of a stream flow which comes from GROUND WATER, as distinct from the surface flow after rain has fallen. G.

base-level, base level the lowest level to which a running stream can erode its bed under stable conditions of the earth's crust; or the level below which a land surface cannot be reduced by running water. The general and ultimate base-level for the land surface would be sea-level, but other local or temporary base-levels may exist, e.g. as presented by a resistant rock. G.

base level of deposition the highest level up to which a deposit can be built. G.

base line, baseline 1. an accurately measured line on the earth's surface which is used as a base in trigonometric observations (TRIANGULATION), and so in the mapping of land **2.** a reference parallel used in the United States Land Office Survey. G.

base map a map of any kind on which additional information may be plotted for specific purposes. G.

basement, basement complex, basal complex a term loosely applied to the assemblage of ancient IGNEOUS and METAMORPHIC ROCKS which usually, but not always, underly the PRECAMBRIAN stratified rocks in any particular region. G.

base metal a METAL that is worthless or of comparatively little value, being neither NOBLE nor PRECIOUS, or that is alloyed with a less valuable metal.

base status of soils the rank of the quality of a soil as measured by the PERCENTAGE BASE SATURATION (PBS). High base status soils have a PBS above 35 per cent, soils with a percentage below 35 per cent being of low base status. BASE CATION.

basic *adj.* **1.** in chemistry, of the nature of a BASE-2, the opposite of ACIDIC,

reacting chemically with acids (ACID) to form salts (pH) **2.** BASAL.

basic activity, basic function, basic industry in urban development, a manufacturing or service activity within a city or urban area which provides goods and services and thus earns revenue from outside the city or urban area, i.e. in this context a 'primary' or 'export' industry. BASIC INDUSTRY, ECONOMIC BASE THEORY, EXPORT BASE THEORY, LOWRY MODEL, NON-BASIC ACTIVITY, URBAN ECONOMIC BASE. G.

basic (Bessemer) process a method used to remove PHOSPHORUS from PIG IRON in the process of making cast STEEL, introduced in 1890 in the USA. A Bessemer converter (BESSEMER PROCESS) was lined with a material incorporating lime to serve as a 'base' with which the phosphorus, escaping from the iron, could combine (hence the name). If the proportion of phosphorus was too high to be removed in that way, more lime was added. The product was known as basic steel. G.

basic grassland grassland of chalk and limestone in which *Festuca ovina* and *Festuca rubra* are dominant.

basic industry 1. a heavy industry of national economic importance, or an industry fundamental to other industries (e.g. iron and steel, or the manufacture of sulphuric acid) **2.** in the UK, sometimes officially applied to mining and quarrying; gas, electricity and water; transport and communication; agriculture and fishing **3.** in urban development, BASIC ACTIVITY. G.

basic lava, basaltic lava the most common type of LAVA, molten IGNEOUS ROCK, high in IRON, MAGNESIUM and other metallic element content, low in SILICA, and with a low melting point, which pours easily and quietly from a volcanic vent, spreading widely before hardening (commonly as BASALT) to form a SHIELD VOLCANO or broad plateau. BASIC ROCK-1, HAWAIIAN VOLCANIC ERUPTION.

basic rock a term applied loosely to **1.** an IGNEOUS ROCK with a SILICA content somewhat below 52 per cent and with more than 45 per cent of basic oxides (of ALUMINIUM, CALCIUM, IRON, SODIUM, MAGNESIUM, POTASSIUM), ULTRABASIC ROCK having the lowest percentage of silica. The typical basic LAVA is BASALT, the HYPABYSSAL is DOLERITE, the PLUTONIC is GABBRO. ACID ROCK, BASIC LAVA **2.** sometimes applied even more loosely to any igneous rock with dark-coloured minerals. Both these uses are obsolescent: neither is precise, and either can be misleading. More precise terms have been suggested in the USA, i.e. subsilic (relating to content of silica), mafic (base content), melanocratic (dark-coloured component minerals). G.

basic slag waste material from blast furnaces, rich in minerals, especially phosphorus, used as a fertilizer.

basil *Ocimum basilicum*, an ANNUAL HERB native to tropical Asia, Africa and the Pacific islands, tolerant of midlatitude climates, valued for its aromatic leaves, used widely in flavouring cooked vegetable and meat dishes, and sauces, salads, drinks, etc.

basin a term applied loosely to some form of natural or artificial depression, varying in extent, in the earth's crust, e.g. **1.** the total area of land drained by a river and its tributaries (DRAINAGE BASIN, RIVER BASIN), termed WATERSHED in USA **2.** in geology, a circumscribed area where the strata dip inward towards the centre (SYNCLINE), or a stratified deposit (e.g. COAL) lying therein (CUVETTE) **3.** a hollow in the ground formed by surface settlement following the natural or artificial removal of underground deposits of salt or gypsum in solution (BASINING) **4.** a large or small depression occupied by a lake (lake basin) or pond **5.** a depression in the ocean floor (DEEP) **6.** an extensive depression occupied by an OCEANIC BASIN **7.** a hollow classified according to origin (TECTONIC basin, GLACIAL basin) **8.** a large area surrounded by high land, with or without access to the sea (Great Basin, USA) **9.** a dock or part of a canal or river widened for navigation and lined with wharfs, etc. **10.** a dock subject to tidal movement, or a depression (natural or artificial) filled with water at high tide (TIDAL BASIN). G.

basin and range a tract of country with a series of asymmetrical ridges separated by basins, the ridge consisting of tilted FAULT BLOCKS. The Basin and Range country of USA lies between the Sierra Nevada and Wasatch mountains.

basin cultivation the practice of dividing land by low earth ridges to form small or large basins where water can be retained and rapid run-off prevented. This technique is used especially near the equatorial margins of the tropics in Africa (e.g. in Nigeria and Ghana) to stem soil erosion by heavy rainfall. G.

basining the settlement of the earth's surface in the form of depressions, varying greatly in extent, and due in many cases to the solution and removal of underground deposits of salt and gypsum, e.g. in the high plains east of the Rocky Mountains. G.

basin irrigation a type of IRRIGATION in which floodwater from a river that annually and for a short time overflows on to its FLOODPLAIN is led off into prepared basins, varying greatly in size, and separated from one another by earth banks. PERENNIAL IRRIGATION. G.

basin of deposition CUVETTE. G.

basisol a very dark, tropical CALCAREOUS soil formed from the weathering of BASALT in areas of high rainfall and high temperatures, the dark colour being due to the TITANIUM content, the organic content being low. It bakes hard and cracks deeply, the surface crumbling to dark grey dust in the dry season; but it swells, blackens and becomes sticky in the wet, holding moisture well. It is especially suitable for COTTON growing, e.g. in the northwest Deccan (REGUR), Kenya, and parts of the West Indies. G.

basket of eggs relief rounded sandy mounds or DRUMLINS arranged in such a pattern that from a distance they resemble eggs in a basket, occurring in glaciated regions in valleys formerly occupied by ice. G.

bass 1. a fibre (RAFFIA) 2. Russian abbreviation of basseyn (basin), usually referring to a coal basin, and used as a suffix, e.g. Donbass. G.

bastard fallow pin fallow. GREEN FALLOW. G.

bastion, rock ROCK BASTION.

batāi (Indian subcontinent: Urdu-Hindi) a type of METAYAGE whereby the tenant (hari) keeps half the produce of the land but personally provides the seeds, fertilizers, implements, and cattle. G.

bat furan, bat hiddan (Arabic bat, sea; furan, open; hiddan, closed) sea conditions affecting local shipping in the Arabian Sea. The former corresponds to winter (with its northeast monsoon) when light winds suit shipping, the latter to summer (with its southwest monsoon) when storms bring danger. G.

batholith, bathylith, batholite a very large dome-shaped mass of IGNEOUS ROCK, usually of GRANITE, formed by a large-scale, deep-seated intrusion of MAGMA, the sides of which plunge down steeply to unknown depths, the base never being seen. The domed upper surface may be exposed by DENUDATION over a long period, to form uplands, e.g. Dartmoor in southwest England. A batholith may be surrounded by a METAMORPHIC AUREOLE. XENOLITH. G.

bathyal *adj.* applied loosely, of, or pertaining to, the deeper parts of the ocean, according to some authors between 180 and 1830 m (100 and 1000 fathoms: 600

to 6000 ft) or more, i.e. very broadly the CONTINENTAL SLOPE between the CONTINENTAL SHELF and the ABYSSAL ZONE; and the deposits and organic life present there. G.

bathymetric *adj.* relating to the measuring of the depth of a body of water.

bathymetry the science of measuring the depth of a body of water, usually the depths of the oceans and seas.

bathyorographical *adj.* applied to maps showing both the relief of the land (OROGRAPHY) and the depths of the ocean. G.

bathysphere 1. deep sphere, applied, incorrectly, by some writers to the interior mass of the earth as an alternative to BARYSPHERE or CENTROSPHERE 2. a diving apparatus, a large spherical chamber, able to withstand great pressure, used for deep sea observations. G.

battery system in poultry farming, a CAPITAL-INTENSIVE system of husbandry common in industrialized countries, in which birds are reared and kept under cover in a carefully controlled environment (ENVIRONMENTAL CONTROL). The birds are confined to cages, fed and watered by automatic devices. They drop their eggs into channels running under the cages; the killing of the birds for meat is automated. FACTORY FARM.

battue ice thick ice, forming particularly on the St Lawrence river in winter, hindering the passage of vessels.

baulk BALK.

bauxite (derived from deposits at Baux, near Arles, France) a naturally occurring ALUMINIUM oxide, $Al_2O_32H_2O$, an amorphous, earthy, granular mineral containing IRON oxides and PHOSPHATE, the colour ranging from grey-white through yellow, brown, red. Now the principal commercial ore of ALUMINIUM, it occurs widely in FELDSPARS and other SILICATES which readily break down in tropical conditions; hence bauxite is found as a surface crust in tropical lands. GIBBSITE. G.

bay 1. a term applied loosely to a wide, curved indentation of the sea or of a lake into the land, especially one with a wide opening, or greater in width than in depth, considered by some authors to be larger than a COVE, smaller than a GULF 2. in law, more precisely defined in the 1958 convention on the delimitation of TERRITORIAL WATERS. Under the terms of this convention the existence of a bay is established by drawing a straight line to connect the seaward extremities of the natural PROMONTORIES on each side of the indentation. This line acts as the diameter of a semicircle which is drawn landwards: if the area of the water in the indentation limited by the straight line is as great as or greater than the area of the semicircle, it qualifies as a bay. If the length of the diameter is 38.6 km (24 mi) or less the waters 'enclosed' by the line can be treated as territorial. The waters in certain 'historic' bays (e.g. Hudson Bay, Canada) which do not meet these criteria are nevertheless considered to be territorial 3. an elliptical, shallow depression in the coastal plain of the eastern USA (the Carolina Bays), origin uncertain 4. in Germany (translation of Bucht), the spread of the lowland into an upland area along a river valley, e.g. Kölnische Bucht 5. a recess in a range of hill or mountains. G.

bay bar, bay barrier (American, baymouth bar) a ridge of sand, mud or shingle extending across a BAY-1 linking the two headlands, caused by the lengthening of one SPIT or the convergence of two, or the moving of an OFFSHORE BAR towards the coast. G.

bay-head beach, pocket beach a crescent-shaped accumulation of sand and shingle piled up at the head of a small COVE on a sea coast.

bay-head delta a DELTA occurring at the head of a BAY-1.

bay ice FAST ICE of more than one winter's growth, possibly fed by snow,

the total thickness of ice and snow rising to about 2 m (6.5 ft) above sea level; if thicker it is termed an ICE SHELF. G.

bay laurel *Laurus nobilis*, an EVERGREEN shrub or tree, native to the Mediterranean region, but now grown in cooler regions, the aromatic leaves being used as flavouring in cooking. In classical times they formed the wreaths for crowning the victorious.

baymouth bar (American) BAY BAR.

bayou (southern USA term derived from French) a sluggish stream or a stagnant body of water such as an OXBOW LAKE or swampy backwater connected or associated with the lower Mississippi and its delta. G.

bazaar economy a commercial system in which a large number of buyers and sellers meet personally to transact their business without the aid of an intermediary such as a retail outlet. The transactions, usually unrelated one to another, are commonly centred on, but not necessarily restricted to, a MARKET PLACE.

beach the accumulation of loose material (mud, sand, shingle, pebbles) on the shore of a lake or of the sea at or near the limits of wave action, mainly between the low water SPRING TIDE line and the highest point reached by storm waves at HIGH TIDE. Beach material is classified by size, i.e. SAND, GRAVEL, PEBBLE and BOULDER. BEACH RIDGES, GROYNE, PROGRADATION, RAISED BEACH.

beach cusp a cone-shaped deposit of sand and gravel with the apex pointing seawards, alternating with bay-like depressions, usually one of a series along a straight, open beach, resulting from the SWASH and BACKWASH of waves breaking at right angles to the coast. The distance between the points, increasing generally with wave height, varies from 9 to 60 m (30 to 200 ft). G.

beach ridges low sandy ridges on a coast representing a successive series of

BERMS-1 produced in the PROGRADA-TION of a BEACH.

beacon 1. a conspicuous hill, used especially in place-names in Britain, e.g. Ivinghoe Beacon 2. a conspicuous object (e.g. a fire, a light, equipment emitting radio waves) used as a signal to indicate direction to the observer. G.

beaded drainage small pools joined by little streams caused by the melting of the ground surface in regions of PERMAFROST.

beaded esker an ESKER with a succession of mounds strung out along the ridge, like beads on a string, indicating pauses in the retreat of the GLACIER that fed the stream which formed the esker. G.

beaded lakes strings of long, narrow lakes between sand dunes. G.

beaded valley a VALLEY with alternating narrow and wide sections.

beaker folk STONE AGE people who spread from Spain to central Europe and Britain c.2000 BC. They made distinctively decorated beakers and bowls, commonly found in their graves.

bean any of the erect leguminous (LEGUMINOSAE) plants of the genus *Phaseolus*, widely grown in all climatic regions, producing edible seeds (and pods in some cases) rich in protein. Those grown in tropical areas are commonly termed PULSES (GRAMS in the Indian subcontinent). Of some varieties both pod and seeds are used as cooked vegetables (e.g. RUNNER BEANS, *P. coccineous*); of others, especially the tropical pulses (e.g. BUTTER BEAN, *P. lunatus*; BLACK GRAM, *P. mungo*) and, in cooler climates, the HARICOT BEAN, *P. vulgaris*, the mature seeds are dried and stored. The sprouts (shoots from newly germinated seeds) of some varieties (e.g. MUNG BEAN, *P. aureus*) are used for food, e.g. especially in Chinese cooking. CHICK PEA, BROAD BEAN, GREEN GRAM, JACK BEAN, KIDNEY (FRENCH) BEAN.

bearing the horizontal angle measured clockwise between a specific reference line and a point viewed by the observer **1.** for a true bearing the reference line is the MERIDIAN, so a true bearing is measured clockwise from TRUE NORTH **2.** a magnetic bearing is measured clockwise from MAGNETIC NORTH **3.** a compass bearing is measured clockwise from the north indicated by the COMPASS **4.** a grid bearing is measured from the north–south GRID-1 lines on a map **5.** a reverse (reciprocal) bearing is the reverse or reciprocal of a given bearing, i.e. a line drawn 180° from any bearing. AZIMUTH, GYROCOMPASS. G.

beastgate CATTLEGATE.

Beaufort Notation a code using letters of the alphabet, devised by Admiral Sir Francis Beaufort, RN, in the early nineteenth century, to show the state of the weather (e.g. *b* blue sky, *q* squalls, *kq* line squall) now superseded by the international system of meteorological symbols listed in the *International Cloud Atlas*, World Meteorological Organization, Geneva, and commonly used on WEATHER CHARTS.

beaver dam, beaver meadow a dam lying across a stream, built by beavers. These semi-aquatic rodents gnaw the trees along the banks of the stream so that they fall across it and become part of the beavers' complex dens (termed lodges), while maintaining a water level suited to the beavers. This may make an area of land near the river bank soft and moist, termed a beaver meadow in Canada.

beck (Old Norse bekk) in those parts of northern and eastern England occupied by Danes and Norwegians, a small stream, especially a winding, swiftly flowing one with a rocky bed. G.

bed 1. the floor, the land at the bottom of a body of water (sea, lake, river, canal, pond), usually permanently covered by the water but possibly intermittently dried out 2. in geology, a layer or STRATUM of rock, a feature of a SEDIMENTARY ROCK-1 distin-

Beaufort scale a scale widely used for measuring and recording the strength of the wind, based on estimated velocity as 10 m (33 ft) above the ground, devised by Admiral Sir Francis Beaufort, RN, in 1805, used internationally since 1874, slightly modified 1926.

	Scale number	Description	Force	
			km/h	mi/
	0	calm	0	0
light winds	1	light air	1.5–5	1–3
	2	light breeze	6–12	4–7
	3	gentle breeze	13–20	8–12
moderate winds	4	moderate breeze	21–29	13–18
	5	fresh breeze	30–39	19–24
	6	strong breeze	40–50	25–31
	7	moderate gale	51–61	32–38
gales	8	fresh gale	62–75	39–46
	9	strong gale	76–87	47–54
	10	whole gale	88–102	55–63
	11	storm	103–121	64–75
	12	hurricane	above 121	above 75

Notice that moderate gale, 7, is for statistical purposes classified under moderate winds. G.

guished from adjacent layers by its composition, structure or texture, and separated from the overlying and underlying layers by well marked BEDDING PLANES. INTERBEDDED. G.

bedding the arrangement of rock strata in bands of various thickness and character. FALSE-BEDDING, STRATIFICATION. G.

bedding cave, bedding-cave a wide, low cave passage with a flat roof formed along a BEDDING PLANE. In Yorkshire, England, it is often associated with washed-out SHALE bands. G.

bedding plane, bedding-plane the plane of stratification, the surface separating the successive distinctive layers of SEDIMENTARY ROCK, in many cases forming a line of weakness. CLINOMETER. G.

bed load, bed-load, bedload traction load, the solid material, e.g. sand and gravel, and sometimes large boulders in time of flood, pushed or rolled by a

STREAM-2 (TRACTION), or bouncing (SALTATION) along the BED-1 of a STREAM-1, as distinct from the material carried in SUSPENSION (SUSPENDED LOAD) or SOLUTION-I (DISSOLVED LOAD).

bed rock, bed-rock, bedrock 1. the unweathered rock underlying the weathered superficial deposits (i.e. underlying the soil, subsoil and other loose unconsolidated rock, the REGOLITH) **2.** more specifically, the solid rock beneath PLACER deposits of GOLD or TIN. G.

beech-hanger HANGER.

beef the flesh from the carcass of a bull, cow or ox. CATTLE.

beeswax 1. the wax secreted by bees, used in building the honeycomb **2.** a product made from this, used mainly in polishes and in modelling. WAX.

beer an alcoholic beverage made from fermented malt (BARLEY) flavoured with HOPS.

beetroot *Beta vulgaris*, an ANNUAL or

BIENNIAL plant with a fleshy, crimson, edible root, possibly derived from the PERENNIAL wild species native to coasts of Europe, north Africa and the eastern Mediterranean. Related to SUGAR BEET, it is grown in midlatitudes, eaten hot or cold when cooked.

beet sugar SUGAR.

behaviour 1. the way in which an organism or a group of organisms (including a person or a group of people) reacts or responds to stimuli in the environment, e.g. to light, sound, touch, chemicals, the presence and activities of other organisms (including a person or people), or to a particular object, or to a particular event 2. the way in which an organ, an organism, or a machine works, in terms of its efficiency. ADAPTIVE BEHAVIOUR, ADOPTIVE BEHAVIOUR, BEHAVIOURAL SCIENCES, BEHAVOURISM, COGNITIVE BEHAVIOURALISM, NORMATIVE EXPLANATION, SATISFICING BEHAVIOUR.

behavioural *adj.* concerned with, or a part of, BEHAVIOUR-1,2.

behavioural approach in psychology 1. the study of humans and ANIMALS-2 in terms of their BEHAVIOUR, the concepts of 'mental' or 'subjective' processes being considered of little importance, and usually excluded 2. a synonym for BEHAVIOURISM.

behavioural environment the part of the ENVIRONMENT perceived (PERCEPTION) by the individual, to which the individual responds or to which BEHAVIOUR is directed. It is the environment in which rational human behaviour begins, in which decisions are taken which may or may not result in conscious use or alteration of the PHENOMENAL ENVIRONMENT, or in a change in the individual's relationship with, or exposure to, that environment. G.

behavioural geography an approach in HUMAN GEOGRAPHY which uses the assumptions and methods of BEHAV-

IOURISM to determine the cognitive processes (COGNITION) involved in an individual's perception of, response and reaction to, his/her environment. The cognitive processes include the construction of mental maps (COGNITIVE MAPS) and the assessments of locations in the individual's ACTION SPACE (PLACE UTILITY), etc. G.

behavioural sciences the scientific studies concerned with the BEHAVIOUR of human beings and other animals, e.g. ANTHROPOLOGY, PSYCHOLOGY, the SOCIAL SCIENCES. BEHAVIOURISM, EARTH SCIENCES, HUMANITIES, NATURAL SCIENCES, PHYSICAL SCIENCES, SCIENCE.

behaviourism, behavioural approach a school of psychology based on the principle that psychological studies should, in order to be scientific, be confined to the observable, and preferably the measurable, reaction of human beings (and ANIMALS-2) to external stimuli, the study of 'mental' or 'subjective' processes (e.g. consciousness, introspection, freewill etc) being excluded because these cannot be directly observed and measured. BEHAVIOUR.

beheading, of river RIVER CAPTURE.

bel, bhel (Indian subcontinent: Panjabi) a sandy island in a river bed, especially that of the River Jhelum. G.

belief 1. the feeling that something is true, or that something actually exists 2. something accepted as true.

bell metal an ALLOY of about four parts COPPER to one of TIN, a type of BRONZE used in the making of bells and ornaments.

bell pit an early type of mine in which a seam of chalk (e.g. in Buckinghamshire), or coal (e.g. in the Black Country and North Staffordshire, England), or other deposit, was worked from the base of a shallow shaft, the pit being abandond when the roof became unsafe. Subsidence around the shafts later resulted in the formation of shallow depressions. G.

bellwether region (bellwether, the sheep with a bell hung round its neck, the leader of the flock) a leading area within a country, i.e. one which displays trends earlier than any other part of that country.

belt **1.** a district with particular, distinctive characteristic(s) e.g. of climate (belt of calms), of vegetation (tundra), of prevalence of a mineral (coal), of a crop (cotton), of land use (green belt), etc. It is generally in the form of a broad, long strip which may or may not encircle something; but in some cases the term is used as a synonym for REGION-1, the shape being disregarded **2.** a long narrow stretch of water (Great Belt in the Baltic sea) **3.** in ice terminology, a long strip of PACK ICE. G.

belted coastal plain, belted outcrop plain an EROSION SURFACE, a PENEPLAIN, with a series of rock outcrops lying in parallel strips, where parallel CUESTAS and lowlands may develop. G.

belt of calms ITCZ.

belt of no erosion the part of a hillslope, near the crest of the hill, where the accumulation of precipitation is too small to cause EROSION.

belukar (Malay) the scrub which grows up after land has been cleared, as distinct from LALANG, which also becomes established on old clearings. G.

ben (from Gaelic beinn or beann) a mountain peak, used in names of Scottish and Irish mountains. PIN. G.

bench a natural step or terrace, usually narrow and backed by a steep slope, produced by structural change, natural (e.g. by EROSION, as is a WAVE-CUT BENCH) or artificial (e.g. by quarrying, mining). RIVER TERRACE. G.

bench mark BM, a surveyor's mark cut in some durable fixed material such as a rock, or the wall or face of a building, for which the height above the DATUM LEVEL (in Britain the Ordnance Datum (OD) at Newlyn, Cornwall) is accurately determined. Bench marks are recorded on British Ordnance Survey maps as BM, with height in metres/feet to one place of decimals. G.

beneficiation the first step in the removal of a commercially valuable mineral from the COUNTRY ROCK or GANGUE surrounding it after extraction from the ground. The process is usually a simple one (e.g. crushing, magnetic separation, flotation) carried out at or near the site of the mine or other working. The aim is to concentrate the ore to keep down the cost of its transport to the works where it is to be further processed.

Benelux countries Belgium, Netherlands and Luxembourg, the countries which formed a customs union on 1 January 1948. They joined the EEC in 1959, and the full economic union of the three came into force on 1 November 1960.

Benioff zone the seismically (SEISMIC) active zone at the bottom of an OCEAN TRENCH where an oceanic plate dives into the MANTLE. PLATE TECTONICS.

benthic *adj.* of, or relating to, the BENTHOS. DEMERSAL.

benthic division, benthic zone one of the two chief divisions of the aquatic environment based on depth of water (the other being the PELAGIC DIVISION), consisting of all the floor of the ocean or lake where BENTHOS-1 live, irrespective of the depth of the floor. It is commonly divided into two systems, the NERITIC and the ABYSSAL, the division being at the edge of the CONTINENTAL SHELF. G.

benthos **1.** the plants or animals living at or near the floor of the ocean or a lake, irrespective of the depth of the floor. The organisms are usually divided into LITTORAL and deep water. ABYSSAL BENTHOS, GEOBENTHOS, NEKTON, PHYTOBENTHOS, PLANKTON, POTAMOBENTHOS, ZOOBENTHOS **2.** the deep zones of water, the bottom of the ocean or a lake. G.

bentonite any of the valuable, very fine

CLAYS, similar to FULLER'S EARTH, formed by changes in deposits of VOL-CANIC ASH, used as a water softener, in blocking of seepages in mines, in the mud used in drilling.

benzene a colourless, inflammable, toxic, liquid HYDROCARBON, distilled from COAL and PETROLEUM and used as a fuel and solvent. BENZOLE.

benzine a mixture of the paraffin series of HYDROCARBONS obtained during the distillation of PETROLEUM. It is used in dry cleaning, and as a solvent of fats.

benzoin an aromatic, resinous substance produced by *Styrax benzoin*, a tree of moderate size, native to Sumatra, but cultivated elsewhere, e.g. in temperate lands. Benzoin is used in medicines (e.g. as an ingredient in 'Friar's balsam', an aromatic inhalant), and in perfumes, etc.

benzol unrefined BENZENE.

berg (Afrikaans and German, a mountain) a term used as an element in various combinations, e.g. Bergschrund, iceberg, Drakensberge (Dragon mountains, a mountain range in South Africa). Notice that the last should not be anglicized as Drakensberg mountains (mountains being indicated by the suffix berge). G.

bergamot *Citrus bergamia*, a small evergreen tree, producing oranges with acid but fragrant pulp. The ESSENTIAL OIL expressed from the skin is used as a flavouring and in perfume-making, especially in Eau de Cologne. CITRUS FRUIT.

Bergschrund (German; French rimaye) a wide CREVASSE or series of crevasses occurring between the rocky mountain wall of a CIRQUE and the mass of ice which occupies it. As the ice, which will become a GLACIER, begins to move down its valley it pulls away from the wall and the ICE APRON attached to it, creating a crevasse with each wall of ice, i.e. the Bergschrund; but if there is not an ice apron on the rock wall, the

gap is termed a RANDKLUFT. SNOW-BRIDGE. G.

Bergwind, berg wind (Afrikaans and German) in general a mountain wind, but specifically in South Africa a hot, dry, FOHN-like wind blowing mainly in winter down from the plateau towards the coast, and thus warming adiabatically. ADIABATIC. G.

bergy bit a heavy, compact piece of sea ice or of HUMMOCKED ICE; or a piece of floating glacier ice. It is usually under 10 m (33 ft) across at its widest measure and under 5 m (16 ft) above sea-level. G.

berm 1. a narrow ledge, shelf or terrace formed by material thrown up on the beach by storm waves to make a horizontal shelf above the FORESHORE 2. a remnant flat surface, part of an earlier, broad valley floor, occurring above the present level of a river, originating from an interrupted CYCLE OF EROSION 3. in military terminology, the bank between a road and the ditch, or the strip of land in a fortification between the ditch and the base of the parapet 4. in USA, the bank of a canal opposite the towpath; or the shoulder of a road. G.

berry a succulent FRUIT with many seeds. The outer part of the wall consists of a skin, the central part is usually fleshy and juicy, and the innermost part is membranous, e.g. a gooseberry. DRUPE, NUT.

bersim (Arabic) *Medicago sativa*, a tropical leguminous plant (LEGUMINO-SAE) grown for FODDER in tropical areas, particularly in Egypt and Sudan.

beryl a beryllium aluminium silicate, a PRECIOUS or SEMI-PRECIOUS stone, varying in colour from white to yellow (termed heliodor), through blue-green (AQUAMARINE), and green (EMERALD) to pink (morganite).

beryllium a hard, poisonous, metallic element, the lightest ALKALINE EARTH, occurring in the atmosphere in minute particles, stable in moist air, resistant to corrosion, used in very hard non-ferrous alloys, in X ray tubes, and in slowing down neutrons emitted in nuclear fission.

Bessemer process a method of producing cast steel (Bessemer steel) devised by Sir Henry Bessemer in 1860. In his original method molten PIG IRON was poured into a vessel known as a converter, lined with a highly refractory material (usually GANISTER), arranged so that cold air could be blown through the molten mass to burn away the carbon and silicon. The due proportion of CARBON was then added and mixed with the fused metal by a repetition of the blowing, resulting in a very brittle steel. R. F. Mushet improved the process by adding the carbon in a compound containing manganese (spiegeleison or ferromanganese), unnecessary if the iron ore used was itself sufficiently rich in manganese. Neither the original nor the improved process (nor the OPEN HEARTH PROCESS) removed any PHOSPHORUS present in the pig iron (phosphorus makes steel brittle), so the Bessemer process was satisfactory only if the iron ore used (which came to be known as Bessemer ore) lacked phosphorus or contained it only in minute quantity. In the BASIC BESSEMER PROCESS, introduced in 1890, the phosphorus was extracted. The last Bessemer converter in the UK was closed in 1974. STEEL. C.

bet, bét (Indian subcontinent: Panjabi) a FLOODPLAIN, usually anglicized as bet lands. G.

betel nut the kernel of the fruit (a DRUPE) of the betel palm, *Areca catechu*. In Africa and parts of Asia it is mixed with an aromatic paste, sandwiched in betel leaves, and chewed.

betrunked river a river affected by subsidence which causes the main stream to be replaced by an indentation of the coast, and its tributaries to commence a separate existence.

betterment the fortuitous increase in the value of land which accrues to the owner as a result of the operation of a planning system or of public or private investment, sometimes termed unearned increment. BLIGHT.

bevel, bevelled surface, bevelling **1.** bevelling, the smoothing of a sharp edge between a horizontal and a vertical surface, specifically the smoothing of the sharp edge of a ledge (the horizontal surface) and the face of a sea cliff (the vertical surface) **2.** a summit which has been smoothed off, e.g. a levelled-off hill top, or summit of an escarpment, to form a flat surface. G.

bhadoi (Indian subcontinent: Bengali) the rainy season, April to August. HEMANTIC, KHARIF, RABI. G.

bhabbar, bhābar (Indian subcontinent: Urdu-Hindi; Panjabi, porous) a band of gravelly deposits fringing the TERAI. G.

bhāngar BANGAR.

bhīl, bīl, bheel, jhīl, jhāor (Indian subcontinent: Bengali) an OXBOW or cut-off lake occupied by stagnant water, sometimes saline, especially one in the Ganga delta. JHIL. G.

bhit (Indian subcontinent: Sindhi) a sand-hill or ridge. G.

B horizon the soil layer underlying the A HORIZON, an ILLUVIAL horizon into which minerals etc. from the A horizon are washed. It is sometimes divided into an upper layer, B_1, high organic content; B_2, the main depositional zone; and B_3, grading into the C HORIZON. The entry on SOIL HORIZON shows refinements in the classification of the B horizon. SOIL, SOIL PROFILE.

bhūr (Indian subcontinent: Urdu-Hindi; Sindhi; Panjabi) **1.** a patch of sandy soil **2.** hills of wind-blown sand capping the high banks of rivers. G.

bhūra (Indian subcontinent: Panjabi) a RAVINE scoured solely by surface drainage, lacking a torrent-bed. G.

bias **1.** in statistics, the distorting effect produced by a sample which does not accurately reflect the characteristics of the POPULATION-4 from which it is drawn owing to SYSTEMATIC ERROR rather than to RANDOM ERROR **2.** in sociological survey, the distorting effect produced either by questions which are framed in such a way that respondents are led to give particular answers, or by the researcher in interpreting and coding the answers given.

bid price curve, bid rent curve a curve on a GRAPH-1 relating price (rent) to distance, showing the price a land user would be willing to pay for a given area of land at various distances from a given point, especially from the city centre. Activities which depend on contacts and need to be located in the most accessible places (e.g. head offices of banks) must have a central location, and can afford the high prices and rents of the centre: they will have a steep curve. Those activities not much affected by their location, and which need to avoid high rents will have curves with a gentle slope. Under those simple terms if the bid rent curve is superimposed on the actual rent curve for a given city, the best location (OPTIMUM LOCATION) for a particular activity will be where the actual rent curve just touches the lowest possible part of the bid rent curve, i.e. where the actual rent curve equals the bid rent curve. The bid price and the value and use of land are interrlated, mutually determining. ALONSO MODEL, TRADE-OFF THEORY.

bield (Anglo-Saxon) a sheep enclosure with dry stone walls, common in the Lake District, England. A term commonly used in place-names.

biennial *adj.* as applied to plants, a plant which after seed germination vegetates for one year, storing food for the second year, when it flowers, fruits and dies, thus taking two years to complete its life-cycle. ANNUAL, EPHEMERAL, PERENNIAL.

bifurcation ratio the ratio between the number of streams of an order of magnitude and the number of streams in the next higher order of magnitude

(STREAM ORDER). The term has also been applied to CENTRAL PLACES, but has not been generally adopted. G.

bight 1. a crescent-shaped indentation of the coastline, usually of considerable extent, normally wider and with a shallower indentation than a BAY **2.** an extensive crescent-shaped indentation in the ICE EDGE, shaped by wind or current. G.

bilberry, blaeberry, whortleberry *Vaccinium myrtillus*, a low shrub of heathland and moorland of European and northern Asia, bearing juicy black fruits (BERRY) which are unpleasantly acid when raw but palatable when cooked.

bill a local term in southern England, a long narrow promontory, almost detached from the mainland (Portland Bill), but not in every case (Selsea Bill). G.

billabong (Australia: Aboriginal, dead river) an elongated waterhole in the bed of an intermittent stream, or a cut-off or OXBOW lake, equated with ANABRANCH by some authors. G.

biltong (Afrikaans) salted and dried meat. G.

bimodal MODE-3.

binary in astronomy, a double star, consisting of two stars revolving round each other.

binary *adj.* consisting of two.

binary connectivity matrix INCIDENCE MATRIX.

binary scale in mathematics, a scale of numbers based on two digits, 0 and 1, other numbers being powers of 1, e.g. the decimal scale 1, 2, 3, 4, 5 in binary scale reads 1, 10, 11, 100, 101.

bing (Scottish) a heap of waste material from a mine. G.

binodal tidal unit an AMPHIDROMIC (TIDAL) SYSTEM in which there are two NODES-1. AMPHIDROMIC POINT.

binomial distribution in statistics, a theoretical DISTRIBUTION-4 which predicts the PROBABILITY of a particular result occurring in a sample when the characteristics of the parent population are known and there are only two possible outcomes.

binomial nomenclature the universally accepted method of naming animals and plants which avoids the confusion arising from the use of local names. The generic name, designating the GENUS to which the animal or plant belongs, is written first, with a capital (upper case) initial letter; second is the specific (or trivial) name, that is the name peculiar to the SPECIES, printed with a small (lower case) initial letter. Those two names, the generic and the specific, are usually printed in italic. The author who named and described the species follows, not in italic; if a species originally allocated to one genus is later transferred to another, the name of the original author is put in brackets. In refinement, one of the species is commonly designated as the 'type specimen' (HOLOTYPE) and in later splitting of the species this type specimen is always included, followed by the name of the new specimen which resembles it. CLASSIFICATION OF ORGANISMS.

biochemical *adj.* of, or relating to, BIOCHEMISTRY.

biochemistry the scientific study of the chemical processes and substances (CHEMISTRY) of living matter. BIOPHYSICS.

biochore BIOSPHERE-4.

biochrome any natural colouring matter of plants or animals.

biochronology 1. the study of the evolution of organisms in relation to geological time 2. a geological timescale or the dating of events in geological time based on FOSSIL assemblages. BIOSTRATIGRAPHY.

bioclastic *adj.* applied to a rock composed of fragmental organic remains. CLASTIC. G.

bioclimatic frontier the boundary beyond which climatic conditions make it impossible for a SPECIES to survive.

bioclimatology the study of climate as it

affects the life and health of animals, plants and people. G.

biocoen in ecology, all the living parts of the environment. ABIOCOEN, HOLOCOEN.

biocoenosis, biocenosis 1. in biology, the participation of diverse organisms in all the resources of their environment; a BIOTIC community 2. MUTUALISM between plants and animals 3. in geology, a group of fossils consisting of the remains of organisms that once lived together, in contrast with THANATOCOENOSIS. G.

biocontent the total energy content of an organism or a community.

biocycle BIOSPHERE.

biodegradable *adj.* applied to a substance or material that can be decomposed by BIODEGRADATION.

biodegradation, biodeterioration the breaking down of a substance or material by the action of living organisms (mainly by AEROBIC BACTERIA).

bioengineering the use of BIOCHEMICAL processes on a large, industrial scale, particularly in the recycling of waste materials to produce foodstuffs for people or livestock.

biogenesis 1. the theory that all living things are descended from previously living things, that a living organism can originate only from a parent or parents generally similar to itself 2. the theory concerned with the biochemical evolution of living things, with the development of living matter from inanimate, complex substances.

biogenetic *adj.* conforming to the principle of BIOGENESIS. ABIOGENESIS.

biogenetic law the theory that the evolutionary stages of an animal species are recapitulated in the life of the individual. PALINGENESIS.

biogenic *adj.* produced by living or once-living organisms.

biogenic sediment a sedimentary deposit, e.g. shelly LIMESTONE, formed by once-living organisms.

biogeochemical anomaly the concentration of minor elements or trace elements in the soil revealed by the chemical analysis of plants growing in it.

biogeochemical cycle the circulation of an ELEMENT-6 within ECOSYSTEMS. CARBON CYCLE, NITROGEN CYCLE, PHOSPHORUS CYCLE.

biogeography the geography of organic life, the study of the spatial distribution of animate nature, i.e. of plants and animals (but usually considered to exclude human beings) and the processes that produce the patterns of distribution. Biogeography combines PHYTOGEOGRAPHY and ZOOGEOGRAPHY. G.

biogeosphere the outer part of the LITHOSPHERE down to the depth at which there is no organic life.

bioherm 1. an ancient rock mass built up by sedentary organisms such as CORAL, MOLLUSCS, ALGAE and/or their remains, e.g. an ancient coral reef (BIOSTROME), especially one surrounded by rocks of different origin (REEF KNOLL) 2. the sedentary organisms which formed it. G.

biohermal *adj.* of, or relating to, a BIOHERM.

biological amplification, biological magnification the concentration of a persistent chemical substance by the organisms at each successive TROPHIC LEVEL-1 of a FOOD CHAIN. The result is that at each successive trophic level the amount of the substance relative to the BIOMASS-1 is increased.

biological control the control of the POPULATION-1 of a PEST (including a PARASITE) by the use of its natural living enemies, e.g. by the introduction of a predator, or a virus, or of sterilized males of the pest or parasite.

biological indicator an organism, usually (but not always) a micro-organism, used to show the level of chemical activity, e.g. LICHEN, used to measure the level of sulphur dioxide in the atmosphere. PRESENCE INDICATOR.

biological magnification BIOLOGICAL AMPLIFICATION.

biology the scientific study of living things, of life in all its manifestations. The biological sciences are usually regarded as including ANATOMY, BIOCHEMISTRY, BIOPHYSICS, BIOSYSTEMATICS, BOTANY, ETHOLOGY, MICROBIOLOGY, PHYSIOLOGY, ZOOLOGY and all the other sciences with 'bio' as a prefix or with 'biology' incorporated in the title (e.g. population biology). BIOTECHNOLOGY, ECOLOGY.

bioluminescence the production of light of various colours, without heat, by some living organisms, e.g. glow-worms and fireflies, many BACTERIA and FUNGI, some PLANKTON and marine fishes, resulting from a biochemical reaction in the organism catalyzed by an ENZYME. NOCTILUCA, PHOSPHORESCENCE.

biomass in ecology 1. the total weight of organisms under consideration, e.g. in a specified area, or making up a particular TROPHIC LEVEL-1 or POPULATION-3 2. the total weight of a SPECIES per unit area. G.

biome a major ecological community (BIOTIC COMMUNITY) of living plants and animals, occupying an extensive area, e.g. DESERT, GRASSLAND-1, RAIN FOREST, TUNDRA. The plants of land biomes comprise FORMATIONS. ECOSYSTEM. G.

biometeorology the science of the relationship between living organisms and the weather.

biometry in ecology, the science of applying statistical techniques to biological knowledge, particularly to the differences and resemblances between groups of related organisms.

bionomics the study of the relation of a population of organisms or an organism to its enviroment, living or inanimate, a term used by some authors as a synonym for ECOLOGY.

biophysics the scientific study of biological material and processes by physical methods (e.g. by tracer techniques using radioactive compounds, ultrasonics, etc.), and the study of physical phenomena in living organisms. BIO-CHEMISTRY, PHYSICS.

bioplex biological complex, a biological system in which the waste products of each stage form the raw materials for a succeeding stage.

biosphere 1. the parts of the earth's crust and atmosphere (LITHOSPHERE, HYDROSPHERE, ATMOSPHERE) occupied or penetrated by living organisms **2.** only the living organisms. ECO-SPHERE **3.** the living organisms together with the parts of the earth's crust and atmosphere that they occupy or penetrate **4.** the part of the earth's crust and atmosphere favourable to at least some form of life, divided into three biocycles (salt water, freshwater and land). Following this last definition, in biogeography the term biochore has been applied to the geographical environment with a distinctive plant and animal life adapted particularly to climatic factors, each biochore thus being characterized by a major type of vegetation. The biochore is subdivided, the smallest division being the NICHE-1. BIOTOPE. G.

biostratigraphy STRATIGRAPHY.

biostrome a modern CORAL REEF, in the course of formation. BIOHERM. G.

biosynthesis the chemical synthesis produced by and within a living organism.

biosystem the interacting plants and animals in a given location.

biosystematics the classification of living things. CLASSIFICATION OF OR-GANISMS, TAXONOMY.

biota a collective term for the animal and plant life of a specific area and/or period of time. G.

biotechnology the application of biological knowedge (BIOLOGY) to industrial processes.

biotic *adj.* of, pertaining to, or relating to, living organisms.

biotic community a local association of interdependent plants and animals living in an area.

biotic factor an influence arising from the activities of living organisms, including people, which affects the environment, as distinct from such factors as the climatic (CLIMATIC ELEMENTS) and the EDAPHIC. G.

biotic index a system of classifying bodies of freshwater, usually rivers and streams, based on the type of INVERTE-BRATE organisms able to survive in them, thus indicating the quantity of dissolved OXYGEN present in the water and, in turn, the degree of POLLUTION.

biotic potential an estimate of the highest possible rate of increase of a SPECIES if there were maximum reproduction and minimum mortality in the absence of parasites and predators.

biotic province in ecology, a major ecological region of a continent.

biotic pyramid a graph showing the number of individuals at each TRO-PHIC LEVEL-1 in a stable FOOD CHAIN, from the primary producers, to the primary, secondary and tertiary consumers. It is inevitably shaped like a pyramid because the number at each trophic level decreases, as explained under food chain.

biotite a common, rock-forming mineral, a silicate (SILICATION) of iron, magnesium, potassium and aluminium, a form of MICA commonly occurring in IGNEOUS and METAMORPHIC rocks, especially as black crystals in GRANITE. It is glassy and transparent, the colour ranging from dark green through to brown-black. FERROMAGNESIAN MINERAL.

biotope 1. a term used by some ecologists to define a small community of plants and animals characteristic of a NICHE-1, the smallest division of an ECOSYSTEM. BIOSPHERE-4 **2.** a HABITAT in which there is uniformity in the main climatic, soil and BIOTIC conditions.

bipolar *adj.* pertaining to, occurring at, associated with, two poles or with the two polar regions. G.

bipolar distribution the distribution of certain species found in areas to the north and south of a median zone, but not in the intervening median zone itself. G.

bird a warm-blooded VERTEBRATE of the class Aves, with forelimbs modified to form wings and, apart from the feet (in some cases the legs also), covered with feathers. AVIAN, CHICKEN, DUCK, GOOSE, ORNITHOLOGY, TURKEY.

bird's foot delta, birdfoot delta a digitate or lobate DELTA with distributaries flanked by relatively narrow borders of sediments, projecting seawards in the pattern of a bird's foot, e.g. the Mississippi delta. DIGITATE MARGIN. G.

birth-rate, birth rate a FERTILITY-3 measure, in human population, the ratio of births to population within a given period, commonly measured by the average number of live births per 1000 of the population. This is the crude birth-rate, or natural increase, and currently ranges from 12 to 50, countries with a high standard of living recording between 15 and 20. The crude death-rate is similarly measured in deaths per 1000 of population, and varies from 6 to 25 or more. At present death-rates continue to fall in most countries, owing to improved medical skill and services; but there is only a slight tendency for birth-rates to fall, despite the spread of birth control (the avoidance of unwanted pregnancies by prevention of fertilization). As a result the world population is still (1980s) increasing at 1.7 per cent per annum, having fallen from over 2 per cent per annum in the 1960s. FECUNDITY, FERTILITY, NATURAL CHANGE, RE-PRODUCTION RATE. G.

biscuit-board relief, biscuit-board topography a rolling upland cut into by CIRQUES resembling large bites. G.

bise, bize (French) a dry, cold, penetrating wind, blowing from the north or northeast across southern France, northern Italy and parts of Switzerland in spring and winter, resembling the MISTRAL and the TRAMONTAINE. G.

biserial correlation coefficient POINT BISERIAL CORRELATION COEFFICIENT.

bismuth a metal, used particularly with TIN, LEAD, CADMIUM to produce ALLOYS which have low melting points and expand on cooling, making good castings; it is also used in making some colouring materials.

bit in mathematics, abbreviation of binary digit, i.e. one of the digits (0 or 1) in the BINARY CODE.

bitumen 1. a general name for various viscous or solid mixtures of native HYDROCARBONS which have lost much of their gaseous material, e.g. ASPHALT 2. tar, the residue from the distillation of COAL.

bituminous coal (American soft coal) humic coal, a COAL containing from 75 to 92 per cent CARBON, 4.5 to 5.6 per cent HYDROGEN and yielding from 15 to 45 per cent volatile matter when heated out of contact with air. The adjective bituminous is erroneous, based on the long-standing incorrect assumption that these coals contain BITUMEN. G.

blackband ironstone a variety of clay ironstone, an ore of IRON containing sufficient coaly material for it to be smelted without the addition of more fuel. It was formerly famous in the Staffordshire coalfields in England.

bivariate frequency distribution FRE-QUENCY DISTRIBUTION.

black body, blackbody a hypothetical perfect absorber and perfect emitter of radiant energy. RADIATION.

black box approach an approach in SYSTEMS ANALYSIS which ignores the internal structure and functioning present within the system under study and deals only with the nature of the output resulting from identified inputs. GREY BOX APPROACH, WHITE BOX APPROACH.

black cotton soil REGUR. G.

blackcurrant *Ribes nigrum*, a small shrub, native to Europe, bearing small black, edible, succulent BERRIES (classified as soft fruit), cultivated for its fruit in most of Europe and northern Asia.

blackearth, black earth a general term covering CHERNOZEM and the dark plastic CLAYS of tropical regions. G.

black frost a hard FROST without RIME. GLAZE.

black gram *Phaseolus mungo*, urd in the Indian subcontinent, woolly pyrol in West Indies, a leguminous (LEGUMINOSAE) food plant, a pulse, probably native to India, still mainly grown in the Indian subcontinent for its seeds, which are dried and eaten cooked, or for the pods which can be cooked and eaten whole; or (as in the West Indies) for its value as GREEN MANURE.

black ice a layer of GLAZE formed on roads, the ice being so clear that it is invisible and extremely dangerous to traffic.

black pepper PEPPER.

black sand a coarse SEDIMENT-2 of HEAVY MINERALS, a PLACER deposit, usually consisting of grains of MAGNETITE, ILMENITE AND HAEMATITE with small quantities of RUTILE, CASSITERITE, CHROMITE, FERROMAGNESIAN MINERALS, GARNET, GOLD, MONAZITE, ZIRCON.

black turf, black turf soil in South Africa, a very dark CLAY soil, 1 to 1.5 m (3 to 5 ft) thick, derived from BASIC IGNEOUS ROCKS, low in organic material, fairly fertile after rain, occurring in parts of the Transvaal. G.

black walnut *Juglans nigra*, a tall DECIDUOUS tree of temperate lands, native to North America, bearing large nuts with thick, hard, shells and strongly-flavoured KERNELS-2 which are used in confectionary, etc. in the USA. WALNUT.

blaen (Welsh, pl. blaenan) head, end, source of river, upland. It is used to contrast with BRO in south Wales, i.e.

the upland and the lowland, the hills and the vale. G.

blanket bog, blanket peat a BOG occurring on a relatively horizontal land surface in regions of high rainfall and low evaporation, and covering the countryside like a blanket, except on steep slopes and rock outcrops. It is common in Ireland. RAISED BOG, VALLEY BOG.

blast furnace a furnace used to produce molten IRON. Poor quality iron ores are subjected to a preliminary roasting to remove volatile impurities. The furnace, a large, vertical steel shell lined with refractory bricks, is then charged with the ore, with COKE and LIMESTONE, through which hot air is blasted. Carbon monoxide from ignited coke reduces the oxides to iron, and the limestone acts as FLUX, so that molten iron flows to the furnace bottom, to be run off and cast into blocks known as pigs, hence the term PIG IRON. The process is now superseded by electric smelting. ARC FURNACE, STEEL.

blight in physical planning, the lowering of the value of land and buildings brought about by official planning proposals which indicate a change of land use or a shortening of the life of the existing buildings. The condition of property affected by planning blight may deteriorate as owners cease to be concerned about its upkeep; and it may become unsaleable from the time when plans are first discussed or from the official designation, to the time of redevelopment, despite the fact that in Britain compensation becomes payable on official acquisition. BETTERMENT.

blijver TREKKER.

blind creek an obsolescent term, a creek that is dry except in winter, termed a draw in USA. G.

blind valley a valley in LIMESTONE country, dry or with a stream, which ends in a steep wall, into the base of which the surface flow of water disappears underground. G.

blizzard originally a very cold, strong, northwesterly wind with fine, drifting snow, associated in the USA in winter with the passage of a DEPRESSION-3; later applied to any very cold, strong wind accompanied by falling or drifting snow, sometimes involving a WHITE-OUT. G.

bloc a group formed of like-minded governments, or political parties, etc. associating together for mutual benefit.

block diagram 1. a perspective drawing giving a three-dimensional impression, used particularly to show landforms 2. a diagram showing the relationship between the surface form of the ground and the underlying geological structure by representing an imaginary block cut out of the earth's crust. G.

block disintegration the mechanical breaking-up of bedded, jointed rocks, usually by FROST action along lines of weakness.

block faulting, block-faulting faulting (FAULT) in which part of the earth's crust is divided into a number of small blocks by a series of faults, in many cases two sets roughly at right angles, some of the blocks being moved up, others moved down, others tilted. FAULT BLOCK, TILT BLOCK. G.

block field FELSENMEER.

blocking *adj.* applied to an atmospheric high pressure system (ANTICYCLONE) which is nearly stationary and, lying in the path of approaching DEPRES-SIONS-3, blocks their passage.

block lava a term used by some authors as synonymous with AA, restricted by others to angular blocks of congealed lava, more acid than the true aa. G.

block mountain a mountain which is structurally an uplifted FAULT BLOCK, prominent because it has been thrown up by earth movements, or because the surrounding land has sunk. HORST, TILT BLOCK. G.

block slumping MASS MOVEMENT down the steep face of an escarpment or sea cliff of well-jointed rocks, e.g. of chalk or limestone slipping on an underlying clay stratum which has become wet through water seepage. The block usually breaks away sharply from the steep face and slumps downward with a rotational movement.

blood rain RAIN coloured by red dust particles in the atmosphere that have been carried by wind from desert regions, in some cases over great distances, e.g. in Italy the red dust washed down during precipitation from the Sahara. G.

blossom shower MANGO SHOWER, a rain shower occurring in March, April, May in southeast Asia, i.e. before the bursting of the monsoon. G.

blowhole, blow-hole, blow hole souffleur, a nearly vertical, smallish hole on land near the seashore, the land aperture of a funnel-shaped CAVE-1. Sea waves force air and water up from the cave through the small opening, so that a spout of spray is carried high in the air. A blowhole is formed when erosion occurs along a vertical or nearly vertical JOINT which passes from the land surface to the cave roof. G.

blowout, blow-out 1. a hollow (deflation hollow) made by eddying wind in tracts of light or sandy soil, occurring especially in a coastal sand dune area or in an arid plain if vegetation cover is lacking (DEFLATION) 2. a sudden, violent escape of gas or steam. G.

blow-well, blowing well 1. a fountain in eastern Lincolnshire and eastern Yorkshire, England, made by boring through a clay AQUICLUDE to the underlying chalk AQUIFER 2. a natural SPRING on the warpland (WARP) of Holderness and the Humber estuary, eastern England. G.

blubber the subcutaneous fat of a whale or other marine mammal from which TRAIN OIL is obtained.

blue-band a layer of clear blue ice without air bubbles in a GLACIER, formed from frozen melt-water.

blueberry a shrub native to the cool

regions of the northern hemisphere, producing juicy, blue-black fruit used in making the traditional blueberry pie of North America. Those in cultivation are Highbush blueberry, *Vaccinium corymbosum*, a large bushy shrub thriving best on moist, acid, peaty soils; and the hardier Lowbush blueberry, *V. angustifolium*, a low shrub grown commercially in northeastern USA and Canada.

blue-collar worker a person who is engaged in and paid a wage for manual work. PROFESSIONAL-1, STEEL-WORKER COLLAR WORKER, WHITE-COLLAR WORKER.

bluegrass any of the grasses of the genus *Poa*, with bluish-green jointed stems, a good pasture-grass, e.g. Kentucky bluegrass.

blue ground (South Africa) the softer zone in diamond-bearing KIMBERLITE, overlying HARDEBANK, and crumbling on exposure to the atmosphere to form YELLOW GROUND. G.

Blue John a blue variety of fluorite (FLUORSPAR), CaF_2, occurring in Derbyshire, England, used ornamentally.

Blue Mud a very fine plastic TERRIGENOUS DEPOSIT, coloured by iron sulphide (IRON), occurring on the CONTINENTAL SLOPE. GREEN MUD, RED MUD.

blue sky the apparent colour of the cloudless sky in daylight, due to the scattering of sunlight (with frequencies corresponding to the blue region of the visible solar SPECTRUM-2) by obstructing MOLECULES in the air.

bluff a steeply rising slope marking the outer margins of the FLOODPLAIN of a river, especially the almost perpendicular, steep slope cut by the stream as it erodes the concave side of a MEANDER. RIVER-CLIFF. G.

boar an uncastrated male PIG.

boca (South America: Spanish) **1.** the mouth of a river or gorge, etc. **2.** the point at which a stream leaves a gorge and enters the plain. G.

bocage (French, woodland) a landscape of small fields bounded by low banks surmounted by hedges with patches of trees (especially pollarded oak and ash) or, in some areas, by drystone walls: the traditional rural landscape, resembling a chessboard, typical of northwestern France. G.

bochorno (Spanish) a warm wind blowing up the Ebro valley. G.

Bodden (German) an irregularly shaped inlet of the southern Baltic coast formed when the sea level rose and flooded the former land surface. SUBMERGED COAST. G.

Boerde, Börde (German, pl. Boerden) the sub-HERCYNIAN loess zone in northern Germany, or individual regions within that zone, e.g. Magdeburg boerde. LOESS. G.

bog **1.** loosely applied to any soft, wet, spongy soil or ground into which the foot sinks **2.** properly applied to an area of wet acid peat and the vegetation complex associated with its poorly drained or undrained surface, i.e. the natural group of wet peat-forming and peat-inhabiting plants (PEAT), sphagnum being characteristic. Bogs occur in areas of poor drainage where lack of oxygen in the waterlogged soil inhibits the decomposition of dead plants, leading to the build-up of humic and other acids which modify plant structure and function. BLANKET BOG, FEN, MARSH, PALSA BOG, QUAGMIRE, RAISED BOG, SCHWINGMOOR, SWAMP, VALLEY BOG.

bogaz (Slavic) a long, narrow chasm in LIMESTONE (KARST), formed by CARBONATION-SOLUTION along a JOINT. STRUGA. G.

boghead coal a type of dark brown coal, revealing remains of algae, with a carbon content lower than that of BITUMINOUS COAL. It yields some oil.

bog iron ore a loose, porous, earthy form of LIMONITE formed in a peat bog or swamp, probably by oxidation (by algae, bacteria, or by the atmosphere) of solutions containing salts of IRON. G.

bog moss SPHAGNUM.

bog peat, moss peat acid, brown PEAT. The plant structure is visible, the cellulose content high, and it supports a vegetation of bog moss (SPHAGNUM) or SEDGES. FEN PEAT.

bohorok (Sumatra) a FOHN-type wind blowing down from the Karo mountains during the northeast monsoon. G.

boli (Sierra Leone) wide inland marshes, flooded in the rainy season and occurring where major rivers cross a belt of old SEDIMENTARY ROCKS.

boiling point steam point. TEMPERATURE SCALE.

bolson (American) **1.** a desert basin surrounded by mountains or high ground **2.** a flat-floored desert valley draining to a PLAYA, especially one in southwest USA and in Mexico.

boma (east Africa: Swahili) **1.** a place protected by a stockade, strong hedge or earthwork **2.** a cattle enclosure **3.** a fort **4.** government offices, particularly the administrative headquarters. G.

bond in chemistry, any of the several forces which hold ATOMS, groups of atoms, or IONS together in a molecular or crystal structure. MOLECULE, VALENCY.

bone ash a whitish ash, mainly CALCIUM PHOSPHATE, obtained by heating animal bones in air. Bone ash is used industrially, especially in industrial chemistry and in making PORCELAIN.

bone bed a STRATUM rich in CALCIUM PHOSPHATE, containing fossilized fragments of the skeletons (e.g. bones, scales of fish) and teeth of VERTEBRATES.

bonemeal the finely crushed or ground bones of animals. It is fed to animals or used as a FERTILIZER.

bonitative map a map indicating land suitable or unsuitable for some specific economic development.

booley, booly, bouille (Irish; other spellings) **1.** a temporary fold or enclosure used by a herdsman in summer **2.** a company of people and their herds wandering in summer, hence booling, applied to this practice.

boolyunyakh, bulgun(n)yakh (Russian: Yakut) a HYDROLACCOLITH, a term sometimes used in American translations of Russian works. G.

bora, borino (Italian) a very cold, often dry, violent north or northeasterly wind (but sometimes accompanied by rain or snow) blowing mainly in winter down from the mountains on to the eastern coast of the Adriatic (comparable with the MISTRAL). The borino is a weaker form blowing in summer. The term bora-type is applied to winds similarly blowing down moderately high mountains from a cold, continental high pressure area towards a low pressure area over warm sea shores or lowlands. G.

borax a mineral, SODIUM tetraborate, $Na_2B_4O_7$, occurring as a surface incrustation in many deserts, notably in Chile and Peru. It is used as a cleansing agent, a mild antiseptic, a flux in soldering, and in PORCELAIN and glass making.

border 1. the district lying along the margin of a country **2.** the actual boundary line, the FRONTIER separating one country from another **3.** specifically, The Border, the boundary or boundary lands between England and Scotland **4.** in USA, in pioneer days, the frontier between the occupied and the unoccupied zones of the country.

bore 1. a tidal wave of some considerable height which regularly or occasionally rushes up certain rivers or narrowing gulfs. EAGRE **2.** a deep hole drilled in exploration for oil or water. G.

boreal (Boreas, Greek god of the north wind) *adj.* belonging to the north, applied especially to **1.** the northern CONIFEROUS FORESTS **2.** the climatic zone with snowy winters and short summers **3.** the climatic period from 7500 to 5500 BC. PREBOREAL.

boric *adj.* of, relating to, or derived from, BORON.

boric acid H_3BO_3, a white crystalline, naturally occurring weak acid. BORON.

Bornhardt (German) a steep-sided, residual hill rising from a plain, a term used by some authors as synonymous with INSELBERG; but more precisely applied by others, particularly by some English authors, to a GRANITE-GNEISS inselberg, a feature of the second erosion cycle in rejuvenated desert or semi-desert country. SHIELD INSELBERG. G.

bornite Cu_5FeS_4, a compound of COPPER sulphide with IRON sulphide, a major ore of copper, occurring in HYDROTHERMAL veins and in the zone of SECONDARY ENRICHMENT.

bōrō (Indian subcontinent: Bengali) summer rice or rice-crop, grown on low-lying land close to irrigation water. G.

boron a non-metallic chemical element occurring as BORAX and BORIC ACID. It has a high melting point, is used in hardening steel, in making enamels and glass, in NUCLEAR REACTORS (it absorbs slow neutrons) and, combined with CARBON, as an abrasive. It is also an essential MICRONUTRIENT for plants. TRACE ELEMENT.

borough 1. in England and Wales, historically, a fortified town or place with a MUNICIPAL organization **2.** a town with a municipal corporation and special privileges given by royal charter. Though the organization may differ, the Scottish burgh (also in older English usage) is roughly similar. Both borough and burgh appear in place-names. COUNTY BOROUGH, LOCAL GOVERNMENT IN BRITAIN, MUNICIPAL BOROUGH, PARLIAMENTARY BOROUGH, ROTTEN BOROUGH **3.** in some States of the USA, a municipal corporation corresponding to an incorporated town or village of other States

4. in New South Wales, Australia, a town incorporated by parliamentary Act of 1857, or a town holding a royal charter; in other Australian states, a municipal area of certain minimum size and population **5.** in New Zealand, a village, township, town with a special governing body (borough council). G.

boss a small BATHOLITH, a STOCK with an upper surface, when exposed by denudation, roughly circular in cross-section.

bosveld BUSH VELD.

botanical garden an enclosed piece of land where plant collections are grown for scientific study and display.

botany the scientific study of plant life and all its manifestations.

botany wool (from Botany Bay, east coast of Australia) fine WOOL from MERINO sheep, mainly from those in Australia.

bote ESTOVERS.

bottom 1. the floor of a sea, lake, pond, river, etc. **2.** the low land or alluvial plains bordering a river, especially those of the FLOODPLAIN of the lower Mississippi river **3.** a dry valley floor, especially in chalk and other limestone regions in southern England, where bottom appears in place-names. G.

bottom ice ANCHOR ICE.

bottom load BED LOAD.

bottomset beds, bottom-set beds layers of fine-grained silt carried out and deposited on the bottom of a sea or lake in front of (and later to be covered by) the foreset beds (DELTA STRUCTURE) of an advancing DELTA. G.

boulder loosely applied to any large, detached, generally rounded mass of rock, larger than a COBBLE, especially one transported by ice, river, or sea, from its original home, but also in some cases one weathered by DISINTEGRATION or EXFOLIATION in situ (MOUNTAIN TOP DETRITUS), specifically exceeding in diameter 200 mm (8 in) in UK, 265 mm (10.5 in) in USA. G.

boulder clay, boulder-clay unstratified,

unconsolidated GROUND MORAINE of mixed rock debris transported by ice and deposited when a former ice sheet or glacier has melted. Whilst usually defined as consisting of stiff clay enclosing boulders of various sizes, in some examples the matrix may be mainly sand instead of clay, and boulders may be few or even absent. For this reason the term has now been dropped in favour of TILL, because till does not specify the constituent materials. Past references indicate boulder clays as yielding soil of varied character: that derived from CHALKY BOULDER CLAY (with pebbles of chalk) providing good agricultural land, that derived from stiff clay with many boulders being almost useless for agriculture. G.

boulder field FELSENMEER.

boulder-train a series of ERRATICS originating from a common bed rock, transported and deposited by a GLACIER either in line or in the shape of a fan with its apex pointing to their source. In either case the erratics signpost the route taken by the former glacier.

boulevard (French) a broad avenue or street, usually laid out with trees or other plants, including grass, properly one occupying the site of a former rampart or town wall. But the application has been extended, especially in Britain and the USA, to various wide, new roads, e.g. ring roads encircling cities, or scenic highways.

boundary a line of demarcation, real or understood, visible or invisible, natural or artificial, of legal or of no legal significance, which may be perceived from either side (or both sides) of it, e.g. between countries (synonymous with FRONTIER) or administrative areas, between regions of various types, between market areas, between service areas. ANTECEDENT BOUNDARY, BORDER, LIMIT, RELICT BOUNDARY, SUBSEQUENT BOUNDARY, SUPERIMPOSED BOUNDARY. G.

boundary current an OCEANIC CURRENT with margins (boundaries) indicated by abrupt changes in temperature and salinity, flowing along the western margins of the deep ocean and forming part of the circulation of deep oceanic water, i.e. southwards in the Atlantic from the deep cold water off south Greenland, northwards in the Pacific from the depths south of New Zealand, and northwards from the cold water off southern Africa towards the Arabian sea.

bounded rationality the concept that a person cannot be completely rational, however hard that individual tries, because no-one can have perfect knowledge or a perfect ability to calculate. Thus the concept of ECONOMIC MAN becomes unreal.

bourbon whisky distilled from corn (MAIZE) mash.

bourgois, bourgoisie (French) **1.** broadly, the middle classes **2.** in Marxism, the capitalist, property-owning class. MODE OF PRODUCTION, PETITE BOURGOISIE, PROLETARIAT, SOCIAL CLASS.

bourne a temporary or intermittent stream which may flow in a DRY VALLEY in chalklands, the intermittent character being due to variations in the level of the WATER TABLE. In winter, when the water table rises above the height of the valley floor, there may be a surface stream, hence the term winterbourne. In summer, when the water table sinks below the level of the valley floor, the stream bed becomes dry. Bourne is commonly incorporated in place-names on the chalklands of southern England, e.g. Bournemouth, Eastbourne. G.

bovate YARDLAND.

bovine *adj.* of, or pertaining to, an OX or a COW.

box-canyon a term applied in the western USA to a CANYON with more or less vertical walls, to distinguish it from canyon, a term commonly applied there to every young valley. G.

BP before the present day, used as a measure of time to avoid the necessity of using BC (before Christ, i.e. in the year before the reputed date of the birth of Christ) or AD (Latin Anno Domini, in the year of our Lord, i.e. in the year since the reputed date of the birth of Christ).

brachycephalic sometimes abbreviated to brakeph, *adj.* short-headed, broad-headed. CEPHALIC INDEX.

bracken a member of a large genus of fern, *Pteridium*, widely distributed from the Arctic to the Tropics, particularly on acid grasslands and heaths, where it becomes a weed difficult to eradicate. Domestic and wild animals will not eat it and it recovers swiftly from burning.

brackish *adj.* applied to water which is slightly salt.

brae (Scottish or Northern Ireland dialect) **1.** a steep bank bounding a river **2.** a hillside. BROW **3.** an upland district, e.g. Braemar. In Scotland, brae may be included in street names if the road has a steep gradient, e.g. Windmill Brae, Aberdeen. G.

braided river course (to braid, to twist in and out, to interweave) an anastomosing river course (ANASTOMOSIS), a stream with a wide, shallow channel split into many small, shallow, interlaced channels separated by bars of alluvial material, visible when the water is low. It occurs particularly when a heavily laden shallow stream deposits so much sediment in its channel that the channel becomes too small, and part of the stream breaks out to follow a new course on the flat land of the valley. This process, which is most likely to happen if the banks of the main stream channel consist of easily erodible deposits, may be repeated; and the diverging streams may or may not return to the main channel. G.

brake a dense clump of bushes, a thicket. G.

brakeph abbreviation of BRACHYCEPHALIC. G.

brandy a spirit distilled from WINE, the finest being made from the white wine of Cognac, France.

Brandt Report *North–South: a Programme for Survival*, the title of the report of the Independent Commission on International Development Issues, published 1980. The Commission was set up in December 1977 (at the invitation of Robert MacNamara, Chairman of the World Bank) under the chairmanship of Willy Brandt, German statesman and winner of the Nobel peace prize 1971. It consisted of people with varied political and professional experience outside the countries with a communist form of government. Eight members represented the North, ten the South. Under the terms of reference of the Commission 'global issues arising from economic and social disparities of the world community' were to be studied, and 'ways of promoting adequate solutions to the problems involved in development and in attacking absolute poverty' were to be suggested. The term North–South is accepted as a misnomer, a very broad generalization to stress the great social and economic imbalance between the rich, developed countries of the North, the northern hemisphere (i.e. North America, excluding Mexico, the countries of Europe, the USSR, China, Japan, to which are added Australia and New Zealand from the southern hemisphere) and the poor, developing countries of the South (very broadly, the rest of the world). There are of course anomalies in each, e.g. the South under that definition includes the developing but rich oil-exporting countries of Arabia. Briefly to summarize the recommendations of the Commission, the members advised the setting-up of a five-year emergency programme to promote food production for the world's rapidly increasing population; to find new sources of energy; to deal with the transnational companies; to transfer financial resources from the rich to the poor countries; to start to reorganize the international institutions with the aim of establishing a reformed economic system.

brash a mass or heap of fragments, especially loose, broken rubbly rock underlying the soil and subsoil, partially weathered in situ. The term is rarely used except in CORNBRASH, a brash or brashy soil. G.

brass ALLOY.

brassica a genus of edible plants of the Cruciferae family (e.g. BROCCOLI, BRUSSELS SPROUTS, CABBAGE, CAULIFLOWER, KALE, KOHLRABI in Europe, PAK-CHOI, PE-TSAI, SHUN-GIKU in eastern Asia) much used in temperate lands as green vegetables for human consumption and as winter fodder for cattle and sheep.

Braunderde BROWN FOREST SOIL.

Brave West Winds the westerlies, the planetary west or northwest winds blowing over the oceans of the southern hemisphere in midlatitudes (40°S to 65°S) where they blow with considerable force and regularity, swinging to north or south under the influence of seasonal change of world ATMOSPHERIC PRESSURE belts. ROARING FORTIES. G.

Brazil nut the seed of *Bertholletia excelsa*, a tall forest tree native to tropical South America. The NUT has a highly nutritious oily, edible, white-fleshed KERNEL (66 per cent fat, 14 per cent protein). Commercial quantities are obtained from wild trees, mainly in Brazil and Venezuela, but the felling of forests is greatly reducing the supply.

breached anticline an ANTICLINE in which the drainage, developed along the ridge (the axis) of the anticline, has eroded the overlying rocks along this line of weakness, revealing the underlying older rocks, and thus creating an ANTICLINAL VALLEY with escarpments facing inwards. G.

breadcrust bomb a VOLCANIC BOMB

with cracks in its glassy crust caused by the shrinkage of the lava as it cooled.

breadfruit *Artocarpus communis*, a large tree, native to the Pacific islands and southeast Asia, now established in most parts of lowland tropical regions, bearing starchy, edible fruit, usually eaten roasted.

breaker a mass of turbulent water and foam, breaking violently against a rocky shore or passing over a reef or shallows, formed when a heavy ocean WAVE-3 rushes from deep to shallow water, so that its CREST steepens, rolls over, and breaks. This occurs particularly when the ratio of wave height to wave length is greater than 1:7. G.

break of bulk the stage at which a shipment is divided into parts, typically at a port where it is transferred from water to land transport.

break of profile a KNICKPOINT. A term used by some authors who wish to avoid the term knickpoint, of mixed German–English origin, derived from Knickpunkt. G.

break of slope any more or less sudden change in a slope, e.g. of a hillside. G.

break-point bar an OFFSHORE BAR-1 formed where the waves first break.

breccia (Italian, gravel, or rubble of broken walls) a rock consisting of angular fragments of other rocks cemented together by some finer material (CRUSH BRECCIA, DESICCATION BRECCIA, FAULT BRECCIA, GASH BRECCIA). The term is not applied in English to a CONGLOMERATE in which the fragments are rounded. AGGLOMERATE, AUTOCLASTIC. G.

breck, breckland 1. a tract of heathland (HEATH) with thickets 2. a tract of land supporting such vegetation, cleared for cultivation from time to time, then allowed to revert. G.

breed a particular group of domestic animals or of plants with similar characteristics inherited from common ancestors. Hence animal breeding, plant breeding, the process of getting animals or plants to reproduce offspring from selected parents, by providing advantageous conditions, in order to produce new types or varieties with characteristics most useful to human beings.

breeze a light wind. BEAUFORT SCALE, BREVA, LAND BREEZE, SEA BREEZE, VALLEY WIND.

breva (Italian) a warm, damp valley breeze blowing upward from the Italian lakes in morning and afternoon.

brickearth, brick-earth 1. originally any earth, usually a loamy CLAY from which bricks could be made 2. in current use, a fine-grained deposit overlying the gravels on river terrances, e.g. on certain of the Thames terraces. Originating from wind-blown material that has been re-worked, re-sorted and re-deposited by water, it has been likened to LOESS. It forms a fertile, friable soil. G.

brickfielder, brick fielder a hot, dry, dusty, squally wind, blowing in southeastern Australia in summer southwards from the interior in front of a DEPRESSION-3.

bridge a structure built over a river, ravine, road, or railway, entrance to a bay, or connecting two points high above general ground level, and carrying a path, road, railway, according to need. LAND BRIDGE.

bridgehead, bridge-head in military terminology, a fortification at that end of a bridge which is nearer to the enemy, hence an advanced position held in territory occupied by the enemy.

bridge-point, bridging point a point at which a river is or could be bridged. It is applied especially to the lowest downstream point at which this is possible, often an important factor in the original location of a settlement.

bridlepath, bridleway 1. a path fit for the passage of a horse or a pedestrian, but not for a vehicle 2. a path which may in English law have a right of way for pedestrians and riders on horseback, but not for wheeled vehicles. G.

brig, brigg a headland or landing place formed by a scarp, consisting of hard rock, specifically the oolitic (OOLITE) limestone outcrops at or near tide marks along the coast of Yorkshire and Northumberland, England. G.

brigalow in Australia, scrub, mainly of *Acacia* species, especially *A. harpophylla*, bordering the MULGA in dry areas of Australia.

Brillouin region, first order DIRICHLET POLYGON.

brimstone SULPHUR.

brine a very salt solution, commonly containing a higher proportion of a dissolved salt than that occurring in sea-water.

brine pan a shallow pit or vessel used in the process of extracting salt from salt water by evaporation.

British Summer Time BST.

briquette, briquet a brick-shaped block of compressed coal fragments, usually of BROWN COAL or LIGNITE, the calorific value being high because water has been expelled in the compression.

bro (Welsh) a region, vale, or lowland. BLAEN. G.

broad bean *Vicia faba*, a hardy ANNUAL plant of the pea family, grown since the IRON AGE in Europe for its broad, flat, edible seed, the shelled beans from mature pods being cooked and eaten, hot or cold, the immature pods being cooked whole, complete with beans. Large quantities are grown for canning and freezing; and some varieties provide FODDER-1.

Broad, Broadlands, the Broads a local term applied in East Anglia, England, to shallow fresh water lakes formed by the broadening out of a sluggish river, the sites where peat for fuel was dug out in the middle ages. G.

broadleaved trees any tree of Dicotyledonae, many with a leaf form generally wide in relation to length. Most are DECIDUOUS, but some are EVERGREEN.

broadtail the skin or fur of a premature

KARAKUL lamb, very flat and wavy in appearance.

broccoli, sprouting a hardy, edible variety of CAULIFLOWER, with loose terminal clusters of purple or white flowerheads on the branches and smaller flowerheads in leaf axils near the base. Green sprouting broccoli, or calabrese, is a tender Italian variety. All types are usually eaten cooked.

Brockenspectre, Brocken spectre (from Brocken, a peak in the Hartz mountains, Germany) an atmospheric phenomenon in mountains in which an observer standing on a mountain peak lying between the sun and a cloud bank sees his/her much enlarged shadow projected on the cloud, encircled by rings of coloured lights.

brockram a SEDIMENTARY ROCK, consisting of angular blocks, occurring in PERMIAN strata west of the Pennines.

broiler a CHICKEN for killing up to ten weeks of age, and therefore sufficiently tender for fast cooking, e.g. by grilling.

bromide any compound of metal with BROMINE. POTASSIUM bromide is used as a sedative in medicine.

bromine a corrosive chemical element, dark red liquid at room temperature, boiling at 58.8°C (138°F), obtained from MAGNESIUM bromide in the Stassfurt deposits, occurring in seawater as BROMIDES. It is used as a catalyst, and in making pharmaceuticals, dyestuffs, photographic materials.

bronze a metal, an ALLOY of COPPER and TIN, special types also with other elements, hard and resistant to moisture and weathering. It expands when solidifying, and thus makes good castings. It ante-dates iron smelting. BRONZE AGE.

Bronze Age an era in human development (succeeding the PALAEOLITHIC, MESOLITHIC and NEOLITHIC and preceding the IRON AGE) when BRONZE was used for tools and weapons. In Europe the COPPER used for personal ornaments in Mesopotamia c.4500 BC

came to be alloyed with TIN by 3000 BC, and the use of this alloy, bronze, had spread widely by 1500 BC, to be gradually superseded by IRON by about 1000 BC. Writing and arithmetic developed in the Bronze Age, the plough, wheeled vehicles, and animals for riding and pulling, came to be used; towns were formed, work became specialized, there was trading and shipping.

brood parasite an animal that depends on another for the rearing of its young, e.g. the cuckoo.

brook 1. a small stream, a rivulet 2. in USA, a natural stream smaller than a river or CREEEK-2, especially one that breaks directly out of the ground (as from a spring). G.

brow 1. the upper part of a hill 2. the projecting edge of a cliff 3. in northern England, synonymous with BRAE in Scotland, a hillside, a slope or ascent. G.

brown alluvial soils, brown calcareous soils BROWN SOILS.

brown coal a brown, fibrous deposit, intermediate between PEAT and BITUMINOUS COAL, a term sometimes used as a synonym for LIGNITE, but more precisely is restricted to a stage nearer bituminous coal when the vegetable origin is no longer apparent because the plant fragments have been changed into an amorphous mass. Brown coal is usually worked OPENCAST. Heavy and soft, it is used mainly in thermal power stations near to the minehead. For domestic and industrial use it is made into BRIQUETTES.

brown forest soil, brown earth, Braunerde the rather unsatisfactory name often applied to a world group of ZONAL SOILS with merging horizons, generally associated with the lands in midlatitudes formerly covered with DECIDUOUS woodland, i.e. the region south of the BOREAL coniferous forest or TAIGA, in northeast USA, northern China, central Japan, northwestern and central Europe. There the humid climate and

MULL-2 lead to the formation of a slightly leached, slightly acid A HORIZON, with a grey-brown lower layer (less leached than that of a PODZOL) and a B HORIZON that is granular, thick, dark brown, with BASES-2 and COLLOIDS from the A horizon. CALCAREOUS brown forest soils are included. The acid brown forest soils of the FAO classification (SOIL ASSOCIATION-3) are identified as having A (B) C profiles (SOIL PROFILE), with merging horizons, and very little CLAY ILLUVIATION. The humus form is usually MULL and the structure of the B horizon is weakly developed. INCEPTISOLS, SOIL, SOIL ASSOCIATION. G.

brown Mediterranean soil RED MEDITERRANEAN SOIL.

brown ores soils leached (LEACHING) of the more soluble minerals but holding IRON oxides and hydroxides.

brown podzolic soils one of the subdivisions of podzolic soils in the 1973 SOIL CLASSIFICATION of England and Wales. PODZOLIC SOILS.

brown sands BROWN SOILS.

brown soils one of the seven groups in the 1973 SOIL CLASSIFICATION of England and Wales. It includes argillic brown earths (brown or reddish, with a loamy horizon overlying a clay layer), brown alluvial soils (non-calcareous, developed on new alluvium), brown calcareous soils (deep, organic, fertile soils of high agricultural quality, developed particularly on limestone, the A HORIZON being reddish-brown overlying a lighter B HORIZON (MOLLISOLS), BROWN EARTHS, and brown sands (a group of brown earths developed on freely drained, non-alluvial deposits of sand and gravel). BROWN PODZOLIC SOILS.

Brückner cycle a 35-year cycle of climatic change, known before E. Brückner drew attention to it, but reiterated by him in 1890, supported by his study of rainfall and temperature records going back to the eighteenth century. He identified irregular alternations of cold, wet periods with hot, dry periods, indicated by the advance and retreat of Alpine glaciers, variations in the dates of opening and closing of Russian rivers, in the level of the Caspian Sea and its inflowing rivers, in the date of the grape harvest, in the price of grain, and in other phenomena.

brumbie, brumby (Australian) the FERAL-1 horse of Australia.

brunizem PRAIRIE SOIL.

brush 1. SCRUB-1 or BUSH-3, or a thicket of small trees and shrubs **2.** vegetation of low, woody plants, especially SAGEBRUSH, in USA.

brushwood 1. small trees and shrubs, undergrowth, thicket **2.** the wood from small branches broken off trees and shrubs.

brussels sprout, Brussels sprout *Brassica oleracea gemmifera*, a cultivar of CABBAGE, with dense, compact axillary buds (also termed Brussels sprouts) resembling small cabbages, suitable for freezing, commonly used as a cooked vegetable, particularly in northern Europe. BRASSICA.

BST British Summer Time, one hour in advance of GMT (Greenwich Mean Time).

buckwheat a low ANNUAL herb (not a true CEREAL-1,2), a member of *Fagopyrum*, family Polygonaceae, especially *F. esculentum* and *F. tataricum*, which will grow on poor soil in temperate lands. It is free from disease and matures quickly, so it is sometimes planted as a CATCH CROP. The seed is fed to livestock and, in North America, ground for flour. It is a source of RUTIN.

buffalo properly, a kind of ANTELOPE, but generally the term is applied to wild oxen of several species, especially *Bos bubalus*, native to the Indian subcontinent but inhabiting most of Asia, southern Europe and parts of north Africa, domesticated and used as a draught animal and beast of burden.

The term is also, but incorrectly, applied to the American bison.

buffalo grass *Sesleria dactyloides*, a grass growing in the PRAIRIES.

buffer state a relatively small and weak state which, lying between two large and powerful ones, may serve to lessen the likelihood of a direct clash between them. LANDLOCKED. G.

building stones rocks of many types used in building, the best being tough and resistant to weathering, e.g. GRANITE, SANDSTONES, LIMESTONES. If either of the latter two split freely in any direction, a much valued characteristic, they are termed FREESTONES.

built-up area in physical planning, that part of a town where the land is so covered with buildings and roads, etc. that there is space for further similar development only if existing structures are demolished. Very small plots of land not built over (e.g. small gardens, school playgrounds, etc.) and derelict land awaiting redevelopment may be included. DEVELOPMENT-2. G.

bull the male of any BOVINE animal (especially, when uncastrated, of the domesticated species, *Bos taurus*) and of certain other large animals, e.g. elephant, sea-lion. BULLOCK, COW.

bullock a castrated BULL; an OX, a STEER.

bulrush a member of *Typha*, family Typhaceae, a genus of reed-like marsh plants with fibrous stems (used in thatching, paper-making, rush mats and chair seats), a dense spike of flowers, and downy-haired seeds; or a member of *Scirpus*, family Cyperaceae, the common bulrush, ANNUAL or PERENNIAL, growing in a similar habitat. It too has stems used in making rush mats and chair seats, but it carries many flowering spikelets gathered in terminal tufts.

bulrush millet *Pennisetum typhoideum*, in white-seeded form sometimes termed pearl millet, a tall, drought-resistant MILLET with stems bearing long cylindrical ears (seed-heads), generally resembling a bulrush, more widely grown than any other food crop in tropical areas with a low rainfall. It may have originated in Africa, and is now an important food crop in Sudan, northern Nigeria and other countries on the southern Saharan border, as well as in the driest areas of the Indian subcontinent.

bunch grass, bunch-grass, bunchgrass any of the coarse grasses which grow in clumps or bunches (instead of forming a continuous cover of matted turf), in many cases separated by bare ground, e.g. in the semi-arid western plains of North America. It is also termed tussock grass, e.g. in New Zealand. G.

bund, bundis (Persian origin, extended and used particularly in the Indian subcontinent) any artificial embankment, large (a dam, dike, causeway, and especially a river embankment) or small (a low earth ridge between ricefields). G.

Bund (German) a union or confederation of states, especially of German states. G.

bunkering the fuelling of ships or vehicles.

buran a strong northeast wind blowing in central Asia at all seasons but most frequently and fiercely in winter (then termed white buran or poorga) when it lifts and carries the snow, and ice particles. In the TUNDRA, especially in southern Russia and Siberia, it is termed purga. KARABURAN. G.

bureaucracy 1. government by officials who are appointed, not elected. It usually incorporates a fixed official area of jurisdiction for the employed officials, who also have fixed duties, are full-time, and whose personal affairs (particularly financial affairs) must be separated from their official function. There is commonly a hierarchical system of command, a centralized authority, a central filing system 2. the rigidity, red tape and emphasis on procedure resulting from such a system **3.** government officials, collectively **4.** in the social sciences, as an ideal type of organization, an organization based on principles that are relevant to various administrative systems, characterized by rationality in decision making, impersonality in social relations, tasks performed according to routine, centralization of authority.

burgage **1.** in England, historically, a form of medieval TENURE whereby lands or tenements in a city or BOROUGH-1,2 were held of the king or lord for an annual, fixed, money rent **2.** the property (and the freehold property) so held.

burgess historically, **1.** loosely, an inhabitant of a BOROUGH-1,2 **2.** a freeman of a borough **3.** the holder of a freehold BURGAGE-2 **4.** a member of the governing body of a town **5.** one elected to represent his fellows in parliament.

Burgess's concentric ring model CONCENTRIC ZONE GROWTH THEORY.

burgh (Scottish) originally equivalent to a BOROUGH-1,2 in England; in Scotland now restricted to a town with a charter. G.

burgundy WINE.

burn (Scotland, Northern Ireland, northern England) a small stream or brook, formerly also a spring or fountain. G.

burrstone, buhrstone **1.** a coarse rock with QUARTZ crystals, occurring mainly in France and North America, commonly used for MILLSTONES **2.** such a millstone.

bush **1.** a SHRUB or small TREE, especially one with branches arising near the ground **2.** uncleared or uncultivated country, especially that covered with trees of this type **3.** widely and variously used locally, e.g. natural vegetation of low woody plants, such as CREOSOTE BUSH (USA); wilder countryside as opposed to cultivated land (Africa); and further extended to

the countryside as opposed to the town. BUSHVELD, VELD.

bushel a measure of capacity which varies for different commodities and in different countries. In British dry and liquid measure it is in general equal to 36.6 LITRES or 2219.36 cu in or 8 GALLONS; in American dry measures it is equal to 35.23 litres or 2150.42 cu in. CONVERSION TABLES (capacity and yield).

bush fallowing a farming practice common in equatorial forest areas in Africa, a modified form of SHIFTING CULTIVATION. A small part of the forest is cleared by cutting and burning, and crops are planted. When the fertility of the soil in that plot is exhausted, another clearing is made and the farmers cultivate it, but they continue to live in their village, they do not themselves move. The abandoned plot quickly becomes covered with such plants as bamboo and eventually, if left untouched, with trees; but very often the plot is recultivated after a lapse of time. FALLOW.

bush veld, bushveld (Afrikaans, bosveld) the SAVANNA of tropical and subtropical south Africa, sometimes open grassland with scattered trees (PARKLAND), grading to close woodland. G.

business park OFFICE PARK.

busti, bustee (Indian subcontinent: Bengali) strictly a village, but commonly applied (with spelling bustee) to the slums of Calcutta with their one-room huts. SHANTY-TOWN. G.

butane a hydrocarbon gas of the PARAFFIN-1 family, obtained in the refining of PETROLEUM, liquefied by pressure for use as a fuel gas.

butt **1.** a large barrel for wine or beer **2.** a section in the common arable field in the OPEN FIELD SYSTEM, shorter than others in the SHOTT owing to an irregularity in the shape of the field. G.

butte **1.** a small, flat-topped, isolated hill with steep sides (its upper layers consisting of resistant rock overlying weaker layers. CAP-ROCK) which remains after partial denudation of the surrounding TABLELAND. It may be a small, isolated part of a MESA 2. in western USA, any isolated flat-topped hill with steep sides. BUTTE TEMOIN. G.

butter 1. a fatty substance obtained by churning cream skimmed from MILK, used at table and in cooking. It enters international trade under refrigeration 2. substances resembling butter in appearance or consistence.

butter bean *Phaseolus lunatus*, Lima or Madagascar bean, a leguminous (LEGUMINOSAE) ANNUAL or BIENNIAL food plant, native to South America, now widely grown but on a small scale in tropical and subtropical lands for its large, protein-rich white seeds, usually dried for storing and cooking.

butternut, white walnut *Juglans cinerea*, a tall DECIDUOUS tree of temperate climates, smaller than the BLACK WALNUT, and bearing fruits with relatively less hard shells, the edible KERNELS-2 being milder and more pleasing in flavour than those of the black walnut. WALNUT.

butte témoin (French) the flat-topped outlier of an escarpment or plateau, of which it was once part, its height, being about the same as that of the escarpment or plateau, bearing witness (témoin) to its origin. Strictly every BUTTE is a butte témoin, but because the term butte has slipped into common usage in western USA the distinction of butte témoin becomes necessary. G.

buyers' market a market in which prices are low because the goods concerned are in plentiful supply and the demand for them is low. SELLERS' MARKET.

Buys Ballot's Law a law postulated by C. H. D. Buys Ballot, Dutch climatologist, 1857, that if an observer in the northern hemisphere stands with back to the wind, the ATMOSPHERIC PRESSURE will be less to that individual's left than to the right, the reverse in the southern hemisphere. CORIOLIS FORCE, FERREL'S LAW. G.

bypass a road which skirts a place, especially a road designed to divert through-traffic from roads in a congested area, e.g. a town centre.

by-product 1. an additional secondary product obtained during a specific process, of greater or less value than the product which is the primary objective of the operation 2. an additional result, which may or may not have been intended or expected.

bysmalith a large IGNEOUS INTRUSION in the shape of a cone or cylinder, ascending from great depths to the earth's surface, and making an arch of the COUNTRY ROCK above it in the manner of a LACCOLITH. Vertical displacement with faulting round its margins is characteristic of this type of intrusion. In some cases the arched overlying strata is denuded and the bysmalith appears as an upland mass. G.

C

C 14 dating   carbon 14 dating.   RADIO-CARBON DATING.

caatinga, catingas   (Brazil: Portuguese) **1.** thorny, DECIDUOUS drought-resistant woodland with XEROPHILOUS plants in northeast Brazil **2.** low, EVERGREEN forest (similar to the Wallaba forest of Guyana) growing on the lowland tropical PODZOLS of the Rio Negro region of Amazonas, Brazil, where there is no marked dry season. G.

cabbage   **1.** the wild cabbage, *Brassica oleracea* (or *B. sylvestris*, identified as a separate species by some botanists), family Cruciferae, native to southern England and Wales, the northern coastlands of the Mediterranean and the western coasts of the Adriatic, is the ancestor of the cultivated varieties used for food by the ancient Greeks, the Romans, the Saxons and Celts in northern Europe, and of the improved cultivars of their descendants. The wild cabbage is a BIENNNIAL or PERENNIAL plant, with a woody stem and a few greyish, blue-green leaves. The vegetative growth of the great variety of modern cultivars is quite different, but if any is allowed to run to seed it will reveal its ancestry by its pale yellow flowerheads and fruits (narrow capsules), characteristic of the wild cabbage **2.** the main cultivated varieties grown in temperate lands for human consumption, cooked or shredded raw, with a much enlarged terminal bud. Spring cabbage is the young form of any cabbage, the savoy has crinkly leaves, red cabbage (commonly cooked slowly with spices or pickled in spiced vinegar) has dark red leaves. The BRUSSELS SPROUT is also a cabbage. BRASSICA.

cabinet wood   wood especially suited to the skilled craftsmanship of the cabinetmaker (maker of furniture), the best type being HARDWOOD, e.g. ebony, mahogany, walnut.

cable, cable length   one tenth of a NAUTICAL MILE, 185 m (approximately 608 ft), about 100 fathoms.

cacao, cocoa   *Theobroma cacao*, a small tree native to tropical America, now widely grown in warm tropical and equatorial regions with fairly high rainfall and a fertile soil. The flowers are borne directly on the main trunk and on the branches, the fruit (the pod) carrying seeds (beans) which after fermentation, drying and roasting, are used in the manufacture of cocoa powder. Cocoa beans contain 50 per cent or more fat (cocoa butter) which is augmented (with the addition of extra cocoa butter) in the making of chocolate, but removed in the production of the cocoa powder used in beverages. They also contain caffeine and THEOBROMINE. Strictly the term cocoa should be restricted to the beans and their products, but in popular usage the term is now also applied to the tree (cocoa tree).

cacimbo   (Angola) fog and low cloud, sometimes giving drizzle on the shorelands of southwestern Africa (Angola) and spreading inwards to the plateau, particularly frequent in July and August, occurring when west winds carry shorewards the cool, damp air lying over the cold offshore Benguela current. CAMANCHACA. G.

CACM Central American Common Market Organización. ODECA.

cadastral (French cadastre, a register of property) *adj.* of, or pertaining to, a CADASTRE.

cadastral map 1. loosely applied to any map on a scale large enough to show every field or plot of land and building, e.g. the 1:2500 map of Britain 2. specifically, a large scale map sufficiently accurate for exact boundaries and (if necessary) the ownership of real property (CADASTRE) to be shown on it. G.

cadastral survey a survey carried out for the purpose of providing information for the drawing of a CADASTRAL MAP-2. G.

cadastre (French, from Latin capistratum, register of the polltax, or the unit of territorial taxation used in the provinces in Roman land tax) occasionally anglicized as cadaster, an official register of the extent and ownership (and usually also the value) of real property (i.e. land and buildings).

cadmium a malleable, ductile, toxic, silvery-white, metallic ELEMENT-6 associated with ZINC ores, used in ALLOYS, in electroplating, sometimes for making control rods for NUCLEAR REACTORS, some of its compounds being used as PIGMENTS.

caesium a white, ductile element of the ALKALI-1 metals. The most electropositive element, it has photoelectric properties and is used in defining the SECOND-2, one of the basic units in SI.

caingin, kaiñgin (Philippines) SHIFTING CULTIVATION, LADANG of Indonesia. G.

Cainozoic, Cenozoic, Kainozoic (Greek kainos, new; zōon, animal) *adj.* of, or pertaining to, the third of the main geological eras (GEOLOGICAL TIMESCALE), the era marked by the rapid evolution of mammals, subsequent to

the Precambrian era. It is still termed the Tertiary era by some geologists. G.

caique 1. a skiff, a light rowing boat characteristic of the Bosporus 2. a sailing ship characteristic of the coast of Turkey-in-Asia.

cairn (Gaelic carn, heap of stones) a pyramid of rough stones piled up as a monument or landmark of some kind. G.

cairngorm a brownish SEMI-PRECIOUS STONE, a variety of QUARTZ or false TOPAZ, used in jewellery or ornamentation.

cake ice an accumulation of ICE CAKES. G.

calabrese SPROUTING BROCCOLI.

calamine (in Britain, the former name for zinc carbonate, $ZnCO_3$, now commonly termed smithsonite; in USA, the former name for basic zinc silicate) a pink powder made from zinc oxide and ferric oxide, used in soothing lotions and ointments, etc.

calamus a tropical climbing plant, member of *Calamus*, family Palmae, the long stems of which (up to 183 m: 600 feet), when stripped, form the RATTAN canes used in furniture and basket making, etc.

calcalkali, calcalkaline ALKALI ROCK.

calcareous, calcarious *adj.* 1. of, pertaining to, consisting of, or containing CALCIUM CARBONATE, $CaCO_3$, or limestone 2. having the character of CHALK or LIMESTONE. G.

calcareous crust in soil science, indurated (INDURATION) crust cemented with CALCIUM CARBONATE. G.

calcareous dam a natural barrier consisting of deposited CALCITE (CALCIUM CARBONATE) behind which a lake forms.

calcic *adj.* applied to a soil containing CALCAREOUS material. CALCIC HORIZON.

calcic horizon a SOIL HORIZON in which there is an accumulation of CALCIUM CARBONATE and/or possibly MAGNESIUM carbonate.

calcicole   a plant which needs or seeks a CALCAREOUS soil, i.e. a CHALK or LIMESTONE soil. CALCIFUGE, CALCIPHOBE, CALCIPHYTE. G.

calcicolous   adj. of, relating to, or characteristic, of a CALCICOLE.

calciferous   adj. containing or producing CALCIUM, CALCIUM CARBONATE or other calcium compounds. G.

calcification   1. generally, the changing into CALCIUM CARBONATE or into a CALCIFEROUS state by the reaction of calcium salts 2. of soil, the deposition of calcium carbonate near the surface of the soil (usually in the B horizon or C horizon) in ARID and semi-arid regions, the result of the rise, by CAPILLARY ACTIVITY, of CALCIUM salts in SOLUTION, followed by the evaporation of the water 3. in geology, the replacement of organic or inorganic material in rocks by the calcium minerals CALCITE and DOLOMITE 4. the hardening of plant or animal tissue by the deposition of calcium salts in it (FOSSIL).

calcifuge   a plant which flees from a CALCAREOUS soil, which can survive on such a soil but which flourishes only on an ACID SOIL. CALCIPHOBE. G.

calcimorphic soil   a soil with an excessive amount of CALCIUM CARBONATE in it.

calcination   the process or action in which a physical change is brought about (usually in inorganic materials) by heating to a high temperature without fusing. It is used in oxidizing (especially metals), in converting a substance to powder form, or in releasing volatile constituents or products. SMELTING.

calciphobe   a plant so intolerant of CALCIUM CARBONATE that it cannot survive on a CALCAREOUS soil. CALCIFUGE. G.

calciphyte   a plant that needs or tolerates a large quantity of CALCIUM in the soil. CALCICOLE.

calcite   a crystalline form of CALCIUM CARBONATE, $CaCO_3$, the hexagonal crystals being colourless (unless coloured by impurities), the main constituent of all LIMESTONES (including Iceland spar, the purest variety of calcite; and CHALK, MARBLE, STALACTITES, STALAGMITES), a GANGUE mineral in some HYDROTHERMAL deposits, a common cementing material in many coarse-grained SEDIMENTARY ROCKS. It forms when material from some weathered IGNEOUS ROCK is transported as a CALCIUM BICARBONATE solution, the bicarbonate decomposes, and the calcite remains as a deposit. ARAGONITE, CALCIFICATION.

calcium   Ca, a soft, white, divalent (VALENCY) element of the ALKALINE EARTH group, occurring mainly as CARBONATE (chalk, limestone, marble, coral). It is used in ALLOYS, is widely used in industry, and is an essential nutrient for plants and animals. ALKALINE ROCK, FELDSPAR, PEDALFER.

calcium bicarbonate $Ca(HCO_3)_2$, a soluble salt formed when carbon dioxide from the air forms a solution of CARBONIC ACID with water, and this solution comes into contact with one of the forms of CALCIUM CARBONATE. Calcium bicarbonate causes the temporary hardness of water (HARD WATER) and acts as a bone-builder in vertebrates.

calcium carbonate $CaCO_3$, an insoluble salt, known as ARAGONITE or CALCITE in its two crystalline forms, occurring in CARBONATITE, CHALK, CORAL, LIMESTONE, MARBLE. It dissolves in water containing CARBON DIOXIDE to form soluble CALCIUM BICARBONATE, which decomposes to form ARAGONITE at temperatures above 30°C (86°F), CALCITE at temperatures below that level. It is used in making quicklime and cement. PEDALFER, PEDOCAL.

calcium cyanamide   a compound which releases ammonia slowly when in contact with water, used as a fertilizer. It is produced artificially by heating calcium carbide in the presence of nitrogen.

calcium hydroxide $Ca(OH_2)$, slaked lime, a white solid alkali produced by the action of water on CALCIUM OXIDE, used agricuturally as a soil conditioner, but also used in the purification of sugar, in glass-making, and in mortars and plasters.

calcium oxide CaO, quicklime, a white solid, the main constituent of lime, which reacts with water to produce CALCIUM HYDROXIDE (slaked lime). It is used in paper-making, metallurgy, the processing of petroleum and in the food industries.

calcium phosphate any of the phosphates of calcium used in manufacturing industry (including the pharmaceutical) and as a fertilizer.

calcium sulphate $CaSO_4$, a white salt, occurring naturally as ANHYDRITE and in hydrated form as GYPSUM.

calcrete, calcicrete any coarse SEDIMENTARY ROCK of which the fragments are cemented, in some cases only loosely, by CALCIUM CARBONATE. HARD PAN, SILCRETE. G.

calculus a branch of mathematics involving or leading to calculations.

caldera (Spanish, a cauldron; Portuguese caldeirão) **1.** a broad, shallow volcanic CRATER, formed by the blowing off of the top of a crater by PAROXYSMAL ERUPTION, or by subsidence, or by combined explosion and subsidence **2.** a large circular or amphitheatre-shaped DEPRESSION-2 of volcanic origin. G.

Caledonian folds, Caledonian orogeny, Caledonides, Caledonoid the great mountain-building movements and associated geological phenomena of the late Silurian–early Devonian periods (GEOLOGICAL TIMESCALE), indicated by the northeast to southwest trend of folds, faults, hills, mountains and valleys, etc. in northwest Europe. G.

calendar a system of dividing time into fixed periods, the natural units being the day (the revolution of the earth on its axis) and the year (the revolution of the earth round the sun). The month (revolution of the moon round the earth) and the week are conventional divisions. There are difficulties if the month is regarded as a natural unit, a natural division of the year, because 12 lunar cycles represent 354 days, but the SOLAR YEAR consists of 365 days. GREGORIAN CALENDAR, JULIAN CALENDAR.

calf **1.** the young of an OX and related genera **2.** leather made from the skin of a young cow of domesticated cattle (i.e. of *Bos taurus*, European; *Bos indicus*, Zebu) **3.** (Norse kalv) an islet lying off a small island, e.g. the Calf of Man, off the Isle of Man. G.

calf ice a piece of glacier ice, smaller than an iceberg, detached directly from a GLACIER or produced by the breakdown of an ICEBERG. G.

calibration **1.** the indication of the scale on a measuring instrument **2.** the checking of the accuracy of a measuring instrument by comparison with an independent standard **3.** the fitting of a mathematical MODEL-4 to a particular data set, e.g. by the weighting of variables.

caliche (Spanish) **1.** in southwest USA, a surface encrustation of CALCIUM CARBONATE which, in solution in the soil in ARID regions, is drawn to the surface by CAPILLARITY and there evaporates. DURICRUST **2.** impure sodium nitrate, $NaNO_3$, occurring in the soils of arid areas, e.g. northern Chile, formerly the main source of NITROGENOUS fertilizer and compounds. HARD PAN. G.

calina (Spanish la calina or la calima; Catalan calitzia) a warm, lead-coloured, dry, dust HAZE characteristic of hot summer days in July and August on Mediterranean coastlands. G.

calisaya bark the bark of *Cinchona calisaya*, a tropical tree of South

America, from which QUININE is extracted.

calm, calms a state of the atmosphere in which there is an absence of appreciable wind, such movement as there is registering Force 0 on the BEAUFORT SCALE. Such a state may occur at any time, anywhere in anticyclonic (ANTICYCLONE) conditions; but periods of calm are prevalent throughout the year in certain latitudes, i.e. the belt of calms, between latitudes 5°N and 5°S (DOLDRUMS) and in the HORSE LATITUDES. G.

calorie, calory a unit of heat energy, the amount of heat needed at sea-level pressure (1 ATMOSPHERE-2) to raise the temperature of 1 cc (1 gram) of water through 1°C (usually from 14.5°C to 15.5°C), a measure now replaced by the JOULE (J) (1 calorie = 4.186 J).

calorific value the heat produced by complete burning of a unit weight of fuel.

calvados an alcoholic liquor distilled from CIDER, originally in Normandy, France. APPLEJACK.

calving of ice, the breaking away of a mass of ice from an ICE FRONT, ICEBERG or GLACIER. CALF ICE. G.

camanchaca (South America: Spanish) a layer of dense fog that occurs twice daily on the coasts of northern Chile and Peru at an altitude of about 1000 m (3280 ft). CACIMBO. G.

Cambrian *adj.* of, or pertaining to, the first geological period or system of rocks of the Palaeozoic era (GEOLOGICAL TIMESCALE) when such rocks as LIMESTONES, SANDSTONES, SHALES, QUARTZITES were formed under shallow seas, and the INVERTEBRATE was the characteristic form of life.

camel a humpbacked QUADRUPED RUMINANT of *Camelus*, family Camelidae, *C. dromedarius* (dromedary) of Arabia having a single hump, *C. bactrianus* of Turkestan having two humps. The animal is physically adapted to withstand hot arid conditions, surviving long periods of time without frequent intakes of water, eating rough herbage, and travelling long distances in hot deserts, its broad spreading feet preventing sinking into the loose surface. It is used in hot desert lands for riding, as a beast of burden, and as a source of CAMEL HAIR.

camel hair, camel's hair the hair of the haunch and underpart of a CAMEL, used to make durable, warm clothing textiles (also termed camel hair, fawn in colour unless dyed), the coarser hairs being used in the manufacture of blankets, carpets and rugs, now usually mixed with other yarns. It was formerly used in making paintbrushes. C.

campagne, campagna CHAMPAIGN.

Campbell-Stokes recorder an instrument for measuring and recording the duration of bright sunshine by means of a graduated, sensitized card on to which the sun's rays are focused by a lens. As the sun and the position of the image move, a line burnt on the card records periods of continuous sunlight.

camphor a white, crystalline, volatile solid, extracted by distillation from the wood and leaves of certain trees, particularly from *Cinnamonum camphora*, native to southeast Asia and China. Used in the production of celluloid, it is repellent to moths and other insects, as is the unprocessed wood.

campo (Brazil: Portuguese) level, open grassland with scattered trees in Brazil, comparable with SAVANNA, probably not a natural CLIMAX vegetation but one arising from human activities, especially burning. Various types are distinguished: campo cerrado (closed grassland) with scrub woodland dominant; campo sujo (dirty grassland) with scattered trees or patches of forest; and campo limpo (clean grassland) open grassland without trees. G.

cañada (Spanish) **1.** a small CANYON, a RAVINE, especially in western USA **2.** a TRANSHUMANCE route for sheep in Spain. G.

Canadian high an ANTICYCLONE centred over central North America in winter.

canal old French term, adopted in English and formerly applied to any narrow piece of water or channel, surviving in this sense in the Lynn Canal of Alaska leading to Skagway. More commonly applied to an artificial watercourse constructed **1.** to unite rivers, lakes, etc., for purposes of inland transport **2.** for water supply and irrigation **3.** to make a SHIP CANAL or seaway, available to ocean-going vessels. G.

canale (Italian, pl. canali) a LONGITU-DINAL VALLEY of the Dalmatian coast, drowned by the rise in sea level which transformed the valleys lying parallel to the coast into gulfs of the sea. G.

candela cd, the basic SI unit of intensity of light, being the luminous intensity, in the perpendicular direction, of a surface of 1/600 000 sq m of a black body at the temperature of freezing PLATINUM under pressure of 101 325 NEWTONS per sq m.

cane sugar SUGAR.

cannel coal a hard, dull grey BITUMIN-OUS COAL, with a high content of volatile material and ash, 77 to 80 per cent CARBON, burning with a yellow flame, like a candle. It is used in making gas.

cañon CANYON.

canopy the high, leafy, continuous, up-permost layer formed by the crowns of trees of approximately the same height, e.g. in RAIN FOREST. G.

canton an administrative division of a country, e.g. in Switzerland.

cantonment a permanent encampment or winter quarters for troops, a term applied especially in the Indian subcontinent to the military stations (usually adjoining towns and cities) which were built during the British period. G.

canyon (Spanish cañon, a tube, a hollow) **1.** a deep valley with very steep sides, with a stream flowing at the bottom, common in arid and semi-arid lands where the downward cutting power of the stream exceeds the rate of WEATHERING of the rocks of the valley sides. The form becomes exaggerated if uplifting of the land occurs at the same rate as the down-cutting of the river. BOX CANYON **2.** a submarine canyon, a deep, steep-sided trough in the ocean floor, in some cases very wide, in some winding. G.

CAP the Common Agricultural Policy of the European Economic Community (EEC). The basic features were adopted in January 1962, the aims being to achieve more efficient agricultural pro-duction, a fair return for farmers, reasonable prices for consumers, stable market conditions; common price levels were to be agreed and national protec-tion systems were to be replaced by a Community system incorporating vari-able levies on imports of some farm products. Most arrangements were op-erating by July 1968. GREEN POUND, UNITS OF ACCOUNT.

capacitated network a NETWORK-2 in which the maximum CAPACITIES-2 of links and nodes handling the flow are known.

capacity **1.** the ability to contain, ac-commodate **2.** the amount so contained or accommodated **3.** the ability of a factory, society, etc. to manufacture or process its product, especially this as a maximum **4.** a measure of the ability of energy to do work. CAPACITY OF A STREAM, CARRYING CAPACITY, CONGESTION, INSTALLED CAPACITY.

capacity of a stream the maximum load of stones, pebbles, sand, etc. a stream can carry, measured in grams per second. COMPETENCE. G.

cape a piece of land jutting into the sea; a prominent headland or promontory. 'The Cape' refers to the Cape of Good Hope, South Africa. G.

Cape coloured a term applied in the Republic of South Africa to people of mixed white and non-white descent who live in Cape Province.

Cape Doctor a strong, southeast, summer wind experienced in Cape Town, South Africa. It may reach hurricane strength (BEAUFORT SCALE) but is advantageous in that it prevents the stagnation of air in the mountain amphitheatre in which Cape Town is built. G.

caper the unopened flower buds of *Capparis spinosa*, a straggling spiny shrub, native to the Mediterranean area. The buds are usually pickled in spiced vinegar, then used in cooking, e.g. in sauces, relishes.

capillarity capillary action, a phenomenon occurring when the surface of a liquid touches a solid. The surface of the liquid is either raised or depressed, depending on the difference between intermolecular attraction in the liquid and between the liquid and the solid. CAPILLARY FLOW, CAPILLARY FRINGE, CAPILLARY MOISTURE, MOLECULAR ATTRACTION.

capillary flow in soil, the rise of water through the soil spaces above the WATER TABLE by means of pore-surface attraction. CAPILLARITY.

capillary fringe the soil layer lying immediately over the WATER TABLE in which water drawn up from the ground water level is held by CAPILLARITY.

capillary moisture in soil, the water held by SURFACE TENSION in pores around soil particles and available to plant roots. CAPILLARY FRINGE, FIELD CAPACITY, HYGROSCOPIC MOISTURE.

capital 1. the head, the chief town of a country, state or province, and usually the seat of government. The term is often used loosely in the sense of the chief town or city, as in the title 'commercial capital' 2. in architecture, the top of a pillar or column 3. in economics, real or financial assets with a money value (FIXED CAPITAL) 4. accumulated wealth used to finance production, or any form of wealth used to help in producing more wealth 5. the stock of goods and COMMODITIES in a country. G.

capital appreciation the increase in the capital value of something (e.g. land, buildings) over a period of time.

capital goods the machinery and equipment, and primary and partly-processed raw materials, which are used in the manufacture of other goods, contrasting with CONSUMER GOODS, which will sooner or later be used up (consumed) by individual purchasers. PRODUCER GOODS. G.

capital-intensive *adj.* needing a large investment of CAPITAL-4 for higher earnings or increased productivity, as opposed to LABOUR-INTENSIVE.

capitalism 1. an economic system characterized by private ownership of, and private investment in, the production of goods; and by private enterprise, competition, profit-making and a MARKET ECONOMY, the allocation of resources and wealth being dependent on market forces. LIBERAL CAPITALISM, NEO-CAPITALISM, STATE CAPITALISM 2. in Marxism, an economic and political system, a MODE OF PRODUCTION in which the property-owning BOURGOISIE (the capitalists) own the factories and other MEANS OF PRODUCTION while the proletarians (the workers), whom they exploit (EXPLOITATION-3), have only the power of their labour (SOCIAL CLASS) which they are free to sell for a wage.

capon a castrated male bird of the common domestic fowl. POULTRY.

cap-rock 1. a layer of resistant rock covering another or others of less-resistant material. BUTTE 2. an impermeable layer overlying an AQUIFER or SALT-DOME 3. unproductive rock covering valuable ore. G.

capsicum a group of tropical plants, genus *Capsicum*, native to tropical America and the West Indies, now grown outside the tropics where sufficient heat is available. They bear pungent, edible fruit known as PEPPER.

capture by river, RIVER CAPTURE.

carapace **1.** a bony shield, the exoskeleton, covering part of the body of some animals, e.g. crab **2.** the back part of the shell-like covering of some animals, e.g. the turtle, formed from plates of the exoskeleton fused with the vertebral column and the ribs.

carapace latéritique (French) in soil science, lateritic crust or HARD PAN found at or near the surface of many tropical soils. LATERITE. G.

carat **1.** the international measure of weight used for PRECIOUS STONES, GEMSTONES equivalent to 1/142 oz or 200 milligrams **2.** a measure of purity of GOLD, pure gold being 24 carat; 22 carat having 22 parts gold, 2 parts of alloy; 18 carat having 18 parts gold, 6 parts of alloy, etc.

caravan **1.** a company of people, especially in north Africa and other hot desert regions, travelling together for security **2.** a vehicle (American trailer) containing cooking, living and sleeping facilities, and usually built to be drawn by a motor vehicle. G.

caravanserai a type of inn in eastern countries, built round a large court, where CARAVANS-1 can stay. SERAI. G.

caraway seeds aromatic seeds of a BIENNIAL plant of the Umbelliferae family, *Carum carvi*, grown in temperate lands, used as flavouring in cooking and in kümmel, a liqueur.

carbo-electricity, carboelectricity electricity produced by the use of CARBONACEOUS fuels (especially COAL). HYDROELECTRICITY. G.

carbohydrate one of a group of neutral organic chemicals, consisting of CARBON, HYDROGEN and OXYGEN and including sugar, starch and cellulose. Produced mainly by green plants, it is important not only in animal diets but in the metabolism of all organisms in providing energy for growth and other functions. FATS, PROTEINS.

carbon an ELEMENT-6 which, combined with other elements, occurs in all living things and in CARBONATES in the earth's crust, the crystalline forms being GRAPHITE and DIAMOND. Organic chemistry is the study of carbon compounds. The radioactive isotope, carbon 14, is used in RADIOCARBON DATING. CARBON CYCLE, HYDROCARBON.

carbonaceous *adj.* containing CARBON, applied to rocks (e.g. COAL, SHALE) or other sedimentary material (e.g. PEAT) consisting largely of carbon usually derived from organic matter. G.

carbonate *adj.* applied in geology to a rock consisting mainly of carbonate minerals, i.e. minerals containing the carbonate group CO_3, e.g. CALCIUM CARBONATE, DOLOMITE etc.

carbonation saturated with or reaction with CARBON DIOXIDE.

carbonation-solution the WEATHERING of rocks by a chemical process in which rainwater charged with CARBON DIOXIDE (forming CARBONIC ACID, H_2CO_3) reacts with and dissolves LIMESTONE and rocks with other basic (BASE-2) oxides. CALCIUM BICARBONATE, CORROSION, DECALCIFICATION, GRIKE.

carbonatite an IGNEOUS ROCK consisting mainly of carbonate minerals (e.g. CALCIUM CARBONATE), in many cases containing RARE EARTH ACCESSORY MINERALS. DOLOMITE.

carbon cycle the movement of CARBON in ECOSYSTEMS. The carbon occurring in the atmosphere as CARBON DIOXIDE is absorbed and stored by plants (PHOTOSYNTHESIS). The plants, some bacteria and animals then oxidize these photosynthetic products, having obtained nourishment from them directly or indirectly, thus giving back some carbon dioxide to the amosphere. Decay and the burning of organic matter (especially FOSSIL FUELS) also contribute to carbon dioxide in the atmosphere. BIOGEOCHEMICAL CYCLE.

carbon-dating RADIOCARBON DATING.

carbon dioxide CO_2, a colourless, heavy gas, present in the ATMOSPHERE and in solution in the HYDROSPHERE, formed by the OXIDATION of compounds containing carbon and by the action of acid on CARBONATES. It does not burn and dissolves in water to form CARBONIC ACID, H_2CO_3. PHOTOSYNTHESIS.

carbonic acid H_2CO_3, a weak acid formed by the solution of CARBON DIOXIDE from the atmosphere and water. CARBONATION, CARBON DIOXIDE, DECALCIFICATION.

Carboniferous *adj.* carbon-bearing, i.e. coal-bearing, applied in Britain to the fifth period (of time) and system (of rock) of the Palaeozoic era (GEOLOGICAL TIMESCALE), between the Devonian and the Permian, lasting some 90 million years, the three main groups of rock being the Carboniferous Limestone (the lowest, not normally carrying coals), the Millstone Grit and the Coal Measures. In the USA the Carboniferous is divided into the (lower) Mississippian and the (upper) Pennsylvanian. G.

carbonification the process of converting into COAL.

carbonization 1. the process of carbonizing, i.e. of converting into CARBON, e.g. the conversion of wood to charcoal by slow combustion 2. the process of combining chemically with carbon 3. charging with carbon or CARBONIC ACID.

carbon monoxide CO, a colourless, odourless, very toxic gas that may be produced by the incomplete combustion of CARBON.

Carbo-Permian *adj.* of, or pertaining to, the end of the Carboniferous and the beginning of the Permian period (of time) or system (of rocks), i.e. the ARMORICAN (HERCYNIAN or VARISCAN) phase of mountain building. GEOLOGICAL TIMESCALE.

carcass 1. the dead body of an animal 2. in trade statistics, the dead animal's body (without head, entrails, etc.) as it reaches the butcher who cuts it up for sale.

cardamom the fruit of *Elettaria cardamomum* of the Indian subcontinent, or of *Amomum cardamon* of Indonesia, or of several other PERENNIAL HERBACEOUS plants of the family Zingiberaceae, native to Asia, Africa and the Pacific Islands. The spicy essence obtained from the seeds is used in confectionery, in curries etc., in flavouring some liqueurs, and in pharmacy.

cardinal points the four main points of the compass: north, south, east, west.

cardoon *Cynara cardunculus*, an HERBACEOUS plant native to the Mediterranean region, cultivated for its leaf stalks which, blanched, are eaten raw or cooked. The roots also are edible, cooked.

careenage 1. a place such as a sandy beach where a wooden ship can be turned on its side (careened) for cleaning, caulking and repair 2. the expense of careening, cleaning, caulking etc.

carfax CARREFOUR.

cargo the FREIGHT-3 consisting of goods, luggage, etc. carried by an aircraft, ship, etc.

caribou a QUADRUPED, a member of the genus *Rangifer*, related to the REINDEER, and native to Canada, Alaska and Greenland. LIVESTOCK.

CARICOM Caribbean Community, an organization of Caribbean states established August 1973 by Barbados, Guyana, Jamaica and Trinidad and Tobago, joined by the six less developed countries of CARIFTA, i.e. Belize, Dominica, Grenada, St Lucia, St Vincent and Montserrat in May 1974 and by Antigua (4 July 1974) and Associated State of St Kitts–Nevis–Anguilla (26 July 1974) with the aim of achieving economic integration through the Caribbean Common Market (which replaced CARIFTA), cooperation in non-economic areas, the operation of certain common services and the co-ordination of the foreign policies of the independent member states. C.

CARIFTA Caribbean Free Trade Area, an organization established in May 1968 by Antigua, Barbados, Guyana, Trinidad and Tobago with the aim of increasing and regulating trade between them, joined by Dominica, Grenada, St Kitts–Nevis–Anguilla, St Lucia and St Vincent in July 1968, by Jamaica and Montserrat in August 1968, by the Bahamas and British Honduras (Belize) in May 1971, becoming the Caribbean Common Market in May 1973, now part of the Caribbean Community, CARICOM. C.

carn (Cornwall, England) a small hill or knoll. G.

carnallite a white or reddish salt, potassium magnesium chloride, a major source of POTASSIUM and MAGNESIUM salts, used as a FERTILIZER.

carnauba wax WAX PALM, WAX.

carnelian, cornelian a variety of translucent CHALCEDONY, composed of SILICA with oxide of iron, varying in colour from wax-yellow to brownish red, distinguished as feminine (pale red), sard (brown), carnelian onyx (red stripes into white), used in signet rings, etc.

carnivore 1. a flesh-eating animal or plant, a secondary consumer in a FOOD CHAIN **2.** a member of Carnivora, an order of mammals with well developed canine teeth, sharp molars and small incisors, which are mostly flesh-eating, though some are omniverous. OMNI-VORE.

carob *Ceratonia siliqua*, a tree grown in the Mediterranean area, producing nourishing pods which are rich in sugar and protein, yield carob gum, and are used for fodder as well as being fermented for beverages.

carpedolith in soil science, a stone line. G.

carr, carr-lands 1. wet, boggy ground

or FEN **2.** fen woodland, the EDAPHIC CLIMAX woodland of the FENLAND in which the alder, *Alnus*, and willow, *Salix*, are the main dominants. G.

carrefour, carfour, carfax (from ancient French quarrefour) a place where four roads meet, formerly anglicized as carfour and carfax but now obsolete and if used at all is regarded as French. Survives as Carfax, the crossroads in the heart of the City of Oxford. G.

carrot *Daucus carota*, a BIENNIAL herb used by Greeks and Romans, now cultivated almost everywhere in the world for its fleshy, orange or reddish, nutritious tap root, rich in sugar and carotene (an unsaturated crystalline hydrocarbon from which VERTE-BRATES make vitamin A), eaten raw or cooked.

carrying capacity 1. the maximum BIOMASS which an area can support for an indefinite period **2.** the maximum number of species that an area can provide food for during the annual period when conditions (e.g. of weather) are hostile **3.** the maximum POPULATION-1 of people or of a given species for which an area can provide food **4.** of agricultural land, the maximum number of grazing animals and/or the maximum amount of food crops that the land can support under a given level of management without suffering deterioration **5.** in planning, the maximum use or number of users that a natural or artificial RESOURCE can sustain under a given level of management without the character and quality of the resource suffering unacceptable deterioration, e.g. the maximum human population that a particular area can carry or support without suffering unacceptable deterioration. When, in such an area, the number of people exactly equals this carrying capacity, the area is said to have reached SATURATION LEVEL, i.e. to be completely filled.

carse (Scottish) a stretch of low, usually fertile, alluvial land flanking the estuary of some rivers, notably the Carse of Gowrie or the Carse of Stirling in Scotland. G.

carso (Italian) KARST. G.

carst, carstification KARST. G.

cart-bote ESTOVERS.

Cartesian coordinate system GRID-1.

cartogram a simplified map presenting statistical information in a diagrammatic form by the use of symbols such as dots, circles, shading, range of colours, etc. G.

cartography 1. the science and art of drawing maps and charts 2. in the broadest application, all the processes involved in a system of communicating spatial information, the making of maps, charts, plans, sections, three-dimensional models and globes representing aspects of the earth or any celestial body at any scale, from the gathering of data (DATUM), e.g. by aerial photography (AERIAL PHOTO-GRAPH) and other REMOTE SENSING processes, or by SURVEY-3,4, through data processing to publication. ICA. G.

cartouche a panel on a map, especially if decorative, displaying the title, legend, scale, etc. G.

carucate HIDE, YARDLAND.

casbah (Arabic qasbah, qasabah) the part of a north African town near the ruler's palace or the CITADEL, or the citadel or the palace itself.

cascade 1. a rush of water falling from a height 2. a waterfall or section of a large waterfall 3. a waterfall in which the water tumbles naturally over rocks, or down a series of artificial shallow steps, a feature of landscape gardening 4. steep RAPIDS. G.

cascade diffusion DIFFUSION-2 in which the dispersal progresses rapidly from larger to smaller centres, from higher to lower levels, as distinct from HIERARCHICAL DIFFUSION in which the movement may be up or down.

cascading system a type of SYSTEM-1,2,3 identified by R. J. Chorley and B. A. Kennedy as consisting of a chain of subsystems dynamically linked by the transfer of mass or energy from one individual subsystem to another, the output from one becoming the input for the one adjacent, the FEEDBACK between components being derived from their sequence, e.g. as in the HYDROLOGICAL CYCLE. CONTROL SYSTEM, MORPHOLOGICAL SYSTEM, PROCESS-RESPONSE SYSTEM. G.

case in statistics, a member of a set of objects or observations. OBSERVATION.

casein an insoluble PROTEIN present in MILK, the basis and chief protein of CHEESE, consumed as food in milk and milk products, and used industrially in the making of some plastic materials, paint, glues, foodstuffs.

cash crop a crop grown primarily for sale, as contrasted with a SUBSISTENCE CROP, grown for the use of the grower and/or the grower's family. G.

cashew nut the valuable fruit with edible KERNEL of *Anacardium occidentale*, a tropical tree, native to Brazil, now widely cultivated there and in the West Indies, the Indian subcontinent and east Africa. A single NUT (the true fruit) hangs from a pseudo-fruit which is fleshy and apple-like. The kernel of the nut is rich in fat and protein, and is eaten after being roasted and shelled. The irritant oil in the shells is used for waterproofing and as a preservative. The pseudo-fruit can be fermented to make liquor.

cashmere a very soft, fine wool YARN or woollen textile made from the fleece of the Kashmir goat.

cassava MANIOC.

cassia member of *Cassia*, family Papilionaceae, a genus of trees cultivated in the Indian subcontinent, the dried leaves producing senna, used medicinally, and the fruits (pods) yielding pectin, used particularly as a setting agent in jam-making.

cassiterite stannic oxide, SnO_2, the

main ore of TIN, brown or black in colour, occurring in acid IGNEOUS ROCKS, in contact metamorphic zones (THERMAL METAMORPHISM), in HYDROTHERMAL veins and, most of all, in PLACER DEPOSITS. STREAM TIN.

caste (Spanish and Portuguese casta, race) **1.** the traditional hereditary, socio-religious rank in Hinduism (Brahmin, priest; Ksatriya, noble warrior; Vaisya, merchant or farmer; Sudra, worker or servant; and the untouchable, one without caste) into which a person is born and remains throughout life, governed by strict rules of behaviour,the ascent from one caste to another being possible only through reincarnation (STRATIFIED SOCIETY). The system in India has gradually become less rigid with the advance of the twentieth century, particularly in relation to the untouchables **2.** in ecology, a class of insect within an insect COLONY, the member of the class being anatomically specialized for specific work within the colony, e.g. in a honey-bee hive there are the queen (fertile female), the workers (sterile females) and the drones (fertile males). POLYMORPHISM.

castellanus cloud (formerly castellatus) a CLOUD formation which presents a mass of turrets when viewed from the side.

castellated *adj.* like a castle, with battlements, turrets.

castellated iceberg a mass of ice with a ragged, high superstructure with pinnacles, broken off from the Greenland ice and floating southwards to the Atlantic. ICEBERG.

cast iron iron-carbon ALLOY, 4 per cent carbon, produced in a BLAST FURNACE. It is brittle, but easily fused.

castor-oil plant *Ricinus communis*, a tropical plant, widely grown in tropical regions for its seed, which yields oil used medicinally and as a lubricant.

cata- (Greek, down) the transliteration kata- is now commonly preferred in scientific literature.

cataclastic KATACLASTIC. G.

cataclinal *adj.* (a term now rarely used) in the direction of the DIP of the surface rocks, applied particularly to a stream or its valley which runs in the same direction as that of the dip, the opposite of ANACLINAL. DIACLINAL. G.

catalyst a substance capable of increasing the rate of chemical reaction without itself suffering permanent chemical change, e.g. an ENZYME, PLATINUM.

cataract formerly applied to a large waterfall, now mainly to a series of rapids of the type occurring in the river Nile. G.

catastrophe theory a theory concerned with the relationship between QUALITATIVE and QUANTITATIVE change within a SYSTEM-1,2,3, with the fact that a sudden qualitative change within the system can abruptly disrupt, and change the form of, a hitherto smooth continuous process produced by a quantitative change. SYSTEMS ANALYSIS.

catastrophism **1.** convulsionism, an old theory (the opposite of UNIFORMITARIANISM) that the earth's crust owed its main features to sudden catastrophes or convulsions rather than to continuous, slow processes such as EROSION and DEPOSITION. NEOCATASTROPHISM **2.** catastrophe theory, a theory concerned with the relationship between QUALITATIVE and QUANTITATIVE change within a SYSTEM-1,2,3, with the fact that a sudden qualitative change within the system can abruptly disrupt, and change the form of, a hitherto smooth continuous process produced by quantitative change. SYSTEMS ANALYSIS. G.

catch crop a fast-maturing crop grown when the ground would otherwise be lying fallow or idle, i.e. between two main crops in a rotation (ROTATION OF CROPS), or as a substitute for a regular crop which has failed, or between the rows of a main crop. G.

catch meadow, catch-meadow a meadow irrigated with water collected by catch-drains from an adjoining hillslope. G.

catchment area, catchment basin strictly, the area over which rain falls and is caught to serve a natural drainage area, a river basin. WATERSHED-2 (American usage).

categorical data analysis the statistical methods used in the analysis of data measured on a NOMINAL SCALE, resembling those used in REGRESSION ANALYSIS, but unlike that technique in that either the DEPENDENT or the INDEPENDENT variables are measured at the categorical (nominal) level, or they both are, not (as in regression analysis) at the interval or ratio level. MEASUREMENT IN STATISTICS.

categorical imperative KANT, IMMANUEL.

categorization the act of putting into categories, of classifying. CATEGORY.

category 1. any division which serves to classify 2. any one of the divisions in a system of classification, e.g. genus in the CLASSIFICATION OF ORGANISMS 3. in philosophy, a division which serves to classify (as in 1.), but only in certain general classes of things or ideas, these classes varying according to the personal theory of the philosopher 4. in statistics, a homogeneous CLASS-1 or group of a POPULATION-4 of objects or measurements. If the category is given an identifying number or letter it is usually termed a code. CELL, CLASSIFICATION, CLASS INTERVAL, COLLAPSE, MODAL CATEGORY.

catena, catenary complex (Latin catena, a chain) in soil science, a sequence of different soils, usually formed from similar parent material, but varying with relief and drainage. The term was originally applied to the succession of soils frequently seen in east Africa to be following a line from a valley bottom to the neighbouring hill top, in the form of a CATENARY CURVE. G.

catenary curve a curve made by a slack flexible chain hanging freely between two suspension points, not necessarily points of the same height.

cation an ION which, having lost one or more ELECTRONS, has a positive charge. ANION.

cation exchange the replacement of certain CATIONS in the soil by other cations on the surfaces of colloidal clay mineral particles (COLLOID, CLAY-2). BASE EXCHANGE, CATION EXCHANGE CAPACITY.

cation exchange capacity CEC, the capacity of a given amount of soil to hold and to exchange CATIONS, measured in milliequivalents. BASE STATUS OF SOILS, PERCENTAGE BASE SATURATION.

catstep (American) a small, backward-tilting terrace. LYNCHET, TERRACETTE. G.

cattle RUMINANT, bovine MAMMALS, the domesticated species being *Bos taurus* (European) and *Bos indicus* (Zebu cattle, Indian subcontinent and the Far East). They are kept in all temperate and tropical lands, except in areas which are too closely forested, too rugged or too dry for the adequate growth of fodder, or elsewhere where disease makes their keeping impossible (e.g. in the parts of Africa infested by TSETSE FLY). In many parts of the tropics, especially in humid areas, water buffalo, *Bubalus bubalis* or *Bos bubalis* are kept instead of *Bos taurus* or *indicus*. Cattle are used as working animals (drawing ploughs, and carts for local transport in tropical lands), for the production of meat (termed beef if taken from the CARCASS of a mature animal, veal if from an immature animal), and for the production of milk (for liquid consumption or for the making of cream, butter, cheese). The distiction between beef and dairy breeds, traditional in midlatitude areas, is steadily disappearing. The tendency is to produce rapidly maturing dual-purpose animals, the males providing meat, the cows producing a high yield of medium quality MILK.

cattle gate, cattlegate, cattle gait **1.** the sole or exclusive right acquired by grant from the owner of the soil or by prescription to graze a particular number of beasts on land in which the holder of the right has no interest **2.** the right to graze a particular number of beasts, fixed according to the extent of the holder's gate, on land which the holder of the right owns in common with the possessors of other gates. Each of these forms of gate may be bought and sold. The term is common in northeast England. There, and in some places elsewhere, beastgate, pasturegate and STINT are sometimes used as alternative terms. G.

cauliflower *Brassica oleracia capitata*, an edible variety of cabbage, widely grown in Europe, bearing a round flower-head with tightly packed, undeveloped, cream-coloured buds, eaten raw, cooked or as an ingredient in pickle.

cauldron subsidence the foundering and sinking of part of a cylindrical block of COUNTRY ROCK enclosed in a vertical RING-DYKE into the underlying MAGMA, e.g. in Glen Coe, western Scotland, where part of the roof of a BATHOLITH sank along peripheral faults. G.

causality **1.** the operation or relation between two events, states of affairs, objects, in which one brings forth, produces, the other, i.e. one is the cause, the other the effect, an essential concept in DETERMINISM **2.** the state of being that which brings forth a result, i.e. of being a cause. MULTIPLE CAUSATION. G.

causal variable INDEPENDENT VARIABLE.

cause and effect analysis a form of explanation concerned with establishing a cause or causes for an event or classes of events. CAUSALITY.

causse, causses (French) **1.** the limestone district, Grands Causses, in the southwest of the Central Massif of France, where relatively pure, thick LIMESTONE is carved by deep valleys into a series of plateaus, the plateau surfaces being of generally low relief, crossed by shallow, dry valleys **2.** a general term for KARST regions, or for a type of karst transitional between HOLOKARST and MEROKARST, i.e. the characteristic karst of the Grands Causses **3.** locally, in southern France, a limestone soil. G.

cave **1.** a natural cavity, recess or chamber in the earth's crust, with an entrance (sometimes restricted to one with a horizontal entrance) from the surface, caused by water erosion or volcanic action. Caves occur particularly in weak areas of seashore cliffs, caused by the eroding action of waves or their load of pebbles, etc. (STACK, BLOWHOLE), or in LIMESTONE regions when water charged with CARBON DIOXIDE dissolves underground channels along a bedding joint to produce a BEDDING-CAVE, or a much larger chamber **2.** an artificial cavity, such as that caused by quarrying, or deliberately constructed for wine storage. CAVE-IN LAKE, SPELEOLOGY. G.

cave breccia, cave deposits, cave earth BRECCIA, earth and other deposits found in caves. Owing to the special features of CAVE formation the debris on the cave floor may show special characteristics, e.g. in a limestone cave, water charged with CALCIUM BICARBONATE on evaporation leaves STALACTITES and STALAGMITES. The deposits in caves inhabited by primitive human beings are of especial interest. G.

cavern **1.** most commonly, a large CAVE **2.** synonym for cave **3.** a large chamber within a cave **4.** (American) a cave formed in limestone country by solution by underground water and streams. G.

cavernous weathering a process of differential SAPPING-2 which creates small and large recesses in rock faces, termed TAFONI by some authors. G.

caving (American) the slumping of river banks.

cave-in lake kettle lake, a lake formed in a depression caused by the melting of GROUND ICE. G.

cavitation a process in which bubbles in a liquid are formed and then collapse in the path of a fast-moving body. Cavitation caused by the sudden increase of velocity in a fast-running stream results in the erosion of rocks because the collapsing bubbles make little shock waves which strike the bed and banks of the stream.

cay (Spanish cayo, a shoal or reef) a bank or reef of sand. KEY. G.

cayenne pepper PEPPER.

CBA COST BENEFIT ANALYSIS.

CBD CENTRAL BUSINESS DISTRICT.

cedar-tree laccolith LACCOLITH. G.

ceja (Peru: Spanish) MONTANE FOREST, consisting of a very dense growth of BROADLEAVED, evergreen species, e.g. in Peru. G.

celery *Apium graveoleus*, a BIENNIAL plant of midlatitudes and their southern tropical margins where sufficient water is available, the cultivated variety being grown for its edible leaf stalks which, blanched in growth (usually by 'earthing up'), are eaten raw or cooked. The seeds are used for flavouring savoury dishes etc.

celestial *adj.* pertaining to the sky, or the heavens.

celestial equator the circle in which the CELESTIAL SPHERE is intersected by the plane of the EQUATOR of the earth.

celestial map a map showing the positions of the planets, stars, etc.

celestial meridian a great circle in the heavens, passing through the ZENITH of a given place and the CELESTIAL POLES (CELESTIAL SPHERE). When the sun crosses the celestial meridian it is twelve o'clock or midday LOCAL TIME. G.

celestial pole one of the hypothetical points at which the extended axis of the earth intersects the CELESTIAL SPHERE.

celestial sphere the 'bowl' of the heavens, an imaginary sphere of infinite radius, with the earth (and the terrestrial observer) at its centre, on the interior surface of which heavenly bodies appear to be placed or may be projected (PLANETARIUM). The plane of the earth's equator, when produced, crosses the celestial sphere at the CELESTIAL EQUATOR. Similarly when the axis of the earth is extended it touches the celestial sphere at its North and South poles (CELESTIAL POLE). ASTROLABE, ZENITH.

cell **1.** in biology, the smallest individual structural unit of every living organism, consisting of translucent, jelly-like, granular material (protoplasm) surrounded by a thin membrane (plasma membrane), in plants surrounded by a cell wall (usually of CELLULOSE); and containing a NUCLEUS or nuclei (CYTOPLASM). A cell may have all the characteristics of a living organism; or it may be highly specialized for a particular function, e.g. the cells of multicellular organisms, which are not only highly specialized but also vary greatly in structure. Many MICROORGANISMS are unicellular (consisting of one cell). ATMOSPHERIC CELL **2.** in statistics, a CATEGORY defined by specific values on several variables simultaneously.

cell frequency in statistics, the frequency with which observations fall into a particular CELL-2, i.e. the number in a particular cell.

cellulose the fibrous constituent of the CELL-1 wall in higher plants, many ALGAE and some FUNGI. For industrial purposes cellulose is obtained mainly from wood pulp, cotton and flax, to make paper, rayon, plastics, explosives, etc.

Celsius scale the internationally ac-

cepted name for the Centigrade TEM-
PERATURE SCALE with 99 divisions
between the ice point (ABSOLUTE
ZERO), the freezing point of pure water
(0°C), and STEAM POINT, the boiling
point of pure water at sea-level with a
standard pressure of atmosphere of 760
mm (100°C). Thus one Celsius degree is
1/100 of the temperature interval between ice point and steam point. The
Celsius scale, with reverse numbering,
was invented by the Swedish astronomer, Anders Celsius, in 1742. CENTI-
GRADE SCALE, FAHRENHEIT SCALE,
KELVIN SCALE, REAUMUR SCALE. G.

Celt, Celtic, Kelt, Keltic 1. a member of
one of the ancient peoples who originated in western Europe and began to
spread westwards about 700BC 2. later
people who spoke, or who now speak,
the languages akin to those of the
ancient Galli, in Brittany, Cornwall,
Wales, Ireland, Isle of Man, western
Scotland. G.

celt a prehistoric tool with a chisel-
shaped edge, made from stone or metal.
G.

Celtic field a roughly square field common in southern England before the
Saxons introduced the strip field, still
identifiable from the air by banks or
earth markings on the chalk downlands. G.

CEMA CMEA.

cement 1. a manufactured substance
widely used in building to bind together
other building materials such as bricks
or stones, to cover floors, to make walls
etc. It is produced by heating together
and then grinding chalk or limestone
with clay or shale, the resultant grey,
powdery material consisting of silicates
of calcium and aluminates which, when
mixed with water, crystallize to a dry
solid. Cement mixed with stones and
water makes concrete, which may be
reinforced by the introduction of steel
rods or bars. Some types of cement
(hydraulic cement) harden under water.
A standard cement made from chalk

and clay mainly in southeast England,
is known as Portland Cement **2.** a
natural SILICEOUS, CALCAREOUS or
FERRUGINOUS material, deposited
from circulating water (CEMENTATION
OF SEDIMENTS) which has converted
loose deposits (e.g. sand, gravel) into a
hard compact rock.

cementation of sediments a natural process that leads to the cohesion of
sediments. Loose deposits such as silt,
sand and gravel can be converted into
hard compact rocks (such as CON-
GLOMERATE, QUARTZITE, SANDS-
TONE etc) by the deposition of some
cementing material such as CALCITE,
QUARTZ or LIMONITE from circulating
waters.

cemetery a place (other than a
CHURCHYARD in Britain) where the
dead are buried.

cenosis COENOSIS.

cenote (Spanish, from Mayan tzonet or
dzonot) a steep-walled natural well in
KARST extending below the WATER
TABLE, usually resulting from the collapse of a cave roof.

Cenozoic CAINOZOIC.

census 1. all the processes involved in
an official complete counting of the
total number of persons inhabiting a
given area at a particular time on a
given day, usually conducted at stated
(commonly decennial) intervals. Such a
census usually incorporates social data
relating to the persons counted **2.** a
similar count conducted through
SAMPLING procedure **3.** a similar complete (or by sampling) count of items in
some other field, e.g. traffic in a
particular area, production at a particular time **4.** the data so collected **5.** the
published results of the count.

centi- c, prefix, one hundredth, attached to SI units to denote the unit
$\times 10^{-2}$, e.g. centigram (one hundredth
part of a GRAM), CENTIMETRE (cm).
CONVERSION TABLES, HECTO-,
KILO-, MILLI-.

Centigrade scale the name formerly

applied to the CELSIUS SCALE, still used in meteorology, but not now used in SI units.

centilitre a unit of capacity in metric measurement, one hundredth part of a LITRE. CONVERSION TABLES.

centimetre cm, a unit of length in the metric system, one hundredth part of a METRE. CONVERSION TABLES.

CENTO Central Treaty Organization (formerly Baghdad Pact to Central Treaty Organization) a pact for mutual defence signed by Turkey and Iraq, joined by the UK, Pakistan and Iran in 1955. The USA became a member of the economic and counter-subversion committees in 1956, of the military committee in 1957, of the scientific council in 1961, and is represented by observers at council meetings. Bilateral defence agreements between the USA and Turkey, Iran and Pakistan were signed in 1959. Iraq ceased to participate in 1958, withdrew in 1959. The headquarters are in Ankara. The members participate in economic development programmes, research and technical assistance; but industrial development projects are still being considered.

centography the determination of centres of distribution and their plotting on maps.

central *adj.* **1.** situated near or at the middle point or CENTRE-1,2,3 **2.** principal, chief (CENTRE-4,5) **3.** having a position between extremes (CENTRE-6). CENTRALITY.

Central American Common Market ODECA.

central area CENTRAL BUSINESS DISTRICT.

central business district CBD (origin in USA, a term not always applicable elsewhere) the heart of a city (commonly termed DOWNTOWN in the USA), the part in which there is the greatest concentration of financial and professional services and major retail outlets, the focus of transport lines, where land use is the most dense and land values are at their highest. It is characterized by tall buildings, a high daytime population and high traffic densities. The importance of many central business districts has declined with the spread of the city and the implementation of the policy of DE-CENTRALIZATION. It has therefore been suggested that in Britain the term central area is more appropriate than central business district. BID PRICE CURVE, INNER CITY. G.

central eruption a volcanic eruption from a single vent or from a tight group of vents, producing a CONE, in contrast to a FISSURE ERUPTION.

central good, central service, central function any good sold or service offered or function performed at any CENTRAL PLACE. ORDER OF GOODS.

centrality **1.** the quality or state of being central **2.** a CENTRAL TENDENCY or central position **3.** in CENTRAL PLACE THEORY, the relative importance of a place with regard to its surrounding area, or the degree to which a centre serves its surrounding area. Christaller applied the term to the 'surplus importance' of a place (i.e. of a town), expressing the centrality of a town as the ratio between all the services provided there (for its own residents and visitors from its COMPLEMENTARY REGION) and the services needed by its own residents only. Centres with a high degree of centrality provided many services per resident; those with low centrality only a few services per resident.

centralization **1.** a CONCENTRATION at one central point **2.** the bringing or putting (e.g. of administration, of a country, of an institution, or of a firm, etc.) under central control. This often puts the minor units on the periphery at a disadvantage, and may hasten the economic decline of units lying at a distance from the central control point. CENTRIFUGAL AND CENTRIPETAL FORCES, DECENTRALIZATION.

centrally planned economy CONTROLLED POLITICAL SYSTEM, PLANNED ECONOMY, PUBLIC SECTOR.

central observation the observation which is in the central position in an ordered ARRAY of data.

central place any LOCATION-1 which provides goods, services, administrative functions for the consuming population of its HINTERLAND, i.e. the surrounding area (termed the COMPLEMENTARY REGION, trade area, or tributary area). The CENTRALITY-3 of the central place is determined by its various localized, specialized functions. CENTRAL PLACE HIERARCHY, CENTRAL PLACE SYSTEM, CENTRAL PLACE THEORY, ORDER OF GOODS.

central place hierarchy in CENTRAL PLACE THEORY, the arrangement of CENTRAL PLACES in a series of discrete classes, the rank of each being determined by the level of specialization of functions. The central places in each class perform all the functions of centres in the classes below them in the hierarchy (i.e.the lower order centres) but in addition perform a group of functions that differentiate them from, and place them above, those lower order centres. Higher order centres stock a wide array of goods and services, and provide specialist goods and services to a wide area; lower order centres stock a limited part of the array of the higher order centres and provide day-to-day goods and services to a smaller area. CENTRAL PLACE THEORY, COMPLEMENTARY REGION, K-VALUE, ORDER OF GOODS, THRESHOLD POPULATION.

central place system the spatial distribution of any set of CENTRAL PLACES which are of different sizes and different spacing and which satisfy the daily, weekly, monthly or yearly needs of the general consuming population. The pattern of this distribution is usually termed a network of central places. ADMINISTRATIVE PRINCIPLE, CENTRAL PLACE THEORY, COMPLEMENTARY REGION, ISOTROPIC SURFACE, MARKETING PRINCIPLE, ORDER OF GOODS, TRAFFIC PRINCIPLE.

central place theory a deductive (DEDUCTION), NORMATIVE theory expounded by W. Christaller in 1933, which asserts that the numbers, sizes, and patterns of spatial distribution of CENTRAL PLACES can be explained by the operation of the forces of supply and demand, by the way in which and the extent to which these centres provide goods and services to their surrounding areas. Christaller's theory, concentrating on the retailing of goods and services, assumes that both suppliers and consumers alike wish to derive the greatest economic benefit from their decisions. The suppliers (the profit maximizers) wish to earn the maximum profit from the sale of goods and services. The consumers (the distance minimizers) wish to satisfy their needs by obtaining goods and services with the minimum of effort and cost. The suppliers, having confirmed that the THRESHOLD POPULATION is sufficient to make their enterprise economically viable, therefore locate their establishments as close as possible to the consumers, taking into account the threshold of success as well as range. Threshold of success is the smallest volume of sales necessary for an establishment to be economically viable; range is the greatest distance consumers are willing to travel to obtain a good or service. Establishments can be classified according to this threshold and range: low order establishments, with a low threshold and range and a fairly compact sphere of influence, meet a daily need for which the consumer is not prepared to travel far; middle order, with a medium threshold and range and a more extensive sphere of influence, supply goods and services less frequently in demand, for which the consumer is willing to make greater effort; and high order establishments, with a high threshold and range and the most extensive sphere of influence, provide specialist goods and services even less frequently in demand, for which the consumer is prepared to travel a considerable distance. In any given area there will thus be many centres with low order establishments, fewer with middle order, fewer still with high order, giving rise to a CENTRAL PLACE HIERARCHY of higher and lower order centres (K-VALUE). In any network of central places (CENTRAL PLACE SYSTEM) there will thus be many smaller, lower order centres, forming a dense network close together, but fewer larger and more widely-spaced higher order centres. Christaller used three principles (ADMINISTRATVE, MARKETING, TRAFFIC) to account for the varying levels and distribution of central places in a central place system. CENTRAL GOOD, CENTRALITY, CONVENIENCE GOOD, ISOTROPIC SURFACE, NESTING, ORDER OF GOODS, SPHERE OF INFLUENCE, ZONE OF INDIFFERENCE.

August Lösch in 1939 presented a central place theory based on the economic forces of supply and demand, to explain the spatial distribution of manufacturing enterprises, which was very similar to Christaller's theory, and is equally applicable to the retailing of goods and services. Like Christaller he used an ISOTROPIC SURFACE to explain his theory (COMPLEMENTARY REGION). But he differed from Christaller in that his approach did not include a rigid CENTRAL PLACE HIERARCHY with a well-defined ranking; instead he saw a continuous sequence of centres. And his larger central places (unlike Christaller's high order central places) do not necessarily perform all the functions of the smaller central places.

Central Powers the countries (Germany, Austria-Hungary, Bulgaria, Turkey) grouped against the ALLIES in the First World War.

central tendency in statistics, the tendency of observations to cluster around a particular value, or to pile up in a particular category, the position of the central value being determined by one of the measures of location, i.e. the MEAN, MEDIAN or MODE, each of which makes different assumptions about the data, is calculated in different ways, and for different reasons. AVERAGE.

Central Treaty Organization CENTO.

centre of its many applications those most common in geographical writing are to **1.** the approximate middle part or point of something **2.** a point equally distant from all sides, the point of maximum accessibility **3.** an exact point in the middle of a circle or globe **4.** an axis, pivot, or point around which an object moves **5.** an area, person, place or thing of most importance in an interest, activity, condition **6.** a place in which an activity is concentrated **7.** the middle (moderate) position between extremes, e.g. in politics. CENTRAL.

centre of gravity the point in a body at which the weight of the body appears to act, i.e. the point of application on the body of the earth's attractive force. If the body is supported at that point it is balanced. An analogy may be drawn in statistics between the centre of gravity of a physical body and the ARITHMETIC MEAN of a FREQUENCY DISTRIBUTION. GRAVITATION, GRAVITY.

centrifugal *adj.* acting, moving, or tending to move away from a centre, the opposite of CENTRIPETAL.

centrifugal force an outward force acting on a body moving along a curved path, directed away from the centre of the curvature of the path. It is generally considered to be equal in magnitude but opposite in direction to the CENTRIPETAL FORCE.

centrifuge an apparatus which, by spinning round at high speed a fluid holding finely divided particles, raises the rate of SEDIMENTATION of the suspended particles.

centripetal *adj.* acting, moving, or tending to move towards a centre, the opposite of CENTRIFUGAL.

centripetal and centrifugal forces two counteracting forces which are said to cause changes in the pattern of land use in urban areas. The CENTRIPETAL FORCE causes CENTRALIZATION, attracting establishments to the central area where they may benefit from the advantages of ACCESSIBILITY and AGGLOMERATION ECONOMIES; but the CENTRIFUGAL FORCE causes DECENTRALIZATION and URBAN SPRAWL as it pushes dwellings and businesses away from congested, expensive inner city areas towards the SUBURBS. CIRCULAR AND CUMULATIVE GROWTH.

centripetal drainage a drainage pattern in which streams flow from many directions into a lake or meet the major stream at a focal area. DRAINAGE.

centripetal force an inward net force acting on a body moving along a curved path, directed towards the centre of the curvature of the path. CENTRIFUGAL FORCE.

centrocline a term sometimes used in the USA instead of PERICLINE. G.

centrography the study concerned with the descriptive statistics used to measure CENTRAL TENDENCY and DISPERSION, e.g. applied to the determination and study of central points of spatial distributions such as the mean centre of a population distribution. G.

centroid the centre of gravity, or the centre of mass.

centrosphere the nucleus or central portion of the earth, termed BARYSPHERE by those who believe it to be a heavy nickel-iron sphere. G.

cephalic index a measure of cranium shape used in studies in physical anthropology. The index is obtained by measuring the maximum width of the cranium and dividing it by the maximum length, multiplied by 100. Long-headed or DOLICHOCEPHALIC peoples have an index less than 75, broadheaded or BRACHYCEPHALIC over 83. Between these limits are the MESOCEPHALIC peoples. G.

ceramics the art of pottery.

cereal 1. any of the cultivated flowering plants of the family Gramineae (GRASS, WHEAT, BARLEY, OATS, RYE, MAIZE, RICE, MILLETS) of which the seed (grain) is used for human and animal food 2. loosely, a member of one of the other plant orders, e.g. BUCKWHEAT, BEANS, PEAS.

cereal *adj.* pertaining to grain used for human or animal food.

cerium a metallic element in the RARE EARTH group. In alloys it has the useful property of sparking when struck or scraped, hence its use in flints for cigarette lighters.

cerrado, cerradão (Brazil: Portuguese) Brazilian savanna. CAATINGA, CAMPO, SAVANNA. G.

chachār, kachār (Indian subcontinent: Urdu-Hindi, Bengali) historically, land left fallow for three or four years. BANJAR, PARANTI, POLAJ. G.

chaco (South America: Spanish, hunting ground, chase; el gran chaco, the great hunting ground) the vast almost level alluvial plain with poor soil supporting grassy SAVANNA or DECIDUOUS scrub woodland in South America north of the PAMPAS of Argentina between the Andes and the river Paraguay. There is heavy rainfall in some places, and the chaco is marshy nearly everywhere. The scrub woodland, where QUEBRACHO grows, is characteristic of the eastern part, grassy savanna of the western. G.

chāhi, chāi (Indian subcontinent: Urdu-Hindi) land or crop irrigated by wells; chāhi-sailābā, such land flooded by river; chāhi-abī such land also irrigated by basket or Persian wheel (SAQIA). G.

chain 1. a number of connected events, things, a series or linear sequence, e.g.

of like physical features such as islands, or lakes, or mountains. When applied to mountains the term implies a complexity of several roughly parallel ranges, e.g. the Andean chain (the term RANGE-2 being applied to a single line of mountains) **2.** a unit of length used in surveying, based on Gunter's chain of 100 joined links, defined in 1644, 22 yards (66 feet or 4 poles) or 20 metres. CONVERSION TABLES. G.

chain migration MIGRATION-1,2 in which the individual worker (usually male) moves and his/her dependants follow once any initial loan given for travel has been repaid and the worker has found a secure job. Other members of the family and/or of the home community follow, using as a base the home of the first migrant, to be followed by their dependants, and so on.

chain reaction a process of change in which the result (the product) of the change itself causes a further change, so that the process is repeated again and again.

chain store one of a number of retail shops that are under the same ownership, each selling the same range of goods.

chak (Indian subcontinent: Panjabi) **1.** a distinct area of land, supplied by a single outlet from the main distributary, in a canal-irrigated district **2.** a village in that area. G.

chalcedony CRYPTOCRYSTALLINE silica occurring in sediments, as GANGUE in HYDROTHERMAL deposits, as filling in AMYGDALES. Some varieties (e.g. AGATE, CARNELIAN, JASPER, ONYX) are classified as SEMI-PRECIOUS STONES, the cut, polished surface appearing waxy because the very small crystals are so tightly packed together.

chalcophile an ELEMENT-6 occurring usually as a sulphide, e.g. ARSENIC, COPPER, LEAD, MERCURY, SILVER, ZINC.

chalk soft, white, friable, fine-grained, pure LIMESTONE, composed mainly of CALCIUM CARBONATE, especially thick and extensive in southeast England. Laid down in the CRETACEOUS period in shallow water, it was once thought to be entirely organic in origin, consisting of shells of foramanifera, coccoliths and other marine MICROORGANISMS. Current thought suggests that chemical precipitation has played a large part in its formation, and that TERRIGENOUS DEPOSITS such as sand from the floor of the seashore are also components.

Chalk (in STRATIGRAPHY it should be spelled with capital C) the upper series of rocks in the CRETACEOUS period (GEOLOGICAL TIMESCALE). In England the Chalk is divided into Upper Chalk (white chalk with abundant flints overlying CHALK ROCK), Middle Chalk (soft, white chalk with fewer flints and some CHALK MARL) and Lower Chalk (grey chalk with more Chalk marl). G.

Chalk marl, Chalk-marl a stratigraphical term restricted to the lowest division of the English CHALK, a stratigraphical horizon consisting of CALCAREOUS material mixed with up to 30 per cent muddy sediments. In referring generally to a MARL rich in chalk (i.e. not specifically to the stratigraphical horizon) the term chalky marl is used. G.

Chalk rock a bed of hard, nodular CHALK, sometimes with green-coated CALCAREOUS or phosphatic nodules, occurring in southern England at or near the base of the Upper Chalk. It has been used in building (CLUNCH). The term should be restricted to this specific bed. G.

chalky boulder clay BOULDER CLAY rich in FLINTS and containing some CHALK, occurring especially in East Anglia, England.

chalybeate *adj.* impregnated with iron salts, applied especially to a mineral spring so impregnated, the water having a bitter metallic taste and reputedly medicinal qualities. G.

chalybite siderite, natural ferrous carbonate ($FeCO_3$), one of the main sources of IRON, occurring in the COAL MEASURES and JURASSIC limestones in England.

chamaeophytes a class of RAUNKIAER'S LIFE FORMS, woody or HERBACEOUS plants with perennating parts from soil level to 25 cm (10 in).

chamois *Antelope rupicapra*, the European, goat-like, horned antelope, a RUMINANT MAMMAL resembling a deer, native to the high Alps and other high mountains in Europe, the skin providing very soft leather, used in clothing etc.

champaign, champain, champagne, champian (French), **campagna** (Italian) (variations of words derived from the Latin campania, a plain, or level country; also connected with modern French campagne, country as opposed to town) open, gently undulating type of country characteristic of large parts of northeastern France, in contrast to the 'close' country or BOCAGE of small fields. The term is extended to the sparkling white WINE produced in the area. G.

chañaral (Spanish) scrub consisting mainly of thorny bushes occurring in South America, particularly in northern Argentina and central Chile.

channel 1. a course for running water, either artificial, as in a canal or irrigation ditch; or natural, as in the deepest part of a river or stream 2. a narrow stretch of water, wider than a STRAIT, connecting two larger stretches of water (e.g. two seas) or two land areas 3. a deep, navigable waterway (natural or dredged) in a BAY, ESTUARY or SHALLOWS, which affords a safe passage for vessels. G.

chapada (Brazil: Portuguese) 1. loosely applied to an elevated ridge or plateau, usually wooded, crossing the CAMPOS of Brazil 2. tableland in Matto Grosso, Brazil, with sandstone strata overlying the crystalline basement rocks. G.

chapparal (Spanish) xerophitic (XEROPHYTE) vegetation, dominated by evergreen dwarf broadleaved trees (e.g. evergreen oak), thorny shrubs, etc., resembling the MAQUIS of Mediterranean Europe, occurring in areas of Mediterranean climate in the southwestern USA and northwestern Mexico. G.

char (Indian subcontinent: Bengali; Hindi car) 1. a newly formed alluvial tract 2. a large, newly formed island in the bed of a deltaic river 3. synonymous with diara, an island on a floodplain. G.

characteristic sheet a reference sheet, a sheet providing the key to the system of CONVENTIONAL SIGNS used on a map. SYMBOL. G.

charcoal an amorphous form of CARBON, the solid residue obtained by the imperfect combustion (DISTILLATION) of animal or vegetable matter (usually wood) in a restricted supply of air. It is useful as a fuel, giving out intense head; it is used in many scientific and industrial processes and, in stick form, in sketching and drawing.

Charnian, Charnoid *adj.* 1. of, pertaining to, a system of folding or mountain building in the late PRECAMBRIAN era, taking its name from Charnwood Forest, Leicestershire, where there are Precambrian folds of northwest to southeast trend (folds with the same trend but of later date are termed Charnoid) 2. of, or pertaining to a stratigraphic division consisting mainly of volcanic rocks of the Precambrian. G.

chart 1. short for sea-chart, a map for use of navigators, indicating coasts, depths of water, positions of rocks, sandbanks, channels, etc. PORTOLAN CHART 2. a graphical representation of varied data, e.g. meteorological data. G.

chase 1. originally, in medieval England, an unenclosed private hunting ground held by a subject of the Crown and therefore subject to Common Law, as distinct from a Royal Forest, held by the King, and subject to Forest Law (FOREST-3, FREE WARREN, PARK-1) **2.** surviving in place-names, indicating a former use, e.g. Cannock Chase. G.

chatter mark, chattermark a crescent-shaped mark, consisting of a series of minute cracks, so finely packed together that they resemble a bruise on the underside of firm but brittle rocks (e.g. granite) insecurely embedded in a GLACIER, the 'horns' of the crescent pointing to the direction of movement of the ice. It is caused by COMPRESSION and by the vibration arising from the looseness of the rocks. G.

chaung (Burmese; also -yaung, -young as a suffix in place-names) a watercourse, whether occupied permanently by a stream or not, thus comparable with NALA (Hindi) or WADI (Arabic). G.

chaur (Indian subcontinent: Hindi) long, semi-circular marshes which during the rainy season form a chain of temporary lakes. G.

chayote *Sechium edule*, an HERBACEOUS PERENNIAL vine, native to central America, grown there and in the West Indies for the sake of its fleshy fruit, cooked and used in a wide range of savoury and sweet dishes.

cheese a solid food with a high PROTEIN content, made from the pressed CURDS of MILK, mainly the milk of cows but also of goats and ewes.

cheesewring a mushroom-shaped rock, a TOR of granite on the southeastern margin of the Bodmin Moor granite mass, Cornwall. The term is occasionally applied to a similar mass elsewhere. Cheesewring is derived from a fancied resemblance of shape to that of the inverted bag into which milk curds are put for wringing out moisture in home cheese-making. G.

chelation the process whereby organisms or organic substances bring about the decomposition and disintegration of soils and rocks.

chelogenic *adj.* shield forming, applied to major cycles in the geological history of the earth. G.

cheluviation the process by which organically rich water in the soil combines with metallic CATIONS and carries aluminium and iron down to lower horizons (commonly to the B HORIZON). CHELATION, ELUVIATION, LEACHING.

chemical reaction the process by which chemical compounds transform one another into different compounds.

chemical weathering CORROSION, HYDROLYSIS, MECHANICAL WEATHERING, ORGANIC WEATHERING, WEATHERING.

chemiotaxis CHEMOTAXIS.

chemistry the scientific study of the components, structure and properties of substances, and of the laws that regulate them when they combine to make compound bodies, and of the way they react to change and behave under different physical conditions. Chemistry is traditionally subdivided into physical chemistry (concerned with the physical laws governing chemical behaviour), organic chemistry (concerned with substances containing CARBON), and inorganic chemistry (concerned with substances containing elements other than carbon), as well as theoretical, analytical and micro-chemistry. BIOCHEMISTRY, GEOCHEMISTRY.

chemoautotrophic *adj.* applied to an organism which is able to produce organic material from inorganic compounds, the energy being supplied by simple inorganic reactions, e.g. by the oxidation of ferrous salts to FERRIC form, the energy source used by some BACTERIA. AUTOTROPHIC.

chemotaxis, chemiotaxis TAXIS in which the gradient of chemical concentration is the source of stimulus. CHEMOTROPISM.

chemotrophic *adj.* applied to an organism that obtains energy from a source other than light (PHOTOTROPHIC), i.e.

by chemical reactions associated with organic or with inorganic substances. AUTOTROPHIC, CHEMOAUTOTROPHIC, HETEROTROPHIC.

chemotropism 1. in botany, direction of growth in response to the stimulus of a gradient of chemical concentration 2. in zoology, sometimes used as a synonym for CHEMOTAXIS. TROPISM.

chena (Sri Lanka: Sinhalese) a form of SHIFTING CULTIVATION practised in Sri Lanka. G.

chenier a beach ridge built on swamp deposits. G.

cheri,   chāri (Indian subcontinent: Urdu-Hindi) a sub-village for segretated untouchables, set up at some distance from the main village . CASTE. G.

chernozem,   tschernosem (Russian, black earth) a zonal group (ZONAL SOIL) of fertile soils, suited to the culivation of cereals. It is granular, well-drained, rich in humus and BASES-2, and develops under tall and mixed grasses in a temperate to cool, sub-humid climate, i.e. in MIDLATITUDES where grassland is/was the natural vegetation, e.g. the Ukraine or central Canada. Typically the dark coloured, nearly black A HORIZON is thick with MULL or mull-like humus, grading to a B HORIZON (if it is present) which is lighter brown, with or without a concentration of clay, beneath which lies the CALCAREOUS C HORIZON. MOLLISOLS. G.

cherry a tree native to western Asia, genus *Prunus*, bearing small edible fruit with a single stone (DRUPE, MARASCA), pale yellow through to dark red in colour. The fruit of *Prunus avium* is sweet, *P.cerasus* or *acida* is sour; both sweet and sour are grown in midlatitudes.

chert a rock which is a variety of SILICA, composed in part of hardened CHALCEDONY, with or without the remains of siliceous and other organisms, opaque and dark-coloured, not unlike FLINT, but with a splintering fracture. It occurs in irregular beds or nodules in calcareous formations other than the CHALK, having been precipitated chemically or biologically on a sea floor.

chervil *Anthriscus cerefolium*, an ANNUAL herb, native to western Asia, but naturalized in temperate lands, and grown for its aromatic leaves, used uncooked in salads, soups, garnishes, etc.

chestnut SWEET CHESTNUT.

chestnut soils a group of ZONAL SOILS, usually dark brown over lighter coloured soil, overlying a calcareous horizon. Such soils are friable, formed under conditions drier than those resulting in CHERNOZEM, and are thus less leached than chernozem, but similarly cover wide areas of land where grassland (but of a drier type) was/is the natural vegetation, e.g. the High Plains of USA, part of the PAMPAS of Argentina, the south African VELD, Hungary, the STEPPE to the south of the chernozem in Russia. The A HORIZON is dark brown and, becoming paler in colour, lies over a B HORIZON in which there is an accumulation of lime, overlying the calcareous C HORIZON. In some places there is little or no B horizon. SOIL, SOIL ASSOCIATION, SOIL HORIZON. G.

chetoi (Egypt) the natural winter harvest of Egypt from seed sown in the Nile mud about November and harvested in February to May. NILI. G.

chicken *Gallus gallus* 1. the domestic fowl kept nearly everywhere in the world for the sake of its eggs and flesh, used for human food 2. the flesh as food. BATTERY SYSTEM, BIRD, BROILER. G.

chick-pea *Cicer arietinum*, a PULSE native to tropical regions, a leguminous plant, cultivated in the Indian subcontinent and other parts of Asia, westward to southern Europe, for the sake of its edible, nutritious seed (also termed chick-pea), rich in protein, and known as gram in India where it is cooked to make a dish known as dhal.

chicle gum a gum which is the basic material of chewing gum, made from latex obtained from zapote, *Achras sapota*, commonly named sapodilla, a tree native to central America but planted elsewhere in the tropics. The tree also yields edible, sugary, pulpy fruit.

chicory *Cichorium intybus*, a herb native to northern Europe, western Asia, and central Russia. Some varieties are grown for the sake of their large roots which, dried roasted and ground, can be blended with coffee or used as coffee substitute, others for their leaves which, blanched in cultivation, are eaten cooked or uncooked.

chili the hot, dry, southerly SIROCCO wind of Tunisia, north Africa. G.

chili, chilli *Capsicum frutescens*, a hot, red PEPPER, a PERENNIAL plant native to tropical America and the West Indies, grown in most tropical regions for the sake of its small, long, pointed, hot, pungent fruit, commonly sun-dried, used as a flavouring in many savoury dishes, e.g. in curries, in pickles, in making tabasco sauce and, ground, to make chili powder or cayenne pepper.

Chiltern Hundreds a manor in Buckinghamshire, England, which is Crown property. To 'apply for the Chiltern Hundreds' is to resign from a parliamentary seat in Britain. A British Member of Parliament (MP) may not resign from his/her seat in Parliament, nor may an MP hold a place of honour or profit from the Crown. The stewardship of the Chiltern Hundreds is nominally a Crown appointment. An MP wishing to resign may therefore apply for the stewardship, resign from Parliament, then resign the stewardship, a practice followed since 1750.

chimney 1. a narrow cleft in a vertical rock wall that can be used by rock-climbers 2. a volcanic vent 3. in USA, a STACK. The application to rock-climbing is preferred.

china 1. fine-grained, hard, semi-transparent ceramic ware used for table utensils, ornaments, etc. 2. an object made from this material. KAOLIN, PORCELAIN.

china clay KAOLIN.

China grass, ramie, rhea (the first version is European, the second Malay, the third is used in the Indian subcontinent) a bushy plant, *Boehmeria nivea*, native to eastern Asia, now grown in southeastern USA. It provides strong, absorbent fibres which can be made into soft textiles, similar to fine linen.

china stone partly kaolinized GRANITE which is whitish but still hard, not so soft as KAOLIN. It is used as a flux in glass and glaze making. KAOLINIZATION.

chine a local term applied to a narrow RAVINE or cleft in soft rocks, usually cut by a short stream flowing to the sea, especially in the Isle of Wight and the Hampshire coasts, England. Chines usually consist of a steep-sided inner valley within a broader, open valley. It has been suggested that processes other than stream action have been involved in this double structure, e.g. land-sliding, enlargement after rainfall, wind erosion, SPRING SAPPING, sea-level changes, cliff retreat. G.

Chinese water chestnut, pi-tsi *Eleocharis tuberosa*, a member of the sedge family, a herb growing in warm shallow water on the margins of lakes and in marshes, cultivated especially in Indonesia and the Far East for the edible crisp-textured basal tuber or corm. It is marketed fresh or canned, and is usually sliced for use in a variety of dishes, including soups. It is not related to the WATER CHESTNUT.

chinook a warm dry, southwest wind, similar to the FOHN, which blows down the eastern side of the Rocky mountains, having been warmed adiabatically (ADIABATIC) as it blew from the west across the mountains. It usually occurs suddenly, accompanied by a rapid rise in temperature which melts the snow in winter. G.

chi-squared test a NONPARAMETRIC statistical test used to measure the extent of the agreement between observed and expected frequencies. Being nonparametric it can be used with nominal or ordinal data (NOMINAL SCALE, ORDINAL SCALE), and being 'distribution free' it does not assume that the data being analysed is normally distributed (NORMAL DISTRIBUTION). The formula for the measure, termed chi-squared, is:

$$\chi^2 = \sum_{i=1}^{k} \frac{(o_i - e_i)^2}{e_i}$$

k denotes the number of CELLS o_i and e_i the observed and expected frequencies respectively, for the ith cell. The value of χ^2 will be 0 if there is perfect agreement between observation and expectation; and the value increases with increasing differences between the observed and expected frequencies. SIGNIFICANCE TEST.

chive *Alium schoenoprasum*, a hardy PERENNIAL small plant, widespread in temperate regions of the northern hemisphere, cultivated for its slender, cylindrical leaves, mildly onion-flavoured, used uncooked in salads, garnishes, etc.

chlorine (Greek chlōros, green) a very toxic, very reactive greenish-yellow gaseous element, widespread in combination with metals and as HELITE in nature, used as a bleaching, oxidizing, and disinfecting agent.

chlorite a greenish ferromagnesian sheet silicate mineral (SILICATES), closely related to MICA, commonly found as an alternative product of other FERROMAGNESIAN MINERALS (e.g. as a result of weathering in the presence of water) and in low grade METAMORPHIC ROCKS.

chlorophyll a green pigment found in the cells of all ALGAE and higher plants (apart from a few PARASITES and SAPROPHYTES), formed only in the presence of light. Several types occur, with small differences in chemical structure, but each contains MAGNESIUM and IRON, and each is essential in absorbing light energy in PHOTOSYNTHESIS. In some plants, e.g. some SEAWEEDS, the green of the chlorophyll is masked by other pigments. AUTOTROPH, AUTOTROPHIC, ZOOLOGY.

C horizon a distinct layer in the SOIL, underlying the A or B HORIZONS, or the organic or mineral horizons (O HORIZONS, MINERAL HORIZONS, SOIL HORIZON), consisting of the PARENT MATERIAL, little altered but weathered bed rock, transported glacial or alluvial material, or an earlier soil, from which the soil is formed. Some soil scientists identify a G layer in it, to indicate much GLEYING; and the C_{ca}(accumulation of CALCIUM CARBONATE) and C_{cs} (accumulation of CALCIUM SULPHATE) as a further refinement. Below the C horizon lies the D horizon, the unaltered bed rock. SOIL PROFILE.

chorochromatic map a map differentiating areas by the use of different colours, or shades of colour.

chorography (Greek chōra, a place, a district) a term much used in the seventeenth and eighteenth centuries to make a distinction between a (geographical) study of a special region or district (chorography) and geography, a study dealing with the earth in general. It is applied today by some authors to the identification of, or to a general account of, a large regional area (hence chorographic map) as distinct from TOPOGRAPHY, which deals with a detailed study of a small area. But some American authors use the term chorographic as relating to a very large area (say, a subcontinent) and a chorographic map as a map on a scale of between 1:500 000 and 1:5 mn. TOPOGRAPHIC MAP. G.

chorology the study of the causal relations of the phenomena present in a region, an explanatory study of a region. G.

chorometrics the statistical study of spatial distributions.

choromorphographic map a map showing a classification of areas of land on the basis of surface configuration.

choropleth map a quantity in area map, presenting the subject under study in terms of average value per unit area within specific boundaries, e.g. density of population per sq km shown within local, regional or national administrative areas; or by dividing the unit area into squares or hexagons and calculating a mean value for each. Sometimes a range of stippling, shading or colouring is used to show orders of density. DASYMETRIC METHOD, DEMOPLETH MAP. G.

chott (French) SHOTT.

-chow (Chinese) as a suffix in place-names denotes the chief town of a district. G.

Christian socialism SOCIALISM.

chromite $FeCr_2O_4$, a mineral in the SPINEL group, a major ore mineral of CHROMIUM, occurring in layers in ULTRABASIC IGNEOUS bodies, in PLACERS, occasionally in HYDROTHERMAL veins.

chromium a hard metallic ELEMENT-6, resistant to corrosion, not occurring freely in nature but derived mainly from CHROMITE, $FeCr_2O_4$ (chrome iron ore or chromate of iron). It is used as an ALLOY with STEEL to make chrome steel, stainless steel; and in chromium plating, a hard stainless surface which does not tarnish.

chromosome a thread-shaped body consisting mainly of DNA and PROTEIN, present in numbers in the NUCLEUS of every animal or plant

CELL-1, and carrying the hereditary factors. GENES.

chrono-isopleth diagram a graph showing hourly values of such phenomena as atmospheric pressure, temperature, etc. as ABSCISSAE and the times of their occurrence in a month as ORDINATES, the points of equal value being joined by isopleths (ISO-).

chronology the science of measuring and adjusting time or periods of time, of recording and arranging events in order of time, and of assigning events to dates considered to be correct in the light of contemporary knowledge. BIOCHRONOLOGY, CHRONOSTRATIGRAPHY, GEOCHRONOLOGY. G.

chronometer an instrument that measures the passing of time very accurately, used in navigation to determine LONGITUDE, and in other exact observations (SEXTANT). Very small intervals of time are measured by a chronoscope. G.

chronometry the science of measuring time.

chronoscope CHRONOMETER.

chronostratigraphy STRATIGRAPHY.

chronotaxis similarity in age geologically, in contrast to HOMOTAXIS. G.

chrysoberyl BERYLLIUM aluminate, yellow or pale green in colour, or brownish-yellow to red. If it is transparent or translucent it is used as a GEMSTONE.

chrysolite any of the magnesium iron SILICATES which are olive-green or yellow in colour, e.g. OLIVINE, some forms being used as gems. PERIDOTITE.

chrysoprase an apple-green variety of CHALCEDONY, used as a gemstone. It is transparent and takes a high polish, but it is affected by heat and sunshine, and gradually loses its colour.

churchyard an enclosed piece of land in which a church stands, commonly used as a burial ground in the past, and still so used in rural areas in Britain.

chute a sloping channel or trough in a CAVE in KARST. G.

cider a beverage (alcoholic or non-alcoholic according to manufacture) made from apple juice. APPLE, APPLE-JACK.

cinder the residue of incompletely burnt coal or similar combustible material. CINDER CONE.

cinder cone a cone formed round the vent of a VOLCANO by debris, usually of volcanic origin, cast up during eruption. As in the ASH CONE the fragmentary material has not been burnt. The majority of the particles exceed 4 mm in diameter or long dimension, the larger fragments being known as LAPILLI or SCORIAE. The slopes of a cinder cone are steeper than those of an ash cone because the angle of repose of its larger constituent fragments is greater. G.

cinglos (Spanish; Catalan cingles) the face of a CUESTA if formed by high abrupt cliffs. G.

cinnabar MERCURY.

cinnamon a fragrant, sweet spice made from the dried inner part of the bark of a small tree, *Cinnamomum zeylanicum*, native to south India and Sri Lanka, the main producer. It is used in pharmaceutical products and in cooking. C.

Cinque Port one of a group of ports, originally five in number (whence the name), along the southeastern coast of England, once important in defence. They were Hastings, Sandwich, Dover, Romney and Hythe, to which Rye and Winchelsea were added in early times. 'Warden of the Cinque Ports' is now an honorific post.

circadian rhythm a term usually applied to the DIURNAL RHYTHM of a human being.

circle GREAT CIRCLE, SMALL CIRCLE.

circular and cumulative growth, circular and cumulative feedback a process whereby growth feeds on and reinforces itself by the creation of new demands for goods and services, etc. It tends to reinforce major cities and favoured regions at the expense of less advantaged areas (CUMULATIVE UPWARD CAUSATION). As the central area prospers, the periphery (reinforcing the centre) suffers the backwash effect of the flow of skilled manpower, investment and locally generated capital to the centre; the periphery becomes poorer than the centre in social services and amenities, etc., and products from the centre flow to the periphery, flooding the market and inhibiting local enterprise. Thus unequal development is maintained. But ultimately the centre's successful development combined with the establishment of an efficient transport and communication network cause a spread effect, a CENTRIFUGAL FORCE (CENTRIPETAL AND CENTRIFUGAL FORCE), leading to decentralization accompanied by development in the periphery, thereby spreading development and reducing regional inequality. CORE-PERIPHERY MODEL, GENERATIVE AND COMPETITIVE GROWTH.

circulation of people, short-term movements, many of which are repetitive, including daily commuting (COMMUTER), weekly movements, the following of seasonal work, travelling employment (e.g. of sales representatives) and short-term changes of residence (e.g. of employees in firms with works and offices in various locations).

circumdenudation, circumerosion 1. in general, DENUDATION all round **2.** the process by which hills or mountains, by DENUDATION all round them, were isolated from an original parent mass such as a plateau. MOUNTAIN CLASSIFICATION, RELIC MOUNTAIN. G.

cirque (Fː nch; Gaelic coire; Scottish corrie; Welsh cwm) a steep-walled amphitheatre, or basin, of glacial origin at the head of a mountain valley (in some cases containing a small lake), seen as eating into the mountain mass, resulting from frost and glacial action (BASAL SAPPING, NIVATION, ROTATIONAL SLIP). At the meeting of two cirques a knife edge or ARETE is formed. G.

cirque glacier a short-tongued GLACIER which fills a separate, rounded basin which it has itself formed on a mountainside.

cirrocumulus a type of high CLOUD (above 6000 m: 20 000 ft), usually formed by ice crystals, appearing as lines of small round puffs interspersed with blue sky, i.e. as a MACKEREL SKY.

cirrostratus a layer of milky, fibrous-looking high sheet CLOUD (above 6000 m: 20 000 ft), lightly veiling the sun, heralding the approach of a WARM FRONT. If it thickens it develops into ALTOSTRATUS.

cirrus high, wispy, fibrous CLOUD (6000 to 12 000 m: 20 000 to 40 000 ft), composed of tiny ice crystals, through which sunlight or moonlight may penetrate. Strong winds in the upper atmosphere may draw out the 'fibres' to form 'mare's tails' or 'stringers'. It is usually associated with fair weather, but if it thickens to CIRROSTRATUS it may signal the approach of a DEPRESSION-3. FALSE CIRRUS.

cisalpine (Latin cisalpinus, on this side of the Alps, i.e. near to Italy) *adj.* on, or relating to, the south side of the Alps. TRANSALPINE.

citadel a strongly armed defensive fortress, usually dominating and protecting a town, built to be a place of refuge in time of war.

citizen 1. a native or naturalized member of a country, enjoying rights conferred by that country and owing allegiance to it 2. a person who lives in a particular city or town, especially one who has voting or other rights there. CITIZENSHIP.

citizenship 1. the state of being a CITIZEN-1,2 2. the legal relationship between an individual and the country which confers rights, and to whom the individual (the citizen) owes allegiance.

citron *Citrus medica*, a small CITRUS tree, apparently introduced to the Mediterranean region from Asia about 300 BC, now cultivated mainly in areas with a MEDITERRANEAN CLIMATE for the sake of its oval fruit with yellow or greenish-yellow skin, the small amount of flesh being sour, but the skin with its thick, white pithy layer being preserved in sugar to make 'candied peel', used in cooking, especially in baking and confectionary.

citronella, citronella oil *Cymbopogon nardus*, a sweet-smelling grass native to southern Asia, the leaves of which yield a fragrant ESSENTIAL OIL used in perfumery and in insect repellents.

citrous *adj.* of or pertaining to the trees (or the fruit) of the genus CITRUS.

citrus a member of *Citrus*, family Rutaceae, a genus of small trees and shrubs, originating in China and southeast Asia, but now widely grown in tropical and subtropical regions (especially in areas of MEDITERRANEAN CLIMATE) for the sake of their juicy fruit, mostly with a high vitamin C content, for the essential oils in the flowers, leaves and fruit skin, and for their pectin (used as a 'setting' agent in jam making). The main species cultivated are the ORANGE, GRAPEFRUIT (an improved form of POMELO or SHADDOCK), TANGERINE, CITRON, LEMON, SMALL LIME. The last is tropical, the others being grown mainly in the areas specified above. The fruits travel well on account of their moisture-retentive, thick oily skins: they are therefore important in international trade.

city 1. in general, in Britain, a large TOWN-1 2. more strictly, a town of any size which is or has been the seat of a bishop and has a cathedral 3. in areas influenced by Britain the term implies a conferment of a definite status, but no special privileges, or a selected town, e.g. Nairobi 4. in USA, city implies the incorporation of a settlement and the establishment of some form of local government, but a city may in this sense be only a few hundred people, and the word is used loosely as being synonymous with town. CENTRAL PLACE THEORY, CITY REGION, CITY STATE, CONCENTRIC ZONE GROWTH THEORY, ECUMENOPOLIS, GATEWAY CITY, INDUSTRIAL CITY, METROPOLIS, POST-INDUSTRIAL CITY, PRE-INDUSTRIAL CITY, PRIMATE CITY, RANK-SIZE RULE. G.

city region a city or large town and the area surrounding it (HINTERLAND-2), the surrounding area being so functionally linked to and dominated by that city or town that the two are interdependent and function together as one unit. CORE-PERIPHERY RELATIONSHIP, FUNCTIONAL REGION.

city state a city which has sufficient POWER-1,2 to constitute an independent, sovereign state, usually exerting its authority over the surrounding region, e.g. Athens, one of the city states of classical Greece.

civil day a MEAN SOLAR DAY of 24 hours, from midnight to midnight. EQUATION OF TIME, MEAN SOLAR TIME.

Civil Law 1. the body of law concerned with private disputes between people and the private rights of individuals, rather than with criminal or military affairs 2. the law of a particular state as distinct from other kinds of law, e.g. international law. COMMON LAW.

civil lines in the Indian subcontinent during the British RAJ, the European section of the larger cities, reserved for the official residences of the local bureaucracy and usually distinct from the CANTONMENT. G.

civil twilight TWILIGHT.

civil year GREGORIAN CALENDAR YEAR.

clachan (Gaelic, Highland and Ayrshire Scots, and recent northern Irish) 1. in northern Ireland, the equivalent of a HAMLET or group of farms in England 2. in Scotland the use is similar, but some authors insist there must be a church. FERMTOUN, KIRKTOUN, RATH. G.

clade a group of organisms that have evolved from a common ancestor. CLADISTIC, CLADISTICS.

cladistic *adj.* of, or pertaining to, a CLADE or clades, sometimes applied to plants to indicate recent origin from a common ancestor. CLADISTICS.

cladistics a systematic method of classification used by some taxonomists (TAXONOMY), based on the theory of evolution by descent with modification, but excluding consideration of the speed or of the mechanism of evolution. DARWIN, NATURAL SELECTION, PUNCTUATED EQUILIBRIUM.

clan 1. a tribe, a group of people descended from a common ancestral family, especially one in Scotland 2. a distinctive small group of plants composed of secondary, subordinate species in local or limited small scattered areas, and generally permanent in a CLIMAX community.

claret red Bordeaux WINE.

class 1. in general, a division based on quality or grade 2. in biology, one of the groups used in the CLASSIFICATION OF ORGANISMS, consisting of a number of similar ORDERS-2, but sometimes of only one order 3. a group of people of similar status, rank, or culture in a community. SOCIAL CLASS 4. a concept or system of social division. CASTE, SOCIAL CLASS, SOCIO-ECONOMIC GROUPING 5. in statistics, a group of occurrences with a common characteristic or set of characteristics formed when data is divided, each group being mutually exclusive. The variate values determining the upper and lower limits of a class are class boundaries; the interval between them is the CLASS INTERVAL; the frequency falling into the class is the class frequency.

class boundary in statistics, CLASS-5.

class-for-itself in Marxism, a CLASS-3 whose members recognize themselves as members of that class and are prepared to organize in their own class interest. CLASS-IN-ITSELF.

class frequency in statistics, CLASS-5.

classical *adj.* applied to **1.** the ancient civilizations of Greece and Rome and to the typically formal and emotionally controlled art and literature of those civilizations **2.** orthodox, traditional, as distinct from experimental.

classical economic theory the body of theory propounded by Adam Smith, David Ricardo, John Stuart Mill and others in the late eighteenth and the nineteenth centuries. To generalize broadly, the theory assumes that in a capitalist society a policy of individualism, of LAISSEZ-FAIRE, combined with the free operation of price in the market place (MARKET PLACE THEORY), the investment of capital to promote economic growth, freedom from government intervention and freedom of international trade, would result in economic benefit to the whole community. The classical period of economics came to be dominated by Ricardo, who expounded his LABOUR THEORY OF VALUE. COMMODITY, NEOCLASSICAL ECONOMIC THEORY.

classical geography a term sometimes applied to the nineteenth century approach to geography promoted by A. von Humboldt and Karl Ritter, in which the study of the relationship of human beings with their environment, based on factual information gathered from the physical and social sciences, was paramount. Too great a reliance on what was considered its rigid DETERMINISM brought this approach into disfavour.

classification **1.** in general, arrangement in CLASSES-1,2, putting into groups systematically, on the criteria of common characteristics or properties **2.** in statistics, the process of putting raw data into CATEGORIES-3 or codes.

classification of lakes the systematic designation of lakes, commonly based on the origin of the depression in which the water accumulated, i.e. erosion (glaciation, solution, wind); enclosure by deposition or barrier (including deltaic and morainic deposits, sand bars, dams of ice or vegetation); structural (sagging of the earth's crust, rift-valley, down-faulted basins); volcanic (crater lakes, lava dam); artificial (especially by great dams built for hydroelectric power and irrigation schemes; and tanks in India). CAVE-IN LAKE, CRATER LAKE, ETANG, FINGER LAKE, LOCH, LOCHAN, LOUGH, MERE, POND, POOL, SHOTTS, TANK, TARN.

classification of organisms the systematic designation of animals and plants (some authorities add PROTISTA) based on groups which reflect evolutionary relationships, the rank of the group being measured by the number of organisms that belong to it. To qualify for group membership the organisms must resemble each other in some property or characteristic more than they resemble any other organism. The smallest group is the SPECIES-1, though in some cases subspecies and VARIETIES-2 are identified; species that resemble each other more closely than they resemble other species are put together to form the GENUS (plural genera); genera that are similarly alike form the FAMILY-6; families form an ORDER-2, orders a CLASS-2; classes together form a PHYLUM (botany or zoology) or a DIVISION (botany); and phyla or divisions form a KINGDOM-3, the highest rank. BINOMIAL NOMENCLATURE, BIOSYSTEMATICS, PARATYPE, TAXONOMY.

class-in-itself in Marxism, a CLASS-3 (identifiable by a social scientist) with members (apparent to the social scientist) who may not see themselves as being members of that class. CLASS-FOR-ITSELF.

class interval in statistics, the size of the interval used in the division of data into CLASSES-5 or CATEGORIES, e.g. in the division of a group of people into age groups, the age groups selected being 0 to 10 years, 11 to 20 years, and so on, the size of the interval is 10 years. CLASS MARK, CUMULATIVE FREQUENCY DISTRIBUTION, FREQUENCY DISTRIBUTION, FREQUENCY POLYGON, GROUPED FREQUENCY DISTRIBUTION, HISTOGRAM.

class mark in statistics, the mid-point of a CLASS INTERVAL, e.g. in the interval 0 to 10 years the class mark is 5.

clast a sedimentary particle, commonly a rock fragment or mineral grain. CLASTIC.

clastic *adj.* applied to **1.** a rock produced by the disintegration of a larger mass or by volcanic explosion, composed of broken fragments (CLAST) of pre-existing rocks or of shell fragments which have been converted by DIAGENESIS into a consolidated mass (LITHIFICATION), e.g. SEDIMENTARY ROCKS (CLAY, CONGLOMERATE, SANDSTONE, SHALE), PYROCLASTIC rocks (AGGLOMERATE, TUFF, VOLCANIC ASH). Some authors restrict the term to rocks in which the fragments are angular, e.g. BRECCIA. AUTOCLASTIC, BIOCLASTIC, KATACLASTIC ROCK, PSAMMITE **2.** CLASTS which have been eroded, transported and redeposited at a distance from their place of origin. G.

clatter (England, Devonshire dialect) a SCREE or a boulder, as on Dartmoor, southwest England. CLITTER. G.

clay a SEDIMENTARY ROCK of very fine texture, consisting mainly of hydrous silicates of ALUMINA, with FELDSPARS and other SILICATES and QUARTZ, and some carbonates and ferruginous and organic material in varying amount, resulting from the weathering and decomposition of feldspathic rocks. Some of the constituents are usually in a colloidal state (COLLOID) and lubricate the grains and flakes of the non-colloidal constituents. When wet, clay is PLASTIC and IMPERMEABLE, because the water held by SURFACE TENSION round the particles

fills the tiny interstices between them; when dry it loses its plasticity and develops cracks; when heated to a high temperature it becomes brittle and stone-like. The term is variously applied: **1.** in PEDOLOGY, the finest particles in a soil, with a diameter of less than 0.002 mm (0.005 mm in USA), and the resulting soils, a clay soil being one which has at least 30 per cent of such particles **2.** in MINERALOGY, a very complex group of minerals (mainly of aluminium and iron silicates) to the individual members of which specific names have been given. SKELETAL MINERALS **3.** in PETROLOGY, a fine-grained deposit, plastic when wet, not laminated, brittle and shrinking when dry, becoming hard and stone-like when heated to redness **4.** in geology, rocks consisting mainly of clay minerals and fine particles and with certain physical characters; and in stratigraphical geology, certain specific beds at different stratigraphical horizons, such as the London Clay (TERTIARY), Oxford Clay (JURASSIC) etc. ARGILLACEOUS, BOULDER CLAY, KAOLIN, LATERITE. G.

clayband a layer of CLAY-IRONSTONE.

clay-ironstone hard, clayey ferrous carbonate (SIDERITE, $FeCO_3$), a source of iron, associated with Carboniferous strata, especially with the COAL MEASURES. The iron occurs in nodules (CONCRETION) in a band or in a thin seam (such layers being known as claybands), divided by shales and sandstones. G.

clay minerals CLAY-2.

clay pan a layer of stiff CLAY formed below the surface of the soil (HARD PAN), acting as a more or less IMPERMEABLE layer and leading to waterlogging.

clay-slate a SLATE formed from the METAMORPHISM of CLAY, as distinct from SLATE derived from compacted VOLCANIC ASH. G.

clay-with-flints **1.** a term loosely applied

to nearly all clay-flint deposits resting on the CHALK 2. more precisely, the mixed chalk-flints with reddish or brown CLAY, sometimes nearly black at the base, becoming lighter and sandier towards the surface, occurring on surfaces of the CHALK (e.g. in southern England) and also in PIPES-2 or POT-HOLES. It is generally considered to be the residue of the clayey impurities and the FLINTS of the Upper Chalk where the Chalk itself has been removed in solution; but part may have been formed from TERTIARY materials. SUPERFICIAL DEPOSIT. G.

clay soil CLAY, SOIL TEXTURE.

cleat one of the main joints along which COAL splits when mined. G.

cleavage 1. in geology, usually applied to slaty cleavage, the fissile structure developed in certain fine-grained rocks (e.g. CLAY-SLATE or compacted VOLCANIC ASH) especially as a result of DYNAMIC METAMORPHISM. Minute flakes of micaceous minerals, formed as a result of the metamorphism, tend to arrange themselves at right angles to the direction of the pressure, and the rock thus splits in a direction quite different from that of the original bedding. As a result roofing slates can be prepared from rock with a well developed slaty cleavage **2.** in CRYS-TALLOGRAPHY, the splitting of a CRYSTAL along a plane (the cleavage plane) or planes formed in its internal structure, when subjected to TENSION. A blow not in the cleavage plane will shatter a crystal into small pieces. G.

clementine an easily peeled CITRUS fruit, authoritatively regarded as a variety of TANGERINE or as a hybrid of the tangerine and the sweet ORANGE.

cleptoparasite KLEPTOPARASITE, PAR-ASITE.

cleugh, clough, cleuch (Scottish) a rocky, narrow GLEN, a steep-sided valley in southern Scotland and northern England, especially in Derby-shire. The term (variously spelled)

enters frequently in place-names in Scotland and Derbyshire. G.

cliff a high, perpendicular or steep face of rock, e.g. along a sea coast (sea-cliff) or bordering a lake (lake-cliff). G.

climate the average WEATHER condi-tions throughout the seasons over a fairly wide or very extensive area of the earth's surface and considered over many years (usually 30 to 35 years) in terms of CLIMATIC ELEMENTS. CLIMATOLOGY, LOCAL CLIMATE, MACROCLIMATE, MESOCLIMATE, MICROCLIMATE, MICROCLIMATO-LOGY. G.

climatic amelioration AMELIORATION.

climatic climax the CLIMAX developed in a LOCAL CLIMATE which differs from the climate normal to the area.

climatic elements ATMOSPHERIC PRESSURE, HUMIDITY (covering CLOUDS, EVAPORATION, PRECIPITA-TION, WATER), TEMPERATURE (cover-ing RADIATION), and WIND, resulting from the interrelationship of latitude, altitude, the spatial distribution of land and sea, ocean currents, relief, soil and vegetation. CLIMATE.

climatic formations vegetation forma-tions (FORMATION-1) classified according to the climatic factors (CLIMATE, CLIMATIC ELEMENTS) that determine them, as distinct from EDAPHIC FORMATIONS. Some authors maintain that the large vegetation for-mations are determined by climatic factors, the smaller units by EDAPHIC FACTORS. G.

climatic geomorphology CLIMATOMOR-PHOLOGY. G.

climato-isophyte a line connecting places with equal plant growth ability resulting from climatic conditions, ex-pressed by the CVP INDEX. G.

climatology the physical science con-cerned with studying the CLIMATES of the earth, describing and where possible explaining them and the part they play in the natural environment. AGRO-CLIMATOLOGY, COMPLEX CLIMATO-

LOGY, DYNAMIC CLIMATOLOGY, LOCAL CLIMATE, MACROCLIMATE, METEORO-LOGY, MICROCLIMATE, MICRO-CLIMATOLOGY, SYNOPTIC CLIMATO-LOGY. G.

climatomorphology the scientific study of the development of landforms under different climatic conditions.

climax in PHYTOGEOGRAPHY, the final stage in the possible development of the natural vegetation of a locality or region, when the composition of the plant community is relatively stable and in equilibrium with the existing environmental conditions. This is normally determined by climate (CLIMATIC CLIMAX) or soil (EDAPHIC CLIMAX). CLISERE, DISCLIMAX, PLAGIOCLIMAX, POSTCLIMAX, POTENTIAL CLIMAX, PRECLIMAX, PRISERE, POLYCLIMAX, PROCLIMAX, SERCLIMAX, SERE, SUBCLIMAX, SUCCESSION-2.

climograph, climagram, climogram a graphical representation of those features of climate which affect human physiological comfort, devised by T. GRIFFITH TAYLOR, c.1920. Selecting a locality, he used its WET BULB temperatures as ORDINATES, and RELATIVE HUMIDITIES as ABSCISSAE, plotting the average monthly figures and joining up the points to produce a twelve-sided graph, the shape and position of which gave an indication of how comfortable that locality would be for human beings. Others used the technique for other climatic data (e.g. temperature and precipitation) to indicate the general climatic conditions at a place (in this case the term climogram is preferable). HYTHERGRAPH. G.

cline (Greek klinein, to slope) a continuous gradation of differences in form or of characters among the members of a SPECIES or other groups of related organisms. The differences may be correlated with the geographical or ecological distribution of the species or groups. POLYTYPIC.

clinographic curve a graphical representation of the actual variations of average slope in a particular area of the earth's surface. The highest point of the slope is shown on the y-axis, and successive pairs of contours marked from it down to the x-axis which indicates the base and extent of the slope. The average gradient (measured in degrees) between the successive pairs of contours is plotted, and a line (the clinographic curve) is drawn to link these points. G.

clinometer an instrument for measuring a vertical angle, used in measuring the angle of a slope or other angles of elevation or depression, e.g. a geological feature such as a BEDDING-PLANE, FAULT-PLANE, JOINT. It usually incorporates a pendulum or spirit level. ABNEY LEVEL.

clint, clent a low, flat-topped ridge, sometimes with LAPIES, in a horizontal LIMESTONE surface, parallel to the BEDDING PLANE and separated from another clint by furrows or fissures. GRIKE. G.

clisere a series of CLIMAXES resulting from a major climatic change, e.g. a change from glacial to postglacial conditions.

clitter, clatter (England, Devonshire dialect) **1.** a mass of loose boulders or shattered stones (SCREE) on Dartmoor, southwest England **2.** sub-angular and rounded boulders of GRANITE derived from TORS, streaming down a hillside, probably detached under PERIGLA-CIAL conditions when SOLIFLUXION would have made their movement easier. G.

close 1. an enclosed space, an enclosure. In this sense the term has specialized, restricted application in different parts of Britain, e.g. a small field (Midlands), a farmyard (Kent and Scotland) **2.** an entry or passage **3.** more commonly, the space around a cathedral. HALF-YEAR CLOSE. G.

closed community a COMMUNITY-4

not open to COLONIZATION-2 because every NICHE-1 is occupied.

closed economy the ECONOMY-1 of a society or group in which exchanges take place mainly within the society or group, any exchanges with those outside being strictly limited. OPEN ECONOMY.

closed system an isolated SYSTEM-1,2,3 enclosed by a boundary through which neither energy nor material can pass (unlike an OPEN SYSTEM). Any heterogeneity existing in such a system is soon destroyed, so there is a trend towards maximum ENTROPY, e.g. in an isolated tank holding water with a temperature higher at one level, the difference in temperature will disappear (heat passing from warmer to cooler bodies) with the passage of time; the heat will eventually be distributed evenly, a homogeneous condition established. Because the local concentration of energy represents organization, its even distribution represents disorganization; and the more disordered the system, the higher the entropy. GENERAL SYSTEMS THEORY, RELAXATION TIME.

cloud a visible mass of tiny particles floating in the atmosphere, consisting sometimes of ice crystals, more usually of water formed from CONDENSATION of WATER VAPOUR on nuclei of dust or smoke particles or ionized MOLECULES of the air itself (NUCLEUS-1, IONIZATION). The formation of cloud thus depends on the cooling of moist air, which may result from the rising and expansion of such air, the mixing of warm air with cold air, or by a loss of heat by radiation. Very low cloud is termed FOG or MIST. CLOUD AMOUNT, CLOUD FORMS. G.

cloud amount cloud cover, estimated visually, expressed as the proportion of sky covered either in tenths or, more commonly, in eighths (OKTA), 0 representing a cloudless sky. The lines drawn to show areas of equal cloudiness are termed isonephs.

cloud base the height above the earth's surface of the lowest part of a CLOUD or of a general cloud layer.

cloud belt the altitudinal belt with constant cloud or mist, e.g. on mountains in the TROPICS.

cloudberry *Rubus chamaemorus*, a small HERBACEOUS plant, native to low northern latitudes, bearing small golden berries even within the ARCTIC CIRCLE, used in puddings and jam making, and for flavouring, especially in Scandinavia.

cloudburst a torrential downpour of rain, usually over a small area and of short duration. G.

cloud forest MIST FOREST, CLOUD BELT. G.

cloud forms in the classification of clouds in *The International Cloud Atlas*, WMO, Geneva, CLOUDS are distinguished by height above sea-level (low, up to 2400 m: 8000 ft; medium, 2400 to 6000 m: 8000 to 20 000 ft; high, 6000 to 12 000 m: 20 000 to 40 000 ft) and by form (cirrus, feathery; cumulus, globular or heaped; stratus, sheet or layer). To these three form names alto is added to show height, nimbus to indicate falling rain. These main genera are subdivided into species (distinguished by shape and structure), varieties (arrangement and transparency, with additional features) and accessory cloud formations. The varieties of form include, among others, lenticular, lens-shaped; castellanus (formerly castellatus), turret-shaped; mammatus, breast-shaped; fracto-, ragged; banner, like a banner. Accessory cloud formations include arcus, arched; incus, anvil-shaped; tuba, column- or cone-like.

cloud reflection the reflection into space of short wave SOLAR RADIATION (ELECTROMAGNETIC SPECTRUM) from the upper surfaces of clouds.

cloud seeding 1. the introduction of dry ice, salt particles or silver iodide smoke into CLOUDS in order to promote rainfall. This technique is also being used experimentally to suppress lightning and to change the structure and movement of HURRICANES-2 **2.** a process in which ice crystals falling from the ICE ANVIL of a CUMULONIMBUS cloud form the nuclei of condensation at lower levels. CLOUD.

clough CLEUGH.

clove *Eugenia caryophyllus*, a tropical tree native to Indonesia, cultivated particularly in Zanzibar and Madagascar. The dried, unopened flower buds (clove) yield a pungent spice used as flavouring in cooking. CLOVE OIL.

clove oil an ESSENTIAL OIL extracted from the buds, stalks and leaves of the CLOVE tree, used in medicine and perfumery and for making vanillin, an artificial substitute for VANILLA. C.

clover *Trifolium*, family LEGUMINOSAE, a genus of some 300 species of ANNUAL, BIENNIAL, PERENNIAL herbs, grown in subtropical and mid-latitude regions, mainly in the northern hemisphere, as a FORAGE CROP and for GREEN MANURE.

Club of Rome an unofficial association founded April 1968 by Aurelio Peccei, Italian manager and consultant, restricted to 100 members, comprising social and environmental scientists, educators, economists, civil servants, managers, philosophers, from different countries. The aims were to foster among policy-makers and the public alike a better understanding of the problems faced by the developing and industrialized world; and to promote new policy initiatives and action.

clunch the lower, harder bed of the Upper Chalk, used as a source of building stone. CHALK ROCK.

cluse (French; Spanish congost) a steep-sided TRANSVERSE VALLEY, cutting a mountain ridge,especially in the Jura mountains and the Fore Alps in Haute Savoie, an example of the development of ANTECEDENT DRAINAGE or SUPERIMPOSED DRAINAGE. G.

cluster a number of similar things growing or gathered together, a group of contiguous elements of a statistical POPULATION-4. A cluster may be indicated on a map by means of symbols (DOT MAP) or on a graph by the flattening of a LORENZ CURVE. CLUSTER ANALYSIS, CLUSTER SAMPLING.

cluster analysis in statistics, an analysis which aims to discover whether the individuals in a POPULATION-4 fall into groups or clusters. In the ideal CLUSTER the members should correlate highly with each other, but the cluster itself should have little correlation with items outside it.

cluster sampling SAMPLING in which the POPULATION-4 is regarded as falling into groups (CLUSTERS), and these groups are sampled. If the process stops at the selection of the whole group it is termed single-stage cluster sampling; if sub-samples are selected from the whole groups the term multi-stage cluster sampling is applied. Multi-stage cluster sampling is sometimes termed NESTED SAMPLING because the higher stage units are 'nested' in the lower stage units. In multi-stage cluster sampling the first sets of clusters selected are termed primary sampling units (psu); the clusters produced by subdividing them are termed secondary sampling units (ssu).

CMEA Council for Mutual Economic Assistance, primarily an economic association, initiated 1949 by the USSR, working language Russian. The founder members were the USSR, Bulgaria, Czechoslovakia, Hungary, Poland, Romania. They were joined by Albania (1949-61), Cuba (1972), German Democratic Republic (1950), Mongolia (1962), Vietnam (1978). Yugoslavia agreed to participate partially (1964). Angola, Laos and North Korea participate as observers, and there are cooperative agreements with Finland, Iraq and Mexico. CMEA is the official acronym, but other popular abbreviations are COMECON and CEMA.

coal a carbonaceous FOSSIL FUEL, a brownish-black or black combustible mineral substance found in beds or seams in SEDIMENTARY ROCKS and derived from vegetable material growing in the CARBONIFEROUS era on level, swampy ground, compacted and hardened by pressure and heat arising from earth movements. A series based on percentage of fixed CARBON (from peats and brown coals with less than 55 per cent, through bituminous or humic coals, to anthracite with more than 93 per cent) may be distinguished, with a corresponding limitation in the amount of volatile material and moisture. ALOCHTHONOUS, ANTHRACITE; BITUMINOUS, BROWN, CANNEL and STEAM COAL; COALFIELD, COAL GAS, COAL MEASURES, COKE, LIGNITE. G.

coalfield a tract of land underlain by COAL. If the workable coal is covered by younger deposits the coalfield is described as concealed.

coalescence the mechanism by which things come together, grow together and unite to form one body, group, mass, etc., e.g. the growing together of small water droplets to form raindrops (RAIN).

coal gas a gas (in volume 50 per cent hydrogen, 30 per cent methane, 8 per cent carbon monoxide, 8 per cent carbon dioxide, oxygen and nitrogen, 4 per cent gaseous hydrocarbons other than methane) used as a fuel. It is obtained by heating suitable COAL in closed retorts (i.e. without air) whereby gas is driven off and COKE remains. C.

coaling station a harbour used by steamships for refuelling with coal; now replaced by oil-depots that serve oil-burning vessels. FUELLING STATION.

coal measures, Coal Measures the series of sediments, mainly SANDSTONES and SHALES, in which COAL is found. With initial capital letters, the term is applied specifically to the upper division of the CARBONIFEROUS. GEOLOGICAL TIMESCALE. G.

coal tar a viscous liquid obtained in the process of COAL distillation in retort or coke oven. It yields ANILINE, BENZENE, creosote, NAPHTHALENE, PHENOL and other products useful in manufacturing. SYNTHETIC FIBRE.

coast a term loosely applied to the zone of indeterminate width where land and sea (or other extensive tract of water) meet, considered as the boundary of the land (COASTLINE). Slightly more specifically applied by some authors to the meeting place of land and sea (width not specified) covering **1.** the narrow strip of land immediately landward of HIGH WATER, the line of the mean SPRING TIDES **2.** a more extensive zone stretching inland **3.** a zone which includes the SHORE **4.** a zone which excludes the shore. Various types of coast are identified, e.g. ATLANTIC, CONCORDANT, DISCORDANT, LONGITUDINAL, PACIFIC, TRANSVERSE. G.

coastal plain any comparatively level land of low elevation, sloping gently seaward and bordering the sea or ocean, resulting from the deposition of sediment washed down from the land, or from denudation by the sea, or by the emergence of part of the CONTINENTAL SHELF following a fall in sea-level. G.

coastline a term applied loosely to the continuous edge of the land, or the general appearance of the COAST, as seen from the sea; or to the zone between the BACKSHORE and the coast; or to the landward limit of the BEACH; or used as a synonym for COAST. More precisely it is applied to the line on the land indicated by **1.** the highest storm waves of the SPRING TIDES **2.** the high-water mark of medium tides **3.** the base of the sea-cliffs. G.

cob **1.** a mixture of CLAY (marl or chalk), straw and sometimes gravel, used in the past, especially in southwest England, for building walls, etc. **2.** a type of NUT, a cob nut (HAZEL) **3.** the spike or ear of MAIZE to which the KERNELS are attached **4.** a generally rounded piece of COAL. G.

cobalt a hard, silver-white, magnetic metallic ELEMENT-6, occurring combined with ARSENIC and SULPHUR, the major sources being linnaeite pyrite, cobaltiferous laterite and COBALTITE. It is used in the production of magnetic and hard ALLOYS that withstand high temperatures and are resistant to abrasion and corrosion; and as a source of GAMMA RAYS used in radiology. It provides a green-blue PIGMENT-2 (cobalt blue) and is a TRACE ELEMENT.

cobaltite cobalt arsenic sulphide, CoAsS, occurring in HYDROTHERMAL veins, one of the major ore minerals of COBALT.

cobble, cobblestone a naturally rounded, water-worn stone, larger than a pebble, smaller than a boulder. In grading beach material the British Standards Institution ranks cobbles (60 to 200 mm: 2.4 to 8 in in diameter) between coarse gravel and a boulder. The USA WENTWORTH SCALE defines a cobble (64 to 256 mm: 2.5 to 10 in in diameter) as between a pebble and a boulder. G.

coca a member of *Erythroxylon*, a genus of shrubs native to South America and the West Indies, the dried leaves of which yield cocaine and other ALKALOIDS.

cochineal red colouring matter (carmine) obtained from the dried female bodies of an insect, *Dactylopius coccus*, which feeds on plants of the cactus family. Cochineal is non-toxic and is used in the preparation of red PIGMENTS-2, including those for food colouring.

cock 1. a male bird, especially of the common domestic fowl **2.** a HAYCOCK.

cockpit 1. a pit where gamecocks were set to fight for sport **2.** in KARST, any natural enclosed depression with steep

sides **3.** a SINK-HOLE with steep sides, especially a star-shaped one with a conical or slightly concave floor, as in the Cockpit country of Jamaica, which has an abundance of such pits in a limestone plateau. This type of cockpit is common in KEGELKARST. G.

cockpit karst tropical KARST landscape with many closed depressions surrounded by conical hills. CONE KARST, KEGELKARST.

cocoa CACAO.

coconut the coconut palm, *Cocos nucifera*, native to tropical lands, tolerant of salty, sandy soils, will also bear fruit in some warmer subtropical areas (e.g. the Bahamas). It is cultivated in the lowlands of both those areas for the sake of its edible fruit (a DRUPE), which has a shiny waterproof skin, inside which is a fibrous mass (the fibres yielding coir, used in making coarse string, matting etc). It is this which makes the fruit light and buoyant so that it can drift great distances on the ocean, carried by ocean currents, germinating when it is stranded on a sandy shore. Inside the fibrous mass is the NUT which has a hard shell lined with a thin, white fleshy layer, the meat of the coconut. There is a hollow within the nut which, before the nut is ripe, is partly filled with a nutritious sugary liquid, coconut milk, pleasant to drink. The milk is gradually absorbed into the flesh as ripening progresses. The flesh is the most valuable part of the coconut palm. It can be eaten directly, or dried and flaked to form desiccated coconut, used cooked or uncooked; and the flesh, when dried, forms copra, from which coconut oil is extracted. The oil is used in cooking, in soap making, in making margarine, and in cosmetics. The residue left after the oil has been extracted from the copra is used as cattle cake. The shells of the nuts can be used as fuel. The tree trunks make good building timber, the leaves are used for thatching; the sap of the tree has a high sugar content, and can be evaporated to make crude sugar or fermented to made a drink (toddy), distilled to make the spirit ARRACK. C.

code in statistics, CATEGORY.

coefficient 1. in mathematics, the non-varying factor of a variable product, i.e. the number or quantity usually placed before and multiplying another quantity (e.g. the 2 in $2x$) **2.** in physics, a number expressing the degree to which a process or substance has a given characteristic, e.g. the coefficient of viscosity of a liquid which (measurable only for STREAMLINE flow) depends on the nature and temperature of the liquid.

coefficient of variability the degree of variability of statistics about a MEAN-1 value. EQUIVARIABLE.

coefficient of variation in statistics, the STANDARD DEVIATION of a distribution divided by the ARITHMETIC MEAN (sometimes multiplied by 100 and expressed as a percentage). It is used to compare variation between different VARIABLES, between variables measured in different units, or between variables with different mean values. The coefficient of variation increases with an increase in heterogeneity (HETEROGENEOUS), decreases with an increase in homogeneity (HOMOGEN-EOUS).

coenocline a series of natural communities associated with an environmental gradient.

coenosis a random assemblage of organisms held together by common ecological needs, as distinct from a COMMUNITY-4. G.

coenosite a commensal. COMMENSAL-ISM.

coenospecies a group of species distinguished by the ability of its members to produce fertile hybrids. ECOSPECIES, ECOTYPE.

coesite a dense SILICA-1 occurring particularly under Meteor Crater, Arizona, formed by impact META-

MORPHISM arising from the impact of a METEORITE on the earth's surface.

coffee a member of a genus of small tropical trees or shrubs, *Coffea*, of which there are over 40 species; of these only three are cultivated commercially on a significant scale. They are grown for the sake of their 'berries' which contain aromatic beans (seeds), the beans being roasted, ground and brewed in hot water to produce a stimulating, non-alcoholic drink. The species commonly grown commercially are *arabica*, native to Ethiopia (flourishes at high altitudes, gives the highest yield of beans of the finest quality; grown mostly in the American tropics, especially in Brazil, and on high land in east Africa); *canephora*, which produces beans known as *robusta*, indigenous to west Africa (the plant lives longer than *arabica* and is more resistant to disease, but cannot tolerate such heights; has a higher yield but lower quality beans than those of *arabica*; grown mostly in the African tropics and in Asia); *liberica*, native to Liberia (a tall tree, hardy and disease-resistant, withstands poor soils and some drought; berries are of poor quality; grown mainly in west Africa, Malaysia and Guyana for local consumption). The famous Mocha coffee is grown in small quantities on the seaward slopes of southern Arabia. Coffee needs a rich, well-drained, slightly acid soil; moderate rainfall and equable heat; and protection from the direct rays of the sun; it can withstand slight frost. C.

cognition 1. the act or the faculty of knowing, a collective term covering all the psychological processes involved in the acquisition, organization and use of knowledge, including PERCEPTION, judgement, reasoning, remembering, thinking and imagining (MENTAL MAP) **2.** the product of the act of knowing.

cognitive *adj.* of, or pertaining to, COGNITION, to those aspects of mental life connected with the gaining of knowledge or the forming of beliefs.

cognitive behaviouralism a school of thought which assumes that the impact of the ENVIRONMENT on people partly depends on their COGNITION-1 (perception) of the resources it offers and the barriers it imposes.

cognitive consonance a condition of harmony in a COGNITIVE SYSTEM, when there is consistency and accord among the items of knowledge, ideas and beliefs in the system. COGNITIVE DISSONANCE.

cognitive description a form of explanation involving the collection, ordering and classification of data. It is applicable to a simple observation or to complex accounts of structure and organization. COGNITION.

cognitive dissonance 1. a condition of disharmony in a COGNITIVE SYSTEM, when there are contradictions and a lack of consistency among the items of knowledge, ideas and beliefs in the system **2.** perceived incongruity between the behaviour and attitudes of an individual.

cognitive map a mental map, an image of a place, of an environment, an organized representation (a MODEL-4) of reality developed in the brain of an individual as a result of information's being received, mentally coded, stored, recalled, decoded and interpreted (COGNITION) and, in some cases, combined with sentiment, feelings, associated with the place or environment (TOPOPHILIA). G.

cognitive system the collection of interrelated items of knowledge, ideas and beliefs which an individual holds about other individuals, groups, events, objects, concrete or abstract subjects, etc. Each individual formulates a number of such systems, and these too are interrelated, the extent of the interrelationship varying widely. COGNITIVE CONSONANCE, COGNITIVE DISSONANCE.

cohort in demography, a group of individuals who experience a significant event during the same period of time, who thus have a common statistical characteristic, e.g. belonging to the same age group, entering hospital at the same time, etc. FERTILITY-3.

cohort analysis in demography, longitudinal analysis, the analysis concerned with the study of a COHORT over a long period of time, e.g. people born or married in a particular year who are studied at selected stages throughout their lives. It is used particularly in the study of FERTILITY-3.

cohort fertility FERTILITY-3, REPRODUCTION RATE.

coir COCONUT.

coire (Gaelic) CIRQUE.

coke the hard, porous, combustible residue, almost pure CARBON, produced when COAL is heated in a closed retort or oven so that COAL GAS and other volatile material is driven off. It is used as a fuel, burning with great heat and little smoke. For metallurgical purposes it must be hard and not easily crushed, so it is made from coals with 65 to 80 per cent carbon. The softer gas coke (a by-product of the gas industry) is produced by heating highly volatile coals of lower carbon content in retorts. Coals which afford good coke are termed coking coals. C.

col (French, neck, from Latin collum) **1.** a marked depression on a mountain ridge or range, commonly occurring where opposed CIRQUES meet, thus affording a PASS through the ridge or range **2.** in meteorology, by analogy (higher pressure representing the ridge, lower pressure the valley), a region of relatively low pressure between two adjacent ANTICYCLONES or between two adjacent DEPRESSIONS-3. G.

colatitude the complement of the LATITUDE, i.e. the difference between 90° and the latitude.

cold-blooded *adj.* applied to an animal (e.g. fish, reptile) with a body temperature dependent on the temperature of its environment. POIKILOTHERMIC, WARM-BLOODED.

cold desert a general term for areas with such low temperatures that plant and animal life are inhibited, e.g. POLAR-1 region, TUNDRA. G.

cold front the boundary zone between an advancing mass of cold air and a mass of warm air. The cold heavy air usually acts like a wedge, undercutting and forcing the lighter, warmer air upward, resulting in a drop in temperature, the formation of CLOUDS (especially CUMULONIMBUS and FRACTO-), rain (sometimes falling in heavy showers, sometimes with THUNDERSTORMS) and winds. ANAFRONT, FRONT, KATAFRONT, LINE-SQUALL, OCCLUSION. G.

cold glacier polar glacier, a moving ice-mass, very rarely with surface melting, maintaining a constant temperature at $-20°C$ $(-4°F)$ or lower. WARM GLACIER. G.

cold occlusion an OCCLUSION in which the overtaking cold air is colder than the cold air ahead of it.

cold pole a popular rather than a scientific term applied to the region in Siberia near Verkhoyansk, at 67°33′N, 133°24′E, where it is said the lowest mean winter temperature or the lowest mean annual temperatures are recorded. The January mean is $-50°C$ $(-58°F)$, the January mean minimum is $-64°C$ $(-83°F)$, and temperatures as low as $-70°C$ $(-94°F)$ have been recorded. Colder temperatures are known to occur in Antarctica. G.

cold wall a DISCONTINUITY-2 layer between the cold and warm ocean currents, e.g. between the cold water of the Labrador Current and the warm water of the Gulf Stream in the North Atlantic ocean.

cold-water desert the continental west coast desert strip (e.g. of northern Chile, northwest and southwest Africa, or northwest Australia) where the climate is influenced by cold sea currents flowing towards the equator. The cool air flowing over the sea to the land reduces summer temperatures and produces fogs and heavy dew. FOG DRIP.

cold wave any sudden drop in temperature, or a period of unusually cold weather, particularly one associated with a fall in temperature following the passage of a COLD FRONT. G.

cold woodland TAIGA.

collapse in statistics, the merging of two or more adjacent coding CATEGORIES, with a consequent loss of some information. GROUPED FREQUENCY DISTRIBUTION.

collapse doline DOLINE.

collective a farm where COLLECTIVE FARMING is practised. LAND TENURE.

collective consumption 1. the services which can be consumed only collectively and are thus provided by the state, e.g. defence services 2. the main services provided by the state, e.g. public transport, welfare 3. in neo-Marxism, the collective ways by which the state works to create a labour force and sustain it, i.e. to provide goods and services for most of the population. COLLECTIVISM, COLLECTIVIST.

collective farming a form of agricultural organization in which the farms are subjected to collectivization, i.e. they are acquired and amalgamated, e.g. by a village (as in the KIBBUTZ of Israel), or by the state (as in the KOLKHOZ of the USSR), which assumes ownership of the land but leases it permanently to a large group of shareholder farm workers who cooperate in running the holding as a single unit. The farm workers usually have shares in the produce or in the revenue from sales, generally in proportion to the work accomplished by the individual worker. In many cases the farm workers are allowed a small plot of land for their own, private use and personal benefit. COLLECTIVE, LAND TENURE, MOSHAV, SOVKHOZ, STATE FARMING.

collectivism 1. a general term applied to a theory which, broadly, advocates that the means of production and/or distribution should be collectively owned or managed, that the state should exert comprehensive, central, political control over social and economic arrangements **2.** a politico-economic system based on that theory. COLLECTIVIST. G.

collectivist one who believes and advocates COLLECTIVISM-1 and usually considers that all COLLECTIVE CONSUMPTION-2,3 should be provided by the state. An anti-collectivist is one who is opposed to such a theory or system.

collectivization COLLECTIVE FARMING.

collision zone in the theory of PLATE TECTONICS, a zone where converging lithospheric plates carrying continental crust meet, with the result that the edge of one plate dives under the other but the rocks of the continental crust pile up, crushed and buckled and mixed with material swept up from the floor of any ocean which may formerly have separated the plates, the ocean being squeezed out of existence. Such a collision produces chains of FOLD MOUNTAINS, e.g. the Alpine-Himalayan chain. OCEANIC TRENCH.

colloid a substance (gas, liquid or solid), finely divided and dispersed in a continuous gas, liquid or solid medium, the particles consisting of very large MOLECULES or aggregation of molecules which do not settle at all, or only very slowly. Thus the system is neither a SOLUTION (in which the dispersed particles are single molecules) nor a SUSPENSION (in which the particles are large enough to tend to fall by GRAVITATION and concentrate as a SEDIMENT-1). The electrical forces in the system are important in SOIL SCIENCE: colloids may loosen or dislodge rock particles from surfaces with which they are in contact, attracting IONS of dissolved substances, especially those that are basic (BASE-2); or some constituent particles of the soil may stick to

each other, as in colloidal CLAY, COLLOIDAL PLUCKING, FLOCCULATION.

colloidal *adj.* of, pertaining to, in the nature of, or characteristic of, a COLLOID.

colloidal or colloid plucking a weathering process in which soil COLLOIDS loosen or pull off small fragments of rock from the surfaces with which they come into contact. G.

colluvial soil a soil formed from COLLUVIUM.

colluvium (American slope-wash) a collection of rock debris of varied origin which has accummulated at the base of a slope as a result of the movement of SCREES-2 and mud flows down the slope under gravity. MASS MOVEMENT. G.

Colombo Plan after several meetings in Colombo, Sri Lanka, and elsewhere, in 1950 the Commonwealth countries of Australia, Canada, Ceylon (Sri Lanka), India, Malaya (Malaysia), New Zealand, Pakistan and the UK drew up the 'Colombo Plan for Cooperative Economic Development in South and South-East Asia', operative from July 1951, with the aim of improving living standards by reviewing development plans and coordinating development assistance. In addition to the originators, the USA, Afghanistan, Bangladesh, Bhutan, Burma, Fiji, Indonesia, Iran, Kampuchea, Republic of Korea, Laos, Republic of Maldives, Nepal, Papua New Guinea, Philippines, Singapore, Thailand and Vietnam are now members.

colonial an inhabitant of a COLONY-1,2, frequently used in a derogatory sense.

colonial *adj.* **1.** of, or pertaining to, a COLONY **2.** in USA, of or belonging to the thirteen British colonies which became the United States, or to the period of time (seventeenth and eighteenth centuries) when they were still colonies, applied especially to the works of art, artifacts, furniture and architecture of that period.

colonial animal an animal which is a member of an association (COLONY-5) of incompletely separated individuals, e.g. CORAL.

colonialism 1. the principle or practice of having or keeping colonies (COLONY-1,2) 2. the economic, political and social policies by which colonies are governed by the sovereign METROPOLITAN-2 country (the colonial power), usually based on the maintenance of a marked distinction between the governing country and the subordinate (colonial) population 3. in a derogatory sense, an alleged policy of exploitation of weak peoples by a large, strong power, which has the effect of perpetuating the economic differences between the colonies and the governing power, the former supplying the raw materials for the latter's manufacturing industry 4. the belief that a colonial system (for policies see 2.) benefits and promotes the welfare of the state colonized.

colonist 1. a person who, or an animal or plant which, helps to establish a COLONY 2. an inhabitant of a colony.

colonization 1. the act or policy of bringing human settlers into a locality, or to what is to them a foreign country, of establishing a COLONY-1, of forming that territory into a COLONY-2. DECOLONIZATION 2. the spread of a group of animals, or of a plant species, into an area. CLOSED COMMUNITY. G.

colony (Latin colonus, a pioneer cultivator or settler; colonia, a farm or settlement, especially a public settlement of Roman citizens in hostile or newly-conquered country) 1. a human settlement formed in a territory by people from other territory, usually from another country (to the government of which it,the colony, becomes in some degree subject) 2. the territory so occupied 3. the people who carry out

the occupying and their descendants. COLONIALISM, COLONIZATION 4. a body of people settling in a new locality within the home territory, and the area so settled 5. people of like occupation (e.g. artists) or of the same foreign nationality living in a place in an unorganized group (e.g. the Chinese colony in London) 6. in ecology, loosely applied to any collection of animals or of plants living together in one place (e.g. a SOCIETY-2), an isolated group, a group of individuals of a plant species migrant in a new habitat, a group of COLONIAL ANIMALS, or a culture of MICROORGANISMS. G.

colt the young male of a horse or similar QUADRUPED.

column in statistics, a vertical line of entries in a frequency table. MATRIX, ROW.

columnar structure in geology, a structure comprising hexagonal columns formed in the cooling of IGNEOUS ROCKS, especially of BASALT. The contraction associated with the cooling results in a series of regular JOINTS at right angles to the surfaces of cooling, thereby producing the columns, e.g. as in the Giant's Causeway, Northern Ireland. Very similar hexagonal cracks develop when mud dries. G.

combe, coombe, coomb, coom 1. in southern England, a deep hollow or valley, especially if short and steep at the head, or closed in, common in CHALK country 2. in southwestern England, a short steep valley opening to the sea 3. a CIRQUE in the English Lake District 4. in the Jura mountains, a narrow, LONGITUDINAL VALLEY developed along the crest of an ANTICLINE; or a SYNCLINAL VALLEY; or a valley drained by a stream which disappears underground at the lower end. G.

combe rock, Coombe Deposit a mass of unsorted rock debris containing angular rock fragments, the product of SOLIFLUCTION in PERIGLACIAL conditions, partly filling a valley bottom or covering the lower slopes. As Coombe Deposit it occurs specifically as hardened CHALK mud with sand and FLINTS in the coastal chalklands of southern England. HEAD-10. G.

combine harvester a farm machine which reaps and threshes grain crops (CEREAL-1) as it moves along, being suitable only for fairly large, flat fields.

comb-ridge an ARETE.

COMECON CMEA.

comet a celestial body consisting of a gaseous cloud enveloping a bright nucleus, moving around the SUN in an elliptical or parabolic ORBIT so eccentrically that some comets escape from the SOLAR SYSTEM. On nearing the sun the pressure of the sun's RADIATION forces the gas of a comet into a tail, pointing away from the sun.

comfort zone the range of TEMPERATURE-2 and RELATIVE HUMIDITY in a climate within which human beings feel comfortable. Common standards are 20° to 21°C (68° to 72°F) and 55 to 60 per cent relative humidity,the latter preferably falling as temperature rises. CLIMOGRAPH, SENSIBLE TEMPERATURE. G.

comfrey *Symphytum officinale*, family Boraginaceae, a genus of some 25 species of erect, sometimes tuberous-rooted herbs, native to Europe and western Asia, reputed to have medicinal properties, including the power to heal wounds. The leaves are edible.

commensalism 1. in ecology, the close association between organisms of different species from which one benefits but the other is unharmed. COENOSITE, INQUILINE, MUTUALISM, PARASITISM, SYMBIOSIS 2. in urban geography, the association between an individual and a group of similar individuals operating in close proximity, the individual cooperating with the other members of the group and benefiting from the advantages derived from group activities and group membership while

competing with the other members, e.g. specialized commercial or professional enterprises (e.g. the clothing industry, lawyers) in a particular district of a town or city.

commercial agriculture the growing of agricultural produce for sale. SUBSISTENCE AGRICULTURE.

commercial crops INDUSTRIAL CROPS.

commercial geography the study of products, their distribution and consumption. Some authors also include a consideration of the geographical and other factors influencing the productivity of people. C.

comminution breaking down into small fragments by crushing or grinding, e.g. of rocks progressively broken down to smaller particles by the agents of erosion and weathering, or by earth movements; or of rock by mechanical means (for road metal or aggregate).

commodity 1. in general, a GOOD which results from a production process (i.e. it is the product of labour), meets human needs, and has an exchange value, being sold on the MARKET-1. Difficulties in definition arise from the inclusion (or not) of FREE GOODS, and of services. Some authors include free goods if they are useful; and, for the sake of clarity, most economists tend to use the term 'goods and services' rather than 'commodity', although some may still use commodity to cover goods and services 2. a synonym for ECONOMIC GOOD, i.e. a GOOD which has a price 3. in early CLASSICAL ECONOMIC THEORY the term commodity excluded services because (according to the LABOUR THEORY OF VALUE) labour was considered to be the source of value, it did not have exchange value 4. in the business community, raw materials, as in commodity exchange.

common, common land in England and Wales, land, usually unenclosed, over which certain persons or groups of people have various COMMON RIGHTS, or rights of common, though they do not own the land, the owner normally being the 'lord of the manor'. At the present day the lord of manor is frequently the local administrative authority. In general common land represents the poorer quality land of the MANOR which, when INCLOSURE of lands took place, was left unenclosed and provided grazing for the villagers or peasants who would otherwise have been left landless. Under later legislation, the general public has been given rights of access on certain commons, including those within boroughs or urban areas, but unless there is specific legislation the public has no rights on rural commons. Today the term common land is commonly applied to MANORIAL WASTE as well as to other land over which common rights exist. COMMONABLE LAND, COMMON RIGHTS, HEAF, LAMMAS LAND, LOT MEADOW, METROPOLITAN COMMON.

commonable animal an animal which a commoner or holder of COMMON RIGHTS may turn out to graze on a common. Usually this includes horses, cattle and sheep, but sometimes sheep are excluded; and in some cases geese (whence goose common) and asses are included. Under the principle known as LEVANT AND COUCHANT the commoner was usually restricted to turning out on the common only the number of animals that could be wintered on that commoner's holding.

commonable land at present, certain land other than MANORIAL WASTE over which COMMON RIGHTS exist at particular times of the year.

Common Agricultural Policy CAP.

commoner one who enjoys COMMON RIGHTS.

common field, common arable in England and parts of western Europe until inclosure in the fifteenth to eighteenth centuries, one of the large, open arable fields worked by the village community. FIELD SYSTEM.

Common Law 1. the unwritten law, especially of England, based on CUSTOM and court decisions rather than on laws made by parliament (STATUTE LAW) **2.** the body of law originating in England and the modern systems of law based on it. CIVIL LAW.

common market an association of nation states formed as a trading group in order to provide its members with better economic prospects than could be achieved by a member trading separately as an individual. The objective is usually to abolish restrictions on trade, movement of labour and of capital, etc. between the members; and to establish a common trading policy with other nation states or groups of nation states. EEC.

Common Market EEC.

common millet *Panicum miliaceum*, a nutritious MILLET, widely grown in temperate regions in Asia and southern Europe since prehistoric times, subsequently in North America, the grain being used for human and livestock food.

common property resource applied by some authors to a flow or continuous resource (NATURAL RESOURES) jointly owned and used by society at large, to which any member has free access, e.g. water, landscape.

common rights, rights of common the rights held by certain persons on COMMON LAND which is the property of another. The chief rights are to pasture certain animals, usually limited in number, stinted (STINT) or gaited, which is known as common of pasture; of digging peat for fuel (common of TURBARY); of gathering, sometimes cutting, wood for fuel or house repairs (common of ESTOVERS); of fishing (common of PISCARY); of digging for sand and stone (common of SOIL AND STONE). Common rights may belong to an individual such as the tenant of a farm (right of common appendant), or may be attached to the dwelling itself (common appurtenant) or by a grant to an individual descending to that individual's heirs (common in gross). ALLOTMENTS (including FUEL ALLOTMENTS), CATTLEGATE, COMMONABLE ANIMAL, FIELD GARDEN, LEVANT AND COUCHANT, MANOR, MANORIAL WASTE, PANNAGE, SOLE PASTURE, SOLE VESTURE.

common salt SALT.

common variance in statistics, in FACTOR ANALYSIS, that part of the total VARIANCE which correlates with other VARIABLES. ERROR VARIANCE, RESIDUAL VARIANCE, SPECIFIC VARIANCE, TRUE VARIANCE.

commonwealth, the Commonwealth 1. a free association of self-governing, individual territories organized in a federation, the federation government taking responsibility for certain common matters, such as defence, e.g. Australia 2. the Commonwealth, a free association of Britain and certain independent SOVEREIGN STATES, each of which was formerly a DEPENDENCY or COLONY-2 within the British Empire. IMPERIAL PREFERENCE.

commune 1. the smallest administrative unit in some countries, e.g. Belgium, France, Italy, Spain 2. a group of people, not of the same family, sharing possessions, accommodation, work and living expenses 3. the place where such a group lives 4. in communist countries (COMMUNISM-2), a group of people working as a team for the general good, e.g. in raising crops and animals, the group usually owning and controlling the MEANS OF PRODUCTION 5. the tract of land owned and used by such a group 6. a group of people working together to protect and advance local interests.

communications 1. travel and transport links between places 2. the means by which people make contact with, exchange information or ideas with, or trade with, others.

communism 1. historically, the common ownership of all property in a society, e.g. as in a non-literate society, extended today to some monastic establishments 2. since 1848, a theory or practice linked especially to the ideas developed by Karl Marx from his interpretation of history. He advocated a classless society, organized on the basis of common ownership of property and the means of production, distribution and supply, the individual members contributing 'each according to his ability' and receiving 'each according to his need'.

community 1. a group of people living in a particular area 2. a group of people living near one another, with distinct social relationships 3. a group of people sharing a common faith, culture, profession, life-style 4. in ecology, a general term applied to any naurally occurring group of different organisms occupying a common environment, interacting with each other, particularly through food relationships, but relatively independent of other groups. The size may vary and the larger communities may contain smaller ones. ASSEMBLY, ASSOCIATION, CLOSED COMMUNITY, CONSOCIATION, SOCIETY.

community council in Scotland, LOCAL GOVERNMENT IN BRITAIN.

commuter one who commutes, i.e. one who travels regularly, usually daily (but also at other regular intervals, e.g. weekly or monthly) from residence to place of work. Historically the greatest number of commuting trips were inward to central employment areas; but outward (or reverse) commuting is now common where the speed of DECENTRALIZATION has outpaced the shift of population to new locations. G.

commuter village dormitory village, a village in a rural area, formerly inhabited by people who worked in, or who had worked in, the village or close to it, now inhabited mainly by people who travel regularly to work in a nearby town. COMMUTER, DORMITORY TOWN.

commuter zone the area in which commuting takes place, from which COMMUTERS are drawn to work in a nearby town.

commuting hinterland DAILY URBAN SYSTEM.

compaction 1. in geology, the process in which fine rock particles, e.g. of silt or clay, are combined tightly together by pressure of earth movements or weight of later overlying deposits. LITHIFICATION 2. of soils, the pressing together of soil particles (e.g. by torrential rain, or by heavy mechanical equipment especially in wet conditions) so that the voids between them are reduced, with consequent loss of air, to the detriment of soil fertility.

compage (obsolete, a means of joining, connecting matter) all the features of the physical, biotic and societal environments functionally associated with the occupance of the earth by human beings. The term was revived by Derwent Whittlesey in his attempt to give greater precision to an aspect of REGION. G.

comparative advantage a concept used in NEOCLASSICAL ECONOMIC THEORY to explain why certain areas or countries specialize in the production of and the trade in certain items. The notion is that an area produces that which it can produce most efficiently and economically, i.e. the items for which it has the greatest ratio of advantage, or the least ratio of disadvantage, in comparison with other areas, assuming conditions of free trade between all areas. COMPLEMENTARITY.

comparative cost analysis an evaluation of the advantages or disadvantages of alternative locations, based on the cost of production at those locations. COST BENEFIT ANALYSIS, VARIABLE COST ANALYSIS, VARIABLE REVENUE ANALYSIS.

comparative method a term variously used in sociology, but generally most frequently applied to 1. the comparison of one set of facts with another 2. the procedures which, by clarifying the differences and resemblances exhibited by various phenomena or classes of phenomena (e.g. social groups, cultures, events) which have on various criteria already been established as comparable, try to find out the causes for the emergence and development of the phenomena as well as the patterns of interrelationship within and between such phenomena 3. a procedure in which a HYPOTHESIS is tested for reliability against actual known facts. CONSTANT COMPARATIVE METHOD.

compass an instrument used to find direction. In a MAGNETIC compass a free-swinging magnetized needle is fixed to, and swings freely over, a dial which is graduated in degrees and shows the cardinal points (north, east, south and west). Under the influence of the local line of magnetic force, the needle indicates the NORTH and SOUTH MAGNETIC POLES. A non-magnetic compass (GYROCOMPASS) points TRUE NORTH. BEARING-3, PRISMATIC COMPASS.

competence of rocks, competent bed the relative strength of a bed of stratum when subjected to folding. If strong enough to bend without distortion when subjected to the stress of folding, it is said to be competent; if weak and thus liable to distortion, it is incompetent. G.

competence of a stream the ability of a stream to transport debris, measured in terms of the size (not the weight) of the largest pebble or boulder it can move. CAPACITY. G.

complementarity one of ULLMAN'S BASES FOR INTERACTION, the mutual benefits derived from the close proximity of one producer or land user to another, the one supplying, or having the potential to supply, goods etc. which the other lacks, or potentially lacks. Such INTERACTION-1 may arise from AREAL DIFFERENTIATION or

from the effects of ECONOMIES OF
SCALE. AGGLOMERATION-4, COM-
PARATIVE ADVANTAGE, EXTERNAL
ECONOMIES, EXTERNALITY, INTER-
VENING OPPORTUNITY, SPECIFIC
COMPLEMENTARITY, TRANSFERABIL-
ITY.

complementary investment in Marxism,
the part of SOCIAL INVESTMENT
which, by socializing (SOCIALIZA-
TION-2) parts of the cost of private
investment, renders profitable some
projects which would otherwise not be
profitable. DISCRETIONARY INVEST-
MENT.

complementary region a trade area, a
tributary area, the area served by a
CENTRAL PLACE, that of a higher
order centre being large and in many
cases overlapping the smaller area
served by a lower order centre. CEN-
TRAL PLACE HIERARCHY, ISOTROPIC
SURFACE.

complex climatology the analysis of the
CLIMATE of a place as shown by the
frequency of the types of WEATHER
occurring there, defined in terms of the
CLIMATIC ELEMENTS.

components of change approach an approach to the study of the changing
pattern of employment (usually employment in manufacturing) in a region
or urban area (a conurbation, town or
city). The changes that have occurred
during a defined time period in the
study area are broken down into four
components, i.e. changes caused by (A)
birth (the formation of new firms); by
(B) death (the closing down of existing
firms); by migration, i.e. movement of
some firms into (C), and others out of
(D), the area; or by (E) in situ change
(the growth or decline of employment
in firms existing in the area at the start
of the period of study). The net change
in employment in the area during the
defined time period can then be calculated: A (birth) minus B (death) plus E
(net change in employment in firms
surviving through the period) plus C

(immigrant firms) minus D (emigrant
firms).

composite preference rating a rating of
PREFERENCE-1 used by an individual
in the process of choosing from alternatives, reached by first defining a
subjective value for each ATTRIBUTE-1
of each alternative, then weighting
these subjective values according to the
importance of the attribute. These
weighted values are then combined to
give the composite preference rating,
the individual choosing the alternative
with the higher ranking.

composite profile termed zonal profile
by some authors, equated with
PROJECTED PROFILE by others, a
PROFILE built up to show the surface
of relief viewed in the horizontal plane
of summit levels from an infinite distance, masking the lower summit levels
of a series of parallel profiles by
including only the highest summits.
COMPRESSED PROFILE, PROJECTED
PROFILE, SUPERIMPOSED PROFILE.

composite volcanic cone a VOLCANIC
CONE composed of layers of ash, cinder
and lava built up over a long period of
time by a series of ERUPTIONS through
the main PIPE which is topped by a
CRATER-1, e.g. Vesuvius. STRATO-
VOLCANIC CONE, VOLCANO.

compost a soil conditioner and FERTIL-
IZER produced by the planned decomposition of organic material, such as
vegetable remains.

compound in chemistry, a substance
formed of two or more ingredients of
constant proportion by weight.

compound *adj.* **1.** composed of separate
substances or parts **2.** in botany, composed of a number of similar parts,
forming a common whole.

Comprehensive Soil Classification System
USA, CSCS, SOIL CLASSIFICATION.

compressed profile a PROFILE in which
a series of COMPOSITE PROFILES are
arranged in such a way that, when
viewed at right angles, only those
features can be seen which are not

hidden by higher ones in the foreground.

compression forcing into smaller compass, reducing in volume, condensation by pressure, pressing together. The effect of compression on the rocks of the earth's surface contributes to FAULTING, FOLDING, LITHIFICATION. CHATTER MARK, TENSION.

compressional wave PUSH WAVE.

conacre, con-acre (Ireland) in Irish LAND TENURE, the letting by a tenant, for the season, of small tracts of land, ploughed and prepared for a crop. G.

concentration coming together, being brought together in a mass, an AGGLOMERATION, e.g. the localization of a particular economic activity in areas favourable to it. CENTRALIZATION, CONGESTION.

concentric zone growth theory a theory introduced by E. W. Burgess in 1927 based on his studies of urban growth specific to the Chicago area. He saw Chicago as a city in an industrialized country, expanding radially from its centre in a series of concentric zones. He suggested that the expansion and the formation of these concentric zones were created by SUCCESSION AND INVASION, as the occupiers of each inner zone, wishing to avoid the invading negative externalities (EXTERNALITY), moved outwards to colonize the next outer zone. From the centre outwards he identified the concentric zones as (a) the inner CENTRAL BUSINESS DISTRICT, (b) a transition zone (INNER CITY) with residential areas invaded by business and industry from the CORE-2, the run-down dwellings being subdivided and overcrowded and inhabited by poor immigrants, especially ETHNIC MINORITIES, (c) a working class residential zone with second generation immigrant dwellings, (d) a higher class (middle class) residential zone with one-family dwellings, and (e) an outer commuting zone with higher class dwellings in suburban areas and satellite towns. He acknowledged that this general, simplified pattern would be modified if applied to other cities (e.g. by terrain, routes, and other constraints); but he suggested that radial expansion along a broad front, stimulated by invasion and succession, was a dominant process in the shaping of the pattern of a city. INDUSTRIAL CITY, MULTIPLE NUCLEI MODEL, SECTOR THEORY, SEED BED, ZONE OF ASSIMILATION, ZONE OF DISCARD. G.

concept 1. a general notion, idea, or understanding, especially one constructed by generalization from particular examples **2.** the meaning of a term, the smallest unit of thought (CONSTRUCT).

concordance of summit-levels ACCORDANCE OF SUMMIT LEVELS.

concordant *adj.* agreeing or consistent with, thus lying or running parallel to the structural trend lines of the relief, of the general strata. DISCORDANT.

concordant coast a coast lying approximately parallel to the structural trend lines of the land, also termed a LONGITUDINAL or PACIFIC COAST. It is usually straight, but if it is drowned by the sea a line of islands (the peaks of a former mountain range) may be formed, separated from the mainland by the drowned parallel valleys (SOUND), e.g. the Adriatic coast of Yugoslavia, the DALMATION COAST. G.

concordant drainage the pattern of DRAINAGE-2 which arises from and closely follows the trends of the underlying STRATA.

concordant intrusion an INTRUSION of IGNEOUS material lying parallel to the trend of the rock strata that it penetrates.

concrete a manufactured building material consisting of CEMENT, SAND, GRAVEL or crushed stone, mixed with water. When used wet, it sets in the shape and position required. In large structures it may be reinforced by having wire mesh, metal bars, etc. embedded in it, the product being termed reinforced concrete.

concretion a nodule, a small roughly rounded mass of rock occurring in a mass of different rock, e.g. iron nodules in claybands (CLAY-IRONSTONE), a useful source of iron ore. DOGGER.

condensation the physical process of the transition of a substance from the VAPOUR to the LIQUID state, e.g. as a result of cooling or increase of pressure. It occurs in the atmosphere when the air is SATURATED (by evaporation into it), or when it is cooled. CLOUD, DEW, DEW-POINT, FOG, MIST, RAINFALL, SNOW.

condensation trail contrail, the white, ribbon-like, cloud-like phenomenon seen behind an aircraft flying at high altitudes in cold, clear but humid air, caused by the CONDENSATION of the water vapour (the product of fuel combustion) coming from the engine's exhaust, and by the lowered pressure behind the wing-tips. Such trails persist only if they become frozen and the ICE EVAPORATION LEVEL favours them.

conditional instability the state of an air mass with an ENVIRONMENTAL LAPSE RATE greater than the SATURATED ADIABATIC LAPSE RATE but less than the DRY ADIABATIC LAPSE RATE, the instability depending on the amount of water vapour held in the air mass. A pocket of unsaturated air, if pushed upward from ground level, will gradually become cooler than the surrounding air and descend; but a pocket of saturated air will remain relatively warmer, and thus will continue to ascend. ABSOLUTE INSTABILITY, LATENT INSTABILITY, POTENTIAL INSTABILITY.

condominium 1. joint rule or sovereignty **2.** territory ruled jointly by two or more countries, as was Sudan by Egypt and Britain, 1898 to 1953 **3.** a residential community in which each individual or family in addition to

owning or leasing a private housing unit (in some cases used as a second home) also holds an equal share with others in the facilities available to that residential community. In North America, e.g. Florida, most condominiums (some occupying HIGH-RISE blocks) are managed by an association comprising elected residents and/or developers. They particularly attract older people (e.g. the retired) or young professional people. G.

conductive THERMAL CONDUCTION.

conductivity the ability of a substance to conduct, i.e. to transmit or be capable of transmitting, heat or electricity.

cone 1. a volcanic peak, with a roughly circular base tapering to a point. ADVENTIVE CONE, ALLUVIAL CONE, ASH CONE, CENTRAL ERUPTION, CINDER CONE, DOME VOLCANO, PARASITIC CONE, SPATTER CONE **2.** the fruit of some trees, e.g. pine trees. CONIFEROUS.

cone karst a type of KARST common in tropical areas, with star-shaped depressions occurring at the base of many steep-sided conical hills, a variety of KEGELKARST. COCKPIT KARST.

cone of dejection an ALLUVIAL CONE or fan, especially if of coarse material, fanning out from a point where a mountain torrent disgorges on to a plain. The term is now rarely used. G.

cone of depression, cone of exhaustion a lowering of the WATER TABLE round a well, occurring when the speed of pumping exceeds the speed of replenishment of the well by AQUIFERS.

cone sheet a zone of dykes (DIKE) or FISSURES-1 sloping inwards and surrounding a circular or dome-shaped IGNEOUS INTRUSION which, in exerting pressure, has fractured the COUNTRY-ROCK. RING DYKE.

Confederate States the eleven states which seceded from the United States in 1860-1. The first six formed a government under the title Confederate

States of America on 4 February 1861. They were, in order of secession, South Carolina, Mississippi, Florida, Alabama, Georgia, Louisiana; joined later by Texas, Virginia, Arkansas, Tennessee, North Carolina. They were defeated in the Civil War which followed. DEEP SOUTH, FEDERAL.

confidence interval in statistics, an INTERVAL ESTIMATE with a very limited range of values within which, as estimated from sample data, a particular POPULATION PARAMETER (e.g. the STANDARD DEVIATION) has a specified probability of lying.

configuration map a simplified relief map on which prominent features are shown by bold, generalized lines.

confluence 1. the place where two streams, about equal in size, converge and unite **2.** the place at which a stream flows into another **3.** the body of water so produced **4.** by analogy, a junction of routeways. G.

confluent a stream that joins another of approximately equal size. AFFLUENT.

conformable *adj.* applied in geology to STRATA deposited one on another in parallel planes in proper geological sequence, without breaks or interruption caused by denudation or earth movements, in contrast to UNCONFORMABLE. UNCONFORMITY. G.

confounded variable a term applied to two VARIABLES-2 which vary with each other so systematically that it is almost impossible to identify which of them is affecting a third variable.

congelifluction, congelifluxion a flow of earth under PERIGLACIAL conditions. SOLIFLUCTION.

congelifract an individual fragment produced by frost splitting. CONGELIFRACTION. G.

congelifractate a mass of material of any grain size resulting from CONGELIFRACTION. G.

congelifraction the splitting of rocks etc. by frost action. G.

congeliturbate a body of material disturbed by frost action. CONGELITURBATION. G.

congeliturbation a general term applied to frost action, including FROSTHEAVING, churning of the ground and SOLIFLUCTION, affecting both the soil and subsoil, in some cases producing PATTERNED GROUND. G.

congestion the state of being packed closely together, clogged by overcrowding, e.g. the result of the use of some facility, such as a road network, in excess of its CAPACITY-2.

conglomerate a SEDIMENTARY ROCK, e.g. PUDDINGSTONE, consisting of round, waterworn pebbles in a matrix of natural cementing material such as CALCIUM CARBONATE, SILICA or IRON oxide. AGGLOMERATE, BRECCIA.

congost (Spain) CLUSE.

congressional township in USA, PUBLIC LAND SURVEY, USA.

coniferous *adj.* applied to trees belonging to Coniferales, a large order of trees and shrubs, commonly EVERGREEN, with slender leaves, and reproducing by means of seeds contained in a cone (a reproductive structure consisting of woody carpels closely grouped around a central axis, bearing pollen or ovules).

coniferous forest forest which occurs under many conditions of soil, climate, aspect, elevation, the species of the order Coniferales being hygrophytic (hygrophilous), mesophytic, xerophytic (xerophilous) (HYGROPHYTE, MESOPHYTE, XEROPHYTE). The largest continuous area is that of the northern coniferous forest (TAIGA), also known as the BOREAL forest. Away from very low temperatures, coniferous trees generally grow rapidly, so are much favoured there for afforestation and as a commercial crop, producing SOFTWOODS of varying quality for widely ranging purposes (wood-pulp for paper-making, timber for house construction, etc. as well as RESIN, TAR, TURPENTINE). Most of the timber consumed commercially is that of coniferous trees. FOREST. C.

conjunction in astronomy, the position of two planets or other heavenly bodies which, when viewed from the earth, lie in line in the same direction (SYZYGY). The 'new moon' is the result of the earth, moon and sun being in conjunction, when forces bringing about tidal action are at maximum strength, giving rise to SPRING TIDES. MOON, OPPOSITION, QUADRATURE. G.

connate water fossil water, water trapped in SEDIMENTARY ROCK at the time of its deposition. G.

conquistador any of the sixteenth century Spanish conquerors of Mexico, of central and of South America.

consequent drainage, consequent river or stream, consequent valley a natural water flow or valley directly related to the original slope of the land surface; also a river or stream flowing in the same direction as the DIP of the underlying rocks. A secondary consequent stream is a tributary of a SUBSEQUENT STREAM; it flows parallel to the main consequent stream. DRAINAGE, OBSEQUENT.

consequent river or stream CONSEQUENT DRAINAGE.

conservancy a body, especially an official body, charged with the task of conservation, e.g. the Nature Conservancy in Britain. G.

conservation protection from destructive influences. A term applied in general to the positive work of maintenance, enhancement and wise management, of reducing the rate of consumption to avoid irrevocable depletion, in order to benefit posterity, as in the conservation of nature, or of natural resources, or of buildings or works of art of special merit, etc. Specifically it is applied to the work of protecting and maintaining the soil (soil conservation) and wild life (nature conservation). PRESERVATION, RESOURCE CONSERVATION.

conservative plate margin PLATE TECTONICS, TRANSFORM FAULT.

consociation a CLIMAX community of plants dominated by a single SPECIES, usually named by adding -etum to the stem of the Latin name of the GENUS of the dominant, e.g.Fagetum, woodland dominated by beech, *Fagus*. ASSOCIATION.

consocies a seral community (SERE) with a single dominant species.

constant in mathematics and physics, a quantity or factor that does not change (VARIABLE), being universal, or applicable solely to a particular operation or circumstance, or characteristic of a substance or instrument.

constant *adj*. continual, uniform, not subject to variation. INVARIABLE, VARIABLE (*adjs*.).

constant capital in Marxism, labour from the past (dead or expended labour) incorporated in the MEANS OF PRODUCTION, qualified as constant because the value is fixed (only living labour being able to create new value). SOCIAL INVESTMENT, VARIABLE CAPITAL.

constant comparative method a process in which segments of data are repeatedly compared with each other. COMPARATIVE METHOD.

constant slope part of a SLOPE profile, the straight slope of the lower hillside, below the FREE FACE and above the WANING SLOPE.

constellation a group of fixed stars commonly given a name relating to a fanciful outline enclosing them.

construct a CONCEPT-2 devised to be used as an element in a theoretical SYSTEM-4.

constructive plate margin OCEANIC RIDGE, PLATE TECTONICS.

constructive wave one of a series of waves rolling regularly and gently on a coast, the SWASH of the wave being more powerful than the BACKWASH, with the result that shingle, etc. is pushed up the beach to form ridges.

consumer 1. one who uses a COM-MODITY-1 or service (SERVICES-3) **2.** one of the organisms using energy in a FOOD CHAIN.

consumer durables those CONSUMER GOODS which are multiple assets, i.e. which can be used many times over a period of time and are thus used up gradually (e.g. cars, clothes, furniture). Consumer goods which can be stored but used only once (e.g. canned or frozen food) are not classified as durable.

consumer goods, consumers' goods goods and services which directly satisfy the needs and desires of the individual person, e.g. food items. CAPITAL GOODS, CONSUMER DURABLES, ECONOMIC GOODS, FREE GOODS, PRODUCER GOODS.

contact field in DIFFUSION-2 of INNO-VATION, the spatial distribution of the contacts, i.e. the friends and acquaintances etc., of an individual or a group of people to whom information may pass from the sender. The density of contacts usually decreases with increasing distance. The probability of anyone receiving information from a sender is high close to the sender, but it progressively and gradually weakens with increased distance. Thus the probability of any person receiving information from the sender is inversely proportional to the distance between them. DIFFUSION WAVE, DISTANCE DECAY PHENOMENON, MEAN INFORMATION FIELD, NEIGH-BOURHOOD EFFECT, PRIVATE INFOR-MATION, PRIVATE INFORMATION FIELD.

contact metamorphism THERMAL MET-AMORPHISM; and IGNEOUS, MAGMA, METAMORPHIC AUREOLE, METAMOR-PHISM.

contagious diffusion EXPANSION DIF-FUSION which depends on direct contact (as contagious diseases depend on direct contact in order to spread), in which the process of spread is CENTRI-FUGAL and strongly influenced by distance, individuals or areas near to the source having a higher probability of contact than those farther away. CASCADE DIFFUSION, HIERARCHIC DIFFUSION.

container, container transport a special large, strong, durable packing case of standard size, usually with internal volume exceeding 1 cu m (35.3 cu ft), made of steel, aluminium-alloy etc., suitable for repeated use and for mechanical handling, into which goods are easily packed and locked for transport by road, sea, air, thus designed for easy switching from one form of transport to another. For example containers at a port may be moved from quay to ship by lift-on/lift-off (Lo-Lo) equipment; or they may be directly transferred by the practice of roll-on/roll-off (Ro-Ro), a wheeled vehicle, complete with containers, being driven along a ramp or link-span pontoon bridge joining the quay with the ship.

contextual effect in an election, the effect on voting behaviour of purely local (as distinct from national) forces. Such local forces become apparent when they do not coincide with the national. NEIGHBOURHOOD EFFECT.

contiguity the state of being **1.** in contact, touching, neighbouring **2.** next in order or in time. CONTIGUOUS ZONE, TOPOLOGY.

Contiguous Zone TERRITORIAL WATERS.

continent one of the large continuous masses of land on the earth's surface. Seven are usually distinguished: North and South America, Europe, Asia, Africa, Australia, and Antarctica. Australia and New Zealand together are often referred to as 'Australasia'. 'Oceania' covers the Pacific islands. 'The Continent' in British writings denotes the mainland of Europe (CON-TINENTAL). Some authors insist that Eurasia is one continent. The traditional boundary between Europe and Asia along the Ural Mountains and the

Ural River does not coincide with any existing administrative boundary within the USSR.

continental a person whose birthplace is in Europe but not in the British Isles.

continental *adj.* of, relating to, or characteristic of, a very large land mass, i.e. of a CONTINENT.

continental air mass an AIR MASS, usually of low humidity, the source of which is a HIGH pressure region over the interior of a continent. Distinguished by the symbol c it may be of high latitude, i.e. polar (Pc) or low latitude, i.e. tropical (Tc).

continental apron, continental rise terms applied by some authors to the slope with a very low gradient stretching up from the deep sea floor to the foot of the CONTINENTAL SHELF.

continental basin a region, usually of inland drainage, in the interior of a CONTINENT. G.

continental borderland a CONTINENTAL SHELF with a series of submarine basins and ridges similar in origin to the faulted structures landward from it. G.

continental climate the climate associated with the interior of a CONTINENT or other places protected from or unaffected by the moderating influence of the sea, and so characterized by great extremes of temperature between summer and winter, low, variable, precipitation, occurring mainly in early summer, and low humidity. The effect of CONTINENTALITY is most marked in midlatitudes, but it is also significant in high and low latitudes. MARITIME CLIMATE, OCEANIC CLIMATE, TROPICAL AIR MASS, TROPICAL CLIMATE.

continental crust PLATE TECTONICS.

continental divide the main waterparting in a continent, e.g. in North America, where the streams flow on one side of the divide to the Pacific and on the other to the Atlantic. G.

continental drift the theory or hypothesis first postulated in 1858, re-stated by A. Wegener in 1911, that the present distribution of the continental masses is the result of fragmentation of one or more pre-existing masses which have drifted apart, the intervening hollows having become occupied by the oceans. PLATE TECTONICS. G.

continental glacier an ice sheet or ice cap covering a considerable area of a continent. GLACIER. G.

continental ice sheet an ice sheet of continental extent, e.g. in Antarctica today, or the ice sheets of the QUATERNARY glaciation, covering the north of Europe and North America.

continental island an island near to and closely associated structurally with the neighbouring land mass. OCEANIC ISLAND. G.

continentality the conditions of being CONTINENTAL (*adj.*) as opposed to OCEANIC, applied especially to the measure of the extent to which the climate of a place is influenced by its distance from the sea. CONTINENTAL CLIMATE. G.

continental margin a zone comprising the CONTINENTAL SHELF and the CONTINENTAL SLOPE, extending from the coastline to depths of approximately 2000 m (1095 fathoms: 6560 ft), but distinct from the deep sea floor. TERRITORIAL WATERS.

continental period in geological time, a period when the area under consideration formed part of a continent or was above sea-level. G.

continental platform a continent and its CONTINENTAL SHELF to the edge of the CONTINENTAL SLOPE, i.e. the part of the earth's crust carrying SIAL, as distinct from the oceanic parts (SIMA). G.

continental rise CONTINENTAL APRON.

continental river a river lacking an outlet to the sea, the water being lost by percolation or evaporation. The term is now rarely used. ENDOREIC, INLAND DRAINAGE. G.

continental sea a partly enclosed sea

lying on continental crust (PLATE TECTONICS) and linked with the ocean, e.g. the Baltic sea. G.

continental shelf a gently sloping submarine plain, usually of 1° slope or less, and of variable width (scarcely present along some CONCORDANT COASTS where FOLD MOUNTAINS lie close to the ocean), forming a border to nearly every CONTINENT, stretching from the coast to the CONTINENTAL SLOPE, i.e. to the point where the seaward slope inclines markedly to the ocean floor. The depth of this outer edge has been defined as lying approximately between 120 m (65 fathoms: 395 ft) and 370 m (200 fathoms: 1215 ft); but the continental shelf itself is, in general, considered to be covered by seawater usually less than 183 m (100 fathoms: 600 ft) deep. The precise definition and delimitation of the continental shelf assume increasing importance in international law in connexion with the ownership of minerals and other resources lying on or under the shelf. CONTINENTAL MARGIN, CONTINENTAL TERRACE, TERRITORIAL WATERS. G.

continental slope the marked slope, commonly with an angle between 2° and 5°, steep in relation to the slope of the CONTINENTAL SHELF, lying between the edge of the continental shelf and the deep ocean floor (ABYSSAL PLAIN), i.e. from approximately 180 to 3600 m (100 to 2000 fathoms: 600 to 12 000 ft). CONTINENTAL MARGIN, BATHYAL. G.

continental subdivision the major distinctive unit in D. L. Linton's hierarchy of MORPHOLOGICAL REGIONS. G.

continental terrace in geomorphology, the marine-built terrace consisting of material removed in the cutting of the marine-cut terrace which lies to landward. The ABRASION PLATFORM and the marine-built terrace together constitute the CONTINENTAL SHELF. G.

continental resources NATURAL RESOURCES.

continuous variable in statistics, a VARIABLE which may take any value between two extremes. DISCRETE VARIABLE, NOMINAL VARIABLE.

contorted drift TILL which has been folded and generally twisted, probably by the pressure from ice, the distortions being dragged out in the direction of movement of the ice.

contour, contour line an imaginary line joining all the points on the ground that are at the same height above sea-level or, for submarine contours, below sea-level, usually based on an instrumental survey. Its representation on a map is commonly termed a contour line or isohypse (ISO-), but some authors insist that a contour is an imaginary line on the earth or its representation on a map, and argue that contour line is tautologous. FORM-LINE, RELIEF MAP. G.

contour interval the vertical distance between the CONTOURS shown on any given map. If the contours shown are at 50, 100, 150, 200 m or ft above sea-level, the contour interval would be 50 m or ft; if at 100, 200, 300 m or ft, the contour interval would be 100 m or ft. The contour interval is often varied on a map. G.

contour ploughing, contour cultivation the farming practice of cutting furrows across a hillslope, following the CONTOURS rather than ploughing up and down the slope, the object being to reduce SOIL EROSION caused by the run-off of rainwater. It is widely practised in the drier parts of the USA. G.

contraction hypothesis a popular nineteenth century hypothesis suggesting that the earth's interior shrank in cooling, thereby causing the crust to contract; and that this contraction (by compression) formed the fold mountains.

contrail CONDENSATION TRAIL.

controlled environment the state within a building (or sometimes within a group of totally enclosed buildings) where the air temperature, the humidity, the rate of movement of and the particle content in the air, is completely controlled, and lighting is artificial.

controlled political system a centrally planned economy, a system in which the government controls nearly all the economy, and profit and competition are normally officially absent. PUBLIC SECTOR.

control system a PROCESS-RESPONSE SYSTEM formed by decision-making. Decisions made at some critical points in the system alter the pattern of the throughputs in the original CASCADING SYSTEM, and this in turn affects the balance of the components in the original MORPHOLOGICAL SYSTEM.

conurbation a continuously urban area formed by the expansion and consequent coalescence of previously separate urban areas. In some cases it may include enclaves of rural land in agricultural use. ECUMENOPOLIS, LOCAL GOVERNMENT IN BRITAIN, MEGALOPOLIS, RANDSTAD. G.

convection the process of heat transfer from place to place within a FLUID (i.e. a gas or liquid) caused by the circulatory movement of the fluid itself (due to differences in temperature and hence in density) and the pull of GRAVITY. Convection produces vertical movement (in contrast to ADVECTION, horizontal movement), as in the upward welling of cold water in the oceans or in CONVECTION RAIN. RADIATION, THERMAL CONDUCTION, UNSTABLE AIR MASS. G.

convection current a stream of FLUID, e.g. in the atmosphere or ocean, produced by CONVECTION.

convection rain precipitation caused by the warming of air (moist with water taken up from the ground and its vegetation) by THERMAL CONDUCTION from the heated land surface. The warmed, moist air expands, rises and cools (ADIABATIC) to DEW-POINT, to form CUMULUS and CUMULONIMBUS CLOUDS, which drop very heavy, torrential rain, often accompanied by thunder. High temperatures and high humidity cause convection rain to fall in most afternoons near the equator; in areas at a distance from the equator it is a feature of the hot season (when the land surface is warmed) if the temperature is high enough and the air sufficiently moist. G.

convenience good a good of low order (ORDER OF GOODS) which needs to be bought frequently but for which people are not willing to travel a great distance, and for which the THRESHOLD POPULATION is fairly low, e.g. bread. CENTRAL PLACE THEORY, NEIGHBOURHOOD UNIT.

conventional name an exonym, the name given by the people of one linguistic group or nation to important towns, localities, physical features etc. of another. Thus in English the exonyms Warsaw, Vienna, Rome are used for cities known to the inhabitants of the countries concerned as Warszawa, Wien, Roma. There is now international agreement that local names should be used.

conventional sign a standard SYMBOL used on a map, and explained in the LEGEND, to convey a definite meaning, e.g. dots of different size may represent towns of varying size. G.

convergence in general, the act of coming together to meet in a common result or at a point of operation (DIVERGENCE). There are many specialized applications: **1.** in biology, the evolutionary development of similarity of some characteristic(s) among groups and organisms (initially different) living in the same environment. CONVERGENT EVOLUTION, POLYPHYLETIC **2.** in climatology, with reference to air flow, ITCZ **3.** in geology, the thinning of rock formations so that the upper and lower horizons draw closer together **4.** in oceanography, the movement of waters of different salinity and temperature **5.** in PLATE TECTONICS, the movement of plates. COLLISION ZONE, OCEANIC TRENCH. G.

conversion tables APPENDIX 2.

convulsionism CATASTROPHISM. G.

coolie, cooly a name given by Europeans to an unskilled Oriental (ORIENT) porter or hired labourer, and adopted in other areas. G.

coombe, Coombe Deposit COMBE, COMBE ROCK.

cooperative farm, co-operative farm a farming enterprise in which individual farmers combine in a group for mutual benefit. Usually the group pays for the bulk purchase of machinery, seeds, fertilizers etc. and for services, and markets the produce, the costs and profits being shared by the members.

coordinate, co-ordinate in mathematics, each of a system of two or more magnitudes specifying the position of a point on a line or a surface or in space, e.g. LATITUDE and LONGITUDE are the coordinates of a point on the earth's surface.

cop a little round-topped hill in north and central England, e.g. Mow Cop. G.

copal a RESIN-1 obtained from the tissues of some tropical plants, especially from *Copaifera copallifera*, used in varnishes and lacquers. C.

copper Cu, a red, ductile, metallic ELEMENT-6, unaffected by water or steam, a CHALCOPHILE, a TRACE ELEMENT, found native in nature, but mainly obtained from a variety of ores, copper pyrites (the sulphide $CuFeS_2$) and the carbonates (malachite and azurite) in particular. The ores occur especially in veins or scattered through METAMORPHIC ROCKS. It was the first metal to be used for ornaments and vessels, later alloyed with tin to form BRONZE. A good conductor of electricity and heat, it is now widely used in alloys, some of its salts being used as fungicides. The adjectives derived from copper vary in application: cupreous, containing or resembling copper;

cupric, of or containing copper, or of compounds of bivalent copper; cupriferous, containing copper; cuprous, of or containing copper, being a compound of univalent copper. VALENCY.

coppice a small WOOD-2 in which the trees are coppiced, i.e. cut off close to the ground and so encouraged to send up several shoots, each to serve in due course for fencing posts or poles. Trees used in this way include the oak, hazel, sweet chestnut. G.

coppice-with-standards a type of woodland common in southeastern England where a few trees are left to grow to full size, the rest being coppiced. COPPICE. G.

coprolite fossil dung or excrement, generally composed of calcium phosphate, hence the extended application of the term to any kind of phosphatic nodule. G.

copse a small wood consisting of trees and undergrowth. COPPICE. G.

coral a hard, calcareous, rock-like substance formed either by the continuous skeleton or fused skeletons of members of a group of sedentary marine animals (coelenterate polyps) that live in COLONIES-6 only in clear, warm, shallow seas, or by the skeleton or fused skeletons together with the polyps that secrete it/them. The colour and form of the coral depend on the habits of the species that builds it. NEGRO HEAD, MADREPORE, MILLEPORE. G.

coral mud an accumulation of minute fragments of CORAL occurring in the BATHYAL area of the CONTINENTAL SHELF near CORAL REEFS.

coral reef an extensive REEF formed by CORAL, CORAL SANDS and other coral derivatives, commonly known as a BARRIER REEF if separated from the shore by a LAGOON, or a fringing reef if bordering the land. ATOLL. G.

coral sand sand composed of comminuted fragments of coral.

cordillera a mountain chain, a term applied particularly to the parallel ranges of the Andes in South America and their continuation through central America and Mexico, and to the ranges of the great western mountain system of North America, including the Rockies, Sierra Nevada, Cascades, Coast Ranges etc. G.

core the innermost part of anything: **1.** of the earth, the central part of the earth's interior (radius approximately 3476 km: 2160 mi) consisting of the outer layer of dense material (probably nickel-iron, NIFE, with a density of about 12.0) which behaves as a liquid, and within which the EARTH'S MAGNETIC FIELD is generated by circulation, and containing the solid inner core (radius approximately 1380 to 1450 km: 860 to 900 mi, density about 17.0) at the centre. Between the outer core and the MANTLE that envelops it is the GUTENBERG DISCONTINUITY **2.** of a city or town, the functionally specialized centre, DOWNTOWN (CENTRAL BUSINESS DISTRICT), usually lacking residences. CORE AREA, CORE-FRAME CONCEPT, CORE-PERIPHERY MODEL.

core area of a CULTURE-1, REGION-1, state, or city, a term loosely applied to the central area, the heart, the nucleus or CRADLE district, the place of birth and nurture of a culture, region, nation, or city, from which the culture/region/state/city expanded and spread. CULTURAL HEARTH, DOMAIN-2, MITTELEUROPA, OUTLIER-2, SPHERE-3,5. G.

core-frame concept in the structure of the central area of a city, the notion that within the central area there is an inner CORE-2 (with high land values, tall buildings, a concentrated daytime population, where strong functional links between various offices, and between offices and shops, create distinct clusters of functions which form particular small land use zones) and a 'frame' (the less intensively developed area surrounding the core, with rela-

tively lower land values and widely distributed functions such as wholesaling, off-street parking, light manufacturing, etc. which have only their location in common). PEAK LAND VALUE INTERSECTION. G.

core-periphery model, centre-periphery model a MODEL-4 in which peripheral areas are defined in terms of their dependence on a relevant core region. It is used particularly in an approach to modeling the spatial pattern of economic development proposed by J. Friedmann, 1973, an American planner who saw the world economy in terms of a rapidly growing, dynamic central region with a slower growing or stagnating periphery. He envisaged four regions, from the centre to the periphery. Core regions are the concentrated urban-industrial economies, with a high potential for INNOVATION and growth, recognizable on several levels, i.e. the international, comprising the urban-industrial belt of eastern North America combined with that of western Europe; and the national, the metropolis, the regional core, the subregional centre, the local service centre. The peripheral zones are first the upward-transition regions, growth areas peripheral to the core regions and much affected by the core, but with their own valuable natural resources (development corridors linking two core cities are a special type). Second and peripheral to the upward-transition regions the resource-frontier regions are zones of origin territory which are being settled and made productive. Third are the downward-transition regions, the old peripheral areas, old settlements with a declining or stagnant economy, a failing rural economy with low agricultural output, or areas characterized by the loss of a primary resource base (e.g. by the exhaustion of a mineral resource), or by out-of-date industrial complexes; innovation and productivity are low and there is a marked inability to adapt, to take advantage of new circumstances. Thus, Friedmann asserted, economic and social power, technical progress and high productivity are concentrated at the core, to the disadvantage of the dependent periphery. CENTRIFUGAL AND CENTRIPETAL FORCES, CIRCULAR AND CUMULATIVE GROWTH.

core region   CORE PERIPHERY MODEL.

core sampling   a method of obtaining a specimen of ice, soil, or rock, by the insertion of a hollow tube into the matter, which fills the tube. The tube is then withdrawn complete with the sample. GEOLOGY.

coriander   *Coriandrum sativum*, an annual plant, probably native to the eastern Mediterranean area, but now widespread, even occasionally on waste land in the British Isles. The unpleasant smell of its fruit disappears when it is dried. It is used to give a sweet-sour flavour to various dishes and to various alcoholic beverages. It may be used medicinally to remedy flatulence.

Coriolis force   (Corioli's is incorrect spelling) the effect of the force produced by the earth's rotation on a body moving on its surface. The body is deflected to the right of the path of movement in the northern hemisphere, to the left in the southern (BUYS BALLOT'S LAW, FERREL'S LAW). This phenomenon was discussed by G. G. de Coriolis, French engineer and mathematician, in 1835 (the force being given his name), and the concept was developed by W. Ferrel in 1855. G.

cork   the impermeable layer of dead cells forming the outermost protective layer of stems or roots, applied particularly to the thick bark of the cork oak, *Quercus suber*, a small tree growing in Mediterranean lands. As soon as the tree is large enough the bark (cork) can be cut off the trunk and main branches. It soon begins to form again, so that after ten or twelve years it can again be cut. In nature the purpose of the bark is to prevent loss of moisture in the hot, dry Mediterranean summer. C.

corn   in North America applied to Indian corn or MAIZE (*Zea mais*); but in Britain (and Europe) applied either to the CEREALS wheat, rye, oats, barley etc. collectively, or to the most important cereal of an area, e.g. barley in some districts, wheat in others.

Corn Belt   the region in the USA, south and southwest of the Great Lakes, where corn (maize) is or was the dominant cultivated plant. G.

cornbrash   a bed of impure limestone in the Middle and Upper JURASSIC of the English lowlands, which affords a stony soil (brash soil) rich in lime and yielding excellent crops of corn. BRASH. G.

cornelian   CARNELIAN.

cornice   (French) **1.** in architecture, a projecting horizontal strip of stone, wood, plaster that crowns a building, especially part of the entablature (architrave, frieze and cornice) above the frieze **2.** an overhanging accumulation of ice and wind-blown snow on the edge of a ridge or cliff face, usually occurring on the lee side of a steep mountain.

corniche, corniche road   from the architectural term, CORNICE-1, the road from Nice to Genova, high up on steep hills overlooking the Mediterranean, a term now applied to other scenic coast roads.

corona   **1.** the upper part of the solar atmosphere, extent unknown, consisting of highly ionized gases, temperature probably exceeding 1 mn°C, visible as a ring round the sun at total eclipse **2.** a series of concentric rings of light seen from the earth to be encircling the sun, moon or a star, caused by the diffraction of light by water-drops, ranging from blue (inner circle), through green and yellow to red (outermost). HALO.

corozo   *Phytelephas macrocarpa*, a South American tree of which the seeds (corozo nuts), when hardened, are used as an ivory substitue, for buttons, carved ornaments, etc.

corporate state a state in which the representation of the individual citizen or the participation of the individual in government is not based on the territorial location of the individual's home, but on the functional group of which by job or profession the individual is a member.

corporatism a form of government in which functional groups in society (e.g. people working in specific industries, commercial institutions, occupational groups, professions) have a legitimate right of access to, and may even participate in, government, being regularly consulted by government officials when the particular interests of the group are involved in decision-making. Corporatism can be especially useful to the government in the economic sphere. The government can use the shared understanding and working relationships established in day-to-day contacts in controlling any activities it so wishes, while leaving these activities nominally in private hands, thereby eliminating the need for taking key sectors of the economy into public ownership. STATE CAPITALISM.

corral (Spanish) in southwestern USA, an enclosure or pen for cattle, horses, etc. G.

corrasion the process of mechanical EROSION of a rock surface by the friction of rock material with the surface, the rock material being moved under gravity, or carried by running water (streams, rivers and waves), by ice (glaciers), or by wind (wind-blown sand). The result of corrasion (the process) is ABRASION. TRANSPORTATION-2. G.

correlation 1. in general, the condition of two or more things mutually or reciprocally related, or the act of bringing two or more things into mutual or reciprocal relationship 2. in biology, mutual relationship 3. in statistics, the degree of relationship, the extent to which two measurable VARIABLES vary together, linear correlation being commonly measured by a correlation COEFFICIENT-1, of which there are several types, most varying between $+1.0$ and -1.0. The term positive correlation indicates that the variables have a tendency to increase or decrease together (i.e. high scores on one tend to be associated with high scores on the other, similarly with low scores). Negative or inverse correlation indicates that there is a tendency for one variable to increase as the other decreases, and vice versa (i.e. high scores on one variable tend to be associated with low scores on the other, and vice versa). If changes in one variable are negatively or positively proportional to changes in the other, the correlation is said to be linear, and non-linear if the changes are not proportional. A correlation of $+1.0$ is termed perfect positive correlation, one exceeding about $+0.7$ indicating a high positive correlation. Zero correlation indicates an absence of relationship; -1.0 indicates perfect negative correlation, a value lying between about -0.4 and zero indicating a slight tendency for high scores on one variable to be associated with low scores on the other, and vice versa. In the condition of perfect positive correlation ($+1.0$) or perfect negative correlation (-1.0) the value of one variable can be predicted with certainty and precision from the given value of the other; but between the values $+1.0$ and -1.0 the certainty and the precision diminish with approaching proximity to zero. Two commonly used measurements of correlation which test the degree to which any two sets of data are correlated are the PEARSON PRODUCT MOMENT COEFFICIENT and the SPEARMAN'S RANK CORRELATION COEFFICIENT. COVARIATION, POINT BISERIAL

CORRELATION COEFFICIENT, SPURIOUS CORRELATION.

corridor a strip of territory belonging to one country but running through the territory of another country in order to give the country owning the strip access to an international waterway, the sea, etc. The term has been extended to air corridor, an AIRWAY-1 allowing a country or countries right of access by air over the territory of a foreign country. G.

corrie (Scottish) CIRQUE. G.

corrosion 1. the wearing away of rock or soil by chemical and solvent action, i.e. by CARBONATION, HYDRATION, HYDROLYSIS, OXIDATION, SOLUTION **2.** the changes in the crystals of a rock produced by the solvent action of residual magma. G.

cors (Welsh gors) a BOG, sometimes specifically SPHAGNUM bog. G.

corsair historical, a pirate, a privateer, especially one of the Barbary coast in the period 1550 to 1850. Barbary, renowned for its pirates, extended in north Africa from Egypt to the Atlantic and from the northern Sahara to the Mediterranean.

corundum a hard (second only to diamond) crystallized mineral, aluminium oxide, Al_2O_3, occurring in some PEGMATITES, in some metamorphosed rocks, or in PLACER deposits. The forms include the GEMSTONES ruby and sapphire; an impure form, emery, is commonly used as an abrasive in polishing powder. Fused BAUXITE is an artificial replacement for corundum.

co-seismal line, co-seismic line homoseismal line, a line connecting points on the earth's surface simultaneously experiencing the arrival of an earthquake wave, not to be confused with an isoseismal line (ISO-) which is related to the amount of earthquake intensity.

cosmic *adj.* of, or pertaining to, the COSMOS-1 (the universe seen as an orderly whole), applied to any phenom-

enon of space outside the atmosphere of the earth.

cosmic dust very small particles of solid matter present everywhere in the universe.

cosmic rays a continuous shower of very high-frequency subatomic particles that, originating in outer space (at that stage termed primary cosmic rays), collide with atoms and molecules of the earth's upper atmosphere and, as secondary cosmic rays, reach the earth's surface, sometimes penetrating it. The subatomic particles of the primary cosmic rays consist mainly of the nuclei of atoms, certainly those of hydrogen (PROTONS), and possibly of silicon and iron.

cosmography (adapted from Greek, writing about the kosmos) **1.** a description or representation of the general features of the earth and/or the universe (COSMOS) **2.** the science concerned with the structure of the earth and/or the universe, including astronomy, geography and geology **3.** historically, a synonym for physical geography. A term with a long, complex history, it is now rarely used. G.

cosmology 1. the science or theory concerned with the physical universe viewed as an ordered entity and with the general laws (PHYSICS) that govern it **2.** in philosophy, a branch of METAPHYSICS concerned with the idea of the universe as a systematic order, a totality of all phenomena in space and through time. ONTOLOGY.

cosmonaut an ASTRONAUT, especially one from the USSR.

cosmopolitan one who is COSMOPOLITAN (*adj.*).

cosmopolitan *adj.* belonging to all parts of the world and free from national limitations. G.

cosmopolite 1. a person of international outlook, without national attachment or prejudices **2.** a plant or animal at home in almost any part of the world. G.

cosmos (Greek kosmos, order, world or universe) **1.** the world or the universe as an orderly whole, a harmonious system **2.** an ordered system of ideas, complete and harmonious.

costa (Spanish) coast, applied particularly to the the Spanish coasts bordering the Mediterranean which have been developed with holiday resorts, e.g. Costa Brava, Costa del Sol etc. G.

cost benefit analysis CBA, the evaluation in monetary terms of the costs and benefits accruing from a scheme or from alternative schemes, taking into account all the factors involved, i.e. not only the commercial but also the social and environmental factors for which value judgements must be made (e.g. social benefits, loss of AMENITY, traffic hazards, mental or physical strain, etc.). COMPARATIVE COST ANALYSIS, NON-MARKET COST, VARIABLE COST ANALYSIS, VARIABLE REVENUE ANALYSIS.

cost curve a line on a two-dimensional graph which shows the relationship between cost of production and volume of output. SPACE COST CURVE.

cost space, cost distance, time space, time distance in the explanation of human spatial structure, relative space as opposed to absolute space (e.g. people travelling or moving goods between places are more concerned with the cost and the time involved than they are with the actual distance to be covered). Some authors prefer the term cost (or time) distance to cost (or time) space as being more precise, maintaining that the term 'space' encompasses too many dimensions. ACCESSIBILITY, EFFECTIVE ACCESSIBILITY. G.

cost structure the total cost of production separated into its component parts, i.e. the cost of individual inputs.

cost surface a three dimensional representation of an area, resembling a CONTOUR map, but constructed to show spatial variations in production costs at varying distances from various locations, the contours joining, for example, points with equal total costs, or the cost of a single item (e.g. land, labour, transport, power, materials, or a particular individual material involved in an operation). A cost surface with its cost contours therefore shows not only the total costs an enterprise at a given location may incur in obtaining land, labour, power, materials etc. and in distributing its products to customers at any and all points throughout an area: it also identifies the LEAST-COST LOCATION. SPACE COST CURVE.

côte (French, coast) **1.** applied particularly to stretches of coast developed with holiday resorts, e.g. the Côte d'Azur along the Mediterranean **2.** an ESCARPMENT in France, e.g. Côte d'Or in Burgundy. G.

coteau (North America: French) a term applied to a prominent ESCARPMENT or sharp-edged ridge of hills by French explorers in North America.

co-tidal line a line drawn on a tidal chart linking points with a common time of high water. Co-tidal lines radiate from the AMPHIDROMIC POINT (OSCILLATORY WAVE THEORY OF TIDES) where the water remains relatively level, the height of the tidal rise growing as the co-tidal lines spread to their extremities. STANDING WAVE.

cottage industry manufacturing wholly or partly carried on in the home of the worker.

cottar, cotter (Scottish; Irish cottier) a rural dweller who occupies a rented cottage usually with a small plot of land. COTTIER TENURE. G.

cottier tenure a system of LAND TENURE (now obsolete) in Ireland in which land was let out annually in small parcels directly to labourers, the rent being fixed by public competition. G.

cotton a small, shrubby plant of the genus *Gossypium*, family Malvaceae, native to tropical and subtropical regions, cultivated for the sake of its seed hairs, i.e. the fibres (used for textiles) attached to its seeds, the seeds and white fibres being contained by the 'boll' or seed case. According to the species and variety cultivated, the fibres vary in length (termed staple) from about 22 to 63 mm (0.9 to 2.5 in), and in texture from harsh and coarse to fine and silky. The many varieties cultivated seem to stem from three original species: *Gossypium herbaceum* and the tree cotton *G. arboreum* (natives of the Old World) and *G. barbadense* the Sea Island cotton of the New World, with very long silky hairs. The ordinary American cotton, *G. hirsutum* is probably only a variety of *G. herbaceum*. The tree cotton (*caravonica*) grown in Queensland, Australia and elsewhere, is said to be a hybrid between Sea Island and rough Peruvian cotton. Being a tropical or subtropical crop, cotton needs annually 200 frost-free days and a low rainfall, especially at picking time when rain would damage the bolls, which are open when ripe. The crop can be gathered by machine, as in the USA, but picking is still mainly by hand. The hairs are separated from the seeds by mechanical 'ginning' (once done by hand). The cotton seed is crushed and the edible oil extracted for use in cooking and food preparations, such as margarine; the residue provides meal or oilcake, used as cattle feed or fertilizer. The cotton fibres (lint) are combed out into slivers, twisted to form yarn (spinning), then woven into fabric (weaving). In many manufacturing areas in Britain spinning and weaving were carried out in separate MILLS-2, even in separate towns. In clothing manufacture cotton gradually replaced wool in the eighteenth century, expanded greatly in the nineteenth century, declined with the advent of synthetic fibres; but cotton fibres, valued for their absorbent qualities, can satisfactorily be woven with synthetic fibres to make fabrics comfortable to wear, so cotton is still in demand. C.

Cotton Belt that part of the south-eastern USA where cotton was the dominant crop. DEEP SOUTH. G.

coulée, coulee (French, flow) **1.** a solidified lava flow **2.** a steep-sided, trench-like valley in the western USA, e.g. Grand Coulee **3.** an abandoned glacial melt-water channel in the USA **4.** a tongue of debris in a periglacial area. G.

couloir (French) a steep, narrow gulley on a mountainside, particularly in the French Alps. G.

Council for Mutual Economic Assistance CMEA, popularly abbreviated to COMECON.

counterdrift in physical planning, a process in which people living on the fringe of a large GROWTH AREA participate in the benefits created by that growth. G.

counterfactual a construction of a hypothetical event, process, or state of affairs, which does not accord with the facts known about an actual event, process, or state of affairs. A counterfactual statement is a statement concerned with a hypothetical event, process, or state of affairs that does not accord with known facts, i.e. which has not in fact taken place or existed, and is not doing so now. A counterfactual conditional statement states what would have happened if something had been the case or occurred; and a counterfactual question queries what might have happened in a non-existent state of affairs. Counterfactual explanation, as an extension of the COMPARATIVE METHOD, has been used in HISTORICAL GEOGRAPHY. ECONOMETRIC HISTORY.

counterfactual *adj.* pertaining to, expressing, that which has not in fact happened or existed but might have happened or existed in different circumstances or conditions. COUNTERFACTUAL.

counter-trade wind an obsolete term. ANTI-TRADE WIND, TRADE WIND.

counter-urbanization the movement of people and industry away from major towns and cities. DE-INDUSTRIAL-IZATION, URBANIZATION.

country 1. a tract of land, a region, lacking precisely defined boundaries, viewed geographically or aesthetically 2. the territory of a nation 3. the native land of an individual person 4. a rural area as opposed to an urban area. G.

country house 1. a large house or MANSION in the country, in many cases standing in its own PARK-2, especially the residence of one who may also own a TOWN HOUSE. COUNTRY-SEAT 2. any fairly large house in the COUN-TRY-4.

country park in Britain, a tract of land, in some cases land and water, commonly of about 10 ha (25 acres), i.e. much smaller than a NATIONAL PARK, designated for the recreational use (without payment) by the public. Signposted nature trails are usually a feature, and in most country parks information about the plant and animal life etc. is supplied.

country rock, country-rock 1. a mass of rock traversed by later INTRUSIONS of IGNEOUS ROCK or penetrated by a mineral VEIN 2. (American) a synonym for bed rock. G.

country-seat the country MANSION and DEMESNE of a wealthy landowner.

county (derived from Latin comitatus, the term came by tortuous paths eventually to denote the domain or territory of a count. Many English counties were of this nature, and although the title Count has been dropped in favour of Earl, an earl's wife is still called a Countess) 1. an administrative or historical unit representing a major territorial division of Britain and of some other countries, e.g. the USA 2. in the USA each STATE-5 is divided into counties, the administrative authorities of which are usually responsible for roads, public health and welfare, law and order. LOCAL GOVERNMENT IN BRITAIN, LOCAL GOVERNMENT IN USA. G.

county borough formerly a large town in England, Wales and Northern Ireland, with administrative status similar to that of a COUNTY-1. BOROUGH, LOCAL GOVERNMENT IN BRITAIN.

coupole (French) a hill with a rounded, almost hemispherical, top, a characteristic feature of KARST. KEGELKARST.

courgette (French; Italian zucchini) a variety of MARROW, bred to be cropped when small.

course 1. the movement from one point to another 2. the direction of movement and the path taken 3. a channel in which water flows 4. a continuous layer of rock, stone, bricks, cement etc (as in a wall) 5. golf course, GOLF LINKS, the ground on which the game of golf is played 6. racecourse, a ground with a track where horses run in competition.

court leet MANORIAL COURTS.

covalency VALENCY.

covariance in statistics, a measure of the extent to which values on two variables vary together, i.e. high values on one being associated with high values on the other, or low with low. It is not, as is a CORRELATION-3 coefficient, restricted to values between $+1.0$ and -1.0. The size of the covariance increases or decreases with the increasing or decreasing strength of the relationship. A zero value indicates no LINEAR-5 relationship; and positive or negative values indicate data showing evidence of corresponding to a positive or negative linear trend. VARIANCE.

covariation the extent to which the SCORES on two VARIABLES vary together, indicated by COVARIANCE and various CORRELATION-3 coefficients.

cove 1. a steep-sided, rounded hollow or recess in a rock 2. a small inlet in a rocky sea coast with a narrow opening and a small curved bay 3. a recess with precipitous sides on a steep mountainside 4. a small rounded hollow at the head of a valley 5. (American) a basin or hollow on the land surface resulting from caving-in, e.g. by CARBONA-TION-SOLUTION. G.

cover crop a fast-growing crop, planted on cleared land between main crops, to form a blanket of vegetation and protect the soil from erosion. G.

covered shield an area of a SHIELD overlain by a thin layer of SEDIMEN-TARY ROCK. EXPOSED SHIELD.

covert something which covers or forms a refuge for animals and birds in hunting, applied especially to a small WOOD-2 with undergrowth, carefully managed, which provides a retreat for foxes in fox-hunting country, or for game birds. G.

cow 1. the fully mature female of the ox family, usually kept by farmers for milk. DAIRY FARMING, HEIFER **2.** the female of any animal of which the male is known as a BULL, e.g. the elephant, the whale.

cowrie, cowry *Cypraea moneta*, a small gastropod, native to the Indian Ocean, the shell of which has been used for currency in Africa and Asia. Similar gastropods of the same genus occur on British coasts.

coypu *Myocastor coypus*, a large aquatic rodent, native to South America, the pelt of which provides the fur known as nutria. Introduced into England, many escaped from fur farms in the 1930s, and their descendants, firmly established in eastern England (East Anglia), now damage crops and river banks in the area.

crab hole (Australia) a hollow resembling a shallow grave, occurring in a clay plain.

cracking HYDROCARBON CRACKING.

cradle the place where something begins or is nurtured. CULTURAL HEARTH. G.

crag 1. steep, rugged rock jutting out from a mountainside **2.** in eastern England, deposits of compacted shelly sand, e.g. Coralline Crag of PLIOCENE age. GEOLOGICAL TIMESCALE. G.

crag and tail a hill or CRAG-1 showing an abrupt and often precipitous face on one side (the stoss side or STOSSEND) which faced a pre-existing glacier and a long gentle slope or tail on the other where the hard rock of the 'crag' protected loose material so that it was not swept away by the ice. A good example is the rock on which Edinburgh Castle stands, with a long gentle slope to the east. G.

cranberry *Vaccinium oxycoceus*, an evergreen, prostrate shrub, native to Europe, northern Asia and North America, with red, acid fruits which are eaten cooked.

crater 1. a bowl-shaped depression or cavity in the earth's surface, especially that around the orifice of a VOLCANO **2.** a depression made by the impact of a meteorite or artificial explosive, e.g. a bomb **3.** a flaring or bowl-shaped opening of a GEYSER. ADVENTIVE CRATER. G.

crater lake a lake occupying a crater, usually the crater of an extinct VOLCANO. G.

craton, kratogen a resistant, very large, stable block of the earth's crust which has remained relatively unaffected by orogenic activity (OROGENESIS) for a very long period of time. G.

cream that part of milk which, richest in fats, rises to the surface of undisturbed milk. The term is also applied to substances resembling cream in appearance or consistence.

creamery a building where BUTTER and CHEESE are made.

cream of tartar TARTAR.

creek diverse uses, but most commonly applied to **1.** a comparatively narrow inlet of fresh or salt water which is tidal **2.** in USA and elsewhere, a branch of a main river bigger than a brook but smaller than a river **3.** in Australia, an intermittent stream. G.

creep the gradual movement downhill of rock debris and soil, primarily due to gravity helped by the presence of water, alternate freezing and thawing or wetting and drying, the growth of plant roots, the work of burrowing animals. G.

creole (French; Spanish criolle) **1.** originally a person of European (usually French or Spanish) descent born in the tropics in the West Indies, Spanish America, Mauritius etc.. Later applied to an African, or to a person of mixed European and African blood, born in those areas. The tendency now is for the term to be used (if at all) in relation to persons of unspecified mixed descent born in those areas **2.** any of the indigenous languages which has developed from the languages of the dominant settlers in the West Indies. G.

creosote bush *Larrea tridentata*, a low XEROPHYTIC shrub characteristic of the hot desert areas of western USA.

crepuscular *adj.* of or pertaining to TWILIGHT; in zoology applied to animals that are active before sunrise or at dusk.

crescentic dune BARCHAN.

crescent moon the MOON in its first or last quarter.

crest of a wave, the position in a WAVE where the displacement or disturbance of particles is at its maximum. TROUGH.

Cretaceous (Latin creta, chalk) *adj.* applied to the geological period (of time) or system (of rocks) which ended the Mesozoic era. During the Cretaceous period the beds of CHALK of southern England were deposited, reptiles were dominant on land and in the sea, and ANGIOSPERMS became the dominant plants. GEOLOGICAL TIMESCALE. G.

crevasse 1. a fissure or chasm in the ice of a GLACIER. LONGITUDINAL CREVASSE, TRANSVERSE CREVASSE **2.** in USA, a break in the natural LEVEE or bank of rivers such as the Mississippi. G.

crevasse filling a straight ridge consisting of layers of sand and gravel formed in a CREVASSE-1 by the melting of a stagnant ice sheet, resembling (but usually smaller than) an ESKER. G.

crib 1. in New Zealand, BACH **2.** (Welsh) a crest or summit in mountains.

criterion pl. criteria, an established rule, a standard or principle by which something is judged.

critical *adj.* applied in physics or in mathematics to the point or state at which an important or fundamental change in properties or conditions takes place. CRITICAL CASE ANALYSIS, CRITICAL PATH ANALYSIS.

critical case analysis the examination of something which is extreme, deviant, or atypical, in order to determine the extent to which an HYPOTHESIS is apt.

critical isodapane the isodapane (ISO-) which indicates the point where additional transport costs balance savings in labour costs.

critical path analysis a technique used to discover the most efficient sequence of phases necessary to carry out and complete a complex task speedily. A schedule of phases is drawn up, and this reveals those places where a slight delay will have a knock-on effect, upsetting the whole schedule. These are the phases which mark the critical path, the identification of which is CRUCIAL to the smooth, speedy running of the operation.

critical temperature 1. the level of temperature above which a gas cannot be liquefied by pressure alone, the highest temperature at which a liquid and its vapour can co-exist **2.** a temperature of vital importance to a plant, e.g. the temperature below which growth cannot take place, in MIDLATITUDES 6°C (43°F) for most food crops. ACCUMULATED TEMPERATURE.

crocus SAFFRON.

croft a small agricultural holding worked for subsistence by an hereditary tenant, especially in the Highlands of Scotland. TOWNSHIP-2. G.

crofter one who works a croft. G.

crofting a system of small hereditary tenant holdings, in which the CROFTER holds the arable land with a right of grazing, with others, on unenclosed hillsides. TOWNSHIP-2. G.

crofting counties officially, seven counties in Scotland: Shetland, Orkney, Caithness, Sutherland, Ross and Cromarty, Inverness and Argyll. G.

Cro-Magnon people AURIGNACIAN EPOCH.

cromlech a megalithic burial chamber, consisting of a large flat horizontal stone supported by vertical stones (synonymous with DOLMEN).

crop 1. the annually or seasonally harvested produce resulting from the cultivation of grain, grass, fruit etc. 2. cultivated produce while growing. CROP FARMING, ROTATION OF CROPS.

crop *verb* 1. to cultivate, sow, plant, reap or bear a crop 2. by animals, to eat the tops of plants, grasses, to graze 3. in geology, to crop out, to come to the surface.

crop farming arable farming, the growing of INDUSTRIAL CROPS, or of food crops such as cereals, vegetables etc. on ARABLE LAND (i.e. land which has been tilled), as distinct from LIVESTOCK FARMING or MIXED FARMING.

cross bedding CURRENT BEDDING, FALSE BEDDING.

crossbreed an animal or plant produced by breeding from parents of different breeds, or of two varieties of the same species.

crosscut in mining, a horizontal passage driven at right angles to the general trend of a coal seam or a mineral vein.

cross-cutting relationships, law of a law which states that an IGNEOUS ROCK is younger than any rock across which it cuts.

cross-faulting two intersecting series of FAULTS.

crosshatching intersecting sets of parallel lines used as a conventional symbol to indicate particular areas on diagrams, maps etc.

cross profile of a valley a transverse profile of a valley drawn approximately at right angles to the stream flowing through it. RIVER PROFILE.

cross section 1. a transverse SECTION-1, the surface resulting from the cutting of a solid at right angles to its length or to the axis of a cylinder. PROFILE-3 2. a sample that is typical or representative of the whole.

cross sectional study LONGITUDINAL STUDY.

cross section approach an approach in HISTORICAL GEOGRAPHY which is concerned with the reconstruction of the geography (the study of a society and its landscape) at a particular point or period in time, i.e. a 'horizontal' approach typical of SYNCHRONIC ANALYSIS as distinct from a VERTICAL THEME of DIACHRONIC ANALYSIS.

cross-wind a wind blowing nearly or exactly at right angles to the course of a moving object, i.e. across the flight path of an aircraft.

crotovine KROTOVINA.

crucial *adj.* critical, decisive, fundamental.

crude oil PETROLEUM.

crumb in soil science, a PED, a rounded, porous AGGREGATE-3 of soil particles, up to 10 mm in maximum dimension. CRUMB STRUCTURE, SOIL STRUCTURE.

crumb structure a soil in which the constituent particles are aggregated into CRUMBS which permit the percolation of air and water between them. SOIL STRUCTURE. G.

crush breccia FAULT BRECCIA, a rock consisting of angular fragments, usually cemented together, produced by crushing and shearing stresses during faulting or folding. FAULT, FOLD-2. AUTOCLASTIC BRECCIA. G.

crush conglomerate a rock consisting of rounded fragments produced during faulting or folding in much the same way that CRUSH BRECCIA is produced. G.

crustacean an INVERTEBRATE of Crustacea, a class of mainly aquatic Anthropoda, the skeleton of which (exoskeleton) covers the outside of the body, or lies within the skin.

crust of the earth the LITHOSPHERE, the outermost shell of the earth, consisting of the surface granitic SIAL and the intermediate basic SIMA layers, separated from the underlying MANTLE by the MOHOROVICIC DISCONTINUITY. There are two kinds of crust: continental, which has an average density 2.7, average thickness 35 to 40 km (22 to 25 mi) but under high mountain chains ranging between 60 and 70 km (37 and 44 mi), with large areas older than 1500 mn years (some exceeding 3500 mn years), a complicated structure, and variable composition; and oceanic, which is heavier than continental, average density 3.0, average thickness only 6 km (3.7 mi), nowhere older than 200 mn years, with a simple layered structure of uniform composition. The junction of the two types of crust is usually obscured by recent SEDIMENTATION. PLATE TECTONICS.

cryergic physical phenomena arising from very cold conditions. G.

cryoconite (Greek kruos, ice; konis, dust) powdered dust-like ice found in cylindrical holes (cryoconite holes) in some ice caps, e.g. in parts of Greenland. G.

cryogenics the science of extreme cold, the branch of chemistry concerned with the effects and products of extremely low temperatures. G.

cryogenic system a system with temperature lower than that of its surroundings, e.g. a refrigerated container.

cryolaccolith PINGO.

cryolite a mineral, Na_3AlF_6, a natural fluoride of aluminium and sodium, occurring in PEGMATITE veins in Greenland, used in the preparation of soda and pure aluminium.

cryology 1. the scientific study of ice and snow 2. (American) the study of refrigeration. GEOCRYOLOGY. G.

cryopedology the scientific study of frost action and of permanently frozen ground, including studies of the processes and their occurrence and also of the engineering devices and techniques which may be employed to overcome the physical problems in such conditions. G.

cryophyte a plant growing on snow or ice, usually a micro-plant such as ALGAE, but it may also be a BACTERIUM, FUNGUS or MOSS. The algae produce RED SNOW.

cryoplanation land reduction by the processes of intensive frost action, i.e. CONGELITURBATION including SOLIFLUCTION and the work of rivers in transporting the material produced by such processes. EQUIPLANATION. G.

cryoplankton microscopic organisms living on the surface layers of snow and ice.

cryosphere the permanently frozen region of the earth's surface.

cryostatic hypothesis the HYPOTHESIS-1 that debris, in freezing, is squeezed between downward freezing ground and the PERMAFROST table, upward injection taking place in areas where relief is easiest. MUD CIRCLE. G.

cryoturbation all phases of weathering under very cold conditions, a term usually now replaced by the more precise, scientific term, CONGELITURBATION. G.

cryovegetation a general term applied to plant communities growing on or in the surface layers of permanent snow or ice. CRYOPHYTE.

cryptocrystalline *adj.* applied to a mineral in which the crystalline structure is concealed, the crystals being imperfectly formed or too small to be resolved by an optical microsope.

cryptophyte a plant with PERENNATING parts covered by soil or water, i.e. a GEOPHYTE, HELOPHYTE or HYDROPHYTE of RAUNKIAER'S LIFE FORMS.

cryptovolcano a fairly rare example of volcanic activity, an area where the sudden release of gases at great depth has shattered BASEMENT rocks, led to the formation of PIPES through FLUID-IZATION, and caused much pyroclastic material (PYROCLAST) to be ejected.

cryptozoic (Greek, concealed life) *adj.* applied to **1.** in geology, the AEON before the CAMBRIAN revealing few visible traces of the remains of primitive life forms. PHANEROZOIC **2.** in zoology, an animal living in a concealed place, e.g. in a crevice, under a stone or a log of wood.

crystal the solid state of many simple substances and their natural compounds, created naturally in the aggregation of their constituent atoms in an ordered, regular pattern. Usually the external, symmetrical plane faces of the solid meet at angles peculiar to the substance. CRYSTALLOGRAPHY.

crystalliferous *adj.* producing or bearing CRYSTALS.

crystalline *adj.* consisting of, or resembling, a CRYSTAL.

crystalline rocks rocks composed partly or wholly of CRYSTALS. The term applies both to IGNEOUS ROCKS (crystallized as a result of cooling of molten rock) and to METAMORPHIC ROCKS (in which the crystals have developed as a result of METAMORPHISM). G.

crystallization the solidification of MINERALS as CRYSTALS either through the cooling of molten MAGMA or as a result of METAMORPHISM. CRYSTALLINE ROCK.

crystallography the scientific study of the external form, the symmetry and structure, and the classification of CRYSTALLINE substances.

crystocrene surface masses of ice formed each winter by the overflow of SPRINGS-2.

crystophene a mass or sheet of ice developed between beds of other material, e.g. rock or alluvial deposits, and wedging them apart. G.

CSCS Comprehensive Soil Classification System, USA. SOIL CLASSIFICATION.

cucumber *Cucumis sativus*, a trailing or climbing plant, probably originating in southern Asia, cultivated for the sake of its long, green, fleshy fruit, eaten raw, pickled or cooked.

cuesta (Spanish) a landform consising of an asymmetrical ridge with an abrupt cliff or steep slope, the ESCARP-MENT (the inface) and a gentle BACK SLOPE (dip slope) which lies almost parallel to the dip of the strata, the result of differential denudation of gently inclined strata where resistant beds overlie weaker layers. The term covers the whole landform and is preferable to the ambiguous terms formerly used, i.e. ESCARPMENT and SCARP, which are better applied not to the whole of the landform but only to the steep slope. G.

culm coal dust or small coal. CULM MEASURES.

Culm Measures a FORMATION-2, approximately of the age of the COAL MEASURES, but consisting mainly of shales and sandstones with only impure anthracite, occurring mainly in south-western England. G.

cultigen a cultivar, abbreviation cv, a cultivated plant produced by human action. G.

cultural determinism the theory that the interaction of people with their environment is largely conditioned by their cultural MILEU.

cultural geography a branch of geography that concentrates on the study of the human activities which have been determined culturally (CULTURE-1), of CULTURAL AREAS, and of the impact of different culture groups and the succession of cultural groups on the landscape (CULTURAL LANDSCAPE) and on other NATURAL RESOURCES. Some authors have used the term broadly, to cover all geographical studies other than the physical. CULTURAL HEARTH, DO-MAIN-2, OUTLIER-2, SPHERE-5. G.

cultural hearth the centre, the CRADLE of a culture or of a cultural group from which, if conditions are favourable, the successful culture or cultural group will spread, effecting the changes in the NATURAL LANDSCAPE which bring into being the particular CULTURAL LANDSCAPE of the culture or group. DOMAIN-2, HEARTH, OUTLIER-2, SPHERE-5.

cultural landscape the NATURAL LANDSCAPE as modified by human activities, i.e. most of the present landscape, there being very few parts of the world now unaffected by such activity. CULTURAL GEOGRAPHY, CULTURAL HEARTH, CULTURE, CULTURE AREA, LANDSCAPE. G.

cultural region CULTURE AREA.

culture 1. the collective mental and spiritual manifestations (aesthetic perception, beliefs, ideas, symbols, values etc.), the forms of behaviour and social structures (modes of organization, rituals, groupings, institutions etc.), together with material and artistic manifestations (tools, buildings, works of art etc.), formulated and created by people according to the conditions of their lives, characterizing a SOCIETY-2,3 and transmitted as a social heritage from one generation to the next, undergoing modification and change in the process (CULTURE AREA, CULTURE CONTACT, SOCIAL STRUCTURE). Culture exerts a strong influence on the way in which the ENVIRONMENT is perceived. COGNITIVE BEHAVIOURALISM, HUXLEY'S MODEL, PERCEPTION **2.** the rearing of fish, silkworms etc. **3.** the growing of microorganisms, tissues etc. in a prepared media, or the product of such cultivation.

culture area, cultural region a REGION-1 identified by the existence within it of a single, distinctive CULTURE-1 or of cultures similar to one another. CULTURAL HEARTH, CULTURAL LANDSCAPE, DOMAIN-2, SPHERE-5.

culture contact the meeting or mingling of groups with differing cultural traditions, brought about by migration, displacement, or enlargement of territory etc. Results vary: a dominant, technologically superior, stronger, more successful culture may swamp the weaker; the weaker may form itself into a self-sufficient, tightly-knit ETHNIC MINORITY within the larger group; cultural groups of equal strength may merge; or, as a result of the interaction of diverse cultural groups and traditions, a new society with its own particular cultural traditions may emerge. ACCULTURATION, ENCULTURATION, PLURALISTIC INTEGRATION, PLURALISM, PLURAL SOCIETY.

cumec CUSEC.

cumin, cummin *Cuminum cyminum*, a small ANNUAL herb, native to the Mediterranean region, long cultivated in Europe, the Indian subcontinent and China for its aromatic seeds, used as a spice in cooking, and reputed to have medicinal properties as a stimulant and sedative.

cumquat KUMQUAT.

cumulative frequency distribution in statistics, a distribution produced by starting at one end of a range of scores and for each successive CLASS INTERVAL adding the frequency in that class interval to all the preceding class intervals. If percentages of the total cases falling in each class interval are used, it is termed a cumulative relative frequency distribution. FREQUENCY DISTRIBUTION, OGIVE-2.

cumulative upward causation the process by which economic activity leading to prosperity and increasing economic development tends to concentrate in an area with an initial advantage, draining investment and skilled labour from the peripheral area (part of the backwash effect). CENTRIPETAL AND CENTRIFUGAL FORCES, CIRCULAR AND CUMULATIVE GROWTH.

cumulo-dome (French mamélon) a

round-topped, apparently craterless volcano formed by a succession of viscous acid LAVA flows.

cumulonimbus a low-based mass of CUMULUS cloud, dark grey when viewed from below, shining white from the side, developed to a great vertical height, the upper part spreading into the shape of an anvil (ANVIL CLOUD, ICE ANVIL, INCUS), usually associated with THUNDERSTORMS, heavy rain or HAIL. CLOUD.

cumulose deposit a superficial layer of decaying organic material such as PEAT.

cumulostratus STRATOCUMULUS.

cumulus a heaped mass of low-lying, rounded convection CLOUD with a large, white, domed crown, which develops vertically from a flat base, in some cases to a considerable height. It may disperse as CONVECTION currents die away, or develop into CUMULONIMBUS if convection currents grow in power. CYCLOGENESIS.

cupola 1. a small, rounded mass of rock projecting upwards from the roof of a BATHOLITH 2. a small isolated mass of INTRUSIVE ROCK occurring near a batholith. G.

cupreous, cupric, cupriferous, cuprous *adj.* COPPER.

cuprite cuprous oxide, Cu_2O, red COPPER ore.

cupronickel an ALLOY of COPPER (usually 70 per cent) and NICKEL.

curd the smooth solid substance formed from MILK as a result of the action of acid, naturally as the milk becomes sour, or artificially when rennet is added. Curd is used in making CHEESE. WHEY.

curie point the temperature, specific to each substance, above which a ferromagnetic material loses its FERROMAGNETISM.

curragh (Irish; Isle of Man) wet land, ill-drained land. G.

currants BLACK, RED, WHITE CURRANTS. GRAPE.

current a body of water, air or other fluid moving vertically or horizontally in a definite direction, e.g. **1.** the vertical movement of fluid material within the earth, of water in the ocean, of air in an air mass. CONVECTION CURRENT **2.** the horizontal movement of water in certain channels of a river **3.** the permanent or semi-permanent, horizontal movement of the surface water of the OCEAN-1 (DRIFT) caused mainly by the dragging action of the PLANETARY WINDS **4.** the horizontal movement of water through a STRAIT due to differences in temperature and salinity at each end. DENSITY CURRENT **5.** the horizontal movement of water through a restricted channel due to differing tidal regimes at each end, i.e. a TIDAL CURRENT (not to be confused with a TIDAL STREAM). LITTORAL CURRENT, LONGSHORE CURRENT. G.

current bedding, current-bedding, cross bedding, cross-bedding LAMINAE that lie transversley or at varying angles oblique to the general stratification of the BEDDING PLANES. This occurs only in granular sediments (particularly in SANDSTONE) and results from a change in the direction of the currents in the water and wind depositing the sand grains. If such a change was very marked the existing beds may be truncated; and laminae deposited thereafter may lie at a different angle. FALSE BEDDING.

current fertility FERTILITY-3, REPRODUCTION RATE.

curry a dish flavoured by a blend of many spices (e.g. CHILLI, CORIANDER, CUMIN, FENUGREEK, GINGER, TURMERIC) and herbs. Curried meat, fish, eggs, vegetables are part of the basic diet in most of MONSOON Asia.

curvilinear *adj.* having the form of a curved line, consisting of, bounded by, a curved line or lines. LINEAR-4.

curvilinear relationship in statistics, the relationship between two VARIABLES which is shown by some kind of curved line rather than by a straight line when one is plotted against the other. LINEAR-4, SCATTERGRAM.

cusec the unit of measurement of flow of a fluid, an abbreviation of cubic feet per second (1 cusec = 102 cu m per hour), commonly used as a measure of river flow (the number of cubic feet per second passing a particular REACH). The abbreviation of cubic metres per second is cumec. DISCHARGE OF A STREAM. G.

cuspate delta a triangular-shaped DELTA with the apex projecting seawards, formed on a straight coast where strong wave action distributes the sediments evenly on each side of the river mouth, e.g. the delta of the River Tevere (Tiber), Italy.

cuspate foreland a large, approximately triangular-shaped FORELAND with the apex pointing seawards, usually formed by the convergence of two curved SPITS of shingle and sand built up by the action of opposing, powerful constructive waves and currents. G.

custard apple any of the species of the family Anonaceae, small trees native to the American tropics, bearing sweet, succulent fruits which are eaten raw.

custom 1. a generally accepted practice, a convention or a habit 2. in law, broadly, a practice so long-established that it has the force of law. COMMON LAW, CUSTOMARY.

customary *adj.* according to CUSTOM.

customs union a group of nation states, united in order to trade freely among themselves while applying, as individuals, the same duties and other trade regulations as other members of the group to states outside the group. COMMON MARKET.

cutan a thin film or skin (e.g. a CLAY skin) coating a PED or a coarse MINERAL-1 grain.

cutch the TANNIN extracted from the bark of some MANGROVE trees in Borneo and Malaysia.

cutcha KUCHA.

cut-off OXBOW.

cuvette (French; German Sammelmude) a large basin in which SEDIMENTATION-1 has been or is taking place. G.

CVP index an index of climate, vegetation and productivity, used to measure the plant growth ability of a climate, taking into account the temperature of the warmest month, the annual TEMPERATURE-2 range, PRECIPITATION, EVAPOTRANSPIRATION and the length of the GROWING SEASON. CLIMATO-ISOPHYTE. G.

cwm (Welsh) CIRQUE.

cwt HUNDREDWEIGHT.

cybernation the condition of control by machines, or the theory or practice of such control. CYBERNETICS.

cybernetics the science of effective organization, the scientific study of the way in which information is moved about and controlled in systems, both in biological systems (e.g. the brain, the nervous system) and in machines.

cycle 1. a series of events or phenomena recurring in the same order 2. a period of time occupied by such events or phenomena 3. an ordered series of phenomena in which a process is completed, e.g. a LIFE CYCLE.

cycle of erosion the geomorphic cycle defined by W. M.Davis, 1850 to 1934, American geologist and meteorologist, the hypothetical sequence of changes or stages through which an uplifted land surface would pass in its reduction to base level by the action of natural agencies in the processes of erosion. YOUTH streams flow through steep-sided V-shaped valleys; in ADOLESCENCE the characteristic features of maturity start to appear; in maturity (MATURE) or middle age valleys are broader with gentler slopes; in OLD AGE valleys are broad and flat, rivers sluggish, and the land surface becomes a PENEPLAIN, ready for REJUVENATION. This hypothesis is no longer uncritically accepted. RIVER. G.

cycle of poverty   POVERTY CYCLE.

cycle of sedimentation   cycle of deposition, the laying down of material by the invading sea (TRANGRESSION-1) in a land basin, in the sequence: dry land, shallow water, deep water, shallow water (each with its associated deposits), dry land. CYCLOTHERM. G.

cyclic time   the approximate length of time needed for the completion of erosion cycles, assessed in terms of millions of years; one of the categories in Schumm and Lichty's GEOMORPHOLOGICAL TIMESCALE.

cyclogenesis   an atmospheric process in which a local heat force causes horizontal convergence and violent vortical disturbance, the rotation of the air being intensified and upward motion speeded up, resulting in the formation of towering CUMULUS clouds and the development of intense tropical storms over the ocean.

cyclone   a system of winds (circulating anticlockwise in the northern hemisphere, clockwise in the southern) round a centre of low barometric pressure, especially applied to a fast-moving tropical revolving storm where the LOW PRESSURE SYSTEM is small (diameter 80 to 400 km: 50 to 250 mi)

and the barometric gradient steep, associated with strong winds, thunderstorms, heavy rainfall, known as a HURRICANE, TYPHOON or BAGUIO. The term used to be applied to any small, travelling low pressure system, but in midlatitudes such a system is now termed a DEPRESSION-3, low, or cyclonic disturbance. FILLING. G.

cyclonic (frontal) rain   precipitation associated with the passage of a DEPRESSION-3 in middle and high latitudes, as the warm moist air mass of the depression meets and overrides colder, heavier air. Characteristically, widespread drizzle is followed by heavy rain and squalls as the COLD FRONT passes.

cyclothem   in geology, a stratigraphical unit consisting of a series of beds deposited during a single sedimentary cycle, beginning with shallow water deposits as the land sank, passing into deeper water and again into shallow water or coastal deposits as the basins became infilled or the land rose. G.

cymotrichous *adj.*   having naturally wavy hair, as opposed to LEIOTRICHOUS or ULOTRICHOUS.

cytoplasm   all the PROTOPLASM of a CELL except the nucleoplasm contained in the NUCLEUS.

D

dacite a fine-grained EXTRUSIVE acid IGNEOUS ROCK of the same composition as QUARTZ-DIORITE, having a SILICA content between that of ANDESITE and RHYOLITE.

dahabiya, dahabeeyah, dahabiah (Egypt: Arabic, other spellings) a large sailing boat, now usually propelled wholly or partly by an engine, on the Nile. G.

dahannah (Arabia: Arabic) in Arabian desert, a hard gravel plain covered at intervals with sand belts of varying width. G.

daïa in north Africa, a basin depression in a sandy desert, smaller than a SHOTT and supporting some pasture. G.

daily urban system a concept which attempts to define the SPHERE OF INFLUENCE-2, in its widest context, of an urban centre (the link between an urban centre and its HINTERLAND-2) by reference to the daily movements of people (especially of COMMUTERS) and the extent of the area from which they are drawn (sometimes termed the commuting hinterland) to travel to the centre. METROPOLITAN LABOUR AREA.

dairy cattle cattle reared specifically for milk production.

dairy farm a farm specializing in DAIRY FARMING.

dairy farming farming devoted primarily to the keeping of cows for their yield of milk, whether for consumption as such, or conversion to butter, cheese and other milk products.

dak-bungalow, dāk-bungalow (Anglo-Indian, formerly dawk) a house for the use of travellers on a dāk route (post or transport route), spaced at approximately 25 km (15 mi) intervals. G.

dale mainly in northern England, a broad, open valley, a term commonly used in place-names, e.g. in the Pennines, Wensleydale; or in the English Lake District, Borrowdale. HILL-AND-DALE. G.

dalesman one who is a native of, who lives in, a DALE. G.

dallol (Nigeria) on the southeast margin of the Sahara, a flat-bottomed, wide valley with steep sides, in form resembling a WADI. G.

Dalmation coast a CONCORDANT COASTLINE, a name derived from the Adriatic coast of Yugoslavia where the coastline lies approximately parallel to the trend of the relief, the tops of former mountain ranges appearing as lines of islands, the parallel valleys having been drowned when the sea level rose. SUBMERGED COAST.

dam a barrier of earth, rock, masonry or concrete built across the course of a river to hold back or restrict the flow of the water for a specific purpose. BARRAGE, BEAVER DAM. G.

dāman, dāmān (Indian subcontinent: Urdu-Hindi) TALUS slope. G.

dambo (Bantu) **1.** the floodplain of a river in central Africa, swampy in the wet season, but dry for most of the year, supporting long grass **2.** a shallow depression, lacking distinct drainage channels, at the head of a drainage system, a term originally applied to such a depression in tropical Africa, now extended to other areas. G.

damson *Prunus damascena*, a small DE-CIDUOUS tree, native to the cooler parts of the northern hemisphere, a species of PLUM, bearing small, oval, black or bluish, not very fleshy, rather acid-flavoured fruit (a DRUPE), eaten cooked or used in jam-making.

Danelaw 1. the code of laws established in north and east England by Norse invaders in the ninth and tenth centuries **2.** the part of England governed by those laws.

Daniglacial stage one of the stages in the retreat of the QUATERNARY (GEO-LOGICAL TIMESCALE) ice sheets in northwest Europe, c.18 000 to 15 000 BC. Norway and Sweden were still under ice, but Denmark became clear, hence the name.

dans (Afrikaans) a broad shallow valley in South Africa, e.g. Leeuwens Dans. G.

Dark Ages a period in European history believed to be one of intellectual darkness, lacking knowledge, enlightenment and artistic expression, considered by some authors to span the five centuries following the fall of Rome (c.500 AD) to c.1000 AD, but by others equated with the MIDDLE AGES. RE-NAISSANCE.

darreh (Iran) a valley, narrower than a DASHT, with a periodic water flow. G.

Darwin, Charles Robert 1809-82, English naturalist, who expounded his theory of EVOLUTION (which came to be termed Darwinism) in his *On the Origin of Species by Means of Natural Selection* in 1859. CLADISTICS, NATURAL SELECTION.

daryākhurdi (Indian subcontinent: Sindhi) the alluvial land on each side of the lower Indus, within the protecting BUNDS, and flooded annually. G.

dasht (Iran) **1.** a valley with a flat floor, normally dry, but with some water in springtime. DARREH **2.** a stretch of firm sand or stones in which SAND DUNES may occur. G.

dasymetric method a method used in drawing a density map (e.g. of population) which does not use average figures related to administrative units (which produces an unrealistic map with sharp contrasts between one area and another, CHOROPLETH MAP) but instead draws on all available geographical knowledge to draw up realistic categories for which densities can be estimated. Such dasymetric plotting gives a more realistic representation.

data in statistics, **1.** the observations originally collected which provide the raw material for statistical analysis, termed raw data until grouped, manipulated or summarized in any way **2.** the figures produced from a structure imposed by the observer on the original observations.

data set in statistics, a collection of observations of several different VARI-ABLES on the same individuals or units, the individual occurrences within the data set being termed VARIATES. AVERAGE, MODE, SET, SUBSCRIPT.

Date Line INTERNATIONAL DATE LINE.

date palm *Phoenix dactylifera*, a long-lived tall tree of dry subtropical areas, characteristic of the oases of hot deserts, cultivated since c.3000 BC in the Middle East. It is usually grown without irrigation, and bears a nutritious fruit with high sugar content, an important food crop in many Arab countries. There are three categories of fruit: soft, grown on a large scale in Iraq for export to other Arab states, often sold pressed; semi-dry, sold boxed, imported to Europe mainly from north Africa; dry, keeps for a long time, can be ground into flour, soaked in water for eating, considerable trade among Arab countries, but few exported to the rest of the world. The crown of the palm can be tapped for its sugary sap, which can be boiled down for sugar, or allowed to ferment to produce palm wine (toddy). C.

dating the determining, so far as is

possible, of dates for structures, events, ARTIFACTS. The techniques used include DENDROCHRONOLOGY, LICHENOMETRY, PALAEOMAGNETISM (ARCHAEOMAGNETISM) investigation, POTASSIUM ARGON DATING, RADIOCARBON DATING, VARVE investigation.

datum (Latin, pl. data) a thing given, something known and made the basis of reasoning or calculation. Purists maintain that the plural, data, should always be used with a plural verb. DATA, DATA SET.

datum level, datum line, datum plane the zero altitude base from which the measurement of elevation starts. The Ordnance Datum (OD) from which heights for British official (Ordnance Survey) maps are calculated is the mean sea level at Newlyn, Cornwall, England. G.

dawn 1. SUNRISE 2. a poetic term for the first sign of something, e.g. the dawn of civilization.

day the time during which the SUN is above the horizon, the opposite of NIGHT.

day degree the measure of the duration of TEMPERATURES-2 above or below a selected basal temperature in a period of 24 hours. ACCUMULATED TEMPERATURE, CRITICAL TEMPERATURE. G.

daylight saving a system in which time in an area is advanced, usually by one hour, in relation to the STANDARD TIME of that area in order to extend the period of daylight at the end of a normal working day.

dead *adj.* 1. having no life 2. inactive, in contrast to ACTIVE.

dead cave a cave in which excavation and deposition have finished. ACTIVE CAVE. G.

dead cliff a sea CLIFF no longer subject to EROSION by waves owing to the build-up of protecting beach material or to a fall in sea level. WEATHERING reduces the angle of slope, and ultimately the dead cliff is colonized by vegetation.

dead ground a tract of land hidden from the observer because the form of the intervening land surface obscures the view. G.

dead ice stagnant ice, usually covered with rock debris, at the edges of a motionless GLACIER or ice sheet. G.

dead reckoning an estimation of a ship's position based on the distance travelled and the course steered by the COMPASS, with corrections for currents, leeway etc. but without the aid of radio or astronomical observations. G.

dead valley a term sometimes applied to a DRY VALLEY. G.

deadweight tonnage SHIPPING TONNAGE.

death-rate BIRTH-RATE.

debris (French débris) DETRITUS, an accumulation of rock waste consisting of disintegrated rocks, sand, clay, moved from their place of origin and redeposited. DENUDATION. G.

debris avalanche a rapid flow of rock debris, sliding in narrow tracks down a steep slope under the influence of gravity. MASS MOVEMENT.

debris fall the precipitate, nearly free, fall of earth debris from a vertical or overhanging face under the influence of gravity. MASS MOVEMENT. G.

debris glacier a GLACIER composed of fragments of ice broken off from a larger, higher glacier. G.

debris slide the rapid downward rolling or sliding of unconsolidated earth debris, without backward rotation of the mass, under the influence of gravity. MASS MOVEMENT. G.

decalcification the removal of CALCIUM CARBONATE from a SOIL HORIZON or horizons as CARBONIC ACID reacts with the CARBONATE mineral material. CARBONATION.

decentralization 1. the action or fact of moving away from a concentration at a central point 2. the diminishing of central control or authority in administration in order to increase the authority of groups at places, branches etc. distant from the centre, or the pursuance of that policy. CENTRAL BUSINESS DISTRICT, CENTRIPETAL AND CENTRIFUGAL FORCES.

decentralized phase URBAN GROWTH PHASES, USA.

deciduous *adj.* applied to trees or shrubs which shed all their leaves at a certain season every year, as opposed to evergreen, applied to trees or shrubs which carry green leaves throughout the year. Although certain CONIFEROUS trees (notably the larch) are deciduous, deciduous is commonly used as synonymous with BROADLEAVED, applied to trees which shed their leaves in the AUTUMN or fall in MIDLATITUDES as the temperature falls. Deciduous trees of the MONSOON FORESTS drop their leaves in the hot dry season as a protection against excessive TRANSPIRATION. DECIDUOUS FRUIT. G.

deciduous fruit the edible fruit of DECIDUOUS trees, such as the APPLE, CHERRY, PEAR, PLUM.

Decke (German) NAPPE.

Deckenschotter (German) sheet gravel, notably in the northern Alpine Foreland. G.

decken structure (German Deckenstruktur) a series of RECUMBENT FOLDS or OVERTHRUST masses lying one above another.

declination 1. of a celestial body, the angular distance of a heavenly body north or south of the celestial equator (CELESTIAL SPHERE), measured along a GREAT CIRCLE passing through both celestial poles and the body, i.e. celestial latitude 2. of the sun, the latitude of the parallel on which the SUBSOLAR POINT is located at any given time. It ranges from about 23°30′N (about 21 June) to 23°30′S (about 22 December) and is published in the *Nautical Almanac* (SOLSTICES). The complement of the declination (the difference between 90° and the declination) is termed the

POLAR DISTANCE. AGONIC LINE, ANALEMMA, MAGNETIC DECLINATION, RIGHT ASCENSION.

declivity a downward slope.

décollement (French) superficial earth folding in which the overlying STRATA ride easily over a BASEMENT surface.

decolonization the process whereby a METROPOLITAN-2 country gives up its authority over a dependent territory so that the dependent territory becomes a SOVEREIGN STATE.

decomposer an agent of DECOMPOSITION.

decomposition decay, disintegration, a breaking down into component parts or elements 1. in biology, the breaking down, the separation, of organic matter into simpler compounds. FOOD CHAIN 2. in geology, the breaking down of rocks, commencing with CORROSION of the component minerals.

deduction 1. reasoning from the general to the particular, i.e. in which the conclusion necessarily follows from the given premises 2. the conclusion reached in this way. ECOLOGICAL FALLACY, INDUCTION, SCIENCE.

deductive *adj.* of reasoning by DEDUCTION.

deep in the ocean, a trough-like depression or trench in the sea floor, of limited extent and great depth (over 5500 m: 3000 fathoms: 18 050 ft). Deeps occur mostly at the convergence of plates in SUBDUCTION ZONES (PLATE TECTONICS). Thus there are many near the ISLAND ARCS in the Pacific, e.g. the Mariana Trench near Guam (11 033 m: 6000 fathoms: 36 198 ft), or the Emden Deep near the Philippines (10 794 m: 5900 fathoms: 35 413 ft). G.

deepening in meteorology, the decreasing atmospheric pressure at the centre of a low pressure system (DEPRESSION-3), as opposed to FILLING. Deepening usually gives rise to greater windspeed and PRECIPITATION.

deep focus of an earthquake, a SEISMIC

FOCUS occurring at a depth below some 300 km (185 mi).

deep-sea plain the generally level area comprising most of the ocean floor, the depth varying between 3600 to 5500 m (2000 to 3000 fathoms: 11 810 to 18 045 ft), but with DEEPS, plateaus, ridges, volcanic islands.

Deep South those southeastern states of the USA coinciding largely with the old cotton states (COTTON BELT) where much slave labour was employed and where there is still a large population of African descent. The states of Mississippi and Alabama may be considered its core. CONFEDERATE STATES.

deer any of the hoofed, RUMINANT MAMMALS of the family Cervidae (e.g. REINDEER, CARIBOU, MOOSE), the males of most species having solid, branched horns which are shed annually. Deer are found in most parts of the world except in the major part of Africa and of Australia. LIVESTOCK.

deer forest FOREST-4.

deer park PARK-1.

defended neighbourhood a NEIGHBOURHOOD-2 with residents who, feeling threatened by outsiders who are seen as competitors for scarce resources (e.g. housing), try to exclude those outsiders from the neighbourhood, e.g. by raising house prices, racial attack, threatening graffiti (GRAFFITO).

defensive restructuring the reorganization and rationalizing of operations by a firm, especially by a large national or multinational company, as a reaction to declining profits. Production is concentrated in the most efficient works (often the largest and most up-to-date) and at the most profitable locations, the least efficient and least profitable being closed down. There is also in many cases investment in new technology with the object of achieving the same level of output from fewer workers.

deferred junction of a tributary with a river, a phenomenon occurring when a tributary is prevented by LEVEES from joining the river, and flows parallel with it for a considerable distance before the ultimate CONFLUENCE-2 which in many cases takes place on the convex side of a large MEANDER. A deferred junction usually occurs on a FLOODPLAIN, e.g. the River Yazoo flows parallel to the Mississippi for nearly 300 km (186 mi) before the two merge. G.

defile a narrow pass or gorge. Formerly, in military terms, a path so narrow that only a single file of men could move along in it. G.

deflation 1. in geomorphology, the removal of fine rock debris by wind, especially likely to occur in ARID or semi-arid areas lacking the protection of vegetation (DUST BOWL). Specifically in discussing arid and semi-arid conditions some authors write of the WINNOWING action of the wind in deflation, which carries the fine, light rock particles high up for great distances, while swirling the coarser near the ground and tossing about the coarsest on the surface. BLOWOUT 2. in economic geography and economics, the reduction of the value of currency or of prices from an inflated condition. INFLATION. G.

deflation hollow a BLOWOUT-1.

deforest, deforestation 1. the permanent removal of FOREST-1,2 and its undergrowth 2. historically, the release of an area from the strict Forest Laws (FOREST-3). G.

deformation in geology 1. a general term applied to the change in the shape, volume or structure of some region or feature of the earth's crust caused by STRESS-1 arising from tectonic activity (e.g. of rocks: COMPRESSION, FAULT, FOLD-2, SHEARING) 2. the process producing those results. G.

dega (Ethiopia) the highest of the altitudinal zones, above 2449 or 2750 m (8000 to 9000 ft) and extending to about 4300 m (14 000 ft), equivalent to ALPINE pasture. The other altitudinal zones are the kolla, the lowest, extending up to approximately 1700 m (5500 ft), above which lies the woina-dega, stretching from the upward limit of the kolla to about 2440 m (8000 ft), the lower limit of the dega. G.

degenerate amphidromic point AMPHIDROMIC POINT.

deglaciation, deglacierization the withdrawal of an ice sheet from an area. Some authors apply the term deglaciation to former times only, preferring deglacierization when referring to the present.

deglomeration the setting-up of productive enterprises at a distance from the centre of an established AGGLOMERATION-4, occurring when the advantages or economies of the agglomeration (AGGLOMERATION ECONOMIES) begin to diminish with its increasing size and land values, etc.

degradation 1. the process of lowering a surface by EROSION and the removal of rock waste 2. the general lowering of the surface of the land by erosive processes 3. in soil science, a change in the soil due to increased LEACHING. G.

degree (symbol °) 1. the measure of temperature on any thermometric scale. TEMPERATURE 2. 1/360th of the angle that the radius of a circle describes in a full revolution; thus a right angle is 90° 3. the unit of angular measurement of latitude and longitude, 1/360th of the earth's circumference, measured for LATITUDE from the EQUATOR, for LONGITUDE from the PRIME MERIDIAN. In latitude and longitude each degree is divided into 60 minutes (symbol ′) and each minute into 60 seconds (symbol ″). A degree of latitude is roughly 111 km (69 statute mi). G.

degrees of freedom in statistics, the highest number of VARIATES it is possible to assign freely before all of those which remain are completely determined.

dehiscence in biology, the spontaneous bursting open of a structure in order to disperse its contents, e.g. of fruits in order to release SEEDS. INDEHISCENCE.

de-industrialization the process in which there is a marked movement of employment away from the production of goods (i.e. from SECONDARY INDUSTRY) usually to the provision of services (i.e. to TERTIARY INDUSTRY). COUNTER-URBANIZATION, INDUSTRIALIZATION.

dejection, cone of CONE OF DEJECTION.

dell a natural hollow or small valley, usually with trees on the slopes, a literary term with no precise definition. G.

delta originally the tract of alluvial land traversed by the distributaries of the River Nile below Cairo and so applied to the more or less triangular terminal floodplain of other rivers. A delta is formed when a river deposits solid material at its mouth at a rate faster than that by which it can be moved by tidal and other currents. As the deposits grow the river splits to make new channels and the distributaries thus formed divide and subdivide, each stream depositing its load. Small deltas may form where a river meets a lake (LACUSTRINE DELTA), or at the confluence of rivers where a swiftly flowing stream, carrying its load of deposits, meets a sluggish older, broader stream. Delta forms are classified as ARCUATE (fan-shaped, as the Nile delta), BIRD'S FOOT (lobate or DIGITATE, as the Mississippi) and CUSPATE (as the Tavere). DELTA STRUCTURE. G.

deltalogy, deltology the study of DELTAS and delta formation.

delta structure a structure produced by the three sets of beds frequently present in a DELTA, i.e. bottomset (consisting of the finer materials carried farthest out to sea), foreset (of the coarser materials, showing the advancing front of the delta), and topset beds (lying above the foreset, forming a continuation of the alluvial plain bordering the delta). G.

demand in economics, the quantity of a commodity which a potential purchaser would buy at a certain price at a certain time, taking personal preferences, money income and the price of other commodities as given. When allowance has been made for those given factors the dependence of demand on the price of the COMMODITY is commonly shown by a DEMAND CURVE.

demand curve a plot on a graph showing the DEMAND for a COMMODITY in relation to its price, the demand appearing on the horizontal axis, price on the vertical. The demand usually falls as the price rises.

demb land (Indian subcontinent: Kashmir) new land on shallow lake margins made by planting willows in the water and infilling the compartments so formed with lake mud and weeds. G.

demersal *adj.* sometimes used as a synonym for benthic (BENTHOS-2), living in the lowest layer of a sea or lake, e.g. demersal fish, such as cod, haddock, halibut, sole, plaice, living near the bottom of a sea in comparatively shallow water and caught mainly by trawling. SEINE. G.

demesne that part of a medieval MANOR which, according to the *Report* of the Royal Commission on Common Land (1955-8), the lord did not grant out, but normally retained for his own occupation and use or that of his servants, as distinct from the manorial land farmed by the villagers. COUNTRY SEAT, MANORIAL WASTE.

democracy 1. a form of government in which political power lies in the hands of the people through elected representatives. The popular interpretation of democracy in western societies is that it is government for the people, of the people and by the people; and that it therefore gives rise to policies and

decisions which accord with the general will of the people **2.** a state so governed. AUTOCRACY, SOCIAL DEMOCRACY.

democratic socialism SOCIALISM.

demographic transition the gradual change in the manner of population growth occurring over a period of time, particularly that associated with the effects of the spread of INDUSTRIALIZATION and URBANIZATION on FERTILITY-3 and MORTALITY. The classical stylized model shows four phases. The first is termed high stationary (high birth and death rates, a fluctuating but low population growth due to famine, disease, war). The second phase is termed early expanding (continuing high birth rate, but declining death rate, and increase in life expectancy resulting from better nutrition, sanitation and medical care, leading to an expansion of population). The third phase, termed late expanding, is characterized by the stabilization of the death rate at a low level, and a decline in the birth rate linked to the growth of an urban-industrial society with its high cost of child-rearing and the ready availability and use of BIRTH CONTROL techniques. The fourth phase, low stationary, is a period of very slow population growth, with birth and death rates stabilized at a low level, the former being more likely to fluctuate than the latter. In this it differs from the first phase, when the death rate is more likely to fluctuate than the birth rate. BIRTH RATE, POPULATION PYRAMID.

demography the scientific study of human populations primarily in respect of their size, structure and development. It is not only concerned with statistics of birth, disease, death, marriage, life expectancy, migration, the division of population into groups on the basis of sex, age, marital status and the changes in those structures, but also with all aspects of population studies, including relationships with social and economic factors. POPULATION GEOGRAPHY.

demoiselle (French) an earth pillar, weathered from VOLCANIC BRECCIA or similar material, but capped by a large boulder which has protected the material underlying it. HOODOO. G.

demopleth map a distinct type of CHOROPLETH MAP, showing distribution by civil divisions. G.

den 1. in Scotland, loosely, particularly in the lowlands around Dundee and Glasgow, a deep hollow between hills, especially if wooded. DENE-1 2. in England, a shortened form of Old English DENN, common in place-names in Kent and Sussex. G.

denationalization PRIVATIZATION.

dendritic drainage (Greek dendron, a tree) a drainage pattern resembling the branching of a tree, developed especially on a gentle, nearly uniform slope where no control is exercised by the underlying structure, so that INSEQUENT STREAMS develop, and as each insequent cuts its own valley it receives its own insequent tributaries. DRAINAGE. G.

dendrochronology in PALAEOBOTANY, dating by the process of counting the ANNUAL GROWTH RINGS of a tree.

dendrology the study of trees.

dene, dean 1. a narrow, steep-sided wooded valley, locally with varied applications 2. a bare, sandy tract by the sea. G.

dene hole, dane-hole a deep excavation in the chalk lands of eastern England from which flints may have been obtained. The name seems to have been derived from the belief that Danish invaders used them as hiding places for plunder, or alternatively that they were used as hiding places from the Danes. G.

denitrification the process in which NITRATES are broken down by BACTERIA in the soil, resulting in the release of free NITROGEN (FREE-2) and a reduction in soil fertility. It usually occurs in ANAEROBIC conditions, e.g. in a WATERLOGGED soil. NITROGEN CYCLE.

denn swine pasture, especially a clearing in the former great forest of the Weald, southeast England. Denn survives as -den in English place-names. G.

density 1. the quantity of anything per unit area, hence the density of persons, of houses, or of habitable rooms per sq km, per ha, per acre etc. 2. the relation of mass (the amount of matter) to the space it occupies (its volume) expressed as gm per cu cm, the unit of measurement used being the density of water at 0°C (at that temperature 1 cc of water weighs nearly 1 gm). Thus density is an absolute quantity, unlike SPECIFIC GRAVITY which, while numerically the same, is a relative quantity. The density of water depends on temperature, salinity, particles in solution; the density of a GAS-1 is proportional to its molecular weight. MOLECULE.

density current an ocean CURRENT resulting from the differences in DENSITY-2 of water caused by variations in salinity and temperature, cold or very saline water being more dense, and therefore sinking and flowing under less dense, warm, less saline water.

density gradient the pattern of the DISTANCE-DECAY PHENOMENON, i.e. the rate of decrease of intensity etc. with distance from a central point.

denudation (Latin denudare, to uncover, to lay bare) very broadly, the uncovering of deeper rocks by any natural agency, i.e. by any agent of EROSION as well as by WEATHERING and MASS-MOVEMENT, and therefore not to be confused with the term erosion, which excludes weathering and mass-movement. Denudation is also applied generally to the lowering of a land surface by erosion and the removal of rock waste, and is thus synonymous with DEGRADATION; but some authors prefer to restrict the term denudation to the process, considering degradation to be the result of denuding activity. Rocks vary in resistance to denudation, hence the term differential denudation is applied to areas where this is apparent. CIRCUM-DENUDA-TION. G.

denudation chronology in geomorphology, the study of the sequence of events leading to the evolution of an existing physical landscape. G.

dependency, dependent territory a territory relying on or subject to the control of another country of which it does not form an integral part.

dependency ratio in population, the ratio of the number of people who cannot be gainfully employed in a population (the dependants) to the number who are actively or potentially active (the employed or employable). The dependant population is sometimes classified as those in the age groups 0 to 14 and 65 years of age and over.

dependent variable a VARIABLE which is to be explained or predicted. It is dependent on one or more other variables which may control it or relate to it. Thus it is not under the control of the experimenter; and it will be affected by other variables which are being manipulated. INDEPENDENT VARIABLE.

depopulation a marked reduction in the number of inhabitants of an area. POPULATION-2.

deposit material laid down, a natural accumulation, a SEDIMENT. DEPOSITION.

deposition the action of laying down of material, especially of the debris transported mechanically by wind, running water, tides and currents in the ocean and seas; of the materials transported in solution, subject to evaporation and chemical precipitation (e.g. ROCK SALT) or to the intervention of living organisms (e.g. CORAL); or of organic matter, mainly the remains of vegetation (e.g. PEAT). Deposition is thus the opposite of DENUDATION, the two processes together acting on the earth's crust at or very near its surface. G.

depositional landform a SEQUENTIAL LANDFORM-1 created by the deposition of SEDIMENT from a fluid medium.

depressed area an area in economic decline, with a high level of unemployment over a long period of time. DEVELOPMENT AREA.

depression 1. in general, the process of sinking, the action of pressing down, or the fact of being pressed down, the condition of being lowered in position, of being less active than usual 2. any hollow or relatively sunken area, especially one enclosed by higher land, without an outlet for surface drainage 3. in meteorology, applied specifically to a region of the atmosphere where the atmospheric pressure is lower than that of its surroundings, i.e. a low pressure system, a 'low' or 'disturbance' in midlatitudes and high latitudes, replacing the term CYCLONE (WARM FRONT). A deep depression is one in which the pressure at the centre is considerably lower than that at the edges; a shallow depression is one in which there is little difference between those two pressures. G.

deprivation the state of being prevented from using, of being taken away from, of lacking something necessary or desirable. MULTIPLE DEPRIVATION, POVERTY CYCLE.

deranged drainage a confused DRAINAGE-2 system which produces a mosaic of small lakes, streams, marshes, small islands (as in Finland) caused by the haphazard distribution of (glacial) DRIFT-1.

derelict land land damaged by some process (e.g. by extractive or other industry) and/or neglect, abandoned and left to fall to ruin, incapable of being used in its present condition.

desalination, desalinization the process of removing dissolved salts from water, especially from sea water, or from the soil.

descriptive statistics 1. statistical data of a descriptive kind 2. the methods used in describing or summarizing such data to reveal the patterns and relationships within the data. INFERENTIAL STATISTICS, STATISTICS.

desegregation the process of abandoning the practice of SEGREGATION, e.g. of bringing to an end the provision of separate facilities, such as educational facilities, for different ethnic or social groups.

desert originally applied to an uninhabited, i.e. deserted, region, occasionally still used by some authors in that sense, e.g. in reference to a 'tropical forest desert'; but most commonly applied to a region in which evaporation exceeds precipitation, from whatever cause, so that the moisture present is insufficient to support any but the scantiest vegetation. ARID, COLD-WATER DESERT, OECUMENE. G.

desertification the spread of desert-like conditions in semi-arid lands and the outward spread of the DESERT fringes there, brought about by the activities of people and their livestock and/or by climatic change.

desert patina, desert varnish a thin, hard, polished, red, brown or black skin formed on rock surfaces in hot DESERT conditions when minute quantities of matter in solution are brought to the surface by CAPILLARITY and there precipitated by EVAPORATION. The loose salts are blown away, but oxides of IRON, accompanied by traces of MANGANESE and other similar oxides, remain to form the coloured 'varnish'. It is sometimes difficult to distinguish desert patina or varnish from the effect of polishing by wind-blown sand. G.

desert pavement, desert mosaic in a hot DESERT, an exposure of bedrock or of pebbles, closely packed after the removal of finer rock material, polished or smoothed by blown sand so that eventually the upper surfaces of the bedrock or pebbles are ground flat. The pebbles are often bonded together by salts drawn to the surface in solution by CAPILLARITY and precipitated by EVAPORATION, which act as a CEMENT-2.

desert soils SOILS of ARID regions where there is a net deficiency of rainfall (commonly areas with rainfall under 250 mm: 10 in), hence a lack of vegetation and a thin or discontinuous organic layer. There is commonly a surface layer of pebbles, the leached layer being only about 15 cm (6 in) thick, underlain by a CARBONATE layer. ARIDISOL, SOIL CLASSIFICATION.

desert varnish DESERT PATINA.

desiccation 1. the action of drying up, or of becoming dried up, applied especially to the loss of moisture from the pore spaces in a soil or sediment 2. the condition of being dried up.

desiccation breccia a rocky mass consisting of the angular fragments produced from wet mud or clay which has been dried out by the sun into irregularly shaped polygonal slabs, floated away by floodwater, redeposited, and compacted.

design disadvantagement the disadvantagement originating from buildings, the design of which fails to meet the needs of people who cannot avoid living or working in them, to such an extent that it contributes to social malaise. The term has been applied particularly to large (especially HIGH RISE) blocks of low-rent flats in densely built-up urban areas.

desilication the removal of SILICA, SiO_2, from the soil in heavy rainfall regions, e.g. in tropical RAIN FOREST, where soils are usually deficient in BASES-2 as well as in silica.

desire line a straight line drawn on a map between the point of origin and the point of destination of a TRIP, i.e. the shortest distance between these two points, indicating the route a person would like to follow if it were possible. TRIP GENERATION ANALYSIS.

de-skilling the replacement of moderately skilled workers by labour-saving machinery, etc., leading to a growth in highly skilled, technical employment and even more growth in less skilled employment, mainly in the distributive and service sector, in which the moderately skilled may find work. MULTIPLIER EFFECT, UNEMPLOYMENT.

desquamation 1. the removal of scales from a surface, exfoliation 2. of rocks, a type of WEATHERING by the removal of 'scales' of rock from the surface, common in the dry tropics; an obsolescent term but in common use in French. EXFOLIATION. G.

destructive wave one of a rapid succession of strong storm waves which drop almost vertically on to the beach, the BACKWASH of the wave being so much more powerful than the SWASH that the beach material is dragged seaward. CONSTRUCTIVE WAVE.

destructive plate margin COLLISION ZONE, OCEANIC TRENCH, PLATE TECTONICS, SUBDUCTION ZONE.

determinism 1. the theory that the world, or nature, or event, or human action is subject to causal law. CAUSALITY, SCIENTIFIC LAW 2. applied in geography to environmental, geographical or physical determinism, the belief that the ENVIRONMENT (particularly its physical factors) dominate, even determine, the pattern of human life and human behaviour, that people are largely conditioned by environmental factors. Scientific determinism expresses the same belief, but the justification for it is statistically based, proceeding from statistical analysis of data sets, rather than from individual case studies. POSSIBILISM, PROBABILISM, STOP-AND-GO DETERMINISM. G.

deterministic model in statistical analysis, a MODEL-4 containing no random elements, the future course of the system being determined by its position, velocities, and so on, at a fixed point in time. DETERMINISM, PROBABILISTIC MODEL, STOCHASTIC MODEL.

deterministic process in statistical analysis, a STOCHASTIC PROCESS with a zero error of prediction, one in which the past totally determines the future of the system. DETERMINISM.

detritus 1. fragmented rock material, formed by the breaking up and wearing away of rocks, that has been transported from the place of origin to a site elsewhere 2. an accumulation of such material. MOUNTAIN-TOP DETRITUS. The term DEBRIS has now generally superseded detritus.

detritus chain FOOD CHAIN.

Deuterozoic *adj.* applied to the younger PALAEOZOIC systems, i.e. the Devonian, Carboniferous and Permian (PROTEROZOIC, PROTOZOIC, GEOGRAPHICAL TIMESCALE). The term is now obsolete or obsolescent.

devaluation of currency, the lowering of the legal international value of currency achieved by increasing the number of units of domestic currency needed to buy a unit of foreign currency.

developed *adj.* applied particularly to a country or a region frequently in connexion with the economy of the area, implying that the area is culturally and socially advanced, and that full use is being made of the natural and economic resources, skills, machinery etc. present there, the necessary capital being available, and that what had formerly been potential is being realized. DEVELOPING, DEVELOPMENT, THIRD WORLD. The gross national product (GNP) is commonly used to measure the degree of such development. For a broader view see UNDERDEVELOPMENT.

developer's profit the profit earned by the person or firm carrying out a DEVELOPMENT-2 scheme. DEVELOPMENT PROFIT.

developing *adj.* applied to a country or a region, formerly UNDERDEVELOPED, now in the process of becoming DE-VELOPED. BRANDT REPORT.

development 1. the act of causing to grow, to expand, to realize what had formerly been potential 2. in British land use planning, according to the Town and Country Planning Act 1971, the carrying out of building, engineering, mining or other operations in, on, over or under land or the making of any material change of use in any building or of the land 3. in WELFARE APPROACH, any progress or improvement in WELFARE facilities and their distribution. BRANDT REPORT, UNDERDEVELOPMENT.

developmental psychology the study of the changes in human behaviour typically taking place with increasing age, together with an analysis of their causes.

development area in British legislation, certain parts of the country, particularly those suffering industrial decline, where DEVELOPMENT-2, especially of new industry, is to be encouraged by the government. In the years between the two World Wars some of the older industrial areas suffered from serious unemployment and decay and were successively designated 'depressed' or 'distressed' areas, 'special' areas, and finally 'development areas'. G.

development control the process by which planning authorities control land use and DEVELOPMENT-2.

development profit in property development, the difference between the cost of development and the market value of the completed scheme. The DEVELOPER'S PROFIT is included in the cost. DEVELOPMENT-2, PROPERTY-2.

Devensian in Britain, the final GLACIAL STAGE of the PLEISTOCENE, characterized by fluctuating advances of the ice interspersed with warmer conditions, when birch and conifers became temporarily, and early human beings became strongly, established. In the Devensian the extent and duration of the ice-cover varied from one area to

another: nearly all of southern England and East Anglia were ice-free; but western Scotland lay permanently under ice for most of this stage. FLANDRIAN.

deviate in statistics, the value of a VARIATE measured from some standard point of location, usually the MEAN-1.

deviation 1. in general, a turning away from. MAGNETIC DEVIATION 2. specifically, the departure from AREAL DIFFERENTIATION (the traditional view of geography), three deviations being defined as the landscape, the ecological and the locational schools 3. in statistics, the difference between a measurement and an average for the set of measurements. If the average is expressed by a regression line (REGRESSION ANALYSIS) and the deviation is measured from this line, the term RESIDUAL is commonly applied to this deviation. MEAN DEVIATION, STANDARD DEVIATION. G.

devolution the passing on of the power or authority of one person or body to another, sometimes to one in a subordinate position.

Devonian *adj.* of the fourth geological period (of time) or system (of rocks) of the PALAEOZOIC era (GEOLOGICAL TIMESCALE) when SANDSTONES, GRITS, SLATES and LIMESTONE were laid down in the sea (as apparent in southwestern England) and red and brown sandstones, CONGLOMERATES, MARLS-3 and LIMESTONE (the OLD RED SANDSTONE) in lakes, and the characteristic life forms were ferns and lower fishes.

dew droplets of water deposited on any cool surface by CONDENSATION of WATER VAPOUR in the atmosphere, especially at night after a hot day. DEW-POINT, PRECIPITATION-1. G.

dew mound, dew-mound a pile of soil put by the cultivator around the base of the trunk of fruit trees (e.g. olives, citrus trees etc.) and capped with flat stones on which dew forms and drops down to irrigate the plant, a device used in ARID lands, especially in the Middle East.

dew-point the temperature at which air, on cooling, becomes saturated with WATER VAPOUR, and below which CONDENSATION begins and DEW forms. G.

dew pond, dew-pond a shallow artificial POND-1, lined with PUDDLED CLAY and straw, made especially on the chalk downlands of southern England, the lining preventing water from percolating downwards. It was long believed that these ponds were fed with dew, hence the name, but dew contributes very little to the water held: some of it comes from the condensation of sea mist, but most is derived from PRECIPITATION-1. G.

dhānd (Indian subcontinent: Sindhi) a salt or alkaline lake in Sind. G.

dhāyā (Indian subcontinent: Panjabi) **1.** Panjabi term for KHADAR **2.** a BLUFF bounding a floodplain of KHADAR (new alluvium). G.

D horizon C HORIZON, SOIL HORIZON.

dhōrō, dhōrū (Indian subcontinent: Sindhi) a dry water channel. G.

dhow (Arabic ?) a general term for a wide variety of locally-built Arab sailing vessels. G.

diabase **1.** a term formerly applied to various dark coloured BASIC IGNEOUS ROCKS, especially those partly metamorphosed (METAMORPHISM) in which some of the constituents have been altered. As PETROLOGY has progressed, these rocks have been separated into different types, but the term is still sometimes applied in Britain to DOLERITE **2.** in USA, a partly CRYSTALLINE rock, a general term for DOLERITE. G.

diabatic *adj.* of the thermodynamic process in which loss or gain of heat occurs (e.g. in an AIR MASS), the opposite of ADIABATIC.

diachronic *adj.* lasting through time or during the existing period, applied particularly to an approach to LINGUISTICS. DIACHRONISM, SYNCHRONOUS.

diachronic analysis the study of the processes by which changes in one component or subsystem of a SYSTEM-1,2,3 pass to and spread among the other component parts. SYNCHRONIC ANALYSIS, VERTICAL THEME.

diachronism in geology, the existence of a lithological unit which is of varying age in different areas, and thus transgresses the palaeontological zones (PALAEONTOLOGY). The *adj.* applied to such a lithological unit is diachronous. DIACHRONIC. G.

diaclinal *adj.* applied to rivers and valleys that have a direction at right angles to the STRIKE, that pass through the AXIS OF A FOLD and are thus in part ANACLINAL, in part CATACLINAL. The term is now rarely used. G.

diagenesis the physical and chemical processes (excluding METAMORPHISM) by which sediments are compacted and cemented (COMPACTION-1, CEMENTATION OF SEDIMENTS) under temperatures and pressures normal at the earth's surface, without crustal movement being directly involved, leading to LITHIFICATION. With rising temperature and pressure diagenesis passes into metamorphism. LITHOGENESIS. G.

diageotropism the growth response of part of a plant (e.g. part of the root system) which is at variance with the stimulus of GRAVITY-2, so that the axis of the part lies at right angles to the direction of the gravitational force. PLAGIOTROPISM, TROPISM.

diagnostic horizon a strictly defined SOIL HORIZON which clearly shows the soil-forming processes at work in an area, and is used as a basis for classifying soils in a SOIL CLASSIFICATION.

dialectic **1.** the art of logical disputation, of critically examining the truth of an opinion or theory by question and answer **2.** in Hegelian philosophy, broadly, the logical subjective development in thought from thesis through antithesis to SYNTHESIS-3 (DIALECTICAL MATERIALISM), or logical objective development in history by the continuous reconciliation, the unification, of parts, of opposites. Marx saw this process at work in his interpretation of the historical succession of MODES OF PRODUCTION. In DIALECTICAL MATERIALISM dialectic is not only equated with the way reality changes, it is declared to be the method of discovering the 'laws of motion', a method, it is asserted, which is applicable to all scientific disciplines. HISTORICAL MATERIALISM, MARXISM.

dialectical *adj.* of, pertaining to, or skilled in, logical argumentation or disputation.

dialectical materialism the philosophy underlying Marxist theory, first formulated by Engels (MARXISM), based broadly on a modification of the standard theory of MATERIALISM-1 combined with a development of Hegelian philosophy (DIALECTIC-2), i.e. that everything that exists can be shown to derive ultimately from matter and that (using Hegelian philosophy) the development of nature, society and thought occurs by conflict between an original direction (the thesis), its direct opposite (the antithesis) and ultimately by SYNTHESIS-2 (the reconciliation and unification of parts of these two extremes). Dialectical materialism maintains that dialectic is not only the way reality changes, it is the method to be used in discovering the 'laws of motion', a method applicable to all scientific disciplines. HISTORICAL MATERIALISM, MARXISM.

dialectics 1. any logical disputation **2.** the gathering together of opposites or apparently contradictory ideas or theories with the object of trying to resolve them. DIALECTIC.

diamond a PRECIOUS STONE, the

CRYSTALLINE form of pure CARBON, the hardest mineral (HARDNESS), usually occurring embedded in PIPES in IGNEOUS ROCK or washed out and redeposited in PLACERS, measured in CARATS. The better quality diamonds are used as GEMSTONES, the poorer in industry for cutting and as abrasives.

diapir a vertical intrusion, cylindrical or dome-shaped (e.g. a SALT DOME) formed by the rising of less dense material through COUNTRY ROCK, e.g. as when the overlying rocks of an ANTICLINAL fold are cracked open by the pressure of mobile material (e.g. IGNEOUS material, salt, mud) forcing its way upwards from below.

diara CHAR.

diastrophic *adj.* of or pertaining to DIASTROPHISM.

diastrophic eustatism, deformational eustatism a global change of ocean level (NEGATIVE MOVEMENT, POSITIVE MOVEMENT) due to a variation in the capacity of ocean beds, caused by filling in by sedimentation or movements of the ocean floors, a change which often leads to REJUVENATION. EUSTATISM, GLACIO-EUSTATISM. G.

diastrophism a general term for the action of forces which have deformed or disturbed the earth's crust. Some authors equate diastrophism with OROGENESIS. But others make a distinction between two types of diastrophic processes: orogenic, covering mountain-building with deformation; and EPEIROGENETIC (EPEIROGENIC), applied to movements of regional uplift without major deformation. G.

diatectic varve VARVE.

diathermancy the property of being diathermic or diathermous, i.e. being pervious to radiant heat (e.g. the property of the atmosphere in allowing the passage of radiant heat).

diatom any of the class of ALGAE in the division Chrysophyta, microscopic, unicellular plants, yellow-brown, with SILICA in the cell walls, occurring singly or in colonies (COLONY-6) which, with other divisions of algae, form part of marine and freshwater PLANKTON (NANOPLANKTON). The deposited remains of dead diatoms of the past appear today as diatomaceous earth (DIATOMITE) or, with other decomposed organisms, as PETROLEUM or as SAPROBEL.

diatomite a dried diatomaceous earth, formed by deposition in the past of silicified cell walls of dead DIATOMS. Chalk-like, light, friable and porous, chemically inert, with low thermal conductivity, diatomite is used industrially as a filler, an absorbent, insulator, refractor or filter, etc.

diatom ooze OOZE consisting of the siliceous skeletons of DIATOMS deposited in the ABYSSAL zone of cold ocean water, occurring particularly in a belt around the earth in the Southern Ocean in latitudes 50°S to 60°S and in the North Pacific Ocean.

diatreme a vent in the earth's crust made by explosive volcanic activity.

dichotomous model a theory or MODEL-4 in which only two, usually competing, categories are considered, sometimes termed a conflict model.

dichotomous variable ATTRIBUTE-2.

dieback a progressive withering in plants (especially in trees and shrubs) from the tips of shoots back to the root, sometimes causing the death of the plant. It is commonly due to bad environmental conditions (lack of soil nutrients, waterlogging, etc.), or to some change in physical conditions, or to infestation by hostile BACTERIA or (wholly or in part) by FUNGI. It may lead to the dying off at the same time of a high proportion of individuals, or of the individuals of one species in a large tract of vegetation (e.g. a forest). G.

differential ablation ABLATION.

differential denudation DENUDATION.

differential disequilibrium, dynamic disequilibrium the state of flux characteristic of an urban system at any one

point in time as different parts of it adjust to changes within it at different rates (e.g. groups of well-educated people with financial resources adapt to change in a system more rapidly than do less privileged groups). There are thus at all times substantial differentials in the disequilibrium in any urban system. G.

diffluence, glacial GLACIAL DIFFLUENCE. G.

diffraction the bending of the direction of propagation of electromagnetic or other waves (e.g. light) around objects that obstruct their passage.

diffusion 1. the spreading out, the propagation, the dissemination through time of a PHENOMENON or phenomena (e.g. plants, animals, ideas, CULTURE-1, languages, knowledge, INNOVATION, techniques) over an ever-extending surface or over SPACE-2 from a single source (termed mononuclear diffusion) or from many sources (polynuclear diffusion). AGE AND AREA HYPOTHESIS, BARRIERS AND DIFFUSION WAVES, CASCADE DIFFUSION, DIFFUSION WAVE, DISTANCE-DECAY PHENOMENON, EXPANSION DIFFUSION, HIERARCHICAL DIFFUSION, DIFFUSIONISM, HORIZONTAL DIFFUSION, INFORMATION FIELDS, NEIGHBOURHOOD EFFECT, RELOCATION DIFFUSION, URBAN SYSTEM 2. in meteorology, the seemingly random mixing of air bodies brought about by a slow process of mixing (termed molecular diffusion) or by TURBULENCE (eddy diffusion). The term is also applied to similar processes in liquids and light. G.

diffusionism the belief that cultural similarities occur mainly (or, in extreme view, solely) as a result of DIFFUSION-1.

diffusion wave, innovation wave the movement of DIFFUSION-1 of INNOVATION, termed the INNOVATION WAVE by T. Hägerstrand, who identified four stages in its progress, i.e. the primary (the beginning when the centres adopting the innovation are established and the contrast between them and remote areas is great); the diffusion (the start of the diffusion process characterized by the creation of new, rapidly-expanding innovation centres distant from the source, and a dimming of the contrast seen in the primary stage); the condensing (marked by a relative increase in the number of acceptances, equal in all locations irrespective of distance from the innovation source); and finally the SATURATION STAGE (characterized by a slowing down and eventual ending of the process, with apparent overall acceptance without regional variation). If the technique of drawing a TREND SURFACE MAP is applied to this diffusion wave profile, it can be shown that at first the wave has limited height (showing limited rate of acceptance), then it increases in height and extent, thereafter decreasing in height but increasing in the total area involved. The gradual weakening of the wave is due to the passage of time (evidenced by the simultaneous slackening of acceptance rates) and the effects of the hazards of space, e.g. inhospitable territory, or barriers (BARRIERS AND DIFFUSION WAVES), or competition from other diffusion waves. The wave will speed up or slow down according to the nature of the medium through which it moves; and if it meets another diffusion wave coming from another direction from another centre it may completely lose its identity.

digitate margin the seaward extension of a DELTA into a finger-like form, e.g. a BIRD'S FOOT DELTA.

digitigrade in zoology, an animal that walks on the lower surface of the digits only, e.g. the dog. ORTHOGRADE, PLANTIGRADE, UNGULIGRADE.

dike, dyke (the spelling dyke is common but etymologically incorrect) **1.** a ditch, a wall, an embankment, a ridge **2.** in geology, an INTRUSION where the molten rock (MAGMA) has ascended through an approximately vertical fissure to solidify as a wall of rock often harder than the rocks of the surrounding strata. DIKE-SPRING, RING-DYKE. G.

dike (dyke) phase the closing episode in a volcanic cycle, when minor INTRUSIONS, especially DIKES-2, are injected. G.

dike- (dyke-) spring a SPRING-2 issuing along the line of a DIKE-2 where water from an AQUIFEROUS, PERMEABLE or PERVIOUS rock meets a dike of IMPERMEABLE rock which is penetrating the aquiferous surrounding strata.

dike (dyke) swarm in geology, a collection of DIKES-2 of the same age, usually with a common trend over a wide area, sometimes radiating from a common centre. G.

dilatancy the state of a SEDIMENT-2 which, when wet and shaken, exudes water. G.

dilatation of rocks, the release of pressure effected within a rock mass when overlying layers are removed by DENUDATION, causing the rock to expand and split along expansion joints (dilatation joints) and concentric layers at right angles to the direction of the pressure release to split away from the upper surface, from which they are commonly removed by WEATHERING.

dill *Anethum graveoleus*, an ANNUAL or BIENNIAL herb, native to Europe excluding Britain, widely naturalized in North America and the West Indies. The aromatic leaves and seeds have been used in cooking and medicine since early times, in the former in pickling and in flavouring cakes, fish, meat and a variety of dishes, in the latter (as dill oil from the seed) as a remedy for flatulence, especially in infants.

diluvion the loss of land by river erosion after flooding. ALLUVION. G.

diluvium (Latin, flood, inundation, deluge) a nearly obsolete term, originally applied to superficial deposits that seemed to have been formed not by the normal slow operation of water but by some extraordinary phenomenon. Therefore applied to the DRIFT-1 deposits (now known to have resulted from the melting ice sheets of the Great ICE AGE) thought to have been laid down in Noah's flood. The term is still sometimes applied (especially by German geologists) to the older of the Quaternary deposits, i.e. to the PLEISTOCENE (GEOLOGICAL TIMESCALE). The term is perhaps best avoided except in historical discussion. ALLUVIUM. G.

dimple a small, usually circular depression, some 4.5 to 14 m (15 to 45 ft) across and up to 7.5 m (25 ft) in depth occurring on steep slopes, especially on valley sides, in chalk terrain, probably an abandoned swallet (SWALLOW) formed in a period when surface water was more abundant. G.

dingo *Canis dingo*, the wild dog of Australia, probably introduced by ABORIGINES-1.

diorite a coarse-grained INTRUSIVE IGNEOUS ROCK belonging to the intermediate group (SILICA 55 to 65 per cent) so that free QUARTZ is usually absent, consisting of PLAGIOCLASE FELDSPAR and a dark mineral, HORNBLENDE or AUGITE and biotite MICA. With an increasing proportion of silica it passes through GRANODIORITE into GRANITE. GNEISS. G.

dip 1. the angle of maximum slope of an inclined surface 2. in geology, true dip, the angle of maximum slope (i.e. maximum INCLINATION-2) of SEDIMENTARY ROCKS (or of rocks bedded with them) at a certain point. The term should not be applied to the INCLINATION-2 of land surfaces. APPARENT DIP, STRIKE. G.

diphotic PHOTIC.

dip slip, dip-slip SLIP.

dip slope, dip-slope the surface slope of the ground where it inclines in approximately the same direction as the DIP-2 of the underlying rocks. Some authors

prefer to use the term BACK SLOPE if the parallelism between the two is very slight. CUESTA. G.

dip stream, dip-stream a stream flowing roughly parallel to the DIP-2 of the underlying rocks. G.

direct correlation in statistics, CORRELATION COEFFICIENT.

Dirichlet polygon a polygon that contains within it areas which are nearer to the point around which they are constructed than to any other points, named after P. G. L. Dirichlet, 1805-59, German mathematician; also known as a Thiessen polygon or a first-order Brillouin region. G.

dirt band a dark band of ice, demarcated by light bands, formed within the ice of a GLACIER between the annual accumulation layers of FIRN. The dark and light ice bands may form a type of OGIVE-1 if they are exposed at the glacier surface. The almost bubble-free dirt band is formed when melt-water containing dirt is re-frozen. The light ice band, a mass of bubbles, may be the result of winter freezing of snow.

dirt road a road with an earth surface.

dirty-boot farmer (American dirt farmer) a FARMER who personally works on the land over which he/she owns property rights. GENTLEMAN FARMER.

disappearing stream a stream that flows on the surface for some distance before vanishing underground, usually down a SWALLOW-HOLE-1 in LIMESTONE (or other pervious rock), working its way down through JOINTS to the base of the limestone, where it may emerge. RESURGENCE.

discharge of a stream the quantity of water passing through any cross section of a stream in a given unit of time. It is usually measured either in cubic feet per second (CUSEC) or cubic metres per second (cumec). River discharge is measured at a gauging station with a current meter and an instrument for measuring the water depth. G.

discipline a branch of learning.

disclimax a SUBCLIMAX which is long-lasting because human or animal activities prevent its reaching a true CLIMAX. PLAGIO-CLIMAX.

disconformity in geology **1.** an UNCONFORMITY where the bedding planes of one set of rocks do not parallel those of another set below, but are not in sharp contrast to them in DIP-2 **2.** occasionally used as a less definite term than unconformity. NON-SEQUENCE. G.

discontinuity a break in continuity **1.** in GUTENBERG DISCONTINUITY and MOHOROVICIC DISCONTINUITY, a sudden change of character in the structure of the earth's interior **2.** in geomorphology, a definite change of slope **3.** in meteorology, the FRONT (or frontal zone) between air masses of differing temperature and humidity.

discontinuous distribution in biology, a distribution pattern in which similar species occur in widely separated areas in the world, usually indicating that the group, once generally and widely distributed, now survives only in parts of its original range. POLYTOPIC.

discordant _adj._ at variance, incongruous; not in accord, not harmoniously connected. CONCORDANT.

discordant coast a coast where the coastline cuts across the FOLDS-2 and FAULTS of the geological structure, i.e. across the 'grain' of the country; a transverse or ATLANTIC TYPE COASTLINE. G.

discordant drainage the condition of drainage when the surface drainage does not directly relate to the DIP-2 of the underlying strata.

discordant intrusion an INTRUSION of IGNEOUS ROCK that cuts across the bedding of the rocks through which it intrudes. DIKE-2.

discordant junction a river junction in which a tributary stream falls abruptly into the main stream, e.g. from a HANGING VALLEY. ACCORDANT JUNCTION. G.

discrete _adj_ separate, individually distinct, detached, not continuous, not connected to other parts.

discrete variable an ATTRIBUTE-2 which is derived from an enumeration or count, which can take only whole numbers or distinct values or SCORES. CONTINUOUS VARIABLE, NOMINAL VARIABLE.

discretionary investment in Marxism, that part of SOCIAL INVESTMENT which, by providing new incentives for investment, has the effect of stimulating further capital accumulation. COMPLEMENTARY INVESTMENT.

diseconomies of scale an increase in unit costs arising from an increased scale of production. This rise in unit costs may be brought about by internal diseconomies (e.g. the need for a large administrative organization, the loss of contact between staff and management), or by the diseconomies experienced when an urban area grows so large that it becomes congested, transport costs rise and staff is no longer readily available. DISECONOMIES OF URBANIZATION, DISECONOMY, ECONOMIES OF SCALE.

diseconomies of urbanization the DISECONOMIES OF SCALE associated with large cities, with their high costs of labour, land, and transport.

diseconomy any unfavourable effect which results from an increased scale of production, e.g. diminishing returns or profitability. It is termed internal diseconomy if those responsible for the increase suffer the unfavourable effect, external diseconomy if the unfavourable effect is suffered by others. DISECONOMIES OF SCALE, DISECONOMIES OF URBANIZATION.

disintegration a breaking-up, separating into fragments, e.g. by FROST.

dismembered drainage the DRAINAGE pattern created when the lower part of a drainage system is drowned by the sea. The lower part of the main river disappears and the tributaries are left to enter the sea as independent streams. G.

dispersal 1. a scattering, breaking-up 2. going away in different directions 3. putting in position at selected points.

dispersed city 1. a group of cities existing at a similar functional level each with its own administration and, in some cases, separated from its neighbours by agricultural land, but functioning as a single unit, the higher order functions being spread among lower order places despite the fact that the population of the whole area is large enough to support a higher order place. CENTRAL PLACE THEORY, ORDER OF GOODS 2. a rural area into which the urban population has spread, most of whom commute to work in the nearest urban centre. The effect of the spread of this urban population with its urban way of life is to change the pattern of rural society so that it becomes urban without any physical urbanization occurring. G.

dispersed settlement a settlement pattern in which farmhouses and rural dwellings are scattered instead of being grouped together in a HAMLET, NUCLEATED SETTLEMENT, VILLAGE. G.

dispersion in statistics, the degree of scatter shown by observations, usually measured as an average DEVIATION about some central value, e.g. MEAN DEVIATION, STANDARD DEVIATION, or by an order statistic, e.g. quartile deviation (INTERQUARTILE RANGE); but it may also be a mean of deviations of values amongst themselves. VARIANCE.

dispersion diagram a diagram used to show the distribution of any quantity for any unit of time over a selected period.

disphotic zone the zone of ocean depth between 100 and 800 m (55 and 437 fathoms: 325 and 2625 ft) where the penetration of sunlight is sufficient for biological responses, between the

APHOTIC ZONE and the EUPHOTIC ZONE. PELAGIC, PHOTIC. G.

displacement tonnage SHIPPING TONNAGE.

dissected *adj.* applied to a land surface cut up by EROSION (DISSECTION), especially to a plateau (dissected plateau) where flat-topped remnants lie between deep irregular valleys. UNDISSECTED.

dissection the cutting of a land surface by EROSION, especially by eroding streams, into numerous valleys. DISSECTED, UNDISSECTED. G.

dissolved load the organic and inorganic material in SOLUTION-1 carried by a STREAM-1, as distinct from the BED LOAD and the SUSPENDED LOAD. The total amount of dissolved material in the water is usually assessed by evaporating a known volume of filtered water and weighing the dry residue.

distance-decay phenomenon the weakening, the fading, of process or pattern with increasing distance. It is apparent, for example, in transport flows in that as the distance between the point of origin and the point of destination increases the intensity of the flow tends to decrease. DENSITY GRADIENT, INTERVENING OPPORTUNITY, NEIGHBOURHOOD EFFECT, NODE-SPECIFIC FLOW DATA, PRIVATE INFORMATION FIELD.

distillation a process in which a LIQUID or a SOLID is subjected to EVAPORATION and CONDENSATION in order to purify it or separate it into smaller parts with different properties. GAS COAL.

distributary 1. a branch of a river which flows away from the main stream and does not return to it, as in a DELTA 2. a branch canal distributing water from a main canal in an irrigation system 3. an ice stream flowing from an ice sheet or ice cap. G.

distribution 1. in general, the action of apportioning, of dealing out, of allocating to distinct places, of dispersing to (or over) all parts of an area or space;

or the condition of being so divided, allocated, dispersed **2.** the dispersal of COMMODITIES among consumers **3.** in biology, the geographical range of an organism or group of organisms **4.** in statistics, a classification or arrangement, especially of statistical information; or the way in which VARIATE values are apportioned. CUMULATIVE FREQUENCY DISTRIBUTION, EXTREME VALUE, FREQUENCY DISTRIBUTION, MEASURE OF LOCATION, NORMAL DISTRIBUTION, POISSON DISTRIBUTION.

distribution free method in statistics, a statistical method (e.g. of testing an hypothesis) which does not depend on the form of the underlying DISTRIBUTION-4. Distribution free tests are sometimes termed NONPARAMETRIC TESTS, but some statisticians prefer to restrict the term nonparametric to tests which do not involve an explicit statement about a parameter.

disturbance in meteorology, a DEPRESSION-3 or LOW (formerly termed a CYCLONE) of no great intensity.

diurnal *adj.* **1.** in general, of, or belonging to, each day, completed once in one day **2.** in zoology, applied specifically to animals active mainly in the daytime in contrast to NOCTURNAL animals, those active mainly at night.

diurnal range the difference between minimum and maximum values in 24 hours, e.g. as applied to air temperature.

diurnal rhythm, circadian rhythm the rhythmic physiological changes that, originating within an organism, occur in every 24 hours even when the organism is isolated from the daily rhythmic changes in its environment, e.g. sleep rhythm in animals (including humans), or leaf movements in plants.

diurnal tide a tidal regime in some parts of the ocean (e.g. in the Gulf of Mexico) where, due to the shape of the water body, there is only one high and one low tide in each 24 hours.

divagating meander a MEANDER which is liable to variation from time to time because the surface on which it occurs approaches the condition of a PENEPLAIN.

divagation of a river, the lateral shifting of the course of a river due to the extensive deposition of ALLUVIUM in its bed. G.

divergence in general, the action of starting off from a point or source, and continuing in separate directions, with the result that the degree of separation increases with distance (e.g. as of two or more paths, real or figurative) (CONVERGENCE). Some specialized applications **1.** in biology, the continuing separate courses of evolution of species with a common ancestor so that with the passage of time the life forms of the species progressively decrease in similarity **2.** in climatology, a type of airflow in which in a certain area at a given altitude the outflow is greater than the inflow, resulting in a decrease in the air contained **3.** in oceanography, the movement of surface water away from a zone, brought about by winddrift, resulting in the rise of water from the depths.

diversivore an animal (e.g. a human being) that eats any type of food. CARNIVORE, HERBIVORE, OMNIVORE.

divide the line of separation, a ridge or stretch of high ground, between drainage basins. CONTINENTAL DIVIDE. G.

divided circle diagram popularly termed a pie diagram or pie graph, a diagram in which a circle, representing the total of the values, is divided into sectors, each sector being proportional to the value it represents.

division a primary group in the CLASSIFICATION OF ORGANISMS, consisting of one class or a number of classes. The term division is commonly used instead of PHYLUM in the classification of plants.

DNA deoxyribonucleic acid, the nucleic acid (inherited material) present in the nuclei of the cells of all living things. NUCLEUS.

do (Indian subcontinent: Urdu-Hindi; Hindustani) two; hence do-fasli-har-sala, land yielding two crops a year; do-fasli-do-sala, a two-year rotation, in contrast with ek, one, and ek-fasli-harsala, land yielding one crop a year. G.

doab (Indian subcontinent: Urdu and Persian, two waters) the land between two rivers, applied specifically in the northern part of the subcontinent to the areas of alluvium with low relief lying between rivers, particularly that lying between the two great confluent rivers, the Ganga and the Jamuna. The term is applied to similar areas of alluvium with low relief lying between rivers in other parts of the world. G.

dock an enclosure or artificial basin, fitted with floodgates, in a harbour or river, in which vessels are loaded, unloaded, refitted, repaired. The pl. docks, denotes the dock basins with adjoining wharfs, warehouses, work-shops and yards, offices. DRY (GRAV-ING) DOCK, FLOATING DOCK, TIDAL DOCK, WET DOCK.

dockyard an enclosure with docks and equipment for building and repairing ships, especially naval vessels (American navy yard).

Doctor the HARMATTAN wind of west Africa. CAPE DOCTOR.

doctrine a principle or a body of prin-ciples of a particular subject, a branch of knowledge or a system of belief, taught and laid down as true.

dod, dodd in northwest England and south Scotland, a rounded hill or summit. G.

doe a female DEER, hare or rabbit.

dog days a popular name for the hot period between early July and mid-August, particularly in the USA, when Sirius (the Dog Star) rises and sets with the sun.

dogger a large concretion or mass of

consolidated material found in certain sedimentary rocks, notably in the JURASSIC SANDSTONES in Dorset and Yorkshire. G.

dokeph DOLICHOCEPHALIC.

doldrums the region of small pressure gradient, the belt of calms with high humidity and high temperatures occur-ring near the equator, approximately between 5°N and 5°S, especially over the eastern part of the oceans. ITCZ.

dolerite 1. a typical medium-grained BASIC, HYPABYSSAL, IGNEOUS ROCK (SILICA 45 to 55 per cent), usually dark in colour, corresponding in composi-tion to the lava BASALT and the plutonic GABBRO, occurring in minor intrusions (e.g. in DIKES-2, SILLS) **2.** in USA, DIABASE. G.

dolichocephalic *adj.* long-headed, some-times abbreviated to dokeph. CE-PHALIC INDEX.

doline (French; derived from Slavic and Italian dolina) a shallow basin or funnel-shaped depression typical of KARST landscape. It usually has a flat floor, sometimes cultivated, linked to the underlying drainage system by a vertical shaft. The size and form vary, the diameter from a few metres to a kilometre, the depth from a few to several hundred metres. If formed mainly by direct solution of surface limestone it is termed a solution doline; if by the collapse of a cave roof following subterranean solution, a col-lapse doline. SOTCH, UVALA. G.

doline lake a body of freshwater occu-pying a DOLINE.

dolmen a Neolithic burial chamber. CROMLECH.

dolomite 1. strictly, a yellow or brown-ish mineral with the formula $CaCO_3Mg_2CO_3$, i.e. consisting of equal molecules of calcium carbonate and magnesium carbonate, commonly occurring in evaporite deposits, e.g. from sea-water; or as a replacement in LIMESTONE, some of the calcium having been replaced by magnesium

(DOLOMITIZATION); as a CEMENT; as a GANGUE mineral in HYDROTHERMAL deposits; and in CARBONATITES **2.** commonly applied to a rock consisting predominantly of that mineral, hence dolomitic limestone, a limestone with some dolomite. Dolomite rock is sometimes termed MAGNESIAN LIMESTONE. CALCIFICATION. G.

dolomitization the alteration of calcite limestones by percolating magnesium carbonate solutions. DOLOMITE.

dolostone a SEDIMENTARY ROCK composed of fragmental, concretionary or precipitated DOLOMITE of organic or inorganic origin. G.

domain **1.** the estate or territory within defined limits over which DOMINION-1, control or influence is exerted **2.** in CULTURAL GEOGRAPHY, the zone which immediately adjoins the CORE AREA of a CULTURE-1 (CULTURAL HEARTH) and into which the culture spreads. OUTLIER-2, SPHERE-5.

dome in general, loosely applied to any dome-shaped (hemispherical) mass of rock or dome-shaped landform. More precisely applied to a structural feature where the underlying rocks form a dome, i.e. the strata dip away in all directions from a central, rounded area. BATHOLITH, DOME VOLCANO, LACCOLITH, OIL DOME, SALT DOME. G.

Domesday Book a documentary, detailed survey of England on a county basis compiled in 1086-7 on the orders of William I (the Conqueror), King of England, recording the extent, value, ownership of estates, census of householders, local customs, in two volumes, one covering Essex, Suffolk, Norfolk, the other the remainder of England apart from Northumberland, Durham, Cumberland and north Westmorland, which were excluded.

domestic *adj.* **1.** belonging to the home or house, to one's own home **2.** of or pertaining to the home country as opposed to a foreign country.

domestic animal a tame animal living under and dependent on human care, usually near a human habitation. G.

domestic market the home MARKET as opposed to a foreign market. G.

domestic port a port serving internal or coastwise trade, as opposed to one serving predominantly international trade. G.

domestic trade internal trade as opposed to international trade. G.

dome volcano a VOLCANO composed of highly viscous LAVA which, on eruption, congeals above and around the orifice instead of flowing away, the older lava sometimes being raised by pressure of the lava welling up from below. G.

dominant *adj.* controlling or ruling, most noticeable, commanding on account of strength or position.

dominant animal the leader of a group of animals.

dominant species of plant or animal present in an ecological community, the characteristic species, the one most thriving and prevalent and exerting the greatest influence on the character of the community, e.g. oak in an oakwood.

dominant wave the largest, most powerful wave rolling on part of the coast.

dominant wind the WIND that blows with the most effect. It may, or may not, be the PREVAILING WIND.

dominion, Dominion **1.** sovereignty, supreme authority **2.** in English law, the right of possession **3.** historically, the lands belonging to a feudal landlord (FEUDAL SYSTEM) **4.** Dominion, the title of a particular self-governing independent country within the COMMONWEALTH-2, formerly a COLONY-2 of Britain, with powers and status as laid down in the Statute of Westminster 1931.

donga **1.** (Bantu) in South Africa, a steep-sided gulley formed as a result of soil erosion **2.** in Australia, a shallow circular depression in the Nullabor Plain resulting from the collapse of the roof of an underground chamber. G.

donkey ASS.

dorbank (South Africa: Afrikaans) a concretion of lime and silica occurring under the surface layer of sandy loam in the Little Karoo. G.

dore (northwest England, local term) an opening or fissure between rocks, often a widened JOINT, especially in a ridge between rock masses. G.

dormant *adj.* sleeping, quiescent, applied specifically to a VOLCANO which has not erupted in historic time, but is not regarded as extinct. G.

dormitory town (Latin dormitorium, a sleeping place) a town from which residents travel daily to work in an accessible nearby larger town or CONURBATION. COMMUTER.

dormitory village COMMUTER VILLAGE.

dorp (South Africa: Afrikaans; Netherlands dorp; German dorf) a relatively small town. G.

dot map a map showing spatial distribution (commonly based on statistical data for an administrative unit) by the use of dots, usually of uniform size, each representing a specific number of the objects concerned. The value of the dot must be carefully chosen, bearing in mind the high and low quantities to be represented and their location. If statistical data only are available the dots have to be spaced evenly within the administrative unit; but with a knowledge of local conditions they can be placed more precisely to give a more realistic representation, less misleading than even spacing, but involving subjective judgment. G.

double tide a tidal regime in which there is a double high tide (the first falling a little before rising again to a second maximum, e.g. in Southampton Water, southern England) or a double low tide (termed a gulder near Portland, Dorset, England). It is due to the effects of the shape of the coast or of shallow water, which deforms a PROGRESSIVE WAVE. TIDE.

double-unilineal UNILINEAL.

doubling time EXPONENTIAL GROWTH.

doup (northern England) a rounded depression or cavity in a rock or hillside. G.

down, downs, downland 1. an open expanse of gently undulating, elevated land, usually of chalk and supporting PASTURE, typically the treeless CHALK uplands of south and southeastern England **2.** in Australia and New Zealand, midlatitude grasslands **3.** The Downs, the name given to part of the North Sea near the Goodwin Sands, off the east coast of Kent, England. G.

downstream 1. *adj.* at a location relatively nearer to the outlet of a stream **2.** *adv.* moving in the direction of the flow of, towards the outlet of, a stream.

downthrow in geology, the subsidence of rock strata on one side of a fault, the strata being lowered on the downthrow side. THROW OF A FAULT.

downtown (American) the main business district of a town or city. CENTRAL BUSINESS DISTRICT.

downward-transition region CORE-PERIPHERY MODEL.

down-warping a smooth, downward deformation or sagging of the earth's crust caused by the pressure of weight of a widespread and great mass of material, such as a continental ice sheet (e.g. as in the Great Lakes margin of the Canadian Shield) or sediments (e.g. as underneath the Mississippi delta). When the great weight is removed the crust recoils and in many cases large shallow lakes result. In North America down-warping and recoil contributed to the formation of the Great Lakes. GEOSYNCLINE.

downwind 1. *adj.* situated to leeward. LEE **2.** *adv.* on the leeward side, in the same direction as the wind.

dråg (Swedish; German Rüllen) a narrow strip of FEN running between MOSSES-2 or crossing a moss.

drain 1. an artificial channel (in some

cases open, in others consisting of a pipe) for carrying off liquid, especially excess water from land, or sewage from a building **2.** in Fenland, a wide, canal-like navigation channel **3.** in WATER-MEADOWS, water is led on to the land by small channels termed drains and taken off by those termed drawns. G.

drainage **1.** the act of taking off excess water from the land by artificial channels. DRAIN **2.** the natural run-off of water from an area by streams, rivers etc. CONSEQUENT, OBSEQUENT, SUBSEQUENT DRAINAGE. The terms applied to the drainage pattern, system or network, i.e. to the arrangement of the main river and its tributaries, include ACCORDANT, ANTECEDENT, CENTRIPETAL, CONCORDANT, DENDRITIC, DERANGED, INCONSEQUENT, INLAND, INSEQUENT, PARALLEL, PINNATE, RADIAL, RECTANGULAR, RESEQUENT SUPERIMPOSED, TRELLIS. G.

drainage area all the land with a common outlet for its surface water, synonymous with RIVER BASIN if the river flows into the ocean; but if several rivers flow into an inland sea the whole area draining to that sea may be included. G.

drainage basin the tract of land drained by a sole river system.

drainge wind a KATABATIC WIND.

draw (American) a blind CREEK. G.

drawn (southern England, local term) a small channel which drains, or draws, off the water from a WATER-MEADOW. DRAIN-3. G.

dray horse a large, strong HORSE, harnessed to pull a dray (a low flat cart with or without sides), used for carrying heavy loads.

Dreikanter (German, three cornered) a pebble that is polished by wind-driven sand blown from different directions in a desert, so that it develops three distinct faces. EINKANTER, VENTIFACT, ZWEIKANTER. G.

driblet cone SPATTER CONE.

drift **1.** in geology, transported super-ficial deposits, especially those transported and deposited by ice, the two main types being stratified drift (FLUVIOGLACIAL DEPOSITION) and unstratified (BOULDER CLAY, TILL). In the British Geological Survey 'drift' maps cover all superficial deposits; the 'solid' edition maps cover the solid BED ROCK. SOLID GEOLOGY **2.** slow movement, e.g. of surface waters in the ocean under the influence of prevailing winds, less distinct than a CURRENT **3.** the movement, and accumulation, of loose material such as snow (SNOWDRIFT) or sand (SAND DRIFT, SAND DUNE) caused by wind **4.** in mining, a passage underground, usually horizontal, but commonly applied to one at any angle other than the vertical (if vertical it is termed a shaft), that follows a coal seam, etc.; also loosely applied to the main exploratory workings of a given level **5.** in many parts of Africa, especially in the south, a stream in flood, especially if flowing over a ford or sudden dip in the road. CONTINENTAL DRIFT. G.

drift-ice pieces of detached floating ice, moved by the wind or sea currents and separated by water, thus easily navigable. ICE EDGE. G.

drift net a large fishing net, held down and open by weights at the bottom and floats at the top, that moves with the tide.

driftway, drift-way a term applied in parts of rural England to a lane or road along which cattle used to be (are) driven. DROVE-ROAD. G.

drizzle a very fine rainfall, with raindrops less than 0.5 mm (0.02 in) in diameter, falling continuously, especially associated with a WARM FRONT. MIZZLE.

dromedary *Camelus dromedarius*, the one-humped CAMEL.

drought a prolonged, continuous period of dry weather, classified in British meteorology as ABSOLUTE DROUGHT, PARTIAL DROUGHT and DRY SPELL. G.

drove-road, drove-way a driftway, driftway, an ancient road or track along which there is free right of way for cattle but which is not necessarily kept in order by any authority. Hence drover, one who drives droves of cattle, sheep etc. to a distant market, and is thus a dealer in cattle. G.

drowned valley a valley which was excavated in a land surface but owing to a change in sea-level has been partly or wholly drowned by the sea. CONCORDANT COAST, FJORD, RIA. G.

drumlin (Irish) a smooth, oval, low hill or mound composed mainly of BOULDER CLAY or glacial sands and gravels, occurring in a once-glaciated region, the long axis in line with the movement of the ice that deposited it. Some authors include rock drumlins, smoothed mounds of rock with or without their veneer of boulder clay. Drumlins often occur in groups (swarms) as a drumlin field or BASKET OF EGGS RELIEF. FALSE DRUMLIN. G.

drumlinoid *adj.* resembling a DRUMLIN.

drupe a succulent fruit in which the wall consists of an outer skin containing a fleshy layer of varying thickness enclosing a hard stony layer, within which lies a single seed. It is commonly called a stone-fruit (e.g. the fruit of CHERRY, PEACH, PLUM). BERRY, NUT.

dry *adj.* lacking moisture, specifically defined when applied to air (less than 60 per cent RELATIVE HUMIDITY) and to climate (generally when evaporation exceeds precipitation, but more detailed specifications for 'dry' climates have been used). ARID. G.

dry adiabatic lapse rate the rate of loss of temperature with increasing height occurring in an unsaturated body of air as it ascends adiabatically (ADIABATIC), about 1°C in 100 m (5.4°F in 1000 ft) of ascent. ENVIRONMENTAL LAPSE RATE, LAPSE RATE, SATURATED ADIABATIC LAPSE RATE.

dry-bulb thermometer an ordinary mercury THERMOMETER used together with a WET-BULB THERMOMETER to discover RELATIVE HUMIDITY.

dry delta an ALLUVIAL CONE or ALLUVIAL FAN. ALLUVIUM. G.

dry (graving) dock a narrow basin into which a vessel passes and from which water is then pumped, leaving the vessel out of the water, dry, for repair. DOCK.

dry farming a farming practice that involves special treatment of the land to overcome a shortage of water. One method is to crop the land only every two years, conserving at least part of the rainfall of one year to add to that received in the next by pulverizing the soil surface or by protecting it by a mulch (a layer of straw or decaying plant leaves etc.).

dry gap WIND GAP.

drying oil an oil that has the property of drying and forming a thin elastic film on exposure to air, e.g. LINSEED OIL, used industrially.

dry point settlement a settlement on a site not liable to flooding in a flood region, or on a patch of dry soil in a wet soil region. G.

dry spell 1. in UK, any period of DROUGHT **2.** formerly precisely defined as 15 consecutive days on none of which more than 1.0 mm (0.04 in) of rain has fallen (definition now obsolete) **3.** in USA, a period of 14 days without measurable precipitation. ABSOLUTE DROUGHT.

dry stone wall a wall, usually of natural stone, built without mortar, to mark boundaries, especially in southwestern and northern Britain.

dry valley a valley, originally carved by water (especially in CHALK and LIMESTONE), which no longer has a running stream, though a BOURNE may flow after heavy precipitation. There are many theories about the origin of dry valleys, including a slow lowering of the WATER TABLE resulting from lowered precipitation; or the divergence of a stream that formed the valley (RIVER CAPTURE); or a change in climatic conditions (e.g. in Pleistocene glaciations); or the cutting back of an ESCARPMENT with resultant lowering of the SPRING-LINE; or surface erosion under PERIGLACIAL conditions; or SPRING SAPPING; or, in limestone, the disappearance of a former surface stream down a JOINT, or the collapse of an underground cavern. It is also possible that some small dry valleys in chalk were formed not by stream erosion but by the enlargement of lines of structural weakness, e.g. some joints enlarged by frost, the debris being moved away by SOLIFLUCTION. A small dry valley formed in this way is termed a vallon de gélivation in France. G.

dualism the quality or state of consisting of two distinct parts, e.g. any theory which considers the ultimate nature of the universe to consist of two irreducible elements (such as mind and matter); or the condition in the economy of a country where a relatively small group of well-educated, affluent, socially and economically advanced controlling elite live in the central city or in the larger towns, where the economy is dynamic and growing, industry (supported by large injections of capital) uses modern production techniques and management and is capital intensive, labour is specialized, commercial exchange is extensive and complex, the professions are gathered together and salaries are high, while the majority of the population (who are much poorer) live in the countryside where the economy is static, industry is labour intensive, techniques are traditional, trade with other areas is limited and services are inadequate.

dual labour market a concept of an economy as being divided into two parts, the primary sector comprising a highly and specifically trained, highly skilled, well paid workforce, the secondary sector comprising a low paid workforce with a low level of skills and training, vulnerable to unemployment.

dual relationship the dual role of a commodity (e.g. housing) in relation to the production process. Taking the example of housing and the production process in an industrial society where some housing is produced for profit, the suggestion is that housing has to be produced not only for the accumulation of profit but also for consumption by workers so that they stay fit and healthy and are able to play their part in the production process.

duār TERAI.

dub (northern England) a deep pool of still water in the course of a swift stream. The plural, dubs, is applied to a resurgence in limestone. G.

duck any of the small web-footed water BIRDS of the family Anatidae (the female of the species being termed a duck, the male a drake), widely distributed, often migratory in the wild, the domestic variety being descended from the mallard. The flesh and eggs provide food for humans, the feathers being used in quilts etc. or processed for animal feed, etc.

duck-billed platypus PLATYPUS.

ductile *adj.* of metals, capable of being pressed or drawn into shape without the aid of heat. MALLEABLE.

dude ranch a RANCH organized for the entertainment of tourists, as a place for a holiday, mainly in the USA.

duff 1. in USA soil science, the surface layer of decomposing vegetable matter resting on mineral soil or on the forest floor 2. in the UK, raw humus. G.

dulse, dulce *Rhodymenia palmata*, edible seaweed, used fresh in salad or washed and dried for later use, to be cooked or chewed. It is sometimes used to make an alcoholic beverage.

dumb-bell island an island consisting of two parts, often rocky, joined by a narrow isthmus, often of sand, which is never in any part of its length below high water mark. TOMBOLO. G.

dump moraine   PUSH MORAINE.

Dumpy level   an instrument consisting of a spirit level with a short telescope and a compass, used for LEVELLING in survey work.

dūn, dhun, dhoon   (Indian subcontinent: Urdu-Hindi from Pahari dialect) a longitudinal intermontane valley. The term, with various spellings, also appears in place-names. G.

dune   a hill or ridge of sand piled up by the wind in dry regions (desert dunes) or along sandy coasts, often independently of any fixed surface feature which might form an obstacle. The form depends on the presence of such an obstacle (which may provide a nucleus), the type and quantity of sand, the characteristics of the land surface over which the sand is moved, the strength and direction of the wind, the presence or absence of ground water, and of vegetation which 'fixes' the sand. Coastal dunes are particularly affected by the presence or absence of vegetation and of welling-up ground water as well as by erosion by the sea. They are identified, in sequence from the sea inland, as FORE-DUNES, MOBILE DUNES, STABILIZED DUNES. For other dune details see ADVANCED DUNE, ANTIDUNE, ATTACHED DUNE, BARCHAN, HEAD DUNE, ICE BARCHAN, LATERAL DUNE, LONGITUDINAL DUNE, PHYTOGENIC DUNE, PLINTH, SEIF DUNE, SLIP-FACE, TAIL DUNE, TRANSVERSE DUNE, WAKE DUNE. G.

dunite   ULTRABASIC ROCK consisting mainly of OLIVINE.

duopoly   the exclusive control of the supply of a product or service in a particular market by two suppliers, who thus dominate the market and between them control the price and scale of the supply. MONOPOLY, OLIGOPOLY, PERFECT COMPETITION.

dura   SORGHUM.

durable goods   goods that are not likely to wear out or decay for a long time,

e.g. carpets, furniture, to buy which the consumer is prepared to travel some distance. CENTRAL PLACE THEORY, CONVENIENCE GOOD, ORDER OF GOODS.

durian   *Durio zibethinus*, a large tree, native to Malaysia, grown there and elsewhere in southeast Asia for the sake of its large fruits, renowned for their sewage-like smell but the delicious taste of the creamy pulp (the aril) surrounding the seed, which is eaten raw and deteriorates quickly.

duricrust   a hard crust covering the soil surface in semi-arid, flat areas with a short rainy season and a long, hot dry season. It consists of aluminous, calcareous, siliceous, ferruginous and magnesian materials, drawn to the surface by CAPILLARITY, which brings to the upper soil during the dry season the minerals dissolved during the wet season. At depth it forms duripan (CALICHE, HARD PAN). FERRICRETE.

duripan   CALICHE, HARD PAN.

durra   *Sorghum vulgare*, a MILLET grown as a food crop in north Africa and southern Asia.

dusk   TWILIGHT.

dust   minute particles of any comminuted dry matter, so fine that they can float in air. AEROSOLS-1, COSMIC DUST, HYGROSCOPIC DUST, VOLCANIC DUST. G.

dust bowl, dust-bowl   a semi-arid tract of land from which the surface soil, exposed by the unwise removal of the covering grassy vegetation by ploughing or overgrazing, has been blown away by wind (DEFLATION). The term became widely used after two or three very dry years (especially 1934-5) in southwestern USA, when strong winds raised huge DUST-STORMS in areas where grassy vegetation, formerly protecting the soil, had been removed by ploughing and the land had been cultivated without the necessary protection of WINDBREAKS. SOIL EROSION. G.

dust devil, dust-devil a local swirl of wind, laden with DUST, forming a fast-moving pillar of dust, varying in breadth and height, common in most arid lands, especially in hot deserts. It is created by extreme, localized heating of the land surface, leading to strong CONVECTION currents, which gather up the dust. WHIRLWIND.

dust storm, dust-storm a broad, general term applied to a strong dust-laden wind in arid and semi-arid regions, arising when the air is very hot, excessively dry and accompanied by high electrical tension. The turbulent wind gathers DUST from the dry surface and carries it to heights up to 3000 m (10 000 ft), sometimes producing a wall of dust, sometimes a dust-laden WHIRLWIND, larger than a DUST DEVIL. BRICKFIELDER, DUST BOWL, SANDSTORM, SIMOON. G.

dust-well a small depression on the surface of a GLACIER, caused by a collection of DUST. The dust particles absorb heat from the sun, melt the ice that encircles them, and sink into the glacier, thus forming a small hollow.

dy (Swedish) muddy material consisting of plant residues deposited from water poor in nutrients. G.

dyestuff any substance used as, or yielding, a dye (a substance capable of colouring materials, e.g. textiles, paper, plastics). At present most consist of the products of the chemical industry, a large range being derived from COAL TAR. Of the age-old natural dyes the dark red comes from logwood, a large central American tree; FUSTIC is a yellow dye from the mora-wood of Nicaragua; indigo comes from the madder plant once very important in the Indian subcontinent; COCHINEAL comes from the dried bodies of an insect.

dyewood any wood from which dye can be obtained. DYESTUFF.

dyke, dyke-phase, dyke-spring, dyke swarm DIKE, RING-DYKE.

dynamic *adj.* of or related to motion or force. DYNAMICS.

dynamic climatology the study of climate based on its relationship with the circulation of the atmosphere, energy processes and the DYNAMICS and THERMODYNAMICS of the atmosphere. CLIMATOLOGY, METEOROLOGY, SYNOPTIC CLIMATOLOGY.

dynamic differential disequilibrium DIFFERENTIAL DISEQUILIBRIUM.

dynamic equilibrium a state in which balance is maintained despite continual change, e.g. on a slope where the rate of weathering of the rock is balanced by the rate of removal of the weathered rock material. STEADY STATE.

dynamic geology a subdivision of PHYSICAL GEOLOGY, concerned with the causes and processes of geological change. STRUCTURAL GEOLOGY.

dynamic lapse rate a rate of adiabatic temperature change associated with rising parcels of air. ADIABATIC, ADIABATIC GRADIENT, ENVIRONMENTAL LAPSE RATE.

dynamic metamorphism, dynamo-metamorphism the alteration of pre-existing rocks by intense pressure associated with earth movements, usually on a relatively small scale and without a great rise in temperature, so that new, well-defined rock is formed. CATACLASTIC, FOLIATION, LAMINA, METAMORPHIC ROCK, METAMORPHISM, THERMAL (CONTACT) METAMORPHISM. G.

dynamic rejuvenation REJUVENATION caused by EPEIROGENIC uplift of a landmass with accompanying tilting and warping. EUSTATIC REJUVENATION, STATIC REJUVENATION.

dynamics 1. the branch of mechanics that deals with matter in motion (kinematics) and the forces that produce or change such motion (kinetics) **2.** the branch of any science concerned with forces.

dynamic spatial model a MODEL-4 which seeks to improve the under standing of the evolution of phenomena (PHENOMENON) through time under various degrees of spatial constraint.

dyne an absolute unit of force: that force which, when applied to a mass of one GRAM, produces an acceleration of one CENTIMETRE per second per second. On SI 10^5 dynes = 1 NEWTON. At sea-level at 45°N and 45°S a mass of 1 gram is subjected to a gravitational force of 980.616 dynes.

dyngja (Iceland) a volcano with gentle slopes, formed by successive outpourings of fluid lava unaccompanied by accumulations due to violent ejection. G.

dysgeogenous *adj.* applied to rocks that do not easily decompose into soil, or yield only a little DETRITUS, as opposed to EUGEOGENOUS. G.

dysphotic DISPHOTIC.

dystrophic *adj.* applied to a body of freshwater poor in plant nutrients and low in calcium, occurring typically in acid peat areas, the bed of the water being covered with undecomposed plant remains. EUTROPHIC, MESOTROPHIC, OLIGOTROPHIC, TROPHIC LEVEL.

E

eagre, egre a tidal wave or BORE.

earth 1. the PLANET on which we live, a flattened sphere (OBLATE SPHEROID) in orbit round the sun, fifth in size and third in order from the sun of the nine planets of the SOLAR SYSTEM. The polar diameter is 12 712 km (7899 mi), the equatorial diameter 12 755 km (7926 mi); the polar circumference 40 008 km (24 860 mi), the equatorial circumference 40 076 km (24 902 mi). It is generally agreed that the surface area is 510 100 448 sq km (196 949 980 sq mi), of which 361 059 266 sq km (139 405 122 sq mi) is water (70.78 per cent of the total surface) and 149 041 182 sq km (57 544 858 sq mi) is land (29.22 per cent of the surface); but some authorities give the surface area as 509 610 000 sq km (196 836 000 sq mi), of which 148 065 120 sq km (57 168 000 sq mi) is land. The mean density is 5.517; the mass is 5.882×10^{21} tonnes. ATMOSPHERE, BARYSPHERE, CORE, CRUST, HYDROSPHERE, HYDROLOGICAL CYCLE, LITHOSPHERE, MANTLE, ORBIT OF THE EARTH, PLATE TECTONICS, ROTATION OF THE EARTH, TERRESTRIAL MAGNETISM **2.** the solid material of that planet, as distinct from air and water **3.** the disintegrated, loose material on the surface of it, the soil as distinct from the solid rock **4.** specifically, the proper name for some amorphous, fine-grained deposits, e.g. FULLER'S EARTH. G.

earthflow a slipping downwards of unconsolidated rock material on the earth's surface, due to its saturation by water and the influence of gravity, occurring on steep and shallow slopes. MASS MOVEMENT, MUDFLOW, SOLIFLUCTION. G.

earth-movement 1. a movement of the earth's crust arising from disturbances in the earth's interior (COMPRESSION, DEPRESSION-1, FAULTING, FOLDING, TENSION, UPLIFT), including both the slow (secular) movements and the sudden (EARTHQUAKES and volcanic activity) **2.** a synonym for OROGENESIS. DIASTROPHISM. G.

earth pillar a high pillar of earthy material or soft rock, sandstone etc. capped by a stone or boulder which protects the underlying soft, easily eroded material. Earth pillars are common in dry regions which are subject to occasionally heavy downpours. DEMOISELLE, HOODOO. G.

earthquake a shaking of the ground caused by deep-seated disturbances, producing a series of elastic shock waves spreading outwards from the EPICENTRE. An earthquake usually originates from sudden adjustments in the crust of the earth, notably by movement along FAULTS (and thus of tectonic origin), or as a result of volcanic activity. Most severe earthquakes are associated with fault lines where there are no VOLCANOES to act as safety valves. The shock waves are classified as P (primary, PUSH WAVE), a body wave within the earth; S (secondary, or SHAKE WAVE); and L (surface, LONGITUDINAL WAVE). The degree of magnitude is usually measured on the RICHTER SCALE and of intensity (related to the effects of waves at the surface) on the MODIFIED MERCALLI SCALE. ASTHENOSPHERE, ELASTIC REBOUND, HOMOSEISMAL LINE, PLATE TECTONICS, SEISMIC FOCUS. G.

Earth Resources Technology Satellite ERTS. LANDSAT.

earth sciences a collective term covering the scientific studies concerned with the earth as a planet (i.e. excluding biological sciences), including ARCHAEOLOGY, ASTRONOMY, CLIMATOLOGY, GEOCHEMISTRY, GEODESY, GEOGRAPHY, GEOLOGY, GEOMORPHOLOGY, GEOPHYSICS, HYDROGRAPHY, HYDROLOGY, METEOROLOGY, MINERALOGY, OCEANOGRAPHY, PALAEONTOLOGY, helped by the techniques of CARTOGRAPHY, REMOTE SENSING, SURVEY and MATHEMATICS. See also BEHAVIOURAL SCIENCES, HUMANITIES, NATURAL SCIENCES, PHYSICAL SCIENCES, SOCIAL SCIENCES.

earth's crust CRUST.

earth's magnetic field TERRESTRIAL MAGNETISM.

earth tremor a small, low intensity EARTHQUAKE.

east 1. one of the four CARDINAL POINTS of the COMPASS, the direction at which the sun rises at the EQUINOX **2.** *adv.* towards the area lying in that direction from the observer, hence, from Europe, the NEAR EAST, MIDDLE EAST, FAR EAST **3.** *adj.* of, pertaining to, belonging to, coming from, or situated towards, the east, e.g. of winds blowing from that direction.

eastern hemisphere HEMISPHERE.

Eastern Question a term applied to the international political problems arising from the breaking-up of the Ottoman Empire in the BALKANS and eastern Mediterranean in the nineteenth century.

easting 1. the first part of a grid reference (GRID), the distance east on a map as measured from a point fixed in its southwest corner. NORTHING **2.** nautical, a sailing towards the east, or

the distance so travelled since the last reckoning point.

ebb the drawing back of tidal water from the shore.

ebb and flow the backward and forward movement of tidal water in relation to the shore. SLACK-2.

ebb channel 1. the channel in which the tide flows out most strongly in a river estuary **2.** the route taken by a TIDAL STREAM as it flows seaward after high tide, in some cases differing from the flood channel, the route taken by the flood tidal stream. G.

ebb tide, ebb-tide the receding or outward movement of tidal water, after high tide and before low tide, in contrast to the FLOOD TIDE. SLACK-2. G.

ebony the wood of any of the trees of the family Ebanaceae, especially of the genus *Diospyros*, native to the tropical forests of Asia and Africa, which is blackened by deposits of gum resin in the heartwood. Heavy and durable, it makes fine CABINET WOOD. IRONWOOD.

éboulis (French) a SCREE. G.

ecesis the successful establishment of a pioneer plant community, an early stage in a PRISERE.

echo sounder an instrument used especially for calculating the depth of water by measuring the time taken for a sound (sonic or ultra-sonic vibration) generated at the surface to return after being reflected from the sea floor (SOUNDING). It is also used to measure the thickness of ice and the depth of different densities of rock or of soil.

eclipse the total or partial cutting off of light received by one celestial body from another by the interception of a third celestial body in passing between the other two. From the earth the eclipse of the sun (solar eclipse) occurs when the moon comes into line between the sun and the earth and casts a shadow on the earth; the eclipse of the moon (lunar eclipse) when the earth intercepts light from the sun to the moon and casts a shadow on the moon.

ecliptic the apparent path of the sun in the CELESTIAL SPHERE during the period of one year as seen from the earth, and relative to the fixed stars. The path makes a great circle, at an angle of 23°27′ with the CELESTIAL EQUATOR; and it is divided into 12 imaginary sections, each identified by a sign of the ZODIAC.

ecliptic *adj.* **1.** of or pertaining to the ECLIPSE of a celestial body **2.** relating to the ECLIPTIC.

eclogite a rather coarse-grained green and red gabbroid rock (GABBRO) composed chiefly of PYROXENE and GARNET which has been affected by DYNAMIC METAMORPHISM, occurring as XENOLITHS in BLUE GROUND and in certain METAMORPHIC belts.

ecoclimate climate considered in relation to plant and animal life. G.

ecological analysis an approach to geography in which the interrelationship of human and environmental VARIABLES are studied and their links interpreted. REGIONAL COMPLEX ANALYSIS, SPATIAL ANALYSIS.

ecological area in urban geography, the association of a particular class, or ethnic, religious or other cultural group, with a particular residential area of a town.

ecological balance the balance of nature, the balance maintained in a stable ECOSYSTEM by the gradual readjustments in the composition of a balanced community in response to natural SUCCESSION, changes in climate, or other influences. Such a delicate balance is easily upset by human activities, e.g. by the introduction or elimination of plants or animals, by pollution of the environment, by the destruction of habitats etc.

ecological efficiency the ability of an organism in a FOOD CHAIN to convert its energy intake into living material.

ecological factor any environmental factor which affects a living organism.

ecological fallacy the false assumption that characteristics or relationships observed in aggregated (AGGREGATE-1) data are also present in the individual data from which the aggregated data is produced. HISTORICAL FALLACY.

ecological validity a research finding held to be true in a range of natural settings or conditions, as distinct from one that is true in artificial settings or conditions only.

ecology the scientific study of the interrelationships between living organisms and the ENVIRONMENT (including the other living organisms present) in which they live. Without qualification the term ecology tends to be confined to plant ecology. Ecologists use many special terms in their studies, particularly in connexion with plant communities (ASSOCIATION, CONSOCIATION) which have developed in harmony with the environment, the idea of development (CLIMAX, SERE, SUCCESSION) being fundamental. ECOTONE, ECOTOPE, ECOSYSTEM. Geography has been defined as HUMAN ECOLOGY, the interrelationship between people and place. AUTECOLOGY, FACTORIAL ECOLOGY, SYNECOLOGY, URBAN ECOLOGY. G.

econometric history an approach to the study of economies in the past (ECONOMIC HISTORY) in which the collection of data, QUANTITATIVE ANALYSIS and COUNTERFACTUALS are used to test the validity of hypotheses.

econometrics economic studies based on observations, the techniques employed in statistical estimation, and the mathematical treatment of economic theories.

economic base theory a theory based on the assumption that urban and regional growth can be explained in terms of the numbers employed in the BASIC ACTIVITIES, responsive to external demand, and the NON-BASIC ACTIVITIES, meeting the internal demands of the city or region itself. LOWRY MODEL.

economic determinism in Marxism, the theory that the economic processes of a society have a determining effect on its other processes, particularly, for example, on the political processes. TECHNOLOGICAL DETERMINISM.

economic good in economics, a good of economic value, i.e. it is useful, scarce and marketable. Economic goods are usually classified in two categories, CONSUMER GOODS and services, and PRODUCER GOODS and services. CAPITAL GOODS, CONSUMER DURABLES, COMMODITY, FREE GOODS.

economic geography the branch of geography dealing with the interaction of geographical and economic conditions, with the production, spatial distribution, exchange and consumption of wealth, and with the study of the economic factors affecting the AREAL DIFFERENTIATION of the earth's surface. INDUSTRIAL GEOGRAPHY. G.

economic growth the increase in real national income or (more commonly) in real national income per head over a long period of time.

economic history the history of economies in the past. ECONOMETRIC HISTORY.

economic man in classical economics, a person who is motivated solely by economic considerations, who manages personal income and expenditure strictly in accordance with personal, material interest, profit being the only objective. This concept assumes perfect knowledge of relevant circumstances and a perfect ability to use that knowledge in order to take the greatest advantage of it; and the gift of totally accurate prediction in order to achieve the lowest possible costs and the highest profits. BOUNDED RATIONALITY, CLASSICAL ECONOMIC THEORY.

economic rent 1. the net surplus available to any factor of production (e.g. capital, labour, land) after the deduction (a) of all the costs, including interest on invested capital, involved in keeping it in its present use; and (b) the returns possible from an alternative use of the factor of production (termed opportunity costs). This concept of economic rent is used particularly in agricultural geography to account for the farmer's decision to use land in one way rather than another (based on the assumption that the farmer will opt for the use yielding the higher economic rent) 2. a variation of that application, used in studies of agricultural location, to account for the fact that the land is farmed at all. In such studies opportunity costs are not taken into account, ignored on the assumption that only one use of the resource is possible, i.e. agriculture. The economic rent then becomes the net income earned by the farmer in excess of the net income which might be earned from producing at the margin. The margin is the point where the level of net income justifies the use of land for agriculture, and where a true economic rent (defined above) is non-existent. The position of the margin is usually governed by the level of transport costs involved in marketing, which increase with increasing distance from the market. David Ricardo, the nineteenth century classical economist, held this same concept of economic rent, but he considered variations in soil fertility resulted in variations of land use, resulting in differences in economic rent 3. a synonym for LAND RENT. RENT, VON THUNEN MODEL. G.

economics a term with a wide range of applications, broadly summarized as 1. the scientific study relating to the production and distribution of material wealth and to human behaviour working within the limits imposed by the relative scarcity of the resources available. MACRO-ECONOMICS, MICRO-ECONOMICS, SOCIAL SCIENCES 2. the science as applied to an organization, industry, etc. 3. the principles of making profit, saving money, producing wealth.

economies of localization ECONOMIES OF SCALE.

economies of scale the lowering of unit costs achieved by increasing the scale of production. The reduction in unit costs is brought about by internal economies, i.e. economies within the enterprise (e.g. greater specialization and division of labour, the spreading of research, development and other fixed costs over more production units) and from the external economies (economies of localization) which arise when firms in the same or similar industries are located close together, thereby benefiting from the availability of a skilled labour force, specialist services, supplies, infrastructure, marketing etc. (AGGLOMERATION-3), or when the growth of an entire industry reduces the costs in each individual firm. DISECONOMIES OF SCALE, ECONOMIES OF URBANIZATION.

economies of urbanization the ECONOMIES OF SCALE achieved by a wide range of industries from the circumstances of URBANIZATION-1, i.e. the well-developed physical structure and services, large labour force with diverse skills, large potential market. DISECONOMIES OF URBANIZATION.

economy **1.** a SYSTEM-5 of production and distribution designed to meet the material needs of a country, region or society. CLOSED ECONOMY, MARKET ECONOMY, OPEN ECONOMY, SUBSISTENCE ECONOMY **2.** a part of such a system, e.g. agricultural economy. G.

ecoparasite a PARASITE which is adapted to live on a specific host or on a group allied to that host.

ecospecies one or more ECOTYPES in one COENOSPECIES.

ecosphere the BIOSPHERE-2 and all the ECOLOGICAL FACTORS which affect organisms. G.

ecosystem ecological system, a SYSTEM-2 formed by the interaction of all living organisms (plants, animals, bacteria etc.) with each other and with the chemical and physical factors of the environment in which they live, all being linked by the transfer of energy and materials (FOOD CHAIN). The boundary of an ecosystem is difficult to define (the whole world may be considered as an ecosystem), but the term is usually applied to a small system where the net transfer of energy and materials across the boundary is low, e.g. a pond, a forest, a small oceanic island. An ecosystem is never totally self-contained or closed: solar energy received crosses the boundary, as does a foraging animal. The part of the world which forms the home for an ecosystem is termed an ecotope; but ecotope is sometimes used as a synonym for ecosystem. G.

ecosystematics the study of ECOSYSTEMS. G.

ecotone a transitional zone between two HABITATS where different plant associations merge. MICTIUM. G.

ecotope ECOSYSTEM.

ecotype a sub-specific group (sometimes classified as a SPECIES) genetically adapted to a particular HABITAT but able to breed with other ecotypes or ECOSPECIES of the same species or COENOSPECIES and still remain fertile.

ectoparasite a PARASITE which lives on the outside of its host for periods of long or short duration.

ecumene, oecumene, ekumene, oikoumene **1.** the habitable world known to the ancient Greeks **2.** the part of the earth's surface suitable, through its climatic conditions, for permanent human settlement. ANECUMENE. G.

ecumenopolis the city of the future, covering most of the habitable surface of the earth as a continuous system, forming a universal settlement. The term was introduced in 1961 by C. A. Doxiadis, who saw the natural hierarchy of large urban settlements as large CITY, METROPOLIS-2, CONURBATION, MEGALOPIS, ecumenopolis. G.

edaphic *adj.* related to, due to, dependendent on, or having characteristics due to, the nature of the soil. G.

edaphic climax the vegetation CLIMAX produced by EDAPHIC FACTORS. G.

edaphic factors soil factors, the biological, chemical and physical properties of a soil. EDAPHIC CLIMAX. G.

edaphic formations vegetation formations classified according to the soil types that determine them, as distinct from CLIMATIC FORMATIONS. G.

edaphology the branch of SOIL SCIENCE concerned with the scientific study of soil in relation to plant growth.

edaphon the whole living community, the plant and animal life, of a soil. G.

eddy a swirling movement of fluid within a larger mass of fluid, in a direction contrary to that of the main flow, e.g. as in depressions or highs in the atmosphere, or in the water of a river as it encounters an obstruction in its flow. HYDRAULIC FORCE.

edge 1. loosely, a sharp ridge, especially one with an exposure of rock, a topographic term not generally used specifically in geographical writing, but commonly used in place-names to indicate an ARETE, ridge or mountain crest, e.g. Cross Fell Edge in the Pennines, Wenlock Edge in Shropshire, England **2.** in mathematics, a link, a route in a TOPOLOGICAL diagram. GRAPH-2, GRAPH THEORY.

EEC, European Economic Community, Common Market a group of European countries established 1 January 1958 on the basis of a treaty signed in Rome 25 March 1957. The aims included increased productivity, free mobility of labour, coordinated transport and commercial policies, and control of restrictive practices among members.The original six members, Belgium, France, Federal Republic of Germany, Italy, Luxembourg and Netherlands, were joined 1 January 1973 by the UK, Denmark (with Greenland) and the Irish Republic. Greenland, on gaining independence

from Denmark, withdrew on 23 February 1982. Customs duties on trade between the six were phased out by 1 July 1968, between the nine by 1 July 1977. Greece joined 1981; Spain and Portugal 1986. Some developing countries have associate status in the EEC, others have trading agreements with the Community. The aims, operation and membership of the EEC are detailed in the *Statesman's Year-Book*, Macmillan. CAP.

effective accessibility the extent to which a place or service is actually accessible, governed not only by the distance to be travelled (LOCATIONAL ACCESSIBILITY) but also by whether or not the means of transport, the time available, and social circumstances make an approach possible. ACCESSIBILITY, COST SPACE.

effective precipitation 1. that part of total precipitation which is of use to plants **2.** in hydrology, that part of total precipitation which flows into a stream channel. G.

efficiency of a stream the measure of a stream's ability to transport debris, taking into account its CAPACITY and COMPETENCE. A formula sometimes used is E = C/QS (C = capacity for stream traction in grams per second; Q = discharge in feet per second; S = slope of channel bed per cent).

efflorescence the surface crust of previously dissolved material left after the evaporation of water which has risen in the soil in arid regions. CAPILLARY FLOW. G.

effluent a pouring out, flowing away, hence a stream flowing out of a lake or out of a reservoir, or from land after irrigation; or the flow of waste liquid from sewage works or from a factory. G.

effluent cave an OUTFLOW CAVE. G.

EFTA, European Free Trade Association a group consisting originally of seven European countries (sometimes known as the outer seven in contrast to the original six countries of the EEC) which linked themselves together effectively in 1960 for the purposes of trade, aiming to abolish tariffs on imports of goods originating in the group. The original members were the UK, Norway, Sweden, Denmark, Portugal, Austria and Switzerland, joined by Iceland 1970, Finland as an associate member 1961. The UK and Denmark left 1972 on joining the EUROPEAN ECONOMIC COMMUNITY.

egalitarian society a SOCIETY-3 of equals, in which the members perform tasks for each other on the basis of need or ability, exchanging goods on a BARTER-1 system. It is considered that such a society, unable to accumulate a surplus, is economically and functionally incapable of becoming urbanized (URBANIZATION). RANK SOCIETY, STRATIFIED SOCIETY.

egocentrism the state or quality of being egocentric, of perceiving the world with the self as centre. In the ordering of the world from the egocentric viewpoint (i.e. with the individual perceiving self as centre), the value of its components, the perceived objects, declines rapidly with increasing distance from the individual. ANTHROPO-CENTRIC, ETHNOCENTRISM. G.

egre EAGRE. G.

E horizon SOIL HORIZON.

eider a sea duck of the genus *Somateria* and related genera, native to northern latitudes. The female has fine downy breast feathers which she plucks to cover the eggs and which, termed down, are used for a filling in bed covers etc.

Einkanter (German) a pebble on a desert surface polished by sand driven by wind blowing constantly from one direction, so that only one facet is cut. DREIKANTER, VENTIFACT, ZWEI-KANTER.

Einzelhof (German) an isolated farm, a term commonly used in describing rural landscapes. G.

Eis (German) ice. Some authors prefer to retain the German spelling in compounds such as Eisblink, instead of using what is really a partial translation in ICE BLINK. G.

eiscir (Irish) ESKER. G.

ejido (Spanish) **1.** a system of land tenure reform under which HACIENDAS were transferred to communal ownership in Mexico **2.** an agrarian property, held and worked in common, and belonging to Mexican villagers.

ek-fasli-harsala DO. G.

ekistics the science of human settlements, covering the settlement itself, its evolution and formation. G.

elastic *adj.* **1.** of or pertaining to any substance that returns to its original shape or size after being deformed by stress so long as the stress does not exceed the elastic character of the substance and the substance itself is not over-fatigued. INELASTIC, PLASTIC **2.** springy; having the power to bend or give without snapping **3.** adaptable to circumstances.

elasticity the state or quality of being ELASTIC.

elastic rebound the recoil of rocks to a position near to that of the original after they have been forced apart by stress and the strain has relaxed, as in FAULTS and EARTHQUAKES.

E-layer HEAVISIDE-KENNELLY LAYER.

elbasin an elevated basin, a term introduced by Griffith Taylor who deplored the fact that such a landform was often wrongly termed a plateau. G.

elbow of capture of a river, RIVER CAPTURE.

El Dorado a legendary place in South America, between the rivers Amazon and Orinoco, reputed to be rich in fabulous treasure, sought by the CON-QUISTADORES. The name is derived from el dorado (Spanish), the gilded man, the reputed ruler.

electoral geography the branch of HUMAN GEOGRAPHY concerned with the study of election statistics and of the geographical aspects of the organization, results and consequences of elections, valuable for studies in POLITICAL or SOCIAL GEOGRAPHY. G.

electric *adj.* of or pertaining to electricity, or producing, produced, or operated by, ELECTRICITY.

electric arc a luminous, continual electrical discharge producing very high temperatures (usually exceeding 3000°C (5420°F), occurring when an electric current is carried by the vapour of the ELECTRODE (or by ionized gas) between two separated electrodes. ARC FURNACE.

electricity, electrical energy, electrical power POWER-3 generated in dynamos in which energy is derived from turbines of two types: **1.** steam-turbines (producing thermal electricity) driven by heat from COAL (especially low-grade bituminous, lignite and brown coal), OIL, NATURAL GAS, PEAT, NUCLEAR ENERGY, GEOTHERMAL ENERGY (and, experimentally, SOLAR energy) **2.** hydraulic turbines (producing HYDROELECTRICITY) driven by WATER POWER, including tidal power (TIDAL POWER STATION). The output of electrical energy is commonly measured in kilowatt-hours (kWh).

electrode either of the two conductors (i.e. the anode or the cathode) by which an electric current is passed by an electrical circuit in an apparatus such as a discharge tube or an electrolytic cell. ELECTRIC ARC.

electromagnet a temporary MAGNET made by winding a coil of wire around magnetic material, e.g. a soft iron core, and passing an electric current through the coil. The MAGNETIC FIELD persists only so long as the electric current flows through the wire.

electromagnetic *adj.* of or pertaining to ELECTROMAGNETISM-1 or to an ELECTROMAGNET.

electromagnetic spectrum the range of wavelength (or of frequencies) of

ENERGY radiated by ELECTROMAGNETIC WAVES. The spectrum is divided into bands on the basis of the type of wave, i.e., in order of decreasing wavelength, RADIOWAVES (the longest), RADAR-1 waves, INFRA-RED RADIATION, VISIBLE LIGHT, ULTRAVIOLET RADIATION, X RAYS and GAMMA RAYS. See also COSMIC RAYS, ELECTROMAGNETISM, LIGHT, NEAR-INFRARED REGION.

electromagnetic wave a wave propagated through space or a medium by simultaneous periodic variation in the electric and magnetic field intensity at right angles to each other and to the direction of propagation. ELECTROMAGNETIC SPECTRUM, ENERGY.

electromagnetism **1.** the magnetic force produced by an electric current, e.g. in an ELECTROMAGNET **2.** the scientific study concerned with the interrelation of ELECTRICITY and MAGNETISM.

electron a fundamental particle, negatively charged, a constituent of the ATOM, orbiting in the NUCLEUS, in number equal to the atomic number of the element. Its antiparticle is the POSITRON. ANION, CATION, FREE RADICAL, NEUTRON, PROTON.

electronic *adj.* of or pertaining to an ELECTRON or electrons, or connected with apparatus that works or is produced by the action of electrons.

electronics the branch of physics concerned with the motion of free electrons and of electrons in motion in electrical circuits.

electrovalency VALENCY.

element **1.** generally, a small quantity of something, present in a larger whole **2.** a basic component of a branch of study **3.** in ancient and medieval philosophy, one of what were considered to be the basic constituents of the universe, i.e. earth, air, fire, water **4.** rhetorically, 'to brave the elements', to go outside in unpleasant weather, e.g. in a heavy snowstorm **5.** in biology, the natural habitat of an organism **6.** in chemistry, a substance that has never been separated into simpler substances by ordinary chemical means because all its atoms have the same atomic number. There are over 100 such substances known, any of which, individually or together, are present in all matter, the first ninety-two (up to and including uranium) occurring naturally.

elevation 1. the vertical distance above a specific level, e.g. above sea-level 2. the vertical angle between the horizontal and a high point, e.g. between the horizon and a star, or between a hill base and a hill top. ALTITUDE 3. in architecture, the view of one of the sides of a building, or a drawing of this view. G.

elfin forest FOREST-2 consisting of dwarf trees suppporting MOSSES-1, occurring at high altitudes in the zones of TROPICAL and EQUATORIAL climates.

elite 1. the few who are considered to be more skilled than, or socially, intellectually or professionally superior to, the rest of the people in a society or a group, such qualities of excellence or distinction being actual, claimed or presumed 2. a more-or-less coherent group of people who know and communicate with each other and, as a group, have access to power. ARISTOCRACY, RANK SOCIETY.

elite theory a theory postulated in the nineteenth century, which suggests that political power lies with a small, ELITE group, the rest of society being seen to be undifferentiated, with no effective influence on policy.

ellipsoid a SPHEROID with a form that is regularly oval. GEOID, OBLATE ELLIPSOID.

el Niño a warm ocean current, originating in the warm equatorial current, which every seven to fourteen years temporarily replaces the cold current (Humboldt current) off the Peruvian coast. This occurs when the southeast trade winds in the Pacific lose their strength; it leads to a fall in the quantity of PLANKTON associated with the cold Humboldt current, and thus to a decrease in fish numbers.

elongated *adj.* 1. made longer 2. long and slender.

elongation 1. a lengthening or being lengthened 2. something that has been ELONGATED 3. in astronomy, the angular distance of a planet from the sun or of a satellite from its primary. SCHUMM'S INDEX OF SHAPE.

elutriation 1. the separation of lighter particles of powder from the heavier by an upward current of a fluid, especially by air 2. the mechanical separation of variously sized grains of a SEDIMENT, for the purpose of analysis, achieved by passing them through water currents of differing velocity.

eluvial horizon the soil layer (broadly the A HORIZON) that is subject to ELUVIATION. HARD PAN, ILLUVIAL, ILLUVIATION, SOIL, SOIL HORIZON.

eluviation in SOIL SCIENCE, the process by which material (especially that consisting of BASES-2) is removed in SOLUTION-1 (LEACHING), or mechanically in water suspension, from the upper horizon or horizons of a SOIL by downward or lateral percolation of the water. ILLUVIATION, SUB-SURFACE WASH. In American usage eluviation is confined to the mechanical transport, leaching being the term applied to the process in which particles are moved in solution. G.

eluvium rock debris produced by weathering in situ, a term used more by American and continental European authors than by British. In speech the distinction between eluvium and alluvium is not always clear. G.

elvan an INTRUSIVE IGNEOUS ROCK, commonly quartz-porphyry, occurring in the GRANITE of southwestern England. The term is sometimes applied, as 'blue elvan', to DOLERITE. Elvan is used as an AGGREGATE-4 and as a building stone.

embayment 1. a widespread dip of

rocks on the margin of a continental mass, in some cases supporting a thick layer of sediments **2.** a projection of sedimentary rocks into a crystalline rock mass **3.** the action of forming a bay in a coast **4.** a well-rounded BAY in a coast.

embouchment, embouchure (French) **1.** the mouth of a river or creek **2.** the place where a river valley opens out on to a plain.

emerald green BERYL of gem quality, a PRECIOUS STONE.

emergence the process of coming forth from concealment, hence applied particularly to **1.** the rise of the level of the land in relation to the sea, so that land formerly under the sea becomes dry, e.g. a RAISED BEACH **2.** the point at which an underground stream comes to the surface.

emery a very hard, granular mineral consisting of a mixture of CORUNDUM, MAGNETITE and SPINEL, used in non-skid road surfaces, in non-slip paints and as an abrasive.

emigrant one who migrates (moves) voluntarily away from the land of birth to take up permanent residence in another country. EXILE, EXPATRIATE, IMMIGRANT, MIGRANT, MIGRATION, REFUGEE.

emir (Arabic amīr, a commander) an independent chief or ruler of a province, especially one in west Africa. G.

emirate the territory ruled by an EMIR. G.

empire **1.** a sovereign state which has enlarged the territory under its control by military or economic conquest, colonization or federation of territories formerly independent of it **2.** an aggregate of subject territories, contiguous or not, governed or dominated by a sovereign state. The term formal empire has been applied to the aggregate of such subject territories which come under the direct sovereignty of the dominant state; and informal empire to the aggregate of those territo-ries in which the more powerful sovereign state lacks formal political power, but dominates commerce and trade **3.** a commercial, industrial or financial organization which has great wealth, power and influence **4.** the sphere of influence of such an organization.

empirical *adj.* **1.** pertaining to, derived from, based on, or making use of, experience, trial and error, observation or experiment rather than theory or knowledge **2.** in mathematics, applied to a formula reached by inductive reasoning, not verified by deductive proof. DEDUCTION, INDUCTION.

empiricism **1.** the theory that all concepts are derived from experience and that all statements claiming to express knowledge depend on experience for their justification **2.** the philosophical system which considers true knowledge to be that which can be perceived and rejects that which cannot be verified (e.g. theoretical statements) **3.** the scientific method of proceeding by inductive reasoning (INDUCTION) from observation to the formulation of a general principle which is then checked by experiment. EMPIRICAL, LOGICAL POSITIVISM, POSITIVISM, SCIENTIFIC LAW.

empolder *verb* to reclaim land by the creation of POLDERS. G.

enclave (French) **1.** an outlying part of a nation state lying within the territory of another nation state, as seen from the point of view of the territory in which it lies. EXCLAVE **2.** a small cultural or linguistic group surrounded by another cultural or linguistic group which is dominant.

enclosure **1.** a place shut in, fenced or delimited in some other way, for a special purpose **2.** the fence etc. that performs this function. INCLOSURE.

enculturation the process by which individuals are brought up to be members of a CULTURE or SOCIETY-3. ACCULTURATION.

endemic *adj.* restricted to a certain

region or people, having originated there, applied especially to a disease that is normally confined and always likely to occur in certain areas, sometimes reaching EPIDEMIC proportions. PANDEMIC.

endemic relic RELIC DISTRIBUTION.

endive *Cichorium endiva*, an annual or biennial plant with a dense rosette of leaves, divided and curled in most varieties, native to southern Asia and northern China but now widely grown also in Europe, used especially in salads. The green leaves are bitter, so to make them palatable they are blanched in growth, by the exclusion of light.

end moraine TERMINAL MORAINE.

endogenetic, endogenic *adj.* arising from within, having an internal cause or origin, e.g. applied to the geological processes which originate from within the earth (e.g. DIASTROPHISM, VULCANICITY) and landforms arising therefrom, in contrast to EXOGENETIC, GENETIC. Endogenetic is the usual English form, endogenic the American; but American authors sometimes differentiate endogenetic (applied to the product, e.g. the rocks formed) from endogenic (applied to the process). G.

endogenous *adj.* loosely and figuratively, formed inside, in contrast to EXOGENOUS (similarly used), formed outside; specifically **1.** in biology, originating within an organism, growing from within, developing from a deepseated layer or within a cell **2.** in geology, formed within a rock mass or within the earth's surface, or proceeding from the interior of the earth. G.

endoparasite a PARASITE that lives inside its host.

endoreic *adj.* in-flowing, applied particularly to basins of inland drainage. AREIC, EXOREIC. G.

endothermic *adj.* having the quality of absorbing heat. EXOTHERMIC.

Endrumpf (German) the final stage of denudation of the land surface, not so much the PENEPLAIN (as defined by

the originator of the term, W. Penck) but the PEDIPLAIN of L. C. King. The term is not in general use. G.

energetics the study of the production of energy and its transformations.

energy in physics, the capacity to do work possessed by a physical body or system of physical bodies, i.e. the capacity to move a force a certain distance, measured in JOULES. The two main types are KINETIC and POTENTIAL, other forms being mechanical (of moving bodies, stretched springs), chemical (energy stored in molecules of compounds), thermal (energy of heat, including geothermal, derived from the interior of the earth; and from the internal random movement of molecules), nuclear (of the nucleus of the atom) and radiant (e.g. solar), carried by ELECTROMAGNETIC WAVES, the only form of energy existing in free space, i.e. in the absence of matter. Each form of energy can be transferred to another form by suitable means, but the transformation is possible only in the presence of matter. ELECTRICITY, NUCLEAR ENERGY, POWER, RADIATION.

energy flow in ecology, the passage of energy through the TROPHIC LEVELS-1 of a FOOD CHAIN.

energy pyramid in ecology, a BAR GRAPH in the form of a pyramid showing the energy lost and retained in the different TROPHIC LEVELS-1 of a FOOD CHAIN. Each trophic level is represented by a horizontal bar representing the total energy received by the organisms in that level; and each bar is divided to show how much of that energy received is retained by and lost by the organisms in that level. Such a graph is particularly useful in providing a quantitative picture of an ECOSYSTEM.

Engels' law a theory put forward by Friedrich Engels, 1820-95, German political philosopher and socialist, that as people become richer they spend proportionately less of their income on food and proportionately more on consumer goods and services.

englacial *adj.* embedded in a GLACIER, as distinct from SUBGLACIAL, SUPERGLACIAL. G.

englacial drift debris enclosed in and carried by the mass of a GLACIER. G.

englacial river a stream of melt-water flowing inside a mass of land ice. SUPERGLACIAL STREAM. G.

engulfment a process in which the walls of a volcanic CONE collapse inwards, occurring when the molten lava lying beneath the cone escapes not through the PIPE but by alternative routes. It is one of the processes by which a CALDERA is formed.

enhancement the processes and techniques used to sharpen the IMAGES-3 received by SENSORS in order to ease the interpretation of them.

Enterprise Zone one of the zones, designated according to the UK Finance Act 1980, in economically derelict urban areas. Various concessions, designed to attract industry to such areas include, e.g. a ten-year exemption from rates, exemption from Development Land Tax, 100 per cent capital allowances, simplified planning controls, few government requests for statistical information.

entisols in SOIL CLASSIFICATION, USA, young MINERAL-1 soils, lacking developed horizons, occurring in all climates, e.g. a FLUVISOL, LITHOSOL, REGOSOL.

entomology the branch of zoology dealing with the study of insects.

entrepôt (French) anglicized **entrepot**, a place to which goods are brought to be stored temporarily while awaiting transfer to another country, and where they are not liable to customs duties. Commonly used as an *adj.*, e.g. entrepot port, entrepot trade. G.

entrepreneur a person who wholly or partly undertakes, manages and controls an enterprise and, in the case of a business, bears the financial risks involved.

entropy **1.** a measure of the amount of disorder in a SYSTEM-1,2,3. The more disordered the system the higher the entropy. NEGENTROPY **2.** a measure of the amount of uncertainty or likelihood in a PROBABILITY distribution or in a SYSTEM-1,2,3 subject to constraints. ENTROPY-MAXIMIZING MODEL.

entropy-maximizing model a statistical model used to find the most probable spatial allocation pattern in a SYSTEM-1,2,3 subject to constraints. ENTROPY-2.

environment that which surrounds, the sum total of the conditions within which an organism, or group, or an object, exists (including the NATURAL conditions, the natural as modified by human activity, and the ARTIFICIAL-1). The term is used broadly in geography, especially in human geography, where the emphasis is on the economic, cultural and social conditions of the surroundings. BEHAVIOURAL ENVIRONMENT, CONTROLLED ENVIRONMENT, CULTURE, ENVIRONMENTAL HAZARD, ENVIRONMENTALISM, ENVIRONMENTAL STUDIES, PHENOMENAL ENVIRONMENT. G.

environmental control the regulation by mechanical means of the temperature, humidity, particle content, and the rate of movement of the air, within a building, combined with the provision of artificial lighting. BATTERY SYSTEM.

environmental determinism DETERMINISM.

environmental geoscience the scientific study concerned with the interrelationship of people and the natural systems of the earth.

environmental hazard a risk (usually to human beings) associated with the physical ENVIRONMENT. Such a risk may be natural (e.g. FLASH-FLOOD, LANDSLIDE, VOLCANIC ERUPTION) or produced by human activity (e.g. POLLUTION).

environmental impact the impression, particularly the undesirable or unpleasant impression, made on an ENVIRONMENT by the introduction of something alien to it, e.g. grazing animals in an area of sparse vegetation.

environmentalism the philosophical concept which stresses the influence of all the items and conditions of the ENVIRONMENT on the life and activities of people. In its extreme form it becomes ENVIRONMENTAL DETERMINISM. PERCEPTION GEOGRAPHY, POSSIBILISM. G.

environmental lapse rate static lapse rate, the actual rate of loss of temperature with increasing height at a specific location at a specific time, averaging some 0.6°C per 100 m (1°F per 300 ft). LAPSE RATE, PROCESS LAPSE RATE.

environmental studies a collective term for the studies which aim to make people aware of the conditions of the world they inhabit, and of the interrelationship of people with the physical, cultural and social items and conditions of their surroundings, i.e. the study of NATURAL HISTORY, ARCHITECTURE, ECOLOGY, METEOROLOGY and many of the other studies commonly included in GEOGRAPHY.

environs the area surrounding a place (not a person). ENVIRONMENT.

enzyme an organic CATALYST, one of a very large class of protein-containing compounds formed in and produced by living matter, which promote the chemical reactions on which life depends (e.g. digestion, reproduction). Each enzyme is usually responsible for one, or a few, of these reactions, the enzyme combining with a specific substance to bring about a chemical change in that substance without itself suffering permanent change. The process is usually dependent on temperature, pH, the presence of co-enzymes, and activators such as vitamins, metallic salts, etc.

eo- (Greek eos, dawn) a prefix signifying dawn (of), hence eolithic, dawn of the Stone Age. ARCHAEOLOGY, EOCENE, EOLITHIC AGE.

Eocene *adj.* of the second epoch (of time) or series (of rocks) of the Cainozoic era or Tertiary period or system, but sometimes applied to the whole division of the Tertiary period or system preceding the Miocene. GEOLOGICAL TIMESCALE.

Eogene in USA, synonymous with PALAEOGENE.

eohypse a reconstructed (imaginary) contour of a former land surface, drawn by plotting the parts that survive and filling in the lost connections by EXTRAPOLATION. G.

eolian AEOLIAN.

eolith a crude stone tool showing some chipping believed, but not by all archaeologists, to be the result of human work. In England and France flint was the stone most commonly used in the EOLITHIC (dawn of Stone Age) and in the PALAEOLITHIC (old Stone Age); in the NEOLITHIC (new Stone Age) stone implements were polished and refined.

Eolithic Age the earliest period of the STONE AGE when EOLITHS first came into use.

eolomotion a fairly slow downwind movement or downhill creep of rock particles caused directly or indirectly by wind action on surface rock particles. G.

eon AEON.

eosere **1.** the development of plant communities during an AEON or era **2.** the main developmental series within the climatic climax of a geological period.

eotechnic *adj.* denoting or belonging to the first phase of industrial and urban development, when transport is little developed, power resources fixed, and technology simple. PALEOTECHNIC, NEOTECHNIC.

Eozoic *adj.* **1.** of or pertaining to all the earliest era and associated rocks (i.e. applied to what other authors term the PRECAMBRIAN) **2.** of or pertaining to the earliest era and rocks of the Precambrian era, succeeded by the ARCHAEOZOIC and PROTEROZOIC. GEOLOGICAL TIMESCALE.

epeiric sea epicontinental sea, a sea on the CONTINENTAL SHELF, shallow in relation to the ocean, created by EPEIROGENIC MOVEMENT.

epeirogenetic, epeirogenic *adj.* of or pertaining to the formation of continents, applied to the type of mass earth movements which result in changes of level over large areas (e.g. continental uplift and depression, which may involve TILTING or WARPING but not intense FOLDING) in contrast to orogenic movement (OROGENESIS). G.

ephemeral *adj.* applied to a plant with a short life cycle (from germination to production of seed). It may produce several generations in a year, e.g. groundsel; or have a brief life when environmental conditions are suitable, e.g. after heavy rain in a hot desert region. ANNUAL, BIENNIAL, PERENNIAL.

epicentre the point on the earth's surface immediately above the centre of origin (the SEISMIC FOCUS) of an earthquake. The point at the antipodes of the epicentre is termed the anticentre. G.

epicontinental *adj.* near a continent, applied especially to seas on the CONTINENTAL SHELF. EPEIRIC SEA. G.

epicycle 1. applied by some authors to a subdivision of a major cycle of erosion initiated by a small change in base-level (sub-cycle and hemi-cycle have also been proposed) 2. a small cycle having its centre on the circumference of a greater cycle. G.

epidemic a disease that becomes widespread in a particular place at a particular time. ENDEMIC, MEDICAL GEOGRAPHY, PANDEMIC.

epidemic *adj.* widespread, applied particularly to a disease. ENDEMIC, PANDEMIC.

epidemiology the branch of medicine

concerned with the study of the factors influencing the frequency and spread of diseases. It is concerned with the features and the prevalence of a disease in a population, rather than in an individual. MEDICAL GEOGRAPHY.

epigene, epigenic *adj.* occasionally applied in geology to the geological processes operating at or on the earth's surface or to the rocks formed by them. HYPOGENE. G.

epigenesis 1. superimposition. EPIGENETIC 2. in biology, a theory that an embryo is formed gradually by the differentiation of cells which were originally not specialized, rather than by the development of preformed structures. PREFORMATION. G.

epigenetic *adj.* superimposed.

epigenetic drainage SUPERIMPOSED DRAINAGE. G.

epigenetic mineral an ore mineral of later origin than the rock that contains it, in contrast to one that is SYNGENETIC. G.

epilimnion the warmer top layer of water in a lake or the ocean, liable to be disturbed by wind and convection currents, lying above the THERMOCLINE. THERMAL STRATIFICATION.

epiphyte a plant which grows on the outside of another plant, using that plant just for support (i.e. not a PARASITE or a SAPROPHYTE), e.g. LICHENS, MOSSES, orchids.

epistemology the branch of philosophy concerned with the study of the nature, origin, foundations, limits and validity of knowledge. IDEALISM-3, METAPHYSICS, ONTOLOGY.

epithermal *adj.* applied to ore deposits, usually in veins, pipes, fissures, formed from an ascending aqueous solution with an approximate temperature of 50°C to 200°C (122°F to 390°F) moving within some 1000 m (3300 ft) of the surface. HYPOTHERMAL, MESOTHERMAL.

epoch the third in the subdivision of geological time. The rocks laid down in an epoch constitute a series. GEOLOGICAL TIMESCALE. G.

equation of time the difference at noon between MEAN SOLAR TIME and APPARENT TIME, LOCAL TIME. The two coincide only on the approximate dates 24 December, 15 April, 14 June, 1 September. The maximum positive value is on 11 February, when mean solar time is in advance of local time (sun fast) by 14 minutes 25 seconds; and the maximum negative value, when mean solar time lags behind local time (sun slow) by 16 minutes 22 seconds, is on 2 November (ANALEMMA). The daily value is published in the *Nautical Almanac*. The inconstant velocity of the earth (ROTATION OF THE EARTH), its elliptical path around the sun (ORBIT OF THE EARTH) and the inclination of the ECLIPTIC to the equator cause the differences of the interval between successive crossings of the MERIDIAN by the sun.

equator an imaginary GREAT CIRCLE round the earth in a plane perpendicular to the earth's axis and equidistant between the north and south poles, thus dividing the earth into the northern and the southern hemispheres; also used by analogy in such expressions as THERMAL EQUATOR. G.

equatorial *adj.* on, near to, or pertaining to, the EQUATOR.

equatorial air mass a warm AIR MASS of high humidity, the source of which lies over the ocean in the zone of the EQUATORIAL CLIMATE.

equatorial belt the zone lying approximately between the latitudes 10°N and 10°S, the area of the EQUATORIAL CLIMATE.

equatorial climate the climate occurring in a belt on each side of the EQUATOR, near sea-level and between approximately 10°N and 10°S, characterized by constant high temperature (about 27°C: 80°F) and humidity with little range throughout day and night and the year, and approximately a 12-hour day, 12-hour night. It is subject to CONVECTIONAL RAIN, the maxima corresponding to the EQUINOXES.

equatorial current, equatorial counter-current the surface movement of ocean water in EQUATORIAL latitudes. In the northern hemisphere the north equatorial current moves towards the southwest or west; in the southern hemisphere the southern equatorial current moves towards the northwest or west; between the two the equatorial counter-current flows eastwards. EL NINO. G.

equatorial forest the luxuriant FOREST-2 mostly of evergreen hygrophilous (HYGROPHYTE) species, growing at lower altitudes in a belt from approximately 7°N to 7°S in areas of EQUATORIAL CLIMATE, particularly in the Amazon and Zaire basins and also, with modifications, in parts of Indonesia, Malaysia, Sri Lanka and west Africa. The constant moisture and heat encourage rapid growth of tall, mostly HARDWOOD, trees, the spreading crowns of which form a thick canopy, and up which LIANAS and other woody climbers struggle to reach the light. In the modified areas, where tree growth is less dense, enough light penetrates to allow smaller plants to grow and form UNDERGROWTH. BUSH FALLOWING, RAIN FOREST, TIMBER. G.

equatorial front equatorial trough, ITCZ. G.

equi- prefix, equal.

equifinality the end state, one of similarity, achieved by SYSTEMS-1,2,3 founded on different initial conditions and undergoing change. The term is applied especially to landforms, e.g. a PENEPLAIN, and in SYSTEMS ANALYSIS. MULTIFINALITY.

equilibrium a state of balance between opposing forces or effects. A body is said to be in a state of stable equilibrium if it returns to its original position after being moved by a small impulse; unstable if it continues to move away from its original position in the direction given to it by a small impulse; neutral if it comes immediately to rest and remains stationary in its new position after being moved by a small impulse. In geography the term equilibrium is usually applied to a static condition, i.e. one of no change; or a dynamic condition, i.e. one where a balance is maintained by continued adjustments in reaction of the opposing forces (LOWRY MODEL), e.g. in slopes where the rate of rock weathering and the rate of removal of the rock debris reach a state of balance, the angle of the slope continually accommodating to changing factors of weathering and removal. DYNAMIC EQUILIBRIUM, RELAXATION TIME. G.

equilibrium line in glaciology, FIRN EQUILIBRIUM LINE. G.

equinoctial, **equinoctial** **circle** the CELESTIAL EQUATOR.

equinoctial *adj.* of, pertaining to, happening at, the time of the EQUINOX. G.

equinoctial point one of the two points at which the CELESTIAL EQUATOR and the ECLIPTIC-1 intersect.

equinoctial year SOLAR YEAR.

equinox day and night of equal length. The moment or point when the sun crosses the EQUATOR during its apparent annual movement from north to south and is directly overhead at noon along the equator, i.e. 21 March and 22 September, respectively the spring (vernal) equinox and the autumnal equinox in the northern hemisphere. G.

equiplanation the reduction of relief to a plain in a cold climate, without involving any loss or gain of material in the affected area. CRYOPLANATION. G.

equiplanation terrace ALTIPLANATION. G.

equipluve a line on a map linking places with the same PLUVIOMETRIC COEFFICIENT.

equity in general, fairness, justice, especially common fairness that is in line with the spirit rather than the letter of justice, e.g. the provision of a good or service in accordance with the needs of the individual.

equivariable an isopleth (ISO-) inserted to link places with equal COEFFICIENTS OF VARIABILITY.

era one of the major divisions of geological time. GEOLOGICAL TIME-SCALE.

erg (north Africa: Arabic) 1. the sandy desert of the Sahara, the KUM of central Asia. HAMADA, REG 2. in southern Africa, a system of fixed sand dunes in a sandy desert, e.g. in western Zimbabwe. G.

ergograph a graph designed to show the character of seasonal human activity and the time spent in each activity. Usually the graph is circular, with months indicated on the circumferences and the proportion of time spent in each activity in each month shown by concentric lines.

ergot *Claviceps purpurea*, a FUNGUS, parasitic on cereals, but especially on RYE, poisonous to animals and human beings, causing the disease which is also termed Claviceps purpurea (popular name Holy Fire), the symptoms being gangrene and violent nervous disorders. At one time the disease was EPIDEMIC in Europe, due to the eating of ergot-infected rye or bread made from the infected flour. A drug made from ergot is used medicinally to control bleeding and induce muscle-contraction.

ericaceous soil an acid, lime-free soil, one on which plants of the genus *Erica*, family Ericaceae, flourish.

erosion the processes of the wearing away of the land surface by natural agents (running water, ice, wave action, wind, and including CORRASION and CORROSION) and the transport of the rock debris that results. This does not include the WEATHERING of rocks in situ or MASS MOVEMENT caused by the force of gravity. Erosion is used frequently, but incorrectly, as synonymous with DENUDATION. ACCELERATED EROSION, THERMAL EROSION. G.

erosion platform a platform or relatively level surface of limited size, formed by EROSION, in contrast to a platform or terrace formed by DEPOSITION or AGGRADATION. EROSION SURFACE. G.

erosion surface a level surface of any size produced by EROSION, not by DEPOSITION or AGGRADATION. If the surface is almost horizontal the term PLANATION SURFACE is preferred. BELTED COASTAL PLAIN.

erratic, erratic block a mass of rock or a boulder transported by a GLACIER or ice sheet (which usually has since disappeared) and deposited in an area remote from its place of origin. The erratics of VALLEY GLACIERS often perch precariously on the valley sides, and are thus termed perched blocks; far-travelled boulders are sometimes distinguished by the term 'exotic'. The former course of the glacier or ice sheet can be traced by tracking erratic blocks back to their sources. G.

error in statistics, the extent to which a set of observations fails to reflect that which the observer wishes it to reflect, possibly because the collection of data is inadequate, the sampling procedure or manipulation of data inappropriate, or the definition of the variable as measured does not match the observer's definition. The HYPOTHESIS under test may be true, but the set of observations indicate it to be false; or the hypothesis under test may be false although conclusions drawn from the set of observations suggest it is true.

error variance in statistics, that part of the VARIANCE in a test which can be accounted for only by attributing it to ERROR. COMMON VARIANCE, RESIDUAL VARIANCE, SPECIFIC VARIANCE, TRUE VARIANCE.

ERTS Earth Resources Technology Satellite, LANDSAT.

eruption a violent outbreak, a pouring forth from restraint or as if from restraint, a term applied especially to the process by which solid, liquid or gaseous material pours forth gently or explosively from the vent of a VOLCANO or from fissures in the earth's crust as a result of volcanic activity; and to the process by which hot water and steam pours forth from a GEYSER.

escarpment 1. generally, any more or less continuous line of cliffs or steep slopes resulting from the differential EROSION of gently inclined strata or from faulting. FAULT **2.** specifically, the abrupt cliff face or steep slope of a CUESTA. The term escarpment should not be applied to the cuesta as a whole, and is preferable to the term SCARP which is sometimes applied to the steep face of a cuesta. G.

esker (formerly also eskir, escar, eskar from old Irish eiscir) a long, continuous, sinuous, steep-sided ridge of glacial and fluvioglacial sands and gravels, deposited by the melt-water of a glacier or ice sheet, termed AS (Sweden), ESGAIR (Welsh), KAME (Scotland). BEADED ESKER. G.

esparto (Spanish) the commercial name of various grasses, mainly *Stipa tenacissima*, but also *Lygeum spartum* and *Ampelodesma tenax*. They have tufts of long, flat grey-green leaves, up to 1 m (3 ft) in height, are grown in Algeria, Tunisia and southern Spain and are used in making ropes, cordage, baskets, high grade paper, cellulose. C.

espiñal (South America) a tract of thorny scrub.

esplanade historically, an open level tract of land separating the CITADEL from the town; hence a levelled, raised piece of land bordering the sea front in a seaside resort, supporting a wide road and serving as a PROMENADE.

essence 1. the most significant part of the nature of something **2.** in philosophy, PHENOMENOLOGY **3.** a concentrated extract, e.g. of VANILLA.

essential minerals the mineral constituents, not necessarily the major constituents, of an IGNEOUS or METAMORPHIC rock which determine its mineralogical classification and from which the rock is named. ACCESSORY MINERAL.

essential oil a volatile, aromatic substance obtained by distillation or extraction from various plants, used especially in medicine, the manufacture of perfume, or in cooking, e.g. ATTAR OF ROSES, CITRUS oils, CITRONELLA etc.

estancia (South America: Spanish) a large farm in South America on which cattle are reared on a large scale. HACIENDA, LATIFUNDIA, RANCH. G.

ester one of a class of compounds which on HYDROLYSIS yield one or more molecules of an acid and one or more molecules of an alcohol. Esters are often fragrant liquids, used in perfumery and as flavouring substances.

estivation AESTIVATION.

estovers (Norman-French estouffer, to furnish) the COMMON RIGHT of cutting and taking the tree loppings, or the gorse, furze, bushes, or underwood, heather or fern of a COMMON for fuel to burn in the commoner's house or for the repair of the house and farm buildings, hedges, fences and farm implements. The Early English equivalent was 'bote', hence cart-bote, fire-bote, hey- or hedge-bote, house-bote, plough-bote etc. G.

estuarine *adj.* of or pertaining to an ESTUARY, applied especially to the deposits laid down in brackish water and to the environmental conditions of an estuary. G.

estuary the tidal mouth of a river where the saltwater of the tide meets the freshwater of the river current. According to circumstances, the mouth of a river may broaden out and flow into its estuary, or divide and deposit material to form a DELTA. G.

étang (French) a shallow pool or lake caused by the ponding back of water draining from the land by beach material thrown up by the sea (an étang salé being one which communicates with the sea). Those on the southwestern coast of France, on the Landes coast, are good examples, as is the Etang de Berre, northwest of Marseille. G.

etchplain a small EROSION PLATFORM occurring in savanna climatic conditions, where the rock, weathered and 'etched', is subjected to seasonal streams that remove debris and maintain a steady, consistent process of wearing downwards. G.

etesian wind (Turkish meltemi) a strong north to northwesterly wind blowing in Greece and other parts of the eastern Mediterranean, caused by a steep pressure gradient associated with the thermal low pressure lying over the Sahara. Strongest in the afternoon, dying out in the evening, it gives rise to rough seas, fog in coastal areas, and clouds of dust; and it may blow with such force and dryness in some localities that it prohibits tree growth. G.

ethane a gaseous paraffin hydrocarbon, C_2H_6, occurring in natural gas, formed as a by-product in the cracking of PETROLEUM. Colourless and odourless, it burns with a clear flame and is used as a fuel and a source of ETHYLENE.

ether in the wave theory of light, a hypothetical substance which forms the medium of propagation of ELECTROMAGNETIC WAVES.

ethics 1. the branch of philosophy concerned with morality, the principles of right or wrong in human conduct 2. the moral principles which determine whether particular acts or activities are right or wrong. KANT, UTILITARIANISM.

Ethiopian Region a ZOOGEOGRAPHICAL REGION.

ethnic a member of an ETHNIC group, especially of a minority ethnic group. ETHNIC MINORITY.

ethnic *adj*. **1.** historically, pertaining to nations not Jewish or Christian, thus Gentile, heathen, pagan **2.** now, of or relating to people unified by common geographic origin, skin pigmentation, cultural, religious or linguistic ties, or deriving from or belonging to such ties. ETHNIC MINORITY.

ethnicity ETHNIC character or peculiarity.

ethnic minority, minority ethnic group a small ETHNIC-2 group, usually composed of IMMIGRANTS or the descendants of past immigrants, commonly living in large towns, sometimes in GHETTOES. Such a group is minor not only in the number of its members but also in the relationship of the group to the structures of political and economic power in the society in which it lives, which are controlled by bigger, more dominant groups. CULTURE CONTACT, HOST SOCIETY, NATIVISM.

ethnobotany the branch of botany concerned with the scientific study or description of the relationship between people and plants, particularly the study of the knowledge and customs, and the understanding, of indigenous people of the plants in their own environment; and the uses to which plants and plant products are put by different cultural groups. G.

ethnocentrism the state or quality of being ethnocentric, i.e. regarding one's own ETHNIC group as being of superior, or supreme, importance, the human attributes of other groups decreasing with increased distance (physical, or the distance created by cultural differences, etc.) from one's own group. An extreme form of ethnocentrism is exhibited in RACISM. ANTHROPOCENTRISM, EGOCENTRISM. G.

ethnography **1.** the scientific description of the distribution of human RACES-1 and cultures **2.** the study of the institutions and customs in small, defined communities in simple societies **3.** a detailed study of small groups of people

(e.g. in factories, hospitals etc.) within a complex society. ANTHROPOLOGY, ETHNOLOGY, ETHNOSCIENCE. G.

ethnohistory the study of the history of human RACES-1 or cultures, especially those of eastern (EAST-2) origin.

ethnology the scientific study of human RACES-1 and cultures, including their distribution, relationships and activities. There is a tendency to drop the terms ETHNOGRAPHY and ethnology in favour of ANTHROPOLOGY. ETHNOSCIENCE. G.

ethnoscience the study of human RACES-1 or cultures, a synonym for ETHNOGRAPHY.

ethnozoology the branch of zoology dealing with the uses made of animals and animal products by people of different cultures.

ethology the study of the actions and habits (especially the instinctive behaviour) of animals and their reaction to their environment, natural or ARTIFICIAL-1.

ethylene a colourless, inflammable, gaseous hydrocarbon obtained from PETROLEUM and used as a raw material in the making of alcohol and PLASTICS. HYDROCARBON CRACKING.

etiology, aetiology **1.** the branch of philosophy which deals with factors of causation **2.** the science, doctrine, demonstration of causes, applied especially in medicine to the study of the factors associated with the causation of disease. EPIDEMIOLOGY, MEDICAL GEOGRAPHY. G.

eugeogenous *adj*. applied to rocks that yield much debris on WEATHERING, in contrast to DYSGEOGENOUS. G.

eugeosyncline a GEOSYNCLINE associated with VULCANICITY during the period of its infilling by SEDIMENTATION. MIOGEOSYNCLINE.

euphotic zone the uppermost zone of ocean or lake depths where penetration by sunlight is sufficient for PHOTOSYNTHESIS to take place, rarely extending to depths below 100 m (55 fathoms: 325 ft). PHOTIC, PELAGIC. G.

European Economic Community   EEC.

European Free Trade Association   EFTA.

euryhaline   *adj.* applied to **1.** a marine organism tolerant of a wide range of salinity, a characteristic of those living in coastal areas and estuaries **2.** a plant tolerant of a wide variation in osmotic pressure (OSMOSIS) of soil water. STENOHALINE, STENOTHERMIC. G.

eurythermic, eurythermous   *adj.* applied to an organism tolerant of large fluctuations of temperature. EURYHALINE, STENOTHERMIC. G.

eurytopic   *adj.* applied to an organism with a widespread distribution. STENOTOPIC.

eustatism, eustasy   a world-wide and simultaneous change of ocean level due to a rise or fall of the ocean, not to that of the land (ISOSTASY). This change of level may be due to GLACIO-EUSTATISM, with a consequent decrease or increase in the amount of ocean water, to ocean floor spreading (PLATE TECTONICS) or (according to E. Suess, who introduced the term in 1888) to changes in the capacity of ocean basins due, e.g., to infilling by sedimentation. Some authors restrict the term to the change caused by the waxing and waning of ice sheets, i.e. to glacio-eustatism. DIASTROPHIC EUSTATISM, NEGATIVE MOVEMENT, POSITIVE MOVEMENT, REJUVENATION. G.

eutrophic   *adj.* applied to a swamp or lake which, having been over-supplied with organic or mineral nutrients, promotes excessive growth of ALGAE and other plants which draw on so much oxygen that little (or none in extreme cases) remains to support animal life, which is accordingly depleted or destroyed. AMMONIUM NITRATE, DYSTROPHIC, MESOTROPHIC, OLIGOTROPHIC, POLLUTION, TROPHIC LEVEL. G.

eutrophication   a change in status to higher productivity in a body of fresh water, often caused by mild POLLU-

TION-1, the process of becoming EUTROPHIC. G.

euxinic   *adj.* applied to water stagnant at great depth, e.g. conditions such as those in the Black Sea (Euxine Sea of the Greeks). G.

evaporation   the process by which a SOLID or LIQUID is converted to a gaseous state, to VAPOUR; or the action or process of driving off the liquid part of a substance in the form of vapour by means of heat. Water vapour in the atmosphere is the result of the evaporation of water from the earth's surface (HYDROLOGICAL CYCLE), a continual process dependent on air temperature, the quantity of water vapour already in the atmosphere, nature of the water or the land surface, and wind. The highest rate of evaporation occurs in hot DESERTS in conditions of great heat, dry atmosphere, lack of surface cover (plant or soil); the lowest rate in equatorial regions, with very high humidity (EQUATORIAL CLIMATE) and much surface cover. LATENT HEAT.

evaporimeter   **1.** an instrument that measures the quantity and rate of EVAPORATION **2.** (American) an atmometer.

evaporite   a SEDIMENTARY ROCK, such as GYPSUM, consisting of minerals which, having been precipitated from a solution, remain after EVAPORATION.

evaporitic   *adj.* characteristic of, pertaining to, an EVAPORITE.

evapotranspiration   the total return of water from the land to the atmosphere, i.e. the combined EVAPORATION from the soil surface and TRANSPIRATION from plants. POTENTIAL EVAPOTRANSPIRATION.

everglade   a low marshy region, usually under water, with hummocks and small islands, and overgrown with tall grass, canes and trees sometimes hung with SPANISH MOSS. The term is specifically applied to a large area of this nature, the Everglades, a National Park in Florida, USA. G.

evergreen a plant that bears green leaves throughout the year, in contrast to a DECIDUOUS plant.

evorsion erosion by vortices or eddies in the rocks of the bed of a river or stream. G.

evorsion hollow a POTHOLE-2 in a stream bed. G.

evolution 1. gradual change and development 2. the process of gradual, cumulative change from a simple to a more complex form. CHARLES DARWIN saw such a process at work in the progressive diversification in the characteristics of organisms or populations in the course of successive generations descended (according to his theory of evolution) from related ancestors, leading to the development of different species and sub-species. He based his theory of evolution on NATURAL SELECTION, the survival of the fittest. To summarize his theory, organisms in each generation produce many, variant offspring, but only the successful variants, i.e. those most fitted to their environmental conditions (ENVIRONMENT), survive and breed, transmitting their advantageous characteristics to their offspring. Thus when environmental conditions change, or organisms spread to new areas (ADAPTIVE RADIATION), new, strong, successful forms emerge, each well adapted to its particular environment. CLADISTICS, PUNCTUATED EQUILIBRIUM.

ewe a female SHEEP.

exaration glacial erosion effected solely by ice and not by ice carrying abrasive debris. PLUCKING.

exceptionalism 1. a methodological approach to geography which, while regarding the discipline as a pure science, considers it differs from other scientific studies in that in its principles of AREAL DIFFERENTIATION it is concerned with the unique (term introduced by F. K. Shaeffer 1953) 2. according to P. Haggett 1965, a quantitative approach in which theoretical MODELS-4 are devised from which exceptions and deviations can be measured. G.

exchange value the VALUE-2 at which a COMMODITY can be exchanged for another commodity. USE VALUE.

exclave a portion of an administrative or other kind of area, e.g. of a state, which is separated from the main part and surrounded by alien territory (and regarded in that territory as an ENCLAVE). If the portion is not physically separated, but is conveniently approachable only through alien territory, it is termed a pene-exclave. If, in practice, an exclave has ceased to be treated legally as such, it is in some cases termed a quasi-exclave. A temporary exclave is one created as a result of some territorial arrangement which has not been concluded, e.g. West Berlin, following the Second World War. G.

Exclusive Economic Zone TERRITORIAL WATERS.

exclusive species a SPECIES confined absolutely, or almost so, to one COMMUNITY-4.

exfoliation (American spalling) onion weathering, the weathering process in which strains lead to the splitting off of the surface of rocks in scales or layers. It is common in hot DESERT and SEMI-ARID lands and in MONSOON lands with a marked dry season, especially in rocks composed of a variety of metals, expanding and contracting at varied rates; and was formerly thought to be the result of the heating of the rock surface by the sun in the day and speedy cooling at night. But this theory is not supported by laboratory tests. The current theory is that water is a necessary element in the process, that exfoliation is due mainly to HYDRATION, resulting in the expansion of salt crystals in the pores of the rock surface. Ground water with its dissolved salts is drawn to the surface by CAPILLARITY where it is subjected to EVAPORATION, salt crystals being precipitated to form a film. HYDROLYSIS, SPHEROIDAL WEATHERING. G.

exhumation the action or process of bringing to the surface something buried in the ground, a term widely applied in geographical literature to previously existing surfaces uncovered by erosion, e.g. mountains, peneplained surfaces and plateaus buried by later deposits. FOSSIL LANDSCAPE. G.

exhumed landscape FOSSIL LANDSCAPE. G.

exile 1. banishment, expulsion from home or native land **2.** a person banished from, compelled to live away from, home or native land **3.** paradoxically, a voluntary exile, one who chooses to go into exile. EMIGRANT, EXPATRIATE, REFUGEE. G.

existentialism a body of philosophical doctrine (inaugurated by the Danish philosopher, S. A. Kierkegaard, 1813-55) concerned with the gulf between the existence of human beings (born with will and consciousness) and the kind of existence of natural objects (possessed of neither). Human beings are thus seen not to be part of an ordered, all-embracing, metaphysical scheme, but to exist in an alien world of objects from which they are estranged. Later existentialists, e.g. Jean-Paul Sartre, held that each individual, being free and responsible but not endowed at birth with character or goals, is self-creating in the situation and environment specific to him/her, through acts of will, of decision, and the self-development of his/her own 'essence'.

exogenetic, exogenic *adj.* arising from without, having an external cause or origin, hence in geology applied to the external processes (i.e. those at or near the surface of the earth), e.g. DENUDATION and DEPOSITION, and to the rocks and landforms arising therefrom. Exogenetic is the usual English form, exogenic the American. ENDOGENETIC, ENDOGENIC. G.

exogenous *adj.* **1.** in biology, originating outside the organism **2.** in geology, formed or occurring outside, proceeding from or acting on the outside of the crust of the earth. G.

exonym the CONVENTIONAL NAME given by people of one linguistic group or nation to the place-names etc. of another. G.

exoreic *adj.* flowing to the outside, applied to normal drainage, i.e. flowing to the sea. AREIC.

exoskeleton CRUSTACEAN.

exosphere the uppermost zone in the upper ATMOSPHERE-1 of the earth, in the IONOSPHERE, above c. 700 km (435 mi), from which neutral particles escape into space. The upper atmosphere may be defined as the atmosphere below the exosphere and above the TROPOPAUSE.

exothermic *adj.* in chemistry, pertaining to the production of heat, as opposed to the ABSORPTION of heat. ENDOTHERMIC.

exotic *adj.* **1.** brought in from a foreign country or from a foreign language; hence applied to plants introduced in a country from outside it **2.** like, or imitating, something foreign.

exotic river a river deriving most of its discharge from headstreams which flow in another, different region, e.g. a river such as the Colorado River in southwest USA which flows across an arid region, having been fed by heavy precipitation on distant uplands.

expanded-foot glacier a GLACIER spreading out from the mouth of a valley to form a broad tongue of ice on the plain, a small PIEDMONT GLACIER.

expanded town in Britain, a town enlarged under the Town Development Act 1952 by accommodating the OVERSPILL population and industry from a major city. NEW TOWN.

expansion diffusion DIFFUSION-1 in which phemonena (or a phenomenon) spread from one place to another but in the process remain and often intensify

in the place or places of origin. CON-
TAGIOUS DIFFUSION, RELOCATIVE
DIFFUSION.

expatriate one who voluntarily, for
personal reasons, chooses to live (and
in some cases work) away from his/her
native country, but not necessarily
permanently. EMIGRANT, EXILE, REFU-
GEE. G.

explanation the process or the end-
product of giving an account of, of
accounting for, or of interpreting,
something, or making it clear or intelligible. FUNCTIONAL EXPLANATION,
STATISTICAL EXPLANATION.

explanatory variable INDEPENDENT
VARIABLE.

exploitable *adj*. having the capability of
being made good use of, of being
developed, e.g. an exploitable seam of
coal. EXPLOITATION, RESERVES.

exploitation the act of **1.** developing the
use of, making the best use of **2.**
deriving unjust profit from the work of
another **3.** in Marxism, the process by
which SURPLUS VALUE is extracted by
the capitalist from the workers.

exponent in mathematics, an index, a
symbol indicating the power to which a
quantity is to be raised. It is printed as a
superior figure or letter following the
number, e.g. in 2^3, 3 is the exponent.

exponential *adj*. applied to an expression in which the EXPONENTS are
variable quantities.

exponential curve a curve which may be
assumed by a surface in accordance
with a particular mathematical formula. G.

exponential growth geometric growth,
compound interest, the growth of a
value over a period by a fixed percentage of the principal, the sum of the
original principal plus the interest accumulated in earlier periods. Thus while
the rate of increase is constant, the
amount of increase grows in successive
periods. The time needed to double the
original value (termed the doubling
time) is approximately 70 divided by

the percentage rate of growth (e.g. a
growth rate of 10 per cent doubles the
value in 7 periods). GEOMETRIC PROG-
RESSION, LINEAR GROWTH.

export something sent out from one
country to another, the sender receiving
in return money, goods or services in
payment (IMPORT). Invisible exports
consist of payments for services, which
include interest on overseas investments and loans, earnings from transport, shipping, banking, insurance, and
tourism. Visible exports are foodstuffs
and raw materials, natural or partly
processed, and manufactured goods.
BALANCE OF PAYMENTS, BALANCE
OF TRADE.

export base theory a theory which sees
the development of a city or region as
being linked solely to its export performance, i.e. to the magnitude of its
exports to areas outside the city or
region. BASIC ACTIVITY, NON-BASIC
ACTIVITY.

exposed shield an area of a SHIELD in
which the BASEMENT rock, usually
PRECAMBRIAN, appears at the surface.
COVERED SHIELD.

exposure 1. the state of being left
uncovered, bare, without protection **2.**
in geology, a place where the solid
rocks reach the surface and are not
obscured by soil etc., the act of uncovering being natural or artificial (a
rock reaching the surface through a
covering of soil produced by its own
disintegration is best termed an OUT-
CROP) **3.** the position of a place with
regard to compass direction or to
climatic influences. ADRET, ASPECT.
G.

expressway in USA, a wide road built
for fast through traffic. It is divided into
a number of lanes, the access to it is
limited, and it crosses other main roads
at a different level.

expropriation NATIONALIZATION.

extended family a FAMILY-3 group
comprised not only of parents and
children (NUCLEAR FAMILY) but also

of blood relatives and relatives by marriage, all living close to one another.

extensive agriculture farming practised in large units in which the amount of capital and the labour employed is small in relation to the size of the unit. The work is sometimes highly mechanized, so that the yield per worker is high though the yield per ha (per acre) is low (due to low inputs), resulting in a high total output from the unit. The term is often, and incorrectly, applied to farming in any large unit. INTENSIVE AGRICULTURE. G.

extent the space between limits, or the time between limits or events.

external diseconomy DISECONOMY.

external economies cost advantages which benefit a commercial activity but which are not produced within it, being derived from a source outside the activity, e.g. as in an AGGLOMERATION-4. EXTERNALITY.

externality a concept concerned with the social and economic costs or benefits caused by the activity of an individual (or an institution) which do not enter the internal production costs of that activity but which affect the activity of another (or others) and over which the latter has/have no control. Thus the costs or benefits are externalized and fall on others. Externalities are often fortuitous and may be negative (creating costs) or positive (beneficial) to the recipient. AGGLOMERATION-4, COMPLEMENTARITY, EXTERNAL ECONOMIES, NEIGHBOURHOOD EFFECT-2.

exterritoriality the immunities granted to a diplomatic envoy and staff in accordance with international law. Not to be confused with EXTRATERRITORIALITY. G.

extinct volcano VOLCANO.

extraction the process of obtaining by pressure, distillation, evaporation, treatment with a solvent etc., e.g. of a metal from its ore. EXTRACTION RATE.

extraction rate the proportion of a mineral or metal obtained from an ore.

extractive agriculture an agricultural practice which is more concerned with immediate short-term benefits than with conserving resources for future agricultural use.

extractive industry a PRIMARY INDUSTRY in which non-replaceable materials are removed in their natural state, e.g. mining, quarrying. Strictly, the term excludes forestry and agriculture, but some authors include forestry. G.

extraneous variable a VARIABLE which might, as well as the explanatory variable (INDEPENDENT VARIABLE), produce or cause changes in the DEPENDENT VARIABLE, and which must therefore be eliminated as a possible explanation for the phenomenon under survey.

extrapolation in mathematical and scientific calculations **1.** the estimation of values outside a range of known values based on the assumption that trends within the known existing range will be continued outside it **2.** drawing a conclusion about a future or hypothetical situation by using observed tendencies as a basis, e.g. drawing an EOHYPSE.

extraterrestrial *adj.* outside the earth or its atmosphere.

extraterritoriality the extension of jurisdiction beyond the borders of the state, involving the partial exemption from local law and jurisdiction enjoyed by diplomatic agents and others (such as troops) stationed in foreign countries, who thus remain under the laws of their own country, usually with the agreement of the foreign country in which they are living. G.

extreme climate a climate with a great range of temperature between the coldest and the warmest months, e.g. as occurring in the interior of continents, away from the moderating influence of the ocean.

extremes of temperature the highest and the lowest shade temperatures, or the mean highest and the mean lowest temperatures, recorded during any selected period (day, month, year) at a meteorological station.

extreme value in statistics, the value at either the top end or the bottom end of a DISTRIBUTION-4.

extrusion the action of pushing out, thrusting out by mechanical force, emergence, e.g. the discharge of solid, liquid or gaseous material to the earth's surface during a volcanic ERUPTION. EXTRUSIVE ROCK.

extrusion flow of ice the horizontal and outward movement of ice in an extensive ice sheet due to the pressure exerted by the great depth of ice, probably helped by PLASTIC DEFORMATION in the basal layers.

extrusive rock VOLCANIC ROCK, an IGNEOUS ROCK resulting from the pouring forth or extrusion of molten material (MAGMA) at the earth's surface as LAVA and its consolidation there. It usually has a glassy texture, and contains CRYSTALS smaller than those occurring in an INTRUSIVE ROCK. EXTRUSION, INTRUSION, PLUTONIC ROCK. G.

eye of a hurricane or other tropical revolving storm, the still, calm area, limited in extent, at the centre of the storm, where the atmospheric pressure is at its lowest.

eyot AIT, a small island, a term surviving in many English place-names in its abbreviated form, -ey. G.

F

fabric structural material, especially the physical composition (textural and structural) of a compound, e.g. TILL, which may be statistically analysed in several ways, including an analysis of the orientation and dip of the stones within the assemblage of clay, stones and sand. PETROFABRIC ANALYSIS.

facet (French, a little face) from analogy with one of the small planes which constitute the surface of a crystal or cut gemstone, a face produced on a pebble by wind action (DREIKANTER, EINKANTER, ZWEIKANTER), or a surface produced as a result of erosion in a complex land surface. G.

facies 1. in botany, the general appearance and form of a plant, or the composition of a natural group 2. in geology, the total character of any part of a FORMATION-2, shown by the FOSSILS it contains, the composition, colour, texture, form of STRATIFICATION and nature of the constituent rocks, or by other special features; hence used in STRATIGRAPHY in differentiating one rock STRATUM from another.

facies fossil, facies (fossil) fauna FOSSILS which are characteristic of a certain FACIES-2, revealing the environmental conditions present when the sedimentary beds in which they lie were formed.

facies mapping the drawing of a map to show the areal differentiation of the lithological character (LITHOLOGY), fossil content, or any other variable relating to the composition of a rock formation that can be expressed quantitatively. Facies maps may be classified according to the number of variables considered: univariate, such as percentage maps and isolith maps, and multivariate. Ratio maps cover the relationship between two variables; and combined ratio maps, entropy-function maps (ENTROPY) and distance distribution maps cover the relationship among three variables. G.

facilities the things (aids, equipment, structures, favourable conditions, opportunities) that make easier some action, specified activity or task.

factor 1. a circumstance, fact, agent or influence contributing to a result, effect or condition 2. in arithmetic, a whole number which, when multiplied by one or more whole numbers, produces a given number (e.g. 2, 3, 4, 6 are all factors of 12) 3. in algebra, an expression which will divide into a given expression 4. in statistics, a quantity under examination in an experiment as a possible cause of variation, e.g. in a FACTORIAL experiment; or in MULTIVARIATE ANALYSIS a function of the observed (OBSERVATION-2) VARIATES (usually LINEAR-5) which may be considered to be part of those variates (and hence as a factor of the variation); or a constituent item in an average or index number (FACTOR ANALYSIS).

factor analysis a branch of MULTIVARIATE ANALYSIS, a statistical technique (similar to PRINCIPAL COMPONENTS ANALYSIS) which ignores the uniqueness of a number of variables (or test items) in a set of observations and aims to describe them in terms of a smaller number of more basic, hypothetical components or FACTORS-4. A factor which to varying degree underlies all items in a test is termed a general factor; one which underlies only a specific group of items is termed a group factor. A factor loading of an item on a factor is the correlation between the factor and the item. FACTORIAL ECOLOGY.

factorial 1. in mathematics, the function of a positive integer n given by the product of the integer and all the integers between it and unity, denoted by $n!$, and with the convention that $0! = 1$ 2. *adj.* relating to a FACTOR-4 or a factorial.

factorial ecology the application of certain multivariate statistical techniques to demographic, socio-economic and housing census data in order to discover, as economically as possible, the variation in neighbourhood characteristics in an urban area. A large number of census variables are subjected to FACTOR ANALYSIS, PRINCIPAL COMPONENTS ANALYSIS or CLUSTER ANALYSIS in order to condense the data into a smaller number of summary, hypothetical variables (FACTORS-4). These variables are not 'real', they are simplifying abstractions which compress and summarize a large part of the variation in the original body of data, extracting the basic pattern.

factorial experiment in statistics 1. an experiment designed to examine the effect of one or more FACTORS-4, each selected factor being applied at two levels at least, to show differential effects 2. an experiment investigating all possible treatment combinations which may be formed from the factors under investigation, the 'level' of a factor denoting the intensity with which it is brought to bear.

factor loading FACTOR ANALYSIS.

factors of production PRODUCTION FACTORS.

factory farm a CAPITAL-INTENSIVE livestock farm in which the stock (e.g. pigs for meat, poultry for meat and eggs, cows for milk) is reared and tended under cover in carefully controlled conditions (ENVIRONMENTAL CONTROL), the food intake, rate of growth, egg production, milk-yield etc. being carefully monitored. The disposal of the effluent, e.g. from a pig unit, often presents difficulties. BATTERY SYSTEM.

facultative parasite a PARASITE that is able to live as a SAPROPHYTE but which sometimes lives as a parasite.

fadama (west Africa: Hausa) a flood-plain in a wide, fairly flat, river valley, subject to annual inundation, common in the SAVANNA zones of Sudan and Guinea, supporting typical savanna vegetation (grasses, sedges, and the tree species *Mitragyna inermis* and *Borassus*). G.

Fahrenheit scale a TEMPERATURE SCALE established 1715 by Daniel Gabriel Fahrenheit, 1686-1736, German physicist, on which the ice point, the freezing point of pure water, is 32° (32°F) and the steam point, the boiling point of pure water at sea-level with a standard pressure of the atmosphere of 760 mm, is 212° (212°F). (0°F represents the melting-point of ice in a sal-ammoniac and water mixture, but 32°F represents the melting-point of ice in water.) The difference between ice point and steam point is thus 180°, and one Fahrenheit degree is 1/180 of the temperature interval between the two points; so 1° CELSIUS or CENTIGRADE equals 1.8° Fahrenheit. CONVERSION TABLES, KELVIN SCALE, REAUMUR SCALE. G.

fairway 1. the main navigable channel in a river or estuary, usually with buoys, and leading to a harbour 2. on a golfcourse, the turf between the tee and the green. G.

fako (west Africa: Hausa) in the Sudan and Sahel zones, land from which the surface layer of the soil has been eroded to the level of the infertile subsoil which, supporting little vegetation even in the rains, is never cultivated. G.

falaise (French) a low cliff. G.

fall 1. in USA, the autumn **2.** a WATERFALL.

fall line, fall-line, fall zone a line or narrow zone where a number of nearly parallel rivers plunge over the edge of a plateau to the lowland below, marking the change in the character of their courses from the rocky channels with swift currents of the upper courses, to the more placid courses on the plain. In the USA applied particularly to the boundary between the ancient crystalline rocks of the Appalachian plateau and the younger, softer rocks of the Atlantic coastal plain. G.

fallow ploughed or cultivated land which is being allowed to rest, uncropped or partially cropped, for one or more seasons, or sometimes for a shorter period. BARE FALLOW, BUSH FALLOWING, GREEN FALLOW, LAND ROTATION, SHIFTING CULTIVATION, SWIDDEN FARMING. G.

false bedding, false-bedding LAMINAE, especially in sandstone, deposited under the influence of changing currents in shallow water areas, lying parallel to each other for short distances, but inclined at varying oblique angles to the general stratification. False bedding (a term synonymous with CURRENT BEDDING or cross bedding), is caused by swift local currents and swirling gusts of wind, thus studies of it provide much information regarding current direction and conditions of deposition. G.

false cirrus thick, grey CIRRUS cloud associated with the top of a thundercloud, and usually heralding bad weather.

false drumlin a formation that resembles a true DRUMLIN but is actually a rock mass overlain by a thin coating of DRIFT deposited by ice.

false origin a selected point in a GRID

SYSTEM used on maps from which the position of any place can be expressed in terms of its COORDINATES from the selected point. The false origin differs from the TRUE ORIGIN (the intersection of the projection axes) in order to exclude negative values. NATIONAL GRID.

falsifiability a principle expounded by Karl Popper, Austrian-born (1902) British philosopher, that theory holds good until it is disproved and that falsification, not verification, is the proper objective of scientific procedures; that a theory which is not falsifiable is not scientific. Thus any scientific HYPOTHESIS must be expressed in terms that make it capable of being submitted to rigorous, sustained testing in an attempt to show it is wrong. For Popper falsifiability distinguishes science from pseudo-science. LOGICAL POSITIVISM.

family (Latin familia, household) **1.** a group of people consisting of parents and children, living together or not. FAMILY LIFE CYCLE **2.** a person's children **3.** a unity formed by people who are, or who are nearly, connected by blood or by affinity. EXTENDED FAMILY, NUCLEAR FAMILY **4.** a household denoting a group of people living in one dwelling, or with one head of household, including parents, children, servants **5.** persons descended or claiming descent from a common ancestor **6.** in the CLASSIFICATION OF ORGANISMS, a group consisting of a number of genera or sometimes just one GENUS. Families grouped together constitute an ORDER. Families of flowering plants used to be termed natural orders.

family life cycle the progressive series of changes in size and composition which a family undergoes over time, e.g. for a NUCLEAR FAMILY (i.e. a married couple with dependent children) the stages usually identified are marriage and the formation of the household, child rearing, child launching, contraction in size as children leave home, post-child (often associated with retirement). The patterns of the life cycle of other types of FAMILY differ with the composition of the family.

family reconstitution in demography, a method of disaggregating demographic data by linking names recorded in parish registers, etc. Information about significant events (baptisms, marriages, burials) in the lives of individuals in a given FAMILY-1,2,3 is collected from such registers etc. and linked together.

fan the alluvial (ALLUVIUM) or stony deposit of a stream where it issues from a ravine or canyon and drops its load on to a plain (ALLUVIAL FAN). Some DELTAS are fan-shaped (ARCUATE DELTA). G.

fan-folding, fan-structure a symmetrical ANTICLINORIUM in which the axes of the folds dip inwards from both sides to the centre. In cross section a belt of country exhibiting fan structure suggests an open fan, hence the term. It was formerly believed that a geological section across a great mountain chain would show an anticlinorial structure with numerous individual folds suggesting an open fan; but it is now known that few if any mountain chains have such a structure. G.

fanglomerate the consolidated material of an alluvial FAN, consisting of rock fragments varying in size compacted and/or cemented together. G.

FAO Food and Agriculture Organization of the United Nations, a specialist organization of the United Nations established on 16 October 1945, secretariat in Rome, with the aims of giving international support to national programmes designed to increase the efficiency of agriculture, forestry and fisheries, and of improving the conditions of people engaged in relevant activities.

Far East far to the east of Europe, i.e. east and southeast Asia, including China, Korea, Japan and, usually,

Malaysia and the Indochinese peninsula. EAST.

farinaceous (Latin farina, flour) *adj.* rich in starch, the main nutritional constituent of CEREAL grains and edible roots, the common STAPLE foodstuffs.

farm originally a piece of land held on a lease for the purpose of cultivation, now applied regardless of the nature of the TENURE to any tract or tracts of land or of water, varying greatly in extent, worked as a unit, used for the cultivation of crops or for the rearing of LIVESTOCK or fish, under individual or collective management. Specific names such as DAIRY FARM, FISH FARM, FUR FARM indicate the purpose of the farm. AGRICULTURE, AGRIBUSINESS, BATTERY SYSTEM, COLLECTIVE FARMING, COOPERATIVE FARM, EXTENSIVE AGRICULTURE, FACTORY FARM, FARMER, FARMING, FRAGMENTATION OF A FARM, HACIENDA, HUERTA, INTENSIVE AGRICULTURE, LADDER FARM, LATIFUNDIA, RANCH, SMALL HOLDING. G.

farmer a term used as loosely as is FARM, broadly one who, whether as tenant or landlord, is occupied in AGRICULTURE, AQUACULTURE or PISCICULTURE. Types of farmer include the DIRTY BOOT FARMER, GENTLEMAN FARMER, PART-TIME FARMER, SIDEWALK FARMER, SUITCASE FARMER. G.

farming the activity of a FARMER. The main types of farming include CROP FARMING, DAIRY FARMING, MIXED FARMING, STOCK FARMING. Examples of some specific farming activities are listed under FARM, but see AGRIBUSINESS, AGRICULTURE, COLLECTIVE FARMING, COMMERCIAL AGRICULTURE, COOPERATIVE FARM, EXTENSIVE AGRICULTURE, INTENSIVE AGRICULTURE, PEASANT FARMING, SHIFTING CULTIVATION, SUBSISTENCE AGRICULTURE; and FARMSCAPE.

farmscape a land use category, an area of land in which viable AGRICULTURE dominates and which includes non-agricultural elements (e.g. agricultural villages) if these do not upset the dominance of the agriculture. LAND-SCAPE, TOWNSCAPE, WATERSCAPE, WILDSCAPE. G.

farmstead the land and buildings of a small farm.

Far West that part of the USA west of the Great Plains.

fast ice sea ice, varying in width, firmly fixed along the coast, where it is attached to the shore, to an ICE FRONT or to an ICE WALL, or lies over SHOALS-1,2, usually in the position of its original formation. It may stretch over 400 km (250 mi) from the coast. ICE EDGE, ICE FOOT. G.

fat 1. any of the neutral compounds, solid or liquid, insoluble in water, containing fatty acids, and extractable from tissue by a solvent such as ether or hot alcohol, true (neutral) fat being a compound of glycerol and fatty acids **2.** a term sometimes applied more widely to cover any substance, not necessarily containing a fatty acid, extractable by a fat solvent. Fats are an important source of energy in the diet of all warm-blooded animals, having the highest calorific value of all classes of food (the other classes being CARBOHYDRATES and PROTEINS). They are stored for food in the bodies of animals and in plant seeds. In industry they are used mainly in the making of soaps and lubricants.

fathom a nautical measurement of the depth of water, based on the span of the outstretched human arms, i.e. 1.829 m (6 ft); 100 fathoms = 1 cable. CONVERSION TABLES.

fault in geology, a fracture or break in a series of rocks along which there has been vertical or lateral movement or both as a result of excessive strain (ELASTIC REBOUND). NORMAL FAULTS are those in which the rocks on one side have slipped down relative to the other; REVERSE FAULTS where they have been pushed up (a reverse fault of very low angle, the upper beds pushed far over the lower, is a THRUST FAULT), OVERTHRUST FAULTS where the plane is near the horizontal. Faults may be nearly parallel to the DIP of the rocks (dip faults), to the STRIKE (longitudinal or STRIKE FAULTS, STRIKE-SLIP or TEAR FAULT), oblique (with planes between dip and strike); and there are groups of faults forming a complex fault or fault zone, STEP FAULTS, TROUGH FAULTS (forming a GRABEN), ridge faults (forming a HORST); and IMBRICATE STRUCTURE. For some of the many other terms applied to faults see those that follow as well as COMPRESSION, FOOT WALL, HADE, HANGING WALL, HEAVE, LAG FAULT, MYLONITE, SLICKENSIDE, NAPPE, THROW, TRANSFORM FAULT. G.

fault apron an accumulation of rock debris at the base of a (fault) scarp resulting from rapid dissection, and formed by the coalescence of many ALLUVIAL CONES. G.

fault block a block of country, often of considerable size, bounded by FAULTS, in many cases resulting from vertical movements in the earth's crust. HORST. G.

fault breccia a rock composed of shattered angular fragments resulting from movement along a FAULT. G.

fault-line scarp a SCARP formed when, by faulting, a hard rock confronts a soft rock and the soft rock is subsequently eroded, leaving the hard rock standing as a cliff or scarp; but this scarp is not necessarily the original FOOTWALL-1 of the FAULT, as it is in a true FAULT SCARP. OBSEQUENT FAULT-LINE SCARP, RESEQUENT FAULT-LINE SCARP, WATERFALL.

fault plane the surface of movement of a FAULT, where material may be shattered and broken off to form FAULT

BRECCIA, or ground to ROCK FLOUR which may be consolidated to form MYLONITE; or where the surface is polished and sometimes scratched (SLICKENSIDE). CLINOMETER.

fault scarp a steep face of rock resulting from the uplift along one side of a recently active FAULT. An early landform, it is rapidly affected by WEATHERING and EROSION, which may result in its disappearance, or its later development into a FAULT-LINE SCARP. ESCARPMENT, SCARP. G.

fault spring a SPRING-2 issuing along a FAULT where a PERMEABLE bed encounters IMPERMEABLE rock.

fauna 1. the animal life, considered collectively, of any given period, environment or region. BIOTA, FLORA 2. the classification of the animals of a period, environment or region. BINOMIAL NOMENCLATURE.

favela (Brazil: Portuguese) a shack, a slum. The plural, favelas, is applied to a SHANTY-TOWN, with shelters improvised from scrap materials, forming a SLUM settlement on the outskirts of a large town, e.g. of Rio de Janeiro.

favelado (Brazil: Portuguese) a person who lives in a FAVELA.

feather-edge, featheredge 1. the thinnest edge of a SEDIMENTARY ROCK as it pinches-out (PINCHING-OUT) in a horizontal plane and vanishes, e.g. of the surface of a BACK SLOPE 2. the thinnest edge of an INTRUSIVE ROCK.

fecundity FERTILITY-2.

Federal in American history, a supporter or soldier of the Union of Northern States in the Civil War, 1861-5. CONFEDERATE STATES.

federal *adj.* 1. of or pertaining to a form of government in which two or more states unite, but have some independence in internal affairs 2. in USA, relating to the central government. FEDERAL DISTRICT 3. in USA, historically, of or supporting the Union of Northern States in the Civil War, 1861-5. CONFEDERATE STATES, FEDERAL.

federal district an area allocated as the seat of the capital of a country, e.g. District of Columbia (for Washington) in USA.

federal system of government, a political system which varies in detail from one country to another, but in which the main authoritative control is usually divided between a central or national government and a regional or state government, with local government being directly responsible to the regional or state government authority, not to the central or national authority. UNITARY SYSTEM.

federation 1. the act of entering a league or covenant for a common purpose, especially when a number of separate states combine to form a political unit, each retaining control of its internal affairs 2. the name given to the sovereign state thus formed, or a group of states thus united.

feedback the action by which the output of a process is coupled to the input, i.e. the return to a system, process or device of part of its output. Normal (termed negative) feedback has a modifying effect in that it controls and corrects discrepancies in the working of the system, process or device, i.e. the response to a stimulus tends to counteract or inhibit the repetition of the stimulus (e.g. there is a lack of the feeling of thirst in an animal which has quenched its thirst: an animal in this situation will not drink any more even if water is available). Positive feedback, which is less common, does not have a modifying effect, but a snowballing effect in that it strengthens the stimulus (e.g. ever greater traffic flow on an ever-widened road: a growing volume of traffic encourages the widening of a road, which in turn attracts a greater volume of traffic).

feeding ground the area where a group of animals, e.g. a herd, commonly feeds.

feldspar, felspar any of a wide-ranging

group of crystalline minerals, white or pink in colour, consisting mainly of aluminosilicates of POTASSIUM, SODIUM, CALCIUM, BARIUM. Usually classified as ALKALI or PLAGIOCLASE, feldspars are abundant in METAMORPHIC ROCKS and in ARKOSES, are used in the manufacture of porcelain and glass, and are a source of SEMIPRECIOUS stones, e.g. MOONSTONE. Alkali feldspar, characteristic of all alkali igneous rocks (ALKALI ROCK) includes ORTHOCLASE (mainly potassium), soda-orthoclase (with a little sodium), anorthoclase (mainly sodium). The plagioclase feldspars range in composition from ALBITE (sodium) to anorthite (CALCIUM) and form constituents of such basic rocks as BASALT and GABBRO (in calcic form) and of sialic rocks (SIAL) such as ANDESITE and DIORITE. FELSIC, LEUCOCRATIC, MAFIC, MELANOCRATIC.

feldspathoid any of the group of silicate minerals (SILICA) that contain SODIUM and/or POTASSIUM, are related to, but contain less silica than, FELDSPARS, and occur in alkali igneous rocks (ALKALI ROCK) but never in rocks containing QUARTZ. FELSIC, MAFIC, NEPHELINE.

fell (Norwegian) in northern England **1.** a wild elevated stretch of rough grazing or moorland, especially if interrupted by boulders or rock outcrops **2.** a summit, e.g. Bowfell, English Lake District. G.

fella (Arabic; pl.**fellaheen**) a peasant in Arabic-speaking countries, especially in Egypt. G.

Felsenmeer (German) a boulder field, block field, rock-block field, rock river, an area on top of a flat-topped mountain or high plateau in a temperate climate, sometimes on lowland in much colder regions, covered with angular blocks of rock created in situ by frost action in JOINTS. G.

felsic *adj.* applied to pale-coloured

minerals, FELDSPATHOIDS and QUARTZ in IGNEOUS ROCK. LEUCRATIC, MAFIC, MELANOCRATIC.

felsite an IGNEOUS ROCK consisting of a compact mass of FELDSPAR and QUARTZ, used as an AGGREGATE-3 and in road making.

felt a fabric made from wool, hair, or fur, by pressing and rolling the fibres together.

felucca (probably of Arabic origin) a small coastal vessel of the eastern Mediterranean, Red Sea and Nile, lateen rigged, sometimes equipped with oars.

fen a lowland covered wholly or partly with shallow water with water-loving vegetation which decays to form PEAT. In contrast to a BOG, where the organic soil is very acid, the soil in a fen is alkaline, neutral or only slightly acid. ACID SOIL, ALKALINE SOIL, FEN PEAT, MARSH, SWAMP. G.

fenêtre (French, window; German Fenster) an opening (a window) worn by erosion in the upper strata of rocks which have been overfolded (OVERFOLD or NAPPE) so that patches of the underlying younger rocks, which form the floor, are exposed, the frame of the window consisting of the older strata of the overfolded rocks. The term should be applied only to this phenomenon and not to the 'normal' result of DENUDATION when younger strata are eroded to expose the underlying older strata. G.

feng-shui, fong-choui, fung-shui (Chinese feng, wind; shui, water) in China, the spirit of wind and water of a locality, the good in a system of good and evil influences that inhabit natural features in the landscape, thus important in the siting of buildings and graves. GEOMANCY. G.

fenland 1. a tract of land occupied by FEN. CARR **2.** with an initial capital letter, Fenland, a synonym for the FENS in eastern England, around the Wash.

fennel *Foeniculum vulgare*, a perennial

herb with aromatic leaves and.seeds, native to the Mediterranean region, but naturalized elsewhere in Europe. The finely-divided leaves are used in flavouring sauces, soups, salad dressing, and various dishes, the oil from the anise-flavoured seeds is used in baking, condiments, liqueurs, cordials, and medicinally as a remedy for flatulence. Florence fennel, *Foeniculum vulgare*, variety *dulce*, is different in character and use, being a short plant with very swollen leaf-bases that give the appearance of a bulb. Blanched by earthing-up in growth, it is that bulb-like part, resembling celery in flavour, which is eaten, raw or cooked.

fen peat ALKALINE or neutral black PEAT in which the plant structure is not apparent, cellulose is absent, but there is a high content of ash and proteins. It supports typical FEN vegetation in which sedges, grasses and rushes predominate. BOG PEAT.

Fens, Fenland in eastern England, a flat lowland below 15 m (50 ft) extending around the Wash in the lower basins of the rivers Witham, Welland, Nene and Great Ouse, consisting of naturally marshy land built up by the deposition of sediment by sea and rivers, forming alluvial flats near the coast and FEN PEAT further inland. Except for Wicken Fen, National Trust property, the area has been largely reclaimed (notably by Vermuyden and the fifth Earl of Bedford) and now provides land valuable for cultivation. G.

Fenster (German) FENETRE.

fenugreek *Trigonella foenum-graecum*, an ANNUAL herb of the pea family (LEGUMINOSAE), native to the Mediterranean region, now widely grown there and in the Indian subcontinent not only for the sake of its edible leaves, but also for its aromatic seeds used in cooking and, with other aromatic ingredients, in making artificial maple flavouring.

feral *adj.* like a wild beast, applied especially to an animal (once domesticated) or to a plant (once cultivated) which has now established itself in the wild. G.

ferik (Sudan: Arabic) in northern Sudan, huts occupied for the whole or part of the year by a group of nomad or semi-nomad pastoralists whose animals (usually cattle or camels) are treated as a common herd. NOMADISM. G.

ferling YARDLAND. G.

fermentation a chemical change brought about by the action of ENZYMES on organic substances, with the release of energy, especially the ANAEROBIC breakdown by yeasts and bacteria of carbohydrates, yielding alcohols, acids and carbon dioxide. ANAEROBIC RESPIRATION.

fermtoun (Lowland Scots, farm town, farm hamlet) a hamlet consisting of a small group of farms without a church, comparable with a baile (bal) in the Highlands. If a church is present it becomes a KIRKTOUN or CLACHAN-2. G.

Ferrel's Law a law postulated in 1856 by the American scientist, W. Ferrel, who developed the concept of CORIOLIS FORCE. The law states that due to the force produced by the rotation of the earth, a body moving over its surface will be deflected to the right in the northern hemisphere, to the left in the southern, the force at the equator being zero, and increasing progressively with distance from the equator. The effects are especially noticeable in water and in air. BUYS BALLOT'S LAW.

ferric *adj.* pertaining to or containing IRON, in chemical terms applied to those compounds having a higher valency (usually 3) of iron, i.e. a trivalent iron compound, e.g. HAEMATITE (ferric oxide, Fe_2O_3). FERROUS, OXIDATION, VALENCY.

ferricrete a DURICRUST in which the cementing agent is formed from iron oxides. CALCRETE, HARD PAN, SILCRETE.

ferriferous *adj.* iron-bearing, applied especially to rocks. IRON.

ferrite any of the IRON ores.

ferromagnesian mineral a dark, dense mineral, containing high proportions of IRON and MAGNESIUM, mainly of the AMPHIBOLE, OLIVINE and PYROXENE groups, e.g. AUGITE, BIOTITE, HORN-BLENDE. MAFIC.

ferromagnetic *adj.* possessing FERRO-MAGNETISM.

ferromagnetism the property possessed by iron, nickel and cobalt, as well as by some alloys, of having a very great magnetic permeability, of holding a residual magnetism in the absence of an external MAGNETIC FIELD and of re-sponding characteristically to a mag-netic field. Horseshoe and bar magnets are usually FERROMAGNETIC. CURIE POINT, ELECTROMAGNET, MAGNET, MAGNETISM.

ferromanganese an alloy of IRON and MANGANESE used in steel making.

ferrous *adj.* of or containing IRON, applied especially to compounds con-taining or derived from divalent iron, i.e. iron compounds not saturated with oxygen. They are usually formed by the reduction of FERRIC compounds. OXI-DATION, VALENCY.

ferruginous *adj.* of or containing IRON or iron RUST-1, or resembling the colour of rust, applied especially to reddish coloured rocks containing some iron.

ferry **1.** a boat (ferryboat) or an air-craft (air ferry) carrying people, goods, vehicles from one point to another on a regular route **2.** a place where people, goods, vehicles are carried, or the service carrying them.

ferry port a port serving ferries. FERRY, PACKET PORT.

fertile *adj.* **1.** highly productive, e.g. a fertile soil **2.** capable of breeding or reproducing **3.** capable of developing, e.g. a fertile egg **4.** in demography, applied to a woman or man who has produced at least one live-born child.

Fertile Crescent the approximately crescent-shaped area of fertile land bounded by the rivers Tigris and Euphrates that was the site of the Sumerian, Babylonian, Assyrian, Phoenician and Hebrew civilizations, and where agriculture was practised with the help of elaborate irrigation and drainage systems. HYDRAULIC CIVILIZATION.

fertility **1.** the state or condition of being FERTILE **2.** in general, sometimes loosely applied to the potential capacity to reproduce (to produce live-born chil-dren), but fecundity is the more precise term for this **3.** in demography, actual reproduction performance (FERTILE-4). In measuring fertility in a human population a broad distinction is made between current fertility (measured by the number of births of a particular year) and period fertility or cohort fertility (the total number of children born to a COHORT throughout the reproductive lives of the members). REPRODUCTION RATE.

fertility ratio the number of children under 3 years of age in a population related to the number of women of child-bearing age (usually taken as between the ages of 15 and 50).

fertilizer a substance consisting of one or a mixture of chemical elements essential for plant growth (NITROGEN, PHOSPHORUS, POTASSIUM) added to the soil to enrich it or make good its deficiencies. Most plant fertilizers are now artificial, made in factories, but natural mineral fertilizers include POT-ASH salts, such as the famous Stassfurt deposits in East Germany, NITRATE of soda from the deserts of Chile, PHOS-PHATES of lime, notably from north America, and GUANO from the drop-pings of birds accumulated in arid lands, especially tropical islands. Natural organic fertilizers include fish meal, BONEMEAL, dried blood, dung, composted vegetable remains (COM-POST).

fetch the distance of open water over which a wind blows and over which a sea WAVE-3, blown by the wind, travels, the length being an important factor in determining the height and energy of the wave and thus its effect on the beach and coast. G.

feud historically, an estate (LAND TENURE) held by a tenant (a vassal) on condition of services being rendered to an overlord. FEUDAL SYSTEM.

feudal system an economic, political and social system, a pre-capitalist MODE OF PRODUCTION common in western Europe between the ninth and sixteenth centuries, and elsewhere (e.g. in Japan to the nineteenth century), taking a wide variety of forms. Commonly the system involved a social hierarchy based on land held in FEUD, and the relationship of landlord to tenant, the former having jurisdiction over the latter. Neither labour nor the products of labour were COMMODITIES-1. The tenant, legally tied to the land, could own some of the MEANS OF PRODUCTION but the land and a set part of the product was the property of the landlord. For example, in England the MANOR was the economic unit. The lord of the manor held the land from the King in return for military service and homage. Tenants who occupied and cultivated the land held it from the lord in return for dues (services or the products of their labour). Bound in service to the lord or to the manor were the serfs, labourers 'attached to the soil', whose mobility was restricted and who were transferred from one lord to another if the land changed hands.

ffridd, ffrydd (Welsh, pl.friddoedd) 1. the fenced-in land near a farmstead in hill-farming country 2. rough grazing (enclosed from a mountain), a sheepwalk, a wood. RHOS. G.

F horizon SOIL HORIZON.

fiard FJARD.

fibre 1. a fine thread-like structure occurring in animal or plant tissue,

being extracted from some plants and used in the making of ropes (HEMP), PAPER, LINEN and other fabrics (CHINA GRASS); or produced by some animals (SILK) 2. such a structure made artificially (SYNTHETIC FIBRE) which, like a natural fibre, can be spun, woven, felted etc. 3. a substance composed of threads, e.g. cotton fibre 4. a fibrous structure.

fibric layer a soil layer consisting of organic material so weakly decomposed that the separate species of its origin can be identified. HUMIC LAYER, HUMUS, MESIC LAYER.

field 1. an area of land, usually enclosed, under pasture or crops 2. an area of land yielding mineral resources, e.g. COALFIELD, OILFIELD, or allocated to special use, e.g. playing field 3. in physics, a region in which force is exerted, originating from a source at the point in the field, e.g. electric field, MAGNETIC FIELD.

field capacity in soil science, storage capacity, the amount of water held in a well-drained soil by CAPILLARITY after excess water has drained away by gravity and the rate of downward movement has materially decreased. INFILTRATION CAPACITY, WILTING POINT. G.

field garden allotments that part of a COMMON inclosed under the Inclosure Acts which was required to be set aside for letting in gardens not exceeding a quarter of an acre (0.1 ha) to the poor inhabitants of the parish or district. Fuel allotments, for the taking of turf, peat etc. for fuel had a similar origin.

field system the medieval system of husbandry in England and parts of western Europe. The arable land of the village consisted of unenclosed strips (STRIP CULTIVATION) held by different owners or cultivators, and was subject to use as common pasture for a certain period in each year. Where the common arable was divided in two (one part cultivated, the other lying fallow) the system is known as Two-Field. More usual was the division into three (two cultivated, one part lying fallow), termed the Three-Field system. OPEN FIELD. G.

fig *Ficus carica*, a deciduous shrub or small tree, probably native to Asia, now widespread in warm midlatitude and subtropical regions. It is grown for the sake of its sweet, nutritious fruit, an unusual multiple fruit consisting of the swollen fleshy axis within which the flowers are borne and fertilized by the female fig wasp which enters through a small hole at the end of the fig. Some varieties, e.g. those grown in the British Isles, bear fruit without being pollinated. The fruit is eaten fresh, or is dried, tinned or preserved.

filbert *Corylus maxima*, a small tree, native to southeast Europe, bearing oval edible NUTS, often crossbred with cob or HAZEL.

filling in meteorology, the condition in a low pressure system (DEPRESSION-3) when there is an increase of atmospheric pressure in the central area so that there is a filling-up, and dying away of the depression, in contrast to a DEEPENING.

film water in soil science, water retained in layers thicker than one or two molecules on the surface of particles in unsaturated soil. G.

financial institution an institution capable of providing finance capital, e.g. pension funds, assurance companies, banks.

financial land ownership the ownership of land by FINANCIAL INSTITUTIONS.

finger lake a long narrow lake occupying a U-shaped glacial valley. G.

finger millet *Eleusine coracana*, a short-stemmed MILLET with the ear consisting of five spikes radiating from a central point, probably originating in the Indian subcontinent, now grown as a food grain there, in the drier parts of Sri Lanka, and in semi-arid areas of Africa, where it is often the staple food.

The grain stores well for as long as five years, and is thus important as famine reserve food. Very little enters trade outside the growing areas.

fiord FJORD.

fire-bote ESTOVERS.

firebreak a wide strip of land cleared of trees in a forest to act as a barrier to the spread of fire.

fireclay a CLAY, almost pure ALUMINIUM SILICATE, with an ability to withstand high temperatures, commonly occurring under coal seams in the Lower Coal Measures, as in many British coalfields. It is used for the lining of furnaces and in making fire-resistant bricks, crucibles etc. GANISTER, SEAT-EARTH.

firedamp 1. a combustible mixture of gases, mainly METHANE, occurring in coalmines, the product of the decomposition of COAL 2. an explosive mixture formed of this and air.

fire ecology the study of the effect of fire on ECOSYSTEMS, the cause of the fire being natural (e.g. lightning strike or spontaneous combustion) or artificial (e.g. deliberate, in the burning of bush to improve pasture growth; or accidental, as in a forest fire started by a burning cigarette butt etc.).

fire succession a plant SUCCESSION-2 which follows a fire. FIRE ECOLOGY.

firiki, firki (west Africa: Kanuri) a tract of very flat land, with a fine-particled, completely IMPERVIOUS clay soil, occurring particularly in the Chad basin and in parts of northern Nigeria. G.

firka (Indian subcontinent: Hindi) a subdivision of a THANA.

firn (German, last year's snow; French NEVE) snow which has been partly consolidated by alternate thawing and freezing but is not yet at the stage of being glacier-ice because, although the particles are partially joined together (which distinguishes it from snow), the small air spaces between the particles communicate with each other (which distinguishes it from ice). G.

firn equilibrium line in glaciology, the line between the net ablation zone (allowing for refreezing) of a glacier and the area of accumulation, i.e. the level at which ABLATION, allowing for refreezing, balances ALIMENTATION.

firnification the process by which snow is transformed to FIRN.

firn line, firn limit the highest level reached by the snowline during the year, or the boundary of the snow-cover at the end of summer. G.

First Point of Aries RIGHT ASCENSION.

First World the countries which have some form of capitalist, market-orientated economy, including the countries of northwestern Europe and of Australasia, and the USA, Canada, Japan. Spain, Portugal, Greece, South Africa and Argentina are sometimes qualified as borderline. SECOND WORLD, THIRD WORLD.

firth (Scottish) loosely applied to an area of coastal water, e.g. an arm of the sea, the lower part of an estuary, a fjord, a strait. G.

fiscal *adj.* **1.** of or pertaining to public money, taxes, debts etc. **2.** financial.

fish a member of Pisces, a class of VERTEBRATE aquatic animals, many yielding nutritious flesh, some yielding oil for food or industrial use. Some (e.g. bream) live in freshwater, others in the ocean, where the PELAGIC, free-swimming species (e.g. herring) swim in shoals in surface waters, the demersal fish (e.g. plaice, sole, halibut) living near the sandy sea bed in shallow waters. The salmon, in nature, spends part of its life cycle in the ocean, part in a river. FISH FARM.

fish farm a FARM where fish are bred and reared for market, under carefully controlled conditions in tanks or in protected parts of coastal waters, lakes, rivers and their estuaries.

fishing port a PORT used by fishery vessels for anchorage, fuelling, repair etc. and for the landing of fish, where these are important among the other activities of the port.

fissile *adj.* able to be split, applied particularly to rocks which are easily split along their well-developed BEDDING PLANES, CLEAVAGE, LAMINAE or JOINTS, e.g. shale, slate.

fission the act or process of splitting into parts. In nuclear fission (NUCLEAR ENERGY) the atomic nucleus is split into roughly equal parts by bombardment with neutrons, with an overall loss in mass (FUSION), thereby releasing great energy.

fissure **1.** an extensive crack, a narrow opening cleft or fracture made by splitting, especially in a rock **2.** a linear volcanic VENT. FISSURE ERUPTION, VOLCANO.

fissure eruption linear eruption, a steady, non-explosive outpouring of LAVA, usually BASIC, from the depths of the earth to the earth's surface, occurring along a weak line (sometimes many kilometres in length) in the crust, often resulting in a BASALT plateau. CENTRAL ERUPTION, VOLCANO. G.

fixation **1.** the conversion of a free element, such as atmospheric NITROGEN, into a solid compound. LEGUMINOSAE, NITROGEN CYCLE **2.** the conversion of any substance to a less volatile form.

fixed capital the buildings and machinery of a firm. CAPITAL-3.

fixed costs **1.** costs that do not vary with volume of output within a certain range (unlike VARIABLE COSTS), i.e. capital investment in plant and machinery, rent, rates and certain other overhead costs incurred irrespective of the volume of production **2.** in spatial economic analysis, costs that are constant in space, e.g. the cost of finance, labour costs if fixed rates apply in the particular area, any components or materials sold at a uniform delivered price.

fjärd, fiard (Swedish) a term applied by some English authors to an inlet of the sea with low banks, the result of a rise in sea-level (SUBMERGED COAST), in contrast with FJORD. This does not conform to the original Swedish application, which was simply to a large continuous area of water surrounded by SKERRYGUARD islands. G.

fjeld (Norwegian, a field) widely applied in Norwegian to any upland rocky area, but restricted in English to an elevated, rocky plateau above the TREELINE, snow-covered in winter. FELL. G.

fjord, fiord (Norwegian, a sea inlet; fiord in old Norwegian) a term widely applied in Norway and Denmark to almost any sea inlet, but restricted in English use to a deep narrow inlet in the coast with high rocky parallel sides smoothed by ice action, HANGING VALLEYS and an irregular rocky floor, the floor being frequently deeper than the sea floor offshore, from which it is separated by a submerged sill near the entrance. It is caused by the submergence of a deep glacial valley. SUBMERGED COAST. G.

flagstone, flag (Old Norse flaga, a slab) a natural hard stone used for paving, capping walls etc., usually cut from SANDSTONE or some types of LIMESTONE which split conveniently and easily along BEDDING PLANES. FISSILE.

Flandrian the post-glacial stage of the QUATERNARY in northwest Europe, succeeding the Late-Glacial (Devensian, GLACIAL STAGES), when the climate became less severe, woodland began to spread, and the sea-level rose (FLANDRIAN TRANSGRESSION). It has been divided into four zones. The first (Fl.I) is termed the protocratic, when trees began to appear as a result of higher temperatures; the second (Fl.II), the mediocratic (5500 to c.3200 BC), the climatic optimum (ATLANTIC STAGE), when temperatures were at their highest; the third (Fl.III), the present, the terminocratic, when temperatures

begin to fall, there is less woodland, and rainfall increases after c.1200 BC. The fourth (Fl.IV), the cryocratic, the future, is purely hypothetical.

Flandrian Transgression the global rise in sea-level of post-glacial times, as applied to the coastal areas of northwest Europe. As the ice sheets retreated great expanses of land became parts of the CONTINENTAL SHELF, the North Sea came into existence and Britain was cut off from Europe (c.5000 BC). ATLANTIC STAGE, FLANDRIAN.

flark (Swedish) a limited wet area occupied by weakly peat-forming FEN vegetation alternating with drier areas which act as dams. G.

flash 1. in Cheshire, England, a subsidence of the land surface due to the working of rock-salt and the pumping of brine; or a small lake, pool or marsh occupying such a depression. SWAG 2. a sudden rise of the water level in a river or stream. FLUSH. G.

flash-flood a sudden violent FLOOD caused by exceptionally heavy rain in a normally dry valley in a semi-arid area, the torrential stream sometimes being laden with debris; or by the collapse or breach of a dam or sea wall. ENVIRONMENTAL HAZARD.

flat 1. a stretch of level ground without marked hollows or elevations, or of country without hills 2. any nearly level stretch of land within hilly country 3. a low-lying tract of marshy land 4. the low land through which a river flows 5. mud-flat, a bank of mud over which the tide flows, exposed at low tide 6. an alluvial deposit yielding tin or gold in a stream bed 7. a horizontal deposit of ore in a BEDDING PLANE. G.

flatt an alternative name in some areas for FURLONG, a unit in the OPEN FIELD system.

flax a plant of the genus *Linum*, family Linaceae, yielding fibre, very resistant to decay, and probably the first of all vegetable fibres to be used by people (in Neolithic times), and seeds containing oil. There are some 90 species, those providing good fibre producing poor oilseed, and vice versa. It grows in a wide variety of climates, but flourishes best in a firm moist soil. *Linum usitatissimum*, is the most widely grown. In northern Europe and the USSR it yields good fibre, termed flax, derived from the inner bark (bast) of the slender stems, up to 127 cm (50 in) in length. It is spun and woven into a cool, absorbent fabric, linen, the shorter fibres being used for twine and cord; and linen rags provide high quality paper. But flax is expensive to produce, needing much labour, and it faces competition from synthetic fibres. Flax for oilseed is grown mainly in the Indian subcontinent and Argentina. Oil from the seed (linseed) is used in making paints and varnish; the crushed cake, remaining after the extraction of oil, makes good cattle feed; the seeds when ground produce medicinal linseed-meal. C.

F layer 1. in the ATMOSPHERE-1, a layer in the IONOSPHERE which reflects high frequency radio waves. It lies at a height of some 250 km (155 mi) above the earth's surface.

fleece 1. the wool coat of a sheep, goat, etc. **2.** the wool taken from a sheep etc. at one shearing. WETHER.

fleet a creek, inlet, run of water, small lagoon. The term is obsolete or of local use only, but common in English place-names, e.g. Fleet Street, London; Northfleet, Kent. G.

flexure 1. the state or process of being bent **2.** the bending of strata etc. under pressure.

flint a heavy, very hard, grey or grey-black nodule or mass of a pure fine-grained SILICA, encrusted with white, occurring in the BEDDING PLANES or JOINTS in the Upper Chalk, and probably formed when silica-rich biogenic OOZE became mixed with chalk mud on the ocean bed, the silica being carried downwards in solution and precipitated

round a core (e.g. the skeletal remains of a sponge, sea-urchin etc.), the flint nodule gradually enlarging within the chalk mud. Easily chipped to a cutting edge, flint was the main source of tools and weapons of the Stone Ages (EOLITH). It has long been used as a building stone; and when struck with steel it emits a spark, hence its use in a flintlock gun.

floatation, flotation a technique used to separate minerals, based on their ability to float or sink, e.g. the separation of minerals from crushed ore by stirring with heavy oil, the great differences in surface tension bringing about the separation. SALINOMETER.

floating dock a large floating structure into which a vessel may pass and which can be used as a DRY DOCK. DOCK.

floating ice a general term applied to all types of ice (apart from ICEBERGS) wholly floating on water.

floating value a potential value for land created by the possibility of future DEVELOPMENT-2.

flocculation (Latin flocculus, a tuft) **1.** in soils, the process by which fine-grained particles come into contact and gather together into tufted masses to form CRUMBS, thereby improving the soil texture, especially the aggregation of colloidal (COLLOID) particles (clay) into tufted masses in the presence of an alkali (lime) **2.** the process of the aggregation of water-borne colloidal particles into small lumps which are able to settle out, occurring when a river carrying electrically charged colloidal clays in its fresh water meets and mixes with the sea, which carries electrically charged particles in solution in its salt water.

floe ICE FLOW. G.

flood the overwhelming of usually dry land by a large amount of water that comes from an overflowing river or lake, an exceptionally high tide, melting snow, or sudden excessive rainfall.

flood basalt basaltic LAVA, very fluid at

221

high temperatures, which has spread over a very extensive area as the result of an eruption or series of eruptions.

flood control 1. the regulation of excessive RUN OFF of water in order to prevent inundation of the land, e.g. by the building of river BARRAGES; the deepening of existing and the cutting of new channels to speed river flow; the making of temporary storage basins, such as TANKS; the conservation or planting of vegetation in a drainage area to slow down run off 2. the building of sea walls etc. to prevent inundation by exceptionally high tides.

flood hazard the dangerous chance of the inundation of usually dry land by water from an overflowing river or lake (e.g. caused by a break in a DAM), melting snow, sudden excessive rainfall, or an exceptionally high tide.

floodplain, flood-plain the relatively level part of the valley bordering a river resulting from alluvium deposited by the river in times of flood. G.

floodplain splays materials spread over the FLOODPLAIN through breaks in natural LEVEES, through restricted low sections in the river course, or along distributary channels. G.

flood stage of a river, the stage when a river begins to overflow its banks. BANKFULL STAGE, OVERBANK STAGE, STREAM STAGE.

flood tide, flood-tide the advancing or rising tide, starting at low tide, ending at high tide, in contrast to EBB TIDE. G.

flora 1. the plant life considered collectively of a region or age 2. a list of plant species of a particular area arranged in families and genera (CLASSIFICATION OF ORGANISMS), with descriptions and a key to aid identification. BIOTA, FAUNA. G.

flotation FLOATATION.

flotsam the wreckage of a ship or its cargo found floating on the sea. JETSAM.

flour 1. the finely-ground meal (the edible part of coarsely-ground GRAIN-5)

of CEREALS such as WHEAT, BARLEY, RYE 2. the finely-ground edible parts of other food crops, e.g. of POTATOES or CASSAVA. ROCK FLOUR.

flow 1. of a fluid, or something that behaves comparably, the smooth movement with a continuous change in shape, e.g. of lava, or of a stream confined between banks, or of grain, under the influence of gravity; or of water gushing forth from a spring; of air; of solid rocks under stress, without fracturing (i.e. plastic flow); of ice in a glacier 2. a steady movement of ideas, goods etc. (e.g. of international trade).

flow diagram, flow chart a diagram showing a sequence of interconnected events, actions or items to indicate the progressive development or evolution of a theme, product, or other objective.

flow-line map, flow map a map with lines which indicate movement of freight and/or of passengers, the direction of movement being shown by arrowheads, the amount of traffic by the thickness of the lines, drawn to scale. In some flow-line maps the scale used is arithmetical, i.e. the width of the line is directly proportional to the intensity of flow. But this scale cannot be used if the range of flows is great, with a few disproportionately large values. It is replaced by a geometrical scale which exaggerates variations in traffic, e.g. by making the width of the flow-line equivalent to the square root of the traffic value, so that a line which is twice the width of another indicates four times the traffic. This allows only relative, not absolute, comparisons between flows to be made from the map. NETWORK-SPECIFIC FLOW DATA, NODE-SPECIFIC FLOW DATA.

flow meadow WATER MEADOW.

flow resources NATURAL RESOURCES.

flow till TILL that is moved by glacier ice after it has been deposited, the structures formed by the parent ice being destroyed. The sliding may be subglacial (due to gravity or shearing stress) or superglacial, down the surface of the ice or the slopes of a ridge of dead ice.

fluid   matter that does not have a fixed shape, i.e. a GAS or a LIQUID, being able to flow along a channel or tube. In general, and in biology, the term is often restricted to a liquid. SOLID, VISCOSITY, VISCOUS.

fluidization   a process in IGNEOUS activity by which hot gases, moving through fine-grained material, cause it to flow, thereby cracking and enlarging existing cracks in the rock by physical and chemical action. In some cases the enlargement of the crack is cylindrical, so that a PIPE forms.

flume   1. an artificial channel made to carry water over some distance for irrigation or for industrial use, e.g. for power, for transport (e.g. of logs), or in PLACER mining 2. (American) a ravine or gorge with a stream flowing through it. G.

fluorine   a nonmetallic element, normally a yellow-green gas, highly reactive and a strong oxidizing agent.

fluorine dating   a method of assessing the age of the remains of VERTEBRATES by studying their FLUORINE content, based on the fact that the remains of bones and teeth absorb fluorine from percolating ground water at a constant rate. As the proportion of fluorine in water varies from one district to another, the method is useful only in establishing that the remains in one locality are of the same age.

fluorspar, fluorite   a native calcium fluoride, CaF_2, occurring in clusters of cubic glassy crystals, commonly as a GANGUE in a vein of metallic mineral, especially of LEAD or ZINC. A source of fluorine compounds, it is colourless or coloured by impurities, the dark blue variety mined in Derbyshire, England, being known as Blue John which, like some other varieties, is used ornamentally. Fluorspar is also used in industry as a flux in the basic OPEN HEARTH PROCESS in steel making, in CERAMICS, and in the production of hydrofluoric acid. It was formerly used in lead smelting.

flush   a sudden growth in the volume of a stream, creating a rush of water that does not quite overflow the banks. G.

fluting and grooving   small ridges and depressions caused by differential erosion, especially by wave action, on an exposed rock-surface with marked jointing. JOINT. G.

fluvial, fluviatile   *adj.* **1.** of or pertaining to a river **2.** found or living in a river. It is usual to apply fluvial to the action of the river (flow and erosive activity); and fluviatile to the deposits laid down by the river, or to the flora and fauna of a river. G.

fluvial landform   a landform shaped by running water.

fluvioglacial   *adj.* produced by or due to the action of streams of water derived from the melting of the ice of the SNOUT of a GLACIER or the margin of an ICE SHEET, applied particularly to the deposits of an OUTWASH FAN. Some glaciologists, especially in the USA, prefer the term glaciofluvial because the action of the glacier precedes the flow of the streams. G.

fluvioglacial deposition   the laying down of stratified DRIFT-1 by melt-water, especially in an OUTWASH APRON, a VARVE, or a PROGLACIAL LAKE. Stratified drift consists of rounded, washed and sorted sand and gravel, unlike TILL, which is angular or subangular and not sorted.

fluvioglacial erosion   **1.** the cutting of a channel or trough by melt-water along the edge of an ice sheet **2.** the cutting of an OVERFLOW CHANNEL across a preglacial watershed.

fluviokarst   KARST landforms that are formed by the joint action of true karst and FLUVIAL-1 processes.

fluviraption   the sweeping away of loose material by moving water. G.

fluvisol   an ENTISOL, an ALLUVIAL

SOIL on a FLOODPLAIN periodically buried as new layers of alluvium are deposited, increasing its depth.

flux a substance added to another in order to help melting and promote fusion, e.g. in a BLAST FURNACE.

fly (Norwegian) the fairly steep slope lying between a SCARP face and the VIDDA below. G.

Flysch (German) a FORMATION-2 term applied specifically to the deposits of coarse sandstone, calcareous shale, conglomerate, marl and clay of early Tertiary (Eocene) and probably late Cretaceous age (GEOLOGICAL TIME-SCALE) laid down on the margins of the Alpine ranges, consisting of material eroded from these rising fold ranges before the maximum building stage in Miocene times (in contrast with MOLASSE, deposited after the maximum building stage). The term is extended to apply to similar deposits elsewhere. G.

focus SEISMIC FOCUS.

fodder 1. loosely, any food for cattle, sheep, horses 2. specifically, dried hay, straw etc. used as food for some livestock, e.g. cattle, sheep, horses.

fog an opaque CLOUD in the ground surface layers of the atmosphere, consisting of condensed water vapour with smoke and dust particles held in suspension, that obscures vision for any distance up to one kilometre (International Meteorological Code). ADVECTION, ARCTIC, COLD-WATER DESERT, FRONTAL FOG, HAZE, ICE FOG, INVERSION, MIST, RADIATION FOG, SMOG, STEAM FOG. G.

fogbow an arc of light seen when the sun is behind an observer and fog is in front. The small size of the droplets in the fog prevents the refraction and reflection of light into the colours of the SPECTRUM, so the arc of light, actually colourless, appears white.

fog drip, fog precipitation moisture deposited from fog especially in COLD-WATER DESERT coastal areas, where low CLOUD or FOG encounters trees,

the water droplets saturate the branches, twigs, leaves, and then drip to the ground. G.

foggara (Sahara: Arabic) in northwest Africa, especially in Mauritania, a gently inclined underground water channel bringing water for irrigation from AQUIFERS near the foot of mountains and carrying it, in some cases over long distances, to dry areas. A foggara is comparable with a KAREZ of Baluchistan, or a QANAT of Iran. G.

Föhn, Foehn, Fön (German) a warm dry wind blowing down slopes on the leeward side of a ridge of mountains, the warmth being due to the compression of the descending air, as in BERGWIND, CHINOOK, NOR'WESTER (New Zealand), SAMUN, SANTA ANA, ZONDA. ADIABATIC. G.

Föhrde, Förde (German) a long, straight-sided, narrow inlet of the sea in BOULDER CLAY country, formed by the drowning of a valley along a low-lying coast when the sea-level rose. These valleys are believed to have been formed originally by rivers flowing beneath an ice sheet. G.

fold 1. in farming, a pen or enclosure, especially a temporary one, often made with movable hurdles, for sheep or other domestic animals 2. the condition when one part of something is made to lie on another, hence, in geology, used as a general term for bending (FLEXURE) in the rocks (STRATA) the earth's crust resulting from COMPRESSION or GRAVITATION-1. The fold may be a simple arch or upfold (ANTICLINE) or a hollow or downfold (SYNCLINE), either of which may be SYMMETRICAL or ASYMMETRICAL, the axis of an asymmetrical anticline being so slanted that an OVERFOLD or RECUMBENT FOLD may result. A very large anticline is a GEANTICLINE or geoanticline; a big syncline a GEOSYNCLINE. A complex anticline is an ANTICLINORIUM, a complex syncline a SYNCLINORIUM. A fold with only one limb is a MONOCLINE (English usage); and repeated tight folds form ISO-CLINAL strata. BREACHED ANTI-CLINE, HETEROCLINAL, NAPPE, OR-OGENY, PERICLINE. G.

foldage in medieval farming, the practice of feeding sheep in movable FOLDS-1. G.

fold course, fold-course, foldcourse in Norfolk, England, fifteenth to eighteenth centuries, an area comprising, usually, open field arable land, permanent pasture (heath, marsh) and sometimes closes (HALF-YEAR CLOSE) within which sheep were confined, eating a different type of feed as it became available. G.

foldland 1. land carrying the right of FOLDAGE **2.** the right of foldage.

fold mountain, folded mountain a mountain resulting from the folding (flexuring) of the earth's crust, in contrast to a BLOCK MOUNTAIN. COL-LISION ZONE, YOUNG FOLD MOUN-TAIN. G.

foliation (Latin folia, leaves) **1.** the more or less parallel, wavy layers or bands of minerals found in some METAMORPHIC ROCKS, e.g. SCHISTS, GNEISS (METAMORPHISM) or in some IGNEOUS ROCKS which flowed as they cooled **2.** the wavy bands of ice seen in the depths of a GLACIER.

folk 1. the great mass of common people of a nation **2.** the unsophisticated, less educated common people of a nation, who maintain and pass on traditional culture (FOLK CULTURE) to succeeding generations **3.** people as belonging to a class or distinct group.

folk culture the social heritage (institutions, customs, conventions, collective wisdom, arts and skills, dress, way of life etc.) of a close group of people (FOLK-2) who maintain a deep attachment to it, and who hand it down orally, by ritual or by behavioural habits from one generation to the next. Such a culture develops indigenously and is relatively static.

Food and Agriculture Organization FAO.

food chain a chain of organisms, existing in any natural community, along which energy passes, the organisms in each link obtaining energy by preying on or parasitizing those in the preceding link, and being so treated in turn by those in the succeeding link. At the first energy level, the first TROPHIC LEVEL (termed T) there is nearly always a green PLANT-1 or other AUTOTROPHIC organism using energy from the sun and making energy available to the consumer levels which follow it. At the second trophic level (T_2) are the primary consumers, the HERBIVORES, which supply energy to the secondary consumers, the smaller followed by the larger CARNIVORES (T_3 and T_4). At each trophic level much of the energy obtained is lost in RESPIRA-TION (SECONDARY PRODUCTION), so that progressively fewer organisms can be supported. Thus in a balanced community the BIOMASS of each trophic level is always greater than that of the succeeding level. DIVERSIVORES will eat anything; BACTERIA, FUNGI and various microorganisms are consumers operating at all levels. All organisms, including predators, die naturally if not eaten, their dead tissues being decomposed by microorganisms (decomposers) which release CARBON DIOXIDE, AMMONIA and mineral salts to the environment. The sequence up to the death of an organism is termed the predator chain, the energy flow after death being termed the detritus chain. Any natural community will have many interlinked food chains, making up a food cycle or FOOD WEB. BIOTIC PYRAMID, BIO-LOGICAL AMPLIFICATION, CARBON CYCLE, ECOLOGICAL EFFICIENCY, ENERGY PYRAMID, NITROGEN CYCLE, PHOSPHORUS, PHOTOSYNTHESIS, PRIMARY PRODUCTION, PRODUC-TION-2, PRODUCTION RATE.

food web all the interconnected FOOD CHAINS in an ECOSYSTEM.

foot ft, pl.feet, a unit of length in

English-speaking countries, one third of a standard YARD, i.e. 12 inches, equal to 30.48 centimetres. CONVERSION TABLES, CUSEC.

foothills a belt of hills aligned approximately parallel to a mountain range and lying between it and the plain below.

footloose industry, footloose activity a mobile enterprise, usually LIGHT INDUSTRY, which, by the nature of the raw materials used and its labour, market and transport needs, is not tied to a particular location. INDUSTRY. G.

footrail in the North Staffordshire coalfield, England, an ADIT MINE. G.

foot wall, foot-wall, footwall 1. the lower side of a FAULT **2.** in mining, the solid rock underlying an ore body or vein.

forage crops in Britain, all CROPS-1 and grass (including grass from ROUGH GRAZING) grown on a farm specifically for grazing LIVESTOCK, but not PULSES or crops harvested as GRAIN-5.

foraminifer a protozoan of the order Foraminifera, class Rhizopoda, a microorganism (mostly marine) with a CALCAREOUS shell, whose skeletal remains form a major part of CHALK and of OOZE. Some shells are microscopic, others (in colonial form) are as much as 1.25 cm (0.5 in) in diameter. COLONIAL ANIMAL.

Forbes' bands DIRT BANDS. OGIVE. G.

force (Old Norse fors) in northern England, a waterfall or cascade. G.

ford a part of a river shallow enough to be crossed by wading, hence common in riverside place-names, e.g. Bedford, Guildford, Oxford.

forecast a prediction based on scientific observation, on experience, or by an estimate of probability, as in a weather forecast.

foredeep 1. a relatively narrow, deep, elongated, steep-sided trough in the ocean floor, near or parallel to a mountainous land area, or associated with an ISLAND ARC **2.** such a trough when infilled with sediment, e.g. the Himalayan foredeep, now in the Indo-Gangetic plain. DEEP, OCEAN. G.

foredune a DUNE or line of dunes most close to the sea on a sandy shore, comparable with an ADVANCED DUNE in a hot desert.

foreland land or territory lying in front, a term widely applied, e.g. **1.** a low PROMONTORY such as a CAPE or headland jutting seawards, e.g. North Foreland, Kent, England **2.** a stable continental mass on the margin of a GEOSYNCLINE **3.** in the study of ports, the seaward trading area connected with the port through maritime organization. CUSPATE FORELAND, GLACIAL FORELAND, MEDIAN MASS. G.

foreset beds BOTTOMSET BEDS, DELTA, DELTA STRUCTURE. G.

foreshore a term loosely applied to the part of the SEASHORE lying between the lowest LOW WATER line and the average HIGH WATER line. BACKSHORE.

forest 1. an extensive, continuous area of land dominated by trees (some authors would specify large trees in order to contrast forest with WOODLAND, implying that woodland consists of smaller trees associated with much undergrowth or scrub – a doubtful distinction, difficult to maintain), sometimes including patches of pasture, economically exploited or not and (FAO qualification) capable of producing wood, or of influencing local climate or the water regime, or providing shelter for livestock and wildlife **2.** in botany, an extensive plant society of shrubs and trees with a closed CANOPY, capable of self-perpetuation (thus forming climax forest) or of development to that CLIMAX (PRIMARY FOREST). Mixed forest is composed of two or more tree species with at least one-fifth of the canopy comprising species other than the dominant one. BOREAL, CONIFEROUS, DECIDUOUS, ELFIN, EQUATORIAL, GALLERY, MONSOON,

MOSSY, RAIN and TROPICAL FOREST and FORESTRY, LUMBER, NATURAL RESOURCES, PATCH CUTTING, SUSTAINED YIELD FORESTRY, TIMBER 3. historically in medieval England, a legal term, an area of land, not necessarily tree-covered, set aside for a royal hunting ground (the hunting rights being the property of the Crown), subject to the strict Forest Laws and outside the Common Law, surviving in such place-names as Dartmoor Forest, Bowland Forest (CHASE, FREE WARREN, PARK-1) 4. the association with hunting similarly survives in the term deer forest, an unenclosed hunting ground, devoid wholly or partly of sheep, cattle, crops, used for deer stalking. G.

forestal *adj.* of or pertaining to a FOREST-1,2. AGRO-FORESTAL.

Forest Law FOREST-3.

forestry the science and art of planting, cultivating and managing FORESTS-1, usually conveying the idea of management with a view to economic development, whereas WOODLAND management may not imply that. PATCH CUTTING, RESOURCE CONSERVATION, RESOURCE MANAGEMENT, SUSTAINED YIELD FORESTRY, TIMBER. G.

forked lightning a type of LIGHTNING taking the form of branching, zigzag flashes from cloud to ground.

form 1. in botany, the smallest sub-specific group in the classification of plants (CLASSIFICATION OF ORGANISMS), characterized by some minor variation from type 2. in zoology, a neutral term, loosely applied to some minor grouping of animals which do not fit readily into the classes species, sub-species or other minor grouping.

formal region a REGION differentiated by uniformity of characteristics (e.g. a desert), as distinct from a FUNCTIONAL REGION, which is characterized by activity.

formal theory THEORY generated by logical DEDUCTION from a priori assumptions. GROUNDED THEORY.

formation 1. in ecology, the largest natural vegetation type, the plants of a land BIOME 2. in geology, sometimes termed a stage, the fourth division in the hierarchy of rocks in GEOLOGICAL TIME, a subdivision of a series, corresponding in the time interval to an AGE (GEOLOGICAL TIMESCALE). It consists of a STRATUM or series of strata which are similar in character (homogeneous), having some property which makes them distinctive. G.

Formkreis, Formenkreis (German, pl. Formenkreise, circle of forms) related to landforms which owe their existence to the same action, e.g. of running water, or ice action, hence glazialer Formenkreis, i.e. a MORPHOGENETIC REGION.

form-line a CONTOUR based on general observations and sketched in, rather than one based on instrumental survey and determined by exact measurement. G.

form ratio, form-ratio of a stream, the relation of depth to width, expressed as a ratio. G.

förna (Swedish) LITTER, leaves and other similar material not yet decomposed and lying on the soil. G.

fors (Swedish, pl. forsar) a waterfall that is not steeply inclined; rapids, or cataract. G.

foss in northern England, probably derived from FORCE, a waterfall. G.

fosse (Latin fossa, a ditch; French fosse, an ocean deep) 1. historically, a trench for protection or drainage, surviving in place-names, e.g. the Fosse Way 2. a depression between a GLACIER and the sides of its trough, along which a KAME TERRACE may develop. G.

fossil 1. specifically, the remains, or an impression left by the remains, or the traces, of a plant or animal, preserved in the earth's crust by natural processes, usually in SEDIMENTARY

ROCKS, sometimes in metamorphic if the METAMORPHISM has not been too violent. The hard parts of the organism may be preserved intact, or they may be replaced by another mineral (CALCIFICATION, MINERALIZATION-2, PETRIFICATION). There may be internal or external casts of the organism in the rock, or impressions of soft parts, or carbon residues of the decomposed organism, or traces of plant and animal activities (e.g. tracks, burrows of animals, root passages of plants); or the organism may be preserved intact, e.g. mammoths in frozen ground in Siberia, insects in AMBER. PALAEONTOLOGY 2. loosely applied to any feature of geological age buried in the earth's surface naturally or by geological agencies, e.g. fossil cliff, FOSSIL FUEL, FOSSIL LANDSCAPE, FOSSIL WATER. G.

fossil erosion surface an exhumed EROSION SURFACE, i.e. a surface which was worn down, buried by deposits, and revealed when DENUDATION removed the deposits. G.

fossil fuel combustible material derived from the fossilized remains of plants and animals, i.e. COAL, NATURAL GAS, OIL.

fossiliferous *adj.* containing, rich in, FOSSILS.

fossil landscape a LANDSCAPE-3 of a former geological age, usually buried by deposits and revealed only when these have been removed by DENUDATION, i.e. when it is exhumed. The alternative term, exhumed landscape, is perhaps preferable. EXHUMATION, FOSSIL EROSION SURFACE.

fossil water connate water, water trapped in SEDIMENTARY ROCKS during their deposition. G.

Foucault's pendulum a pendulum designed in 1851 by J. B. L. Foucault, 1819-68, French physicist, to demonstrate the rotation of the earth on its axis. A heavy ball is suspended on a long wire and set swinging freely in space in a certain direction. The direction of the swing of the ball gradually moves round to the right until it returns to the original direction of the swing. The cause of this rotation is the earth's rotation. At either Pole Foucault's pendulum would make a complete rotation in 24 hours (15° per hour); but at the equator there would not be any rotation.

Fourth World THIRD WORLD.

foxtail millet *Setaria italica*, a strong, cultivated millet of subtropical and warm temperate regions, probably derived from *Setaria viridis*. The grain is used for human food, the remainder of the plant making good hay and silage.

fracture 1. a split, break or state of being broken 2. in geology, a clean break in STRATUM brought about by DEFORMATION under the strain of COMPRESSION or TENSION.

fractus cloud, fracto- ragged, shredded CLOUD indicating strong winds and stormy conditions in the upper atmosphere. The prefix fracto- is used to indicate a similar ragged condition in other cloud forms, e.g. fractocumulus, fractonimbus, fractostratus, etc.

fragipan HARD PAN.

fragmental rock CLASTIC rock, a rock composed of recognizable fragments of other rocks, combined by CEMENTATION or COMPACTION, e.g. clay, sandstone.

fragmentation 1. the breaking of a whole into separate, isolated parts 2. of a farm, the division of the land of a FARM into separate, isolated parts. If the farm remains a single holding it may become difficult to work as a unit if the land is divided (e.g. if the construction of a MOTORWAY cuts through the farm). If the fragmentation is the result of the law of equal inheritance, the land being parcelled out to family members, a number of small farm units of doubtful economic viability may be created. LAND REFORM.

fragmented *adj.* composed of isolated parts broken off and separated from the whole, e.g. a country with an EXCLAVE.

francium a radioactive ALKALI-1 metal element produced by the disintegration of actinium (a radioactive element occurring in URANIUM ores).

frankincense an aromatic gum resin burnt as incense, obtained from various trees of the genus *Boswellia*, family Burseraceae.

frazil ice fine needles or minute plates of ice suspended in water, the first stage of freezing, giving the surface of the water an oily, opaque appearance. SLUDGE. G.

free *adj* **1.** unhampered, unimpeded, at liberty, without restrictions **2.** in chemistry, not combined, e.g. free nitrogen, FREE OXYGEN.

freeboard a narrow, irregular belt of land marking the boundary between medieval parishes and counties, giving access to arable fields on each side, the origin of many of the winding lanes characteristic of the Midlands of England after inclosure. G.

free economy MARKET ECONOMY.

free face a vertical facet cut in bare rock, part of the slope profile of a hillside, below the WAXING SLOPE, above the CONSTANT SLOPE, too steep for debris to rest on it (the debris falls downwards to form SCREE). SLOPE. G.

free goods in economics, goods (GOOD) which are so abundant that at any given time and place they are obtainable without cost or, being scarce enough to warrant a price, they are impossible to sell. CAPITAL GOODS, COMMODITY, CONSUMER DURABLES, CONSUMER GOODS, ECONOMIC GOODS, PRODUCER GOODS.

freeholder, freehold ownership ownership in perpetuity. LAND TENURE, LEASEHOLD OWNERSHIP.

free oxygen gaseous OXYGEN or oxygen dissolved in a medium such as water. ANAEROBE, ANAEROBIC.

free port, free city an enclosed guarded city, usually a PORT or part of a port, where goods may be loaded or unloaded free from customs dues, most customs regulations and similar restrictions. It usually develops a large ENTREPOT trade. G.

free radical in chemistry, an ATOM or group of atoms with unpaired ELECTRONS.

freestone **1.** a rock that can easily be cut in any direction, thus a fine-grained rock of uniform particles occurring in thick beds, e.g. sandstone, limestone. BUILDING STONES **2.** in botany, a fruit in which the stone does not stick to the flesh, e.g. some varieties of PLUM or PEACH; or the stone of such fruit.

free trade the commercial interchange between nation states or an economic group of nations and others, in which there are no tariff barriers or other protective measures such as quota restrictions.

free trade area an area within which a group of nation states has agreed to operate a policy of FREE TRADE while each member maintains its separate trade policies with states outside the area.

free warren in medieval England, a right granted by the Crown to the lord of a manor to hunt smaller game (e.g. the fox, hare, rabbit, wild cat, pheasant and partridge) over his estate. CHASE, FOREST-3, PARK-1.

freeway in USA, a motor road, an EXPRESSWAY, usually free of TOLL.

freeze-thaw a form of WEATHERING in PERIGLACIAL areas where the temperature hovers around freezing point, below which frost breaks up the rock and above which the ice melts, so that water flows and carries away the rock fragments. NIVATION, PATTERNED GROUND, SOLIFLUCTION, TEMPERATURE SCALE, THAW.

freezing point the temperature at which a substance in the liquid state changes to the solid state, e.g. at which pure water changes to ice, 0°C (32°F). FROST, ICE POINT.

F region the outer region of the IONO-SPHERE, lying 200 to 400 km (125 to 250 mi) above the earth's surface.

freight 1. the hire of a vessel or vehicle to carry goods 2. the service of carrying goods, originally by ship but now extended to land vehicles and aircraft 3. the goods so carried 4. the payment for that service, sometimes being specifically applied to goods carried in bulk, more slowly than the normal service and at lower cost.

freight car (American) goods wagon.

freight container CONTAINER.

freighter 1. a ship or aircraft that carries cargo, i.e. FREIGHT-3 2. one who loads a ship or aircraft with cargo 3. one whose business is the receiving and forwarding of freight.

freight rates the money paid to a carrier for the loading, transporting and unloading of goods.

freight train (American) goods train.

French bean BEAN, KIDNEY BEAN.

French chalk a variety of TALC used as an absorbent or grease remover, and in drawing a line on cloth, etc.

frequency 1. in general, the condition or quality of repeatedly occurring 2. the number of times something, some event, or some process occurs in a given unit of time 3. in statistics, the number of members of a POPULATION-4, or a number of occurrences of a given type of event, falling into a single class in a statistical survey of the variation of specified characteristics. ARRAY, FREQUENCY DISTRIBUTION, RELATIVE FREQUENCY and the following entries qualified by frequency.

frequency curve a graph of a continuous FREQUENCY DISTRIBUTION. The variate (the class interval) is shown on the horizontal axis in order of increasing value from the intersection of the two axes, the class frequency on the vertical axis in order of increasing frequency from the point of intersec-

tion (FREQUENCY POLYGON). The data may group around the values in the centre of a range of values (CENTRAL TENDENCY), with a few low values (to the left) and about an equal number of high values (to the right). This forms a symmetrical curve, the curve of the theoretical NORMAL DISTRIBUTION. If the curve is asymmetrical it is said to be skewed. If most of the data group around relatively low values and there is a long tail of very high values, the slope on the right flank of the curve becomes shallower and longer than that on the left, and the curve is said to be positively skewed. But if most of the data group around relatively high values, so that the slope on the left flank of the curve is shallower and longer than that on the right, it is said to be negatively skewed. The peak of the curve (KURTOSIS) is described as leptokurtic if it is high and tapering, platykurtic if it is flattened and broad, mesokurtic (the curve of the NORMAL DISTRIBUTION) if it lies midway between the two extremes. If the curve of a positively skewed frequency distribution lacks a short tail and ends abruptly at the peak, it is termed a truncated positively skewed frequency distribution curve (POISSON DISTRIBUTION). FREQUENCY POLYGON, MOMENT MEASURES.

frequency distribution in statistics, the organization of statistical data brought about by separating the range of values covered by a set of observations on a single variable into CLASS INTERVALS, arranging the class intervals consecutively in increasing or decreasing order, and tabulating the number of observations (the frequencies) falling in each. That is the most common frequency distribution, termed a univariate frequency distribution. If the frequency distribution is drawn up on more than one variable at the same time, the result is termed a bivariate frequency distribution. ARRAY, AVERAGE, BINOMIAL

DISTRIBUTION, CUMULATIVE FREQUENCY DISTRIBUTION, EXTREME VALUE, GROUPED FREQUENCY DISTRIBUTION, INTERQUARTILE RANGE, NORMAL DISTRIBUTION, POISSON DISTRIBUTION, RELATIVE FREQUENCY DISTRIBUTION.

frequency polygon in statistics, a graph showing the form of a FREQUENCY DISTRIBUTION, the class frequencies being the ORDINATES, the values of the VARIATE the ABSCISSAE, i.e. a HISTOGRAM. The tops of the columns erected from the horizontal axis are joined up. If the diagram is used to show the frequencies of a continuous distribution, the ordinates are erected at the CLASS MARKS, i.e. at the mid-points of the class intervals. If the resultant polygon is smoothed out, the resultant line is termed a FREQUENCY CURVE, but for this to be effectively achieved the number of observations tend to be infinitely large and the class intervals excessively small. If the frequencies of each class are expressed as a percentage of all the classes, the resultant diagram is termed a relative frequency curve, or a relative frequency polygon. FREQUENCY DISTRIBUTION.

freshet 1. a sudden flood or overflowing of a stream, due to heavy rainfall or the melting of snow in the drainage area upstream 2. a clear stream. G.

freshwater water that is not salt, registering less than 0.2 per cent salinity.

freshwater *adj.* of, pertaining to, or living in water that is not salt.

fret SEA-FRET.

friable *adj.* easily crumbled, hence applied to rocks having that characteristic and to soils with a good CRUMB STRUCTURE, a desirable quality for cultivation.

friagem SURAZO.

friction the force which offers resistance to the relative movement of one surface sliding or rolling over another with which it is in contact.

friction layer the planetary boundary layer, the atmospheric layer extending from the earth's surface up to about 600 m (2000 ft) in which the movement of the air is markedly affected by surface FRICTION.

friction of distance, friction of space the retarding effect of distance on human interactions, hindering perfect or immediate ACCESSIBILITY. Efficient transport partly offsets this friction; but site rentals and transport costs represent the charge for the friction remaining.

frigid *adj.* extremely cold, sometimes applied to PERMAFROST areas covered with snow for most of the year.

frigid zone a term applied in classical times to the coldest of the latitudinal temperature zones, i.e. the very cold, frozen zones lying within the polar circles, the Arctic and the Antarctic. ZONE-2.

fringing reef CORAL REEF.

front the surface of separation between two large or small AIR MASSES of markedly different temperature and humidity, termed a frontal zone if the air masses are widely separated. ANAFRONT, COLD FRONT, KATAFRONT, OCCLUDED FRONT, OCCLUSION, WARM FRONT. G.

frontal *adj.* 1. relating to a meteorological FRONT 2. in, at, or relating to the forward part.

frontal apron OUTWASH APRON, VALLEY TRAIN. G.

frontal fog FOG formed when very fine warm drizzle resulting from the passing of the WARM FRONT of a DEPRESSION encounters cold air near the ground, which becomes saturated.

frontal rain CYCLONIC RAIN.

frontier 1. that part of a country or other political unit which fronts or faces another country or other political unit, sometimes applied to the actual BOUNDARY, sometimes to a zone. MARCH 2. in USA, the border or advance area of settlement, carrying few people. G.

frontogenesis the processes in the atmosphere that lead to the formation or intensification of FRONTS. G.

frontolysis the dispersal of a FRONT or frontal zone.

frost 1. the temperature of the atmosphere, at about 1.2 m (4 ft) above the ground surface, at or below the FREEZING POINT of water, expressed in the number of degrees C or F that the temperature falls below that freezing point. AIR FROST, GROUND FROST 2. frozen dew, fog or water vapour, termed HOAR FROST or RIME, appearing as white ice-crystals on exposed surfaces. BLACK FROST, GLAZE 3. a powerful, MECHANICAL WEATHERING, disintegrating agent effective when water that has penetrated cracks and gaps in rocks and soil freezes, expands, and splits apart the rock or soil particles, e.g. in jointed rocks in regular beds, shattered by frost along lines of weakness to produce rectangular blocks, a process known as BLOCK DISINTEGRATION. G.

frost action CONGELIFRACTION, CONGELITURBATION, FREEZE-THAW, FROST HEAVE, FROST THRUST.

frost boundary FROST LINE.

frost climate a climate characterized by perpetual FROST-1. G.

frost heave, frost heaving the raising of the soil surface by FROST-1 formed in it. CONGELITURBATION, FROST THRUST. G.

frost line a boundary indicating the incidence or the seasonal limit of FROST-1, revealing areas 1. that never experience temperatures below the FREEZING POINT of water, e.g. the altitudinal limit in tropical countries below which frost never occurs 2. with a mean minimum temperature or the lowest mean monthly temperature higher than freezing point, 0°C (32°F) 3. that do not experience frost at certain times of the year, e.g. with a frost-free period of a specific number of days, important to certain crops, such as

maize or cotton, which need a frost-free growing season 4. lacking a month with a mean temperature above 0°C (32°F), the boundary of a FROST CLIMATE. G.

frost pocket a small, low-lying area into which cold air rolls down by gravity from a hillslope, the cold air having been created by swift radiation of heat from the hillslope (KATABATIC WIND). The cold air is trapped in the pocket and may stay at FREEZING POINT after the temperature of the air over the hillslope has risen. Thus fruit-growers, anxious to avoid frost damage to blossom, do not plant in frost pockets.

frost smoke fog-like clouds formed when cold air comes into contact with relatively warm sea-water, e.g. sometimes occurring over newly-formed LEADS-2, or pools on an ice sheet, or leeward of the ice edge. G.

frost thrust the sideways pressure of freezing ground water. FROST HEAVE.

fruit the enlarged reproductive part of a SEED plant. Fertilization of the egg cell in flowering plants leads to changes in growth of parts of the flower and other tissue, the result being termed the fruit. If only the carpel (part of the female reproductive organ) is enlarged the result is a true fruit; but if other parts of the flower head are enlarged the result is known botanically as a false fruit (pseudocarp). A fruit when ripe may be dry (NUT), or succulent (BERRY, DRUPE), dehiscent or indehiscent (DEHISCENCE, INDEHISCENCE), and it is often important in human diet, supplying carbohydrates, minerals, vitamins. See references to individual fruits.

fruitwood the wood of some fruit trees, e.g. pear, used in making furniture.

fu, -fu (Chinese) 1. a department or prefecture in China 2. the chief town of that administrative division. As a suffix in place-names, -fu indicates a city or town. G.

fuel a combustible or fissionable material used as a source of energy or heat. Gaseous fuel comes from natural or artificial sources (GAS), the main liquid fuels, i.e.diesel oil, petrol (gasolene) and paraffin (kerosene), are obtained from PETROLEUM. Solid fuels are COAL in all its forms, PEAT, WOOD and CHARCOAL. The fissionable material used as a source of NUCLEAR ENERGY is mainly THORIUM or URANIUM.

fuel allotments   FIELD GARDEN ALLOTMENTS.

fuelling station   a port with an oil depot, regularly used for refuelling by those vessels that cannot carry enough fuel oil to complete their journey, replacing the COALING STATION.

fula   (Sudan: Arabic) a shallow pool formed in the rainy season in northern Sudan, usually capable of retaining water until the middle of the dry season. G.

fulgurite   a tube of glassy rock caused by the fusing of sand grains where lightning strikes the ground, occurring especially in a desert region or on an exposed mountain. G.

fulji   (northern Arabia: Arabic) a hollow between DUNES in northern Arabia. G.

full   1. the rounded part of a sandbank 2. a ridge of sand or shingle created by wave action, usually lying almost parallel to the coastline. Depressions between fulls are known as SWALES, slashes or furrows. G.

fuller's earth   a clayey material consisting mainly of hydrated aluminium, magnesium, calcium SILICATES, used in processes of absorbing grease (e.g. from a FLEECE-2), in the refining of oils and fats, and formerly medicinally as a soothing paste when mixed with water.

fulling   the process in which cloth (especially woollen cloth) is shrunk and thickened by dampening and heating and the application of friction and pressure. PANDY.

fumarole   (Italian) a vent in the ground in volcanic regions emitting steam and other gases (e.g. AMMONIUM CHLORIDE, CARBON DIOXIDE, SULPHUR DIOXIDE) in the form of powerful jets, a manifestation of dying or extinct volcanic activity. MOFETTE, SOFFIONI, SOLFATARA. G.

function   1. a characteristic or special activity of a person or thing 2. the activity for which something exists, the purpose of a person or thing, the consequence of some kind of existence and/or action of persons or things 3. any quantity, trait or fact that depends on and varies in accordance with another, or is determined by another.

functional   adj. 1. of or pertaining to a FUNCTION 2. made for or having a special purpose, or designed in accordance with criteria determined by use 3. in medicine, having an effect on the working but not the substance of an organ.

functional analysis   an analysis of phenomena on the basis of the part they play within a particular, wider organization, e.g. of rivers in terms of their role in DENUDATION, of towns in terms of the part they play within an economy. FUNCTIONAL EXPLANATION, FUNCTIONALISM. G.

functional classification of towns   a classification of cities and towns based on their characteristic economic activities.

functional explanation   EXPLANATION based on the identification of the purpose for which such-and-such exists, what it is (they are) for. Functional explanation does not try to account for the existence of such-and-such; or why it/they acted in a certain way or took some decision. FUNCTIONAL ANALYSIS, FUNCTIONALISM.

functional integration   the state of one and another of being so strongly linked and interdependent by the way each acts to carry out its purpose as to become one unit, e.g. areas beyond the boundary of a town or city so strongly linked to, and interdependent with, that urban centre on account of commuting, administrative, and social networks that they become part of that centre, and they with the centre are best considered as a whole.

functional interdependence the interdependence of different activities or functions, e.g. the interdependence of a vehicle assembly works and the suppliers of the vehicle parts.

functionalism 1. broadly, a concept of a culture or a society regarded as an entity, a system comprising smaller, differentiated and interdependent systems, all parts of which function to maintain one another and the whole, any change occurring in one part leading to readjustment in the others 2. in design, the theory that the FUNC-TION-2 of the thing to be designed should dictate the form of the design, extended in the 1920s and 1930s to the conviction that the form most closely concerned with function was the most beautiful, the most desirable.

functional region a REGION-1 differentiated by the activity within it, i.e. by the interdependence and organization of its features (CITY REGION). In some cases this may be a single feature (e.g. as in a river system), but more often different kinds of interrelated features are involved, operating at or from differing locations, the concept of unity being based on their interplay. FORMAL REGION, NODAL REGION. G.

functional segregation the spatial separation of different types of urban functions and land uses, e.g. the separation of commercial and industrial areas from those which are primarily residential and commercial. SEGREGATION, SOCIAL SEGREGATION, SPATIAL SEG-REGATION.

fundamental complex a term formerly applied to the areas of Precambrian rocks which were once believed to be the fundamental or original crust of the earth. BASEMENT COMPLEX (basal complex) is a better term and now more often used. G.

fungus pl. fungi, a member of Eumyco-phyta, a phylum of thallophytes (THALLOPHYTA), a parasitic or saprophytic plant, lacking chlorophyll, having a world-wide distribution. Some are important in the breakdown of the remains of dead plants and animals, especially in highly acid soils where they largely take over the role of bacteria; some, e.g. the mushroom, are valued for food; some for the organic fermentation they produce, e.g. some fungi are the source of certain antibiotics such as penicillin. MYCOLOGY, PHYTOPATHOLOGY, PROTISTA.

funicular *adj.* pertaining to a rope or cable, hence funicular railway, a railway that is worked by cables, the ascending rail car balancing the one descending.

funnel cloud a TUBA CLOUD, a spinning cone of dark grey cloud descending from the base of a low-lying CUMULONIMBUS cloud until it reaches the surface of the sea, where it may become part of a WATERSPOUT. It may also be associated with a TORNADO-2.

fur 1. the soft thick hair (the coat) covering many animals, e.g. cats, dogs 2. the dressed pelt of some animals, used for clothing; most pelts are taken from animals especially bred for the purpose, killed when their coats are in the best condition (FUR FARM). Common in trade are rabbit (fur treated in various ways and given trade names), squirrel, muskrat (musquash), coypu (nutria), fox, seal, mink, ermine, sable; young lamb (especially karakul), kid, pony. The sale of fur of some protected wild animals is now prohibited.

fur farm a FARM concerned with the rearing of certain fur-bearing animals under controlled conditions with the objective of producing pelts for sale. FUR.

furfural an oily, sweet-smelling, liquid aldehyde obtained by distillation from bran (e.g. from oats) or wood, used in making synthetic fibres, solvents, dyes, lacquers, resins.

furlong (Old English furlang, the length of a furrow) a unit of distance, originally the length of a FURROW in the COMMON FIELD, the common field varying in size, but usually covering a square of ten ACRES. The furlong was also regarded as being equal to a Roman stadium, i.e. one-eighth of the Roman mile. Thus it came to be standardized as 220 yards (201.17 metres). It is still used in horse racing and surveying. BUTT, SHOTT-3, CONVERSION TABLES. G.

furrow a narrow trench made in the earth by a plough, into which seed is sown. The ridge of earth turned up by the plough in making the furrow is known as the furrow slice. FURLONG. G.

fusain 1. in banded coal, a friable, dusty layer, with flakes like charcoal 2. fine CHARCOAL used by artists, obtained from spindlewood 3. a drawing made with that charcoal.

fusion a melting or blending together, e.g. of metals, to form a fused mass, the components becoming indistinguishable. In nuclear fusion (NUCLEAR ENERGY) light atomic nuclei are exposed to extreme temperature and extreme pressure to form heavier atomic nuclei, without loss in mass, resulting in a great output of energy. LATENT HEAT.

fusion point melting point.

fustic 1. a yellow-olive green coloured dye obtained from the heartwood of a tropical American tree, *Chlorophora tinctoria*, family Moraceae 2. the wood of that tree. DYESTUFFS.

G

gabbro a coarsely crystalline BASIC, PLUTONIC ROCK, dark in colour, consisting essentially of calcium-rich PLAGIOCLASE FELDSPAR and a FERROMAGNESIAN MINERAL (commonly AUGITE) and ACCESSORY MINERALS, commonly occurring as a LOPOLITH or RING-DYKE. GNEISS. G.

gage *Prunus italica*, a variety of PLUM, a small DECIDUOUS tree, originating in Turkey, cultivated in midlatitudes for its succulent yellow or green fruit, used raw or cooked, e.g. in jam-making.

gait CATTLE GATE.

galaxy one of the great number of systems in the universe, consisting of stars (individual and in clusters), nebular and intersteller particles etc., and classified by shape (amorphous, irregular, ellipsoidal, elliptical, spheroidal, spherical, spiral). The earth lies in a spiral galaxy, the Milky Way.

gale a strong wind. BEAUFORT SCALE. G.

galena the chief ore of LEAD, lead sulphide, PbS, which is also an important source of SILVER, occurring as dull, bluish-grey cubic crystals, commonly in CARBONIFEROUS LIMESTONE.

gallery forest (Italian and Spanish galeria, a tunnel) the forest fringing both river banks in otherwise open country. The foliage may meet in an arch over narrow streams, hence the association with a tunnel. G.

gallon a measure of capacity, in liquid British measures 1 gallon of water at 15°C (59°F) weighs 10 lb and is equal to 4.545 litres (277.5 cubic inches), in American liquid measures equal to 3.785 litres (231 cubic inches). One British gallon equals 1.20094 American gallon; one American gallon equals 0.83268 British gallon. Both the British and the American gallon are divided into four quarts. In dry measures one gallon equals one-eighth of a BUSHEL. CONVERSION TABLES.

galvanizing the process of coating with metal by electrolysis, especially of iron or steel with ZINC to form galvanized iron.

gamboge a gum resin obtained from the trees of the genus *Garcinia*, especially from *G. cambogia* and *G. hanburyi*, native to southeast Asia, a powerful cathartic and a source of yellow pigment. Some *Garcinia* produce refreshing, palatable fruit.

gambo hemp, Deccan hemp *Hibiscus cannabinus*, a plant of the same family as the cotton plant, grown in the Indian subcontinent for the sake of its fibre, obtained from its stems. C.

game theory the application of mathematical logic to the strategy, tactics and fluctuating odds in situations where two or more people (the players) are totally or partly in conflict, and where each opponent is free to adopt several courses of action in order to select the OPTIMUM strategy and achieve the desired goal of minimizing the maximum loss the other(s) can impose on him/her (the MINIMAX solution).

gamma rays extremely short wavelength (high frequency), high energy radiation in the ELECTROMAGNETIC SPECTRUM.

gangue the worthless mineral material,

e.g. QUARTZ, surrounding or accompanying a metallic ore in a LODE-1 or a VEIN, of little value in relation to the ore itself. The term expresses a judgement of economic value: mineral material regarded as gangue at one place or at a certain period of time may be valued highly in another place or time where or when economic conditions are different; or it may be of higher value in places where the concentration of the mineral is greater.

ganister, gannister a fine-grained, hard, very pure ARENACEOUS-1, SILICEOUS-1 rock underlying some coal-seams in the Lower COAL MEASURES, used in making heat-refractory bricks in furnace linings. SEAT-EARTH.

gap a break in a ridge or belt of hills. If the break is just an indent and no stream is present, it is termed a dry gap or WIND-GAP; if the cutting is deep and a stream flows through it, it is termed a WATER-GAP. G.

gap town a town situated in or at the entrance to a GAP, a natural centre for lines of communication, holding a commanding position. G.

garden, garden cultivation 1. an enclosed piece of ground, usually near the home, where flowers, fruit, vegetables can be grown (GROUND-4); hence garden cultivation, implying intensity of cultivation and a rich, well cultivated tract of land **2.** often pl., enclosed GROUND-3, usually for public use, provided for pleasure or study, e.g. botanical gardens, zoological gardens.

garden city 1. specifically, based on the definition of Ebenezer Howard who used the term in 1898 (apparently unaware of its use by A. T. Stewart in Long Island, USA, in 1869): a planned, self-contained, industrial-residential settlement, rural in character, with open spaces and trees, catering for inhabitants of various grades of society, providing properly planned facilities for cultural and recreational opportunities, with properly planned facilities for

industry, and with a relatively low housing density (10 units per acre, or 35 persons per acre), the whole of the land being in public ownership or held in trust for the community, being of a size that makes possible a full measure of social life, but not larger, and surrounded by a rural belt. Such a settlement is exemplified in England by Letchworth (1903) and Welwyn Garden City (1920) **2.** now loosely applied to an open density, planned, industrial-residential settlement, similar in character to Howard's specification, but not necessarily so tightly restricted, e.g. with regard to land ownership, or an encircling green belt. G.

garden suburb a suburb with a well-planned, open layout (not to be confused with a GARDEN CITY), e.g. Hampstead Garden Suburb, northwest London. G.

garden village in England, a model estate developed by a manufacturing enterprise to house its employees, e.g. Bournville, Port Sunlight. G.

garigue, garrigue (French, in Provence, garroulia, scrub oak) scrub vegetation, especially of aromatic XEROPHYTES and stunted evergreen oak, *Quercus coccifera*, occurring on LIMESTONE in drier areas of Mediterranean climate. It is ANTHROPOGENIC-2 in origin, resulting from the effects of human settlements, fire and the browsing of DOMESTIC and FERAL goats, on the earlier vegetation, i.e. forest of tall oaks, *Quercus ilex*. In southern France the term includes the uncultivated land with CALCAREOUS soil supporting such vegetation. MAQUIS. G.

garlic any of the several plants of the genus *Allium*, but especially *A. sativum*, native to Central Asia, the bulb of which consists of 'cloves', enclosed in a paperlike covering. Garlic is cultivated for its pungency, and is used in flavouring savoury dishes etc.

garnet a group of minerals consisting of a compound of iron, magnesium, manganese or calcium with aluminium or chromium, hard and brittle, without cleavage, occurring in a wide range of rocks, mainly in METAMORPHIC ROCKS but also, as an ACCESSORY MINERAL, in IGNEOUS ROCKS, especially in ECLOGITE, GNEISS and mica-schist (SCHIST). The PRECIOUS garnet, transparent or semi-transparent, a gemstone sometimes termed almandine, is a rich dark red; the pyrope or Bohemian garnet is blood-red; uvarovite or uwarowite is emerald green; spessartite, yellow, red or brownish-red; and in essonite the colour ranges from yellow to brown. The much coarser common garnet is used as an abrasive.

garnierite a SERPENTINE mineral occurring in LATERITE, formed from ULTRABASIC IGNEOUS ROCKS, and one of the two major ores of NICKEL. PENTLANDITE.

garth (derived from Icelandic) **1.** a small piece of enclosed ground, usually near a building, used as a garden or paddock, sometimes defined (e.g. apple-garth), a suffix in place-names in northern England, e.g. Gatesgarth in the English Lake District, but apparently obsolete in Scotland **2.** abbreviation for cloister garth, an open court bounded by a cloister (i.e. by the covered walk, usually walled on one side, open to the open court on the other), especially of a religious building, e.g. of a cathedral.

garúa (South America) a heavy, dense mist or drizzle occurring in winter on the Pacific slopes of the Coastal Range or the lower foothills of the Andes, soaking the land and promoting the growth of quick-flowering plants and grasses. LOMA. G.

gas 1. a fluid substance with neither definite volume nor definite shape, having a fixed mass but no fixed volume, the volume changing with temperature and pressure, filling and taking the shape of its container. FLUID, LIQUID, SOLID **2.** a fluid

substance at a temperature above its CRITICAL TEMPERATURE. VAPOUR **3.** a substance or mixture of substances in a gaseous state used for some specific purpose, e.g. provision of light and heat, an anaesthetic etc. COAL GAS, FUEL, NATURAL GAS, POWER, RARE GAS, WATER GAS.

gas coal a COAL, usually bituminous, used for making GAS-3 by DISTILLATION.

gas coke the COKE resulting from the DISTILLATION of COAL for GAS-3.

gaseous *adj.* of, pertaining to, or in the form of, a GAS-1,2,3.

gash breccia a rock consisting of angular fragments of DOLOMITE and LIMESTONE in a MATRIX-3 of CALCITE and CLAY, peculiar to the CARBONIFEROUS LIMESTONE of southwest Wales.

gat 1. a natural channel or strait through shoals, sandbanks or offshore islands, or connecting a lagoon or a coastal plain with the sea (GRAU in French) **2.** in Kent, England, a natural or artificial opening in cliffs, serving as a landing place and, as gate, used in place-names, e.g. Margate, Ramsgate, Sandgate. G.

gate 1. an opening or passageway through a wall or barrier **2.** a mountain pass, or an opening wider than a GAP in hilly areas **3.** a broad gap forming an entrance to a particular area or a region **4.** the entrance to a harbour, between promontories **5.** a restricted section in a river valley, e.g. the Iron Gates on the Danube **6.** in English towns, an opening or passageway through city walls; or, occasionally (from Swedish gata), a street in a town, e.g. in York. CATTLE GATE. G.

gateway city a settlement, usually occupying a favourable, commanding site, which acts as a link between two areas and in many cases becomes a PRIMATE CITY. MERCANTALIST MODEL.

gathering ground the area over which water is collected from precipitation, sometimes combined with percolation from springs, a term commonly applied if the water is stored in a natural or artificial lake or reservoir and used for an urban water supply, or for other purposes. CATCHMENT AREA, WATERSHED. G.

GATT General Agreement on Tariffs and Trade, a treaty signed in 1947 by 23 nations. By 1981 there were 85 contracting parties with others participating, aiming to reduce barriers to international trade, to overcome trade problems, and to expand world trade by negotiation and consultation. Headquarters, Geneva.

gauge 1. an instrument for measuring the amount of rainfall, wind force etc. **2.** RAILWAY GAUGE.

gault in eastern England, a stiff, dark CLAY.

gauthānā (Indian subcontinent: Marathi) the village common.

gazetteer a list of places and/or named geographical features, arranged in alphabetical order, with references to their location, with or without brief descriptions or notes. When issued with an ATLAS, it usually includes GRID-1 references to locate positions on specific maps. G.

GDP gross domestic product, the total market value of commodities (goods and services) produced in a country in a given period of time, usually a year. No allowance is made for capital consumption and depreciation; the use of market prices ensures the value of indirect taxes and subsidies are incorporated. The value of intermediate goods used in the production of other goods is excluded, being incorporated in the market price of the final goods. The distinction between intermediate and final products is an arbitrary one, varying from one country (or one economic statistician) to another. COMMODITY, GNP.

geanticline, geoanticline a very large ANTICLINE, commonly extending over many kilometres. G.

Geest (German, dry or sandy soil) an infertile, dry area of FLUVIOGLACIAL sand and gravel, supporting HEATH vegetation, occurring particularly in north Germany and inland of coastal marshes in Denmark, the Netherlands, Poland. G.

gelatin, gelatine 1. a tasteless, colloidal protein, soluble in water, extracted from the collagen of animal hides, skin, bones and used in foodstuffs, photography, adhesives, the making of size and of plastic compounds **2.** a substance having physical properties resembling those of true gelatin, e.g. AGAR-AGAR.

gelation (Latin gelare, to freeze) the act or process of freezing.

gelifluction MASS MOVEMENT associated with PERMAFROST.

geltozem a yellow soil developed under Mediterranean conditions. G.

gem, gemstone any PRECIOUS or SEMIPRECIOUS stone cut and polished, in some cases engraved, for use as an ornament, e.g. in jewellery.

Gemeinschaft, Gesellschaft (German, community, society) terms used to contrast one with the other, Gemeinschaft being the long-lasting, intimate social relationship between individuals, based on affection, kinship or membership of a close community such as a family, or a group of friends, or a religious sect (PRIMARY GROUP); Gesellschaft being a loose, rational association of isolated individuals concerned with personal self-interest, each of whom enters the association for limited purposes and with limited commitments. Gemeinschaft is considered to be typical of pre-industrial, pre-capitalist, pre-urban society; Gesellschaft to be typical of industrial, capitalist, urban society.

gemma (Sudan: Arabic) **1.** a shallow well close to the Nile **2.** a shallow well dug at the foot of rocky hills in Blue Nile Province, Sudan, yielding water throughout the dry season. G.

gendarme (French) a sharp rock pinnacle arising from an ARETE.

gene the basic unit of the material of inheritance, part of a CHROMOSOME, passed on from parent to offspring and responsible for controlling the processes of growth, development and reproduction which distinguish each species.

genecology the study of the GENETIC differences of related species and of the GENETICS of populations of a species in relation to their environments.

General Agreement on Tariffs and Trade GATT.

general circulation of the atmosphere, ATMOSPHERIC CIRCULATION.

general factor FACTOR ANALYSIS.

general land use model a hierarchical model devised by Alice Coleman in 1969 to ease the interpretation of land use maps, consisting of three stable 'scape' territories (wildscape, farmscape, townscape) with intervening less-stable territories (MARGINAL FRINGE and RURBAN FRINGE). G.

general regional system GENERIC REGIONAL SYSTEM, SPECIFIC REGIONAL SYSTEM.

general systems theory a concept developed by Ludwig von Bertalanffy in the 1940s – 1950s as a framework for a science of 'wholeness' and organization. By making theoretical generalizations about the properties common to different types of SYSTEM-1,2,3,4,5 an attempt is made to discover ISO-MORPHISMS between systems, to liken a system drawn from one discipline to that of another or others, and thus to produce a doctrine of 'wholeness'. This approach emphasizes the relationship between form and process and draws attention to the multivariate character of phenomena; so it has come to be applied to geographical systems where isomorphism may often be found, e.g. in systems displaying ALLOMETRIC GROWTH, ENTROPY, or the HIERARCHICAL structure of the processes involved in the filling of space. SYSTEMS ANALYSIS.

generation 1. in general, a producing, or being produced, procreation 2. all the members of any developing class of things (e.g. computers) at a certain stage 3. in biology, a whole group of animals or plants which have in common their degree of distance from an ancestor 4. a whole group of organisms, particularly of people, regarded as being born about the same time, thus of roughly the same age group 5. in demography, a period of time, some 25 to 30 years, roughly correspondening to the age of parents at the time of birth of their children.

generative and competitive growth two different types of growth in urban development identified by some authors, who relate generative growth to the linkages which transmit (generate) growth from one area to another; and competitive growth to the mechanisms which result in growth at one place only, at the expense of (in competition with) less-favoured places. CIRCULAR AND CUMULATIVE GROWTH.

generator in transport studies, a land use which generates a great demand for transport facilities, e.g. a SHOPPING CENTRE or an INDUSTRIAL COMPLEX.

generic *adj.* 1. in biology, of or pertaining to a GENUS. CLASSIFICATION OF ORGANISMS 2. of or pertaining to a group or class 3. not specific, having a general application, being shared by or 'typical' of, a whole class of things.

generic (general) regional system a regional system in which types of places resemble one another according to a combination of attributes (e.g. climatic, cultural) chosen to suit the purpose of the classification, but which are separate and possibly widely spaced, as distinct from those in a SPECIFIC REGIONAL SYSTEM. REGION-1. G.

genetic *adj.* 1. of or relating to the origin or the development of something 2. of or relating to GENETICS.

genetic description of landforms, the systematic study of landforms, covering their origin, the processes involved in their development, and their present stage.

genetics the branch of BIOLOGY concerned with the heredity and variation of organisms.

genocide the deliberate destruction of an ETHNIC-2 group by those outside the group, whether by murder or by extreme persecution.

genotype 1. the constitution of an organism inherited through its GENES. PHENOTYPE **2.** a group of organisms with identical GENETIC-1 constitution **3.** the type species of a genus. CLASSIFICATION OF ORGANISMS.

genre de vie (French) the attitudes, traditions, purposes, institutions and technology of a particular cultural group.

gentleman farmer a person of fairly high SOCIAL STATUS and usually of independent means, engaged in farming, commonly on his own estate. FARMER.

gentrification a process in which wealthier people move into, renovate and restore run-down housing in an inner city or other neglected area, housing formerly inhabited by poorer people, the tenure usually shifting from private rented to owner occupation. It occurs particularly in inner city areas as a result of the wishes of the wealthier people to have easy access to their jobs and the recreational facilities (e.g. theatres) in the central area. Once started in a district, the process spreads rapidly until most of the poorer inhabitants are displaced, so that the social character of, and the value of the property in, the neighbourhood markedly changes. G.

genus (pl. genera) one of the groups used in the CLASSIFICATION OF ORGANISMS, consisting of a number of similar SPECIES-1, sometimes of only one species, a group of similar genera constituting a FAMILY-6. BINOMIAL NOMENCLATURE.

genya (Japanese) pastures, plains. G.

geo, gio (Norse gya, a creek) in northern Scotland and Faeroes **1.** a long narrow inlet of the sea, penetrating cliffs, especially in areas of well-jointed OLD RED SANDSTONE **2.** in geomorphology, a coastal cleft, often marking JOINTS, FAULTS or DIKES from which material has been removed by wave action. G.

geo-, ge- (Greek, the earth) a prefix used in many compounds, it usually implies a relationship with the EARTH-1 as a whole, as in geology; but it sometimes implies something on a large, or world scale, e.g. geosyncline. The prefix before a vowel is usually ge, e.g. geoid. G.

geoanthropolitics the study of anthropological POLITICAL GEOGRAPHY, the political geography of people living in a simple, 'tribal', PRE-INDUSTRIAL SOCIETY. G.

geoanticline GEANTICLINE.

geobenthos that part of the floor of a freshwater lake which does not support plants with roots.

geobiocenosis in ecology, the combination of the geocenosis (the physical habitat) and the biocenosis (BIOCOENOSIS), a term first used by the Russian school of ecologists who in this application equated biocenosis with the BIOME. G.

geobotany the branch of botany, comprising plant ecology and plant geography, covering the interrelationship of plants and the earth's surface.

geochemistry the scientific study of the distribution of, and chemical changes in, the elements in the earth's crust and atmosphere (including the LITHOSPHERE, HYDROSPHERE, BIOSPHERE, ATMOSPHERE-1).

geochronology 1. in geology, the science of using geological data to establish the duration of geological periods, thus enabling a timescale to be drawn up for the earth's history before the time of recorded history. ABSOLUTE AGE, CHRONOLOGY, GEOLOGY, GEOLOGICAL TIMESCALE **2.** in archaeology, all the methods of dating based on measurable changes in natural substances, e.g. DENDROCHRONOLOGY, RADIOCARBON DATING, the study of VARVES etc. G.

geochronometry the precise measurement of geological time.

geo-code a world reference system used mainly by biologists to identify the location of information (in a hierarchical form) on species. GEOREF. G.

geocratic *adj.* applied by T. Griffith Taylor, 1951, to the control of human beings by Nature. WE-OCRATIC. G.

geocryology the scientific study of frozen ground, including not only PERMAFROST but also ground seasonally frozen. CRYOLOGY. G.

geode a hollow, rounded, rock nodule, its interior walls lined with inward-pointing crystals, common in SEDIMENTARY ROCK, especially in LIMESTONE.

geodesic line the shortest distance between two points on a surface, e.g. a straight line on a plane, or an arc of a GREAT CIRCLE on a sphere.

geodesy, geodetics the branch of applied mathematics dealing with the external shape, size and gravity of all the earth, or of very large parts of it, and the use of this information in surveying large areas when such factors must be taken into account. G.

geodetic distance the shortest distance ('as the crow flies' distance) between two places on the earth's surface.

geodetic line a GEODESIC LINE on the earth's surface.

geodetic position the position of a point on the earth's surface expressed in LATITUDE and LONGITUDE determined by triangulation from an initial station the position of which has been arbitrarily adopted after consideration of astronomical data. G.

geodimeter a surveying instrument used for estimating the distance between two points. A light signal flashed from one point is sent back by a reflector stationed at the other, the time interval between the flash and the observed return being measured.

geodynamics the scientific study of processes or deformation forces within the earth. G.

geo-econometrics the geographical study based on the combination of ideas and methods taken from geography, economics, statistics. ECONOMETRICS. G.

geognostic *adj.* of or pertaining to GEOGNOSY.

geognosy the geological study dealing with the materials and structure of the earth.

geographic, geographical *adj.* of, pertaining to, or relating to GEOGRAPHY. On the whole, geographic is used more frequently by American authors than by British.

geographical determinism DETERMINISM.

geographical inertia the tendency of a place with established installations and services to maintain its size and its importance as a focus of activity after the conditions originally influencing its development have appreciably altered, ceased to be relevant, or disappeared. GEOGRAPHICAL MOMENTUM, INDUSTRIAL INERTIA, INDUSTRIAL MOMENTUM. G.

geographical mile a NAUTICAL MILE, a mile theoretically equal to one minute (1′) of LATITUDE but, owing to the fact that the earth is a sphere flattened at the poles (GEOID), a minute of latitude varies slightly in length. The geographical or nautical mile has therefore been standardized internationally as being equal to 1.852 kilometres (1 km = 0.539 international nautical mile). CONVERSION TABLES, KNOT, MILE. G.

geographical momentum the tendency of a place with established installations and services not only to maintain (GEOGRAPHICAL INERTIA) but also to increase its size and its importance as a focus of activity after the conditions originally influencing its development have appreciably altered, ceased to be relevant, or disappeared. INDUSTRIAL INERTIA, INDUSTRIAL MOMENTUM. G.

geographical name the proper name or geographical expression by which a particular geographical entity is known. TOPONYMY.

geographical region REGION.

geography 1. the branch of knowledge concerned with the study of the material and human phenomena in the space accessible to human beings and their instruments (SPACE-2,3,4), especially the patterns of, and variation in, their distribution in that space, on all scales, in the past or present. It involves description, classification, analysis, synthesis, explanation. That is a very broad concept of geography as a whole. But in the subject's evolution, advances in knowledge of the phenomena studied have induced geographers to draw on other disciplines to an ever greater extent, while focusing the ideas and material so derived on the study of spatial variations on the earth's surface (AREAL DIFFERENTIATION). As a consequence geography has developed a number of specialist studies, each separately defined in the *Dictionary*.

Within traditional academic organization, geography (as an academic discipline and an applied science) straddles the HUMANITIES, the NATURAL SCIENCES and the SOCIAL SCIENCES; and partly for convenience in study, partly to take into account the particular interests of those who study it, the subject is commonly separated in two parts, human geography and physical geography.

HUMAN GEOGRAPHY, concerned with human activities (of individuals and of groups) and organization in so far as these relate to the interaction (past or present) of people with their physical environment and with the environments created by human beings themselves, and the consequences of these interrelationships, draws on studies in ARCHAEOLOGY, ANTHROPOLOGY, DEMOGRAPHY, ECONOMICS, HISTORY, LANGUAGE, LAW-1,2, LITERATURE, MEDICINE, PHILOSOPHY, POLITICAL SCIENCE, PSYCHOLOGY, SOCIOLOGY and their branches; and to some extent on BIOLOGY, BOTANY, ECOLOGY and their branches, as well as the specialist geographical studies, BIOGEOGRAPHY, PHYTOGEOGRAPHY, ZOOGEOGRAPHY. It has many branches of its own, appearing in this *Dictionary* as GEOGRAPHY qualified by BEHAVIOURAL, CULTURAL, COMMERCIAL, ECONOMIC, ELECTORAL, HISTORICAL, HUMANISTIC, MEDICAL, PERCEPTION, POLITICAL, SOCIAL, URBAN, WELFARE.

PHYSICAL GEOGRAPHY is concerned with the physical characteristics and processes of the ATMOSPHERE-1, BIOSPHERE, HYDROSPHERE and LITHOSPHERE. If restricted to inanimate matter and to energy, it draws on studies in such related disciplines as CHEMISTRY, CLIMATOLOGY, GEOLOGY and GEOMORPHOLOGY, HYDROGRAPHY and HYDROLOGY, METEOROLOGY, OCEANOGRAPHY, PEDOLOGY, PHYSICS and their branches (to which ASTRONOMY may be added), and information provided by REMOTE SENSING and land SURVEY. If not so restricted it (like human geography) draws on the related disciplines of BIOLOGY, BOTANY, ECOLOGY, ZOOLOGY; and on the specialist geographical studies, BIOGEOGRAPHY, PHYTOGEOGRAPHY, ZOOGEOGRAPHY.

Both human and physical geography make use of the techniques of CARTOGRAPHY, SURVEY, MATHEMATICS and STATISICS **2.** the data with which geography, as a branch of knowledge, is concerned **3.** a textbook or treatise concerned with that branch of knowledge. G.

geoid 1. an earth-shaped body **2.** the earth in geometrical terms, i.e. an OBLATE spheroid or ELLIPSOID (a SPHERE, SPHEROID or ELLIPSOID flattened at the poles), regarded as a mean sea-level surface (i.e. supposing all mountains and ocean basins to be levelled to the mean sea-level surface), or as an undulating surface related to gravitational force (GRAVITY-2), higher than the actual surface of the spheroid under the continents, lower under the oceans. G.

geoisotherm surfaces in the LITHO-SPHERE, below ground level, having the same temperature at all points. G.

geologic, geological *adj.* of, pertaining to, derived from GEOLOGY. Geological is the usual English usage; both forms appear in American usage, but geologic is sometimes restricted to that which is part of the subject-matter of geology. G.

geological inversion STRATA which have the normal sequence of layers, but in reverse order (i.e. upside down, as in the lower limb of an OVERFOLD), occurring in regions of intense folding (FOLD-2).

geological structure STRUCTURE-3.

geological time the chronology of the history of the earth revealed by its rocks. The hierarchy of time-periods shown in the GEOLOGICAL TIME-SCALE is not always strictly observed: there are variations in the nomenclature and the periods used. An epoch (time) may be divided into AGES-2, during which a FORMATION-2 (of rocks) may be laid down. Some authors refer to the smallest division of time as a HEMERA, the rocks then deposited constituting a ZONE-4. G.

geology 1. the scientific study of the earth's crust, the rocks of which it is composed, and the history of its development and changes. There are many branches of geological study, the broad subdivisions being PHYSICAL GEOLOGY and HISTORICAL GEOLOGY. Physical geology is concerned with the processes that shape the earth's crust. The detailed study of rocks is PETROLOGY and of the minerals which compose them MINERALOGY. The study of FOSSILS or former life is PALAEONTOLOGY. The relationship of water with the structure of the earth is covered by HYDROGEOLOGY. (The study concerned with landforms, GEOMORPHOLOGY, is included as a branch of geology in the USA, but in Britain it is usually regarded as a branch of GEOGRAPHY-1). Historical geology is the study of the sequence of earth history, but because it depends on the study of the strata of the crust it is also termed STRATIGRAPHY. The specialized study of the ages of rocks, which aims to establish the duration of geological periods, is GEOCHRONOLOGY. It is possible to date accurately many rocks of the earth's surface, the various techniques used being the study of the measurement of RADIOACTIVE DECAY in minerals which break down very slowly in the course of time, RADIOCARBON DATING, DENDROCHRONOLOGY, the study of the deposition in SEDIMENTARY ROCKS (LAMINA, VARVE), CORE SAMPLING, FLUORINE DATING. On such evidence a GEOLOGICAL TIMESCALE can be drawn up. ABSOLUTE AGE **2.** the data of that scientific study, i.e. of geology **3.** a treatise on the subject (of geology). G.

geomagnetic *adj.* of or pertaining to the MAGNETISM-1 of the earth.

geomagnetism 1. the MAGNETIC FIELD of the earth. TERRESTRIAL MAGNETISM **2.** the study of the earth's magnetic properties and the phenomena associated with them; hence geomagnetician, geomagnetist, an expert or student engaged in this scientific study.

geomancy divination by means of signs

Era	Period *(time)* or System *(rock)*	Epoch *(time)* Series *(rock)*	Duration in million years (approx.)	Million years ago (approx.)
	Quaternary	Holocene (Recent)		(12 000 years)
		Pleistocene (Glacial)	*c.*1.5	
CAINOZOIC (CENOZOIC)		Pliocene	5.5	— *c.*1.5 —
		Miocene	19	— 7 —
	Tertiary	Oligocene	12	— 26 —
		Eocene	16	— 38 —
		Palaeocene	11	— 54 —
				— 65 —
MESOZOIC (SECONDARY)	Cretaceous		71	— 136 —
	Jurassic		54	— 190 —
	Triassic		35	— 225 —
	Permian		55	— 280 —
PALAEOZOIC (PRIMARY)	Carboniferous	*Am* Pennsylvanian Mississippian	65	— 345 —
	Devonian		50	— 395 —
	Silurian		35	— 430 —
	Ordovician		70	— 500 —
	Cambrian		70	— 570 —
PRE-CAMBRIAN	PROTEROZOIC			
	ARCHEOZOIC			

Geological timescale

derived from random figures, from lines on paper, or from lines on the ground, e.g. cracks, or the pattern made by dropping a handful of earth on the ground. FENG-SHUI.

geometrical scale   FLOW-LINE MAP.

geometric growth   EXPONENTIAL GROWTH.

geometric progression   a series of numbers in which each is multiplied by a fixed number to produce the next, e.g. 1, 3, 9, 27. ARITHMETIC PROGRESSION.

geomorphic threshold in   GEOMOR-PHOLOGY **1.** the point at which there is a significant change in a LANDFORMS without a change occurring in such external controls as climate, base level, or land use **2.** the point at which there is an abrupt change in a landform as a result of progressive change of external controls.

geomorphological   *adj.* of or pertaining to GEOMORPHOLOGY.

geomorphological map   a map showing the origin or history of the LANDFORMS of a selected period of time. MORPHO-LOGICAL MAP.

geomorphological timescale a timescale devised by S.A. Schumm and R. W. Lichty, who identified three categories in GEOMORPHOLOGICAL development: CYCLIC TIME, GRADED TIME, STEADY TIME.

geomorphology (Greek ge, the earth; morphe, form; logos, study) the scientific study concerned with LAND-FORMS, especially the genesis, evolution and processes involved in the formation of the surface forms of the earth. PHYSIOGRAPHY. G.

geonomics ECONOMIC GEOGRAPHY. A term rarely used and not generally adopted. G.

geopacifics a term introduced by T. Griffith Taylor in 1947 as an alternative to GEOPOLITICS which had become identified with the use of geography to promote Nazi political ideas in Germany in the 1930s. Taylor declared that geopacifics was an attempt to base the teaching of freedom and humanity on real geographical deductions, a 'humanized' geography. He defined it as the study of geography to promote peace G.

geophysical *adj.* of or pertaining to GEOPHYSICS.

geophysical survey a survey using the many complex techniques employed in GEOPHYSICS, concerned with the investigation of the earth's crust in order to discover and locate structures associated with minerals of economic importance. G.

geophysics 1. an interdisciplinary scientific study concerned with the PHYSICS of the earth's interior and crust and the application of the techniques of the PHYSICAL SCIENCES to the investigation of the earth, involving the application of precise quantitative techniques to EARTHQUAKE waves (natural and those induced deliberately by explosives), MAGNETISM, GRAVITATIONAL fields, electrical conductivity etc. **2.** such an interdisciplinary study but with wider scope, incorporating all the con-

cerns listed, together with studies of the earth's environment and its evolutionary history, and thus including ASTRO-PHYSICS, CLIMATOLOGY, METEOR-OLOGY, OCEANOGRAPHY. G.

geophytes a class of RAUNKIAER'S LIFE FORMS, HERBACEOUS plants with PERENNATING parts below the surface of the soil.

geopolitics (German Geopolitik) **1.** the study of STATES-2 or NATIONS viewed as organic entities in space, and as such subject to biological laws of growth and decline (in territorial extent and political influence, as well as in economic and social terms), being, like other organisms, engaged in a perpetual struggle for survival, for control over the space occupied. F. Ratzel is credited with the introduction of this concept in the late nineteenth century; it was taken up by the English geographer, Halford Mackinder; by the Swedish political scientist, Rudolf Kjellen; and later by the German geographer, K. Haushofer. But it became particularly prominent (and disreputable) before and during the Second World War when adopted by the Nazi Party in Germany to advance its theory of race superiority and to justify the Third Reich's demand for LEBENSRAUM **2.** the influence, or the study of the influence, of spatial aspects on the political nature, history, institutions, etc. of states or nations, and especially on their political relations wth other states. GEOPACIFICS. G.

georef system a world system devised to locate points on the earth's surface. A rectangular GRATICULE, laid on a world map, shows meridians from 180° eastwards, at 15° intervals, lettered A to Z (I and O being omitted), and parallels from the South Pole, also at 15° intervals, lettered A to M (I omitted). The origin of map reference (FALSE ORIGIN) lies where the meridian of 180° meets the parallel representing the South Pole. The world surface is thus covered by 288 quadrangles (each identified by two letters, that of the meridian being put first), each quadrangle being subdivided into degrees, minutes and hundredths of minutes. G.

geosere a series of CLIMAX formations developed through geological time, the total plant succession of the geological past. SUCCESSION-2.

geosophy the study of the nature and expression of geographical ideas. G.

geosphere the earth, excluding the ATMOSPHERE-1, HYDROSPHERE and BIOSPHERE.

geostrategy the study of the significance of the environment in relation to the understanding of a problem of economic or political welfare, primarily, but not necessarily, of international scope. G.

geostrophic *adj.* of, pertaining to, or caused by, CORIOLIS FORCE, commonly applied to a wind or water current in which the Coriolis force is balanced by a horizontal pressure gradient. G.

geostrophic flow, geostrophic wind the concept of a wind blowing parallel to the ISOBARS as a result of the force exerted by the horizontal ATMOSPHERIC PRESSURE gradient in one direction, balanced by the deflection of the CORIOLIS FORCE in the opposite direction. It is only in the upper atmosphere that winds come near to perfect geostrophic flow, because FRICTION near the earth's surface upsets the balance below 455 m (1500 ft) above sea-level, so that at these lower levels winds blow at an oblique angle to the isobars towards low pressure. FERREL'S LAW, GRADIENT WIND.

geosyncline a very large linear depression or SYNCLINE or down-warping of the earth's crust, filled (especially in the central zone) with a deep layer of sediments derived from the land masses on each side and deposited on the floor of the depression at approximately the same rate as it slowly, continuously subsided during a long period of geological time. G.

geotaxis the locomotory movement of an organism or cell in response to the stimulation of the force of GRAVITY-2. TAXIS.

geotectonic *adj.* in geology **1.** structure **2.** relating to the structure, form, arrangement of the rocks in the earth's crust.

geotectonic plate PLATE TECTONICS.

geothermal energy energy, in the form of heat derived from anomalies in the GEOTHERMAL GRADIENT, which gives rise in nature to the delivery of hot water and steam to the earth's surface by THERMAL SPRINGS and GEYSERS (HYDROTHERMAL). The hot water and steam can be used in generating ELECTRICITY, e.g. as in Iceland and New Zealand. In addition the heat of the rocks in the earth's crust can be used to warm cold water pumped down from the surface through a borehole. The water is passed over the HOT ROCK and rises to the surface by another borehole, where its acquired heat can be tapped locally.

geothermal gradient the rise in temperature in the rocks of the earth's crust with increasing depth. There are indications that the rate is not constant but speeds up with increasing depth. Estimates range from $1°C$ in 28 m ($1°F$ in some 51 ft) to $1°C$ in 40 m ($1°F$ in about 73 ft), but an average in the SIAL could be $1°C$ in about 28.6 m ($1°F$ in about 52 ft). The temperature of the base of the sial may be some $986°C$ (about $1806°F$). HOT ROCK, GEOTHERMAL ENERGY.

geotropism the response of plant growth to the force of GRAVITY-2, main roots being positively geotropic (downward growing), main stems usually negatively geotropic (upward growing). DIAGEOTROPISM, PLAGIOTROPISM.

gerf (Sudan: Arabic) land in Sudan annually left uncovered after the flood waters of the Nile subside. SELUKA. G.

germanium a metallic element occurring in some COPPER and ZINC ores, semi-conducting, used in the making of transistors.

gerontocracy 1. government by old people 2. a governing body composed of the old 3. a government controlled by the old. AUTOCRACY.

gerrymander the drawing of the boundaries of electoral districts in such a way that one political party is given an advantage.

Gestalt a shape, pattern, configuration, structure which can be perceived only as a whole or unity, with qualities different from those of its components: it cannot be expressed as the sum of its parts, because the parts acquire certain characteristics produced by their inclusion in the whole, and the whole has some characteristics belonging to none of the parts, giving it an additional, often indefinable, quality (e.g. a melody cannot be expressed as the sum of its components, the notes). Hence the Gestalt theory of the Gestalt school of psychology, that human perceptions, reactions, responses are all Gestalten, that enquiries about them should examine the whole to discover what are the components, rather than start with the components and try to synthesize them to form the whole. BEHAVIOURAL ENVIRONMENT, PHENOMENAL ENVIRONMENT. G.

Gewann (German, pl. Gewanne) a division of the communal lands of a village. G.

Gewanndorf (German) a nucleated village with strip fields. G.

geyser (Icelandic geysir, gusher or roarer) a hot spring that intermittently, sometimes at regular intervals, throws up a jet of hot water, steam etc. in areas that are or were volcanic. Famous examples are in the volcanic districts of Iceland and the USA. The original 'geyser', the Giant Geyser in Iceland, erupts on occasion to 55 m (180 ft), the Giant Geyser in Yellowstone National Park, USA, to 60 m (200 ft) at irregular intervals, Old Faithful, also in Yellowstone, every 66 minutes to about 45 m (150 ft). The Waimangu Geyser, New Zealand, is said to have erupted to 305 m (1000 ft) in 1909. GEOTHERMAL ENERGY. G.

geyserite SINTER, a deposit of SILICEOUS-1 material that forms around GEYSERS and HOT SPRINGS. TRAVERTINE. G.

gezira (Arabic, island) a tract of land partly or wholly enclosed by water, either by rivers or the sea. The Gezira is the large lowland plain, widely irrigated, renowned for its cotton, lying between the Blue and White Nile and the Sobat rivers in Sudan; but the name is also applied specifically to the much smaller, irrigated area north of Sennar. G.

ghat, ghaut (Anglo-Indian; Hindi ghāt) 1. a path down to a river 2. a flight of steps leading to a landing stage 3. a mountain pass. Europeans transferred the term from the passes to the mountains themselves, so the term Western Ghats was applied to the mountainous edge of the great Indian plateau, the Eastern ghats to the eastern edge. G.

ghee (Hindi ghi) clarified butter, very soft and almost fluid, usually made from the milk of buffaloes, sometimes of cows.

gher (Indian subcontinent) 'sweet water' lands above areas of salt marsh. G.

ghetto 1. historically, the quarter of a town or city in which Jews were required to live in parts of medieval Europe, e.g. Poland, Italy 2. the quarter of a town or city where members of a minority group live on account of economic or social pressure. SEGREGATION. G.

G horizon SOIL HORIZON.

ghost town a TOWN, usually once a mining town, now wholly abandoned, or inhabited by a few people if the town is now a tourist attraction, e.g. in Klondike, USA.

ghyll   GILL.

giant's kettle   KETTLE.

gibber (Australian: aboriginal) a large stone or boulder. G.

gibber plains desert areas strewn with pebbles or boulders, a type of gravel desert. G.

gibbsite $Al(OH)_3$, one of the main ores of aluminium, being a constituent of BAUXITE and LATERITE.

gibli, ghibli the local name for the SIROCCO in Libya and Tunisia. G.

gibbous (Latin gibbus, a hump) *adj.* convex, rounded out, applied especially to the MOON or a planet that appears more than half but less than full.

gilgai (Australian) a low, rounded mound on the surface of the plain in parts of the Murrumbidgee basin, New South Wales, formed when fragments of the A HORIZON fall down cracks in the B HORIZON. When the soil becomes saturated the B horizon expands, and the A horizon is pushed up in a series of mounds, continually eroded until the B horizon is exposed at the surface. G.

gill (ghyll, obsolescent spelling) in northern England, Lake District and Yorkshire **1.** a swiftly flowing, narrow stream **2.** a wooded ravine. G.

gilt a young SOW.

gin an alcoholic spirit, distilled from grain, flavoured with juniper berries.

ginger *Zingiber officinale*, a PEREN-NIAL plant with a tuberous starchy root (rhizome), native to tropical Asia, cultivated widely in tropical areas for its strong 'hot' flavour. The rhizomes formed at shallow depth are dug up after ten months' growth, to be sold as fresh ginger, or processed by washing, soaking, sometimes boiling, peeling and drying. Preserved ginger is made from young fleshy rhizomes, boiled in sugar, packed in syrup. Dried ground ginger is commonly used in cooking.

Gini coefficient, Gini index a measure of concentration, calculated by taking the ratio between the area contained by the diagonal and the LORENZ CURVE and the area in the whole triangle below the diagonal. The value of the index increases with the increasing size of the area between the curve and the diagonal. The index is frequently used as a measure of inequality in the welfare approach (WELFARE GEOGRAPHY).

ginseng *Panax schinseng* of eastern Asia, *P.quinquefolius* of North America **1.** a plant with an aromatic root especially valued in Asia as a panacea **2.** the processed product of the root.

Gipfelflur (German) the surface of uniformity of summit levels in Alpine mountains, independent of any glacial remodelling, or of the folded structure or the nature of the rocks, thought not to be the heritage of a hypothetical PENEPLAIN but connected with the distribution of slope form produced by a regional balance in down wasting, the levels being lowered simultaneously. ACCORDANCE OF SUMMIT LEVELS, SUMMIT PLANE. G.

gipsy, gypsy, gip an intermittent spring or BOURNE. G.

gja, gia (Gaelic, geodha) **1.** a narrow opening in a sea cliff, penetrated by waves, usually an eroded DIKE-1 **2.** a fissure from which volcanic eruptions take place. G.

glacial *adj.* **1.** icy, frozen, used very loosely in connexion with a glacier, ice age, lake etc. **2.** strictly, of or pertaining to a GLACIER **3.** in chemistry, having ice-like crystals. G.

glacial age, epoch, era, period a period of GEOLOGICAL TIME when much of the earth's surface was covered by GLACIERS. ICE AGE. G.

glacial boulder a BOULDER which has been carried by a glacier. ERRATIC.

glacial control theory R. A. Daly's theory to account for the formation of CORAL REEFS at abnormally great depths, which assumes that in tropical latitudes during the Great Ice Age of the QUATERNARY period the sea water was cooler and the sea-level fell. These conditions were hostile to the then existing live coral and led to its destruction. The preglacial reefs and coral islands then suffered marine erosion and were planed to sea-level. As the ice melted, the sea-level rose, the water temperature increased and new coral colonies built up on the platforms created. But recent research indicates that there was insufficient time in glacial periods for such marine erosion to have taken place; and that most erosion was probably subaerial. G.

glacial diffluence the branching of a VALLEY GLACIER, usually due to an obstruction in the channel which causes a build-up of ice in the main glacier and an overflow (often over a COL) into an adjoining valley.

glacial divergence the obstruction of an established drainage pattern by an advancing glacier or ice sheet, leading to a change of course in the drainage.

glacial drainage channel a stream bed cut by the melt-water of a glacier.

glacial drift DRIFT.

glacial foreland a lowland area on the flanks of a mountain range which supported glaciers in the QUATERNARY ice age, on to which the glaciers spread (PIEDMONT GLACIER), e.g. the Bavarian Foreland.

glacial lake a body of water contained by the wall of a valley and the margin of a glacier.

glacial outburst a sudden flood occurring when melt-water contained within or by a glacier is released.

glacial stages the final glacial stage of the PLEISTOCENE in Britain was the Devensian (the later period of which is sometimes termed Late-Glacial). It lasted from about 68 000 to 8000 BC, and was contemporaneous with the Weichsel stage in the North European Plain, the Würm of the Alps, the Wisconsin of North America. This glacial stage was preceded by the interglacial Ipswichian (Britain), Eemian (North European Plain), Riss-Würm (Alps), Sangamon (North America); preceded by the glacial stage Wolstonian (Britain), Saale (North European Plain), Riss (Alps), Illinoian (North America); before that was the interglacial Hoxnian (Britain), Holstein (North European Plain), Mindel Riss (Alps), Yarmouth (North America). The glacial stage which preceded that interglacial was the Anglian (Britain), Elster (North European Plain), Mindel (Alps), Kansan (North America). An earlier glacial stage in the Alps has been identified as being contemporary with the Nebraskan in North America (which was separated from the Kansan glaciation by the Aftonian interglacial); and an even earlier glacial stage in the Alps has been termed the Donau.

glacial trough a deep, U-shaped rock trench with steep sides formed by the erosion of an ALPINE GLACIER. TROUGH LAKE.

glaciated valley U-SHAPED VALLEY.

glaciation 1. the occupation of an area by an ICE SHEET or by GLACIERS 2. the action of ice on rocks over which it has passed 3. the time when such took place (e.g. first and second glaciations). Some authors restrict glaciation to evidence of past ice cover, applying glacierization to the present ice cover. G.

glacier (French) originally a river of ice moving down a valley (now termed a VALLEY GLACIER, ALPINE GLACIER or mountain glacier). Now applied to a mass of snow and ice formed by the consolidation (under pressure) of snow falling on high ground (FIRN, NEVE), moving outward and downward from the zone of accumulation to lower ground (or, if afloat, continuously spreading), in some cases carving a broad, steep-sided valley on its way. It may be active, moving quickly and carrying much material; or passive (the rates of accumulation and ablation being low, the slope gentle), moving little if at all, and carrying little or no material. ABLATION, ALIMENTATION,

GLACIERET, GLACIER MILK, ICE CAP, ICE PIEDMONT, ICE SHEET, ICE SHELF, INTERGRANULAR TRANSLATION; and glacier qualified by the terms ACTIVE, APPOSED, CIRQUE, COLD, EXPANDED-FOOT, HANGING, PASSIVE, PIEDMONT, TIDAL, WALL-SIDED, WARM. G.

glacier band OGIVE.

glacier breeze a cold, KATABATIC type breeze, blowing down the course of a glacier, the result of air being chilled as it touches the ice. G.

glacière (French) an ice cave. G.

glacieret a small GLACIER.

glacier ice any ice, on land or floating in the sea, originating from a GLACIER. G.

glacierization the occupation of an area by a glacier or ice sheet at the present day. GLACIATION. G.

glacier karst the KARST-like appearance of an ice mass caused by the action of melt-water lying under a covering of morainic and glaciofluvial material which, varying in thickness, leads to variations in the rate of ABLATION. As in true karst, underground streams, POLJES etc. are produced. GLACIOKARST.

glacier mice GLACIER MOUSE.

glacier milk a murky, greenish-white stream of melt-water containing ROCK FLOUR, issuing from a GLACIER SNOUT. G.

glacier mill MOULIN.

glacier mouse (Icelandic jökla mýs, glacier mice) a small, rounded, moss-covered stone occurring on some glaciers in Iceland. POLSTER. G.

glacier retreat the stage of a glacier when the ice is moving forward but the snout is receding, reached when the rate of ABLATION exceeds that of ACCUMULATION.

glacier snout the cavernous end of a VALLEY GLACIER from which GLACIER MILK flows. G.

glacier table a large block of stone resting on a supporting pedestal of ice

on a glacier. The stone insulates the ice beneath it from the heat of the sun, with the result that the supporting ice pedestal remains as the surrounding ice melts. G.

glacier tongue the part of a glacier extending to the sea, and usually afloat. G.

glacio-eustatism a global change of ocean level caused by the abstraction of water by the growth of ICE SHEETS, or the return of water to the ocean when these melt. DIASTROPHIC EUSTATISM, EUSTATISM, GLACIO-ISOSTASY, NEGATIVE MOVEMENT, POSITIVE MOVEMENT, REJUVENATION. G.

glaciofluvial, glacifluvial FLUVIOGLACIAL. G.

glacio-isostasy the deformation of part of the earth's crust caused by the weight of an extensive ICE SHEET. The great weight of the ice depresses the crust but the land surface recoils as the ice melts, in many cases leading to TILTING and WARPING-2. GLACIO-EUSTASY, ISOSTASY.

glaciokarst, nival karst karstic landforms (KARST) in a limestone area resulting from the effects of GLACIATION and subsequent melting, the ice subjecting the area to differential erosion (ABRASION), and the melt-water from the ice and snow, charged with CARBON DIOXIDE, dissolving the CALCIUM CARBONATE in the limestone. GLACIER KARST.

glaciology the scientific study of all forms of natural ice and its action. G.

glacis (French, slippery place) **1.** in medieval times, a bank sloping away from fortifications and thus vulnerable to firing from above **2.** a gentle slope or bank. G.

glaçon ICE FLOE.

glade an open space or passage in woodland or forest, natural or produced by the felling of trees. G.

glam (Malay) an inland freshwater swamp in Malaysia. G.

glass an artificial substance, amorphous, transparent or translucent, hard and brittle. Common glass is made by the fusion by melting of a SILICA (sand) with LIMESTONE and SODIUM carbonate. The components are varied to produce glass of special quality, e.g. if the limestone is replaced by oxides of ALUMINIUM and BORON a heat-resistant glass results; and the addition of LEAD oxide produces a clear, highly refractive glass.

glasshouse GREENHOUSE.

glauconite a green coloured mineral, hydrous silicate of IRON and POTASSIUM, occurring in rocks of marine origin. It was formerly used as a pigment by painters. GREEN MUD, GREENSAND.

glaze, glazed frost (American silver thaw) a generally HOMOGENEOUS-2, transparent clear ice formed when **1.** raindrops or non-supercooled DRIZZLE droplets fall on an exposed surface with a temperature well below 0°C (32°F) (e.g. on a road, forming BLACK ICE) **2.** supercooled raindrops or drizzle droplets fall on objects with a surface temperature below or slightly above 0°F (32°F) (e.g. on tree branches) **3.** FROST recurs after some thawing. The weight of ice of a glazed frost can be very damaging, especially to the branches of trees. SUPERCOOLING.

glebe the portion of land assigned to a rector or vicar of the Church of England as part of his benefice. G.

glen (Scottish and Gaelic; Irish glean; Welsh glyn) a steep-sided, usually glaciated, valley characterized by a narrow floor, sometimes contrasted with a broad, open and cultivated valley or STRATH. G.

gley, glei in soil science, a sticky yellow and grey mottling in the soil, or the horizon thus mottled. It is caused by poor drainage, intermittent waterlogging reducing OXIDATION or causing the deoxidation of FERRIC compounds. The FERROUS compounds which result are bluish-grey, mottled, sticky, clayey

and compact. PODZOLIC SOILS, WATERLOGGED. G.

gleying the development of a GLEY structure in soil. G.

gley podzol PODZOLIC SOILS.

gley (glei) soils, gleysols one of the seven major groups in the 1973 SOIL CLASSIFICATION of England and Wales, covering soils subject to periodic or permanent waterlogging (WATERLOGGED), not artificially drained, and to some extent displaying the MOTTLED horizon below the surface layer typical of GLEYING. There may or may not be a humic, peaty topsoil. They are characterized by the extent to which humus, clay, sand, peat are present, and to the severity of waterlogging. Those with a peaty surface horizon overlying a clayey, MOTTLED impermeable horizon are termed humic gley soils or stagnohumic gley soils: they are excessively wet and typically non-calcareous. Those that lack the peaty surface horizon and have a subsurface horizon that is slightly less impermeable, are termed stagnogley soils: in general these are non-alluvial and non-calcareous and are classified as loamy or clayey. Other gley soils lacking the peaty surface horizon are termed argillaceous (clayey) or sandy. PODZOLIC SOILS.

glint a steep cliff, steep terrace, steep edge of a plateau. KLINT. G.

glint line an erosion escarpment, the conspicuous edge of a stretch of denuded rock, particularly the edge between an ancient SHIELD (such as the Laurentian shield) and younger rocks. G.

glint-line lake a lake, frequently one of a string, formed along a GLINT LINE. G.

glitter (Northumberland) a SCREE.

global adj. **1.** of or pertaining to the earth as a whole **2.** spherical **3.** comprehensive, total, encompassing all or nearly all considerations, categories, items etc.

globate adj. shaped like a GLOBE.

globe 1. a spherical body 2. the EARTH-1 3. a spherical model with the configuration of the earth (terrestrial globe) or the heavens (celestial globe) shown on it. G.

globe artichoke ARTICHOKE.

globigerina a member of *Globigerina*, family Globigerinidae, a genus of FORAMINIFERA the shells of which occur as fossils in TERTIARY formations, and form the main constituent of the CALCAREOUS mud that covers about one-third of the whole ocean floor. OOZE.

globule a small rounded particle or drop.

gloup, gloap (Scottish) BLOWHOLE. G.

glyders, glydrs (north Wales) SCREE.

glyptolith VENTIFACT.

GMT, Greenwich Mean Time the LOCAL TIME at the meridian of Greenwich England (0° longitude, passing through the former Royal Observatory at Greenwich), the STANDARD TIME for the British Isles and the standard from which most of the countries of the world reckon their standard times. MEAN SOLAR TIME.

gneiss a foliated (FOLIATION), coarsely crystalline rock with characteristic alternating dark and light bands composed of dissimilar minerals, having the composition of GRANITE or DIORITE but produced by the METAMORPHISM and recrystallization of both IGNEOUS rocks and sediments. Those resulting from the DYNAMIC METAMORPHISM of igneous rocks are termed orthogneiss; those from metamorphism of SEDIMENTARY ROCKS are termed paragneiss. Gneiss is the most common of crystalline rocks (the term covering a wide range of crystalline rocks derived from DIORITE, GABBRO, GRANITE, SYENITE), and gneisses make up a large part of the areas of ancient rocks, e.g. of the Laurentian shield. G.

gneissose structure a structure shown by coarsely crystalline rocks, characterized by discontinuous FOLIATION. G.

GNP gross national product, a measure of the total flow of output in an economy during any specified period of time (excluding goods or services used as inputs in the production of further goods or services), i.e. the GDP (gross domestic product) of a country combined with income derived from overseas investments less profits generated by production within the country but due to foreigners abroad.

goal an objective, the object of efforts or ambition. ATTITUDE-2.

goat a horned RUMINANT of the genus *Capra*, family Bovidae, wild and domesticated, feeding on scanty, coarse vegetation, a source of rather poor meat, skins for leather, and hair (ANGORA, CASHMERE). Goats are bred mainly in hot dry countries because they are able to thrive on the poorest of fodder which no other animal will touch; but they are very destructive of vegetation, and it is often claimed that, if uncontrolled, they are more of a liability than an asset.

Gobi, gobi 1. Gobi, the great arid basin in central Asia containing minor basins. TALA 2. gobi, a small inland basin containing late MESOZOIC or TERTIARY sediments, occurring in a TALA in the great basin of the Gobi. G.

godown (Malay gadong, godong) a warehouse in the Far East.

gold an indestructible, yellow, MALLEABLE, DUCTILE PRECIOUS METAL, almost resistant to chemical attack (NOBLE METAL) but dissolving in AQUA REGIA. It occurs in alluvial deposits (PAN) or in quartz veins in a free state. It is sometimes associated with metallic sulphides in the quartz veins or in the other forms in which these occur, thus gold is sometimes mined directly or won as a byproduct from other metals. In use in industry, works of art, jewellery etc. it is alloyed with other metals to achieve hardness or to change its natural colour. CARAT. C.

golf course, golf links terms used interchangeably (COURSE-5), but strictly golf links (LINKS) should be restricted to a tract of land lying near the sea coast, on which the game of golf is played.

gombolala (Uganda: Luganda language) an administrative division, smaller than a SAZA, sometimes termed a sub-county. G.

Gondwanaland 1. the southern part of the great PRECAMBRIAN land mass, PANGAEA. Current scientific evidence suggests that it included today's Africa, Madagascar, Australia, part of South America, Antarctica and the Indian subcontinent, and began to break up in PALAEOZOIC times. CONTINENTAL DRIFT, LAURASIA, PLATE TECTONICS, TETHYS OCEAN 2. the land of the Gond tribe in India. G.

good in economics, anything (morally good or morally bad, material or intangible) which is capable of satisfying a human want. CAPITAL GOODS, CONSUMER DURABLES, CONSUMER GOODS, ECONOMIC GOODS, FREE GOODS.

goodness of fit in statistics, the extent to which a model or theory fits the data which are to be explained.

goods train a railway train consisting of wagons designed to carry FREIGHT.

goose pl. geese, a large waterfowl of the family Anatidae, wild or domesticated (DOMESTIC ANIMAL). The domesticated varieties are reared on land where they feed on herbage. They are valued for their meat, feathers and eggs. BIRD.

gooseberry a member of the species of *Ribes*, the only genus in the order Grossulariaceae, small shrubs native to cool, moist climates, *R. grossularia*, grown in Europe, being a spreading, pendulous or upright shrub derived from a plant native to most of Europe. It bears juicy, fleshy fruit (BERRY). Some varieties are sweet, eaten raw; others are sour, used in jam making and other cooking.

gorge a rocky-walled, steep-sided, deep, narrow river valley. G.

gorich (Indian subcontinent: Baluchi) the dry, steady, strong northwest wind in Baluchistan. G.

gossan Cornish miners' term applied to a cellular mass filling the uppermost part of a mineral vein, near the surface of the ground, consisting of hydrated iron oxide (LIMONITE), usually with QUARTZ and other GANGUE minerals, from which sulphide minerals have been oxidized, leached and carried down in percolating solutions (PERCOLATION) to enrich the ore vein at greater depths. An alternative term is iron hat.

gouffre (French) a large vertical shaft in LIMESTONE, synonymous with AVEN, sometimes appearing in place-names, e.g. Gouffre de Padirac, southern France.

gouge 1. clay or other fine material on a fault plane, as distinct from FAULT BRECCIA 2. the soft material which in some cases surrounds the containing wall of a mineral vein, and is removed (gouged out) without difficulty, thereby easing the mining of the vein.

gouging a term sometimes, but rarely, applied to the local basining of BED ROCK surfaces brought about by glacial EROSION. PLUCKING. G.

gourd a plant of the Cucurbitaceae family, native to tropical areas, some varieties being grown for the sake of their fleshy, succulent fruit, others, e.g. of the genus *Lagenaria*, for the dry hard shells of their fruit, used as containers for liquids etc.

goz (Sudan: Arabic) the gently undulating tracts of accumulated sand, lacking steep-sided dunes, in Darfur and Kordofan, Sudan. G.

Graben (German) a long, narrow, depressed tract of the earth's crust, a trough (TROUGH FAULT) between parallel NORMAL FAULTS, the THROWS of which face in opposite directions. The term is sometimes applied to a RIFT

VALLEY, but such usage is unacceptable to some authors on the grounds that a Graben is a structural feature and not necessarily a valley. G.

gradation the bringing of the surface of the land to a uniform GRADE by the processes of AGGRADATION and DEGRADATION.

grade 1. a class of persons or things distinguished by having the same rank, quality, value 2. in geomorphology, the concept of a system in EQUILIBRIUM, e.g. a GRADED RIVER, or a ground slope where erosion, transport and deposition are in equilibrium 3. sometimes used as a synonym for GRADIENT 4. the degree of METAMORPHISM undergone by rocks. HIGH GRADE METAMORPHISM, LOW GRADE METAMORPHISM. G.

graded river, graded stream a river which has by EROSION and DEPOSITION adjusted its channel and the slope of its bed so that the rate of its flow is exactly that needed for the carrying of the material supplied by the drainage basin, a delicately balanced condition which may be upset at any time.

graded sediments loose or cemented sediments with particles sorted by natural means, according to size. Various systems of grain size are used in the study of sediments and soils (SOIL TEXTURE, WENTWORTH SCALE), e.g.

Grade	International diameter (mm)	American diameter (mm)
stones	over 2.00	over 2.00
coarse sand	0.2–2.00	0.2–2.00
fine sand	0.02–0.2	0.05–0.2
silt	0.002–0.02	0.002–0.05
clay	below 0.002	below 0.002

graded time one of the categories in Schumm and Lichty's GEOMORPHOLOGICAL TIMESCALE, the length of time needed (e.g. by a stream) for the development of a graded profile.

gradient 1. the degree of slope of a land surface, stream, road, railway, away from the horizontal, expressed as a percentage, or an angular measurement from the horizontal, or as a proportion between its VERTICAL INTERVAL, reduced to unity, and its HORIZONTAL EQUIVALENT. SLOPE LENGTH 2. the degree of variation in certain phenomena, e.g. atmospheric pressure, temperature, density etc. GEOTHERMAL GRADIENT.

gradient wind a wind resembling the hypothetical GEOSTROPHIC WIND in relation to the effects of ATMOSPHERIC PRESSURE force and CORIOLIS FORCE, but also taking into account the CENTRIFUGAL effect due to the curvature of the isobars (ISO-) and the effects of FRICTION near the earth's surface.

graffito (Italian, pl. graffiti) 1. unauthorized drawing or writing on a wall or other surface (the term originally implied scratching, as on the ancient walls of Pompeii and Rome etc.) 2. an incised decoration, e.g. on a pot, produced by scratching through a superficial layer of plaster, glaze, etc. to reveal a GROUND-5 of different colour. G.

grain 1. the general trend of the geological structure, folds, faults etc. reflected in the dominant direction of mountain ranges, river valleys etc. 2. the natural arrangement of strata in rock, coal etc. 3. the direction and pattern in which wood fibres grow 4. a very small hard particle, especially of a mineral 5. the seed of a CEREAL; or harvested cereal crops in general. FLOUR, MILLSTONE 6. the smallest unit of weight in British measures. In apothecaries and TROY 480 grains = 1 ounce; in AVOIRDUPOIS 7000 grains = 1 lb (POUND-3); 15.4323 grains = 1 GRAM-2. CONVERSION TABLES.

gram 1. a pulse (CHICK PEA) 2. gram, gramme, gm, the unit of mass in the metric measurement system, one thousandth part of a kilogram. CONVERSION TABLES, KILO-.

255

granadilla *Passiflora quadrangularis*, a PERENNIAL climbing plant, native to Brazil, widely grown in tropical areas for its juicy, sweet fruit (passion fruit), eaten raw or pressed for its juice. *P. edulis*, with its purple, wrinkled ripe fruit, is the preferred variety.

granary 1. a storehouse for threshed CEREAL (grain) crops 2. a region producing or exporting much grain.

Gran Chaco CHACO.

grange 1. a country house with farm buildings, usually the residence of a GENTLEMAN FARMER 2. historically, an outlying farmhouse with farm buildings, attached to a monastery or a feudal lord 3. a building for storing threshed grain, a GRANARY-1. G.

granite 1. specifically, a granular, coarsely-crystalline acid PLUTONIC ROCK, the most common and widespread of plutonic rocks, consisting essentially of ORTHOCLASE FELDSPAR with free QUARTZ and a MICA, formed by the slow cooling of a large INTRUSION of MAGMA or possibly by the transformation of pre-existing COUNTRY ROCK by a form of METASOMATISM 2. loosely, any coarsely-crystalline, pale, IGNEOUS ROCK. Granite is used as a building stone, especially polished for facings, and as an AGGREGATE-4. ACID ROCK, BATHOLITH, GNEISS, MICROGRANITE. G.

granitization a process by which pre-existing rocks may be formed into GRANITE. METASOMATISM.

granodiorite a coarsely-crystalline acid IGNEOUS ROCK with QUARTZ, PLAGIOCLASE and ORTHOCLASE FELDSPARS (the plagioclase predominating) and, usually, MICA, the main PLUTONIC igneous rock occurring in most of the BATHOLITHS. DIORITE.

granophyre a very finely-grained MICROGRANITE or QUARTZ-PORPHYRY, the QUARTZ and FELDSPAR being very closely intergrown, occurring as a stage in the GRANITIZATION of another rock.

granular *adj.* 1. in general, consisting of GRAINS-4 (GRANULE) 2. in soil science, GRANULAR STRUCTURE.

granular disintegration a type of MECHANICAL WEATHERING, the breaking down of porous rocks (POROSITY) into fragments to form a granular mass, caused either by the freezing and thawing of water absorbed in the rock pores (which dislodges rock particles), or by differential contraction and expansion caused by INSOLATION-1,2 (which results in the disintegration of the grains).

granular structure in soil science, a soil arrangement consisting of GRANULES-2. SOIL STRUCTURE.

granule 1. in general, a small hard particle, a grain 2. in soil science, a friable, rounded, irregularly shaped AGGREGATE-1 up to 10 mm in diameter. SOIL STRUCTURE.

granulite a high grade METAMORPHIC ROCK with a GRANULAR-1 texture.

grape the juicy edible green or purple BERRY growing in bunches on the vine of the genus *Vitis*. *V. vinifera* (and its cultivars) is the species that produces grapes for dessert, wine, and for drying as currants, sultanas, raisins. Probably native to western Asia, this species is now grown widely in Mediterranean conditions and in mild midlatitude areas (including southern England) in the open air, or under glass in cooler areas (VINERY). The finest quality dessert grapes have to be grown under glass in Britain. Dried fruits are derived mainly from wine grapes, currants from a small-fruited variety grown around Patras in Greece since classical times; sultanas from a small seedless variety; raisins, with or without seeds, from a variety with larger berries. The traditional drying method is to spread out the crop and expose it to sunlight, but muscatels (a high-quality dessert grape and raisin) are traditionally dried on the vine by partial cutting of the stalk. VINEYARD, VITICULTURE.

grapefruit *Citrus paradisi*, a tree some-what larger than the orange tree, the most important producer of CITRUS FRUIT to have originated outside Asia. It is native to the West Indies and considered to be either a POMELO – SWEET ORANGE hybrid, or a sport from a pomelo. It is widely grown in tropical and subtropical areas for the sake of its large, yellow-skinned, acid, slightly bitter fruit, which is canned, pressed for juice, made into marmalade or eaten fresh.

graph 1. a geometric representation of a function, i.e. a diagram indicating the relationship of one variable quantity to one or more others by showing their values as distances from two axes (occasionally three) usually placed at right angles to one another, one of the VARIABLES to be plotted being scaled along each of the axes. By convention, in a two-dimensional graph showing a FREQUENCY DISTRIBUTION, the DE-PENDENT VARIABLE is plotted along the vertical axis (i.e. the *y*-axis), repre-senting the FREQUENCY of occurrence, the INDEPENDENT VARIABLE along the horizontal axis (i.e. the *x*-axis) 2. a mathematical structure, a TOPO-LOGICAL diagram, consisting of a set of objects (which may also be termed access points, junctions, nodes, points, terminals, or vertices, sing. vertex) some of which are connected to each other by edges (which may also be termed links or routes), thus establish-ing a binary relation between the objects (access points, junctions etc.). Such a graph, being topolological, ignores the spacing of the objects and the lengths of the connecting edges (links, routes). It is used to represent a NETWORK and is basic to GRAPH THEORY. BAR GRAPH, DIVIDED CIRCLE DIAGRAM, HISTOGRAM, IN-CIDENCE MATRIX, INTERCEPT.

graphic *adj.* 1. relating to GRAPHS or diagrams. GRAPHICACY 2. concerned with decoration and representation on a flat surface 3. in geology and mineral-ogy, showing marks resembling written characters on the surface or in a section of a rock, e.g. as in some GRANITES, MICROGRANITES, PEGMATITES with intergrown QUARTZ and alkali FELD-SPAR crystals.

graphicacy the state or condition of being able to form an idea of (to conceptualize), interpret and express relationships (e.g. two- or three-dimen-sional spatial relationships) that cannot be expressed in words and/or mathe-matical terms alone (i.e. not in the languages of LITERACY and/or NUM-ERACY alone) by the use of maps, diagrams, graphs (CARTOGRAPHY in its widest sense) and illustrative mater-ial such as photographs, supported by the skills of literacy and of the various branches of MATHEMATICS. ARTICU-LACY, INGRAPHICATE. G.

graphite 1. an almost pure, crystalline form of natural CARBON, soft and black, commonly occurring in veins or lenticular masses, used in the manufac-ture of 'lead' pencils, paints, dry lubri-cants, electrical apparatus, e.g. dry batteries 2. a crystalline form of pure carbon manufactured from coke and pitch to form electrodes, or to serve as blocks used as moderators in nuclear reactors.

graph theory the branch of mathemat-ics concerned with the properties of GRAPHS-2, i.e. with the vertices (access points, junctions, nodes, points, termi-nals) and edges (links, routes). Graph theory is used to describe NETWORKS, to indicate ACCESSIBILITY, etc.

grass a member of Gramineae, a large family of loose HERBACEOUS plants with shallow, fibrous roots, and with a world-wide distribution in areas with sufficient moisture. Among the mem-bers are CEREALS, reeds, bamboos, pasture grasses. Grasses provide valu-able food for people and their livestock, through their starchy seeds, rich in protein, and their nutritious foliage.

The leaves of some species are used in paper-making (ESPARTO), the stems of others (BAMBOO) provide structural timber. BLUEGRASS, SUGAR CANE.

grassland land covered with grass **1.** occurring naturally in areas with sufficient moisture to stimulate and maintain seasonal growth, e.g. tropical (SAVANNA), midlatitude (DOWNS, PRAIRIE, PAMPAS, PUSZTA, STEPPE, VELD), mountain (above tree line, ALP) **2.** cultivated for pasture or hay, either permanent or short LEY (grown in rotation with arable crops), usually sown and fertilized.

graticule the network of meridians and parallels on a map. GRID. G.

grau (French) a strait connecting a lagoon of a coastal plain with the sea. GAT. G.

graupel granular snow, not so hard as HAIL.

Grauwacke GRAYWACKE.

gravel a loose, water-worn sediment in which small rounded stones, variously defined as 2 to 10 mm (0.08 to 0.4 in), or 2 to 20 mm (0.08 to 0.8 in), or 2 to 50 mm (0.08 to 2 in) in 'diameter', predominate. It is used especially in making CONCRETE. When consolidated it becomes a CONGLOMERATE. BOULDER, COBBLE, PEBBLE and RUDACEOUS ROCK, SOIL TEXTURE. G.

gravel pit a surface working from which GRAVEL is extracted.

gravel train a VALLEY TRAIN in which the dominant material is GRAVEL. G.

graveyard a cemetery, a burial ground.

gravitation **1.** the act or process of moving under the force of GRAVITY-2 **2.** the attractive force between any two pieces of matter, between two bodies, that is independent of the chemical nature of those bodies, or the presence of bodies between them **3.** the law (NEWTON'S LAW OF UNIVERSAL GRAVITATION) stating that any two pieces of matter attract one another with a force which is directly proportional to the produce of their masses,

and inversely proportional to the square of the distance between them.

gravitational, gravity *adj.* of or pertaining to GRAVITATION.

gravitational slumping GRAVITY SLUMPING, MASS MOVEMENT.

gravitational transfer MASS MOVEMENT.

gravitational water GRAVITY WATER.

gravity **1.** weight, heaviness **2.** the gravitational force (GRAVITATION) between the earth and a body on the earth's surface or in the earth's gravitational field. The force decreases with increasing distance from the earth, the force of gravity being inversely proportional to the square of the distance. CENTRE OF GRAVITY **3.** *adj.* synonymous with GRAVITATIONAL.

gravity anomaly a deviation from the predicted value of the GRAVITY of rocks, termed POSITIVE ANOMALY if exceeding expectation, NEGATIVE ANOMALY if below. The deviation may be due to the existence of rocks of unexpectedly different densities or magnetic characters in the earth's crust, so the detection of gravity anomaly is important in GEOPHYSICAL surveying and in the study of the structure of the EARTH'S CRUST. ANOMALY, ISOSTATIC ANOMALY, ISOSTASY.

gravity collapse structure the falling apart and sliding down of the strata on each side of an ANTICLINE, particularly likely to occur if incompetent beds (COMPETENCE OF ROCK) lie between resistant strata and erosion has occurred.

gravity model a mathematical MODEL-4 inspired by NEWTON'S LAW OF UNIVERSAL GRAVITATION, i.e. the force of attraction of two masses, or bodies, is proportional to the product of their masses and inversely proportional to the square of the distance between them. By analogy with this physical law a mathematical gravity model may be used to predict the attraction of places, and thus the potential movement of people between them.

Thus, in a crude analogy, the force of attraction becomes the transactions (T) between two places, i and j (T_{ij}); the masses become the two places represented by the numbers of their inhabitants (P), P_i representing the population of one place, P_j the other; and the distance between them D_{ij}:

$$T_{ij} = \frac{P_i P_j}{D_{ij}^2}$$

Using this equation the relative transactions between towns A and B (T_{AB}) and A and C (T_{AC}) can be calculated:

$$T_{AB} = \frac{Population\ of\ A \times population\ of\ B}{Distance\ between\ A\ and\ B^2}$$

$$T_{AC} = \frac{Population\ of\ A \times population\ of\ C}{Distance\ between\ A\ and\ C^2}$$

This version of a gravity model is rarely applicable in such a simplified form, so is usually much refined in practice. REILLY'S LAW OF RETAIL GRAVITATION.

gravity slope that part of the slope of a hillside which is steep in comparison with the gentler slope, the HALDENHANG, which occurs lower down. G.

gravity slumping, gravitational slumping, sliding the downward slippping of rock masses or sediments due to GRAVITY-2. MASS MOVEMENT, SLUMP.

gravity water, gravitational water water in the soil, moving by the force of GRAVITY-2 towards the WATER TABLE. Either term may appear as a synonym for VADOSE WATER. G.

gray-brown podzolic soil GREY-BROWN PODZOLIC SOIL.

graywacke,   greywacke (German Grauwacke) an old term (now superseded by TURBIDITE) loosely applied to the dark, coarse SANDSTONES or gritstones (GRIT) composed of angular or subangular rock fragments firmly cemented with ARGILLACEOUS mater-

ial, characteristic of some older FORMATIONS-2, especially of the earlier PALAEOZOIC, commonly associated with sediments deposited in a GEOSYNCLINE. G.

grazing a PASTURE area where livestock feed on growing grass or other HERBS-1. A distinction is often made between land used for grazing and land which is precisely similar, but from which the herbage is cut for hay or drying. MEADOW, ROUGH GRAZING. G.

great circle any hypothetical circle on the earth's surface the plane of which passes through the earth's centre, and thus divides the earth into two HEMISPHERES. The EQUATOR is a great circle, and each MERIDIAN is half of a great circle. The number of great circles which can be drawn on a sphere is limitless. The shortest distance between any two points on the earth's surface is the arc of the great circle on which they lie, hence the use of the great circle routes by aircraft, e.g. over the north polar regions. G.

Great Ice Age ICE AGE.

greenbelt, green belt a tract of open country of varying width, and not necessarily continuous, with farmland, open recreational areas, woodland etc. surrounding a large built-up area and, by planning regulations, protected from further building in order to prevent URBAN SPRAWL and to provide an AMENITY. G.

green fallow PARTIAL FALLOW, fallow land from which a quickly maturing green crop (e.g. turnips, potatoes) is taken, the crop being planted in rows and weeded by hoeing. Bastard fallow (pin fallow) is a type of green fallow, the land being sown with a quickly maturing crop (e.g. vetches) in autumn, fed-off in spring. FALLOW. G.

green field, greenfield site a planning term for a plot of land, previously undeveloped, for which DEVELOPMENT-2 is proposed, or on which it is in progress.

greengage GAGE.

green gram *Phaseolus aureus*, a PULSE (MUNG in the Indian subcontinent, to which it is probably native), a leguminous food plant (LEGUMINOSAE) bearing green, brown or mottled nutritious seed, widely grown in tropical and subtropical areas not only for its seed but for 'bean sprouts', obtained by germinating seeds in the dark until shoots reach the required length. The sprouts are high in vitamin C, the seeds are rich in protein.

greenhouse glasshouse, a building with glass or clear plastic roof and sides, sometimes heated artifically, used to provide the protection from air turbulence and warmth needed in some areas by certain growing plants. The glass or clear plastic allows incoming short-wave SOLAR RADIATION to enter the structure, but acts as a barrier to outoing long-wave radiation from the earth. GREENHOUSE EFFECT.

greenhouse effect the phenomenon in which the ATMOSPHERE near the earth's surface holds heat. Incoming short-wave SOLAR RADIATION (which includes visible light and heat) has no difficulty in passing through the atmosphere to the earth's surface. It is absorbed there by materials acting as BLACK BODIES and these bodies re-radiate on longer wavelengths. But this outgoing returning long-wave RADIATION from the earth does have difficulty in passing through the atmosphere: it is inhibited especially in the presence of cloud, being absorbed and radiated back by water vapour, particles and carbon dioxide. Thus the atmosphere near the earth's surface is warmed by long-wave radiation and, like the glass of a greenhouse, restrains the rising heat, an effect enhanced by cloud cover. ATMOSPHERIC WINDOW.

green labour people who for the first time go outside the home to work for payment. COTTAGE INDUSTRY.

green lane an unmetalled road, still bearing a turf or grass surface, but constituting a right-of-way. Often identical with DRIFTWAY.

green manure a green crop, especially one rich in nitrogen, which is, while still green, ploughed directly into the soil to increase soil fertility and improve structure. CLOVER.

Green Mud a fine-grained deposit with a high content of CALCIUM CARBONATE occurring on some CONTINENTAL SLOPES, e.g. of west Africa and North and South America, the colour being derived from GLAUCONITE. BLUE MUD, RED MUD.

green pepper *Capsicum annum*, PEPPER-1.

green pound in the CAP of the EEC, the unit (of varying value) used to represent the rate of exchange for agricultural purposes between the UK pound sterling and the UNITS OF ACCOUNT.

green revolution an AGRICULTURAL REVOLUTION in less developed countries, especially in Asia in the 1960s and 1970s, which gave rise to increased food production, brought about mainly by the rapid improvement of yields of cereal crops (particularly of RICE, MAIZE and WHEAT) resulting from the introduction of new high-yielding varieties and the techniques necessary to grow them.

greensand a CRETACEOUS rock series similar to a SANDSTONE, consisting largely of GLAUCONITE, hence the green colour, the denominations Upper and Lower relating to the position above or below the GAULT.

green village a small nucleated settlement in which the homesteads surround an open space or village green. G.

Greenwich Mean Time GMT.

Greenwich meridian the prime MERIDIAN (0° longitude) passing through the former Royal Observatory at Greenwich, England, from which other meridians are calculated, expressed east or west, to 180° longitude.

gregale the 'wind from Greece' of Malta and its neighbourhood, a strong northeast wind blowing mainly in winter in the central Mediterranean area, associated with a DEPRESSION-3 over north Africa and a large high pressure system (ANTICYCLONE) over the Balkans. G.

Gregorian calendar year, civil year a YEAR consisting of 365 MEAN SOLAR DAYS of 24 hours. In compensation for the additional 0.2422 solar days, every fourth year (leap year) has 366 days, the extra day being added to the second month, February. There is still an inbuilt inaccuracy of one day per century (25 leap years) and in compensation the last year of a century becomes a leap year only if the first two figures are divisible by 4. The Gregorian calendar (not immediately adopted by Protestant Europe) was introduced by Pope Gregory XIII in 1582 to correct the error in the JULIAN CALENDAR year (11 minutes 10 seconds too much). The accumulated deficit of the Julian calendar (11 days) was taken into account when the Gregorian calendar was adopted in England and the English colonies in America in 1752. CALENDAR, INTERCALATION.

grey box approach an approach in SYSTEMS ANALYSIS which considers some of the sub-systems present in the system under study, and is not concerned with the internal structure and the functioning of those sub-systems. BLACK BOX APPROACH, WHITE BOX APPROACH.

greisen a crystalline IGNEOUS ROCK consisting of QUARTZ and white MICA.

grey-brown (gray-brown) podzol (podzolic soil) an acidic soil, transitional between a PODZOL and BROWN FOREST SOIL, less leached and with a greater organic content than a podzol. It occurs mainly in western Europe (including the UK) and northeast USA, where humid conditions lead to some leaching, but less than in the climatic conditions resulting in a true podzol. It supports good quality grassland, and is suitable for mixed farming. It is defined by FAO as a soil with A B C horizons (SOIL HORIZON), characterized by the marked ILLUVIATION of clay in the B HORIZON, appearing as clay coatings on the soil particles or clay linings to pores or cavities, the B horizon also containing small, round FERRUGINOUS concretions. The upper layer of the A HORIZON, i.e. the A_0 horizon, has a MULL-HUMUS form; the A_1 horizon may be only weakly developed; the A_2 horizon is not strongly leached. SOIL, SOIL ASSOCIATION, SOIL PROFILE. G.

grey earth a soil in arid areas of midlatitudes, deficient in organic matter and nitrogen, commonly salt encrusted, because the lack of rainfall discourages plant growth and LEACHING is rare. SEROZEM. G.

greywacke GRAYWACKE.

greywether (WETHER, a male sheep) a SARSEN, a grey, rounded block of SANDSTONE or QUARTZITE remaining as a residual boulder when the less resistant surrounding material has been eroded away. From a distance it appears to resemble a grazing sheep, hence the name. G.

grid 1. an arbitrary network drawn on a map so that an exact reference can be made to any point on the map. In reference grids for small areas the Cartesian coordinate system is commonly used, the location of a point being identified by its distance from two reference lines intersecting at right angles. In this system the grid usually consists of two sets of parallel lines at right angles, the lines being such a distance apart as to represent a fixed distance (e.g. 1 km, 10 km). The lines are numbered eastwards and northwards from a point fixed in the southwest corner of the whole area covered (FALSE ORIGIN); thus any point can be expressed exactly in terms of its EASTING and NORTHING, without the cumbersome use of LONGITUDE and LATI-TUDE. Grid north is the direction of the approximately north to south grid lines, but coincides with TRUE NORTH only along the meridian of origin. BEARING, GRATICULE, NATIONAL GRID 2. any uniform pattern laid on a surface for the purpose of mapping (and subsequently spatially analysing) data, or for calculating values at the NODE-5 of each unit of the pattern so that ISOPLETHS may be drawn 3. a NETWORK-2, e.g. of electricity cables or gas pipelines for a power supply. G.

grike, gryke a deep cranny or fissure lying between ridges (CLINTS) and traversing LIMESTONE pavements, caused by CARBONATION SOLUTION along well-defined joints. G.

grit, gritstone an ambiguous term applied to 1. small, usually angular, particles of rock 2. coarse-grained SANDSTONE, or sandstone (coarse or fine) made up of angular grains 3. sandstone of grains of conspicuously unequal size 4. stratigraphically, in names of such FORMATIONS-2 as Millstone Grit. G.

grivation the angle between grid north (GRID-1) and MAGNETIC NORTH on a map with a grid.

groin GROYNE.

gross domestic product GDP.

gross migration a measure of the total flow of people into and out of an area. MIGRATION-1,2, NET MIGRATION.

gross national product GNP.

gross production rate in ecology, PRO-DUCTION RATE.

gross reproduction rate REPRODUC-TION RATE.

gross tonnage SHIPPING TONNAGE.

grotto a small picturesque cavern, especially one with an attractive feature, e.g. a waterfall, or natural or simulated stalactites etc. Artificial grottos were very fashionable in landscape gardening in Europe in the eighteenth and nineteenth centuries.

ground 1. the surface of the earth 2. the uppermost layer of the soil 3. an enclosed area of land set aside for a particular purpose, e.g. camping ground, playground 4. grounds, enclosed land with lawns, trees, shrubs, flower beds etc. larger than a GARDEN-1, but similarly attached to a house 5. the basic surface or foundation colour of pottery, painting etc. revealed by GRAFFITO-2.

grounded theory THEORY-1 purposefully generated and based on DATA, as distinct from FORMAL THEORY.

ground fog RADIATION FOG.

ground frost a FROST occurring on the surface of the ground or in the upper soil layer when the ground surface temperature drops to $-1°C$ (30.2°F), a condition likely to have a marked effect on plant tissues. AIR FROST.

ground ice 1. ice formed on the bed of a river, lake or shallow sea when the water as a whole remains unfrozen (also termed ANCHOR ICE or bottom ice) 2. fossil ice, a mass of clear ice associated with PERMAFROST. G.

ground icing AUFEIS.

ground information in REMOTE SENS-ING, information concerning the state of the physical environment derived from aerial photographs and measurements (based on sample estimates) of soil moisture, temperature, biomass etc. Ideally this information is recorded at the time when a remote sensor, also collecting data, is passing over the area concerned. GROUND TRUTH.

ground-mass, groundmass the compact basal element of an IGNEOUS ROCK, in which the crystals are embedded. MATRIX-3.

ground moraine debris carried by the underside of a glacier or ice sheet and deposited as TILL. BOULDER CLAY, MORAINE. G.

groundnut, ground nut, monkey nut, peanut *Arachis hypogaea*, family LEGUM-INOSAE, native to South America, now an important food plant on light soils in tropical and subtropical areas. The plant is erect in some varieties, prostrate in others. The stalk bearing the flowers lengthens after germination and forces the seed pods into the soil, where they develop (underground). Harvesting is carried out by hand or machine. The nuts (seeds) are high in protein, oil and vitamins B and E. They are cooked, eaten raw or roasted, roasted and ground to form peanut butter. The oil extracted from them is valuable as a salad or cooking oil, and in the making of margarine, soap and toilet preparations. After oil extraction the residue is made into oilcake for animal feed.

ground radiation   RADIATION.

ground rent   the rent paid by a leaseholder to the freeholder for the privilege of using or leasing the ground on which a building stands. Ground rent is termed unsecured until the building is actually erected; when the building has been erected it is termed secured. LAND TENURE.

ground swell   the deep, slow rolling of the sea (sometimes caused by a distant storm or a SEISMIC disturbance) which in passing into shallow water raises the height of the waves.

ground truth   an inaccurate synonym for GROUND INFORMATION. Ground information is the term preferred because 'truth' implies an accuracy not achieved by measurements of soil moisture etc. based on sample estimates. REMOTE SENSING.

ground water, groundwater   PHREATIC WATER, all the water derived from PERCOLATION of rainwater, from water trapped in a sediment at its time of deposition (CONNATE WATER) and from magmatic sources (JUVENILE WATER), lying under the surface of the ground above an IMPERMEABLE layer, but excluding underground streams. VADOSE WATER, WATER TABLE. G.

group   1. a number of animals (including human beings), plants or things gathered closely together and considered as a whole 2. a number of things (e.g. languages), persons or other organisms classed together. CLASSIFICATION OF ORGANISMS 3. an organized body of people with a common purpose. PRESSURE GROUP 4. in geology, a SYSTEM of rocks formed in one of the eras, e.g. in the Precambrian, Palaeozoic etc. GEOLOGICAL TIME-SCALE 5. in statistics, a set of elements, individuals or observations all of which have one characteristic, or more, in common. CLASS-5.

grouped frequency distribution   a FREQUENCY DISTRIBUTION in which some of the original CLASS INTERVALS have been COLLAPSED (joined together) in order to emphasize the pattern in the data.

group factor   FACTOR ANALYSIS.

grove   a small group of trees or a small wood, natural or planted. If scattered, such groups may mark the transition zone between FOREST-2 and GRASSLAND-1, hence the term grove belt applied to the zone between the coniferous forests and the prairies of Canada. Sacred groves, associated with mountains and temples in China, Japan and elsewhere, were planted to honour gods or serve as places for worship. G.

growan   (southwest England, Dartmoor) local term, a coarse-grained deposit consisting of decayed GRANITE. GRUS.

growing season   that part of the year when plant growth is active on account of favourable temperature, availability of moisture, and sufficient hours of daylight. The duration of the season usually decreases with increasing distance from the equator. The crucial temperature for many plants in midlatitudes is 5°C (42°F), below which vegetative growth does not take place and the plant is dormant; but the number of continuously frost-free days (FROST CLIMATE) while the plant is growing to maturity is sometimes an overriding factor. COTTON, MAIZE, VERNALIZA-TION. G.

growler a piece of HUMMOCKED ICE, or a very small ICEBERG, or a mass of ice smaller than a BERGY BIT, almost awash and large enough to be a hazard to shipping. G.

growth area, growth centre, growth point in planning, a district in which economic growth starts or is deliberately concentrated as part of regional development strategy (GROWTH POLE-2, REGIONAL PLANNING) and continues if stimulated, thereby affecting the surrounding region. COUNTERDRIFT.

growth pole a concept introduced by Francois Perroux, French economist, in 1949, who related it only to economic space. It has since been related by others to various indistinct, ill-defined concepts and ideas. In geographical literature growth pole usually refers to **1.** a concentration of integrated, vigorous industries associated with a dynamic, leading industry or industrial sector, the whole being capable not only of growing but of generating growth, especially by the MULTIPLIER EFFECT, in the rest of the economy **2.** an urban centre that may with official support gain strength and stimulate economic development in its HINTERLAND-2. G.

groyne, groin a construction of timber, concrete, or stone jutting seawards, usually at right angles to the coast, to combat LONGSHORE DRIFTING of sand and shingle and maintain a level BEACH. G.

grumusol in SOIL CLASSIFICATION, a heavy texture, black, tropical and subtropical clay soil such as REGUR, in which the CLAY minerals expand and cause the soil to crack in alternating wet and dry seasons, so that the soil becomes churned up. It may contain some CALCIUM CARBONATE. VERTISOLS. G.

grus a coarse-grained FELSIC IGNEOUS ROCK, such as GRANITE, that is partially decomposed. GROWAN.

G-scale a geographical measurement scale proposed by Haggett, Chorley and Stoddart, 1965, the standard geographical measurement being the earth's surface area (Ga), and the scale of measurement (the G-scale) derived by successive subdivisions of this standard area by the power of ten, the G-value rising with decreasing area, thus:

G-value	Earth's surface	Area in sq miles
0	Ga	1.968×10^8
1	$Ga\ (10)^{-1}$	1.968×10^7
2	$Ga\ (10)^{-2}$	1.968×10^6

guadi WADI.

guano (Spanish, excrement) **1.** the thick deposit formed by the excreta of sea-birds valued, as a source of nitrogen and phosphates, for FERTILIZER, e.g. occurring on the many islands of the dry Chilean and Peruvian coasts where there is little or no rainfall to wash it away **2.** other similar deposits, e.g. bat-guano in the limestone caves of Malaysia. G.

guava *Psidium guajava*, a small tree of the American tropics, grown for its fruit, which has a sharp flavour, pulp full of seeds, is high in vitamin C, and is juicy in some varieties. It is usually cooked or canned in syrup, or used to make jam or jelly.

Guinea corn MILLET.

gulch in western USA, a narrow, deep, rocky ravine. G.

gulder DOUBLE TIDE.

gulf 1. a large inlet of the sea, cutting into the land more deeply than a bay. It is more enclosed by the coast than a bay is; and it may itself contain one or more bays **2.** a deep hollow, a chasm, an abyss, e.g. a steep-walled SINK HOLE with a flat alluvial floor in which an underground stream may sink or rise. G.

gully, gulley a narrow channel worn in

the earth by the action of water, especially a miniature valley resulting from a heavy downpour of rain. GULLY EROSION. G.

gully erosion, gullying SOIL EROSION resulting from a sudden heavy downfall of rain which cuts channels (GULLY), especially in soil or soft rock. G.

gully gravure an erosion surface scored by many gullies. GULLY. G.

gum a sticky liquid exuded from certain shrubs and trees, and some seaweeds, soluble in water, insoluble in alcohol and essential oils, hardening in air, used as an adhesive and in pharmaceutical, cosmetic, and food preparation industries; in lustring of silk fabrics, in calico printing, in lithographic ink. The most important in trade are gum arabic, used in confectionery and as postage stamp adhesive, derived from species of *Acacia*, particularly *A. senegal* or *A. verek* and *A. drepanolibium*; and gum tragacanth, more expensive, used in calico printing and in the pharmaceutical industry, derived from several species of thorny shrub (milk vetches), *Astralagus*. Another is karaya gum, used in icecream and sauces, derived from *Sterculia* species. KAURI-1. C.

gumbo 1. in Africa, central and northern Angola, the Bantu name for OKRA, both the plant and its pods **2.** in western USA, a fine-grained soil which when saturated with water becomes black and sticky. The term is derived from a thick, palatable soup made from the mucilaginous pods of okra. G.

gumbotil 1. a dark, leached, sticky clay, formed by weathering of glacial TILL over a very long period **2.** a fossilized soil beneath later tills. G.

gumland in New Zealand, land from which fossil KAURI gum has been obtained, but commonly applied to land which may or may not have yielded gum, now covered with scrub. G.

gum tree, gums any of the species

Eucalyptus, a tree native to Australia. G.

gun (Japanese) a county. G.

gunmetal a variety of bronze, an alloy of COPPER, TIN, ZINC, sometimes with LEAD and NICKEL, dark grey in colour, hard and resistant to corrosion; and formerly used in making cannon.

gur, goor in the Indian subcontinent, coarse sugar.

gust a sudden, strong, brief rush of wind.

gut a narrow water channel opening into the sea or an estuary, especially in the eastern USA.

Gutenberg Discontinuity the discontinuity occurring between the lower surface area of the MANTLE and the CORE of the earth, where the material of the earth's interior is in transition between the solid and plastic state, and where the speed of the P-waves of EARTHQUAKES slackens and S-waves are absent. In the Gutenberg Channel, lying between 100 and 200 km (60 to 120 mi) beneath the LITHOSPHERE, at the upper surface of the mantle, the material becomes more plastic and both P-waves and S-waves exist but are slowing down. ASTHENOSPHERE, PUSH WAVE, SHAKE WAVE.

gutta-percha a rubber-like exudation obtained from tropical trees of the Sapotaceae family, especially now from *Payena Leerii*, vulcanized when mixed with carbon, the addition of sulphur controlling the degree of hardness. It is used in dentistry and as insulation in some electrical cables.

guttation a process in which droplets of excess moisture form on the leaves of plants, usually in highly humid conditions, occurring when the pressure rises in the vascular tissue of the plant as water is taken up by the roots. Such droplets should not be confused with DEW.

gutter an artificial drainage channel along the side of a road, or a metal or plastic channel fixed to the edges of roofs of buildings to catch and carry away rainwater. Formerly, gutter had specific technical applications (e.g. a submarine valley-like depression on a continental shelf), but these are now all obsolescent, and best avoided. G.

guyot tablemount, a topographical feature of the ocean bed, a flat-topped mountain, volcanic in origin, occurring especially on the floor of the Pacific ocean (as opposed to a SEAMOUNT with its pointed summit). It may rise to a great height, e.g. 3200 m (some 10 000 ft), but the summit usually lies well below the ocean surface, e.g. 800 m (2700 ft) below. The term honours Arnold Guyot, 1807-84, a Swiss-American scientist. G.

gwaun (Welsh) moor, mountain pasture. G.

gwely (Welsh, family) a socio-anthropological term covering an extended family claiming descent from a common ancestor. TYDDYN. G.

gymnosperm a seed-bearing plant of the class Gymnospermae, producing seeds not enclosed in an ovary (e.g. conifers), and thus unlike an ANGIOSPERM. By the end of the CRETACEOUS period angiosperms had succeeded gymnosperms as dominant land plants.

gypsey a BOURNE.

gypsum an EVAPORITE mineral, also occurring in SHALES and LIMESTONE, hydrated calcium sulphate, $CaSO_42H_2O$, SELENITE in its fully CRYSTALLINE form, ANHYDRITE without its water of crystallization. In hydrated form it is used in making plaster of Paris, which sets quickly, expanding in the process, to provide clean, clear, sharp casts; as a conditioner in soils that have become saline; as an insulator; and in cement manufacture. Anhydrite is used in the chemical industry.

gyrocompass a COMPASS that makes use of the properties of a continuously driven gyroscope (a fast-spinning wheel mounted in such a way that its plane of rotation can vary, but which, unless subject to a very great disturbing force, keeps the plane constant in space). In the gyrocompass the spinning axis of the continuously driven gyroscope can move only in the horizontal plane and, due to the rotation of the earth, this axis aligns itself with the axis of the earth, thus pointing TRUE NORTH (unlike a magnetic compass, which points to MAGNETIC NORTH).

gyttja (Swedish, pl. gyttjor) a HUMIC grey-brown to blackish A C soil (SOIL HORIZON) of sedimentary origin, rich in organic matter, the humus consisting mainly of plant and animal residues precipitated from standing water. G.

H

haar a cold sea fog accompanying an easterly drift of air on the east coast of Britain, especially in summer from Lincolnshire northwards, notably in the Firth of Forth. G.

habit 1. a tendency to repeat an act again and again without consideration of alternatives **2.** a pattern of behaviour that is automatic to some degree and cannot easily be abandoned **3.** in biology, a characteristic mode of growth and appearance or occurrence **4.** the general appearance of a plant, e.g. prostrate.

habitant (French, inhabitant) originally a peasant settler of French descent (as distinct from one of noble birth) in Canada and Louisiana. Now applied especially to the rural population of the Canadian province of Québec whose native language (NATIVE-3, *adj.*) is French. G.

habitat a place that provides a particular set of environmental conditions for the organism or organisms inhabiting it. Some geographers use the term as a synonym for ENVIRONMENT. NICHE. G.

habitat analysis the study concerned with the evaluation of the non-living factors in a HABITAT.

habūb, habob (Sudan: Arabic, to blow violently) a local strong wind in northern Sudan, especially near Khartoum. Accompanied by thick dust and, in the rainy season, preceding a thunder-shower, it usually blows in the afternoon and evening, mainly between May and September, but it may occur at any time of the year. G.

hachure (French) short lines of shading on a map drawn at right angles to the CONTOURS to represent slopes, the lines being thicker and more closely spaced at areas of greatest slope, but not giving specific information about altitude. G.

hacienda (Spanish) in Spain and former Spanish colonies, e.g. those in South America or the Philippines, a large agricultural estate with a dwelling house. Some authors use the term to cover a country house and its associated activities, e.g. in monoculture, cash crop farming, sheep or cattle ranching (and some would add mining or manufacturing conducted on the estate). Others restrict the application of the term to the country house itself. FARMING, LATIFUNDIA. G.

hade the angle made by a mineral vein, lode or fault plane with the vertical, i.e. the slope of the FAULT as contrasted with the DIP of the beds. G.

Hadley cell a thermally driven vertical circulation cell (ATMOSPHERIC CELL-2) forming that part of the general circulation of the ATMOSPHERE-1 between the equator and approximately 30°N and 30°S. Warmed air rises from the equatorial area to high altitudes, spreads polewards and, cooled, descends at about 30°N and 30°S, flowing towards the equator in the lower levels of the atmosphere. The cell takes its name from G. Hadley who, in his explanation of trade winds in 1735, issued his modification of the original theory put forward by Edmund Halley in 1686. Edmund Halley, 1656-1742, was the English astronomer who theorized on the circulation of the atmosphere, produced the first record of stars visible in the southern hemisphere from personal observation, recorded the comet of 1682 (Halley's comet) and correctly calculated its orbit and its return in 1758.

haematite, hematite an important iron ore, Fe_2O_3, ferric oxide, grey, black or reddish in colour, abundant and widely distributed, valued for its lack of phosphorus, occurring in CRYSTALLINE, MASSIVE or GRANULAR-1 form, sometimes in kidney-shaped nodules (kidney ore), as a CEMENT-2 in SANDSTONE, as an ACCESSORY MINERAL in IGNEOUS ROCK, and in HYDROTHERMAL veins and replacement deposits. BANDED IRONSTONE, LIMONITE.

Haff (German, pl.Haffe) a coastal lagoon of fresh or brackish water, fed by a stream which is blocked by a NEHRUNG, through which it is linked to the sea by a channel, typical of the Baltic coast of East Germany, Poland and the USSR. G.

hafir (Sudan: Arabic) a reservoir in northern Sudan made by mechanical means or by deepening a FULA. G.

hafod, hafoty (Welsh, a summer dwelling, a SHIELING) upland pasture in Wales to which TRANSHUMANCE took place in summer. HENDREF, MEIFOD. G.

hag, hagg PEAT-HAG.

haggs, hagges HAYES. G.

hagiocracy 1. government by a group of people considered to be holy 2. a state governed by such a group 3. the members of such a group. ARISTOCRACY.

ha-ha a hidden barrier marking and protecting the boundary of a garden, park or estate, usually consisting of a trench with a perpendicular inner-side lined with stone or brick, and an outer-side sloping and turfed, the aim usually being to keep out grazing animals while ensuring an uninterrupted view of the landscape from within the barrier. G.

hail, hailstone a small ball or piece of ice with a concentric layered structure, the diameter usually ranging between 5 and 50 mm (0.2 to 2 in), falling separately or agglomerated from CUMULONIMBUS cloud at the passing of a COLD FRONT, sometimes damaging crops, trees, glasshouses and injuring livestock and people. Hailstones are caused by the fast ascent of moist air in which frozen droplets of ice are carried ever higher by the force of the updraught, their size growing as additional WATER VAPOUR freezes on them, until their weight overcomes the force of the ascending air current and they fall to earth. On the way they sometimes gather more ice from supercooled (SUPERCOOLING) water droplets in the moist air. PRECIPITATION-1. G.

hailstorm a violent fall of HAIL.

hained cow pasture (England, Midlands) pasture from which cattle are temporarily excluded to allow the grass to grow freely. G.

Haldenhang (German) wash slope, the part of a slope that is less steep than the part above it, occurring at the foot of a rock wall, usually beneath an accumulation of TALUS. G.

half-hardy *adj.* applied to a plant that is able to tolerate some FROST-1 but is killed by a severe one.

half-life 1. in biology, the time needed for one half of the amount of a substance to be metabolized, eliminated etc. after it has entered an organism 2. in physics, the time needed for one half of the ATOMS of a sample of a radioactive element (RADIOACTIVITY) to disintegrate or change into the ISOTOPE of another element. The half-life of a radioactive element is an important and constant characteristic (ISOTOPE), e.g. the half-life of carbon 14 is 5600 years (but see RADIOCARBON DATING); of radium 226 is 1620 years; of uranium 238, U^{238}, is 4.51 × 10^9 years; decaying to lead 206; of U^{235}

is 7.13 × 10^{18} decaying to lead 207; potassium 40 has a half-life of 47 000 mn years decaying to strontium 87. LEAD-RATIO.

half-year close (eastern England) an enclosure opened for sheep pasturage during the winter half-year. G.

half-year land LAMMAS LAND.

halite NaCl, ROCK SALT, sodium chloride, usually occurring in association with SANDSTONE or SHALE. It tends to flow under the pressure of overburden, giving rise to SALT DOMES, over the tops and on the flanks of which it traps OIL.

hällanalys (Swedish) the analysis of the glacial surfaces of outcropping rocks. G.

Halley, Edmund HADLEY CELL.

Hallstatt *adj.* applied to the transitional cultural period in Europe, about 700 to 450 BC, when the use of bronze gave way to that of iron. The name is derived from Hallstatt, a village in Austria. BRONZE AGE, IRON AGE.

halo a ring or rings of diffused light seen from the earth to encircle the sun or moon in conditions of thin cloud, caused by the refraction of light by water-drops or ice crystals in cloud. It usually appears to be white, but if it is sharply defined, the innermost ring appears red, the outermost blue. CORONA-2.

halobiont an organism that lives in a saline habitat. SALINE, *adj.*

halobiotic *adj.* living in the sea.

haloclasty the disintegration of rocks common in hot deserts resulting from periodic wetting which gives rise to the crystallizing of salts, to swelling and thus to stresses in the rocks. CHEMICAL WEATHERING, HYDRATION, THERMO-CLASTY.

halocline the boundary between two masses of water with differing SALINITY.

halo effect of a BOUNDARY, the detrimental effect of a boundary on locations close to it, making them un-attractive to people, e.g. of a political boundary in disputed territory, resulting in the emigration of the population. See INTERVENING OPPORTUNITY EFFECT.

halolimnic organism an organism living in freshwater but having an affinity with salt water forms.

halomorphic soil saline soil, a soil with properties determined by the presence of salts. SALINE, *adj.*, SOLONCHAK, SOLONETZ. G.

halophyte a plant that tolerates soil impregnated with salt, e.g. the soil of a salt marsh, the seashore, a salt desert.

halosere halarch succession, the stages in a plant SUCCESSION-2 in which the pioneer community develops in saline conditions. HYDROSERE, MESOSERE, SALINE, *adj.*, SERE, XEROSERE.

ham in southern England, a local term **1.** a plot of meadow land, especially a tract of rich pasture by a river **2.** (Old English ham, home) a settlement ranging in size from a single homestead to a town, common in place-names in the parts of England influenced by Anglo-Saxons, surviving in HAMLET. G.

hamada, hamāda, hammada (north Africa: Arabic) the rocky desert of the plateaus of the Sahara, stripped of sand and dust by air currents, the surface smoothed by ABRASION. ERG, REG. G.

hamlet a small group of dwellings in the English countryside, usually smaller than a village and lacking a church. HAM-2. G.

hammer-pond a small body of water, artificially impounded, used to drive the hammer in the early iron smelting industry in the Weald of southeast England and in the Sheffield district.

hamūn (Indian subcontinent: Pashto) a PLAYA or lake in Baluchistan. G.

handicraft an art or craft in which skilled work is carried out by hand, e.g. weaving at a handloom, pottery, needlework, wood carving.

hanger a WOOD-2, most commonly of beech trees (hence beech-hanger), on the steep chalk slopes of southern England. Some authors have transferred the term to the slope itself. G.

hanging garden a garden on a steep slope.

hanging glacier a truncated GLACIER projecting from a basin or shelf high on a mountainside, from which ice may break off and fall as an ice AVALANCHE.

hanging valley a tributary valley, the lower end of which is well above the bed of the main valley (and from which a WATERFALL may descend), commonly found where the main valley has been deepened by a GLACIER which has since disappeared. U-SHAPED VALLEY. G.

hanging wall 1. the upper surface of a FAULT or VEIN 2. the limit of the ore on the upper side of an inclined ore-body.

haor, haur, hrād (Indian subcontinent: Bengali) a depressed water-filled area, a BHIL, especially in the Sylhet plain. G.

haptotropism THIGMOTROPISM.

harbour 1. an anchorage, haven, stretch of water, close to and sheltered by the shore and protected from the sea and swell by artificial or natural walls which allow access for vessels through a narrow entrance (the harbour mouth). In the harbour vessels can lie at anchor, secure to buoys or alongside wharves, piers, etc. The facilities of a PORT may or may not be provided. ISLAND HARBOUR 2. a term used in place-names in England, especially applied to farms in sheltered situations. G.

harbourage 1. a HARBOUR-1 2. the shelter provided by a harbour.

hardebank the lower hard layer of KIMBERLITE, the dark, diamond-bearing rock of South Africa, changing nearer to the earth's surface to the softer BLUE GROUND and at the surface to YELLOW GROUND. G.

hardeveld (Afrikaans) VELD where a thin layer of soil is underlain by hard gravel or rock, making ploughing impossible.

hardness 1. the quality of being hard, in solid objects generally, of being resistant to cutting, cracking or crushing 2. in metals, resistance to indentation or deformation 3. in minerals, resistance to ABRASION. The hardness of minerals is indicated on a scale devised by Friedrich Mohs (Mohs' Scale), a German mineralogist, 1773-1839, in which the softest (talc) was represented by 1 and the hardest (diamond) by 10 (subsequently revised, the diamond now being 15) 4. of water, HARD WATER.

hard pan, hard-pan, hardpan an indurated (INDURATION) or cemented layer of soil of varying character found at varying distances below, or sometimes at, the upper surface of the soil. It is usually formed from material carried down mechanically or in solution by rainwater percolating from the surface, and later deposited. The term is applied to the layer or the material, e.g. clay pan, a dense subsoil formed by the washing down of CLAY or syntheses of clay; duripan or silcrete, cemented SILICEOUS minerals; fragipan, an acid, cemented, platey layer between the parent material and the upper soil layers; iron pan, a layer of sand or fine gravel cemented with IRON oxides; laterite or plinthite, a layer cemented with FERRIC oxide; lime pan (CALICHE, petrocalcic), a thick layer of CALCIUM CARBONATE; moorpan, compact redeposited HUMUS compounds. CALCRETE, CARAPACE, LATERITIQUE, LATERITE. G.

hardware 1. ironmongery, utensils and other articles made of iron and other metals 2. the physical equipment used for data collection and handling, including REMOTE SENSING equipment, computers etc. and their parts. SOFTWARE 3. weapons, machinery.

hard water water which does not easily form a lather with soap or detergent on account of the inhibiting effect of dissolved calcium, magnesium and iron compounds present in it, derived from the rocks over or through which the water has passed (e.g. water flowing over LIMESTONE). The greater the quantity of these salts in it, the harder the water.

hard wheat   WHEAT.

hardwood   1. any DECIDUOUS or evergreen BROADLEAVED tree, with vessels in its wood, that produces close-grained wood **2.** the TIMBER of such a tree. Most hardwoods, e.g. oak, beech, maple, walnut, grow in temperate regions. Other examples are the evergreen oak, native to the region of Mediterranean climate; the eucalyptus, native to eastern Australia; the hard, heavy teak of the regions of monsoon climate; and ebony and mahogany, growing in tropical forests. All are used mainly in furniture making. CABINET WOOD, SOFTWOOD. C.

haricot bean   *Phaseolus vulgaris*, a tender ANNUAL plant, probably native to South America, now cultivated in warm midlatitude, tropical and subtropical climates for its edible seeds which, when dried, can be successfully stored for very long periods. BEAN.

harmattan, hamattan   a strong, dry wind blowing from a northeasterly or sometimes easterly direction over northwest Africa, from the Sahara to the northwest African coast, the southern limit averaging 5°N in January (mid-winter) and 18°N in July. Heavily dust-laden and parching in the interior, it helps to evaporate the high humidity of the Guinea Coast, and thus seems a relatively cool and healthy wind in that area, hence its local name there, the Doctor. G.

harrah   (Arabia: Arabic) a tract of rough lava with a surface sharp enough to cut the feet. G.

harratin   (Sahara: Arabic harāthin) freed slaves, especially the black agricultural serfs in the Sahara. G.

ha-ta   (Japanese) dry cultivation of rice. G.

harvest   1. the gathering in of ripe crops

or FODDER **2.** ripe crops **3.** the season's yield of any natural product **4.** the time of year when crops etc. are gathered in.

Haufendor   (German, pl. Haufendörfer) a nucleated village, irregular in ground plan. NUCLEATED SETTLEMENT-1. G.

haugh, haughland   in northern England and Scotland **1.** low-lying land beside a river **2.** the floodplain of a river. G.

haven   a small sheltered bay or inlet providing a safe, sheltered anchorage for ships.

havsband   (Swedish) the outermost SKERRIES, lying seaward in the skerries that enclose a SKERRY-GUARD. G.

Hawaiian high   one of the persistent atmospheric high pressure cells stationed over the central and eastern sectors of the North Pacific ocean in HORSE LATITUDES, stronger in summer than in winter. ANTICYCLONE.

Hawaiian volcanic eruption   a nonexplosive volcanic eruption in which basic (BASE-2) and highly fluid LAVA flows over a large area and hardens to form a volcano shaped like a shield (SHIELD VOLCANO), e.g. Mauna Loa in Hawaii. PELEAN, STROMBOLIAN, VULCANIAN ERUPTION. G.

hay   the stems and leaves of grasses, cut and dried, and used as FODDER.

haycock   a loose, small pile of HAY set up in a MEADOW-1 to dry.

hayes   in Leicestershire, England, a local term applied to land won from forest or woodland, synonymous with ASSART, a term used in old deeds. In neighbouring areas the terms haggs or hagges, and heage in Derbyshire, had the same application. G.

hayfield   a MEADOW-1 in which grass is grown for HAY.

hayrick, haystack   a large pile of compressed hay built in a regular shape out-of-doors, formerly with a thatched, protecting top, now commonly covered with plastic sheeting or the roof of an open-sided barn, i.e. a Dutch barn.

haystack hill   MOGOTE.

hazard   1. a risk, a chance associated

271

with danger or with a damaging effect. ENVIRONMENTAL HAZARD, FLOOD HAZARD **2.** an obstacle in a path.

hazardous *adj.* risky.

haze a somewhat vague term applied to the condition in the atmosphere near the earth's surface when visibility is restricted to more than 1 km (0.6 mi) but less than 2 km (1.2 mi), commonly resulting from the suspension in the air of solid matter such as dust or smoke particles, or from shimmering caused by intense heat, which produces irregularities and changes in density in the layers of the atmosphere. FOG, MIST. G.

hazel, cob *Corylus avellana*, a bush or small tree native to Europe and Turkey-in-Asia, cultivated for its nutritious NUTS, used in baking and confectionery and for dessert, and known to have been eaten by people since MESO-LITHIC times.

HDC highly developed country, a country which is well advanced in realizing its full potential. LDC, MDC, UNDERDEVELOPMENT.

head **1.** the source of a river **2.** the limit of navigation on a river **3.** a body of water at some height above an outfall or held at a height to supply water to a water mill, etc. through a HEAD-RACE **4.** the height of such a body of water, or the force of its fall **5.** the leading part of a BORE-1 **6.** a projecting, high coastal feature such as a promontory with a steep face, or a cape (HEADLAND-1 is the more usual term) **7.** the inner part of a bay, creek, etc. **8.** head sea, a strong movement of the sea, a heavy swell or wave coming from the direction to which a vessel is steering (heading) **9.** in geology, the attitude or direction of a set of parallel planes in a mass of crystalline rock along which fracture is most difficult **10.** a layer of angular debris of adjacent strata which generally overlies the raised beaches of England **11.** The Head, a compacted PERIGLACIAL deposit of PLEISTO-

CENE age, or a product of SOLIFLUCT-ION, mantling the slopes and filling the valley bottoms, e.g. in southwest England **12.** in mining, an underground tunnel where coal is being worked. G.

head dune a sand DUNE formed on the windward side of an obstacle, where the air is stagnant.

head-dyke in Scotland, a dry stone wall built originally to mark off the farm area, with its arable and meadow land, from the open moorland beyond. G.

heading a horizontal tunnel made in an AQUIFER to carry to a reservoir or well the GROUND WATER that has penetrated the fissures.

headland **1.** a comparatively high promontory with a steep face, projecting into the sea or a lake **2.** the unploughed strip between the ploughed land of the OPEN FIELD in medieval England on which ploughs were turned and which formed a winding routeway between the fields. G.

head-race the channel through which water passes to a WATER MILL or generating station. RACE. G.

headship rate the percentage of individuals in particular age groups who are heads of households.

headstream the stream at the source of a river.

headwall the steep wall that forms the back of a CIRQUE.

headward erosion the action of a stream in extending its valley upstream. AB-STRACTION-4. G.

headwater, head-water, headwaters a stream or streams forming the source or sources and upper parts of a river, especially a large river. G.

head wind a wind blowing head-on, directly against the face of a person or the forepart of a vehicle, aircraft or vessel, etc.

headwork apparatus for controlling the water flow in a river or canal.

heaf **1.** north of England dialect, pasture, usually on part of a FELL, common, MOOR, or mountainside, unenclosed but recognized as the grazing land of a specific flock of sheep. Flocks of many breeds of hill sheep become so closely attached to a particular tract of grazing (their 'heaf') that they pine and lose condition if moved. Consequently if a farm changes hands the 'hefted' flocks are sold with the farm **2.** a small plot of land, somewhat similar to a CATTLE GATE, the pasture of which may be leased or sold as private property. G.

heage HAYES. G.

hearth the place of origin of agriculture, of cultivated plants, of a CULTURE-1. CRADLE is the more usual term. CULTURAL HEARTH, SAUER HYPOTHESIS, URBAN HEARTH. G.

heartland, heart-land **1.** the central part of a land mass inaccessible to a seapower. The term was introduced by Sir Halford Mackinder, 1904, with reference to the heart of the Euro-Asian continental mass, and subsequently developed by him. WORLD ISLAND **2.** the heartland of the USA, usually defined as the area within the rectangle Boston-Washington-St Louis-Chicago. URBAN GROWTH PHASES, USA.

heartland-periphery phase URBAN GROWTH PHASES, USA.

heat balance the condition of equilibrium in which radiation reaching the earth and its atmosphere from the sun (INSOLATION) is approximately equalled by radiation and reflection (ALBEDO) from the earth. HEAT ENGINE.

heat engine a mechanical system set in motion by heat ENERGY, e.g. as in the ATMOSPHERE-1, where differences in net RADIATION balance (HEAT BALANCE) provide the thermal power, and the various movements in the atmosphere distribute the energy (the heat energy being converted to KINETIC ENERGY).

heat equator THERMAL EQUATOR.

heat gradient GEOTHERMAL GRADIENT.

heath, heathland, heather moor an uncultivated, open tract of land, with poor acid soils, supporting shrubby plants of which the dominant are of the family Ericaceae, such as the common heather or ling, *Caluna vulgaris* and species of heath, *Erica*. The term heath or heathland may persist in place-names where such vegetation has almost or completely disappeared. G.

heat island, heat-island the persistent warmth of the densely built-over part of a large town, the overlying air having an average temperature higher than that of the air overlying the more open surrounding area. This is due in part to the storage of solar heat in roads and the mass of building etc. and its slow release, in comparison with the heat stored and released more quickly from the surrounding, more open area; in part to the effect of buildings in reducing wind speed; in part to the heat derived from the internal heating of buildings, from industry, transport, power generation, human metabolism; in part to the blanket effect of polluted air overlying large towns. The effects of a heat island spread to areas lying at some distance from it. G.

heat wave, heat-wave a relatively long unbroken spell of abnormally hot weather.

heave of a fault the forward, lateral displacement of STRATA on an inclined NORMAL FAULT, expressed as the horizontal distance between the ends of the surfaces of the displaced strata. THROW OF A FAULT. G.

Heaviside Layer the E layer of the ATMOSPHERE-1, lying in the IONOSPHERE at about 100 to 120 km (60 to 72 mi), and reflecting medium and long RADIOWAVES (ELECTROMAGNETIC SPECTRUM) back to the earth while allowing short radiowaves from the earth to penetrate until they reach the APPLETON LAYER. The name is derived from Oliver Heaviside, 1850-

1915, who suggested the existence of such a layer; but it is also known as the Kennelly-Heaviside Layer, Arthur Edwin Kennelly, 1861-1939, having made the same prediction. G.

heavy industry 1. loosely, a SECONDARY INDUSTRY in which large weights of materials are handled **2.** more specifically, industry in which the weight of materials used per worker is high, the MATERIAL INDEX is high, and the finished products have a low value in relation to their weight, e.g. ship building. LIGHT INDUSTRY. G.

heavy land, heavy soil land that is difficult to plough, especially in wet conditions, e.g. a thick CLAY soil.

heavy minerals the ACCESSORY MINERALS with a high specific gravity present in some SEDIMENTS-2 or SEDIMENTARY ROCKS, and concentrated in PLACER deposits. BLACK SAND.

heavy oil an oil with a high specific gravity obtained by distilling COAL TAR.

heavy water water with a high content of deuterium oxide, D_2O, used as a moderator in nuclear reactors. NUCLEAR ENERGY.

hectare ha, a metric unit of area, equivalent to 10 000 sq m or 2.471 ACRES. 100 hectares equal 1 square KILOMETRE. CONVERSION TABLES.

hecto- h, prefix, a hundred, attached to SI units to denote the unit $\times 10^2$, e.g. hectolitre (100 litres). CENTI-, CONVERSION TABLES, KILO-, MILLI-.

hedge a continuous row of bushes and/or low trees, closely planted to mark off or protect a particular piece of land, usually cut to a desired height, traditionally planted in lowland Britain to mark a boundary, e.g. of a field or garden.

hedge-bote ESTOVERS.

hedgerow a row of bushes forming a HEDGE, with occasional tall trees in it; or the line of a hedge.

hegemon a leading or a paramount power.

hegemony (originally used with reference to the states of ancient Greece) predominance, leadership **1.** the predominance of, the leadership exercised by, one state in relation to others in a group, e.g. in a union or confederacy of states **2.** in Marxism, the predominance of one SOCIAL CLASS over others.

Heide (German) in northern Germany, heathland characterized by heather, *Calluna vulgaris*. STEPPENHEIDE. G.

heifer a young COW-1 that has not given birth to any young.

hekistotherm, hecistotherm 1. a plant able to withstand prolonged very low temperatures, growing in a brief summer, e.g. the Arctic and Antarctic LICHENS and MOSSES **2.** a climate with very low temperatures, characterized by such plants, i.e. one of the E climates of KOPPEN, where for twelve months the temperature does not exceed 10°C (50°F). MEGATHERM, MESOTHERM, MICROTHERM.

heliophobe a plant growing best in shaded places. HELIOPHYTE.

heliophyte a plant growing best in full sunlight. HELIOPHOBE.

heliotaxis PHOTOTAXIS.

heliotropism PHOTOTROPISM.

helium a light gaseous element, colourless, odourless, chemically inert, small quantities being present in the ATMOSPHERE-1 and in a few minerals (PITCHBLENDE), larger quantities in some natural gases. It is used in meteorological and other research balloons, in medicine and in industry.

Helm wind a strong, steady, cold wind blowing from the east or northeast down from the summit of hills in northern England, especially down Cross Fell to the Eden valley, creating an isolated, stationary, cloud (BANNER CLOUD), (termed helm cloud or Helm) over the summit and, a few kilometres distant to the west, a thin parallel line of cloud (the Helm Bar). It commonly occurs in spring and winter and is caused mainly by LEE WAVES combined with eddies, turbulence and an inversion layer, leading to continuous condensation on the windward side of the summit and continuous evaporation on the leeward. LENTICULAR CLOUD. G.

helophytes a class of RAUNKIAER'S LIFE FORMS, HERBACEOUS plants of marshes, with PERENNATING parts lying in the mud.

hemantic season (Indian subcontinent) the principal cropping season of Bangladesh, extending from the end of the rains, about October, to the cool dry season, November-December. BHADOI, RABI, KHARIF. G.

hematite HAEMATITE.

hemera the short period of geological time corresponding to a ZONE-4 as determined by an assemblage of FOSSILS in the rocks.

hemicryptophytes a class of RAUNKIAER'S LIFE FORMS, HERBS-1 with PERENNATING parts at soil level, protected by the soil and by the dry debris of the plant itself.

hemiparasite a PARASITE that draws only part of its food from its host.

hemisphere the half of a sphere formed on each side of a plane passing through its centre. The earth's surface is commonly considered to be bisected by the EQUATOR, producing the northern hemisphere and the southern hemisphere; and the eastern hemisphere (the Old World, i.e. Asia, Africa, Europe, Australia and New Zealand) is separated from the western hemisphere (the New World, i.e. North and South America) by the meridians 20°W and 160°E. GREAT CIRCLE, LAND HEMISPHERE, WATER HEMISPHERE. G.

hemp a term applied to a number of plant fibres. True hemp is a soft fibre from *Cannabis sativa*, an ANNUAL plant of the stinging nettle family (Moraceae), native to central Asia, widespread in cultivation. It is used for making twine, rope, coarse fabric, and is the source of the narcotics known as hemp, i.e. bhang, hashish, marijuana. Manila hemp (ABACA) is a harsh pale coloured fibre obtained from *Musa textilis*, a plant of the Musaceae family (of which the BANANA is a member), grown mainly in the Philippine Islands, from the capital of which, Manila, it takes its name. Sisal hemp or sisal is also a harsh pale fibre, used mainly for making twine, obtained from the huge hard pointed leaves of various species of *Agave*, chiefly *A. rigida*, native of Yucatan, Mexico, introduced into other tropical areas, particularly in Tanzania and Brazil. The Mexican HENEQUEN hemp is similar. NEW ZEALAND HEMP or flax is finer and obtained in small quantities from a marsh plant, *Phormium tenax*, native to New Zealand. GAMBO HEMP, INDIAN HEMP, SUNN HEMP. G.

hendref, hendre (Welsh, old home) a winter dwelling, a permanent home, contrasted with HAFOD, a summer dwelling. MEIFOD. G.

henequen a strong fibre used for making twine or cord, obtained from the leaves of *Agave fourcroydes*, a tropical plant native to Yucatan, Mexico. AGAVE, HEMP.

herb 1. in botany, a non-woody vascular plant lacking parts that persist above the ground (ANNUAL, BIENNIAL, PERENNIAL) as distinct from a SHRUB or a TREE 2. in general, a plant that is not necessarily a herb in the botanical sense, but is valued for its fragrant, medicinal or flavouring properties.

herbaceous *adj.* of, pertaining to, or resembling a HERB-1, i.e. a herb in the botanical application.

herbage growing HERBACEOUS plants, especially if grazed. GRAZING.

herbicide a chemical agent used to kill plants that are considered to be WEEDS. Many of the herbicides now in use are selective, formulated to attack only specified plants.

herbivore a plant-eating animal, a primary consumer in the FOOD CHAIN. CARNIVORE, OMNIVORE.

Hercynian (Latin Hercynia (silva), the Hercynian wood) *adj.* associated with the Harz mountains of Germany and originally applied (in Germany) to any mountains, varying in age, formed by folding and faulting; but now restricted to the earth building movements and associated mountain remnants of Upper Carboniferous to Permian times (GEOLOGICAL TIMESCALE) in Europe, variously and confusingly named. Hercynian is applied by some authors to the whole mountain system of central Europe (also termed the ALTAIDES); others use it as a synonym for VARISCAN; others restrict the use of Variscan to the eastern sector of the Hercynian earth movements, applying ARMORICAN to the western, i.e. to earth movements in Brittany and southwestern Britain. There are remnants of mountains of similar age in Asia (the Urals, T'ien Shan, Nan Shan), in North America (the Appalachians) and in South America (the foothills of the Andes). G.

herd a number of herbivorous animals (HERBIVORE) of one kind, moving and feeding together in order (in natural conditions) to have some protection from PREDATORS, lone individuals being more likely to fall prey to CARNIVORES.

herd instinct the tendency in gregarious animals for most individuals to respond in the same way to a particular stimulus, or automatically to follow a lead. The phenomenon is sometimes associated with the tendency for MOBILE firms, especially those from the same area of origin, to select the same new locations as those selected by earlier movers in the same industry, for no apparent reason other than that of gregariousness. This particularly applies to foreign firms setting-up factories in countries new to them.

hermeneutic *adj.* explaining, interpreting.

hermeneutics the art, science or skill of interpretation, of the classification of meaning, of understanding the significance of human actions, statements, institutions, products.

heteroclinal fold a FOLD-3 of which one side slopes at an angle steeper than that of the other. G.

heteroecious parasite heteroxenous parasite, a PARASITE which does not need a specific host, or which needs more than one host in its life cycle.

heterogeneity the quality of being HETEROGENEOUS.

heterogeneous *adj.* **1.** in general, diverse or dissimilar in kind or character **2.** composed of different or disparate ingredients or elements **3.** in geology, a rock composed of diverse materials **4.** rocks varying in nature or kind and adjacent to each other. HOMOGENEOUS.

heterogenesis **1.** ABIOGENESIS **2.** alternation of generations. HOMOGENESIS.

heteromorphic *adj.* having a form different or dissimilar from that of another. HOMOMORPHIC, ISOMORPHIC.

heteropic a term applied to two FORMATIONS-2 deposited at the same time but of different FACIES. ISOPIC. G.

heterotroph an organism which is HETEROTROPHIC.

heterotrophic *adj.* applied to an organism which needs an organic food supply from outside itself in order to produce its own constituents and (with a very few minor, PHOTOTROPHIC, exceptions) to obtain energy. All animals and fungi, most bacteria, and some flowering plants are heterotrophic, relying on organic substances provided by the activity of AUTOTROPHIC organisms, most being CHEMOTROPHIC, a few being PHOTOTROPHIC. HETEROTROPH, MIXOTROPHIC.

heuristic *adj.* **1.** useful for discovery or for solving problems, e.g. applied to a method of education which encourages the student to gain knowledge by personal investigation; or to a procedure (heuristic method) for discovering an unknown goal by a progressive sequence of operations or investigations based on a known criterion, e.g. a person put against a wall in a totally dark room and asked to find the door will discover it by moving systematically in the same direction round the walls **2.** of practical, though perhaps unexplained, use in invention or discovery, e.g. in social science, applied to conceptual devices such as MODELS-4 and working hypotheses the aim of which is not to describe or explain the facts but to suggest or to eliminate possible EXPLANATIONS.

H horizon O HORIZON, SOIL HORIZON.

hibernation the dormancy of some animals during the coldest season, the winter, e.g. many mammals, most reptiles and amphibians, and many vertebrates of temperate and arctic regions. The metabolic rate is much slowed, and in mammals the temperature drops to that of the surroundings. Some animals hibernate completely through the cold winter, but others (e.g. bats, hedgehogs) wake and feed if there is a warm spell. AESTIVATION.

hide 1. the raw or dressed skin of dead cattle, used especially in shoe making. The term usually appears linked to SKIN in trade statistics **2.** (Anglo-Saxon) an inexact measure of land in the DOMESDAY BOOK, reckoned at four YARDLANDS (especially sulung), but the term sometimes refers to an area of land capable of supporting one family. G.

hierarchical *adj.* of or relating to a HIERARCHY.

hierarchical (hierarchic) diffusion DIFFUSION-2 in which the dispersal progresses up or down through a regular sequence of order, classes or HIERARCHIES, usually moving down

from higher to lower levels (e.g. from the METROPOLIS-1,2 to remote rural villages), but in some cases beginning at a lower point and moving up. Upward dispersal is usually slower than downward. CASCADE, CONTAGIOUS, EXPANSION, HORIZONTAL DIFFUSION.

hierarchic model, hierarchic theory a MODEL-4 or THEORY-2 expressed in terms of a set of levels in ascending or descending order.

hierarchy the organization of a SYSTEM-6 into successive ranks, in ascending or descending order, e.g. CENTRAL PLACE HIERARCHY, URBAN HIERARCHY.

high an ANTICYCLONE, an area of high ATMOSPHERIC PRESSURE.

highest and best use 1. of land, the most profitable use of land at a particular location at a particular time, measured only by economic considerations, e.g. profit maximization in a free market (MARKET ECONOMY), any social considerations (e.g. SOCIAL COSTS OF PRODUCTION and EXTERNALITIES) being ignored **2.** in development of REAL ESTATE, the use which can pay the highest rent and derive the greatest economic benefit from the site.

high farming a term applied to the period in England in the 1850s and 1860s when the large landowners adopted new farming techniques and spent considerable amounts of money on underdraining their land, improving its fertility, raising the quality of their livestock and crops, and improving farm buildings.

high grade metamorphism METAMORPHISM by high temperatures or high pressure, as distinct from LOW GRADE METAMORPHISM.

highland, highlands in general, any tract of high or elevated land or the more mountainous parts of any country. Used specifically, usually in plural, in proper names, e.g. the Highlands of Scotland, the Kenya Highlands. LOWLAND. G.

high latitudes LATITUDE.

highly developed country HDC.

high plains **1.** PLAINS lying at an elevation above some 600 m (2000 ft) **2.** the section of the Great Plains in the USA which is relatively UNDISSECTED.

high positive correlation CORRELATION COEFFICIENT.

high pressure HIGH.

high rise, high-rise building a tall building with many storeys. CENTRAL BUSINESS DISTRICT, CONDOMINIUM.

high seas the open sea or ocean beyond TERRITORIAL WATERS.

high technology the application of the knowledge of and/or the use of advanced techniques, complex equipment, and materials drawn from any convenient source to a task or an industrial proces, or to the solution of a problem arising from the interaction of people with their ENVIRONMENT. LOW TECHNOLOGY, TECHNOLOGY-3.

high tide **1.** the TIDE at highest flood **2.** the level of the sea at, or the time of, the highest flood. FLOOD TIDE, LOW TIDE, NEAP TIDE, SPRING TIDE.

high water the state of the TIDE when the water is at its highest for any given tide. The level varies through the year so that the high water ordinary spring tide (HWOST) is higher than high water ordinary neap tide (HWONT), both of which may appear on maps. HIGH-WATER MARK, LOW WATER, NEAP TIDE, SPRING TIDE.

high-water mark the mark left by the TIDE at HIGH WATER, or at the highest level ever reached, similarly applied to the water level in a river or lake. On British Ordnance Survey maps the high-water mark indicates the high-water mark of medium tides (HWMMT).

highway **1.** in Britain, a main, principal public road **2.** in USA, a public road, especially one that is direct, wide and well surfaced; or a main route by land or sea or air. G.

hill loosely, a natural elevation of the earth's surface, not so high as a MOUNTAIN. It is sometimes defined as an elevation under 300 m (1000 ft), but such exactness is misleading: some ranges termed hills exceed 2680 m (8800 ft). HILL STATION. G.

hill-and-dale **1.** a literary term applied to the hill and valley landscape of northern England, particularly of Yorkshire, Lancashire and the Lake District. DALE, HILL **2.** the alternating ridges and hollows of waste rock or overburden remaining as the result of surface mining or QUARRYING carried out by large machines. G.

hill-billy, hillbilly (American) originally a poor white settler in the hills of Alabama, now a local, country-bred person living in the mountains of the southeastern USA. G.

hill farming generally, farming in hill country, but given a more precise meaning in British legislation on account of the subsidies designed to encourage the continued farming of poor hill land. LESS FAVOURED LAND. G.

hill fog low CLOUD covering the higher levels of HILLS as seen by the observer standing below.

hill-island BAKKEOER. G.

hillock a small HILL. G.

hill-shading a method of indicating relief on maps, usually by shading the slopes only on the south and eastern sides of hills and mountains, the shading darkening with increasing steepness.

hillslope SLOPE, STANDARD HILLSLOPE.

hill station, hill-station a mountain resort in the tropics, e.g. in Burma, Java, the Indian subcontinent, where elevation ensures pleasantly cool temperatures in the hot season. The term is associated particularly with the British administration in the Indian subcontinent, where provincial and central governments moved to hill stations (e.g. Simla, Darjeeling) in the hot season to escape the heat of the plains, a movement known as 'going to the hills'. G.

hill village   a village, a hamlet, or even a small market town or other non-agricultural settlement occupying a site from which the ground slopes markedly downwards. G.

hindcast   extrapolation backwards in the study of a sequence of events (as opposed to FORECAST, extrapolation forwards), based on recorded events or on tendencies and probabilities as well as on accepted laws.

hinderland   a rarely used English translation of German HINTERLAND.

hinge of fold   that part of a FOLD where the STRATA is under maximum STRESS.

hinge-line   a real or imaginary line on the earth's surface representing a boundary between a stable area and one moving upwards or downwards. G.

hinny   ASS.

hinterland   (German) **1.** originally the district behind a settlement on a coast, especially the area serving and served by a PORT-1 **2.** the area influenced by or tributary to any settlement, but for this application some authors prefer the term UMLAND **3.** the sphere of influence of an establishment within a settlement.   COMMUTING HINTERLAND,   MOVEMENT   HINTERLAND,   URBAN HINTERLAND. G.

hirsel   (Scottish) a natural division of land grazed by sheep, delimited by the hefting instinct (HEAF) of a sheep, whereby it tends to stay in the area in which it was reared. G.

hirst   HURST. G.

histic   *adj.* applied to soil surface layers that have a high content of organic CARBON and are seasonally saturated with water.

histogram   a two-dimensional graph showing a FREQUENCY DISTRIBUTION (of data measured on an INTERVAL or RATIO SCALE) by means of rectangles, the widths of the rectangles being proportional to the CLASS INTER-

VALS or CATEGORIES (shown on the horizontal axis), the heights (on the vertical axis) showing the frequencies of occurrence in each CLASS. If the frequency of each class is expressed as a percentage of all classes (i.e. not by the absolute frequency of each class), the resultant diagram is termed a relative frequency histogram (RELATIVE FREQUENCY   DISTRIBUTION).   BAR GRAPH, FREQUENCY-3, FREQUENCY CURVE, FREQUENCY POLYGON.

historical fallacy   the false assumption that a relationship observed in a CROSS SECTIONAL STUDY will be present in a similar LONGITUDINAL STUDY, or vice versa. ECOLOGICAL FALLACY.

historical geography   the geography, physical and human, real, perceived or theoretical, of the past. Defining the limits of historical geography is difficult because the content of the studies involved and the methodology used range widely, but two approaches should be mentioned: the horizonatal (CROSS-SECTION APPROACH) and the vertical (VERTICAL THEME). G.

historical geology   one of the two main divisions of GEOLOGY, concerned with the sequence of changes in or on the earth, especially as revealed by the STRATA of the CRUST. GEOCHRONOLOGY, STRATIGRAPHY. G.

historical materialism   a materialist conception of history, the Marxist theory of history, seen as a natural process related to human material needs. Marx believed that this historical human evolution was the product of the class struggle arising between and amongst the exploiting and exploited classes throughout the succession of the MODES OF PRODUCTION, each mode of production determining the general character of the social, political, intellectual and spiritual processes in a society (social existence being determined by human consciousness, not vice versa). Marx and Engels came to modify their views that such economic determinism governed all aspects of historical development; and Marxists have since variously interpreted the theory and its concepts. DIALECTICAL MATERIALISM, MARXISM, MATERIALISM.

historicism 1. the belief that historical change is governed by laws, that the course of history is predictable and unalterable 2. the belief (held especially in the nineteenth century) that all historical phenomena are unique, that each age should be explained in terms of its particular ideas and principles, that past actions should not be explained by reference to the beliefs, motives, valuations of the present.

history 1. the study of past events, economic, social and political conditions, ideas, cultural manifestations etc. 2. a record of such past events etc. 3. the study of the development of anything in time.

history of geography the history of geographical knowledge and ideas. G.

histosols in SOIL CLASSIFICATION, USA, an order of soils developed mainly by the accumulation of organic matter in a waterlogged area, and thus including BOG soil and soils rich in PEAT.

hithe, hythe a PORT or HAVEN, particularly a small haven or landing place on a river; a term now rarely used except historically and in place-names. G.

hoar frost, hoar-frost, hoarfrost a white deposit of ice with a crystalline appearance, formed directly from the cooling of water vapour on surfaces with a temperature below that of DEW-POINT when the dew-point is below 0°C (32°F). FROST, RIME. G.

hobby farmer a PART-TIME FARMER who owns the freehold of, or rents, a FARM, who may or may not use any building on it as a main residence, who does not rely on the farm output for a livelihood, whose main income comes from another source, e.g. from an urban occupation, and who farms mainly for pleasure.

hoe a projecting ridge of land, a height ending abruptly or steeply. The term is now obsolete except in place-names, e.g. Plymouth Hoe, Ivinghoe. G.

hog a PIG reared for slaughter, especially a castrated male pig.

hogback, hog back, hogsback, hog's back an elongated narrow ridge shaped like the back of a hog (PIG), the result of unequal erosion on alternating hard and soft layers of steeply inclined rocks. It differs from a CUESTA because both slopes of the ridge (i.e. the dip slope and the scarp slope) are steep and more nearly equal. G.

holding in agriculture, land held or occupied by legal right for purposes of farming. The term is sometimes used as a synonym for FARM, but several farms may be combined and worked as one holding. SMALL HOLDING. G.

holism a philosophy so named by J. C. Smuts, that there is a tendency in nature to produce 'wholes' (whole bodies or whole organisms) from an ordered grouping of unit structures, the whole being greater than the sum of the properties (PROPERTY-3) and relationships of its parts (GESTALT, ORGANICISM). This phenomenon can be recognized in other 'wholes', leading to the doctrine that a functioning whole (e.g. an organization, institution, society) affects its component parts, is inimical to analysis, and that therefore the parts should not be studied in isolation (e.g. in ETHNOGRAPHY that individual actions can be fully understood only in relation to the whole social context). It has also led to a theory of science which sees science not as a collection of disparate parts but as an integrated system. G.

holisopic ISOPIC.

holistic *adj.* of or relating to HOLISM.

holm 1. a small island in a river, estuary or lake, or near the mainland (SKAR) 2. low, flat land along the banks of a river, liable to flood, often more or less surrounded by water, affording a landing place for small boats (howm in Scotland), especially in southern Scotland and northern England, common in place-names **3.** in Germany, a SAND DUNE. G.

Holocene the period in which we are living, termed Recent or Postglacial by some authors, i.e. the most recent of the geological periods, the youngest of the QUATERNARY (GEOLOGICAL TIME-SCALE), and the rocks of that period, dating from the end of the last ICE AGE.

holocoen in ecology, the whole environment, i.e. including all the living (BIO-COEN) and the non-living (ABIOCOEN) components.

holokarst wholly karst, a region of KARST with karst characteristics developed to the full and little or no surface drainage. MEROKARST. G.

holophytic *adj.* feeding in the manner of a green plant, i.e. by means of CHLOROPHYLL, using the energy of sunlight to synthesize organic compounds from inorganic components (PHOTOSYNTHESIS). AUTOTROPHIC, FOOD CHAIN, HOLOZOIC.

holotype in biology, the type specimen, i.e. the original specimen(s) forming the basis of identification of a new SPE-CIES. BINOMIAL NOMENCLATURE, CLASSIFICATION OF ORGANISMS, PARATYPE, SYNTYPE.

holozoic *adj.* feeding in the manner of an animal, i.e. by eating other organisms or organic matter changed by them. FOOD CHAIN, HOLOPHYTIC.

holt **1.** a wood, a copse or, locally, a plantation of osiers, a wooded hill, a term frequently appearing in place-names **2.** the lair of an animal, especially of the otter (a carnivorous aquatic mammal).

homalographic, homolographic *adj.* applied to a map or diagram delineating parts of the whole in correct proportions and relationships. TOPOLOGICAL DIAGRAM.

home farm a FARM used personally by

the owner of an estate on which there are other farms.

homegrown *adj.* applied to agricultural produce, including animals, grown or reared by the consumer or by the direct seller.

homeostasis, homoiostasis the maintenance of a constant, balanced internal environment within a system (OPEN SYSTEM), person, group etc. MORPHO-STASIS, STEADY STATE.

home range the whole territory occupied for some time by an animal or group of animals, the part where it spends most time being termed the core area. G.

homestead **1.** a house or home, especially a farm with its associated buildings **2.** in USA, a plot of land adequate for the residence and maintenance of a family, given special and legal meaning, notably under the 1862 Homestead Act, a grant of 65 ha (160 acres) being regarded as sufficient to support a family, rising to some 255 ha (640 acres) in mountainous or semi-arid areas **3.** a rural settlement of dispersed farms. G.

homio-osmotic,   homoeo-osmotic see POIKILOSMOTIC.

homiothermy, homoeothermy the state of being WARM-BLOODED. POIKILO-THERMIC.

homoclimes places having similar climates, shown by a comparison of their CLIMOGRAPHS. G.

homoclinal ridge, homoclinal valley a ridge or valley formed by denudation from one side of a folded sructure. G.

homocline in geology, a structure in which the STRATA dip evenly, continuously, in one direction. Some authors prefer the synonym UNICLINE. MONO-CLINE. G.

homogamy inbreeding resulting from isolation.

homogeneous *adj.* **1.** similar in kind, character or nature **2.** having at all points the same composition and properties, the opposite of HETERO-

GENEOUS **3.** in mathematics, an expression in which all the terms are of the same degree, or represent commensurable quantities.

homogenesis in biology, reproduction in which like begets like. HETEROGENESIS.

homogenous *adj.* **1.** in biology, exhibiting structural correspondence of individuals or of parts, due to descent from a common ancestor **2.** HOMOGENEOUS-1,2, in contrast to HETEROGENEOUS.

homolographic HOMALOGRAPHIC.

homologue that which is similar or corresponds in type to an original or to others in respect of structure, value, relationship, type etc., e.g. a climatic homologue or HOMOCLIME. G.

homomorphic *adj.* **1.** having a form similar to that of another despite differences in structure or origin **2.** in botany, bearing perfect flowers of only one kind. HETEROMORPHIC, ISOMORPHIC.

homoseismal line CO-SEISMAL LINE.

homotaxis similarity in order of arrangement, especially the similarity of position of rock strata and FOSSILS in the stratigraphic sequence at widely separated localities. CHRONOTAXIS. G.

honeycomb weathering a type of WEATHERING which causes a rock surface to resemble that of a honeycomb. The current theory is that hard material filling some of the joints in the rock is left standing as softer material is worn down more speedily by wind and water. The same pattern sometimes also occurs on a shore, where wind and water act together on small pools in the rocks, enlarging the hollows in which they lie to such an extent that they are eventually separated only by steep, sharp ridges.

hoodoo in North America, a rock pinnacle of the nature of an EARTH PILLAR or DEMOISELLE, formed by weathering in a semi-arid region where the rare rain, flowing down the sides of the column, accumulates near the base to effect chemical and physical undercutting there when the upper part of the column is already dry. The name is presumably due to the imagined resemblance of such an earth pillar to embodied evil spirits, e.g. in the groups in Banff and Jasper National Parks, Canada. G.

hook **1.** a sharp bend or angle in a length or course, e.g. the bend in a river, used in proper names **2.** (Dutch Hoek van Holland) a projecting point, a spit of land **3.** a recurved spit resembling a fish-hook. SPIT. G.

hop *Humulus lupulus*, a PERENNIAL HERBACEOUS climber of the family Cannabiaceae, native to Europe and western Asia, now cultivated there and in other midlatitude areas for the sake of the essential oils and soft resins contained in the lupulin which is produced in resin glands, used in flavouring beer in Europe since the Middle Ages.

hope in southern Scotland, the midlands and northeast England, a small enclosed valley; occurs in place-names. G.

hope value that part of the value of a parcel of land which reflects the probability of the parcel of land's being granted (under planning legislation) a change of use from that existing to one more profitable to a developer. FLOATING VALUE.

horizon **1.** visible horizon (also termed apparent, geographical, natural, physical, or sensible, horizon) the line at which the earth or sea and the sky appear to meet when seen from any given viewpoint, excluding anything interrupting or obstructing the view. In clear visibility, to a person standing at an elevation of 3 m (10 ft) from the horizontal the horizon is nearly 6.5 km (just over 4 mi) distant; at 30 m (100 ft) it is 21 km (13 mi) away **2.** in astronomy, a circle with its plane parallel to the sensible horizon and passing through the centre of the earth (also termed the true celestial or geometrical horizon) **3.** in geology, a plane in a series of geological STRATA, or the level at which a particular FOSSIL occurs **4.** in soil science, a distinct soil layer with more or less well-defined characteristics produced by soil-forming processes. SOIL, SOIL HORIZON. G.

horizonation the degree of development of SOIL HORIZONS.

horizontal *adj.* having a direction perpendicular to a VERTICAL direction, e.g. the surface of static water in a container.

horizontal diffusion DIFFUSION-2 in which the dispersal spreads to units of similar size, e.g. to settlements of similar size. CASCADE DIFFUSION, CONTAGIOUS DIFFUSION, HIERARCHICAL DIFFUSION.

horizontal equivalent HE, the distance between two points on the land surface when projected on to a horizontal plane, as on a map. If two points on a hillside, say 75 m (245 ft) apart measured down the actual SLOPE LENGTH, are projected on to a map they will be a shorter distance apart in the horizontal plane: this is the HE or Horizontal Equivalent, say 60 m (195 ft) in the example. A rise of 1° in an HE of 30 m (100 ft) represents a VERTICAL INTERVAL of about 0.5 m (1.74 ft). GRADIENT. G.

Horn (German) a pyramidal peak in a mountain range occurring when several CIRQUES are formed back to back, thereby leaving a high central mass, an unreduced part of the original mountain range, with marked faces and sharp ridges (ARETE). The term is common in mountain names, e.g. the MATTERHORN. G.

hornblende a rock-forming mineral of the AMPHIBOLE group, consisting mainly of mixed silicates of CALCIUM, IRON and MAGNESIUM occurring as a

crystalline mass in IGNEOUS and METAMORPHIC rocks, e.g. GABBRO.

horse *Equus caballus*, family Equidae, a large herbivorous animal (HERBIVORE) with undivided hoof, domesticated since prehistoric times as a draught animal, a beast of burden and a mount for riding. ASS, DRAY HORSE, PONY.

horse gram, Madras gram *Dolichos biflorus*, a leguminous plant (LEGUMINOSAE) closely related and similar to LABLAB, grown in the Indian subcontinent and Sri Lanka for animal feed or as GREEN MANURE, and as a PULSE-2 for human consumption.

horse latitudes the belts of calms (CALM), the subtropical belts of high atmospheric pressure (ANTICYCLONE), moving north and south with the sun, lying north and south of the equator from about 30° to 35°N and 30° to 35°S (but interrupted by the land and sea pattern), regions of the descending air which flows towards the equator and the poles to produce calm, stable, dry weather conditions with light variable winds. The origin of the name is uncertain. G.

horse power, horsepower 1. a British standard unit of power equal to 550 footpound per second, 75.9 kg per second, nearly equal to 746 watts **2.** the power of an engine measured in such a unit.

horseradish, horse-radish *Armoracia rusticana*, family Cruciferae, a PERENNIAL herb with a long cream-coloured fleshy tap root which is crushed, minced, grated or powdered and used as a pungent condiment.

horseyculture the practice of pasturing horses and ponies for riding on relatively small plots of land, usually on town outskirts.

horst (German) a structural feature, an elevated block of the earth's crust, usually with a level summit, standing prominently above parallel NORMAL FAULTS, formed either by the sinking of the crust on each side outside the faults, or by the uplift of a block between the faults. In some cases it is denuded to the extent that it is no longer upstanding. GRABEN, RIFT VALLEY. G.

horticulture originally the cultivation of a garden, now applied to the intensive cultivation of vegetables, fruit and flower crops on relatively small plots, in a MARKET GARDEN, GLASSHOUSE or NURSERY. G.

Horton's Law the theory relating to streams and their order (STREAM ORDER) and the area of the basin they drain, i.e. that the number of streams of a given order decreases sharply with increasing order, and that the total length of a drainage net increases regularly with order. Generally, in most drainage basins the number of streams of a given order is about three times the number in the next higher order.

hosier bails a portable milking apparatus, used for milking cows in the field, popular in Britain in the Second World War. G.

host in biology **1.** any organism in which a PARASITE spends the whole or part of its life cycle **2.** an organism into which a graft is transplanted.

host society the society within which ETHNIC MINORITIES or IMMIGRANTS come to live.

Hotelling model a MODEL-4 used to account for the locations of two firms in competition with each other, based on the assumptions that these firms are producing identical goods, that their production costs are the same at all locations, that the price of the goods covers the cost of transport to the consumer, that the market to be satisfied is LINEAR-2, with customers evenly distributed along its length, and that the demand for the goods is inelastic.

hot rock a deep-seated rock in the earth's crust which has a temperature higher than might be expected from the normal GEOTHERMAL GRADIENT. GEOTHERMAL ENERGY.

hot spot PLUME.

hot spring THERMAL SPRING.

house-bote ESTOVERS.

housing class a group of people who have access to a particular type of housing. HOUSING CLASS THEORY.

housing class theory the theory that housing can be ranked in a hierarchy of desirability and that only the richest and most powerful members have access to the highest, most desired levels, the poorer and least powerful being restricted to lower levels. HOUSING CLASS, SOCIAL SEGREGATION.

how (Old Norse haug-r, mound, cairn) **1.** a low hill, a hillock, now used only in place-names in northern England **2.** an artificial mound, tumulus, barrow.

howe (Scottish) a hollow, a depression on the earth's surface, especially in eastern Scotland, e.g. Howe of Fife. G.

Hoyt's sector model SECTOR MODEL, SECTOR THEORY.

hpoongyi, hpongyi (Burmese) a Buddhist monk. G.

hpoongyi-kaung (Burma) a Buddhist monastery providing a village school and shelter for travellers in Burma.

-hsien (Chinese) as a suffix to place-names, a county town. G.

huasipungo (highland Ecuador: Quechua) a plot of land which a landless labourer attached to a HACIENDA in highland Ecuador is allowed to cultivate personally in return for work given to the landowner. G.

huerta (Spanish, derived from Latin hortus, a garden) a highly cultivated, irrigated area (originally developed by Muslims) along the eastern coastlands of Spain which may yield several crops a year of vegetables, fruit, etc. INTENSIVE AGRICULTURE, VEGA. G.

hum a residual hill in KARST country, analogous to a MONADNOCK, resembling a haystack. Similar features are termed pepino hills or haystack hills in Puerto Rico, MOGOTES in Cuba, BUTTES TEMOINES in the Causse region of France. G.

human agency the capacity of human beings to act, in the light of their experience and creativity, e.g. to improve a soil; or to reproduce and reinforce SOCIAL STRUCTURES by repeatedly conforming to the rules, constraints and conventions of their social system (SOCIALIZATION); or to change their circumstances, or an even wider area of the social structure, e.g. by coming together as a group on the basis of some shared experience (such as place of residence, social class, gender) and as members of that group pursuing an agreed goal. HUMANISTIC GEOGRAPHY, STRUCTURATION.

human ecology 1. the study of the interrelationships between human beings and the ENVIRONMENT (physical and social) in which they live. Some authors maintain that GEOGRAPHY is synonymous with human ecology **2.** in sociology, the study of the interrelationship between human beings. Most studies in human ecology try to link the structure and organization of a human community to interactions with its local environment. ECOLOGY.

human geography one of the two parts into which GEOGRAPHY is often separated, the other being PHYSICAL GEOGRAPHY. Human geography is concerned with the study of those features and phenomena in the space accessible to human beings which relate directly to or are due to people (as individuals or in groups), their activities and organization, past or present. It concentrates on the interrelationship of people (as individuals or in groups) with space, with their physical ('natural') environment and with their social (societal) environment, covering spatial and temporal distribution, the organization of society and social processes etc. on a local to global scale. It draws on ARCHAEOLOGY, ANTHROPOLOGY, DEMOGRAPHY, ECONOMICS, HISTORY, LANGUAGE, LAW-1,2, LITERATURE, MEDICINE, PHILOS-OPHY, POLITICAL SCIENCE, PSYCHOLOGY, SOCIOLOGY and their branches; on BIOLOGY, BOTANY, ECOLOGY, ZOOLOGY and their branches; and on the specialist geographical studies, BIOGEOGRAPHY, PHYTOGEOGRAPHY, ZOOGEOGRAPHY. The various branches of human geography are covered in this *Dictionary* by GEOGRAPHY qualified by BEHAVIOURAL, CULTURAL, COMMERCIAL, ECONOMIC, ELECTORAL, HISTORICAL, HUMANISTIC, MEDICAL, PERCEPTION, POLITICAL, SOCIAL, URBAN, WELFARE. See also HUMAN ECOLOGY, REGIONAL GEOGRAPHY, SYSTEMATIC GEOGRAPHY. G.

humanism any theory, or doctrine, or movement, which is concerned with the primacy of human beings and their interests (as distinct from the primacy of a divine being, of nature, of STRUCTURES-1, or of SYSTEMS); or with the human race in general (as distinct from an individual member of it); or with the studies of human culture and creativity, particularly those exemplified by the cultures of classical Greece and Rome, the prime concern of the HUMANISTS of the RENAISSANCE. HUMANISTIC.

humanist a person who adheres to one of the theories, doctrines or movements of HUMANISM, e.g. one of the scholars of the RENAISSANCE whose main concerns were the studies and the promotion of study of the language, history, literature and the arts of classical Greece and Rome.

humanistic *adj.* relating to, characteristic of, HUMANISM or of a HUMANIST, e.g. one of the humanists of the RENAISSANCE who were primarily concerned with the products of human effort as revealed in the history, language, literature and the arts of classical Greece and Rome.

humanistic geography humanist geography, an approach in HUMAN GEOGRAPHY which emphasizes the SUBJECTIVE as distinct from the OBJECTIVE in that it stresses the importance of perception, creativity (e.g. in literature and landscape painting), thinking and beliefs as well as human experience and values in formulating people's attitudes to their environment and in affecting their relationships with it. It gained attention in the 1970s largely as the result of the reaction of some geographers to POSITIVISM-1,3 and to what they considered to be the excesses of MECHANISM displayed in the QUANTITATIVE REVOLUTION. HUMAN AGENCY, HUMANISM, HUMANIST, HUMANISTIC, ICONOGRAPHY.

humanities a term covering the disciplines of art, history, languages, literature, music, philosophy and theology. BEHAVIOURAL SCIENCES, EARTH SCIENCES, NATURAL SCIENCES, PHYSICAL SCIENCES, SOCIAL SCIENCES, SCIENCE.

human regions the seven regions identified by H. J. Fleure on the criteria of physical characters related to human activities, i.e. the regions of hunger (isolated or extremely cold: hunting and plant collecting); of debilitation (equatorial rain, TIERRA CALIENTE, excessively hot islands: hunting and plant collecting, some gardening and fishing); of increment (climate with adequate alternating sun and rain, higher land in hot areas, some oases: fruit and rice growing, gardening, commerce, cities, engineering); of effort (temperate climate without extremes, all-year moisture: corn growing, organization for defence, privileges of property, available energy for resource exploitation); of industrialization (up to mid-twentieth century, temperate climate, some natural source of industrial power, facilities for communication: manufacturing, usually specialized, agriculture threatened but usually maintained by special effort, invention, organization, financial institutions); of lasting difficulty (high valleys of temperate regions, plateaus with cold win-

ters: small farming, stock raising, herding with transhumance, export of men for manual labour and the mercantile marine); of wandering (great variations of temperature, seasonal drought: herding and, in some areas, hunting).

humic *adj.* of or derived from HUMUS.

humic acid a complex organic acid found in soils and resulting from the partial decay of organic matter. HUMUS.

humic coal BITUMINOUS COAL, COAL.

humic gley soil GLEY SOILS.

humic layer, humose layer in a soil, a layer consisting of organic material so well decomposed that very little fibrous material remains, in contrast to FIBRIC LAYER. HUMIFICATION, HUMUS, MESIC LAYER.

humidity the state of the ATMOSPHERE-1 with respect to its content of WATER VAPOUR, warm air being able to hold more water vapour than cold air. When the air is holding its maximum amount of water vapour it is said to be SATURATED. ABSOLUTE HUMIDITY, RELATIVE HUMIDITY, MIXING RATIO, SATURATION DEFICIT, SPECIFIC HUMIDITY. G.

humid tropicality, humid tropics the climatic condition relative to a standard period of time (e.g. a month) in which the RELATIVE HUMIDITY exceeds 65 per cent, pressure 20 mb, mean temperature 20°C (68°F). To this is sometimes added rainfall exceeding or equalling evaporation for the period, approximating to 75 mm (3 in) per month. The term humid tropics is applied to tropical areas in which the condition of humid tropicality prevails for a minimum of nine months of the year. TROPICAL CLIMATE, TROPICS. G.

humification in soil science, the transformation of organic material into HUMUS by slow DECOMPOSITION-2 and OXIDATION.

hummock a low knoll or hillock of earth, ice or rock.

hummocked ice an untidy pile of sea ice, which may be WEATHERED. A piece rising over 1.5 m (5 ft) above the sea surface is termed a GROWLER. G.

hummocking of ice a process in which LEVEL ICE breaks up under pressure into humps and ridges. G.

humus loosely, organic matter (of vegetable or animal origin) in the soil, but the term is better restricted to such organic matter as has decomposed and been converted into an AMORPHOUS COLLOIDAL substance by the process of HUMIFICATION. HUMIC, LITTER, MOR, MULL, O HORIZON. G.

hundred 1. in England, an old subdivision of a county or shire with its own court, still occasionally used (CHILTERN HUNDREDS). The origin of this subdivision is obscure: it may originally have been the area inhabited by a 100 families, or equivalent to 100 HIDES, but these suppositions are doubtful. WAPENTAKE, YARDLAND 2. in USA, a similar subdivision of a county, originally in the British American colonies, now surviving only in Delaware. G.

hundredweight cwt, a measure of weight equal in British measures to 112 lb (50.802 kg), in US to 100 lb (45.359 kg). CONVERSION TABLES, POUND-3.

hungry rice *Digitaria exilis*, a tropical MILLET locally important as a food crop in the dry areas of west Africa, e.g. the Jos plateau in Nigeria.

hurdle a short, rectangular, light, portable section of fencing, usually made of WATTLE-1, used as a gate or to form an enclosure, e.g. a temporary pen for sheep.

Huronian (from Lake Huron) *adj.* applied to 1. the PRECAMBRIAN orogenesis responsible for part of the folding in the rocks of the Canadian Shield 2. one of the main rock systems of the Shield 3. a division of Precambrian time. G.

hurricane 1. in the BEAUFORT SCALE, wind of force 12, with velocity equivalent at a mean velocity exceeding 34 m/sec or 121 km (75 mi) per hour 2. a violent cyclonic storm with torrential

rain and thunderstorms and wind velocity over 117 km (73 mi) per hour (often exceeding 160 km: 100 mi per hour), originating in latitudes 5° to 20°N over the west Atlantic, moving west-northwest over the Caribbean Sea and the Gulf of Mexico to Florida, then northeast at about 30°N along the eastern coast of the USA. BAGUIO, CYCLONE, TYPHOOON. G.

hurst, hirst 1. a hillock, knoll or bank, especially a sandy one 2. a grove of trees, a copse or wood. Almost obsolete as a separate term, it appears in placenames, especially in southeast England, e.g. in the Weald.

husbandry, husbandman farming, an old term, surviving in animal husbandry (LIVESTOCK FARMING), or in such phrases as 'according to the rules of good husbandry' (i.e. good farming practice). The term husbandman, a farmer, is archaic and almost obsolete. G.

husk the dry outer covering layer of various fruits and seeds, e.g. of CEREALS.

Huxley's model the classification of the components of a CULTURE devised by Julian Huxley, 1887-1975, British biologist. He identified three main categories: mentifacts (the central, durable elements, e.g. language, religion, folklore and magic, artistic traditions); sociofacts (the links between individuals and groups, e.g. family structure, reproductive and sexual behaviour, child rearing and, at a group level, political and educational systems); artifacts (the material manifestations, e.g. technology, systems of land use and agriculture as well as tools and clothing).

HWMMT, HWONT, HWOST HIGH WATER.

hybrid a plant or animal resulting from crossbreeding between two parents which are genetically unlike, sometimes restricted to the offspring of parents of different species, subspecies, or distinct varieties. In many cases hybrids are sterile; the more distant the genetic relationship between the parents, the higher is the probability of sterility.

hydrarch HYDROSPHERE, SUBSERE.

hydrate a compound or other chemical SPECIES-2 containing HYDROGEN and HYDROXYL, either in the proportion in which they form water, or are associated with one or more molecules of water; a solid salt which contains water molecules in its molecules as water of crystallization.

hydrated *adj.* formed into a HYDRATE, i.e. chemically combined with water or its elements. ANHYDROUS.

hydration the action of hydrating, or the condition of being hydrated, or the chemical addition of water to a compound (HYDRATE, SLAKED LIME, SLAKING). As involved in the MECHANICAL WEATHERING of rocks, the term refers to the process in which minerals combine with water and expand, thereby exerting pressure within the rock pores (EXFOLIATION). Minerals that have undergone hydration are very likely to be affected by CHEMICAL WEATHERING.

hydraulic *adj.* of or pertaining to water in motion or to the pressure exerted by water when carried through pipes, or to mechanical devices operated by moving fluids. WATER POWER. G.

hydraulic cement a CEMENT-1 capable of hardening under water. POZZOLANA.

hydraulic civilization a rural-urban agrarian civilization depending on big, productive waterworks for irrigation, flood control and power, as distinct from a rural-urban civilization depending on rainfall (termed non-hydraulic). FERTILE CRESCENT, HYDRAULIC HYPOTHESIS. G.

hydraulic force the eroding power of water on rocks, by turbulence, eddying, wave action. If the water carries a load of material the process of EROSION is termed ABRASION.

hydraulic gradient the slope of the WATER TABLE, expressed as the ratio between the HEAD-3 of water and the length of the flow.

hydraulic hypothesis the proposition that the problems attached to the development of agriculture in broad, seasonally flooded river valleys could have been tackled in prehistoric times only by the integrated, collective efforts of many small groups, the members of which would have built and maintained large scale irrigation works. This large labour force would have needed a supervising authority which could ensure a fair distribution of water to the groups over space and time; and that this led to the emergence of URBAN centres. HYDRAULIC CIVILIZATION, URBAN HEARTH.

hydraulic limestone a LIMESTONE containing some SILICA and ALUMINA and yielding a QUICKLIME or HYDRAULIC CEMENT. G.

hydraulic radius of a STREAM, the area of cross-section of the channel divided by the length of the wetted perimeter. If the stream is shallow and wide, with the height of the channel walls negligible in comparison with the channel width, the hydraulic radius equals the mean depth. G.

hydraulic tidal current a tidal current which surges through a long, narrow, confined stretch of water to compensate for the difference in the height of the water when the high tide at one end of the stretch occurs at a time different from that at the other, thereby producing water of differing heights.

hydraulics the scientific study of the movement of water or other liquids through artificial channels and of the engineering applications of the force produced by the pressure of moving fluids.

hydrazine NH_2NH_2, a colourless reducing base liquid used in the production of antioxidants, explosives, photographic chemicals and in rocket and jet engine fuels.

hydric *adj.* having abundant water. MESIC, XERIC.

hydrocarbon an organic compound, solid, liquid or gaseous, consisting primarily of CARBON and HYDROGEN, e.g. petroleum, coal, natural gas. ACID RAIN, HYDROCARBON CRACKING.

hydrocarbon cracking the process of breaking down PETROLEUM or heavy petroleum fractions into finer products (e.g.motor or domestic fuel, or the basic materials for the chemical industry such as ETHYLENE and PROPYLENE for the making of PLASTICS) by the use of heat, with or without catalysis (CATALYST) and/or pressure, and usually taking place in a refinery.

hydrochloric acid HCl, a strong acid consisting of hydrogen chloride dissolved in water, widely used in laboratories and in industrial processes.

hydrochore a plant of which the seeds or other reproductive parts are dispersed by water, e.g. the COCONUT.

hydrodynamics the scientific study of the flow of liquids.

hydroelectricity, hydro-electricity ELECTRICITY produced by HYDROELECTRIC POWER. CARBO-ELECTRICITY, THERMAL ELECTRICITY. G.

hydroelectric power HEP, electric POWER generated in dynamos moved by the energy of falling water. ELECTRICITY, WATER POWER.

hydrogen the lightest element, having the simplest structure (one PROTON, one ELECTRON), gaseous, inflammable, occurring in very many compounds, but not normally in the uncombined state on earth, used in many chemical processes, e.g. in syntheses (e.g. of HYDROCHLORIC ACID), in HYDROGENATION, as a component in rocket fuel, in producing (by combustion with OXYGEN) high-temperature flames and (in liquid form) as a cooling agent. Heavy hydrogen, when used in the hydrogen bomb, is converted to HELIUM with the release of great heat and other radiation. HYDROCARBON.

hydrogenation a chemical process in which HYDROGEN is incorporated by organic compounds either by the addition of, or reduction with, molecular hydrogen, usually directly in the presence of a CATALYST and in suitable conditions of temperature and pressure, e.g. in the production of margarine, where the hydrogen hardens unsaturated vegetable oils; or in the production of mineral oil from coal (hydrogenation of coal), where the hydrogen destroys the highly aromatic molecules of coal.

hydrogenic rock a SEDIMENTARY ROCK formed as a result of the precipitation of chemicals (especially CALCIUM) in solution. CALCIUM CARBONATE.

hydrogen ion the positive ION of HYDROGEN, affecting the properties of ACIDS. In soil science the hydrogen ion activity in a SOLUTION is used to assess the acidity and alkanity of soils. pH.

hydrogeology a branch of geology which deals with the relations of water with the structure of the earth, both on the surface and underground. G.

hydrogeomorphology the study of landforms resulting from the action of water, especially that of rivers.

hydrograph a graph indicating the rate of flow or the level of water in a STREAM or an AQUIFER measured at a given point during a selected period of time.

hydrography the science and art concerned with the study of all bodies of water on the earth's surface, and especially with charting oceans and seas, their beds and coastlines, and the tides, currents, winds, with particular regard to safe navigation. The terms hydrography and HYDROLOGY overlap, and the tendency now in Britain is to restrict hydrography to cover survey and mapping etc. with an emphasis on marine waters, and to apply hydrology to the study of the water of land areas. G.

hydrolaccolith a dome-like hummock of unconsolidated material resembling in form a LACCOLITH, occurring in the soil in cold lands when water, including the upwelling of water from an ARTESIAN BASIN, freezes and is trapped by an underlying layer of PERMAFROST. The ensuing ice core expands and raises the surface to form a dome. The top of the dome may later collapse to form a crater which partially fills with water. A hydrolaccolith differs from a PINGO in that it occurs only in the presence of permafrost. G.

hydrological cycle the continuous circulation of water from the earth's surface to the ATMOSPHERE-1 to the earth's surface, brought about by EVAPORATION from the surface water and the land and by EVAPOTRANSPIRATION from vegetation, giving rise to WATER VAPOUR in the atmosphere which condenses (CONDENSATION), forms CLOUDS, and returns to the earth's surface as PRECIPITATION, swelling the oceans, seas, lakes and rivers, or becoming GROUND WATER. HYDROGRAPHY, HYDROLOGY.

hydrology 1. the scientific study of the occurrence, movement, properties and use of water and ice on or under the earth's surface, from its precipitation to its discharge to the sea or its return to the atmosphere, with a particular emphasis in Britain on inland water. HYDROGRAPHY 2. specifically, in USA Geological Survey, the scientific study of underground water resources as distinct from hydrography, there defined as applying to surface water supplies and resources. HYDROLOGICAL CYCLE. G.

hydrolysis a chemical reaction of water in which the reagent (i.e. the substances creating the chemical reaction of the water) is decomposed and HYDROGEN and HYDROXYL and, commonly, other new compounds are added. This process is involved in 1. the chemical WEATHERING of rocks, in which the salt constituents of the rock combine with water, and an ACID and a BASE-2 are formed. CORROSION 2. in the decomposition of organic compounds by interaction with water, ALCOHOLS and acids from ESTERS being formed.

hydrometeor in meteorology, any weather phenomenon in the ATMOSPHERE-1 produced by the CONDENSATION of WATER VAPOUR, e.g. FOG, HAIL, RAIN, SLEET, SNOW.

hydrometer an instrument used in measuring the SPECIFIC GRAVITY of a liquid by means of the principle of FLOATATION. SALINOMETER.

hydromorphic soil a soil containing excess water, e.g. GLEY.

hydrophilous HYGROPHILOUS.

hydrophyte 1. an aquatic (water) plant, one growing only in water or a saturated soil. HYGROPHYTE, MESOPHYTE, TROPOPHYTE, XEROPHYTE 2. one of a class of RAUNKIAER'S LIFE FORMS, an HERBACEOUS plant with PERENNATING parts lying in water. G.

hydroponics the cultivation of plants without soil, i.e. in water to which essential plant nutrients are added, the roots developing in inert material irrigated by the water.

hydrosere hydrarch succession of plants, the stages in a plant SUCCESSION-2 in which the pioneer community develops in freshwater or a wet habitat. HALOSERE, MESOSERE, SERE, XEROSERE.

hydrosphere the water sphere, all the waters (liquid or solid) of the surface of the earth collectively, including soil and ground water, in comparison and contrast with the LITHOSPHERE and the ATMOSPHERE-1. The HYDROLOGICAL CYCLE blurs the distinction between the atmosphere and the hydrosphere. G.

hydrostatic equation in the ATMOSPHERE-1, the basic interrelationship of density, pressure, gravity and altitude.

hydrostatic equilibrium the state of a FLUID in which the various forces acting on it (including GRAVITY and pressure) balance one another. ISOSTATIC.

hydrostatic pressure the pressure exerted by water at rest equally at any point within that body of water. ARTESIAN WELL.

hydrostatics a branch of hydromechanics, the study of the characteristics (especially the pressure) of FLUIDS, and of immersed solid bodies, at rest.

hydrotaxis the movement of an organism or cell in response to the stimulus of moisture. TAXIS.

hydrothermal *adj.* **1.** relating to hot water **2.** in geology, applied to the combined action of heat and water that brings about changes in the earth's crust, by making strong solutions, by altering processes in minerals (KAOLINIZATION) and by depositing minerals in VEINS and on the earth's surface around GEYSERS **3.** applied to the rocks (hydrothermal deposits), ore-deposits and springs (hydrothermal springs) so produced. GEOTHERMAL ENERGY, HOT ROCK, METAMORPHISM, METASOMATISM, SINTER, TRAVERTINE. G.

hydrotropism the growth response in plants and sedentary animals to the stimulus of water, shown by the growth curvature (especially of roots) to the direction of the stimulus. TROPISM.

hydrous *adj.* containing water.

hydroxyl (hydrogen, oxygen) the monovalent (VALENCY) OH group or radical characteristic of hydroxides, alcohols, oxygen acids etc.

hyetal *adj.* related to rain.

hyetograph **1.** a tipping-bucket gauge, a self-recording RAIN GAUGE in which an inked pen, attached to a float, draws a line on a rotating roll of paper to indicate the quantity of water collected in the gauge **2.** a graph showing the monthly maximum and minimum rainfall and, usually, the standard deviation and probable deviation from the accepted mean.

hyetography the study and mapping of rainfall distribution.

hyetology the scientific study of PRECIPITATION.

hygric *adj.* of, relating to, or containing, moisture.

hygrograph an instrument which continuously records changes in the RELATIVE HUMIDITY of the ATMOSPHERE-1 on a rotating roll of paper, the line being traced by an inked pen attached to a hair HYGROMETER.

hygrometer an instrument used to determine the HUMIDITY or RELATIVE HUMIDITY of the atmosphere or a gas, some types making use of a human hair (which stretches and contracts), others a lithium-chloride strip (which has a varying resistance to moisture), the very slight reaction of the moisture-sensitive component being enlarged and registered on a graded scale. PSYCHROMETER.

hygropetric *adj.* inhabiting the wet surface of rocks.

hygrophilous, hydrophilous *adj.* inhabiting wet places.

hygrophilous coniferous forest the natural FOREST-1,2 of the west and southeast of North America, southern Chile, western Europe, parts of China and Japan, southeastern Australia and New Zealand. It includes the species Douglas fir, sequoia, red cedar, Sitka spruce, hemlock, white and yellow pine. MESOPHYTE, XEROPHYTE. G.

hygrophyte a water-loving plant, one which thrives in moisture (e.g. a tree of the tropical RAIN FOREST) but is not an aquatic plant, i.e. not a HYDROPHYTE. HYGROPHILOUS. G.

hygroscope an instrument that indicates, without actually measuring, HUMIDITY.

hygroscopic *adj.* the quality of having an affinity with water, applied to substances which absorb and retain moisture, e.g. salt particles in the atmosphere, which act as nuclei for the CONDENSATION of WATER VAPOUR.

hygroscopic moisture water in the soil held by such strong SURFACE TENSION that it is unavailable to plants. CAPILLARY MOISTURE, FIELD CAPACITY.

hypabyssal *adj.* half-abyssal, applied especially to an IGNEOUS ROCK or to an IGNEOUS INTRUSION which has risen towards the surface, but which has crystallized below it. They are thus intermediate in physical form between the rocks which are deep seated, having cooled from MAGMA deep in the earth's crust (PLUTONIC ROCK) and those which have been poured out and solidified on the surface (VOLCANIC ROCK). Hypabyssal rocks are found as DYKES, SILLS, small intrusions etc. ACIDIC ROCK, BASIC ROCK, INTERMEDIATE ROCK, INTRUSIVE ROCK. G.

hypermarket a very big, self-service store which is larger than a SUPERMARKET, sells a wide variety of foodstuffs, household goods etc., has an extensive car park and is usually situated on the outskirts of a town.

hyperparasite, superparasite a PARASITE that lives in or on another parasite.

hypogene *adj.* (the alternative, hypogenic, is not generally accepted or used) applied **1.** by C. Lyell to the rocks formed, and by A. Geikie to the geological processes originating, in the depths below the earth's crust **2.** to the action or force operating in the interior of the earth, in contrast to epigenic forces (EPIGENE) acting on the earth's crust **3.** to mineral deposits formed by an ascending aqueous solution, in contrast to SUPERGENE. G.

hypolimnion the non-circulating, coldest layer of water, with the least oxygen, in a deep LAKE or the ocean, lying below the THERMOCLINE. THERMAL STRATIFICATION.

hypothermal *adj.* applied to a deposit formed from an ascending aqueous solution at very high temperatures (300° to 500°C: 575° to 930°F) and at great depths. EPITHERMAL, MESOTHERMAL.

hypothesis pl. hypotheses **1.** an idea or proposition that is not the outcome of experience, but is formulated and used, as an untested assertion, to explain certain facts, or the relationships between two or more concepts (ALTERNATIVE HYPOTHESIS, FALSIFIABILITY, NULL HYPOTHESIS) **2.** the primary assumption of an argument. LAW, SCIENTIFIC LAW, THEORY.

hypsographic curve, hypsometric curve a graph used to indicate the proportions of the area of the earth's surface at various elevations above or depths below a given datum, such as sea-level, the height/depth being shown on the vertical axis, the land area on the horizontal. CLINOGRAPHIC CURVE. G.

hypsography, hypsometry a branch of geography observing, determining, describing, mapping etc. variations in the height of the earth's surface above sea level, a term now rarely used except adjectivally. HYPSOGRAPHIC CURVE.

hypsometer an instrument used for determining altitude in terms of atmospheric pressure, by measuring very accurately the temperature at which water boils at the particular height (e.g. the boiling point of water at 1013 mb is 100°C (212°F), at 728 mb is 90°C (195°F).

hypsometric curve HYPSOGRAPHIC CURVE.

hypsometry HYPSOGRAPHY.

hythe HITHE.

hythergraph a twelve-sided graph of temperature and humidity or precipitation, giving a broad picture of climatic conditions. The mean monthly temperatures are plotted as ORDINATES, and the mean monthly precipitation for each of the twelve months as ABSCISAE, the points being joined to produce a closed, twelve-sided polygon, the hythergraph. CLIMOGRAPH. G.

I

IBP International Biological Programme, an organization established by the International Council of Scientific Unions (ICSU) to study biological productivity and human adaptability on a world-wide scale from 1964 to 1974. More than 40 nation states participated; many of the studies undertaken have been extended and continue.

ICA International Cartographic Association, an international association of cartographers, formed in 1964, affiliated to the IGU (International Geographical Union) and having a broadly similar constitution. The meetings are usually held quadrennially in the country then acting as host to the IGC (International Geographical Congress).

ICAO International Civil Aviation Organization, a United Nations agency established in 1947 to promote safety and cooperation in international air transport.

ice the solid form of water, formed in nature by the freezing of water (as in a river or sea ice), by the condensation of atmospheric WATER VAPOUR at temperatures below freezing point direct into ICE CRYSTALS, by the compaction of SNOW (with or without the movement of a GLACIER), by the seepage of water into snow masses and its subsequent freezing. The DENSITY-2 of ice is lower than that of water, thus ice floats. G.

ice age in the long course of geological time, from Precambrian time onward (GEOLOGICAL TIMESCALE) there have been several periods when extensive GLACIERS covered large parts of the land surface in the northern and southern hemispheres; but the greatest and best known of these ice ages is the last or Great Ice Age, also known as the Glacial Epoch, which occurred in the PLEISTOCENE epoch when human beings had already appeared on the earth's surface. In Europe and North America there were at least four fluctuations in the Pleistocene glaciation, periods with extending ICE SHEETS being interrupted by warmer interglacial episodes. Many local names have been given to these periods of ice expansion. In the European Alps four were distinguished and named, from the oldest to the youngest: Günz, Mindel, Riss and Würm. In North America the four periods corresponding to those of Europe are, from the oldest, Nebraskan, Kansan, Illinoian, Wisconsin, with the Iowan as the earliest stage of the Wisconsin. LITTLE ICE AGE. G.

ice anvil the flattened head of a CUMULONIMBUS cloud, formed by very small snow and ice crystals falling down below the spreading layer. ANVIL CLOUD, INCUS CLOUD.

ice apron a thin mass of ice and snow adhering to a mountainside. BERGSCHRUND, RANDKLUFT. G.

ice barchan a crescentic dune resembling a SAND DUNE but formed of tiny, very cold ICE CRYSTALS.

ice barrier the edge of the Antarctic icefield or Ross ice barrier, now generally called an ICE FRONT or ICE WALL. G.

iceberg a large mass of land ice, often of great height, broken off from a GLACIER or from an ICE SHELF and floating in the sea, at the mercy of winds and currents. The ratio of ice below the water to that above is some three or four to one. Arctic icebergs (CASTELLATED ICEBERGS) have pinnacles in their superstructures; coming from Greenland, they are carried south by the East Greenland and Labrador currents to the Atlantic (a few passing through the narrow, shallow Bering Strait to the Pacific). The superstructures of Antarctic icebergs, which come from the Ross ice shelf, are flat (TABULAR ICEBERGS), and these icebergs float northwards to about 60°S in the Pacific. G.

ice blink, iceblink a yellowish white luminous glare in the sky produced by reflection on clouds from an ice surface which may be distant and out of sight. EISBLINK, ISBLINK. G.

icebreaker a high-powered ship with strongly reinforced bow that breaks its way through floating ice by means of a downward thrust.

ice cake a general term applied to flat fragments of floating ice, sometimes defined as an ICE FLOE smaller than 10 m (33 ft) across. G.

ice cap a permanent covering of ice, a dome-shaped GLACIER, smaller than an ICE SHEET, covering a highland area or an island in high latitudes. INLAND ICE. G.

ice cluster a concentration of sea ice, extending hundreds of square kilometres, present in the same area annually in summer. G.

ice column, ice pillar a column of ice capped by a boulder.

ice cover the amount of sea ice, measured in tenths of the visible sea surface that is covered with ice.

ice-covered *adj.* applied to land at present concealed by an extensive GLACIER. G.

ice crystal a single ice particle with regular structure. G.

ice dam a river dam formed of blocks of ice.

ice-dam lake, ice-dammed lake a lake formed by a barrier of ice stretching across a valley mouth.

ice edge the boundary between sea ice (DRIFT or FAST), and the open sea, termed compacted ice edge when the wind or the sea swell compact the ice edge, or open ice edge when they disperse the ice along the edge. ICE LIMIT. G.

ice evaporation level the level in the ATMOSPHERE-1 at which the temperature is low enough for the transition of water from the gaseous (GAS-2) to the solid state (occasionally with a very brief intervening LIQUID phase). The direct formation of ice from saturated air occurs at any temperature below −40°C (−40°F); and if ICE CRYSTALS in air at such low temperatures meet very dry air they will change from the solid to the gaseous state without the intervening liquid phase (a process termed SUBLIMATION). CONDENSATION TRAIL.

ice fall, icefall a cataract of ice, comparable to a WATERFALL or a CATARACT in a river, usually marked by a heavily crevassed (CREVASSE) area in a GLACIER and a marked, sudden drop in the glacier's surface level.

ice field, ice-field, icefield 1. generally, an extensive area of LAND ICE 2. specifically, a large continuous area of PACK ICE or sea ice, more than 8 km (5 mi) across, defined as large (over 20 km: 12.5 mi across), medium (15 to 20 km: 9 to 12.5 mi across), or small (10 to 15 km: 6 to 9 mi across). G.

ice floe, ice-floe, floe any piece of floating sea ice, level or hummocked, termed light (up to 1 m: 3.3 ft thick) or heavy (exceeding 1 m: 3.3 ft thick). Floes are described as vast (over 10 km: 6.2 mi across), large (between 1 and 10 km: 0.6 to 6.2 mi across), medium (200 to 1000 m: 655 to 3280 ft across), small (10 to 200 m: 33 to 655 ft across).

A floe under 10 m (33 ft) across is termed an ICE CAKE, under 200 m (655 ft) across a glaçon. G.

ice fog a formation of ICE CRYSTALS suspended in the air, so numerous that they restrict visibility at the earth's surface and reflect sunshine to produce optical phenomena such as haloes and luminous pillars. The sun shining above an ice fog can produce an effect so dazzling that it can be dangerous to the naked eye. ICE PRISM, STEAM FOG. G.

ice foot, icefoot 1. a narrow strip of ice formed along the shore between LOW and HIGH WATER marks, attached to the coast, unmoved by the tide and staying in position after FAST ICE has broken off 2. the ice forming the front part of a GLACIER. G.

ice front the floating vertical cliff formed by the seaward face of floating ice, e.g. of an ICE SHELF or other glacier entering the sea. ICE WALL.

ice island a rare form of Arctic ICE-BERG, having a tabular, regularly undulating superstructure, and originating from an ICE SHELF in northern Ellesmere Island or northern Greenland. G.

ice jam a great quantity of broken river, lake or sea ice caught in a narrow channel or piled up on a shore. G.

Icelandic low the mean atmospheric sub-polar low pressure area (LOW) lying over the North Atlantic ocean between Iceland and Greenland, particularly in winter, comprising swiftly moving areas of low pressure interspersed with occasional periods of higher pressure. ALEUTIAN LOW.

ice limit the average position of the ICE EDGE in any given month or period based on observations recorded over a number of years. G.

ice needle a very small, thin, sharp-pointed piece of ice (a spikelet of ice). PIPKRAKE. G.

ice pedestal SERAC.

ice piedmont an ice belt covering a coastal strip of low land backed by mountains, the ice shelving seawards, extending to any width between 50 m and 50 km (160 ft to 30 mi) and forming ice cliffs along stretches of the coastline, frequently passing to an ICE SHELF. G.

ice point the melting point of pure ice at standard pressure. ABSOLUTE ZERO.

iceport a bay in an ICE FRONT, usually temporary, where vessels can moor. G.

ice prism a very small unbranched ICE CRYSTAL in the form of a needle, column or a plate, which may fall from a cloud or from a cloudless sky, seen to glitter in the sunshine (termed diamond dust), and sometimes producing a luminous pillar or a halo. ICE FOG. G.

ice rind a thin layer (less than 5 cm: 2 in) of elastic, shining ice on a quiet sea surface, formed by the freezing of SLUDGE. G.

ice rise a mass of ice, frequently dome-shaped, sometimes including SNOW, resting on rock, surrounded wholly or partly by an ICE SHELF, partly by sea and/or ice-free land. The underlying rock may occasionally be exposed. G.

ice segregation a process or formation in which layers of clear ice form within the soil from the freezing and upwelling of GROUND WATER and of water within the soil, thereby causing the soil to rise. FROST HEAVING.

ice sheet a continuous mass of ice and snow of considerable thickness and covering a large area of rock or water, such as those ice masses occupying a major part of the Antarctic continent and Greenland at the present day. An ice sheet of under some 50 000 sq km (20 000 sq mi) and overlying rock is termed an ICE CAP. ICE AGE.

ice shelf a thick ICE SHEET of great extent and with a level or undulating surface, fed by snow and sometimes by glaciers, which has reached the sea and is floating, although parts may be aground. BAY ICE, ICE FRONT.

ice storm a storm in which falling rain

freezes as soon as it touches any object on the surface of the earth which has a temperature below freezing point. GLAZED FROST. G.

ice stream part of an ICE SHEET in which the ice flows more swiftly than, and not necessarily in the direction of flow of, the surrounding ice. G.

ice wall an ice cliff based on rock and not floating (unlike an ICE FRONT, which is floating), forming the seaward margin of an inland ICE SHEET, ICE PIEDMONT or ICE RISE. The rock base may be at or below sea-level. FAST ICE.

ice wedge a mass of ice in the ground, shaped like a wedge with a point facing down, formed when melt-water, penetrating cracks in the ground in summer, freezes in winter under PERIGLACIAL conditions. Repeated melting and refreezing breaks up the material around the wedge, thereby increasing its width and penetration.

icicle a hanging spike of clear ice formed by the freezing of dripping water. G.

icing the accumulation of a deposit of ice (ranging from dense and transparent to white and opaque) on exposed objects, produced by the deposition of WATER VAPOUR as FROST, or by the freezing on impact of droplets suspended in air. CONDENSATION TRAIL, GLAZE, RIME.

iconic model a simple scaled-down representation of reality in which the phenomena selected are shown in the form in which they exist, reduced to scale, e.g. people shown as reduced-to-scale people. ANALOGUE MODEL, SYMBOLIC MODEL. G.

iconography the study of the sources and meanings of images used in the visual arts and literature and by society, particularly as applied in HISTORICAL GEOGRAPHY and HUMANISTIC GEOGRAPHY.

ICSU International Council of Scientific Unions, the non-governmental organization to which international scientific unions (including the International Geographical Union, IGU) adhere. It

aims to facilitate the exchange of scientific knowledge, issues reports on problems encountered by scientists in their work, and initiates international research programmes, e.g. the International Biological Programme (IBP). Headquarters in Paris.

ideal 1. a MODEL-2 of perfection 2. a belief in high or perfect standards.

ideal *adj.* 1. pertaining to an IDEAL-1 2. expressing possible perfection which is unlikely to exist in the real world 3. existing only in thought. IDEALISM-3.

idealism 1. the belief that a system or a standard conceived as perfect or nearly perfect (but which is unlikely to exist in the real world) exists 2. a system which consists of forming IDEALS-1,2 and attempting to realize them 3. in philosophy, one of the various approaches which, in general, maintain that ideas are the only real things, i.e. that the object of external perception consists (either in itself or as perceived) as an idea, a notion, that the only things which exist in reality are minds or mental states, or both. For example, the metaphysical doctrine (METAPHYSICS, ONTOLOGY) maintains that ultimate reality is either mental or spiritual; and the epistomological doctrine (EPISTOMOLOGY) holds that either the objects of perception or ideas are the only knowable entities. MATERIALISM, REALISM.

ideal-type a term introduced and applied by Max Weber to a freely created hypothetical construction (representing a type of society, institution, or activity etc.) built-up by an investigator from empirically observable or historically recognizable elements, which is useful in making comparisons, in classifying, and in developing theoretical explanations. It is 'ideal' in the logical sense: it is not an 'average type' (either in the statistical sense or in the sense of representing a common denominator of a number of empirical phenomena); and it is not a model of perfection (IDEAL-1).

ideology **1.** the science of ideas, concerned with their origin and nature, particularly those springing from sensory stimulation **2.** ideal or abstract speculation **3.** a set of ideas adopted as a whole, held implicitly and maintained regardless of the course of events, used in support of, justifying, an economic, social or political theory, or the conduct of a class or group **4.** the way of thinking of an individual class or culture.

-ides a suffix frequently attached to the names of major mountain ranges or chains, e.g. Alps, Alpides; Altai, Altaides. The purpose is to indicate the broad geological or structural units of which the present mountains form only part.

idiographic approach explanation concerned with individual cases or situations (explanation by case history) rather than with general types or theories (a law-seeking approach). NO-MOTHETIC APPROACH. G.

idiomorphic *adj.* applied to the minerals in IGNEOUS ROCK which have the characteristic crystalline form. ALLO-TRIOMORPHIC.

IGC International Geographical Congress, the main international meeting of geographers, held every three or four years since the first at Anvers in 1871 (planned for 1870 but delayed by the Franco-Prussian war). The earlier congresses were organized by committees of the host countries and before the close of one congress the meeting would accept the invitation of a host country for the next. But in 1922 a permanent body, the International Geographical Union (IGU), was formed, and it is the Executive of this body which now accepts the invitations from countries willing to organize a congress. There is thus a distinction between the President of the Union and the Chairman of the Organizing Committee of the Congress. Countries which are members of the Union send official delegations, but participation in congresses is open to any individual geographer on payment of the appropriate fee. The organization of each congress is in the hands of the host country, subject to the agreement of the Executive Committee of the Union. Meetings, divided into various sections, are usually held over several days in a main centre, preceded and followed by study tours and symposia in local centres. Regional meetings are sometimes arranged by other hosts in their countries, and take place about midway between congresses.

igneous *adj.* of, pertaining to, containing, resembling, or emitting, fire.

igneous rock a rock which has originated from the cooling and solidification of MAGMA from the heated lower layers of the earth's crust. The chemical composition of igneous rocks depends on the nature of the magma. Arranged in order of their SILICA content they are classified as ACID ROCK (e.g. GRANITE, OBSIDIAN), INTERMEDIATE ROCK (e.g. ANDESITE), BASIC ROCK (e.g. GAB-BRO), ULTRABASIC ROCK (e.g. PERI-DOTITE). Their character is also affected by the mode of cooling. When solidifed slowly at depth they are coarsely crystalline (e.g. GRANITE), and termed INTRU-SIVE or PLUTONIC; when rapidly, at the surface, they have fine crystals and are termed extrusive (EXTRUSIVE ROCK) or volcanic (VOLCANIC ROCK) (e.g. AN-DESITE, BASALT); between the two are HYPABYSSAL (e.g. DOLERITE). A few igneous rocks (e.g. TUFF) are formed from compacted or cemented fragments of pre-existing igneous rocks. POIKILI-TIC.

ignimbrite NUEE ARDENTE, WELDED TUFF.

IGU International Geographical Union, an organization established in 1922 to serve as a permanent body linking geographers all over the world who had previously come together only for the international congresses (IGC). The Union comprises three main elements: the Executive Committee and the Secretariat; the Member Countries (a small country with only a few geographers may become an Associate Member); the Commissions. The Executive Committee consists of a President, seven Vice-Presidents chosen as widely as possible from member countries throughout the world, and a Secretary-Treasurer who is in charge of a small secretariat of paid officers. The Executive Committee holds office from the end of one international congress to the end of the next, and makes recommendations for its continuance. These are put before the General Assembly of the Union, consisting of the delegates of member countries, who vote by countries, each member country being required to set up a National Committee to represent it (usually the Academy where such exists). Between congresses scientific and other work is carried on by Commissions and Working Groups appointed by the General Assembly at each congress. A Commission may co-opt corresponding members and hold its own meetings. Each reports to the next congress and may be reappointed for a further period if its work is incomplete. The secretariat publishes the *IGU Bulletin* in English and French. ICA.

IGY International Geophysical Year, the most recent being 1 July 1957 to 31 December 1958 when 70 countries cooperated in coordinated astronomical and geophysical research programmes.

IHO International Hydrographic Association, an intergovernmental consultative and technical organization, established in 1970, headquarters (IHB, the International Hydrographic Bureau) in Monaco, the membership consisting of 45 maritime states in 1982. Its objectives are to bring about the coordination of the activities of national hydrographic offices, the greatest possible uniformity in nautical charts and documents, the adoption of reliable

and efficient methods of conducting and exploiting hydrographic surveys, and the development of the sciences linked to HYDROGRAPHY as well as the techniques used in descriptive OCEANOGRAPHY.

illuvial horizon in SOIL SCIENCE, the SOIL HORIZON which has received material in SOLUTION or SUSPENSION from the overlying soil layer(s). ARGILLIC HORIZON, ELUVIAL HORIZON, ELUVIATION, ILLUVIATION. G.

illuviation in SOIL SCIENCE, the process by which material removed in SUSPENSION or SOLUTION from the upper part of a soil (the ELUVIAL HORIZON) is washed down and deposited in the lower layers or ILLUVIAL HORIZONS (usually the B HORIZON). ARGILLIC HORIZON, HARD PAN. G.

ilmenite an oxide of IRON and TITANIUM, $FeTiO_3$, a black, crystalline ore, occurring in detrital SAND, BASIC IGNEOUS ROCK and METAMORPHIC ROCK.

image 1. a concept, a mental picture. COGNITIVE MAP, PERCEPTION **2.** something that represents or is taken to represent something else **3.** a counterpart, a copy (solid or optical, or the product of REMOTE SENSING) of an object, the same size as the object, or diminished or magnified, erect or inverted (e.g. in optics a copy produced by a mirror or lens).

imbricate *adj.* having parts which overlap, as the tiles on a roof, e.g. the overlapping of flat pebbles in a stream, the tapering ends pointing upstream. G.

imbricate(d) structure a term applied by some geologists to a structure produced by intense pressures of mountain building activity, as a result of which many individual blocks of rock have been thrust one over another. G.

immature soil in soil science, a young, imperfectly developed soil occurring on recently laid deposits or where EROSION keeps pace with the development of a SOIL PROFILE. AZONAL SOIL. G.

immature town URBAN HIERARCHY. G.

immigrant 1. a person who voluntarily comes from the home country to settle in another, especially for the purpose of permanent residence 2. a plant or animal which comes from its native habitat into another. EMIGRANT, EXILE, EXPATRIATE, HOST SOCIETY, MIGRANT, MIGRATION, REFUGEE.

imperialism 1. the making of an EMPIRE through the extension by one sovereign state of its authority over other territories, by military conquest or by political or economic means, thereby creating a relationship in which those territories contribute resources to, and become dependent on, the dominant sovereign state (IMPERIAL POWER), usually to the economic benefit of the latter. COLONIALISM is a form of imperialism 2. the policy or the doctrine of such an extension of authority.

imperial preference, Commonwealth preference a tariff system of the British Empire (later the COMMONWEALTH) under which goods imported by one member from another were favoured either by lower (sometimes zero) import duties than those imposed on similar goods imported from elsewhere, or by some other arrangement. Similar tariff systems have operated in trade relations between other sovereign states and their dependent territories.

impermeable *adj.* not PERMEABLE, not permitting the passage of fluids, especially of water. IMPERVIOUS ROCK. G.

impermeable rock a rock which does not allow water to soak into and through it because it is IMPERVIOUS or non-porous (or practically non-porous) or both. IMPERVIOUS ROCK, PERMEABLE ROCK, PERVIOUS, PERVIOUS ROCK, POROSITY. G.

impervious *adj.* not PERVIOUS, impenetrable, that which cannot be entered or passed through.

impervious rock rock through which

water cannot pass freely, e.g. CLAY (POROSITY), unfissured GRANITE. Some authors use impervious as synonymous with impermeable, but some draw a certain distinction, outlined in PERMEABLE ROCK, PERVIOUS ROCK. (PERMEABILITY should not be confused with porosity.) G.

import something brought in, introduced from a foreign or external source, or from one use, connexion, or relation, into another, especially goods or merchandise brought into a country from a foreign source in the course of international trade. BALANCE OF TRADE, EXPORT.

import substitution the replacement of previously imported goods by home produced goods in order to improve the BALANCE OF PAYMENTS.

improved *adj.* made better in quality, more productive, more valuable.

improved land frequently used as a technical term, but not always with the same meaning. In the Agricultural Statistics of most countries it refers to farm land where by ploughing, cultivation, manuring, or some form of management, the natural condition of the land has been improved. In Britain it covers ploughland and grassland in fields, but not open moorland; similarly in the USA, land in farms but not open range land. G.

in-by land (from Scandinavian term by, a farm) in northern England, the fenced-in land nearest the homestead in those holdings where the farm may include tracts of unenclosed grazing. G.

incense the aromatic gum of some plants, or the fragrant product of others, or a mixture of the two, burnt to produce a sweet-smelling vapour used particularly in religious ceremonies. It is obtained from various trees, especially of the genera *Boswellia* and the *Icica*, and species of *Pittosporum*. Incense-wood is the wood of *Icica heptaphylla*, a tree native to South America.

inceptisols in SOIL CLASSIFICATION,

USA, an order of young soils with weakly developed horizons, occurring in variable climates, e.g. brown earths (BROWN FOREST SOIL) and TUNDRA SOILS. G.

inch **1.** (Scottish and northern Ireland, from Gaelic innis) a small rocky island **2.** a flat area of alluvium by a river which may become an island in flood time **3.** a measure of length, one-twelfth of a FOOT, 2.54 centimetres or 25.4 millimetres. CONVERSION TABLES.

incidence matrix a square or rectangular table which displays the elements of two sets and shows whether or not the elements of one relate to those of the other. The rows and columns are the elements of the sets; a 1 (positive) or an O (negative) in each row and column intersection indicates the relationship. As a binary connectivity matrix it may be used in the interpretation of a NETWORK-2. GRAPH-2, GRAPH THEORY, NETWORK CONNECTIVITY.

incised meander a MEANDER deeply sunk into the general level of the surrounding country. This may be the result when a mature river, characterized by extensive meanders, is rejuvenated by an uplift of the land and begins to cut down its bed. Most authors use incised and INTRENCHED (or entrenched) as synonymous; but some give a narrower meaning to intrenched, applying that term to a meander in which the valley sides are roughly of the same slope, and using the term ingrown meander when they are not of similar slope. G.

incision the deepening of its channel by a river. G.

inclination **1.** the angle of approach of one line or plane to another **2.** a slant, a slope, a deviation from the vertical or horizontal. DIP.

inclination-dip the angle of DIP of a FAULT, STRATUM or VEIN, measured from the horizontal. DIP SLOPE.

inclosure in England, the legal act, permitted by an Act of Parliament,

either general or special, whereby open land or land formerly worked in common (OPEN FIELD) is cut up into individual fields surrounded by fences, walls, hedges, etc. In the process the COMMON RIGHTS over a piece of land are extinguished, the land being turned into 'ordinary freehold' (LAND TENURE). The spelling of the physical process is usually ENCLOSURE, and although that of the legal Act should be inclosure, the two spellings are commonly used interchangeably. G.

inclosure in severalty the parcelling out of a COMMON or COMMONABLE LAND among all the persons legally interested, usually in proportion to the value of the respective interest of each.

incoming *adj.* taking the place formerly held or occupied by someone, something, else.

incompetent, incompetence (of rocks), incompetent bed COMPETENCE.

inconstancy of climate climatic change considered as the norm rather than the exception.

inconsequent drainage a drainage pattern not conditioned by (not consequent on) the present structure, being either ANTECEDENT or SUPERIMPOSED. ACCORDANT, DISCORDANT, DRAINAGE, INSEQUENT DRAINAGE. G.

incus cloud the ANVIL CLOUD spreading above a CUMULONIMBUS.

indaing (Burmese) the dry deciduous woodland of Burma, on the margin of the Burmese dry belt, consisting mainly of in or eng, *Dipterocarpus tuberculatus*, co-dominant with ingyin, *Pentacme suavis*. G.

indehiscence in botany, the quality of being INDEHISCENT.

indehiscent *adj.* in botany, applied to FRUITS that do not burst open spontaneously in order to release their SEEDS or SPORES. DEHISCENCE.

indelta inland delta, an inland area where a river subdivides. G.

indented *adj.* notched, jagged, cut into,

having deep recesses, e.g. as in a coastline.

indenture a contract that binds one person to work for another for a prescribed period of time.

independent variable a VARIABLE which produces or causes (or is thought to produce or cause), changes in another variable (termed the DEPENDENT VARIABLE) and, in an experiment, is manipulated by the experimenter (EXTRANEOUS VARIABLE, REGRESSION ANALYSIS). Synonyms include causal variable and explanatory variable.

index 1. an indicator, a sign **2.** in mathemetics, a numerical ratio or other number deduced from observations and used as an indicator or measure of a process or condition, e.g. cost of living index. INDEX NUMBER.

index fossil a FOSSIL typical of, and giving its name to, a particular rock FORMATION-2.

index number in economics, a number used to show relative change (if any) in cost, in price, or in quantity of a heterogeneous collection of economic objects (e.g. commodity prices, wages, interest rates, industrial or agricultural production, business activity, provision of services etc.) either between one point in time (or one point in space) and the point in time (or in space) selected as the base, to which the index number 100 is usually assigned.

index of circuity NETWORK CONNECTIVITY.

index of decentralization a measure of the degree to which some activity (e.g. manufacturing employment) or other variable is centrally located within an area (town, city, conurbation or region). Maximum concentration within the inner area has a value of zero, grading to maximum concentration on the periphery, value 100.

index of dissimilarity a measure of the difference between the areal distributions of occupation groups, used in the study of social stratification and residential differentiation in urban areas. Using the occupations recorded in census data, the percentage of all those working in each occupation group living in each unit area is calculated. The index of dissimilarity between two occupation groups is then computed as being one half the sum of the absolute values of the differences between the respective distributions, taken area by area. INDEX OF SEGREGATION. G.

index of segregation a measure of the degree of residential separation of sub-groups (e.g. occupation groups, ethnic groups) within a population, used particularly in the study of social stratification and residential differentiation in urban areas. The SEGREGATION may be shown graphically by a LORENZ CURVE, the cumulative percentage of each sub-group being shown on the x-axis, the cumulative percentage of the remaining groups over the sub-areas of the whole area on the y-axis, the diagonal straight line indicating no segregation. The distance of the curved line from the diagonal indicates the degree of segregation. Or two simple indices may be employed. In the first an INDEX OF DISSIMILARITY is computed, not between one sub-group and another, but between one sub-group and all the other sub-groups combined (e.g. in occupation groups, the total employed except those in the given occupation group). This indicates the percentage difference between the distribution of one group and that of the rest of the population. Or a simple LOCATION QUOTIENT may be used, showing the ratio between the percentage of one population and the percentage of another population in a given area. G.

index of shape a measure used to indicate the form (e.g. round, elongated) of a geographical phenomenon (e.g. a nation state) as determined by its BOUNDARY, i.e. $1.27A$ divided by L, A

being the area of the whole phenomenon, *L* the length of the long axis. A circle is represented by the value of 1.0, and as the shape becomes more elongated the value moves progressively towards 0.

index species an organism adapted only to a very restricted range of environmental conditions, and used to characterize such conditions. CLASSIFICATION OF ORGANISMS, SPECIES-1.

Indian corn MAIZE.

Indian hemp *Apocynum cannabinum*, a North American plant, dog-bane, providing fibre from which cordage used to be made. HEMP.

Indian rice AMERICAN WILD RICE.

Indian summer a period of calm, dry, mild weather, with cloudless sky but hazy atmospheric conditions, occurring fairly regularly in late autumn (fall) or early winter in the USA and the UK. The origin of the term is uncertain. G.

indicator plant, indicator species a plant or SPECIES-1 which shows by its presence in a locality the existence of a particular environmental factor or certain environmental conditions, e.g. the OLIVE, an indicator of the MEDITERRANEAN CLIMATE.

indifference theory TRADE-OFF THEORY.

indifferent equilibrium NEUTRAL STABILITY.

indifferent species a SPECIES-1 lacking marked affinities for any community.

indigenous *adj.* originating in, native to, a particular area, not introduced, applied to 1. plants, animals, human population 2. a rock, mineral or ore in its place of origin, not carried in from somewhere else.

indigo a blue dye obtained from plants, especially from the genus INDIGOFERA.

indirect contact space the part of an urban area perceived by an individual from secondhand contact, e.g. from information given by acquaintances, the MASS MEDIA, etc. AWARENESS SPACE.

indivisible services services (high order goods), e.g. hospitals, theatres, which can be provided only by large centres (ORDER OF GOODS). TIME-DIVISIBLE SERVICES.

induction 1. the drawing of a general conclusion from a number of known facts 2. the conclusion reached in this way. DEDUCTION, SCIENTIFIC, SCIENTIFIC LAW.

inductive *adj.* based on or pertaining to INDUCTION-1.

induration the hardening of a rock by heat, pressure or cementation, or of SOIL HORIZONS by chemical action (HARD PAN).

industrial archaeology the study of past industrial processes and methods, especially of the early period of the INDUSTRIAL REVOLUTION, based primarily on examination of the physical remains, e.g. of buildings (including housing associated with manufacturing), machinery, mines, means of transport, and their equipment.

industrial centre INDUSTRIAL COMPLEX.

industrial city the type of city commonly found in industrialized countries today, broadly conforming to the CONCENTRIC ZONE THEORY of Burgess. INDUSTRIAL SOCIETY, POST-INDUSTRIAL CITY, PRE-INDUSTRIAL CITY.

industrial complex a large assemblage of manufacturing enterprises concentrated in a relatively restricted area (thus differeng from an industrial region, where widespread industry covers an extensive area), served by good transport, commercial and financial facilities. It usually comprises one or more basic manufacturing industries combined with diverse other manufacturing enterprises, technically and economically interdependent (thus differing from an industrial centre, which is small, with less diversified manufacturing enterprises). GENERATOR. G.

industrial crop, commercial crop a

CROP-1 grown not for food but as a raw material for manufacturing industry, e.g. COTTON.

industrial geography a branch of ECONOMIC GEOGRAPHY concerned with manufacturing industry, particularly its location and spatial distribution.

industrial inertia the tendency of an industry or firm to remain in a location or site after the conditions originally influencing the choice of that location or site have altered, ceased to be relevant, or disappeared. GEOGRAPHICAL INERTIA, GEOGRAPHICAL MOMENTUM, INDUSTRIAL MOMENTUM. G.

industrialization 1. the process of growth of large-scale machine production (MECHANIZATION) and the factory system **2.** the process of setting-up such organizations, especially in the introduction of MANUFACTURING INDUSTRY (SECONDARY INDUSTRY) in countries or regions where people are engaged mainly in agricultural activities (PRIMARY INDUSTRY). It is usually accompanied by the establishment of service industry (TERTIARY INDUSTRY) and by social change, e.g. in patterns of consumption, and in migration of people from rural areas to the growing urban settlements. AIC, COTTAGE INDUSTRY, DE-INDUSTRIALIZATION, INTERNATIONAL DIVISION OF LABOUR, NIC, UNDERDEVELOPMENT. G.

industrialized country generally, a country in which the contribution of manufacturing industry (SECONDARY INDUSTRY) and the service industry (TERTIARY INDUSTRY) to the economy is greater than that of PRIMARY INDUSTRY, e.g. of agriculture.

industrial land ownership the ownership of land by industrial companies. FINANCIAL LAND OWNERSHIP.

industrial linkage all the exchanges between an industrial enterprise and the factors, material and non-material, influencing its location, including, for example, the exchange of information with other firms engaged in similar work in the locality.

industrial location theory a body of theory which seeks to account for the location of industry, including the identification of the factors which make a particular site more advantageous than another for a particular enterprise. LOCATION THEORY.

industrial momentum the tendency of an industry or firm in a given locality not only to maintain its activity (INDUSTRIAL INERTIA) but also to increase its importance after the conditions originally influencing its establishment in that locality have appreciably altered, ceased to be relevant, or disappeared. GEOGRAPHICAL INERTIA, GEOGRAPHICAL MOMENTUM. G.

industrial movement the setting-up by an existing manufacturing firm of a new factory in a location new to the firm, whether or not this involves only the setting-up of a new branch factory (the firm continuing its activities in its existing location) or the complete transfer of all the activities of the firm to the new location.

industrial park a tract of land, in some cases park-like (PARK-2), planned and officially designated for the accommodation of clean, relatively small industrial enterprises. OFFICE PARK, PARK-6, RESEARCH AND SCIENCE PARK.

industrial region INDUSTRIAL COMPLEX.

industrial reserve army the body of workers who, according to production needs, are employed or dismissed. Frequently unemployed, when they are working they are paid low wages and given little job security.

industrial revolution the changes generated by the MECHANIZATION of MANUFACTURING INDUSTRY, i.e. the change from domestic industry to the factory system, which leads to the mass production of goods, with the consequent great changes in social, economic and technical structures. The term Industrial Revolution is commonly applied to the period in Britain in the eighteenth and early nineteenth centuries when the mechanization of the textile industry produced such changes.

industrial society POST-INDUSTRIAL SOCIETY.

industry any work performed for economic gain, but popularly applied especially to MANUFACTURING INDUSTRY (SECONDARY INDUSTRY). Industries are variously described as BASIC, EXTRACTIVE, FOOTLOOSE, HEAVY, LIGHT, LINKED, PRIMARY, SECONDARY, TERTIARY and QUATERNARY. See terms qualified by industrial; and INDUSTRIALIZATION, LOCALIZATION OF INDUSTRY, LOCATION OF INDUSTRY, LOCATION THEORY, LOCATIONAL TRIANGLE, MARKET ORIENTATION, RESOURCE ORIENTATION. G.

inelastic *adj.* **1.** not ELASTIC, not adaptable **2.** in economics, of supply and demand, not readily conditioned by, or responsive to, changing conditions or fluctuations.

inface the scarp face of a CUESTA. G.

infantile landform an early landform developing on a PENEPLAIN as it is slowly uplifted, as distinct from a senile landform, displayed before the uplifting. G.

infantile town URBAN HIERARCHY.

infant mortality the average number of deaths of infants under one year of age per 1000 live births. BIRTH-RATE.

inferential statistics a statistical approach in which appropriate statistical techniques are used to make estimates or predictions, or to draw conclusions, from a set of DATA, as distinct from DESCRIPTIVE STATISTICS.

infield-outfield a system of farming which confines intensive cultivation, manuring etc. to enclosed fields near the farm (infield), the outer parts of the farm (outfield) being used for grazing or periodically cropped on a SHIFTING CULTIVATION basis. G.

infiltration the ABSORPTION and downward movement of PRECIPITATION in the REGOLITH. INFILTRATION CAPACITY.

infiltration capacity, infiltration rate the maximum rate at which water can be absorbed and seep downwards through the soil. Infiltration rate is now the preferred term. FIELD CAPACITY, OVERLAND FLOW.

infiltration excess overland flow OVERLAND FLOW.

inflation a general increase in prices produced by an increase in the proportion of currency and credit to the goods available.

influent stream a stream, common in CHALK and LIMESTONE country, that has its bed higher than the WATER TABLE, and which flows into a cave, etc. The bed is usually IMPERMEABLE, being lined with fine silt; but part of the water inevitably disappears down cracks or joints in the chalk and limestone. G.

information fields AGGREGATE INFORMATION FIELD, CONTACT FIELD, MEAN INFORMATION FIELD, PRIVATE INFORMATION FIELD.

infra-red, infrared *adj.* relating to, producing, or produced by INFRA-RED RADIATION.

infra-red radiation, infrared radiation ELECTROMAGNETIC WAVES in the wavelength range of 0.7 to about 200 microns, invisible, perceived as heat, with wavelengths just a little longer than those of red light but less than short MICROWAVES, used especially in some cooking devices and in photography with infra-red or colour infra-red film. Infra-red photographic devices used in REMOTE SENSING successfully distinguish features on the surface of the earth through darkness or cloud. ACTINIC RAYS, ATMOSPHERIC WINDOW, NEAR-INFRA-RED, THERMAL INFRA-RED SENSING, ELECTROMAGNETIC SPECTRUM.

infrastructure the basic structure, the

framework, the system which supports the operation of an organization (e.g. the power and water supplies, the transport and communications facilities, the drainage system), which makes economic development possible, the basic capital investment of a country or enterprise.

ingraphicate *adj.* not versed in GRAPHICACY. G.

ingrown meander INCISED MEANDER, INTRENCHED MEANDER. G.

ings in Yorkshire, England, a tract of alluvial soil bordering a river and subject to flooding in winter and summer. G.

inherited valley a valley cut by an old, obsolete drainage system and subsequently occupied by a new drainage system. G.

initial advantage the advantage gained by a city, region or nation by being the first in some respect, e.g. in establishing a market area, or in adopting a new technique. Some authors maintain that urban centres which are first in establishing industry or in developing new techniques build up a self-generating lead over other centres in terms of the size of their population and in the volume and variety of their industries. URBAN SIZE RATCHET.

initial landform 1. a landform produced directly by vulcanism (VULCANICITY) and TECTONIC activity, with its original features almost intact, having been only slightly modified by DENUDATION, unlike a SEQUENTIAL LANDFORM-1 2. the term applied by W. M. Davis to the ideal form resulting from ideal advance in the series of developmental change, as opposed to the SEQUENTIAL LANDFORM-2.

inland basin a DEPRESSION-2 entirely surrounded by higher land, with or without an outlet to the ocean. BASIN-8. G.

inland drainage INTERNAL DRAINAGE. G.

inland ice, inland ice sheet an extensive

ICE SHEET (exceeding 50 000 sq km: 19 300 sq mi) of great thickness and covering rock, e.g. in Greenland. It may merge into an ICE SHELF if it lies near sea-level. G.

inland sea a large, isolated expanse of water, without a link to the open sea. G.

inlet 1. a narrow opening by which the water of the sea, of a lake, or river, penetrates the land 2. the passage between islands into a lagoon. G.

inlier an exposed rock formation entirely surrounded by geologically younger rocks, hence the opposite of an OUTLIER. G.

inner city a term loosely applied to an undefined area with a wide range of economic and social problems, lying within a long-established, generally large, urban area. It may be applied to such an undefined area, usually in economic decline, lying between the commercial centre of the city (CENTRAL BUSINESS DISTRICT) and its suburbs, i.e. to the transition zone of CONCENTRIC ZONE GROWTH THEORY. That is a zone usually characterized by aged, run-down housing in multiple occupation by people with low incomes (especially new immigrants), people who stay for a relatively short period of time, and a dwindling number of aged 'local' people, an area which may still provide a SEED BED. GENTRIFICATION, INDUSTRIAL CITY, VILLAGE-2.

inner lead the calm water lying between the mainland and the skerries. SKERRY, SKERRY-GUARD.

innings lands taken in, enclosed, reclaimed, especially reclaimed alluvial land, e.g. INGS. G.

innovation the making of changes, the introduction of practices, processes, etc. which are new in a particular context.

innovation diffusion the spread or movement of INNOVATION through space and time. DIFFUSION-2, DIFFUSION WAVE, INNOVATION WAVE.

innovation wave a concept of the DIF-FUSION of INNOVATION as being analogous to the onward movement of an ocean wave. DIFFUSION WAVE. G.

inorganic *adj.* **1.** of or relating to substances not composed of plant or animal material. ORGANIC **2.** in chemistry, of or pertaining to substances that are not organic, hence inorganic chemistry, the branch of chemistry concerned with the scientific study of chemical elements and their compounds other than the CARBON compounds (but including the simpler compounds of carbon, e.g. CARBON DIOXIDE). ORGANIC CHEMISTRY.

inorganic chemistry INORGANIC-2.

input-output analysis a method used in ECONOMICS to trace the connexions between products and services (the output) and the resources needed to produce them (the input). During any period of time the output of one sector of the productive system may become the input of another. The quantities, commonly expressed in money values, are displayed in a matrix as an input-output table, each sector of the productive system being assigned a row (showing the destination of the outputs) and a column (showing the provenance of the inputs). Outputs which are absorbed in the productive system are termed intermediate outputs, those which pass out of it into final demand are final outputs; inputs derived from the system are intermediate inputs, those that come in from outside (e.g. land, labour, capital) are primary inputs. For each sector the sum of the output entries (i.e. total revenue), displayed in the row, equals the sum of the input entries (i.e. total costs) displayed in the column. The interdependence of different sectors of the productive system is thus revealed and can be analysed; and the effects of changes in one sector on those in another traced.

inquiline any animal which lives in the abode of an animal of a different species and shares its food. COMMENSALISM-1.

insectivore **1.** an animal or plant that eats insects **2.** a member of the order Insectivora, a group of primitive, placental MAMMALS, e.g. a shrew, hedgehog, mole.

Inselberg (German, island mountain) an isolated hill of circumdenudation, e.g. a steep-sided, isolated residual hill, common in semi-arid and savanna lands, rising from a plain which is, in many cases, monotonously flat (PEDIMENT). BORNHARDT.

insequent drainage, insequent stream **1.** a DRAINAGE-2 pattern developed on the present land surface (especially on horizontal strata), bearing no direct relation to the underlying structure, and seemingly determined by accident. DENDRITIC DRAINAGE, INCONSEQUENT DRAINAGE **2.** (American) inconsequent drainage. G.

inshore *adj.* an imprecise term, moving towards the shore (e.g. inshore breeze); or applied loosely to that which is close to the shore; or shorewards of a position in contrast to seawards of it. G.

in situ (Latin, in place, in position) *adv.* associated with the occurrence of a fossil, mineral, rock or soil in its original place of deposition or formation, e.g. a RESIDUAL DEPOSIT.

insolation **1.** exposure to the rays of the SUN **2.** SOLAR RADIATION received, applied especially to that reaching the surface of the earth, greatest at the equator for the year as a whole, but polewards decreasing, at first slowly, then more rapidly, then again slowly, the variation throughout the year being least at the equator, most at the poles. It is an important climatic factor and has a (sometimes disputed) role in atmospheric weathering. The interrelationship of relief and insolation is important in human geography (ADRET) **3.** the rate at which solar radiation reaches a specified area. G.

installed capacity all the potential CA-PACITY-4 of plant (especially of hydro-electric plant) or a machine as distinct from the capacity used. The percentage relationship between used and installed capacity is the capacity factor (or plant factor) when average load is taken. The term is applied especially to WATER POWER and sometimes contrasted with potential capacity, e.g. of a WATER-FALL. G.

institution 1. an organization established with the purpose of advancing learning, public welfare, etc. 2. the building or group of buildings used by such an establishment 3. in sociology, activities which are continuous or repeated within a regularized pattern accepted as the NORM, usually classified in four groups: political (regulating competition for power), economic (concerned with the production and distribution of goods and services), cultural (concerned with religious, artistic and expressive activities and traditions in the society), and kinship (concentrating on marriage, the family, the rearing of young). FINANCIAL INSTITUTIONS, SOCIAL STRUCTURE.

instrumentalism a pragmatic, philosophical theory that thought is an instrument of adjustment to the environment, and that the validity or truth of ideas and meaning is determined by their utility.

insular *adj.* of, inhabiting, situated on, forming, characteristic of, an island.

insular climate a climatic regime with little range of temperature between summer and winter, characteristic of many islands and some sheltered coastal areas. G.

intake 1. a parcel of land taken in (enclosed) from moorland or common 2. a quantity or amount of something taken in 3. a place where liquid or gas is taken into a pipe or channel.

integration in society, the process by which a sub-group, e.g. an ETHNIC MINORITY, adapts to, fits into, and participates fully in the social and economic structure in which it finds itself, while keeping its identity, its individuality and cultural distinction. ACCOMMODATION, ASSIMILATION, PLURAL SOCIETY.

intensive agriculture a farming system in which large amounts of capital and/or labour are applied to a relatively small area of land to achieve high yields per unit area. EXTENSIVE AGRICULTURE. G.

interactance, interactance hypothesis an INTERACTION hypothesis postulated by S. C. Dodd, 1950, concerned with the interrelationship of two contiguous, separated groups of people (e.g. those parted by a political boundary) and the activities they generate, expressed by the GRAVITY MODEL

$$I = \frac{KTP_A P_B A_A A_B}{D}$$

K is a constant; T = time over which interactions are measured; P_A, P_B = populations of the two interacting groups; A_A, A_B = specific indices per capita activity of populations; D = space dimension. G.

interaction 1. in general, reciprocal action, the action or influence of forces, objects, or persons on each other. SOCIAL INTERACTION, SPATIAL INTERACTION 2. in statistics, the effect produced if the various levels of categories of one INDEPENDENT VARIABLE fail to affect the DEPENDENT VARIABLE in the same way within all levels of categories of another independent variable.

interbedded *adj.* applied to a layer of rock deposited in sequence between two other BEDS-2.

intercalation insertion, especially 1. the insertion of a day or month in a CALENDAR so that the calendar year corresponds to the SOLAR YEAR 2. an interpositioning between rock STRATA.

intercept on the axis of a GRAPH-1, the value on the axis at the point where a line crosses it.

interception the capture of rainwater by the foliage of trees and other plants before it reaches the ground and its return to the ATMOSPHERE-1 by evaporation.

intercision by a stream, the capture of a STREAM-1 by sideways swinging of a neighbouring mature stream. G.

intercommoning the pasturing of stock of two or more villages on COMMON LAND which was never part of a particular MANOR but which stayed common to certain specified villages which lacked adequate common pasture of their own. COMMON RIGHTS.

interface 1. the surface that constitutes the common boundary between two bodies, two systems, two spaces, two different contiguous parts of the same substance, or between phases in a heterogeneous system (e.g. the surface formed between a liquid and a solid), extended to cover other boundaries, e.g. between related disciplines, especially if such boundaries are ill-defined (e.g. between geography and a related study) 2. the connexion between two pieces of equipment, by analogy extended to cover the intercommunication between different social groups.

interfluve the tract of land between two adjacent rivers, regardless of its character. DOAB. G.

interglacial *adj.* of a period of time between two glacial periods. The term is sometimes used as a noun, referring to an INTERGLACIAL PERIOD and/or to a deposit formed in that period. G.

interglacial period a period of time with a relatively warm climate, when the ice sheets retreated, occurring between two periods of glacial cold, e.g. as in the ICE AGE, which was not a period of unremitting cold. G.

intergranular translation the slipping of ice grains over each other within glacier ice, a factor assisting in the gravity flow of a GLACIER.

interior basin INLAND BASIN.

interior drainage INTERNAL DRAINAGE.

interlocking spur one of the series of protrusions (SPUR) of land that, lying between bends in the winding course of a young river in its V-shaped valley, juts into a concave bend and interdigitates with its opposing neighbours lying upstream and downstream. Interlocking spurs thus obscure the upstream or downstream view of the river. They are caused initially by the stream's flowing swiftly round an obstacle in its course, undercutting the bank opposite the obstruction, and thereby making the concave bend. MEANDER. G.

intermediate rock an IGNEOUS ROCK classified chemically as being between acid and basic in its composition (ACID ROCK, BASIC ROCK), that is it has a SILICA-2 content lying between 52 and 66 per cent and no free QUARTZ. A plutonic (intrusive) representative is DIORITE, a volcanic (extrusive) is ANDESITE. EXTRUSIVE ROCK, HYPABYSSAL, INTRUSIVE ROCK, LAVA, PLUTONIC ROCK, VOLCANIC ROCK. G.

intermittent saturation zone in soil science, a layer, lying below the surface soil, which may hold VADOSE WATER in a period of prolonged rainfall, but which quickly dries out in even a short-lived DROUGHT.

intermittent spring a spring that flows from time to time, usually depending on the height of the WATER TABLE (itself affected by fluctuations in precipitation), but also occasionally caused by a SIPHON in a cave system.

intermittent stream a stream which does not flow continuously but dries up from time to time, e.g. a BOURNE.

intermontane, intermontain (intermontane is more common) *adj.* lying between mountains, e.g. the intermontane plateaus, the high plateaus, lying between the east and west ranges of the Andes. G.

internal diseconomy DISECONOMY.

internal drainage interior drainage, EN-DOREIC drainage, a drainage system in which the waters do not reach the sea. AREIC, EXOREIC.

internal migration the movement of people within a country, e.g. in search of employment. G.

international airport an AIRPORT with facilities suitable for handling international traffic and meeting the needs of international AIRLINES-1.

International Cartographic Association ICA.

International Civil Aviation Organization ICAO.

International Council of Scientific Unions ICSU.

international date line an imaginary line agreed internationally which follows the meridian of 180°, with some deviations to accommodate certain land areas. In crossing the line from west to east a day is repeated; in crossing it from east to west a whole day is lost. Thus an aircraft flying from Japan to Honolulu arrives at an earlier time on the same day; an aircraft leaving Honolulu late on Monday evening would not reach Japan until Wednesday morning, although the duration of the flight is only a few hours. G.

international division of labour 1. the separation of employment into parts on an international scale, a feature of the 'old' international division of labour (OIDL) (i.e. before c.1940) when there was specialization on a territorial basis in the tasks performed to supply world markets. To summarize, broadly, subject territories of a dominant country provided it with primary products (foodstuffs and other RAW MATERIALS) to augment the food supply of that dominant country and meet the needs of its manufacturing industry. The principal exports of the dominant country were manufactured goods, sold to the subject territories and to other countries. The subject territories bought these manufactures with earn-ings from the exports of their primary products. In the process the most powerful controlling powers became dominant in the international organization of production and trade, and in world markets **2.** in the 'new' international division of labour (NIDL) (i.e. after the mid-1940s) the production of manufactured goods is widespread throughout the world (not concentrated, as formerly, in the home territories of dominant countries), countries which were formerly predominantly exporters of primary products having developed their own (in many cases substantial) manufacturing industry and, benefiting from improved transport and communications as well as from investment from external sources, entered international trade as exporters of manufactured goods. In many cases their production costs are lower than those current in the formerly dominant countries; their prices are highly competitive in world markets; and many of them specialize in particular products. Such developments have far-reaching social, economic and political effects; and are especially important to the MULTINATIONALS, with their world-wide interests and ability to finance technical advance. AIC, INDUSTRIALIZATION, NIC, UNDERDEVELOPMENT.

International Geographical Congress IGC.

International Geographical Union IGU.

International Geophysical Year IGY.

International Hydrographic Association IHO.

international region a group of states considered to form REGION-1 because they have one or more features or characteristics in common, e.g. the states forming MITTELEUROPA, lying centrally in the continent of Europe.

interpluvial period a period of time, a stage, in low latitudes resembling an INTERGLACIAL PERIOD of higher latitudes, but not necessarily occurring at exactly the same time.

interquartile range in statistics, a measure of DISPERSION of a FREQUENCY DISTRIBUTION. A quartile is produced by splitting a distribution into four equal parts, the quartiles being those values of the VARIABLE below which lie 25 per cent, 50 per cent and 75 per cent of the distribution. The interquartile range is the distance between the 75 per cent (upper quartile) and the 25 per cent (lower quartile). It thus contains one half of the total frequency and provides a simple measure of dispersion which is useful in DESCRIPTIVE STATISTICS. Half the interquartile range is termed the quartile deviation, or the semi-interquartile range. INTERQUARTILE RATIO, MEDIAN.

interquartile ratio synonymous with quartile dispersion coefficient, a COEFFICIENT-1 used to measure the HETEROGENEITY of data. It relates the INTERQUARTILE RANGE (a measure of dispersion) to an appropriate measure of location.

interstadial *adj.* between stages, applied to the period or the deposits laid down between two stages or phases in the retreat of glacial ice in the Great Ice Age, not so important, distinct or prolonged as an INTERGLACIAL PERIOD. Interstadial (like interglacial) is sometimes used as a noun. ICE AGE, STADIAL MORAINE.

Intertropical Convergence Zone (ITCZ), Intertropical Front (ITF) ITCZ.

interval 1. a period of time between two events or two actions 2. a space between two points 3. in statistics, the range between two extremes over which a variable can have any real number value. CLASS INTERVAL, INTERVAL SCALE, NOMINAL, ORDINAL.

intervale (American) a tract of low-lying land, especially between hills or bordering a river.

interval estimate in statistics, a range of values within which a particular POPULATION PARAMETER has a specified probability of lying. CONFIDENCE INTERVAL, POINT ESTIMATE.

interval scale in statistics, a MEASUREMENT scale which lacks an absolute zero, but in which the INTERVALS-3 between the scale points are equal. A VARIABLE measured on this scale is termed an interval variable.

interval variable a VARIABLE measured on the INTERVAL SCALE, not to be confused with an INTERVENING VARIABLE.

intervening location effect the effect of its intermediate location on a place which occupies such a location on a route which is well served for reasons other than those of meeting the needs of the place itself. The result is that the place enjoys better services than its own characteristics would justify. ACCESSIBILITY, SHADOW EFFECT.

intervening opportunity a concept (one of ULLMAN'S BASES FOR INTERACTION) which proposes that the number of movements, e.g. of commodities, of people, of traffic etc., from a place of origin to a destination is directly proportional to the number of opportunities at that destination and inversely proportional to the number of opportunities between the place of origin and the destination (the intervening opportunities). It is claimed that distance is not a deterrent and that the observed decline in movement with increased distance (DISTANCE DECAY) is caused by the rise in the number of intervening opportunities.

intervening opportunity effect of a boundary, the beneficial effect of a BOUNDARY on locations close to it, making such locations attractive to people, e.g. in giving shoppers from one side of the boundary access to goods sold at lower prices on the other. HALO EFFECT.

intervening variable a VARIABLE not necessarily directly measured but postulated as an aid in explaining an observed association.

intratelluric water JUVENILE WATER. G.

intrazonal soil a well-developed soil with the form and structure reflecting the influence of some local factor of relief, parent material or age, rather than of climate and vegetation. AN-THROPOMORPHIC SOIL, AZONAL SOIL, SOIL, SOIL CLASSIFICATION, ZONAL SOIL.

intrenched meander an INCISED MEAN-DER with steep, symmetrical valley sides, produced by swift, vertical erosion. G.

intrusion the forceful entry of a mass of molten rock (MAGMA) in the pre-existing rocks of the earth's crust, sometimes CONCORDANT as a sheet or SILL along BEDDING PLANES, or as a lens-shaped mass (LACCOLITH and PHACOLITH), sometimes DISCOR-DANT, i.e. across the beds (DYKE). HYPABYSSAL, INTRUSIVE ROCK.

intrusive rock PLUTONIC ROCK, an IGNEOUS ROCK with CRYSTALS larger than those occurring in an EXTRUSIVE ROCK, formed from the consolidation of molten material (MAGMA) which has penetrated or been forced into pre-existing solid rocks at depth in the earth's crust. EXTRUSION, HYPABYS-SAL, INTRUSION, VOLCANIC ROCK. G.

invar a NICKEL-IRON alloy with about 36 per cent nickel, which has a very low coefficient of linear expansion, used in the making of precision instruments, such as a surveyor's tape.

invariable *adj.* **1.** having the quality of being unchanging, unable to change **2.** in statistics, a quantity which cannot change in magnitude. VARIABLE.

invasion and succession SUCCESSION AND INVASION.

inverse correlation see CORRELATION COEFFICIENT.

inversion the reversal of the normal or expected order of position. GEO-LOGICAL INVERSION, INVERSION OF TEMPERATURE. G.

inversion of structure GEOLOGICAL IN-VERSION.

inversion of temperature a phenomenon in which there is in the air a temperature increase with increasing height instead of the normal decrease (LAPSE RATE). It can occur at high altitudes (e.g. when a cold air mass flows under a warm one, as at a COLD FRONT, or when a warm air mass flows over a cold one, as at a WARM FRONT, or when an OCCLUSION develops); or near the earth's surface (e.g. in temperate latitudes on a calm, clear night when radiation of heat from the ground at night is rapid, or when warm air flows over a cold surface; or in mountainous regions when radiation of heat from the upper slopes is rapid, and cold dense air behaves like cold water and flows down the valley). FRONT, RADIATION FOG. G.

invertebrate in zoology, a member of Invertebrata, i.e. all the animals which are not members of the Vertebrata. CAMBRIAN, VERTEBRATE.

invertebrate *adj.* lacking a backbone. VERTEBRATE.

inverted lapse rate INVERSION OF TEMPERATURE, LAPSE RATE.

inverted relief inversion of relief, a landscape displaying the effects of a long period of DENUDATION and ERO-SION, the upfolds in the strata (ANTI-CLINES) corresponding with low ground and the SYNCLINES with high ground, the opposite of UNINVERTED RELIEF. G.

invierno in intertropical America, the rainy season. VERANO. G.

invisible exports the income-earning items in the international trade of a country, representing not the sale or transfer of goods, but payments made by foreign countries to that country, for services provided (banking, insurance etc.), for shipping, air freight, for interest on investments; and also expenditure by tourists visiting the country (TOURISM) and remittances home by migrants. Invisible imports cover similar items, but for these payments are made by the country to foreign countries. G.

involution **1.** the results of frost action in the upper soil layers, i.e. CRYOTURBATION **2.** the partial change in an existing NAPPE occurring when a younger nappe is forced into the older one, or when two nappes are re-folded together.

iodine a non-metallic grey-black ELEMENT-6, very volatile, giving off a violet coloured vapour, obtained from sodium iodate, $NaIO_3$, occurring in Chile saltpetre and some SEAWEEDS. Quickly soluble in ALCOHOL and slightly so in water, it is a TRACE ELEMENT essential for the proper functioning of the thyroid gland. It is used in chemical analysis, in photography; and in medicine in diagnosis, in treatment of disorder of the thyroid gland, and as an antiseptic.

ion an ATOM or group of atoms with either an excess (CATION, a positive ion) or a deficiency (ANION, a negative ion) of ELECTRONS, which is therefore electrically unbalanced and electrically charged. An ion may be formed in a gas or in a solution and carry current through either. ACID, BASE, COLLOID, pH.

ionic *adj.* pertaining to IONS.

ionic bond an electrovalent bond. VALENCY.

ionization the production of IONS, converting to ions, or being converted to ions, by the addition or removal of ELECTRONS from ATOMS, e.g. by addition to an ionizing solvent or by means of high-energy RADIATION, as in the IONOSPHERE.

ionosphere thermosphere, the outermost zone of the ATMOSPHERE-1, above the MESOPAUSE, at a height of about 65 km (40 mi), the lower level dropping to some 55 km (35 mi) in daylight, rising to some 105 km (65 mi) at night, so named because ULTRA-VIOLET and X RAYS radiated by the sun ionize its gases to a degree determined by the solar cycle, season and time of day. Particles arising from this IONIZATION concentrate in distinct layers (distinguished by the letters D, E, F2, F1) and refract radiowaves back to earth. APPLETON LAYER, HEAVISIDE LAYER, SUNSPOT.

ipecacuanha *Cephaelis ipecacuanha*, a creeping plant native to tropical South America, its dried rhizome and roots being used medicinally in the production of an emetic.

iridium a metallic ELEMENT-6 similar to PLATINUM (PLATINOID), hard and chemically resistant, used in the manufacture of scientific instruments, for the tips of pen nibs, and in ALLOYS.

iron a widely occurring heavy, MALLEABLE, DUCTILE, MAGNETIC METALLIC ELEMENT-6, the second most widespread MINERAL-1 (ALUMINIUM being the first), estimated to constitute chemically some 5 per cent of the earth's crust by weight. It does not occur NATIVE-1 (*adj.*) in nature, rusts readily in moist air (i.e. it combines with the oxygen of the atmosphere to form an oxide) and is chemically active, forming FERRIC or FERROUS compounds. The chief ores are (a) haematite or red kidney ore (red ferric oxide, Fe_2O_3) and magnetite or magnetic iron ore (Fe_3O_4), the purest and richest ores, having up to 70 and 72.4 per cent respectively weight of iron. They tend to occur in large masses associated with IGNEOUS or METAMORPHIC ROCKS; (b) the bedded ores of hydrated oxide of iron which include limonite (hydrated ferric oxide, $2Fe_2O_3 3H_2O$) and are usually very impure; (c) siderite (ferrous carbonate, $FeCO_3$); (d) sulphide ores of which iron pyrites (FeS_2) and copper pyrites ($CuFeS_2$) are the chief, although neither is an important source of metal, iron pyrites being more important as a source of sulphur and copper pyrites of copper. The impurities in the ore are important in processing, e.g. phosphoric iron ore needs special treatmet, because PHOSPHORUS makes the iron brittle. PIG IRON, STEEL.

Iron Age the era in human development (succeeding the BRONZE AGE) when IRON was smelted and used for tools, utensils and weapons. It probably began among the Hittites c.1400 BC, reaching southern Europe by c.1000 BC and Britain by 500 BC.

Iron Curtain a term introduced by Winston Churchill in a speech in 1946 describing the divide between the USSR and its associated communist states in eastern Europe on the one hand and the countries of western Europe on the other. The USSR, Poland, Czechoslovakia, Hungary, Romania, Bulgaria and East Germany were considered to be within the Iron Curtain, and sometimes Yugoslavia and Albania were included.

iron pan, ironpan HARD PAN.

ironwood a very hard, dense, heavy timber obtained from various trees and shrubs, e.g. EBONY, LIGNUM VITAE, QUEBRACHO.

irradiance the amount of radiant power (RADIANT FLUX) per unit area falling on a surface or object.

irredentism the policy or programme advocated by irredentists, i.e. those who wish to incorporate within their country the areas that are inhabited by people having linguistic or ethnic links with that country, but which lie beyond its borders in a neighbouring country; e.g. in 1937-8 the claim of the German irredentists that Sudetenland (a region with a largely German-speaking population, bordering Germany and lying within Czechoslovakia), should be incorporated in Germany.

irreversible transformation TRANSFORMATION.

irrigation the action of artificially supplying land with water to help the growth and productivity of plants. In addition to methods long used of bringing water by KAREZ, by CANAL-2 from a river or reservoir, and by BASIN IRRIGATION, more up-to-date methods include closed pipes (preventing loss by evaporation) and sprinklers producing artificial rain. PERENNIAL IRRIGATION.

isabnormal line term superseded by isanomalous line (ISANOMAL).

isallobar a line drawn on a map to join places undergoing equal change in ATMOSPHERIC PRESSURE during a given period, plotted to indicate the development and progress of a PRESSURE SYSTEM.

isanakatabar a line drawn on a map joining places with equal ATMOSPHERIC PRESSURE amplitudes. G.

isanomal, isanomalous line a line drawn on a map joining places with equal difference from the normal or average of any climatic element. ANTIPLEION, PLEION.

isarithm any line drawn on a map to link places having the same value or quantity. G.

isblink (Danish) in Greenland, the seaward splayed-out end of a stream of inland ice. G.

ishinna (Swedish) ice-film, ice-scum, the initial freezing on the sea surface, a thin film of ice, oil-like, formed in calm and even windy frosty weather. G.

island a piece of land entirely surrounded by water, a small island sometimes being termed an ISLE, very small an ISLET. By analogy applied to many other phenomena that are isolated, as an island, e.g. island site, a building site surrounded by roads; or a HEAT ISLAND. G.

island arc the disposition of an island chain in the form of an ARC, e.g. as in the Pacific ocean. DEEP, OCEANIC TRENCH, OPHIOLITE, PLATE TECTONICS. G.

island harbour a HARBOUR formed or mainly protected by islands. G.

islands authorities LOCAL GOVERNMENT IN BRITAIN.

isle a poetic, romantic form of the term ISLAND, rarely found in scientific writing, but retained in place-names, e.g. British Isles, Isles of Scilly. G.

islet a small ISLAND, AIT or EYOT. SKAR. G.

iso- (Greek, equal) a prefix (sometimes is- before a vowel) used in very many scientific terms, especially in geography. Over 70 are listed in Stamp and Clark (eds). *A Glossary of Geographical Terms*, 3rd edn., Longman, applying to various lines drawn on a map to link points having similar values or similar quantities, to which the general term applied is isopleth, isogram, isontic line or isoline. On the whole **isopleth** (ISO-PLETH-1) is preferred; isoline, a hybrid term, is disliked by some geographers. Individual entries for some terms with the prefix is- or iso- (ISALLOBAR, ISANOMAL, ISOCLINE, ISOMETRIC, ISOMORPHIC, ISONOET, ISOPIC, ISO-STASY, ISOSTATIC, ISOTHERMAL, ISOTIM, ISOTYPE) appear below.

Some standard isopleths (with the factor of similarity of their points specified) are: **isobar** barometric pressure; **isobase** elevation or depression of land; **isobath** depth of body of water; **isobathytherm** or **isothermobath** temperatures in a vertical section of sea water; **isobront** places experiencing thunderstorms at the same time; **isochime** or **isochimenal line** mean winter temperatures; **isochrone** distance travelled in equal time from a common starting point; **isocryme** coldest period of time; **isodapane** transport costs; **isodynamic line** intensity of terrestrial magnetism; **isoflor** plant species; **isogeotherm** temperature of subterranean points; **isogloss** vocabulary or pronunciation; **isogon** or **isogonic line** magnetic variation (AGONIC LINE); **isograd** rocks of the same FACIES-2; **isohaline** or **isohalsine** salinity; **isohel** amounts of sunshine; **isohyet** amounts of rainfall; **isohydrics** hydrogen-ion concentration; **isohyomene** wet months; **isohypse** or **isohyp**

contour; **isoikete** degree of habitation; **isokeph** cranial variation; **isokeraunic** frequency or intensity of thunderstorms; **isokinetic** wind speed; **isokrymene** minimum temperature at marine stations; **isomer** mean monthly amount as a percentage of average annual amount of rainfall; **isomesic** or **isomeisic** sediments formed under the same conditions; **isoneph** amount of cloudiness; **isonif** amount of snow; **isonotide** rain factors; **isopath** or **isopachyte** thickness of a selected geological bed; **isophene** seasonal biological phenomena, e.g. flowering date; **isophode** cost contour, e.g. transport costs; **isophore** load rates, transport; **isophotic** emission of quantity of light; **isophyte** height of vegetation; **isophytochrone** long growing season; **isopore** annual change of magnetic variation; **isopotential level** surface to which artesian water may rise; **isopract** population; **isopycnic** density; **isorad** radiation from rock; **isoryme** frost; **isoseismal** earthquake activity or intensity; **isoseismic** (CO-SEISMAL) or **isoseist** phase of earthquake wave at the same instant; **isostade** significant dates; **isotach** distance travelled in a specified period of time; **isotalantose** range of temperature between the mean of the hottest and that of the coldest months of the year; **isoterp** comfort; **isothere** summer temperature; **isotherm** temperature; **isothermombrose** summer rainfall; **isovol** ratio of fixed to volatile carbon in coal; **isoxeromene** atmospheric aridity. G.

isocline, isoclinal folding a fold which is so intense that the two limbs now incline or dip in the same direction and to an equal amount, i.e. at approximately the same angle. Where such a fold coincides with a ridge or valley, the terms isoclinal ridge or isoclinal valley may be used.

isolated state term applied by Von Thünen to a notional state serving as the basis for his model (VON THUNEN MODEL), a state completely cut off from the rest of the world, dominated by a very large town which serves as the sole market and is situated at the centre of a broad, featureless, uniform plain (an ISOTROPIC SURFACE), bounded by an uncultivated wilderness which prevents communication between the state and the outside world. The plain has a uniformly fertile soil, is not crossed by a navigable river or canal, and the ease of movement over it is uniform. Production and transport costs on the plain are everywhere the same. The farmers provide the large town with agricultural produce in exchange for the manufactured goods produced in the town. They themselves haul their produce to market along a close, dense network of converging roads of equal quality, at a cost directly proportional to the distance covered. All the farmers wish to maximize their profits, so automatically adjust the output of crops to the needs of the central market.

isometric *adj.* having equal measure.

isometric projection a drawing showing the three main axes of the subject inclined equally to the drawing surface, giving the appearance of a relief model viewed obliquely, but not in perspective. It is used in BLOCK DIAGRAMS, especially to illustrate geomorphological (GEOMORPHOLOGY) features. An isometric projection is true to scale in the horizontal axes, but the scale of the vertical is usually exaggerated in relation to them. Isometric graph paper is printed with one vertical and two diagonal lines, so that the drawing surface is divided into equilateral triangles. G.

isomorphic *adj.* in general, and in biology **1.** of the same or an analogous form **2.** in botany, applied to ALGAE and some FUNGI that have alternating generations which are vegetatively identical **3.** in chemistry, mineralogy, having shape or structure similar to that of another, usually due to a

similarity of composition **4.** in mathemetics, applied to two data sets, or two theories, which are precisely equivalent in form and in the nature and product of their operations, the elements of one corresponding with those of the other. GENERAL SYSTEMS THEORY, HOMO-MORPHIC, ISOMORPHISM.

isomorphism the state or quality of being ISOMORPHIC, e.g. specifically in biology, the apparent similarity of individuals of different races or species; and in mathematics, a one-to-one correspondence between data sets.

isonoet, isonetic an ISOPLETH-1 showing the incidence of a given level of intelligence.

isopic *adj.* applied in PETROLOGY to **1.** FORMATIONS-1 having the same fauna and flora although occurring in different provinces or in the same province at different times (if the LITHOLOGY and the general characteristics are also similar, the term HOLISOPIC is applied) **2.** two contemporaneous formations which are of the same FACIES. G.

isopleth 1. common use in geography, ISO- **2.** a graph showing the frequency of any phenomenon as the function of two variables **3.** in mathematics, a straight line on a graph joining corresponding values of the variables when one of the variables has a constant value.

isostasy, isostatic theory, isostatic adjustment (Greek, equal standing) isostasy, the condition of relative equilibrium of the earth's crust, accounted for by the theory that the surface features of the earth have a tendency to reach a condition of gravitational equilibrium. Isostatic theory maintains that where equilibrium exists on the surface of the earth, equal mass must underlie equal surface area. Thus under an elevated plateau there would be rocks of low density, e.g. GRANITE; under ocean basins the rocks would be of high density, e.g. BASALT. The instability of continental margins where high mountains are found close to ocean deeps is explained by underground movement of magma to effect the necessary adjustment (isostatic adjustment). GLACIO-ISOSTASY, ISOSTATIC, ISOSTATIC ANOMALY, PLATE TECTONICS. G.

isostatic *adj.* **1.** under equal pressure from all sides **2.** in HYDROSTATIC EQUILIBRIUM **3.** of, relating to, or characterized by, ISOSTASY.

isostatic anomaly a GRAVITY ANOMALY resulting from horizontal and vertical variations of density below the surface of the GEOID. G.

isosteric surface a surface with constant density in the ATMOSPHERE-1.

isothermal *adj.* **1.** having the same temperature **2.** without change of temperature **3.** relating to or showing a change in pressure or in volume at a constant temperature.

isothermal expansion the expansion of a given mass of GAS without change of temperature.

isothermal zone a layer in the ATMOSPHERE-1 now commonly termed the STRATOSPHERE.

isotim one of a series of equally spaced, concentric circular contours linking points where transport costs for a single element in a manufacturing process (e.g. delivery or procurement) are equal from a resource supply site, assuming equal transport costs per unit weight. COST SURFACE. G.

isotope one of two or more forms of an ELEMENT-6, having the same atomic number as the other forms but identified by small differences in atomic weight and, usually, by minute differences (due to mass) in chemical and physical properties. An isotope is named by the mass number with the name or symbol of the element, e.g. carbon 14, or ^{14}C, a radioactive carbon. In NATURAL RADIOACTIVITY one isotope changes very slowly but at a known rate into another that is more stable (HALF-LIFE), thus the proportion of one to the other present in a

sample (e.g. of organic remains, or in a geological sample) which is being investigated gives a measure of the age of the material which can be expressed in years (RADIOMETRIC AGE). RADIO-ACTIVITY, RADIOCARBON DATING, RADIOGENIC ISOTOPE, RADIO-ISOTOPE. G.

isotropic (Greek iso, equal; tropic, direction) *adj.* showing physical properties or actions equal in all directions. If the properties or actions shown are unequal, they are termed anisotropic. G.

isotropic surface a notional, unbounded, uniformly flat plain on which population density, purchasing power, transport costs, ACCESSIBILITY in all directions, etc. are kept uniform and unvarying. Christaller used this plain to show the theoretical distribution of CENTRAL PLACES in CENTRAL PLACE THEORY, specifying that CENTRAL GOODS should be bought from the nearest central place, all parts of the plain should be served by a central place (thus the COMPLEMENTARY REGIONS should cover the whole of the plain), there should be minimum movement of consumers, and no central place should make excess profits. In order to fulfil all these specifications the plain has to be divided into hexagonal complementary regions. URBAN HINTERLAND.

isotype a duplicate of the TYPE SPECIMEN of an animal or plant.

isthmus a narrow strip of land, with water on each side, connecting two larger land masses, e.g. two continental land masses, or a mainland and a peninsula. G.

ITCZ, ITF Intertropical Convergence Zone, Intertropical Front (sometimes termed the equatorial front or equatorial trough) a broad trough of low pressure, a zone rather than a front, defined more sharply over land than over the ocean, where the tropical maritime air masses converge, i.e. where the northeast trade winds and the southeast trade winds meet, broadly in the region of the equator, but moving north and south according to season. The air masses may be almost stagnant, the winds light and variable, hence the old name of belt of calms or DOLDRUMS. The air is very unstable, a factor in the heavy CONVECTION RAIN of the equatorial belt; and in this area shallow, slow-moving DEPRESSIONS-3 develop which may stray from the zone towards the poles, intensify and become fast-moving tropical revolving storms (CYCLONE). G.

iteration in mathematics, the repeating of an operation on the product of the operation, e.g. in LINEAR PROGRAM- MING an approximate solution to the given problem is drawn up, and this is repeatedly adjusted by the repeated application of a formula devised to obtain a successively closer approxima- tion, thus leading to the optimum solution.

ivory **1.** the dentine of any teeth **2.** the hard, cream-coloured dentine forming the tusks (long curved sharp teeth) of elephant and the canine teeth of hippo- potamus, walrus or narwhal. Ivory is still worked for ornaments and jewel- lery, but has been replaced by plastic materials for commercial use (e.g. for piano keys, formerly made from ivory).

ivory nut   VEGETABLE IVORY.

J

jaborandi the small dried leaflets of *Pilocarpus jaborandi* or the leaves of *P. microphyllus*, or the root of *Piper jaborandi*, plants native to South America, all of which yield alkaloids, including PILOCARPINE, used medicinally.

jack bean *Canavalia ensiformis*, an ANNUAL leguminous plant (LEGUMINOSAE) native to tropical America, widely grown in most tropical areas for GREEN MANURE or FODDER-1, but the seeds (white beans) are used for human food when other food is scarce. SWORD BEAN.

jack (jak) fruit the fruit of *Atrocarpus integrifolia* or *A. heterophyllus*, large trees native to southeast Asia and the Pacific islands. The fruit, very large and strong-smelling, eaten raw or cooked, is similar to BREADFRUIT.

jade a general term for three distinct, hard, translucent minerals, used for jewellery and ornaments. The most highly valued is jadeite (a PYROXENE, a silicate of calcium, magnesium and aluminium etc.). The second is nephrite (an AMPHIBOLE, a silicate of calcium and magnesium once worn as a remedy for kidney disease: Greek nephros, kidney), the third, the dark green chloromelanite, being the least prized. In their rarest and purest forms both jadeite and nephrite are white with a tinge of colour, the colours (green, rose, blue, brown, red-brown) being due to admixture of other minerals.

jama (Slavic) a pothole, a hole or cavity in limestone of which the bottom cannot be seen. It may pass, but not necessarily so, into a subterranean cavern. ABIME. G.

Japanese millet *Echinochloa frumentacea*, a MILLET widely grown in the warm parts of temperate regions, the grain being used for human food in Japan and Korea, the whole plant as a FORAGE CROP in North America.

Japan pepper PEPPER.

jasper an opaque QUARTZ, a form of CHALCEDONY, yellow, green, dark red or brown, used as an ornamental stone.

jebel, jabal, djebel (Arabic) a mountain, commonly used in place-names. G.

jessero a marshy or lake-filled DOLINA; a term now rarely used. G.

Jerusalem artichoke ARTICHOKE.

jet a very hard, black form of natural CARBON (LIGNITE) which occurs in England in the Upper Lias in Yorkshire, near Whitby. It polishes to a high lustre, and is used in jewellery and ornaments.

jetsam the cargo etc. thrown overboard from a ship in distress to lighten the load. FLOTSAM.

jet stream, jet-stream a high altitude, fast-moving air current, a few thousand kilometres in length, a few hundred kilometres wide, a few kilometres in depth, more or less horizontal, flattened, tubular, occurring in the vicinity of the TROPOPAUSE, usually blowing more strongly in winter than in summer. POLAR FRONT JET STREAM, SUBTROPICAL JET STREAM, TROPICAL EASTERLY JET STREAM, WESTERLIES.

jetty a structure built to project from the shore into a sea, lake or river to break currents or waves, to shelter a

HARBOUR, or to provide a landing stage.

jhil, jhīl, jheel, jhaor (Indian subcontinent: Bengali; Urdu-Hindi jhīl) a lake, swamp or marsh, especially a BHIL. G.

jilla, jila, zila (Indian subcontinent: Urdu; Hindi; Bengali) a district.

Job's tears (adlay) *Coix lachryma-jobi*, a millet-like tropical cereal, the local food grain grown in the Philippines and other parts of southeast Asia. MILLET.

joint in geology, a crack or fissure in a rock following a dominant direction along a line of weakness, usually transverse to the BEDDING, and produced by tearing apart under TENSION or shearing under COMPRESSION, but without dislocation (as in a FAULT). In stratified rocks (STRATIFICATION) the joints are usually at right angles to the bedding. If the BEDDING PLANES are well marked a number of parallel joints will divide up the rock bed into more or less regular blocks. There may be two or more sets of joints following different directions, in which case the most strongly marked set constitutes the master joints. A joint coinciding with a STRIKE is termed a strike-joint, with a DIP a dip-joint. EROSION and WEATHERING are helped by the weak surfaces of well-developed joints, which are also useful to the quarryman in stone extraction. CLINOMETER. G.

joint plane the plane, the flat surface, of a JOINT.

joint valley a valley course or part of a valley course controlled by major JOINT systems. G.

jökla mýs (Icelandic) glacier mice (GLACIER MOUSE). POLSTER. G.

jökula (Icelandic) a river flowing from a GLACIER. G.

jökull (Icelandic, pl. jöklar) a small ICE CAP, an ice-covered mountain. G.

joran a cold, dry wind, blowing at night from the Jura mountains towards the Lake of Geneva. G.

joule j **1.** the unit of energy and work in SI, the work done when the point of application of a force of one NEWTON is displaced in the direction of the force through a distance of one metre **2.** a unit of heat. KILOCALORIE.

Julian calendar the solar CALENDAR introduced by Julius Caesar, 45 BC, based on the calendar of ancient Egypt. It established a year consisting of 365 days divided into twelve months of 28, 30 or 31 days; an extra day was added every fourth year, making that year 366 days. This incorporated an error which amounted to 8 days in 1000 years (the Julian year being 11 minutes 10 seconds too long). This was taken into account when the GREGORIAN CALENDAR was introduced in 1582. INTERCALATION.

junction of rivers, ACCORDANT JUNCTION, DISCORDANT JUNCTION.

jungle a word brought home by the British from India where as jangala (Sanskrit) and jangal (Hindi and Marathi) it meant waste or uncultivated ground as opposed to cultivated land. Frequently such land was covered with scrub and tangled vegetation, including long grass, and so was the haunt of wild animals. It therefore has no precise meaning, and is best avoided in scientific literature (especially the term 'jungle-forests' applied to equatorial or hot wet forest). G.

juniper a member of *Juniperus*, family Pinaceae, a genus of hardy or nearly hardy EVERGREEN trees or shrubs native to cold or warm temperate regions of the northern hemisphere, usually with fragrant red wood (*J. virginiana* being used for lead pencil casing). In some species the wood yields fragrant essential oil, a medicinal diuretic oil is obtained from the leaves and shoots of others. The berries of some species are used as flavouring in cooking and in GIN.

junk a term of uncertain oriental origin, introduced into many European languages and applied to various types of wooden sailing vessels, especially those in the China seas. G.

Jurassic *adj* of or relating to the middle geological period of the Mesozoic era and the system of its rocks (GEOLOGICAL TIMESCALE) dating from some 180 to 135 mn years ago, when sediments of CLAYS and SANDS and CORAL REEFS were laid down in shallow seas, dinosaurs were at their peak, birds began to appear, and the flora included ferns and conifers. The rocks include CLAYS, LIMESTONES, SANDSTONES.

jute the coarse fibre of the bark of *Corchorus olitorius* or *C. capsularis*, family Tiliaceae, plants native to tropical Asia, cultivated especially in Bangladesh, used in making coarse canvas (hessian) for wrapping, sacking, cordage, paper, and carpet-backing. The fibre will not bleach, but can be dyed. C.

juvenile *adj.* youthful, immature.

juvenile relief a landscape with steep-sided valleys characteristic of the early stages of a CYCLE OF EROSION. G.

juvenile town URBAN HIERARCHY.

juvenile water magmatic water, intratelluric water, water from great depths of the earth reaching the earth's surface for the first time, as a result of volcanic activity, i.e. not the METEORIC water which is already present in the ATMOSPHERE-1 and HYDROSPHERE. G.

K

K **1.** KELVIN, the basic SI unit of temperature **2.** usually in italic, the symbol for a constant, e.g. in statistics **3.** K-VALUE.

kaatinga CAATINGA.

kabouk (Sri Lanka: Sinhalese) LATERITE, especially when cut out in blocks, hardened in the air, and used in building or the construction of minor roads. G.

kabuli gram CHICK PEA, PULSE.

kachchi, katchchi (Indian subcontinent: Panjabi; Urdu katchchi) SAILABA a flood plain, the tract of land actually flooded by a river. G.

kachha KUCHA.

Kaffir, Kafir originally Arabic, kaffer, an infidel, applied by Arabs to non-Muslim Africans of the east coast of Africa; later applied by the Dutch, the English and the Portuguese to the Bantu peoples of southeast Africa and to their language. The term is much disliked by African peoples, and is now rarely, if ever, used. G.

kaffir corn SORGHUM.

kainga PA.

kale *Brassica oleracea acephala*, family Cruciferae, a variety of cultivated cabbage, a hardy green vegetable of temperate climates, some kales being strongly flavoured. Curly kale is preferred for human consumption, the kale with large, erect juicy stem bearing large leaves being used for livestock feed. BRASSICA.

kallar (Indian subcontinent: Panjabi) saline soil, REH. G.

kame an old Scottish term, perhaps connected with comb, applied to a long ridge suggesting a cock's comb. Now an imprecise, unspecific term applied to any ridge or mound of poorly sorted water-laid materials (glacial sands and gravels) associated with former ICE FRONTS. ESKER, MORAINE, PERFORATION DEPOSIT. G.

kame-and-kettle country an undulating landscape consisting of KAME MORAINES and shallow depressions. KETTLE.

kame complex an assemblage of KAMES. G.

kame moraine an end or TERMINAL MORAINE that includes many KAMES, the kames being more likely to consist of TILL than of water-laid materials. G.

kame terrace a terrace formed of sand and gravel deposited by a stream of MELT-WATER in the depressions between a GLACIER and the sides of its trough. FOSSE. G.

kampong (Malay) **1.** in Malaysia, a cluster of buildings making a large homestead or small hamlet, together with the surrounding gardens and, sometimes, fish-breeding ponds. Hence kampong horticulture, cultivation carried on in such a setting **2.** in Java, a village. G.

kanat KAREZ, QANAT. G.

kankar, kunkur (Indian subcontinent: Urdu-Hindi) the nodules or concretions of CALCIUM CARBONATE occuring in LIMESTONE in the older alluvium of the Indo-Gangetic plain. They are sometimes collected and used for surfacing paths or secondary roads, or for burning to make LIME-1. G.

Kant, Immanuel German philospher,

1724-1804. To sketch broadly some of his theories/beliefs: in METAPHYSICS he asserted that the ultimate nature of reality (things in themselves) is inaccessible to the human mind and that phenomena (PHENOMENON-3) are the only knowable things (PHENOMENAL-ISM); that the mind itself synthesizes and impresses its forms of sensibility (space and time) on the original data of the senses and orders them in categories of thought, e.g. substance and cause (Kant uses the term 'categories' in the sense of metaphysical concepts). He disagreed with transcendent metaphysics on the grounds that *a priori* thought provides knowledge about reality only if it is applied to the data of experience. He rejected POSITIVISM on the grounds of the crudeness of EMPIRICISM. In ETHICS he maintained that an absolutely valid law by which specific moral values can be tested can be determined by the principle of categorical imperative, i.e. the obligation to discover for every case of conscience a solution which constitutes a universal law, i.e. which will hold good for all people for all time.

kaoliang a general term for grain SOR-GHUM, extensively grown in northern China, a staple food (with SOYA BEANS) of humans and livestock. A strong spirit is prepared from the grain, and the large stalks of the plant provide thatching material. MILLET. G.

kaolin (Chinese kao, high; ling, hill: Kaoling, a mountain in northern China) china clay, a fine, white CLAY of which KAOLINITE is the main constituent, occurring especially in pockets in GRANITE masses, resulting from the decomposition of FELDSPARS (especially the ORTHOCLASE) by HYDRO-LYSIS and by ascending gases and vapours (mainly CARBON DIOXIDE and superheated steam) from a deep-seated MAGMA. It is used in making ceramic ware (CHINA, PORCELAIN), paper, pharmaceuticals, rubber and

cosmetics. The term is derived from the name of the mountain in China (Kaol-ing) from which it seems originally to have been obtained. KAOLINIZATION, SUPERHEATING.

kaolinite $Al_2Si_2O_5(OH)_4$, a fine crystalline form of hydrated aluminium silicate, formed by the decomposition of FELDSPAR. KAOLIN, KAOLINIZA-TION. G.

kaolinization, kaolinisation, kaolisation the process by which GRANITE is attacked and its constituent feldspars with other aluminium silicates altered by heated gases (PNEUMATOLOYSIS) and waters (HYDROTHERMAL, SUPER-HEATING) to a soft, white CLAY (china clay, KAOLIN). G.

kapok vegetable down, the silky fibres of the seed pods of silk cotton trees, *Bombax ceiba*, native to tropical America, *B. malabaricum* to the Indian subcontinent, *Eriodendron anfractuo-sum* to the Indian subcontinent and to southeast Asia. The hollow fibres, too short for spinning, are buoyant, resil-ient and waterproof, and are thus useful in making life-jackets, and in filling cushions, mattresses etc.; but in those uses they face competition from synthetic fibres. The oil from the seed is used in soap-making. C.

Kar (German, pl. Kare) a CIRQUE; not to be confused with KARRE or KAR-RENFELD. G.

karaburan black buran, a strong north-east wind laden with dust and sand blowing in daytime in the arid Tarim basin in central Asia, darkening the sky, changing river courses by deposit-ing sand, and carrying dust particles great distances to settle as LOESS. G.

Karakul, karakul 1. a breed of broad-tailed, brown-haired sheep originating from the arid regions of central Asia and introduced into the semi-desert areas of south and southwest Africa 2. the skin of the newborn Karakul lamb, glossy black, with tightly curled hair, yielding valuable furs (ASTRAKHAN,

BROADTAIL, PERSIAN LAMB). The ewes survive under exceptionally hard conditions if they do not suckle their lambs. C.

kārez term applied in Baluchistan to the QANAT (kanat) of Iran and the FOGGARA of north Africa: an almost horizontal underground, hand-engineered irrigation channel or tunnel dug from the arid plains to tap water at the foot of a nearby hill range, the water flowing through by gravity. G.

karite *Butyrospermum parkii*, SHEA BUTTER-NUT.

karling a term, not in general use, applied by T.Griffith Taylor to a dome into which CIRQUES have cut deeply, e.g. Snowdon, in Wales; Mount Anne, in Tasmania. G.

karoo, karroo (Afrikaans karoo, English spelling sometimes karroo) **1.** a plateau in southern Africa between the Swartberge and Nuweveldberge, covered with semi-desert vegetation of small shrubs **2.** the vegetation in this region, extending into the Little Karoo, south of the Swartberge, and the Northern Karoo, Upper Karoo, or Karroid plateau, north of the Nuweveld range. KARROID VEGETATION. G.

Karre (German, usually in pl., Karren) a channel or furrow varying in depth from a few millimetres to over a metre, and separated from others by ridges, caused by solution on limestone surfaces. CLINT, KARST, LAPIE. G.

Karrenfeld (German) a surface cut with and dominated by Karren (KARRE). G.

karroid vegetation (South Africa) in botanical literature and on some maps, the vegetation of the country north of the Nuweveld range in South Africa, generally termed karooveld by local farmers. G.

karst (German form of Slavic kras, a bleak waterless place; French causses; Italian carso) originally the barren limestone plateau of Istria, between Carniola and the Adriatic, where nearly all the natural drainage is underground and where there are bare, limestone ridges, caverns, sinks and underground drainage caused by rainwater which, charged with CARBON DIOXIDE from the atmosphere, dissolves the CALCIUM CARBONATE in the porous LIMESTONE, producing an uneven landscape. Karst is now applied to any area of similar limestone or dolomite country. Only a few of the many specialized terms applied to karst from the Slavic, French, German and English languages have been included in this dictionary. COCKPIT KARST, CONE KARST, FLUVIOKARST, GLACIOKARST (NIVALKARST), HOLOKARST, KEGELKARST, MEROKARST, MICROKARST, POLYGONAL KARST, TOWER KARST, TURMKARST, TROPICAL KARST. G.

karstic *adj* of or pertaining to karst country. The more usual adjective is karst. G.

kās (Indian subcontinent: Panjabi) a ravine carrying water after the rains. G.

kasba (Arabic, pl. ksabi) **1.** a town or small city **2.** in Algeria, the citadel (fort) in the town **3.** in Morocco, al-qasba (Arabic) the citadel or fortified refuge of a city; the old Muslim quarter of a modern city. QASR. G.

kashmir wool CASHMERE.

katabatic wind a drainage wind, a cold wind that blows down a valley or slope, especially at night, when dense cold air, cooled by radiation at higher levels, flows downhill by gravity, behaving much like a stream of water. ANABATIC, FROST POCKET. G.

kataclastic rock, cataclastic rock a CLASTIC rock produced by the fracture of pre-existing rocks as a result of earth-stresses, e.g. CRUSH-BRECCIA. G.

kataclastic structure, cataclastic structure structure in rock caused by great mechanical STRESS-1, the constituent minerals generally being deformed and granulated. G.

katafront a COLD FRONT in which

warm air flows down over a wedge of cold air. FRONT.

katamorphism the destructive processes of METAMORPHISM in contrast to anamorphism, the constructive processes. G.

kauri 1. a member of the genus *Agathis*, EVERGREEN trees native to Australasia and parts of southeast Asia. *A. australis* is of economic importance: it is a magnificent timber tree with a columnar trunk of almost uniform diameter, free from branches when fully mature, a source of Agathis resin (kauri gum, also termed copal), used for varnishes **2.** the timber or the resin of *A. australis*.

kava *Piper methystricum*, a shrub native to and now commonly cultivated in the Pacific islands, the root being used to make a fermented, intoxicating alkaloid liquor.

kavir (Iran) a salt marsh, a PLAYA. G.

kayak an Eskimo canoe, traditionally made of sealskin, now usually of canvas, stretched over a light frame to enclose and cover the user.

Kegelkarst (German) the term now applied internationally to several types of tropical humid KARST characterized by COUPOLES, PITONS and TOURELLES with pits on the surface, sometimes termed CONE KARST in translation. COCKPIT, COCKPIT KARST. G.

keld (England: Cumbria and Yorkshire dialect, common in place-names) **1.** a spring, fountain or well **2.** a deep, still stretch of a river. G.

kelp any of the large brown SEAWEEDS, especially those used as a fertilizer, and from which IODINE used to be obtained. ALGIN is extracted from the giant kelp.

kelvin K, the basic SI unit of temperature, defined from the triple point of water (the point at which water, ice and water vapour are in equilibrium), valued at 273.16K, ice point (ABSOLUTE ZERO) being 273.15K. The value of the degree in the KELVIN SCALE is

the same as that of the degree in CENTIGRADE. A temperature expressed in K (the symbol °K and the term degree kelvin have been superseded by K or kelvin) is equal to the temperature in °C less 273.15°C (commonly rounded to 273°C); and to express °C in K it is necessary only to add 273°C to the Centigrade value (e.g. $-3°C = 270K$).

Kelvin (K) scale a TEMPERATURE SCALE with 1K (one KELVIN) equal to 1°C (CELSIUS, CENTIGRADE) but with an ABSOLUTE ZERO temperature calculated to be $-273.15°C$ or $-459.4°F$ (FAHRENHEIT). The advantage of the Kelvin scale is that it has no negative quantities. It is thus especially valuable if one is dealing with very low temperatures.

Kelvin wave a tidal system in an approximately rectangular sea area, the tidal range being greater on the right of the direction of a PROGRESSIVE WAVE, decreasing on the left, e.g. in the English Channel the tidal range on the south coast of England is less than that on the north coast of France. AMPHIDROMIC POINT, AMPHIDROMIC SYSTEM, OSCILLATORY WAVE THEORY OF TIDES.

kenaf *Hibiscus cannabinus*, a fibrous plant native to southeast Asia, tolerant of temperate climatic conditions, widely cultivated in southeast Asia for its fibre, used in canvas and cordage.

kerangas a term originally applied to a type of podzolic soil (PODZOL) in Borneo, but extended to apply to the heath-like vegetation growing on it. PADANG. G.

kermes a member of the family Kermesidae, a scale insect occurring on the kermes oak, *Quercus coccinea*, a small evergreen tree of the Mediterranean region. The dried bodies of the females of this insect provide a red dye.

kernel 1. the soft, innermost part of a SEED, usually edible. NUT.

kerogen insoluble organic material present in sedimentary rocks, apparently derived from plant remains, but having a content of oxygen and nitrogen higher than that of PETROLEUM.

kerosene PARAFFIN.

kettle, kettle-hole, kettle-lake kettle seems originally to have been a local term applied to a pothole in a river, then, as a 'giant's kettle', to a pothole formed by whirling stones in a stream under a glacier. Later it came to be applied to a circular hollow in a stretch of glacial sands, gravels and clays, caused by the former presence of a great detached block of ice which had eventually melted. Such hollows became filled with water to form kettle-lakes (CAVE-IN LAKE), and the drifts in which they occurred became known as kettle-drift or kettle-moraine. G.

kettle-drift a mound or ridge of gravelly DRIFT-1 formed by water at or beyond the margin of the ice. G.

kettle-moraine a TERMINAL MORAINE pitted with many kettle-holes. G.

key, kay, cay (Spanish cayo, a shoal or reef) a low sand and coral island, sandbank or reef, lying a little above high tide, dry at low tide, a term used especially in the West Indies and Florida, sometimes appearing in place-names, e.g. Key West. SAND KEY. G.

Keynesianism the economic theory of John Maynard Keynes, 1883-1946, in part of which he states that a condition of unemployment may continue for long periods, even indefinitely, unless a government steps in to remedy it by spending to suppplement a deficient private sector demand. AGGREGATE DEMAND, MONETARISM.

key village, king village a minor centre with facilities (e.g. a primary school, a village hall) to serve even smaller villages and hamlets, but reliant on major centres for other facilities. G.

khad (Indian subcontinent: Panjabi) a torrent in the hills. G.

khādar, khaddar, khuddar (Indian subcontinent: Urdu-Hindi) 1. new allu-vium, in contrast to BANGAR 2. a low-lying area of new alluvium liable to river flooding. DHAYA. G.

khaderā, khuddera (Indian subcontinent: Panjabi) a deep ravine, eroded by rainwater. G.

khāl (Indian subcontinent: Bengali) 1. a narrow natural water channel 2. a sluggish creek of the lower Ganga delta. G.

khamsin, khamseen (Arabic, fifty) a hot, dry, often dust-laden southerly wind known elsewhere as ghibli, samiel or leveche, which blows intermittently for some fifty days in March, April, May from the deserts of the south across Egypt and the southeast Mediterranean, commonly after a HEAT WAVE, and often becoming humid in passing over the Mediterranean sea. The SIROCCO is similar, but warmer. G.

khari (Indian subcontinent: Bengali) a small deep stream of local origin flowing in Barind, Bangladesh. G.

kharif, khareef (Indian subcontinent: Urdu-Hindi; Sudan: Arabic) 1. kharif, the rainy season of the northern Indian subcontinent 2. the crops planted during the monsoon or rainy season and harvested at the end of the rains in the autumn; hence applied to the autumn or winter harvest in contrast to RABI, the spring harvest 3. khareef (Arabic, autumn), the rainy season in northern Sudan. G.

khoai (Indian subcontinent: Bengali) the lateritic areas of west Bengal. G.

khor (Sudan: Arabic) an intermittent stream. G.

khud kasht (Indian subcontinent: Urdu) land cultivated by the owner himself. G.

khushkābā (Indian subcontinent: Panjabi) dry unirrigated land. G.

kibbutz, kibutz (Israel: Hebrew, a gathering, pl. kibbutzim) a form of COL-LECTIVE or communal rural settlement in modern Israel, with egalitarian communal ownership of the land and collective economic and social organization. The family is not recognized as an important social and economic unit, the children being considered to be the responsibility of the whole community. Originally the kibbutzim were devoted solely to farming, but many now produce manufactures. LAND TENURE, KVUTZA, MOSHAV. G.

kibbutznik a member of a KIBBUTZ.

kid a young GOAT or ANTELOPE-1, yielding fine skin for leather, used in clothing, baggage manufacture etc.

kidney bean, French bean a general term for any of the varieties of *Phaseolus vulgaris*, a tender annual plant (some climbing) probably native to South America, now widely grown in temperate, subtropical and tropical areas for the sake of its edible pods and seeds, usually cooked. BEAN.

kidney iron ore HAEMATITE.

killas an imprecise term applied by Cornish miners to low-grade slates, micaceous schists, resulting from the METAMORPHIM of CLAYS or SLATES near GRANITE masses. G.

kilo k, prefix, a thousand, attached to SI units to denote the unit $\times 10^3$, i.e. KILOCALORIE, KILOGRAM (one thousand GRAMS), kilometre (one thousand METRES). CENTI-, CONVERSION TABLES, HECTO-, MILLI-.

kilocalorie a unit of 1000 calories or 1 Calorie, now replaced by the JOULE. CALORIE.

kilogram, kilogramme kg, the basic SI unit of weight, 1000 GRAMS (2.2046 lb) being defined as the mass of a standard piece of platinum-iridium alloy kept in the Bureau International des Poids et Mesures at Sèvres, near Paris.

kilometre km, an SI unit of length, 1000 METRES (0.62 mile, or 3280.84 feet). In measures of area one square kilometre equals 100 HECTARES each of 10 000 square metres. CONVERSION TABLES.

kimberlite a DIAMOND-bearing ULTRABASIC IGNEOUS ROCK containing MICA and OLIVINE, filling volcanic PIPES-3

in the Kimberley district of South Africa, termed HARDEBANK at depth, changing near the surface to the softer BLUE GROUND which oxidizes at the surface to YELLOW GROUND. G.

kinesis a random movement of an organism in response to a stimulus, the direction of the locomotion being unspecific, unrelated to the position of the stimulus. TAXIS.

kinetic energy the energy of a moving mass associated with its speed and equal to half the product of the mass and the square of its velocity.

kingdom 1. the territory over which a MONARCH has authority **2.** a state with a monarchical government. MONARCHY-2 **3.** one of the highest ranking groups in the CLASSIFICATION OF ORGANISMS, i.e. ANIMAL-1, PLANT-1, and some authorities add PROTISTA; and also of MINERALS.

king village KEY VILLAGE.

kinship in anthropology, the condition of being tied by close relationship, by descent from a common ancestor (defined as someone standing in the social position of a mother or a father, and not necessarily a forbear in the biological sense).

kinship system a social system, varying in form, based on the relationship ties traditionally recognized and accepted in a culture, and the rights of, and reciprocal obligations between, its members.

kipuka (Hawaiian) an island of the old land, frequently with vegetation, left within a lava flow; a STEPTOE. G.

kirktoun (lowland Scots, church town) a hamlet or small village consisting of a group of farms with a church. CLACHAN, FERMTOUN. G.

kitchen-midden, kitchen midden a mound containing the domestic refuse (mainly shells of edible molluscs, animal bones, tools, etc.) of early people, an obsolescent term superseded by shell mound. G.

kivas STRAATE.

kiwi (Maori) **1.** an apteryx, order Aptergiformes, a goose-sized flightless New Zealand bird **2.** a familiar term for a New Zealander. G.

kiwi fruit Chinese gooseberry, the brown-skinned, egg-shaped, succulent fruit of *Actinidia chinensis*, a vigorous climbing plant, successfully cultivated by commercial growers in New Zealand, who gave the fruit the name kiwi, and marketed it as such. KIWI-2.

kizshi (Baluchistan) a nomad's tent. G.

kleptoparasite an animal which obtains food by stealing it from the catches of animals of another species.

klint (Swedish) **1.** a steep cliff or steep terrace or steep edge of a plateau, a GLINT, e.g. a limestone cliff extending from the coast of the Baltic sea to the southern margin of Lake Ladoga. In countries bordering the Baltic sea the term is also applied to a nearly-vertical, free mountain-wall or abrasion precipice at least some metres in height and a hundred or more long **2.** an exhumed BIOHERM or CORAL REEF. G.

Klippe (German, crag or granite tor) a rock outlier representing a remnant, isolated by DENUDATION, of an overthrust rock mass or NAPPE. G.

klong (Thailand: Thai) a waterway, partly natural, partly artificial, in Thailand, especially in Bangkok. G.

kloof (Afrikaans) a ravine or gorge, sometimes applied to a pass. G.

knickpoint, knick-point, knick point, nickpoint (German Knickpunkt, pl.Knickpunkte) a break of slope, particularly one in the long PROFILE of a river valley, occurring when a relative lowering of the sea-level (NEGATIVE MOVEMENT)compels the river to regrade its course to the new sea-level. In doing this the river makes a new curve which, by headward EROSION, cuts across an earlier one. The junction is marked by a break of slope (the knickpoint, or rejuvenation head) which, owing to continued erosion, moves progressively upstream. REJUVENATION. G.

knob a rounded hill or mountain summit. G.

knob and basin topography, knob and kettle the irregular surface of a TERMINAL MORAINE, marked by low KNOBS and lake-filled hollows (KETTLES), more common on terminal moraines formed by ICE CAPS than those formed by ICE STREAMS.

knoll, knowe, know, knowle (Old English cnoll) a more or less rounded small hill. G.

knot **1.** a unit of speed, one nautical mile (standardized at 1.852 km) per hour, derived from the original use of pieces of knotted string fastened to the logline trailed from sailing vessels, the number of knots being measured against a period of time indicated by a sand-glass. It is tautological to refer to so many 'knots per hour'. LOG, MILE **2.** a complex of mountains, especially one where several ranges meet and the arrangement is irregular, e.g. the Pamir Knot. G.

knotted wrack *Ascophyllum nodosum*, a species of SEAWEED commonly occurring on sheltered beaches of the British Isles, important as a source of ALGINATES, used for thickening soups, as an edible emulsifer, as a gelling agent (e.g. in confectionery) and, made into film, as sausage 'skin'.

koa *Acacia koa*, a local name for a Hawaiian tree with close-grained wood, highly valued for cabinet making.

koembang KUMBANG.

kohlrabi any of the varieties of CABBAGE (BRASSICA) with the stem base swollen and resembling a root, hence the popular name, turnip-rooted cabbage, used for livestock feed and, cooked, for human consumption. G.

kola tree, cola tree a member of *Cola*, family Styerculiaceae, a genus of trees native to the drier parts of tropical west Africa, the fruit of which (kola or cola nut), containing CAFFEINE and THEOBROMINE, is used as a tonic, an aid to digestion and an ingredient of soft drinks.

kolkhoz (Russian contraction of kollektivnoe khoziaistvo) a collective farm in the USSR. COLLECTIVE FARMING, SOVKHOZ. G.

kolla DEGA

kop, koppie (Afrikaans kopje, diminutive of kop; German Kopf, head) an isolated prominent hill (INSELBERG), or a row of hills (pl. koppies) often composed of old volcanic rock, characteristic of the interior parts of South Africa. Kop, without the diminutive ending, is in some cases applied to a large mountain, especially as part of its proper name.

Köppen's climatic classification a classification devised by Vladimir Peter Köppen, 1846-1940, born at St Petersburg, but lived from 1875 to 1919 in Hamburg, hence the initial W (for Wladimir) he is commonly given. He was an official of the Deutsche Seewarte, but he devoted his life to climatic-vegetation studies, the first version of his World Climatic Regions being published in 1900, later revised in collaboration with R. Geiger. He based his climatic classification on the climatic needs of certain types of vegetation, and identified five major groups, A to E, to which he added H, the mountain zone. The major groups are: A, tropical zone (12 months with a temperature exceeding 20°C); B, subtropical zone (4 to 11 months with temperature exceeding 20°C, and 1 to 8 months ranging between 10°C and 20°C); C, temperate zone (4 to 12 months with temperature between 10°C and 20°C); D, cold zone (1 to 4 months with temperature between 10°C and 20°C, and 8 to 11 months below 10°C); E, polar zone (12 months with temperature below 10°C). These major climatic groups were subdivided to take account of refinements of rainfall and temperature characteristics, expressed by lower case letters (10°C = 50°F; 20°C = 68°F). HEKISTOTHERM, MEGATHERM, MESOTHERM, MICRO-THERM, THORNTHWAITE'S CLIMATIC CLASSIFICATION.

koum KUM.

koustar, kustar industries (Russian koustar, kustar, one earning a living by HANDICRAFT in the home) the old peasant industries (COTTAGE INDUSTRY) of Russia. G.

kraal (Afrikaans derived from Portuguese corral) **1.** an enclosure for cattle **2.** an African village. G.

kraaling (South Africa) the putting of sheep or cattle into a KRAAL-1 at night as a protection against wild animals. G.

krans, krantz, krantzes (Afrikaans) precipitous rock-face on a mountain. G.

krasnozem a red, ZONAL SOIL developed under Mediterranean conditions, not to be confused with TERRA ROSSA, an AZONAL SOIL. G.

kratogen CRATON.

kremlin a fortress or CITADEL of a Russian city, specifically the Kremlin, the twelfth century citadel of Moscow, seat of the government of the USSR, with residences, administrative offices, cathedrals and gardens within its walls.

krotovina, crotovine (Russian) in soil science, a filled-in animal burrow in the soil, e.g. a filled-in worm burrow. G.

kuala (Malay) a confluence or estuary; common in place-names, e.g. Kuala Lumpur. G.

kucha, cutcha, kachha, kacha (Indian subcontinent: Hindi, various other Anglicized spellings) *adj.* poor, weak, temporary, built of mud or earth, as opposed to PUKKA, e.g. a kucha well, one not lined with masonry. G.

kum (Tadzhik and many Turkish languages, sand; French koum) any of the sandy deserts of central Asia, equivalent to the Saharan ERG. In Russia kum appears in place-names. G.

kumbang, koembang a southeast FOHN-type wind of Java. G.

kumquat, cumquat a member of *Fortunella*, a genus of Asian shrubs, not a true CITRUS but bearing acid fruit resembling an ORANGE. It is resistant to cold and can be grown beyond the northern limit of true *Citrus* trees. The fruit is used mainly in conserves and pickles, etc.

kumri SHIFTING CULTIVATION in Kanara, India. G.

kunkur KANKAR. G.

kursaī (Indian subcontinent: Pashto) a shepherd's hut. G.

kurtosis in statistics, a shape characteristic of a FREQUENCY DISTRIBUTION that reflects the pointedness of the peak and the length of the tails. FREQUENCY CURVE.

kusam *Schleichera oleosa*, a HARDWOOD tree native to southeast Asia, with red timber and seeds that yield macassar oil.

K-value in CENTRAL PLACE THEORY, a value given by Christaller to a CENTRAL PLACE, to describe its place in, and the nature of, the hierarchy (CENTRAL PLACE HIERARCHY). The K-value expresses the total number of central places at a certain level in the central place hierarchy served by a central place at the next highest order in the system. The value includes the higher order place itself, e.g. in a K-3 hierarchy, the higher order place serves two adjacent lower order places (i.e. two places and the place itself). ADMINISTRATIVE PRINCIPLE, MARKETING PRINCIPLE, TRAFFIC (TRANSPORTATION) PRINCIPLE. G.

kvutza, kvutzah (Israel: Hebrew, a group) a collective village, an agricultural group in Israel, a term now superseded by KIBBUTZ. COLLECTIVE FARMING, MOSHAV. G.

kwin (Burmese) a small division of land. G.

kyaung (Burmese) a monastery, more correctly HPOONGYI-KYAUNG. G.

kyle (Scottish, from Gaelic cael) a narrow channel or strait between an island and the mainland, or between two islands, e.g. the Kyles of Bute. G.

L

laagte, laegte (Afrikaans) broad shallow hollows between widely rising ground in a gently undulating landscape. G.

labelling a social process by which individuals or groups classify and categorize social behaviour in others, e.g. a particular group or area may be reputed to have socially undesirable characteristics (e.g. being characterized by criminality) and is stigmatized, given a disparaging name. People who are not 'socially undesirable' but who live in such a stigmatized area may have difficulty in finding work or obtaining credit. It has been suggested that labelling affects the behaviour of the labelled, in that people who are labelled come to see themselves in terms of the label, and behave accordingly.

lablab (Arabic lubia) *Dolichos lablab*, the bonavist bean or hyacinth bean, a PERENNIAL leguminous (LEGUMINOSAE) plant native to Asia, grown usually as an annual in the Indian subcontinent, Egypt, Sudan and southeast Asia for its pods (cooked) and for its ripe seeds (beans), poisonous when raw, cooked as a split pulse. The plant leaves are used for animal feed.

labour 1. workforce 2. in economics, work as a factor of production. METROPOLITAN LABOUR AREA, PRODUCTION FACTORS.

labour-extensive *adj.* needing a very small work force to achieve a very high output. CAPITAL INTENSIVE, LABOUR INTENSIVE.

labour hoarding the action of employers in holding on to their skilled work people in times of recession, even if there is insufficient work for them to do, as a form of 'insurance' for the future when trade picks up and competition for skilled workers makes recruitment difficult.

labour-intensive *adj.* needing the efforts of a large work force for increased productivity or higher earnings, as opposed to CAPITAL-INTENSIVE. LABOUR-EXTENSIVE.

labour market inefficiency the simultaneous existence of job vacancies and unemployment.

labour power in Marxism, the ability to work, or the commodity that workers sell, the exchange value being determined by the socially necessary labour (LABOUR THEORY OF VALUE-2) needed for subsistence, i.e. the cost of production and reproduction of labour itself. CONSTANT CAPITAL, MEANS OF PRODUCTION, PRODUCTIVE FORCES, REPRODUCTION OF LABOUR POWER, SURPLUS VALUE.

labour reserve the pool of under-used labour, often immobile and existing, for example, in depressed industrial areas.

labour theory of value 1. in CLASSICAL ECONOMIC THEORY, a theory proclaimed by Adam Smith and David Ricardo, that any two products will exchange one with another in proportion to the amounts of labour needed to make them, i.e. that VALUE-1 is the product of the expenditure of labour. The part played by capital in production is explained by assuming that the amount of capital used per unit of

330

labour in making every product is constant, or by treating capital equipment as stored-up labour (interest and SCARCITY RENTS being ignored) **2.** in Marxian economics the theory is similar, one of the main tenets of Marxism being that value can be created only by the expenditure of human labour. Thus the price of a commodity should be the amount of labour time needed to produce it under normal conditions. The labour so required is termed socially necessary labour. CONSTANT CAPITAL, LABOUR POWER, VARIABLE CAPITAL **3.** in NEOCLASSICAL ECONOMICS it is asserted that capital and land as well as labour contribute to the production process, so that they also are entitled to a return and should be reflected in the price of a commodity.

lac a resin secreted by the lac insect, *Laccifer lacca*, a parasite on a number of host-trees of the Indian subcontinent. It is melted and made into sheets of shellac, used in the manufacture of varnishes and lacquers, and formerly of gramophone records.

laccolith, laccolite a mass of IGNEOUS ROCK intruded along the BEDDING PLANES of SEDIMENTARY ROCKS, like a SILL but swelling out to form a lens-shaped mass, the flat base being concordant with the strata into which it is intruded, the upper surface swelling out in the shape of a dome so as to cause the overlying strata to arch over it. Sometimes a laccolith may be more complex, with several masses one above the other, so that a section through the whole looks like a cedar tree with spreading branches; hence the term cedar-tree laccolith. PHACOLITH. G.

lacquer LAC.

lacustrine *adj.* of or pertaining to a LAKE, hence applied, e.g. to deposits laid down in a lake; or to terraces on lake margins left when the area of the lake diminishes. G.

Lacustrine civilization, Lacustrine period LAKE-DWELLING.

lacustrine delta a DELTA spreading into a LAKE, built by a stream flowing into the lake.

lacustrine plain a plain occupying the site of a former LAKE. G.

ladang (Indonesia) SHIFTING CULTIVATION in the Malay archipelago and, particularly, in Indonesia. CAINGIN. G.

ladder farm striped farm, a FARM with a succession of small, ribbon-like fields. G.

LAFTA Latin American Free Trade Association, headquarters Montevideo, an organization of some South American states, established in February 1961 by Argentina, Brazil, Chile, Mexico, Paraguay, Peru and Uruguay, with Colombia and Ecuador (October 1961) and Venezuela (September 1966) associated by treaty, to promote economic cooperation. It has two subgroups: the Andean Group, established May 1969, comprising Bolivia, Chile, Colombia, Ecuador and Peru (Venezuela, expressing interest, has not yet signed); and the River Plate Association, comprising Argentina, Brazil, Paraguay and Uruguay.

lag deposits, lag gravel coarse residual materials, sorted and remaining **1.** on a stream bed **2.** on the surface of a hot desert after wind-dispersal of finer materials. G.

lag fault a low-angled FAULT resulting from movement in a series of rocks, those nearer the top moving less than (lagging behind) those nearer the bottom of the series.

lagg (Swedish) marginal FEN. G.

lagoon 1. a shallow area of salt or brackish coastal water completely or partly separated from the open sea by some more or less effective obstacle, such as a low sandbank (BARRIER BEACH, BARRIER ISLAND) or a CORAL REEF **2.** the sheet of water enclosed in an ATOLL.

lahar (Indonesia) a flow of mud arising from volcanic activity, the fine-grained volcanic material being impregnated with water derived from heavy rainfall during the eruption; or from the sudden emptying of a CRATER LAKE; or from melted snow (e.g. during the eruption of a snow-capped volcano). G.

laissez-faire, laisser-faire (French, let act, i.e. let things alone) a term originated by the PHYSIOCRATS, the philosophy or the practice of the avoidance of planning, particularly, in economic affairs, as expressed in the avoidance of government control. The doctrine is based on the theory that general good and harmony will ensue if each individual is allowed to work for his or her economic advantage in a freely competitive economic system, a theory supported by Adam Smith, David Ricardo and others. CLASSICAL ECONOMIC THEORY.

lak (Indian subcontinent: Pashto) a pass. G.

lake a broad, general term applied to an accumulation of water lying in a depression in the earth's surface, normally to a sheet of water of considerable size, but sometimes to a small artificial ornamental feature, e.g. in a PARK-2,3; if very large, natural and saline, the term sea is commonly used (e.g. Caspian Sea, Sea of Aral, Dead Sea), if very small and natural, POND or POOL are used. An inflowing and/or outflowing river may or may not be present; and a lake may not necessarily be a permanent body of water (e.g. Lake Eyre, PLAYA). The CLASSIFICATION OF LAKES is usually based on the origin of the depression which they occupy (e.g. BARRIER LAKE, CRATER LAKE, GLACIAL LAKE, TROUGH LAKE). LACUSTRINE, LIMNOLOGY. The term is also applied to a fairly large accumulation of ASPHALT, a viscous substance, lying in a depression in the earth's surface. G.

lake breeze SEA BREEZE.

lake-dwelling a dwelling built on piles driven into a marsh or the bed of a shallow lake, common in some tropical areas today, and characteristic of certain periods of NEOLITHIC times in Switzerland, France and central Europe, to which the terms Lacustrine period or Lacustrine civilization are applied. G.

lake rampart a conspicuous ridge or ridges on the shores of a lake, caused by the freezing of the lake in winter with consequent expansion of the ice which presses or shoves (ice-shove) against the lake shores, making ridges of the lakeshore deposits. G.

lalang (Malay) a coarse grass, *Imperator cylindrica*, that infests deserted plantations and clearings in Malaysia, occupies large areas, and appears as a specific category on land use maps. G.

lamb the young of sheep, and its flesh used as food.

lambskin the dressed skin of LAMB, used in the making of clothing, as leather or complete with its wool.

lamella (pl. lamellae or lamellas, diminutive of Latin lamina, a plate) , a very thin plate or layer, or a scale, especially of bone or TISSUE.

lamellar *adj.* consisting of, characterized by, or arranged in, lamellae, sometimes applied to some minerals, e.g. MICA.

lamina pl. laminae, any thin plate, scale or layer. LAMINATION.

laminar *adj.* having the nature of a thin plate, scale or layer. LAMINA.

laminar flow 1. non-turbulent flow of a FLUID closely following the streamlined surface of a solid object in the fluid (TURBULENCE) so that adjoining levels do not mix 2. the movement of a GLACIER along a slope caused by the thrust of the weight of solid ice in the upper part, in some cases so powerful that the SNOUT of the glacier moves uphill.

laminate, laminated *adj.* having or consisting of a LAMINA or laminae. LAMINATION.

lamination in geology, STRATIFICA-

TION on the finest scale, the usual definition being that each layer should be under 1 cm (0.39 in) in thickness, typically occurring in fine-grained SANDSTONES and SHALES. G.

lammas land, half-year land a class of COMMON land, usually under grass but sometimes arable, formerly important and still existing in England, used or cultivated by individuals severally for part of the year but thrown open to the severalty owners and other classes of COMMONERS after the gathering of the crop. The usual date for throwing open is Lammas Day (1 August) or Old Lammas Day (12 August), the land generally being used for common grazing from then until some date in the spring. LOT MEADOW. G.

land 1. the solid surface of the earth where it is not covered by water (the permanent ice of Antarctica is usually considered to be 'land' in world statistics) 2. a part of this solid surface distinct from other parts naturally or for political, economic, or cultural reasons 3. a part of this solid surface in relation to ownership of rights 4. in occasional use, the soil, especially in respect of its quality 5. the countryside, particularly farmland, as opposed to town 6. a strip of ploughed open field divided by water furrows 7. as suffix, BADLANDS, heathland (HEATH) etc. G.

Land (German, pl. Länder) an administratrive unit in Germany that replaced the former kingdoms, grand-duchies, principalities and other units which made up the old Federation. When the Federal Republic of Germany became a soverign independent country on 5 May 1955 it retained the Länder; but they were abandoned by the German Democratic Republic in favour of districts. G.

land agent in the UK, a person who manages an estate; or one who acts for the sale of private or public land.

land breeze a cool breeze (BEAUFORT SCALE) that blows at night from the land to the sea in coastal regions (or from the land surrounding a large lake to the lake), due to the differential heating of land and water. At night air cooled relatively quickly by radiation over the land descends, the atmospheric pressure over the land is slightly higher than that over the water, and the air over the land flows away from the land towards the warmer water. Land breezes are particularly likely to occur in equatorial latitudes; and in other areas where temperature changes are regular in calm, settled weather. SEA BREEZE. G.

land bridge 1. in geology, a land link between continents 2. in anthropology and biology, an ancient route used by migrating land animals 3. in transport, an overland route lying between and connecting two sea routes.

land capability the potential usefulness of land for agriculture (including forestry) based on environmental factors, e.g. soil and climatic factors. LAND CLASSIFICATION.

land classification a systematic classification of land, usually devised for a specific purpose, e.g. as a basis for land use planning and the conservation of land resources, and designed in most cases to indicate the quality, the relative fertility of the land for different types of farming or for some other land use. In some cases 'land capability' or 'potential land use' classes are favoured; but potential is a matter of judgment and may be radically changed by research findings and technological progress (e.g. in Australia, where land previously thought to be worthless was transformed by the addition of TRACE ELEMENTS). In a land classification system devised by L. Dudley Stamp in Britain (fully described in L. Dudley Stamp. *The Land of Britain: Its Use and Misuse*, Longman), ten types of land (1 to 4 good; 5 to 6 medium; 7 to 10 poor) were identified, based on site and soil and history of land use. In Britain, as in

much of Europe, land has been cultivated for two or three thousand years,
and he considered the history of its use
over that long period gave a guide to
potential. More recently other systems
have been devised, drawing particularly
on available economic data. G.

lande (French from Celtic landa, probably free open land) land on which only
brushwood and wild plants (e.g. gorse,
broom, heath) grow. G.

landed *adj.* 1. owning land 2. consisting
of land.

landed gentry people of good birth and
breeding, below the rank of nobility
(the peerage) whose PROPERTY-1 includes land. YEOMAN.

landed property in Marxism, landlords,
as a class.

Landes, Les (French) a region of lowland with sand dunes, lagoons and pine
forests in southwest France, bordering
the Bay of Biscay. G.

landfall the sighting of, or arrival on,
land after a voyage by sea or air.

landform, land form the shape, form,
nature of a specific physical feature of
the earth's surface (e.g. a hill, a plateau)
produced by the natural processes of
DENUDATION and DEPOSITION (including WEATHERING, GLACIATION-1
etc) and by TECTONIC processes.
GEOMORPHOLOGY. G.

land hemisphere the half of the earth's
surface that includes most of the land,
i.e. the part lying north of the equator,
centred on Paris. WATER HEMISPHERE.
G.

landholder one who has a proprietary
interest in, or who occupies, land;
sometimes, specifically, a TENANT who
holds land from a LANDOWNER (proprietor). LAND TENURE.

land ice ice formed from fresh water
lying inland.

landlocked *adj.* applied to an area (particularly a state) which lacks a sea coast
and thus does not have direct access to
the sea. Some landlocked states are
BUFFER STATES.

landlord 1. a proprietor who lets land
or a building to a TENANT 2. a person
in whose house one boards for payment
3. an innkeeper.

landlord capital in an agricultural tenancy system, the PROPERTIES-1 owned
and contributed by the LANDLORD-1.
They usually comprise the fixed assets
on the farm, i.e. the land, drainage
system, buildings. Normally the landlord is responsible for their repair and
upkeep, but this arrangement is by
agreement with the tenant, and is
sometimes varied. LAND TENURE,
SHARECROPPING, TENANT CAPITAL.

landmark 1. a conspicuous feature in
the landscape, serving as a guide in
checking the direction of a course, or
marking a boundary 2. a significant
event or object associated with a turning point in a process or period of time.

landmass a very large area of continental crust (PLATE TECTONICS) lying
above sea-level.

landowner a proprietor of land, one
who has an entitlement to certain
property rights in land. In England and
Wales these may include a surface right
(to enjoy the use of the land), productive right (to make a profit), development right (DEVELOPMENT-1,2),
pecuniary right (to benefit from actual
or anticipated development), restrictive
right (the right not to develop), disposal
right (the right to sell). LAND TENURE.

land power a nation with military
strength on land. SEA POWER.

land reclamation a term applied
broadly to cover not only the winning
back, the recovering, of land that has
been spoilt for agricultural use, but also
the improvement of land so that it can
be made useful, or more useful, for
economic (including agricultural) or
social purposes. Some of the types of
land and the techniques employed are:
land under water or waterlogged (by
drainage or by the filling-in of a water-
filled depression); arid land (by irrigation and, if saline, by chemical treat-

ment); unstable slopes and loose soil (by planting of vegetation cover); land subject to water erosion (by planting of vegetation cover, by terracing, by embankment); land subject to wind erosion (by planting of vegetation cover including shelter belts of trees); land spoiled by quarrying, mining or industrial activity (by filling-in of quarries and pits, levelling, planting of spoil tips, restoration of soil profile); land impregnated with salt or industrial effluent (by chemical treatment); land covered with undesirable trees and/or scrub (by clearance).

land reform changes in a system of LAND TENURE, commonly brought about by government intervention and usually aimed at removing what is considered to be unfairness in the system, or at improving agricultural efficiency, etc., e.g. by the breaking up of big estates and the redistribution of the land as small holdings to farmers who become owner-occupiers; or by the consolidation of fragmented holdings (FRAGMENTATION-2) to form larger, more efficient, farming units. AGRARIAN REFORM.

land rent the concept of economic rent (similar to the economic rent of English CLASSICAL ECONOMIC THEORY) developed by J. H. Von Thünen for his model (VON THUNEN MODEL). He defined it as that part of the total (gross) product of land which remains as a surplus after the deduction of all costs, including interest on invested capital, i.e. the net profit earned by a farmer from his/her chosen productive system, opportunity costs (ECONOMIC RENT) being ignored. The net profit is governed by production costs and market price per unit of product, transport rate per distance unit for each product, the yield per unit of land, and the distance from the point of origin of the product to the market centre. RENT.

land rotation a regular system of land management in which land is cultivated for a few years and then allowed to rest, perhaps for a considerable period, usually by simply allowing scrub or bush to grow up over it. In due course it is cleared and cultivated again, the farms on which, or settlements from which, cultivation takes place being fixed. This is a type of SHIFTING CULTIVATION common in Africa. Land rotation should not be confused with the ROTATION OF CROPS. G.

LANDSAT one of the US satellites orbiting the earth without a crew, at a height of over 12 km (7.5 mi) and by REMOTE SENSING surveying the natural resources, land use, environmental conditions of the earth (e.g. crop disease, water pollution etc). LANDSAT 1 was launched on 23 July 1972 by the National Aeronautics and Space Administration (NASA) as the Earth Resources Technology Satellite (ERTS), its name being changed on 22 January 1975 when LANDSAT 2 was launched. LANDSAT 3 was launched in 1978. LANDSAT passes round the earth in a sun-synchronous, near-polar orbit, completing 14 orbits a day, and achieving global coverage every 18 days. It records images by means of two systems. One consists of television-like cameras, the other of four-channel multispectral line-scanner devices, each system operating in various bands of the green, red and two near-infra-red wavebands. SEASAT 1, SKYLAB, SPACE SHUTTLE.

landscape (Dutch landschap, the representation in painting of inland natural scenery) **1.** still used in the Dutch sense, e.g. a landscape by Constable **2.** the scenery itself **3.** an area of the earth's surface characterized by a certain type of scenery, comprising a distinct association of physical and cultural forms. From this came the separation of NATURAL LANDSCAPE from CULTURAL LANDSCAPE, and the terms 'exhumed' or 'fossil' landscape (EXHUMATION, FOSSIL LANDSCAPE). G.

landscape architecture LANDSCAPE GARDENING on a large scale, involving the harmonization of groups of buildings, factories, roads etc. with the landscape as a whole. G.

landscape evaluation the assessment of the qualities of the components of a LANDSCAPE-2,3.

landscape gardening the art and practice of laying out a garden or estate associated with houses or buildings to form a harmonious whole.

landscapist approach an approach to the study of the origin and spread of *homo sapiens* and of cultures that concentrates on discoveries of material remains (animal, plant, buildings, tools etc.), on features visible today in the landscape, and on historical records, as distinct from a locationist approach, i.e. the approach of SPATIAL ANALYSIS applied to DIFFUSION.

Landschaft a German term often translated as landscape. But in German it has many diverse applications, the most common being specifically to a region delimited by its appearance. G.

landskap (Norwegian) a province in Norway. G.

landslide, landslip **1.** the sliding down under the force of gravity of a mass of land on a mountain or hillside **2.** the part which has so fallen. ENVIRONMENTAL HAZARD, MASS MOVEMENT, SLIP-2. G.

landslip terrace a short terrace with a rough surface, resulting from the slip of part of a hill. G.

land survey system in USA, PUBLIC LAND SURVEY.

land tenure the rules and regulations governing the rights of holding, disposing and using land, i.e. the conditions on which land is held (TENURE), varying with the social and economic organization of the country concerned. In England and Wales where the terms estate or interest indicate the period of time for which the holder may enjoy the land, the largest and highest estate that can exist in land is the fee simple estate (popularly termed the freehold). The natural and proprietary rights flowing from the ownership of that estate may be enjoyed in perpetuity. The owner of the freehold may make any number of grants of the enjoyment of the land to others for limited periods. Such rights to the present possession and use of land for these lesser periods are also termed estates or interests, their duration being indicated by a specific name (e.g. life interest, or leasehold, i.e. term of years). The grant of a lease (a tenancy for a term certain) gives the leaseholder exclusive possession of the land for a certain period, usually in consideration of the payment of contract rent. The leaseholder may, unless the terms of the lease prevent it, grant an underlease. The holding of land under a lease (or sublease) is termed a tenancy, the term tenancy being also applied to the duration of a tenure (GROUND RENT, RACK RENT, TENANT).

In other systems of ownership the tenant may, instead of paying money rent, pay for the right to use land by providing labour on the land which the owner has kept for personal use, or by giving the owner a share of the crop produced on his/her (the tenant's) holding (SHARE CROPPING). In COLLECTIVIST systems the land is owned by a collective interest, e.g. the state (COLLECTIVE, STATE FARMING), or a small community (KIBBUTZ) etc. In SHIFTING CULTIVATION an individual farmer or a group of farmers establishes a right merely by using the land. LANDLORD CAPITAL, LAND REFORM, TENANT CAPITAL.

land use, land utilisation, land utilization- terms commonly applied interchangeably to the use made by human beings of the surface of the land. A land use survey, though literally surveying and mapping the use of the land surface, usually also includes in sparsely populated countries the natural and semi-natural vegetation. LAND UTILISATION SURVEY OF BRITAIN, GENERAL LAND USE MODEL. G.

land use conflict the clash of interests in the use of a piece of land, e.g. between a user or a potential user over its current or potential use; or between a user or a potential user and the objectives of social or land use planning schemes.

land use displacement the displacement of lower value, lower yielding uses by those which give higher yields.

land use intensification the DEVELOP-MENT-1 of a site by the owner and/or the developer with the object of achieving the highest possible rental income and/or capital values commensurate with MARKET-3 demand.

land use planning the demarcation of land for specific uses, usually (but not necessarily) over an extensive area, based on environmental, social and economic criteria, which takes into account present and possible future needs. PLANNED ECONOMY, PLAN-NING, REGIONAL PLANNING.

land use segregation FUNCTIONAL SEG-REGATION.

Land Utilisation Survey of Britain the First survey, started in 1930, field work carried out mainly in 1931-33, was a voluntary organization established by L. Dudley Stamp with E. C. Willatts as organizing secretary. Mainly with the help of volunteers from educational institutions, the use of every acre of land in England, Wales, Scotland, the Isle of Man and the Channel Islands (Northern Ireland was separately surveyed later) was recorded on over 15 000 6-inch (1 : 10 560) field sheets, the results being reduced to the scale of 1 in to 1 mi and published in 150 sheets for all of England and Wales and for the more populous parts of Scotland. The published maps were accompanied by county monographs under the title *The Land of Britain*. The whole work is summarized in L. Dudley Stamp. *The Land of Britain: Its Use and Misuse*, Longman.

Basically the land use was classified in six categories, with some subdivisions: arable (brown), permanent pasture (light green), rough pasture (yellow), woodland (dark green, with subdivisions), gardens, orchards, nurseries (purple, with subdivisions), and land agriculturally unproductive (red). The picture shows Britain at a time of agricultural depression.

The Second Land Utilisation Survey of Britain, far more detailed than the original Survey, was started in 1960 under the direction of Alice Coleman of King's College, London. The 52 categories identified in the fieldwork were recorded on the scale of 1 : 10 000, the results being published on the scale of 1 : 25 000.

In 1949 the IGU set up a Commission on a World Inventory of Land Use, under the auspices of which the World Land Use Survey was establised to encourage countries all over the world to carry out studies on a comparable basis. The broad, simple categories distinguished were settlements, horticulture, perennial crops, cropland, improved grass, rough grazing (divided on a floristic basis), woodland, swamps and marshes, unproductive. Land use surveys have now been undertaken or are in progress in most countries of the world. REMOTE SENSING is important in general land use surveys.

landward 1. *adj.* located towards the land 2. *adv.* towards the land.

landward population in Scotland, the part of the population not resident in a BURGH. G.

lane 1. a narrow road in a rural area, commonly edged with hedges, trees, fencing and sometimes unmetalled 2. a narrow street, an alley in a town, usually bordered by walls or buildings 3. an established route for sea or air traffic 4. one of the parallel, marked-out sections of a wide road **5.** a channel of clear water through an ice field.

language **1.** the organized system of words, of speech, by which human beings communicate with each other **2.** such a system differentiated and used by a certain group of the human race **3.** such a system adapted to meet the needs of a special purpose (e.g. of diplomacy), or of a particular group, science or profession **4.** an organized system of communication (e.g. using signs, symbols), as in mathematics **5.** any seemingly organized system of communication using gestures, movements, etc. (e.g. the language of animals).

lanolin, lanoline WOOL.

lapiaz an area of LAPIE, equivalent to German KARRENFELD. G.

lapié (French) an exposed limestone surface in a KARST region, with etching, pitting, grooving, fluting, caused by CARBONATION, mainly, some authors maintain, from small free-flowing streams; equivalent to German Karren (KARRE). There are also very small forms, microlapiés, German Rillensteine, perhaps best termed rock-rills. CLINT. G.

lapilli (Italian lapillo, pl.lapilli) small pyroclastic fragments ejected from a volcano, commonly classified as having a long dimension or diameter of between 5 and 10 mm. PYROCLAST.

lapis lazuli a brittle SEMI-PRECIOUS STONE, a sodium aluminium silicate with some sulphur, varying in colour from pale to deep blue, flecked with white or yellow, with little lustre, losing its polish by constant use, used in making jewellery and ornaments, the deep blue variety being ground for pigment. ULTRAMARINE.

lapse rate the rate of decrease of air temperature normally occurring with height (vertical temperature gradient), but varying with time and place. The average lapse rate in the atmosphere, termed ENVIRONMENTAL LAPSE RATE, is 0.6°C per 100 m (about 3.5°F per 1000 ft, or sometimes stated as 1°F per 300 ft). This continues up to the TROPOPAUSE unless an INVERSION OF TEMPERATURE occurs, causing an increase of temperature with height (termed an inverted lapse rate and indicated by a minus sign). ADIABATIC LAPSE RATE, DRY ADIABATIC LAPSE RATE, DYNAMIC LAPSE RATE, ENVIRONMENTAL LAPSE RATE, PROCESS LAPSE RATE, SATURATED ADIABATIC LAPSE RATE. G.

Laramide orogeny, Laramide revolution a period of earth movement, a mountain-building movement in the early TERTIARY, i.e. before the main ALPINE OROGENY, important, for example, in the development of the Rocky mountains. G.

latent heat **1.** the amount of heat energy (thermal energy) needed to bring about an isothermal chemical or physical change of a body without making it hotter, e.g. from a SOLID to a LIQUID (the latent heat of FUSION) or from a liquid to a GAS-1 (the latent heat of EVAPORATION or vapourization) or from a solid to a gaseous state (the latent heat of SUBLIMATION) **2.** the thermal energy released in such a process. SENSIBLE TEMPERATURE.

latent instability the state of the upper part of an air mass of CONDITIONAL INSTABILITY that lies above the level where convection freely occurs.

lateral dune a small DUNE lying beside a major dune within a dune pattern caused by an obstacle in a desert.

lateral erosion the wearing away of its banks by a stream. BANK CAVING, MEANDER.

lateral moraine the rock debris from valley slopes that lies on the surface of a GLACIER, making a low ridge along each side. It may form an embankment along the valley wall as the glacier melts. MORAINE. G.

laterite (Latin later, a brick) a subsoil product of WEATHERING in humid tropical areas (HUMID TROPICALITY) where there are alternating wet and dry seasons which lead to the formation of a mottled red-yellow and grey mass, sufficiently soft to be cut out with a spade but hardening on exposure to the atmosphere. In humid tropical conditions the soft grey clayey or sandy matter is leached of SILICA and ALKALI-1, leaving a concentration of sesquioxides of ALUMINIUM and IRON. When cut and exposed to the air a sponge-like red rock is formed, hard enough to be used for building, especially as foundations for light structures, for paths or secondary roads. The process of formation is called laterization. Sometimes the rock is sufficiently rich in iron to be usable as an iron ore. In other cases it is rich in alumina and grades into BAUXITE (GIBBSITE), the chief ore from which ALUMINIUM is extracted.

The soils derived from laterites can obviously be called lateritic soils, but in the past almost any red soils have been called 'lateritic'; thus the more specific term LATOSOL was proposed, but not widely accepted outside the USA. G.

laterite soil, lateritic soil a ZONAL SOIL formed on LATERITE, porous and leached, and of little agricultural value. LATOSOL. G.

laterization the process of WEATHERING in humid tropical areas (HUMID TROPICALITY) that leads to the formation of LATERITE. G.

latex the milky fluid exuding from the cut surface of some flowering plants and trees, coagulating on exposure to the air. Some of these fluids are commercially useful, providing, e.g. the raw material of RUBBER from *Hevea* species, or of chewing gum (CHICLE GUM) from the SAPODILLA TREE.

latifundia (pl. of Latin latifundium, a large estate) originally large landed properties in South America cultivated by peons (agricultural labourers) for the Spanish Crown, now applied to the extensively farmed large estates or ranches in Spain and South America in contrast to the intensively farmed HUERTAS. In some cases very small holdings on the estate are leased to tenants (LAND TENURE), the rest of the land being farmed by the landlord, who employs day labourers. The comparable Italian term is latifondo, but this is applied to an agricultural area with extensive cereal cultivation and grazing which includes large estates and peasant holdings. EXTENSIVE AGRICULTURE, FARMING, INTENSIVE AGRICULTURE. G.

Latin America those countries of the NEW WORLD which were discovered, explored, or conquered by the Spaniards (or the Portuguese in the case of Brazil), i.e. that part of the New World where Spanish is spoken together with Portuguese-speaking Brazil, comprising the whole of mainland South America (except Guyana, French Guiana, Surinam), all the countries of central America (except Belize), as well as Mexico and the islands of the West Indies, Cuba and Dominica, where Spanish is spoken.

Latin American Free Trade Association LAFTA.

latitude the angular distance of any point on the earth's surface north or south of the equator, as measured from the centre of the earth, in degrees, minutes and seconds. The equator itself is $0°$, the NORTH POLE is $90°N$, the SOUTH POLE is $90°S$. Low latitudes are broadly those between the TROPIC OF CANCER ($23°30'N$) and the TROPIC OF CAPRICORN ($23°30'S$). Midlatitudes extend from the Tropics to the ARCTIC CIRCLE ($66°30'N$) and ANTARCTIC CIRCLE ($66°30'S$). High latitudes lie within the Arctic and Antarctic circles, i.e. from the Arctic circle to the North Pole, from the Antarctic circle to the South Pole. PARALLEL OF LATITUDE. G.

latosol a soil with thin A_0 and A_1 layers over reddish or red deeply weathered material which is low in silica and high in sesquioxides. A term not widely accepted outside the USA. LATERITE. G.

Laurasia the northern part of the great PRECAMBRIAN landmass, PANGAEA. Current scientific evidence suggests that it included today's North America, Europe, Asia (the Indian subcontinent excluded) and the Arctic. CONTINENTAL DRIFT, GONDWANALAND, PLATE TECTONICS, TETHYS OCEAN.

lava molten rock (MAGMA) that issues from a volcanic VENT or FISSURE to the surface of the ground, where it solidifies. Chemically it is divided into acid (with excess silica so that some crystallizes out as quartz, SiO_2), intermediate, basic, ultrabasic (ACID LAVA, BASIC LAVA, INTERMEDIATE ROCK, ULTRABASIC ROCK). Some lavas cool quickly and are glassy, others more slowly, giving crystals of minerals time to grow, so that they are CRYSTALLINE. AA, BASALTIC LAVA, PILLOW LAVA, VOLCANO, VOLCANIC ROCK.

lavant BOURNE.

lava tube a tunnel formed when a river of LAVA solidifies where it touches cold ground and at the same time crusts over at the upper surface. The hot fluid lava within continues to flow and drains out, leaving an enclosed passageway, a tunnel, which may be large enough to walk through, and may be occupied eventually by an underground stream. G.

laver *Porphyra umbilicalis*, edible SEAWEED, rich in protein and vitamins B and C, common on rocks on the coasts of temperate North Atlantic countries. The genus *Porphyra* has a world-wide distribution, and is cultivated in Japanese waters. It is cooked, used in making laver bread, soups etc., pickles, preserves and sweetmeats etc.

law 1. a custom or practice recognized as binding by the members of a society, governing their behaviour and supported by the power of government 2.

the whole body of such customs and practices, and obedience to it 3. a relationship between cause and effect, based on observation and experiment, and accepted as true, e.g. Newton's law of cooling, the law of supply and demand. It may be affirmative or conditional (HYPOTHESIS). SCIENTIFIC LAW, THEORY 4. Scottish, a rounded or conical hill, e.g. Traprain Law. G.

lawn 1. an open, grassy area in woodland, especially in the New Forest, Hampshire, England 2. an area of land covered with closely-cut grass in a garden or the grounds of a house or pleasure garden etc. 3. the cultivated terraces on the Portlandean (Upper JURASSIC) limestone in the Isle of Portland, Dorset, England (LYNCHET). G.

law of crosscutting relationships CROSSCUTTING RELATIONSHIPS, LAW OF.

law of included fragments the assumption that identifiable fragments of rock enclosed within another rock are older than the enclosing rock.

law of retail gravitation REILLY'S LAW OF RETAIL GRAVITATION.

law of superposition the assumption that in a sequence of stratified or sedimentary rocks (and in extrusive igneous rocks) the oldest bed is at the bottom, unless the sequence has been reversed by intense folding, faulting or other disturbance. G.

law of unequal slopes the assumption that if the opposing slopes of a ridge are dissimilar, one being steep, the other gently sloping, the steep one will be eroded more quickly than the other, so that the ridge retreats on that side, e.g. as in an ESCARPMENT.

layer colouring, layer tinting a mapping technique in which different colours or tints are used to emphasize differences, e.g. on relief maps, where land up to a selected height is shown in one colour or tint, from that height to the next limit in a different colour or tint, and so on.

lazy-bed a piece of land deeply dug for growing potatoes, the seed tubers being placed in straight lines on the surface and covered with soil taken from between the rows. The potatoes thus grow on in raised ridges separated by trenches which help drainage. The system is favoured in the Hebrides and western highlands of Scotland. G.

LDC less developed country. HDC, MDC, NIC, UNDERDEVELOPMENT.

lea 1. in poetic use, an expanse of open land, most commonly applied to grassland 2. obsolescent spelling of ley or lay. LEY FARMING. G.

leaching the process whereby percolating water removes materials from the upper layers of a rock, soil or ore, and carries them away in SOLUTION or SUSPENSION. It brings about the secondary enrichment of ores and POROSITY in LIMESTONE, and is especially important in the formation of soils in that it removes soluble salts from the A HORIZON to the zone of accumulation, the B HORIZON. G.

lead a blue-white, tarnishing to dark grey, soft, dense, MALLEABLE, DUCTILE divalent or tetravalent (VALENCY) metallic ELEMENT-6, a CHALCOPHILE, not occurring native in nature, most being obtained from an ore in its sulphide form, galena, PbS, which often occurs with zinc sulphide in veins in igneous rocks or as irregular masses in some limestones; also in an oxidized form, cerussite, $PbCO_3$. It is used in pipework (a declining use), protective coverings and linings, or as a shield against radioactivity. Some compounds are used in glass making and other manufactures, in accumulators; and as pigments. LEAD-RATIO.

lead ratio the ratio of LEAD to URANIUM and/or THORIUM in a rock containing those elements, based on knowledge of the progressive radioactive breakdown in some of the urani-

um/thorium (HALF-LIFE). It is used in estimating the age of a rock. ISOTOPE, RADIOACTIVITY,   RADIOCARBON DATING.

lead 1. an artificial water-course, especially one leading to a mill. LEAT 2. a channel in an ICE FIELD or in sea ice 3. in mining, a lode 4. in Australia, an old river bed in which gold is found. G.

leap year GREGORIAN CALENDAR YEAR.

leasehold, leasehold ownership, leaseholder control or ownership of rights to land or buildings over a certain period. FREEHOLD, LAND TENURE.

leasow (Old English laes, laeswe) 1. pasture or meadow land 2. in English Midlands, an individual allotment or field, often arable. G.

least cost location the siting of industry in a place where the costs of transport (assessed on weight and distance) and labour, and the advantages of industrial AGGLOMERATION-3 or DEGLOMERATION are most favourable in economic terms. LOCATION THEORY, LOCATIONAL TRIANGLE. G.

least squares criterion, least squares method LINEAR MODEL, REGRESSION ANALYSIS.

leat an open water-course taking water to a mine, mill, house or reservoir. G.

leather the dressed skin of an animal, tanned to make it flexible and long-lasting, used in the making of shoes, garments, luggage, harness etc. TANNERY, TANNIN.

Lebensraum (German, living space, habitat) 1. the area inhabited by or habitable by (the area of distribution of) a living organism 2. a term used by Nazi geopoliticians in Germany in the sense of a claim to what they considered adequate living space for Germans especially at the time of the Third Reich and the Nazi party's dominance. GEOPOLITICS. G.

lectotype a specimen chosen from original plant or animal material to be used as the TYPE SPECIMEN when this is missing or was not designated on publication. NEOTYPE, PARATYPE.

ledge 1. a narrow, horizontal, shelf-like projection from a vertical or steeply sloping surface, e.g. from a cliff or a mountainside 2. an underwater ridge of rock.

lee, lee side, leeward the sheltered side, i.e. the side opposite to that against which the wind blows (WINDWARD) or oncoming ice impinges (STOSS); but see LEE-SHORE.

lee depression an OROGRAPHIC LOW, a non-frontal low pressure system occurring on the sheltered side of a mountain range due to the eddy made by the passing air-stream.

leek *Allium ampeloprasum*, an onion-flavoured plant of the Alliaceae family, in its wild state native to the Mediterranean area and possibly to southern England and Wales, cultivated in the eastern Mediterranean region for three or four thousand years. The modern leek in cultivation is usually *Allium porrum*, the lower part of the plant (the elongated bulb) being blanched by earthing-up to exclude light; usually eaten cooked.

lee-shore the shore towards which the wind is blowing, thus dangerous to shipping. LEE. G.

leet, court leet MANORIAL COURTS.

leeward LEE.

lee wave, lee-wave a wave formation in an air-stream caused by a relief barrier which forces air to rise. In slight wind the air closely follows the form of the ground, and the flow on the lee side of the barrier is LAMINAR; but if the air speed is greater a STANDING WAVE forms; and if greater still a lee wave develops in the stable part of the air-stream and clouds frequently form in the cool crest of the lee wave (BANNER CLOUD, LENTICULAR CLOUD) under which the air becomes turbulent, a revolving phenomenon termed a rotor. HELM WIND.

legend an explanation, a key to, the symbols used on a map or diagram.

legitimate function in Marxism, the function of the state in capitalist societies (CAPITALISM) in providing the social conditions, (e.g. by providing economic benefits, defence services, the police, the diffusion of ideas) which facilitate the accumulation of capital.

legume 1. the fruit of a member of the LEGUMINOSAE family (which includes beans, peas, pulses), a pod formed from a single carpel which when ripe splits along each side allowing the seeds to fall out 2. any plant which is a member of the LEGUMINOSAE family, particularly, in agriculture, clover and allied FODDER plants. NORFOLK ROTATION.

Leguminosae a large family of plants, divided into three sub-families, Papilionaceae, Caesalpinoideae, Mimosoideae, many species (e.g. BEANS, BERSIM, CLOVER, LUBIA, PEAS) being important food plants for humans and animals. Most bear root nodules that contain nitrogen-fixing bacteria, and are thus useful in soil management. LEGUME, NITROGEN CYCLE, NITROGEN FIXATION, RHIZOBIUM.

leiotrichous *adj.* having straight hair, as opposed to CYMOTRICHOUS or ULOTRICHOUS.

lembah (Malay) low-lying land in Malaysia. G.

lemon *Citrus limon*, a small CITRUS tree, native to China and southeast Asia, widely grown in Mediterranean and subtropical areas for its acid fruit, rich in vitamin C, used for flavouring in cooking and beverages, with a skin yielding ESSENTIAL OILS (used in perfumery and soap making) and pectin (used in jam making).

lentic *adj.* slow, calm, hence lentic water, the standing water of lakes, ponds, or swamps that lack a continuous flow of water in a definable direction. LOTIC, LOTIC WATER.

lenticular cloud a cloud shaped like a lens, usually produced by an eddying wind (HELM WIND) in mountains or hills. LEE WAVE.

lentil *Lens culinaris*, a leguminous AN-
NUAL plant (LEGUMINOSAE) bearing
lens-shaped seed pods, probably native
to the Mediterranean region, widely
cultivated in Europe and Asia for the
sake of its protein-rich SEEDS, usually
orange-red in colour, which are dried
and used when cooked.

less developed country LDC.

less favoured land an area of infertile
land of limited potential, low economic
performance, and a low, dwindling
population, one of the categories of
agricultural land identified under EEC
regulations (EEC Directive 75/268 on
mountain and hill farming). Most of
the HILL FARMING land of Britain lies
in this category.

lessivage the process by which percolating water carries CLAY MINERALS
downwards through the soil, leading to
ILLUVIATION in a lower horizon (usually the B HORIZON).

leste (Portuguese) a hot, dry, often
dust-laden, easterly to southerly wind
blowing from the Sahara and experienced in Madeira, heralding an advancing DEPRESSION-1. G.

lettuce *Lactuca sativa*, family Compo-
sitae, an ANNUAL plant, origin unknown but possibly native to the
Mediterranean region, now widely
grown with or without protection in
temperate regions for the sake of its
leaves, rich in vitamin A, usually used
in salads, sometimes in cooking, the
leaf shape and crispness varying with
variety.

leucocratic *adj.* light-coloured, applied
especially to IGNEOUS ROCK in which
QUARTZ, FELDSPAR and white MICA
predominate to produce a pale-colourd
rock. FELSIC, MAFIC, MELANOCRA-
TIC.

Levant **1.** formerly applied to the countries of 'the East', later to the eastern
part of the Mediterranean with its
islands and the lands bordering it. Now
rarely used, being partly replaced by
the equally ambiguous term MIDDLE

EAST **2.** sometimes used as a regional
name for southeastern Spain. G.

**levant and couchant, levancy and cou-
chancy** a legal phrase (from medieval
Latin through French, rising up and
lying down, of cattle) applied to the
rule, dating back at least to the thirteenth century, whereby the number of
cattle, sheep and horses which a holder
of COMMON RIGHTS was entitled to
graze on COMMON land in summer was
restricted to the number which could be
supported on the commoner's farm
through the winter. G.

levante, levanter, llevante, llevantades
(Spanish) a usually humid easterly
wind blowing over the western Mediterranean area, taking its name from
the direction from which it blows. It is
due to a DEPRESSION-3 in the region,
which sometimes causes heavy rainfall,
and particularly affects southeast
Spain, the Balearic islands, Gibraltar
and northern Algeria, especially between July and October. When moderate it causes a BANNER CLOUD to
develop from the summit of the Rock
of Gibraltar, when strong it produces
dangerous currents and eddies in the
sea on the LEE side of the Rock. When
especially stormy it is termed llevantades. SOLANO is an alternative term.
G.

Levantine an inhabitant of the LE-
VANT-1, usually restricted to those of
European or mixed parentage who
have no other country and have largely
adopted the languages, customs etc. of
the country in the Levant in which they
live. G.

leveche (Spanish) a hot, dry wind,
sometimes dust-laden, originating in
the Sahara and blowing towards Spain,
due to the eastward movement of a
DEPRESSION-3 in the western Mediterranean area. KHAMSIN, SIROCCO.
G.

levee, levée a natural or artificial embankment of a river which confines the
river within its channel and hinders or

prevents flooding, applied especially to those of the lower Mississippi. G.

level 1. a nearly horizontal tract of land, unbroken by hills and valleys, applied specifically to certain tracts of such land, e.g. the Bedford Level in Fenland, England. There, in a famous experiment, three stakes were driven into the ground so as to stand exactly the same height above a water surface, by which the curvature of the earth could be seen and measured **2.** a surveying instrument (ABNEY LEVEL, DUMPY LEVEL) **3.** a nearly horizontal DRIFT-4 or gallery (a working level) in a mine **4.** the horizon at which an ore-body is worked **5.** a position in a scale of importance **6.** relative position in respect to a NORM in a scale of estimating **7.** the magnitude of a physical quantity. G.

level *adj.* horizontal, having no part higher than another.

level crossing a place where a railway crosses a road at the same level.

level ice smooth-surfaced sea ice that has never been HUMMOCKED.

levelling in surveying, the process of establishing the difference in height between successive pairs of points by means of sighting through various instruments that incorporate a telescope with spirit level, graduated measuring rod, etc.

level of living in WELFARE GEOGRA-PHY, the level of well-being (SOCIAL WELL-BEING) of a group of people in a particular place at a particular time. QUALITY OF LIFE, STANDARD OF LIVING.

ley, lay, lea (ley is the usual spelling) arable land put down to grass and/or clover for a period of years, a short ley applying to two to four years, a long ley to longer periods, even up to twenty years. The character of the ley may be indicated, e.g. grass-ley, clover-ley, etc. G.

ley farming a system of farming in which grass-leys or clover-leys are an essential part of the land management.

liane, liana (possibly from French lier, to bind; liana possibly derived from a mistaken idea of Spanish origin) a climbing woody plant, with roots in the ground, characteristic of tropical forests. G.

lias a bluish-LIMESTONE occurring interbedded with SHALE or CLAY in the Lias, a division of the Lower JURASSIC (GEOLOGICAL TIMESCALE) and in other FORMATIONS-2.

libeccio (Italian) a strong westerly or southwesterly wind blowing across Corsica, most frequently in summer, in winter bringing rain or snow to the western mountainsides.

liberal capitalism sometimes used as a synonym for CAPITALISM, an economic system based on private ownership, competition, private enterprise and profit-making, in which most of the means of production, distribution and exchange as well as the objects of consumption are controlled privately by a large number of competing firms, and sales are for profit in the MARKET-3, 4,5. MARKET ECONOMY, STATE CAPITALISM.

lichen a member of Lichenes, a group of slow-growing dual organisms formed from the symbiotic (SYM-BIOSIS) association of a FUNGUS and a green or blue-green alga (ALGAE), surviving only in unpolluted air, size variable, flat and leaflike or upright and branched in form. Lichen are primary colonizers of bare areas, the dominant flora in mountainous and Arctic regions where few other organisms can survive; but they occur also elsewhere on tree trunks, walls, exposed rock surfaces. A valuable food for animals in Arctic regions (e.g. REINDEER MOSS), they are also a source of dyes (e.g. *Rocella* provides the dye litmus, for litmus paper, which is turned red by the application of an acid or blue by an alkali). LICHENOMETRY, THALLO-PHYTA.

lichenometry an imprecise method of measuring the passage of time (DAT-ING), e.g. the length of time a stone has been lying in situ in a MORAINE, based on the rate of growth of LICHEN (the assumption being that the diameter of the largest lichen growing on the surface under investigation is proportional to the length of time that the surface has been exposed to colonization and growth). It is not a completely satisfactory method because the rate of lichen growth is greatly affected by climatic and other factors, e.g. atmospheric POLLUTION-1.

lidar light detection and ranging (lasar radar), a system that uses laser radiation in the way that RADAR-2 uses microwaves to detect the existence, location, direction of movement of objects. It is used particularly in meteorology.

lido (Italian, a barrier of sand or silt in front of a lagoon) one of the best examples of such a barrier (BARRIER BEACH) is that protecting the lagoon of Venice. This has been converted into a bathing beach and resort. The term lido was transferred and applied to it, and later extended even to a freshwater or artificial lake beach/resort. The term is best avoided in its original, Italian, sense. PLAGE. G.

life cycle the progressive series of changes undergone by an organism from fertilization (the union of gametes, the reproductive cells) to death or, in a lineal succession of organisms, to the death of that stage producing the gametes which begin an identical series of changes. In some organisms (e.g. flukes, parasitic worms) there is a succession of individuals, connected by sexual or asexual reproduction, in the complete cycle. FAMILY LIFE CYCLE.

life expectancy the amount of time, calculated by actuaries, an individual person or a class of people is expected to live.

life form 1. the characteristic form of a plant or animal species at maturity 2. of plants, RAUNKIAER'S LIFE FORMS. G.

life science any science that deals with the description, classification, structure or performance etc. of living organisms, e.g. biology, medicine.

lifespan the period of time from the birth to the death of an organism. AGE, LIFE CYCLE.

life table a mortality table, an actuarial table showing LIFE EXPECTANCY at any given age at a particular time and place.

light 1. the wave band of electromagnetic radiation (ELECTROMAGNETIC SPECTRUM) to which the retina of the eye is sensitive and which is interpreted by the brain 2. the part of the electromagnetic spectrum which includes INFRA-RED, VISIBLE and ULTRA-VIOLET radiation.

light detection and ranging LIDAR.

lighthouse a prominent permanent building housing signalling equipment to warn shipping of navigational hazards, e.g. rocks, the shore.

light industry 1. loosely applied to a SECONDARY INDUSTRY which does not come under the definition of HEAVY INDUSTRY 2. more specifically, a secondary industry in which the weight of materials used per worker is low, the MATERIAL INDEX is low, and the finished products have a high value in relation to their weight 3. in town planning, any industry which may be located in a residential area without detracting from its amenity. FOOT-LOOSE INDUSTRY. G.

lightning an electrical discharge seen as a flash occurring within or between clouds, or between a cloud and the ground. BALL, FORKED, SHEET LIGHTNING, THUNDER.

lightship a permanently anchored vessel, equipped with a signalling device or devices to warn shipping of navigational hazards, such as sandbanks, shoals, etc.

light-year an astronomical measure of

distance, the distance travelled by light in one year at 3.04×10^8 m per second (186 326 mi per second), i.e. 9.7×10^{12} km (6×10^{12} mi).

lignite a low-grade COAL, generally post-CARBONIFEROUS in age, intermediate in properties between the PEATS and BITUMINOUS COAL. The term is sometimes used as a synonym for BROWN COAL, but lignite is darker in colour, the vegetable structure is not so apparent, the carbon content is higher, the moisture content lower that that of brown coal. Lignite is used mainly as a fuel to produce heat in thermal-electric generators. G.

lignum vitae (Latin, wood of life) any of the trees of the genus *Guaiacum*, family Zygophyllaceae, especially *G. officinale*, a tropical American tree producing resin and a very hard, heavy, dense wood. IRONWOOD.

Lima bean (Madagascar) bean PULSE.

liman (Russian)**1.** a LAGOON formed by the barring of the mouth of an estuary by sand **2.** a broad freshwater bay of the sea, especially of the Black and Azov seas, where SPITS block river mouths **3.** a valleyside cultivation terrace in arid USSR. G.

limb of a fold the rock strata on one or the other side of the central line (the AXIS) of a FOLD.

lime a caustic, infusible solid consisting essentially of CALCIUM OXIDE (quicklime), obtained by heating CALCIUM CARBONATE. It is used in agriculture and metallurgy, in building materials, the treatment of sewage, and in various manufacturing processes. CALCIUM HYDROXIDE, KANKAR, LIMEKILN, LIME PIT, LIMESTONE.

lime *Citrus aurantifolia*, a small citrus tree, so branched as to resemble a SHRUB, cultivated in tropical areas and in subtropical (where frost-free), bearing small fruits rich in ascorbic acid (vitamin C), the juice being used in beverages or squeezed over other foods to enhance their flavour, the whole fruit

being used in making marmalade, chutnies, etc.

limekiln a chamber, usually of brick, in which CHALK or LIMESTONE is heated to a high temperature to produce LIME.

lime pan HARD PAN.

lime pit **1.** a LIMESTONE quarry **2.** a pit filled with water and LIME into which HIDES are immersed to remove the hair.

limes (Latin, boundary, pl. limites) a limit, an end. A term rarely used except in historical geography, where it is applied to the limit or boundary of a territory, especially territory of the Roman Empire. G.

limestone a broad, general term for a SEDIMENTARY ROCK consisting mainly of CALCIUM CARBONATE with varying amounts of other minerals. The many types are distinguished by qualifying adjectives, which may refer to texture (e.g. oolitic, pisolitic, crystalline etc., limestone of tiny rounded grains being OOLITE, of larger grains PISOLITE; CRYSTALLINE limestone being MARBLE), to mineral content (dolomitic), to origin (e.g. organic, coral, sedimentary, precipitated, shelly etc.), to geological age (CARBONIFEROUS, JURASSIC), and to other characters. The popular concept of a limestone is that it is relatively hard, hence such references as to 'chalk and limestone', despite the fact that CHALK itself is a limestone, but can be comparatively soft. CALCITE, CAMBRIAN, LIME.

limit **1.** a point, line, surface beyond which it is impossible to go, perceived from one side only (in contrast to BOUNDARY) **2.** the value of a quantity, or of a period of time, that is the largest or smallest possible.

limits of error in statistics, UNCERTAINTY-2.

limnetic *adj.* of, pertaining to, or inhabiting, the open freshwater of a lake or pond, away from the bottom or shore. PELAGIC **2.** inhabiting MARSHES, LAKES or PONDS.

limnic *adj.* applied to **1.** a sediment formed or deposited in inland, standing, freshwater, e.g. in a lake or swamp, in contrast to PARALIC **2.** the environment of such an area of deposition.

limnology the scientific study of the physical, biological and chemical conditions of freshwater and/or freshwater life. G.

limnoplankton microscopic animal and vegetable organisms living in still freshwater, in lakes or ponds.

limon (French) a superficial fine-grained deposit widespread in northern France, spread like a blanket regardless of minor relief, and providing brown, loamy soils. It seems to have been deposited, as wind-sorted, wind-borne material, around the margins of retreating ice sheets in the Great Ice Age, and in that its origin may be compared with that of LOESS. Some authors distinguish between the wholly wind-sorted, wind-borne limon and the wind-sorted, wind-borne limon which may have been reworked and redeposited by later stream action. G.

limonene a liquid terpene, $C_{10}H_{16}$, with lemon fragrance, occurring in many ESSENTIAL OILS.

limonite hydrated ferric oxide, $2Fe_2O_3.3H_2O$, forming a brown or yellowish-brown ore of IRON. It is widespread and important, although of low-grade. It occurs particularly in Lorraine.

limpo (Brazil: Portuguese) a type of SAVANNA in Brazil, grass with few trees.

linch **1.** one of the alternative spellings of LYNCHET, applied to rising ground, or a ridge, or a ledge, especially one on the side of a hill in chalk downland **2.** an unploughed strip making a boundary between fields. G.

Line, The colloquial term for the equator, as in 'crossing The Line'. G.

linear *adj.* **1.** of, or in, lines **2.** long, narrow and of generally uniform width **3.** involving one dimension only in a unit of measure **4.** in mathematics, having the property of being able to be shown as a straight line on a graph, e.g. as in LINEAR MODEL or linear regression analysis (REGRESSION ANALYSIS) **5.** having or involving the property that a change in one quantity corresponds to or is accompanied by a directly proportional change in another quantity.

linear eruption a FISSURE ERUPTION.

linear growth arithmetic growth, simple interest, the growth of a value over a period based on a fixed percentage calculated on the original principal only, so that the sum added at the end of each successive period is constant (unlike that in EXPONENTIAL GROWTH). ARITHMETIC PROGRESSION.

linearity the property of being LINEAR-4, e.g. in the relationship between variables on a graph, when the points tend to fall in a straight line, as exhibited in a LINEAR MODEL.

linear model a MODEL-4 in which the DEPENDENT VARIABLE is assumed to be related in direct proportion to one or more INDEPENDENT VARIABLES. When the variables are shown against each other on a graph, the line of best fit (a line drawn to come as close as possible to the data points) can be drawn. The fitting is accomplished by using the least squares method (REGRESSION ANALYSIS, including linear regression analysis).

linear programming an OPTIMIZATION MODEL devised in such a way that the objective function (the quantity to be maximized or minimized) and the identified constraints are LINEAR-4, the optimum solution being reached by way of ITERATION.

linear regression analysis REGRESSION ANALYSIS.

linear scale SCALE.

linear settlement, linear town an elongated settlement, especially an elongated urban settlement, developed alongside a major routeway on account of either the constriction of the terrain (e.g. valley slopes rising steeply from the routeway) or the advantages of the ease of transport provided by the routeway.

linebreeding the inbreeding of animals with a common ancestor so as to strengthen or maintain a desirable characteristic.

linen a fabric, yarn or thread made from the fibres of FLAX, *Linum usitatissimum*, the fabric varying in coarseness from cambric, the finest, to canvas, the most coarse.

line of best fit REGRESSION ANALYSIS.

liner a large passenger ship or aircraft (airliner) or one of a fleet of trucks belonging to a regular line operator, i.e. the operator of a regular service.

line scanning a process in which a facsimile device such as a cathode-ray tube produces an IMAGE-3 by viewing and recording an object or scene etc. one line at a time.

lines the barracks and huts arranged in lines in a CANTONMENT. In India in the period of the British raj the homes for Europeans, similarly laid out in regular rows, were known as CIVIL LINES. G.

line squall a phenomenon associated with the passage of a COLD FRONT, in which violent storms with strong gusts of wind and heavy precipitation occur simultaneously along a line extending in some cases from 480 to 650 km (300 to 400 mi) in length. SQUALL. G.

ling *Trapa bicornis*, a water plant cultivated in China, Japan, Korea, the edible seeds of which are boiled for various dishes, or preserved in honey and sugar, or ground for flour. WATER CHESTNUT.

lingua franca (Italian) originally applied to a mixed language used in the eastern Mediterranean area (the LEVANT) and based on Italian. Now applied to any language usually, but not always, simplified, and used as a

means of communication between peoples of different tongues.

linguistics the scientific study of LANGUAGE-1 or of LANGUAGES-2, divided into several branches, the major distinction usually being made between diachronic linguistics (concerned with history and language change) and synchronic linguistics (the study of the state of language at any point in time, descriptive and structural). SOCIAL SCIENCES.

link in a NETWORK-2,3,4, a line, a route, an edge (GRAPH THEORY) between NODES-4.

linkage 1. the contact and the flow of information between two individuals, the connexion between and within different types of activities or different functions. FUNCTIONAL INTERDEPENDENCE, INDUSTRIAL LINKAGE 2. in SYSTEMS ANALYSIS, any sequence of behaviour originating and recurring in one system which gives rise to a reaction in another.

linked data PAIRING.

linked industries SECONDARY INDUSTRIES where the final product, e.g. motor vehicles, depends on many separate preparatory processes and materials. INDUSTRY.

links (Scottish, always pl.) a narrow coastal strip with accumulations of blown sand giving rise to sand dunes, supporting coarse grass and low shrubs, so often used for the game of golf that the term golf links, or links, is now regarded as almost synonymous with GOLF COURSE. G.

linn (Scottish and northern England) variously applied to 1. a cascade, waterfall, or a torrent flowing over rocks 2. a pool into which a cataract falls 3. a precipice, a ravine with steep sides. G.

linseed the seed of the FLAX plant, *Linum usitatissimum*, a valuable source of oil (LINSEED OIL). The best seed producers are the plants grown in tropical and subtropical climates, e.g.

in the Indian subcontinent, or in Argentina. LINSEED CAKE, LINSEED MEAL.

linseed cake the residue of linseed after the oil has been expressed, used as cattle food.

linseed meal ground LINSEED, used medicinally.

linseed oil the DRYING OIL expressed from LINSEED, used in the making of paints, varnishes, linoleum, and some printing ink, and as a wood preservative.

liquefaction the action or process of making liquid, of reducing to a liquid condition; or the state of being so reduced. LIQUID.

liquid fluid matter having a definite volume but no definite shape, taking on the shape of its container but, unlike a GAS-1, not expanding to fill the containing vessel, i.e. it keeps its own volume at any given temperature. CONDENSATION, FLUID, GAS, LIQUEFACTION, SOLID.

liquid *adj.* having the form of a LIQUID.

liquorice *Glycyrrhiza glabra* a PERENNIAL herb of the LEGUMINOSAE family, cultivated in Europe, especially in the Mediterranean region. It has an extensive system of rhizomes and roots that are used medicinally and as a flavouring, being pulped and boiled in water to produce liquorice extract, which is concentrated by evaporation to powder and used in cough pastilles and cough mixtures, confectionery and beverages.

lis, liss (Irish; Welsh llys) a circular enclosure protected by an earth wall, commonly used as a fort. G.

litchi, litchee, lychee *Litchi chinensis*, a medium-sized tree of the Sapindaceae family, native to China, cultivated in the drier parts of tropical and subtropical regions for its fruit, which has succulent, slightly acid, firm pulp encased in a brittle, rough skin. It is usually eaten raw, but is also canned and preserved.

literacy the state of condition of being able to read and/or write (and, in some contexts, of having a knowledge of LITERATURE). ARTICULACY, GRAPHICACY, NUMERACY.

literature 1. written works (prose or verse) of artistic value and lasting quality 2. written works produced in a certain country or during as certain period 3. written works dealing with a particular subject.

lithic *adj.* 1. of or pertaining to stone 2. of or pertaining to LITHIUM.

lithification 1. the process of forming into stone 2. in geology, the result of the transformation of an accumulation of loose SEDIMENTS-2 into a rock mass, e.g. SANDSTONE, SILTSTONE or SHALE, by the processes of CEMENTATION, COMPACTION, COMPRESSION. DIAGENESIS, LITHOGENESIS.

lithium a silvery-white element of the alkali metal group (ALKALI), the lightest known metal, used in alloys. It may be used as a fuel in nuclear fusion reactors.

lithogenesis the forming of solid rocks, the accumulation of sediments, especially sand and mud (e.g. in a GEOSYNCLINE), before DIAGENESIS. LITHIFICATION.

lithographic stone an extremely fine-grained limestone, formerly used in lithographic printing, a process in which a smooth surface is prepared so that the image to be printed is receptive to printing ink, the remainder being ink-repellent. The heavy bulky stones were superseded by plates made of lighter materials.

lithology 1. the study of rocks (in contrast to MINERALOGY) 2. the study of the composition, colour, texture and structure of a rock formation (of more concern to the geographer and soil scientist than is the geological age of the rocks) 3. the general physical characteristics of a rock, especially the visible characteristics. G.

lithometeor an assemblage of any solid

particles (apart from water) in the atmosphere. METEOR-2.

lithomorphic soils one of the seven groups in the 1973 SOIL CLASSIFICATION of England and Wales. They include RANKERS and RENDZINAS, and are generally well-drained soils with an organic surface layer lying on bed rock or little altered parent material which underlies the soil layer at a depth of some 30cm (12 in).

lithophyte a plant growing on a rock or stone.

lithosere a PRISERE originating on an exposed rock surface. PSAMMOSERE, XEROSERE.

lithosol a thin, stony, young, AZONAL SOIL or ENTISOL consisting of rock fragments, dominantly mineral, weathered not at all or only in part, lacking a defined B HORIZON, the thin A HORIZON merging into or resting on shattered hard rock near the surface. Some authorities equate a lithosol with a SKELETAL SOIL. G.

lithosphere 1. the earth's CRUST, including the SIAL and SIMA layers above the MOHOROVICIC DISCONTINUITY 2. the sial, sima and upper part of the MANTLE above the Gutenberg Channel (GUTENBERG DISCONTINUITY). ASTHENOSPHERE, ATMOSPHERE-1, BIOGEOSPHERE, GEOTHERMAL GRADIENT, HYDROSPHERE, PLATE TECTONICS. G.

lithostratigraphy STRATIGRAPHY.

litmus LICHEN.

litre a metric unit of capacity, defined as the volume of 1 kilogram of pure, air-free water at 4°C and 760 mm pressure = 1000.027 cubic centimetre, rounded to 1000 cc, equivalent approximately to 0.22 British gallons. CENTI-, CONVERSION TABLES, HECTO-, KILO-, MILLI-.

litter 1. the layer of leaves, twigs and other organic remains lying on the soil surface (especially on the forest floor), which may decompose to form HUMUS (L LAYER). PATABIONT 2. straw and other materials used as bedding for animals or the protection of growing plants 3. the young of a multiparous animal, produced at one birth 4. waste material, refuse, lying about in disorder.

Little Ice Age 1. the period c.3000 to 500 BC, but varying with latitude (SUBBOREAL), when the climate became cooler than in the preceding millenia, and GLACIERS reappeared in parts of Alaska and the Sierra Nevada. NEOGLACIAL 2. the advance of glaciers in the Alps between 1550 and 1850 AD.

little millet *Panicum miliare*, a MILLET widely grown in temperate regions, especially in the Indian subcontinent near the Tropic of Cancer, yielding well even on poor soil, providing an important grain used for human food.

littoral *adj.* of, on, or along, the shore, whether of sea, lake or river. G.

littoral current a CURRENT in the zone of the SURF running parallel to the SEASHORE, caused by waves breaking obliquely to the shore. LONGSHORE CURRENT.

littoral deposit the sand, shells, shingle, deposited in the area between the marks of HIGH WATER and LOW WATER. Some authors include the offshore mud and all the shallow water marine deposits. G.

littoral district a region alongside a shore. G.

littoral drift the movement of material in a LITTORAL CURRENT.

littoral zone variously applied to 1. the aquatic zone between the marks of HIGH WATER and LOW WATER 2. the upper part of the BENTHIC DIVISION, from the water surface to a depth of about 200 m (110 fathoms: 655 ft) 3. the part of the benthic division favourable to the growth of green plants (PHOTIC). In a lake this usually applies to the part extending from the shore down to the limit for rooted plants (SUBLITTORAL); in the ocean to the part which usually has strong waves and currents, the upper part of it forming the tidal zone, the lower (the sublittoral zone) starting at about 40 to 60 m (20 to 30 fathoms: 130 to 200 ft). G.

livestock the DOMESTIC ANIMALS commonly kept on a subsistence or commerical basis to provide food (e.g. eggs, MEAT, MILK) and other raw materials (e.g. BONEMEAL, HIDES, WOOL) for people **1.** in general, applied to cattle, sheep, pigs to which may be added poultry (chickens, ducks, turkeys) (often); and goats and horses (sometimes) **2.** FAO international statistics include cattle, sheep, pigs, poultry (chickens, ducks, turkeys), buffaloes, horses, mules, asses, camels (all, except the mules, asses and camels being slaughtered for meat). FAO also include indigenous animals as livestock, as providers of meat, i.e. beef and mutton as well as buffalo, goat and pig meat. But reindeer (important in semi-NOMADISM in Lapland), caribou and other deer and antelope, all providing meat, skins etc., are not included. See references to all the animals listed, and LIVESTOCK FARMING.

livestock farming, stock farming, pastoral farming, pastoralism the farming activity based on the rearing of animals (LIVESTOCK) for eggs, HIDES and SKINS, MEAT, MILK, WOOL etc. as distinct from CROP FARMING. The methods used and the size of the enterprise vary greatly, from NOMAD-ISM to small scale farming (where, for example, a small herd of dairy cattle may be kept) to large scale RANCHING, or to the large scale, highly organized, scientific rearing of animals under cover in a completely controlled environment (the term PASTORAL being perhaps an inappropriate description of the last). BATTERY SYSTEM, FACTORY FARM, GRAZING, PASTURE.

living space the English translation of LEBENSRAUM, the German being commonly used in geopolitical discussion. G.

liwa (Iraq: Arabic) a province. Each liwa is subdivided into qadhas and nahyah.

llama any of the several DOMESTIC or wild RUMINANT MAMMALS native to South America, closely allied to and resembling a small camel (but without the hump), used as a transport animal, especially as a beast of burden, and prized for its thick, woolly coat, the fibres of which make durable, warm cloth.

llan (Welsh) originally an early religious enclosure, later applied to a church, or an enclosure. G.

llano (Spanish, a plain, level ground) a term applied especially to the extensive plains in the basin of the Orinoco in South America. These are treeless, and the term came to be transferred, in international literature, to the type of vegetation, i.e. tropical grassland, or SAVANNA. G.

L layer the organic LITTER-1 lying on the land surface and not yet incorporated in the soil. O HORIZONS, SOIL HORIZON.

loading in statistics, FACTOR ANALYSIS.

load of a river, load of a stream all the solid matter transported by a river or stream, by being rolled and bounced (SALTATION) along its bed (BED LOAD), or carried in SUSPENSION or in SOLUTION. COMPETENCE OF A STREAM. G.

loadstone LODESTONE.

loam an old term variously applied, now best restricted to a soil having clay and coarser materials in such proportion as to form a PERMEABLE, FRIABLE and easily worked mixture. This is attained, according to standards in the USA, when clay is 7 to 27 per cent, silt 28 to 50 per cent and sand less than 52 per cent (GRADED SEDIMENTS). Loam is termed clay loam if the clay content lies between 27 and 40 per cent and the sand between 20 and 45 per cent; silty loam if clay is below 30 per cent, sand between 20 and 50 per cent, silt between 72 and 80 per cent; sandy loam if clay is between 10 and 20 per cent, sand between 50 and 70 per cent (SOIL TEXTURE). The best agricultural soils are usually loams, hence tracts with loamy soils, or loam-terrains, have always attracted settlers. G.

lobate delta BIRD'S FOOT DELTA, DELTA.

lobe a tongue-shaped mass, applied to **1.** such a mass of ice, or of drift, projecting from a larger mass **2.** wet clay in such a shape, creeping down a steep slope **3.** in USA, a tongue-shaped tract of land enclosed by a well-defined MEANDER.

local climate a general term applied to the climate of a small area (larger than that of a MICROCLIMATE), e.g. a valley with a particular aspect, which differs in one or more elements from the climate of nearby areas, those within, say, a kilometre or less. The difference(s) may be caused by slight variations in slope, aspect, type of soil or vegetation, or the presence (or absence) of water or tall buildings. MACROCLIMATE, MESOCLIMATE. G.

local government in Britain before the reforms of the 1960s and 1970s, local government in Britain was divided broadly between URBAN-1 and RURAL authorities. London was governed by the London County Council, with a second tier of government, metropolitan borough councils, concerned with the small areas within it. Outside London all the largest and some of the smaller towns came under the authority of COUNTY BOROUGH councils, each entirely responsible for all government services within its area. The COUNTIES were divided into second tier authorities, i.e. non-county boroughs (larger towns), urban districts (smaller towns) and rural districts (some rural districts containing some sizeable settlements, RURAL POPULATION). Administration in Scotland was similar, but the main

cities (Aberdeen, Dundee, Edinburgh, Glasgow) had their own special systems of government.

The reforms of the 1960s and 1970s included in their aims the elimination of some of the long-accumulated anomalies of the old system, the achievement of the ECONOMIES OF SCALE presumed to be attached to the proposed new larger units, and the recognition of the interdependence of urban and rural areas (by including both in the proposed new units, i.e. those to be classified as non-metropolitan counties).

The administration in London was reorganized in 1965, the Greater London Council being created, with a second tier, the London boroughs, below it. Administration in the rest of England and Wales was reformed in 1974, in Scotland in 1975. In England and Wales, outside London, the six largest CONURBATIONS became metropolitan counties, with 36 metropolitan districts forming the second tier; and outside London and the metropolitan counties (the conurbations), 47 non-metropolitan counties were formed, comprising urban and rural areas. This involved the merging of some of the smaller counties of the old system and the creation of some new counties. The non-metropolitan counties were subdivided into 333 non-metropolitan districts (plus one in the Isles of Scilly). In Scotland nine regions were created (roughly corresponding to the non-metropolitan counties of England and Wales); and the regions were subdivided into a second tier, 53 districts and 3 islands authorities. In Scotland there were also relatively powerless community councils, forming a third tier. As a result of all this reorganization in Britain, the number of local authorities was reduced from the 1500 of the old system to the 522 of the new, some of them very large. In 1984 the government proposed the abolition of the metropolitan counties and the Greater London Council with effect from 1985.

local government in the USA The organization of local government varies from one STATE-5 to another in the USA, because it is arranged by each individual state, not prescribed by federal government (FEDERAL-2, *adj*). There are however some administrative units common to most states: the COUNTY, the TOWNSHIP-3, the MUNICIPALITY, the SCHOOL DISTRICT, the SPECIAL DISTRICT; and there is a growing number of standard metropolitan statistical areas (METROPOLITAN AREA).

local relic RELIC DISTRIBUTION.

local relief the difference between the highest and the lowest altitudes in a limited area. RELATIVE RELIEF, RELIEF.

local time local solar time, APPARENT TIME, the time expressed with reference to the MERIDIAN of a given place, e.g. it is twelve noon local time when the sun's centre crosses that meridian and the shadows of vertical objects at the given place are at their shortest. It would be inconvenient for every place to use its local time, hence the STANDARD TIME adopted for use in most countries. ANALEMMA, EQUATION OF TIME, MEAN SOLAR TIME, ROTATION OF THE EARTH APPENDIX 3. G.

locality 1. the situation or position of an object in time or space 2. an undefined area, the particular site occupied by certain persons or things, or the scene of certain activities.

localization economies the economies achieved by setting up an enterprise in a particular area where the components needed by the enterprise are readily available in the locality.

localization of industry the concentration of an industry in a certain district or districts. Various means have been proposed to measure the degree of concentration, such as a LOCATIONAL (LOCALIZATIONAL) COEFFICIENT. G.

location 1. geographical situation, a part of SPACE-2 or a point or position in space where objects, organisms, FIELDS-3 or events may be found. ABSOLUTE LOCATION, RELATIVE LOCATION 2. the action of placing or the condition of being placed 3. the fact or condition of occupying a particular place 4. a local position, or a position in a series or succession 5. the action of finding the position of someone or something; or the ability to do this 6. in Australia, a farm or STATION-4. MEASURE OF LOCATION, PLACE, POSITION, SITUATION.

locational *adj.* of or pertaining to LOCATION.

locational accessibility ACCESSIBILITY measured in terms of the actual distance to be travelled to reach a place or a service. EFFECTIVE ACCESSIBILITY.

locational analysis 1. the study of the LOCATION of economic activity 2. in urban studies, a set of theories seeking to explain the spatial distribution of urban land uses in terms of economic factors, particularly those associated with NEOCLASSICAL ECONOMIC THEORY.

location allocation model any OPTIMIZATION MODEL used to find the OPTIMUM location for central facilities in view of the need to keep down to the minimum the costs of transport, etc.

locational (localizational) coefficient a statistical measure expressing the relative amount of a function or of a population present in a part of a large region. In P. Sargent Florence's formula the workers are shown region by region as percentages of the total in all regions, and the coefficient is the sum (divided by 100) of the plus deviations of the regional percentages of workers in the particular industry from the corresponding regional percentages of workers in all industry. Uniform dispersal gives a coefficient of 0, extreme differentiation a coefficient of 1. G.

locational triangle a MODEL-4 suggested by Alfred Weber for determining the OPTIMUM location for an industrial enterprise. He visualized the location of the source of materials and of cheap labour and the location of the market to be fixed, and the cost of movement (not limited to a particular direction) to be constant per unit of distance. The points of the triangle drawn represented a single market point and the sources of two materials. The weight of the two commodities to be moved from their respective sources to the market as well as transport costs per unit distance being given, it is possible to find the LEAST COST LOCATION for the enterprise within the drawn triangle. Total transport costs at different locations can be calculated, and thus contours of equal additional transport costs (isodapanes, ISO-) drawn from the given information on the cost of the given two materials combined with the delivery cost of the products to the market. This makes the model flexible and responsive, allowing factors other than least transport cost (e.g. the location of a cheap labour pool, or AGGLOMERATION ECONOMIES) to be the overriding consideration in the fixing of optimum location. WEBERIAN ANALYSIS.

location factor a measure of any particular industry in any given area provided by comparing the proportion of all occupied persons who were occupied in that industry in the given area with the corresponding proportion for the country as a whole. The technique, introduced by P. Sargent Florence in 1937, can be applied to any particular activity or characteristic in an area compared with a specified norm. LOCATION QUOTIENT.

locationist approach LANDSCAPIST APPROACH.

location of industry the areal distribution of industrial activities. G.

location quotient LQ, a measure used to show the degree of concentration of a particular activity or characteristic in an area compared with a specified norm, e.g. the degree to which a particular industry is concentrated in a particular part of the country. Using this example, the location quotient for an area is the ratio between the area's share of the national total of the activity under consideration and its (the area's share) of all activities. For example, if engineering is the particular activity, the location quotient for the area is calculated by dividing the percentage of all the engineering workers in the country employed in the area by the percentage of the total employed population working in the area. If the resultant location quotient exceeds 1.0 it indicates a 'surplus' in the share of a particular activity or characteristic, if less than 1.0 a shortfall. INDEX OF SEGREGATION, LOCATIONAL (LOCALIZATIONAL) COEFFICIENT, LOCALIZATION OF INDUSTRY. G.

location theory a theory which attempts to explain and predict the location of economic activities. The traditional, classical theory is based on three levels of observation: the location of an enterprise; the location of groups of enterprises judged to be in stable competition with each other; the location of sets of activities (e.g. different kinds of agricultural land use) in relation to each other, so that all activities are in stable competition. More recent theories incorporate BEHAVIOURAL assumptions; and some stress more refined economic concepts. INDUSTRIAL LOCATION THEORY, LEAST COST LOCATION, LOCATION QUOTIENT, MINIMAX LOCATION, OPTIMUM LOCATION, VON THUNEN MODEL, WEBERIAN ANALYSIS. G.

loch (Scottish; Gaelic and Irish) **1.** a lake **2.** an arm of the sea (a sea loch), especially if it is narrow and bounded by steep sides, or partly landlocked. G.

loch, loch-hole (England, Derbyshire, a miner's term) a cavity in a mineral vein. VUG. G.

lochan (Scottish, from Gaelic) a small loch or lake, especially if lying in a CIRQUE. G.

lock a stretch of canal or river confined within gates so that the water level can be controlled to lift or lower a vessel to allow it to pass from one REACH of navigable water to another at a different level.

lockage 1. the system of locks on a waterway **2.** any of the activities connected with a LOCK, i.e. the passing through of a vessel, the toll extracted, the amount of rise or fall of a vessel, the volume of water transferred in the opening and closing of the lock gates.

locust a large insect of several species of Acrididae, a family of grasshoppers, which has long been an agent of devastation. There are breeding grounds in the arid regions of the tropical areas of the Old World from which vast numbers periodically fly in great clouds to invade cultivated fields and eat every green plant on which they settle.

lode 1. a mineral vein or systems of veins in a rock **2.** an old local term in Cambridge and East Anglia applied to a water-course, partly artificial and embanked.

lodestar, loadstar a star that shows direction, e.g. the pole star.

lodestone, loadstone 1. a variety of MAGNETITE, a strongly magnetic oxide of iron, Fe_3O_4 **2.** a piece of this mineral used as a MAGNET.

lodge 1. a small house at the gates of the park or of the grounds of a large COUNTRY HOUSE **2.** a house occupied only during the sporting season (hunting lodge) **3.** the den of a beaver or otter **4.** the TEPEE or WIGWAM of an American Indian.

lodgement till 1. in general, GROUND MORAINE **2.** more precisely, debris deposited under ice while the ice is moving, the sticky clay being lodged on the under surface of the ice and the longer axes of the larger stones being aligned with the direction of movement of the ice, often resulting in DRUMLINS. ABLATION TILL, TILL (includes deformation till, deformed lodgement till).

loess, löss (from German dialect lösz; spelling loëss is erroneous) originally applied to a fine-grained yellowish loam occurring in the valley of the Rhine and elsewhere in Germany, comparable with the LIMON of France. Internationally the term has come to be applied to the fine-grained AEOLIAN deposit, the PERMEABLE, wind-sorted and wind-deposited morainic material, unconsolidated and unstratified, laid down away from the margins of the great ice sheets of PLEISTOCENE times, and covering vast areas in central Asia, Europe, North America and elsewhere. It ranges from clay to sand (GRADED SEDIMENTS), is buff or brownish in colour, is usually calcareous and contains concretions of CALCIUM CARBONATE and in some cases iron oxide. It has been suggested that increased rainfall helped to wash the fine-grained material down from the air to the ground; and also that much may have been reworked and redeposited by later stream action. The soils derived from loess are of high quality, being fine in texture, well-drained, fertile, deep and easily worked. G.

log 1. a long and heavy piece of a branch or trunk of a tree, trimmed, but with the bark attached **2.** a nautical device to gauge a ship's speed, consisting traditionally of a float, trailing at the ship's stern, with a line divided into equal parts by knotted cords attached to a reel on the ship (KNOT); now replaced by a mechanical device **3.** in navigation and aeronautics, a log book, a book in which all details of the progress of a voyage or flight are recorded; or the record itself.

loganberry *Rubus loganobaccus*, a hybrid between a blackberry and a raspberry, a climbing, prickly or thornless plant with arching canes, developed in California by James L. Logan, 1841-1928, widely cultivated in the northern hemisphere, sometimes on a commercial scale, for the sake of its juicy, dark red berries, eaten cooked or uncooked, canned or used in jam making. The canes may bear heavy crops for fifteen years or so.

logan stone (southwest England) a rocking stone. The original is a block of GRANITE in the Land's End peninsula, so balanced that it can be rocked by hand, the result of the chemical weathering that produces TORS. G.

logarithmic scale a scale in which an increase of one unit represents a power increase in the quantity involved.

logger (North America) a lumberman, one who fells timber or cuts it into LOGS-1.

logging (North America) **1.** the felling of timber and the cutting of if into LOGS-1 **2.** the quantity of felled timber.

logical positivism a body of philosophical thought developed from the late 1920s under the leadership of Schlick and Carnap in Vienna, the main aim of which was to create a comprehensive philosophy of science stemming from EMPIRICISM and proceeding by INDUCTION. Among its tenets, logical positivism maintained that traditional METAPHYSICS, consisting of propositions that could not be verified by empirical observation, was without meaning (thus religious and moral statements, being metaphysical, were considered by most logical positivists to be meaningless); and that philosophy generally consisted of an analysis of language, of the discovery of verbal forms of expression for complex ideas and propositions, helped by formal logic. Logical positivism differs from POSITIVISM in that it accepts that some statements are verifiable without recourse to experience. FALSIFIABILITY.

Lo-Lo lift-on/lift-off of CONTAINERS. RO-RO.

loma (South America: Spanish)**1.** a small, broad-topped hill, or rising ground on a plain, a term appearing in place-names **2.** according to P. E. James, the Peruvian term for the growth of quick-flowering plants and grasses promoted by the GARUA. G.

longhouse the communal dwelling-house of certain peoples, e.g. the Land Dyaks of Sarawak.

longitude the angular distance between the MERIDIAN passing through a given point and the prime, standard, initial or zero meridian (usually considered to be the meridian passing through the old Observatory at Greenwich, London, England, and numbered 0°). This angular distance, i.e. longitude, is measured in degrees, minutes and seconds east or west of the Greenwich meridian (0°) to 180°, the meridian 180°E thus coinciding with 180°W. All points through which a meridian passes have the same longitude. The alternative term for meridian is line of longitude; and all lines of longitude meet all PARALLELS OF LATITUDE at right angles. The actual distance represented by 1° of longitude becomes less as the meridians converge towards the poles. Thus at the equator the distance is roughly 111 km (69 mi), at latitude 45° it is 78.8 km (nearly 49 mi), and at the poles it is zero. Fifteen degrees of longitude are equivalent to a difference of one hour in LOCAL TIME. LATITUDE. G.

longitudinal *adj.* **1.** of or pertaining to LONGITUDE **2.** of or pertaining to length as a dimension **3.** running lengthwise, e.g. LONGITUDINAL COAST, LONGITUDINAL STUDY.

longitudinal coast a CONCORDANT COAST, one running broadly parallel to the main geological structure or fold lines. Commonly occurring around the Pacific ocean, it is also termed a Pacific coast, in contrast to a transverse or ATLANTIC COAST. G.

longitudinal crevasse a crack or fissure in a GLACIER with a trend nearly parallel to its sides, formed by the lateral spreading of the ice. CREVASSE, ICE FALL, TRANSVERSE CREVASSE. G.

longitudinal dune a SAND DUNE with its crest running parallel to the direction of the prevailing wind. SAND RIDGE DESERT, TRANSVERSE DUNE.

longitudinal profile of a river long profile, RIVER PROFILE.

longitudinal study, longitudinal analysis a study or analysis in which selected variables are studied in the same group or groups of subjects at intervals over a period of time, often over many years, as distinct from a cross sectional study or analysis, in which a cross section of the population is sampled, similar variables being studied at different ages, but on different subjects at each age. Thus in a cross sectional study the subjects are approached only once. ECOLOGICAL FALLACY, HISTORICAL FALLACY.

longitudinal valley a VALLEY with a trend roughly parallel to the STRIKE of the rocks or the grain of the country. G.

longitudinal wave L-wave, a form of shock wave produced by an EARTHQUAKE passing over the land surface and responding to the character of the rocks encountered. LONGITUDINAL WAVE MOTION, SHAKE WAVE, TRANSVERSE WAVE, WAVE.

longitudinal wave motion a form of transference of energy in a material medium, in which the disturbance of particles in the medium displaces the particles in the direction of propagation of the wave, e.g. a LONGITUDINAL WAVE. PUSH WAVE, RAYLEIGH WAVE, SHAKE WAVE, TRANSVERSE WAVE MOTION, WAVE.

long profile of a river the PROFILE-3 of a river bed, from the source of the river to its mouth. PROFILE OF A RIVER, TALWEG.

long-range weather forecast a weather forecast for a period usually longer than five days, based on the handling by computer of great quantities of data,

and the use of a synoptic ANALOGUE-1.

longshore bar BAR.

longshore current a current generated by tides, waves, winds, running generally parallel to the coast. LITTORAL CURRENT, LONGSHORE DRIFT.

longshore drift the movement of material along a beach (not only in the surf zone) by the action of waves that meet the shore at an angle. The SWASH of a BREAKER carries the material up the beach at an oblique angle, the BACKWASH pulls some back seawards at right angles, so that there is a gradual build-up of material along the beach, reinforced if there is a LONGSHORE CURRENT. GROYNE, LITTORAL DRIFT.

lōō, look (Indian subcontinent: Urdu; Sindhi look) a very hot dust-laden wind; or a HEAT WAVE. G.

loop district, the loop originally that part of the DOWNTOWN business district of Chicago, enclosed by a loop of elevated railways; later, by analogy, applied in other towns to what came to be called the CBD, or CENTRAL BUSINESS DISTRICT. G.

lopolith a large INTRUSION of igneous rock CONCORDANT with the strata, allied to a LACCOLITH or PHACOLITH, but having the form of a saucer-shaped basin. GABBRO. G.

loquat *Eriobotrya japonica*, family Rosaceae, a small EVERGREEN tree native to China, cultivated in China, Japan, the north of the Indian subcontinent, parts of the Mediterranean region and parts of tropical areas, for the sake of its sweet-acid, small, pear-shaped fruits, eaten cooked or raw, and used in jam and jelly making.

lord of the manor COMMON, FEUDAL SYSTEM, MANOR.

lōrā (Indian subcontinent: Baluchi) a hill TORRENT (carrying rainwater). G.

Lorenz curve a curved line drawn on a graph to show, by its concavity, the extent to which a distribution (e.g.

concentration of population) differs from a uniform distribution (INDEX OF SEGREGATION). Percentage values are used for each axis, the uniform distribution appearing as a straight line drawn across the graph at 45°. GINI COEFFICIENT.

Lösch's theory CENTRAL PLACE THEORY.

lotic (Latin, washing) *adj.* of or pertaining to organisms or habitats in rapidly moving water, e.g. streams, and lakes subject to wave action. LOTIC WATER.

lotic water flowing water. LENTIC, LOTIC.

lot meadow a form of LAMMAS LAND marked off in separate portions for which the commoners draw lots each year.

lotus of the many plants with this popular name those most frequently used for human food are 1. *Nelumbium niciferum*, a water plant native to Asia, the rhizomes of which, when young, are roasted, steamed or pickled, in China sometimes ground into flour. The seeds are boiled or roasted or eaten raw; and the fruit, flower stem and leaves are also edible, commonly eaten raw 2. any of the species of water lilies, *Nymphaea*, of world-wide distribution, with rhizomes which can be roasted or dried and ground into meal, the seeds roasted and used in sauce, or also ground into meal.

louderback displaced segments of LAVA flow on the sides of a FAULT. G.

lough (Irish; Scottish LOCH) a lake or narrow arm of the sea. G.

lovage any of the several herbs of the Umbelliferae family, especially *Levisticum officinale*, native to southern Europe, formerly used as a vegetable and herbal tea, now more commonly used mainly in flavouring and perfumery.

low, atmospheric a low pressure system or DEPRESSION-3 in the atmosphere, indicated on a weather chart by closed isobars (ISO-), the values of which fall towards the centre. If the strength of the air flow around the system increases, the depression is said to intensify. ANTICYCLONE, CYCLONE, V-SHAPED DEPRESSION.

low, on a beach SWALE.

low grade metamorphism METAMORPHISM by low temperatures and low pressure, as distinct from HIGH GRADE METAMORPHISM.

lowland, lowlands an imprecise term applied broadly to relatively level land at a lower elevation than that of adjoining districts, often used to contrast with highland, and sometimes defined as lying at an elevation under 180 m (600 ft). G.

low latitudes LATITUDE.

low negative correlation in statistics, CORRELATION COEFFICIENT.

low pressure system LOW, ATMOSPHERIC.

Lowry model a computer MODEL-4 devised by I. S. Lowry to portray the spatial organization of human activities in an urban area, so that the impact of public decisions on such an area can be evaluated and changes in urban form predicted in view of anticipated changes in key variables such as the pattern of employment, the efficiency of the transport system, or the growth of population.

Using ECONOMIC BASE THEORY, the distribution of 'primary' and manufacturing employment (BASIC ACTIVITIES) in the area are plotted on one mile square cells (Lowry's measure) on the basis of POPULATION POTENTIAL-2; and service employment (e.g. retail activities) is plotted in proportion to the market or employment potential in each cell. The service activities naturally create further employment opportunities, located to take advantage of the market potentials. Constraints on land use in the cells are incorporated in the model (e.g. the minimum sizes of clusters of services, maximum housing densities etc.). By ITERATION more and more residents come into the cells, until the market potentials are so disturbed that the pattern of retail activities has to be modified. These activities are re-allocated to the limits of the land use constraints, so that eventually a state of EQUIILIBRIUM is reached, the final distribution of the population corre-sponding to that used initially to com-pute potentials. G.

low technology the application of the knowledge of and/or the use of simple methods, simple equipment and readily available inexpensive materials to a task or an industrial process, or to the solution of a problem arising from the interaction of people with their EN-VIRONMENT. HIGH TECHNOLOGY, TECHNOLOGY-3.

low tide 1. the TIDE at lowest ebb 2. the level of the sea or the time of the lowest ebb. EBB TIDE, HIGH TIDE, NEAP TIDE, SPRING TIDE.

low velocity zone (LVZ) ASTHENO-SPHERE.

low water 1. LOW TIDE 2. the low level of water in a lake or river. HIGH WATER.

loxodrome, loxodromic curve RHUMB LINE.

lubia *Dolichos lablab*, a tropical plant of the LEGUMINOSAE family, with edible pods and seeds, grown for human consumption and for fodder in tropical areas, especially in Egypt and Sudan. LABLAB.

lucerne ALFALFA.

lumb (England, Yorkshire) locally in the Sheffield area, a steep-sided valley. G.

lumber (American) TIMBER, especially that recently felled and roughly sawn into LOGS-1 and planks.

lumbering (American) the act of 1. felling and sawing of TIMBER and removing it from the area 2. felling trees and sawing them into LOGS-1.

lumberjack (American) one who cuts TIMBER and prepares it for the sawmill or market.

lumberman (American) one employed in the felling of trees and in preparing them for market, especially as a man-ager.

lunar *adj*. of, relating to, similar to, the MOON.

lunar day the period of time in which the earth rotates once in relation to the MOON, i.e. 24 hours 50 minutes be-tween two successive crossings of the same meridian, despite the fact that the earth rotates once in 24 hours. The discrepancy is due to the orbiting of the moon itself around the centre of gravity of the moon and the earth, causing it to cross each meridian 50 minutes later: hence the interval of some 12 hours 25 minutes between one high tide and the next, i.e. high tide on one day is 50 minutes later than the corresponding high tide of the day before.

lunar month the period from one new moon to the next, averaging 29 days 12 hours 44 minutes. MOON.

lunar year a period of twelve LUNAR MONTHS.

lunate *adj*. in biology, crescent-shaped.

lunette a crescent-shaped dust dune consisting of black loam carried by dust-laden winds blowing over the lakes in Victoria, Australia, and lying on the LEEWARD side of the lakes. G.

lusaka (Africa: Bantu, pl. malusaka) dense bush vegetation in Zambia. G.

lusuku (Africa: Luganda language) a garden plot, averaging 0.5 to 0.8 ha (1 to 2 acres) in Uganda, the banana garden in which the homestead is traditionally set. G.

lutite, lutyte an ARGILLACEOUS sedi-ment or SEDIMENTARY ROCK consist-ing entirely of particles of clay-size (less than 0.002 mm in diameter). GRADED SEDIMENTS. G.

LVZ, low velocity zone ASTHENO-SPHERE.

L-wave LONGITUDINAL WAVE.

lychee LITCHI.

lynchet, linchet (old English dialect LINCH-1 and other spellings) probably originally applied either to a strip of green land between two pieces of ploughed land, or (as in present use) to a narrow terrace on a hillside, especially in the chalk downlands of southern England. In most cases lynchets seem to mark old cultivation strips, perhaps of IRON AGE or earlier, usually lying parallel to the contours to make a level, well-drained strip of land, accidentally or intentionally preventing soil erosion; but sometimes running down the hill-slope; and sometimes retained by a wall of stones. Their origin is much discussed. G.

lysimeter a simple device for measuring the percolation of water through the soil, and so of determining the soluble constituents removed in soil drainage. A container holding the material under investigation is fitted with instruments that measure the quantity or quality of the water that has passed through it, revealing what happens in the field. G.

M

Maar (German) **1.** originally, a small CRATER LAKE, encircled by a low ring of crushed COUNTRY ROCK, resulting from a volcanic eruption which has blown a hole through the surface rocks, unaccompanied by any igneous EXTRUSION **2.** later applied to the crater itself. The term is derived from the Eifel region of western Germany, where many examples occur (e.g. Lachermaar). G.

macadam, macadam road, macadamized a road building method and its material. Early in the nineteenth century a Scottish road engineer, John Loudon McAdam, 1756-1836, discovered that certain types of stone, broken into angular pieces of nearly uniform size, would bind together under pressure to form a hard road surface. The use of such stone under the pressure of specially designed rollers became standard practice in road making. Macadam is the term applied to the broken stone, the roads being termed macadam roads or macadamized roads. The binding of the stone is helped if the pieces are coated with TAR, hence tar-macadam, a later development. G.

macassar oil KUSAM.

macchia (Italian) MAQUIS. G.

mace a spice derived from the fleshy part of the fruit of the NUTMEG TREE. The flesh is flattened, dried and usually sold in the form of powder, used widely in cooking to enhance other flavours.

machair (Scottish) whitish, shelly sand forming a low plain along the western shores of Scotland, and in the Hebrides in South Uist and Tiree, providing useful light arable soils (machair soils). The term is also applied by some authors to the calcareous vegetation growing on it. G.

mackerel sky rows of CIRROCUMULUS CLOUD or of small ALTOCUMULUS CLOUD that resemble the pattern on the skin of the sides of mackerel (a food fish of the North Atlantic).

macro- (Greek) large, as opposed to MICRO-, small.

macroclimate the climate of a large region, in contrast to LOCAL CLIMATE, MESOCLIMATE, MICROCLIMATE.

macro-economics the branch of ECONOMICS concerned with the operation of an economy as a whole and its major component sectors (e.g. public and private sectors, housing, industry, construction, services etc.), including the relationship between the GNP, national economic growth or decline, taxation, inflation, imports/exports, prices, wages and the levels of employment and unemployment, as distinct from MICRO-ECONOMICS. MESO-ECONOMIC SECTORS.

macro-level analysis an approach used in the investigation of INDUSTRIAL MOVEMENT, involving the testing of aggregate statistical patterns by gravity and other models. MICRO-LEVEL APPROACH.

macronutrient an ELEMENT-6 or combination of elements needed in relatively large quantities by an organism in order to maintain health, in contrast to a MICRONUTRIENT. The macronutrients needed in varying amount by plant crops are carbon, hydrogen, oxygen (from the air), and calcium nitrogen, potassium, phosphorus, magnesium and sulphur (from the soil).

macrophagous *adj.* applied to an animal which eats large pieces of food at intervals. MICROPHAGOUS.

macrophyte a large aquatic plant, in contrast to a very small one, such as one of the PHYTOPLANKTON.

macroscopic *adj.* general, comprehensive, concerned with an overall view, with large units, rather than with minutiae, with small, exact details.

Madras gram HORSE GRAM.

madrepore any of the tropical CORALS of the order Madreporaria which, with members of the order Milleporina (MILLEPORE), are the chief constituents of CORAL REEFS.

maelstrom (Dutch) a large dangerous whirlpool, a term derived from the name of a notorious whirlpool caused by tidal currents in the Lofoten Islands, off the west coast of Norway, formerly supposed to suck in and destroy all vessels within a long radius. G.

maerdref (Welsh) a hamlet attached to a chief's court; a lord's DEMESNE. G.

maestrale (Italian) MISTRAL (French). G.

mafic *adj.* applied to a FERROMAGNESIAN MINERAL, dark in colour (the opposite of FELSIC), or to an IGNEOUS ROCK with a relatively high content of ferromagnesian minerals. MALANOCRATIC is synonymous. BASIC ROCK, LEUCOCRATIC, ULTRAMAFIC IGNEOUS ROCK.

Magdalenian epoch (from La Madeleine, a place in southwest France) the last cultural epoch of the PALAEOLITHIC, succeeding the AURIGNACIAN CULTURAL EPOCH. In addition to the refining of the tools already in use, the carving of bone and ivory began, and the art of cave painting advanced.

magma extremely hot, viscous, molten rock material, charged with gas and volatile matter, lying under great pressure below the surface of the earth.

It contains among a wide range of elements (mainly in oxide form) silica (ACID ROCK, ACID LAVA) and basic oxides (BASIC LAVA, BASIC ROCK). It is believed to have accumulated in magma basins, from which it may force its way to the surface; but it often solidifies underground to form INTRUSIVE PLUTONIC ROCKS. Less frequently it reaches the surface as LAVA from which EXTRUSIVE ROCKS, VOLCANIC ROCKS, are formed on solidification; or it stops short of the surface and solidifies in minor INTRUSIONS (e.g. dykes, DIKE), LACCOLITHS, SILLS as HYPABAYSSAL ROCK. When solidifying as a plutonic rock (such as GRANITE) in a great mass or BATHOLITH it heats and alters the surrounding rocks by what is known as CONTACT METAMORPHISM, converting sedimentary and other rocks into METAMORPHIC ROCKS through the distance from the molten mass known as the METAMORPHIC AUREOLE. The magma may dissolve and absorb some of the surrounding rocks (magmatic assimilation), and in particular penetrate fissures, joints, cracks and replace fragments of the adjoining rocks by its own mass (magmatic stoping). In the course of these changes, and in the slow cooling of the mass of the magma, a wide range of different igneous rocks may be formed, originating from one great magma basin (magmatic differentiation). The main mass eventually solidifies as a BATHOLITH or BOSS with associated smaller masses such as DIKES, LACCOLITHS, PHACOLITHS, SILLS. There may be a connexion in the earth's surface through a FISSURE or VOLCANIC NECK. G.

magmagenesis the formation of MAGMA from solid material, either by the partial melting of PERIDOTITE in the earth's MANTLE or by the melting of material of the CRUST above a SUBDUCTION ZONE. PLATE TECTONICS.

magmatic assimilation MAGMA.

magmatic differentiation MAGMA.

magmatic stoping the process by which an igneous MAGMA is believed to penetrate the rocks into which it is intruded by engulfing and absorbing blocks of rock.

magmatic water JUVENILE WATER.

magnesia MgO, a white solid oxide of MAGNESIUM, with a low melting point, occurring naturally or obtained by calcining (CALCINATION) carbonates of magnesium. It is used in refractories, medicinally as an antacid and laxative, and in the making of fertilizers.

magnesian limestone a LIMESTONE containing 5 to 15 per cent MAGNESIUM carbonate (MAGNESITE). As Magnesian Limestone (with initial capitals) it is the name given to the rocks of this character which constitute an important part of the PERMIAN rocks in northeast England. DOLOMITE. G.

magnesite magnesium carbonate, $MgCO_3$, an ore of MAGNESIUM, a white mineral used in making refractories, chemicals, fertilizers.

magnesium a light divalent (VALENCY) silver-white metallic element, Mg, occurring naturally only in combination, especially as the carbonates MAGNESITE and DOLOMITE and as the chloride CARNALLITE, present in seawater, in plants (e.g. in CHLORPHYLL) and in animals (in bones). It is usually obtained by electrolysis for use in alloys (particularly with ALUMINIUM), in photography and in medicine. BROMINE, PEDALFER, PEDOCAL.

magnet an object producing a MAGNETIC FIELD, e.g. a piece of iron, cobalt, nickel or alloy exhibiting FERROMAGNETISM; or the magnetized needle of a COMPASS; or the soft iron core together with the surrounding coil through which an electric current is passed, of an ELECTROMAGNET. MAGNETISM.

magnetic *adj.* 1. of or relating to a MAGNET or to MAGNETISM 2. having the magnetic north pole as a reference.

MAGNETIC POLE 3. producing a MAGNETIC FIELD or being capable of so doing.

magnetic anomaly a deviation from the predicted value of the earth's MAGNETIC FIELD, at a particular point on its surface, due to changes in the internal magnetic field (recorded in rocks at the time of their formation, PALAEOMAGNETISM) or to local concentrations or deficiences of magnetic minerals. The symmetry of the deviations (MAGNETIC REVERSAL) revealed by the rocks on each side of an OCEANIC RIDGE supports the theory of sea floor spreading and PLATE TECTONICS.

magnetic bearing BEARING-2.

magnetic compass COMPASS.

magnetic declination, magnetic variation the angular distance between the magnetic needle of a COMPASS aligned with the MAGNETIC MERIDIAN and pointing to the MAGNETIC NORTH and the TRUE NORTH of the geographical meridian, expressed in degrees east and west of true (geographical) north, at any point on the earth's surface. It varies with place and time owing to irregularities in the earth's MAGNETIC FIELD. AGONIC LINE, MAGNETIC DEVIATION, MAGNETIC INCLINATION, MAGNETIC POLE, TERRESTRIAL MAGNETISM. G.

magnetic deviation the apparent variation or declination of the magnetic needle of a COMPASS due solely to local conditions, e.g. the presence of magnetic iron ore, iron objects etc. MAGNETIC DECLINATION, MAGNETIC INCLINATION. G.

magnetic dip MAGNETIC INCLINATION.

magnetic division of rocks, MAGNETIC REVERSAL.

magnetic equator ACLINIC LINE.

magnetic field a field of force existing at a point if a small MAGNET placed at that point experiences two equal, unlike (i.e. opposing) parallel forces. It is formed by a permanent magnet or by a circuit carrying an electric current. ELECTROMAGNET, FERROMAGNETISM, MAGNETIC, MAGNETISM, TERRESTRIAL MAGNETISM.

magnetic flux the product of the area of a surface and the average normal component of the magnetic intensity over that surface. POLE-3.

magnetic inclination, magnetic dip the angular distance between the vertical and the direction of the MAGNETIC FIELD of the earth (TERRESTRIAL MAGNETISM) in the (horizontal) plane of the MAGNETIC MERIDIAN. MAGNETIC DECLINATION, MAGNETIC DEVIATION.

magnetic interval MAGNETIC REVERSAL.

magnetic meridian any line joining the north and south MAGNETIC POLES along which the free-swinging MAGNETIC NEEDLE of a COMPASS aligns itself.

magnetic needle the thin, magnetized steel rod of a COMPASS, freely swinging so that it comes to rest in line with a MAGNETIC MERIDIAN.

magnetic north the direction to which the MAGNETIC NEEDLE of a COMPASS, swinging freely in a horizontal plane, points in its search for the magnetic north pole. BEARING-2, MAGNETIC DECLINATION, MAGNETIC INCLINATION, MAGNETIC MERIDIAN, MAGNETIC POLE, TRUE NORTH.

magnetic pole either of the two poles (the north or the south) of the earth's MAGNETIC FIELD, indicated by the MAGNETIC NEEDLE of a COMPASS swinging freely in a horizontal plane. The precise locations vary, partly due to irregularities of the earth's magnetic field. At present the magnetic north pole lies in Canada near Prince of Wales Island and the south in South Victoria Land in Antarctica. A diameter drawn to link these two poles does not, however, pass through the centre of the earth: it misses it by some 12000 km (745 mi).

magnetic reversal a 180° swing round of direction of the earth's MAGNETIC FIELD (MAGNETIC ANOMALY), causing the MAGNETIC POLES to change position, so that the north magnetic pole becomes the south magnetic pole, and vice versa. This phenomenon, revealed by studies in PALAEOMAGNETISM, helped to confirm the theory of sea floor spreading and PLATE TECTONICS. The time period during which the reversal occurs is termed the magnetic interval, and the rocks formed at that time are termed a magnetic division.

magnetics the science of MAGNETISM.

magnetic storm a sudden marked, natural disturbance of the earth's MAGNETIC FIELD, seemingly (but not proven to be) related to solar activity such as SUNSPOTS or the AURORA.

magnetism 1. the property of producing or being affected by a MAGNETIC FIELD 2. the study of MAGNETS and the behaviour and effects of magnetic fields. ELECTROMAGNETISM, MAGNETICS, PALAEOMAGNETISM, TERRESTRIAL MAGNETISM.

magnetite a black iron oxide mineral, $FeOFe_2O_3$, one of the SPINEL group, occurring in METAMORPHIC and IGNEOUS ROCKS, constituting a valuable source of IRON and of VANADIUM. It is sometimes naturally magnetic. LODESTONE.

magnetopause the outer boundary surface of the MAGNETOSPHERE.

magnetosphere the zone of influence of the earth's MAGNETIC FIELD, extending in the ATMOSPHERE-1 up to and including the EXOSPHERE and IONOSPHERE. SUNSPOT, TERRESTRIAL MAGNETISM.

mahogany a tropical HARDWOOD of some tall trees of the Meliaceae family, especially of *Swietenia mahogani*, native to tropical America (including the West Indies), providing fine-grained, red-brown wood, highly valued for cabinet making.

maidan (Persian maidān; various other

spellings) **1.** a term variously used and spelled, but applied generally, especially by British residents in the Indian subcontinent, to an open space in or near a town, a parade ground, an esplanade **2.** specifically, a large rolling plateau in peninsular India. G.

main effect in statistics, the overall effect exerted by an INDEPENDENT VARIABLE over the whole sample. INTERACTION.

mai-yu (Chinese) plum rains, the Japanese BAI-U. G.

maize a tall, coarse ANNUAL cereal, *Zea mais*, bearing large 'cobs' on which the edible KERNELS-2 are carried. It is native to the Americas, where it is known as CORN, was brought to Europe by Columbus, taken to Africa by the Portuguese, where it spread to the areas with sufficient rainfall (and where, in central and southern Africa, it is known as mealies). It needs good, deep soil, plenty of moisture and freedom from frost. It can be grown only where there are at least 150 frost-free days, and where the summers are warm. New hybrid strains with short stalks ripen quickly and are pushing the limit of cultivation farther to the north, as well as giving higher yields. The grain does not make good bread, but maize is the staple grain for people in many maize-growing countries (especially in South America and Africa). The sweeter varieties grown for use as a vegetable are termed, appropriately, sweet corn. The grain, when very finely ground, produces cornflour, used in baking and confectionery; it also yields corn oil, used in cooking and salad dressings, and corn syrup, which is sweeter than sugar. But in Europe and North America the larger part of the crop is fed to animals, especially to cattle and pigs, and so leaves the farm on the hoof.

maize rains one of the periods of maximum rainfall, February to May, in east Africa (Nairobi and Entebbe). MILLET RAINS.

malachite a bright green mineral, basic copper carbonate, $Cu_2CO_3(OH)_2$, polished and made into ornaments and jewellery.

malacology the branch of zoology concerned with MOLLUSCS.

malleable *adj.* of metals, capable of taking a permanent change of shape by being beaten, pressed, rolled etc., i.e. by the application of STRESS-1. DUCTILE.

mallee (Australian: aboriginal term) a low, scrubby, EVERGREEN plant, *Eucalyptus dumosa*, growing in arid, subtropical parts of southern Australia. G.

mallee scrub a densely growing, low, EVERGREEN shrub formation of *Eucalyptus* species, including *E. dumosa* and *E. oleosa*, growing in the arid, subtropical parts of southern Australia.

malpais (Spanish) a tract of land covered with rough, bare lava.

malt processed grain, especially of BARLEY, used in brewing and distilling. The grain is put in water to germinate, then heated and dried.

Malthusianism the body of doctrines derived from the writings of Thomas Robert Malthus, 1766-1834, a British economist and demographer, especially from *An Essay on the Principle of Population as it affects the Future Improvement of Society*. His thesis was that the human population, if unchecked, increases at a geometric rate (GEOMETRIC PROGRESSION) while the food and other resources needed for its subsistence increase only at an arithmetic rate (ARITHMETIC PROGRESSION); that population always grows to the limits of the means of its subsistence, checked only by war, famine, pestilence and the influence of miseries derived from a consequent low standard of living. Some French writers have applied the term to any form of restriction on production. NEO-MALTHUSIANISM. G.

Malvernian, Malvernoid *adj.* applied to the earth building movements, part of the ARMORICAN OROGENY, named from the Malvern Hills, a conspicuous ridge on the borders of Worcestershire and Herefordshire in England, which has a north to south trend in contrast to the usual east to west trend of Armorican folding. Some authors distinguish Malvernian folds, initiated in PRECAMBRIAN times, from Malvernoid folds of later date, but also trending north to south. The use of the terms Malvernian and Malvernoid is much discussed by geologists. G.

mamélon (French) CUMULO-DOME.

mamma cloud a CLOUD FORM in which pendulous protruberances hang from the underside of a cloud. MAMMATUS CLOUD.

mammal a member of Mammalia, the highest class (CLASSIFICATION OF ORGANISMS) of VERTEBRATES, which includes human beings. The class is divided into three main groups, Monotremata (rare, egg-laying, with some primitive reptilian characteristics, e.g. duck-billed PLATYPUS, native to the Australian ZOOGEOGRAPHICAL REGION); Marsupialia (no PLACENTA-2, or with an ineffective placenta, in general the imperfectly developed young being nourished and carried by the female in an external abdominal pouch until fully developed, e.g. kangaroo); and Placentalia (with a well-developed PLACENTA-2, the most common mammals now living, includes human beings). The many distinguishing characteristics of mammals include warm blood (WARM-BLOODED), mammary glands producing nourishment (milk) for the young, a heart with four chambers, three small bones in the ear, and different types of teeth adapted for the cutting, tearing and grinding of food. WALLACE'S LINE.

mammatus cloud a breast-shaped CLOUD (CLOUD FORM) usually occurring in the formation of thunderclouds, associated with CONVECTION.

mammilated surface a rock surface smoothed and rounded by various agencies, a term applied to various scales of landform, e.g. large scale rounding due to PERIGLACIAL conditions, small scale due to glacial GOUGING. G.

mandarin orange a CITRUS fruit. ORANGE, TANGERINE.

mandate, mandated territory (mandate, command or order) a territory designated under Article 22 of the League of Nations. When the League of Nations was created after the First World War, in 1919, the former colonial possessions of Germany and Turkey (the losing countries) came under the control of the League, which allocated the administration of these territories (mandated territories) to certain powers, notably Great Britain and France (mandatory powers). The Class A mandates (Iraq, Palestine, Transjordan, Syria, Lebanon) were regarded as likely soon to be ready for independence and fairly soon became independent. Class B (Cameroons, Togoland, Tanganyika, Ruanda) were less developed; Class C (South-West Africa, New Guinea and certain Pacific islands, Samoa) were to be governed as part of the territory of the mandatory power. In 1945 the United Nations took over such responsibilities as remained, and the territories became TRUSTEESHIP TERRITORIES.

man-day, standard smd, STANDARD MAN-DAY.

manganese Mn, a hard, brittle, reddish-white or grey-white metallic ELEMENT-6, occurring in nature usually as an OXIDE, CARBONATE or SILICATE, found in a great variety of compounds, an essential MICRONUTRIENT for plants, and especially important (as FERROMANGANESE) in making certain types of steel.

manganese dioxide MnO_2, a dark brown or grey compound used in manufacturing as a CATALYST or as an oxidizing agent (OXIDATION). PYROLUSITE.

manganese nodule a hydrated MAN-

GANESE OXIDE CONCRETION, with some IRON, COPPER, COBALT and NICKEL, occurring widely on the floor of the ABYSSAL ZONE of the ocean (RED CLAY). Manganese nodules were first discovered about 1873, but research continues into their origin and development.

mango *Mangifera indica*, a medium to large tree, native to Burma-Indian subcontinent, widely cultivated in tropical areas for the sake of the sweet juicy pulp of its fruit (a DRUPE), eaten raw, or canned when ripe, but used in pickles, chutneys, jams etc. before full ripening.

mango-shower a term applied in the Indian subcontinent, especially by Europeans, to one of the heavy showers, usually of thunderstorm origin, occurring before the breaking of the main MONSOON (March, April, May) when mangoes (MANGO) begin to ripen. G.

mangosteen *Garcinia mangostana*, a slow-gowing tall tree native to southeast Asia, difficult to propagate, thus not so widely cultivated as the MANGO, but occasionally grown in hot, humid tropical areas for the sake of the sweet, whitish, segmented pulp of its fruit, each segment containing seeds.

mangrove **1.** a member of a number of genera and species, especially of the families Rhizophoraceae, Verbenaceae and Lythraceae (e.g. of the genera *Rhizophora, Bruguiera* or *Avicennia*), low trees and shrubs which grow and spread quickly on tidal mud in tropical areas, so that their dense root systems (which include adventitious, aerial roots termed pneumatophores which stretch out at a distance from the main stem, bend towards the ground, strike, and send up new trunks) are covered by salt or brackish water at each tide and effectively bind the mud. The roots under the mud have air supplied by aerial roots which rise above the surface. The seeds have a peculiar characteristic: while the fruit is still on the tree they let down a root which in many cases reaches the ground and almost immediately sends up a shoot. Some species provide hard, durable wood, the bark of some yields a substance used in tanning (CULCH), the sweet fruit of the common mangrove (*Rhizophora mangle*) is edible **2.** a plant community dominated by such trees and shrubs. MANGROVE SWAMP, SUNDRI. G.

mangrove swamp the association of trees and shrubs covered by the collective term MANGROVE, growing with members of other families in tidal mud in DELTAS, ESTUARIES and along the coasts in TROPICAL regions. SUNDRI, SWAMP. G.

man-hour the amount of work done by one man in one hour, used as a unit of measurement of labour cost in production statistics. STANDARD MAN-DAY.

manioc *Manihot utilissima*, a shrub native to Brazil but widely cultivated as a staple food plant in tropical areas of South America and Africa, the source of cassava and tapioca, produced by subjecting the poisonous tuberous roots of manioc to heat and pressure, then drying. It is the non-poisonous granular substance produced that is known as cassava or tapioca (according to treatment), high in starch but low in protein. The juice of some bitter varieties is boiled to make cassareep, used in West Indian sauces. The leaves can be boiled and eaten, and cassava is fermented to make alcoholic liquor. Manioc is usually recorded in international agricultural statistics as cassava.

man-made fibre SYNTHETIC FIBRE.

man-made soils the term applied by those who drew up the 1973 SOIL CLASSIFICATION of England and Wales to one of its major soil groups. Made by human endeavour, these are good deep soils, over 40 cm (16 in) thick, rich in HUMUS, with a surface layer which may or may not be totally artificial.

Mann-Whitney test in statistics, an hypothesis test used to analyse two unrelated samples of data in order to compare two population MEDIANS-3 (i.e. the samples are drawn from two different populations). The sample data must be ordinal (ORDINAL SCALE) and the sample sizes small (unlike MOOD'S MEDIAN TEST). It is assumed that both population distributions have the same shape.

manor the extent of the land held by the lord of the manor together with, in feudal times in England (and to some extent in Wales), certain legal and administrative rights, dues and responsibilities (FEUDAL SYSTEM). The lord of the manor, holding land from the king, owned the soil of the whole, and kept for himself the parts termed DEMESNE LANDS (MANORIAL WASTE). The remainder was worked by his tenants under various systems of payment (LAND TENURE). With INCLOSURE the lord of the manor retained the demesne lands and the ownership of the soil of the common grazing (COMMON), subject to COMMON RIGHTS. (The lordship of the manor is now in general little more than an archaic, empty title.) The affairs of the manor were controlled by MANORIAL COURTS from the manor house or mansion of the lord. Manor (from manor house) is now often applied loosely to the principal house of an estate, or to a mansion or COUNTRY HOUSE. G.

manorial courts the courts which a lord of the MANOR has the right to hold. There used to be three: the court baron for freeholders of the manor; the customary court for copyhold or customary tenants; and the court leet for criminal jurisdiction. The surviving manorial courts, now usually termed courts leet, have generally assumed the responsibility for managing any remaining COMMON land. G.

manorial waste part of the DEMESNE LAND of a MANOR left uncultivated and unenclosed over which the freehold and customary tenants might have rights of common (COMMON RIGHTS). Not all manorial waste was subject to common rights. G.

mansion historically the term has many applications, but today is applied mainly to a very large, imposing residence. COUNTRY-SEAT, MANOR.

mantle 1. of the earth, that part of the earth's interior lying between the CORE and the CRUST, consisting of ULTRABASIC ROCKS (including the silicate minerals OLIVINE and PYROXENE as well as DUNITE, ECLOGITE and PERIODOTITE), probably solid unless the pressure of the overlying rocks is relieved, when they become viscous (MAGMA). The mantle is some 2900 km (some 1800 mi) thick, density about 3.0 to 3.4, the lower surface forming the GUTENBERG DISCONTINUITY, the upper the MOHOROVICIC DISCONTINUITY, the uppermost layer forming the ASTHENOSPHERE. Movements in the mantle, presumed to be convectional, are reflected in the structure of the earth's CRUST 2. the accumulation of loose rock debris, consisting of weathered rock and soil, lying on the older solid BEDROCK and covering most of the land of the earth; termed mantle rock by some authors (REGOLITH). G.

manufacturing differential area the change in manufacturing employment which is not caused by changes in the national economy or by changes in the composition of the various industrial sectors in the area.

manufacturing industry SECONDARY INDUSTRY, the making of articles or materials (now usually on a large or relatively large scale) by physical labour or mechanical power, in general terms the processing of raw materials and foodstuffs (or of semi-processed or recycled materials), the working-up of materials into a useful form. FOOTLOOSE INDUSTRY, HEAVY INDUSTRY, INDUSTRY, LIGHT INDUSTRY; and PRIMARY, TERTIARY, QUATERNARY INDUSTRY. G.

manure animal dung, compost or chemicals applied to the soil in order to fertilize it.

map a representation of the earth's surface or a part of it, or of the heavens, delineated on a flat sheet of paper or other material. Generally used loosely, thus diagrammatic maps or CARTOGRAMS are included. CHART, COGNITIVE MAP, MENTAL MAP, PLAN, TOPOLOGICAL DIAGRAM (MAP). G.

maple a member of *Acer*, family Aceraceae, a large genus of HARDWOOD trees and shrubs growing mainly in the MIDLATITUDE ZONE of the northern hemisphere, providing syrup (MAPLE SYRUP) and close-grained wood used in furniture making.

maple sugar sugar obtained from MAPLE SYRUP.

maple syrup the concentrated sap of the sugar MAPLE, collected especially in Canada.

mapping 1. the making of a map 2. in mathematics, transformation, the establishing of correspondence between points in two regions, the points in the second region being the IMAGES-2 of those in the first.

map projection 1. the method or methods and their results of representing part or the whole of the earth's surface on a plane surface 2. the orderly arrangement of parallels and meridians which enables this to be done. It is impossible to map any part of a sphere on a plane, or a flat sheet, with complete accuracy. It is necessary therefore to choose the properties which are desirable for the purpose of the proposed map (correct area, shape, bearing, scale). True shape can be obtained for small areas only, and is incompatible with correct area. A map which shows correct areas cannot give true direction. No projection can give true distances over the whole surface.

Certain classes, e..g. perspective projections, can be obtained by geo-metrical construction: the zenithal or azimuthal (true direction from the centre); the gnomic (shortest distance between two points in a straight line); stereographic (preserves correct shape); and orthographic (produces effect of a globe). These are useful features but the modified, or non-perspective, projection are in greater use since the net (GRATICULE) can be calculated to meet particular requirements.

Two much-used classes are derived from the so-called developable surfaces, those of the cylinder and the cone (Mercator's projection is a modified cylindrical and Bonne's a modified conic). Conventional projections include the Mollweide (giving the whole surface within an elipse) and various interrupted projections. The latter modify the central meridian or meridians to show conveniently the areas of most interest, with the result that the whole surface is not continuous (Goode's projection). Good atlases generally discuss in their introductory section the projections employed, and indicate the correction to be applied to distances. Projections frequently used for atlas maps are Mercator (bearings are straight lines and shapes correct, but areas in high latitudes are greatly exaggerated; also used for navigational charts); Mollweide (equal area but shapes greatly distorted on the margins); Bonne (modified conic, used for great continental areas); zenithal equidistant (correct direction and distance from the centre, used for polar regions). A number of projections have transverse and oblique forms, when in place of the equator as the axis, a suitable GREAT CIRCLE is used.

maquis, macquis, macchia, matorral (French maquis; Italian macchia; Spanish matorral) the low scrub vegetation of EVERGREEN shrubs and small trees (including arbutus, rose laurel, myrtle, heaths, rosemary, holm oak and cork oak) characteristic of SILICE-

OUS soils in Mediterranean lands, the result not only of the summer aridity of the Mediterranean climate in association with the soils, but of human activities in the felling of the natural forest cover (including particularly ilex, the holm oak), of grazing animals (particularly goats) and of fire. The maquis afforded good cover in the days of vendetta and banditry in Corsica, and again for the French Resistance Movement opposed to the German occupation of France during the Second World War, hence the term was applied to the Resistance Movement and its members. GARIGUE. G.

mar (Swedish, pl.marer) a bay or creek with silted-up entrances, so that its water is almost fresh. G.

marasca *Prunus cerasus marasca*, a small CHERRY tree grown especially in Dalmatia, producing black bitter fruit, the source of maraschino, a liqueur distilled from the fruit. Maraschino cherries are preserved in this liqueur.

marble 1. a naturally occurring CAL-CIUM CARBONATE, a CRYSTALLINE LIMESTONE, veined or mottled by the presence of other crystallized minerals, produced by dynamic or thermal METAMORPHISM. It forms a hard rock that takes a high polish. It is used by sculptors, and for decorative purposes in building and in ornaments 2. loosely applied to any other stone that can be highly polished and used in the manner of true marble for similar decorative purposes, especially as a facing stone in buildings.

march, marches, marchlands (French marche) a boundary or frontier, the borderland, now generally applied to the borderland between two states (e.g. the Welsh Marches between Wales and England), or a zone of land of debatable ownership between two states. G.

mare's tails wispy streaks of CIRRUS CLOUD, indicating strong winds in the upper ATMOSPHERE-1.

marg (Indian subcontinent: Kashmiri)

ALP, high pasture above the tree line; commonly appearing in place-names. G.

margin 1. an outer, limiting border 2. a narrow area adjoining the border of something 3. a reserve supply of something 4. in commerce, the difference between net sales and costs, out of which expenses are paid and profits come 5. in economics, the least profit at which a transaction is economically sound.

marginal *adj.* 1. of or related to a MARGIN 2. close to the limit of possibility or acceptability 3. of an electoral seat, being held by a very small majority.

marginal channel the course between ice and rock cut by and along which flows melt-water from a GLACIER or ICE SHEET.

marginal cost the expenditure actually incurred (or saved) in producing (or not producing) the next unit of a product or service. OPPORTUNITY COST, SPATIAL MARGIN (to profitability).

marginal deep a narrow trough in the ocean floor lying parallel to and alongside an ISLAND ARC. DEEP.

marginal depression a linear hollow at the base of an INSELBERG or of an ESCARPMENT at its junction with the PEDIMENT-2.

marginal efficiency of capital the productivity of the last (i.e. the marginal) unit of CAPITAL-4 employed on a given project. MARGINAL COST.

marginal fringe a term applied by Alice Coleman, Second Land Utilisation Survey of Britain, to a zone of poor quality land where improved and unimproved patches are interspersed and where the area of the improved land expands or contracts according to fluctuations in farm prices. Thus neither improved farmland nor natural vegetation is dominant. G.

marginal land land which is just fertile enough to yield an average return from agricultural use, no more than is sufficient to cover the costs of production. It goes in or out of agricultural use according to fluctuations in economic conditions. G.

marginal nunatak a NUNATAK with ice on three sides, the fourth being bounded by the sea, a fjord, or land. G.

marginal producer a producer operating at the MARGIN OF PROFITABILITY who, if costs rise, may be forced out of business.

marginal sea a nearly enclosed sea bordering a continent and lying on a submerged part of the continental mass, e.g. the Mediterranean sea. CONTINENTAL SEA, EPEIRIC SEA. G.

marginal utility the UTILITY or VALUE-1,3 yielded by the last (the marginal) unit of consumption. In PERFECT COMPETITION the price of a product is determined (so far as demand is concerned) by its marginal utility to the consumer, i.e. the price the consumer is generally willing to pay will be equal to the price the consumer is willing to pay for the last unit of the product consumed.

margin of profitability the limit of profitability below which a firm operates at an economic loss.

marin (French) a warm, moist, southeast wind blowing over the coastland of southern France, especially in spring and autumn, caused by low atmospheric pressure in the Golfe du Lion. LOW (atmospheric).

marina any waterside location providing mooring and other facilities for pleasure boating. G.

marine *adj.* 1. of, relating to, found in, or produced by, the sea (examples given under RAISED BEACH) 2. of or relating to shipping or navigation.

marine deposits deposits laid down under the sea, as opposed to TERRESTRIAL DEPOSITS.

marine park a conservation area on a small part of the sea bed, e.g. on the Great Barrier Reef of Australia.

maritime air mass AIR MASS.

maritime climate a climate similar to an INSULAR or OCEANIC CLIMATE, directly influenced by the proximity of the sea, which causes a comparatively cool summer and a comparatively mild winter on account of the different thermal capacities of land and water. There is thus a small daily and seasonal range of temperature, much cloud, fairly uniform precipitation, and humidity higher than that of a CONTINENTAL CLIMATE. This regime may be experienced on islands and in coastal regions in any latitude, but it is particularly prevalent in, and typical of, western midlatitude coastal areas. G.

maritime polar air mass POLAR AIR MASS.

maritime tropical air mass TROPICAL AIR MASS.

marjoram any member of the genera *Origanum*, family Labiateae, especially *Origanum majorana* and *O. onites*, PERENNIAL aromatic herbs native to the Mediterranean region, used as a flavouring in cooking.

market many applications, the most common in geographical writing being 1. a congregation of sellers and buyers of goods 2. a public place where goods are displayed and put on sale. MARKET PLACE 3. the demand for a COMMODITY-1 4. the outlet for a COMMODITY-1 5. the trade, or traffic, in a particular COMMODITY-1 6. the people concerned with buying and selling a particular COMMODITY-1 7. the class of persons to whom a particular COMMODITY-1 can readily be sold. G.

market demand the effective demand for a COMMODITY-1 supported by an ability to pay for it.

market economy, free economy an economic system which operates according to free market forces, i.e. in which there is free enterprise and competition, most of the production, distribution and exchange being in private hands, with government intervention kept to the minimum; and in which goods and services are allocated by the price mechanism, prices being determined by the free exchange of commodities for money (SUPPLY AND DEMAND). Today most countries with a market economy modify to some extent (by STATE INTERVENTION) the forces of the free market. CAPITALISM-1, CLASSICAL-ECONOMIC THEORY, LAISSEZ-FAIRE, LIBERAL CAPITALISM, MARKET PLACE THEORY, MIXED ECONOMY, NEOCLASSICAL ECONOMIC THEORY, PERFECT COMPETITION, STATE CAPITALISM, SUBSISTENCE ECONOMY.

market garden an area of land on which MARKET GARDENING is practised.

market gardening in Britain, the intensive cultivation of vegetable crops, soft fruit or flowers for sale. When organized on a large scale, with a marked concentration on one or two crops, it becomes comparable with TRUCK FARMING in the USA. HORTICULTURE. G.

marketing principle the main principle used by W. Christaller to account for the varying levels and distribution of CENTRAL PLACES in a CENTRAL PLACE SYSTEM. According to the marketing principle the supply of goods and services from a central place should be as close as possible to the place supplied. To achieve this a higher order central place will serve two centres of the next lower order (a K-3 hierarchy). See ADMINISTRATIVE PRINCIPLE, K-VALUE, TRAFFIC OR TRANSPORTATION PRINCIPLE.

market orientation the tendency of a firm or industry to be located close to its MARKET-3. RESOURCE ORIENTATION.

market place, market-place 1. a public open area where a MARKET-2 is held. MARKET TOWN 2. the outlet for a commodity. MARKET-4.

market place theory an economic theory which assumes that production and consumption are kept in balance by the interaction of supply and demand acting through the price mechanism; that price encourages or suppresses demand and stimulates or depresses supply. Thus the quantity of a good produced at a given price will be matched by the demand at that price. It assumes that a large number of independent buyers and sellers together produce this result, despite the fact that they are unable independently to affect either the market price or the supply. CLASSICAL ECONOMIC THEORY, MARKET ECONOMY, NEOCLASSICAL ECONOMIC THEORY, PERFECT COMPETITION.

market price the price at which it is possible to sell or buy a COMMODITY-1 in a MARKET-1,2.

market town in Britain, a town with a legal right to hold a MARKET-2 on certain days, usually a settlement which developed at the intersection of routes. 'Market' commonly occurs in the names of market towns, e.g. Market Harborough.

market value the value a COMMODITY-1 would have if it were offered for sale to the public.

marl 1. a calcareous CLAY or MUDSTONE with an admixture of CALCIUM CARBONATE 2. in agriculture, loosely applied to any friable clayey deposits 3. in geology, used specifically in the proper name of certain types of rock, e.g. Keuper Marl (commonly devoid of calcium carbonate), Chalk Marl (an impure chalk with much clayey matter, but often hard) and the Marlstone (Middle Lias, a hard ferruginous rock worked as an IRON ore, together with its associated beds). G.

marling the old practice of spreading MARL or CLAY on light soil to improve its texture and its capacity to hold water. G.

marrow vegetable, a variety of *Cucurbita pepo*, family Cucurbitaceae, a climbing herb, probably native to America, grown in temperate midlatitudes for its large, fleshy fruit. Some varieties (COURGETTE) have been bred to be cut when young and small (usually well under 30 cm: 12 in); but if grown to maturity they become the normal, large, fleshy marrow. Large varieties are cooked, small varieties are eaten raw or cooked.

Marsch (German, pl. Marschen) the low-lying reclaimed land of the coast and estuaries west of the River Elbe, with good soil providing pasture for cattle. G.

marsh a wet area of mainly mineral (i.e. inorganic) soil commonly flooded periodically or at intervals and covered with water-loving vegetation. It differs from a SWAMP (where the summer water level is usually above the surface of the soil), a BOG (which has a mainly organic, acid peat soil), or a FEN (which has a purely organic soil, typically alkaline in reaction, though occasionally neutral or slightly acid). SALT MARSH. G.

Marshall Plan a popular name for the European Recovery Programme proposed by G. C. Marshall in the USA on 5 June 1947 that materials and financial aid should be provided for the countries of Europe by the United States on condition that the European nations took the initiative in cooperative action. It came into force in 1948, administered by OEEC, terminated 1952.

marsh gas CH_4, an odourless, inflammable gas, formed largely of METHANE, produced by vegetation that is decomposing in wet conditions, e.g. in a MARSH.

Marxism a doctrine based on the political, social and economic views of Karl Marx, 1818-83 (German economist, sociologist and philosopher), and Friedrich Engels, 1820-95 (German socialist philosopher), both of whom were profoundly influenced by German philosophers, particularly by G. W. F. Hegel, 1770-1831. Very broadly, the doctrine's philosophical bases are DIALECTICAL MATERIALISM and HISTORICAL MATERIALISM. The MODE OF PRODUCTION is viewed as the dominant factor governing economic and social interaction in society, the impetus for change in the mode of production being generated by class conflict, until ultimately the harmonious, classless, communist society (COMMUNISM-2) is to be achieved. In economic theory the Marxist view is especially concerned with the determination of the EXCHANGE VALUE of COMMODITIES-1; it broadly accepts David Ricardo's LABOUR THEORY OF VALUE-2. References that help to explain the doctrine will be found elsewhere in the *Dictionary*, labelled 'in Marxism'.

mascaret (French) a local term applied to a BORE-1 on the River Seine, France.

mashlum, maslin 1. a FODDER-1 crop consisting of a mixture of CEREALS-1 and PULSES-1 (usually PEAS or BEANS) **2.** a mixed cereal crop, usually RYE and WHEAT grown together (so that if one fails there will some yield from the other) **3.** bread made from the flour of such mixed GRAINS-5.

mass 1. an aggregation of a quantity of matter **2.** loosely, in general, a large amount or number **3.** the larger part or number **4.** in physics, one of the fundamental properties of matter, i.e. that property of a piece of matter which causes it to be attracted to any other piece of matter by gravitational force (GRAVITATION-3). The force attracting an object to the earth is that object's weight. The mass of an object also measures its inertia, i.e. resistance to a force applied to set it in motion, or change its motion.

massif (French) a term applied widely and somewhat loosely to a compact mass of mountain or highland with relatively uniform characteristics and well-defined boundaries, thus similar to the Massif Central of France. G.

massive *adj.* applied to a thick STRATUM of rock in which STRATIFI-CATION, jointing (JOINT), CLEAVAGE, FOLIATION-1 etc. are almost or completely absent.

mass media the systems or impersonal means used for the transmission of information and entertainment for the benefit of a large number of people, i.e. newspapers, radio, television, etc. MEDIA.

mass movement, mass wasting the spontaneous downward movement by GRAVITATION-1 of rock material, usually helped by rainfall or melt-water from snow or ice, classified as: slow, creeping, usually imperceptible except through extended observation (CREEP, ROCK CREEP, ROCK-GLACIER CREEP, SOIL CREEP, SOLIFLUCTION, TALUS CREEP); rapid (DEBRIS AVALANCHE, EARTHFLOW, MUDFLOW); and landslides, which are perceptible and involve relatively dry masses of earth debris (BLOCK SLUMP, DEBRIS FALL, DEBRIS SLIDE, ROCKFALL, ROTATIONAL SLIP-2, SLIP, SLUMP, SUBSIDENCE-2). COLLUVIUM, SCREE. G.

mass production the making of one article, or type of goods, in large numbers by a standardized process.

mass society a SOCIETY-2,3 with a high level of large-scale manufacturing and urban development, an elaborate and extensive BUREAUCRACY and powerful MEDIA, in which habits, activities, opinions and tastes are shared by most of its members, although some individuals feel estranged from such a society (ALIENATION).

master as prefix, applied in the sense of chief, commanding, dominating, as in master-joint, master-stream, master-cave, etc. G.

master sample in statistics, a SAMPLE-3 drawn from a population for use on a number of future occasions, to avoid repeated ad hoc sampling. The master sample may be large, and if necessary a sub-sample may be drawn from it.

mata, matta (Brazil: Portuguese) forest; a term rarely used internationally. G.

matched samples in statistics, a pair or set of SAMPLES-3 in which each member of one sample is matched by a corresponding member in every other sample by reference to qualities other than those being immediately investigated. The object of matching is to obtain better estimates of differences by eliminating the possible effects of other variables. Assessments of significance may prove difficult if the members of the second sample have to be selected purposively in order to match those of the first, instead of being chosen at random. RANDOM SAMPLE.

maté, yerba maté, Paraguay tea *Ilex paraguensis*, a small tree of tropical regions, cultivated particularly in Brazil and Paraguay for the sake of its leaves, which are picked, dried, ground and used to produce a tea-like beverage containing a small amount of CAFFEINE, drunk mainly in South America.

material index a measure of materials used in manufacturing industry, calculated by the total weight of localized materials used per product divided by the weight of the product. Most manufacturing industries have an index greater than 1.0 and are described as 'weight-losing'. HEAVY INDUSTRY, LIGHT INDUSTRY.

materialism **1.** in ONTOLOGY, the theory that matter is the basic reality of the universe, that nothing exists except matter, that consciousness and will can be shown to be the products of material agencies, and thus that they too are derived ultimately from matter. The possibility of disembodied minds and mental states is therefore excluded unless they are identified with states of the brain and the nervous system. The DIALECTICAL MATERIALISM of Marx and Engels introduces an element of evolution to that standard theory of materialism: it maintains that mind, while originating in matter, is distinct in nature from it. HISTORICAL MATERIALISM, IDEALISM, REALISM, SPIRITUALISM **2.** the belief that material things are more valuable than spiritual things.

material orientation the tendency of a manufacturing industry to be located close to sources of the materials used in its processes. MARKET ORIENTATION.

mathematical geography the aspects of geography deriving from the shape, size and motions of the earth which are capable of mathematical definition or expression, identified as a division of geography by eighteenth century writers. G.

mathematical model a MODEL-4 which expresses relationships between variables in terms of numerical values (e.g. REGRESSION MODEL). SYMBOLIC MODEL.

mathematics the science of studying and expressing the relationships between quantities and magnitudes, as represented by numbers and symbols.

matmura (Sudan: colloquial Arabic) a pit for the storage of harvested grain, dug in the soil on rising ground, free of stagnant rainwater, in the drier, termite-free areas of northern Sudan. G.

matorral (Spanish) equivalent to French MAQUIS, Italian MACCHIA. G.

matriarch a woman who rules a group, especially a mother having authority over her immediate family or larger family group.

matriarchate a community ruled by a MATRIARCH.

matriarchy 1. government by women. PATRIARCHY **2.** a form of social organization in which descent passes through the mothers **3.** a MATRIARCHATE.

matrilineal *adj.* applied to a social system in which descent is traced only through mothers. MATRIARCHY-2, PATRILINEAL, UNILINEAL.

matrix (late Latin, womb; pl. matrices) **1.** a place or medium within which

something is bred, formed, developed, produced; a place or point of origin and growth **2.** in botany, the material on which a FUNGUS or a LICHEN grows **3.** in geology, the fine-grained material of a rock within which something (e.g. coarser particles, FOSSILS, GEMSTONES, METALS, PHENOCRYSTS etc.) is embedded (GANGUE) or to which something adheres. In SEDIMENTARY ROCKS the matrix is usually the cementing (CEMENT-2) material, in IGNEOUS ROCKS it is the GROUND-MASS of CRYSTALLINE or glassy minerals **4.** the impression left in a rock after a FOSSIL etc. has been removed **5.** in mathematics, an ordered ARRAY of mathematical elements conveniently arranged for the carrying out of various operations on it, e.g. in a square or rectangular arrangement of ROWS and COLUMNS of quantities or symbols. MINIMAX.

matroclinous *adj.* in biology, **1.** applied to an offspring which in inherited character or characters resembles the female rather than the male parent **2.** possessing or involving the tendency to inherit a character or the characters of the female parent only. PATROCLINOUS.

Matterhorn (Swiss-German) the name of the famous PYRAMIDAL mountain peak in the Alps in Europe, sometimes applied to a similar peak (horn peak, HORN) elsewhere. G.

mature *adj.* having reached a stage of full natural development, applied to **1.** a landscape exibiting the features resulting from a long CYCLE OF EROSION when dissection of the original surface is complete, little trace of it remaining; such a landscape may be maturely dissected by RIVERS which are still young **2.** a river characteristic of such a landscape, in which erosion is at a minimum because the stream has acquired a normal fall in its bed **3.** a shoreline in a condition of approximate equilibrium between erosion, weathering and transportation **4.** a soil with well-developed characteristics produced by the processes of soil formation, and in equilibrium with its environment, or a soil with well-developed horizons. SOIL HORIZON **5.** a town (URBAN HIERARCHY). MATURITY, OLD AGE, SENILE RIVER, YOUTH. G.

maturity the quality or state of being MATURE-1,2,3,4,5, an additional example being the optimal stage of development in the theory of industrial growth, characterized by a broad, balanced industrial structure, a substantial infrastructure, and the existence of developed skills and techniques.

maximum-minimum thermometer a THERMOMETER used to register the maximum and the minimum ambient air TEMPERATURES reached over a selected period of time. A common type consists of a column of MERCURY housed and partly filling a graduated, U-shaped tube, the rest of each side of the tube being topped-up with a transparent liquid. One side of the U is completely filled; but a small space is left at the top of the other side. Small metal needles rest on top of each end of the mercury, inside the tube; they are mobile but stick to the tube sides when not being actively pushed by the column of mercury. When the temperature rises, the liquid on the side of the U which is completely full expands, exerting pressure on the column of mercury, which accordingly rises on the other side of the U (i.e. the side with the space in it), pushing up the needle, which sticks to the tube to register the maximum temperature reached. When the temperature drops, the liquid in the full side contracts, the mercury rises under the weight of the liquid in the side with the space, the needle in the full side rises and sticks to the tube to register the minimum temperature reached. As the mercury swings up and down, the needles stay in the ultimate positions to which they were pushed. They have to be reunited with the

mercury by a magnet operated by hand or by a push-button device.

maximum thermometer a THERMOMETER used to register the highest ambient air TEMPERATURE reached over a selected period of time. There are many types in use. One consists of a sealed glass, graduated tube containing MERCURY; on the top of the column of mercury lies a metal needle. As the temperature rises the mercury expands and pushes the needle upwards; but the needle is arranged in such a way that it cannot fall with the mercury when the temperature falls and the mercury contracts. It is left standing, stuck to the sides of the tube, and the lower end of it (i.e. the end which touched the mercury) registers the highest temperature reached. The needle remains set in this position until reunited (usually manually, with a magnet) with the column of mercury. MAXIMUM-MINIMUM THERMOMETER, MINIMUM THERMOMETER.

Mayen (Switzerland: German) an intermediate shelf, between ALP-1 and valley floor, where cattle stay for a while on their upward journey in May and their downward trek in September. TRANSHUMANCE, VORALP. G.

mbuga (Tanzania, central plateau: Swahili and other Bantu languages) a seasonal swamp in a wide, shallow valley. It has a dark clay soil which cracks deeply when dry, is very sticky when wet, and supports open grassland. G.

MDC **1.** moderately developed country **2.** more developed country, as defined by United Nations, 1980, with reference to the USSR, Japan, Australia, New Zealand and all the countries in North America and Europe. HDC, LDC, UNDERDEVELOPMENT.

meadow **1.** a piece of land, of any size, permanently covered with grass which is mown for hay **2.** any piece of enclosed grassland but especially one low-lying or close to a river. CATCH MEADOW, GRAZING, PASTURE, WATER

MEADOW **3.** (American) as English usage, but also applied to a wet lowland with an abundant flora and what is called a peaty 'meadow soil' (the soil of SWAMPS or MARSHES). **4.** a part of the ocean where fish feed. G.

meadow soil   MEADOW-3.

mean   1. something occupying (or, as *adj.*, applied to something occupying) a position midway between two extremes in number, quantity, degree, kind, value etc. **2.** an average. ARITHMETIC MEAN.

meander   a loop-like bend, a pronounced bend or loop in the course of a sluggish river (term derived from the River Maiandros, which has many such features) or of a valley (MEANDERING VALLEY). The river itself develops the curve of a meander by LATERAL EROSION, the bank on the concave side (the outside) of the curve being eroded by the current (RIVER-CLIFF), while deposition is taking place on the convex side (the inside) (SLIP-OFF SLOPE). Eventually the meander becomes so well-developed that it is nearly circular. At this stage the river may break across the strip of land (the NECK) separating the stream, a cut-off (OXBOW) may be formed, and the river flows on a straight course. MEANDER CORE, DIVAGATING MEANDER, INCISED MEANDER, MEANDER BAR, MEANDER BELT, POOL AND RIFFLE, SCAR, TERRACE. G.

meander bar   POINT BAR.

meander belt   the part of the flat floor of a valley across which a stream and its channel wind, i.e. the area between the outer banks of successive MEANDERS. G.

meander core   the piece of land in the centre of an INCISED MEANDER nearly encircled by the river, or completely so if the river has broken through the MEANDER NECK. It is usually the remnant of the spur that helped to cause the meander.

meandering valley   a winding valley with large, sweeping curves that cut into the solid rocks that contain it. It has extensive deposits of alluvium on the floor over which existing streams have created meanders smaller than those of the valley as a whole. MEANDER, MISFIT RIVER.

meander neck   the strip of land sparating the stream on each side of a well-developed MEANDER. OXBOW.

meander scar   a discernible depression, infilled with deposits and vegetation, marking the channel of a cut-off meander (OXBOW).

meander scroll   POINT BAR.

meander terrace   a terrace formed on one bank of a river as it meanders in its valley and at the same time (due to REJUVENATION) erodes the valley floor. In flowing across the valley current on the outside of the curve of the meander cuts down into the former level of the river's FLOODPLAIN, so that part of that higher level is left as a terrace.

mean deviation   in statistics, a statistic of DISPERSION, rarely used, derived from the sum of ABSOLUTE DIFFERENCES in a set of observations divided by the number of differences. STANDARD DEVIATION.

mean diurnal range   the mean difference between the highest and lowest rainfall, or temperatures, etc. recorded at a place for each day of a selected period (usually four weeks) over a number of years. RANGE.

mean information field   in SPATIAL DIFFUSION, an area in which contacts might occur. The probability is derived theoretically from the characteristics of PRIVATE INFORMATION FIELDS, the expression of a negative logarithmic relationship between the probability of contact between any pair of individuals and, other things being equal, the distance separating them.

meaningfulness   in statistics, the true significance of a test SCORE, covering the extent to which it relates to other measurements, fits into a theoretical framework, or provides a link with another concept.

mean sea-level MSL, the average level of the surface of the sea (which varies slightly from place to place), calculated from a series of continuous records of tidal oscillation over a long period, the standard level from which all heights are calculated. The British standard is calculated at Newlyn, Cornwall, and forms the Ordnance Datum (OD) from which heights on British maps are measured.

means of production in Marxism, materials and machinery. The materials are seen as the object of labour (either natural raw materials such as minerals, to which labour is to be applied; or objects such as manufactured components, or harvested crops, on which some labour has already been used). The machinery (i.e. a scythe as well as a complicated machine) is seen as the means or tools of labour, used to transform the materials. CONSTANT CAPITAL, MODE OF PRODUCTION, PRODUCTIVE FORCES, SLAVE MODE OF PRODUCTION.

mean solar day the average length of the SOLAR DAY, the period of time between successive returns of the MEAN SUN to the meridian, i.e. 24 hours. CIVIL DAY, EQUATION OF TIME.

mean solar time an average or MEAN SOLAR DAY of 24 hours, a useful measure because the period of time between two successive daily returns of the real sun to the meridian (LOCAL TIME) is not always the same. ANALEMMA, APPARENT TIME, EQUATION OF TIME, ROTATION OF THE EARTH.

mean sun an imaginary sun that travels along the celestial equator at a constant rate that is equal to the average rate of the real sun.

mear, mears (English dialect) **1.** land bordering the cultivated area, presumed to be used for turning the plough **2.** waste or poor quality land

between the cultivated area and a stream, hence Maer Field, The Mears **3.** also mere, ploughland marking headlands between open fields. G.

mearstone, meerstone (English dialect, Midlands) **1.** a boundary stone between holdings in common fields **2.** any boundary stone, especially one marking the limits of waste, common, and wood. Trees have been used for the same purpose, hence mear oak. G.

measurement 1. the magnitude, length, degree etc. of something in terms of a selected unit **2.** in statistics, at its simplest, the classification of individuals, groups or other units and the placing of them in previously defined CATEGORIES-3 (termed categorical or qualitative measurement; if a number can be applied meaningfully to each category it is termed a quantitative measurement). Four scales or levels of measurement are usually employed, each with its particular properties, a particular series of statistical techniques being applied at each level. The scales are the NOMINAL SCALE, ORDINAL SCALE, INTERVAL SCALE, RATIO SCALE (from the simplest to the most precise). CLASS-5, CLASS INTERVAL, VARIABLE.

measurement error in statistics, the deviation of the experiment measurement from a true measurement.

measure of location in statistics, a quality which locates a distribution (or a set of sample values derived from it) by means of a value which is in some way typical or central (e.g. the ARITHMETIC MEAN, the MEDIAN, or the MODE).

meat the flesh of dead animals (excluding fish) used for food. CARCASS, CATTLE, LIVESTOCK, LIVESTOCK FARMING, PIG, POULTRY, SHEEP.

mechanical adj **1.** of, pertaining to, involving, made by, or operated by, a machine **2.** made or operated as if by a machine.

mechanical analysis in soil science, the analysis of the size of soil particles. Dry particles exceeding 0.06 mm in diameter are usually measured by sieving; but the size of smaller particles is calculated by the rate at which they settle when put into water. SOIL TEXTURE.

mechanical weathering the disintegration of rock by the agents of weathering (e.g. EXFOLIATION, FREEZE-THAW, GRANULAR DISINTEGRATION, SPALLING etc.) which cause internal and external stresses without chemical alteration of the rock, as opposed to chemical weathering (CORROSION). HYDRATION, ORGANIC WEATHERING, WEATHERING. G.

mechanism in philosophy, the theory that the workings of the universe can be explained by the SCIENTIFIC LAWS of PHYSICS and CHEMISTRY, that the ocurrence of any event can be deduced (DEDUCTION) from the conditions that preceded it. TELEOLOGY takes the opposing view.

mechanist one who believes in the philosophical theory of MECHANISM.

mechanistic *adj.* of, pertaining to, connected with, or believing in, the philosophical theory of MECHANISM.

mechanization the introduction and use of machines in an industrial process in order to enhance, lighten, or replace human power, an integral part of INDUSTRIALIZATION. INDUSTRIAL REVOLUTION.

media the systems or impersonal means used for the transmission of information and entertainment, including newspapers, magazines, books, advertisement hoardings, films, records, tapes, radio, television. Popular newspapers, television and radio together, reaching a very large, heterogeneous audience, are commonly termed the MASS MEDIA.

medial moraine, median moraine the debris lying centrally in a line along the surface of a GLACIER, occurring when the LATERAL MORAINES of two confluent glaciers meet.

median 1. something situated in the middle **2.** in geometry, a line joining the vertex of a triangle to the middle of the opposite side **3.** in statistics, the centrally occurring value in a DATA SET which is arranged in rank order, i.e. the value above and below which lie 50 per cent of the observations in a distribution. If there is an even number of observations, the median lies midway between the two centrally occurring values. The upper quartile covers the highest 25 per cent of the values, the lower quartile the lowest 25 per cent; and the INTERQUARTILE RANGE is the difference between the lowest number in the upper quartile and the highest number in the lower. ARITHMETIC MEAN, CENTRAL TENDENCY, MODE.

medical geography the study of the geographical aspects of health and the provision of health care. It covers the study of the spatial distribution of human disease and causes of death, together with the factors of the environment conducive to human health and sickness. It includes deaths from disease (MORTALITY), illness not necessarily fatal (MORBIDITY), the diseases permanently located in certain areas (ENDEMIC disease), and the wider spread (PANDEMIC) or sudden outbreak (EPIDEMIC) of disease. It is thus concerned with aetiology (ETIOLOGY) and EPIDEMIOLOGY. G.

medicine the science of understanding, prevention and cure of disease and of the preservation of health.

medieval, mediaeval *adj.* of, relating to, or characteristic of, the MIDDLE AGES.

Mediterranean climate the western margin warm temperate climate (one of KOPPEN'S C climates), occurring on the coastal lands round the western Mediterranean and on the narrow western coastal margins of continents in latitudes 30° to 40° (California, central Chile, South Africa, Australia). It is a climate with wide variations, but is generally characterized by mild wet winters (average temperature for coldest month usually over 6°C: 43°F) and hot dry summers (warmest month usually over 21°C: 70°F) with a high amount of sunshine. Some of the most marked variations, especially in precipitation and temperature, occur in the Mediterranean area itself. The regime is due to the dominance of subtropical high pressure systems (ANTICYCLONE) in summer, and the passage of DEPRESSIONS-3 associated with moist winds from the oceans in winter. G.

medlar *Mespilus germanica*, a small tree native to southern Europe and southwest Asia, but surviving farther north, grown for its fruit, usually used in jam making. The fruit is eaten raw from the tree in warmer lands where it ripens satisfactorily, but partly decayed, when it becomes palatable, in cooler climates.

megalith a large stone used as a monument or in construction. Some cultures have been marked by the use of such stones (e.g. Stonehenge in England). G.

megalithic *adj.* applied to the large monumental constructions termed MEGALITHS, to the people who put them up, and to the period in which they were constructed (Megalithic Age), i.e. the cultures of NEOLITHIC - BRONZE AGE times. G.

megalopolis a very large and spreading urban complex, with some open land, formed when built-up areas, widespread over an extensive area, enlarge to such an extent that they become linked together, as in the northeastern seaboard of the USA, or along the northern shore of the Inland Sea of Japan. ECUMENOPOLIS. G.

megatherm 1. a plant characteristic of warm habitats, needing a continuously high temperature, in general an average monthly temperature exceeding 18°C (65°F) 2. a climate characterized by such plants, one of consistently very high temperatures, i.e. one of the A (tropical) climates of KOPPEN, where for 12 months the temperature exceeds 20°C (72°F).

Megathermal Period ATLANTIC STAGE.

meifod (Welsh) an intermediate dwelling, between a HAFOD and a HENDREF (summer and winter dwellings).

meion THERMOMEION, a climatological station where the TERMPERATURE ANOMALY is markedly negative. ANTIPLEION, ISANOMAL, PLEION, THERMOPLEION.

mélange a mixture of fragments of differing rocks cemented into a SEDIMENTARY ROCK. BRECCIA.

melanocratic *adj.* consisting mainly of dark minerals, i.e. of FERROMAGNESIAN MINERALS. BASIC ROCK, FELSIC, LEUCOCRATIC, MAFIC.

melon *Cucumis melo*, family Cucurbitaceae, an ANNUAL trailing herb native to tropical Africa, of which many varieties are cultivated to suit conditions in warm temperate, subtropical and tropical regions. The main varieties entering trade are the cantaloupe (hardy enough to be grown in Europe, including Britain if protected); ogen (grown particularly in Israel); water melon, *Citrullus vulgaris* (widely grown in warm temperate, subtropical and tropical regions). It produces large, succulent fruit with juicy, sweet pulp, eaten raw; the seeds contain edible oil.

meltemi ETESIAN WIND.

melting the change from SOLID to LIQUID state by the action of heat.

melting point the temperature at which a SOLID becomes LIQUID under normal pressure.

melt-water, meltwater water derived from the melting (MELTING POINT) of snow and ice, e.g. from the SNOUT of a GLACIER. G.

melt-water channel a channel in the solid rock or in drift deposits in a once-glaciated area, unrelated to the trend of the present drainage pattern, and in some cases cutting across it. DRAINAGE-2.

membrane extremely thin TISSUE, pliable and strong, which lines, covers or connects parts of the body of an animal or of a plant. OSMOSIS.

me-nam, mae-nam, menam (Thai, big water) a large river. G.

menhir a free-standing, tall, upright, isolated, single-stone monument of varying prehistoric age; or such a stone in a prehistoric group, e.g. the Hele Stone of Stonehenge in England (MEGALITH). Menhirs are found especially in Brittany, as well as in Africa and Asia.

mental map COGNITIVE MAP.

mentifact HUXLEY'S MODEL.

Mercalli scale, modified a scale formerly used in measuring EARTHQUAKE intensity, based on the observed effects on buildings, etc., ranging from I (detectable only by seismograph reaction) to XII (catastrophic, i.e. the total destruction of buildings). It superseded the ROSSI-FOREL SCALE, and has now in turn been superseded by the RICHTER SCALE.

mercantile *adj.* pertaining to, concerned with, engaged in, trade and commerce.

mercantile model, mercantilist model an approach to the study of urban systems suggested in 1970 by J.E.Vance, American geographer, as an alternative to CENTRAL PLACE THEORY. To summarize, it views wholesaling as the key urban function, and explains the development of an URBAN SYSTEM-1 in terms of trading. His study was particularly related to the evolution of the MERCANTILE cities of the USA Atlantic seaboard, hence the name. MERCANTILE PHASE.

mercantile phase URBAN GROWTH PHASES.

mercantilism 1. trade and commerce 2. a theory popular among European nations during the sixteenth and seventeenth centuries, that the economic and political strength of a country lay in its acquiring large quantities of gold and silver, to be achieved by restricting imports, developing production for ex-

ports, and prohibiting the export of gold and silver.

mercury Hg, a silver-coloured, poisonous, metallic ELEMENT-6, liquid at temperatures above $-38.8°C$ ($-37.8°F$), boiling at $356.9°C$ ($674.4°F$) under normal pressure, obtained mainly from the compound mercuric sulphide in its red form, cinnabar (CHALCOPHILE). It is used in industrial processes and in BAROMETERS and THERMOMETERS.

mercury barometer BAROMETER.

merdeka (Malay) in Malaysia, independence, especially the independence dating from 31 August 1957, when the British administration came to an end. G.

mere a large pond or shallow lake, occupying a hollow in glacial drift, especially in TILL or BOULDER CLAY. The term is used especially in place-names in Cheshire and Shropshire in England, relating not only to water-filled KETTLE HOLES but also to subsidence hollows resulting from the removal of subterranean salt. G.

meridian terrestrial, one of the lines of LONGITUDE which link the North Pole to the South Pole and cut the equator at right angles, i.e. half of one of the GREAT CIRCLES, the other half being termed the antimeridian. The prime, standard, initial or zero meridian, $0°$, is usually considered to be the meridian passing through Greenwich, and meridians are numbered from it to $180°$ east or west of it. CELESTIAL MERIDIAN, MERIDIAN CIRCLE, MERIDIONAL-2, PRINCIPAL MERIDIAN. G.

meridian *adj.* pertaining to noon, especially to the position of the sun at noon.

meridian circle a GREAT CIRCLE, consisting of a MERIDIAN and its complement, the antimeridian.

meridian day 1. the day on which a traveller crosses the INTERNATIONAL DATE LINE 2. the day gained in crossing the INTERNATIONAL DATE LINE from west to east, i.e. in order to avoid having, say, two Wednesdays in one week, the second is entitled meridian day. G.

meridional *adj.* **1.** of, pertaining to, characteristic of the south, especially of the inhabitants of the south, particularly of southern Europe **2.** of or pertaining to a MERIDIAN.

meridional flow an atmospheric circulation in which the dominant flow of air is from north to south, or from south to north, across the PARALLELS OF LATITUDE, in contrast to ZONAL FLOW.

meridional wind a wind that blows across the PARALLELS OF LATITUDE along the MERIDIANS, from north to south or from south to north.

merino 1. a Spanish breed of SHEEP with very fine white wool, extensively crossbred (CROSSBREED) with other breeds of sheep to improve fleeces **2.** a soft, fine woollen fabric originally woven from pure merino wool, later from merino wool and cotton. BOTANY WOOL, WOOL.

merokarst imperfect or partially developed KARST (unlike perfectly formed karst, HOLOKARST), occurring on thin, impure or chalky LIMESTONE where, in addition to karst features, dry valleys and surface drainage are present. G.

meromixis the permanent stratification (chemical or thermal, or a combination of chemical and thermal) of water in lakes, occurring in the absence of complete circulation. MONIMOLIMNION. G.

mesa (Spanish, a table) a high, extensive TABLELAND or an isolated flat-topped hill (broader than a BUTTE), the remnant of a plateau that has been subjected to DENUDATION in a semi-arid region. It consists of horizontal strata capped by a more resistant stratum, and one or all sides slope steeply or form cliffs. MESETA. G.

mesarch SUBSERE.

meseta (Spanish) **1.** the high TABLELAND of the heart of Spain **2.** an alternative to MESA. G.

mesic *adj.* applied to conditions in which MESOPHYTES flourish. HYDRIC, XERIC.

mesic layer in a soil, an organic layer in a state of decomposition between fibric and humic. FIBRIC LAYER, HUMIC LAYER.

mesocephalic a skull shape between long and broad, that is with a CEPHALIC INDEX between 75 and 83.

mesoclimate the CLIMATE of a local area marked by its deviation from the norms of the climate of the region in which it lies, i.e. from the MACROCLIMATE. In area it is larger than a LOCAL CLIMATE or a MICROCLIMATE. G.

meso-economic sector that sector of the economy which does not fall within either of the two traditional economic realms (MACRO-ECONOMICS and MICRO-ECONOMICS) and comprises all large and diversified private or publicly-owned organizations, many of which are MULTINATIONALS.

Mesolithic *adj.* of or pertaining to the culture period midway in the STONE AGE between the PALAEOLITHIC and the NEOLITHIC, from about 8000 to 6000 BC in Europe, when the bow and arrow came into use, the dog was domesticated, pottery came to be made, England was cut off from continental Europe by the rising sea, and there was marked climatic change. ATLANTIC STAGE.

mesopause the layer in the ATMOSPHERE-1 some 80 km (50 mi) above the earth's surface, the boundary between the MESOSPHERE and the THERMOSPHERE, and the layer with the lowest temperature.

mesophile an organism for which the best temperature for growth lies between 20° and 45°C (68° and 113°F). PSYCHROPHILE, THERMOPHILE.

mesophyte a plant growing under medium conditions of moisture, where there is neither an excess nor a deficiency of water. Mesophytes are typical of the CONIFEROUS FOREST growing in high latitudes (TAIGA) and high altitudes, and include pine species, fir, spruce, larch. HYDROPHYTE, HYGROPHYTE, TROPOPHYTE, XEROPHYTE.

mesosaprobic SAPROBIC CLASSIFICATION.

mesosere a SERE, the stages in a plant SUCCESSION-2 in which the pioneer community develops on damp aerial surfaces such as alluvial mud. HALOSERE, HYDROSERE, XEROSERE.

mesosphere the zone in the ATMOSPHERE-1 extending from the STRATOPAUSE to the MESOPAUSE, between some 50 to 80 km (30 to 50 mi) above the earth's surface, in which the temperature, having reached its maximum in the stratopause, decreases with height to the mesopause, where it is at its lowest. G.

mesotherm 1. a plant thriving in regions without extremes of temperature, in warm, temperate conditions, where the average temperature of the coldest month exceeds 6°C (43°F) and the average temperature of the hottest exceeds 22°C (72°F) **2.** a warm, temperate climate (e.g. the MEDITERRANEAN CLIMATE) characterized by such plants, one of the C climates of KOPPEN'S classification, where for most of the year the temperature ranges between about 10°C and 20°C (50°F and 68°F), with short colder and hotter periods. HEKISTOTHERM, MEGATHERM, MICROTHERM.

mesothermal *adj.* applied to an ore deposit formed in conditions between those which produce EPITHERMAL and HYPOTHERMAL deposits, i.e. in the temperature range 200°C to 300°C (400°F to 575°F) at fairly high pressure. G.

mesotrophic *adj.* applied to a body of freshwater with a moderate amount of plant nutrients, which is accordingly generally productive in plant and animal life. DYSTROPHIC, ENTROPHIC, OLIGOTROPHIC, TROPHIC LEVEL.

Mesozoic *adj.* of or pertaining to the (Secondary) geological era that followed the (Primary) PALAEOZOIC era, divided into three periods, the CRETACEOUS, JURASSIC, TRIASSIC. GEOLOGICAL TIMESCALE.

mesquite (Spanish) **1.** *Prosopis juliflora*, a leguminous plant (LEGUMINOSAE) of the family Papilionaceae, a spiny, deep-rooted shrub or small tree of the dry lands of southwestern USA and Mexico, often almost the only vegetation over extensive areas. It yields a gum resembling gum arabic (GUM), and it has large sugary pods (similar to those of the CAROB TREE) which provide food for humans and fodder for animals **2.** a thicket formed by mesquite trees and shrubs. G.

mesta (Spanish) the pastoral organization typical of Castile. G.

mestizo a person of mixed parentage in South America, especially one of European (usually Spanish or Portuguese) and American Indian descent. MULATTO, ZAMBO. G.

metacartography 1. the portrayal of spatial properties on maps as distinct from and opposed to their representation in graphs, language, mathematics, photographs and pictures **2.** the study of the effectiveness of maps in conveying information when compared with that of other means, e.g. graphs, language, mathematics, photographs or pictures. G.

metal 1. an ELEMENT-6 of high specific gravity, with atoms structured in such a way that they easily lose ELECTRONS to form positively charged IONS, an element that is usually DUCTILE, MALLEABLE, and a good conductor of heat and electricity **2.** a compound or ALLOY of such an element. PRECIOUS METAL, ROAD METAL.

metallic *adj.* **1.** of or pertaining to a METAL **2.** resembling a metal **3.** made of a metal.

metalliferous *adj.* containing or yielding a METAL.

metallogenic, metallogenetic *adj.* **1.** in

geology, of, pertaining to, or designating, the origin of ores; or **2.** as applied to a province or epoch, a geographical region or a specific geological epoch characterized by a particular assemblage of mineral deposits. GEOLOGICAL TIMESCALE **3.** applied to a map, a map showing the distribiution of one or more mineral deposits and relating their distribution to the geological formations or periods and to the tectonic features of a region. G.

metalloid 1. an ELEMENT-6 with properties between those characteristic of METALS and those characteristic of nonmetals **2.** a nonmetal which can be combined with a metal to form an ALLOY.

metalloid *adj.* **1.** of, relating to, designating a METALLOID **2.** resembling a METAL.

metallurgy the science and technology concerned with METALS and their ALLOYS, their structure, their extraction from their ores, their purification, their suitability for various uses; and with creating new alloys or devising new methods of treatment to meet a particular demand.

metamorphic *adj.* pertaining to, characterized by, formed by METAMORPHISM, applied especially to rocks and rock formations. G.

metamorphic aureole the zone of COUNTRY ROCK surrounding an IGNEOUS INTRUSION in which the country rock is metamorphosed by heat and migrating fluids (THERMAL or CONTACT METAMORPHISM) from the intrusion. BATHOLITH, MAGMA, SKARN. G.

metamorphic differentiation a process by which the mineral content of some METAMORPHIC ROCKS is graded, some of the minerals moving within the rock to form layers with distinct chemical and mineral characteristics. FOLIATION-1, METAMORPHISM.

metamorphic rock a well-defined, new type of rock derived from pre-existing rock by mineralogical, chemical and structural changes in the pre-existing rock brought about by its contact with a great heated mass of MAGMA (THERMAL or CONTACT METAMORPHISM) or by folding and pressure (DYNAMIC METAMORPHISM). METASOMATISM. G.

metamorphism a change of form, applied in geology to the process of the transformation of pre-existing rock into a new, well-defined type of rock by ENDOGENETIC processes, i.e. by the action of heat (THERMAL or CONTACT METAMORPHISM, METAMORPHIC AUREOLE) or by severe compressional earth movements (DYNAMIC METAMORPHISM, REGIONAL METAMORPHISM) or by both, which modifies or changes the texture, composition, physical or chemical structure of the pre-existing rock. ANATEXIS, AUTOMETAMORPHISM, COESITE (IMPACT METAMORPHISM), FOLIATION, HIGH GRADE METAMORPHISM, LOW GRADE METAMORPHISM, HYDROTHERMAL, MAGMA, METAMORPHIC, METAMORPHIC DIFFERENTIATION, METASOMATISM, PNEUMATOLYSIS. G.

metamorphosis in biology, a period of rapid change of form and structure which some animals (e.g. amphibians, insects) undergo in their development from embryo to adult stage, e.g. the transformation of the tadpole to the adult frog.

metaphysics (Greek, the works of Aristotle that come after the *Physics* in the sequence of his work) **1.** the branch of PHILOSOPHY-1 concerned with the first principles of things, with what really exists in the world (e.g. the concepts of time, space, being, substance). The investigation is usually conducted by rational argument, not by mystical intuition, and may be described as transcendent (regarding reality as being above and beyond normal experience, e.g. as in the view of the universe supplied by supernatural religion) or immanent (regarding reality as consisting only of objects of experience). COSMOLOGY, EPISTEMOLOGY, ETHICS, IDEALISM-3, KANT, ONTOLOGY 2. with 'of', the theoretical principles or philosophical explanation of a particular branch of knowledge.

metaquartzite QUARTZITE.

metasomatism the change in a pre-existing rock brought about by a solution from external sources which percolates through the rock and (introducing material from other series of rocks) causes chemical reactions by which one mineral in the pre-existing rock is partly or wholly altered into other minerals, or is replaced by another mineral of different composition, without a change in texture. PNEUMATOLYSIS is not involved. GRANITIZATION, HYDROTHERMAL, PETRIFICATION. G.

métayage (French) a system of LAND TENURE in western Europe, especially in France, in which the landowner provides seed and stock and the tenant farmer pays a proportion of the produce as rent to the landowner, i.e. the farmer is a share-tenant. Only a very few holdings are now farmed under this system. G.

meteor **1.** a solid body moving rapidly in SPACE-4 glowing when it enters the earth's ATMOSPHERE-1 on account of the heat generated by friction with the atmosphere, and appearing as a 'shooting star' or 'falling star'. The friction causes most meteors to disintegrate to dust (METEORIC DUST), but some solid remnants may reach the earth's surface as METEORITES.

meteoric *adj.* **1.** of, relating to, or originating in, the earth's ATMOSPHERE-1 **2.** of or pertaining to a METEOR.

meteoric dust the dust in the ATMOSPHERE-1 derived from disintegrating METEORS and trapped in the earth's gravitational field, a component of COSMIC DUST.

meteoric water water on the earth's surface which is derived from PRECIPITATION, as distinct from CONNATE WATER, JUVENILE WATER.

meteorite a solid extra-terrestrial body which reaches the earth's surface, typically formed of METALS (iron-nickel), or SILICATES (TEKTITE), or a combination of metals and silicates, described as stony, stony-iron or iron according to composition. Some stony meteorites contain rounded clusters of OLIVINE and PYROXENE crystals. A meteorite may make a meteorite crater as it hits the earth's surface, e.g. Grand Meteor Crater, Arizona. METEOR, METEORIC DUST.

meteorological screen STEVENSON SCREEN, a white painted wooden box on legs, about 1 m (4 ft) above the ground, designed to shelter meteorological instruments from strong wind and solar and terrestrial radiation. The roof is usually insulated, the sides are louvred (one usually hinged to act as a door), so that air can flow through freely; and the THERMOMETERS within (usually dry and wet bulb, maximum and minimum thermometers) are supported on a frame, standing free from the roof, floor and sides, so that they give shade readings as accurately as possible. Some types also house a HYGROGRAPH and a THERMOGRAPH.

meteorology the scientific study of the processes and physical phenomena operating in the earth's ATMOSPHERE-1 in the short term, as distinct from CLIMATOLOGY, which is concerned with the long term. Weather forecasting depends on meteorological studies, and on the whole meteorologists confine their attention to the layers of the earth's atmosphere where the weather that effects the earth's surface is generated, i.e. to the TROPOSPHERE and STRATOSPHERE. The idea that the processes and physical phenomena of the upper zones (the MESOSPHERE and IONOSPHERE) are more the concern of geophysicists (GEOPHYSICS-2) is now

generally accepted. MICROMETEORO-
LOGY, SYNOPTIC METEOROLOGY. G.

meter an instrument for measuring and
recording the flow or amount of something, e.g. HYGROMETER.

methane an odourless, colourless,
inflammable HYDROCARBON, CH_4, a
product of decaying organic matter
(MARSH GAS), found in NATURAL GAS
and in coal mines (FIREDAMP), and one
of the constituents of COAL GAS.

methodological individualism the concept that all explanations of social
phenomena must ultimately consist of
statements about individuals.

metoecious (metoxenous) parasite a
PARASITE which does not have to rely
on a specifc host in order to live.
AMETOECIOUS.

metre m, the basic SI unit of length,
defined in 1960 from a wavelength in
the spectrum of krypton, equivalent to
about 39.37 in. A metre is divided into
10 decimetres (rarely used) or 100
centimetres (cm) or 1000 millimetres
(mm). In measures of area 1 square
metre (10.8 sq ft) equals 10 000 square
centimetres or 1 million square millimetres. In measures of volume 1 cubic
metre (35.3l5 cu ft) equals 1 million
cubic centimetres. CONVERSION
TABLES.

metroland 1. the area immediately surrounding a METROPOLIS-1, especially
that of London, a term derived from
the area served by the Metropolitan
Line of the London underground railway system 2. the people (collectively)
living in such an area, the individual
member being termed a metrolander.

metrology the science of, or the system
of, weights and measures.

metropolis 1. the mother city of a
colony or settlement, hence the capital
or chief city of a country, serving as the
seat of government or of ecclesiastical
authority, or as the main commercial
centre 2. the largest town or agglomeration of people in an extensive area.
ECUMENOPOLIS 3. the see (bishop's

diocese or office) of a metropolitan
bishop. G.

metropolitan 1. the head of an ecclesiastical province, with a rank between
that of an archbishop and a patriarch,
in the Orthodox Eastern Church, an
archbishop in the Western Church 2. a
person who inhabits a METROPOLIS-
1,2.

metropolitan *adj.* 1. of, pertaining to,
characteristic of, or constituting, a
METROPOLIS in the limited senses
defined under (1), but also as applied to
those large urban centres etc. specified
under (2), as in British local government
administration after 1971, e.g. the metropolitan authorities of Manchester,
Birmingham, Liverpool, metropolitan
districts, metropolitan county councils.
LOCAL GOVERNMENT IN BRITAIN 2.
as applied to colonial powers, the
controlling 'mother' country, e.g.
METROPOLITAN FRANCE, as distinct
from the colonies. COLONY-1,2. G.

metropolitan area in USA, the Standard Metropolitan Statistical Area
(SMSA), a very large urban settlement
comprising the continuous built-up
area and its adjoining suburbs
(SUBURB-2) divided administratively
(as are all other urban areas) into two
zones: the inner (the central city) and
the outer (consisting of the independent
MUNICIPALITIES together with unin-
corporated parts of the COUNTIES
lying in the built-up area, and the
SUBURBS-2). Within the suburban
zone there are single-purpose local
government units (SCHOOL DISTRICT,
SPECIAL DISTRICT), usually with
powers of taxation. LOCAL GOVERN-
MENT IN USA.

metropolitan common any COMMON
land lying wholly or partly within the
area of the Metropolitan Police District
on 10 August 1866, the date of the
passing of the Metropolitan Commons
Act 1866.

metropolitan county LOCAL GOVERN-
MENT IN BRITAIN.

Metropolitan France the home (mother) country as distinct from France's overseas territories. G.

metropolitanism the idea or spirit of, the character of, the attachment to, the manifestations of living in (including the social institutions, the fabric of the buildings etc.), a METROPOLIS-1,2.

metropolitan labour area the commuting hinterland of a METROPOLITAN AREA. DAILY URBAN SYSTEM.

metropolitan region a very large urbanized area, distinguished by its great size, and regarded as a FUNCTIONAL REGION, the locations within it being more closely linked to one another than to locations outside it. REGION.

mica a group of transparent SILICATE minerals, common in ACID IGNEOUS ROCKS, in METAMORPHIC ROCKS, and in derived sediments, which, having perfect CLEAVAGE-1, can easily be split into very thin, tough, pliable, lustrous plates, used as an electrical insulator and as a heat-resistant substitute for glass. The main micas are the dark BIOTITE; the light-brown, green or red muscovite (both occurring in GNEISS and SCHIST); and the yellow phlogopite (rich in magnesium and present in MARBLE-1 and PERIDOTITE). LAMELLA.

micro- (Greek) small as opposed to MACRO-, large, e.g. MICROCLIMATE.

microbe a MICROORGANISM, especially one causing disease.

microbial metallurgy the use of BACTERIA in purifying METALS or in separating them from their ores.

microbiology the branch of BIOLOGY concerned with MICROORGANISMS.

microchemistry CHEMISTRY concerned with very small quantities, especially in analysis.

microclimate the CLIMATE of a very small area, the modification of the general climate produced by conditions in the immediate environment of a subject, e.g. of a cereal crop, of a building. The area involved is smaller than that of LOCAL CLIMATE or MESOCLIMATE and is confined to the very shallow layer of the atmosphere close to the ground. CLIMATOLOGY, MACROCLIMATE, MICROCLIMATOLOGY, MICROMETEOROLOGY.

microclimatology the scientific study of MICROCLIMATES. MICROMETEOROLOGY.

microcosm a miniature universe, formerly applied to what is now termed an ECOSYSTEM. G.

microcrystalline *adj.* having CRYSTALS so small that they can be seen only with the aid of a microscope.

micro-economics the branch of ECONOMICS based on the concepts of supply and demand and concerned with the behaviour of the individual firm or consumer and the analysis of aggregate outcomes of the numerous individual responses to changes in price, cost and revenue, and quantity, as distinct from MACRO-ECONOMICS. MESO-ECONOMIC SECTOR.

micro-erosion EROSION processes on a very small scale and affecting depths of just a few centimetres, e.g. the alternate wetting and drying on a surface. RILL.

microgabbro 1. in Britain, dolerite 2. in USA, diabase. DOLERITE.

microgeography a detailed study of a small area.

microgranite an ACIDIC IGNEOUS ROCK with medium size grains, similar to GRANITE in mineralogical and chemical composition and occurring in MINOR INTRUSIONS.

microhabitat a very small HABITAT, e.g. on the bole of a tree, on one side of a hillock, on a pebble. MICROSERE.

microkarst 1. an area of KARST in which all the typical features are very small 2. an area dominated by such features.

micro-level approach an approach used particularly in the study of INDUSTRIAL MOVEMENT, involving the identification and investigation by interview or questionnaire survey, of individual firms which have moved location. MACRO-LEVEL ANALYSIS.

micrometeorology the detailed scientific study of the layer of the ATMOSPHERE-1 nearest to the earth's surface i.e. from ground level up to about 1.2 m (4 ft), important in the study of MICRO-CLIMATES. MICROCLIMATOLOGY.

micron a unit of linear measurement, one-millionth of a METRE, or one-thousandth of a millimetre.

micronutrient an ELEMENT-6 or combination of elements needed by an organism in order to maintain health, but only in very small quantities. The micronutrients needed by plants in varying amount are boron, cobalt, copper, iron, manganese, molybdenum, zinc. MACRONUTRIENT, TRACE ELEMENT, VITAMIN.

microorganism any organism too small to be seen by the unaided eye, being of microscopic or even smaller size (e.g. a BACTERIUM). VIRUSES, having many of the characteristics of living organisms (e.g. they can grow and multiply in living cells), are usually included, although they are not living organisms in the strict sense.

microphagous *adj.* applied to an animal which eats, almost unceasingly, particles of food very small in relation to its size, e.g. a whale, nourished by PLANKTON. MACROPHAGOUS.

microrelief very small variations in form of the surface of the ground originating in the REGOLITH.

microseism a minor earth tremor, such as one of those constantly occurring and due to such natural causes such as rock falls, winds, waves, as well as to heavy road traffic etc. G.

microsere all the stages in a plant SUCCESSION-2 occurring within a MICROHABITAT. SERE.

microtherm 1. a plant characteristic of habitats in a cool temperate to cold climate, able to withstand low winter temperatures, but preferring areas where the coldest average monthly temperature does not fall below 6°C (43°F) and the temperature of the warmest month lies between 10°C and 22°C (50°F and 70°F) **2.** the cool temperate to cold climate characterized by such plants and by generally low temperatures, i.e. one of the D climates of KOPPEN, where for 8 to 11 months the average temperature is generally below 10°C (50°F) and for 1 to 4 months it ranges between 10°C and 20°C (50°F and 68°F).

microwave an electromagnetic wave of wavelength ranging from the wavelengths of very short RADIOWAVES to those of long INFRA-RED (ELECTRO-MAGNETIC SPECTRUM). With the aid of microwave relay stations microwaves can be transmitted over distances between 50 and 80 km (30 to 50 mi). They are used in RADAR-2 and in radio astronomy, and as carrier waves in telephone, telegraph and television transmission. MICROWAVE RADIOMETER.

microwave radiometer a REMOTE SENSING device in which MICROWAVES are used to detect variation in soil moisture.

mictium a mixture of SPECIES-1, e.g. occurring in an ECOTONE.

Middle Ages the period of European history between ancient and modern times. There is general agreement that it terminated with the flowering of the RENAISSANCE between 1400 and 1500 AD, but the onset is variously taken as coinciding with the decline of the Roman Empire (c.500 AD) or (more commonly in England) with 1100 AD. DARK AGES, MEDIEVAL.

Middle East part of the earth's surface EAST of Europe, now usually taken to include southwest Asia and northeast Africa, stretching from Turkey through Iran, Iraq and the countries of Arabia to Sudan and Egypt and including the countries bordering the eastern shores of the Mediterranean. FAR EAST, NEAR EAST.

midlatitude *adj.* applied to the latitudinal zone lying between 23°30′ and 66°30′ in the northern and southern hemispheres. MESOTHERM, TEMPERATE.

midlatitudes LATITUDE.

midnight sun the SUN-2 still to be seen shining above the horizon at midnight in latitudes higher than 63°30′ north or south, in the northern hemisphere between mid-May and the end of July, and in the southern between mid-November and the end of January. SOLSTICE.

mid-ocean ridge OCEANIC RIDGE.

migmatite a METAMORPHIC ROCK in which alternating layers or lenses of SCHIST and GRANITE can in some cases be seen, formed at a stage in the process of GRANITIZATION when the MAGMA, penetrating metamorphosed unmelted rock, cools. MIGMATIZATION.

migmatization the process of the transformation of high grade METAMORPHIC ROCK to GRANITE whereby the composition of the metamorphic rock is changed mainly by a raised SODIUM and POTASSIUM content, the resultant rock being MIGMATITE.

migrant one who migrates. EMIGRANT, EXILE, EXPATRIATE, IMMIGRANT, MIGRATION, REFUGEE.

migrant relic RELIC DISTRIBUTION.

migration 1. the act or process of moving from one place to another with the intent of staying at the destination permanently or for a relatively long period of time 2. of humans, such a movement from one area (usually the home area) to work or settle in another. EMIGRANT, EXILE, EXPATRIATE, GROSS MIGRATION, IMMIGRANT, MIGRATION MOVEMENT, NET MIGRATION, PLACE UTILITY, PUSHPULL THEORY, REFUGEE, RESIDENTIAL MIGRATION, STRENGTH THEORY, WAGE DIFFERENTIAL THEORY 3. of animals, e.g. birds, the seasonal movement from one region to another 4. of plants, the movement to extend habitat 5. in chemistry, the movement of an atom or group of atoms in a molecule from one position to another, the movement of ions from one electrode to another under electromotive force. G.

migration movement (human) the movement of a person that is essentially one-way and relatively permanent, in which the fixed base, the centre of gravity of the weekly movement cycle (i.e. the home), is removed to a new location, in contrast to RECIPROCAL MOVEMENT. G.

migration of divide a shift in position of the divide between river basins as the more forceful river cuts back rapidly and captures the drainage area of the other, weaker, stream; or, in a CUESTA, the down-dip shift in position of the watershed at its crest, resulting from the unequal erosion of the steep slope and the back slope (LAW OF UNEQUAL SLOPES).

migratory agriculture SHIFTING CULTIVATION.

mil a unit of measurement equivalent to one-thousandth of an inch, or 0.0254 mm, used especially for the diameter of wire.

milch *adj.* giving milk, applied to domestic animals kept for this purpose, e.g. milch cow.

mile (Latin mille passus or passuum, a thousand paces) mi, a British unit of linear measurement 1. statute mile, 1760 yards or 5280 feet (1609.35 metres) 2. geographical mile, one minute of arc measured along the equator, 6087.2 ft, rounded to 6080 ft (1852 m) 3. nautical mile, one minute of arc or $\frac{1}{21600}$ of a GREAT CIRCLE, standardized in Britain as one minute of arc at 48°N, 6080 ft, equivalent to 1.1516 statute mile (1853.25 m) 4. international nautical mile (used by the USA and other countries) 6076.1033 ft, approximately 1.15 statute mile (1852 m). KNOT 5. in measures of area, 1 square mile (2.59 square kilometres) equals 640 ACRES each of 4840 square yards or 43 560 square feet.

milestone in Britain, a roadside stone recording the distance in miles from a selected place.

milieu (French) the environment, a term covering all the factors of the environment, human and physical, philosophical and tangible, with special emphasis on the social factors and on the conditions of human life.

milk 1. the watery, creamy-white liquid containing the essential proteins, enzymes, sugar, minerals (apart from iron), vitamins, and fat globules in suspension secreted by female MAMMALS for the nourishment of their young. BUTTER, CHEESE 2. a liquid resembling that secretion, e.g. coconut milk, contained in the fruit of the COCONUT PALM; or GLACIER MILK.

Milky Way the spiral GALAXY in which the SOLAR SYSTEM lies.

mill 1. a machine (in some cases hand-operated) that grinds a solid substance, e.g. that grinds grain into flour; or the building housing such a machine 2. a building housing machinery used in some kind of manufacturing, e.g. a cotton mill. MILLING.

mille map a quantitative distribution map on which dots are marked, each representing one-thousandth (0.1 per cent) of the total commodity represented (e.g. a crop) and located as accurately as possible. G.

millenium 1. a period of one thousand years 2. a thousandth anniversary.

millepore a member of *Millepora*, order Milleporina. MADREPORE.

millet any of several small-grained cultivated cereals, the most widely grown food grains in tropical regions, being drought-resistant, tolerant of poor soils, storing well, some having a higher mineral content that that of other cereals, e.g. BULRUSH MILLET, FINGER MILLET. Locally important are JOB'S TEARS, HUNGRY RICE and TEFF. By some botanists SORGHUM is classified as a millet. The most important millets of temperate regions are

COMMON MILLET, LITTLE MILLET, FOXTAIL MILLET, JAPANESE MILLET.

millet rains one of the periods of maximum rainfall, October to December, in east Africa (Nairobi and Entebbe). MAIZE RAINS. G.

milli- m, prefix, one thousandth, attached to SI units to denote the unit × 10^{-3}, e.g. millimetre (mm), one thousandth part of a METRE. CENTI-, CONVERSION TABLES, HECTO-, KILO-.

millibar mb, a unit of atmospheric pressure indicated by a BAROMETER, equal to 1000 DYNES per square centimetre of MERCURY; 1000 mb equals 1 BAR-3. The formula for conversion is not universally applicable, but at 0°C (32°F) at latitude 45°, 29 in mercury = 982 mb; 30 in = 1016 mb; 31 in = 1049 mb; and 1000 mb = 750.1 mm (29.531 in). Millibars are commonly used in drawing weather charts, isobars (ISO-) being drawn at 2 or 4 millibar intervals.

millimetre CONVERSION TABLES, MICRON, MILLI-.

milling of cloth, the action of shrinking and thickening cloth by the application of damp heat, friction and pressure.

million city a city with a resident population of one million people or more.

millstone one of the two large circular, grooved flat pieces of hard stone (BURSTONE) between which GRAIN-5 is ground in the making of flour. The stones are place horizontally one above the other, the grain being fed through a central hole in the upper stone.

Millstone Grit a hard, coarse-grained, SANDSTONE occurring in Britain in the central Pennines and in Northumberland, under the Coal Measures at the base of the Upper CARBONIFEROUS. It was probably a delta-deposit laid down in a shallow sea.

mine an excavation deep in the ground, made for the purpose of extracting MINERALS, together with its shaft and associated buildings. MINING, OIL

WELL, OPENCAST MINING, PIT, QUARRY.

Minamata disease a disease of the central nervous system caused by MERCURY poisoning. The name is derived from a bay and town in Japan where many people were poisoned in 1959 by eating fish and shellfish which had ingested dimethyl mercury, present in the sediments of the bay as a result of effluent discharged by a nearby factory. POLLUTION.

mineral a naturally occuring inorganic HOMOGENEOUS-2 substance, usually CRYSTALLINE with a definite chemical composition (capable of being expressed by a chemical formula, or varying only within certain limits); hence the distinction from a rock, which is commonly (apart from rock salt) a mixture of minerals. On this definition water (ice) is a mineral, but (being of organic origin) petroleum and natural gas are not, although they are commonly described as such, as are some organically-derived limestones and siliceous rocks **2.** loosely applied to any mineral deposit won by mining, e.g. metallic ore, coal. G.

mineral horizons one of the two major classes of SOIL HORIZON, the other class being the O HORIZONS. The mineral horizons consist mainly of inorganic matter (MINERAL SOIL), and the two groups of minerals present in them are identified as CLAY MINERALS and associated products of weathering and SKELETAL MINERALS. SOIL CLASSIFICATION.

mineralization 1. the process whereby gases and water from the heated lower layers of the earth's crust by passing through fissures, cracks etc. cause changes in the rocks, and the deposition of minerals of economic importance. METASOMATISM **2.** the replacement of the organic parts of a plant or animal by MINERALS-1 during decomposition or fossilization. FOSSIL.

mineralogy the scientific study of MINERALS-1.

mineral oil any oil of so-called MINERAL-1 origin, notably PETROLEUM.

mineral soil a soil low in HUMUS, consisting mainly of material of MINERAL-1 origin (i.e. of inorganic material). MINERAL HORIZONS, ORGANIC SOIL.

mineral spring a SPRING-2 containing a high proportion of mineral salts in solution. G.

mineral water 1. naturally occurring water containing mineral salts or gases, especially if of medicinal value **2.** a term extended to artifically made effervescent soft drinks. G.

minette (French) **1.** a phosphoric low grade IRON ore, with a metal content between 25 and 40 per cent, occurring in vast quantities in Lorraine, France, similar in geological age and chemical composition to the English JURASSIC ores **2.** loosely, any sedimentary iron ore with LIMONITE as the main constituent. G.

ming land (England, Midlands) historically, lands of different proprietors lying intermixed in common fields. G.

minifundio (Spanish) **1.** a very small farm in Latin America **2.** a very small plot of land cultivated by a Spanish settler for personal use in return for military service to the Spanish crown. LATIFUNDIA.

minimax in a mathematical MATRIX-5, a value which is the maximum in a row and a minimum in the column which intersects at the value concerned. GAME THEORY.

minimax location the site for a firm selected by an entrepreneur looking for the minimum cost and the likelihood of maximum profit for the enterprise. LOCATION THEORY.

minimax solution GAME THEORY.

minimum thermometer a THERMOMETER used to register the lowest ambient air TEMPERATURE reached at a particular place over a selected period of time. One type commonly used consists of a horizintal glass tube filled with ALCOHOL with a dumb-bell shaped metal marker held at one end of the alcohol by surface tension. As the temperature falls the alcohol contracts and the marker is drawn to the level of the lowest temperature registered, where, even if the temperature rises, it sticks until reunited with the alcohol in its original position (by the use of a magnet). MAXIMUM THERMOMETER, MAXIMUM-MINIMUM THERMOMETER.

mining 1. deep excavating in the earth in order, by underground workings, to extract MINERALS-1, METALS, METALLIC ORES. A distinction is sometimes made between mining (also termed deep mining) by underground workings and QUARRYING (also termed surface mining), conducted at the surface (PIT, QUARRY). OPENCAST MINING is associated with extensive surface workings, whereas quarrying is more limited in extent 2. the extraction of non-renewable NATURAL RESOURCES.

minor intrusion in geology, an IGNEOUS INTRUSION that is small when compared with a large, deep-seated plutonic intrusion. Examples include DYKES, SILLS, small LACCOLITHS and VEINS. MICROGRANITE. G.

mint 1. any of the aromatic plants of the family Labiatae, especially of the genus *Mentha*, native to Europe, Asia and Australia, some being used as flavouring, others as a source of ESSENTIAL OILS (PEPPERMINT, SPEARMINT) 2. a place where official coins are made.

minute a unit of measurement applied to 1. time, one-sixtieth of an hour 2. an angle, one-sixtieth of an (angular) degree; one-sixtieth of a degree of latitude or longitude. SECOND.

Miocene *adj.* of the geological epoch (of time) or series (of rocks) in the TERTIARY period, between the OLIGOCENE and the PLIOCENE. GEOLOGICAL TIMESCALE.

miogeosyncline a GEOSYNCLINE sub-

jected to only slight VULCANICITY while being infilled by SEDIMENTATION, in contrast to a EUGEOSYNCLINE.

miombo MYOMBO. G.

mirage an optical phenomenon in which distant objects may be seen inverted, as if mirrored in water, or suspended in mid-air. This is due to the unusual distribution of density in the ATMOSPHERE-1. For example, when the air near the ground is greatly heated by conduction and becomes less dense, rays of light from above (approaching at a slight angle) are refracted towards the observer, i.e. they are bent upwards so that the sky appears as a glistening sheet of water on, for example a road surface. In high latitudes, when a warm layer of air rests on a cold layer, the light rays may be bent down from the warm layer, so that an inverted or even double image of a distant object appears.

mire an area of spongy, waterlogged ground; sometimes used in place-names in England, e.g. Great Close Mire, in Yorkshire. G.

misfit river, misfit stream, underfit river a river that appears now to be too small for its valley, due to BEHEADING by another stream, or a change of climate, or the valley's being enlarged and broadened by glaciation, or because the volume of water has been reduced by seepage through the thick alluvium of the floodplain. MEANDERING VALLEY. G.

mist 1. obscurity in the lower layers of the ATMOSPHERE-1 caused by particles of condensed moisture held in suspension, limiting vision (officially) to between one and two kilometres (about 1000 to 2000 yards). FOG, HAZE. G.

mist forest, cloud forest hygrophilous forest (HYGROPHYTE) occurring on mountain slopes in tropical regions, where mist or cloud is constant or frequent. G.

mistral (French) a powerful, cold, dry

northwesterly or northerly wind, blowing from the high Massif Central towards the relatively warm Golfe du Lion, particularly affecting the Rhône delta and the north coast of the Mediterranean. It is particularly common in winter when the cold air from the winter high pressure system lying over central Europe is channelled through the lower Rhône valley to the low pressure area lying over the western Mediterranean. BORA.

Mitteleuropa central Europe, variously used by different authors as a politico-geographical term, e.g. by the promoters of Pan-Germanism, who proposed that central Europe should form one great German-speaking empire; or by other authors who see Mitteleuropa as a CORE AREA in central Europe. INTERNATIONAL REGION. G.

mixed cultivation the growing of two or more crops intermingled on the same field or plot, especially a mixture of tree and ground crops. G.

mixed economy an economic system in which some parts operate according to the forces of the free market (production, distribution and exchange being in private hands), while other parts are in the hands of the government (i.e. there is more STATE INTERVENTION in a mixed economy than there is in a true MARKET ECONOMY). CENTRALLY PLANNED ECONOMY, LAISSEZ-FAIRE, PLANNED ECONOMY, PRIVATE SECTOR, PUBLIC SECTOR, STATE CAPITALISM.

mixed farming AGRICULTURE in which both crops and livestock are produced on an individual farm (not to be confused with MIXED CULTIVATION). ARABLE LAND, CROP FARMING, FARMING, LIVESTOCK FARMING. G.

mixing ratio the ratio of the weight of a parcel of an atmospheric gas to the total weight of the parcel of air with which the gas is mixed, commonly used to express the ratio of the weight of

WATER VAPOUR in a parcel of air, where it is expressed in grams of water vapour per kilogram of dry air. ABSOLUTE HUMIDITY, SPECIFIC HUMIDITY.

mixotroph an organism that is AUTOTROPHIC as well as HETEROTROPHIC.

mizzle (comparable with Dutch dialect, miezelen) misty DRIZZLE, rainfall so fine that it appears to be MIST.

MNE multinational enterprise. MULTINATIONAL.

mobile *adj.* **1.** capable of moving or being moved from place to place **2.** moving or moved with ease, fluid **3.** of organisms, having the power of moving from one place to another. MOTILE.

mobile dune a coastal DUNE partially fixed by vegetation but still liable to DEFLATION-1 and BLOW-OUTS-1, a type transitional between a FOREDUNE and a STABILIZED DUNE.

mobile industry FOOTLOOSE INDUSTRY.

mobility the state or quality of being MOBILE, e.g. of people moving readily from one place to another and from job to job, in search of employment or higher incomes; or of individuals in relation to the degree to which each one has access to travel facilities (personal mobility); or of individuals and households in being able to move between SOCIAL CLASSES and income groups (SOCIAL MOBILITY).

mocha COFFEE.

modal category in statistics, the CATEGORY-3 containing the greatest number of observations.

mode **1.** a way or a manner of doing, of being, of taking place **2.** a fashion **3.** in statistics, in PROBABILITY theory, the value of the VARIABLE occurring most frequently in a set of observations. If the set has one value around which observations cluster, it is termed unimodal; if there are two distinct values around which observations cluster it is said to be bimodal; and if there are more than two, it is multi-modal. The mode is the simplest type of average and the only one that can be used on NOMINAL data. ARITHMETIC MEAN, CENTRAL TENDENCY, MEDIAN, MODAL CATEGORY **4.** in geology, the percentage of each constituent mineral present in a METAMORPHIC or IGNEOUS ROCK.

model 1. a pattern, an exemplar, or an archetypal (ARCHETYPE) image **2.** a thing, person or process considered to be so near to perfection in some respect as to be worthy of imitation **3.** a three-dimensional representation, more or less to scale, of something that exists or is to be constructed, e.g. of a building, which may be to scale in every dimension and detail; or of a landscape, with ground measurements to scale but the vertical scale exaggerated to show ALTITUDE-3; or a working scale model, such as one representing tides, which includes processes **4.** a representation of some aspects of reality, selected and brought together to show certain of its properties, and providing a working HYPOTHESIS against which reality can be tested. On a range of ABSTRACTION-1 these are, at the first level, the ICONIC MODEL, the most realistic, a scaled-down representation of reality (e.g. with real phenomena shown in their characteristic form, but scaled-down). At the second level is the ANALOGUE MODEL in which real phenomena are represented by different but analogous phenomena (e.g. clusters of people shown by clusters of points). The final level of abstraction, the furthest from reality, is reached by the SYMBOLIC MODEL, in which real phenomena are represented by mathematical expressions. ALONSO MODEL, ANALOGUE MODEL, DICHOTOMOUS MODEL, DYNAMIC SPATIAL MODEL, GRAVITY MODEL, HOTELLING MODEL, ICONIC MODEL, REGRESSION MODEL, SIMULATION MODEL, SYMBOLIC MODEL. G.

mode of domestic production the way in which work is organized in the home. It usually involves the unpaid labour of women.

mode of production (one of the central concepts in Marxism) the economic base of a society, the way in which the productive activities in the society are organized, and thereby affect the social as well as the economic relations in that society. Marx identified the historical succession of modes as primitive communal, slave, feudal, capitalist, state capitalist, socialist, communist, the changes of mode being brought about by class conflict, e.g. between landlord and peasant, capitalist (BOURGOISIE-2) and workers (PROLETARIAT-2) as each class tries to gain control of the MEANS OF PRODUCTION, asserting that the other class is inept at providing society with an acceptable level of subsistence. Marx maintained that the proletariat would eventually overcome the capitalist ruling class (DIALECTIC, DIALECTICAL MATERIALISM) and that a classless communist society would emerge. CAPITALISM, COMMUNISM, FEUDALISM, HISTORICAL MATERIALISM, PRIMITIVE COMMUNISM, PRODUCTIVE FORCES, SLAVE MODE OF PRODUCTION, POST-INDUSTRIAL SOCIETY, SOCIALISM, SOCIAL FORMATION, SOCIAL RELATIONS OF PRODUCTION.

moder in soil science, a HUMUS layer intermediate in composition between MOR and MULL-3, occurring where decomposition is greater than in mor, but has not advanced so much as in mull, but where there is some mixing with particles from the underlying MINERAL SOIL, due to the presence of soil fauna.

moderately developed country MDC.

modernization 1. in general, the process or act of making up-to-date, of changing something from the past so that it becomes in harmony with current taste, thinking, technology **2.** in society, a process of social change which commonly accompanies or follows INDUS-

TRIALIZATION and may include an increase in SOCIAL MOBILITY, the blurring of boundaries between SOCIAL CLASSES, the advancement of education, the development of social services, and the adoption of procedures in government more effective than those previously prevailing. MODE OF PRODUCTION.

modified Mercalli scale MERCALLI SCALE.

moela (Spanish) a plateau in the form of an elevated SYNCLINE. G.

mofette (French) an opening in the earth's surface, occurring in regions of former volcanic activity, which emits carbon dioxide, oxygen, nitrogen and, sometimes, water vapour. FUMAROLE, SOFFIONE, SOLFATARA, SOLFATARIC STAGE. G.

mogote (Cuba: Spanish) a karst INSELBERG, a steep-sided limestone residual hill in KARST, rising from a nearly flat alluviated plain, formerly termed a haystack hill in some areas, e.g. in Puerto Rico. The term was originally local to Cuba, where it referred to residual hills of folded limestone, but it is now applied internationally to karst residual hills in tropical regions. HUM, TOWER KARST. G.

mohair 1. the hair of the ANGORA GOAT **2.** the fabric made wholly or partly from that hair, or a fabric of cotton, wool or synthetic fibres woven to resemble true mohair fabric.

Mohorovičić Discontinuity, Moho the boundary surface between the MANTLE of the earth and the rocks of the earth's surface, lying at a depth of some 40 km (25 mi) under the continents but only some 6 to 10 km (4 to 6 mi) under the ocean. In the Discontinuity, owing to the different densities of the crust and the mantle, there is a very sharp change in the rate of travel of EARTHQUAKE waves (PUSH WAVE, SHAKE WAVE): they accelerate. G.

Moh's scale HARDNESS OF MINERALS.

moisture in soil science, the water that

can be removed from soil by heating it to 105°C (221°F). G.

moisture index THORNTHWAITE'S CLIMATIC CLASSIFICATION.

mol MOLE-3.

molasse, mollasse (French) a deposit of soft greenish SANDSTONE and red and grey MARLS-2 mainly of MIOCENE age, derived from material denuded from the Alps in Europe during and after the maximum mountain building stage and laid down under continental or shallow freshwater conditions (in contrast with FLYSCH, laid down before the orogenic maximum). Applied to deposits similar in origin and character laid down elsewhere. G.

molasses (American treacle) **1.** the syrup drained from raw sugar **2.** the thick, dark brown sugary liquid, a by-product in the refining of sugar, left when the crystals of sugar separate out from the juice of the SUGARCANE or SUGAR BEET. It has various uses: it can be refined to produce (British usage) golden syrup and treacle, and is the basis for the distilling of the sugar spirit, rum.

mole 1. a strong masonry wall built out from the coast into the sea to serve as a breakwater **2.** a HARBOUR formed by such a wall **3.** mole, mol, in chemistry, the basic SI unit of amount of substance, the amount being equal to the amount of substance which contains as many elementary units as there are atoms in 0.012 kg of carbon 12. The elementary unit has to be specified. It may be an ATOM, MOLECULE, ELECTRON, ION etc. or a specified group of such entities.

molecular *adj.* pertaining to, involving, or consisting of, MOLECULES.

molecular attraction the attraction of MOLECULES for each other, especially, in HYDROLOGY, the attraction of the molecules of the surfaces of solid rocks for the molecules of water, and of the water molecules for each other, the means by which some GROUND

WATER is held in fine-grained rocks. CAPILLARITY.

molecule two or more ATOMS linked by chemical bonding (BOND) and constituting the smallest group of combined atoms of an ELEMENT-6 or of a COMPOUND which can exist freely (FREE-2) while retaining the characteristic properties of the substance.

mollic *adj.* applied to dark, organically-rich surface layers of a soil, rich in CALCIUM and MAGNESIUM, strongly structured so that the soil does not harden unduly when dried out. MOLLISOLS.

mollisol the seasonally ACTIVE LAYER in the soil.

mollisols in SOIL CLASSIFICATION, USA, an order of soils characteristic of grassland, with a thick, organically-rich surface layer. It includes soils with widely varied profiles, all structurally well developed, e.g. BRUNIZEM, CHERNOZEM, RENDZINA.

mollusc a member of Mollusca, a large phylum of animals (CLASSIFICATION OF ORGANISMS) that includes mostly aquatic INVERTEBRATES (e.g. snails, mussels), unsegmented, with a soft body (often with a calcareous shell), a head, and a muscular foot variously modified for digging, swimming or creeping.

molluscoid a member of Molluscoidea, a large phylum (CLASSIFICATION OF ORGANISMS) of INVERTEBRATE animals having some of the characteristics of a MOLLUSC.

molluscoid *adj.* like a MOLLUSC or a MOLLUSCOID.

molybdenite a mineral, molybdenum disulphide, MoS_2, the chief ore of MOLYBDENUM.

molybdenum a hard white metallic ELEMENT-6, produced by hydrothermal activity associated with acid igneous rocks, mainly occurring in MOLYBDENITE. It is an essential MICRONUTRIENT, and is used in hard STEEL alloys.

moment measures in statistics, terms used to describe the character of a distribution curve (FREQUENCY CURVE), i.e. the KURTOSIS, MEAN, SKEWNESS, STANDARD DEVIATION.

momentum 1. the force built up by a moving body 2. a measure of the ability of a moving body to resist forces acting on it. GEOGRAPHICAL MOMENTUM, INDUSTRIAL MOMENTUM.

monadnock a residual hill, a remnant of erosion, left standing above the general level of a denuded plain (PENEPLAIN). The term is derived from the name of a mountain of this character in New Hampshire, USA. G.

monarch one who holds a hereditary, lifetime right to rule a kingdom, state or people, being invested with either absolute or constitutional power. MONARCHY.

monarchy 1. rule of a kingdom, state or people by a MONARCH 2. a kingdom, state or people ruled by a monarch. AUTOCRACY.

monazite a phosphate mineral, containing compounds of CERIUM, THORIUM and other RARE EARTH METALS, occurring in small quantity in GRANITE and PEGMATITE but more in sand (especially in BLACK SAND). It is used in alloys and as a catalyst, and (when present in pegmatites) in measuring the RADIOMETRIC AGE of rocks.

monetarism the economic theory which asserts that the level of activity of an economy can be controlled by controlling the money supply. KEYNESIANISM.

monimolimnion the lower, stable, unmixed layer of a meromictic lake. EROMIXIS. G.

monoclimax vegetation exhibiting a single CLIMAX formation developed within a climatic region, as distinct from POLYCLIMAX.

monoclinal fold an asymmetrical FOLD-2, with one limb dipping more steeply than the other, due to compression in the earth's crust.

monoclinal shifting UNICLINAL SHIFTING.

monocline 1. in geology, British usage, a FOLD-2 in which the bend is in only one direction (i.e. a fold with one limb, a single bend in horizontal beds), the rock stratum, through TENSION in the earth's crust, changing its dip by increasing the steepness of inclination, and then levelling out again or resuming its original dip. It is termed a monocline because only one fold, or one half of a fold, is presented instead of the two occurring in an arch or trough 2. in American usage, monocline is synonymous with HOMOCLINE (British usage), i.e. a structure of several beds dipping evenly in one direction. UNICLINAL. G.

monoclinic *adj.* applied to a CRYSTAL having one oblique axial intersection.

monoculture cultivation in which a single crop predominates and is planted successively on the same land, in contrast to a ROTATION OF CROPS. G.

monoglacial *adj.* applied especially to the theory that the GREAT ICE AGE was marked by one great ice extension only, without INTERGLACIAL PERIODS of recession followed by advance of the ice. G.

monoglaciation a glacial period during which there was only one major advance of the ice, although in that period the limit of the ice may have advanced and retreated to some small extent, and in the process deposited TILLS.

monoglot 1. one who understands, speaks, writes, only one language 2. written in only one language. POLYGLOT. G.

monolith 1. a single block of stone, especially one shaped into a monument. MEGALITH, MENHIR 2. in soil science, a vertical section taken from the soil in order to study the SOIL PROFILE. G.

mononuclear diffusion DIFFUSION-2.

monophagous *adj.* applied to an animal which feeds on only one kind of food.

MACROPHAGOUS, MICROPHAGOUS, PHYTOPHAGOUS, POLYPHAGOUS, XYLOPHAGOUS.

monophyletic *adj.* applied to a natural, closely related group of species or other taxa descended from a common ancestor which is a member of the same TAXON. POLYPHYLETIC, TAXONOMY.

monopoly 1. the exclusive control of the supply of a product or service in a particular market by a single supplier, who thus dominates the market 2. an exclusive right to conduct a particular business or provide a particular service, granted by a ruler, government, etc. 3. a COMMODITY-1 under exclusive, single control 4. a single supplier who has exclusive control. DUOPOLY, OLIGOPOLY, PERFECT COMPETITION, SPATIAL MONOPOLY.

monopoly capitalism in a capitalist economy, the domination of the economy by a relatively small number of large companies.

monoxenic parasite AMETOECIOUS PARASITE.

monsoon (Arabic mausim, season) originally applied to the regular winds of the Arabian sea, blowing for six months from the northeast, six months from the southwest. Now generally applied to those and some other winds that blow with considerable regularity at definite seasons of the year due to the seasonal reversal of pressure over land masses and their neighbouring oceans. In the typical area of the Indian subcontinent and southeast Asia it is the seasonal inflowing moist winds that bring rain, hence the monsoon season is termed the RAINS, and the term monsoon is applied to the rains without reference to the winds. G.

monsoon forest the FOREST-2 of the tropical monsoon lands where the annual rainfall is between 1000 and 2000 mm (40 and 80 in) and there is a marked dry season. It consists of BROADLEAVED TREES that lose their leaves in the hot dry season (February to May in the Indian subcontinent and Burma). The trees, mainly HARDWOOD, do not grow so closely together as those in EQUATORIAL FOREST, and the number of species is low; teak is the most important commercially. G.

montaña (Spanish) mountain or highland, applied in South America to the forested eastern slopes of the Andes in Peru. G.

montane *adj.* of or pertaining to mountain regions, applied particularly to the vegetation growing on high land below the TREE LINE. G.

montane forest the FOREST-2 of the cool uplands in the zones of tropical and equatorial climates. MOSSY FOREST.

montan wax a bituminous wax present in some LIGNITES and PEATS, extracted for use in making polishes, carbon paper, etc.

monte (Spanish and Italian, mountain) DECIDUOUS, BROADLEAVED low trees and shrubs, XEROPHYTIC scrub, on the foothills in Argentina and elsewhere in South America. There is so much disagreement concerning the location and extent of this type of vegetation that some authors prefer to avoid the term.

Mood's median test in statistics, an hypothesis test used to compare two population MEDIANS-3. The sample data must be ordinal (ORDINAL SCALE) and the overall sample sizes fairly large. This test cannot be used to deal with data values which are exactly equal to the overall median: if some of the data values are equal to the overall median, another test will have to be used. MANN-WHITNEY TEST.

moon 1. in astronomy, the natural satellite of any planet 2. the earth's only natural satellite, appearing to move in the CELESTIAL SPHERE, in relation to the stars, from west to east, responsible with the sun for tidal action on the earth. It revolves round the earth in a slightly elliptical orbit, the distance from the earth varying from 348 285 km (216 420 mi) surface to surface PERIGEE to 398 587 km (247 67 mi) APOGEE. The diameter is about one-quarter of that of the earth, its mass 1/81 of the earth. It has no atmosphere, no water, and shines by reflecting light emitted by the sun. It revolves round the earth every 27 days 7 hours 43 minutes 15 seconds (sidereal month), and one revolution related to the sun on average 29 days 12 hours 44 minutes (lunar or synodic month, i.e. from one 'new' moon to the next); the LUNAR DAY is about 24 hours 50 minutes. Because it rotates on its own axis once in each revolution of its orbit (i.e. the period of rotation is the same as the period of revolution), the same face is always seen from the earth. The changes in aspect seen from the earth are due to changes in the relative position of the earth, moon and sun, and are termed phases, the sequence being: new moon, invisible to faintly visible, the moon lying between the earth and the sun (CONJUNCTION) so that viewed from earth it is not illuminated by the sun's light, although it may reflect a faint glow of light from the earth; the first quarter (QUADRATURE), when the moon has moved through about one-quarter of its orbit round the earth, and appears as a semicircle, having 'grown' from a CRESCENT, bow facing west; GIBBOUS moon, the phase reached when the moon has passed through another eighth of its orbit, so that three-quarters of its face as seen from earth is illuminated by the sun; full moon, when the earth lies between the moon and the sun (OPPOSITION) so that viewed from earth the whole face of the moon is illuminated by light from the sun. From new moon to full moon the moon is said to be waxing; from full moon to new moon it is waning. OSCILLATORY WAVE THEORY OF TIDES, SYZGY, TIDE.

moonstone andularia, a translucent,

lustrous, SEMI-PRECIOUS STONE, a variety of ALBITE, a form of FELD-SPAR, resembling MOTHER-OF-PEARL. It is usually pale grey to milky white in colour; if tinted with red or green it is sometimes termed sunstone.

moor, moorland a term loosely applied in Europe and especially in the British Isles to open, unenclosed land, generally elevated with acid peaty soil, which is not good pasture though used to some extent as ROUGH GRAZING. Different types are distinguished by dominant plants, hence cotton grass moor, heather moor, etc. G.

moorpan HARD PAN.

moose *Alces americana*, a large North American DEER, with very large branched antlers, living in forests in Canada and northern USA.

mopane (Afrikaans) *Copalifera mopane*, a BROADLEAVED tree of the northern BUSH VELD regions of South Africa, adapted to fairly low rainfall, occurring as a shrub or standard tree, the close stands of the latter constituting mopane forest. G.

mor in soil science, raw HUMUS, low in animal life, acidic and crumbly, unmixed with and sharply demarcated from the underlying MINERAL SOIL. MODER, MULL. G.

moraine (French) an accumulation of unstratified (STRATIFIED) debris, especially boulders and coarse material, carried down and deposited by a GLACIER or ICE SHEET. Some of the debris falls from above, from the mountain slopes, on to a glacier (SUPERGLACIAL), some is plucked (PLUCKING, SUBGLACIAL) from the floor beneath a glacier or ice sheet by ice action. On the surface of a VALLEY GLACIER there are usually LATERAL MORAINES on each side, and a MEDIAL MORAINE if two valley glaciers have joined. ENGLACIAL moraines are those enclosed in the ice. The debris may be deposited as GROUND MORAINE when the ice melts, or as a TERMINAL or END MORAINE. The term applies both to the material and to the feature produced. ABLATION MORAINE, PUSH MORAINE, RECESSIONAL MORAINE, SUBGLACIAL MORAINE, STADIAL MORAINE, WASHBOARD MORAINE. G.

morass loosely applied to a wet, swampy tract, a BOG or MARSH. G.

morbidity the incidence of disease in a population.

more developed country MDC, UNDER-DEVELOPMENT.

morello *Prunus acida*, a small bushy, DECIDUOUS tree, cultivated in temperate regions for the sake of its crop of fruit (a DRUPE), dark red, sour CHERRIES, used in making jam and liqueurs.

morfa (Welsh) MARSH, sea FEN, frequently fronting coastal cliff scenery. G.

morocco leather originally high quality leather made in northern Nigeria from goatskin, coloured and roughened by means of a boxwood stamp on one side only, introduced to Europe via Morocco, hence the name. C.

morphochronology the dating of morphological phenomena or landforms. MORPHOLOGY.

morphogenesis the origin and development of forms, especially of LAND-FORMS. FORMKREIS, GEOMORPHO-LOGY, MORPHOGENETIC, MORPHO-GENETIC REGION, MORPHOLOGY.

morphogenetic *adj.* of or pertaining to MORPHOGENESIS.

morphogenetic region a REGION-1 differentiated from other areas by the particular characteristics of its landscape which result from climatically influenced geomorphological processes. FORMKREIS. G.

morphographic map landform map, a small scale map on which the physiographic features are shown by standardized conventional pictorial symbols representing oblique aerial views of them.

morphological map a map which usually covers a small area and on which detailed surface forms, determined by field survey, are shown symbolically, to assist in the interpretation of the relationships between soil and vegetation distribution and landforms. The origin or history of landforms is not shown, as it is in a GEOMORPHOLOGICAL MAP.

morphological region in GEOMORPHOLOGY, a physiographic region, a distinctive unit of landform identified on the basis of form, rock structure, evolutionary history, and varying in scale or order. D. L. Linton postulated a hierarchy of increasing size and complexity, from the smallest, the site, through stow, tract, section, province to continental subdivision.

morphological system a type of SYSTEM-1,2,3 identified by R. J. Chorley and B. A. Kennedy, 1971, consisting solely of the physical properties of its components, the strength and direction of their connectivity being revealed by statistical correlations. CASCADING SYSTEM, CONTROL SYSTEM, PROCESS-REPONSE SYSTEM. G.

morphology 1. the scientific study of the form, structure, origin and development of organisms; or of the external structure of rocks in relation to form; or of landforms or topographic features resulting from erosion 2. the branch of linguistics concerned with word formation, inflectional forms, etc. GEOMORPHOLOGY, URBAN MORPHOLOGY. G.

morphometric analysis analysis concerned with the examination of shapes and forms in space (e.g. the shape and pattern of town locations, structure of networks, etc.), an extension of the mathematical morphometric techniques (MORPHOMETRY) used in geomorphology to various branches of human geography. G.

morphometry 1. the exact measuring of the external form of a substance 2. in geomorphology, the exact measurement of external features of landforms and a mathematical treatment of them. G.

morphosequent *adj.* applied in geomorphology to surface features which do not reflect the underlying geological structure. TECTOSEQUENT. G.

morphostasis the process by which any tendency to change a SYSTEM-1,2,3 is counteracted, so that the STEADY STATE characteristics are perpetuated. GENERAL SYSTEMS THEORY, HOMEOSTASIS.

mortality 1. the number of deaths in a particular period or place 2. the death-rate (BIRTH-RATE). MORBIDITY.

mortlake a dead lake, an old practically obsolete term for an OXBOW LAKE or cut-off. G.

mosaic 1. in aerial photography, a composite photographic representation of an area obtained by joining together individual photographic prints 2. in ecology, a vegetation pattern in which community types are interspersed 3. MOSAIC DISEASE in a plant 4. in geology, the angular and granular structures visible in a rock which has been subjected to DYNAMIC METAMORPHISM. MOSAIC-1, *adj.* G.

mosaic *adj.* 1. of or relating to a surface decoration made by joining and cementing together small pieces of coloured stone, glass etc. 2. of the design or picture so made 3. resembling that design in pattern or structure 4. of a plant, suffering from MOSAIC DISEASE.

mosaic disease any of several VIRUS diseases of certain flowering plants, usually recognizable by the mottled patches on the leaves.

moshav (Israel: Hebrew, pl. moshavin) an agricultural Jewish village in modern Israel, where there is some cooperative enterprise but where each farmer owns and works an area of land, and the importance of the family as a social and economic unit is recognized. KIBBUTZ, KVUTZA. G.

mosore a MONADNOCK which has survived on account of its remoteness from streams. G.

moss 1. any member of Musci, a class

of Bryophyta, primitive land plants distributed throughout the world, growing on rocks, walls, trees, heaths, damp ground **2.** a term applied to a BOG or SWAMP, especially in northern England and parts of Scotland, dominated by bog moss (SPHAGNUM); and also to summit plateaus of the southern Pennines where the moorland is dominated by cotton grass (*Eriophorum vaginatum*) and some *Sphagnum*; and to coastal marshes with FEN PEAT, e.g. Solway Moss. G.

mossy forest FOREST-2 in which the living and dead trees and ground surface support a strong growth of MOSSES-1, occurring particularly on uplands supporting MONTANE FOREST, e.g. in the humid cloudy conditions of the windward side of a high OCEANIC ISLAND in the TROPICS.

mota (Indian subcontinent: Hindi) a clay pan. HARD PAN. G.

motel a roadside hotel designed specifically for motorists, usually with self-contained accommodation and adjacent parking space or garage.

mother country 1. the country of one's birth, or of one's ancestors **2.** a country in relation to its COLONIES or dependencies. METROPOLITAN-2, *adj*. G.

mother-of-pearl the hard, iridescent internal layer of certain shells, e.g. of abalones or oysters, used for jewellery, decoration, furniture inlay etc. and valued for making 'pearl' buttons.

mother-or-pearl cloud nacreous cloud, a rather rare, usually LENTICULAR CLOUD, occurring in the upper layers of the STRATOSPHERE when atmospheric pressure and temperature are low.

motile *adj*. applied to an organism that moves freely by its own energy, usually restricted to MICROORGANISMS which have a means of locomotion and move about aimlessly (e.g. a BACTERIUM). MOBILE.

motivation the stimulus internal to the individual, conscious or subconscious, that leads to action towards a desired

GOAL and gives purpose or direction to BEHAVIOUR.

motorway in UK, a wide road with limited access, carried over or under other mainroads at a different level, accommodating several lanes of high-speed through traffic, and restricted to certain types of vehicle, comparable with an AUTOBAHN, AUTOSTRADA, or EXPRESSWAY, see autobahn for more details. HIGHWAY, TURNPIKE. G.

mottled *adj*. marked with blotches, with patches of differing colours, e.g. in a GLEY SOIL or MOSAIC DISEASE.

mould 1. a soil rich in HUMUS, e.g. leaf mould **2.** the popular name for FUNGUS **3.** more precisely, any surface growth of fungus mycelium (the mass of filaments constituting the reproductive part of a fungus) **4.** any fungus producing a superficial growth of mycelium, e.g. *Penicillium notatum*.

moulin (French, a mill) glacier mill, a steep shaft or vertical circular hole carved through the ice of a GLACIER or ICE SHEET by a stream of melt-water as it swirls, laden with rock debris, down a fissure in the ice. G.

mound 1. a low hillock, natural or artificial **2.** a bank of earth, especially if artificial and (historically) used for defence purposes, i.e.a rampart.

mount mt, a mountain or high hill, commonly used in place-names. SEA-MOUNT.

mountain any natural elevation of the earth's surface with a summit small in proportion to its base, rising more or less abruptly from the surrounding level. In Britain the term is commonly restricted to an elevation exceeding 600 m (2000 ft), the term HILL being applied to a lower elevation; but mountain may be applied to an elevation even under 300 m (1000 ft) if it rises sufficiently abruptly from the surrounding level. G.

mountain building OROGENESIS. G.

mountain chain a complex series of roughly parallel mountain ranges

(RANGE-2) forming a connected system, i.e. consisting of more than one mountain range. CHAIN. G.

mountain classification the old descriptive classification of mountains, i.e. mountains of accumulation (e.g. volcanoes), mountains of circumdenudation, and fold mountains, now little favoured. G.

mountain glacier GLACIER.

mountain sickness SOROCHE, a feeling of nausea, accompanied by sickness, which attacks some unaccustomed or untrained people at high altitudes, caused by breathing air deficient in oxygen at those heights.

mountain-top detritus a BOULDER FIELD formed by the disintegration of bed rock in situ by frost shattering on mountain tops. G.

mountain wave LEE WAVE.

mountain wind ANABATIC, FOHN, KATABATIC.

Mousterian, Moustierian (Le Moustier, a cave in southwest France) *adj.* of or pertaining to the mid-PALAEOLITHIC period, i.e. the cultural period associated with NEANDERTHAL MAN and the use of stone flake tools, preceded by the ACHEULEAN and succeeded by the AURIGNACIAN cultural periods.

mouza (Indian subcontinent: Bengali) a village area, an administrative unit. A number of mouzas constitute a union; and the next larger unit to the union in the administrative hierarchy is the THANA. G.

movement the act or process of changing location or position.

movement block one of the three basic patterns made by the direction of transport flows, a group of points or areas in which there is much movement relative to external movement. MOVEMENT HIERARCHY, MOVEMENT HINTERLAND.

movement frequency the number of times a movement is made for the same purpose within a given period of time. RECIPROCAL MOVEMENT.

movement hierarchy one of the three basic patterns made by the direction of transport flows, a pyramidal structure, with very many small flows and a few large ones. MOVEMENT BLOCK, MOVEMENT HINTERLAND.

movement hinterland one of the three basic patterns made by the direction of transport flows, the major flows converging on and diverging from a single location. MOVEMENT BLOCK, MOVEMENT HIERARCHY.

movement repetitiveness the proportion of movements made to the same locations for the same purpose within a given period of time. MOVEMENT FREQUENCY, RECIPROCAL MOVEMENT.

msitu (Africa: Bantu) a type of dense bush in Zambia.

muck 1. MANURE 2. (American) a dark, moist, HUMUS-rich soil. G.

mud 1. an unconsolidated rock of clay and/or silt grades (GRADED SEDIMENTS) with much water, e.g. as commonly deposited in estuaries, lakes, lagoons and at depths under the ocean. It may be partly consolidated in certain geological formations to form MUDSTONE, resembling a soft shale, but non-plastic. MUD FLAT, MUD-POT 2. a manufactured SLURRY-1 used in the process of sinking BORES-2.

mud circle a roughly circular pattern developed on the ground in very cold lands, e.g. on the gravel beaches of Cornwallis Island, due to alternate freezing and thawing. When young, the mud circle is represented by a plug under a conical pit, but in thawing the mud plug rises (injected upwards), increases in size and eventually breaks out into the pit. The CRYOSTATIC HYPOTHESIS may explain the formation of mud circles.

mud flat an expanse of fine clay or silt (GRADED SEDIMENTS) deposited by FLOCCULATION-2 in estuaries and sheltered coastal areas (e.g. behind a sandspit), covered by water at high tide and sometimes colonized by hygrophilous plants, such as MANGROVE in tropical areas, various grasses in cooler regions. HYGROPHYTE.

mudflow a moving mass of soil made fluid by rain or melting snow, i.e. a mud avalanche. AVALANCHE, EARTH-FLOW, LAHAR, MASS MOVEMENT. G.

mudir (Arabic) an administrator or a governor (according to the country) of an administrative unit varying in extent from a village to a province. G.

mudiriya (Sudan: Arabic) a province in Sudan. G.

mudiriyet (Egypt: Arabic) a province in Egypt. G.

mud-pot a bubbling, boiling pool of mud, usually sulphurous, occurring in a volcanic area, e.g. in Yellowstone National Park, Wyoming, USA.

mudstone a general term applied to an unlaminated, non-plastic, indurated SEDIMENTARY ROCK, consisting of CLAY MINERALS and other clay-grade constituents. ARGILLITE, GRADED SEDIMENTS, INDURATION. G.

mud-volcano 1. a cone of mud associated with escaping gases in the earth's surface, formed when gases, trying to escape through a stratum of wet clay, whisk the clay to a soft slurry which is ejected at the surface with hisses and bubbles, though usually quite cold, to build a cone 2. an ejection of hot mud from a volcanic vent, building a small ephemeral cone. VOLCANO. G.

muirburn (Scottish) the burning of heather or moorland vegetation as part of land management. G.

mukin the smallest administrative unit in Malaysia, corresponding roughly with a parish in the UK. G.

mulatto a person who is a member of the first generation offspring of mixed parentage, especially in South America, properly applied to one of mixed African and European parentage, but loosely applied to an offspring of any white and black parents. MESTIZO, QUADROON, ZAMBO. G.

mulatto *adj.* (American) dark-col-oured, applied particularly to dark-coloured clay, land, loam, soil, prairie (PRAIRIE with dark-coloured soil).

mulberry *Morus nigra*, the common or black mulberry, an irregularly shaped, medium-sized DECIDUOUS tree, probably native to western Asia, but cultivated widely in Asia and Europe since ancient times for the sake of its juicy, purple, multiple fruit, eaten raw when fully ripe or used in jam making. The leaves of the white mulberry, *Morus alba* (the fruit of which lacks the flavour of the black mulberry), are the food of the common SILKWORM.

mule ASS.

mulga (Australia: aboriginal term) a scrubby tree or spiny shrub, *Acacia aneura*, dominant over large areas of arid Australia, hence the term mulga scrub. MALEE, MALLEE SCRUB. G.

mull 1. (Scottish, from Gaelic) a promontory or headland in Scotland 2. (Swedish) in soil science, mild HUMUS derived mainly from leaf-mould, occurring as a surface layer in DECIDUOUS forests 3. in soil science, a loose, crumbly humus layer mixed with the underlying MINERAL SOIL, occurring where decay is rapid in the presence of a plentiful soil fauna, including especially earthworms. MODER, MOR 4. a loosely woven muslin used in book binding and map mounting. MUSLIN. G.

mulola OSHANA. G.

multiband spectral photography photography which makes use of a combination of selected narrow spectral bands in the VISIBLE LIGHT and the NEAR-INFRA-RED wavelengths to give simultaneously two or more images of the same area. ELECTROMAGNETIC SPEC-TRUM, REMOTE SENSING.

multi-cycle landscape, multi-grade valleys POLYCYCLIC. G.

multi-factor analysis FACTOR ANALY-SIS, MULTIVARIATE ANALYSIS.

multifinality the end state, one of dissimilarity, achieved by systems founded on similar initial conditions and undergoing change. EQUIFINALITY.

multi-modal *adj.* MODE-3.

multinational, multinational company, multinational enterprise (MNE) a very large business enterprise which has subsidiary companies, branches, offices, factories etc. in many countries. NEO-CAPITALISM.

multinational *adj.* of, pertaining to, or consisting of many nationalities or ETHNIC-2 groups. MULTINATIONAL.

multiple causation the effect caused by many different factors acting together. CAUSALITY.

multiple deprivation a state in which people (especially a family) who are disadvantaged in one respect are also likely to be disadvantaged in others, e.g. those disadvantaged by poverty and hence a low income, are likely to have poor housing, poor health etc. DEPRIVATION, POVERTY CYCLE.

multiple-feature region a REGION distinguished by the presence within it of several, overlapping sets of features. SINGLE-FEATURE REGION.

multiple land use the use in common of a tract of land for two or more purposes, one of which is in many cases recreational (RECREATION), e.g. as in NATIONAL PARKS in Britain. PARALLEL LAND USE.

multiple nuclei model a model of urban land use introduced by C. D. Harris and E. L. Ullman, 1945, who suggested that in many urban areas the pattern of land use is built around several discrete nuclei, not around a sole centre. These nuclei may date back to the origins of the city, or they may have developed as the growth of the city stimulated migration and specialization. In the latter case nuclei are formed because some activities need specialized facilities (e.g. maximum ACCESSIBILITY for retailing), and these generate an AGGLOMERATION-3 of EXTERNAL ECONOMIES (e.g. theatres, law districts). Other activities are incompatible (e.g.

heavy manufacturing industry and high status housing); and some activities or land uses cannot afford high land values, high rents. Residential areas grow around the nuclei, old and new, and eventually coalesce to form an urban area in which these separate centres still dominate. CONCENTRIC ZONE THEORY, SECTOR THEORY. G.

multiplier in economics, the process (or the index or coefficient measuring such a process) by which initial changes within an economic system have cumulative effects on the system, its components and its equilibrium which are, in principle, measurable. MULTIPLIER EFFECT.

multiplier effect the way in which an increase or decrease in activity acts as a stimulus to the initial effect of that activity, and thereby multiplies its effect, e.g. the opening of a new factory will give rise to employment additional to that in the factory itself by stimulating employment in firms providing goods and services to the factory, and in the local services and shops which meet the needs of those working in the factory. All this additional employment increases purchasing power, which gives rise to further employment in a wide range of other firms, and so on. Reverse effects are generated by the closure of a factory, resulting in an increase in unemployment. DE-SKILLING, MULTIPLIER.

multiracial *adj.* of, pertaining to, or consisting of, people of several ETHNIC-2 or cultural groups, coexisting amicably and cooperatively together, each group having equal rights and opportunities.

multispectral *adj.* applied in REMOTE SENSING to a device that makes use of several wavebands in recording images. ELECTROMAGNETIC SPECTRUM.

multispectral scanner in REMOTE SENSING, a scanning device that operates simultaneously in various wavebands in recording images. ELECTROMAGNETIC SPECTRUM, LANDSAT.

multispectral sensing in REMOTE SENSING, the recording of images by one or more SENSORS operating in several wavebands. ELECTROMAGNETIC SPECTRUM, MULTIBAND SPECTRAL PHOTOGRAPHY.

multi-stage cluster sampling CLUSTER SAMPLING.

multi-story, multi-storey consisting of many floors, applied to buildings, e.g. an office block, a car park.

multivariate analysis 1. a statistical analysis of data in which more than one type of measurement or observation is involved, the number of VARIABLES being greater than two **2.** in REGRESSION ANALYSIS, an explanation of VARIANCE in terms of several variables, taking into account not only the relationship of INDEPENDENT to DEPENDENT variables, but also the interrelationship of independent variables. FACTOR ANALYSIS.

mung bean GREEN GRAM.

municipal *adj.* of, relating to, or carried out by, local self-government, especially of a town or city.

municipal borough (historical) a non-county borough, a BOROUGH in England and Wales included within an administrative COUNTY but having some powers of self-government. LOCAL GOVERNMENT IN BRITAIN.

municipality 1. in Britain and USA, a town or city or other relatively densely populated unit of local government which has powers of self-government. In the USA these powers vary from one state to another. LOCAL GOVERNMENT IN USA **2.** the corporation or council governing such a unit.

municipal socialism in Britain, the control of PUBLIC UTILITIES, provision of education, public health and social welfare services and the systematic regulation of living conditions exercised by urban local authorities under statutory powers in the period 1870 to 1920.

munro (Scotland) a hill of 900 m (3000 ft) or more, separated from another by a dip of 150 m (500 ft) or more. G.

mura-yama (Japanese) waste land. G.

muri evergreen scrub on podzolic soils (PODZOL) in Guyana, similar to PADANG in Malaysia. G.

murram (east Africa, especially Uganda: original language unknown) lateritic ironstone formed in level areas as a result of the vertical movement through a succession of seasons of otherwise stagnant waters. It is often used as a road surfacing material. KABOUK, LATERITE. G.

muscatel GRAPE.

muscovite a pale-coloured MICA.

muskeg (Canada: Algonquian origin) an undrained basin in a subarctic or transition forest region in Canada, filled in with PEAT and bog moss (SPHAGNUM). There are dense stands of tamarack and black spruce around the margins, the trees declining in height towards the centre of the bog. G.

muslin a woven cotton fabric, bleached or unbleached, varying in texture and weight from fine, soft and light for clothing, to coarse and heavy, used in bookbinding. MULL.

mustard ANNUAL or BIENNIAL herbs of the Cruciferae family, grown in temperate lands, the seeds of both the white mustard, *Sinapis alba*, and of the black mustard, *Brassica nigra*, being ground to powder, strong and hot in flavour, used in making condiments, etc. Both species are also used as food in seedling form, at seed-leaf stage (mustard and cress); and both are sometimes sown to be dug in as they begin to flower as a GREEN MANURE crop. The RAPE plant, *Brassica napus*, is used in the same ways. Two other mustards are grown, but only for the sake of their edible green leaves: Chinese mustard, *Brassica juncea* and *Brassica japonica*.

mutation a changing or being changed, especially a sudden change in the CHROMOSOMES of a cell, the changes in the DNA of individual genes (gene-mutation) being the most common. Mutations occurring in the gametes (reproductive cells) or their precursors can produce an inherited change in the characteristics of the organisms that develop from them; thus mutation is a potent force in evolution.

mutualism **1.** broadly, any association of one organism with another of a different species **2.** SYMBIOSIS, pure, but also modified to the extent that neither partner is vitally important to or totally beneficial to the life of the other **3.** COMMENSALISM **4.** the concept that mutual dependence is an essential, basic factor if social well-being is to be achieved.

mutually exclusive categories in statistics, two CATEGORIES characterized by the fact that an observation may fall in either one of them, but not in both.

mycetophagous *adj.* applied to an animal which feeds on fungi (FUNGUS).

mycology **1.** the study of fungi **2.** the fungal life of an area **3.** the LIFE CYCLE of a FUNGUS.

mycorrhiza the association of a FUNGUS with the root of a higher plant. There are two main types: endotrophic, in which the vegetative part of the fungus (i.e. the mycelium) is within the cortex cells of the root; and ectotrophic, in which the mycelium is external, covering the smaller roots completely. The association is not always mutually beneficial; but in some cases it seems to be helpful, or even vital, to the host, e.g. mycorrhizal fungus is essential to the growth of seedling pine trees. MYCOTROPHIC.

mycotrophic *adj.* applied to a plant involved in MYCORRHIZA.

mylonite a hard, streaky or banded rock consisting of compacted ROCK FLOUR, sometimes produced along a THRUST PLANE. FAULT-PLANE.

myo (Burmese) a town; used as a suffix in place-names. G.

myombo (Tanzania: Swahili, pl. miombo) a tree of the open DECIDUOUS woodland in south central Africa, especially in western and southwestern Tanzania. G.

myrrh a resin with a sweet smell and bitter taste obtained mainly from the trees of *Balsamodendron* species in east Africa and Arabia, used in incense, scents and medicine. C.

myth a traditional story, a narrative, an image (of the past, present or future) which cannot be verified by EMPIRICAL data, but which expresses beliefs, perceptions and aspirations, and serves to explain phenomena and events.

myxomatosis a VIRUS disease affecting rabbits and sometimes hares, contagious and usually fatal. ENDEMIC in South America (where it is transmitted by mosquitoes), it was deliberately introduced into Australia in 1950, and reached Britain (where it is transmitted by the rabbit flea) about 1953. It almost eliminated rabbits in many areas in Britain, and this markedly affected the vegetation, especially of the downland (DOWN). But in the late 1950s the rabbit populations in Australia and Britain began slowly to recover, and a growing number of rabbits are now able to survive an outbreak of the disease.

N

nab (England, Yorkshire) **1.** a headland **2.** a spur of an ESCARPMENT. G.

nabob (Indian subcontinent: Urdu nawwāb) a Muslim official under the Mogul emperors in India, extended to a European returning home from the EAST-2 with great wealth.

nacreous cloud (nacre, mother-of-pearl) MOTHER-OF-PEARL CLOUD.

nad (Indian subcontinent: ? Hindi) swampy land, kept permanently moist by the presence of SPRINGS-2. G.

nadir **1.** the lowest point **2.** the point on the CELESTIAL SPHERE directly opposite to the ZENITH. G.

nagana (Africa: Zulu, u-nakane) a fatal disease of cattle and horses in parts of tropical Africa, carried by the TSETSE FLY.

Nagelfluh (German) the conglomerates which accompany the MOLASSE of the Alpine region. G.

nahya (Iraq: Arabic) LIWA. G.

nāi (Indian subcontinent: Sindhi) a hill torrent. G.

nailbourne a temporary stream occupying a dry valley. BOURNE. G.

nālā (Indian subcontinent: Urdu-Hindi) a dry river bed, or one with an intermittent stream; commonly anglicized as nullah. ARROYO, CHAUNG, NULLAH, WADI. G.

naled, naledee (Russian) AUFEIS.

nanism the condition of being, or tendency to become, dwarfed or stunted, e.g. the condition of plants, especially trees, at high altitudes or in high latitudes.

nanization the process of artificially dwarfing plants.

nanoplankton, nannoplankton the extremely small microscopic organisms, both PHYTOPLANKTON and ZOO-PLANKTON, occurring in bodies of fresh or salt water. Fossil nanoplankton are common in PELAGIC sediments. DIATOM, DIATOM OOZE, PLANKTON.

nano-relief very small variations in form of the surface of the ground, smaller than those of MICRORELIEF.

naphtha one of the volatile HYDRO-CARBON mixtures obtained by the distillation of coal, tar or petroleum. It is used in some paints, and as a fuel, a solvent, a thinner for varnish.

naphthalene an aromatic crystalline HYDROCARBON present in coal tar, used in making dyestuffs and as a fumigant, e.g. mothballs.

nappe (French, a tablecloth) **1.** a very large overfold in the earth's crust, an OVERTHRUST mass of rock in a near horizontal fold that has moved forward for many kilometres from its 'roots', covering the formations beneath (as a cloth over a table). It may be either the hanging wall of a low-angled THRUST FAULT or a RECUMBENT FOLD in which the reversed middle limb has been sheared out by great pressure. As a result of later DENUDATION a piece of nappe may be left isolated as a nappe outlier, or KLIPPE **2.** applied less precisely, especially in France, to any overlying sheet of rock, e.g. a LAVA flow, equivalent to the German Decke. FENETRE. G.

nappe outlier NAPPE-1.

narrow gauge of railway, a gauge less

than 143 cm (4 ft 7 in). RAILWAY GAUGE.

narrows a constricted passageway in a STRAIT, in part of a river, in a valley, or in a PASS-1.

NASA National Aeronautics and Space Research, the civilian agency concerned with the space exploration programme of the USA, established by act of Congress, 1958. LANDSAT.

nastic movement in plants, the response of plants to a stimulus, the response being unconnected with the direction of the stimulus. TROPISM. Nastic movement may be triggered by touch (seismonasty), by light (photonasty), by temperature (thermonasty), or by the combined factors of day and night, i.e. by illumination and temperature (nyctinasty).

natal *adj.* of or pertaining to birth.

natality rate birth rate. DEATH RATE.

nation the largest SOCIETY-1,2,3 of people united by common culture and consciousness, generally linked by common descent, historical, ethnic and possibly linguistic ties, having common interests of place and land, and usually recognized as a separate, political entity. It is sometimes defined as an independent political unit, but a nation does not necessarily enjoy statehood (STATE-2) or political autonomy: it may exist as an historical community or be identified by its cultural ties. NATION STATE. G.

national a member of a NATION.

national *adj.* of or pertaining to a NATION.

National Aeronautics and Space Research NASA.

national grid 1. on Ordnance Survey maps in Britain at present, the metric GRID-1 based on the transverse Mercator projection. The axes are 2°W and 49°N, and from their intersection at TRUE ORIGIN the FALSE ORIGIN is transferred 400 km W and 100 km N. Drawn on the metric system, with 500, 100 and 1 km squares, one reference

system covers the whole of Britain, the grid lines corresponding with sheet lines **2.** a network of transmission lines linking the main generating stations to distribution centres in a country or region in order to maintain a constant supply (e.g. of electricity, of water) to the consumer.

nationalism 1. the devotion to one's NATION **2.** national aspiration **3.** the advocacy of national unity or independence **4.** the process whereby a nation has been established as an independent political unit. G.

nationalist one who believes in or who supports NATIONALISM.

nationality 1. membership of a NATION **2.** the legal relationship between a state and an individual entailing reciprocal rights and dues. G.

nationalization the act of putting privately controlled or owned PROPERTY-1 (e.g. an activity, industry, land) under public control or ownership, i.e. under the control or ownership of the state, with or without compensation. The term expropriation is applied to nationalization without compensation. PRIVATIZATION. G.

national park an extensive area of countryside officially designated by government in order to protect and conserve its special natural features (scenic beauty, native flora and fauna, display of geological phenomena, etc.) and in some cases its historical associations, for public enjoyment and for scientific purposes. The concept originated with the Yellowstone National Park, USA, in 1871; and national parks, varying in type, are now to be found in most countries. COUNTRY PARK, NATIONAL PARKS (two examples selected, Britain and the USA). G.

National Parks 1. in Britain (England and Wales). Ten National Parks (Lake District, Yorkshire Dales, North York Moors, Peak District, Snowdonia, Brecon Beacons, Exmoor, Dartmoor, Cheviots, Pembrokeshire Coast) were designated in 1965 under the National Parks and Access to the Countryside Act 1949. They are areas of natural beauty but for the most part the land is farmed or otherwise used, villages and other settlements lie within their boundaries, and normal country life and their use for recreation continues (MULTIPLE LAND USE). Building development is controlled. The same Act set up the Areas of Outstanding Natural Beauty (AONB) **2.** in USA, the home of the first National Park (Yellowstone, 1871) there are now 28 other National Parks, mainly extensive wild areas, largely uninhabited, providing areas for recreation and the conservation of wild life. Some of the most famous, apart from Yellowstone with its GEYSERS, are Crater Lake, Everglades, Glacier, Grand Canyon, Mammoth Cave, Carlsbad Caverns, Mount KcKinley, Mount Rainier, Sequoia, Yosemite, Zion. There are also National Historical Parks, National Recreational Areas, National Parkways, National Monuments, and a range of lesser areas, e.g. State Parks, controlled by the states themselves. G.

nation state a SOVEREIGN STATE most of the members of which constitute a NATION.

native 1. a person born in a given place or country **2.** one of the original inhabitants of a country **3.** a plant or animal originating in an area.

native *adj.* **1.** applied to metals occurring in nature or in a pure state **2.** inherent, belonging to a person or thing by nature **3.** belonging to someone by birth, e.g. native land **4.** of or relating to a NATIVE-1 of a particular place **5.** of or belonging to a plant or animal originating in an area.

nativism a movement which aims to reaffirm the native, tribal culture of a group as a reaction to the pressure of ACCULTURATION.

native state historical, an Indian state ruled by its own prince under British supervision during the period of the British RAJ in the Indian subcontinent.

NATO North Atlantic Treaty Organization, an alliance for mutual defence, the treaty being signed on 4 April 1949 by Belgium, Canada, Denmark, France, Iceland, Italy, Luxembourg, the Netherlands, Norway, Portugal, the UK, the USA. Greece and Turkey were effectively admitted in February 1952, the Federal Republic of Germany in May 1965. Headquarters in Brussels. WARSAW PACT.

natural *adj.* pertaining to, existing in, or formed by NATURE-1, i.e. not ARTIFICIAL or SUPERNATURAL **2.** in chemistry, found in the earth's crust. G.

natural change the net change in the total population of an area arising from the balance of births and deaths. BIRTH-RATE.

natural gas a mixture of combustible gaseous HYDROCARBONS and non-hydrocarbons (ETHANE, METHANE) occurring, frequently with PETROLEUM, in the rocks of the earth's crust, a source of ENERGY, used as a fuel or as a raw material in the PETROCHEMICAL industry. It is transported by pipeline; or in liquid form (LNG, liquefied natural gas) in carriers at a temperature of $-160C(-260F)$. GAS. C.

natural history 1. historically, the systematic study of all natural objects, animal, vegetable, mineral, especially applicable to the broad studies of the nineteenth century **2.** a study of animal, and sometimes of plant, life, with popular appeal.

natural increase of population the rate of population growth shown by subtracting the number of deaths from the number of births. REPRODUCTION RATE.

naturalization 1. the admittance of a foreigner to the citizenship of a country **2.** the introduction of an animal or plant to a habitat in which it is not native, but where it can flourish and reproduce. G.

natural landscape the LANDSCAPE as unaffected by human activities, i.e. the physical landscape (including relief and NATURAL VEGETATION) as opposed to the CULTURAL LANDSCAPE. But human activities have been so widespread that little 'natural landscape' thus defined still exists, and it can be said that nearly all landscape is now cultural; thus it is perhaps preferable to refer to the natural and cultural elements in the landscape.

natural radioactivity the process in which the NUCLEUS of an unstable ISOTOPE of an ELEMENT-6 spontaneously disintegrates to form more stable products. All isotopes of elements with atomic number above 83 display natural radioactivity. RADIOACTIVITY, RADIO-METRIC AGE.

natural region 1. a part of the earth's surface characterized by a comparatively high degree of uniformity of structure, surface form and climate within it 2. a part of the earth's surface possessing a unity based on any significant geographical characteristics, whether physical, biological or human, or any combination of these, in contrast to an area demarcated by a boundary imposed for political or administrative purposes, regardless of any geographical unity . There has been much discussion on whether the natural region is an OBJECTIVE or SUBJECTIVE concept. REGION-1. G.

natural resources the wealth supplied by nature and available for human use, including energy, mineral deposits, soil fertility, timber, water power, fish, wildlife and natural scenery etc., the list being indeterminate because the assessment of what constitutes a RESOURCE-1,2 is constantly changing. Natural resources are now commonly classified as: flow (those that are renewable, being always available but open to human modification in that they may be depleted, sustained or increased by human activity, e.g. amenity landscape,

soils, forests); stock (non-renewable, e.g. minerals); and continuous (always available and independent of human action, e.g. solar and tidal energy). ALTERNATIVE TECHNOLOGY, CONSERVATION, RESOURCE MANAGEMENT. G.

natural sciences the scientific studies concerned with the physical world, thus including biology, chemistry, geology, physics as well as some aspects of philosophy (via mathematics) and psychology. BEHAVIOURAL SCIENCES, EARTH SCIENCES, HUMANITIES, PHYSICAL SCIENCES, SCIENCE, SOCIAL SCIENCES.

natural selection the mechanism of evolutionary change suggested by Charles DARWIN in his theory of evolution (EVOLUTION-2) and also by the English naturalist, Alfred Russell Wallace, 1823-1913 (WALLACE'S LINE), i.e. that of the many variant offspring produced by a generation of organisms only those most fitted to their environmental conditions will survive and breed, transmitting their advantageous characteristics to their offspring, the weaker, less well adapted variants failing in competition with them, and thus not perpetuating their disadvantageous characteristics. MUTATION.

natural succession SUCCESSION.

natural vegetation the plant-association which is primarily due to nature rather than to human activity. But little of the world's vegetation is entirely unmodified by human activities, which include the burning of plant cover, the introducttion of alien species (e.g. rabbits), or the grazing of LIVESTOCK. Thus a large part of the so-called natural vegetation is at best only SEMI-NATURAL. The term is therefore now usually applied to all vegetation not deliberately managed or controlled in farming activities, i.e. it includes the 'natural' as well as the 'semi-natural'. SUCCESSION-2. G.

natural year SOLAR YEAR.

nature **1.** the physical universe, including the laws and forces ruling changes within it, excluding objects made by human beings **2.** the essential, fundamental character of something or someone.

nature reserve an area of land or water managed for the protection and conservation of its animal and plant life and its physical features.

nature versus nurture the controversy current in DEVELOPMENTAL PSYCHOLOGY as to which of two sets of factors (the genetic on the one hand or the environmental on the other) has the greater influence in forming the characteristics of an organism, especially of a person.

nautical mile CONVERSION TABLES, KNOT, MILE.

nautical twilight TWILIGHT.

naze, nose, ness, nore a headland or promontory. NESS.

nazzaz in soil science, a compact, IMPERMEABLE, concretionary pan (HARD PAN) occurring at a slight depth below the surface of the red sandy soils in the LEVANT. G.

Neanderthal man *Homo neanderthalensis*, a species of hominid from the PLEISTOCENE (GEOLOGICAL TIMESCALE), characterized by stocky build, heavy brow ridge, low receding forehead, and a lack of chin, particularly associated with the MOUSTERIAN culture, becoming extinct perhaps only some 50 000 years ago.

neap tide a TIDE in which the difference between HIGH WATER and LOW WATER is small, the high tide being lower and the low tide being higher than usual (TIDAL RANGE). It occurs twice a month, about the time of the first and last quarters of the moon, when the earth, sun and moon lie at right angles to each other (QUADRATURE), with the effect that the gravitational pull of the sun opposes that of the moon instead of reinforcing it. MOON, SPRING TIDE, SUN.

Nearctic Region a ZOOGEOGRAPHICAL REGION.

Near East EAST-2 of western Europe, a term formerly applied to the territory of the Ottoman Empire in the eastern Mediterranean region and to the Balkan states, or to Palestine and the adjacent lands facing the Mediterranean, a term now usually superseded by MIDDLE EAST.

nearest-neighbour analysis a statistical technique used to describe a point pattern, involving measuring the distance between each point and one or more of its neighbours. The observed point pattern is then compared with a theoretically derived random pattern, which allows the non-randomness of the observed point pattern to be judged. The average of the distances between each observed point and its nearest neighbour is divided by the expected random spacing, to give the statistic Rn, with values varying (not in linear progression) from 0, indicating maximum clustering, through 1 (random distribution) to 2.15, indicating that all points are uniformly distributed throughout the area. Problems arise over the spacing, shape and size of areas. It is always important to be consistent in defining the area to be analysed in comparative studies, and in some cases it may be necessary to measure not only the proximity of the nearest neighbour to each observation but also the distance to the second, third or nth nearest neighbour. For example, if observations occur in widely scattered pairs over the specified area, measurement of the distance to the second nearest neighbour (second order nearest-neighbour measurement) will be needed. The technique was originally used by botanists, and later applied to the study of geographical problems. G.

near-infra-red region a narrow band of wavelengths in the INFRA-RED region, adjoining the red band of VISIBLE

LIGHT (ELECTROMAGNETIC SPEC-
TRUM), used in MULTIBAND SPECTRAL
PHOTOGRAPHY.

nebula in astronomy, a cloud of gas
and/or dust, luminous in many cases,
occurring in the interstellar medium of
a GALAXY.

nebular hypothesis an hypothesis postulated by Kant, 1724-1804, developed
by Herschel and Laplace, that the
SOLAR SYSTEM was formed from a
large rotating NEBULA which cooled
and contracted, forming rings of matter
from which the bodies in the solar
system condensed. G.

neck 1. of land, isthmus or promontory, a narrow stretch of land with
water on each side 2. a narrow stretch
of woodland or of ice 3. a high level
pass, especially the narrowest part
(NEK). VOLCANIC NECK. G.

necropolis a large CEMETERY, especially one associated with an ancient
city.

nectarine *Prunus persica nectarina*, a
smooth-skinned variety of PEACH.

needle ice PIPRAKE.

neese a prominent ridge, a spur.

nefūd (Arabia: Arabic) the ERG or
sandy desert in Arabia. G.

negative anomaly a departure, a deviation from the normal predicted or
uniform state, value, etc., being smaller,
less than that value or state etc.
ANOMALY, POSITIVE ANOMALY.

negative area a term sometimes used to
suggest a range of environmental factors which render an area unfit for
human habitation, i.e. ANOECUMENE
as opposed to OECUMENE. G.

negative correlation in statistics, COR-
RELATION-3.

negative externality EXTERNALITY.

negative feedback FEEDBACK.

negative landform a landform that
sinks from general ground level, e.g. a
valley, a basin, an ocean-basin, as
distinct from a POSITIVE LANDFORM.

negative movement of sea level, the
lowering of the sea level in relation to

the land caused by 1. a global lowering
of ocean level (EUSTATISM); or 2. a
local vertical movement such as WARP-
ING, TILTING or isostatic recovery
(ISOSTASY) of the land. KNICKPOINT,
POSITIVE MOVEMENT of sea level,
REJUVENATION.

negative planning POSITIVE PLAN-
NING.

negative skewness in statistics, FRE-
QUENCY CURVE, MOMENT MEASURES.

negentropy a measure of order or
organization in a SYSTEM-1,2,3.
ENTROPY.

'negro head' (Australia, Great Barrier
Reef) a block of CORAL broken off
from the BARRIER REEF, probably by
storm waves, possibly by TSUNAMI,
darkened by a subsequent growth of
LICHEN on it, and commonly now
found lying, thrown by waves, on the
surface of the REEF-FLAT-2.

Nehrung (German) a long sandspit separating a HAFF or LAGOON from the
sea, common along the south coast of
the Baltic sea. LIDO.

neighbourhood 1. a small district inhabited by people, in which there are close,
everyday social contacts and within
which the individual feels secure, 'at
home', i.e. the home territory. The
boundaries are indeterminate and more
readily discernable by those outside
than by those inside the district 2. the
inhabitants of such a district 3. the
relations between the inhabitants of
such a district, and the fact or quality of
their nearness one to another 4. the
friendly relations among the inhabitants
5. in the neighbourhood of, nearness.

neighbourhood effect 1. the influence of
the local residential area on the decisions and behaviour of a person living
in it. CONTEXTUAL EFFECT 2. the
effect of proximity in the DIFFUSION of
INNOVATION, new adopters being
most likely to be near to existing users,
the likelihood of adoption of innovation decreasing with increased distance.
PRIVATE INFORMATION.

neighbourhood unit in town planning, a physical and social unit within a large town, the unit being self-contained in the sense of having its local shops (selling particularly CONVENIENCE GOODS), banking, postal and other service facilities and social amenities, but depending on the main centre of the town for other than daily needs. A neighbourhood unit may arise naturally from the absorption of a village within an expanding town. The term was introduced in the 1920s by the American planner, Clarence Perry, who applied it to his ideal design of a neighbourhood free of through traffic, and with a population sufficient to support an elementary school. He thought of it as a planning means for enhancing the spirit of local community. G.

nek (Afrikaans, a neck) a natural, low tract in a mountain range, a SADDLE or COL. G.

nekton, necton a collective term for animals that live at various depths in the ocean or in lakes, i.e. in the PELAGIC ZONE, and swim actively, in contrast to PLANKTON which float, and BENTHOS which live on the bottom. G.

nematoda a phylum (CLASSIFICATION OF ORGANISMS) of unsegmented worms, free-living in soil and water, many being PARASITES of animals and plants, e.g. hookworm in human beings, eelworm in root crops.

nemoriculture (Latin nemus, a glade or grove) the primitive stage of CULTURE-1 associated with the gathering of fruit, roots etc. in forest glades. G.

neocapitalism a form of CAPITALISM-1 in which large-scale state intervention, corporations, conglomerates and MULTINATIONALS are dominant.

neocatastrophism a theory which, while not totally accepting the doctrine of CATASTROPHE THEORY-2, postulates that sudden convulsive episodes (e.g. earthquakes, volcanic eruptions) in the earth's crust effect speedy change in the natural system, that periods of relative stability during which earth processes bring about slow evolution, are interrupted by these short paroxysmal periods of instability. Thus neocatastrophism also questions the doctrine of UNIFORMITARIANISM, which sees changes in and on the earth's crust as being brought about solely by continuous, slow processes.

neoclassical economic theory a body of economic theory introduced in the latter part of the nineteenth century (e.g. by W. S. Jevons) and forming the dominant economic analysis used (usually with modifications) in capitalist societies today (MARKET ECONOMY). Outlined briefly, in general it accepts CLASSICAL ECONOMIC THEORY (apart from Ricardo's LABOUR THEORY OF VALUE), but refines and extends it. For example, neoclassical economic theory, related to a free-enterprise, capitalist system (CAPITALISM-1), is founded on the idea that the maximization of PROFIT-2 and UTILITY by a large number of small producers and consumers who do not have power to influence to any great extent the operation of the market in which they act (PERFECT COMPETITION), benefits the entire community. It assumes that the whole economic system is regulated by the interaction of supply and demand in the market place (MARKET PLACE THEORY). Business decisions are made primarily on the basis of consideration of production processes to be used and the scale of output, not of the location of plant. Business buys or hires land, labour and capital (PRODUCTION FACTORS) and uses them in production processes in a way designed to maximize profits. The prices of production factors and of the finished goods sold are beyond the control of business. The public offers for sale (to business) labour, land, capital goods; and the interaction of supply and demand for these determines prices paid as wages, rent, interest (i.e. the distribution of income). The public uses income to buy goods and services chosen to maximize personal satisfaction or UTILITY; and this all-powerful consumer demand, interacting with the costs at which business can supply goods, determines the prices of those goods and services. The theory also assumes that markets are self-regulating, that they automatically adjust to changes and always tend to move towards equilibrium at a price which balances supply and demand. Thus neoclassical economics focuses on individual decisions and the aggregates of those decisions, generally ignoring social costs and benefits, and concentrating on the analysis of cost, profit, revenue and utility.

neoclassical economic land use theory a theory which assumes that the supply of and demand for land in different locations by different users is in equilibrium, that changes in supply are instantly balanced by changes in demand. DIFFERENTIAL DISEQUILIBRIUM.

neoclassical theory of regional development the theory that imbalance in growth and well-being between regions or between cities is temporary and will be resolved eventually by the effect of market forces alone. It assumes perfect price flexibility, and perfect mobility of labour and capital.

neo-colonialism, neocolonialism 1. the situation in which a foreign power intervenes in the economic, and sometimes the political, affairs of another country, in some cases to the resentment and annoyance of some nationals of that country. The intervention does not necessarily stem from a former colonial relationship **2.** the transfer of power from external colonial control to internal control accompanied by the preservation of the trade and investment (sometimes also of military, fiscal and political) relations existing before

independence was gained from the dominant, external colonial power. COLONIALISM.

Neogea ZOOGEOGRAPHICAL REGION.

Neogene *adj.* little used in the UK, applied by various authors to the epoch (time) and series (of rocks) either of the Miocene onwards or of the younger two divisions of the TERTIARY period (i.e. the MIOCENE and PLIOCENE). GEOLOGICAL TIMESCALE, PALAEOGENE.

Neoglacial *adj.* used particularly in the USA, of or pertaining to the (European) Little Ice Age (the Neoglaciation), a short period when glaciers of the Great Ice Age which had retreated in the relatively warm period that succeeded it, began to advance again, particularly in Alaska and the Sierra Nevada. ICE AGE.

neolith (Greek, new stone) a polished stone tool of the last period of the STONE AGE.

Neolithic *adj.* of or pertaining to the last period of the STONE AGE (succeeding the PALAEOLITHIC and the MESOLITHIC) from about 6000 to 3000 BC in Europe and western Asia, when NEOLITHS came into use as well as implements of polished bone and horn, animals came to be domesticated, crops cultivated, weaving undertaken, and the wheel used. The long BARROWS and MEGALITHS of Britain are associated with this cultural period. BRONZE AGE. G.

Neo-Malthusianism the body of doctrines derived from the writings of Thomas Robert Malthus (MALTHUSIANISM) brought up-to-date and applicable to present day conditions, when the rapidly increasing pressure of the human population on land and resources leads to a low standard of living, misery and even starvation, conditions which can be partly relieved by the limitation of population by means of artifical birth control. G.

neotechnic *adj.* denoting or belonging

to the third phase of urban and industrial development, when technical innovations in industry, transport, agriculture and power supplies etc. are the NORM-1. EOTECHNIC.

neotype a plant or animal specimen selected to replace a TYPE SPECIMEN when all the original material has been destroyed or lost. LECTOTYPE, PARATYPE.

nephanalysis the analysis of cloud patterns, especially of the cloud patterns revealed by the wide coverage of REMOTE SENSING, which provide information for weather forecasting. NEPHOSCOPE.

nepheline a FELDSPATHOID mineral which cannot co-exist with QUARTZ and is characteristic of alkali BASALT and nepheline SYENITES. PHONOLITE.

nephelinite a BASIC to ULTRABASIC volcanic rock (BASIC ROCK) consisting of NEPHELINE and a variety of FERRO-MAGNESIAN MINERALS.

nephoscope an optical instrument used in determining the direction of passage of moving clouds, and in measuring their speed of travel.

nephrite (Greek nephros, kidney) a silicate of calcium and magnesium, a low quality JADE, once worn as a remedy for kidney disease.

Neptunism the theory, current some 200 years ago, that water was reponsible for the origin of certain geological formations, in contrast to the view expressed in PLUTONIC THEORY. CATASTROPHISM, NEOCATASTROPHISM, UNIFORMITARIANISM. G.

neritic *adj.* associated with shallow water, especially with shallow coastal water. OCEANIC. G.

neritic province, neritic system, neritic zone one of the zones of the aquatic environment based on depth of water, variously defined, but commonly applied to the LITTORAL and SUB-LITTORAL marine zones between LOW WATER mark and depths of 180 to 365 m (100 to 200 fathoms: 590 to 1200 ft),

or the edge of the CONTINENTAL SHELF. PELAGIC. G.

neroli an ESENTIAL OIL yielded by flowers of the Seville ORANGE, used in flavouring (especially liqueurs) and perfumery (e.g. Eau de Cologne).

ness (Scotland and eastern England, also naze, nore, nose) a headland or cape, a SPUR of a mountain ridge; used especially in place-names, apparently where Scandinavian influence was strong. G.

nested sampling in statistics 1. synonymous with multi-stage CLUSTER SAMPLING 2. SAMPLING in which certain units are embedded in larger units which form part of the whole SAMPLE-3.

nesting the enclosing of objects by a succession of similar, ever-larger ones, e.g. in CENTRAL PLACE THEORY, the pattern displayed when low order trade areas lie within the boundaries of high order trade areas.

net 1. material made of wire, string, thread or other yarns, twisted, tied or woven together with pronounced open spaces between them 2. a NETWORK.

net *adj.* 1. of an amount, clear of all charges and deductions 2. final, when everything has been considered 3. of a price, not allowed to be made lower, e.g. the price of some books.

net aerial production in ecology, the BIOMASS or BIOCONTENT incorporated in the aerial parts (i.e. the stem, leaf, seed and associated organs) of a plant commuity. FOOD CHAIN, PRIMARY PRODUCTION, PRODUCTION, PRODUCTION RATE.

net migration a measure of the difference between inward and outward MIGRATION-1,2 in an area.

net primary production in ecology, PRIMARY PRODUCTION.

net production rate in ecology, PRODUCTION RATE.

net radiation a measure of the difference between incoming and outgoing RADIATION.

net radiometer an instrument used in measuring NET RADIATION.

nets in PERIGLACIAL areas, PATTERNED GROUND.

net weight in transport, weight clear of tare, i.e. the gross (the total) weight less the weight of the container or vehicle in which goods are packed and weighed.

network 1. the actual structure that forms a NET, i.e. the knotted yarn of a fishing net, or the veins of a leaf etc. 2. any set of interlinking lines (links or routes) that cross or meet one another (at NODES, junctions, terminals) in the manner of those in a net, e.g. a railway network, a GRID-3 3. a system with its unit members interlinked in some way, e.g. a SOCIAL NETWORK, an information network 4. a chain of radio or television stations etc. CAPACITATED NETWORK, GRAPH-2, GRAPH THEORY, NETWORK CONNECTIVITY, NETWORK DENSITY, NON-PLANAR NETWORK, PLANAR NETWORK, TOPOLOGY.

network connectivity the extent to which movement is possible between different parts of a NETWORK-2,3 and, if movement occurs, the extent to which it is direct. Directness is expressed as the ratio between route distance and GEODETIC DISTANCE (a ratio termed the route factor, or index of circuity). INCIDENCE MATRIX.

network density the length of links, routes, edges (GRAPH-2) of a NETWORK-2,3,4 per unit area.

network-specific flow data DATA of transport flow along a NETWORK-2 (e.g. as collected by a traffic count) used to indicate variations in the level of traffic flow along the network, but not the origin and destination of the flow. NODE-SPECIFIC FLOW DATA.

neuston a collective term for the microscopic organisms associated with the surface film of a body of fresh, quiet water, their position stabilized by SURFACE TENSION. G.

neutral coast a stable coast, where the level of land in relation to sea level does not alter.

neutrality 1. the quality or state of not taking the side of or assisting either of two opposing sides in a dispute, controversy, war etc. 2. the quality or state of having no distinctive colour or other quality 3. in chemistry, the state of being neither ALKALINE nor ACID (pH) 4. in international law, the status of a state (or a NATION) which has a declared policy of nonparticipation. The state may have adopted this policy of neutrality from choice (e.g. Sweden, Switzerland in the Second World War); or it may be imposed on the state by others (e.g. Austria, following the Second World War).

neutral occlusion OCCLUSION.

neutral stability, indifferent equilibrium the state of a parcel of air in balance with its surroundings, i.e. when a saturated parcel has an ENVIRONMENTAL LAPSE RATE equal to the SATURATED ADIABATIC LAPSE RATE or when an unsaturated parcel has an environmental lapse rate equal to the DRY ADIABATIC LAPSE RATE.

neutron a fundamental particle having no charge, present in all atomic nuclei (ATOM, NUCLEUS-2), apart from that of HYDROGEN. Outside the nucleus it is unstable, and decays. ELECTRON, PROTON, POSITRON.

nevados (Ecuador: Spanish) a cold wind blowing regularly down the valleys of the high Andes in Ecuador, caused partly by radiation at night, partly by the cooling of air on contact with ice and snow. KATABATIC. G.

nevé, névé (French) glaciologists prefer the term FIRN. The correct spelling is nevé. G.

new ice a general, indefinite term applied to newly formed ice, in the form of CRYSTALS, SLUDGE, ICE RIND or PANCAKE ICE, less than 5 cm (2 in) thick. G.

newly industrializing country NIC.

New Red Sandstone the red SAND-

STONE deposited in the Permian and Triassic periods (GEOLOGICAL TIME-SCALE) in Europe. OLD RED SANDSTONE.

newton n, the unit of force in the SI system, defined as the force needed to accelerate a mass of 1 kg by 1 m per second per second. 1 newton = 10^5 dynes, 10 n = 1 bar; 1 n per sq m = 1.4504 × 10^4 lb per sq in; 1 lb per sq in = 6894.8 n per sq m. BAR-3, DYNE, JOULE, PASCAL.

Newton's law of cooling the law stating that a body loses heat at a rate proportional to the difference in temperature between the body and its surroundings if the temperature of the body is higher than that of its surroundings.

Newton's law of universal gravitation the law stated by Sir Isaac Newton, 1642-1727, English mathematician, physicist, astronomer and philosopher, that any two pieces of matter attract one another with a force which is proportional to the product of their masses and inversely proportional to the square of the distance between them. Thus the force of attraction, F, between two masses m_i and m_j respectively, when placed at a distance apart, d, is:

$$F = \frac{Gm_im_j}{d^2}$$

where G is the gravitational constant. The earth attracts the moon strongly enough to hold it in orbit, and the moon is large enough and close enough to tug insistently at the earth's mantle. GRAVITY MODEL.

New Town, new town a town designated in Britain under the New Towns Act 1946, planned as a well-balanced, self-contained unit to include housing, employment, educational facilities, social amenities etc. primarily to relieve population pressure in OVERCROWDED cities and CONURBATIONS. In practice these New Towns have tended to become major urban centres, GROWTH CENTRES, providing employment, shopping facilities and other services for people living outside the designated area of the town. The term new town (without initial capital letters) is sometimes more generally applied to planned, self-contained towns in Britain and elsewhere, which (like the New Towns) have not evolved slowly but start life in a mature state, being built specifically for a particular purpose, e.g. to bring to life a hitherto sparsely-settled, stagnant region by acting as a development centre; for political considerations, to form a new administrative centre (e.g. Brasilia, Canberra); for research and scientific purposes to act as a centre for those engaged in research and development in a particular field of study (the 'academic' towns of the USSR); or to cater for tourists.

New Zealand flax *Phormium tenax*, a marsh plant native to New Zealand and Norfolk Island, with sword-like leaves up to 3 m (10 ft) in length, yielding strong fibres, historically important in the making of binder twine and rope, used by Maoris for cloaks, skirts, mats etc. HEMP. C.

New World a term commonly applied to the continents of North and South America, the western HEMISPHERE. G.

NGO non-governmental organization, a voluntary body, usually with international membership, recognized officially by United Nations, so that it may give evidence, act as consultant, and attend meetings of UN committees.

niai (Indian subcontinent: Panjab) highly manured land, usually near a settlement.

niaye (west Africa: Senegal, a clump of oil palms) a marshy depression lying between DUNES, created by water from intermittent streams and dew, supporting a luxuriant vegetation around its edges. G.

NIC new industrializing country. AIC,

LDC, INDUSTRIALIZATION, UNDER-DEVELOPMENT.

niche 1. in ecology, the specific part of a HABITAT occupied by an organism, where it can exist and develop. RANGE-7 2. the role played by an organism in the ECOSYSTEM 3. in geology, a small recess or shelf in a rock face. G.

niche glacier a small CIRQUE GLA-CIER, lying in a funnel-shaped hollow high on a steep mountain slope. Alternative terms are cascade glacier, cliff glacier, wall-sided glacier. G.

nickel a hard, rust-resistant metallic ELEMENT-6, occurring with copper and iron as sulphide-ore, used as a protective and ornamental coating and, with other metals, in making steel ALLOYS which are hard, strong and non-magnetic, particularly nickel steel (which is tough and strong yet DUCTILE) and stainless steel. C.

nickel silver an ALLOY of nickel, copper, zinc, used in tableware, medical equipment etc.

nickpoint American and British translation of German Knickpunkte, but generally KNICKPOINT is preferred in Britain. G.

nicotine a narcotic, volatile ALKALOID occurring in the leaves of the TOBACCO plant and used as an insecticide.

nieve penitent snow penitent, a term formedly used by writers in English, but not by those in Spanish, in relation to a PENITENT in South America. POCKET-PENITENT. G.

nife an acronym for the mass of nickel (Ni) and iron (Fe), density about 12.0, believed to be the main constituent of the CORE of the earth.

night the time during which the SUN is below the horizon, the opposite of DAY.

night soil manure or excrement, especially human excreta, so-called because of the old custom of removing it from cesspits at night, under cover of darkness. G.

nili (Egypt) the harvest obtained from the higher fields in the Nile valley which have been irrigated by some form of water lift and fertilized to give a harvest later than the CHETOI. G.

nilometer a graduated pillar by which the height of the Nile was measured in Cairo. As more elaborate types were developed and used in the Nile, the term came to be applied to certain self-recording gauges used elsewhere. G.

nimbostratus a low, thick, dark grey mass of CLOUD from which continuous rain or snow usually falls. It is commonly associated with the WARM FRONT of a DEPRESSION-3, having thickened from ALTOSTRATUS.

nimbus CLOUD, CLOUD FORMS.

niobrium a rare metallic ELEMENT-6, usually occurring with TANTALUM and used in alloys.

nitrate 1. a SALT-1 or ESTER of nitric acid, HNO_3 2. sodium nitrate or potassium nitrate used as a FERTILIZER. NITROGEN.

nitre 1. naturally-occurring potassium nitrate 2. sodium nitrate, e.g. CALICHE.

nitric *adj.* of or pertaining to a compound derived from NITROGEN.

nitrification the process by which AEROBIC soil bacteria convert organic compounds of NITROGEN (which cannot be absorbed by green plants) into NITRATES (which can be absorbed by green plants). DENITRIFICATION, NITROGEN CYCLE.

nitrogen a colourless, odourless gaseous ELEMENT-6, the main constituent (some 78 per cent by volume) of the ATMOSPHERE-1, an essential constituent of living organisms, an essential MACRONUTRIENT for plants. Atmospheric nitrogen is the main raw material used in the manufacture of fertilizers, nitric acid and ammonia; and naturally occurring potassium nitrate (saltpetre) is used as FERTILIZER. ACID RAIN, NITROGEN CYCLE, NITROGEN FIXATION.

nitrogen cycle the circulation of NITRO-GEN atoms through ECOSYSTEMS brought about by natural processes in which living organisms play the major part. Inorganic nitrogen compounds, mainly NITRATES, are converted into organic nitrogen compounds by AUTO-TROPHIC plants which either die, decay or are eaten by animals. These organic nitrogen compounds then pass, through the excreta or by the death and decay of the animals or plants, to the soil or water, where they are converted back to inorganic nitrogen compounds by nitrifying bacteria (NITRIFICA-TION), again becoming a necessary nutrient for green plants. But some of the atmospheric nitrogen is processed by nitrogen-fixing bacteria and blue-green algae (NITROGEN FIXATION) to form organic nitrogen compounds; and some nitrogen passes to the atmosphere by means of denitrifying bacteria (DE-NITRIFACTION) which convert some of the nitrates to atmospheric nitrogen. BIOGEOCHEMICAL CYCLE, LEGUMI-NOSAE, RHIZOBIUM.

nitrogen fixation 1. the process by which atmospheric NITROGEN is converted into organic nitrogen compounds by some blue-green ALGAE and by nitrogen-fixing BACTERIA living in the soil. Some of these bacteria live independently in the soil, others symbiotically (SYMBIOSIS) with legumi-nous plants (LEGUMINOSAE) in NODULES-2 on their roots (RHIZO-BIUM), absorbing atmospheric nitrogen and from it forming organic nitroge-nous compounds which enrich the soil. NITROGEN CYCLE **2.** the industrial conversion of atmospheric nitrogen into useful compounds (e.g. ammonia and ammonium compounds) used in fertilizer, explosives.

nitrogenous *adj.* of or relating to NITROGEN, or containing nitrogen.

nitrous *adj.* of, pertaining to, or im-pregnated with NITRE.

nival *adj.* pertaining to snow, growing in snow; snowy.

nivalkarst GLACIOKARST.

nivation 1. a general term for the effects produced by snow and nevé (FIRN) in the weathering and sculpture of rocks in contrast to those produced by glacier ice **2.** the rotting and disintegration of rocks underlying and round the edges of a patch of snow lying in a hollow (nivation hollow or nivation niche) brought about by FREEZE-THAW and CHEMICAL WEATHERING (sometimes termed snow patch erosion). This en-larging of the hollow may lead to the formation of a CIRQUE (nivation cirque). G.

nivation cirque NIVATION-2.

nivation hollow NIVATION-2.

niveo-aeolian *adj.* applied to the work of snow storms or strong winds in a PERIGLACIAL climate, and to the forms made by them (e.g. mounds of snow mixed with blown sand etc.). G.

noble gas an inert gas.

noble metal a METAL resistant to OXIDATION or CORROSION, i.e. GOLD and PLATINUM.

noctiluca any member of the genus *Noctiluca*, marine flagellates with a BIOLUMINESCENCE that causes the PHOSPHORESCENCE of the ocean.

noctilucent cloud a luminous blue-silver cloud occurring mainly in midlatitudes and high latitudes in the STRATO-SPHERE, thought to consist of ice crystals or METEORIC DUST.

nocturnal an early fifteenth century star-clock used at sea to determine time at night.

nocturnal *adj.* of or relating to night; in zoology, active mainly at night.

nodal *adj.* of, like, or situated at a NODE or nodes.

nodality 1. the quality of being a NODE-1, of being NODAL **2.** the extent to which lines, routes or other linear things approach each other or actually converge at a point **3.** a synonym for CENTRALITY, e.g. the degree to which phenomena such as retail sales, major offices, theatres, land values, density or volume of traffic etc. show a concentration in the city centre. NODE.

nodal region originally a REGION-1 differentiated by the fact that it had a central place interconnected with the neighbouring countryside. But some authors, concerned mainly with the functional relationship between the parts, now define a nodal region as a FUNCTIONAL REGION. NYSTUEN-DACEY METHOD. G.

node 1. a central point in any system or complex. NODALITY 2. in astronomy, one of the two points in which the ORBIT of a planet intersects the ECLIPTIC-1, or where the orbit of a satellite intersects the plane of the orbit of its planet 3. in botany, a usually swollen part of a plant stem from which a leaf or branch emerges (the part of the stem between successive nodes is termed the internode) 4. in geometry, the point at which a looped curve cuts itself 5. in a NETWORK-2,3, the point at which routes (edges, links) meet 6. in physics, a point of zero displacement or zero variation in a STANDING WAVE, or the point at which the displacement, or velocity of particles, or the amplitude of pressure, has a minimum value (the term antinode is applied to the point at which particle displacement, or particle velocity, or pressure amplitude, has a maximum value).

node-specific flow data DATA of transport flow between specified points (i.e. NODES-5) on a transport NETWORK-2, sometimes termed an origination-destination flow count. NETWORK-SPECIFIC FLOW DATA, NYSTUEN-DACEY METHOD.

nodular *adj.* of, pertaining to, or having NODULES.

nodule 1. a small rounded mass, e.g. a discrete rounded concretion of minerals in a sedimentary rock 2. in botany, a root nodule, such as a mass on the root of a leguminous plant (LEGUMINOSAE).

nodum a general term for a unit of vegetation of any rank.

noise POLLUTION.

nomad a wanderer, one who practises NOMADISM.

nomadism a type of human life style, now rare, based essentially on constant movement in search of sustenance, especially for grazing animals. Most nomads wander within defined areas, often using regular routes. The term semi-nomadism is sometimes applied to the life style of nomads who use fixed quarters in the wet season but migrate in the dry season to find pasture. SEDENTARIZATION. G.

nomen nudum (Latin, naked name; pl. nomina nuda) the name given to an animal, plant, mineral etc. without a full description. Many soil series have been given such names, without detailed descriptions, so that the type can be recognized elsewhere.

nomenclature the systematic naming used in a particular branch of knowledge, especially the naming used in classification, e.g. the CLASSIFICATION OF ORGANISMS. BINOMIAL NOMENCLATURE.

nominal *adj.* applied in statistics to a classification of cases based on verbal decription (e.g. male, female) rather than on numerical values. INTERVAL-3, ORDINAL.

nominal scale in statistics, a system of CATEGORIZATION, and the simplest form of measurement. It has a low level of differentiation, the observations being put into convenient, MUTUALLY EXCLUSIVE and exhaustive classes according to a particular attribute, without any particular order or preference. BAR GRAPH, CATEGORICAL DATA ANALYSIS, MEASUREMENT IN STATISTICS.

nominal variable a VARIABLE, sometimes termed an attribute, which may be placed only on a NOMINAL SCALE.

nominum conservandum the scientific name of an organism commonly used by general agreement, even it if does not conform to the NOMENCLATURE currently accepted.

nomothetic approach a law-seeking approach, concerned with the search for general laws or theories, not with particular cases or situations (in contrast to the IDIOGRAPHIC APPROACH). The method used can be either deductive or inductive. DEDUCTION, INDUCTION. G.

nonaligned *adj.* applied to a nation state that does not support the policies of either of two opposed groups of powers (e.g. of NATO or of the WARSAW PACT countries). NEUTRALITY.

non-basic activity, non-basic function, non-basic industry in urban development, a manufacturing or service activity within a city or urban area which meets needs within that city or urban area, resulting in an internal exchange of revenue. It is sometimes termed 'secondary' basic industry in contrast with the 'primary' basic industry (BASIC ACTIVITY). ECONOMIC BASE THEORY, URBAN ECONOMIC BASE.

nonconformity a type of UNCONFORMITY where denuded igneous or metamorphic rock is overlain by material which is eventually compacted to form sedimentary rock.

non-governmental organization NGO.

non-hydraulic civilization HYDRAULIC CIVILIZATION.

non-market cost a cost which cannot easily be evaluated in terms of MARKET PRICE, e.g. the destruction of a unique habitat or of a scenic view. COST BENEFIT ANALYSIS.

non-metropolitan county LOCAL GOVERNMENT IN BRITAIN.

non-metropolitan district LOCAL GOVERNMENT IN BRITAIN.

nonnull hypothesis 1. in general, an hypothesis alternative to the one under test, i.e. to the NULL HYPOTHESIS-1 **2.** an hypothesis under test where the effect is not equal to zero (NULL HYPOTHESIS-2).

nonparametric test a statistical test which does not need an estimate of a POPULATION PARAMETER or, generally, any assumptions as to the distribution of the scores. DISTRIBUTION FREE METHOD, MEASUREMENT IN STATISTICS, PARAMETRIC TEST. G.

non-place urban realm a realm comprising the communications between and interactivities of heterogeneous groups of people independent of their spatial distribution, e.g. the web of contacts maintained by specialized, professional people. G.

non-planar network a NETWORK-2 in which the links cross each other without intersecting (without any possibility of traffic exchange) and which cannot be represented by a GRAPH-2 drawn in a single plane. PLANAR NETWORK.

non-probability sample purposive sample, any SAMPLE-3 taken from a POPULATION-4 without the certain knowledge that all the members of that population had a known probability of being selected at the outset. RANDOM SAMPLE.

non-renewable resources any of the NATURAL RESOURCES classified as stock, assessed to be finite and which, once used, could be replaced only after a considerable span of geological time, e.g. FOSSIL FUELS, MINERALS.

non-sampling error in statistics, SAMPLING ERROR.

non-sequence in geology, a small gap in the normal sequence of sedimentary rocks, detected only by a study of successive fauna, representing a short time interval. The time interval in a DISCONFORMITY is longer. G.

Nordic *adj.* of or pertaining to the Germanic, i.e. Teutonic, peoples of northern Europe, especially the SCANDINAVIAN, in FAO usually applied to the countries Denmark, Finland, Iceland, Norway, Sweden.

nore, nose NAZE, NESS.

Norfolk rotation a four-year crop rotation (ROTATION OF CROPS) originating in East Anglia in eastern England in the eighteenth century, comprising wheat (which needs much nitrogen), a root crop, barley, and a leguminous crop (LEGUME-2) which replenishes the NITROGEN in the soil.

noria an arrangement of bucket and chain, worked by an animal, and used to draw water from a well, especially in Mediterranean countries. G.

norm **1.** that which is usual, expected, average, conforming to an accepted standard, measure or pattern. The *adj.* is NORMAL **2.** a standard or ideal to which people think behaviour ought to conform or which is laid down by a legislating authority, *adj.* NORMATIVE.

normal **1.** the usual or average state, level etc. **2.** in mathematics, a line drawn at right angles to the tangent line at a point on a curve.

normal *adj.* **1.** standard, regular, conforming to a NORM-1 **2.** in mathematics, forming a right angle **3.** approximating to an average or statistical NORM-1.

normal distribution normal probability distribution, in statistics, a continuous FREQUENCY DISTRIBUTION of infinite range, represented on a graph by a symmetrical curve (FREQUENCY CURVE) shaped like a bell, and termed a normal curve, most of the values being grouped around the mean, with a generally regular fall in the numbers of values away from the mean as distance increases in both directions. The normal distribution is important in PROBABILITY theory, and is used to estimate the closeness of the results of a sample to the true figure of the POPULATION-4.

normal erosion a term originally applied to land sculpture effected by rain, rivers, weather, water (termed the normal agents) in temperate climatic conditions, i.e. not in desert or glacial conditions where special erosion would be effected by ice and wind (the special agents). The term is now obsolete, the distinction between normal and special agencies being regarded as unreal.

normal fault a FAULT caused by TENSION in which the INCLINATION of the FAULT PLANE is at an angle of between 45° and the vertical and the direction of the DOWNTHROW is the same, i.e. the beds abutting the fault on its upper face (HANGING WALL) are displaced downwards relative to those against the lower face (FOOTWALL). This is not, as the use of 'normal' (NORM-1) might imply, the most usual, the most common type of fault.

normal watershed a WATERSHED in a mountainous region which runs along the crest of the highest range of a mountain chain. ANOMALOUS WATERSHED. G.

normative *adj.* **1.** of or relating to a NORM-2, thus concerned with rules, regulations, proposals **2.** establishing a NORM-2.

normative approach an approach which concentrates on what ought to occur in certain circumstances rather than on what actually occurs.

normative explanation an explanation based on the assumption that the NORM-2 rules BEHAVIOUR, and from that deducing spatial consequences.

norte (Mexico and central America: Spanish) **1.** a cold northerly winter wind, sometimes strong, blowing in Mexico and central America especially over the coast lands (a continuation of the NORTHER), causing a sudden, often great, drop in temperature **2.** applied by some authors to a similar wind blowing in eastern Spain during winter. PAPAGAYO. G.

north **1.** one of the four CARDINAL POINTS of the COMPASS, directly opposite the south, lying on the left side of a person facing due EAST **2.** towards or facing the north, the northern part, especially of a country. BRANDT REPORT.

north *adj.* of, pertaining to, belonging to, situated towards, coming from, the NORTH, e.g. of winds blowing from the north.

North America usually applied geographically to the whole continent northwards from the isthmus of Panama, including the West Indies. In view, however, of the close affinities of the Spanish/Portuguese speaking countries, collectively named LATIN AMERICA, a grouping very commonly used is North America, Central America, South America; and in that case North America is the mainland north of the Mexican-USA border (though sometimes Mexico is included). Greenland is obviously part of North America, but is often excepted on account of its political ties with Europe; and Iceland is normally included with Europe.

North Atlantic Treaty Organization NATO.

northeast trades TRADE WIND.

norther a cold, northerly winter wind, blowing in the rear of a DEPRESSION-3 over the southern USA, especially Texas and the Gulf of Mexico, sometimes violent, in some parts dry and dusty, in others accompanied by thunderstorms and hail, bringing a sudden great fall in temperature that devastates fruit crops. NORTE, PAPAGAYO. G.

northern hemisphere the half of the earth north of the equator.

Northern Lights AURORA.

northing 1. the second part of a grid reference (GRID-1), i.e. the distance north on a map as measured from a point fixed in its southwest corner. EASTING 2. nautical, a sailing northwards, or the distance so travelled since the last reckoning point.

north magnetic pole MAGNETIC POLE.

North Pole the geographical North Pole, the northern extremity of the earth's axis. MAGNETIC NORTH, MAGNETIC POLE, POLE, TRUE NORTH.

North-South BRANDT REPORT.

northwester, nor'wester a strong wind or gale blowing from the northwest, e.g. in South Island, New Zealand

where it is a FOHN-type wind (hot and dry). G.

nose NAZE.

nosographical map, nosological map a map showing the distribution of diseases. NOSOLOGY. G.

nosography the scientific description of diseases. NOSOLOGY.

nosology (Greek nosos, disease) a scientific or systematic classification of diseases. G.

notch 1. the undercutting of sea cliffs seen near HIGH WATER mark 2. in USA, a defile or passageway between mountains, especially the narrow part of such a passageway. G.

notifiable *adj.* applied to a disease that has to be reported to the public health authorities.

Notogea a ZOOGEOGRAPHICAL REGION.

nubbins 1. the last remnants of a mountain range surviving along its crest as the range is worn down by desert erosion 2. a small rounded spur-remnant 3. a small, generally rounded clod of earth on the ground surface in a PERIGLACIAL area, probably caused by the action of ice needles (PIPRAKE) in the surface layer of the ground. G.

nuclear energy the ENERGY released by the fission or fusion of atomic nuclei, a very important POWER-3 source. ATOM, ELECTRICITY, HEAVY WATER, NUCLEAR FISSION, NUCLEAR FUSION, NUCLEUS-3.

nuclear family a FAMILY-1 unit comprising husband, wife and children. EXTENDED FAMILY.

nuclear fission the act or process of splitting an atomic NUCLEUS-3 into smaller nuclei. It is usually achieved by the bombardment of the large nucleus with high energy particles, which releases a great amount of energy, available for industrial or scientific purposes, or for a war weapon. ATOM, NUCLEAR ENERGY, NUCLEAR FUSION, NUCLEAR REACTOR.

nuclear fusion the combination of light

atomic nuclei (NUCLEUS-3) to form heavier ones under very high temperatures and pressure, the loss of mass occurring in the process releasing a great amount of energy as RADIATION, e.g. as in a hydrogen bomb. ATOM, LITHIUM, NUCLEAR ENERGY, NUCLEAR FISSION.

nuclear reactor the equipment in which NUCLEAR FISSION is initiated, carried out and controlled.

nucleated settlement a rural settlement comprising **1.** a cluster of dwellings, in contrast with a DISPERSED SETTLEMENT **2.** a cluster of dwellings with an organizational centre. G.

nucleus pl. **nuclei 1.** the central, identifiable part of a whole around which the rest of the whole accumulates and may grow (e.g. in a CLOUD) **2.** of CELLS-1, a variously-shaped body, bounded by membrane, and containing the CHROMOSOMES present in most cells of plants and animals and probably in bacteria (BACTERIUM), but not in VIRUSES. CYTOPLASM, DNA, PROTOPLASM **3.** of an ATOM, the small positively charged central core containing most of the mass of the atom. NUCLEAR ENERGY, NUCLEAR FISSION, NUCLEAR FUSION, NEUTRON.

nuée ardente (French) a dense mass of exceedingly hot, gas-charged and gas-emitting fragmental lava, with particles separated by compressed gas, usually incandescent but sometimes dark, emitted from some volcanic eruptions, and rolling downwards at high speed in explosive, devastating blasts, e.g. the town of St Pierre, Martinique, was destroyed by such phenomena in 1902 when Mount Pelée erupted (PELEAN). IGNIMBRITE is deposited by nuée ardente. G.

nugget a small lump of a native precious metal, especially of GOLD.

nullah (Anglo-Indian corruption of NALA) **1.** the dry bed of an intermittent stream **2.** in Hong Kong, an artificial water supply channel or a drainage channel, constructed of concrete (presumably so-named because either channel may be periodically dry). G.

null hypothesis 1. in general, an HYPOTHESIS under test, as distinct from alternative hypotheses that are under consideration. NONNULL HYPOTHESIS **2.** an HYPOTHESIS about to be tested statistically which states that no difference between the groups being tested or that no relationships between the variables will be discovered. All hypotheses to be tested statistically are stated in this form. A null hypothesis rejected on the grounds that a difference or correlation does exist provides evidence of an ALTERNATIVE HYPOTHESIS. FALSIFIABILITY, STATISTICAL TEST.

numeracy the state or condition of having a basic knowledge of the principles of mathematics and some mastery of the interpretation of mathematical and statistical evidence. ARTICULACY, GRAPHICACY, LITERACY.

nunakol a rounded rock island rising from a GLACIER, contrasted with a NUNATAK which is characterized by a peak. A term rarely used. ROGNON. G.

nunatak, nunataq (Eskimo; Swedish pl. nunatakker) an island of rock or a mountain peak rising above a glacier or land ice. MARGINAL NUNATAK. G.

nunja (Indian subcontinent: Tamil; Hindi) wet land, as opposed to PUNJA, dry land. G.

nuptiality the frequency of marriage in a population.

nursery in horticulture, a place where young plants are grown and from which they will be transplanted.

nut 1. a dry, dehiscent or indehiscent (DEHISCENCE) single-seeded fruit with a firm, usually edible KERNEL, enclosed in a hard, often brittle, shell **2.** the edible kernel itself. See entries under particular names, e.g. almond, Brazil, cashew, etc.

nutation a slight change of position of the axis of the earth, occurring approximately every nineteen years, caused by the gravitational pull of the moon and the sun on the earth's equatorial region.

nutmeg *Myristica fragrans*, a large tree native to the Molucca Islands, now cultivated there and in the West Indies, especially in Grenada, for the sake of its hard aromatic seed, which is enclosed in a fleshy network (MACE). It is used as a spice in cooking.

nutrient a substance that serves as a food, a term applied especially to such a substance used by a PLANT. BASE CATION.

nutrition 1. the act of providing or the state of being provided with food 2. the branch of PHYSIOLOGY concerned with the nature of foodstuffs and their use in a living organism.

nyctinasty NASTIC MOVEMENT.

nyika (Kenya: Swahili and other Bantu languages) a wilderness, a barren desolate area, e.g. the semi-desert with ACACIA shrubs and BUNCH GRASS lying inland from the Kenya coastal strip. G.

Nystuen-Dacy method a procedure for simplifying NODE-SPECIFIC FLOW DATA to indicate the hierarchy of flow movements in an area. Transport flows between locations in the area are drawn up in matrix form, showing origins and destinations. Then the total flows to any location from all the other centres (termed the in-degree of a location) are added together, to give an indication of the level of attractiveness of each location. The centres are then ranked in order of their in-degree, the largest outflow from each location being noted with the destination of these flows being identified. The final stage is to note whether or not the destination of the largest outflow is ranked higher in terms of in-degree than that of its origin. If the destination has the higher ranking, the flow is considered to be a dominant flow. This technique not only generalizes transport data, it indicates the area over which a particular centre exerts the major influence, and thus divides an area into NODAL REGIONS with some functional independence.

O

OAS Organization of American States, a group of American nation states, the members of which on 30 April 1948 signed the Charter of the Organization of American States, aiming to achieve an order of peace and justice, promote American solidarity, strengthen collaboration among the member states and defend their sovereignty, territorial integrity and independence. The aims were extended on 14 April 1967: to promote Latin American economic integration and foreign trade, to raise agricultural productivity and living standards, and to expand other social development programmes. The member countries are Argentina, Barbados, Bolivia, Brazil, Chile, Colombia, Costa Rica, Dominican Republic, Ecuador, El Salvador, Grenada, Guatemala, Haiti, Honduras, Jamaica, Mexico, Nicaragua, Panama, Paraguay, Peru, Surinam, Trinidad and Tobago, USA, Uruguay, Venezuela.

oasis 1. an area in a hot desert where the presence of water at a suitable level permits sustained plant growth. Such oases vary in extent from a very small area supporting a few palm trees (typical of the Sahara) to a tract of hundreds of square kilometres supporting a large settled agricultural population (e.g. those near the rivers Nile and Euphrates) **2.** a patch of ice-free land in an icebound landscape. G.

oast the kiln in which HOPS are dried.

oasthouse a traditional building housing an OAST and the other equipment necessary for processing HOPS, with a chimney vent that turns in response to wind direction.

oat, oats (pl. is common use) *Avena sativa*, a cultivated member of the family Gramineae, and one of the main CEREAL-1 crops of temperate regions, absent from Mediterranean and tropical lands. Oats existed in Europe in the BRONZE AGE. They grow under conditions similar to those needed by BARLEY and WHEAT, but need less heat and more moisture. Oats are fed to livestock as grain, processed pellets, hay, or in green stage; the grain is used for human food (oatmeal, rolled oats). Oat flour is an antidixant, used as a preservative in the preparation of some foods (e.g. some margarines), and the husks yield FURFURAL. C.

OAU Organization of African Unity, an organization founded by 30 African countries in May 1963 to which all independent African nation states now belong, aiming to advance the unity and solidarity of African countries, to coordinate political, economic, cultural, health, scientific and defence policies, and to eliminate colonialism.

objective *adj.* **1.** not influenced by personal feelings **2.** having a real existence external to an observer, or pertaining to an object or event independent of the feelings or imagination of an observer. HUMANISTIC GEOGRAPHY, SUBJECTIVE.

oblast (Russian) an administrative division in the USSR, surrounding a town of importance (from which it usually takes its name), below the hierarchical level of a republic. G.

oblate *adj.* of a SPHEROID, flattened at the poles. GEOID.

oblate ellipsoid, oblate spheroid GEOID.

obligate parasite a PARASITE which can survive only as a parasite.

oblique angle any angle greater or less than a right angle. Aerial photographs are often taken from an aircraft with the lens of the camera pointing down at such an angle. VERTICAL.

oblique fault a FAULT in which the STRIKE lies at an oblique angle to the strike of the bed it traverses.

obnoxious *adj.* offensive, unpleasant, e.g. obnoxious fumes produced by an industrial process. POLLUTION, TOXIC.

obruk (Turkish) a deep natural SINK-2 in limestone, usually with steep walls and sometimes occupied by a lake or pond. G.

obsequent fault-line scarp a FAULT-LINE SCARP which has 'turned around' so that it is now facing in the reverse direction to that produced by the initial earth movement. It occurs when prolonged denudation takes away the resistant strata from the higher side of the fault-line scarp to expose the underlying less resistant strata; in comparision with this the strata on the DOWN-THROW side may be more resistant, so that it persists (as the surface is generally lowered) along the line of the original fault but facing in the opposite direction to that of the original fault-line scarp. RESEQUENT FAULT-LINE SCARP.

obsequent stream, obsequent river, obsequent drainage a natural water flow, a stream, or a river that flows in the direction opposite to that of the DIP of the rock strata. An obsequent stream flows in an obsequent valley. CONSEQUENT STREAM, SUBSEQUENT STREAM. G.

observation 1. the act of scientifically watching and recording, e.g. a natural phenomena 2. a record so made 3. in statistics, an individual measurement, a

CASE or SCORE made according to certain prescribed rules.

obsidian an EXTRUSIVE IGNEOUS ROCK, rich in silica, glass-like, very dark in colour, used as a gemstone. ACID LAVA, RHYOLITE.

oca *Oxalis tuberosa*, a plant native to tropical and subtropical South America, with edible, branched, groved tubers (usually sun-dried to reduce acidity), and edible leaves, eaten raw or cooked.

Occident, the originally Europe as opposed to Asia (the Orient). Now extended to those parts of the world peopled by Europeans or by those of European descent.

occlusion, occluded front a phenomenon in an atmospheric DEPRESSION-3 when an advancing COLD FRONT overtakes a WARM FRONT, thereby raising the WARM SECTOR of the depression and cutting it off (occluding it) from the earth's surface, to form an occluded front, reduced at the earth's surface to a line, termed the line of the occlusion. The cold air continues to advance and meets the cold air ahead of the warm front. If the overtaking cold air is colder than this cold air ahead, it forms a cold occlusion; if it is not so cold, a warm occlusion; if there is very little difference, a neutral occlusion.

occupational classification a classification of population based on the type of work people do, rather than on the industry in which they work. People are thus classified as employers, managers, professionals, other non-manual, skilled manual etc. REGISTRAR GENERAL'S CLASSIFICATION OF OCCUPATION.

ocean 1. the body of salt water which covers 70.78 per cent of the earth's surface 2. one of the main areas of the whole, i.e. the Atlantic, Pacific, Indian, Arctic, to which some would add the Southern or Antarctic south of 50°S. Sometimes the Atlantic and the Pacific are divided into North and South, the

427

equator forming the conventional limit between the North and South Atlantic and the North and South Pacific. In the South Pacific the ocean extends to within 490 km (305 mi) of the South Pole at the snout of the Scott Glacier. G.

ocean current CURRENT-3.

ocean floor all the bed of the ocean below LOW WATER mark.

ocean floor spreading, sea floor spreading OCEANIC RIDGE, PLATE TECTONICS.

oceanic *adj.* **1.** of, pertaining to, or occurring in, the ocean **2.** associated with the deep sea (OCEANIC PROVINCE) as opposed to NERITIC. G.

oceanic basin the great depression occupied by an OCEAN-2 but excluding the CONTINENTAL SHELF. G.

oceanic climate a climate that is equable owing to the moderating influence of surrounding waters, i.e. with a moderate range of temperature in summer and winter. CONTINENTAL CLIMATE, INSULAR CLIMATE, MARITIME CLIMATE. G.

oceanic crust that part of the crust of the earth which lies under the floor of the OCEAN BASINS. PLATE TECTONICS.

oceanic depths the depths of water in the ocean, generally distinguished as BATHYAL and ABYSSAL (the deeper). DEEP, PELAGIC, PELAGIC DIVISION, PHOTIC ZONE.

oceanic island an island in the ocean, far away from any continent, as opposed to a CONTINENTAL ISLAND. G.

oceanicity, oceanity the trend from an OCEANIC CLIMATE to a CONTINENTAL CLIMATE. Various indices have been drawn up to measure the degree to which an oceanic climate (MARITIME CLIMATE) prevails, an important factor in the understanding of plant distribution. G.

oceanic mud the BATHYAL deposits on the CONTINENTAL SLOPE consisting of Blue, Coral, Green and Red Muds derived from clay particles eroded from the land, which are larger than those of the OOZE of the ocean floor.

oceanic province the part of the PELAGIC DIVISION, PELAGIC ZONE constituting the deep sea, the edge of the CONTINENTAL SHELF forming the boundary between it and the NERITIC PROVINCE.

oceanic ridge a mid-ocean ridge, broadly the symmetrical ridge formed on the OCEAN FLOOR where, according to the theory of PLATE TECTONICS, lithospheric plates drift apart (ocean floor spreading, sea floor spreading), the edges of the plates lift to form the ridge and hot MAGMA surges through the weakened crust, cooling quickly to make new OCEANIC CRUST, each plate being enlarged by an identical amount, the boundary between them being termed constructive. Not all oceanic ridges are now active. Studies in the PALAEOMAGNETISM of rocks on each side of the ridge show that the pattern of the direction of magnetism fossilized in the rocks is exactly the same on each side of the ridge, i.e. on the margin of each plate, the youngest rocks (the outcome of the most recent outpouring of magma) being nearest to the crest of the ridge. Reversals in the direction of the earth's magnetic field have been recorded, and by checking the magnetic pattern of the rocks against this timetable the spreading of the ocean floor, and thus of plate movements, can be calculated. The evidence provided by mid-ocean ridges thus strongly supports the theory of plate tectonics. MAGNETIC ANOMALY.

oceanic trench, ocean trench in the theory of PLATE TECTONICS, a subduction zone, a long narrow depression in the ocean floor (TRENCH-1) where, as lithospheric plates converge under the ocean, one dives under the other, the edge of the descending plate going steeply and deeply into the MANTLE where it is absorbed (BENIOFF ZONE). Friction arising from the grinding together of the two plates leads to EARTHQUAKES. The SEDIMENTARY ROCKS of the diving plate are scraped off, the rocks of the OCEANIC CRUST descend into the mantle, but not so deeply as does the plate: they are less dense than the rocks of the mantle and, deformed by heat, they rise, molten, to the ocean floor, where they erupt as LAVA and combine with the scraped-off sediments to form an ISLAND ARC, an arc of volcanic islands near the weakened margin of the overriding plate, e.g. the Aleutian and the Japanese islands. ANDESITE, COLLISION ZONE.

oceanography the scientific study of the OCEAN-1, covering the distribution, physiography of the floor, phenomena of the water (chemical composition, currents and flora and fauna associated with it) as well as the allied ecological, economic and legal issues. The alternative terms, oceanology or thalassography, have not been widely accepted. G.

ocean wave WAVE-3.

ochre a yellow or red form of natural hydrated ferric oxide, Fe_2O_3, used as a pigment.

octoroon, octaroon a person with one-eighth African blood, i.e. the child of a QUADROON and a white person. MULATTO. G.

OD, Ordnance Datum the mean sea-level at Newlyn, Cornwall, calculated from hourly observation of the tide in the period 1915 to 1921, from which heights shown on British maps are measured. BENCH MARK. G.

ODECA Organización de los Estados Centroamericanos (Organization of Central American States), Central American Common Market (CACM), an organization set up in December 1960 with the aims of furthering economic, cultural and social cooperation among the members, i.e. El Salvador, Guatemala, Honduras and Nicaragua, joined by Costa Rica in 1962. Of the central American countries only Panama has not joined.

OECD Organization for Economic Cooperation and Development. In 1961 the OEEC (Organization for European Economic Cooperation) which had existed since 1947 gave place to OECD, headquarters in Paris. The members, Australia, Austria, Belgium, Canada, Denmark, Federal Republic of Germany, Finland, France, Greece, Iceland, Irish Republic, Italy, Japan, Luxembourg, the Netherlands, New Zealand, Norway, Portugal, Spain, Sweden, Switzerland, Turkey, UK, USA (Yugoslavia, with special status, takes part in activities) aim to develop each other's economic growth and social welfare, coordinate policies, and work together in helping developing countries.

oecumene ECUMENE.

OEEC Organization for European Economic Cooperation, an organization established in 1947-8 by sixteen European countries with the aim of assisting in the distribution and administration of resources sent to Europe under the MARSHALL PLAN, and of coordinating their economic activities. It was superseded in 1961 by the OECD.

office park, business park a tract of land, in some cases park-like in character (PARK-2), with good transport facilities and communications, planned and in some cases officially designated for groups of modern office buildings. INDUSTRIAL PARK, PARK-6, RESEARCH AND SCIENCE PARK.

offshore *adj*. applied to **1.** movement away from the shore towards the sea **2.** located at a point or in an area relatively near to the shore, e.g. offshore island, offshore fishing zone **3.** the zone to seaward of the FORESHORE and the BACKSHORE. G.

offshore bar BAR.

ogive a pointed arch hence **1.** in glaciology, applied generally to any feature that originally stretched from one side to the other across the surface of a GLACIER, or part of a glacier between two MORAINES, and which is drawn by the flow of the glacier (faster in the centre than on the margins) into the shape of a pointed arch, the point aimed downstream. Applied especially to one of the ogival broad bands of dark and light ice (also termed DIRT BANDS or Forbes' bands) that occur on the surface or within a glacier below ice falls, and form ridges, the separation between the successive dark bands being estimated to represent one year's flow; or the wavy surface of a glacier occurring below ice falls, the wavelength representing one year's flow. By analogy with dirt bands these have been termed wave ogives. On glaciers without ice falls, features that resemble Forbes's bands (dirt bands) but lack their regularity have been termed Alaskan bands, but as their occurrence is not confined to Alaska this use of the term is not favoured. (The entry on DIRT BANDS explains the origin of the dark and light ice bands) **2.** in statistics, the curve on a diagram representing a cumulative frequency distribution or a relative cumulative frequency distribution (CUMULATIVE FREQUENCY DISTRIBUTION), the cumulative frequency (or relative cumulative frequency) being plotted on the vertical axis, the value of the variable on the horizontal. FREQUENCY DISTRIBUTION. G.

O horizons organic horizons, one of the two major classes of soil horizon used by some soil scientists in their SOIL CLASSIFICATION, the other class being the MINERAL HORIZONS. O horizons correspond, broadly, to the A_{00} and A_0 horizons. They are formed of accumulated material derived from plants and animals and lie over the predominantly mineral or inorganic horizons. The organic material in the uppermost layer, O_1, is easily recognizable to the unaided eye; but in the layer below, O_2, it has decomposed, is not so easily identifiable, and is termed HUMUS. Some soil scientists identify the O layer

as: L (the LITTER); F (partly decomposed litter); H (decomposed litter). Below the O horizons lie the A_p and A_h horizons (SOIL HORIZON), soil layers roughly corresponding to A_1, and profile grading to A_2.

oil crude oil, mineral oil, PETROLEUM.

oil cake, oil-cake, oilcake a general term for the crushed and pressed seeds remaining after the oil has been expressed from any of the OILSEEDS (VEGETABLE OILS), the 'cake' being used as cattle food and occasionally as fertilizer.

oil crops plants that yield oils used either for food products (e.g. margarine) or for industrial purposes (e.g. soap-making). DRYING OILS, ESSENTIAL OILS, OIL CAKE, VEGETABLE OILS.

oil dome an underground, roughly hemispherical structure in gently flexed sedimentary strata containing an accumulation of PETROLEUM.

oilfield an area containing one or more subterranean pools of PETROLEUM (OIL-POOL) associated with one geological phenomenon, a term applied especially when the petroleum is being exploited.

oil palm *Elaeis guineensis*, a tree native to western tropical Africa, now cultivated there and elsewhere in tropical Africa and in similar climatic conditions in Asia (e.g. Malaysia, Indonesia), the largest producer of vegetable oil measured in yield per ha (per acre), hence valuable commercially. The oil is obtained from the fibrous pulp of the fruit (the pericarp) and from the kernel (palm kernel oil, the more valuable), and is used in the making of industrial products (e.g. soap) and food products (e.g. margarine, for which palm kernel oil is preferred). The sap of the tree is fermented to make alcoholic liquor.

oil-pool a separate reservoir of PETROLEUM or NATURAL GAS occurring, with water, in the pores and fissures of SEDIMENTARY ROCKS.

oil-sand, oilsand, tar-sand, tarsand porous sandstone or sand at or near the surface of the land, impregnated with viscous HYDROCARBONS (BITUMEN-1), occurring particularly in Canada (Athabasca valley). The Athabasca tar-sands are estimated to constitute one of the largest oil reserves in the world, but they are expensive to exploit commercially.

oilseed, oil-seed any of the plant seeds that yield useful oil. DRYING OILS, OIL CROPS, VEGETABLE OILS.

oil-shale bituminous shale, i.e. black, brown or green SHALE containing HYDROCARBONS from which oil and gas can be distilled at very high temperatures.

oil well, oil-well a shaft sunk into the ground, in many cases to a great depth, for the purpose of extracting PETROLEUM.

okra, okro lady's fingers, gumbo, *Hibiscus esculentus*, family Malvaceae, an ANNUAL plant, native to tropical Africa, widely cultivated in tropical and subtropical areas. The palatable edible fruit ('pods') are eaten fresh or cooked (especially to thicken soups, stews etc.), and can be successfully canned.

okta in meteorology, a unit used in the measurement of cloud cover, one okta representing one-eighth of the sky occupied by cloud, e.g. 0 okta, clear sky; 8 oktas total cloud cover, usually shown on weather maps as proportional shading on a disc.

old age SENILITY, the final and declining stage of development in theories of growth, e.g. of a RIVER, the other stages being YOUTH, MATURITY. G.

Older Drift one of the two main groups of glacial deposits occurring in the PLEISTOCENE (GEOLOGICAL TIMESCALE), well weathered, eroded and frequently re-sorted by POSTGLACIAL streams, and partly overlain by YOUNGER DRIFT.

oldland an extensive area of ancient crystalline rocks reduced to low relief by DENUDATION. G.

Old Red Sandstone a series of DEVONIAN (GEOLOGICAL TIMESCALE) SANDSTONES, red, brown, white, with CONGLOMERATES, LIMESTONE, MARLS-1, SHALE, thought to have been deposited in large lakes. NEW RED SANDSTONE.

Old World the eastern HEMISPHERE, especially Europe.

oligarchy 1. a form of government or of administration in which a few people hold the power **2.** a state so governed. AUTOCRACY.

Oligocene *adj.* of the third epoch (time) or series (rocks) of the TERTIARY period or system in the CAINOZOIC era, following the EOCENE. GEOLOGICAL TIMESCALE.

oligoclase a PLAGIOCLASE FELDSPAR, very rich in SODIUM.

oligomict a sedimentary rock consisting of fragments of one kind of material, e.g. a CONGLOMERATE with one kind of stable, hard pebble. POLYMICT.

oligophagous *adj.* applied to an animal which feeds on just a few kinds of food. POLYPHAGOUS.

oligopoly in economics, the control of a market by a few producers, none being dominant. DUOPOLY, MONOPOLY, PERFECT COMPETITION.

oligosaprobic SAPROBIC CLASSIFICATION.

oligotrophic *adj.* applied to a freshwater body in which the plant nutrients are low and the oxygen content is high, especially at lower levels in the summer, and which can therefore support only a few organisms. Most oligotrophic lakes are deep, with steep sides, the water being clear and poor in dissolved minerals. DYSTROPHIC, EUTROPHIC, MESOTROPHIC, TROPHIC LEVEL. G.

olive *Olea europaea*, a small, slow-growing, long-lived, EVERGREEN tree with oval leaves thickly covered with silky hairs to prevent loss of moisture, giving the tree a greyish appearance. Native to the Mediterranean region, it is cultivated there and in other warm

431

temperate and subtropical areas for the sake of its edible nutritious fruit (a DRUPE), green in the first stage, almost black when ripe, which yields oil, used for culinary, medicinal and cosmetic purposes, the inferior grades being used as lubricants and in soap-making. The whole fruit may be pickled in brine, to be used in various dishes, or eaten raw. The olive is sometimes used as an INDICATOR PLANT for the true Mediterranean climate. C.

olivine a group of rock-forming silicate minerals (SILICA-2), usually dark green, consisting of silicates of MAGNESIUM and IRON, occurring in BASIC and ULTRABASIC ROCKS, and probably a major component of the SIMA. CHRYSOLITE, FERROMAGNESIAN MINERAL, PERIDOTITE, SERPENTINE.

ombrogenous, ombrophilous, ombrotrophic *adj.* applied to an organism living in a wet climate.

ombrothermic *adj.* related both to rainfall and temperature, applied to diagrams illustrating climatic types. G.

omnivore an animal, sometimes termed a DIVERSIVORE, that eats plants and animals, as opposed to a CARNIVORE or HERBIVORE.

omuramba (southwest Africa, central and northeastern parts) the clearly defined dry bed of an intermittent stream that carries water for a few days or weeks each year, sometimes only in a series of pools. G.

onion *Allium cepa*, a BIENNIAL bulbous plant, probably native to central Asia, a bulb eaten since very early times, e.g. 3000 BC in Egypt, now widely cultivated in temperate regions, widely used in cooking or eaten raw. The spring onion, a special variety, is grown from seed and harvested before maturity, to be eaten raw. Another variety, the shallot, is grown mainly in gardens, the bulbs multiplying freely to produce clusters of small bulbs, used when mature, in cooking or pickled in spiced vinegar.

onion weathering EXFOLIATION, SPHEROIDAL WEATHERING.

onomastics the scientific study of names as names. TOPONYMY. G.

onset and lee in glaciology, the result of the action of a GLACIER on a rock mass over which it has passed, the rock being smoothed by ABRASION on the part facing upstream (the onset part) and marked by the effect of PLUCKING on the downstream (the lee) side. ROCHE MOUTONNEES.

ontogeny the life history of the growth and development of an individual organism. PHYLOGENY.

ontology the branch of METAPHYSICS concerned with the essence of things, the science or study of being, the theory of existence, or, more precisely, the theory concerned with what really does exist as distinct from that which seems to (but does not) exist or from that which can be correctly said to exist but only if it can be thought of as a whole composed entirely of ELEMENTS-1 that really do exist. COSMOLOGY, IDEALISM, MATERIALISM.

onyx a SEMI-PRECIOUS STONE, QUARTZ or a variety of CHALCEDONY, consisting of layers of different colours, used in making cameos (a design or portrait carved to reveal the difference of the colours), jewellery, ornaments etc.

oolite (Greek oos, egg; lithos, stone) a sedimentary rock, usually CALCAREOUS, e.g. limestone, with very small concretions (OOLITH) resembling the roe of fish, hence the name. Oolite (with capital initial letter) is the formation name applied to the upper part of the JURASSIC (GEOLOGICAL TIMESCALE) in England, on account of the oolitic texture of its limestones.

oolith a small rounded grain of rock (smaller than a PISOLITH) ranging in diameter from some 0.25 to 2 mm, consisting of a particle of shell or mineral (e.g. quartz) enclosed by concentric layers of other material, and resembling a fish-egg. OOLITE.

oolitic limestone OOLITE.

ooze a deposit on the floor of the deep ocean far from land, classified as biogenic or non-biogenic according to its origin. The biogenic oozes consist mainly of minute organic remains (e.g. DIATOMS, GLOBIGERINA, PTERO-PODS, RADIOLARIA, that give the ooze a specific name); at an even lower level lie the non-biogenic oozes (RED CLAY), consisting of wind-blown VOLCANIC ASH that has settled on the ocean surface and sunk, METEORIC DUST, and material carried by ICEBERGS etc. G.

opaco UBAC.

opal an amorphous form of hydrous silica, iridescent, ranging widely in colour, fragile, cut and used ornamentally, especially in jewellery.

OPEC Organization of Petroleum Exporting Countries, an organization established in 1960 by what were then the chief oil-producing countries, with the aim of formulating a common policy in respect of dues paid to them by the oil companies, and of their share in the capitalization of the oil companies.

opencast mining (American open-cut mining or strip mining) the form of excavation used when extensive mineral deposits, e.g. of coal, brown coal or iron ore, lie near the earth's surface, the mineral strata being exposed by the removal of the overlying strata (the overburden) and worked mechanically on a large scale (a QUARRY being more limited in extent). AREA STRIP MINING, MINE, MINING, PIT, QUARRYING. G.

open economy the ECONOMY-1 of a society or group in which exchanges take place within the society or group and with economic systems outside it. CLOSED ECONOMY.

open field common field, a large field which, before the INCLOSURE of the village lands of England (mainly in the seventeenth and eighteenth centuries),

was worked by the villagers in common. Each village usually had two or three open fields. FIELD SYSTEM. G.

open hearth process Siemens-Martin process, a process for STEEL production, an improvement on the BESSEMER PROCESS. It incorporates a reverberatory furnace with a cup-shaped hearth for melting and refining PIG IRON, iron ore and scrap iron, the air employed to remove the carbon being played over the molten metal (instead of being blown through it as in the Bessemer process). Neither the open hearth nor the Bessemer process removes any phosphorous present in the pig iron (phosphorus makes steel brittle).

open system a SYSTEM-1,2,3 characterized by the supply and escape of energy and material across its boundaries (unlike a CLOSED SYSTEM). Such a system regulates itself by HOMEOSTASIS and may eventually reach a STEADY STATE (e.g. a CENTRAL PLACE, NODAL REGION, the concept of the URBAN HIER-ARCHY). ECOSYSTEM, GENERAL SYSTEMS THEORY, RELAXATION TIME.

ophiolite an association of PELAGIC sediments, BASIC ROCK and ULTRABASIC ROCK, probably part of the oceanic lithosphere, placed on an ISLAND ARC or a continental margin by tectonic processes. OCEANIC TRENCH, PLATE TECTONICS.

opisometer a small instrument used in measuring distances on a map. A calibrated dial is attached to and records the revolutions of a toothed wheel which is small enough to be pushed along and negotiate any bends on the specified course (of a river, road, etc.).

opium poppy *Papaver somniferum*, a flowering plant native to eastern Mediterranean lands, cultivated there and in Asia since ancient times for the sake of its content of opium, a habit-forming stimulant or narcotic drug, obtained from the seed-head. The seed-head is scratched before it is ripe, and the exuding juice (LATEX) dried. Formerly opium was widely used medicinally for the relief of pain, but in this it has generally been superseded by derivatives. The cultivation of the opium poppy is now, theoretically, controlled in most of the countries that grow it. Seeds are used in baking, and also yield oil, used in cooking and in soap-making. ALKALOID. C.

opportunity cost in economics, cost expressed in terms of the best of the alternative opportunities foregone, a measure of the loss or sacrifice involved in using a resource or a location for one particular purpose rather than for another. ECONOMIC RENT, MARGINAL COST.

opposition in astronomy, the position of a heavenly body when it is directly opposite another one, i.e. when their longitudes differ by 180°, e.g. when three celestial bodies share a common line, the two outer bodies are in opposition to each other and to the central body. Thus, when viewed from the earth, the position of a heavenly body when its direction is opposite to that of the sun. Full MOON and SPRING TIDES occur when the sun and moon are in opposition to the earth. CONJUNCTION, QUADRATURE, SYZYGY.

opstal (Afrikaans) a group of farm buildings as distinct from the farm land. G.

optimal *adj.* OPTIMUM.

optimization model any MODEL-4 concerned with making the best possible choice from a set of alternatives. OPTIMIZER CONCEPT, OPTIMIZER MODEL.

optimizer concept the theory that people arrange themselves in space in order to make the most favourable use of available resources and demands. G.

optimizer model a MODEL-4 based on the assumption that people organize

themselves, their production and their consumption in space so as to maximize utility or revenue. G.

optimum 1. the most favourable or the best number, quality, etc. 2. in biology, the most favourable conditions (of temperature, moisture, light, food supply) for the life and reproduction of an organism.

optimum, optimal *adj.* the most favourable or the best possible in the given circumstances (e.g. in constraining conditions) or for a particular purpose.

optimum city size a city of a size above which the benefits (the advantages of further growth) would be outweighed by the costs (the disadvantage of such growth). The criteria used may be economic, social, psychological or political, and may refer to the city itself or to the city in relation to a region or to the country in which it lies.

optimum location the site for a firm which is central, minimizes the costs of the necessary spatial relations and (if selling prices for the product vary) maximizes the differences between costs and revenues (the classical theory of firm location formulated by A. Weber in 1909). BID-PRICE CURVE, LOCATION THEORY, WEBERIAN ANALYSIS. G.

optimum population 1. the number of individuals that can be accommodated in an area to the maximum advantage of each individual 2. the number of human inhabitants considered to be the most favourable for the full use of all the resources available in an area, so that the standard of living achieved is adequate or as high as possible 3. the number of human inhabitants who would produce the highest total production or real income per head in an area. OVERPOPULATION, UNDER-POPULATION. G.

orange a small CITRUS tree native to China and southeast Asia, bearing the most popular citrus fruit entering world trade 1. the sweet orange, *Citrus sinensis*, many varieties, now widely grown in tropical and subtropical areas, especially in the Mediterranean region, being able to withstand some low temperatures. Its succulent sweet juicy fruit is eaten fresh or used in orange drinks; and it is produced for home consumption or export. Some of the flowers are used for decoration, their essential oil (NEROLI) being used in perfumery **2.** Seville or bitter orange, *Citrus aurantium*, cultivated mainly in Spain, the sour and bitter fruits being used in making marmalade, especially in Britain. The flowers yield essential oil used in flavouring and perfumery.

orbit (Latin orbis, wheel or circle; orbita, wheel-track) the closed course of a heavenly body, especially the closed path of a planet around the sun or of a satellite (natural or artificial) around another heavenly body, or of one star around the other in a BINARY system.

orbit of the earth the counterclockwise closed elliptical path of the EARTH-1 around the sun, taking 365.26 days to complete. The velocity of the earth is not constant, it varies in some sections of the orbit, the average speed being 106 000 km (66 000 mi) per hour. The APHELION occurs about 4 July when the earth is some 152 mn km (94.5 mn mi) distant from the sun, the PERIHELION about 3 January, when the distance is 147 mn km (9l.5 mn mi). EQUATION OF TIME.

order 1. sequence, systematic ranking, arrangement, the way in which one thing follows another **2.** in biology, one of the groups in the CLASSIFICATION OF ORGANISMS, lying between FAMILY-6 and CLASS-2, and consisting of a number of similar families, but sometimes of only one family. The 'natural orders' of flowering plants are equivalent to families **3.** in the hierarchy of streams, STREAM ORDER.

ordered attribute synonym for ORDINAL VARIABLE. ATTRIBUTE.

order of goods in CENTRAL PLACE THEORY, Christaller's ranking (accord-

ing to their scarcity in a CENTRAL PLACE SYSTEM) of the retail and service facilities (to which he applied the term goods) carried by a CENTRAL PLACE. High order goods are specialized and scarce, middle order goods less specialized, low order goods generally available everywhere. Thus the order of goods is linked to CENTRAL PLACE HIERARCHY, higher order goods being available in higher order centres with larger COMPLEMENTARY REGIONS, lower order goods (e.g. a CONVENIENCE GOOD) being available in lower order centres with a smaller complementary region. INDIVISIBLE SERVICES, RANGE-10, TIME-DIVISIBLE SERVICES.

order of magnitude ORDER-1 of size.

ordinal *adj.* of a number, showing position or order in a series, arrangement according to order or rank. INTERVAL-3, NOMINAL.

ordinal scale in statistics, a MEASUREMENT scale which allows individuals in a data set to be placed in correct relative order even if absolute values are unknown; thus the distance between categories or ranks is unspecified. A VARIABLE measured on this scale is termed an ordinal variable or ordered attribute. BAR GRAPH.

ordinal variable in statistics, an ordered attribute, a VARIABLE measured on the ORDINAL SCALE. ATTRIBUTE.

ordinate in mathematics, the vertical or *x*-coordinate in a plane coordinate system. ABSCISSA.

Ordnance Datum OD.

Ordnance Survey OS.

Ordovician *adj.* of the second oldest period of the PALAEOZOIC era, and of the system of rocks formed at that time. GEOLOGICAL TIMESCALE.

ore an indefinite term applied to any solid naturally-occurring mineral aggregate of economic interest from which one or more valuable constituents (especially metallic minerals) may be obtained by treatment. G.

ore-body a mass of ore capable of being exploited.

oregano ORIGANUM.

organ in biology, a unit which is both structural and functional, is adapted for some specific and essential function, and forms part of an animal or plant, e.g. a heart, a leaf.

organic *adj.* **1.** of or pertaining to an ORGAN of an animal or plant **2.** having the characteristics of, being related to, belonging to, or derived from living animals and plants and their remains **3.** in chemistry, applied to compounds that contain, or are derived from, CARBON (excluding the simpler compounds such as carbon dioxide), the naturally occurring constituents of animals and plants; hence organic chemistry, the branch of chemistry concerned with the scientific study of organic substances, i.e. with the compounds of carbon (excluding the oxides of carbon, carbonic acid and the carbonates) **4.** arising from the natural processes of or the substances produced by plants and animals. INORGANIC.

organic acids acids produced during the first stage of the decomposition of organisms, before HUMIFICATION. They include acetic, lactic and OXALIC acids.

organic chemistry ORGANIC-3.

organic deposit a sediment formed by and consisting of the remains of organisms.

organic farming agriculture practised without the use of 'artificial' (i.e. chemical) fertilizers or pesticides, only ORGANIC-4 manure being added to be soil (and, in some cases, only pesticides such as PYRETHRUM being used).

organic horizons O HORIZONS.

organicism the philosophy of HOLISM applied to complex wholes which have the kind of ordered grouping of unit structures characteristic of living ORGANISMS. Such wholes are distinguished from complex machines or collections of parts on account of the fact that the properties and existence of their parts depends on the position of the parts in the whole. Thus the parts of the whole are unified by their relationship, and the whole has a LIFE CYCLE typical of an organism.

organic soil a soil consisting mainly of ORGANIC-4 material, as distinct from a MINERAL SOIL. O HORIZONS, MINERAL HORIZONS.

organic weathering the disintegration of rocks brought about mechanically by the penetration of plant roots, or chemically by the ORGANIC ACIDS and HUMIC ACID created by plants such as MOSSES which, growing in crevices and hollows, steadily enlarge their NICHES-1 as the acids they produce rot the rock. Burrowing and grazing animals also contribute to these processes. CHEMICAL WEATHERING, MECHANICAL WEATHERING, WEATHERING.

Organization (Organisation) for Economic Cooperation and Development OECD.

organism **1.** a living being or entity with ORGANS which, while functioning independently, are mutually dependent **2.** a living being or its material structure **3.** any complex whole which is comparable to a living organism in that its components interact, are integrated and mutually dependent.

Organization for European Economic Cooperation OEEC, OECD.

Organization of African Unity OAU.

Organization of American States OAS.

Organization of Central American States (Spanish) Organización de los Estados Centroamericanos. ODECA.

Organization of Petroleum Exporting Countries OPEC.

Orient, the (to orient, to cause to face eastwards, especially to build a church with its altar at the eastern end) the countries lying to the east of Europe, especially the FAR EAST, the opposite of OCCIDENT.

Oriental Region a ZOOGEOGRAPHICAL REGION.

orientation (French orienter, to place facing east; Medieval maps have east at the top) **1.** the positioning of, or the position of, someone or something in relation to the points of the compass, especially the positioning of a map or of a surveying instrument in the field so that a north-south line on the map lies parallel to the north-south line on the ground **2.** in chemistry, the relative position of ATOMS or groups in relation to the NUCLEUS or an existing configuration **3.** in geology, the arrangement of particles in a rock or deposit **4.** the adjustment of someone or something to the surroundings or to a situation.

origanum, oregano *Origanum vulgare*, wild marjoram and any of several aromatic fragrant, bushy plants of the Labiateae family, native to the Mediterranean region, the leaves (and in some varieties the flowers) being used as flavouring in cooking.

origin 1. the point in time or in space at which the existence of something begins. CRADLE, FALSE ORIGIN, SEISMIC FOCUS, TRUE ORIGIN **2.** the source, the cause, or the beginning of something.

origination-destination flow count NODE-SPECIFIC FLOW DATA.

origin of grid on a map FALSE ORIGIN, NATIONAL GRID, ORIGIN, TRUE ORIGIN.

origin of species EVOLUTION-2.

ornithology the branch of ZOOLOGY concerned with BIRDS.

orocratic period a period of earth movements and vulcanicity in which earth-building takes place on a vast scale. PEDIOCRATIC PERIOD. G.

orogeny, orogenesis tectonic activity and mountain building. Orogenesis is usually applied to the process; orogeny to the great periods of mountain building, e.g. the ALPINE, ALTAIDES, ARMORICAN, CALEDONIAN, CHARNIAN, HERCYNIAN, HURONIAN, LARAMIDE, MALVERNIAN, VARISCAN periods. EPEIROGENETIC. G.

orogenic, orogenetic *adj.* applied to the forces that cause OROGENESIS (the process). G.

orographic, orographical *adj.* of or pertaining to mountains.

orographical map or model a map or MODEL-3 representing relief.

orographic precipitation, orographic rain relief rainfall, precipitation caused by the cooling of moisture-laden air as it rises over a high relief barrier, occurring particularly on the high ground and windward slopes that face a wind blowing steadily from a warm ocean. It leads to RAINSHADOW on the lee side of the barrier; but the orographic factor may be just additional to other rain-inducing factors, e.g. a high relief barrier may slow down the progress of a DEPRESSION-3, lengthening the duration of the CYCLONIC (FRONTAL) RAIN. THUNDERSTORM.

orography 1. a branch of physical geography concerned with the knowledge and description of the surface relief of the earth **2.** more specifically, the scientific study of the relief of mountains and mountain systems. A term now rarely used. G.

oroide a gold-coloured ALLOY mainly of copper, zinc, tin, used in the manufacture of cheap jewellery.

orrery a working MODEL-3, usually clockwork, that shows the relative position and movement of the planets and sun in the solar system, invented c.1700 by George Graham, 1675-1751.

orris root the fragrant root of *Iris florentina* and other European irises, used in cosmetics and medicine.

ortanique a CITRUS tree, hybrid of the ORANGE, grown particularly in the West Indies, valued for the high quality of its succulent, sweet fruit.

Orterde (German) in soil science, the compacted B HORIZON of a PODZOL. Some authors equate it with clay pan which they distinguish from HARD PAN. ORTSTEIN. G.

orthoclase POTASSIUM FELDSPAR, a

MONOCLINIC mineral occurring in ACID IGNEOUS ROCK and in many METAMORPHIC ROCKS, e.g. in GRANITE, characterized by two CLEAVAGES at right angles to each other. ACID ROCK, FELDSPAR.

orthoclastic *adj.* **1.** of or pertaining to ORTHOCLASE **2.** of crystals having CLEAVAGES at right angles to each other.

orthognathous *adj.* having jaws that are straight, i.e. not projecting. PROGNATHOUS.

orthogneiss a true GNEISS that results from the METAMORPHISM, of an IGNEOUS ROCK (e.g. GRANITE) in contrast to PARAGNEISS. G.

orthogonal *adj.* **1.** right angled, rectangular **2.** applied in statistics to two FACTORS-4 which are not correlated with each other (thus having a geometrical representation as two straight lines at right angles). FACTOR ANALYSIS, PRINCIPAL COMPONENTS ANALYSIS.

orthograde *adj.* applied to an animal that walks erect on two legs. DIGITIGRADE, PLANTIGRADE, UNGULIGRADE.

orthophotomap a PHOTOMAP on which the distortion of perspective has been removed by transforming the central projection of one or more air photographs to an ORTHOGONAL projection. MAP PROJECTION.

orthoquartzite QUARTZITE.

orthorhombic *adj.* applied to crystals having three unequal axes of symmetry at right angles to each other. ARAGONITE.

orthotropic *adj.* applied to a plant growing, or have the tendency to grow, approximately vertically.

Ortstein (German) in soil science, the hard cemented B HORIZON of a PODZOL. Some authors equate it with HARD PAN. ORTERDE. G.

OS Ordnance Survey, the official British map-making authority.

os (pl. osar; Swedish ås, pl. åsar, a ridge, ridges) an ESKER. G.

oscillatory wave theory of tides oscillation wave theory of tides, the hypothesis that the surface of the ocean can be divided into tidal units (amphidromic systems) each with a centre (a node), termed an AMPHIDROMIC POINT. Within each unit the water oscillates in response to its depth, the relative positions of the earth, moon, and sun (CONJUNCTION, OPPOSITION, SYZYGY) and the gyratory movement resulting from the earth's rotation (CORIOLIS FORCE). At the nodes the water stays almost level, rotating in an anticlockwise direction in the northern hemisphere, clockwise in the southern; but from the nodes CO-TIDAL LINES radiate outwards and the height of the tidal rise grows (STANDING WAVE). In the English channel a KELVIN WAVE results, the theoretical amphidromic point, the degenerate amphidromic point, being in Wiltshire. PROGRESSIVE WAVE THEORY OF TIDES.

oshana (Afrikaans; Angola mulola) in Ovamboland, Namibia, one of the numerous periodic river channels in which water flows only at times of flood and falls quickly to chains of standing pools before drying away completely. There is no well-defined main channel, hence the distinction from the OMURAMBA of Heroroland.

osier **1.** any of the varieties of WILLOW, especially *Salix viminalis*, with shoots pliable enough to be woven, thus suitable for basketry **2.** the willow shoot so used.

osmium a hard metallic ELEMENT-6 of the platinum group (PLATINUM METAL, PLATINUM), used in alloys with IRIDIUM and PLATINUM or as a catalyst for gas reactions.

osmosis the process whereby the solvent (e.g. water) from a weak (dilute) SOLUTION (termed the hypotonic solution) separated by a SEMI-PERMEABLE MEMBRANE from a stronger (i.e. more concentrated) solution (termed the hypertonic solution) passes through the membrane so that the two solutions reach a state of equilibrium in concentration (solutions of equal concentration being termed isotonic). Similarly a pure fluid (e.g. water) will pass through the membrane to a solution (e.g. a sugar solution). If pressure is applied to the stronger solution the process stops. This pressure is termed osmotic pressure and the more concentrated the solution the greater is the pressure needed to stop the process. The movement of water in living organisms occurs, in the main, through osmosis, most cell MEMBRANES being semipermeable to some extent.

osmotic pressure OSMOSIS.

osseous *adj.* 1. composed of or similar to bone 2. having many fossilized bones.

Ostpolitik (German, Eastern Policy) a policy inaugurated by Willy Brand, Foreign Minister, later Chancellor of the Federal Republic of Germany 1969-74, aiming to bring about the end of hostility between the Federal Republic of Germany and its eastern communist neighbours.

Ottoman Empire a Muslim state, founded in the late thirteenth century in the peninsula of western Asia, historically known as Asia Minor (now Turkey-in-Asia) that steadily spread its influence in Asia into Europe and north Africa to reach its maximum power in the mid-sixteenth century. It gradually declined after the seventeenth century and was overthrown by Atatürk in 1922. Turkey itself was declared a republic in October 1923, Atatürk becoming the first president.

ouklip (Afrikaans, old rock) a gravel hardened into a kind of CONGLOMERATE. G.

ounce a unit of weight in British measures, equivalent to one-sixteenth of a POUND AVOIRDUPOIS (28.35 grams) or one-twelfth of a pound TROY (31.1 grams). CONVERSION TABLES.

outback in Australia, back country, i.e.

the areas remote from the main settlements.

outbuilding a small building adjoining and subsidiary to a house, serving the needs of the household, e.g. for stabling, kennelling, storage, outside lavatory etc. The pl. (outbuildings) is most common in use. OUTHOUSE.

outcrop in geology, the part of a rock body, STRATUM or VEIN which reaches the earth's surface and is exposed, or covered (e.g. by superficial soil, vegetation or buildings). EXPOSURE. G.

outfall 1. the narrow outlet of a river 2. the vent of a drain.

outfield, out-field INFIELD-OUTFIELD. G.

outflow cave an effluent cave, a CAVE-1 from out of which a stream flows, or formerly flowed.

outhouse an OUTBUILDING, i.e. a small building adjoining and subsidiary to a house, serving the needs of the household in some way, e.g. as a store for tools and garden equipment, fruit and vegetables etc.

outlet 1. an opening that gives access to the outside, e.g. the outlet of a river 2. in trading, a market for goods.

outlet glacier a GLACIER issuing from the margin of ice capping a high plateau.

outlier 1. a mass of rock (not necessarily elevated above the surrounding country) normally of stratified sediments, completely surrounded by older rocks, usually as a result of circumdenudation. By analogy applied to other units separated from the main mass, e.g. to a hill or hills separated from a main highland even if the feature is not an outlier in the geological sense 2. in cultural geography, an outlying area in which a particular CULTURE is dominant but which lies as an island in an extensive area of another culture/other cultures. The outlier is linked to its distant CORE AREA by long-distance routes. DOMAIN-2, SPHERE-5. G.

outport a port situated nearer the sea

and consequently more accessible to vessels (especially those of greater size and draught) than the main port to which it is subordinate. It may eventually supplant the older main port if the latter cannot be modified to meet changing needs in size and port facilities. G.

output a PRODUCT or PRODUCTION which can be measured. INPUT-OUTPUT ANALYSIS.

outskirts parts remote from the centre, e.g. of a town.

outward-bound *adj.* applied to transport (ships, aircraft) travelling away from the home port or airport.

outward commuting COMMUTER.

outwash apron, fan, plain, frontal apron material (clay, sand, gravel) washed out by melt-water flowing from GLACIERS and ICE SHEETS and deposited, beyond the ice, sometimes over wide areas (FLUVIO-GLACIAL). Sorting and re-sorting takes place in this process, the coarser material being deposited near to the ice, the finer being carried farther before settling. VALLEY TRAIN. G.

outwork work subcontracted by a commercial firm or enterprise and carried on away from the business premises, e.g. sewing done at home by machinists for a garment manufacturer.

ouvala UVALA.

ova (Turkey) a sunken basin, occupied by marshes and mud flats. G.

overbank deposit material deposited on the FLOODPLAIN by a river at OVER-BANK STAGE.

overbank stage in river flow, the stage when a river overflows the banks of its normal channel and spreads in flood on to the FLOODPLAIN, usually carrying material which is deposited there and termed OVERBANK DEPOSIT. BANK-FULL STAGE, FLOOD STAGE, STREAM STAGE.

overburden the overlying soil and rock which has to be removed in OPENCAST MINING or STRIP MINING before the seam of coal or bed of ore is exposed. G.

overcast *adj.* cloudy, sometimes applied as a noun to total cloud cover, e.g. in aviation.

overcropping the overplanting and taking of too many harvests from the soil without restoring its fertility, thus leading to the creation of poor soils which lack sufficient plant nutrients. OVER-GRAZING, OVERSTOCKING.

overcrowding the cramming into a given space of more people, more organisms or more things than there is room for, or than is allowed or desirable. The term is applied especially to an excessive number of people living in a specific dwelling, usually measured by the number of persons per room, a matter of judgment varying in one social context to another (e.g. a dwelling in London, a dwelling in Hong Kong). OVERPOPULATION.

overflow channel the channel carved by water escaping from an ice-dammed lake at a period when the water level of the lake was high, e.g. such channels draining lakes dammed by ice in the GREAT ICE AGE. G.

overfold an overturned asymmetrical ANTICLINE or an anticline of which one limb is inverted and lies beneath the other. FENETRE, GEOLOGICAL INVERSION, RECUMBENT FOLD. G.

overgrazing, overstocking the putting of so many animals on land that the pasture or other vegetation cover is damaged, sometimes beyond recovery. OVERCROPPING.

overland *adv.* by land rather than by sea or air.

overland flow surface runoff, the unhampered, unchannelled, downslope movement of a broad expanse of shallow water produced by sudden, heavy precipitation, often leading to SOIL EROSION. If the flow is due to the fact that the rate and amount of rainfall exceeds the rate at which the soil can absorb it (INFILTRATION CAPACITY, INFILTRATION RATE) it is termed infiltration-excess overland flow; but if it is due to the SATURATION of the soil over which it flows, it is termed saturation overland flow.

overlap in stratigraphy, the extension of one bed or stratum beyond that of the bed below, occurring when deposits were being laid down in a basin which was steadily sinking. G.

overlay a transparent sheet laid over a map, photograph, diagram etc. It may bear information to be used in conjunction with that shown on the map, photograph, or diagram, or on another overlay; or it may be used for rough work or for correcting the material beneath. SYNOPTIC MAP.

overpopulation too many people, usually applied to the population in an area where the available RESOURCES are inadequate for the support of the great number of people living there, i.e. when the number exceeds that of the OPTIMUM POPULATION, with the result that the standard of living declines and economic and social aspirations cannot be realized. Some authors relate the term to production as well as to available resources, distinguishing absolute overpopulation (where the absolute limit of production has been reached, but living standards stay low) from relative overpopulation (where present production does not support the population but greater production is feasible). UNDERPOPULATION. G.

oversaturated rock SATURATED-4.

oversea, overseas *adj.* of or pertaining to countries, people, things beyond, on the other side of, the sea.

overspill of population, people in excess of the number than can be properly housed and served in an area, and who accordingly have to be accommodated elsewhere, e.g. the overspill is often housed in NEW TOWNS or, in Britain, in EXPANDED TOWNS.

overthrust in geology, the THRUST of the rocks of the upper limb of a FOLD along a horizontal or near horizontal plane over the lower limb of the fold,

caused by folding so intense that the rocks may fracture. NAPPE.

overurbanization a term applied by some authors to the condition in some THIRD WORLD countries where the proportion of the population living in urban areas is much higher than that living in urban areas in the countries of North America and western Europe when they were at a similar level of DEVELOPMENT-1. URBANIZATION.

ox a member of the genus *Bos*, family Bovidae, and related genera (e.g. BUFFALO), but particularly the castrated male of *Bos taurus*, used as a draft animal or reared for food.

oxalic acid a poisonous acid occurring in some plants, especially in the genus *Oxalis*, and one of the ORGANIC ACIDS formed in the early stages of the decomposition of organisms. It is also produced synthetically, and used industrially, particularly in the textile industry in bleaching, and dyeing and printing.

oxbow, oxbow lake a cut-off or mortlake, a crescent-shaped lake formed when a river breaks across the neck of a well-developed MEANDER. G.

ox-gang in agricultural history, YARDLAND.

oxic *adj.* applied to a SOIL HORIZON which has been highly weathered and leached (LEACHING, WEATHERING) of the SILICA that had been combined with IRON and ALUMINA, leaving it rich in iron and alumina or SESQUIOXIDES of low CATION EXCHANGE CAPACITY.

oxidation the process of combining with, or being combined with, OXYGEN. In the chemical WEATHERING of rocks (CORROSION) oxidation occurs particularly in rocks containing IRON because iron takes up oxygen very readily, thus the FERROUS state changes to the oxidized FERRIC state with a yellowish-brown, crumbly crust, e.g. LIMONITE.

oxide a compound of OXYGEN with another element. OXIDATION.

oxisols in SOIL CLASSIFICATION, USA, an order of soils occurring in tropical and subtropical areas, well-weathered and including most bauxite and lateritic soils. LATERITE.

oxygen a colourless, odourless, invisible gaseous ELEMENT-6, the most abundant in the earth's crust, forming about eight-ninths by weight of water, nearly one-half by weight of all the rocks of the earth's crust, and about 20 per cent of the earth's ATMOSPHERE-1. It is essential to plant and animal life, combines easily with other elements, forming OXIDES. In industry it is used in cutting, welding, in producing flames of high temperature in steel making, and in the manufacture of NITRIC acid from ammonia, etc. FREE OXYGEN.

oxygenation mixture or combination with OXYGEN.

oyster an edible marine bivalve mollusc of the family Ostreidae, with a rough shell, living on the sea floor in shallow coastal waters and estuaries in tropical or temperate regions, farmed in beds on the sea or estuary floor for food and for the production of PEARLS.

ozone an allotropic form of oxygen, O_3, i.e. oxygen with MOLECULES consisting of three ATOMS instead of the usual two. It has a strong smell and is present in very small quantity in the ATMOSPHERE-1 of the earth, the main concentration occurring between heights of 30 and 80 km (20 and 50 mi) above the earth's surface. Its presence there is vital to atmospheric processes and to the existence of life on on earth, because it absorbs harmful short-wave ULTRAVIOLET RADIATION while allowing the beneficial longer ultraviolet radiation from the sun to pass to the earth's surface. Its own existence results from the absorption by OXYGEN of the short-wave ultraviolet radiation from the sun, a process leading to the rise in temperature that occurs at the STRATOPAUSE.

P

pa (New Zealand) a Maori fortified village, as distinct from kainga, an unfortified Maori settlement. G.

Pacific suite a PETROGRAPHIC PROVINCE distinguished by the PACIFIC TYPE OF COAST, intense folding, and rocks rich in calcium. ANDESITE LINE, ATLANTIC SUITE, SPILITIC SUITE. G.

Pacific type of coast CONCORDANT COAST. G.

pack in ecology, a group of mammalian PREDATORS (e.g. hyenas) which act together in hunting their prey. HERD, SOCIETY.

pack animal an animal (usually an ass or mule) that carries goods, fodder etc., usually loaded on a pack saddle (PANNIER); a procession of pack animals is termed a pack train.

packet boat originally a boat or vessel plying at regular intervals between two or more ports (packet ports, packet stations) and carrying 'the packet' of state letters and despatches. The term packet was later applied to the ordinary mail, goods and passengers, hence packet boat. The term is obsolescent, now generally replaced by ferry or ferry-boat. G.

packet port, packet station a FERRY PORT. PACKET BOAT. G.

pack ice any area of sea ice, whatever the form, however distributed, other than FAST ICE. A collection of pack ice with visible limits is termed a patch. A long strip, from a few kilometres to more than about 100km (60 mi), is termed a belt. G.

padang (Malay) treeless waste land in southeast Asia, supporting a scrubby, heath-type vegetation common on leached sandy soils. Often (wrongly, it is suggested) applied to the vegetation itself. KERANGAS, LEACHING. G.

paddock 1. a small grassy enclosure, usually close to the stables, where a few horses or ponies can graze and exercise 2. in a racecourse, the enclosure where racehorses are paraded for inspection by racegoers 3. in New Zealand, a grass, as distinct from an arable, field. G.

paddy PADI.

pa-deng (Thai) dry, deciduous, open forest in eastern Thailand, consisting mainly of *Dipterocarpus tuberculatus* and *Pentacme suavis*, in Burma termed INDAING. G.

padi (Malay; commonly anglicized as paddy) 1. RICE as a plant 2. the grain in the seed-head of the plant 3. the unhusked grain (hence paddy field, paddy cultivation, paddy harvest etc.). The term 'rough rice' was formerly used internationally for unhusked rice. FAO now applies padi to unhusked rice to distinguish it from milled rice (rice grain from which the husk has been removed). HUSK. G.

padiny (Russian) a depression in dry STEPPE lands in which snow, rain and melt-water collects to promote the formation of soil, with HUMUS, giving rise to patches of cultivation. G.

pagoda (Portuguese) a sacred building, a temple in the FAR EAST, now applied particularly to the many Buddhist shrines (Burmese dagon) which are widespread in Burma, nearly all with the characteristic form of a tapering tower with many storeys, each storey with its projecting and upward curving roof. G.

pahoehoe (Hawaiian) a newly consolidated LAVA flow with a smooth, glassy or ropy surface. Its chemical composition is in many cases identical to that of AA. PILLOW LAVA, ROPY LAVA. G.

paint-pot a coloured mud-well asociated with GEYSERS and HOT SPRINGS, as in Yellowstone National Park, USA. Rising water throws out pink, blue, yellow, red and white mud, generally when there is too little water to form a spring and the surrounding rocks have been weathered to clay with compounds of iron. G.

pairing in statistics, a method of control whereby people or things are selected from the whole population and matched in pairs, each pair being characterized by the same quality or qualities. This gives rise to paired, matched or linked data (linked data is the term applied to data from two batches of data where each value in one batch is naturally linked with a unique value in the other). MATCHED SAMPLES.

pakeha (New Zealand: Maori, a stranger) a white New Zealander as contrasted with a Maori. G.

pakihi (New Zealand: Maori) a waterlogged gravel flat. G.

pakka PUKKA.

pāko (Indian subcontinent: Sindhi) alluvial land in the lower Indus valley outside the protection BUNDS. G.

Palaearctic Region a ZOOGRAPHICAL REGION. G.

palaeo- (Greek palaios, ancient) as prefix, ancient.

palaeobotany, paleobotany a branch of PALAEONTOLOGY concerned with the scientific study of plant life of past geological periods based on the evidence of FOSSIL plants. PALAEOETHNOBOTANY.

Palaeocene, Paleocene *adj.* of the earliest of the epochs or of the series of

rocks of the TERTIARY period. GEOLOGICAL TIMESCALE.

palaeoclimatology, paleoclimatology the scientific study of the climate of a period in the geological past.

palaeoecology, paleoecology the scientific study of the environments of past geological times, their landforms, climates, flora, fauna.

palaeoethnobotany a branch of PALAEOBOTANY concerned with the evolution of cultivated plants.

Palaeogene *adj.* of the two earliest epochs or series of rocks of the TERTIARY, i.e. the Palaeocene and the Eocene together. Some authors add the Oligocene. A term used in Europe, but rarely in Britain and the USA. EOGENE, GEOLOGICAL TIMESCALE, NEOGENE.

palaeogeography, paleogeography the scientific study of land and water distribution, i.e. of the geography, of a particular former geological epoch, or of former geological time in general. A palaeogeographical map seeks to show a reconstruction of past geological times. PALEVENT, PALSTAGE. G.

palaeohydrology, paleohydrology the scientific study concerned with the spatial distribution of drainage patterns, ground water etc. at a specific period in geological time, and with the attempt to reconstruct such patterns etc.

palaeolith, paleolith an unpolished, chipped stone implement used in the second period of the STONE AGE.

Palaeolithic, Paleolithic *adj.* of or pertaining to the second period of the STONE AGE, preceded by the EOLITHIC age and followed by the MESOLITHIC age, ending some 8000 years ago, coinciding in the main with the PLEISTOCENE glacial epoch (GEOLOGICAL TIMESCALE). It is the period when, it can be confidently stated, human beings began to shape tools. It is divided into various epochs of cultural development, the most significant being ABBEVILLIAN,

ACHEULEAN, AURIGNACIAN, MAG-
DALENIAN, MOUSTERIAN. G.

palaeomagnetism, paleomagnetism sometimes termed archaeomagnetism, the
fossil MAGNETISM-1 evident today in
IGNEOUS ROCKS. When cooled, igneous rocks (containing iron oxides)
retain the magnetism present in them at
the time of their cooling. This indicates
the direction of the earth's MAGNETIC
FIELD at that time and provides an
historical record of the shifting position
of the MAGNETIC POLES. MAGNETIC
ANOMALY, MAGNETIC REVERSAL,
OCEANIC RIDGE, PLATE TECTONICS,
REMANENT MAGNETISM, TERRES-
TRIAL MAGNETISM. G.

palaeontology, paleontology the scientific study of FOSSIL remains of past
geological periods. The study of fossil
plants is now usually designated
PALAEOBOTANY, of fossil animals,
PALAEOZOOLOGY. G.

palaeopathology, paleopathology the
scientific study of disease, injury and
nutrition indicated in ancient human
remains.

palaeopedology, paleopedology the scientific study of fossil soils.

Palaeosere, Paleosere the EOSERE of
the PALAEOZOIC ERA.

palaeosol, paleosol an ancient soil horizon or fossil soil underlying younger
deposits.

Palaeozoic, Paleozoic (Greek palaios,
ancient; zōē, life) *adj.* or of pertaining
to the first geological era following the
PRECAMBRIAN (GEOLOGICAL TIME-
SCALE), when seed-bearing plants, fish,
reptiles, amphibeans appeared, and invertebrates began to develop (some
becoming extinct towards the end of
the era).

palaeozoology, paleozoology a branch
of PALAEONTOLOGY concerned with
the scientific study of animal life of past
geological periods, based on the evidence of FOSSIL animals. PALAEOBO-
TANY.

pale, The Pale 1. one of the narrow

lengths of wood, or stakes, or rods or
bars of iron, usually with a pointed top,
driven vertically into the ground, and
with others forming a fence. A fence
made of such pales is termed a paling **2.**
the barrier enclosing a medieval PARK-
1, usually consisting of a strong earth
bank topped by a wooden fence (a
paling) or by a stone wall with an inside
ditch **3.** an obsolete general term for a
district or territory subject to a particular rule, especially, as The Pale, that
part of Ireland around Dublin, varying
in extent, over which the English
established jurisdiction between the
fourteenth and sixteenth centuries; or
an area around Calais, France, under
similar jurisdiction, established by the
English, 1346-1558. G.

paleotechnic *adj.* denoting or belonging
to the second phase of urban and
industrial development, when coal is
the source of power, the transport
system relatively primitive, markets are
generally localized and urban settlements compact. EOTECHNIC, NEO-
TECHNIC.

palevent abbreviation for a palaeogeographical event. PALAEOGEOGRAPHY.
G.

pālēz (Indian subcontinent: Pashto) a
generic term for plants of the family
Cucurbitaceae but also applied to the
beds in which they are grown. G.

palichnology the study of FOSSIL tracks
of animals, e.g. burrows or the tracks
left by wriggling animals. G.

paling PALE-1,2.

palingenesis 1. in biology, the recapitu-
lation of adult forms of ancestors in the
early stage of the life of their descendants. BIOGENIC LAW **2.** in geology,
the re-constitution of granitic magma
by melting and re-crystallization.

palisade 1. a high fence **2.** in military
history, a defensive, generally vertical,
structure of wooden, pointed stakes **3.** a
high fence, as used in early European
settlements in North America for
protection against wild animals and

attackers, the term being transferred there to a picturesque, columnar, unscaleable cliff, rising precipitately from the margin of a stream or lake. G.

palladium a metallic ELEMENT-6 resembling and occurring with PLATINUM (PLATINOID), used in some alloys and as a CATALYST.

pallet a portable platform used in the mechanized transport of goods or materials. It usually consists of two decks separated by bearers, or of a single deck supported by feet. A quantity of goods can be assembled on each deck to form a unit load for the purpose of transporting it, or for handling and stacking it with the assistance of mechanical appliances such as pallet trucks, fork-lift trucks (based on the definition of the *European Convention*, 1960, Art.1).

palm kernel, palm kernel oil OIL PALM.

palm oil OIL PALM.

palmyra palm, borassus palm *Borassus flabellifer*, a tree native to the drier parts of tropical regions, the trunks being used for building, the leaves for thatching as well as for mats, hats, umbrellas. The soft kernel of the fruit is eaten fresh, the embryo in germinated fruit is used as a vegetable, and the sap is tapped to be boiled down for sugar or fermented to make an alcoholic beverage (toddy) which is distilled to make arrack, a spirit.

palsa (Swedish pals, from Finnish palsa, elliptical) a dome-like peat hillock, containing several lenses of ice, varying in form and size, up to 3 m (10 ft) or more in height, often surrounded by open water, occurring in northern Sweden and in the TUNDRA elsewhere. The term should not be equated with PINGO. G.

palsa bog a BOG-2 in which ice-heaved hillocks of peaty soil covered with LICHEN and a few metres in height, occur in the wetter areas. G.

palstage an abbreviation for palaeo-geographical stage. PALAEOGEOGRAPHY.

paludal *adj.* of or pertaining to marshes or swamps.

palynology the study and analysis of pollen grains preserved in soil or peat and of FOSSIL POLLEN and other microfossils highly resistant to acids found in sedimentary rocks of all ages, the findings providing valuable environmental indicators and aids to stratigraphic correlation. STRATIGRAPHY.

pamir, Pamirs 1. poor grassland on the high plateaus of central Asia 2. the Pamirs, or the Pamir Knot, the knot of high mountains and lofty plateaus where the Hindu Kush, the Sulaiman, the Himalaya, the Karakoram, the Kunlun-Atyn Tagh and Tien Shan mountains converge.

pampa, pampas (South America: Spanish, adaptation of Quechua bamba) the midlatitude or temperate grassy plains stretching from southern Brazil through Uruguay into the heart of Argentina, the western part (dry pampa) being largely arid and covered with semi-desert grassy vegetation, the eastern part (humid pampa) having a higher rainfall and covered with tall, coarse grass (pampa). Some of the dry pampa and much of the humid pampa has now been ploughed and planted with European grasses or wheat and other crops, and extensive areas are used for cattle rearing. GRASSLAND-1. G.

pampero (South America: Spanish) a strong southwest wind associated with the cold front in the rear of a LOW PRESSURE SYSTEM moving eastwards, blowing in the PAMPAS area, particularly in Argentina and Uruguay, and on the adjoining coasts, sometimes accompanied by rain, thunder, lightning, taking the form of a LINE SQUALL, sweeping up dust from the pampas and causing a great fall in temperature as the storm passes. It occurs most frequently in summer, and results from the northward advance of a polar air mass. G.

pan **1.** in soil science, HARD PAN **2.** a container in which gold-bearing sand or gravel is washed in order to separate the valuable GOLD from the other alluvial deposits. PANNING **3.** (Afrikaans, pl. panne) any shallow, generally rounded hollow occurring in arid and semi-arid regions (e.g. in the southern Kalahari) which may in some cases hold water only in the rainy season, but in others throughout the year **4.** more specifically, the flat central part of such a depression which may be briefly or seasonally flooded. PLAYA **5.** the dried-out deposit in a shallow depression. G.

pān (Indian subcontinent: Hindi) the leaf of the betel palm, or a preparation of BETEL NUT.

panama *adj.* applied to a hat made of finely plaited straw consisting of the dried leaves of jipijapa, *Carludovica palmata*, a tropical plant native to central and South America.

pancake ice pieces of newly-formed sea ice, generally thin and circular with raised rims, less than 2 m (6.5 ft) in diameter. G.

pandemic *adj.* applied to a disease so widespread that it affects the whole world or a whole continent or a whole country. ENDEMIC, EPIDEMIC.

pandy (Welsh) a FULLING mill. G.

panfan the end stage of geomorphological development in an arid area (a desert) in which the ridges have been reduced and eliminated, basins have been filled in, PEDIMENTS-2 are markedly extensive, effected by a cycle of DEGRADATION and AGGRADATION-1. The panfan (associated with an arid climate) can be compared with the PENEPLAIN (the end stage of the general process of degradation in a humid climate). G.

Pangaea the name given by A.Wegener in his theory of CONTINENTAL DRIFT to a great land mass, the supercontinent of PRECAMBRIAN times, probably split

in two parts, GONDWANALAND in the south being separated by a vast ocean, TETHYS, from LAURASIA in the north. It is thought that Pangaea started to break up some 190 mn years ago, perhaps first in what is now the Gulf of Mexico. PLATE TECTONICS.

panhandle a narrow protruding strip of land, especially a narrow strip of the territory of a state protruding into that of another state, or between the territory of another state and the sea, e.g. Alaska Panhandle (US territory) lying between Canada and the Pacific ocean.

pannage historical, in law **1.** the right or privilege of putting swine to pasture (on acorns, beech mast etc.) for a limited period in a forest or woodland owned by another **2.** the feeding of swine in this way **3.** the materials on which the swine feed **4.** the payment to the owner of the forest or woodland for such a right **5.** the accrued profit on such payments. In this context 'owner' is applied to the one who holds superior rights over the forest or woodland. COMMON RIGHTS. G.

panneveld (Afrikaans) a sub-humid region with many PANS-3 which collect the run-off of rainwater and from which it evaporates, thus preventing the formation of organized drainage. It has been suggested that the terms humid AREIC or small-scale ENDOREIC might appropriately be applied to such a region. G.

pannier a large basket, sometimes with a lid, especially one of a pair of baskets which are slung over the back of a PACK ANIMAL to hang down the animal's sides.

panning the act or process of separating a valuable element, especially gold, from alluvial deposits by washing the whole in a PAN-2.

panplain, panplane a plain formed by the joining together of spreading flood-plains, i.e. by long-continued lateral erosion by rivers (as distinct from a

PENEPLAIN, caused by DEGRADA-
TION). G.

panplanation the process of forming a
PANPLAIN.

pantanal (South America: Portuguese)
the vegetation cover of grassland and
trees (MATA), a type of SAVANNA,
growing on the floodplain of the
Paraguay River and its tributaries in
Brazil, flooded during the summer,
suffering drought in other months. In
general the grasses occupy the lower
ground, the trees the low hillocks. G.

pantograph an instrument used in the
mechanical copying of a drawing or
map on any scale selected, consisting of
hinged rods arranged to form a paral-
lelogram and rotating about a fixed
point.

papa (New Zealand: Maori, earth) soft
MUDSTONES, SILTSTONES and SAND-
STONES of TERTIARY age covering
large areas of North Island, New
Zealand. G.

papagayo (Mexico: Spanish) a dry,
strong and cold northerly to northeasterly wind blowing over the plateau of
Mexico in winter, causing the temperature to fall suddenly to a low level. A
continuation of the NORTHER of the
USA, it is due to the movement of air
from the high Mexican plateau to the
low pressure area lying over the Gulf of
Mexico and the Caribbean sea. NORTE.

papaya, papaw, pawpaw *Carica papaya*,
an unbranched, EVERGREEN tree native to tropical America, now widely
grown there and elsewhere in tropical
areas, bearing large succulent fruit
resembling a melon, eaten raw, or
preserved, or (when unripe) boiled as a
vegetable. The fruits and leaves produce the enzyme papain that breaks
down protein and is used as a tender-
izer of meat, in brewing and in other
manufacturing processes, as well as
medicinally.

paper a substance made from CELLU-
LOSE fibres processed in such a way
that they become tightly interwoven,

the fibres being obtained from WOOD
PULP (mechanically ground or chemically digested), from various grasses
(e.g. ESPARTO, PAPYRUS), cloth rags
etc. and waste paper.

paprika term commonly applied to the
large ripe fruit of the European type of
sweet PEPPER, cooked or eaten raw,
and to the mildly spicy powder
obtained from it when dried, used
as a spice and colouring in savoury
dishes. There is also a hot, pungent
pepper, dried and powdered, termed
hot paprika.

papyrus *Cyperus papyrus*, a PEREN-
NIAL, strong, rush-like tall aquatic
herb native to the Nile valley, providing
pithy tissue from its flowering stems,
the raw material from which writing
paper was made in ancient times in
Egypt, Greece, Rome. The term has
been extended to apply to the paper
and to the matter written on it.

parabolic dune a crecent-shaped SAND
DUNE that faces the wind, i.e. the curve
is convex, with a steep face downwind,
the 'horns' pointing into the wind, i.e.
in the opposite direction to those of a
BARCHAN. Parabolic dunes are found
particularly on sandy shores and plateaus inland where sudden wind eddies
and BLOW-OUTS whisk away the central part of the dune and carry it
downwind. G.

paradelta that part of a DELTA where
CORRASION is taking place, as distinct
from that part (the true delta) where
deposition is occurring. G.

paradigm 1. in general, an accepted
example, MODEL-1 or pattern, an
ARCHETYPE 2. specifically, an approach, a school of thought with its
associated methodology, accepted by a
group of leading scholars as being of
particular importance, and used by
them and others for a while in their
field of study. It points to the kinds of
phenomena which should be investigated and to the best means of carrying
out the investigations; and it owes its

pre-eminence to its perceived ability to be more successful than its competitors in solving some problems regarded as acute at a particular time. PARADIGM SHIFT. G.

paradigm shift a significant change of approach in a field of study occurring when a prevailing PARADIGM-2 fails to meet the needs of changing thought or new theories, falls from favour, and a new paradigm emerges which is, by consensus, accepted.

paraffin 1. a complex mixture of saturated HYDROCARBONS, waxy, white and odourless, distilled from wood, coal or petroleum 2. in solid state, as paraffin wax, it is used as a stable, water repellent coating, as a seal, or as candles (WAX) 3. any of the several mixtures of similar hydrocarbons in liquid or semi-solid state, e.g. paraffin or paraffin oil (kerosene in USA) used as a fuel or solvent.

paraffin oil PARAFFIN.

paraffin wax PARAFFIN.

paragneiss a GNEISS that results from the METAMORPHISM of detrital sediments. ORTHOGNEISS.

paralic *adj.* of or pertaining to a coast with shallow water, e.g. applied to 1. a sediment deposited on the landward side of a coast in an area sometimes inundated by salt water 2. the environment of such an area of deposition, in contrast to LIMNIC.

parallax a seeming change of position (or the difference in the change of position) resulting from the real change (or the difference in the real change) of position from the point of observation. For example, if two objects are in line with the eye they coincide, but if the eye is moved to the right the object nearer the eye seems to lie to the left of the one in the rear; if the eye is moved to the left the nearer object seems to lie to the right of the one in the rear. The two objects are said to show parallax with respect to the observer. When the two objects coincide there is said to be a situation of no parallax. In astronomy, parallax is expressed by the angle subtended at a celestial body by a selected base line (e.g. the equatorial radius of the earth). G.

parallel *adj.* applied to 1. lines, curves, planes which are at the same distance from each other at all points and have the same direction or curvature 2. a line, curve, plane equidistant from another at all points and having the same direction or curvature.

parallel drainage a DRAINAGE pattern in which the channels of streams and their tributaries are almost parallel to one another.

parallel land use a type of MULTIPLE LAND USE in which the tract of land is not used in common for two or more purposes, but in which the uses are kept spatially separate, e.g. a limited recreational strip bordering the highway through a commercial FOREST-1.

parallel of latitude a line drawn on a map to link all points on the earth's surface with the same angular distance north or south of the equator (LATITUDE), termed a parallel because each is a line encircling the earth parallel to the equator. Because the earth is a SPHEROID the circles become smaller from the equator towards the poles. Thus each is a SMALL CIRCLE, but the equator (0° latitude) is a GREAT CIRCLE. Parallels of latitude are marked off in ninety divisions, i.e. ninety degrees, from the equator to each of the poles, each degree being subdivided into 60 minutes, each minute into 60 seconds; and the parallels of latitude meet MERIDIANS (lines of LONGITUDE), which are not parallel to one another, at right angles. G.

parallel retreat of slope a progressive backward movement of a slope with very little change of GRADIENT, the result of weathering and erosion. The theory that the form of a slope may alter little despite a long period of erosion is favoured by many slope authorities. REPOSE SLOPE, ZONAL INSELBERG.

parallel roads two series of matching horizontal terrace-like features, facing one another at corresponding levels on each side of a valley, e.g. in Glen Roy, near Fort William, Scotland, where they are considered to indicate the shorelines of lakes dammed by ice at various levels in the GREAT ICE AGE.

parameter in mathematics **1.** a quantity which is constant in the CASE under consideration but which may vary from case to case **2.** an unknown quantity which may vary over a certain set of values **3.** a VARIABLE of which other variables are taken to be FUNCTIONS-3 **5.** a descriptive summary of some characteristic of a POPULATION-4, e.g. MEAN, MEDIAN, STANDARD DEVIATION **6.** any of the factors in a whole that helps to define main characteristics or limits. In statistics the term parameter commonly occurs in expressions defining FREQUENCY DISTRIBUTION (e.g. POPULATION PARAMETER) or in models describing a STOCHASTIC situation (e.g. REGRESSION parameter).

parametric *adj.* of, pertaining to, or concerned with a PARAMETER.

parametric test a statistical test which involves the estimation of a POPULATION PARAMETER and also some assumptions about the data, i.e. that the data are measured on an INTERVAL SCALE or RATIO SCALE and that the POPULATIONS-4 from which the samples are drawn have a NORMAL DISTRIBUTION and equal variances. NONPARAMETRIC TEST.

paramo (Spanish páramo) pl. paramos **1.** the zone of alpine meadows in the Andes, a high treeless plateau above the TREELINE at 3000 m (10 000 ft) but below the snowline, supporting pasture and tundra vegetation, encroached on in places by GLACIERS **2.** the high, dry tableland of the northern MESETA in Spain. G.

paranti (Indian subcontinent: Urdu-Hindi) historically, land uncultivated for some time to allow it to regain its fertility. BANJAR, CHACHAR, POLAJ. G.

paraselene mock moon, an image of the moon appearing near the moon, caused by the refraction of light from ice crystals in a high cloud. PARHELION.

parasite an organism which lives on a living organism of a different species (the host) from which it draws most or all of its food sometimes, but not always, to the detriment of the host. Parasites are classified as AMETOECIOUS (MONOXENIC), AUTOECIOUS, BROOD PARASITE, ECOPARASITE, ECTOPARASITE, ENDOPARASITE, FACULTATIVE, HEMIPARASITE, HETEROECIOUS (HETEROXENOUS), HYPERPARASITE, KLEPTOPARASITE, METOECIOUS (METOXENOUS), OBLIGATE, TROPOPARASITE, XENOPARASITE. See COMENSALISM, EPIPHYTE, MUTUALISM, PARASITOID, SAPROPHYTE, SYMBIOSIS, ZOOBIOTIC.

parasitic *adj.* of, pertaining to, having the characteristics of a PARASITE.

parasitic cone ADVENTIVE CONE.

parasitism the close association between the PARASITE and its host.

parasitoid an animal that is partly a PARASITE, partly a PREDATOR in that it ultimately destroys its host, e.g. an insect that lays its eggs in the eggs or larvae of other insects.

paratonic, aitogenic *adj.* applied to a plant movement stimulated by external stimuli, e.g. NASTIC MOVEMENT, TROPISM. AUTONOMIC.

paratype in the CLASSIFICATION OF ORGANISMS **1.** any specimen, other than the TYPE SPECIMEN or its duplicates, included in the original description of a SPECIES **2.** an abnormal type within a species. HOLOTYPE, SYNTYPE.

parent material in soil science, the little altered but weathered bed rock, or the transported glacial or alluvial material, or an earlier soil, on which soil-forming processes work to make the soil layers. SOIL, SOIL HORIZON.

Pareto optimum a state of maximum general economic efficiency achieved in a SOCIETY-2,3, identified by Vilfredo Pareto, 1848-1923, Italian economist and sociologist, as the state reached when it is impossible by a shift in income to make all its members better off; when those members who would gain by such a shift would be unable sufficiently to recompense the losers without themselves becoming losers. The same principle can be applied to the allocation of RESOURCES-1, i.e. the OPTIMUM allocation being reached when it is impossible by the redistribution of the resources to satisfy the needs of some people without lowering the satisfaction of others. The Pareto optimum is an important concept in NEOCLASSICAL ECONOMICS; and discussions of its lack of ETHICS-2 figure in WELFARE-1 economics.

pargana (Indian subcontinent) the smallest administrative unit, a subdivision of a TAHSIL. G.

parhelion mock sun, an image of the sun appearing as a bright disc on or close to the solar halo, resulting from the refraction of light by ice crystals in a high cloud. PARASELENE.

Parian marble a fine white MARBLE-1 occurring in the Greek island of Paros, used in sculpture and decorative works.

Parian ware a type of white porcelain with a slightly granular surface, resembling PARIAN MARBLE, used mainly for busts, figures etc. in the mid-nineteenth century.

paring and burning a method, formerly used in Britain, of preparing peatland for cultivation. The surface layer of PEAT was removed (i.e. it was pared from the surface), dried and burnt, the crops being grown in the underlying peat or soil enriched by the ash of the burnt peat. The value of the method was much disputed. G.

Paris green a compound of COPPER arsenite and copper acetate used as an insecticide and as a pigment.

parish in the UK, originally an ecclesiastical unit, a subdivision of a COUNTY, consisting of a village with its own church and a clergyman in charge, in England either the rector (the rightful holder of the post) or someone appointed (the vicar) to undertake his duties. The TITHES and other dues were paid by the inhabitants of the parish to the rector or vicar, often in kind, hence TITHE BARNS. In a town the parish became the part of the town attached to a church. Later the parish, not necessarily identical with an ecclesiastical parish, became a unit for local government (civil parish). TOWNSHIP-1. G.

park 1. in medieval England, a piece of land held by the lord of the manor by royal grant for keeping animals (e.g. deer), for a source of food, for hunting, differing from a CHASE or FOREST-3 in being enclosed (and also differing from a forest in not being subject to Forest Laws, FREE-WARREN); later applied to 2. a large, enclosed piece of ground, usually landscaped, comprising grassland and widely planted trees, used for private recreation and the grazing of deer, cattle etc. and attached to a MANSION or COUNTRY HOUSE; and later to 3. a public park, an enclosed recreational area in or near a town, laid out with flowerbeds, paths, lawns, perhaps a small ornamental lake etc. 4. in the concept of a NATIONAL PARK, an extensive area of countryside to be protected and conserved for public benefit, originating in the Yellowstone National Park, USA, in 1871 5. in the mountains of Colorado, Wyoming, USA, a high, enclosed valley with a flat, grassy floor, frequently with a RANCH 6. a tract of land, in some cases park-like (PARK-2) in character, set aside and officially designated (e.g. by planning authorities) for a particular purpose, e.g. a COUNTRY PARK, INDUSTRIAL PARK, OFFICE (BUSINESS)

PARK, RESEARCH AND SCIENCE PARK 7. in Scotland and Ireland, a PADDOCK or field, a small enclosed area of land usually used for pasture, sometimes for crops. G.

parkland 1. see applications PARK-1,2 2. by analogy with a typical English park (consisting of grassland with scattered trees, PARK-2), the tropical SAVANNA lands of grassland with scattered trees, especially in Africa (also termed park savanna). G.

park savanna PARKLAND-2.

parliament the supreme legislative body of certain countries, e.g. of the UK.

parliamentary borough in Britain, a town or urban constituency with a right to send one or more members to PARLIAMENT.

parna (Australia: aboriginal) sandy, dusty ground, consisting of aggregates of wind-borne clay particles. G.

parochial *radj.* 1. or of relating to a PARISH or parishes 2. applied to attitudes, ideas, interests, opinions, or perception limited in scope; narrow, provincial.

paroxysmal volcanic eruption a sudden violent eruption of a VOLCANO, usually of one that has been dormant for a long period. PLINIAN ERUPTION.

parsley *Petroselinum crispum*, family Umbelliferae, a fragrant BIENNIAL herb, native to southern Europe, rich in vitamin C, widely grown for use as flavouring in cooking, or as a garnish.

parsnip *Pastinaca sativa*, family Umbelliferae, a BIENNIAL plant native to Europe, widely cultivated for its long, white taproot, rich in sugar and starch, cooked for human food, or used for cattle feed.

parterre an ornamental, level area in a large garden, usually close to the residence, laid out symmetrically (usually in an intricate pattern) with paths and formal beds planted with low-growing flowering plants or dwarf shrubs, in which garden ornaments,

statuary, or a fountain are, in some cases, incorporated.

partial *adj.* of or pertaining to only a part, as distinct from the whole.

partial drought in Britain, a period of 29 consecutive days on some of which slight rain may fall, but during which the daily average does not exceed 0.25 mm (0.01 in). DROUGHT.

partial eclipse ECLIPSE.

partial fallow land left uncropped in the first half of the summer or in the late summer. FALLOW, GREEN FALLOW.

participation the act or fact of having, with others, a share or a financial interest in, or taking part in, or being a part of, an activity or enterprise. PUBLIC PARTICIPATION.

particle 1. a very small piece of matter 2. in physics, a piece of matter assumed to have mass but to be so small that it does not have dimensions, i.e. it is a point. GRADED SEDIMENTS.

particulate *adj.* 1. existing as a very small, separate particle 2. of or relating to such a very small, separate particle.

part-time farmer a FARMER who has a regular occupation, other than that of FARMING, to provide an income. The two main types are the HOBBY FARMER, who farms primarily for pleasure, not for a living; and the worker-farmer, whose main occupation is in the town (e.g. in MANUFACTURING INDUSTRY) but who relies on the output of a small holding, worked in any spare time, for a food supply and an important auxiliary source of income.

pascal pa, the unit of pressure in SI, equal to one NEWTON per sq METRE.

pass 1. a narrow gap or COL in a mountain range providing a PASSAGE-WAY through the barrier 2. a narrow passageway through a BARRIER REEF.

passageway a way between two points by which it is possible for a person, animal, vehicle or vessel etc. to move from one to the other.

passerine bird a member of Passeriformes, an ORDER of birds that perch. More than half of all living birds have this characteristic and belong to this order. CLASSIFICATION OF ORGAN-ISMS.

passion fruit, purple granadilla *Passiflora edulis*, a tropical PERENNIAL climbing plant, native to Brazil, now widely grown there and elsewhere in the tropics and, in some places warmer than normal, in the Mediterranean area, for its purple-skinned sweet, succulent fruit, eaten fresh or pressed for its juice.

passive glacier a GLACIER with a low rate of accumulation (ALIMENTATION) and ABLATION because it receives only light snowfall and undergoes little melting in summer. It flows very slowly and transports little ice and debris. ACTIVE GLACIER.

pastoral (Latin pastor, a shepherd) *adj.* of, pertaining to, or characterized by, the care of grazing animals. GRAZING, LIVESTOCK FARMING.

pastoralism LIVESTOCK FARMING.

pasturable *adj.* applied to land capable of being used for PASTURE.

pasturage PASTURE.

pasture (Latin pascere, to feed, graze, attend to the feeding of beasts) land covered with growing grass and/or other HERBS-1 on which LIVESTOCK can feed, as distinct from a MEADOW-1 where the vegetation is mown for hay or silage. COMMON RIGHTS, GRAZ-ING, LIVESTOCK FARMING, ROUGH GRAZING.

pasture gate, pasturegate CATTLE GATE.

pat (Indian subcontinent: Sindhi) **1.** an arid clay plain **2.** a small steep-sided plateau in Chota Nagpur. G.

patabiont an animal living in the LIT-TER of a forest floor for the whole of its LIFESPAN.

patch cutting the FORESTRY practice of felling trees in a defined area within a FOREST-1 so that the tree cover may regenerate naturally, inward towards the centre of the cleared area from the

surrounding forest. The aim is to conserve the forest as a FLOW RE-SOURCE and to maintain a sustained yield (SUSTAINED YIELD FORESTRY). CONSERVATION, RESOURCE CONSER-VATION, RESOURCE MANAGEMENT.

patchouli *Pogostemon cablin*, family Labiatae **1.** an aromatic plant native to and cultivated in Asia **2.** an essential oil yielded by its leaves, used in perfumery and as an insect repellent **3.** the scent made from the essential oil.

paternoster lake one of a chain of lakes in a glaciated valley caused by the damming action of morainic ridges or rock bars, so-named because the chain resembles a rosary (in this sense the string of beads that constitute a ro-sary). Such lakes occur especially where the downward slope of the valley floor takes the form of a series of steps. G.

pathetic fallacy the attribution of human characteristics or emotions to natural phenomena, e.g. an angry sea.

pathogen an agent (e.g. a BACTERIUM, a VIRUS) that causes disease.

pathology the scientific study of all aspects of departures from the norm in respect of bodily function and health. PHYTOPATHOLOGY.

patina a coloured hard film on the surface of exposed rocks, formed by WEATHERING. DESERT PATINA, DESERT VARNISH. G.

patrial one who has the right of abode in the UK (e.g. a person born in the UK or one of whose parents or grand-parents were; a naturalized citizen; a former citizen of the Commonwealth already residing in the UK).

patrial *adj.* **1.** having the right of abode in the UK **2.** of or belonging to one's native country.

patriality the eligibility to become or right to be a PATRIAL-1.

patriarchy 1. government by men **2.** a social system in which the chief author-ity is the father or eldest male member of the family or clan **3.** a community characterized by this system. MATRI-ARCHY, PATRILINEAL.

patrilineal *adj.* applied to a social system in which descent is traced only through fathers. MATRILINEAL, PATRIARCHY, UNILINEAL.

patroclinous *adj.* in biology **1.** applied to an offspring which in inherited character or characters resembles the male rather than the female parent **2.** possessing or involving the tendency to inherit a character or the characters of the male parent only. MATROCLINOUS.

patterned ground ground that is embellished with circles, polygons, nets, stripes etc., varying in size and occurring in (but not wholly confined to) regions of present or former PERMA-FROST. The rock fragments may or may not be sorted, e.g. in a polygonal pattern the smaller fragments may appear near the centre and increase in size outwards, or the pattern may be outlined by vegetation, no sorting of fragments having taken place; and on a slope there may be step-like forms, or stripes consisting of lines of stones, graded as the slope flattens. It has been suggested that the processes of freeze-thaw and SOLIFLUCTION, moisture conditions and the extent (or non-existence) of vegetation cover contribute to the development of the patterns. PIPRAKE.

paved surface a hard, artificial, flat surface laid on a road or path, commonly built by the laying of several courses of various materials, usually incorporating ASPHALT, AGGREGATE-4, BITUMEN etc. MACADAM, PAVEMENT, SIDEWALK.

pavement **1.** in geomorphology, applied loosely to bare rock surfaces of varied origin, especially if more or less horizontal, i.e. resembling a PAVED SUR-FACE **2.** a PAVED SURFACE **3.** in Britain (American sidewalk), a path with a paved surface by the side of a ROAD or STREET used by walkers **4.** (American) a SIDEWALK, a paved street or road,

any PAVED SURFACE, or the material used in making it.

pawpaw PAPAYA.

pawindāh, pavindá (Indian subcontinent: Pashto) Afghan migratory traders who travel to Pakistan in the winter. G.

pays (French) a small NATURAL RE-GION in France, of no administrative significance, demarcated by the unity of some physical feature or features (e.g. of geology, relief, land use) and/or of social or cultural features. A term used by geographers writing in English who wish to avoid the problem of defining 'natural'. G.

paysage (French) landscape. A term sometimes used by geographers writing in English in order to avoid the controversies raised by the use of the term LANDSCAPE. G.

pea *Pisum sativum*, family Papilionaceae, a LEGUME-2, a climbing or bushy ANNUAL plant, probably (but not certainly) native to western Asia, now widely grown for the sake of its PODS, containing seeds also termed peas, used as human food since ancient times, the whole plant also being used for animal feed and for GREEN MANURE. Many cultivars with tender pods (e.g. sugar pea, mangetout) are eaten whole when the pods are flat in youth; other cultivars are bred to produce seeds for a specific purpose (e.g. freezing, drying, picking when young and small). CHICK PEA.

peach *Prunus persica*, a small tree native to China, widely cultivated there and in Mediterranean regions since ancient times, and in south-facing, sheltered locations in adjacent regions, for its fruit (a DRUPE), of varying size, with succulent sweet flesh, ranging in colour from near white to deep golden yellow, eaten fresh, or canned, dried, frozen, made into jam. NECTARINE.

peak (pike, a sharp point) **1.** a pointed top or projection **2.** the highest point, the maximum **3.** the more or less pointed prominent summit of a mountain, often used in place-names; but this does not apply to The Peak of Derbyshire, England, which is a high, flat-topped plateau of MILLSTONE GRIT, an area scheduled as a NATIONAL PARK **4.** a high isolated mountain. G.

peak land value intersection PLVI, a point within the CORE-2 of a city or town where land values are at their highest, usually occurring at a major traffic intersection, i.e. at the most accessible area.

peak-plain GIPFELFLUR.

peanut GROUNDNUT.

pear *Pyrus communis*, a small DECIDU-OUS tree of temperate areas, bearing fairly firm, edible, juicy, sweet fruit (a POME) with a generally rounded base, tapering towards the stem. There are many varieties in cultivation, eaten raw, cooked (hard varieties best), canned, dried, used in chutney etc. PERRY is made from the hard varieties with a high TANNIN content. Special care is needed in marketing because pears are perfectly ripe and palatable for a very short time.

pearl a secretion formed by certain MOLLUSCS (especially by OYSTERS), particularly that formed around irritants such as a parasite or a grain of sand that has entered the shell naturally (a natural pearl) or been introduced articially (a cultured pearl).

pearl barley a grain of BARLEY reduced by friction to a small round shape resembling that of a PEARL.

Pearson product moment correlation a measure of the strength of the relationship between variables, the data of each being measured on an interval or ratio scale. It is possible for its value to lie only between $+1$ (perfect positive linear relationship) and -1 (perfect negative linear relationship), zero indicating no linear relationship. CORRELATION-3.

peasant very loosely applied to an agricultural worker, usually (but not always) farming at or near subsistence level and either holding a proprietary right over the land farmed or being an employed labourer. Some specialists in LAND TENURE restrict the term to the former category, i.e. the farmer holding a proprietary right. PEASANT ECON-OMY, PEASANT FARMING, SUBSIS-TENCE AGRICULTURE. G.

peasant economy an economic system in which production depends primarily on small scale, technologically simple agriculture, fishing and craftwork, usually carried out by family groups mainly in rural surroundings. Such a system is usually characteristic of a PRE-INDUSTRIAL SOCIETY or a partly industrialized society (INDUSTRIALI-ZATION-2). PEASANT FARMING, PEASANT SOCIETY.

peasant farming a farming activity carried on by a small family unit at or near subsistence level on a small area of land, e.g. as distinct from that on a large farm with employed labour, organized primarily for commerce. In peasant farming the family provides the labour. Production is primarily for the benefit of the family but in most cases (particularly if the land is leased from a landlord) surplus products are sold in the market (in order to pay the rent). If the land is leased from a landlord the tenancy is usually held by the family group, rather than by an individual. If the land is owned by the peasant the right of ownership is usually vested in the family group, not in an individual. FARMING, LAND TENURE, PEASANT, PEASANT ECONOMY, PEASANT SOCI-ETY, SUBSISTENCE AGRICULTURE.

peasant society a rural sub-society of a large STRATIFIED SOCIETY, the members of the peasant society being engaged in PEASANT FARMING, with the family as the most important social unit, being attached to the soil, the local community and tradition, having a simple culture and low social status, and being in most cases economically interdependent with an urban centre. Such a society is generally seen as characteristic of a pre-industrial or partly industrialized nation. INDUS-TRIALIZATION-2, PRE-INDUSTRIAL SOCIETY.

peat a dense deposit of dead vegetable matter only partially decomposed (DE-COMPOSITION) mainly because it has accumulated in water or in very damp conditions where oxygen is deficient (ANAEROBIC). Dark brown or black, partially CARBONIZED, it forms the first stage in the development of LIGNITE, BROWN COAL and COAL; and may be neutral or alkaline (FEN PEAT) or acid (BOG PEAT). It is burnt as fuel, or applied to the soil to improve the texture or raise the water retaining property of the soil. PARING AND BURNING.

peat hag, peat-hag, peat hagg, hagg a more or less vertical bank cut in upland peat by wind or water erosion. Hag is the more usual spelling. G.

pebble a small stone, naturally rounded by the action of water or wind, diameter between that of GRAVEL and COBBLE. There is some confusion over precise size, but a pebble is commonly defined as having a diameter lying approximately between 10 and 50 mm (0.4 to 2 in); on the WENTWORTH SCALE it lies between 2 to 64 mm (0.08 to 2.5 in). The term is sometimes wrongly applied (as 'angular pebbles') to the fragments in a BRECCIA. G.

pecan *Carya illinoensis*, a large tree native to North America bearing smooth-shelled oval NUTS containing a kernel with a taste resembling that of a mild, sweet WALNUT. A dessert nut, it is used in confectionery etc. and is sometimes salted.

peccary a wild, hoofed MAMMAL native to tropical America, resembling a small PIG. The tanned skin is used in clothing manufacture, especially in glove-making.

pectin a substance present in plant tissue, especially in varying amount in that of fruits. It readily forms a gel (a jelly-like colloidal dispersion) in solution, and is therefore used in jam-making to ensure 'setting', i.e. the formation of a gel. PECTOSE.

pectose a substance present in the CEL-LULOSE of plant tissue, especially in that of fruits, which changes to PECTIN as the fruit ripens.

ped a naturally-formed aggregate of soil particles. CRUMB.

pedalfer in soil science, a soil from which the base compounds CALCIUM CARBON-ATE and MAGNESIUM carbonate have been leached, and which contains accumulations of ALUMINIUM, Al, and IRON, Fe, compounds. Such soils occur especially in a humid climate with annual precipitation over 610 mm (24 in), and include BROWN EARTHS, LATOSOLS, PODZOLS. Pedalfer is applied to one of the two types of ZONAL SOIL, the other being PEDOCAL. SOIL CLASSIFICATION. G.

pedestal, pedestal rock a residual, columnar mass of weak rock capped with a harder rock. Opinions differ as to whether it is formed by differential WEATHERING helped by rainwash, or by wind ABRASION. HOODOO. G.

pediment 1. in architecture (a corruption of periment, itself probably a corruption of pyramid), the gable over the front of a building, triangular in shape in classical design, the apex arched or broken in baroque design 2. in geomorphology (Latin pes, pedem, foot, base, foundation), an eroded rock platform, cut into the BED ROCK, usually slightly concave, resembling an arthitectural pediment in shape, bare or carrying a very small quantity of rock debris, extending over a considerable area at the foot of an abrupt mountain slope or face, the lower edge sloping gently away to a PERIPEDIMENT (WANING SLOPE). Pediments form basal slopes of transport over which occasional rainstorms carry weathered material derived from the steeper slope above, and are characteristic of ARID and semi-arid lands. There is much controversy over their origin. A pediment should not be confused with a BAJADA, formed by deposition. RUWARE, SENESCENT DESERT, SLOPE ELEMENTS. G.

pedimentation the processes collectively involved in the formation of a PEDIMENT-2. ZONAL INSELBERG.

pediocratic period the relatively quiet period of time between two OROCRATIC PERIODS.

pediplain, pediplane a gently undulating or level surface, either exposed or lightly covered with a thin layer of alluvium, possibly formed by the merging of several PEDIMENTS-2 in the course of erosion. G.

pediplanation the process of formation of a PEDIPLAIN. G.

pedocal in soil science, a soil, only slightly leached (LEACHING), containing an accumulation of CALCIUM CARBONATE and MAGNESIUM carbonate, and occurring mainly in climates with an annual precipitation below 610 mm (24 in), insufficient to remove the soluble constituents. Pedocal is applied to one of the two types of ZONAL SOIL, the other being PEDALFER. SOIL CLASSIFICATION. G.

pedogenesis 1. in soil science, the formation of soil by natural processes from parent material 2. in zoology, an alternative spelling for paedogenesis, reproduction by animals sexually mature but otherwise in a pre-adult stage. G.

pedogenic *adj.* of or pertaining to soil formation 2. of or pertaining to the effect caused by soil factors. EDAPHIC.

pedogenics the scientific study of the origins and development of soils. PEDOGENIC, PEDOLOGY. G.

pedogeography the geography of soils, the study of the spatial distribution of soils. G.

pedography the description of soils. G.

pedology the scientific study of the morphology, composition and spatial distribution of soils, including their classification and, in a general way, their uses. G.

pedon a long, narrow column of soil, from the upmost surface down to the PARENT MATERIAL, exhibiting all the SOIL HORIZONS present at a point, small in diameter but large enough for any of the horizons to be studied laterally. G.

pedosphere that part of the earth's crust in which soil-forming processes occur. G.

pegmatite an IGNEOUS ROCK, composition simple or complex, with coarse, very large CRYSTALS, usually occurring in DYKES or VEINS and especially associated with BATHOLITHS. Pegmatite is formed at a late stage in the cooling of the MAGMA and may be rich in rare minerals of such elements as BORON, LITHIUM, NIOBIUM, TANTALUM and the RARE EARTHS. The most common, granite pegmatite, is composed mainly of QUARTZ and FELDSPAR.

pelagic *adj.* of, pertaining to, occurring in, living in, the mass of open water of the ocean or a lake, i.e. away from the land and the floor of the ocean or lake.

pelagic deposit an OOZE deposited on the ABYSSAL floor of the ocean (or of a very deep large lake), derived from minute remains of organisms previously living in the PELAGIC DIVISION. PELAGIC ORGANISMS.

pelagic division, pelagic zone one of the two chief divisions of the aquatic environment based on depth of water (the other being the BENTHIC DIVISION), consisting of the whole mass of open water, including the NERITIC PROVINCE and the OCEANIC PROVINCE, the boundary between those two being at the edge of the CONTINENTAL SHELF. G.

pelagic organisms the animals and plants living in the PELAGIC ZONE, divided into PLANKTON (NANO-PLANKTON, PHYTOPLANKTON, ZOO-PLANKTON), and NEKTON.

Pelean, Peléan *adj.* of or pertaining to a type of volcanic eruption (Pelean or Peléan eruption) characterized by the extrusion of a very acid, very viscous lava which tends to consolidate as a solid plug and by the violent emission of NUEE ARDENTE, causing widespread devastation. The term is derived from Mount Pelée, Martinique, which erupted in that way in 1902. Pelean is applied to one of the four main types of volcanic eruption, the others being HAWAIIAN, STROMBOLIAN, VULCAN-IAN or VESUVIAN G.

Pele's hair (Hawaii) volcanic glass spun out into hair-like form from basaltic lava as it explodes, or from gas bubbles in liquid lava as they burst; not to be confused with Pelée or PELEAN.

Pele's tears (Hawaii: Pele, the fire goddess, whose home was the volcano of Kilauea) congealed lava droplets, small droplets of basaltic volcanic glass ejected and solidifying during a volcanic eruption, formed particularly when a shower of lava falls into the sea.

pelite a rock composed of metamorphosed clay particles. METAMORPHISM, PSAMMITE, PSEPHITE. G.

pelitic *adj.* of or pertaining to PELITE, hence applied to rocks derived from pelite, e.g. pelitic gneiss, pelitic schist. ARGILLACEOUS used to be regarded as synonymous with pelitic by some authors, but pelitic is now applied only to metamorphosed argillaceous rocks, i.e. to pelite. G.

pellicle in biology 1. any very thin skin, MEMBRANE or film 2. any of those forming a protective covering.

pellicular *adj.* of, pertaining to, having the character of, a PELLICLE-1, hence applied to a film of water adhering to a surface. G.

pelosols one of the seven major groups in the 1973 SOIL CLASSIFICATION of England and Wales, clay soils which are argillic (ARGILLACEOUS), CALCAREOUS or non-calcareous, with a brown, grey or red and commonly MOTTLED layer below the surface.

pelt 1. an animal skin with wool or hair on it 2. untanned cattle hide (raw hide) after the hair has been removed from it. FUR.

pen a small enclosure for livestock, usually in the open air, e.g. sheep pen.

pene-exclave EXCLAVE.

peneplain, peneplane peneplain, almost a plain, an almost featureless, gently undulating land surface. The term was introduced in 1889 by the American physical geographer, W. M. Davis, who conceived the idea that the forces of erosion and denudation were destined, unless interrupted, to reduce any surface (regardless of structure) to an almost featureless plain showing little connexion with structure, i.e. to a peneplain, the penultimate stage before the stage of the plain without relief, the end of the Davisian CYCLE OF EROSION or denudation. Later authors used the spelling peneplane (almost a flat surface) because, they contended, the term plain implied a region of horizontal structure. Others introduced the terms PANPLAIN and PEDIPLAIN, landforms which they identified as resulting from processes different from those producing Davis's peneplain. G.

peneplanation the process of forming a PENEPLAIN. The spelling peneplaination is, by general agreement, incorrect. CYCLE OF EROSION. G.

peninsula almost an island, a tract of land, large or small, projecting into a body of water, nearly surrounded by water or having water on at least three sides, so that the greater part of the boundary is coastline. ISTHMUS. G.

peninsular *adj.* of, pertaining to, belonging to, a peninsula. G.

penitent (South America: Spanish penitente) a spike of old compact snow or of glacier ice, as applied to such features appearing in the Andes of Santiago.

pentlandite a sulphide mineral, composed of iron and nickel, one of the two major ores of NICKEL. GARNIERITE.

pentref (Welsh) a village, a homestead. G.

penumbra UMBRA.

peon (South America: Spanish peon, Portuguese peão) a labourer **1.** generally, in Spanish America, a labourer hired daily **2.** in Mexico, formerly, a labourer who works for a master to whom he is indebted, in order to pay off his debt **3.** in India, a foot-soldier, constable, attendant, messenger.

peperite a mixture of lava and sediment, especially such a deposit in the Puy de Dome district, France. G.

pepino hill (Puerto Rico) HUM. G.

pepitos the seeds of PUMPKIN, fried in deep oil and salted to eat, rich in fats and protein.

pepo the fleshy fruit, with firm skin (rind) and many seeds, of any member of the Cucurbitaceae family, e.g. CUCUMBER, MELON, PUMPKIN.

pepper a term applied to the fruits of two different kinds of plants, red or Cayenne pepper (chilli) and sweet pepper being derived from CAPSICUM, but white and black from the climbing vine, *Piper nigrum*. **1.** red pepper and CHILLI (various spellings) are the very hot fruits of *Capsicum frutescens*, a perennial, bushy plant native to tropical America and the West Indies, used only in cooking or as a condiment, the powdered form being known as cayenne. Sweet pepper is the fruit of *Capsicum annum*, a bushy annual plant native to the tropics which can be grown, under adequate heat, beyond the tropical zone. It is consumed when immature (green) or ripe (red), cooked or uncooked; varieties differ in degree of pungency. The term PAPRIKA is applied to the large mild fruit of European type (pimiento in Spanish) (but also, in Hungary, to the powdered

form of a more pungent variety, sometimes specified as hot paprika) **2.** peppercorns, used as flavouring, historically the most important spice in world trade, are the berries, born in long hanging spikes, of a climbing vine, *Piper nigrum*, which has a lifespan of 15 years or more, native to the Indian subcontinent, cultivation restricted to a wet tropical climate. The berries are harvested at different stages of development, and treated in different ways. For black peppercorns they are picked before ripening and dried in the sun; for white peppercorns they are ripened on the vine, the husks being removed before drying. Green peppercorns are the untreated unripe berries, milder in flavour than black or white, sometimes canned or bottled for export. Red or pink peppercorns are the berries of *Schinus terebinthifolius*, a shrub native to Brazil. They are used in medicine and in cooking, but can be mildly toxic. In addition, the pepper-flavoured black seeds of Japan pepper, *Zanthoxylum piperitum*, a deciduous shrub or small tree, hardy in Europe, are ground and used as a condiment in Japan and China.

peppercorns PEPPER-2.

peppermint *Mentha × piperita* (*Mentha aquatica* and *Mentha splicata* hybrid), family Labiateae, an aromatic herb native to Europe, and cultivated widely in temperate regions for its ESSENTIAL OIL, distilled from the leaves, stems, flowers, and used in cordials and confectionery and medicinally.

perambulation **1.** an official walk round a district to establish rights of ownership, the boundaries and so on, an old practice, often annual, still extant in some places in Britain **2.** the boundary so established.

percentage base saturation PBS, the percentage of the exchangeable BASE CATIONS in relation to the total CATION EXCHANGE CAPACITY (CEC) of a given amount of soil. BASE STATUS OF SOILS.

percept a mental image of something perceived, a product of PERCEPTION.

perceptible *adj.* capable of being measured or perceived (PERCEPTION), thus capable of being isolated from a similar object.

perception in general, sensory experience which has acquired meaning or significance. The term perception is used with varying technical meanings in philosophy, psychology and physiology, entering widely into theoretical discussions. It is used in geography mainly as the mental interpretation of the sensations produced by the awareness and appreciation of situations or external objects by the senses (sight, hearing, touch, smell, taste); but, in preference to perception, some geographers use the term COGNITION, with its stress on the mental processes of understanding and interpretation. PECEPTION GEOGRAPHY, PERCEPT, PERCEPTIBLE, PERCEPTIVE, PERCEPTUAL.

perception geography the branch of human geography concerned with the part played by psychological factors in attitudes to, and decision-making in, environmental affairs. COGNITION, MENTAL MAP, PERCEPTION.

perceptive *adj.* capable of perceiving (PERCEPTION), having or showing a marked ability to understand.

perceptual *adj.* of, pertaining to, involving, characteristic of, PERCEPTION.

perched block ERRATIC, ERRATIC BLOCK.

perched water table GROUND WATER (often isolated) separated from an underlying body of ground water (the WATER TABLE proper) by unsaturated rock, forming part of a zone of saturation (VADOSE) different from that occupied by the water table proper. G.

percolation the downward seeping of water through pores, joints and other interstices in soil and rock, sometimes accompanied by LEACHING. G.

percoline a SOIL HORIZON along which water seeps laterally. THROUGHFLOW.

perennating *adj.* applied to those parts of a plant (the renewal organs) which carry it through an unfavourable period, especially the unfavourable period of its annual cycle. PERENNATION, PERENNIAL, RAUNKIAER'S LIFE FORMS.

perennation in botany, the act of tiding over an unfavourable period, e.g. a cold winter, a dry summer. PERENNATING.

perennial *adj.* applied particularly to plants that continue in growth from year to year. The aerial parts of perennial HERBACEOUS plants die down after seed production and are replaced by new shoots in the following year. The new shoots of woody perennials (PHANEROPHYTES) start from the permanent woody stems above ground, hence the great size of some woody perennials, e.g. TREES, SHRUBS. ANNUAL, BIENNIAL, EPHEMERAL, RAUNKIAER'S LIFE FORMS.

perennial irrigation a system of IRRIGATION in which water is artifically made available to plants, not seasonally, as in BASIN IRRIGATION, but throughout the year. The methods used to maintain a regular water supply range from the simple lifting of stream water by ARCHIMEDES' SCREW, SAQIA, SHADUF or water pump, by sinking of wells and by the creation of ponds (TANK) to the building of dams and barrages on rivers to hold back enough water to last from one flood period to the next, thus providing a constant supply for the canals and channels by which the water is led on to the dry land.

perennial stream a stream that flows continuously throughout the year, in contrast to an INTERMITTENT STREAM.

perfect competition a theoretical market

condition in which there are many independent buyers and sellers, each of the sellers holding only part of the good involved, all resources being perfectly divisible and mobile. The sellers are not allowed to collude. Buyers and sellers are fully informed of prices throughout the market, and none of them individually can influence prices. DUOPOLY, MARGINAL UTITLITY, MARKET ECONOMY, MONOPOLY, OLIGOPOLY.

perfect negative correlation CORRELATION-3.

perfect positive correlation CORRELATION-3.

perforation deposit a mound of glacial sand and gravel deposited when a debris-holding pool, lying on the surface of stationary ice, melts and cuts its way through the ice and the ice around the deposited debris is subsequently subjected to ABLATION. G.

pergelation (Latin per, by, through; gelu, frost) the formation of PERMAFROST or the local freezing of soil.

pergelisol PERMAFROST. G.

pericline a small ANTICLINE in which the strata are arched up into a dome so that the beds dip away on all sides from a central point. QUAQUAVERSAL. G.

peridotite a group of coarse-grained ULTRABASIC IGNEOUS ROCKS, usually dark green, consisting mainly of OLIVINE with or without FERROMAGNESIAN MINERALS, and without FELDSPAR. The characteristic variety, peridot, is used as a gem, the darker the green the more valuable. CHRYSOLITE, MAGMAGENESIS, TALC.

perigean *adj.* of or pertaining to the PERIGEE.

perigean tide the tidal condition characteristic of a period when the moon is at its PERIGEE, so that its gravitational pull is great, resulting in a high tide that is higher, a low tide that is lower, and a TIDAL RANGE greater, than usual. APOGEAN TIDE.

perigee the point in the orbit of any of

the earth's planets or satellites when it is nearest to the earth, applied particularly to the orbit of the MOON around the earth. APOGEE, APSIS. G.

periglacial *adj.* applied to areas adjacent to an ICE SHEET or GLACIER of the past or of the present, and to all phenomena associated with such a situation. G.

perihelion the point nearest to the SUN in the orbit of a comet or planet around it. The earth arrives at its perihelion about 3 January, when it is some 147.3 mn km (91.5 mn mi) from the sun. APHELION, APSIS. G.

period in geology, a division of geological time, part of an ERA. The rocks laid down in a period constitute a system. GEOLOGICAL TIMESCALE. G.

period fertility COHORT FERTILITY, FERTILITY-3.

periodic *adj.* **1.** occurring repeatedly at constant intervals **2.** recurring from time to time **3.** characterized by regularly occurring stages or processes.

periodicity the quality or fact of recurring at constant intervals.

peripediment an area of accumulated sands and gravels peripheral to a PEDIMENT in a desert basin, preferably termed a BAJADA.

permafrost permanently (i.e. for a continuous period of at least two years) frozen soil, subsoil and, some authors add, bed rock (ACTIVE LAYER), a term more acceptable than the synonymous pergelisol. GEOCRYOLOGY, TELE, TJALE.

permafrost table the surface between the upper limit of the PERMAFROST and the lower limit of the overlying ACTIVE LAYER.

permanent way in British railway system, a track laid for long-term use.

permatang (Malay) an old sand beach, rising sandy ground, or a sandy ridge in Malaysia. G.

permeability the quality or state of being PERMEABLE.

permeable *adj.* capable of being wholly

penetrated by a fluid, of allowing the passage of a fluid, of being SATU-RATED. The opposite condition is termed IMPERMEABLE. G.

permeable rock a rock that allows the free passage of water through it owing to its POROSITY, e.g. sandstone, oolitic limestone. Some writers include also rock with joints, bedding planes, cracks, fissures etc. that allow the free passage of water, defining the porous rock as being of primary permeability, the rock with joints etc. of secondary permeability. Other authors distinguish the secondary group as being pervious (PERVIOUS ROCK). IMPERMEABLE ROCK, IMPERVIOUS ROCK. G.

Permian *adj.* of the latest period (of time) or system (of rocks) of the PALAEOZOIC era, when amphibians declined, reptiles developed, sandstone strata were deposited. PERMO-TRIASSIC, GEOLOGICAL TIMESCALE.

Permo-Triassic *adj.* of or pertaining to the period (of time) or system (of rocks) transitional between the PERMIAN and the TRIASSIC. GEOLOGICAL TIME-SCALE.

perpendicular *adj.* **1.** vertical, at right angles to the horizontal at any point on the earth's surface **2.** of a line, plane or surface forming an angle of 90° with another line, plane or surface.

perry an alcoholic drink made from pear juice, preferably from only a single variety of PEAR.

Persian lamb the skin of KARAKUL lamb, used in clothing manufacture.

Persian wheel SAQIA.

persistent *adj.* lasting, enduring despite opposition or difficulties, repeatedly occurring, applied to **1.** leaves which, though withered, stay on a plant **2.** an organism which continues to occupy an area in which conditions are now generally hostile to the species to which it belongs **3.** chemicals which, because of their stability, do not readily decom-pose **4.** meteorological conditions which continue for more than the normal or expected length of time, e.g. a persistent high pressure system (HIGH).

personal mobility MOBILITY.

Peruvian bark the bark of CINCHONA.

perverse growth in economics, growth that undermines rather than builds up the potentialities of an economy for long term development.

pervious rock some authors use this term as synonymous with PERMEABLE ROCK, but others restrict it to a rock that allows the free passage of water through it owing to the presence in it of joints, cracks, fissures etc., e.g. chalk, carboni-ferous limestome. IMPERMEABLE ROCK, IMPERVIOUS ROCK, POROSITY. G.

pest an insect or other animal con-sidered by human beings to be harmful to them and their activities, e.g. to the growing of crops, food storage etc. PESTOLOGY.

pesticide a chemical susbstance used in killing insects or other animals re-garded by human beings as being harmful to them and their activities. PEST.

pestology the scientific study of PESTS and of the control of pests.

petite bourgoisie **1.** the middle classes collectively **2.** applied by some authors to self-employed people who own a small amount of capital but do not employ labour (the spelling petit, frequently used, is incorrect). BOURGOISIE.

petrification the turning of organic material into stone or into a substance nearly as hard as stone, the original organic tissue being very slowly re-placed by deposits of minerals such as silica, agate or calcium carbonate carried in solution. FOSSIL.

petrocalcic horizon HARD PAN.

petrochemical a chemical derived from PETROLEUM or NATURAL GAS or from a derivative of either, extensively used industrially in the production of plas-tics, synthetic fibres, drugs, detergents, fertilizers, fungicides, insecticides, pes-ticides etc., e.g. ETHYLENE.

petrofabric analysis the measurement of the extent of, and the nature of the orientation of, mineral grains, particles and rock fragments present in a rock or till, particularly of those in the area of JOINTING or CLEAVING in METAMORPHIC ROCKS. FABRIC.

petrogenesis the branch of PETROLOGY concerned with the scientific study of the origins of rocks, especially of the processes involved, with particular reference to IGNEOUS and METAMORPHIC ROCKS.

petroglyph a carving on a rock face, especially one made by prehistoric people, e.g. on cave walls.

petrographic province a region in which the various IGNEOUS ROCKS are so related by marked specific peculiarities as to differentiate them from other assemblages of igneous rocks of other regions or cycles. The term 'suite' has been applied to rocks showing such affinities with each other, and three suites are usually identified: ATLANTIC, PACIFIC, SPILITIC. G.

petrography a branch of PETROLOGY concerned with the classification and description of rocks, special attention being given to the TEXTURE-1.

petrol (American gasoline) refined PETROLEUM, a refined mixture of hydrocarbons, used as a fuel in the internal-combustion engine.

petrolatum petroleum jelly, a white or yellow gelatinous substance derived from a mixture of hydrocarbons, used as lubricant and as a base in ointments and cosmetics.

petroleum (Latin petra, rock; oleum, oil) crude oil, mineral oil, a mixture of HYDROCARBONS in a solid (BITUMEN), liquid or gaseous state (NATURAL GAS), occurring in porous SEDIMENTARY ROCKS interbedded with SHALES and other rocks. It originates from the alteration of vegetable and animal remains entombed in the sediments when these were deposited, especially under brackish water condi-

tions (SAPROBEL). The term is usually applied specifically to the liquid form, commonly trapped, with natural gas, at the top of the arch (dome) of an ANTICLINE between impervious beds. This petroleum, obtained by drilling into the OIL DOME, is subsequently distilled in a refinery, yielding fractions making PETROL, PARAFFIN OIL, diesel fuel, fuel oil, lubricating oils and heavy fuel oils; ASPHALT and PARFFIN WAX are obtained from the residue. The hydrocarbons in the heavier fractions are further 'cracked' to make them suitable for, and to increase the amount of fuel for, internal combustion engines. HYDROCARBON CRACKING, OIL FIELD, OIL POOL, OIL-SAND, OIL-SHALE, OIL WELL.

petrology the branch of GEOLOGY which includes GEOCHEMISTRY, LITHOLOGY, MINERALOGY, PETROGENESIS and PETROGRAPHY and deals with the scientific study of the origin, occurrence, chemical and mineral structure and composition of rocks. G.

pewter an alloy of TIN and LEAD, sometimes with a small quantity of COPPER or ANTIMONY added for hardening, malleable, does not tarnish, used for table utensils, ornamental objects.

pH potential hydrogen, the standard measure of ACIDITY or alkalinity (BASE-2) of a substance, based on the activity of HYDROGEN IONS in a litre of a solution (or of a pure liquid), expressed in gram equivalent per litre. The pH values range from 0 to 14.7, NEUTRALITY-3 being 7.2 (the pH of pure water at 25°C: 77°F). Numbers lower than neutrality signify increasing acidity, the higher numbers increasing basicity (alkalinity). Knowledge of the pH of the soil under cultivation is most important in agriculture and horticulture. ACID RAIN, ACID SOIL, ALKALI SOIL, ALKALINE SOIL. G.

phacolith, phacolite a lens-shaped intrusion of IGNEOUS ROCK occupying the saddle (the crest) of an ANTICLINE or the keel (the trough) of a SYNCLINE, differing from a LACCOLITH in that it is shallower and the laccolith has a flat base. G.

phanerophytes plants forming a class of RAUNKIAER'S LIFE FORMS, woody plants with PERENNATING buds (renewal buds) more than 25 cm (10 in) above soil level, e.g. many trees and shrubs (PERENNIAL). Shrubs between 25 cm and 2 m (10 in and 6.5 ft) are nanophanerophytes, trees and shrubs between 2 m and 8 m (6.5 ft and 26 ft) are microphanerophytes, trees exceeding 8 m (26 ft) in height are mesophanerophytes and megaphanerophytes. G.

Phanerozoic (Greek phaneros, visible or evident; zoon, animal) *adj.* applied by geologists to the period of the earth's history from the beginning of the CAMBRIAN period to the present, the rocks of which contain FOSSILS, i.e. visible evidence of life. AEON, CRYPTOZOIC, GEOLOGICAL TIMESCALE. G.

phatic communication, phatic communion a process of communication between people (by vocal sound, speech or some other form of symbolization) in which precise information is not imparted but which creates a social relationship and states of feeling which help to produce common attitudes and social solidarity.

phenetic *adj.* in biology, of or pertaining to a relationship or classification based on the maximum similarity that can be seen, i.e. a relationship based on the PHENOTYPE as opposed to the GENOTYPE or to PHYLOGENY. G.

phenocryst a large conspicuous crystal occurring in the finer grained groundmass (MATRIX-3) in porphyritic igneous rock. PORPHYRY.

phenol 1. a white, crystalline, smelly, corrosive, poisonous hydroxy acid, distilled from COAL TAR, used in the manufacture of disinfectants, dyes, plastics 2. any of the hydroxy derivatives of BENZENE.

phenology in biology, the scientific study of periodically recurring natural phenomena in plant and animal life (e.g. flowering, breeding, migration) in relation to seasonal climatic change and other factors. G.

phenomenal environment the ENVIRONMENT seen as including not only natural phenomena (PHENOMENON-1) but also the effects of human activity, i.e. the BIOTIC and physical environment combined with the environments altered by, and in some cases almost entirely created by, human beings: the 'real' world as distinct from the 'perceived' world represented by the BEHAVIOURAL ENVIRONMENT. G.

phenomenalism in philosophy, the theory that phenomena (PHENOMENON-3) are the only things that can be known, that everything else is either nonexistent or inaccessible to the human mind. METAPHYSICS.

phenomenology 1. the observation and description of phenomena. PHENOMENON-1 2. in philosophy, a method of enquiry in which living experience is anlaysed in disregard of scientific knowledge in the search for absolute essences (the sum of the intrinsic properties of a thing or instances of a kind of thing which has made either of these what it is, which are unaffected by accidental modification, and without which the thing or the instances of a kind of thing would cease to be what it is) 3. in sociology, one of the adaptations of the philosophical approach outlined above applied to the investigation of the assumptions involved in everyday life. One such approach studies the way in which commonsense knowledge about society comes, through SOCIAL ACTION, to mould that society. Another approach confines itself to a generalized description of the feelings and thoughts of people in different societies concerning the world and their place in it; another incorporates an analysis of the unconscious routines by which people conduct their inter-personal contacts.

phenomenon pl. phenomena **1.** a fact or event that can be described and explained scientifically **2.** a person, fact, thing, or event that is extraordinary **3.** in philosophy, something known by (sense) perception, i.e. not by thought or intuition. PHENOMENALISM.

phenotype the observable characters of an organism that are the response of its inherited characters (genetic constitution) to the environment. GENOTYPE.

philosophy 1. very broadly indeed, the love, the pursuit of wisdom; the study of the nature of existence, of reality, knowledge, moral values etc. (A. L. Quinton's 'thought about thought'). EPISTEMOLOGY, ETHICS, METAPHYSICS **2.** with 'of', the study of the general principles of a particular subject, branch of knowledge, activity or experience.

phlogopite a form of MICA, yellowish in colour.

phonolite a grey or greenish finegrained PORPHYRITIC EXTRUSIVE IGNEOUS ROCK, consisting mainly of alkali FELDSPAR and NEPHELINE and giving a ringing sound when hit. TRACHYTE.

phosphate 1. any salt or ester of PHOSPHORIC acid **2.** any of the substances obtained naturally from the weathering of PHOSPHATE ROCK or from GUANO, or industrially by processing phosphate rock or basic slag from blast furances. Phosphates, an important source of PHOSPHORUS, are used as a soil FERTILIZER and in industrial processes. SUPERPHOSPHATE.

phosphate rock PHOSPHORITE-2, a sedimentary rock with a high content of APATITE, the main source of PHOSPHORUS for FERTILIZERS etc. (PHOSPHATE, SUPERPHOSPHATE). The rock is almost insoluble and has to be processed with sulphuric acid in order to obtain a more useful soluble substance.

phosphor bronze an exceptionally hard, strong BRONZE containing a small quantity of PHOSPHORUS, used in bearings etc.

phosphorescence the delayed emission of light from a substance which has been subjected to radiation and continues to give off light after the exciting radiation has ceased. BIOLUMINESCENCE, NOCTILUCA.

phosphoric *adj.* of or pertaining to PHOSPHORUS, especially to compounds in which phosphorus has its higher VALENCY. PHOSPHOROUS.

phosphorite 1. in general, all deposits with a high content of PHOSPHORUS **2.** specifically, PHOSPHATE ROCK.

phosphorous *adj.* of or pertaining to PHOSPHORUS, especially to compounds in which phosphorus has its lower VALENCY. PHOSPHORIC.

phosphorus a nonmetallic ELEMENT-6 existing in several structural forms, in nature only in combined state, chiefly in PHOSPHATE ROCK. White or yellowish waxy phosphorus is poisonous, unstable, inflammable, glowing at room temperature, but when white phosphorus is heated it forms a stable less inflammable non-poisonous red powder (red phosphorus). All living cells contain organic compounds of phosphorus (it is an essential MACRONUTRIENT), the inorganic compounds being important constituents of minerals, soil, bones, teeth. It is used in making FERTILIZER, detergents and matches.

phosphorus cycle the circulation of ATOMS of PHOSPHORUS brought about by natural processes, the major part being played by living organisms which contain organic compounds of phosphorus. On their death their tissues decompose, the phosphorus returns to the soil in a form suitable to sustain plants, and the plants provide food rich in phosphorus for animals. BIOGEOCHEMICAL CYCLE.

photic region, photic zone (Greek phōs,

light) the layer of water of a lake or of the ocean indicated by the penetration of light and the distribution of plants. It includes the DISPHOTIC ZONE and the EUPHOTIC ZONE. APHOTIC, APHOTIC ZONE, PELAGIC.

photobiology the branch of biology concerned with the influence of light on biological activity.

photochemical *adj.* **1.** of or relating to PHOTOCHEMISTRY **2.** of or relating to the action of light on chemical properties or reactions. PHOTOSYNTHESIS.

photochemical smog SMOG-2.

photochemistry the branch of chemistry concerned with the effects of electromagnetic radiation (especially the visible region of the ELECTROMAGNETIC SPECTRUM) on chemical reactivity.

photodynamic *adj.* of or pertaining to the toxic effect of sunlight on living organisms.

photogeology the study of vertical air photographs or of images provided by remote sensors on satellites (LANDSAT) as indicators of geological phenomena.

photogrammetry **1.** the technique of using photographs to obtain measurements of the subject of the photograph **2.** the science or art of making a topographical map from air photographs. The various adjustments needed in plotting involve the use of highly complex instruments.

photokinesis movement caused by reaction to light.

photolysis chemical decomposition of molecules caused by ELECTROMAGNETIC radiation.

photomap an aerial photograph or a PHOTOMOSAIC on which symbols, names, boundaries etc. have been added so that it may serve as a map.

photometeor any luminous phenomenon in the atmosphere caused by the reflection, refraction or diffraction of light (e.g. a BROCKENSPECTRE, FOGBOW, HALO, RAINBOW). METEOR-2.

photomosaic MOSAIC-1.

photoperiod the duration of daylight (when direct SOLAR RADIATION reaches the earth's surface). The relative length of dark and light periods affects the behaviour and growth of animals and plants. NASTIC MOVEMENT, PHOTOPERIODISM.

photoperiodism the response of plants or animals to the relative duration of hours of darkness and hours of light, e.g. the timing of flowering of plants, or the breeding season in some VERTEBRATES.

photorelief relief on a map indicated by shading which simulates the shadows appearing on a photograph of an illuminated relief model. The method formerly used involved making a three-dimensional relief model of a tract of country and illuminating it by a light (traditionally from the northwest) to cast shadows which, when the whole was photographed, realistically represented the distribution of hills and valleys. G.

photosynthesis the SYNTHESIS-1 by living cells of complex organic compounds from simple inorganic compounds, with the aid of light energy, the process by which the living cells of green ALGAE and higher green plants manufacture CARBOHYDRATES from CARBON DIOXIDE and water, using energy absorbed from sunlight by CHLOROPHYLL, and releasing OXYGEN (AUTOTROPHIC, PHOTOTROPHIC). The products of photosynthesis directly or indirectly supply all plants and animals with the energy they need for metabolism; thus, with a few minor exceptions, all other forms of life depend for their existence on the photosynthesis carried out by green plants. FOOD CHAIN, HETEROTROPHIC, PRIMARY PRODUCTION.

phototaxis, heliotaxis TAXIS in which light is the source of stimulus.

phototrophic *adj.* applied to an organism that obtains energy from sunlight.

AUTOTROPHIC, CHEMOTROPHIC, HETEROTROPHIC, PHOTOSYNTHESIS.

phototropism, heliotropism the response of an animal (e.g. an insect) or a plant to a source of light, e.g. the growth response of plants to the stimulus of light. TROPISM.

phreatic *adj.* of or pertaining to a well, applied to water obtained from a well. PHREATIC WATER.

phreatic eruption a violent volcanic eruption resulting from the underground contact of hot magma with GROUND WATER. The heat of the magma rapidly converts the ground water to steam, which erupts with great force with the lava at the ground surface.

phreaticolous *adj.* applied to an organism living in subterranean fresh water.

phreatic surface GROUND WATER table. PHREATIC WATER.

phreatic water GROUND WATER below the PHREATIC SURFACE, i.e. water in the zone of saturation. If several wells were sunk and the water levels in them linked by a flat or undulating surface, the ground water table, i.e. the phreatic surface, would be the result. G.

phreatophyte a plant with roots long enough to reach the WATER TABLE.

phrygana scrub vegetation in Greece, corresponding more or less to the GARIGUE of France. G.

phylogenetic, phyletic *adj.* in biology, of or pertaining to a relationship or classification based on closeness in line of descent, i.e. in evolution. PHENETIC, PHYLOGENY (PHYLOGENESIS).

phylogeny, phylogenesis the history of the evolution and relationships of a group of organisms or of species, as distinct from ONTOGENY. SYNPHYLO-GENY.

phylum the primary division in the CLASSIFICATION OF ORGANISMS, consisting of one CLASS or of a number of classes. The term is commonly used in the classification of animals instead

of DIVISION, which is more commonly applied to plants. G.

phyochorology the scientific study of the ecology and the geographic occurrence pattern of areal production. G.

physical determinism DETERMINISM.

physical geography one of the two main divisions of GEOGRAPHY (the other being HUMAN GEOGRAPHY), concerned with the study over time of the characters, processes and distribution of the 'natural' phenomena in the space accessible to human beings and their instruments, i.e. in the ATMOSPHERE-1, BIOSPHERE, HYDROSPHERE, LITHOSPHERE. Thus physical geography (if the definition is restricted to inanimate matter and energy) draws on CHEMISTRY, CLIMATOLOGY, GEOLOGY, GEOMORPHOLOGY, HYDROGRAPHY, HYDROLOGY, METEOROLOGY, OCEANOGRAHY, PEDOLOGY, PHYSICS (some would include ASTRONOMY), using information provided by land SURVEY and REMOTE SENSING and employing the techniques of MATHEMATICS, STATISTICS and CARTOGRAPHY. The concerns of BIOGEOGRAPHY, which draw on BIOLOGY, BOTANY, ECOLOGY, and ZOOLOGY, straddle physical and human geographical studies.

physical geology one of the two main divisions of GEOLOGY, sometimes subdivided into STRUCTURAL GEOLOGY (concerned with the structure of the earth, the materials of which it is made) and DYNAMIC GEOLOGY (concerned with the processes, causes and agencies by which it is modified). It overlaps GEOMORPHOLOGY to some extent. G.

physical landscape NATURAL LANDSCAPE.

physical sciences the DISCIPLINES concerned with inanimate matter or with energy, e.g. including chemistry, geology, physics, astronomy etc. BEHAVIOURAL SCIENCES, EARTH SCIENCES, HUMANITIES, NATURAL SCIENCES, SCIENCE, SOCIAL SCIENCES.

physical weathering mechanical weathering, the disintegration of rocks caused by the weather (e.g. frost and temperature change) without any chemical change taking place. ERO-SION, WEATHERING.

physics 1. the scientific study concerned with matter and energy and their interactions **2.** physical processes or properties. BIOPHYSICS.

physiocrat a believer in the economic theory postulated by Francois Quesnay, 1694-1744, that land, being the source of all wealth, should alone be taxed, that economic law is immutable, and that there should be complete freedom of trade, a theory that influenced Adam Smith. LAISSEZ-FAIRE.

physiognomic, physiognomical *adj.* of or relating to the general appearance of a landscape, of vegetation, of facial features etc.

physiognomic classification a classification of vegetation based on the structural characters of vegetation rather than on its floristic composition. RAUNKIAER'S LIFE FORMS. G.

physiographic pictorial map a map on which relief is shown by a series of conventional pictorial symbols portraying oblique aerial views of the physical features.

physiography a term, becoming obsolete, variously used **1.** originally, a description of nature or the science of natural objects **2.** a synonym for PHYSICAL GEOGRAPHY **3.** the study of the structural features of the earth's surface (now superseded by the term GEOMORPHOLOGY) **4.** the integration of GEOMORPHOLOGY, PHYTOGEO-GRAPHY and PEDOLOGY. G.

physiology 1. the branch of BIOLOGY dealing with the functions rather than the structure and constitution of living organisms **2.** the functions, collectively, of a living organism or its parts. NUTRITION-2.

physique as applied to a region, the relief, physical structure and organiza-

tion of a region (by analogy with the physique of a human body). G.

phytal zone of a lake, the part of the lake water which is shallow enough to allow the growth of rooted green plants. G.

phytobenthos the plants that live on the ocean or lake floor. BENTHOS, GEOBENTHOS, ZOOBENTHOS.

phytoclimate those aspects of climate that affect the life of plants.

phytocoenosis an assemblage of plants living in a specific area.

phytogeocoenosis a plant community and its physical environment.

phytogenic dune a DUNE covered and fixed by covering vegetation.

phytogeography plant geography, the branch of BIOGEOGRAPHY concerned with the scientific study of the distribution of plants, in space and over time, and of the environmental elements influencing this distribution. It is akin to plant ecology, but broader in scope, drawing on chemistry, climatology, genetics, geology, history, palaeontology, physics, taxonomy. PHYTO-SOCIOLOGY. G.

phytopathology 1. the scientific study of plant diseases **2.** the study of diseases caused by vegetable organisms, e.g. by fungi (FUNGUS).

phytophagous *adj.* applied to an animal which feeds on plants. HERBIVORE, MACROPHAGOUS, MICROPHAGOUS, MONOPHAGOUS, OLIGOPHAGOUS, POLYPHAGOUS, XYLOPHAGOUS, ZOO-PHAGOUS.

phytoplankton minute plants, many microscopic, mostly algae (e.g. DIA-TOMS), which float in bodies of fresh or salt water, which are basic in the FOOD CHAIN, and on which nearly all animal life in the open sea depends. MACRO-PHYTE, PLANKTON, ZOOPLANKTON. G.

phytosociology plant sociology, the scientific study concerned with plants as social units and with the phenomena that affect their life-cycles, including the origin, formation, composition and structure of plant communities, their development and change, the relationships among species and between species and their environment, and the classification of communities.

phytotoxic *adj.* poisonous to green plants.

phytotron a building in which plants are cultivated under completely controlled environmental conditions.

piassava, piasava, piassaba, piasaba terms applied both to the plants and to the hard brown fibres obtained from the outside of the stems of the South American palms *Attalea funefera* and *Leopoldinia piassaba* and from the west African *Raphia vinifera*. A softer fibre is obtained from *Vonitra fibrosa*, the only species of the genus, native to Madagascar. All these fibres are used in brushes and brooms. VICUA. C.

piccottah, peccottah (Indian subcontinent: Tamil) a SHADUF in southern India. G.

pidgin a spoken language that incorporates the vocabulary of two or more languages and an extremely modified grammar of one, used especially in trade, e.g. in southeast Asia.

pidgin English the form of PIDGIN used especially in southeast Asia, originally between English- and Chinese-speaking peoples, now the LINGUA FRANCA used in northern Australia and the Pacific islands.

piecework work paid for by the quantity of the output, not by the time taken to produce it.

pie diagram, pie graph DIVIDED CIRCLE DIAGRAM.

piedmont at the foot of a mountain, term used as a noun and applied to the gentle sloping ground lying between the steep slope of a mountain and the plain below, including the PEDIMENT and the BAJADA; or, very widely, as an *adj.*, e.g. piedmont alluvial plain, piedmont depressions etc. G.

piedmont glacier an extensive sheet of ice covering low ground at the foot of a mountain range, formed by the merging of a number of parallel VALLEY GLACIERS, e.g. the Malaspina glacier in Alaska. EXPANDED-FOOT GLACIER. G.

Piedmonttreppen (German) the term applied originally by W. Penck to a succession of benches, like steps, which occur around the flanks of the Black Forest and other mountains. G.

piezometric level, piezometric surface the level to which water from a confined acquifer will rise under its own pressure in a borehole, the underground level to which water will rise under its full head in an acquifer. G.

pig 1. a member of Suidae, genus *Sus*, an omniverous, cloven-hoofed, short legged, stocky, quickly maturing MAMMAL, usually with thick, bristled skin, widely reared in all (except Muslim) countries, the numbers fluctuating from year to year. The CARCASS is used mainly for meat, eaten fresh (pork) or cured (salted and smoked, known as bacon or ham) and for fat (rendered into lard), used in cooking. The bristles are used for brushes, the skin (pigskin) for clothing and leather goods **2.** (American) a young, as distinct from an adult, member of the family Suidae. BATTERY FARMING, BOAR, GILT, HOG, PORCINE, SOW, SWINE.

pigeon pea *Cajanus cajan*, a perennial, deep-rooted, drought resistant, frost-sensitive, shrubby leguminous plant (LEGUMINOSAE), possibly native to Africa, now widely cultivated in tropical areas for its seeds, which are known as red gram in the Indian subcontinent where, split and cooked, they form the protein-rich dish, dhal.

piggery a PIG farm or a PIGSTY.

pig iron cast iron, impure iron with high combined carbon content, sometimes also with sulphur and phosphorus, the product of a BLAST FURNACE. The molten metal from the blast furnace is run into channels or lateral moulds known as pigs, hence the name (in the large modern integrated steel works the molten metal is fed directly to the steel converter). The pig iron (cast iron) is very brittle by reason of its high content of carbon and phosphorus. These impurities are removed by puddling, a process in which the cast iron is re-melted and stirred when molten, so that the carbon escapes, to be burnt in the heated air of the furnace, the product being termed wrought or malleable iron. Wrought iron is tolerably hard, but not flexible, elastic or hard enough to be used in making machinery, construction equipment, cutlery and so on: for those purposes it must be converted into STEEL. C.

pigment 1. the colouring matter in the cells and tissues of plants and animals 2. colouring substances used in paints etc.

pigsty an enclosure with an uncovered area in front of a low building in which pigs are housed.

pike a mountain PEAK-3,4, especially in the Lake District of northern England.

pillow lava LAVA which has consolidated so as to resemble a jumbled mass of pillows (globular or cylindrical in shape) due to the fact that MAGMA (especially BASALTIC magma) was extruded under water on a sea floor, or flowed into the water before consolidation, so that the water cooled the outer skin quickly and the lava formed a rounded mass which partly collapsed. Some pillow lavas have chemical and mineralogicial characters which distinguish them from closely allied BASALTS and make them typical members of the SPILITIC SUITE, in contrast to the ATLANTIC and PACIFIC SUITES (PETROGRAPHICAL PROVINCE). AA, PAHOEHOE. G.

pilocarpine an ALKALOID obtained from JABORANDI, used medicinally, especially to contract the pupil of the eye.

pilot *adj.* 1. serving as a guide 2. serving as an experimental model for others to follow, e.g. in a pilot survey or study research work may be conducted on a small scale first in order to test the feasibility of a projected survey or study. SURVEY ANALYSIS.

pimenta, pimento *Pimenta dioica*, ALLSPICE.

pimiento PEPPER.

pin (Irish dialect, from Gaelic beann, a peak) a mountain PEAK-3, a term rarely used. BEN. G.

pinchbeck an alloy of copper and zinc, imitating gold, used especially in inexpensive costume jewellery.

pinching-out the thinning out of a rock STRATUM in a horizontal plane, leading ultimately to its disappearance. FEATHER-EDGE.

pine 1. a member of *Pinus*, family Pinaceae, a genus of CONIFEROUS trees with many species, the majority being native to the northern TEMPERATE ZONE, cultivated for their easily worked TIMBER (LUMBER) and for their resinous products. CONIFEROUS FOREST, PINE KERNEL, RESIN-1 2. the wood of such a tree.

pineapple *Ananas comosus*, a PERENNIAL tropical plant, native to South America, now widely cultivated on a commercial, plantation basis on light sandy soils, preferably near the sea, in tropical lands (especially in Hawaii); or under glass. Its commercially valuable, firm, sweet juicy yellow fruit, composed of the many fruits of individual flowers, takes two seasons to mature. It is eaten fresh, the poorer fruit being crushed for juice; or it is canned, having been peeled and cut into various shapes.

pine kernel the edible seed of various PINES, the best known internationally being those of the stone pine, *Pinus pinea*, a Mediterranean tree. The seed is eaten raw or roasted and salted, or used in confectionery, savoury dishes etc.

pin fallow bastard fallow (GREEN FALLOW), land left FALLOW not throughout the year but usually during the period between an autumn crop and the spring sowing of a different crop. G.

pingo (Eskimo, a conical hill) an isolated more or less conical mound of unconsolidated gravel or earth, varying greatly in height between 6 and 90 m (20 to 300 ft), generally with a core of clear ice, occurring in TUNDRA lands in Alaska, Arctic Canada, Greenland, Siberia when water under the surface freezes in autumn. The ensuing ice core expands and raises the surface to form a dome. The top of the pingo may later collapse, forming a crater which may become partially filled with water. Some authors distinguish a closed-system pingo (formed from the freezing of an isolated body of water) from an open system pingo (formed from the freezing of ground water at its pressure head). The suggestion that the larger pingos should be termed cryolaccoliths does not seem to have been favourably received. HYDROLAC-COLITH, PALSA.

pinnate *adj.* resembling a feather.

pinnate drainage a distinctive DENDRI-TIC pattern of DRAINAGE-2 that resembles a feather in that the tributaries to the main stream are closely spaced and meet the main stream at acute angles. It is likely to occur if the slopes on which the tributaries develop are unusually steep.

pint any of the various liquid or dry units of capacity equal to one-eighth of a GALLON, especially in British measures a unit equal to 0.568 LITRE; in American 0.473 litre or (dry measure) 0.550 litre. CONVERSION TABLES.

Pinyin TRANSLITERATION OF GEO-GRAPHICAL NAMES.

pioneer boundary ANTECEDENT BOUND-ARY.

pioneer fringe in a 'new country', the undeveloped land bordering the settled area, into which those who are open-ing-up and developing the country (the pioneers) have not yet moved.

pioneer stage the first stage in a SUCCESSION-2.

pipe 1. in a VOLCANO, the vent opening into the CRATER 2. in chalk country, a vertical joint made wider by CARBONATION-SOLUTION and filled with sand and gravel 3. a mass of mineral ore or diamond-rich rock, e.g. KIMBERLITE, shaped lik a column, in some cases formed by FLUIDIZATION 4. a tunnel in the soil, commonly in the interface between horizons (SOIL HORI-ZON), resulting from THROUGHFLOW.

pipeline a long stretch of linked pipes with pumps at intervals, used to carry gases, liquids, or solid material in the form of SLURRY-2.

piprake (Swedish) the layer or layers of perpendicular ICE NEEDLES, varying in length from one or two mm to nearly 0.5 m, formed in PERIGLACIAL conditions on the surface of the ground, i.e. the limit surface between the soil and free atmosphere, or just below the surface, where layers of needles may be separated by a thin MINERAL SOIL layer. They contribute to FROST HEAV-ING and the making of PATTERNED GROUND and NUBBINS-3. G.

piracy of streams RIVER CAPTURE.

piscary the COMMON RIGHT of fishing with others (and sometimes with the owner of the soil also) in water owned by another. It cannot exist in the sea or, generally, in a tidal navigable river.

pisciculture the breeding and rearing of fish for commercial purposes. AQUACULTURE, FARM, FARMER, FISH FARM.

pisolite, pisolith pea stone, a type of LIMESTONE made up of rounded bodies the shape and size of a pea, diameter 2 to 10 mm (0.08 to 0.40 in), i.e. larger than the OOLITHS of an OOLITE. Jurassic pea grit is typical; and other rocks may show pisolitic structure, e.g. some LATERITES and IRON ores. G.

pistachio *Pistacia vera*, a small DE-CIDUOUS tree native to western Asia, bearing ovoid DRUPES, the green-coloured kernel of which (the NUT) is eaten fresh, salted, used in confection-ery etc. for flavouring and decoration.

piste (French, from Italian pista) a track beaten by animals, e.g. as applied to old established routeways across the Sahara, which are now termed pistes automobiles. G.

pit 1. a surface working, a deep hole in the ground from which minerals or other materials are extracted, e.g. clay, gravel 2. a coal mine, including the associated PITHEAD buildings and the shaft. MINE, MINING 3. the stone of the fruit of a cherry, plum etc.

pitch 1. any of the several dark, sticky, resinous substances, solid at low tem-peratures, plastic at high, which is a residue formed in the distillation of TARS, turpentine or fatt oils etc., but especially the naturally occurring nearly black substance termed AS-PHALT. When an oil-sand has been exposed by natural erosion the oil gradually evaporates, leaving such nat-ural pitch as a residue (as in the pitch lake of Trinidad). Pitch is used for road surfacing, proofing wood and fabric and in the manufacture of soaps and plastics. BITUMEN 2. a resin derived from certain CONIFEROUS trees, e.g. PITCH PINE 3. pitch of a fold, the direction in which the AXIS OF A FOLD dips. G.

pitchblende a mineral, a massive form of uraninite, a complex oxide of ura-nium with small quantities of other elements (HELIUM, LEAD, THORIUM) from which URANIUM and RADIUM are extracted.

pitch pine any of the resinous pines, especially *Pinus rigida* of North America, which provide PITCH-2 and timber.

pitchstone a solidified mass of acid LAVA, the surface being predominantly glassy, but not so lustrous as OBSIDIAN. The waxy appearance is caused by the high content of water in the extruded mass.

pitfall (northern England) a subsidence hollow occupied by a pond, caused by underground mining. G.

pithead 1. the top of a mine shaft 2. the buildings and machinery on the sur-face, at ground level, associated with the working of a MINE.

piton (French, in mountaineering, a metal spike driven into rock for sup-port, a step, or hold for a rope) a peak with a sharp point in tropical or subtropical KARST, resulting from the dissection of limestone. One of the features of KEGELKARST. G.

place 1. a particular part of space, an area or volume of SPACE-2, un-occupied or occupied, e.g. by a person, object or organism 2. a particular area in space, e.g. town, village, district etc., real or as perceived (MENTAL MAP), where people and environment interact over time to give it characteristics distinct from those of surrounding places 3. POSITION in a hierarchy, scale, orderly arrangement, or in space. LOCATION, SITUATION.

placenta 1. in botany, part of the ovary wall bearing ovules 2. in zoology, a highly organized vascular organ in most MAMMALS (the Placentalia) which unites the developed embryo (the foetus) with maternal tissue (the wall of the uterus) and through which there is a metabolic exchange, the foetus being able to receive nourishment and oxygen from the maternal system and to get rid of waste products.

placer (Spanish, sandbank) a mining term for an alluvial deposit of sand or gravel with particles of valuable miner-als, especially of GOLD, weathered from rocks or veins and washed down by a stream. STREAM TIN. G.

place utility 1. the measure of approval accorded by an individual to a location in his/her ACTION SPACE. Dissatisfac-tion with place utility, expressed by a low measure, may result in MIGRATION-2 **2.** the VALUE-1,3 of a PLACE-1,2 brought about by efficient TRANSPORT facilities.

pladdy (Northern Ireland) a residual island DRUMLIN awash at high tide. G.

plage (French, a beach, a shore) a beach with special facilities provided for the enjoyment of visitors, a term now commonly superseded by LIDO. G.

plagioclase feldspar (Greek plagios, oblique; klasis, cleavage) a FELDSPAR with an oblique cleavage, containing SODIUM and CALCIUM.

plagioclimax (Greek plagios, oblique) the stage reached by a plant community that is stable and in equilibrium with the existing conditions of its environment but which has been deflected or drawn back by the influence of BIOTIC FACTORS from what would have been its natural CLIMAX. See DISCLIMAX, POSTCLIMAX, PRECLIMAX, PROCLIMAX, SERCLIMAX, SUBCLIMAX.

plagiosere a plant SUCCESSION-2 which has been deflected by BIOTIC FACTORS from its normal course of development, resulting in the formation of a PLAGIOCLIMAX rather than a CLIMAX.

plagiotropism the differing response of groups of cells in the organs or members (e.g. roots, lateral branches) of a plant to the stimuli of gravity and other external forces, causing the longer axes of the organs or members to form an oblique angle or a right angle with the vertical line of gravitational force, DIAGEOTROPISM, GEOTROPISM, TROPISM

plain a term with wide variations in use, especially in compound terms, but generally applied to an unbroken flat or gently rolling land surface, lacking prominent elevations and depressions, especially one of low elevation. Some authors restrict the application to such a feature with a horizontal structure. HIGH PLAINS.

plan **1.** a map of a small area on a large scale on which everything is drawn precisely to scale **2.** a large scale detailed chart **3.** a drawing to scale representing a horizontal section of a solid object **4.** a formulated scheme of action, or the way in which it is proposed or intended to carry out some proceeding or course of action to achieve some goal, e.g. in connexion with economic and social development. PHYSICAL PLANNING, PLANNING, PLANNING REGION, REGIONAL PLANNING. G.

planalto (Brazil: Portuguese) a high plain or plateau, especially the great Brazilian plateau. G.

planar network a NETWORK-2 which can be represented by a GRAPH-2 drawn in a single plane, with links intersecting at vertices (but in which some non-planar features may also appear). NON-PLANAR NETWORK.

planation the making of a plane or flattened surface, the DENUDATION of rocks so as to produce a level surface (the PLANATION SURFACE). PEDIPLANATION, PANPLANATION, PENEPLANATION. G.

planation surface a flat surface produced by PLANATION, a term now used in preference to EROSION SURFACE. G.

planet **1.** a heavenly body that is not a meteor, comet, or artificial satellite, which revolves around a STAR by which it is illuminated **2.** specifically, such a body revolving about and illuminated by the sun in the earth's SOLAR SYSTEM. ASTEROID, EARTH, PLANETISIMAL HYPOTHESIS.

plane table, plane-table a small drawing board mounted on a tripod and used in the field with an ALIDADE in topographical survey.

planetarium **1.** a device for displaying the positions and relative movements of stars and planets by projecting spots of light on to the inner surface of a dome, representing the sky, within which the onlooker sits **2.** the building housing this device.

planetary *adj.* of or pertaining to a planet or to the planets, specifically to the earth as a planet.

planetary boundary layer FRICTION LAYER.

planetary desert an extensive desert area found in arid zones, as distinct from a TOPOGRAPHIC DESERT. G.

planetary winds the major latitudinal winds (TRADE WINDS, WESTERLIES etc.) in the ATMOSPHERIC CIRCULATION in contrast to local winds (LAND BREEZES, SEA BREEZES etc.). G.

planetisimal a small solid planetary body which is considered to be PLANETISIMAL.

planetisimal *adj.* of, pertaining to, or designating a minute solid planetary body thought by some to have been derived from the break-up of a primitive sun and another star as they came together, and to have existed in the earliest stage of the evolution of the SOLAR SYSTEM. PLANETISIMAL HYPOTHESIS.

planetisimal hypothesis the theory that PLANETS were formed by the aggregation of PLANETISIMALS in the evolution of the SOLAR SYSTEM. G.

planetoid ASTEROID.

planèze (French) a type of MESA, a term applied especially to one of the lava-capped plateaus in the Puy de Dome district of France where former CINDER CONES have been denuded, leaving only those parts which were protected by a lava flow. G.

planimeter an instrument for measuring areas on a plane figures, especially on a map. There are many types in use, some of great refinement, but they all incorporate a small tracer wheel or pointer linked in some way to a recording dial; and they are all operated by moving the wheel or pointer along the boundary of the area to be measured. G.

planimetry the measurement of the area of a plane figure, e.g. a selected area on a map. PLANIMETER.

planina (Slavic, pl. planine) a broad featureless limestone plateau of the KARST, the flattened crest of a ridge. G.

plankton the collective term for all forms of minute plant and animal life (especially of the many microscopic organisms freely floating in the ocean or in fresh water), in contrast to NEKTON and BENTHOS. It does not include the larger plants, e.g. floating SEAWEED. All plankton are of great ecological and economic importance, in that they are the essential food for fish, marine mammals, sea birds etc. (FOOD CHAIN). AEROPLANKTON, CRYOPLANKTON, DIATOM, NANOPLANKTON, PELAGIC ORGANISMS, PHYTOPLANKTON, PLEUSTON, POTAMOPLANKTON, PROTOZOA, SAPROBEL, ZOOPLANKTON.

planned economy an economic system in which the central government controls capital, labour, production, distribution of goods etc. to meet the needs of a comprehensive economic plan, instead of allowing free play of market forces (supply and demand) as in a LAISSEZ-FAIRE approach. CONTROLLED POLITICAL SYSTEM, MIXED ECONOMY.

planning 1. a method for outlining or defining goals and ways of achieving them **2.** the drawing-up and implementation of a PLAN-4. LAND USE PLANNING, PLANNING REGION, POSITIVE PLANNING, REGIONAL PLANNING.

planning region a specific unit area for which there is a PLAN-4 for economic and social development. LAND USE PLANNING, REGIONAL PLANNING.

planosol an INTRAZONAL SOIL with a sharply defined clay pan (HARD PAN) compacted between the surface and the bed rock, commonly formed under forest or grassland vegetation in upland areas in humid to semi-arid climate usually, but not always, with a fluctuating water table. The shallow A HORIZON may be leached and the clay pan nearly saturated (GLEY). In SOIL CLAS-

SIFICATION, USA, the planosols are divided into ALFISOLS, MOLLISOLS, ULTISOLS. G.

plant 1. any member of the kingdom Plantae (CLASSIFICATION OF ORGANISMS), the typical plant being a living organism lacking means of independent locomotion, having no central nervous system, the cell walls consisting of CELLULOSE, nutrition depending on simple gaseous or liquid substances (principally carbon dioxide and water) which, with the aid of CHLOROPHYLL functioning in light, the organism builds up into sugars and other complex materials (FOOD CHAIN, PHOTOSYNTHESIS); but many plants, e.g. FUNGUS, do not exhibit those characteristics and some microscopic organisms, e.g. some BACTERIA, seem to be as much ANIMAL-1 as plant. ECOLOGY, PARASITE, RAUNKIAER'S LIFE FORMS, SERE, SUCCESSION-2, TAXIS, TROPISM, VEGETATION 2. such an organism specified as being smaller than a SHRUB, bush or TREE, but large enough to be handled 3. in industry, the buildings, machinery and other equipment used in the manufacture of goods or the production of power.

plantain a species of BANANA with higher starch, lower sugar content than that of the dessert banana, cooked or made into banana flour in banana-growing lands. Specific varieties are grown for steaming or roasting. They are the staple food crop in some tropical areas, especially in parts of Uganda and Tanzania, where special varieties are also grown for beer making. The term is applied less specifically in importing countries to some varieties of dessert banana.

plantation 1. something which is planted, hence applied particularly to a group of growing trees which has been planted to provide timber and wood pulp, and/or to function as part of a land reclamation scheme 2. a farm or estate, especially one in the tropics,

with large scale planting of CASH CROPS, especially on a monocultural basis (MONOCULTURE) under scientific management, often with a large labour force, sometimes with facilities and plant to process the crop, e.g. as on tea or rubber plantations. The term plantation agriculture or plantation system implies rather more: the use of a large labour force (in many cases formerly of slave labour) to produce one or two commercial crops on a large scale, in contrast to PEASANT FARMING 3. historically, a colony or settlement in a new or conquered country. G.

plantation agriculture PLANTATION-2.

planter 1. one who manages a PLANTATION-2 2. historically, a colonist. PLANTATION-3.

plant formation FORMATION-1.

plant geography PHYTOGEOGRAPHY.

plantigrade *adj.* applied to an animal-that walks on all the lower surface of the foot, i.e. of metacarpus or metatarsus and of the digits, e.g. a human being. DIGITIGRADE, ORTHOGRADE, UNGULIGRADE.

plant indicator INDICATOR PLANT.

plant sociology PHYTOSOCIOLOGY.

plant succession SUCCESSION-2.

plastic, plastic material non-cellulosic material, made from the by-products of oil refining (HYDROCARBON CRACKING), coal distillation etc. which was PLASTIC under heat or pressure or both at some stage in its manufacture, but which finally is firm and stable in everyday use. If the final product softens when heated it is termed thermoplastic; if it resists such heat it is termed thermoset.

plastic *adj.* of or pertaining to any substance which changes shape under pressure and keeps the new shape when the pressure is eased. ELASTIC.

plastic deformation of ice the condition when the ice in the lower part of a thick GLACIER or ICE SHEET is subjected to so much pressure that inter-molecular movement occurs in it, it becomes

PLASTIC, and eases the flow of the glacier or ice sheet. EXTRUSION FLOW OF ICE.

plate a geotectonic plate or lithospheric plate, PLATE TECTONICS.

plateau (French) a markedly elevated tract of comparatively flat or level land, a tableland, usually bounded on one or more sides by steep slopes which drop to lower land, or by steep slopes rising to a mountain ridge. Some large plateaus consist of LAVAS, especially of BASALTS, poured out through fissures, hence the term plateau lava (PLATEAU BASALT). Originally French, with plural plateaux, the term is regarded as having been absorbed into the English language, the plural becoming plateaus. INTERMONTANE, PLATEAU GRAVEL, WATERFALL. G.

plateau basalt BASIC LAVA, very fluid at high temperatures, that reaches the surface through fissures in the earth's crust (FISSURE ERUPTION) and spreads evenly over an extensive area. It builds up a lava PLATEAU, especially if there are successive eruptions. FLOOD BASALT. G.

plateau gravel gravel (small stones, sand, grit) occurring either as an extensive sheet on the surface of a PLATEAU or as a capping on hills which represent the dissected remains of former plateaus. Such gravel deposits provide important evidence in the study of the chronology of DENUDATION and of geomorphological history. G.

plateau lava PLATEAU BASALT.

plate tectonics the theory that the EARTH'S CRUST (the LITHOSPHERE) consists of several large and some small, rigid, irregularly-shaped plates (lithospheric geotectonic plates) which carry the continents (continental crust) and the ocean floor (oceanic crust) and float on the ASTHENOSPHERE, moving very slowly, the movement probably resulting from currents in the asthenosphere. The theory is supported by evidence from studies in PALAEOMAG-

NETISM. Boundaries between plates are termed constructive (OCEANIC RIDGE), destructive (COLLISION ZONE, OCEANIC TRENCH, SUBDUCTION ZONE), conservative (TRANSFORM FAULT), EARTHQUAKES being associated with the destructive and conservative. BENIOFF ZONE, PLUME.

platform a general term for a natural or artificial flat, plain, or bench-like surface, widely used with varying meanings, but **1.** in geology and geomorphology, a wave-cut shelf, or any bevelled surface, sometimes applied to a more extensive rigid peneplaned block underlying later sediments, e.g. the Russian platform (some authors prefer the term Russian table) **2.** in REMOTE SENSING, the gyroscopically stabilized mounting which provides a frame for a REMOTE SENSOR carried aboard an aircraft or SATELLITE-1 etc. **3.** in oil and gas industry, a structure that stands on the bed of the sea, lake or estuary and makes a stable base above the water level for the workers and equipment involved in the drilling or regulation of offshore oil or gas wells.

platinoid a member of the PLATINUM group of metals (IRIDIUM, OSMIUM, PALLADIUM, PLATINUM, RHODIUM, RUTHENIUM) occurring native with NICKEL and sulphides.

platinum a silvery-white metallic ELEMENT-6, heavy, ductile, malleable, highly resistant to acids, corrosion, and heat, a good conducor of electricity, used in the manufacture of chemical and electrical apparatus, jewellery and, in the form of black powder (platinum black), as a CATALYST especially in HYDROGENATION or OXYGENATION, e.g. in PETROLEUM refining.

platinum metal any of the six metallic elements listed under PLATINOID (of which platinum is one) resembling or occurring with PLATINUM.

platypus *Ornithorhynchus anatinus*, commonly known as duck-billed platypus, a small Australian aquatic, furry, primitive MAMMAL that has a flat leathery bill in the shape of a duck's bill, webbed feet, a long flat tail, lays eggs, and suckles its young.

plav a reed swamp in the Danube delta. G.

playa (Spanish, shore) **1.** a flat basin in an arid land which may become covered with water periodically, so that a shallow, usually salt, lake forms amidst flats of saline and alkaline mud. PAN **2.** the fluctuating lake so formed. G.

Playfair's Law the law postulated by John Playfair, 1748-1819 (Scottish geologist, mathematician and divine who became successively professor of mathematics and natural philosophy at Edinburgh, and extended the basic geological ideas of James Hutton), that in areas of uniform bed rock and structure which have been subjected to long-continued river erosion, valleys are proportional to the size of the streams they contain, and the level of stream junctions in these valleys is accordant. ACCORDANT JUNCTION.

pleck, plek (England, Midlands) **1.** waste or common **2.** a small individual holding or allotment of indifferent land, usually supporting pasture or meadow. G.

pleion an area with a high POSITIVE ANOMALY in relation to a climatic element, especially in relation to temperature. ANTIPLEION, ISANOMAL, THERMOPLEION.

Pleistocene (Greek, most recent) *adj.* with several applications **1.** the most common and now generally accepted, of or pertaining to the earlier epoch (time) or series (rocks) of the QUATERNARY period or system. GEOLOGICAL TIMESCALE **2.** of the last 1 to 2 mn years (or of the last 600 000 years), including the HOLOCENE, i.e. coinciding with the appearance of human beings on the earth **3.** of the last 1 to 2 mn years (or the last 600 000 years) excluding the

Holocene, i.e. characterized by the formation of glaciers, coinciding with the most recent GREAT ICE AGE.

pleuston hemi-plankton, a plant community consisting of microscopic plants (NEUSTON) floating freely within a body of water as well as on its surface.

Plimsoll line the diagram on the hull of a cargo vessel indicating the safety level to which the ship may be submerged by the weight of the cargo in accordance with the British Merchant Shipping Act of 1876 (promoted by Samuel Plimsoll, 1824-98). The level varies according to season and the salinity of the water.

Plinian eruption a PAROXYSMAL VOLCANIC ERUPTION, the name derived from Pliny the Younger (c.62 to c.114 AD), a Roman author and administrator, who saw and recorded such an eruption of Vesuvius in 79 AD. VULCANIAN ERUPTION.

plinth the lower and outer portion of a SAND DUNE, beyond the SLIP-FACE boundaries, which has never been subjected to sand AVALANCHES. G.

plinthite DURICRUST, HARD PAN.

Pliocene *adj.* of or pertaining to the most recent epoch (time) or series (rocks) of the TERTIARY period or system. GEOLOGICAL TIMESCALE.

plottable error the smallest distance on the ground that can be shown to scale on a MAP, governed by the finest line it is possible to draw, i.e. some 0.25 mm.

plucking **1.** one of the main erosion processes carried out by a GLACIER, effecting the removal of rocks from its valley floor. Water enters cracks in the rocks of the floor, freezes and detaches rock fragments, which become frozen to, and carried away by, the under surface of the glacier as it moves along. COLLOIDAL PLUCKING, SAPPING **2.** applied by some authors to one of the erosion processes carried out by a stream. QUARRYING. G.

plug a roughly cylindrical mass of volcanic rock marking the neck (VOL-

CANIC NECK) of an ancient volcano, sometimes left isolated, exposed by denudation of the rest of the CONE.

plum one of the many species of the genus *Prunus*, family Rosaceae, a small DECIDUOUS tree native to and cultivated in temperate regions for its succulent fruit (a DRUPE), varying in colour and sweetness. It is eaten raw, cooked, made into preserves. The species includes the sloe, bullace, damson, gages, cherry plum.

plumbago GRAPHITE.

plume 1. in PLATE TECTONICS, the upward movement of magma in a convection current caused by a locally very hot area lying between the CORE and the MANTLE of the earth which melts the rocks of the mantle and results in a localized swelling on the earth's surface, followed by a cracking open of rocks. There are certain localities (termed hot spots) on the earth's surface where this volcanic activity is relatively frequent, e.g. Hawaii 2. a stream of effluent consisting of gases or gases and particulates emitted by a chimney, or a larger ribbon-like cloud of similarly polluted air produced by an industrial complex, city etc., drifting downwind from its source, the form varying, shaped by turbulence in the atmosphere. POLLUTION.

plum rains BAI-U, MAI-YU.

plunge pool the deep pool at the base of a WATERFALL into which the water plunges, cut out by the whirling round of boulders and stones to form a large POTHOLE-2. If the stream later deserts its course the plunge pool remains as a nearly circular lake, cut in the bed rock and often very deep. G.

plural *adj.* of or including more than one.

pluralism 1. the institutional arrangements for the distribution of political power 2. an approach to the study of political systems which argues that power should be diffused in a political system, that the system should be open

and accessible to all its members, and that it is possible by democratic processes to reach compromise agreements when disagreement and conflicts arise in society 3. the existence or the toleration of different attitudes and beliefs within a group or an institution etc., or of these as well as different ethnic or cultural groups within a society. DUALISM.

pluralist analysis of political systems, the analysis of the institutional arrangements for the distribution of power in a political system in which there is no dominant ideological, political, cultural, or ethnic group. PLURALISM.

pluralistic integration INTEGRATION in society in which distinct groups coexist within a common framework of political and legal rights.

plural society a society within which there are two or more elements, social orders or cultural groups which, in many areas of social behaviour, do not mix because each group wishes to retain its identity, its beliefs, traditions etc. All societies except the very simplest are pluralist to some extent, containing as they do groups based on CLASS-3 in addition to identifiable local and regional communities. But if in a nation state the demand of a group to retain its separate identity becomes very strong a delicately poised unity may break down under the strain of warring factions, especially if that group demands separate, independent statehood. Such a demand for separate territorial and political sovereignty is termed SEPARATISM. See ACCULTURATION, CULTURE CONTACT, DUALISM, PLURALISM, PLURALISTIC INTEGRATION.

plutocracy 1. plutarchy, rule by the rich 2. a state ruled by the rich. AUTOCRACY.

pluton a cylindrical mass of PLUTONIC ROCK, e.g. a granite BATHOLITH. G.

plutonic *adj.* of or pertaining to any

process associated with great heat deep in the earth's crust.

plutonic rock (Greek Pluto, god of the underworld) an INTRUSIVE IGNEOUS ROCK which, having cooled and solidified slowly at great depth in the earth's crust, is usually coarse in texture, with large CRYSTALS. ABYSSAL ROCK, DIORITE, EXTRUSION, EXTRUSIVE ROCK, GABBRO, GRANITE, INTRUSION, PERIDOTITE, VOLCANIC ROCK.

Plutonic theory, Plutonism an eighteenth century theory which maintained that most geologial phenomena were due to the action of internal heat (CATASTROPHISM, NEOCATASTROPHISM, UNIFORMITARIANISM). The term Plutonism was revived about 1940 and applied to all phenomena associated with deep-seated PLUTONIC or IGNEOUS ROCKS. G.

plutonium an artificial metallic element produced in nuclear reactors from natural URANIUM, used in nuclear weapons and as a nuclear fuel. An excess of plutonium is produced by fast breeder reactors in their generation of power.

pluvial, pluviose *adj.* of, pertaining to, due to, or characterized by, rain. Frequently used in connexion with postglacial climatic fluctuations (a pluvial period being a long period when rainfall was distinctly heavier than that in those relatively dry stages that preceded and followed it), the term came to be used as a noun, i.e. a pluvial period became a pluvial. INTERPLUVIAL PERIOD. G.

pluviometer an instrument for measuring rainfall, a rain gauge.

pluviometric coefficient the value reached by expressing the actual average rainfall of a particular month for a given place as a ratio of the hypothetical amount taken to be equal to each month's rainfall if the total rainfall were evenly distributed through the year. EQUIPLUVE.

ply 1. one of the sheets of wood that make up PLYWOOD 2. a length of fibre, twisted with others to form a rope, cord, sewing or knitting thread.

plywood a light construction material composed of thin layers of wood (PLY) glued or cemented together, the grain of each sheet lying at right angles to that of its neighbouring sheet or sheets.

p.m. post meridiem, the time period after noon (1200 hours) and before midnight (2400 hours). A.M.

pneumatolysis the process in which chemical changes are brought about in rocks by the action of heated gases and vapours from the earth's interior, usually in the later stages of a cycle of IGNEOUS activity, affecting both the IGNEOUS ROCK itself and the surrounding COUNTRY ROCK, and usually resulting in the formation of new minerals, including metalliferous ores. AUTOMETAMORPHISM, HYDROTHERMAL, METASOMATISM. G.

pneumatolytic *adj.* of or pertaining to a rock or ore formed by PNEUMATOLYSIS.

pocket beach BAY HEAD BEACH.

pocket-penitent NIEVE PENTITENTE and allied formations in the Himalaya. G.

pod a dry, dehiscent (DEHISCENCE) fruit with many seeds, applied particularly to the fruit of a LEGUME.

podu (Indian subcontinent: Telegu) SHIFTING CULTIVATION. G.

podzol, podsol (Russian podzol, ash) podzol is now the preferred spelling, a ZONAL SOIL developed under cool moist coniferous forest climatic conditions (in which precipitation is fairly low but the evaporation rate is also low) especially from sandy parent rock, e.g. in the heathlands of western Europe and the CONIFEROUS FOREST region of northern Europe and Canada. Downward leaching produces an acid A HORIZON with three layers, a top thin surface layer of black organic matter, below which lies a partly leached layer and under that a heavily leached, ash-coloured layer (hence the name) devoid of iron compounds and lime. The two layers of the B HORIZON consist of an upper layer of partly cemented leached material and a lower layer rich in iron oxide, clay and humus. The C HORIZON below is of little altered parent rock. Podzols are not generally good agricultural soils, but they can be upgraded by the application of lime and fertilizers. GREY-BROWN PODZOL, LEACHING, PODZOLIC SOILS, RED-YELLOW PODZOLIC SOIL, SOIL, SOIL ASSOCIATION, SOIL HORIZON, SPODOSOLS. G.

podzolic soils one of the seven main groups in the 1973 SOIL CLASSIFICATION of England and Wales, covering soils characterized by a dark B HORIZON rich in aluminium and/or iron oxide and humus. The overlying leached A HORIZON and the organic surface layer typical of a pozdol may be absent. They range from well drained to badly drained and cover true PODZOLS as well as brown podzolic soils (well drained loamy or sandy soils, lacking an organic or bleached A horizon, pH 5 to 6, developed in humid temperate climates, transitional between BROWN EARTHS and PODZOLS), GLEY podzols (poorly drained soils developed on land with intermittent waterlogging, typically MOTTLED in the gleyed grey horizon, with an overlying dark brown to black A horizon which may or may not have the bleached horizon and/or organic layer of the true podzol), and stagnopodzols (soils with the typical thin peaty surface layer of the true podzol, overlying an intermittently waterlogged gleyed bleached horizon, below which lies a thin iron pan layer, rich in iron oxide and/or a more crumbly layer). HARD PAN, LEACHING, WATERLOGGED. G.

podzolization 1. a general term applied to the process or processes by which soils are depleted of BASES-3, become acid, and come to have developed

eluvial A HORIZONS and illuvial B HORIZONS. ELUVIATION, ILLUVIATION 2. specifically, the process by which a PODZOL is developed, including the more rapid removal of IRON and ALUMINA than of SILICA from the surface horizons; but also applied to similar processes at work in the formation of certain other soils of humid regions. G.

poikilitic *adj.* applied in PETROGRAPHY to an IGNEOUS ROCK or to the texture or structure of an igneous rock in which larger crystals of a mineral enclose smaller crystals of another.

poikilosmotic (Greek poikilos, variegated, various) *adj.* applied to an animal the SOLUTE in the body fluids of which varies in accord with variations in the concentration of the solute in the fluid of its surrounding medium. An animal lacking this ability is termed homoeo-osmotic. OSMOSIS.

poikilothermic, poikilothermal, poikilothermous *adj.* applied to an animal the body temperature of which fluctuates in accord with the temperature of its surroundings, i.e. a cold-blooded animal, unlike a homoeothermic (homiothermic) animal, which is WARM-BLOODED.

point 1. a narrow piece of land projecting into the sea, e.g. Start Point, Devon, England 2. any of the 32 direction marks on the circumference of a COMPASS, or any measurable position in a scale 3. a specific place, a location, having a definite position in space but lacking definite size and shape.

point bar the deposit of sand or gravel developing on the inside of a MEANDER bend, growing by the ACCRETIONS which accompany the migration of the meander. It is sometimes termed a meander bar, meander scroll or scroll meander, but point bar is now usually preferred. G.

point biserial coefficient in statistics, a measure of the strength of the relationship between two variables when one is measured on an interval or ratio scale and the other is a simple dichotomy, according to the presence or absence of an attribute. CORRELATION-3.

point estimate in statistics, an estimate of a summary measure, e.g. the point assessed as giving the best estimate of the value of a POPULATION PARAMETER from some sample data, as distinct from an INTERVAL ESTIMATE.

point sampling in statistics, a method of SAMPLING a geographical area by selecting points in it, especially by selecting points at random on a map or aerial photographs.

Poisson distribution a theoretical discontinuous FREQUENCY DISTRIBUTION which is positively skewed and truncated (NORMAL DISTRIBUTION) and in which, if perfect, the MEAN and the VARIANCE are equal, i.e. the square root of the STANDARD DEVIATION is equal to the square root of the variance. The Poisson distribution is most likely to be produced by a random proces in which there is only a small probability of an event occurring, i.e. the probability of the event occurring is very small indeed in comparison with the probability that it will not occur. BINOMIAL DISTRIBUTION.

polaj (Indian subcontinent: Urdu-Hindi) historically, land cultivated annually. BANJAR, CHACHAR, PARANTI. G.

polar *adj.* of or pertaining to those parts of the earth, or of the celestial sphere, close to the POLES-4 **2.** of or pertaining to the POLE-3 of a MAGNET.

polar air mass an AIR MASS with cool temperatures, symbol P, originating in midlatitudes (40° to 60°) over the ocean (polar maritime air mass, Pm) or over the continental interior (polar continental, Pc). The term was introduced originally to stress the difference in temperature with a TROPICAL AIR MASS. It should not be confused with the ARCTIC AIR MASS, symbol A, or

the ANTARCTIC AIR MASS, symbol AA, air masses which originate near the poles.

polar desert soil TUNDRA SOIL.

polar distance the angular distance from the nearest POLE-4 of a point on a sphere. DECLINATION-1.

polar front the FRONT or frontal zone where the polar maritime and tropical maritime (POLAR AIR MASS, TROPICAL AIR MASS) meet, over the North Pacific and North Atlantic oceans. Disturbances along this front, which shifts over a broad zone, play a major part in determining north European weather. ARCTIC FRONT, ATLANTIC POLAR FRONT, POLAR FRONT JET STREAM. G.

polar front jet stream a JET STREAM formed along the POLAR FRONT.

polar glacier COLD GLACIER.

polar high a persistent area of high atmospheric pressure (ANTICYCLONE, HIGH) located over the POLAR region of ANTARCTICA.

polar ice heavy sea ice, up to 3 m (10 ft) or more in thickness, of more than one winter's growth. Much HUMMOCKED, it may ultimately be weathered to a generally level surface. WEATHERED ICE. G.

polarity **1.** the quality of having POLES **2.** the state of having or developing one or the other of two opposite POLAR conditions **3.** the tendency of bodies with MAGNETIC POLES to become aligned with the earth's MAGNETIC POLES **4.** the state of having or developing two opposite poles or opposite qualities.

polarization **1.** the act or process of gathering, concentrating about two opposite, conflicting or contrasting positions or points **2.** the state of being divided into two such opposing groups in such a way **3.** the accentuation of the difference between two things or groups **4.** the formation of something, e.g. public opinion, in a definite direction **5.** the act of giving something, or causing it to have, such a unity of direction **6.** the state of having such a unity of direction.

polar low a persistent area of low atmospheric pressure (LOW) located in the upper atmosphere over high LATITUDES.

polar outbreak the movement of a polar continental air mass (POLAR AIR MASS) from high latitudes and midlatitudes to lower latitudes, accompanied by cool or cold weather and strong winds (NORTHER). It occurs particularly where there is no obstructing east to west mountain barrier, e.g. in North America.

polar wind a very cold wind blowing from the north or south POLAR-1 region.

polder (Dutch) low-lying land, usually below sea-level, reclaimed from the sea, lake, or floodplain of a river by embanking (DIKES-1) and draining, often supported by continued pumping. The work of creating polders is termed empoldering, and occasionally the tract enclosed is termed an empolder. G.

pole **1.** one of two extremes **2.** a point of attraction **3.** one of the two or more points in a MAGNET where most MAGNETIC FLUX is concentrated **4.** either end of the axis of a sphere, e.g. the northern or southern extremity of the axis of rotation of the earth, i.e. the NORTH POLE, SOUTH POLE, the earth's geographical poles. CELESTIAL POLES, COLD POLE, MAGNETIC POLE, TRUE NORTH, TRUE SOUTH. G.

pole of inaccessibility the most difficult point to reach in ANTARCTICA. G.

Pole Star the bright star in the constellation Ursa Minor, seen in the ZENITH at the NORTH POLE. It is used to find TRUE NORTH from any point in the NORTHERN HEMISPHERE, the height of the Pole Star above the horizon seen from that given point being equal to the LATITUDE. SOUTHERN CROSS.

-polis (Greek) city, as in metropolis, sometimes used in the form of -opolis

to form names or nicknames of towns, e.g. Cottonopolis. G.

political *adj.* **1.** of, belonging to, pertaining to, or concerned with, public affairs and/or government (of a STATE-2,3 or a body of people) **2.** of, relating to, or concerned with politics, i.e. with the art or science of government.

political geography the branch of HUMAN GEOGRAPHY concerned with the boundaries, extent, divisions, territories, resources and internal and external relationships of politically organized areas, especially those of nation states. It is also concerned with the effects of political actions on social and economic conditions, and with the significance of geographical factors behind political situations, problems and activities. ELECTORAL GEOGRAPHY, GEOPACIFICS, GEOPOLITICS. G.

political science the study of the organization, functions and conduct of the state and of government. SOCIAL SCIENCES.

polje (Slavic, field or cultivated area) **1.** in KARST, a large or very large (e.g. with an area of some 400 sq km: 155 sq mi) closed depression in limestone, usually elliptical, with a flat floor either of bare limestone or covered by alluvium, sometimes marshy, sometimes with an intermittent lake, generally surrounded by steep limestone walls **2.** in Yugoslavia, any enclosed or nearly-enclosed valley. G.

pollard **1.** a tree cut back to the main trunk. POLLARDED-1 **2.** a hornless variety of goat, sheep, ox etc. **3.** a mixture of fine bran and coarse flour fed to livestock **4.** a coarse wheat bran.

pollarded *adj.* applied to **1.** a living tree cut back to the main trunk (POLLING-2) so that the new growth forms a thick head of many branches, e.g. a pollarded WILLOW, not to be confused with coppiced (COPPICE) **2.** an animal from which the horns have been removed.

pollen the very fine powder (microspores) carrying the male element of seed-producing plants, each grain of pollen (i.e. each microspore) containing a (much reduced) male gametophyte. The pollen grains are carried by wind, insects, birds or water to the female parts of the plant, where pollen tubes containing the male nuclei develop and eventually penetrate the embryo sac.

pollen analysis PALYNOLOGY.

pollen count a measure of the POLLEN in the atmosphere in a given volume of air during 24 hours.

polling 1. the act of removing or cutting short the horns of cattle 2. the act of removing the branches from the main trunk of a growing tree. POLLARD-1,2, POLLARDED-1,2 3. the registering or casting of votes.

pollution 1. the direct or indirect process by which any part of the ENIVIRONMENT is affected in such a way that it is made potentially or actually unhealthy, unsafe, impure or hazardous to the welfare of the organisms which live in it, i.e. the results are harmful (TERATOGENIC POLLUTION). The term is sometimes also applied, loosely, to such processes if they give rise to results which are merely objectionable 2. the state of being so harmfully affected. Pollution usually occurs as a result of the presence of too much of some substance, or the excessive occurrence of a process or action, in an inappropriate place at an unsuitable time, such as oil spillage, sewage outfall, or industrial effluent in a river, lake or sea, e.g. mercury in the sea (MINIMATA DISEASE); sulphates etc. in the atmosphere (ACID RAIN); soil nutrients causing EUTROPHICATION; industrial heat causing THERMAL POLLUTION; noise of flying aircraft in residential areas near airports at night; OBNOXIOUS fumes from an industrial process in a residential area (PLUME-2). Pollution is an example of negative externality (EXTERNALITY). SAPROBIC CLASSIFICATION.

polster a small, spheroidal, silt-packed moss cushion, resembling a GLACIER MOUSE, associated with some glaciers, e.g. Matanuska Glacier, Alaska. G.

polyandry in zoology, an uncommon animal breeding system in which one female mates with several males. POLYGAMY, POLYGYNY.

polycentric hierarchy a many-centred arrangement of command in administration, e.g. in a large, diversified, multi-locational industry in which there are a number of semi-automonous divisional headquarters each with its own hierarchical structure. UNITARY HIERARCHY.

polyclimax many climaxes, vegetation exhibiting several CLIMAXES in a climatic region, probably due to the fact that the influence of EDAPHIC FACTORS is stronger than that of the climatic factors (CLIMATIC ELEMENTS).

polycyclic, multicyclic (polycyclic preferred) *adj.* applied in geomorphology especially to features passing through several erosion cycles or part-cycles, e.g. a stream, the course of which indicates levelling to several former sea levels, or swallow holes that operate at different levels, i.e. under separate cycles, within a limited area. G.

polygamy 1. in botany, the condition of having male, female, hermaphrodite flowers on the same or different plants 2. in zoology, an animal breeding system in which an individual has more than one mate at the same time. POLYANDRY, POLYGYNY.

polygenetic *adj.* of many origins, applied particularly to soils of complex origin, to pebbles in a conglomerate, to mountains formed by different forces at different times, to geomorphological features which have been subjected to several different erosion processes. G.

polyglot 1. a person who understands, speaks or writes several languages. MONOGLOT.

polyglot *adj.* applied to a person, a text or a book, understanding, speaking, writing or containing several languages.

polygon **1.** a plane figure enclosed by five or more straight lines **2.** in soil, a polygonal pattern varying in diameter from a few millimetres to tens of metres, usually with a slightly concave or convex surface, visible on the soil surface not only in areas of PERMA-FROST or of very cold winters (PATTERNED GROUND), but also in other areas where contraction has taken place, e.g. in PLAYAS, in arid areas, in deserts. G.

polygonal karst KARST characterized by a surface completely pitted with closed depressions, the divides between them forming a roughly polygonal network. Such a formation is common in humid tropical CONE KARST terrain, but it also occurs in temperate karst terrain with DOLINES.

polygyny **1.** in zoology, an animal breeding system in which one male mates with and has exclusive breeding rights with several females, which he defends **2.** in botany, the condition of bearing male, female, hermaphrodite flowers on the same plant. POLYAN-DRY, POLYGAMY.

polymict a rock consisting of fragments of different materials, e.g. a CONGLOM-ERATE with pebbles of varying origin. OLIGOMICT. G.

polymorphism the occurrence of something in several different forms **1.** in biology, the occurrence of distinctly different forms of individuals in the same species, the different forms being in fairly constant proportion in a freely interbreeding population, e.g. human blood groups, or the queen, drone, worker CASTES in the honey bee **2.** the occurrence in a single population of more than one genetically distinct type, each occurring too often to be accounted for solely by repeated mutation **3.** the occurrence of distinctly different forms or different forms of organs in the same individual at differ-

ent stages in its life cycle **4.** in geology, the occurrence of separate minerals of the same chemical composition, but differing in physical structures and properties **5.** in crystallography, the capability of crystallizing into various forms.

polynuclear diffusion DIFFUSION-2.

polynya (Russian poluinya) a large area of open water surrounded by sea ice, sometimes bounded on one side by the coast, and occurring annually in the same region, notably off the mouths of big rivers. G.

polyphagous *adj.* applied to an animal that feeds on many kinds of food (unlike one which is OLIGOPHAGOUS). MICROPHAGOUS, MACROPHAGOUS, MONOPHAGOUS, PHYTOPHAGOUS, XYLOPHAGOUS.

polyphyletic *adj.* applied to a group of organisms the members of which are not all descended from a common ancestor which was also a member of the group, some of the members having had distinct evolutionary histories. The group is therefore an artificial one, the similarities of its members resulting from convergent evolution (CONVER-GENCE). MONOPHYLETIC, TAXO-NOMY.

polysaprobic SAPROBIC CLASSIFICA-TION.

polytopic *adj.* of, pertaining to, or characterizing, an organism or group of organisms occurring in more than one district. DISCONTINUOUS DISTRIBU-TION.

polytypic *adj.* applied to a species which has varying forms inhabiting different parts of its range. CLINE.

pome a so-called fruit, most of which is developed not from the ovary, but from the receptacle of the flower, e.g. in the APPLE the core is formed from the ovary, and the fleshy edible part from the receptacle.

pomegranate *Punica granatum*, a DECI-DUOUS shrub or small tree, native to western or southern Asia, now widely cultivated in tropical and subtropical areas for its fruit, a BERRY with a thick skin, containing many seeds surrounded by juicy acid pulp, which is eaten raw or pressed for its juice, used in drinks or in wine-making.

pomelo, pummelo SHADDOCK. CITRUS.

pond an area of still water, smaller than a lake, lying in a natural hollow (POND *verb*) or in a depression formed by digging or by embanking a natural hollow (and specifically named according to use, e.g. fish-pond, HAMMER-POND).

pond *verb* to dam a stream, to check a flow, to form a pond, e.g. when the normal flow of a stream is interrupted (e.g. by an uplift of part of the stream bed or by an obstruction which may include a strong flow of water from a side valley), the water of the main stream is said to be ponded back, forming a lake or large pond. G.

ponente a westerly wind, usually cool, usually bringing dry weather, blowing on the coasts of Corsica and Mediterranean France.

ponor (Slavic, pl. ponore) an AVEN, a vertical or steeply inclined shaft in limestone country, leading from a SWALLOW-HOLE or from the ground surface to an undeground cave, and through which water may pass. G.

pony any of several breeds of small, sturdy HORSE, in Britain up to 13 hands (c.1.32 m) (or polo ponies 13 to 15 hands: c.1.32 to 1.52 m), domesticated for riding.

pool 1. a small body of standing water, artificially impounded or occurring naturally, specifically named to accord with use or formation, e.g. swimming-pool, WHIRLPOOL 2. a still, usually deep, area in a stream or river.

pool and riffle a pattern developed on the bed of a stream by a sequence of alternating scoured pools and shallow gravel bars termed RIFFLES, even if the channel is straight and the bed uniform. The distance between the pools de-

pends on the width of the stream (commonly the distance is 5 to 7 times the width of the stream). If the bed is easily eroded the channel begins to wind, pools develop on the outer (concave) bank, riffles grow on the inner side to become POINT BARS, the beginning of MEANDERS.

poort (Afrikaans, a gate) a water-gap, cut by a river through a transverse ridge, a gorge in which a river breaks through a range of hills or mountains. G.

population 1. the total number or a specified group of people or of animals or of plants living in an area 2. the process of providing an area with inhabitants. DEPOPULATION 3. a GROUP-1 of individuals regarded without consideration of the interrelationships within the group 4. in statistics, a term synonymous with AGGREGATE-1, any finite or infinite collection of individuals under consideration (not necessarily a collection of living organisms) of which the ATTRIBUTES-2 may be estimated by SAMPLING.

population density the number of individuals occupying a particular unit area. DENSITY-1.

population explosion a great, rapid, increase in POPULATION-1,3 in a specific area, or in the world, e.g. as experienced in the human population in the twentieth century.

population geography a division of HUMAN GEOGRAPHY concerned with the scientific study of people in their spatial distribution and DENSITY-1, increase or decrease in numbers, movements and mobility, occupational structure, grouping in settlements, the way they form the geographical character of places, the way in which places in turn react to population phenomena, varying in space and over time. The boundary between population geography and DEMOGRAPHY is becoming more and more blurred. G.

population parameter in statistics, a

descriptive summary measure of some characteristics of a POPULATION-4. INTERVAL ESTIMATE.

population potential 1. the number of people who might be able to live in a certain area at a standard of living that is reasonable in relation to the resources available in that area. CARRYING CAPACITY-4 2. a measure of the accessibility of a particular mass of people to a point. LOWRY MODEL.

population pyramid a BAR GRAPH in the form of a pyramid drawn to express the age and sex composition of a human POPULATION-1. The age groups are shown on the vertical scale, commonly graduated into five-year intervals, youngest at the base, and the number or percentage of males and females within each of the age groups on the horizontal scale, the males lying traditionally to the left, females to the right of a line drawn perpendicular to the horizontal axis, and expressing zero (percentages or numbers increasing to the left for males, to the right for females). The result is shaped more or less like a pyramid, but is rarely symmetrical. If the pyramid has a wide base and tapers to a pointed top, it is termed expansive (denoting an expanding population, with many children and a declining death rate); if its shape resembles a tall dome rather than a pyramid, it is termed stationary (denoting a stable, slowly growing population, with a decline in mortality and a low birth rate); if the shape is broadly oval with a pointed top, the base cutting the oval below its widest part, it is said to be constrictive (denoting a declining population, with a birth rate lower than the death rate). In addition to birth and death rates, migration and the tendency of females to outlive males affect the shape of population pyramids. BIRTH RATE, DEMOGRAPHIC TRANSITION.

poramboke (Indian subcontinent: Urdu) unassessed waste land where cultiva-tion is prohibited, e.g. footpaths, village sites. G.

porcelain a hard, translucent, non-porous ceramic ware, true porcelain (hard paste) being made from a mixture of KAOLIN, FELDSPAR and QUARTZ fired at high temperature (1250° to 1350°C: 2280° to 2470°F), used especially for tableware and ornaments. Its predecessor, soft paste, was made from a wide variety of materials, including BONE ASH, but mainly from white clay and glass fired at temperatures under 1250°C (2280°F).

porcine *adj.* of, relating to, similar to, a PIG.

pore water pressure the pressure on rock and soil particles produced by water contained in the interstices between the particles.

pork PIG.

pororoca (South America) the BORE in the estuary of the River Amazon.

porosimeter an instrument for measuring the POROSITY of rocks by injecting a rock sample with a liquid or gas at a specific pressure. It shows a range of porosity from under 1 per cent to over 50 per cent, the porosity of sandstone lying between 5 to 15 per cent, of loose gravel 45 to 50 per cent.

porosity the quality of being porous, full or abounding in pores. The porosity of rocks, i.e. the ratio or percentage of the total volume of the pore spaces (minute interstices through which liquids or gases can pass) in relation to the total volume of the rock, is measured by a POROSIMETER. Sand, gravel, sandstones, with open textures and coarse grains, are typical porous rocks. Porosity is quite different from perviousness (PERVIOUS ROCK, PERMEABLE ROCK). Dry CLAY, for example, is highly porous and will hold much water in its pores, but when saturated the small spaces between the grains become blocked with water held by SURFACE TENSION, preventing the passage of water. To be an ACQUIFER or source of water a rock must be both porous and pervious. Porosity may be increased by LEACHING or decreased by COMPACTION. IMPERMEABLE. G.

porphyritic *adj.* applied to the texture of an IGNEOUS ROCK with large crystals (PHENOCRYSTS) embedded in a crystalline or glassy ground-mass (MATRIX-3) of finer grain.

porphyry a HYPABYSSAL rock with large crystals (PHENOCRYSTS) in a crystalline or glassy ground-mass (MATRIX-3) of finer grain. The hard Eygptian porphyry composed of red or white FELDSPAR crystals embedded in a red or purple glassy base was much used by the Romans as a decorative material for buildings, vessels, ornaments. ACIDIC ROCK.

port 1. a term loosely applied to a place where ships may anchor to load or unload cargo, or to the HARBOUR itself, or to the harbour and adjoining settlement, quays, handling facilities for cargo, warehouses, transport termini etc. ENTREPOT, FISHING PORT, FREE PORT, OUTPORT, PORT OF CALL, PORT OF ENTRY, RIVER PORT, TREATY PORT and AIRPORT 2. a fortified sweet wine made in Portugal. G.

port *adj.* applied to the side of a ship or aircraft that lies on the left of someone aboard who is facing the forepart (the bow).

portage (French) 1. the practice, especially in the days of the exploration by Europeans of North America, of carrying a boat, goods etc. by land from one navigable waterway to another, or from one section of a navigable waterway to another section in order to avoid obstacles such as rapids and waterfalls, or from one lake to another 2. the place or route over which this is accomplished. G.

port of call a PORT-1 at which a ship regularly stops to replenish supplies, carry out repairs, refuel.

port of entry a place, not necessarily a

PORT-1, where incoming foreign goods are cleared through customs offices.

portolan chart a CHART-1 drawn and used widely in the fourteenth to sixteenth centuries characterized by radiating systems of LOXODROMES or RHUMB LINES. G.

portolano (Italian) a sailing handbook and guide to harbours, illustrated by charts (PORTOLAN CHART), used in the fourteenth to sixteenth centuries.

position 1. the PLACE-1 occupied by a person or an object in relation to another person or object 2. rank in a hierarchy 3. physical posture. LOCATION, SITUATION.

positive anomaly a departure, a deviation from the normal, predicted or uniform state, value etc., being greater, higher than that state, value etc., e.g. of temperature, applied to the mean temperature (reduced to sea-level) at a meteorological station exceeding that for all the stations in the same latitude. ANOMALY, NEGATIVE ANOMALY, PLEION.

positive correlation in statistics, CORRELATION-3, CORRELATION COEFFICIENT.

positive discrimination the practice of singling out for favourable, rather than unfavourable, treatment.

positive externality EXTERNALITY.

positive feedback FEEDBACK.

positive landform a landform that rises from the ground, e.g. a mountain, a plateau, as distinct from a NEGATIVE LANDFORM.

positive movement of sea-level, the raising of the sea-level in relation to the land caused by a global raising of the ocean level (EUSTATISM) or by the subsidence of the land resulting from earth movements or isostatic processes. ISOSTASY.

positive planning PLANNING-2 in which a planning authority initiates development schemes, as opposed to negative planning in which the authority merely reacts to development schemes initiated by others, e.g. by public authorities or private developers.

positive skewness FREQUENCY CURVE.

positivism 1. a system of philosophy named (and elaborated from 1830 onwards) by Auguste Comte, 1798-1857, mathematician and philosopher, based on the assumption that all true knowledge is SCIENTIFIC-2, is based on facts or experience, and completely represented by observable phenomena and scientifically verified facts, their objective relations, and the laws that determine them. EMPIRICISM, IDEALISM **2.** a religious system based on this philosophy, humanity (as a single corporate being) being the object of worship **3.** an abbreviation for LOGICAL POSITIVISM.

positron positive electron, a fundamental particle with the same mass and spin as an ELECTRON but with a positive charge (the electron being negatively charged), the antiparticle of the electron.

possible *adj.* **1.** that may exist, happen or be done **2.** reasonable **3.** potential. POSSIBILISM.

possibilism a philosophical concept which, while accepting that every ENVIRONMENT has its limits which restrict human activity, argues that within those limits there is a set of opportunities which offer human beings freedom of choice of action, in contrast to DETERMINISM or ENVIRONMENTALISM. See PROBABILISM, STOP-AND-GO DETERMINISM. G.

postclimax the stage beyond the CLIMAX reached by a stable plant community the composition of which indicates climatic conditions more favourable than average for the area. CLIMAX, PLAGIOCLIMAX, PLANT SUCCESSION, PRECLIMAX, PRE-CLISERE, POSTCLISERE.

postclisere the series of vegetation formations which develop when the climate becomes wetter. CLISERE,

POSTCLIMAX, PRECLIMAX, PRE-CLISERE.

postdiction a statement or deduction about something in the past, e.g. a forecast based on past and present records which projects a situation at a period before that of the time of the past records taken into account.

postglacial *adj.* occurring after a glacial period, generally applied to all time succeeding the glacial epoch (PLEISTOCENE) in the QUATERNARY period. GEOLOGICAL TIMESCALE.

post-industrial city a city (a large town) which it is thought will become typical of the POST-INDUSTRIAL SOCIETY. It is presumed that employment will be mainly in the TERTIARY and especially in the QUATERNARY INDUSTRIES, population density will be low, differences in living standards and SOCIAL STATUS low, recreational facilities abundant, personal (rather than public) transport the norm. It is suggested that Canberra, Australia, is a prototype. INDUSTRIAL CITY, PRE-INDUSTRIAL CITY.

post-industrial society a form of society which some authors see as evolving in the industrialized societies of western Europe, the USSR, Japan and North America in the latter part of the twentieth century as a result of changes in their SOCIAL STRUCTURE. This new form is variously seen to involve a pronounced shift in employment from manufacturing to service industries (SECONDARY INDUSTRY, TERTIARY INDUSTRY, QUATERNARY INDUSTRY); the rise to pre-eminence of professional and technically qualified people; or the rejection by young people of MATERIALISM-2; or the decline of the role of the PROLETARIAT-2 as the agent of change (MODE OF PRODUCTION), resulting from technological progress. In this concept, the stage preceding the post-industrial is termed the industrial (employment mainly in SECONDARY INDUSTRY) and, preceding that, the pre-industrial (employ-ment mainly in PRIMARY INDUSTRY). POST-INDUSTRIAL CITY.

post meridiem P.M..

pot 1. POT HOLE **2.** used in some place-names in Ireland as a synonym for CIRQUE.

potamobenthos an organism living on the bed of a river. BENTHOS.

potamology the scientific study of rivers and their drainage basins, a term seldom used. G.

potamoplankton minute living organ-isms floating in sluggish rivers and streams. PLANKTON, POTAMOBEN-THOS.

potash 1. loosely applied to any POTASSIUM salt or compound **2.** specifically a potassium carbonate, K_2CO_3, especially an impure form. FERTILIZER.

potassic *adj.* of, relating to, containing POTASSIUM.

potassium a silvery-white, soft metallic ELEMENT-6, rapidly oxidizing in air, occurring in plants and animals, and in combined form in minerals. An essential plant NUTRIENT, it is used in the compound POTASH in mineral fertilizers; its salts are extensively used in medicine and industry. The rate of radioactive decay (RADIOACTIVITY) in potassium-argon is sometimes measured in DATING, especially in dating rocks. BASE CATION, BROMIDE, FELDSPAR, RADIO-GENIC HEAT, SYLVITE.

potassium-argon dating POTASSIUM.

potato *Solanum tuberosum*, a descen-dant of the PERENNIAL herb native to South America, now cultivated in most areas except the arctic and equatorial regions, one of the most extensively grown and important food crops of the world, yielding higher food value per area cultivated than any CEREAL crop. The potato has fibrous roots and many underground stems (rhizomes) culmi-nating in swollen tips that make the edible tubers. The shoots that develop from the 'eyes' of these tubers (termed

seed potatoes when used for planting) generate new growth. The tubers, rich in carbohydrates with some protein and potash, are eaten cooked, used for stock feed, and for producing alcohol.

potential climax a CLIMAX that would occur, replacing the existing climax, if there were a climatic change.

potential energy in physics, one of the two main types of energy, the other being KINETIC. Potential energy is the energy a system has as a result of its position in a field of force, e.g. if a body is raised above the ground to a suffici-ent height it has potential energy, the potential energy being equal to the work the body can do when it goes back to its original position.

potential evapotranspiration the highest amount of water that could be evapo-rated or transpired from plants from a given area if the plants had an unlim-ited water supply. EVAPOTRANSPIRA-TION, THORNTHWAITE'S CLIMATIC CLASSIFICATION.

potential instability the original condi-tion of an AIR MASS before it becomes conditionally unstable (CONDITIONAL INSTABILITY) as a result of being lifted over a relief barrier or over a mass of cooler air at a FRONT.

potential model a mathematical MODEL-4 which measures the force exerted by any defined phenomenon on a point in space by reference to the same phenomenon located at all other points in the area being studied.

potential temperature the TEMPERA-TURE-2 a parcel of air would reach if it were lowered adiabatically (ADIABATIC) to the level of standard atmospheric pressure (1000 millibars) near sea-level.

pothole, pot hole, pot-hole, pot 1. ap-plied loosely, especially by speleologists (SPELEOLOGY), to any deep hole, verti-cal cave system, or underground CAVE-1, in LIMESTONE country, hence the term pot holing for caving or exploring underground caverns, etc., the sport of pot holers **2.** in studies of erosion, a hole, more or less circular, worn in rocks by whirling stones, as in the BED ROCK of the channel of an eddying swift stream **3.** popularly applied to holes in roads developed by traffic **4.** a PONOR. G.

poultry CHICKENS, DUCKS, geese (GOOSE), TURKEYS and other domesticated birds raised for food, i.e. for the eggs they produce as well as for their flesh. BATTERY SYSTEM, LIVESTOCK.

pound **1.** an enclosure in which stray or trespassing animals are detained by a local authority until claimed **2.** a monetary unit, e.g. the pound sterling, the British monetary unit, equal to 100 pence **3.** lb, a British unit of mass equal to 16 ounces (7000 grains) AVOIRDUPOIS, or 12 ounces (5760 grains) TROY, or to 0.45 KILOGRAM. 14 lb = 1 stone (6.350 kg); 28 lb = 1 quarter (of a hudredweight) (12.70 kg); 112 lb = 1 HUNDREDWEIGHT (50.802 kg). CONVERSION TABLES.

poverty **1.** a deficiency in, a lack of, or an inadequate supply of, something **2.** of soil, unproductive **3.** the state or quality of being poor, a relative term without precise meaning. At subsistence level, only those with insufficient food and shelter for survival may be said to be in a state of poverty; but poverty is usually defined in sociology as a concept of relative deprivation, i.e. the absence or inadequacy in the lives of the poverty-stricken of the diets, amenities, standards, services and activities which are common or customary in the society in which they are living. Official definitions of poverty are customarily given in social welfare schemes: such definitions are often governed by political considerations.

poverty cycle the recurrent transmission of POVERTY-3 and DEPRIVATION from one generation to the next. This has been explained in terms of the assumed passing-on from parents to children of negative attitudes towards work, education and the law, as a consequence of which the children lack adequate equipment to compete for jobs. The assumption is that children of poor parents start school at a disadvantage, may not be encouraged and supported by their parents while at school, achieve little, leave school with few or no qualifications, can find only low-paid work, if at all, and stay poor, their children in turn inheriting their disadvantages. MULTIPLE DEPRIVATION.

powder snow loose, very dry snow crystals forming at low temperatures, e.g. in northern Canada and Siberia. SNOW.

power **1.** an ability, faculty, control, controlling influence **2.** in international affairs, a state of sufficient size, wealth, authority and strength to be able to maintain its influence among other nation states **3.** in physics, the rate at which work is done (as distinct from ENERGY, the capacity for doing work), the transfer of energy. GEOTHERMAL POWER, HYDROCARBON, HYDROELECTRIC POWER, NUCLEAR FISSION, NUCLEAR FUSION, SOLAR BATTERY, TIDAL POWER STATION, WATER POWER, WIND POWER **4.** in mathematics, the number of times a quantity is multiplied by itself, or the index showing this, e.g. 10 to the power of 3 (10^3) is 1000. G.

power of exclusion the ability of a social group to bring about and maintain SOCIAL CLOSURE.

power of solidarism the ability of people who have been excluded by SOCIAL CLOSURE from rewards and opportunities to combine and form themselves into a group to reinforce or improve their own status.

pozzolana, pozzuolana (Italian Pozzuoli, a town near Naples) fine VOLCANIC ASH used in HYDRAULIC CEMENT and some mortars.

PPBS programme, planning and budgeting system, a method of determining the allocation of resources to organizations in the PUBLIC SECTOR of an economy, e.g. to education, local authorities, the social services etc.

pradoliny (Polish pr, old; doliny, valley) German URSTROMTAL a large channel carrying glacial melt-water. G.

pragmatic, pragmatical *adj.* **1.** of or relating to PRAGMATISM, concerned with practical results and values **2.** dealing with events in the light of practical lessons or applications, rather than by following a principle **3.** practical, matter-of-fact **4.** dogmatic **5.** of or relating to state affairs.

pragmatism **1.** a system of philosophy formulated in the USA by C. S. Peirce, 1839-1914, developed by William James, John Dewey and others, which in general asserts that the truth or value of a conception or assertion expresses itself in practical consequences and may be tested in the light of these as they affect human interests and purposes. Truth is held to be relative, not attainable by metaphysical speculation **2.** a practical approach to assessing situations and acting on them, a matter-of-fact treatment of things **3.** dogmatism, officiousness, pedantry.

pragmatized, pragmatizing *adj.* represented (or representing) as real or actual something which is imaginary or subjective.

prairie (Latin pratum, a meadow; French prairie, a tract of meadow land) an extensive area of unbroken grassland, generally without trees, occurring in midlatitudes in North America and considered by some authors to be equivalent to the STEPPE of Europe, the PAMPA of South America, the VELD of southern Africa. Prairie occurs where summer rains are light (total annual rainfall 250 to 500 mm: 10 to 20 in, with some local drought) and summer temperatures are high, conditions well suited to grain crops. Thus little remains of the original prairie, most having been ploughed and sown to cereals or grasses

finer than the indigenous. Some authors distinguish long grass prairie from short grass prairie, equating the latter with steppe. Confusion arises from the French common application of the term prairie to grassland or meadow or, in French literature, to a small enclosed field. In FAO statistics 'prairies et pâturages permanents' is translated 'permanent meadows and pastures'. G.

prairie soil **1.** brunizem, a ZONAL SOIL developed in midlatitude subhumid temperate areas that were formerly under PRAIRIE grasses but which are now sown and cultivated. It is very similar to CHERNOZEM, but is dark brown and slightly acidic on the surface; some leaching has occurred but there is no great accumulation of calcium carbonate in the B HORIZON **2.** a general term for all dark soils of treeless plains. MOLLISOLS. G.

praxis **1.** practice, especially customary practice **2.** practical activity, the exercise of or practice of an art, science or skill **3.** practical ability **4.** a function resulting from a particular STRUCTURE-1 **5.** the conscious action by which a theory or a philosophy becomes a reality.

Preboreal, Pre-Boreal a climatic phase, with cold, dry conditions favourable to the growth of birch-pine forest in England, pine forest in Scotland, following the QUATERNARY glacial epoch (GEOLOGICAL TIMESCALE), lasting in eastern England until about 7500 BC. The sea-level was more than 60 m (195 ft) below the present level, thus the British Isles were part of continental Europe. BOREAL, MESOLITHIC, SUB-ATLANTIC, SUB-BOREAL.

Precambrian *adj.* of or pertaining to all geological time and rocks before the Cambrian period (time) or system (rocks). ARCHAEON, ARCHAEOZOIC, GEOLOGICAL TIMESCALE.

precession of the equinoxes the gradual change in the annual occurrence of the EQUINOXES, due to the relative change of position of the ECLIPTIC and the EQUATOR as the axis of the earth's rotation describes a cone-shaped rotation (similar to the movement of a gyroscope). The conical rotation is caused by gravitational forces between the earth and the sun, the earth and the moon. It is calculated that the equinoxes swing round the ecliptic once in 25 800 terrestrial years. ZODIAC.

precinct **1.** commonly, the space within the walls or other boundaries of a building or place, e.g. of an abbey, church, school **2.** the boundaries (real or imaginary) of such a building or place **3.** the immediate surroundings of a place **4.** historically, especially in Norfolk, England, fifteenth to seventeenth centuries, a division (composed of furlongs and strips, and in some cases coterminous with a SHIFT) of the arable field of a TOWNSHIP-1 where the common arable was not divided into 'fields' **5.** in town planning, an area in a town with a specialized and defined function, especially one not crossed by a major road, or one accessibly only on foot **6.** (American) a subdivision of a county, city or city ward, demarcated for police and election purposes.

precious metal a METAL-1, especially GOLD, PLATINUM, SILVER, prized for its high value. TROY WEIGHT.

precious stone one of the GEMSTONES of great value (e.g. DIAMOND, EMERALD, RUBY, SAPPHIRE), qualified as male if it has a particular depth or brilliancy of colour. Precious stones are weighed in CARATS.

precipice (French précipice) the high, steep or overhanging part of a cliff face.

precipitation **1.** in meteorology, the deposition of moisture on the surface of the earth from atmospheric sources (METEORIC WATER), including DEW, HAIL, RAIN, SLEET, SNOW **2.** in chemistry, the formation, the settling out, of solid particles in a SOLUTION. G.

precipitation-day a period of 24 hours, commencing normally at 0900 hours,

on which rainfall exceeds 0.25 mm (0.01 in) or snowfall 0.25 cm (0.1 in). It is a more appropriate and precise term than RAIN-DAY, especially when applied to areas with much snowfall, and is generally now the preferred term. WET-DAY. G.

precipitation efficiency index THORNTHWAITE'S CLIMATIC CLASSIFICATION.

precipitous *adj.* resembling a PRECIPICE or containing precipices.

preclimax a CLIMAX reached by a stable plant community the composition of which indicates climatic conditions less favourable than the average for the area. PLAGIOCLIMAX, PLANT SUCCESSION.

preclisere the series of vegetation formations which develops when the climate becomes drier. CLIMAX, CLISERE, POSTCLISERE, POSTCLIMAX, PRECLIMAX.

pre-Columbian *adj.* of or relating to the period before the discovery of America by Columbus.

predator an ANIMAL-1 that kills another animal for food. FOOD CHAIN.

predator chain FOOD CHAIN.

preference **1.** a liking for, esteeming, or choosing one thing rather than an another. ATTRIBUTE, COMPOSITE PREFERENCE RATING **2.** in international trade, a system in which the import duties levied on goods from selected favoured countries are waived or are lower than those levied on the goods of other countries, a system (colonial preference) much used in the past in the trade between Great Britain and what were then her colonies.

preformation in biology, an old theory that the fertilized egg of an organism contains the entire diversity of structure of the adult, and that embryonic development consists only of the enlargement and manifestation of this structure, brought about by adequate nourishment. EPIGENESIS.

prehensile *adj.* in zoology, capable of

grasping and holding, e.g. the tail of a monkey.

prehistoric *adj.* of or relating to the period before recorded history.

prehistory 1. the period of history before there were written records 2. the study of the past (usually of a given region), before the appearance of written records (relevant to the region).

pre-industrial city a CITY-1 of the past or present serving a population of which most are engaged in agriculture. It has characteristics markedly different from those of an INDUSTRIAL CITY typical of the highly developed industrialized countries of today. For example, the pre-industrial city is usually identified by the non-existence of a CENTRAL BUSINESS DISTRICT; in some cases one may exist but if it does it is not dominant. Usually there is no specialization of land use, the urban layout is relatively un-ordered, the street markets, shops, workshops and homes being mixed together (although a particular craft may be concentrated in a particular district). The poorest live on the periphery, not in the INNER CITY zone. There is a marketing economy based on crafts, little specialization of labour in the large population, and in most cases an extended KINSHIP SYSTEM that ensures rigid social, ethnic, and tribal segregation, often on occupational lines, so that there is little SOCIAL MOBILITY. POST-INDUSTRIAL CITY.

pre-industrial society POST-INDUSTRIAL SOCIETY, PRE-INDUSTRIAL CITY.

presence indicator, plant indicator, species indicator INDICATOR PLANT.

preservation the act of keeping in existence, preventing ruin and decay. The term usually implies the maintenance of something in its present form (or as close to that form as is possible), without change; whereas the work of CONSERVATION, aware of present and future needs, is more positive, forward-looking and flexible, and does not rule out necessary change.

pressure 1. the action of exerting a steady force on something, of trying to force or persuade. PRESSURE GROUP 2. in climatology, meteorology, ATMOSPHERIC PRESSURE 3. in physics, the force per unit area, or the force acting on a surface divided by the area over which it acts.

pressure gradient BAROMETRIC GRADIENT.

pressure group a group of people united by shared attitudes and goals who try to obtain decisions favourable to their interests by gaining access to and influencing the governmental process.

pressure ice a general term applied to sea ice which has been subjected to pressure, squeezed, and in places forced upwards to form RAFTED ICE or HUMMOCKED ICE or a PRESSURE RIDGE. G.

pressure melting point the temperature at which ice, near to melting, needs only slight pressure on it to make it melt. This is normally 0°C (32°F) at the surface of a glacier, but it becomes fractionally lower with increasing depth, due to the increasing pressure of the thickening layer of ice.

pressure plate anemometer ANEMOMETER.

pressure release the outward force of pressure set free from inside a rock mass when the overlying strata is removed by DENUDATION. This commonly occurs in massive, unjointed rock, e.g. in granite.

pressure ridge of ice, a ridge or wall of HUMMOCKED ICE occurring where one FLOE presses against another. PRESSURE ICE. G.

pressure system a distinct atmospheric circulation system of high or low pressure. ANTICYCLONE, COL, DEPRESSION, RIDGE OF HIGH PRESSURE, SECONDARY DEPRESSION, WEDGE.

pressure tube anemometer ANEMO-METER.

pressure wave PUSH WAVE.

prevailing wind a WIND blowing most frequently from a specific direction in a particular area. DOMINANT WIND.

price in economics, a measure in terms of money (or some other agreed COMMODITY-1,2) of the exchange value of a GOOD or service.

Primärrumpf (German) a primary PE-NEPLANE, or an 'old from birth pene-plain', a term introduced by W. Penck and applied to a land mass which is being slowly and progressively uplifted so that it forms a flat-topped, expanding dome; but the slow uplift is matched by DENUDATION, with the result that there is no actual increase in the relief or net rise of the surface, and a low, featureless plain is formed. ENDRUMF. G.

primary *adj.* of the first in order of time of origin or precedence.

Primary *adj.* applied in the nineteenth century to the period of geological time and the associated rocks of what is now termed the PRECAMBRIAN. Later applied first to the Lower Palaeozoic, then to all the PALAEOZOIC. The term Primary is little used now, except by authors who like to keep to the sequence Primary, Secondary (Mesozoic), Tertiary, Quaternary. GEOLOGICAL TIME-SCALE.

primary consumer FOOD CHAIN.

primary forest a FOREST-2 consisting mainly of CLIMAX or subclimax species.

primary group a group of people in which there is an informal, personal, face-to-face relationship between members, e.g. as in an EXTENDED or a NUCLEAR family. GEMEINSCHAFT, SECONDARY GROUP.

primary industry, primary activity, primary sector an activity concerned with the collecting or making available of material provided by nature, i.e. agriculture, fishing, forestry, hunting, mining and quarrying. INDUSTRY, SECONDARY INDUSTRY, TERTIARY INDUSTRY, QUATERNARY INDUSTRY. G.

primary mineral a mineral formed directly from cooling magma, keeping its original identity even in a sedimentary rock. SECONDARY MINERAL.

primary production in ecology, the total quantity of organic matter as it is newly formed by PHOTOSYNTHESIS. It is estimated that only some one-tenth of one per cent of the SOLAR energy reaching the earth is fixed in plants by photosynthesis. The matter remaining after the energy needs of the plants have been met is termed the net primary production. FOOD CHAIN, PRODUCTION, PRODUCTION RATE, SECONDARY PRODUCTION.

primary rock a rock formed directly from magma. SECONDARY ROCK.

primary sampling unit CLUSTER SAMPLING.

primary sere PRISERE.

primary succession plant succession starting on ground surfaces that have not previously supported vegetation, e.g. sand dunes, lava flows. CLIMAX, SECONDRY SUCCESSION, SERE, SUCCESSION-2.

primary wave of earthquakes, PUSH WAVE.

primate in zoology, a member of the order Primates, an order of placental MAMMALS which first appeared at the beginning of TERTIARY. The order, primarily arboreal, includes lemurs, monkeys, apes and human beings, characterized by having nails (rather than claws), a well developed big toe and thumb, the thumb being opposable to the other digits, well developed eyesight (some with binocular vision) and a relatively large brain. CLASSIFICATION OF ORGANISMS, PLACENTA.

primate city a CITY-4 which is first in the rank of city sizes in a country, being larger than any other, its great size giving it an impetus to become even larger, so that the gap between its population total and that of the city or cities of the second rank is exceptionally wide. According to the RANK-SIZE RULE, the primate city will be more than twice the size of the second largest city; according to CENTRAL PLACE THEORY it will be three times the size of the second rank city. Some authors regard the primate city as being equivalent to a METROPOLIS. GATEWAY CITY. G.

prime meridian, initial meridian the MERIDIAN from which LONGITUDE is measured, i.e. 0°, the longitude of Greenwich.

primeur (French, anything new or early) a fruit or vegetable cultivated in such a way that it is ready for market before, or very early in, its normal season. G.

primitive communism a MODE OF PRODUCTION in which the PRODUCTION FACTORS are communally owned and labour and the products of labour are equally shared.

primogeniture 1. the state or fact of being the firstborn of parents **2.** a principle, custom or law by which property or title is passed down to the firstborn child, usually the firstborn son **3.** the feudal rule of inheritance in England whereby the whole of the real estate of one who dies intestate (not having made a valid will) descends to the eldest son.

principal components analysis a statistical technique (similar to FACTOR ANALYSIS) of deriving a smaller number of FACTORS-4 to represent a larger number of tests. It assumes that all the variance of all the tests may be assigned to the common factors to be extracted. The new variables created correlate with the original variables, but are not correlated with each other. No allowance is made for either the ERROR VARIANCE or the SPECIFIC VARIANCE of the tests to be kept out of the factors. The method extracts orthogonal factors. ORTHOGONALITY.

principal meridian in USA public land system, PUBLIC LAND SURVEY.

prisere primary SERE, a natural plant SUCCESSION-2, originating on a bare area and progressing to a CLIMAX. CLISERE, HALOSERE, HYDROSERE, LITHOSERE, MESOSERE, PLAGIO-CLIMAX, PLAGIOSERE, PSAMMOSERE, SECONDARY SUCCESSION, SERE, SUB-SERE, XEROSERE.

prismatic compass a magnetic COMPASS used in surveying, in which a prism is incorporated so that an image of the angle of the bearing can be read by the operator while using sights through a telescope.

private information the information passed from person to person in face-to-face contact, telephone calls, personal letters etc. DIFFUSION, PRIVATE INFORMATION FIELD, PUBLIC INFORMATION.

private information field information field, the probable spatial distribution of PRIVATE INFORMATION passed from person to person during a certain period of time. CONTACT FIELD, DIFFUSION, DIFFUSION WAVE, DISTANCE DECAY PHENOMENON, MEAN INFORMATION FIELD, NEIGHBOUR-HOOD EFFECT, PUBLIC INFORMATION.

private sector one of the two major divisions of a national economy, the other being the PUBLIC SECTOR. The private sector covers consumer expenditure for goods and services and business expenditure for plant and equiment, and thus includes the economic activities of private persons, industrial and commercial companies, institutions (insurance companies, building societies, finance houses), trade unions, charities and churches etc. MIXED ECONOMY, PLANNED ECONOMY.

privatization denationalization, the reverse of NATIONALIZATION, i.e. the removal of state control or ownership of an activity or property in order to allow it to pass into private hands.

probable *adj.* likely to occur or to be true. PROBABILITY.

probabilism 1. in philosophy, the doctrine that no knowledge is absolutely certain and that PROBABILITY provides sufficient grounds for action 2. in geography, a modification of POSSIBILISM postulated by O. H. K. Spate, that everywhere there are possibilities but some are more probable than others. G.

probabilistic model in statistics, a term applied by some authors to a MODEL-4 which, in covering the results of a series of trials in terms of PROBABILITY-2, contains an element of prediction as to the outcome of the individual trials. DETERMINISTIC MODEL, ENTROPY, STOCHASTIC MODEL.

probability 1. generally, loosely applied to the degree of belief that an event may occur 2. in statistics, the likelihood of an event occurring, expressed as a ratio between the number of the actual occurrences of the event and the average number of cases that would favour its occurrence, taken over an infinite series of these cases. The probability of an event occurring lies within the range 0 to 1, zero indicating absolute impossibility, 1 indicating absolute certainty, the probability of the occurrence of most events lying between the two. BINOMIAL DISTRIBUTION, PROBABILISTIC MODEL, SCIENTIFIC LAW, STOCHASTIC.

probability sample NON-PROBABILITY SAMPLE, RANDOM SAMPLE.

probability sampling, probability selection RANDOM SAMPLING.

process in general, a continuous change consisting of a connected and related series of events or acts.

process elements the measurable contributions made by the various forces which give rise to a set of RESPONSE ELEMENTS.

process lapse rate the rate of decrease of temperature in a small parcel of rising air. DRY ADIABATIC LAPSE RATE, ENVIRONMENTAL LAPSE RATE, LAPSE RATE, SATURATED ADIABATIC LAPSE RATE.

process-response system a type of SYSTEM-1,3 formed when at least one CASCADING SYSTEM and one MORPHOLOGICAL SYSTEM link up. CONTROL SYSTEM. G.

proclimax a stable stage of development reached by any plant SUCCESSION-2, similar to the CLIMAX in its permanence, but due to the influence of factors other than the existing climatic. DISCLIMAX, PLAGIOCLIMAX, POSTCLIMAX, PRECLIMAX, SERCLIMAX, SUBCLIMAX.

produce agricultural or horticultural products.

producer gas a mixture of CARBON MONOXIDE, HYDROGEN and NITROGEN made by passing air over COKE heated to a high temperature. It is used as a fuel.

producer goods, producers' goods goods such as tools and raw materials needed by a manufacturer to make other goods. Thus producer goods do not satisfy the needs or desires of the individual person directly (CONSUMER GOODS), but only indirectly, in the production of other, final products. CAPITAL GOODS, CONSUMER DURABLES, CONSUMER GOODS. G.

producer level FOOD CHAIN.

product something grown, manufactured or created, the end result of a process. OUTPUT, PRODUCTIVITY.

production 1. that which results from a process, effort, action. PRODUCT 2. the total OUTPUT, especially of something grown or manufactured, measured in absolute terms. FOOD CHAIN, NET AERIAL PRODUCTION, PRIMARY PRODUCTION, PRODUCTION RATE, SECONDARY PRODUCTION.

production factors the components needed in a production process, conventionally, capital, land, labour (including enterprise, e.g. of an ENTREPRENEUR).

production rate in ecology, the number of organisms formed in an area in a given period of time. Gross production rate is the rate of assimilation shown by organisms of a given TROPHIC LEVEL-1. Net production rate is the gross production rate less loss of matter brought about by respiration, decomposition and predation. FOOD CHAIN, PRIMARY PRODUCTION, PRODUCTION-2.

productive forces the means of PRODUCTION and the LABOUR POWER of a society, which, by their interaction, show the society's CAPACITY-3 to produce.

productivity 1. the ability to produce, i.e. to put or bring forth, cause, create, make or manufacture, grow, something 2. the measured ability to grow things or the calculated rate of making goods. PRODUCTIVITY OF LAND. Thus in economics, the rate of output per unit of input used in measuring capital growth and in assessing the effective use of equipment, materials and labour. INPUT-OUTPUT ANALYSIS.

productivity of land 1. actual productivity, the equivalent of YIELD-1 2. potential productivity, the hypothetical yield under stated conditions, especially under the best conditions possible (land potential is the term sometimes used in this sense, LAND CLASSIFICATION). G.

profession 1. a form of employment regarded by a society as being honourable, of direct personal service to others, and possible for an educated person only after specific training in some special branch of knowledge, e.g. traditionally, religion, the law, medicine 2. the body of people engaged in a particular branch of such service.

professional *adj.* 1. applied loosely to a person who habitually earns a living from a skilled occupation, but usually restricted to an occupation considered to be socially superior in some way. BLUE-COLOUR WORKER, WHITE-COLLAR WORKER, STEEL-COLOUR

WORKER 2. specifically, of or pertaining to a PROFESSION.

profile 1. the shape of something, viewed from the side 2. a short, concise descriptive written sketch 3. a side elevation or section, e.g. the shape shown in outline where the plane of a SECTION cuts vertically the surface of the ground, i.e. producing the surface outline alone, as in a RIVER PROFILE. Other profiles include COMPOSITE PROFILE, PROFILE OF EQUILIBRIUM, PROJECTED PROFILE, SOIL PROFILE, SUPERIMPOSED PROFILE.

profile of equilibrium 1. of a river, the theoretical curve along the course of a fully mature or GRADED RIVER (GRADATION, RIVER PROFILE) 2. of a shore, the sloping shore where the accumulation of beach deposits is balanced by the amount removed.

profit 1. an excess of income over expenditure, especially in a particular transaction over a period of time 2. the surplus money produced by industry after deductions have been made for the cost of wages, raw materials, rents, charges.

proglacial *adj.* before, in advance of, to the front of, a GLACIER, applied to time or position. PROGLACIAL LAKE. G.

proglacial lake a stretch of water ponded during a glacial period between an ice front and rising ground, e.g. a morainic ridge, the sediments in the lake consisting of FLUVIOGLACIAL DEPOSITS. G.

prognathous *adj.* having projecting jaws.

progradation a process in which the shore is extended seaward by the action of waves on a seashore, continuing so long as the current builds up the sea bottom offshore. AGGRADATION, BEACH RIDGES. G.

progressive wave applied to ocean waves, a wave propagated in a channel of (theoretically) infinite length, its wave length being the distance between two successive crests, and its period the time taken to move one wave length. PROGRESSIVE WAVE THEORY OF TIDES.

progressive wave theory of tides 1. a theory formerly held, offering an explanation for tides. The theory suggested that two tidal waves formed in the Southern Ocean, one following and slightly lagging behind the moon, matched by another at the opposite end of the diameter of the earth. From these two tidal waves branches went north into the Atlantic, Indian and Pacific oceans and thence to their adjoining seas, the speeds of these branches being affected by the shape of the oceans and depths of water rather than by the period of the moon. The theory has been superseded by the OSCILLATORY WAVE THEORY OF TIDES.

projected profile one of a series of PROFILES-3 constructed to show relief, using information taken from a physical map at equally spaced intervals. These profiles are plotted on a single diagram, one behind the other. Each profile includes only those parts of the land surface which are not hidden by the high land of the intervening profiles. This produces an outline landscape drawing which shows detail only of the summits; but the effect is panaoramic, with a foreground, middle distance and skyline (unlike a COMPOSITE PROFILE diagram which, in effect, shows only the skyline, or the SUPERIMPOSED PROFILE diagram which shows a mass of detail). G.

projection MAP PROJECTION.

proletariat 1. the lowest CLASS-3 in ancient Rome **2.** the lowest SOCIAL CLASS in a modern society, particularly (in Marxist theory) the wage earners possessing neither property nor capital and living by the sale of their labour. BOURGOISIE, MODE OF PRODUCTION, POST-INDUSTRIAL SOCIETY.

proluvial *adj.* applied in the USSR to dry sediments formed by rivers that exist only temporarily. G.

promenade a wide path, commonly paved, used for walking or strolling, and lying between the road and the seashore in a seaside resort. ESPLANADE.

promontory a headland, a cliff, usually rocky, projecting into the sea, often associated with offshore rocks and STACKS.

property 1. possessions, a thing or things owned **2.** REAL ESTATE or a piece of real estate (especially a building such as a house) **3.** an attribute, a characteristic, a quality.

property market 1. the MARKET-5 in REAL ESTATE, i.e. the relationship between the demand for and the supply of land and buildings and the rents and prices paid **2.** broadly, land ownership, changes in the financial climate in property, the process of development, and shifts in type of development, etc.

property rights in land LANDOWNER.

proportional equality a share for each one in proportion to some measure of worthiness. ARITHMETIC EQUALITY.

proportional representation an electoral system designed in such a way that representatives elected to a legislative body reflect as accurately as possible the divisions of opinion within the electorate. Thus in a parliamentary (PARLIAMENT) DEMOCRACY each party has a number of representatives in the governing body proportional to the votes cast for it.

propylene an inflammable gaseous HYDROCARBON produced from PETROLEUM, used in making detergents etc. HYDROCARBON CRACKING.

protalus rampart an accumulation of coarse angular rock debris, resembling a MORAINE, consisting of material that has slipped down from perennial banks of snow, and lying parallel to the slope that produced it. TALUS.

protein any of the countless number of naturally occurring complex combinations of amino acids, a basic constituent of all living cells, and necessary for growth and maintenance of tissue, the essential nitrogenous food for ANIMALS-1. Plants can synthesize proteins from inorganic nitrogenous material, but animals apparently cannot. NITROGEN CYCLE.

Proterozoic *adj.* **1.** of or pertaining to the more recent of the two PRE-CAMBRIAN geological eras. GEOLOGICAL TIMESCALE **2.** in USA, of or pertaining to all Precambrian time and rocks **3.** by some authors, of or pertaining to the third era of Precambrian time (the others being the EOZOIC and ARCHAEOZOIC). G.

Protista the term now applied to a kingdom(CLASSIFICATION OF ORGANISMS) comprising all the simple organisms (BACTERIA, FUNGI, slime MOULDS, ALGAE as well as the PROTOZOANS). Formerly the term was applied to unicellular organisms only.

proton a positively charged, fundamental particle present in all ATOMS.

protoplasm in popular usage, the living matter of a CELL, in animals and plants differentiated into NUCLEUS and CYTOPLASM.

Protozoa a PHYLUM (CLASSIFICATION OF ORGANISMS) of small, in many cases microscopic, single-celled or COLONIAL ANIMALS (PROTISTA), some members of which are present in PLANKTON.

protozoan a member of the phylum PROTOZOA.

Protozoic *adj.* obsolete, formerly applied to the older period (time) and system (rocks) of the Palaeozoic era, comprising the Cambrian, Ordovician, Silurian periods and systems, the similarly obsolete *adj.* DEUTEROZOIC being applied to the remainder of the Palaeozoic, i.e. the Devonian, Carboniferous, Permian periods and systems. GEOLOGICAL TIMESCALE.

provenance **1.** the place of origin **2.** of SEDIMENTARY ROCKS, the place of origin of the constituent materials (CLASTIC ROCK) or the area from which the sediments came. Sediments that come from one source are termed monogenetic, those from more than one source, polygenetic.

proven reserves RESERVES of which the existence and the location are known and the quantity can be estimated, but which may be exploited only if the demand is great enough to make it economically feasible to do so. RESOURCES.

province **1.** an administrative division within certain countries **2.** part of a country outside the area of the capital **3.** a territory administered by but outside a country, i.e. an overseas administrative division **4.** an administrative area of an ecclesiastical order **5.** a sphere of action or of learning **6.** in biology, a sub-region, smaller than a region, a distinctive area of sea or land populated by a particular assemblage of plants and animals **7.** in geomorphology, a unit larger and more complex than a section, inferior to a continental subdivision, in the hierarchy of MORPHOLOGICAL REGIONS. G.

prune the dried fruit of a PLUM.

psammite (Greek psammos, sand) a term formerly applied to a CLASTIC SEDIMENT composed of sand-sized particles, i.e. an ARENACIOUS ROCK; but now applied only to an arenaceous rock that has been metamorphosed. PSEPHITE. G.

psammitic *adj.* of or pertaining to, having the character of a PSAMMITE.

psammophyte a plant which thrives in sandy soil.

psammosere a PRISERE in which the pioneer community developed on dry sandy soil. LITHOSERE, SERE, XEROSERE.

psephite (Greek psephos, a pebble) a term formerly applied to a rock composed of large rock fragments, pebbles or blocks in a ground-mass varying in kind and amount, e.g. a CONGLOM-

ERATE; but now applied only to a rock composed of metamorphosed particles, e.g. PELITIC or PSAMMITIC rocks. G.

psephitic *adj.* of or pertaining to, having the character of a PSEPHITE.

psephology (Greek psephos, pebble; logos, discourse) a term derived from the Athenians' use of pebbles in voting, hence the study of voting, of elections, and the scientific analysis of election results.

pseudo-cirque an armchair-shaped hollow, resembling a CIRQUE but non-glacial in origin, and occurring in arid lands in limestone, sandstone, granite rocks. G.

pseudogley soil an acid soil with A, B, C horizon sequence (SOIL PROFILE), the B HORIZON having compact texture, and both the A HORIZON and the B horizon showing strong mottling due to a temporary water table lying above the B horizon. The destruction of clay minerals occurs in the upper layer of the B horizon, into which tongues of the A horizon may penetrate. GLEY, SOIL, SOIL ASSOCIATION, SOIL HORIZONS. G.

pseudokarst a terrain that resembles KARST but which has been formed by processes other than the dissolving of rock, e.g. the rough surface above a lava field where the ceilings of LAVA TUBES have collapsed.

pseudomorph a mineral that resembles another in CRYSTALLINE form.

pseudovolcanic features topographic features that resemble volcanic forms, e.g. bomb and meteor craters. G.

psychic income factors which, reflecting personal psychological and mental attitudes, give the decision-maker the greatest comfort, happiness, pleasure, satisfaction.

psychology a term variously defined, but commonly applied to **1.** the study of the mind, the study of human or animal behaviour, the study of people as they interact with their social and physical ENVIRONMENT **2.** the mental

characteristics with which a particular kind of behaviour is associated.

psychosomatic *adj.* of, pertaining to, characteristic of, resulting from, the interaction between the human mind and the human body, particularly applied to ailments (trivial illnesses) and diseases.

psychosphere the human mind, human thought and culture (as studied by psychologists, anthropolgists and other social scientists) as an environmental factor, to be taken into account together with the lithosphere, atmosphere, hydrosphere and biosphere in environmental studies. TECHNO-SPHERE. G.

Psychozoic (Greek psyche, soul or mind) *adj.* applied by some writers to the era of geological time commencing with the appearance of human beings on earth, i.e. the QUATERNARY. GEOLOGICAL TIMESCALE.

psychrometer a type of HYGROMETER, an instrument used to measure RELATIVE HUMIDITY. One type of psychrometer consists of WET and DRY BULB THERMOMETERS, the wet bulb lying in a current of air produced by a small fan to give maximum evaporation. In another type, the sling psychrometer, the thermometers are swung round and round to give maximum evaporation.

psychrophile an organism which grows best at temperatures below 20°C (68°F). MESOPHILE, THERMOPHILE.

pteropod ooze a calcareous deep sea OOZE occurring locally in small areas in the North and South Atlantic oceans, formed from the shells of minute Pteropoda (various MOLLUSCS which have the foot spread into a wing-shaped paddle for swimming).

public housing housing units financed directly by government agencies or local authorities and rented to TENANTS.

public information the information received by an individual through the MEDIA. PRIVATE INFORMATION.

public land survey, public land system in USA, a system used in surveying by the US Land Office Survey, the land being parcelled into congressional townships (TOWNSHIPS-3) of 6 mi square; 32 major zones were drawn up, with a principal north to south meridian (the PRINCIPAL MERIDIAN) drawn to meet and cut a parallel of latitude (the BASELINE-2). The zones were then divided into 6 mi squares by measuring off north to south rows (RANGES-9) at 6 mi intervals to the east and to the west of the principal meridian to provide the eastern and western boundaries, the northern and southern boundaries being provided by measuring-off 6 mi intervals (townships) along the principal meridian to the north and to the south of the baseline.

public participation PARTICIPATION by the public in political decision-making, especially in relation to policies which have a direct effect, e.g. land use planning policy.

public sector, government sector one of the two major divisions of a national economy, the other being the PRIVATE SECTOR. The public sector covers the official economic activity of central and local government, the nationalized industries and any corporations partly financed by the government, but excludes the consumer expenditure of government employees. In a MIXED ECONOMY it usually represents some 25 per cent of all economic activity; in countries with a CENTRALLY PLANNED ECONOMY, e.g. the USSR or China, it usually represents 100 per cent.

public utility a business concern which provides and administers a public service, e.g. by providing gas, electricity, water, telephones, railways etc., which requires special wayleaves or other rights or compulsory powers over the use of land.

puddingstone a country name in Britain for a CONGLOMERATE which, in structure, resembles a plum pudding, e.g. Hertfordshire puddingstone.

puddling the kneading or mixing of CLAY and SAND (or CONCRETE) with water in order to make an IMPERVIOUS mass. DEW POND.

pueblo (Spanish, people, population, town, village) a term so variously and loosely used that it is best avoided **1.** a town or village in Spain, Spanish-speaking America, or the Philippines **2.** a native Indian settlement of Arizona, New Mexico and adjoining parts of Mexico and Texas, with houses often of more than one storey in height, sometimes forming terraces, built of stone or ADOBE-2 **3.** a native Indian inhabitant of one of these settlements **4.** any Indian village in the southwest USA.

pukka, pakka (Indian subcontinent: Urdu-Hindi) *adj.* built of stone or masonry, hence strong and good. e.g. a pukka well, a well lined with masonry or bricks, as opposed to KUCHA. G.

pulpwood TIMBER, usually from SOFTWOOD trees, suitable for the making of pulp for PAPER manufacture.

pulque a Mexican fermented drink prepared from various species of *Agave*, especially *Agave atrovirens*.

pulsar a STAR that gives out strong but very brief pulses of light, radio or X rays at precise intervals.

pulse 1. a leguminous food plant (LEGUMINOSAE). BUTTER BEAN, CHICK PEA, BLACK AND GREEN GRAMS, LABLAB, LENTIL, PEA **2.** the edible seeds of these plants, an important source of PROTEIN. BEAN, PEA.

pulverulent *adj.* applied to a rock that easily disintegrates to form a fine powder.

pumice solidified froth on the surface of an ACID LAVA flow, formed by the escape of gases and vapours from the cooling lava, to give a very light, fine-grained, cellular acid rock, capable of floating in water and used as an abrasive in smoothing, cleaning and polishing. Pumice dust forms one of the components of RED CLAY-1 on the deep ocean floor. G.

pumpkin any of the members of *Cucurbita*, family Cucurbitaceae, a climbing plant cultivated in temperate regions for its very large, round, fleshy, edible fruit, cooked as a savoury, or sweet dessert dish. Vegetable oil is obtained from the seeds, which are rich in protein and fat. PEPO, PEPITOS.

puna (South America, Peru: Spanish from Quechua language) **1.** a high bleak intermontane plateau stretching from Peru into Bolivia **2.** the vegetation of the area, cacti, coarse grass and other XEROPHYTIC herbacious plants **3.** the cold PREVAILING WIND blowing in the area. G.

punctuated equilibrium the term popularly applied to the theory of allopatric speciation which suggests that the evolution of species occurs not as a steady, continuous process (DARWIN, EVOLUTION) but by a process of sudden leaps forward, followed by a calm period with litle change; that new species develop rapidly from a small subpopulation of ancestors, often in an isolated area at the limit of the ancestral range, thereby splitting the lineage. CLADISTICS.

punctured *adj.* pierced, perforated, e.g. applied to the shape of a country in which there are ENCLAVES.

punja (Indian subcontinent: Tamil; Hindi) dry land as opposed to NUNJA, wet land. G.

purdah (Persian: Urdu) a curtain, especially one screening women from men or strangers, hence applied to the traditional practice of secluding Indian and some other eastern women of rank. G.

purga (Russian from Finnish purku) a violent snowstorm or blizzard in Siberia. G.

purposive sample synonym for NON-PROBABILITY SAMPLE.

push moraine, push-moraine (less frequently push-ridge moraine, or shoved moraine; French moraine de poussée; Dutch stuwwal) mounds of sand and gravel pushed into broad, smooth, massive, parallel ridges, frequently arc-shaped, on the margin of ice as it advanced over glacial drift from an earlier glaciation. It may be composed of SUPERGLACIAL and ENGLACIAL material dumped in front of the ice (dump moraine); or of local glacial and non-glacial debris which the advancing ice piled up in its path. CONTORTED DRIFT, MORAINE. G.

push-pull theory of MIGRATION-1,2, a theory which suggests that people are pushed by adverse conditions (e.g. overpopulation, poverty, political repression, war, dislike of a development scheme) to leave an area, and are at the same time attracted to another area by what are perceived as favourable conditions (e.g. the likelihood of a better job, higher wages, freedom of movement) in another area. STRENGTH THEORY, WAGE DIFFERENTIAL THEORY.

push wave P-wave, a primary wave, a pressure or compressional wave, a shock wave produced by an EARTHQUAKE, a compressional vibration resembling a sound wave, which passes through solids, liquids and gases. It is termed push wave because each particle is displaced by the wave along the direction of its movement through the earth's crust, mantle and core. GUTENBERG DISCONTINUITY, LONGITUDINAL WAVE, MOHOROVICIC DISCONTINUITY, RAYLEIGH WAVE, SHAKE WAVE, TRANSVERSE WAVE MOTION, WAVE.

puszta (Hungarian, waste) Hungarian STEPPE region, the open treeless plains of temperate grassland in the heart of Hungary, also termed alföld. G.

P-wave PUSH WAVE.

puy (French) a small hill, the cone of an extinct volcano, rising from a plateau in the Auvergne, France. The term does not have a precise definition: the puy may be composed of ash or cinder, of acid lava in the shape of a dome, or it may have a double cone; and there may be evidence of PELEAN or STROMBOLIAN volcanic activity. The term is applied elsewhere to similar hills. G.

pygmy a member of the very small Negrillo people of equatorial Africa, commonly under 137 cm (4 ft 6 in) in height.

pyramid peak, pyramidal peak a HORN, a sharp mountain peak formed when three or more CIRQUES cut back into the original upland or mountain, the faces being bounded by steep ARETES, e.g. the MATTERHORN.

pyrethrum *Chrysanthemum coccineum* **1.** a PERENNIAL plant bearing flowers containing a substance poisonous to insects **2.** an insecticide made from the dried powdered flowers. ORGANIC FARMING.

pyrheliometer an instrument used in measuring SOLAR RADIATION.

pyrite an iron disulphide mineral, FeS_2, popular name fool's gold, occurring as an ACCESSORY MINERAL in IGNEOUS ROCK and in contact METAMORPHIC ROCK, and in HYDROTHERMAL and replacement deposits.

pyrochemical *adj.* of or pertaining to chemical change at high temperature.

pyroclast fragmental material, derived from MAGMA as well as from the wall of the VENT of a VOLCANO, ejected by the explosion of rapidly freed gases in an explosive volcanic eruption. It consists of ash, bombs, cinders, dust, lapilli, fragments of older rocks etc. CLAST, CLASTIC, PUMICE, PYRO-CLASTIC, SCORIA, TEPHRA, and entries under VOLCANIC. G.

pyroclastic *adj.* applied to a rock formed from the debris of an explosive volcanic eruption. CLAST, CLASTIC, PYROCLAST.

pyroligneous acid an acid fluid obtained by the distillation of wood, the main source of acetic acid.

pyrolusite a naturally occurring MANGANESE DIOXIDE, one of the main ores of MANGANESE, found in sedimentary rock, as a residual deposit of LEACHING, and in deep sea MANGANESE NODULES.

pyromagma a fluid, very hot basaltic LAVA, highly charged with gas, as found in Hawaii. G.

pyrometamorphism THERMAL (CONTACT) METAMORPHISM. G.

pyrometer an instrument used to measure temperatures higher than those measured by a normal mercury thermometer, e.g. as used in measuring molten LAVA. It may use as an indicator the degree of intensity of the light being emitted by the subject; or the changes in electrical resistance resulting from the heat being generated and the expansion of gases.

pyrometry the measuring of very high temperatures. PYROMETER.

pyrope a fiery, dark red variety of GARNET, used as a gemstone.

pyrophyte a plant adapted to withstand fire, in many cases able to reproduce only after being subjected to fire.

pyroxene a group of silicates (SILICA-2) varying widely in composition, occurring in BASIC and ULTRABASIC IGNEOUS ROCK. The group includes jadite (JADE), AUGITE and other complex minerals. ALKALI ROCK, ECLOGITE, FERROMAGNESIAN MINERAL, SERPENTINE.

Q

qadha (Iraq: Arabic) LIWA. G.

qanat, kanat (Persian from Arabic) an underground conduit, karez or foggara, carrying water from higher land to the plains in Iran, in many cases over very long distances. G.

qasr (Arabic) a fortress, usually one in a rural area. G.

qoz (Sudan: Arabic, popular term in northern Sudan, a sand dune) an extensive area of DUNES and undulating sheets of sand deposited during an arid period of the PLEISTOCENE in Kordofan and Darfur, and now fixed by varied forms of SAVANNA. G.

quadrant 1. one quarter of the circumference of a circle 2. one quarter of the area of a circle, i.e. the part bounded by the quarter of the circumference and the radii drawn from each end of it to the centre of the circle; or of something having this shape 3. one quarter of a spherical body 4. an obsolete instrument, usually with a calibrated 90° arc, used for measuring altitude and angles. SEXTANT.

quadrat one of the equal-sized sample areas (usually a square) used in QUADRAT ANALYSIS. In plant ecology a quadrat of one metre square is commonly used for sampling in order to gain an accurate statistical record of the composition of low plant cover, but for tree cover the quadrat has to be much larger.

quadrat analysis a statistical technique derived from one used in plant ecology (QUADRAT). The area covered by a point pattern is divided into equal-sized sample areas, and the number of points occurring within each is counted. The observed distribution can then be tested against a theoretical pattern of distribution.

quadrature 1. in astronomy, the position when a celestial body lies at 90° or 270° to another, e.g. when the sun, earth and moon form a right angle, the earth being the apex. This occurs twice each LUNAR MONTH with the result that the tide-producing effects of the moon and sun are in OPPOSITION, producing tides of low range. NEAP TIDES 2. one of the two points on the orbit of a celestial body midway between the syzgies (SYZGY). CONJUNCTION, MOON.

quadroon a person whose blood is one-quarter African, the child of white and MULATTO parentage, especially in South America. OCTOROON. G.

quadruped four-footed, especially a four-footed animal, e.g. a MAMMAL such as an OX.

quagmire in general applied to any soft, wet ground, but specifically a quaking bog, an area of soft, wet ground so soft that it quakes or trembles when trodden on. BOG. G.

quake an EARTHQUAKE.

quaking bog QUAGMIRE. G.

qualitative *adj.* relating to, concerned with, or involving, QUALITY.

qualitative variable in statistics, usually, a VARIABLE-2 which may be measured on nominal or ordinal scales, commonly termed an ATTRIBUTE. Some statisticians however apply the term qualitative variable to a variable measured on the nominal scale only, maintaining that the use of ordinal, interval or ratio measurement (MEASUREMENT IN STATISTICS) to display increasing quantities of an attribute involves QUANTIFICATION. Thus a quantitative variable may refer to a variable measured on ordinal, interval or ratio scales, or a variable measured on interval or ratio scale only. This difference in approach arises from the fact that all variables can be represented by numbers.

quality 1. grade, degree of goodness, worth **2.** an attribute, trait, characteristic.

quality of life the degree of goodness of the conditions of life and of the life-style of a person. Objectively it may be possible to assess the degree of goodness by the extent to which SOCIAL WELL-BEING is achieved. But 'one person's meat is another's poison', and judgement as to what should be included among the factors contributing to social well-being (apart from the basic sufficiences of shelter, food, water, clothing etc.), let alone the evaluation of the importance of each factor, can only be subjective, varying from one individual to another, from one society to another, from one time to another. Perhaps the term quality of life may therefore best be interpreted as people's subjective feelings of satisfaction with their living conditions and life-style.

quango a quasi-autonomous non-government organization, a government appointed body or agency in the UK which is financed by, but is not part of, a government department.

quantifiable *adj.* measurable in terms of quantity, conceivable or treatable as a QUANTITY.

quantification the action of measuring QUANTITY, the expression of a property or QUALITY in numerical terms.

quantitative *adj.* relating to, concerned

with, quantity or with the measurement of QUANTITY.

quantitative analysis the numerical manipulation of data to test HYPOTHESES and to produce summary statistics.

quantitative revolution a movement in approach to geographical studies in the 1950s and 1960s concerned with the use of statistical and mathematical methods and techniques (e.g. the use of explanatory MODELS-4 and REGRESSION, CORRELATION, VARIANCE and COVARIANCE ANALYSIS) to analyse associations in the attempt to produce objective systems of classification and theories of spatial organization. G.

quantitative variable QUALITATIVE VARIABLE.

quantity 1. an amount, sum or number **2.** a great deal, very many **3.** the property of things that can be measured **4.** in mathematics, anything which is measurable, or a figure or symbol used to represent this.

quaquaversal, quâquâversal, quâ-qua-versal *adj.* formerly applied by geologists to strata dipping away in all directions from a central point, a structure now termed a structural dome of PERICLINE. G.

quarry an open excavation on the surface of the earth, worked usually for the extraction of rocks and certain non-metallic minerals. Until the mid-nineteenth century the term was applied to a mine as well as to surface working. MINE, MINING, OPENCAST MINING, PIT. G.

quarrying 1. the extracting activity conducted in a QUARRY, limited in extent when compared with OPENCAST MINING **2.** the eroding of its channel by a young stream by the lifting effect of water as it penetrates cracks in rocks, termed PLUCKING-2 by some authors. G.

quart any of various units of liquid or dry capacity, especially in British measures a unit equal to one-quarter of a British GALLON or 69.355 cubic inches, or 1.1361 LITRES; or, in American, a unit equal to one-quarter of a liquid gallon or 57.75 cubic inches, or 0.9461 litres; or, in dry capacity (American) equal to 1/32 of a bushel or 67.200 cubic inches, or 1.101 litres. CONVERSION TABLES.

quartile one of the four equal parts of a data distribution. INTERQUARTILE RANGE, MEDIAN.

quartile deviation half the INTER-QUARTILE RANGE. MEDIAN.

quartile dispersion coefficient INTER-QUARTILE RATIO.

quartz silicon dioxide, SiO_2, one of the commonest minerals in nature. It is characteristic of ACID IGNEOUS ROCKS, is resistant to chemical WEATH-ERING, is abundant in SEDIMENTARY and METAMORPHIC ROCKS, often fills JOINTS, VEINS and cavities, sometimes mixed with other MINERALS, as well as ORES, so stable that it usually consti-tutes the majority of sand grains in a sand or SANDSTONE. When pure (ROCK CRYSTAL) it is like clear glass, but much harder (no 7 on the HARD-NESS scale), and occurs in hexagonal crystals and massive form, usually colourless and transparent, but also in coloured (AMETHYST), translucent and opaque forms. Rigid, it does not easily expand and is therefore used in making heat-resistant apparatus. It is also used in radio-transmitters and astronomical clocks because the opposing faces of the crystals are able, under pressure, to take up opposite electrical charges; conversely there is a change of volume if an electromotive force is applied. FELSIC.

quartzite a very hard, resistant and IMPERMEABLE rock, consisting mainly of QUARTZ re-cemented by SILICA, some forms being metamorphic in origin (METAMORPHISM) and termed metaquartzite, others sedimentary (orthoquartzite). Many quartzites are quarried for AGGREGATE-3. CAM-BRIAN.

quartzose sandstone orthoquartzite, QUARTZITE.

quasi-exclave EXCLAVE.

quassia a drug extracted from the roots and heartwood of several members of Simaronbaceae, a family of tropical trees. It has long been used as an insecticide and in medicaments.

Quaternary *adj.* **1.** of the most recent period (time) or system (rocks) of the CAINOZOIC era, i.e. following the PLIOCENE of the TERTIARY period or system (GEOLOGICAL TIMESCALE), and including the PLEISTOCENE and HOLOCENE **2.** applied by some geolo-gists to the fourth era in the sequence of geological time, i.e. post-Cainozoic, post-PLIOCENE, commencing about the time of the onset of the most recent ICE AGE. G.

quaternary industry, quaternary activity, quaternary sector the activities in the TERTIARY INDUSTRY which are con-cerned with research, with the assem-bly, processing and transmission of information, and with administration, including the control of other industrial sectors. INDUSTRY, PRIMARY INDUS-TRY, SECONDARY INDUSTRY. G.

quebracho (South America: Spanish from quebrar, to break; hacha, axe) any of several EVERGREEEN tropical trees of South America with exception-ally hard wood, yielding useful sub-stances and providing durable timber for posts etc. The bark of the white quebracho, *Aspidosperma quebracho*, family Apocynaceae, of Chile and Argentina, contains substances used medicinally; and that of the red que-bracho, *Schinopsis Lorentzii*, family Anacardiaceae of Argentina, yields TANNIN used in dyeing and leather processing. CHACO, IRONWOOD.

quicklime CALCIUM OXIDE.

quicksand a thick mass of unstable, fine loose sand, sometimes mixed with mud, supersaturated with water, occurring on some coasts and near river mouths, liable to suck down any heavy object that comes to rest on it. SUPERSATU-RATION. G.

quicksilver MERCURY.

quince *Cydonia vulgaris*, family Rosa-ceae, a small tree, native to western Asia, now cultivated in temperate cli-mates, bearing hard, acid pear-shaped yellow fruit, cooked, used especially in making jam, jelly and as a flavouring. The tree itself is used as dwarfing rootstock for pears. A quince-like fruit, even more acid than the true quince, is born by the *Japonicas* (now classed as *Chaenomeles*) and used in making jam, jelly, and especially in flavouring dishes made with apples or pears.

quinine an ALKALOID obtained from the bark of CINCHONA CALISAYA, long used medicinally in the treatment of malaria and heart diseases, now mainly for flavouring tonic water or quinine water.

quinta (Spanish: Portuguese) a country mansion, especially in Spanish-speak-ing South America. G.

quintal a unit of weight, equal to 100 kilograms, or 100 pounds (lb), or a hundredweight (112 lb). CONVERSION TABLES.

quograph, quo-graph a term applied by T. Griffith Taylor, 1938, to a graph giving the same results as a slide-rule. He defined it as 'a device for determin-ing quotients (e.g. densities), by the intersection at the diagonals of the ordinates for dividends (e.g. crops) and divisors (e.g. areas). It is a simple form of nomograph, and serves the same purpose as a slide rule.' G.

quotient the result obtained after divid-ing one number by another.

qurer (Sudan: Arabic) silt land along the river bank immediately above the highest flood level attained by the Nile, and irrigated by SHADUF or SAQIA. G.

R

ra (Norwegian) a ridge of gravel and clay with a surface layer of large stones, a moraine, occurring generally in or near the sea, especially in southern Norway. G.

rabi (Indian subcontinent: Urdu-Hindi) **1.** the cool dry or winter season in the north of the Indian subcontinent **2.** a winter-sown crop harvested in the spring. BHADOI, KHARIF, HEMANTIC. G.

race 1. classificatory term, in broad terms equivalent to a SUBSPECIES, but sometimes specifically and unscientifically applied to a distinct group of the single species *Homo sapiens* (from which all persons living today are descended). Such a group is identified as possessing well-developed and primarily heritable physical characteristics which differ from those of other groups (e.g. head shape, hair character, skin colour, facial features etc.), a biological concept of little value when applied to the world population of today, there having been much migration and inter-breeding since the time of the appearance of *Homo sapiens* on earth. RACISM **2.** in biology, a subdivision of a species, a permanent variety or a particular breed **3.** swiftly flowing water in a narrow channel, natural (as in a river) or artificial, controlled to provide power, as in the channel leading river water to the wheel of a watermill (HEAD-RACE) or from the mill (TAIL-RACE) **4.** tidal race, a rush of sea water through a restricted channel occurring where there are considerable differences in the tides at each end of it **5.** an offshore current flowing strongly round a headland. G.

racecourse a track, often with associated buildings, laid out and prepared for racing, especially for horse racing.

racism 1. the assumption that the abilities and characteristics of a person are determined by RACE-1, and that biologically one race is inherently superior to another. 'Race' in this context is often arbitrarily extended to include religious sects, as well as national, linguistic and cultural groups. ETHNOCENTRISM **2.** a political programme or social system based on such assumptions.

rack rent the rent charged to the actual user of a property by the freeholder or head leaseholder, as distinct from GROUND RENT. LAND TENURE.

radar acronym of Radio Detection and Ranging, a device for determining the presence, distance or speed of movement of an object by means of transmitting MICROWAVES at it and measuring by electronic devices the speed of the microwaves' return after reflection from the object. It is used especially in air and sea navigation, in tracking satellites and missiles and in automatic guidance; in land use and geological studies, and in measuring and locating atmospheric phenomena (thus of great benefit to weather forecasters). LIDAR, REMOTE SENSING, SLAR, SONAR.

Radburn layout in town planning and urban studies, a planned urban layout, developed by Clarence Stein, applied in Radburn, New Jersey, USA in 1928, which separates pedestrians from vehicles by arranging that 'superblocks' of housing, shops, offices, schools etc. enclose a central green or pedestrian space. Each superblock has its peripheral ring roads, off which come service cul-de-sacs. The central green or pedestrian space has pedestrian access only, by underground passages or surface walks.

radial drainage a pattern of DRAINAGE produced when streams flow down and outward from a central dome or cone-shaped upland, occurring especially in mountainous areas or on volcanic cones. G.

radian a unit of angular measurement, the angle subtended at the centre of a circle by an arc equal in length to the radius; 1 radian = 57.29578°.

radiance in REMOTE SENSING, the spatial distribution of RADIANT power density.

radiant in physics, a point from which radiant energy is emitted. RADIATION.

radiant *adj.* emitting energy in the form of ELECTROMAGNETIC WAVES. ENERGY, RADIATION.

radiant energy the ENERGY originally transferred from the sun.

radiant exitance in REMOTE SENSING, the measure of RADIANT ENERGY (ENERGY) per unit area leaving the surface or object which is undergoing sensing.

radiant flux, radiant power in REMOTE SENSING, the time-rate of the flow of RADIANT ENERGY (ENERGY) measured in watts. RADIANCE, IRRADIANCE, RADIANT EXITANCE.

radiate *verb* 1. to send out something in all directions from a central point 2. to emit RADIATION.

radiation the act or process of radiating, i.e. of propagating or transmitting energy in the form of particles or ELECTROMAGNETIC WAVES, e.g. an electric light bulb emits radiation in the VISIBLE and INFRA-RED regions of the ELECTROMAGNETIC SPECTRUM (ELECTROMAGNETIC WAVE); radiant

ENERGY is emitted by the SUN (INSOLATION, SOLAR CONSTANT, SOLAR RADIATION); and there is loss of heat from the surface of the earth by ground radiation, i.e. long-wave radiation emitted by land or water surfaces and passing up into the ATMOSPHERE (TERRESTRIAL RADIATION), especially on a clear night. ACTINOMETER, ATMOSPHERIC WINDOW, BLACK BODY, CONVECTION, COSMIC RAY (cosmic radiation), GREENHOUSE EFFECT, THERMAL CONDUCTION.

radiation fog, ground fog a layer of white FOG occurring particularly in low-lying areas when the weather is settled, the sky is clear, the air near the earth's surface is calm and moist and there is little wind, e.g. in Britain, in spring and autumn. In such conditions the ground surface cools quickly at night by RADIATION, cooling the overlying moist air, which flows downward by gravity to the hollows where, cooled to DEW POINT, it is condensed. Such fog usually disappears when the sun rises, but it may persist if the layer is thick and it lies under a layer of TEMPERATURE INVERSION. It may also lead to SMOG.

radical in chemistry, an ATOM or group of atoms that retains its identity despite chemical changes in the rest of the molecule. VALENCY.

radioactivity, radioactive decay the property possessed by some natural elements (e.g. URANIUM) and many synthetic elements of spontaneously decaying and in the process emitting ionizing particles which may consist of an alpha particle (thus reducing the atomic number of the element by 2 and mass number by 4) or a beta particle (increasing atomic number by 1, mass number unaffected), accompanied by electromagnetic radiation from the nuclei of individual atoms, usually at a constant known rate (HALF-LIFE) to form ISOTOPES, e.g. uranium^{238} to lead^{206} (LEAD-RATIO). ARTIFICIAL

RADIOACTIVITY, DATING, POTAS-
SIUM, NATURAL RADIOACTIVITY,
RADIOCARBON DATING, RADIO-
METRIC AGE.

radio astronomy the branch of astronomy concerned with radio frequency
radiation (i.e. electromagnetic radiation with frequences of RADIOWAVES)
from bodies and areas beyond the
earth's atmosphere using such instruments as radio telescopes and radio
interferometers. Maps of space can be
constructed to show the source of the
radio noise located by these devices,
and such maps compared with others
based on optical observation. By using
the techniques of radio astronomy it
has been possible to identify radio
sources (most of which are galaxies)
outside the range of optical observation.

radiocarbon dating a method of
DATING proposed by W. F. Libby in
1946 to determine the age of prehistoric
organic remains, such as wood, bone,
shells etc. (effective for material up to
3000 years in age, some would say
fairly accurate up to 20 000 years, the
accuracy diminishing beyond 30 000
years). Very broadly, the method is
based on the assumption that carbon
14, a rare radioactive ISOTOPE of
CARBON incorporated in organic matter, diminishes at a known rate after the
death of the organism. Carbon 14, with
a HALF-LIFE of about 5600 years, is
formed in the upper atmosphere by the
bombardment of nitrogen by particles
from COSMIC RAYS, oxidizes to carbon
dioxide, enters the CARBON CYCLE and
is absorbed by all living organisms. The
proportion of carbon 14 to carbon 12 is
constant in the atmosphere and in
living organisms. But when an organism dies it ceases to absorb carbon
dioxide and the proportion of carbon
14 to carbon 12 decreases with the
decay of the unstable radioactive isotope. With the knowledge of the half-
life of carbon 14 and the ratio of

carbon 14 to carbon 12 in a sample, it is
thus theoretically possible to calculate
the age of the sample at the time of
death. But recent research suggests that
to improve accuracy the 'old half-life'
of some 5600 years should be replaced
by the recalculated half-life of 5730
years (the 'preferred half-life'); and
comparison between radiocarbon dates
and known historical dates has revealed
discrepancies. These discrepancies have
been confirmed by comparing the age
of tree rings as revealed by radiocarbon
dating with their known age (DENDRO-
CHRONOLOGY). It seems that the production of carbon 14 is not constant
through time, and that to reach a
more satisfactory measure radiocarbon
years should be compared with the
known tree ring dates, thus giving 'real
years'. Radiocarbon years are now
sometimes expressed as bc, and the
recalibrated approximation to 'real
years', rather confusingly, as BC. Dates
are usually given as BP (before the
present day).

radiogenic heat the natural heat produced in the upper levels of the earth's
continental crust by the radioactive
decay (RADIOACTIVITY) of POTAS-
SIUM, THORIUM and URANIUM.

radiogenic isotope an ISOTOPE formed
from the breakdown of a natural
parent element, particularly important
in RADIOCARBON DATING, RADIO-
METRIC AGE. RADIOACTIVITY.

radioisotope any ISOTOPE of an element which is radioactive. RADIO-
ACTIVITY.

Radiolarian ooze a type of OOZE occurring in the deeper parts of the tropical
oceans, especially the Pacific, characterized by an abundance of the minute
siliceous shells of Radiolaria which are
less soluble in sea water than the
calcareous shells of Foraminifera which
are apt to be dissolved by sea water
before they reach the greatest depths.
But in the deepest parts even the
siliceous Radiolarian shells are dis-

solved, and there the bottom deposit is RED CLAY.

radiometer in REMOTE SENSING, any device that measures/records RADIANCE from, or IRRADIANCE on to, a surface.

radiometric age of rocks, the age of rocks measured in years before the present (i.e. before 1950) as revealed by the time taken for a particular ratio of daughter atoms to parent atoms to be formed by natural radioactive decay of the parent atom (NATURAL RADIOACTIVITY, RADIOACTIVITY). The HALF-LIFE of a particular natural radioactive element (commonly URANIUM 238 or 235, or POTASSIUM) is used in the calculation, which is based on the assumption that the system is a closed one (CLOSED SYSTEM) and that none of the daughter atoms existed in the original material. ISOTOPE, LEAD-RATIO, RADIOCARBON DATING.

radiosonde (French) a self-recording, radiowave-transmitting instrument used in meteorology for measuring atmospheric pressure, temperature, humidity, and winds at various levels in the upper atmosphere, carried aloft by a balloon filled with hydrogen. BALLOON-SONDE, RAWINSONDE, ROCKETSONDE.

radiowaves waves in the ELECTROMAGNETIC SPECTRUM with wavelengths ranging from 0.1mm (extremely high frequency) to 30 km (very low frequency), and including the RADAR band. HEAVISIDE LAYER.

radium a strongly radioactive white divalent (VALENCY) metallic element which disintegrates into radon (a radioactive gaseous element). First isolated by Pierre and Marie Curie, it is used in radiography and radiotherapy, and in luminous paints. RADIOACTIVITY.

radish *Raphanus sativus*, a quickly maturing annual plant, known to have been grown by the Egyptians over 2000 years ago, now cultivated in most parts of the world. It has an edible, crisp, pungent tuberous taproot, eaten raw.

raffia bass, the fibre obtained from the leaves of *Raphia ruffia*, a low-growing raffia palm native to Madagascar. Raffia is used for tying up small plants, weaving into matting, etc. C.

rafting of ice, the overriding of one ice floe on another, a mild form of pressure, producing rafted ice. PRESSURE ICE. G.

rag, ragstone in Britain, a hard, coarse and rough stone, notably the calcareous sandstone of the Lower Greensand in Kent, known as Kentish Rag, of Lower CRETACEOUS age, used as a building and road stone. G.

rail (railway) gauge the width in the clear between the top flanges of the rails, the parallel metal lines on which trains run. The standard gauge used through Europe (except Spain, Portugal, the USSR and certain European countries linked to the USSR) as well as throughout North America and parts of Australia is 1435 mm (4 ft 8½ in). Broad gauges of 1600 mm (5 ft 3 in) and 1650 mm (5 ft 6 in) are used in Spain, Portugal, USSR, parts of the Indian subcontinent, Australia and South America. Narrower gauges, especially 1066 mm (3 ft 6 in) or less, are used in South Africa and parts of Australia; and the metre gauge (3 ft 3⅜ in) in many parts of the world, including the Indian subcontinent. G.

railway (American railroad) 1. a system of transport in which trains run on parallel metal lines laid on easy gradients 2. the equipment, organization and people operating such a system 3. the tracks for the trains of such a system. RAIL GAUGE.

railway yard an area of land with a complex system of tracks, usually near a railway station or terminus, where ROLLING STOCK (especially carriages and wagons) is shunted and parked and trains are assembled.

rain drops of water, large enough to fall under the influence of gravity from CLOUDS to the earth's surface, formed by the COALESCENCE of water droplets produced by condensation of WATER VAPOUR in the atmosphere. BAI-U, BLOOD RAIN, CONVECTIONAL RAIN, CYCLONIC RAIN, DRIZZLE, MAI-YU, MIST, OROGRAPHIC PRECIPITATION, PRECIPITATION-1, RAINS. G.

rainbow an arc of concentric bands of light in the colours of the SPECTRUM seen in the sky when the sun is behind and RAIN is in front of an observer, caused by the reflection and refraction of sunlight in the water drops. The larger the drops, the brighter the colours. FOGBOW.

rain-day, rainy day, day of rain in UK, a period of 24 hours, commencing normally at 0900 hours, on which 0.25 mm (0.01 in) or more of RAIN is recorded. PRECIPITATION-DAY. G.

raindrop RAIN.

raindrop erosion soil erosion caused by large raindrops falling on bare earth, occurring especially in tropical and semi-arid areas. The raindrops dislodge soil particles, or cause SOIL COMPACTION, which leads to increased surface RUNOFF.

rain factor the ratio between mean annual temperature in °C and annual PRECIPITATION in mm, used to indicate climatic ARIDITY, an index devised by R. Lang as an aid in delimiting climatic regions.

rainfall the quantity of RAIN falling in a certain time within a given area, usually expressed in millimetres (mm) or inches (in). Unless otherwise stated in statistics, snow and hail (converted to water equivalent) are included. For calculation of averages a period of 35 years is commonly used, though many records are based on a shorter period. DROUGHT, PRECIPITATION-1, RAIN-DAY, RAIN SPELL. G.

rain forest, rain-forest, rainforest FOREST-2 composed mainly of EVER-GREEN hygrophilous trees (HYGROPHYTE). If the term is strictly applied it should be restricted to such a forest growing in moist TROPICAL lowlands only in non-seasonal tropical climates, i.e. with evenly distributed rainfall (EQUATORIAL FOREST), e.g. of the Amazon and the Zaire basins, the islands of Borneo and New Guinea. But if more generally applied, it may include the somewhat less luxuriant EVERGREEN forest with some DECIDUOUS species growing at low altitudes on tropical mountains, or the evergreen forests of OCEANIC subtropical climates in southwestern China, southern Africa, eastern extra-tropical Australia, and New Zealand. CANOPY, TROPICAL FOREST. G.

rain gauge an instrument used in measuring rainfall, consisting of a funnel with the diameter of the mouth 12.5 cm (5 in) or 20 cm (8 in), the rim ideally 30 cm (12 in) from the ground, fitting closely into a collecting container, the water collected being periodically measured in a vessel graduated to the area of the mouth of the funnel. It should be positioned at a distance from the nearest building, ideally at a distance equal to twice the height of that building.

rains, the the rainy season or wet season, also termed the MONSOON season in India and other monsoon countries. In India the rains 'break' between early June and early July, especially about 15 June, and last until September - October. G.

rainshadow an area with a relatively small average rainfall occurring on the LEE side of a high land barrier, e.g. a mountain barrier. The high land gives rise to OROGRAPHIC PRECIPITATION on the windward slopes, thereby reducing the moisture content of the airstream on the leeward, which is warmed and dried further as it descends. CHINOOK, FOHN. G.

rain spell in Britain, a period of at least

fifteen consecutive RAIN-DAYS on each of which at least 0.25 mm (0.01 in) of rain falls. G.

rain-wash, rainwash   hill wash, hill-wash **1.** the surface creep of soil and weathered rock down a slope under the influence of gravity aided by rainwater. MASS MOVEMENT **2.** material which originates in this way. G.

rainy day   RAIN-DAY.

raised beach   **1.** a former beach of a sea (or lake), sometimes with a cliff at the rear and a wave-cut platform in front, covered with ancient beach deposits, now situated above the present sea (or lake) level as a result of NEGATIVE MOVEMENT of sea-level or of ISOSTASY. If it lies 40 to 45 m (130 to 150 ft) above the present sea-level, it is usually termed a marine erosion surface, marine platform, marine terrace. The term raised beach does not apply to a beach abandoned by a lake which is drying up or being drained **2.** the ancient beach deposits themselves. G.

raised bog   a deep, lens-shaped BOG occuring in a shallow basin having grown above the WATER TABLE and therefore dependent on the precipitation-evaporation rate. The growth thickens towards the centre, e.g. as in central Ireland. It may provide PEAT for fuel, used domestically and in power stations.   BLANKET BOG, VALLEY BOG.

raisin   the dried fruit of any of several varieties of GRAPE.

raj   (Indian subcontinent: Hindi) rule, reign.

rake   (Scottish) **1.** a well-defined group of patches of different types of vegetation over which hill sheep move in a fairly regular daily pattern **2.** a sloping terrace on a mountain side or rock face. G.

ram   **1.** the underwater projection of ice from an iceberg, ice floe, ice front or ice wall **2.** a male SHEEP. WETHER. G.

ramie grass   CHINA GRASS.

raml, ramla   (north Africa: Arabic,

sand) **1.** el raml, the sand, the desert **2.** in Spain, ramla, ramlah, the bed of an ephemeral river when dry. G.

rañas   (Spanish) FANGLOMERATE (REG in the Sahara) consisting of a surface of rolled pebbles. G.

ranch   a FARM where cattle, horses or sheep are bred and reared, especially on a large scale in areas once covered with grassland in southern and western USA,   southwest   central   Canada, Argentina, Uruguay. The animals used to roam to find food on the open RANGE-3, but now they are mainly confined to enclosures and fed on fodder crops grown on the farm. DUDE RANCH.

ranching   the activity of breeding and rearing animals on a large scale on a RANCH.

rand   (Afrikaans rant, pl. rante) **1.** historically in South Africa rand was applied to a line of hills, especially when forming an escarpment, thus Witwatersrand is White Water Ridge; and although linguists prefer the spelling rant, the gold-bearing reef near Johannesburg has become universally known as The Rand **2.** the South African unit of currency. G.

Randkluft   (German) a gap (with one wall of rock, the other of ice) between the rocky mountain wall of a CIRQUE and the ice mass which occupies it, created when the ice, later to become a GLACIER, begins to move away and downwards. Radiation of heat from the rock wall makes the ice melt, so there is no ICE APRON left attached to it, in contrast to a BERGSCHRUND with its two ice walls. G.

random   *adj.* applied in mathematics to numbers as likely to come up as any others in a set.

random error   in statistics, a deviation from observed true value, occurring as if it had been chosen at random from a PROBABILITY-2 distribution of such errors. Over a large number of cases random errors balance out because, unlike SYSTEMATIC ERRORS, they do not occur predominantly in one direction.

random numbers sets of numbers drawn purely by chance, e.g. by rolling a dice. APPENDIX 3, RANDOM SAMPLING.

random sample a probability sample, a SAMPLE-3 which has been selected by a method of random selection (RANDOM SAMPLING). SIMPLE RANDOM SAMPLE.

random sampling, random selection probability sampling, probability selection, a method of selecting SAMPLE UNITS based on the theory of PROBABILITY, so that each sample unit has a fixed and known chance of selection. Random selection is usually carried out by reference to tables of RANDOM NUMBERS which are generated by a computer in such a way that all the numbers appear an equal number of times in the overall table.

random sampling error the SAMPLING ERROR occurring in cases where the SAMPLE-3 has been drawn by random selection (RANDOM SAMPLING), termed sampling error where it is assumed or understood that random selection has been used.

random variable VARIATE.

Randstad (Dutch) in the Netherlands, a ring-city, a circular CONURBATION consisting of Amsterdam, Haarlem, Leiden, 's Gravenhage, Rotterdam and Utrecht, interspersed with agricultural land devoted mainly to market gardening and glasshouse cultivation and containing within the central area agricultural land cultivated to meet the food needs of the urban ring. G.

range 1. a row or line of things 2. a single line of mountains forming a connected system. CHAIN, MOUNTAIN CHAIN 3. a natural or semi-natural grazing area, usually unenclosed, over which animals may roam in search of food. RANCH 4. the difference between the least and the greatest of a series of numerical values, e.g. of pressure, ele-

vation, temperature, the difference made clear by the context, e.g. the difference between average day temperatures and average night temperatures (the daily range), or the difference between average winter and summer temperatures (annual range). ABSOLUTE RANGE, MEAN DIURNAL RANGE 5. the maximum attainable distance, e.g. of a missile 6. the maximum distance a vehicle can travel without refuelling 7. in ecology, the limited area within which an organism is distributed, i.e. the limit of its habitat which affords it an appropriate NICHE; or the period within which it occurred, e.g. in PALAEONTOLOGY 8. in mining, a mineral belt, e.g. the Mesabi range 9. specifically in the USA PUBLIC LAND SURVEY, the row or line of congressional townships between two successive meridians 9.6 km (6 mi) apart, the rows being identified by numbering in sequence from east or west from the PRINCIPAL MERIDIAN 10. of a good, in CENTRAL PLACE THEORY, the maximum distance consumers are willing to travel to obtain a good or service. ORDER OF GOODS. G.

ranger 1. a person living on a RANGE-3 2. an official, such as a gamewarden or officer, who looks after expanses of woodland or forest 3. in USA, an official who patrols a public forest. G.

rank 1. position in a hierarchy. RANKING, RANK-SIZE RULE 2. the stage reached by COAL in the progress of its CARBONIFICATION and change in physical composition, from LIGNITE (with the lowest CARBON content) to ANTHRACITE (with the highest).

ranker in soil science, soils with an A C horizon developed on silicate (SILICA) rocks or sediments, the content of the organic matter in the A HORIZON being much higher than in the A horizon of non-calcareous LITHOSOLS or REGOSOLS. SOIL, SOIL ASSOCIATION, SOIL HORIZON. G.

ranking the action of arranging, assigning RANK, thus the arrangement of numerical data in sequence, in rank order, according to a specific quality, without making any assumptions about the intervals between the ranks.

rank-size rule   an empirical rule describing the distribution of town or city sizes in an area. It states that if a set of towns in an area are ranked in descending order of size of population, the population of any given town will be inversely proportional to its rank in the list, i.e. the population of any given town tends to be equal to the population of the largest town in the set divided by the rank of the given town; e.g. if the population of the largest town numbers 100 000, the population of the fifth largest town will be 20 000 (100 000 divided by 5). PRIMATE CITY. G.

rank society   a pre-capitalist SOCIETY-3 (CAPITALISM-1) in which there are more people qualified to perform high-level jobs than there are positions available. This creates a privileged ELITE, a ranked social structure, but the elite has no greater claim on the basic resources on which life depends than the mass of the population. A redistribution process replaces the barter system of the EGALITARIAN SOCIETY, and there is a flow of goods, sustained by the use of force, coercion or religious ideology, towards the elite who live or take part in rituals in a centre. Thus, it is considered, an economic surplus may accumulate in the centre, and URBANIZATION becomes possible. STRATIFIED SOCIETY.

rant   (Afrikaans) RAND. G.

rantjie   (Afrikaans) a narrow ridge of rocky hills, a diminutive form of rant. RAND. G.

rape   1. *Brassica napus*, an ANNUAL herb of the Cruciferae family, origin uncertain, grown widely in Europe, North America and Asia for cattle feed, for spring 'greens' for cooking, in seedling form for salad; or dug in as a GREEN MANURE (MUSTARD). The

seeds are pressed for their oil, used in cooking and as a lubricant, the residue being pressed to form nutritious cattle cake (OIL CAKE) 2. one of the six administrative divisions of Sussex, England, recorded in the DOMESDAY BOOK.

rapids   (rarely rapid) part of a stream where the water is relatively shallow and the rate of flow is accelerated by continuous and unbroken increased slope of the stream bed or by a gently dipping outcrop of hard rocks, causing the flow of water to become swift, turbulent and broken. CASCADE, WATERFALL.

rare   *adj.* applied to 1. some thing or event infrequently occurring, seen, experienced 2. a gas, air etc. which is not dense, lacking tight packing of its components. RARE GAS.

rare earth, rare element   any of the widely distributed but scarce group of oxides of the lanthanide elements (with atomic numbers fom 57 to 71, atomic weight 138.92 to 175.00), e.g. CERIUM, THORIUM. ACCESSORY MINERAL.

rare earth metal   a RARE EARTH element. THORIUM.

rare gas   an inert gas, i.e. a gas which does not normally form chemical compounds.

raspberry   *Rubus idaeus*, a tall growing shrub of annually renewed prickly canes, native to temperate regions of the northern hemisphere, hardy, widely cultivated for its sweet juicy fruit (a BERRY), consisting of many round, one-seeded small DRUPES attached to a conical receptacle. It is eaten fresh or cooked, is used in making jams, jellies, drinks, and commonly canned or frozen.

rasputitsa   (Russian) in Siberia and northern Russia, the period of spring characterized by floods and mud resulting from the thaw, making roads difficult or impossible to use.

ratafia   a liqueur flavoured with almonds or the kernels of the peach, apricot, cherry.

rate 1. the amount of something in relation to some other thing, e.g. BIRTH RATE **2.** a fixed ratio between two things, quantities etc., e.g. rate of exchange **3.** the speed of motion or change **4.** in Britain, a local property tax.

rath (Irish) **1.** historically, a fortified enclosure where the chief of a tribe lived, or a hill-fort **2.** a mayor's settlement **3.** an isolated farm settlement (as contrasted with a CLACHAN). G.

ratio the relation between two quantities of the same kind, indicated by a colon (e.g. 3 : 2), and expressed by dividing the magnitude of one by that of the other. Thus 3 : 2 where the division is implied but not carried out, or as $3/2 = 1.5$, indicating that for each unit in the denominator there are 1.5 in the numerator.

ratio scale in statistics, a MEASURE-MENT scale which is the same as, but more precise than, the INTERVAL SCALE, because it has a true, fixed zero point. Thus equal proportional variations in the data correspond to equal absolute variations on the scale. A VARIABLE measured on this scale is termed a ratio variable.

ratio variable a VARIABLE measured on the RATIO SCALE.

rattan, ratan any of several climbing palms native to the Indian subcontinent and southeast Asia, especially of the genera *Calamus* and *Daemonothops*, providing long stems used in wicker-work, and the making of ropes, chair seats, nets etc. CALAMUS.

Raunkiaer's life forms a system of classification of plants based on their morphological adjustment to local conditions, specifically on the type of organs present in the plant which indicates how it survives the unfavourable season (i.e. the PERENNATING parts) and the relation of the position of those organs to soil level (CHAMEO-

PHYTES, CRYPTOPHYTES, GEO-PHYTES, HELOPHYTES, HEMI-CRYPTOPHYTES, HYDROPHYTES, PHANEROPHYTES, THEROPHYTES). The life form, or growth form, of a plant is the form which its vegetative body produces in response to all the life processes, including those that are induced by the environment within the lifetime of the plant and are not heritable. Different individuals of the same species, growing in different environments, may therefore exhibit different life forms and belong to different groups of life forms, because the groups of life forms merely consist of plants of similar habit, plants which in their entirety show similar morphological adjustments. Life forms may therefore be useful indicators of local environmental conditions. PHYSIOGNOMIC CLASSIFICATION.

ravine a small, narrow valley with steep sides, larger than a GULLY, smaller than a CANYON. G.

ravinement (French, gullying)**1.** the act of ravining, of making a gully **2.** a small scale geological UNCONFORMITY in which the overlying beds have scooped down into (ravined) the underlying beds, although the two beds are parallel to each other. It occurs when the sea, in depositing the upper beds, slightly erodes the top layer of the lower, i.e. the underlying, beds. G.

raw data DATA.

rawhide untanned cattle hide.

rawinsonde (French, radar wind-sounding) a meteorological device carried up into the atmosphere by a hydrogen-filled balloon to record atmospheric conditions. It is equipped with self-recording and transmitting instruments, a parachute and a radar target to enable the ground staff to plot its flight course and recover the instruments when they are parachuted down to earth. BALLOON-SONDE, RADIO-SONDE, ROCKETSONDE, SONDE.

raw material the basic commodity or

commodities, natural (e.g. a plant product such as cotton, or a mineral) or partially processed, i.e. the product of another activity (e.g. wood pulp, wheat flour), which are to be transformed by some industrial or manufacturing process into some further product before being used. The term is sometimes loosely applied to include the source of energy employed, e.g. coal, petroleum. G.

Rayleigh wave a ground surface SEISMIC WAVE-1 in which the particles move in elliptical orbits in the plane of the direction of movement of the wave. EARTHQUAKE, LONGITUDINAL WAVE, PUSH WAVE, SHAKE WAVE, TRANSVERSE WAVE, TRANSVERSE WAVE MOTION, WAVE.

rayon **1.** an artificial fibre made from solutions of natural CELLULOSE forced through a fine mesh and solidified in a chemical bath or in warm air **2.** a textile made from such a fibre.

reach an uninterrupted stretch of water, e.g. in a straight section of a river, especially if navigable, between two bends, or between locks on a canal. G.

reafforestation REFORESTATION.

real estate in law, immovable property, e.g. land or buildings, as opposed to movable personal PROPERTY-1.

realism **1.** an attitude based on facts and reality, as opposed to one founded in imagining and emotions etc. **2.** in philosophy, as opposed to IDEALISM, the doctrine which maintains that material things, the object of sense perception, have a real existence, i.e. that they exist outside the mind; or the doctrine which holds that ideas, or universals (a general proposition, concept or idea) have an absolute existence outside the mind. MATERIALISM-1.

rean, rein (England, Northumbria) an old terrace or ridge produced by cultivation. G.

Réaumur scale a TEMPERATURE SCALE, now obsolete, introduced by René-Antoine Ferchault de Réaumur, French physicist, 1683-1757, in which the ice point, the freezing point, of water at one atmosphere is 0° and the steam point, the boiling point, is 80°; thus $1°C = 33.8°F = 0.8°R$.

reave a BRONZE AGE stone boundary wall found particularly on Dartmoor, southwest England.

Recent *adj* of or pertaining to the period of time since the end of the PLEISTOCENE (GEOLOGICAL TIME-SCALE), i.e. equivalent to the HOLOCENE epoch (time) or series (rocks), broadly the time since the last Great Ice Age; but regarded by some geologists as the last subdivision of the Pleistocene. G.

recessional moraine stadial moraine, one of a succession of TERMINAL MORAINES which mark the temporary limit of an ice sheet, formed as it pauses in retreat, or sometimes as it slightly readvances and again retreats. MORAINE.

recharge a process in which water in an AQUIFER, e.g. in an ARTESIAN BASIN, is replenished by the sinking of precipitation into the land surface. ARTIFICIAL RECHARGE.

reciprocal bearing BEARING-4.

reciprocal movement the movement of an individual involving both an outward and a return journey, beginning from a fixed base, proceeding to one or more stopping places, and returning to base (e.g. everyday movements from home to work or to shop and then return home), as distinct from a MIGRATION MOVEMENT. A reciprocal movement cycle is determined by aggregating all the reciprocal movements of an individual over a period of time (e.g. daily, weekly, monthly). MOVEMENT FREQUENCY, MOVEMENT REPETITIVENESS. G.

reclamation the act of reclaiming, winning back. LAND RECLAMATION.

recovery rate in DIFFUSION-1 process, the time taken for the straight line of the DIFFUSION WAVE front to re-form. BARRIERS AND DIFFUSION WAVES.

recreation a leisure-time activity undertaken for the sake of refreshment or entertainment (e.g. in TOURISM), in many cases away from home, and in the countryside bringing about MULTIPLE LAND USE.

rectangular drainage a DRAINAGE-2 pattern formed usually under the influence of a rectilinear joint pattern, the tributaries meeting larger streams mainly at right angles, and all streams having sections of approximately the same length between junctions.

recumbent fold an OVERFOLD. NAPPE, THRUST FAULT. G.

recycling the long-established practice of collecting and purifying waste materials and converting them to new and useful products, e.g. as in the making and use of SHODDY, the reconstitution of glass bottles, the recovery of metals from scrap, the re-use of waste paper and rags in paper making.

red clay 1. a fine-grained, soft deposit, mainly of hydrated silicate of ALUMINA, rich in IRON oxides, occurring on the floor of the deepest parts of the ocean (ABYSSAL ZONE, OOZE) where the water is so deep that the calcareous shells of FORAMINIFERA and even the siliceous shells of Radiolaria (RADIOLARIAN OOZE) are dissolved before they reach the bottom. Thus the red clay, in which MANGANESE NODULES occur, is derived mainly from volcanic (PUMICE) and meteoric dust, material carried by icebergs, and insoluble remains of marine life (e.g. sharks' teeth) 2. applied loosely to any red-coloured clay. G.

red currant *Ribes sativum*, a cultivated bush derived from a wild species of *Ribes*, the only genus in the order Grossulariaceae, native to stream banks and wet woodlands of western Europe, grown for the sake of its edible succulent, sharp-flavoured, round red BERRIES, eaten cooked, used in making jam, jelly. BLACK CURRANT.

reddish-brown soil in soil science, FAO, a soil with an A C or A B C profile, the upper part of the A HORIZON being reddish-brown to red in colour, the B HORIZON red or reddish-brown. The C HORIZON, normally present, is often hardened. BROWN SOILS have the same horizon sequence and similar properties but are brown in colour. SOIL, SOIL ASSOCIATION, SOIL CLASSIFICATION, SOIL PROFILE. G.

red earth a term loosely and unscientifically applied to a red-coloured ZONAL SOIL with clay, quartz and iron compounds, occurring in tropical areas (e.g. in Brazil, Guyana, eastern Africa, southern Deccan in India, Sri Lanka), resulting from chemical WEATHERING in conditions of high humidity and high temperatures in a region with a markedly seasonal rainfall. It can be 15 m (50 ft) in thickness. The term should not be used as a translation of TERRA ROSSA. LATERITE, LATOSOL. G.

redevelopment in urban areas, URBAN RENEWAL.

redir (Arabic, pl. redair) 1. a pool of temporary water left in a WADI after the torrent has ceased to flow 2. a natural reservoir of water. G.

red-line district in an urban area, an area delimited on a map by a red boundary line and regarded by mortgage controllers as being in decline, unstable in social and economic terms, a poor security risk, in which property is therefore considered to be unacceptable as a loan security. The practice of such delimiting is termed redlining.

red Mediterranean soil in soil science, FAO, a soil with an A B C horizon sequence (SOIL PROFILE), formed on CALCAREOUS and noncalcareous parent materials, the B HORIZON being red or yellow with thick clay. The brown Mediterranean soil is the same apart from the colour of the B horizon, which is brown, and the parent material, which is frequently noncalcareous. SOIL, SOIL ASSOCIATION, SOIL CLASSIFICATION. G.

Red Mud a TERRIGENOUS DEPOSIT coloured by FERRIC oxide and occurring on some CONTINENTAL SLOPES. BLUE MUD, GREEN MUD.

red ochre red HAEMATITE, used as a pigment.

red pepper PEPPER.

red rain rain coloured by red dust carried by high-level winds from an arid to a more humid area, e.g. from the Sahara to southern Europe.

red snow 1. snow coloured red by the presence of various organisms, found in snowfields throughout the world 2. specifically, snow coloured by the presence of the cryophytic red algae, *Chlamydomonas nivalis*, which occurs in acidic alpine and arctic environments. CRYOPHYTE.

red-yellow podzolic soil in soil science, FAO, a soil with an A B C horizon (SOIL HORIZON) sequence. The profile has an A_2 horizon unless it has been removed by erosion. The clayey B HORIZON is 'blocky' in structure. PODZOL, SOIL, SOIL ASSOCIATION, SOIL CLASSIFICATION.

reed any of several tall erect grasses growing in swamps, used for thatching roofs etc.

reef 1. a mass of rock or coral, sometimes of shingle or sand, occurring in the sea, usually covered at high tide, but often partly exposed at low tide. ATOLL, BARRIER REEF, CORAL REEF (includes fringing reef), STACK 2. a bed or vein of metal ore or metal, particularly of gold-bearing QUARTZ. G.

reef flat 1. a flat area of coral sand or coral fragments, accumulated on the landward side of a fringing CORAL REEF and fixed by vegetation 2. a generally flat, fairly extensive coral platform, with patches of coral debris and small rock pools on the surface, uncovered at low tide, formed by the erosion of an old coral reef.

reef knoll a low, rounded or conical hill or knoll, varying in diameter from a few to hundreds of metres, occurring in certain LIMESTONES. It consists of a compacted mass of material originally forming part of an elevated section of a CORAL REEF which, being more resistant to WEATHERING than the sediments that surrounded it, has been exposed by denudation, e.g. in the Carboniferous limestone district of the Craven district of Yorkshire, England. BIOHERM. G.

reentrant, re-entrant *adj.* pointing inward 1. a markedly angled inlet into a coastline 2. a marked indentation in a landform, e.g. one formed at the side of the main valley where it is joined by a transverse valley.

reference net an arbitrary reference GRID-1 put on to a map with letters along one side, numerals along the other, no further subdivision, to ease identification of the general location of a place. Thus a place can be located from the information that it lies, e.g., in square C3, but no more precisely than that. Such a net is frequently used in atlases, the references being conveniently brief for the index or gazetteer. GRATICULE, NATIONAL GRID.

reference sheet see CHARACTERISTIC SHEET.

reflectance REFLECTION.

reflection the process by which a beam of particles or a wave (e.g. VISIBLE LIGHT), in collision with an opaque surface, may be deviated or reversed in direction. Reflection is regular from a smooth (especially a polished) surface, diffuse (not coherent) from a rough surface. The proportion of radiant energy incident upon a surface which is reflected or scattered by it is termed reflectance.

reforestation, reafforestation the planting of trees on land previously forested but from which the trees have been removed by natural causes or by cut-

ting, burning or other means. AFFOR-
ESTATION. G.

refraction the change in direction of the path of a ray occurring when ELEC-TROMAGNETIC WAVES (e.g. VISIBLE LIGHT) or other energy-bearing waves, pass obliquely from a less dense to a denser medium. REFLECTION.

refugee a person who, owing to religious persecution or political troubles, seeks shelter or protection from danger in a foreign country. EMIGRANT, EXILE, EXPATRIATE, IMMIGRANT, MIGRANT, MIGRATION. G.

refugium in ecology, a locality which has escaped the great changes suffered everywhere else in the region in which it lies, and thus commonly supports RELIC SPECIES.

reg (north Africa: Arabic) a stony desert, a desert plain covered with tightly packed, wind-scoured gravel, e.g. as in Algeria, the SERIR of Libya and Egypt. The gravel may be cemented by salts drawn to the surface in solution by CAPILLARITY and precipitated by evaporation, thereby forming a DESERT PAVEMENT. ERG, HAMADA. G.

regelation the re-freezing of ice which has melted under pressure, the re-freezing taking place as pressure is released (the melting point of ice is lowered by pressure, but it rises when pressure is released, PRESSURE MELT-ING POINT). The process is a contributary factor to the flow of a glacier: the melt-water flows down to a place where pressure is less, where it again freezes. G.

regeneration 1. the act of giving (or being given) new life and vigour 2. in biology, the renewal or regrowth by an organism of tissue, organs, or substances lost or damaged.

regime (French régime) in the sense of a recurring pattern 1. the seasonal fluctuation in respect of precipitation, or of the volume of a glacier, or of the volume of water in a river 2. recurring seasonal pattern of climatic changes. G.

regimen as applied to rivers, or a sloping shore, a term now rarely used, superseded by PROFILE OF EQUILIB-RIUM-1,2. G.

region 1. an area of the earth's surface with one or more features or characteristics (natural or the result of human activity) which give it a measure of unity and make it differ from the areas surrounding it. According to the criteria used in the differentiation, a region may be termed cultural, economic, morphological, natural, physiographic, political etc.; and a region may be identified by single, multiple or 'total' attributes (COMPAGE). See REGION qualified by BELLWETHER, COMPLE-MENTARY, FORMAL, FUNCTIONAL, INTERNATIONAL, MORPHOGENETIC, MULTIPLE-FEATURE, NATURAL, NODAL, SINGLE-FEATURE, UNIFORM; and GENERAL REGIONAL SYSTEM, SPECIFIC REGIONAL SYSTEM 2. the area or space surrounding a specific place, e.g. the London region 3. an area which is a unit of administration, e.g. a planning region 4. a local government division in Scotland. LOCAL GOVERN-MENT IN BRITAIN. G.

regional *adj.* of, pertaining to, characteristic of, derived from, a REGION. GENERIC REGIONAL SYSTEM, SPE-CIFIC REGIONAL SYSTEM.

regional complex analysis an approach to geography in which the results of ECOLOGICAL ANALYSIS and SPATIAL ANALYSIS are combined, appropriate regional units being identified through AREAL DIFFERENTIATION, and the flows and links between pairs of REGIONS-1 established.

regional development the economic and cultural growth in a REGION-3, large or small, especially in one suffering serious economic problems, usually stimulated and organized, sometimes directed, sometimes financed, by direct or indirect government action. REGIONAL PLANNING.

regional geography a geographical approach concerned with AREAL DIFFERENTIATION, with the detailed study of REGIONS-1, i.e. of particular defined areas of the earth's surface identified as differing in some respect from adjoining areas. The regional study takes into account the physical and human (including social and economic) features operating in space and over time within the region, and their interrelationship with each other, the EXPLANATIONS of all these aspects being integrated in order to produce an overall view. SYNTHESIS, SYSTEMATIC GEOGRAPHY. G.

regional geology the study of the geology of a large spatial area, rather than a small, localized one.

regionalism 1. the local feeling of group consciousness associated with a particular geographical area, e.g. the South, the West, the Middle West 2. the French movement of the late nineteenth century directed to the revival of REGIONAL identifies and feelings, sometimes associated with political overtones 3. the movement to decentralize central government, placing it at a level intermediate between that of the state and the small local government units 4. in economic and social planning (PLAN-3), the selection of a REGION-3 to serve as the basic area for future development. REGIONAL PLANNING. G.

regionalization 1. a division or classification into REGIONS-1 2. regional organization, i.e. REGIONAL SYSTEMS 3. the making, the development of regional character. REGIONALISM-1,2. G.

regional metamorphism the alteration of pre-existing rocks by a combination of pressure (DYNAMIC METAMORPHISM) and heat (THERMAL METAMORPHISM) over a very extensive area, associated with an OROGENY, and leading to the formation of a wide range of new rocks and minerals. G.

regional multiplier concept the concept that a rise in income, production or employment in one group of economic activities in a REGION-1 stimulates the expansion of other groups through the increased demand of the first group for the goods and services provided by the others, a rise often due to changes outside the region. MULTIPLIER EFFECT. G.

regional planning 1. comprehensive PLANNING (i.e. concerned with socio-economic and political affairs) on a spatial basis, the area concerned ranging from a CITY-1 and its surrounding rural HINTERLAND-2 or several cities and their overlapping hinterlands, as distinct from town planning, which is localized, concerned with small areas 2. such comprehensive planning within such an area by central government with the aim of reducing inequalities between REGIONS-3 in a nation state 3. sometimes applied as equivalent to economic planning in REGIONAL DEVELOPMENT, although the economic region may differ in design and size from the region identified or specified by the city regional planners. LAND USE PLANNING. G.

regional science an interdisciplinary field of study within the SOCIAL SCIENCES, linking ECONOMICS, GEOGRAPHY and PLANNING, concerned with economic and social phenomena in a regional setting (REGION-1), making use especially of mathematical MODELS-4 in the forming of (regional science) theories. G.

regional sea a term applied by United Nations to a landlocked sea, or to one with water that does not speedily mix with that of the ocean, and which is consequently liable to accumulate polluting substances, e.g. the Baltic, the Mediterranean.

regional systems systems concerned with the areal classification of the earth's surface according to the criteria outlined under REGION-1. Some authors identify two types of regional system: the general (or generic), in which places resembling one another according to characteristics selected on account of their appropriateness to the classification, are widely separated; and the specific, in which the places so selected lie next to each other, touching. G.

Registrar General's classification of occupation, UK a publication giving brief descriptions of every type of job in the UK, classifying these occupations in various ways, e.g. by the many socio-economic groups (SOCIO-ECONOMIC GROUPING) or by SOCIAL CLASS (five recognized). OCCUPATIONAL CLASSIFICATION.

regolith (Greek rhegos, blanket; lithos, stone) a general term applied to the mantle of loose material (soil, sediments, broken rock, volcanic ash, wind-blown material etc., i.e. the soil and weathered rock) overlying the solid BED ROCK. MANTLE ROCK, WEATHERING.

regosol in soil science, an AZONAL, immature soil, lacking definite horizons, developed from deep unconsolidated CALCAREOUS or noncalcareous, coarse-textured rock, or from soft MINERAL-1 deposits. Dominantly mineral, but not stony, it may form a moderately thick layer, but it is usually thin. LITHOSOL, SOIL HORIZON, SOIL ASSOCIATION, SOIL CLASSIFICATION. G.

regression 1. in general, the act of moving back, the tendency to move back 2. in astronomy, the movement of a heavenly body, e.g. a planet, in a direction opposite to the normal, e.g. as in the PRECESSION OF THE EQUINOXES 3. in biology, the return to an earlier or less complex form 4. in statistics, the relationship between the mean value of a random variable and the corresponding values of one or more other variables. REGRESSION ANALYSIS.

regression analysis a statistical technique, using data measured on the interval or ratio scale (MEASUREMENT in statistics), which aims to explain the variation in an observed quantity in terms of the dependence of one VARIABLE (the DEPENDENT VARIABLE) on one or more other variables (the INDEPENDENT VARIABLES). If only one independent variable is involved the technique is termed univariate analysis; if more than one it is termed MULTIVARIATE ANALYSIS (CATEGORICAL DATA ANLYSIS).

In LINEAR-4 regression analysis the data are plotted on a graph, the dependent variable on the vertical axis, the independent variable(s) on the horizontal axis, the trend of the points on the graph indicating the relationship between dependent and independent variables. A straight line is fitted as close to the data points as possible to give the best description of the trend (termed the regression line, or the line of best fit, the distances between the points and the line being termed regression RESIDUALS). The line is placed so that the sum of the squares of the deviations of the points from the line is the smallest possible (termed the least squares method). The line can then be used to predict expected values of one variable given the value of the second variable. REGRESSION MODEL, SLOPE-3.

regression model regression equation, a mathematical model, used in REGRESSION ANALYSIS.

$$\hat{Y} = b_1 X_1 + b_2 X_2 + \ldots b_n X_n + \text{constant}$$

$\hat{Y}$ is the prediction of the SCORE-1 on a DEPENDENT VARIABLE (Y), $X_1 X_2$ etc. the observed scores on a set of INDEPENDENT VARIABLES, $b_1 b_2$ etc. are the constants, termed regression weights or REGRESSION COEFFICIENTS. This is an example of an ADDITIVE model, so-termed because it is assumed that $\hat{Y}$ may be derived by adding a proportion of X_1 and another proportion of X_2 etc.

regur (Indian subcontinent) a black cotton soil (BASISOL, TIRS) occurring in the northwest Deccan. G.

reh (Indian subcontinent: Panjabi; Hindi) **1.** a SALINE SOIL **2.** saline lands. G.

reification the act of regarding an ABSTRACTION as a material thing.

Reilly's law of retail gravitation a law postulated by W. J. Reilly, New York, in 1931, that two cities attract retail trade from an intermediate city or town in the vicinity of the breaking point, approximately in direct proportion to the populations of the two cities and in inverse proportion to the square of the distances (distance via most direct improved automobile highway) from these two cities to the intermediate town. GRAVITY MODEL. G.

reindeer any of the large arctic or subarctic deer of the GENUS *Rangifer*, native to northern Europe, Asia and America. Both male and female bear antlers. They are half-tamed and domesticated especially in Lapland, drawing sleighs, and providing meat and leather. LIVESTOCK, REINDEER MOSS.

reindeer moss *Cladonia rangiferina*, a grey, tufted LICHEN, the chief winter food of REINDEER and CARIBOU.

rejuvenation becoming again youthful; in geology and geomorphology, the development of younger surface forms, appropriate to the earlier stages of the CYCLE OF EROSION, occurring when a comparatively well-advanced cycle is interrupted by an increase in the rate of erosion, especially erosion by a RIVER, due to causes defined as dynamic, eustatic (DIASTROPHIC EUSTATISM, GLACIO-EUSTATISM) or STATIC, resulting in such river features as ALLUVIAL TERRACES, INCISED MEANDERS, KNICKPOINTS, TERRACES. G.

rejuvenation head KNICKPOINT.

relations of production SOCIAL RELATIONS OF PRODUCTION.

relative age of rocks, the determination of the age of rocks by their relationship to each other, or by some particular characteristic, e.g. an igneous rock is younger than the rock it cuts across (CROSS-CUTTING RELATIONSHIPS, LAW OF); a sedimentary rock is younger than the rocks of which it is composed, and the lowest stratum in a series of sedimentary rocks is presumed to be the oldest, the overlying strata being progressively younger (but see SUPERPOSITION, LAW OF); and rocks with like fossil assemblages are of like age. ABSOLUTE AGE.

relative frequency curve FREQUENCY POLYGON, RELATIVE FREQUENCY DISTRIBUTION.

relative frequency distribution in statistics, a FREQUENCY DISTRIBUTION in which the frequency of each class is expressed as a percentage of the total of all classes. CUMULATIVE FREQUENCY DISTRIBUTION (includes cumulative relative frequency distribution), FREQUENCY POLYGON.

relative humidity RH, the ratio between the actual amount of moisture in the air and that which would be present if the air were saturated at the same temperature, expressed as a percentage. Air with an RH of 60 may be considered as approximately separating DRY from 'moist' atmosphere. ABSOLUTE HUMIDITY, DRY-BULB THERMOMETER, HUMIDITY, HYGROMETER, SPECIFIC HUMIDITY, WATER VAPOUR, WET-BULB THERMOMETER.

relative location the location of a point in relation to another point or points. ABSOLUTE LOCATION.

relative relief relative altitude, the relation of the highest and lowest points in a land area, the difference between the two extremes being termed the amplitude of relative relief. AVAILABLE RELIEF, RELIEF.

relaxation time the lapse of time during

which a physical SYSTEM-1,2,3, having been upset by changes in the factors controlling or influencing it, readjusts to those changes and reaches a new state of EQUILIBRIUM.

relevance 1. the quality of being connected with, pertinent to, the subject 2. specifically, the degree to which something known or being discussed or studied has a bearing, particularly a practical bearing, on some current issue in society, e.g. the contribution geographers may make in helping to find practical solutions to current environmental or social problems.

relic, relict 1. the material evidence of a thing that no longer exists 2. in ecology, an organism, population or community the remnants of which still exist but which was at an earlier time common in, or charactersitic of, an area. REFUGIUM.

relic distribution the RANGE-7 of a RELIC-2 population of plants and/or animals surviving in an area, being the remains of a much wider range of an earlier time. If such a population occupies an area throughout its range it may be termed an absolute relic; if only an isolated part of the area, a local relic; if restricted to a single region, an endemic relic; if its area has been restricted by human activity, an anthropogenic relic; and if it establishes a secondary distribution by occupying a suitable habitat it is termed a migrant relic.

relic mountain, relict mountain an obsolescent term applied to a mountain of CIRCUMDENUDATION, a mountain representing the remains or RELIC-1 of a pre-existing plateau or range in an area of denudation. INSELBERG, MONADNOCK. G.

relict boundary a BOUNDARY that can still be discerned in the CULTURAL LANDSCAPE despite the fact that it has been abandoned and no longer serves a political purpose. G.

relict sediments sediments deposited

under conditions different from those of the present.

relief 1. the physical shape of the surface of the earth, its mountains and valleys, plains and plateaus, the physical landscape 2. often applied loosely to indicate inequalities or variations in shapes and forms of the earth's surface. The use of the terms 'high relief' and 'low relief' is best restricted to areas which respectively show a great or little variation in altitude. TOPOGRAPHY should not be confused with relief. AVAILABLE RELIEF, RELATIVE RELIEF. G.

relief map a map depicting the surface configuration of an area, e.g. a CONTOUR map. Other types include PHOTO-RELIEF maps, which are either photographs of a model or diagrammatic maps simulating a photograph. G.

relief model a three-dimensional MODEL-3 depicting the surface relief of an area, but not necessarily true to scale. Usually the vertical scale is exaggerated in comparison with the horizontal so as to accentuate mountains and plateaus. G.

relief rainfall OROGRAPHIC PRECIPITATION, rainfall resulting from the RELIEF-1 of the land. Hills and mountains cause air to rise, resulting in cooling, condensation of moisture, and rain. G.

relocation diffusion DIFFUSION-2 in which a phenomenon spreads from place to place but leaves the place or places from which it originated as it moves on to new areas. EXPANSION DIFFUSION.

remanence the residual MAGNETISM present in a ferromagnetic substance after the magnetizing field has been reduced to zero.

remanent magnetism the MAGNETISM retained in a rock or mineral from the time when it cooled through the CURIE POINT, providing information about the MAGNETIC FIELD of the earth at that time. PALAEOMAGNETISM.

remote sensing the examination of, the obtaining of information about, an object or phenomenon at a distance from it, without physical contact with it, particularly by devices based on the ground, by SENSORS carried aboard ships or aircraft, or by spacecraft or satellites orbiting the earth, which gather data at a distance from their source (LANDSAT, SKYLAB, SPACE SHUTTLE, SPOT). The data collected by such equipment may be based on measurements of variations in electro-magnetic radiation, in acoustic energy, or in gravitational or magnetic force fields. Computers are commonly used in the retrieval and storage processes. The information obtained from remote sensors orbiting the earth in this way has greatly advanced the understanding and knowledge of the earth's surface, its geological structure and mineral resources, the circulation of the ocean, and atmospheric phenomena. AERIAL PHOTOGRAPH, ELECTROMAGNETIC SPECTRUM, GROUND INFORMATION, GROUND TRUTH, IMAGE, INFRA-RED RADIATION, LIDAR, MICROWAVE, MULTI-BAND SPECTRAL PHOTOG-RAPHY, MULTISPECTRAL SENSING, RADAR-2, RADIOMETER, REFLEC-TANCE, REFLECTION, NEPHANALY-SIS, SIGNATURE, SURVEY-4, SLAR, SONAR, SYNOPTIC IMAGE, THERMAL INFRA-RED SENSING.

Renaissance, Renascence the artistic and literary revival which began in Italy in the fourteenth century and pro-foundly affected thought and the arts in Europe for some two centuries there-after. Broadly, it was characterized by the spread of HUMANISM, a return to classical values (which weakened the grip of the medieval church), realism in the visual arts, an architecture radically different from that typical of the MIDDLE AGES or DARK AGES, a spirit of objective scientific enquiry, the birth of printing, and the founding of centres of learning. It opened the way for the development of the modern world. HUMANIST, HUMANISTIC.

rendzina, rendsina (Polish) in soil sci-ence, a group of INTRAZONAL SOILS, shallow and calcareous, with an A C horizon, usually with a brown or black friable surface, humus content MULL-2,3, underlain by light grey or yellow calcareous material, CALCIUM CAR-BONATE being distributed throughout the profile. It develops from relatively soft parent rock, under grass and forest vegetaion in humid and semi-arid regions. It is well developed in England on Chalk, but it also occurs on harder limestones. MOLLISOLS. G.

renewal in urban areas, URBAN RE-NEWAL, REDEVELOPMENT.

rent 1. an item of income or payment to any factor of production which is limited in supply 2. any surplus earned on account of superior quality or ability 3. popularly, the price paid for the use of any durable GOOD (usually of land or buildings) at certain specified or customary times, or for a certain period of time. ECONOMIC RENT, LAND RENT, LAND TENURE, SCAR-CITY RENT.

replat (French) 1. a bench, a shoulder or tilted terrace, wider than a ledge, above the steep side of a U-shaped valley 2. a stretch of more or less horizontal land interrupting a slope. G.

repose slope a slope, usually steep, with its steepness governed by the ANGLE OF REPOSE of the superficial layer of debris, which maintains its angle in its retreat. PARALLEL RETREAT OF SLOPE.

representative fraction RF, the fraction expressing the ratio between the dis-tance measured between two points on a map and the corresponding distance measured on the ground, the usual way of expressing the scale of a map. The numerator is one (indicating one unit, such as an inch or centimetre), the denominator is the number of units on the ground which this represents. Thus an RF of 1/1 000 000 (often written

1 : 1 000 000, or 1 : 1 mn) means that 1 in or 1 cm on the map represents one million in or cm on the ground. Since a mile is 63 360 in, this is approximately 16 miles, and a map of 1 : 1 000 000 is roughly 16 miles to one inch. G.

representative sample in statistics, a SAMPLE which truly reflects all the important characteristics of a POPULA-TION-4.

reproduction of labour power the process by which the supply of capacity to work (LABOUR POWER) is generally ensured, including the provision of services for existing workers and the production of future workers (children). The provision of services includes housework and the preparation of meals within the home (undertaken mainly by women) as well as communal and collective services, e.g. health services; and child care in relation to 'future workers'.

reproduction of workers the part of the REPRODUCTION OF LABOUR POWER concerned with the domestic work involved in the rearing of children (as future workers) and in feeding and caring for current workers.

reproduction rate any of the calculations used to indicate trends in numbers of human population. They vary from simple and crude to complex and refined. The simplest is the gross reproduction rate, calculated by the number of all females of reproductive age (15 to 50 years) expressed as a ratio of the actual number of females born; or by the number of daughters a woman would produce if throughout her life she gave birth in accord with current (age-specific) fertility rates (FERTILITY-3). Net reproduction rate involves a more refined calculation, the factors modifying the calculated gross reproduction rate being taken into account, e.g. deaths of females before and during reproductive age, infertility, un-mated state, numbers of reproductive males. If continued over a period of time a net reproducton rate of less than one leads to population decline; of one, to a stationarypopulation(STABLEPOPULA-TION); and if more than one, to an increasing population. BIRTH-RATE, NATURAL INCREASE.

reptile a member of Reptilia, a class of COLD-BLOODED land VERTEBRATES (e.g. snake, lizard, crocodile), characteristically having a skin covered with strong scales or plates.

republic 1. a form of government without a MONARCH, in which sovereign power is widely vested in the people through some form of elective process 2. a form of government in which the head of state is an elected president, not a monarch (as in a MONARCHY) 3. a state without either of those two forms of government. AUTOCRACY.

resam (Malay) vegetation consisting of fern and bracken in Malaysia. G.

research and science park a tract of land, commonly park-like (PARK-2,6), planned and officially designated for the accommodation of individuals and organizations engaged in research, development and scientific production. In some cases these activities (especially the last) overlap with those found in an INDUSTRIAL PARK, so the difference between these two types of park is not everywhere distinct.

resequent, re-consequent drainage a DRAINAGE-2 pattern in which streams lie approximately along the line of former CONSEQUENT streams, after long-term denudation of a folded area. They thus appear to be consequent but, due to some agency such as RIVER CAPTURE, they have in fact developed (with synclinal valleys and anticlinal ridges) from a SUBSEQUENT drainage pattern. G.

resequent fault-line scarp a FAULT-LINE SCARP facing in the original direction of the DOWNTHROW, produced by erosion along the OBSE-QUENT fault-line scarp. G.

reservation an area of land demarcated for a special use. RESERVE-2.

reserve 1. something held back, kept in store for future use. RESERVES, RESOURCES 2. reservation, an area of land demarcated for a special use, e.g. to protect the habitat of rare species, a nature reserve.

reserves part of a RESOURCE considered to be usable in current conditions of technical skill and economic and social needs. The existence and location of some reserves is known (PROVEN RESERVES), but of others it may be hypothetical (UNDISCOVERED RESERVES).

reservoir 1. a container or receptacle, natural or artificial, in which a liquid or gas collects and is stored 2. a natural or artificial lake, the water of which is used in irrigation, in producing hydroelectricity, in manufacturing processes and for domestic purposes 3. a highly porous (POROSITY) PERMEABLE ROCK mass which is able to transmit as well as to hold a fluid 4. a reserve supply of something, e.g. a relict species. RELIC 5. in a plant or animal, a sac or cavity in which a fluid collects or is secreted.

residence 1. the act or fact of living in a particular place for some considerable period of time 2. the period of time during which one lives at a place 3. the place, the dwelling, in which a person lives.

residential *adj.* 1. of or relating to RESIDENCE 2. occupied mainly by housing.

residential migration the change of permanent RESIDENCE-3 from one dwelling to another.

residual in statistics, a quantity remaining after another quantity has been subtracted, e.g. if the computed value of a variable is subtracted from an observed value and there is a difference, the difference is termed the residual; or if a mathematical MODEL-4 is fitted to data, the values by which the observations differ from the model values are termed residuals. DEVIATION, REGRESSION ANALYSIS.

residual *adj.* 1. remaining, left over 2. of or relating to something remaining, left over.

residual deposit an accumulation of rock waste, from clays to boulders in grain-size, caused by the WEATHERING, the disintegration, of rocks in situ. RESIDUAL SOIL. G.

residual soil a SEDENTARY SOIL, a soil resting on the material from which it was formed. SECONDARY SOIL. G.

residual variance in statistics, the part of the VARIANCE of a set of data which remains after the removal of the effect of certain systematic elements. It measures the variability due to unexplained causes or experimental error. ERROR VARIANCE, TRUE VARIANCE.

resin 1. a general name for certain fluids (secretions) in the tissues of some plants (e.g. in PINES-1), which become solid on exposure to the air. They are used in the manufacture of varnishes, printing inks etc., or burnt as incense. FRANKINCENSE, MYRRH, ROSIN 2. one of the many synthetic resins which have some physical properties similar to the natural resin, but which are chemically dissimilar, used as varnishes and in adhesives, etc.

resistant *adj.* the quality of being opposed to physical force or chemical change.

resistant rock a rock with physical and chemical properties which make it able to withstand the processes of EROSION and WEATHERING. G.

resolution in REMOTE SENSING 1. the ability of a SENSOR to make a distinction between objects which have similar temperatures, or between objects which have similar spectral characters 2. the minimum distance (linear or angular) between two objects at which they continue to appear distinct and separate on an image or photograph.

resource, resources 1. a source of supply or support, the means (or collective means) of meeting a need or deficiency, especially an economic or social need or deficiency **2.** a stock (or stocks) which can be used if necessary, the total stock consisting of the total amount of a substance, i.e. the accessible, the inaccessible, and the unusable, at the current stage of technology. NATURAL RESOURCES, RESERVE, RESERVES, RESOURCE CONSERVATION, RESOURCE MANAGEMENT. G.

resource conservation the careful management and maintenance of NATURAL RESOURCES. CONSERVATION, RESOURCE, RESOURCE MANAGEMENT. G.

resource development the bringing of a RESOURCE-2 into production, the realization of its potential.

resource-frontier region CORE-PERIPHERY MODEL.

resource management the skilful control of a RESOURCE-1,2 by those who ensure that it is used economically and with forethought, who determine the present need for and the value of such a use, balance benefits and costs, take into account environmental constraints, social, economic and political implications, technological inventiveness, national policy; and possible future needs, technology and uses. A decision not to develop a resource at a particular time is also part of management. CONSERVATION, NATURAL RESOURCES, RESOURCE CONSERVATION, RESOURCE DEVELOPMENT.

resource orientation the tendency of a firm or a particular industry to be located close to the source of its raw material(s) when the source of the raw material(s) is localized, not widespread like that of UBIQUITOUS MATERIALS. MARKET ORIENTATION.

respiration any of the processes by which an organism takes in air or dissolved gases, uses one or more of them in chemical reactions which produce energy, and expels the unused parts of the air or gases together with

the by-products of the chemical changes. Plants and animals take in OXYGEN and expel CARBON DIOXIDE produced by the oxidation of CARBON compounds (e.g. glucose) in the system. In daylight green plants use carbon dioxide from the air to form starch and expel oxygen as a by-product. AEROBIC RESPIRATION, ANAEROBIC RESPIRATION, PHOTOSYNTHESIS.

respiratory quotient RQ, the ratio of the volume of CARBON DIOXIDE expelled to the volume of OXYGEN used by an organism during RESPIRATION. It varies with the compounds being oxidized and is used as an indicator of AEROBIC or ANAEROBIC RESPIRATION.

response elements the features resulting from a series of processes (PROCESS ELEMENTS) acting together.

response surface TREND SURFACE MAP.

resurgence the emergence of an underground stream (DISAPPEARING STREAM) from a cave, commonly occurring near the point where an IMPERMEABLE layer underlying the PERMEABLE layer through which the stream has passed intersects the surface. VAUCLUSIAN SPRING.

retrogressive method a method of approaching an understanding of the past which begins by a careful, detailed analysis of the identifiable factors which have created the present (particularly applied to past landscapes). RETROSPECTIVE METHOD.

retrospective method a method of approaching an understanding of the present based on a study of the past which, it is maintained, illuminates present conditions. HISTORICAL GEOGRAPHY has been described as retrospective geography. RETROGRESSIVE APPROACH.

retting the initial process in which the fibre is extracted from the leaves of some plants, e.g. of FLAX or JUTE. The leaf is soaked in water so that the outer sheaf is partly decomposed by bacterial action. That process is termed the retting, and following it the leaves are washed and dried, the outer parts (now brittle) are scutched (broken into pieces without damage to the fibre) and subsequently hackled (drawn through metal combs) until only the fibres remain.

reverse (reciprocal) bearing BEARING.

reverse commuting COMMUTER.

reversed drainage RIVER CAPTURE.

reverse(d) fault a thrust fault, a FAULT in which the FAULT PLANE dips to the upthrow side (THROW OF FAULT), the result of older beds on one side of the fault plane being compressed and THRUST over younger beds on the other side. G.

reversible transformation TRANSFORMATION.

RF REPRESENTATIVE FRACTION.

RH RELATIVE HUMIDITY.

Rhaetic *adj.* of or pertaining to a rock series transitional between the TRIASSIC and the JURASSIC system of rocks. GEOLOGICAL TIMESCALE.

rhea CHINA GRASS.

rheology (Greek rheos, stream; logos, word) the study of the flow and deformation of matter, particularly if it is PLASTIC, important, for example, in studies of the folding of rocks or the movement of glacier ice. PLASTIC DEFORMATION OF ICE.

rheophyte a plant growing in flowing water.

rheotaxis TAXIS in which flowing movement in the surroundings (e.g. as in the current of a stream) is the source of stimulus.

rheotropism curvature of growth in plants or sedentary animals in response to the influence of a current of water or air. TROPISM.

rhine, rhyne, rhene (southwest England) a wide, open, artificial watercourse, a drainage channel, sometimes marking a field boundary.

rhinn (Scotland, Southern Uplands) a rugged ridge.

rhizobium nitrogen-fixing BACTERIA living in the root nodules of leguminous (LEGUME, LEGUMINOSAE) and a few other plants. NITROGEN CYCLE, NITROGEN FIXATION.

rhizosphere the part of the soil immediately surrounding the roots of plants. G.

rhodium a metallic element, resembling PLATINUM (PLATINOID), used in hard ALLOYS and as a CATALYST.

rhodonite a red silicate of MANGANESE used as an ornamental stone.

rhos (Welsh, pl. rhysodd) rough moorland, in contrast with FFRIDD. G.

rhyolite a fine-grained to glassy, ACIDIC extruded IGNEOUS ROCK of QUARTZ and alkali FELDSPARS, with FERROMAGNESIAN MINERALS, corresponding in chemical composition to PLUTONIC granite. The banding in some cases indicates the flow of the rhyolite as molten MAGMA which, even at very high temperatures, is much less fluid than basic lavas (BASALT), tending to consolidate in masses. Rhyolite may contain PORPHYRITIC quartz crystals and SPHERALITES. The surface may be glassy (OBSIDIAN) or wax-like (PITCHSTONE). ACIDIC ROCK.

rhubarb *Rheum rhaponticum*, a large-leaved PERENNIAL plant, probably derived from wild Siberian species, cultivated particularly in Europe. The long, fleshy, edible, usually red leaf-stalks are used as 'fruit', canned, stewed, used in jams and preserves and in home wine-making. The leaves have a high content of OXALIC ACID.

rhumb line a loxodrome, a line of constant bearing, cutting all MERIDIANS at the same angle. On Mercator's projection (MAP PROJECTIONS) it is a straight line. G.

ria (Spanish) a funnel-shaped indentation on a coast, narrowing inland, increasing in depth seaward, a drowned river valley, occurring particularly along coasts of the ATLANTIC TYPE as a result of a rise in sea level (EUSTATISM). A ria differs from a FJORD in being shorter and lacking the irregularities of depth characteristic of the fjord. The stream that made the original valley, and which flows into the head of the ria, is clearly too narrow in relation to the present size of the inlet. G.

ribbon development the building of houses etc. along each side of the main roads extending outwards from a built-up area, typical in Britain between 1920 and 1939.

rice *Oryza sativa*, a CEREAL-1, native to Asia, cultivated in wet tropical areas (especially in monsoon Asia, the large producing region) and in some subtropical areas for its seed (PADI), used for human food, the staple cereal in the growing countries. There are many different varieties in cultivation, the high-yielding strains produced by hybridization playing an important part in the GREEN REVOLUTION. Upland or hill rice (dry rice) is grown without irrigation on tropical hillslopes; but most rice is grown in flat fields which can be flooded, either through heavy rainfall or by irrigation water. In Asian countries the seed is usually germinated in a nursery field from which the young plants are transplanted to larger fields when they are some 15 to 25 cm (6 to 10 in) in height (other crops may be growing in the larger fields whilst the seeds are germinating in the nurseries). Transplanting, by hand, is usually done into the mud of the flooded fields. As the plants grow, rapidly, the water is absorbed and by harvest time the fields are almost dry. Reaping is still mainly by hand. In some parts of the tropics three, or even more, crops a year are obtained, two are common, and in many parts of the monsoon lands rice as the summer crop can be grown on the same land as wheat or barley as winter crops. With intensive cultivation under these methods yields may be very high; but very high yields are also obtained from rice grown under mechanized conditions in Calfornia, Spain and northern Italy. Poorer qualities of rice are used for making starch; the hulls are fed to cattle, as are the stalks, which are also used in making paper, thatching, footwear; fermented rice kernels provide alcoholic beverages. AMERICAN WILD RICE. C.

Richter scale a scale, in Arabic numerals from 0 to over 8, used in measuring the magnitude of EARTHQUAKES, based on instrumental recordings of a standard SEISMOGRAPH 100 km (62 mi) from the EPICENTRE, the larger numbers being applied to the larger disturbances, 7 being a major earthquake. The Richter scale superseded the MERCALLI SCALE and the ROSSI-FOREL SCALE.

ride a path for horseback riding, especially one in a forest.

ridge a term loosely applied to any long, narrow, steep-sided rise in the land, sometimes to a small feature in a mountain RANGE-2, but infrequently to the range itself and never to a mountain chain. OCEANIC RIDGE. G.

ridge and furrow a method of working heavy soils, attributed to the Anglo-Saxons in England, but continuing after INCLOSURE, to promote surface drainage. The land was ploughed so as to form broad ridges separated by furrows which served as drains to draw off water. Where such land is now in grass, the ridge and furrow can, in many cases, still be seen. G.

ridge and valley a form of relief in which ridges and valleys lie close together and almost parallel, e.g. the ridge and valley region of the Appalachians, where the RESISTANT ROCKS (sandstones, quartzites and conglomerates) form the ridges and the valleys have been carved through the more easily eroded shales and limestones.

ridge of high pressure a long, narrow region of HIGH atmospheric pressure, broader than a WEDGE, lying between two LOW pressure areas, and responsible for a brief period of fine weather in a generally rainy period. ATMOSPHERIC PRESSURE.

riding (Old Norse thrithjungr, third part) historically one of the three former administrative divisions of Yorkshire, England.

Ried a marshy floodplain, especially that of the Rhine, commonly flooded in spring and early summer, and supporting hygrophilous vegetation. HYGROPHYTE.

Riegel (German) an outcrop of resistant rock forming a barrier across a glacially eroded valley. G.

riffle 1. in gold washing, a bar fixed across a sluice etc., or a channel or space between two bars in the bottom of a sluice etc. to catch gold carried by the water 2. a cross bar in a fish ladder 3. a rocky obstruction in a river bed, or the riffled water resulting from such an obstruction 4. a shallow gravel bar over which water flows rapidly with a ruffled surface, formed in a stream with a gravel bed which, with intervening pools, gives rise to a POOL AND RIFFLE pattern.

rift a crack or fissure in the earth's surface. G.

rift valley a long, narrow section of the earth's crust let down between two parallel series of FAULTS with THROWS in opposite directions, thus appearing as a long, flat-floored valley with steep sides, as in the rift valleys of east Africa, or in the Midland Valley of Scotland. It is a relief feature, as opposed to a GRABEN (a structural feature which may or may not coincide with a valley). G.

right ascension in astronomy, one of the two published references (the other being DECLINATION-1) used to locate the position of a heavenly body in the CELESTIAL SPHERE, equivalent to terrestrial longitude. Right ascension lines, or hour lines (analogous to lines of terrestrial longitude) pass through the celestial poles and cut at right angles the celestial equator and all the parallels of declination (analogous to parallels of latitude). Right ascension is the angular distance measured eastward along the celestial equator from a point known as the First Point of Aries (analogous to the meridian of Greenwich, i.e.0°) and the point where the circle of declination of the heavenly body cuts the celestial equator. (The First Point of Aries was chosen some 2000 years ago when it did in fact lie in Aries; now, owing to the PRECESSION OF THE EQUINOXES, it lies in Pisces. Nevertheless it retains its name as the First Point of Aries.) ZODIAC. G.

right-of-way 1. the right of using a path or thoroughfare over another person's property, i.e. over land in which another person or persons has/have a superior interest 2. the path or thoroughfare so used.

rights of common COMMON, COMMON RIGHTS.

rill 1. a small natural STREAM-1 of water 2. in soil erosion, a small erosion channel. G

rill channel a channel made by a RILL-1.

Rillenstein (German) solution-grooved rocks. KARRE. G.

rill erosion the removal of soil by RILLS which, if persistent, may enlarge the rill channels so much that they unite to form GULLIES, leading to gully erosion. SOIL EROSION.

rill marks 1. small furrows made by small streams on sand or mud surfaces as the tide recedes, or flow from a SPRING-2 declines 2. fossil features of this type. G.

rillstone VENTIFACT.

rimaye (French) BERGSCHRUND.

rime an accumulation of white, opaque, granular ice tufts formed on the windward side of objects with a temperature below that of freezing point when supercooled droplets (e.g. in cloud or freezing fog) are blown against them. It resembles HOAR FROST. SUPERCOOLING. G.

rimstone 1. a CALCIUM CARBONATE deposit formed around the rim of a still pool (a rimstone pool) on a cave floor in KARST **2.** term applied by W. M. Davis to a CALCAREOUS-1 deposit formed around the rim of an overflowing basin. G.

ring-city RANDSTAD.

ring-dyke a dyke (DIKE) in the area surrounding a circular or dome-shaped igneous INTRUSION. The vertical and outward pressure of the MAGMA forming the intrusion causes fractures in the COUNTRY ROCK-1, into which the magma flows and solidifies, forming ring-dykes if the fractures are vertical. CAULDRON SUBSIDENCE, GABBRO.

ring fence a fence which completely encloses a piece of land, especially to protect the grazing and to contain the grazing farm animals.

ring road a road encircling a built-up area, used as a bypass or as a service road for the area.

rio (Portuguese and Spanish) a river, usually applied to a permanent stream. In Spanish it contrasts with ARROYO. G.

rip a disturbance and turbuluence, rough water, sometimes in a river, more commonly in the sea, occurring when two tidal streams flowing from different directions meet; or a strong outflowing surface or near-surface current meets the incoming surf; or when a tidal stream suddenly flows into shallow water; or when a tidal or river current flows over an irregular floor, especially one with abrupt changes in depth. G.

riparian *adj.* of, pertaining to, situated on, or associated with, a river bank, applied especially in legal terms, e.g. riparian rights, riparian land, riparian states. RIVERINE, RIVERSIDE. G.

ripple, ripple-mark, sand ripple a series of small more or less parallel ridges, produced especially on sand by the wind, the current of a stream, or by waves on a shore. G.

ripple-till, ribble till patterned till-sheets with sinuous, smooth-topped ridges resembling RIPPLES, 180 m to 3 km (200 yd to 2 mi) long, some 6 to 15 m (20 to 50 ft) higher than the intervening depressions, running across and very roughly at right angles to the direction of ice movement, occurring in North America, especially in Canada. TILL. G.

rise, swell on the deep sea floor, a gently sloping, long, broad elevation rising from the deep sea floor, its summit far below the surface of the water. G.

risk 1. the chance of danger, injury, loss etc. **2.** a situation in which there is the possibility of several possible outcomes occurring as a consequence of a decision or action taken when the probabilities of the possible outcomes are known and can be calculated for each individual outcome (e.g. as in insurance risk). UNCERTAINTY.

river a general term applied to a natural STREAM-1 of water flowing regularly or intermittently over a bed, usually in a definite channel, towards the sea, a lake, or an inland depression in a desert basin, or a marsh or another river. Stages of development may be recognized, in youth with swift streams actively eroding steep-sided V-shaped valleys; MATURITY, middle age, with broad open valleys with gentler slopes and the beginning of MEANDERS; and old age (SENILE RIVER) with broad valleys and a sluggish flow of water. Not all rivers fit this ideal model, e.g. the flow may be sluggish but the stream not old (CYCLE OF EROSION). The point or points of origin are termed the SOURCE of a river, the path it follows is its course, the part where it enters a sea or lake its mouth. Most rivers have an upper mountain course, a middle or plains course, and a lower or estuarine-delta course. Rivers are especially active as eroding agents in REJUVENATION, particularly in STATIC REJUVENATION.

See BED LOAD, DELTA, DISSOLVED LOAD, DRAINAGE-2, STREAM-1, STREAM ORDER, SUSPENDED LOAD and entries qualified by river. G.

river bank the rising land bordering a river.

river basin all the area of land drained by a river and its tributaries.

river bed, riverbed the channel in which a river flows or once flowed.

river capture, river piracy the action by which a river, by rapid headward erosion, captures and diverts to itself the headwaters of another stream, thereby enlarging its own drainage area and diminishing that of the other. The stream of which the headwaters have been captured is said to be beheaded; the point at which the capture occurs is the elbow of capture; self-shortening of the course of a river is AUTOPIRACY. River capture may give rise to MISFIT STREAMS and WIND GAPS. ABSTRACTION, INTERCISION. G.

river cliff in a MEANDER of a river in a young river valley, the steep, concave (outer) curve cut by the stream current, facing the gentle SLIP-OFF SLOPE of the spur opposite.

river icing AUFEIS.

riverine, riverain *adj* **1.** of, pertaining to, situated on, living on, the banks of a river **2.** of or relating to a river or its vicinity. The application is less restricted than that of RIPARIAN.

river port a PORT-1 situated on a river, usually at the point farthest from the mouth where the water is deep enough for navigation by trading vessels.

river profile a section of a river valley, either longitudinal (along the course of the river, showing the slope from source to mouth, RIVER) or transverse (across the valley at right angles to the stream). The long PROFILE-3 ('long' is preferred to 'longitudinal') usually takes in the actual length of the centre of the stream, and the height of the surface at mean level, measurements being adjusted for minor variations of level. Some authors prefer to take the height of the floodplain (especially useful in comparing the present profile with reconstructed profiles); in this case minor windings are ignored because such features cannot be reconstructed for past phases of the river at higher levels. TALWEG. G.

riverside a general, imprecise term, the land alongside a river. RIPARIAN, RIVERINE.

river terrace a part of the former floodplain of a river, left on the side of a river valley as the stream cut down its bed and now appearing as a generally flat, step-like strip on the side of the valley, at a level higher that of the present channel. Such a terrace is usually built up of gravel, coarse sand and alluvium deposited by the river when it was flowing at the level of the terrace. Thus the terrace represents a part of the valley floor at that time, and may not be perfectly flat, still preserving BARS-2, SWALES-2 etc. The term is applied to a rock BENCH as well as to a gravel-covered terrace. ALLUVIAL TERRACE. G.

Riviera 1. the coastal strip with its numerous resorts, bordering the Mediterranean from Marseille in southern France to Genova in Italy **2.** by analogy, applied to resort coasts elsewhere, e.g. Cornish Riviera, the southern coast of Cornwall, England. G.

rivulet a small river. G.

road a cleared strip of land with some form of hard surface, made to make a passageway for walkers and vehicles.

road metal not metal, but broken, tough, hard stone which breaks into angular pieces without splintering or making much dust, used for surfacing roads. MACADAM. G.

roadstead, roads, road an open, relatively sheltered anchorage near the shore which may in some places be protected from the open sea by shoals, reefs etc. G.

Roaring Forties the latitudes between 40° and 50°S where the westerly winds are not obstructed by land as they blow strongly and regularly over the ocean, bringing gales, rough seas and rain associated with the regular procession of DEPRESSIONS-3 moving from west to east. The term is occasionally applied to the winds. WESTERLIES. G.

robber economy the working of RESOURCES-2, especially of stock resources which, once used up, are not renewed, in contrast to the development of renewable resources (NATURAL RESOURCES). Sometimes also applied to the needless destruction of resources for the sake of quick profits, especially if it is unlikely that the resource will recover in the future. G.

roche moutonnée (French, sheep-like rock) a rock mass forming a hillock, resulting from ice action, seen in most glaciated valleys. The upstream side is smoothed and rounded owing to the effect of ABRASION and commonly shows glacial striation (STRIAE); the downstream side is steeper and rougher, owing to PLUCKING. The term was coined by de Saussure, 1779-96, who saw a resemblance between this form of rock and a thick fleece, or between it and the wavy wig (moutonnée, slicked down with mutton tallow) fashionable in his day. ONSET AND LEE. G.

rock 1. in general use, a large, hard, consolidated, compact part of the earth's crust, also a large piece of this material protruding from the land or sea 2. in geology, any naturally formed aggregate of mineral particles, whether it is hard, relatively soft, unconsolidated or incompact, which constitutes an integral part of the lithosphere (thus includes mud, clay, sand, coral etc.). Rocks are classified by the manner of formation (IGNEOUS, METAMORPHIC, SEDIMENTARY) or by age (GEOLOGICAL TIMESCALE, STRATIGRAPHY). COUNTRY ROCK. G.

rock bastion a solid rock mass projecting from the outlet of a hanging valley into the main valley at the point where a tributary glacier joins the main glacier. It indicates the weakening power of the main glacier due to the thrust exerted on it at right angles by the tributary glacier, or to the reduced rate of flow brought about by an accumulation of debris at the confluence of the glaciers or at a DISCORDANT JUNCTION.

rock creep the slow movement of rock blocks down a slope under the influence of gravity. MASS MOVEMENT. G.

rock crystal the purest, most transparent form of QUARTZ.

rock drumlin, rockdrumlin, rocdrumlin also drumlinoid or false drumlin, a mound in glaciated country, resembling a DRUMLIN but consisting of solid rock. G.

rocketsonde (French, rocket sounding) a collection of self-recording and transmitting meteorological instruments, including photographic equipment, carried by rocket into the atmosphere and ejected at a selected height, to return to earth by parachute tracked by radar. BALLOON-SONDE, RADIOSONDE, RAWINSONDE, SONDE.

rock fall a free fall of individual boulders or blocks of bed rock down any steep slope under the force of gravity. MASS MOVEMENT, ROCK SLIDE. G.

rock flour, rock-flour 1. very finely powdered rock material produced by the grinding action of a glacier when, with rocks frozen into its mass, it abrades its bed. The process is mechanical, there is little or no chemical action, hence rock flour has the same mineralogical composition as that of the rocks from which it is formed. It is a major constituent of BOULDER CLAY. ABRASION, GLACIER MILK 2. very finely ground, powdered rock material mechanically produced, with FAULT BRECCIA, by movement along a FAULT.

rock glacier a tongue of angular rock

waste, with an ice core overlain by fine rock debris, in turn overlain by large blocks of rock raised into concentric ridges, resembling a glacier and occurring at the head of a CIRQUE or similar landform. It commonly occurs where part of a former glacier is covered by morainic material or falling rock debris. ROCK GLACIER CREEP.

rock glacier creep the slow movement of a ROCK GLACIER down a slope under the force of gravity. MASS MOVEMENT. G.

rocking stone a large boulder, formed as a result of CIRCUMDENUDATION of an ERRATIC, so poised that it can be rocked by hand. LOGAN STONE. G.

rock pavement RUWARE.

rock-rill LAPIE. G.

rock salt NaCl, HALITE, sodium chloride, a mineral occurring in a clear or white to brownish crystalline mass and forming, with BRINE, the source of SALT-2 used in commerce and industry.

rock slide, rock-slide the sliding of an individual rock mass down a gentle gradient, e.g. down a bedding joint or fault surface, under the force of gravity. MASS MOVEMENT, ROCK FALL.

rock wool a fibrous material, used in heat and sound insulation, made by directing a jet of steam at limestone, siliceous rock etc.

rognon a rounded NUNATAK, i.e. a NUNAKOL. G.

role in society, SOCIAL ROLE.

roller a term popularly appllied to the ocean swell which even in calm weather gives rise to very large BREAKERS along coasts, due to the length of the FETCH.

rolling stock 1. the collective term applied to the wheeled vehicles of a railway 2. (American) the fleet of vehicles of a road carrier.

Roman Empire the European empire based on Rome which, encircling the Mediterranean, spread its control outwards into Europe, and into southwest Asia, reaching its maximum extent between the time Augustus took power in 27 BC until c. 300 AD. Many of the Roman names for the provinces and territories of the Empire continued for centuries and are often referred to at the present day, e.g. Britannia (Britain as far north as the lowlands of Scotland or Hadrian's Wall, Tyne to Solway); Gallia (Gaul or France); Aquitana (part of southwest France); Rhaetia (roughly Switzerland); Hispania (Spain and Portugal); Cyrenaica (eastern Libya); Aegyptus (Egypt); Nubia (southern part of what is now Egypt); Arabia Petraea (Sinai); and Asia, which was applied to a small part of western Asia Minor, i.e. Turkey.

rondawel, rondavel (Afrikaans) a cylindrical hut with a conical roof, as built by many Bantu people in Africa.

roof pendant a downward-pointing mass of COUNTRY ROCK penetrating the top of an igneous intrusion such as a BATHOLITH.

rooikalk (Afrikaans, red lime) a layer of hard sandy loam cemented by siliceous material underlying a thin, light brown sandy loam occurring in areas in southern Africa where the rainfall is between 305 and 380 mm (12 to 15 in). G.

root crop a term applied particularly to plants with a root system incorporating swollen fleshy parts which serve as a store of food for animals as well as the plant, grown especially for the feeding of animals and, because they can be stored, much used for winter feed for cattle and sheep, e.g. in midlatitudes turnips, Swedish turnips or swedes and mangolds (formerly called mangelwurzels). In midlatitudes root crops feature in the ROTATION OF CROPS (NORFOLK ROTATION) and those grown mainly for human consumption include SUGAR BEET, POTATOES, CARROTS, PARSNIPS, TURNIPS. Tropical root crops include ARROWROOT, CASSAVA, SWEET POTATOES, YAMS.

ropy lava, corded lava LAVA which has

solidified so that the surface of the flow is glassy and smooth with surface shapes resembling ropes or cords. PAHOEHOE. G.

Ro-Ro roll-on/roll-off technique, CONTAINER.

rosemary *Rosmarinus officinalis*, an aromatic EVERGREEN bushy shrub, native to the Mediterranean region, cultivated there and in cooler areas for its pungent leaves, used as flavouring in cooking, and from which ESSENTIAL OIL is extracted for use in medicines and cosmetics.

rose oil ATTAR OF ROSES.

rose quartz QUARTZ coloured pink by slight impurities.

rosin the substance left after oil of turpentine has been distilled from the RESIN-1 of some coniferous trees. G.

Rossby waves long waves with a wavelength of some 2000 km (1250 mi), very large-scale westerly movements in the upper ATMOSPHERE-1, stretching from sub-polar to tropical latitudes. There are generally about four such waves in the northern hemisphere, four in the southern.

Rossi-Forel scale a scale formerly used in the measuring of the intensity of EARTHQUAKE shocks, superseded by the modified MERCALI SCALE c.1931, now superseded by the RICHTER SCALE.

rotational slip the downward movement of a mass of rock or ice on a slip-plane, in which the solid mass appears to rotate on a pivot as it descends (an action which may, in the case of ice, contribute to the basin shape of a CIRQUE), and which usually results in the solid mass presenting a well-defined, uphill-facing back-slope.

rotation grass, rotation grassland, temporary grassland grass which is sown and grown for one or more years in rotation with other crops. LEY, ROTATION OF CROPS. G.

rotation of crops a farming system in which a systematic succession of differ-

ent crops is grown on the same piece of land so that the maximum use is made of soil nutrients but soil fertility is not exhausted. LAND ROTATION, MONOCULTURE, NORFOLK ROTATION. G.

rotation of the earth the revolving movement of the earth on its axis from west to east, from which it appears (from the earth) that the sun, moon and stars move round the earth from east to west. The average period of rotation of the earth in relation to the sun, i.e. the time interval between two successive crossings of the sun over a meridian, is 24 hours (MEAN SOLAR TIME, EQUATION OF TIME); but the rotation period measured by two successive crossings of a meridian by a selected star is less, being 23 hours 56 minutes 4.09 seconds (SIDEREAL DAY). The velocity of the earth's rotation is some 1690 km (1050 mi) per hour at the equator, some 845 km (525 mi) per hour at 60°N and S, and zero at the poles.

rotor flow LEE WAVE.

rotten borough in Britain before 1832, one of the parliamentary constituencies which still held the right to elect members of parliament even though the population was much reduced (and thus open to pressure from the unscrupulous) or even non-existent. Rotten boroughs were abolished by the Reform Bill 1832.

rotten ice disintegrating sea ice, resembling a honeycomb as it melts. G.

rottenstone a general term for a well-weathered rock, usually calcareous but in some cases fossiliferous sandstone, which is greatly leached but has not disintegrated. Friable siliceous limestone in this form is used in polishing metals.

rough grazing unimproved, usually unenclosed grazing, including many different types of natural and semi-natural vegetation, e.g. moorland, scrubland, salt marsh, mountain pasture. In Britain the distinction between rough grazing (usually unenclosed) and improved grazing (usually enclosed in fields) is clear, the two categories being shown separately in official statistics. In other countries enclosed grassland for grazing does not exist, and statistics merely separate pasture from arable: thus statistics for grassland, pasture, grazing are not internationally comparable. G.

route 1. a COURSE-1 of travel from one place to another 2. a regularly followed COURSE-2 3. in a GRAPH-2 the link (edge) between access points (vertices).

route factor NETWORK CONNECTIVITY.

row in statistics a horizontal line of entries in a frequency table. COLUMN, MATRIX.

rubber 1. an elastic substance made from the sap (LATEX) of a number of different tropical trees and climbers. Formerly, 'wild rubber' was collected from such plants in the equatorial forests of South America and tropical Africa; now most of the world's rubber comes from plantations of the Brazilian rubber tree, *Hevea brasiliensis*, in Malaysia (some from those in Indonesia, Thailand, Sri Lanka, Nigeria). The trees are 'tapped' by cutting a thin slice of bark, the white sap is collected, curdled by the adition of a litle acid, and stickiness countered by the addition of sulphur. It was this treatment of rubber with sulphur (vulcanizing) which opened up the possibility of its extensive use in manufacturing. BALATA 2. a synthetic substitute for the natural product.

rubidium a soft, silvery metallic ELEMENT-6, very similar chemically and physically to SODIUM.

rubrozem in soil science, FAO, a soil with an A B C horizon (SOIL PROFILE), the upper A HORIZON being dark with low BASE-2 saturation, overlying a B HORIZON which is reddish-brown or red. SOIL, SOIL ASSOCIATION, SOIL CLASSIFICATION. G.

ruby red CORUNDUM, a PRECIOUS STONE, used mainly in jewellery and art objects, but also in precision instruments such as watch movements. SPINEL.

rudaceous rock a sedimentary rock in which the grain is coarser than that of sand, both compacted (e.g. BRECCIA, CONGLOMERATE) and unconsolidated (e.g. BOULDER CLAY, GRAVEL). GRADED SEDIMENTS, PSEPHITE.

ruderal *adj.* growing on or among stone rubbish, hence applied to a plant inhabiting waste places (e.g. the site of a demolished or abandoned human settlement), waste land on road verges, waste tips near human habitations etc.

rug (Afrikaans, the back of an animal) in pl. rûens applied to a landscape of low plateaus cut by river erosion into numerous rounded ridges of equal height, resembling the backs of a herd of large animals. G.

rum an alcoholic drink, a spirit obtained by fermenting SUGARCANE, molasses and distilling the product.

ruminant any MAMMAL of the order Artiodactyla, hoofed and even-toed, no upper incisor teeth, usually with a complicated stomach consisting of several chambers, an animal that chews the cud, e.g. a camel, cow, deer, giraffe, sheep.

rundale (Ireland) 1. the joint occupation of land in Ireland, in which strips or patches of land, not necessarily contiguous, are occupied and cultivated by each holder 2. the land occupied in this way. RUNRIG. G.

rundale mountain a mountain or hill pasture grazed in common. RUNDALE.

Runddorf, Rundling (German, round, circular village) a village typical of parts of central Germany, possibly of Slav origin, with farmhouses encircling a central green, originally with only one means of access. G.

rundhäll (Swedish) the Swedish term for ROCHE MOUTONNEE.

runnel 1. a small stream 2. the small channel in which a small stream flows.

runner bean, Scarlet runner *Phaseolus*

coccineus, a PERENNIAL, vigorously-climbing BEAN native to South America, grown as an ANNUAL in the British Isles, producing long bean pods, usually cooked complete with seeds when young, before the seeds ripen.

running mean, running median a moving average, a VALUE-4 produced in a statistical procedure which aims to smooth out irregularities in a time series of values, especially when there is a marked fluctuation in adjoining values (e.g. agricultural output, annual precipitation), so that the moving average may be plotted for each time-interval on a graph. Each value is in sequence converted to the mean of the sum of itself and its two nearest neighbours (or more neighbours as required). Thus on a three-year moving average in a series $x_1 - x_n$, the value

for year 2 would be $\dfrac{x_1 + x_2 + x_3}{3}$,

for year 3 would be $\dfrac{x_2 + x_3 + x_4}{3}$.

runoff, run-off, run off a term best restricted to that part of the PRECIPITATION which runs off the land surface into streams, in contrast to those parts which either soak into the ground or evaporate, although some authors include water which returns to the surface by seepage and from springs. OVERLAND FLOW, THROUGHFLOW. G.

runrig, run-rig (Scottish) Scottish term corresponding to RUNDALE in Ireland. G.

rural (Latin *rus*, country) *adj.* of, belonging to, relating to, characteristic of, the country or country life, in contrast to the town or urban life. It is now recognized that in industrialized countries the distinction between rural and URBAN is blurred. Some authors write of a rural-urban continuum, or URBAN-RURAL CONTINUUM, and the creation of a rural-urban complex as

'rurbanization'. RURBAN, RURBAN FRINGE. G.

rural district LOCAL GOVERNMENT IN BRITAIN.

rural population 1. broadly, the population living in the countryside, not in a town 2. more specifically, that part of the population defined on the basis of various criteria, e.g. functional, those maintained by the exploitation of the intrinsic resource of the land (agrarian, or agrarian and mining); landscape-sociological, those living in a 'non-built-up' area; statistical, the size of the agglomeration, or the density of population; socio-psychological, many 'primary' social contacts 3. in Britain, before the reform of local administration in 1974, the population living in the administrative units known as rural districts, many of which included fairly large towns or suburbs of towns. LOCAL GOVERNMENT IN BRITAIN. G.

rural-urban continuum RURAL, URBAN-RURAL CONTINUUM.

rurban *adj.* having the characteristics of country and town combined.

rurban fringe, rural-urban fringe an indeterminate transitional zone around a town where urban functions and activities impinge on those that are agricultural, rural. URBAN-RURAL CONTINUUM. G.

rurbanization RURAL.

Russia leather a durable leather with a very agreeable smell, made by tanning a variety of skins with TANNIN obtained from birch, willow or oak bark, and then steeping them in birch-tar oil.

rust 1. a hydrated oxide of IRON, $Fe_2O_3H_2O$, formed when iron is exposed to air and moisture 2. the red-brown colour of that substance 3. any of the several diseases of plants caused by an order of parasitic fungi, or the parasitic fungus (e.g. *Puccinia graminis* causing black stem rust in cereals) responsible for the disease.

ruthenium a rare metallic ELEMENT-6, a platinum metal. PLATINOID.

rutile T_1O_2, titanium oxide, a lustrous, crystalline mineral, commonly reddish-brown, used in making white pigment in high quality paint, and in glass and porcelain manufacture. It is a source of TITANIUM.

rutin a glycoside, present in plants, especially in BUCKWHEAT and TOBACCO leaves, used in the treatment of radiation injuries and of hypertension.

ruware a rock pavement, a low, rounded, in some cases elongated exposure of rock, like a WHALEBACK-2, rising from a plain in a tropical area. Opinions differ as to whether it is an early stage in the formation of a SHIELD INSELBERG, or the result of the wearing down of an inselberg combined with the merging of surrounding PEDIMENTS-2.

rye *Secale cereale*, a tall-growing CEREAL with dark grain, probably native to Turkey-in-Asia, cultivated in the colder parts of Europe as far north as the Arctic Circle, and in mountainous areas up to 4270 m (14 000 ft). It tolerates the poor soils unsuited to most of the other midlatitude grain crops. The grain yields a dark-coloured flour, used in making bread which is heavy, with a rather sour flavour, and in making crispbread (a thin, hard biscuit) which keeps well and is favoured in low-calorie diets. The grain is also used in making whisky, gin, beer. Young plants are fed to livestock, the straw is used in thatching, bedding, paper-making. ERGOT. C.

ryot, rayat (Indian subcontinent: Urdu-Hindi; various other spellings) a tenant, a cultivator; hence ryotwar, ryotwary, the system of tenant farming in the subcontinent characterized by direct settlement between the government and cultivator without the intervention of a ZEMINDAR or landlord. G.

S

saaidam, zaaidam (Afrikaans, sowing dam) a basin with low earthen walls into which flood waters are diverted, used in irrigation especially along the Sak river in southern Africa. G.

sabzbār (Indian subcontinent: Pashto) an autumn crop.

saddle 1. a COL or PASS or any landform resembling in shape a saddle, especially a broad flat col in a ridge between two mountains 2. a mining term applied to an ANTICLINE and, in Australia, to a SADDLE-REEF. G.

saddle-reef a lens-shaped ore-body, usually of QUARTZ (e.g. gold-quartz veins) adjoining beds near the axis of an ANTICLINE; not to be confused with a PHACOLITH, an igneous intrusion similar in form. G.

saeter, seater, seter, setr, setter (Norwegian) formerly saeter, current spelling seter, but see SETER 1. a mountain pasture in Norway grazed by cattle in summer, comparable to ALP in Switzerland 2. a farm high up in mountain districts, used only in summer, which includes such pasture 3. in Orkney and the Shetlands, pasturage attached to a farm. G.

saffron 1. *Crocus sativus*, family Iridaceae, a low-growing herbaceous plant, probably native to Turkey, cultivated for its flowers, the three-branched style of which, when dried, yields an orange-coloured substance used since ancient times as spice, dye, cosmetic and medicine 2. the substance thus obtained 3. the colour of the substance.

sag and swell topography TILL sheets with an undulating surface as seen in the Middle West of USA. SWALE-2, SWELL AND SWALE. G.

sage *Salvia officinalis*, an aromatic, low-growing herb, native to southern Europe, now cultivated there and farther north, in midlatitudes, the leaves used as a flavouring in savoury dishes.

sagebrush a semi-desert type of vegetation in western North America dominated by sagebrush, *Artemisia tridentata*, in association with other small-leaved shrubs. G.

sago the starch obtained from the pith of some tropical palms, especially of the sago palm, *Metroxylon sagu* (of the freshwater swamps in southeast Asia); of *Caryota urens* (Indian subcontinent and Malaysia); of *Metroxylon rumphii* (Indonesia); of *Phoenix acaulis* (Indian subcontinent and Burma); of *Arenga saccharifera* (southeast Asia); of *Oreodoxa oleracea* (American tropics); and of the two palm-like plants, *Cycas circinalis* (Sri Lanka and Indian subcontinent) and *Cycas revoluta* (Japan). The palm or other plant is felled when the starch content in the pith of the trunk is at it highest. In the case of the sago palm this is just before its only flowering, at the age of about 15 years. The pith is ground and washed to extract the starch, the solution being dried to form sago flour. The granular pearl sago, used in puddings and sweet dishes in Europe, is produced by passing the starchy paste through a fine sieve, the resulting granules being dried on a hot surface.

Sahara (Arabic sahra) a desert, a plain, a term applied to the greatest of all

540

deserts, the Sahara, northern Africa (which should therefore not have 'desert' added to its name).

Sahel (Arabic shore) a term adopted from French authors, applied to the phytogeographical semi-arid zone (PHYTOGEOGRAPHY) lying between the SAHARA and the savanna lands to the south in west Africa, covering large parts of Mauritania, Sénégal, Mali, Upper Volta, Niger and Chad. The Sahel, which is prone to long periods of drought, e.g. 1968-73, is characterized by a short uncertain rainy season and a long dry season. The original CLIMAX was thorn woodland, but the vegetation now consists of patches of poor wiry and tussocky grass, acacia and thornbushes. G.

sailābā (Indian subcontinent: Panjabi) **1.** KACHCHI, flood plain, the area actually flooded by a river **2.** the area irrigated by flood waters. ABI-SAILABA. BARANI. G.

St Elmo's fire small flickering flames around the tops of tall objects, such as mastheads, visible at night, occurring in stormy weather, associated with the passing of a FRONT. The phenomenon constitutes a brush discharge, occurring when a stream of molecules of air, electrically charged, is repelled by a sharp-pointed, charged conductor.

saké, sake, saki an alcoholic liquor made, especially in Japan, from fermented RICE, usually drunk warm.

sakiya, sakya, sakiyeh SAQIA.

sal *Shorea robusta*, a valuable hardwood timber tree of tropical DECIDUOUS forests, especially of the eastern Indian subcontinent and Burma. The RESIN is used in varnishes.

salar a PLAYA, an inland drainage basin, often with a salt lake or SALT FLAT, in southwestern USA.

salic horizon a SOIL HORIZON enriched by water-soluble salts (GYPSUM excluded).

salience flow a transport flow which is

greater than might be expected from the trend of general regional flows.

salience score a measure used in the study of SALIENCE FLOWS, calculated by dividing the actual flow value by the expected flow value and subtracting 1.0 from the result. The salience score will be zero if the actual flow equals expected flow. Salience scores above zero indicate a higher than expected level of movement, those below zero the reverse.

saliferous *adj.* impregnated with salt.

salina (Spanish) a PLAYA with a high concentration of salts, enclosed from the sea. G.

saline **1.** an anglicized form of SALINA **2.** a salt spring.

saline *adj.* salt, or of, containing, or tasting SALT-2; or relating to or being characteristic of chemical SALTS-1.

saline soil an INTRAZONAL SOIL with a high concentration of soluble salts, occurring in areas with high evaporation, e.g. in hot deserts or in dry areas with high summer temperatures (as in cool temperate continental interiors), and especially resulting from the drying of salt lakes that once lay in inland drainage basins. Irrigation, unless carefully managed in such areas, may increase salinity. The salt solution is drawn upwards in the soil by CAPILLARITY, it dries out and forms a surface crust over a salt-impregnated more granular layer. SOIL, SOIL CLASSIFICATION.

salinity the degree of concentration of common salt in a solution, determined by measuring the density of the solution, and usually expressed in parts per thousand by mass. SALINOMETER.

salinity anomaly of ocean, the difference between the SALINITY at a certain location in the ocean and the average salinity of all the ocean.

salinization the PRECIPITATION-2 of soluble salts within the soil.

salinometer an instrument used in cal-

culating SALINITY, a type of HYDRO-
METER designed to measure the density
of a salt solution using the principle of
FLOATATION.

salpausseklä (Finnish) a RECESSIONAL
MORAINE or STADIAL MORAINE. G.

salsify *Tragopogon porrifolius*, vegetable oyster, oyster plant, a BIENNIAL
plant native to southern Europe, grown
for its white swollen root, eaten
cooked. The young leaves are suitable
for salads.

salt 1. in chemistry, a chemical compound derived from acids formed when
all or part of the replaceable HYDRO-
GEN ATOMS in a MOLECULE of the
ACID are replaced by a METAL, directly
or indirectly. Classified as normal salt
when all the replaceable hydrogen
atoms have been replaced by a metal,
acid salt when only part of all replaceable hydrogen atoms have been replaced **2.** common salt, NaCl, sodium
chloride, a white crystalline compound,
widespread in nature as a solid (HAL-
ITE, ROCK SALT) around margins of
salt lakes, in SALT DOMES, or in
solution in seawater, and present in all
animal fluids. It is obtained commercially by the evaporation of BRINE from
seawater, from brine wells and salt
lakes; or from the solid deposits in rock
salt mines. It is used as a food seasoning and preservative, and widely as a
raw material in industrial processes and
manufacturing industry, e.g. the chemical industry, in glass and soap making
etc.

salt *adj.* **1.** impregnated with, preserved
with tasting of, SALT-2.

saltation (Latin *saltare*, to leap) **1.**
leaping, jumping **2.** the mode of transportation of sediments bouncing along
a surface, e.g. of material by a river,
whereby particles such as small pebbles
make intermittent leaps from the bed of
the stream. BED LOAD **3.** the similar
movement of grains of sand propelled
by the wind in a hot desert. G.

salt corrie a cirque-like hollow in a

salt plug (SALT DOME), the result of
solution, resembling a CRATER or
CALDERA. G.

salt dome, salt-dome, salt plug an almost circular mass of ROCK SALT or of
other salt forced upwards from a great
depth in the earth's crust and, being
plastic under pressure (HALITE),
squeezed towards the weak part of the
sedimentary cover, thus often topped
by limestone cap rock. Oil and gas
fields are in many cases associated with
salt domes, which extend to great
depths in the earth's crust. DIAPIR. G.

salt flat a stretch of unbroken, salt-
encrusted horizontal land, the bed of a
former salt lake, now dried out, usually
permanently but sometimes only temporarily.

salt-field a tract of land underlain by
EXPLOITABLE salt deposits.

salt glacier a tongue of salt extending
down the slope of a SALT DOME. G.

saltigrade *adj.* applied to an animal
having legs adapted for leaping, e.g.
some spiders.

saltings a slightly elevated natural area
of salt marsh, with muddy channels,
supporting HALOPHYTES, and covered
by the sea at high water. Some authors
do not make the distinction between
saltings and SALT MARSH. G.

salt lick a place where animals come to
lick salt, either a natural deposit or a
salt block provided for them.

salt marsh, salt-marsh a natural coastal
marsh, supporting HALOPHYTES, regularly covered by the sea at high water
(HALOSERE). Some authors use the
term to include the elevated area
(SALTING), others exclude it. Salt
marshes may be enclosed for grazing,
for LAND RECLAMATION, for the recovery of salt, and when protected in
this way the term sea-marsh is sometimes applied. G.

salt mine a MINE where a natural
deposit of rock salt is worked. SALT-2.

salt pan, salt-pan 1. a small undrained
natural basin in which water evapo-

rates, leaving a deposit of salt **2.** a shallow vessel in which salt water accumulates and from which SALT-2 is obtained by evaporation.

salt plug   SALT DOME.

sample   1. a small part taken from the whole by which the characteristics of the whole can be deduced **2.** an individual portion, a specimen, by which the quality of more of the same sort can be judged **3.** in statistics, a part of a POPULATION-4 or a subset from a set of units, deliberately selected with the object of investigating the properties of the parent POPULATION-4 or set. MASTER SAMPLE, MATCHED SAMPLES, RANDOM SAMPLE, REPRESENTATIVE SAMPLE, SAMPLING, SIGNIFICANCE TEST, STRATIFICATION, SUBSAMPLE, SYSTEMATIC SAMPLE.

sample design   in statistics, a set of rules or a specification for the drawing of a SAMPLE-3.

sample estimate   SAMPLING THEORY.

sample size   the number of SAMPLE UNITS which are to be included in the SAMPLE-3.

sample survey   a survey carried out by using a sampling method, i.e. by surveying a representative part only, not the whole.

sample unit   any one of the units constituting a specified SAMPLE-3.

sampling   1. in general, the judging of the quality etc. of the whole by examining a part **2.** in statistics, the process of selecting a part or a subset in order to judge the quality, characteristics etc. of the whole by investigating the properties of the part. CLUSTER SAMPLING, NESTED SAMPLING, POINT SAMPLING, QUADRAT, RANDOM SAMPLE; and SAMPLE, with all its cross references.

sampling accuracy   the extent to which the SAMPLE-3 is free of all ERRORS, including BIAS-1, MEASUREMENT ERROR, SAMPLING ERROR, SYSTEMATIC ERROR.

sampling error   in statistics, the difference between a POPULATION-4 value and an estimate of it derived from a SAMPLE-3. No sample, however carefully selected, can be a perfect representation of the population from which it is drawn; but if the sample units are selected at random from the population (RANDOM SAMPLE) the sampling error can be calculated. The greater the precision of the sample, the less the sampling error. Sampling error does not include errors due to imperfect selection, BIAS-2 in response or estimation, mistakes in observation and recording etc. Errors arising after the sampling has been done are termed non-sampling errors. RANDOM SAMPLING, RANDOM SAMPLING ERROR, SAMPLING ACCURACY.

sampling fraction   the proportion of the total number of the SAMPLING UNITS in the whole (i.e. the population, stratum or higher stage unit) within which simple RANDOM SAMPLING (RANDOM SAMPLE) is made. The sampling fraction determines the sampling interval (SYSTEMATIC SAMPLE). SAMPLE-3.

sampling frame   a record or set of records which displays and identifies the whole (the POPULATION-4) from which a SAMPLE-3 is to be drawn.

sampling interval   SYSTEMATIC SAMPLE.

sampling procedure   any one of a number of methods used for drawing a SAMPLE-3.

sampling theory   the set of mathematical deductions and assumptions underlying RANDOM SAMPLING procedures which allows an estimate (the sample estimate) to be made from sample statistics to POPULATION-4 values or PARAMETERS.

sampling unit   one of the units into which an aggregate is divided or considered to be divided for the purposes of SAMPLING-2, each unit being regarded as individual and indivisible when the selection is made. SAMPLE UNIT.

sampling with partial replacement SAMPLING WITH REPLACEMENT-2.

sampling with replacement, without replacement, with partial replacement 1. a sampling method in which a SAMPLE UNIT is drawn from a finite population and replaced in the population after its characteristics have been noted so that it may be selected again. If it is not put back, the sampling is said to be without replacement **2.** in sample surveys conducted on successive occasions, if the same members are used for successive samples, there is said to be no replacement; but if some are retained and others replaced by new members there is said to be partial replacement.

sampling without replacement SAMPLING WITH REPLACEMENT-1.

samsam (Malay) a person of mixed Thai-Malay origin. G.

samun, samoon (Iran) a warm, dry FOHN-like wind in Iran, descending from the mountains of Kurdistan; not to be confused with SIMOOM, SIMOON. G.

sand 1. comminuted rock or mineral fragments of small size commonly, but not necessarily, SILICEOUS-1, ranging from 2 to 0.2 mm diameter, classification internationally agreed as coarse (2 to 0.2 mm), fine (0.02 to 0.002 mm) **2.** a soil of which 90 per cent or more is sand. ARENACEOUS-1, GRADED SEDIMENTS. G.

sandalwood 1. *Santalum album*, an EVERGREEN tree, parasitic on the roots of other plants, native to southeast Asia, the Pacific islands and Australia, with fragrant, pale-coloured wood (sandalwood) used in cabinet making, and yielding an aromatic oil (sandalwood oil), used commercially and in perfumery **2.** the fragrant wood of some other trees, also popularly termed sandalwood, resembling *Santalum album*.

sandbank an accumulation of sand, piled up by the action of waves or currents, occurring in a river (POINT BAR) or by the sea and exposed at low water.

sand bar BAR, SPIT. G.

sand drift a formation of blown sand, occurring in the lee of a gap between two obstacles, caused by funneling of the wind or by the concentration of the sand stream on the windward side from a broad to a narrower front. G.

sand dune, sand-dune a general term for a mound or ridge of loose, well sorted sand, piled up by the action of wind on sea coasts or in hot deserts. DUNE. G.

sand key, sandkey, sand cay a small, elongated sand island, lying parallel to the shore. KEY. G.

sand levée a WHALEBACK-1.

sandplain (Australia) a large, sand-covered plain of uncerain origin, a term used especially in western Australia. G.

sandr SANDUR.

sand ribbon a long, very slender but marked line of sandy material deposited on the sea floor, shaped by, and running parallel to the direction of, a strong TIDAL STREAM.

sand ridge desert a sandy DESERT in which most of the DUNES are longitudinal. LONGITUDINAL DUNE.

sand sea an area dominated by TRANSVERSE DUNES.

sand shadow an accumulation of sand lying in the shelter of and immediately behind a fixed obstruction lying in the path of sand-driving wind. DUNE. G.

sand sheet a sand area with a very flat surface. G.

sandstone a porous, ARENACEOUS SEDIMENTARY ROCK, widespread and laid down throughout geological time, formed mainly from rounded grains of QUARTZ and various minerals, varying considerably in colour, laid down in shallow seas, estuaries and deltas, along shallow coasts, in hot deserts, to be consolidated, compacted, and cemented by such substances as CLAY or SILICA which affect the colour of the sandstone and can be used as criteria in classification, e.g. calcareous, siliceous, ferruginous, or dolomitic sandstone. CAMBRIAN, NEW RED SANDSTONE, OLD RED SANDSTONE.

sandstorm a phenomenon in arid or semi-arid regions caused by a very turbulent wind which, passing over sandy soil, lifts and carries clouds of sand for a relatively short distance and rarely above a height of 15 to 30 m (50 to 100 ft). DUST STORM.

sandur, sandr (Icelandic) **1.** a general term for sandy ground, sand flat, sand-bank **2.** an alluvial outwash sand plain (OUTWASH APRON) formed by glacier streams flowing from a glacier edge to the sea. The spelling sandur is pre-ferred, sandr being out-of-date.

sandveld (Afrikaans) a sandy soil covered with vegetation, of some use for grazing. G.

sandy soil SAND, SOIL TEXTURE.

sanindo, sanyōdō (Japanese) regional names in central Japan, sanindo being applied to the shady, i.e. the wet, cloudy western side, contrasted with sanyōdō applied to the sunny, i.e. the drier eastern side. G.

Santa Ana, Santa Anna (Spanish: Cali-fornia) a hot, dry wind, commonly dust-laden, blowing from the north and northeast down the Sierra Nevada and over the south Californian deserts. It blows mainly in winter, but also occurs in spring, when it may damage the blossom or young fruit of fruit trees.

sapodilla *Achras sapota*, a medium-sized EVERGREEN tree, native to cen-tral America, now cultivated in many tropical areas, bearing succulent, fleshy fruit with black seeds and brown skin, the flesh, eaten when fully ripe, tasting like brown sugar. It also yields LATEX, coagulated by heating to make CHICLE GUM, the basis of chewing gum.

sapphire a PRECIOUS STONE, transpar-ent CORUNDUM, colour range white, yellow, green, blue, purple. The blue stones are the most highly valued, the colour being due to traces of cobalt.

sapping **1.** a term commonly used syn-onymously with glacial PLUCKING, but some authors restrict it to the detachment of rocks occurring at the bottom of a crevasse **2.** the undermining of slopes by chemical WEATHERING, water seepage, erosion by swiftly flowing water, especially in tropical regions (BASAL SAPPING). SPRING SAPPING. G.

saprobe an organism which feeds on decaying or dead organic matter.

saprobic classification a classification of river organisms based on their capacity to withstand organic POLLUTION, from the most tolerant group, polysa-probic (surviving in water so polluted that decomposition is ANAEROBIC), alpha mesosaprobic (surviving in partly AEROBIC, partly anaerobic decomposi-tion), beta mesosaprobic (tolerant of mild pollution), to the least tolerant group, the oligosaprobic (surviving only in non-polluted water). BIOTIC INDEX.

saprolite a deeply weathered rock formed from the disintegration of BED ROCK and lying in situ. It occurs particularly in the warm humid tropics. G.

sapropel sludge or mud which collects in swamps or shallow marine basins, rich in organic matter, formed by slow ANAEROBIC decomposition of remains of small organisms, e.g. of DIATOMS and PLANKTON, which, if compressed by accumulated sediments, may form PETROLEUM compounds.

saprophagous *adj.* applied to an animal which feeds on dead or decaying organic (plant or animal) material. SAPROPHYTE.

saprophyte an organism, usually a plant, obtaining organic matter in solu-tion from decaying or decayed organic matter, e.g. a FUNGUS living on dead wood; or YEASTS, which include among their activities the production of alcohol. EPIPHYTE, PARASITE.

saprophytic *adj.* of, pertaining to, or characteristic of, a SAPROPHYTE.

saqia, saqiya, sakya, sakiya, sakiyeh (Sudan: Arabic) a simple animal-powered mechanical device, also known as the Persian wheel, used to raise water from a river to irrigate the land above. Types vary, but a common, simple one, resembling the hand-powered SHADUF, incorporates a bucket suspended from the end of a pole, the pole pivoting on the top of a support mounted vertically on the ground and attached to a system of cog-wheels operated by a circling animal. The bucket is dipped in the water and when full swung round by the cog-wheel system to be tipped and emptied, usually into a channel carrying water to the cultivated land. In the Persian wheel a series of buckets is fixed round the circumference of a vertically mounted wheel, similarly operated by a cog-wheel, animal-powered system. QURER. G.

sardar SIRDAR.

sargasso (Portuguese sargaço) *Sargussum*, gulfweed, a tropical brown seaweed which gives it name to the calm Sargasso sea, characterized by the mass of sargasso which floats on it, supporting a variety of marine organisms, some of which are unique and peculiar to it.

sarn (Welsh, pl. sarnau) a causeway. G.

sarsen a mass of hard SANDSTONE, varying in constituents, especially one formed in the EOCENE, left as a residual mass on the surface of southern England. The alternative term, GREY-WETHER, indicates the resemblance in form to a sheep seen from a distance. It is widely used as a building stone, and in prehistoric times in megalithic monuments. MEGALITH. G.

sastrugi, zastrugi (Russian, pl.) preferred spelling zastrugi, wavelike ridges of hard snow formed by the action of wind carrying ice particles, occurring on a level surface (e.g. of a snow field or ice field), the axes of the ridges lying at right angles to the wind. The term now generally supersedes SKAVLE. G.

satellite 1. a natural or artificial celestial body constrained by GRAVITATION and moving in ORBIT around another more massive heavenly body, e.g. one of the artificial satellites in orbit around the earth, particularly one used in REMOTE SENSING 2. a state which depends economically and politically on another, more powerful, state.

satellite town a self-contained town, i.e. with its own industry, etc., especially a NEW TOWN, in some cases in the style of a GARDEN CITY, associated with a major city with which it has good communications. G.

satisficing behaviour behaviour in which the individual does not strive for the ultimate level of perfection, but is content to settle for a level which is personally satisfying and more likely to be achieved. BOUNDED RATIONALITY. G.

saturated *adj.* applied to 1. a SOLUTION having the highest possible amount of a SOLUTE in a specified amount of the SOLVENT at a given temperature 2. the ATMOSPHERE when it cannot hold any more WATER VAPOUR, i.e. when the number of molecules of water going in coincides with the number going out. If cooling occurs at that stage CONDENSATION results, giving MIST, CLOUD or RAIN 3. a rock holding in its interstices the maximum possible amount of water. WATER TABLE 4. a rock having the maximum possible amount of combined SILICA (if the rock is oversaturated the excess silica occurs as free QUARTZ). SATURATION LEVEL.

saturated (or wet or moist) adiabatic lapse rate the rate of loss of TEMPERATURE-2 with increasing height occurring in a moist or SATURATED-2 body of air as it ascends ADIABATICALLY. The rate is governed by the WATER VAPOUR content (itself governed by temperature) because latent heat is given up as CONDENSATION takes place (the higher the temperature the greater the content of water vapour, the greater the release of latent heat). Thus the rate can vary between 0.4°C and 0.9°C per 100 m (20°F and 5.3°F per 1000 ft). CONDITIONAL INSTABILITY, DRY ADIABATIC LAPSE RATE, ENVIRONMENTAL LAPSE RATE, LAPSE RATE, NEUTRAL STABILITY, PROCESS LAPSE RATE.

saturation the act of completely filling, or of completely satisfying, or the result of those acts. SATURATED.

saturation deficit the quantity of WATER VAPOUR needed to bring to saturation point a mass of non-saturated air at a given temperature and pressure. SATURATED-2.

saturation level, saturation point, saturation stage the level or point or stage at which a SOLUTION, the ATMOSPHERE-1, or a substance, or an area etc. reaches a stage of SATURATION, e.g. when the POPULATION-1,3 of an area equals its CARRYING CAPACITY-1,2,3,5, or when the DIFFUSION-1 of INNOVATION is complete. DIFFUSION WAVE.

saturation overland flow OVERLAND FLOW.

Sauer hypothesis the proposition of Carl Orwin Sauer, 1889-1875, American geographer, that the location of the first agricultural communities (agricultural HEARTHS) would have been in areas where food was sufficiently abundant, where there was a great variety of plants and animals for experiments and hybridization, and where the vegetation was woodland (trees being easily felled and burnt whereas grassland is difficult to dig with primitive tools). He considered wide river valleys, needing advanced techniques in water control, were unlikely locations; and that the inhabitants of the agricultural hearth areas, the original group of cultivators, were SEDENTARY.

savanna, savana, savannah (from Carib zavana, via Spanish) a term with a wide variety of applications, but best restricted to the natural, open, tropical grassland with scattered trees and bushes (mainly XEROPHYTES) covering vast areas in Africa, South America and northern Australia between the EQUATORIAL FORESTS and the hot deserts, a region with its particular soil conditions and a regular climatic regime. Rain falls in the hot summer and leads to a sudden luxuriant growth, but this withers in the drying winds of the winter with its low rainfall, or is scorched in frequent extensive fires. CAMPO, LLANO. G.

savanna woodland park-like woodland with xerophytic undergrowth. PARKLAND-2, XEROPHYTE. G.

savory *Satureja*, a fragrant herb native to the Mediterranean region, used as flavouring in cooking. Summer savory, *Satureja hortensis* is a bushy ANNUAL; but winter savory, *Satureja montana*, is a hardy PERENNIAL low shrub adapted to north African conditions.

sawah (Indonesia) a flat field enclosed by small embankments which prevent the runoff of water, used in rice cultivation. G.

saylo (Indian subcontinent: Kashmir) UBAC. TAILO. G.

saxicolous, saxatile, saxicoline *adj.* applied to an organism which lives among rocks.

saza (Uganda: Luganda language) an administrative division corresponding to a county in Buganda, the alternative term, county, sometimes being preferred. GOMBOLALA. G.

scabland a landscape with flat-topped hills or plateaus, bare or covered with a thin soil consisting of angular debris formed in situ and supporting a sparse vegetation, formed by the glacial erosion of a hard basalt surface, occurring particularly in northwestern USA. Compare BADLANDS, associated with soft sediments. G.

scalded flat (Australia) 1. in Australia, a low plain with soil impregnated with salt **2.** in soil science, specifically a red-brown soil which has lost part of its A HORIZON. G.

scale **1.** a level of representation of reality, the proportion of a representation to the object it represents. FLOW-LINE MAP **2.** linear scale, the indication on a MAP or PLAN-1,2,3, of the ratio between a given distance on the map or plan and the corresponding distance on the earth's surface. This is shown by a graduated line, or by the REPRESENTA-TIVE FRACTION; or it is expressed in words **3.** an arrangement of marks spaced at intervals to represent a series of numerical values and used in measuring temperature (THERMO-METER), length, angles etc. TEMPER-ATURE SCALE **4.** a measuring instrument bearing such marks **5.** an ordered series of graduated quantities, values, degrees **6.** relative magnitude, e.g. small scale, large scale. MEASUREMENT, in statistics. G.

scallop (American flute) one of several oval-shaped hollows which form patterns on stream beds and cave walls. Asymmetric in cross-section along the main axis, these scallops are steeper on the upstream side, and thus indicate the direction of flow of turbulent water. G.

scar, scaur (northern England) a crag, cliff, hilltop, where bare rock is exposed and soil or weathered debris is absent, seen especially in the bare limestone faces of the Great Scar Limestone (CARBONIFEROUS Limestone) in northern England. G.

scarcity rent a payment made in respect of any of the factors of production (land, labour, equipment) which are temporarily or permanently in short supply. RENT.

scarp, scarp-face, scarp-slope the abrupt, sometimes cliff-like face or slope terminating an elevated surface of low relief, i.e. an ESCARPMENT-2, the steep slope of a CUESTA. Escarpment is the preferred term. G.

scarp-foot spring a SPRING-2 near the foot of an ESCARPMENT, occurring particularly where chalk, limestone or sandstone overlies clay.

scarp land, scarpland, scarplands a region characterized by a number of parallel or sub-parallel scarped ridges separated by VALES, as in the scarplands of England, also termed scarp-and-vale terrrain. G.

scattergram a graph showing the way in which a DEPENDENT VARIABLE, plotted on the ordinate or vertical axis (the y-axis), relates to an INDEPEN-DENT VARIABLE plotted on the abscissa or horizontal axis (the x-axis). The scattered points may lie in such a way that they form, for example, a LINEAR-4 or a CURVILINEAR RELA-TIONSHIP; or they may not form a pattern at all, indicating an absence of any relationship. Scattergrams are therefore often used to discover if there are relationships which would not be revealed in an ordinary CORRELATION COEFFICIENT.

scattergraph in statistics, a visual presentation of a series of values along a numerically continuous scale, the scale covering the whole range of DATA and each observation being plotted according to its size along it.

Schattenseite (German) the shaded side of a valley, synonymous with UBAC. ADRET, SONNENSEITE. G.

scheelite calcium tungstate, $CaWO_4$, an ore of TUNGSTEN, occurring in veins and contact metamorphic deposits. CONTACT METAMORPHISM.

schist a foliated METAMORPHIC ROCK in which the various minerals have crystallized or been recrystallized into thin layers, lying parallel to each other, a rock which will split into more or less irregular flakes, owing to the occurrence in it of LAMELLAR minerals such as mica, chlorite etc. The texture is independent of the BEDDING PLANES of the original rock, the grains are of medium size, not so coarse as those of

GNEISS; and the type of schist is distinguished by the dominant mineral, e.g. mica-schist. Nearly all types of SEDIMENTARY and IGNEOUS rocks will become schists if subjected to sufficient heat and pressure.

schiste (French) a term applied in French to SHALES and SLATES as well as to SCHISTS, and thus not necessarily indicating a METAMORPHIC ROCK as does the English term schist. The French equivalent of the English schist is 'schiste crystallin'; for mica-schist it is 'schiste lustré'. G.

school district in USA, a single-purpose unit of local government responsible for providing junior and secondary education in its territory. LOCAL GOVERNMENT IN USA.

Schratten (German) Karren, KARRE. G.

Schumm's index of shape an index of measuring shape which compares the area of a shape with the length of its longest axis, and thereby attempts to measure the level of ELONGATION-1,2: $4A^2/\pi L$, where A is the area of the shape, L the length of the longest axis, π a constant value equal to 3.14.

Schuppenstruktur (German) IMBRICATE STRUCTURE, the Schuppen being the individual thrust masses in between the thrust planes in the imbricate structure, which is caused by a large number of small thrusts. G.

Schwingmoor (German) a term adopted by many British ecologists and applied to BOGS or MOSSES-2 largely of SPHAGNUM replacing former glacial lakes or MERES, as in the Cheshire plain in England; a quaking bog (QUAGMIRE). G.

science (Latin scientia, knowledge) **1.** the condition or fact of knowing **2.** knowledge gained by detailed observation, by DEDUCTION of the laws governing changes and conditions and by testing these deductions by experiment. SCIENTIFIC LAW **3.** a branch of study, especially one concerned with facts, principles and methods, e.g. BEHAVIOURAL SCIENCES, EARTH SCIENCES, PHYSICAL SCIENCES, SOCIAL SCIENCES. SCIENTIFIC.

scientific *adj.* **1.** of, pertaining to, used in, SCIENCE-1,2 **2.** of or using methods based on well-established facts and conforming to well-established laws **3.** using knowledge made available by scientists.

scientific determinism DETERMINISM.

scientific law a general statement of fact methodically established (according to the orthodox view) by INDUCTION, on the basis of observation and experiment (EMPIRICISM). Scientific laws are usually rooted in DETERMINISM-1, and are universal in their application in that they usually make statements which cover all members of a particular class of things. But they stray into PROBABILITY in statistics in making statements about a methodically estimated proportion of the class of things under observation. Some philosophers of science do not agree that the method of establishing a scientific law is wholly inductive. They see two phases in the scientific method, the first comprising inspired guesswork, leading to the formation of an HYPOTHESIS; and the second, the confirmation of the hypothesis by induction.

sciophilous *adj.* applied to an organism adapted to life in shaded conditions. SCIOPHYTE.

sciophyte a plant which flourishes in shaded conditions. HELIOPHOBE, HELIOPHYTE, SCIOPHILOUS.

scirocco SIROCCO.

sclerophyll a plant with hard, leathery, commonly EVERGREEN leaves which have few stomata (STOMA), thereby resisting loss of water by TRANSPIRATION, e.g. the OLIVE, characteristic of Mediterranean lands with hot dry summers. G.

sclerophyllous *adj.* of, pertaining to, in the nature of a SCLEROPHYLL.

score in statistics **1.** a number yielded

by a test item, test, or series of tests etc.
2. a quantitative assessment of an individual on a scale, related to performance in a test, or derived from the individual's reaction to certain stimuli.

scoria pl. scoriae **1.** a mass of volcanic rock, fine-grained and resembling clinker from a furnace, the holes being caused by the expansion of gases and steam imprisoned in the LAVA and the rapid cooling of its surface **2.** an accumulation of similar clinkery material which has been ejected from a volcano as PYROCLASTS. G.

scoriaceus *adj.* having the nature of, resembling, SCORIA.

scorzonera *Scorzonera hispanica*, a hardy PERENNIAL plant native to central and southern Europe, usually cultivated as an ANNUAL or PERENNIAL for its swollen, nutritious root containing inulin, a sugar which most diabetics can tolerate. Normally cooked, it can be used as a coffee-substitute; and the young leaves can be eaten raw.

Scotch mist MIZZLE, precipitation resembling mist and drizzle, occurring when CLOUD lies near the ground. It is especially common in hilly or mountainous areas, e.g. in Scotland, hence the name.

scour the strong, erosive action of a current or flow of water, e.g. of the tide, of a river, in clearing away deposits such as mud and sand; or of boulders frozen into the base of an ice sheet or glacier. ABRASION. G.

scree 1. a slope consisting of an accumulation of loose angular rock debris of any size and commonly formed by frost action from the parent rock, lying at a uniform angle (commonly of some 35°) at or near the foot of a steep cliff, rock-buttress, mountain etc. **2.** the angular rock debris itself **3.** a synonym for TALUS, especially in the USA. ANGLE OF REPOSE, MASS MOVEMENT, REPOSE SLOPE. G.

scroll 1. on a FLOODPLAIN, a narrow stretch of floodplain added to the outer end and downstream side of spurs between enclosed meanders **2.** a type of POINT BAR, sometimes termed a meander bar or meander scroll, a low, narrow ridge running in line with the curve of a MEANDER, formed when the river overflows its banks. OVERBANK DEPOSIT, OVERBANK STAGE. G.

scrub 1. vegetation consisting of dwarf or stunted trees and shrubs, often very thick, XEROPHILOUS and growing on poor soil, or in a semi-arid area, or in an exposed position. It may be a natural CLIMAX or a transitional stage in a PLANT SUCCESSION, leading to woodland **2.** the land covered with this type of vegetation **3.** applied by some authors to RAIN FOREST in Queensland, Australia. G.

scrub forest FOREST-2 consisting of malformed, small or stunted trees and shrubs. G.

scrub woodland an open cover of the trees and shrubs characteristic of SCRUB FOREST. G.

scud fragmented low, thin CLOUD moving quickly underneath the main cloud cover of NIMBOSTRATUS, blown by strong wind and indicating stormy weather.

sea 1. in general, applied to the great body of salt water on the earth's surface (the OCEAN-1), i.e. as opposed to land **2.** one of the smaller bodies of salt water of the ocean with a proper name, e.g. Mediterranean sea, China sea **3.** a large body of inland salt water, e.g. Sea of Aral, Dead Sea, Salton sea. TERRITORIAL WATERS.

sea breeze, lake breeze a local breeze (BEAUFORT SCALE) which blows usually during the afternoon from the sea to the land, owing to the differential heating and cooling of land and water. The heating of the land by day causes the ascent of warmed air in a small low pressure area, and cooler air from the sea flows inland for a short distance to take its place. A sea breeze occurs in calm, settled weather where temperature changes are regular, and especially in equatorial latitudes. A similar breeze is associated with large lakes. LAND BREEZE. G.

sea floor spreading OCEANIC RIDGE, PLATE TECTONICS.

sea-fret (southwestern England) a salt mist moving inland from the sea, often very destructive of vegetation. G.

seakale *Crambe maritima*, a PERENNIAL plant native to the sea cliffs of westernEurope, cultivated for its leaf stalks which, when blanched, are eaten cooked.

seakale beet, chard *Beta vulgaris*, a plant resembling SPINACH BEET, producing edible leaves, the white stalks being cut and eaten cooked, the green parts, when cooked, tasting rather like SPINACH.

sea-level, sea level MEAN SEA-LEVEL.

sea loch LOCH.

sea-marsh SALT MARSH.

sea mile NAUTICAL MILE. KNOT, MILE.

sea-mill tide mill, a mill powered by the force of seawater, e.g. from the tide or, rarely, from a swallow-hole close to the sea. The sluices are closed at high water so that a head of water to drive the mill wheel is available on the ebb tide, the same principle as that applied in a TIDAL POWER STATION. G.

seamount a topographical feature rising from the ocean floor, an isolated peak, usually a volcano, with a pointed summit (as opposed to a flat-topped GUYOT), the summit usually lying well below the ocean surface, e.g. sometimes 3000 m (nearly 10 000 ft) below. G.

sea power a nation with naval strength, as distinct from a LAND POWER.

search behaviour the way in which an individual or group of individuals (e.g. a firm) looks for information to meet particular needs, interprets and assesses it, and reaches a decision on a course of action. ASPIRATION REGION, SEARCH SPACE. G.

search space the locations within an area where an individual searches in order to meet a specific need or specific needs (e.g. for housing), based on information from that individual's current AWARENESS SPACE.

SEASAT 1 the first of the NASA SATELLITES-1 to be launched (1978), equipped with RADAR sensors, to obtain information for oceanographic research. LANDSAT, REMOTE SENSING.

seascarp an ESCARPMENT on the ocean floor, caused by faulting.

seashore 1. a general term applied to land immediately adjoining the sea 2. the land between the lowest water line of the SPRING TIDE and the highest limit of storm waves.

season 1. a division of the year associated with the duration of daylight and/or characteristic climatic conditions related to changes in the intensity of SOLAR RADIATION brought about by the inclination of the earth's axis to the plane of the ECLIPTIC and the elliptical nature of the ORBIT OF THE EARTH around the sun. Defined astronomically there are four divisions in the year, lasting from EQUINOX to SOLSTICE, from solstice to equinox (sequence repeated). In midlatitudes the seasons are associated with the life cycle of plants, winter (dormant), spring (sowing), summer (growing and ripening), autumn (harvesting), conventionally and arbitrarily defined as WINTER (December to February), SPRING (March to May), SUMMER (June to August), AUTUMN (September to November) in the northern hemisphere, the reverse in the southern. In tropical regions the seasons are linked to rainfall, the year being commonly divided into two, the wet (rainy) season and the dry season; in monsoon regions there are commonly three seasons, termed cold, hot, rainy; in polar regions the periods of change between winter and summer are so brief as to be scarcely noticeable, so there are in effect only two seasons; and in equatorial regions there is little differentiation of season **2.** a period of time most favourable to something, e.g. strawberry season, when that fruit is at its best; or to some activity, e.g. the football season **3.** in season, as applied to plants, fruit and vegetables harvested after natural growth in the open, thus available fresh and in peak condition; as applied to sexually mature females of many species of MAMMAL, oestrus, the brief period in the reproductive cycle when ovulation takes place and the female is willing to copulate with the male. G.

seat-earth a mining term, a layer consisting mainly of FIRECLAY or GANISTER underlying coal-seams.

seaway 1. in general, a way across the open ocean used regularly by shipping **2.** a SHIP CANAL large enough for ocean-going vessels, notably the St Lawrence Seaway in Canada.

seaweed any marine plant, but especially the red, brown, or green ALGAE living in or by the sea, many edible and used directly as human food, particularly in Japan and China; or as a fertilizer; or as a source of useful substances (e.g. ALGIN, IODINE, or the gelatinous AGAR-AGAR, used in making jellies and as the base of media for bacteriological culture). Some seaweeds popular for food include carrageen, *Chondus crispus*, from which carrageenin, an emulsifier used in food preparation, is obtained; dulse, *Rhodymenia palmata*, used fresh in salads, dried for cooking or for chewing and in making alcoholic liquor; knotted wrack, *Ascophyllum nodusum*, providing ALGINATES for thickening, gelling and emulsifying in food preparation, seaweed meal for livestock feed, and thin film for sausage skins; laver, *Porphyra umbilicalis*, cooked and widely eaten in China and Japan and elsewhere, e.g.Wales; oarweed, *Laminaria digitata*, also a source of alginates, boiled as food for humans or livestock. SARGASSO, THALLOPHYTA.

sebka, sebhka (north Africa: Arabic) a salt-encrusted mud flat, a closed depression in an arid area which becomes temporarily marshy after the (rare) rain. PLAYA. G.

second a unit of measurement **1.** of time, one-sixtieth of a minute or $\frac{1}{86400}$ of a MEAN SOLAR DAY **2.** the SI basic unit of time, defined from a frequency in the spectrum of CAESIUM **3.** of angle, one-sixtieth of one minute of an arc, used in the measurement of latitude and longitude. MINUTE.

Secondary *adj.* formerly applied to the second era of geological time (GEOLOGICAL TIMESCALE), i.e. subsequent to the PRECAMBRIAN, corresponding to MESOZOIC, the term now in general use.

secondary consequent stream a tributary to a SUBSEQUENT STREAM which, although formed after the subsequent stream, flows parallel to the original CONSEQUENT STREAM.

secondary consumer FOOD CHAIN.

secondary depression in meteorology, a relatively small area of low atmospheric pressure associated with a main primary DEPRESSION, encircling the latter in an anti-clockwise direction in the northern hemisphere (clockwise in the southern) as it moves along its course. It may be linked to the primary depression, appearing in the isobars (ISO-) as a protruberance, or it may be self-contained, with closed isobars; in either case its pressure may be lower than that of the primary, which it may eventually absorb.

secondary enrichment the natural augmentation of an ORE by material of later origin held in an aqueous solution and deposited by downward percolation. SUPERGENE.

secondary group large, freely associating groups of people brought together by shared beliefs and common goals but lacking the intimate interrelationship of the PRIMARY GROUP.

secondary industry, secondary activity, secondary sector MANUFACTURING INDUSTRY, i.e. INDUSTRY concerned with transforming material provided by PRIMARY INDUSTRY into something more directly useful to people, e.g. manufactured goods, construction work, electric power production (but some countries include the last in primary industry). QUATERNARY INDUSTRY, TERTIARY INDUSTRY. G.

secondary mineral a MINERAL-1 formed **1.** in consolidated MAGMA as it cools, the result of reactions within the rock itself or of the reaction with the circulating ground water present **2.** by METAMORPHISM **3.** by WEATHERING of IGNEOUS ROCK **4.** during DIAGENESIS of SEDIMENTARY ROCK. AUTHIGENIC. G.

secondary production in ecology, the energy lost in the RESPIRATION of the consumer levels in a FOOD CHAIN. PRIMARY PRODUCTION, PRODUCTION.

secondary rock rock formed by the alteration of pre-existing rock (PRIMARY ROCK) or from material derived from the disintegration of rocks, e.g. CLASTIC sediments, METAMORPHIC ROCK.

secondary sampling unit CLUSTER SAMPLING (multi-stage).

secondary sere SUBSERE.

secondary soil a transported soil, soil formed on transported material, as opposed to a RESIDUAL SOIL. SEDENTARY SOIL. G.

secondary succession a PLANT SUCCESSION which follows the destruction of part or all of the original vegetation (PRIMARY SUCCESSION) of an area and gives rise to a SUBSERE. CLIMAX, SERE.

secondary vegetation a general term applied to the natural plant-cover growing on land that was once cleared of vegetation, used for a while, and then abandoned. NATURAL VEGETATION, SEMI-NATURAL VEGETATION.

secondary wave of EARTHQUAKE. SHAKE WAVE.

second home weekend cottage, a dwelling of a freeholder or leaseholder (LAND TENURE) whose usual, commonly larger, residence is elsewhere, and who occupies the second home, which is usually in a rural area or near the sea, at weekends and/or holiday periods.

Second World the countries which have adopted a centrally planned communist system, including the countries of eastern Europe and the USSR. FIRST WORLD, THIRD WORLD.

section 1. in geometry, the plane figure resulting from the cutting of a solid by a plane; hence **2.** the formation revealed by a cut, or representation of a cut, made vertically through a landform, rock or soil so as to show the surface and subsurface layers, e.g. in a geological section, the surface layer and the underlying strata. In that example the representation of such a cut may be small scale and generalized (diagrammatic section), or have an accurate surface profile with the strata in diagrammatic form (semi-diagrammatic). A totally accurate representation, without exaggeration of the vertical scale, would give a misleading impression of the dip of the strata (VERTICAL EXAGGERATION). CROSS-SECTION, SOIL PROFILE **3.** a unit in the hierarchy of MORPHOLOGICAL REGIONS **4.** a unit of land, one mile square, in the USA Land Office Survey, forming 1/36 of a township. PUBLIC LAND SURVEY, USA **5.** a thin slice of something, e.g. of TISSUE, prepared for study under a microscope.

sector model sectoral model, of land use, a MODEL-4 developed on the assumptions of Hoyt's SECTOR THEORY, that the arrangement of routes radiating from a city centre condition the structure of the city and its growth;

and that differences in accessibility between the radial routes lead to variations in land value and consequently of land use in the sectors created by these routes, the outer arc of each sector tending to repeat the pattern of its earlier growth.

sector theory a theory introduced by H. Hoyt in 1939 to explain land use patterns and urban growth, based on his studies of the structure and growth of residential neighbourhoods in American cities. The theory asserts that the wealthy occupy the most desirable areas, the less wealthy the areas around the most desirable, the least desirable areas being those lying adjacent to industrial development; that the high status residential area, identified as the central agent of change over time, combined with the disposition of the radial routes from the city centre, determine urban residential structure and growth. The high status residential areas move out radially along the most rapid, convenient transport lines, and the new urban growth tends to repeat the pattern of the established earlier growth (i.e. the wealthy occupy the most desirable areas, etc). Differences in accessibility between the radial routes lead to variations in land value and consequently of land use in the sectors created by these routes, the outer arc of each sector tending to repeat the pattern of its earlier growth. Sector theory thus takes into account not only distance but also direction from the city centre (unlike CONCENTRIC ZONE GROWTH THEORY). MULTIPLE NUCLEI MODEL. G.

secular *adj.* **1.** of a change or event occurring very rarely, e.g. once in a century or other very long period of time **2.** continuing, lasting over such a long period **3.** concerned with temporal, worldly matters rather than with religion.

sedentarization the stage of transition

fron NOMADISM or semi-nomadism to a settled way of life. G.

sedentary (Latin sedere, to sit) *adj.* being established in, staying in, one place, i.e. not migratory, not transported.

sedentary agriculture farming as practised by a settled farmer in one place, as distinct from SHIFTING CULTIVATION.

sedentary soil a soil formed from the decay and decomposition of the solid rocks on which it lies. SECONDARY SOIL. G.

sedge any member of Cyperaceae, a very large family of coarse grass-like herbs, especially of the genus *Carex*, usually with solid, three-sided stems, commonly growing in wet places, although some are XEROPHILOUS. BOG PEAT.

sediment **1.** matter which, owing to its greater density, naturally sinks by gravitation to the bottom of any undisturbed liquid with which it was formerly mixed, e.g. as in a SOLUTION (COLLOID) **2.** in geology, unconsolidated particles or grains of rocks deposited by river, ocean, ice, wind. ELUTRIATION, GRADED SEDIMENTS, SEDIMENTARY ROCKS, SEDIMENTATION. G.

sedimentary rock a rock consisting of material derived from pre-existing rocks (i.e. from SEDIMENTS-2) or from organic debris, laid down in layers, in some cases FOSSILIFEROUS, some being consolidated (CEMENTATION OF SEDIMENTS, COMPACTION-1, DIAGENESIS), others unconsolidated. They may be mechanically formed (CLASTIC rock) and ARENACEOUS (e.g. SANDSTONE), ARGILLACEOUS (e.g. SHALE), or RUDACEOUS (e.g. CONGLOMERATE). They may be organically formed and CALCAREOUS, FERRUGINOUS, SILICEOUS-1 (e.g. DIATOMITE) or CARBONACEOUS (e.g. the COALS). Those formed by chemical processes include carbonates such as DOLOMITE, or silicates such as FLINT, or ironstone such as LIMONITE; those formed by DESICCATION include EVAPORITES and sulphates such as GYPSUM, or chlorides such as ROCK SALT. The layers vary greatly in thickness and may lie horizontally or at an angle; considerable tilting reveals that they were disturbed after the period of deposition. In all sequences of sedimentary rocks the oldest bed lies at the bottom, the youngest at the top, unless the order is upset by folding, faulting or other disturbance. LAW OF SUPERPOSITION.

sedimentation 1. the downward movement of finely divided solid particles through a fluid under the influence of GRAVITATION, a natural process which can, in a laboratory, be speeded up if the suspending medium is put in a CENTRIFUGE 2. the act or process of deposition, of settling as a SEDIMENT-2.

sediment discharge rating in hydrology, the ratio between the discharge of SEDIMENT-2 carried by a stream and the total discharge of the stream.

sedimentology the scientific study of SEDIMENTS-2 and SEDIMENTARY ROCKS and the processes forming them.

sediment yield the mean load of SEDIMENT-2 carried by a stream measured by the weight of the sediment per unit area of DRAINAGE BASIN.

seed the product of a fertilized ovule of a plant containing within a protective coating an embryonic plant and sufficient food for its development. FRUIT.

seed bed an industrial area near the CENTRAL BUSINESS DISTRICT where rents are low and premises cheap and easily converted to small factories, conditions which favour the setting-up of new small manufacturing enterprises. Those that flourish transplant themselves to larger premises farther from the central area. They and those that fail are soon replaced by new, hopeful, industrialists. There is thus a rapid turnover of manufacturing establishments in the seed bed area; but such areas are squeezed out by policies of URBAN RENEWAL and CENTRALIZATION-1.

seeding of clouds the dropping of chemical particles (e.g. solid carbon dioxide, silver iodine) from aircraft on to clouds in order to stimulate CONDENSATION and lead to rainfall.

seepage the very slow percolation or oozing out of a fluid through a porous body (POROSITY), or along a fault or joint-plane, e.g. the slight oozing out of PETROLEUM on the ground surface, an important indication of the presence of oil in rocks below; or the percolation of surface water into the soil; or the slow oozing out of ground water at the surface when the flow of water is insufficient and the pressure not high enough to form a SPRING-2. G.

segregation 1. separation from others of a group, especially the establishment by law or custom of separate facilities for different social or ethnic groups, as for whites and blacks in the Republic of South Africa. APARTHEID 2. the process by which individuals and groups settle in areas already housing people with tastes, preferences, or social characteristics similar to their own. DESEGREGATION, GHETTO, INDEX OF SEGREGATION, SOCIAL SEGREGATION, SPATIAL SEGREGATION.

seiche a periodic or occasional, brief, undulation of the water in a restricted area, e.g. in a lake, estuary or bay, apparently caused by abrupt changes in ATMOSPHERIC PRESSURE, or by wind, or EARTHQUAKE. G.

seif-dune (Arabic seif, sword) a SAND DUNE piled up longitudinally as a steep-sided ridge, sometimes stretching over many kilometres, and lying parallel to the direction of the prevailing wind. The origin of seif-dunes is uncertain; some authors maintain that they form when a strong wind is funnelled through lines of small crescent-shaped dunes (BARCHAN), sweeping away the horns, leaving the ridges, and causing the small dunes to coalesce; others that they are just an extended sand drift formed behind an obstacle. G.

seine a large fishing net held vertically in the water by floats fixed to one edge and weights to the other. The fish (especially DEMERSAL fish) are trapped as the ends are drawn together.

seismic *adj.* relating to, characteristic of, produced by, movement within the earth, e.g. an EARTHQUAKE.

seismic focus, seismic origin (Greek seismos, earthquake) the place in the earth's crust, under the surface, from which an EARTHQUAKE shock originates. DEEP FOCUS, EPICENTRE.

seismic wave **1.** a shock wave generated by an underground explosion or EARTHQUAKE **2.** a TSUNAMI.

seismograph a scientific instrument used in recording the duration, magnitude and direction (horizontal and vertical) of earth tremors, natural (EARTHQUAKE) or ARTIFICIAL-1.

seismology the scientific study of EARTHQUAKES and of other movements of the solid earth, including earth tremors produced artificially. SEISMO-GRAPH.

seismonasty NASTIC MOVEMENT.

seistan (Iran, Seistan, a former province in east Iran) a strong northerly wind, sometimes reaching hurricane force (BEAUFORT SCALE), blowing in the four summer months in eastern Iran. G.

selenite a transparent, colourless, fully crystalline variety of GYPSUM.

selenium a non-metallic element used in its silver-grey crystalline form in photoelectric cells, photographic light meters, etc. because its electrical resistance varies according to the intensity of its received light.

selenography the scientific study of the physical features of the MOON.

selenology the scientific study of the MOON, including its movement and astronomical associations, and the formation of its crust.

selenomorphology the study of landforms on the MOON'S surface. G.

selion historical, a ridge or narrow strip of land of indeterminate width lying between the furrows, formed in the division of an OPEN FIELD, used as the basic unit of measure of ploughing in the common arable field; synonymous with stitch, stretche, ridge, rig (and also, sometimes, strip). G.

sellers' market a MARKET in which prices are in general high because there is a great demand for scarce goods. BUYERS' MARKET, MARKET ECONOMY.

seluka (Sudan: Arabic) **1.** a digging stick with a foot-rest, used especially on GERF land in Sudan **2.** the land cultivated with such a tool. G.

selva (South America: Portuguese and Spanish) **1.** dense EQUATORIAL FOREST of the Amazon region **2.** such equatorial forest growing elsewhere **3.** the Amazon region in which it grows. G.

semi-arid climate, semi-desert the transitional zone lying between SAVANNA and the true hot desert (SAHEL), or between the hot desert and a Mediterranean climatic region, characteristrically supporting patchy XEROPHILOUS vegetation.

semi-natural vegetation vegetation not actually planted by human hand, but resulting directly or indirectly from the activities of human beings or their livestock, e.g. SECONDARY VEGETATION. NATURAL VEGETATION.

semi-permeable membrane, semipermeable membrane a MEMBRANE through which only a solvent may pass, i.e. not the dissolved or colloidal substances contained in it. COLLOID, OSMOSIS, SOLUTE.

semi-precious stone a GEMSTONE valued for use in jewellery or ornaments, but not so commercially valuable as a PRECIOUS STONE.

semolina the small hard grains remaining after the sieving of a hard WHEAT, especially of *Triticum durum*, used in cooking, for puddings, the making of pasta etc.

senescent desert a desert in which the undulating profiles of recently coalesced sloping PEDIMENTS can be seen. G.

senile *adj.* of old age, arising from old age, e.g. a SENILE RIVER. SENILITY.

senile landform INFANTILE LANDFORM.

senile river a RIVER system in old age as defined by W. M. Davis in his CYCLE OF EROSION, i.e. when all slopes have been worn down and the products of decomposition accumulated, covering any irregularities in the surface, so that the FLOODPLAIN may become a marsh without regular outflow: thus the river lies on a completely formed PENEPLAIN. G.

senile town URBAN HIERARCHY. G.

senility the condition of being SENILE.

sensible temperature cold or heat as felt by the human body, depending not only on actual temperature but also on RELATIVE HUMIDITY and wind. LATENT HEAT. G.

sensor any apparatus used to detect variations in electromagnetic radiation, acoustic energy or force fields associated with gravity and magnetism at a distance from their source, especially such a device used to gather information about distant objects or phenomena in the ATMOSPHERE-1, BIOSPHERE, HYDROSPHERE, LITHOSPHERE (as in REMOTE SENSING). ELECTROMAGNETIC SPECTRUM, PLATFORM, REFLECTANCE, REFLECTION.

separatism the demand of a particular group of people that a particular piece of territory should become separate in territory and political sovereignty from the state within which it lies. NATIONALISM-2, PLURAL SOCIETY.

sequanian type of river a river which does not rise in mountains and which diminishes in volume in summer owing to evaporation and the demands of vegetation; a term rarely used. G.

sequence the orderly following in space or time of one thing, happening or idea after another to which it is in some way connected.

sequential *adj.* **1.** constituting a SEQUENCE **2.** characterized by a sequence **3.** following as a result, as a sequel to **4.** following in order of time or place.

sequential landform 1. a landform so modified by DENUDATION that little of the structural form of the INITIAL LANDFORM-1 can be identified **2.** a term applied by W. M. Davis to the actual landform resulting from greater or less advance in the sequence of developmental change (as opposed to his ideal INITIAL LANDFORM-2). DEPOSITIONAL LANDFORM. G.

sequent occupance a term applied by Derwent S. Whittlesey to a chronological series of CROSS-SECTIONS-2 of the geography of an area with special emphasis on its human occupants and the effects of their various activities. He likened it to a PLANT SUCCESSION in botany. G.

serac, sérac (French) a pinnacle of ice formed in the part of a glacier where crevasses intersect, usually at the point where the glacier breaks on reaching a steep slope. G.

serai, sarai (originally Persian; in many eastern languages) a building for the accommodation of travellers, an inn. CARAVANSERAI. G.

seral community SERE-1.

seration a series of communities within an ecological FORMATION-1 or ECOTONE.

serclimax the stage in a PLANT SUCCESSION preceding the SUBCLIMAX (the stage at which a plant community may become stable and in equilibrium with existing environmental conditions). PLAGIOCLIMAX, PROCLIMAX.

sere **1.** a developmental series of plant communities resulting from the process of SUCCESSION-2. A group of plants representing a stage in the process is termed a SERAL COMMUNITY. HALOSERE, HYDROSERE, LITHOSERE, MESOSERE, PLAGIOSERE, PRISERE, PSAMMOSERE, SECONDARY SERE, XEROSERE, CLISERE, SUBSERE (HYDRARCH, MESARCH, XERARCH) **2.** any stage in a plant SUCCESSION-2.

serendipity the gift of being able to make delightful, unexpected discoveries accidentaly, of looking for something and unexpectedly and pleasingly finding something else. A term coined by Horace Walpole, 1754, from *The Three Princes of Serendip*, a fairy tale.

sericulture the rearing of silkworms and the production of raw SILK. G.

series **1.** a general term applied to a number of similar things, events etc. occurring in an orderly way, one after another in space or time, each being in some respect like the one preceding it **2.** in geology, a stratigraphic unit, the major subdivision of a system, corresponding to an epoch. GEOLOGICAL TIMESCALE, STAGE, STRATIGRAPHY.

serir (Egypt and Sahara: Arabic) a gravel desert, a stony desert in Egypt and Libya, synonymous with REG, a plain covered with numerous pebbles of all kinds, the sizes varying from that of a pea to that of a fist. G.

serozem (Russian) a desert soil with pale grey upper layers. GREY EARTH, SIEROZEM. G.

serpentine a group of silicate (SILICA) rock-forming minerals, widespread, in some places occurring in large masses, produced mainly by the alteration of OLIVINE and PYROXENE by hydrothermal alteration (termed serpentinization). The group includes the fibrous chrysotile (ASBESTOS) and the nickel-bearing GARNIERITE. SERPENTINITE.

serpentinite a SERPENTINE rock formed by the METASOMATISM of various ULTRABASIC ROCKS. Usually a dark, mottled green, sometimes with red streaks, it cuts and polishes well and is used in making ornaments etc.

serra (Portuguese) **1.** a range of mountains (Spanish SIERRA) **2.** applied specifically in northeast Brazil to the elevated mountain zones with comparatively luxuriant tropical vegetation, supporting a relatively advanced agriculture. G.

sertão (Brazil: Portuguese) **1.** in northeast Brazil, parched uplands with XEROPHILOUS plants, brushwood and grasses interspersed with patches of CAATINGA **2.** the back country, a wilderness supporting a sparse population, beyond the frontiers of concentrated settlement. G.

serule a succession of very small or microscopic organisms devoted to the disintegration of organic matter to its simple constituent parts.

service industry TERTIARY INDUSTRY.

service pipe a pipe linking a main pipe (e.g. a gas or water main) to a building.

service road a road branching from and in many cases running parallel to a main road, providing access to shops, housing etc.

services **1.** the products of the paid activities of an employee or professional person **2.** in Britain, utilities (gas, water, electricity) as supplied to a consumer **3.** the product of human activity intended to satisfy a human need or needs but not constituting an item of goods.

sesame, sesamum, sim-sim, benniseed, till, gingelly and various other names, *Sesamum indicum*, an ANNUAL plant native to Africa now, despite its relatively low yield, cultivated there and in tropical and subtropical regions elsewhere for its seeds, which provide edible oil used in cooking etc. Whole seeds are used in confectionery and baking and other cooking; and the residue of the seed after the extraction of oil is used as livestock feed.

sesquioxide an oxide with three oxygen

and two metallic elements, in soil mainly iron (Fe_2O_3) and alumina (Al_2O_3).

seston the bioseston (living organisms) and abioseston (inanimate matter, also termed tripton) swimming or floating in a body water.

set 1. a generally square block of GRANITE or other hard stone used in paving paths, roads etc. **2.** the direction of flow of a current, tide, wind **3.** the hardening of a liquid, paste or soft solid **4.** a group of things similar or complementary to one another in some respect, considered as a whole, regardless of orderly arrangement in space or time **5.** in mathematics, a definable collection of objects (which can itself be considered as a single abstract object) the members (elements) of which are characterized by some common property. DATA SET.

seter (Norwegian sete, a bench; pl. seter) a rock terrace, the base of a RAISED BEACH that has been worn away. The term is best avoided for two reasons: it is a plural form applied to the singular; and it is likely to be confused with SAETER, for which the alternative spelling is seter. G.

settlement 1. any form of human habitation, even a single dwelling, although the term is usually applied to a group of dwellings **2.** the act of peopling a formerly uninhabited or under-populated land **3.** a decision or choice made to put an end to a controversy. G.

settlement hierarchy the ranking of urban places with their associated trade areas, graded according to their functional importance. CENTRAL PLACE HIERARCHY, CENTRAL PLACE THEORY.

settlement net the settlement structure of a region, including rural and urban elements, covering pattern and form of AGGLOMERATION-3, FUNCTION, POPULATION-1 etc.

settler one who settles, a COLONIST, SETTLEMENT-2.

settling velocity the steady, uniform speed of a particle sinking in a liquid or gas.

Seven Seas the Arctic, Antarctic, North and South Atlantic, North and South Pacific and Indian oceans (OCEAN). In classical literature the term was applied to the seven supposed salt water lagoons on the east coast of Italy, including the lagoon of Venice (cut off from the Adriatic by the LIDO). G.

Seventh Approximation SOIL CLASSIFICATION.

sex ratio the ratio of males to females (i.e. the number of males divided by the number of females) in any given group, at birth or at any other age.

sextant (Latin sextans, sixth part) an instrument with a 60 degree arc (one-sixth of a circle) used in measuring the angle subtended by two distant objects. Looking through the eyepiece the observer is able to see simultaneously the horizon and the image of the sun, moon or a star in a mirror, and is then able to read off the angle. In navigation the sextant is used in ascertaining LATITUDE by measuring the apparent altitude of the sun, moon or a star above the horizon. This information, combined with reference to a CHRONOMETER and the *Nautical Almanac*, allows the position of the observer to be determined. QUADRANT.

sexual division of labour a distinction based on the idea that unpaid labour in the home is normally carried out by women, paid labour at a place of employment, by men.

shachiang (Chinese) a poorly drained, tightly packed clay soil in northern China, resembling some of the soils with a high lime content in the Indus and Ganga lowlands, the B HORIZON having lime concretions derived from calcium in the ground water.

shaddock pomelo, *Citrus grandis*, a large CITRUS fruit, from which the GRAPEFRUIT is descended. It has a thick skin and fairly bitter pulp, some cultivars being more palatable than others.

shade temperature the temperature shown on a THERMOMETER sheltered from the sun's rays, from radiation from surrounding objects (including the ground), from strong wind, and from precipitation, best achieved by putting it in a screen, a white-painted box, raised on legs so that it is 1.25 m (4 ft) above the ground, the sides louvred for ventilation (STEVENSON SCREEN). In climatic statistics the temperature given is shade temperature, unless sun temperature is specified.

shadow effect the effect of a large, well-served urban centre on the transport services of a nearby small centre, the smaller place being relatively ill-provided with direct services. ACCESSIBILITY, INTERVENING LOCATION EFFECT.

shādūf, shadouf, shadoof (Arabic) a simple hand-operated irrigation device used for raising water from a river or shallow well. It consists of a long pole with a bucket suspended from a rod, chain or rope at one end and a weight (a stone or pieces of iron etc.) at the other. The pole is mounted and pivots on a vertical support which is fixed firmly in the ground. The pivot point is nearer to the weighted than the bucket end of the pole; the weight acts as a counterbalance to the bucket. The bucket is dipped in the water and the counterpoising weight takes over most of the effort of raising and swinging it round so that the bucket may be emptied into a trough or into a channel by which the water is carried to cultivated land. These, or similar dippers or sweeps, have been/are used in many parts of the world, including southeast Europe and as far north as Norway (the Swedish term is svängel). QURER, SAQIA.

shaft DRIFT-4.

shakehole, shackhole (England, Derbyshire term, transferred to Yorkshire and spreading to other areas; shack or shackhole in Yorkshire) a collapsed DOLINE (American collapsed sink), a round funnel-shaped depression, usually dry, with debris on the floor, common in limestone (sometimes in Millstone Grit) areas, caused by solution down vertical pipes or the collapse of a cave roof. SINK,SWALLOW-HOLE. G.

shake wave S-wave, the secondary wave (also termed shear wave or transverse wave) produced by an EARTHQUAKE, a body wave in the earth which passes through solids but not liquids, resembling a light wave in that it is a transverse wave, i.e. it displaces particles at right angles to the direction of its own movement. LONGITUDINAL WAVE, MOHOROVICIC DISCONTINUITY, PUSH WAVE, RAYLEIGH WAVE, TRANSVERSE WAVE MOTION, WAVE.

shale a fine-grained ARGILLACEOUS SEDIMENTARY ROCK, formed from particles of CLAY minerals compressed by overlying rocks, very finely LAMINATED in the direction of the BED ROCK, the thin layers easily splitting apart and disintegrating (compare SLATE). CAMBRIAN, SHALE OIL.

shale oil an oil distilled from bituminous SHALE.

shallot ONION.

shallows an area of little depth of water in a sea, lake, river.

shamal, shamels (Arabic) a strong, dry, violent, dust-laden northwest wind blowing across Iraq and the Gulf of Oman, regularly in summer, intermittently in winter, and producing HAZE. G.

shamba (east Africa: Swahili) **1.** a general term for a plantation, estate, farm, garden, or any plot of cultivated land **2.** specifically, a cultivated plot, sometimes temporary, in contrast to the permanent LUSUKU. G.

shamilat (Indian subcontinent: Urdu) village common land. G.

shanty-town, squatter settlement a

settlement, lacking services, which consists of a collection of small, crude shacks made of discarded materials and serving as habitations for poor people on the outskirts of towns, especially in South America and parts of Africa, variously termed (South America) FAVELA or rancho; (central America) barrio; (Asia) BUSTI or kampong; (Africa) bidonville or shanty-town.

shape index   INDEX OF SHAPE.

sharaf, sharav   (Israel: Hebrew) a hot desert wind blowing especially in spring in Israel. G.

sharecropping   (French métayage) an agricultural tenancy system in which the tenant renders rent to the landlord in the form of produce rather than cash. The systems vary, but usually the landlord in addition to providing and being responsible for the land, buildings, drainage and farm roads (LANDLORD CAPITAL) also provides the sharecropper (the tenant) with machinery, stock, seeds and fertilizer. In return the landlord receives an agreed proportion of the farm produce. LAND TENURE.

sharp sand   SAND-1 with angular, as opposed to rounded, GRAINS-4.

shatter belt, shatter-belt   a zone of movement in the earth's crust where rocks have been broken into angular fragments, i.e. into FAULT-BRECCIA. It occurs where FAULTS are ragged and extensive, so that a line of weakness develops in the crust, along which WEATHERING and EROSION are facilitated. G.

shattering   a form of physical WEATHERING in which strong mechanical stresses produce fresh fractures in the rocks.

shaw   (England, Yorkshire and Lancashire) **1.** a small wood or grove on a hillside **2.** a thicket **3.** a strip of woodland and undergrowth bordering a field.

shea butter-nut   karite, *Butyrospermum parkii*, a tree of tropical Africa, bearing seeds rich in edible oil, used in making shea butter.

shear   **1.** in physics, STRESS applied to a body but along one face only of the body (termed shearing stress), or the STRAIN-2 produced by shearing stress, producing a change of shape but not of volume **2.** in geology, a change in the direction of a STRATUM due to lateral pressure. SHEARING.

shearing   in geology, the bending, twisting or drawing out, sometimes accompanied by crushing or shattering, of a rock near a FAULT or THRUST-PLANE, due to STRESS with resultant slipping (hence shear-fault, shear-plane, shear-cleavage). The volume of the rock does not alter, but its form does as the two neighbouring parts slide past each other, in some cases causing crushing and shattering along the line of SHEAR-1.

shear wave   SHAKE WAVE.

sheep   a member of *Ovis*, family Bovidae, a gregarious RUMINANT animal (a MAMMAL), domesticated and crossbred for a very long time to produce animals suitable for specific purposes, i.e. for supplying MEAT or WOOL of varying quality, and milk (especially in the Mediterranean region) for CHEESE. Sheep flourish on land poorer than that required for cattle provided it is not too wet underfoot, those bred for meat (lamb and mutton) needing better fodder than those kept for wool. The fine wool breeds (MERINO) thrive in dry, warm climates, those with medium quality wool in cooler midlatitudes where they provide meat as well as wool. RAM, WETHER.

sheepskin   **1.** the skin of sheep, especially complete with wool **2.** leather prepared from the skin after the wool has been removed **3.** parchment made from the skin.

sheet erosion   very slow EROSION of soil from an extensive, flat, gently sloping area, the result of RUNOFF, most likely to occur in areas where the soil layer is thin. SOIL EROSION. G.

sheetflood, sheetflow an unhampered, broad expanse of water derived from PRECIPITATION flowing down a slope, occurring where CHANNELS are not present or when the RUNOFF is so great and fast that the existing channels, RILLS, etc. cannot carry it, and thus it overflows. OVERLAND FLOW.

sheeting the splitting away of shells of rock from the upper surface of a massive rock (particularly an IGNEOUS ROCK) resulting from the expansion of the rock by the release of pressure (DILATATION); not to be confused with EXFOLIATION.

sheet lightning a discharge of LIGHTNING within a cloud or between clouds, the brilliance of the flash being diffused by the clouds so that it appears as a sheet of illumination.

sheet metal metal flattened out to form a thin sheet.

sheikh, sheik (Arabic, old man; many other spellings) **1.** in general, a title of respect used by Arabs **2.** in particular, an Arab chief (of family or tribe), a headman of a village, a governor, a son of the royal line. G.

sheikhdom **1.** the office or standing of a SHEIKH **2.** the territory controlled by a sheikh. G.

shelf a LEDGE or PLATFORM of rock. CONTINENTAL SHELF. G.

shelf-ice a floating ice-mass formed by the coalescence of glaciers bordering the margin of Antarctica, terminating in an ICE FRONT.

shellfish a term applied by traders in fish to aquatic invertebrates with a shell, i.e. a MOLLUSC or CRUSTACEAN, e.g. oyster, crab.

shell sand beach sand consisting mainly of comminuted shell fragments, and therefore highly CALCAREOUS, e.g. MACHAIR.

shelterbelt a windbreak, usually a stand of trees planted to act as a screen

against the wind, especially in areas subject to wind EROSION.

sherry WINE.

shield in geology, a very large rigid mass of PRECAMBRIAN rock, forming a major continental block, relatively stable over a long period of geological time, disturbed only by some slight WARPING, e.g. the Laurentian Shield. COVERED SHIELD, GLINT-LINE.

shield inselberg a dome-shaped INSELBERG which some authors suggest is a high point in the BASAL SURFACE OF WEATHERING exposed by the removal of weathered, overlying rock layers. BORNHARDT, RUWARE, ZONAL INSELBERG.

shield volcano a volcano shaped like a shield, i.e. a broad dome, the diameter of the base being large, the angle of slope small, basic LAVA forming the cone, e.g. Mauna Loa in Hawaii. BASALTIC LAVA, HAWAIIAN VOLCANIC ERUPTION, SINK.

shieling (Scottish; many spellings) summer pasture of the Scottish hills asociated with a simple dwelling, inhabited only in summer, where butter making and other farm pursuits were carried on. It was an essential feature in TRANSHUMANCE in the Scottish highlands and islands. G.

shift **1.** in Norfolk, England, fifteenth to seventeenth centuries, a division of the common arable where it was not divided into fields, the basis of the ROTATION OF CROPS and fallowing. FALLOW, PRECINCT-4 **2.** in geology, the maximum or total displacement of rocks possible on opposite sides of a FAULT at a distance from the dislocated zone itself. G.

shifting cultivation loosely applied to any of the many systems of cultivation where land is cropped and after a few years, with the initial fertility exhausted, abandoned in favour of a new patch. A distinction can be made between the true shifting cultivation of nomadic peoples who do not practice a

LAND ROTATION but move on when the soil fertility is exhausted; a regular system of land rotation or BUSH FALLOWING practised by people who usually have a fixed central village; and shifting cultivation associated with certain cash crops whereby land is abandoned when yields begin to drop below an economic level. Some of the 150 or so vernacular terms applied to shifting cultivation are given in G. FARMING, LAND TENURE, SWIDDEN FARMING. G.

shift-share analysis an analysis which seeks to assess the relative importance of different components in the decline or growth of employment in a region or city. The decline in jobs may be due to the fact that the study area has a disproportionately large share of industries declining nationally in terms of employment. The growth in jobs may be due to the presence of a high concentration of industries that are growing rapidly nationally, or to the shift of location within an industry; or it may be quite unrelated to national growth rates. The employment change in the region or city is disaggregated into three components: national (the change which would have occurred if employment in the area had changed at the same rate as the national rate); structural (the change which would have occurred if employment in each sector in the area had changed at the same rate as that sector had changed nationally, less the national component); and differential (the difference between actual change in employment in the area and the sum of the other two components). If the value of the differential component is positive, the region or city is considered to have performed better than expected; if negative, it is deemed to be a poor location for most industries. The analysis does not offer an explanation for the growth or decline in jobs in the region or city, or in the different sectors of industry

there, or for the shifts in location taking place.

shingle an accumulation of coarse stones, rounded by water. The term is usually restricted to cover only such an accumulation on a BEACH. G.

ship canal an artificial waterway large enough for the passage of ocean-going vessels, e.g. the Manchester Ship Canal. SEAWAY. G.

shipping tonnage a measure of capacity of ships, calculated as follows: **Gross tonnage**, the capacity of the permanently enclosed space between the frame of the vessel and the deck together with any closed-in space above the deck, 2.83 cu m (100 cu ft) being reckoned as 1 ton. **Net or registered tonnage**, gross tonnage less the space occupied by engines, gear, crew's and officers' quarters, i.e. the space available for cargo and passengers, calculated on the same basis as gross tonnage. Dues are usually paid on net or registered tonnage. **Cargo tonnage**, the weight of the cargo carried, calculated by volume; in UK 1.19 cu m (42 cu ft), in USA 1.1 cu m (40 cu ft), being equal to 1 ton. **Deadweight tonnage**, dwt, the total load carried at maximum loadline, including the total weight of cargo, fuel and passengers etc. measured in tonnes. **Displacement tonnage**, the weight of water displaced by the vessel when fully laden, i.e. the weight of the vessel and its contents when calculated on the basis that 0.99 cu m (35 cu ft) of water equals 1 ton. As a rough conversion for a mixed fleet, consisting of tankers and cargo vessels, gross registered tons plus 50 per cent equals deadweight tons; for giant tankers the dwt may be 120 per cent higher than gross tonnage.

shippon, shippen (English dialect) a cattle shed, cowhouse. G.

shipyard a large enclosed area on the shore or riverside equipped for the building and repair of ships.

shire Old English term for an administrative unit, usually part of a kingdom, of the same status and age as a COUNTY and consisting of a number of smaller units (HUNDRED, WAPENTAKE), under the joint rule of an ealdorman and a sheriff. Those English counties which do not carry the suffix -shire (Essex, Kent, Sussex, East Anglia or Norfolk and Suffolk) were in most cases separate kingdoms, whereas the the larger of the old Anglo-Saxon kingdoms, notably Mercia and Wessex, came to include a number of counties or shires. The Shires is the term applied (especially by those living elsewhere in Britain) to the counties carrying the suffix -shire; the grassy shires is a term loosely applied to the Midland counties of England. G.

shitwi (Sudan: Arabic shita, winter) **1.** wintry **2.** in northern Sudan, the cooler season, December to January. G.

shoal **1.** a shallow part of a river, sea, lake **2.** an accumulation of sand, mud, pebbles creating such shallow water and in many cases dangerous to navigation **3.** a group of fish.

shoal *verb* to shoal, to become shallower, to move (a ship) into shallower water, to gather together to form a shoal.

shoaly *adj.* characterized by the presence of many SHOALS-1,2.

shock waves of earthquakes EARTHQUAKE.

shoddy **1.** a yarn made from the shredded and reconstructed fibre of fabric or fabrics which have already been used **2.** a fabric made from such reconstituted yarn. RECYCLING.

shoot **1.** the land used for shooting game **2.** people shooting game over such land.

shooting box SHOOTING LODGE.

shooting lodge a shooting box, a small house or cabin used by those shooting game in season.

shooting star METEOR.

shore **1.** loosely applied to the land immediately bordering the sea or other large expanse of water **2.** the meeting of sea and land considered as a boundary of the sea, thus the land as seen from the sea **3.** the area between the lowest water of a SPRING TIDE and the highest point reached by unusually strong waves in a STORM **4.** in law, the ground between the ordinary low and high water marks (LOW WATER, HIGH WATER). BEACH, COAST, FORESHORE, SHOREFACE, SHORELINE, SHORE ZONES. G.

shoreface the zone between the SHORE and the OFFSHORE region. G.

shoreface terrace the outer margin of a marine ABRASION PLATFORM where, in deep still water, wave-worn material is deposited.

shoreline the line where the SHORE meets the water, an imprecise term sometimes regarded as synonymous with COASTLINE (equally imprecise), sometimes applied to the line reached by an ordinary low tide. There is a tendency to regard coastline as the landward limit fixed in position for considerable periods of time, shoreline as a moving phenomenon. G.

shore zones in ecology, the division of the SEA SHORE based on its relationship with the SEA and TIDE, each zone supporting characteristic fauna and flora. From the land seawards the zones are: splash (above HIGH WATER level of SPRING TIDES); upper shore (between the levels of the average ordinary high tide and the high spring tide, covered by sea occasionally and briefly); middle shore (between the levels of the average ordinary LOW TIDE and average ordinary high tide, covered regularly twice a day by sea); lower shore (between levels of low spring tide and average ordinary low tide, occasionally and briefly uncovered by sea); sublittoral fringe (under level of low spring tide, never uncovered, but with a greater range in water temperature than that of the OCEANIC PROV-

INCE, see NERITIC PROVINCE, PELAGIC DIVISION). LITTORAL ZONE.

shott (north Africa: Arabic; French chott) **1.** a fluctuating shallow brackish or saltwater lake in north Africa, especially in Tunisia and Algeria, dry for much of the year, water-filled in winter **2.** the depression holding such a lake. PLAYA, SALINA **3.** an alternative term for FURLONG in the OPEN FIELD system. BUTT. G.

shoulder 1. a rounded spur on a mountainside **2.** a BENCH on the side of a valley, most likely to occur on the side of valley deepened by a glacier at the point where the gentle slope of the upper part (unaffected by glacial erosion) changes abruptly to the steep slope of the inner, glaciated valley side. ALP, U-SHAPED VALLEY. G.

shower a fall of RAIN, HAIL, SLEET or SNOW of brief duration.

showery *adj.* characterized by frequent, brief falls of rain.

shrub a PERENNIAL plant with many persistent woody stems branching from or near the base. SUB-SHRUB, TREE.

SI, Système Internationale d'Unités a simplified metric system based on seven basic units, agreed in 1960 by an international committee and now adopted by most countries using the metric system. The seven basic units, from which all other SI units are derived, are the METRE (m), KILOGRAM (kg), SECOND (s), AMPERE (A), the KELVIN (K), MOLE (mol) and CANDELA (cd). Multiples and submultiples preferably separated by the factor of 1000 are used with these basic units, i.e. 10^{12} (prefix tera-, T), 10^9 (giga-, G), 10^6 (mega-, M); 10^3 (kilo-, k); 10^{-3} (milli-, m); 10^{-6} (micro-, u); 10^{-9} (nano-, n); 10^{-12} (pico-, p); 10^{-15} (femto-, f); 10^{-18} (atto-, a). CENTI-, HECTO-, JOULE, KILO-, MILLI-, NEWTON, PASCAL. CONVERSION TABLES.

sial *si*lica and *al*umina, granitic rocks (GRANITE-1) of the surface of the earth's continental crust (PLATE TECTONICS), composed mainly of SILICA and ALUMINA, light in colour and density (between 2.65 and 2.70). There is a tendency for the term to be replaced by the less specific term upper crust. GEOTHERMAL GRADIENT, ISOSTASY, SIMA. G.

Siberian high a persistent anticyclone situated over north central Asia in winter.

sidereal *adj.* of or relating to the fixed constellations or stars, especially as they are used in measures of time, e.g. SIDEREAL DAY.

sidereal day the time interval between two successive transits of a star over the same meridian, i.e. the rotation of the earth on its axis in relation to the stars, equal to 23 hours 56 minutes 4.099 seconds of solar time, thus almost 4 minutes shorter than a MEAN SOLAR DAY.

sidereal month MOON.

sidereal year the time taken for the earth to make a complete revolution in its orbit round the sun, in relation to fixed stars, equal to 365 days 6 hours 9 minutes 9.54 seconds, i.e. 365.2564 MEAN SOLAR DAYS.

siderite 1. native ferrous carbonate, $FeCO_3$, a valuable ore of IRON present in the Coal Measures and Jurassic limestones of England **2.** a METEORITE consisting wholly of metal (nickel-iron).

siderolite a METEORITE with high proportions of metal and stone.

sidewalk (American) a hard-surfaced path for pedestrians. PAVEMENT.

sidewalk farmer in USA, a person who lives in an urban area and cultivates land distant in the COUNTRY-4, the farm equipment being housed on the farm land. The crop raised is usually one which needs little attention in growth, e.g. a CEREAL. SUITCASE FARMER.

sienna a clay coloured by iron and manganese, used as a pigment.

Side Looking Airborne Radar SLAR.

sierozem, sierosem (Russian) a grey desert soil of arid and semi-arid zones, the weekly developed upper layer of the A HORIZON being greyish-brown or grey, with little organic material, grading into a generally hardened calcareous layer (lime pan). The preferred Russian spelling is SEROZEM. GREY EARTH, SOIL, SOIL ASSOCIATION, SOIL HORIZON. G.

sierra (Spanish, a saw; Portuguese serra) a high range of mountains with jagged peaks resembling the teeth of a saw. The term was originally applied to such mountains in Spain and Spanish-speaking South America, but it is now extended and applied in Spanish to almost any high mountain range; and in English generally to 'the mountains' or a mountain region. G.

sieve map, sieve method a series of maps drawn on transparent material (OVER-LAY), each showing the distribution of a selected factor. By superimposing the transparencies the factors wanted or not wanted for a particular purpose can be 'sieved out'. G.

signature the unique pattern of wavebands (ELECTROMAGNETIC SPECTRUM) peculiar to and emitted by an object on the earth's surface. REMOTE SENSING.

significant *adj.* in statistics, unlikely to have occurred by chance. SIGNIFICANCE TEST.

significance test a statistic calculated to indicate the likelihood that a characteristic in a SAMPLE-3 reflects accurately the characteristic of the parent POPULATION-4 of that sample, and that it has not occurred by chance in the sampling, e.g. CHI-SQUARED TEST.

silage green fodder (e.g. grass, clover, alfalfa, maize plant) packed into a silo, usually with molasses, fermented by ANAEROBIC bacteria to preserve it, and cut into blocks for animal feed when needed.

silcrete HARD PAN.

silica silicon dioxide, SiO_2 **1.** the mineral of that composition, e.g. QUARTZ, CHALCEDONY **2.** the silicate mineral content of a rock, commonly expressed chemically as the percentage of silica by weight. Silicate minerals are the largest group of compounds in the earth's crust. SILICATES.

silicate magma MAGMA from which silicate minerals are formed. SILICA-2, SILICATES.

silicates silicate minerals, a group of minerals based around the highly stable SiO_4, constituting (with the SILICA-1 group) some 95 per cent of the earth's crust, and including CLAY minerals, FELDSPAR, GARNET, OLIVINE, PYROXENE, QUARTZ etc. as members. SILICA, SILICATION, SILICON.

silication the process whereby the proportion of SILICA in a SOIL HORIZON rises as other materials are removed from it.

siliceous *adj.* **1.** of, pertaining to, containing, resembling SILICA-1 **2.** growing in or needing a soil containing silica.

siliceous sinter GEYSERITE.

silicify *verb* to impregnate with, to turn into, to become impregnated with, or to be turned into SILICA-1.

silicon Si, a nonmetallic element occurring, as a brown powder or dark grey crystals, abundantly in nature, always in compounds. It is the second main element in the earth's crust (the first being OXYGEN), comprising by volume some 28 per cent (SIAL). Combined with oxygen it forms SILICA and, with various other oxides, a large group of rocks termed SILICATES, e.g. FELDSPAR, HORNBLENDE, MICA, OLIVINE etc. It is used in glass-making, the manufacture of very hard alloys and, in SILICONE compounds, in lacquers, lubricants, water-repellent finishes etc.

silicone any of the large class of synthetic SILICON-containing compounds in which the atoms of SILICON are held together by bonds to OXYGEN atoms which act as bridges, each silicon atom being attached to at least one organic atom, which forms the radical, the building block, e.g. R_2SiO (R, the radical, being a hydrocarbon) which generates oils and polymers, used as lubricants, heat-resisting resin and varnishes, water repellent film etc. These silicon-containing oxygen-linked compounds have an advantage over carbon-linked compounds: the polymers are more heat-resistant, and the viscosity of the oil is constant over a wide range of temperature.

silk 1. a fine, strong, protein, thread-like structure secreted by some insects 2. such a substance secreted by the caterpillar (silkworm) of the moth *Bombyx mori*, so called because it feeds on the leaves of the white mulberry, *Morus alba*. When the caterpillar is about to pass into the chrysalis stage it sends out streams of a jelly-like substance through two minute apertures in the head. These harden on exposure to air and unite to form a single thread of silk which the creature winds round itself to form the cocoon. If the chrysalis is allowed to hatch into a moth, the cocoon is destroyed. If the silk thread is needed for manufacturing, the cocoon is destroyed by steam, hot water or gas and the silk unwound, 45 kg (100 lb) of cocoons yielding about 4 kg (9 lb) of silk thread 3. fabric woven from silk threads, correctly applied only to those woven from threads of natural (i.e. not artificial or synthetic) silk. SERICULTURE, SYNTHETIC FIBRE.

silk cotton soft, silky short fibre, too short for spinning, obtained mainly from the tropical trees *Bombax ceiba*, native to tropical America, *Bombax malabaricum* of the Indian subcontinent, *Eriodendron anfractuosum* of the Indian subcontinent and Indonesia. The fibres are marketed under the name kapok or vegetable down, and are used in filling cushions etc. C.

sill 1. an INTRUSION of IGNEOUS ROCK of tabular form, as when a very fluid MAGMA is forced between the bedding planes of sedimentary or volcanic formations, i.e. it is CONCORDANT with the STRATA. DOLERITE, of the same composition as basalt, is often found in sills 2. a submarine ridge between ocean basins, or between a sea and an ocean or, termed submerged sill, near the entrance to a FJORD. G.

silt fine particles, larger than those of clay, finer than those of fine sand, diameter 0.002 to 0.02 mm, suspended in, carried or deposited by, water. GRADED SEDIMENTS.

siltstone a CLASTIC SEDIMENTARY ROCK formed by the LITHIFICATION of particles of SILT grade. GRADED SEDIMENTS.

Silurian *adj.* of or pertaining to the third period (time) or system (rock) of the PALAEOZOIC era, preceded by the Ordovician, succeeded by the Devonian. GEOLOGICAL TIMESCALE.

silver a white, stable, malleable, ductile metallic ELEMENT-6, a CHALCOPHILE, a precious metal, some occurring in silver ores, and a little NATIVE-4 in nature, but more as an impurity in lead ores, particularly the lead sulphide, galena, which is considered to be argentiferous (silver-bearing) if it has more than 0.1 per cent of silver. A good conductor of heat and electricity, silver is resistant to oxidation. It is used in electrical apparatus, coins, photography, electroplating, backing mirrors, jewellery, silverware etc. C.

silver thaw GLAZE.

silviculture, sylviculture (Latin silva, a wood) a branch of the science of forestry concerned with the breeding, development and cultivation of forest trees. G.

sima *si*lica and *ma*gnesium, BASALTIC rocks composed of SILICA and MAGNESIUM, forming part of the earth's crust, relatively heavier in density than the SIAL of the CONTINENTAL CRUST which overlies it in places. In areas without sial, sima forms most of the ocean floor. There is a thus a tendency for the term sima to be replaced by OCEANIC CRUST. ISOSTASY, PLATE TECTONICS. G.

simoom, simoon (Arabic) a scorching-hot, heavily dust-laden, swirling wind occurring in the hottest months in the northern Sahara, usually associated with the northward passage of a LOW pressure system. It may carry so much dust and sand that visibility is reduced to zero; and it greatly affects the shape of SAND DUNES in its path. G.

simple random sample a RANDOM SAMPLE is qualified as simple if every member of the POPULATION-4 has an equal chance of being selected and successive drawings are independent, e.g. in SAMPLING WITH REPLACEMENT.

simulation model a physical or mathematical MODEL-3,4 which imitates, represents a real system and is used for experimental purposes. If mathematical, it may be deterministic (DETERMINISTIC MODEL) (based on direct cause and effect, consisting of a set of mathematical assertions from which consequences can be derived by logical, mathematic argument) or STOCHASTIC (based on PROBABILITY rather than on mathematical certainty). If physical, a variety of techniques may be used, e.g. by using floating magnets in a representation of CENTRAL PLACE THEORY.

singhara nut WATER CHESTNUT.

single-feature region a REGION-1 distinguished by the presence within it of one feature, a single attribute. MULTIPLE-FEATURE REGION.

single-stage cluster sampling CLUSTER SAMPLING.

singularity in climatology, the tendency of a type of weather to recur fairly regularly at about the same date annually.

sink, sinkhole 1. in general, a hollow in which surface water collects and escapes through a shaft, i.e. a SWALLOW HOLE 2. specifically, a feature characteristic of limestone country (KARST), a closed depression which is dry or through which water seeps downwards, resembling in shape a basin, funnel or cylinder, comparable with a SHAKE-HOLE in Britain. DOLINE, PONOR 3. a large depression in a SHIELD VOLCANO or lava dome formed when the surface has cooled and solidified but subsides as the underlying molten lava flows away. G.

sinter 1. a chemical deposit of SILICA (siliceous sinter or GEYSERITE) formed around a GEYSER or hot spring, the material having been previously held in solution in the water (HYDROTHERMAL). The term calcareous sinter (applied to a similar deposit, but consisting of calcium carbonate) is sometimes used, but is best avoided, TRAVERTINE being preferable 2. a substance heated, without complete melting, to the point where it becomes a coherent, solid mass (like a cinder). Metal particles, glass, ceramics and some other non-metallic substances may be sintered in this way. G.

siphon in speleology, an underground passage of inverted U-shape which allows the passage of water when the level of the water in the adjoining cave to the rear reaches the level of the loop at the top of the inverted U.

sirdar, sardar (Persian origin) in the Indian subcontinent, a military chief or leader, a term adopted by the British army and applied elsewhere, e.g. in Egypt. Currently applied in the Indian subcontinent to an old-style landlord. G.

sirocco, scirrocco a hot south or south-easterly wind, sometimes dust-laden, dry as it blows over north Africa, sometimes humid by the time it meets the south Italian shore, blowing from the Sahara over the Mediterranean to Malta, Sicily, Italy, ahead of a depression moving eastwards over the Mediterranean. It is common in spring, when it may damage crops, especially blossoming vines and olives. KHAMSIN. G.

sisal 1. a strong fibre, used for making twine and cord, obtained from the leaves of *Agave*, particularly *rigida* and *sisalana*, especially in Tanzania and Brazil 2. HENEQUEN.

site 1. in general, a fixed position where an object, structure or tissue is placed or where something occurs, e.g. the position on the ground of a place, town, building etc. in the past, present or future. SITUATION 2. in law, the whole space occupied by a house, building or other erection between the level of the bottom of the foundations and the level of the base of the walls 3. in geomorphology, the smallest distinctive unit in the hierarchy of MORPHOLOGICAL REGIONS, a number of similar or related sites forming a STOW.

site factors in soil science, the elevation, slope and aspect in relation to soil formation. G.

Site of Special Scientific Interest SSSI.

situation 1. the PLACE, POSITION or LOCATION of something, e.g. a house, a town, in relation to its surroundings or to another thing 2. a state of affairs. SITE.

size-density rule in urban land use, a maxim stating that as the population of an URBAN SETTLEMENT increases the density of development increases, but at a declining rate, i.e. as the population increases the rate of increase of the density of development decreases.

skär (Swedish; Norwegian skjer, skjaer, a skerry, rocky islet, har) a general term for a small island, the smallest islets being termed bada, kobb, klabb. In the inner SKERRY-GUARD and in lakes the preferred term for a small island is holme. HOLM. G.

skare (Swedish) crust on the SNOW. G.

skärgård (Swedish, skär, skerry; gård, garden, enclosure, yard; Norwegian

skjergaard, skjaergaard) enclosure by a line of skerries, SKERRY, SKERRY-GUARD. G.

skarn (Swedish) 1. GANGUE or COUNTRY ROCK of contact megnetite ore. METAMORPHIC AUREOLE 2. an impure LIMESTONE or DOLOMITE which has been subjected to METAMORPHISM and METASOMATISM, commonly containing calcium silicates and borosilicates, worked for manganese silicates and sulphide minerals. G.

skärtråg (Swedish; German Sichelwanne) a sickle-shaped flat rock basin characteristic of some glaciated regions. G.

skauk an extensive field of crevasses. G.

skavle (Norwegian, pl. skavler) SASTRUGI. Although skavle was preferred by some scientists, it has now generally been superseded by sastrugi. G.

skeletal minerals the mineral particles, mainly silt and sand (GRADED SEDIMENTS) constituting the inactive part of a soil as distinct from clay minerals. CLAY-2,3,4.

skeletal soil formerly applied to an AZONAL SOIL, a newly formed soil consisting of almost unweathered rock fragments and lacking a B HORIZON; the term preferred now is ENTISOL. G.

skerry (Scotland and Northern Ireland: Scottish, derived from Swedish SKAR) a small islet, sometimes one of a series lying parallel to the main trend of the coast, usually rocky, sometimes composed of morainic material, over which the sea may break at high tide or in stormy weather. HAVSBAND. G.

skerry-guard the area of calm water between a line of skerries (SKERRY) and the mainland. The term should not be applied to the line of skerries itself. HAVSBAND. G.

skewness in statistics, asymmetry. FREQUENCY CURVE, MOMENT MEASURES.

skid row, skidrow in USA, the dilapidated district in a North American city which has become the resort or refuge of the 'down and out', people who are temporarily or permanently destitute or nearly destitute. G.

skin 1. the membrane or one of its layers forming the outer cover of an animal (including human) body 2. such a membrane when stripped from the body of a small animal (e.g. sheep, goat), with or without the hair, and used particularly in the clothing and associated industries 3. the outer covering of something, e.g. of a fruit.

skjer, skjaer (Norwegian) SKAR, SKERRY. G.

skjergaard, skaergaard (Norwegian) SKARGARD. G.

sky the atmosphere enveloping the earth, with or without clouds. Its blue appearance if cloudless in the daytime is due to the scattering of sunlight by the molecules of air. At high altitudes the sky appears to be deep blue because there the short waves of the blue-violet end of the spectrum of solar light are easily scattered by the fine molecules of air present.

Skylab a U.S. SATELLITE-1 launched in 1973 with a crew of three, whose members carried out various experiments and had control over the REMOTE SENSING equipment aboard. This included cameras as well as instruments for electronic imagery. LANDSAT, SPACE SHUTTLE, SPOT.

slack 1. a shallow hollow among coastal sand dunes or mud banks 2. the state of the TIDE when tidal currents are almost still, commonly about high or low water when there is neither EBB nor FLOW 3. small pieces of coal, refuse coal. G.

slade 1. a hollow in the side of a valley, sometimes well-wooded, corresponding to the CWM at its head, often marking the early stages of the rejuvenation of a tributary of the main valley 2. sometimes applied to a PLAIN. G.

slag the nonmetallic residue resulting

from the smelting of metallic ore. BASIC SLAG.

slaked lime HYDRATED lime, calcium hydroxide, $Ca(OH)_2$.

slaking 1. quenching of thirst. HYDRATE, HYDRATION 2. the crumbling and disintegration of earth resulting from alternating saturation with water and drying-out 3. in particular, the disintegration of clay-rich sedimentary rocks by such a process when exposed to air; or of dry clay when covered with water. G.

slar Side Looking Airborne Radar, a development of RADAR used in REMOTE SENSING in which the radar sensing systems direct their impulses toward each side of the aircraft and provide clear images of the earth's surface even through cloud or darkness, penetrating dry soil to give a picture of the underlying BED ROCK.

slash 1. locally in south and southeast USA, a low, wet, swampy, boggy area, overgrown with bushes, cane etc. favourable for the growth of any one of the slash pines, e.g. *Pinus caribaea* yielding gum and turpentine 2. the debris of felled trees 3. part of a forest strewn with such debris. SLASH AND BURN. G.

slash-and-burn a method of clearing land, as in SHIFTING CULTIVATION, by felling trees and burning the SLASH. SWIDDEN FARMING.

slate a fine-grained, LAMINATED, dark grey, METAMORPHIC ROCK derived from SHALES or MUDSTONES subjected to pressure by earth movements. It has the property of being fissile into thin slabs (slates) by the development of minerals such as MICA, the thin flakes of which lie at right angles to the pressure in DYNAMIC METAMORPHISM. Thus the splitting is along the lines of CLEAVAGE, independent of the original BEDDING PLANES, differing from the splitting of SHALE, which takes place along the bedding planes.

Slate is used in roofing, walling etc. CLAY-SLATE. G.

slave a person who is the property of and absolutely subject to another.

slave mode of production in Marxism, a MODE OF PRODUCTION in which the labourer is regarded as a component in the MEANS OF PRODUCTION, being bought, sold and owned by the slave owner as any other piece of machinery.

slavery a very old institution in which a master forces a subordinate (a slave) to labour or render other services for the master's benefit, the extent of domination and subordination ranging from the extreme, carrying the right of life and death of the master over the slave, to agreed mutual rights and privileges established by a legally binding agreement. SLAVE MODE OF PRODUCTION.

sleet 1. in UK, PRECIPITATION-1 consisting of snow and rain mixed, or of partially thawed snow. GLAZED FROST 2. (American) precipitation consisting of raindrops which have frozen and then partially melted.

slickenside a polished and sometimes scratched, sometimes finely fluted, surface of a FAULT PLANE resulting from friction and pressure along the divisional planes, caused by the movement of the fault. G.

slide 1. a mass of rock or earth falling as a whole, rapidly and sometimes catastrophically down a BEDDING PLANE or JOINT through the force of gravity 2. the mark on the hillside, caused by such a slide. MASS MOVEMENT, ROCK SLIDE, SLUMP.

sling psychrometer PSYCHROMETER.

slip 1. in a FAULT, the actual relative movement along the FAULT PLANE, either in the direction of the STRIKE, termed strike-slip; or in the direction of the DIP of the fault plane, termed dip-slip 2. a LANDSLIDE in which a mass of rock or surface debris moves as a whole down a slip-plane. MASS MOVEMENT,

ROTATIONAL SLIP 3. of a glacier, BASAL SLIP.

slip-face the leeward side of a SAND DUNE, steeper than the windward side from which sand is blown. DUNE, PLINTH.

slip-off slope of a MEANDER, the gentle slope of the spur on the convex (inside) curve opposite the steep bank or RIVER CLIFF on the concave curve (outer side). POINT BAR.

slobland (Northern Ireland, especially Belfast) muddy ground, especially if it has been reclaimed. G.

sloe, blackthorn *Prunus spinosa*, wild plum, a small DECIDUOUS tree native to western Europe, bearing blue-black, hard, sharp-flavoured DRUPES with little flesh, used only in making sloe gin or sloe wine.

sloot, sluit (Afrikaans) a narrow water channel, artificial (for irrigation or drainage) or natural (a shallow erosion gully). G.

slope 1. the upward or downward inclination of a natural or artificial surface, a deviation from the perpendicular or horizontal 2. the degree or nature of such an incline. The study of the development of slopes on the earth's surface is a complex one, theories abound (SLOPE RETREAT) and many specialized terms are in use apart from those which follow and the entries CONSTANT SLOPE, FREE FACE, GRAVITY SLOPE, HALDENHANG, PARALLEL RETREAT OF SLOPE, SLIP-OFF SLOPE, SLOPE ELEMENTS, SLOPE RETREAT, TRANSPORTATION SLOPE, WANING SLOPE, WAXING SLOPE 3. on a straight line on a graph, the amount by which the dependent variable (on the vertical axis) increases/decreases for each unit of the independent variable (on the horizontal axis), the slope of a regression line for two variables being the REGRESSION COEFFICIENT. REGRESSION ANALYSIS. G.

slope elements of hillside slope, the component parts of the hillside slope profile as defined by W. Penck and A. Wood, the WAXING SLOPE being relatively concave and at the top, succeeded downwards by the nearly vertical FREE FACE, the CONSTANT SLOPE, rectilinear in profile, and the WANING SLOPE, relatively concave, at the base. STANDARD HILLSLOPE.

slope length the actual length of the surface of a SLOPE-1, from its highest to its lowest point, not the length projected on to a plane, termed the HORIZONTAL EQUIVALENT, which measures less. GRADIENT.

slope profile analysis the division of a surveyed slope PROFILE-3 into its component parts, identified by their individual form, often expressed statistically. SLOPE.

slope retreat the progressive wearing back of a slope profile by erosion. W. M. Davis suggested this would lead to a flattening of the slope; W. Penck that the slope would be replaced by others of different angle; L. C. King that (due to wholesale PARALLEL RETREAT OF SLOPES) the form would always be maintained until the end of the cycle of erosion. SLOPE-1.

slough 1. a piece of soft, muddy, boggy, waterlogged ground, a term used only in a literary or poetic sense 2. in USA, a sluggish side channel of a river, a comparatively narrow stretch of backwater, a BAYOU.

sludge in ice terminology, the stage in freezing following that of FRAZIL ICE when needles and plates of the ice coagulate to form a gluey, opaque, lead-coloured surface layer, the dull appearance being due to the lack of reflection of light. G.

sludging SOLIFLUCTION, the downward movement of a thawed mud layer over ground that is still frozen.

sluggy, slugga (Ireland) a SWALLOWHOLE waterlogged because it lies at the WATER TABLE.

sluice 1. in water control, an artificial waterway equipped with a device such as a sluice gate, which controls the flow and level of the water 2. the device itself, or the stream beside the device 3. any channel effective in draining-off water 4. a sloping trough carrying water, used in mining to wash ores, or in lumbering to float logs downstream.

sluiceway an artificial channel controlled by a sluice gate. SLUICE.

slum a rundown settlement or part of a settlement, usually in or near an urban area and characterized by dilapidated buildings or shacks (FAVELA), the poverty of its inhabitants, squalor, the presence of refuse, and overpopulation.

slump block the mass of rock involved in SLUMPING.

slump-fold a FOLD in strata produced not by earth building movements but by the sliding or SLUMPING of soft sediments (especially of MUDSTONES) down a slope, e.g. down the edge of the CONTINENTAL SHELF. G.

slumping the downward, usually distinctly rotational, slipping under gravity of a mass of rock, torn away from its base, over a curved slip fault (SLIP-1), leaving a scar on the slope surface. It commonly occurs where more massive rocks overlie a weaker layer, e.g. limestone over clay, along an escarpment. It is also believed to occur extensively down the steep slope of the CONTINENTAL SLOPE from the CONTINENTAL PLATFORM, termed submarine slumping. CAVING, MASS MOVEMENT, SLUMP FOLD. G.

slurry 1. very wet, mobile MUD-1,2 2. a mixture of water and insoluble matter, e.g. chalk, clay, coal, lime, sometimes transported by pipeline.

slush zone a zone above the FIRN LINE of a glacier from which a mass of snow, melted by the summer thaw, may pour down the ice surface in an AVALANCHE.

small area a small district in a city, within which it is considered particular problems can be identified and policy solutions tried out, a process termed small area approaches.

small circle any hypothetical circle on the earth's surface the plane of which does not pass through the earth's centre, in contrast to the plane of a GREAT CIRCLE. Thus all parallels of LATITUDE to north and south of the equator are small circles, decreasing in size polewards; but the EQUATOR itself is a great circle. G.

small fruit (American) SOFT FRUIT.

small holding, small-holding, smallholding written as two words, with or without a hyphen, applied generally to any small farm or holding. But as one word given a special legal meaning in Britain under the Agriculture Act 1947, i.e. a holding of less than 20.25 ha (50 acres), having a rental value below a certain level. G.

smelting the extraction of metal from its ore, usually by a heat process which reduces the oxide of the metal with carbon in a furnace, separating out the metal in a molten state; or by CALCINATION of sulphide ores.

smithsonite CALAMINE, ZINC, ZINC CARBONATE.

smog (term derived from smoke and fog, 1905) **1.** originally applied to thick, yellow RADIATION FOG, injurious to health, occurring over a built-up area where sooty particles from smoky fuels (SMOKE) formed the nuclei for condensation in the atmosphere, and SULPHUR DIOXIDE added to the POLLUTION. A very dense smog in London in 1952 stimulated a campaign for smoke abatement and was so successful that the use of smoky fuels within specified areas was banned under the clean Air Acts of 1956 and 1968 **2.** the term has since been applied to other foggy air pollution, not necessarily caused by smoke but by the pollution of the air by NITROGEN oxides and HYDROCARBONS from the exhausts of motor vehicles, combined with the chemical change brought about by the action of

sunlight, as in the common Los Angeles PHOTOCHEMICAL smog. G.

smoke fine particles suspended in the ATMOSPHERE-1 and carried by air currents, usually consisting of carbon particles formed from incomplete combustion. SMOG.

smonitza in soil science, a HYDROMORPHIC black or dark grey soil of Yugoslavia, usually derived from calcareous clay overlying sand, the surface layer leached of lime. G.

SMSA STANDARD METROPOLITAN STATISTICAL AREA.

snag a tree or branch embedded in a river or lake bottom, not visible at the surface, and therefore hazardous to boats. G.

snout of a GLACIER, the lower extremity of a VALLEY-GLACIER, sometimes partially hidden by morainic material, but commonly featuring a cave from which MELT-WATER flows. MORAINE. G.

snow PRECIPITATION-1 in the form of delicate, feather-light, hexagonal, variously patterned, individual ice crystals aggregated to form snowflakes. The ice crystals are formed when water vapour in the atmosphere condenses quickly at a temperature below freezing point, does not liquefy but passes immediately to the solid state, the sparkling whiteness releasing LATENT HEAT, causing a rise in air temperature. Sometimes snow melts in descending, to reach the ground surface as rain; it arrives as snow only if the lower atmosphere is cold enough to prevent melting. Air is trapped between the crystals in snowflakes causing internal reflection of light at the crystal surfaces and giving snow its sparkling whiteness. This trapped air, combined with the air between the flakes, makes snow a good insulator, preventing heat loss by radiation from the surfaces on which it collects. Snow can be dry and powdery, and therefore great in volume (as in cold temperatures, e.g. in the Antarctic); so to make for uniformity in meteorological recording, it is collected (usually in a special cylindrical gauge) and melted, the amount being expressed as the rainfall equivalent, and usually added to the precipitation total. AVALANCHE, RED SNOW.

snow avalanche a swift fall of a mass of snow down a slope (AVALANCHE, STAUBLAWINE), distinguished as dry, consisting of newly-fallen snow in winter; wet, caused by spring thaw; wind slab, where the surface layer of the snow is compacted and hard.

snow-bridge a mass of compacted or partly compacted snow connecting one side of a BERGSCHRUND or CREVASSE with the other.

snowdrift a pile or bank of snow, heaped by the wind against an obstacle.

snow-field,snowfield a stretch of permanent snow with a relatively level, smooth surface, occurring in shallow depressions in mountainous areas or on high plateaus.

snow limit, snow-limit the limit north and south from the equator indicating a zone within which no snow falls and stays unmelted, varying with physical conditions (elevation, proximity to the ocean etc.); not in general use as a technical term. G.

snow line, snow-line the variable lowest level on mountains above which snow never completely disappears, considered to be a permanent level (varying with latitude, altitude, temperature, humidity, precipitation, aspect, steepness of slope) if summer warmth does not melt and remove the winter accumulation. The snow line in winter is commonly lower than this so-called permanent snow line. G.

snow niche NIVATION.

snow patch erosion NIVATION.

soak 1. loosely applied to a depression holding moisture after rain 2. in west and central Australia, a damp or swampy spot occurring round the base of a granite rock. G.

soapstone steatite, a cream to brown or grey-green coloured TALC, soft and easily carved. It feels like soap, resembles the much harder JADE when carved, the surface being waxlike; and it has long been used in China and Japan and by Eskimos for making ornaments and jewellery.

socage a form of LAND TENURE held by a sokeman in the feudal system, whereby the sokeman originally paid in produce and labour to his lord (later partly in money), but was not liable to military service. G.

social *adj.* 1. relating to human SOCIETY-1 2. as applied to human beings, of any BEHAVIOUR or attitude that is influenced or created by experience of the behaviour of other people; or any behaviour or attitude directed towards other people 3. applied to an animal which is part of a SOCIETY-5 4. applied to a plant growing in clumps. SOCIETAL, SOLITARY; and entries qualified by social.

social action 1. in politics, activity by an interested group concerned with securing a particular reform, or support for some cause 2. in sociology, human activity looked at from the point of view of its social content, i.e. the subject-matter of SOCIOLOGY.

social area a social region, an area usually identified by the homogeneity of the social character of its inhabitants (age group, class, ethnic group etc.) or (less commonly) by the strength of social interdependence present.

social area analysis a technique used to link social structure and urban residential patterns. Widely ranging data, e.g. concerning rank in SOCIAL CLASS, occupation, fertility, size of families, racial and ethnic grouping, are analysed and classified in order to make distinctions between small areas within a city. Sometimes termed SOCIAL ECOLOGY.

social capital in Marxism, state expenditure which, by providing resources that firms themselves would otherwise have to provide, contributes to the profitability of the private sector of an economy. SOCIAL CONSUMPTION, SOCIAL EXPENSES, SOCIAL INVESTMENT, VARIABLE CAPITAL.

social city a planned cluster of GARDEN CITIES, the whole designed to be big enough to provide city-scale social and economic facilities without sacrificing the small-scale advantages of each individual garden city.

social class a problematic, disputed concept, widely used in the social sciences, applied to a group of people of similar rank or status in a community (CLASS-3), the basis for the grouping being variable, e.g. determined by education, power, income, wealth, prestige, occupation, or relationship to the MEANS OF PRODUCTION. A distinction may be made between social class (distinguished in relation to the means of production) and SOCIAL STATUS (distinguiished on the basis of consumption of goods, of particular life-styles). The REGISTRAR GENERAL'S CLASSIFICATION, UK, distinguishes five categories, from higher managerial or professional through skilled manual to unskilled manual workers. But the term social class is often used loosely in the UK to distinguish upper, middle (also stratified) and WORKING CLASSES without any precise definition of the criteria used. In Marxist theory, social class is related to the ownership or non-ownership of the means of production (MODE OF PRODUCTION), a class comprising individuals with a common behavioural pattern, sharing a common relationship to property and power, the classes being the capitalist, the BOURGEOISIE (the capitalist class which owns and controls property) and the PROLETARIAT (the propertyless workers, dependent on selling their labour to the owners of capital for wages in the market). SOCIAL STATUS.

social closure a process by which a social group tries to maximize rewards by restricting access to rewards and opportunities to those whom the group judges to be eligible. It may be effected by the POWER OF EXCLUSION or the POWER OF SOLIDARISM.

social consumption in Marxism, state expenditure which acts mainly as VARIABLE CAPITAL during the process of capitalist production, i.e. state investment in the production and reproduction of LABOUR POWER. SOCIAL CAPITAL, SOCIAL EXPENSES, SOCIAL INVESTMENT.

social costs of production the costs imposed on local residents by a production process, e.g. roads, housing, community services. SOCIAL OVERHEAD CAPITAL.

social democracy a political concept which accepts the democratic form of government (DEMOCRACY) but wishes to make social change (by means of reforms rather than by revolution). SOCIALISM.

social distance 1. the voluntary or enforced separation of distinct social groups for most of their activities 2. the distance as perceived by individuals or small groups between themselves and other individuals or social groups.

social ecology the study of the distribution of social groups in urban areas. SOCIAL AREA ANALYSIS.

social expenses state expenditure which does not directly contribute to the profitability of the private sector of the economy but which is necessary to maintain the economic conditions for that profitability. SOCIAL CAPITAL, SOCIAL INVESTMENT.

social formation the collection of various types of SOCIAL RELATIONS OF PRODUCTION characteristic of different MODES OF PRODUCTION and present in a particular society at a specific time, commonly comprising some types characteristic of a previous mode (or modes), those (the main group) characteristic of the current mode, and an identifiable few indicating a mode of production yet to come.

social geography a branch of geography which has been defined broadly as the analysis of social phenomena in space, but also equated with HUMAN GEOGRAPHY, CULTURAL GEOGRAPHY (in USA), and many of the aspects of URBAN GEOGRAPHY, HUMANISTIC GEOGRAPHY and WELFARE GEOGRAPHY. It deals generally with the interrelationship of people with their environment. But SOCIAL-1,2 implies an individual living and functioning with others in a group or a SOCIETY-1,3; thus social geography emphasizes the importance of studies of population, urban and rural settlements, social activities and problems (including such considerations as SOCIAL JUSTICE, SOCIAL STRUCTURE, SOCIAL NETWORK, SOCIAL WELL-BEING).

social indicator, social welfare indicator in WELFARE GEOGRAPHY, a measure of SOCIAL WELL-BEING (i.e. a measure of 'worse off' or 'better off'), used in differentiating groups of people, territories, or periods of time. TERRITORIAL SOCIAL INDICATOR.

social interaction the reciprocal influence of the social actions of people on each other.

social investment in Marxism, state expenditure which acts mainly as CONSTANT CAPITAL during capitalist production, thereby lowering the costs of private sector investment in the MEANS OF PRODUCTION, i.e. raising the productivity of labour. COMPLEMENTARY INVESTMENT, SOCIAL CAPITAL, SOCIAL CONSUMPTION, SOCIAL EXPENSES.

socialism a political, social and economic concept which takes various forms. Each form, in general, is opposed to uncontrolled CAPITALISM-1, seeks equality of opportunity for each person, advocates that there should be collective ownership of the means of production and control of distribution, and maintains that in return for contributing to the community, the individual should be entitled to receive the care and protection of that community. To generalize, it may be said that Marxian socialism is concerned largely with economic issues, and stresses the importance of communal ownership and control of the means of production, distribution, exchange. Christian socialism, more concerned with social aspects, sees socialism as a way of life. Democratic socialism stresses the political aspect, and compromises in the economic field between state and private enterprise. SOCIAL DEMOCRACY.

socialization 1. the process by which the accepted values, rules and ways of operating of a society or a group (the social patterns) are passed to and absorbed by the offspring of its members or by other new entrants to it, a process which continues throughout their lives as the social patterns change **2.** the act or fact of establishing in terms of SOCIALISM, e.g. the putting of medical services under state control.

social justice in WELFARE GEOGRAPHY, the fair distribution of benefits and burdens among the members of a SOCIETY-1.

socially necessary labour LABOUR THEORY OF VALUE.

social mobility the movement of people between SOCIAL CLASSES. People moving from unskilled and manual occupations to those that are skilled, non-manual and professional are often considered to be going up (upwardly mobile), those moving in the reverse direction, from what may be considered superior occupations to the inferior, going down (downwardly mobile) in social class position; such mobility is said to be vertical. The term horizontal mobility is applied to movement which involves a change of status (SOCIAL STATUS) and role (particularly in occupation, SOCIAL ROLE) without a change in social class position. MOBILITY, MODERNIZATION, PRE-INDUSTRIAL CITY.

social network the relatives, friends, neighbours with whom an individual person or a family is linked (the persons being represented by nodes, the relationships by connecting links, NETWORK-3).

social overhead capital the public investment in roads, housing, community services necessary for production to take place. SOCIAL COSTS OF PRODUCTION, SOCIAL EXPENSES.

social pathology an approach to social problems which concentrates on the characteristics of problem individuals and communities. It suggests, for example, that the problem of poverty can best be understood by studying any physical and social inadequacies which may be present in the poor themselves.

social physics a mechanistic approach to the study of human geography which uses analogy with physical laws in analysing human behaviour. Introduced in the mid-nineteenth century, it is represented today by, e.g., the GRAVITY MODEL, REILLY'S LAW OF RETAIL GRAVITATION.

social relations of production, relations of production in Marxism, the relationships between social groups which are generated by, and form the basis of, a particular MODE OF PRODUCTION. SOCIAL FORMATION.

social role the pattern of behaviour expected by others from an individual in a particular social position, of a particular SOCIAL STATUS, e.g. a mother, doctor, employer, schoolteacher.

social sciences those DISCIPLINES that try in a generally systematic way to study human society, its organization, and the relationship of individual members to it or to groups within it, e.g. ANTHROPOLOGY, ECONOMICS, POLITICAL SCIENCE, and some aspects of PSYCHOLOGY, SOCIOLOGY and

LINGUISTICS. See also BEHAVIOURAL SCIENCES, EARTH SCIENCES, HUMANITIES, NATURAL SCIENCES, PHYSICAL SCIENCES, SCIENCE, SOCIOLOGY.

social segregation the residential grouping, the spatial separation of people within an area, on the basis of social distinctions. FUNCTIONAL SEGREGATION, HOUSING CLASS THEORY, SEGREGATION.

social space the space perceived to be homogeneous by those living in it who, in using it, give it its special character, the space itself reflecting their activities, preferences, aspirations, and thereby becoming separate, identifiable by the social group inhabiting it.

social status the social standing of a person, based on life-style, consumption of goods, the esteem in which that person is held by others, differing from SOCIAL CLASS, which is more concerned with the production and division of labour. SOCIAL MOBILITY.

social structure the form, shape, pattern, framework of the interrelationships of people in a SOCIETY-2,3, in a social system, which can be analysed by identifying the roles and sets of roles played by the individual in that society, as well as the rules, constraints, conventions, resources and facilitations which underpin them, any of which may be considered as a social structure in its own right, with structures of its own. That, broadly, is one of the traditional applications of the term social structure in anthropology and analytic (formal) sociology. In human geography in the mid-1980s it is more commonly applied (as social structures) to the social rules and resources etc. which underly a social system. CULTURE-1, INSTITUTION-3, POST-INDUSTRIAL SOCIETY, STRUCTURATION, STRUCTURE-1,2, as well as many of the entries qualified by 'social', especially SOCIAL ROLE, SOCIAL STATUS.

social welfare indicator SOCIAL INDI-CATOR.

social well-being, human well-being a state in which the needs and wants of a POPULATION-1 are, in general, satisfied. The identification of these needs and wants is subjective and in many cases historically determined, varying from one CULTURE to another and changing with the passage of time. In western industrialized societies in the twentieth century, it may be said that it is generally accepted that in order to be in a state of ideal social well-being a population should be in good health, have sufficient income for basic needs and thus be free from want, be well fed and clothed, housed with sufficient space in a benign, POLLUTION-free environment; should have command over goods and services, receive all the education desired, be protected by the administration of justice, have social and economic mobility, as well as time and facilities for recreation and leisure; and be able to participate in social affairs in a stable (preferably democratic) administration. LEVEL OF LIVING, QUALITY OF LIFE, SOCIAL INDICATOR, STANDARD OF LIVING, WELFARE.

sociation a HOMOGENOUS-1 plant community, each part being character-istic of the whole.

socies a group of subdominant plants in a stage of SUCCESSION-2 preceding that of the CLIMAX.

societal, societary *adj*. of or pertaining to, concerned with, SOCIETY-2,3 or with social conditions. Societal is com-monly used as a synonyum for SOCIAL, but some authors restrict the applica-tion to the attributes of society as a whole, i.e. to its structure or to the changes within it.

society 1. the state of living in organ-ized groups 2. people living and work-ing together and considered as a whole 3. a large group of people associated together geographically, culturally or otherwise, with collective interests, shared laws, customs etc. and with a particular organization. EGALITARIAN SOCIETY, INDUSTRIAL SOCIETY, MASS SOCIETY, POST-INDUSTRIAL SOCIETY, PRE-INDUSTRIAL SOCIETY, RANK SOCIETY, SOCIAL STRUCTURE, STRATIFIED SOCIETY 4. an association of people with some special interest, some central discipline 5. in ecology, a group of animals living and working together, organized on a CASTE-2 system and forming a COMMUNITY (in contrast to a PACK, in which all the members carry out all forms of labour) 6. a community of plants forming a minor CLIMAX within a CONSOCIA-TION, but in which the dominant SPECIES differs from that of the consociation.

socioeconomic *adj*. of or relating to social and economic conditions.

socioeconomic grouping a classification of people into groups, by industry, intended to contain people whose social, cultural and recreational standards are similar. The REGISTRAR GENERAL'S CLASSIFICATION OF OCCUPATIONS, UK, has seventeen socioeconomic groups, based on em-ployment status and occupation.

sociofact HUXLEY'S MODEL.

sociology the study of the SOCIAL BEHAVIOUR or SOCIAL ACTION of human beings, of the origin, the history and the structure of human SOCIETY-2,3 and its institutions. SOCIAL SCIENCES.

soda strictly, crystalline sodium car-bonate or sodium hydroxide, but com-monly applied to any form of SODIUM present in a rock or mineral.

sodium a soft, white metallic ELEMENT-6, oxidizing rapidly in air, reacting with water to liberate hydrogen and pro-ducing a solution of sodium hydroxide. It is widely distributed in many compounds, the most common being common SALT-2 (sodium chloride, NaCl), an essential MICRONUTRIENT.

Sodium salts are important in industrial processes. See FELDSPAR, SODA.

soffione, suffione, soffoni (Italian, pl. suffioni) **1.** an orifice in an old volcanic area emitting steam and sulphurous vapours **2.** the emission itself. FUMAROLE, MOFETTE, SOLFATARA, SOLFATARIC STAGE. G.

soft coal BITUMINOUS COAL, COAL.

soft fruit 1. black, red and white currants, gooseberries, raspberries, strawberries etc., small succulent fruits of temperate midlatitude lands, forming a distinct category in the agricultural statistics of the UK **2.** (American) small fruit, British soft fruit but including cranberries and blueberries.

soft layer of mantle ASTHENOSPHERE.

software the set of systems, in the form of programs, which controls the operation of a computer, simplifying and linking the work of computer and user. HARDWARE.

softwood any tree of Gymnospermae, or its TIMBER which is soft, without vessels in its wood, and relatively light, with an open texture. Most commercial softwoods are CONIFEROUS trees (the timber being known commercially as pine, fir, deal), taken from the great northern forests of Canada, Scandinavia and the USSR; others are grown in plantations elsewhere in Europe and in New Zealand. Softwood is used especially for pulp, cellulose and wood products, and in construction work. HARDWOOD.

soil (Latin solum, the ground) loosely, the earth or ground, but specifically the loose material of the earth's surface in which terrestrial plants grow, usually formed from weathered rock or REGOLITH changed by chemical, physical and biological processes. Thus the soil may be considered as an entity, quite apart from the rocks below it. It consists partly of mineral particles and partly, to a varying extent, of organic matter (HUMUS). The mineral particles can be graded according to size (GRADED SEDIMENTS); and according to the proportion of the grade present the terms clay soil, sandy soil etc. are applied. A soil is said to be MATURE if it has a fully developed profile (SOIL PROFILE); immature if it lacks a well-developed profile; truncated if it has lost all or part of the upper horizons (SOIL HORIZON). Human beings, by their cultivating activities, have affected the development of many soils (especially MAN-MADE SOILS), and led to the destruction of others (SOIL EROSION). PEDOGENESIS, PEDOLOGY, SOIL ASSOCIATION, SOIL CLASSIFICATION, SOIL SCIENCE and other entries qualified by soil.

soil acidity ACID SOIL, pH.

soil and stone COMMON RIGHT in, the right to dig for sand, stone, coals or minerals for the household use of the COMMONER.

soil association a term used by some soil scientists but not by others, usually applied to soils, not necessarily with the same profiles (SOIL PROFILE), lying close to each other, but also **1.** in UK, a group of SOIL SERIES developed on parent material derived from similar rocks or combinations of rocks (1959 Soil Survey Research Board) **2.** (American) an area in which different soils occur in a characteristic pattern, or a landscape which has characteristic kinds, proportions and distribution of component soils **3.** FAO identifies 21 soil units to form the 33 associations shown on the 1:2.5 mn Soil Map of Europe, the soil units including lithosols, regosols, alluvial soils, organic soils, rankers, rendzinas, brown forest soils and acid brown forest soils, greybrown podzolic soil, podzolized soil (PODZOL, PODZOLIZATION), redyellow podzolic soils, red Mediterranean soils, brown Mediterranean soils, chestnut soils, reddish-brown and brown soils, sierozem soils, rubrozem soils, pseudogley soils. See references to those, and SOIL CLASSIFICATION. G.

soil atmosphere the mixture of gases occupying the gaps between soil particles or crumbs with their enveloping films of water, tending to have a lower proportion of oxygen and a much higher proportion of carbon dioxide than the air above ground, and normally to be saturated with water (except in the surface layers of very dry soils). Waterlogged soils are liable to be deficient in oxygen.

soil classification a systematic grouping of SOILS. Most soils which develop in the SOLUM fall into one or other of two great groups, the lime-rich PEDOCALS, containing an accumulation of CALCIUM CARBONATE and the lime-poor PEDALFERS containing accumulations of ALUMINIUM, Al, and IRON, Fe, compounds. (The terms pedocal and pedalfer are now little used by soil scientists.) From another viewpoint soils fall into three world groups: ZONAL, soils with profiles which show a dominant influence of climate and vegetation in their development; AZONAL, or skeletal, soils lacking such a profile; and INTRAZONAL, well developed soils with profiles reflecting the influence of some local factor of relief, parent material, or age, rather those of climate and vegetation (SOIL PROFILE). Most soil scientists recognize the existence of world soil groups at one end of the scale and SOIL SERIES or soil types as the units for description and mapping, but the intermediate soil families and SOIL ASSOCATIONS are differently interpreted.

In 1960 the Soil Conservation Service of the US Department of Agriculture in *Soil Classification: A Comprehensive System, Seventh Approximation* (commonly termed the Seventh Approximation), later entitled the *Comprehensive Soil Classification System* (CSCS), drew up another classification in which ten major orders were based on the present state of development of soils, divided into sub-orders, great groups, subgroups, families and soil series. The ten major orders are ALFISOLS, ARIDISOLS, ENTISOLS, HISTOSOLS, INCEPTISOLS, MOLLISOLS, OXISOLS, SPODOSOLS, ULTISOLS, VERTISOLS.

FAO desribes the main soils groups of Europe by indicating whether or not all three horizons (SOIL HORIZON) are present, e.g. CHERNOZEM is typical A C, a LITHOSOL is (A) C, neither showing a B HORIZON.

A new soil classification in England and Wales, 1973, groups soils on a consideration of their land use capability. Taking landform, geology, climate and natural vegetation into account, seven major groups emerge: peat soils, and six groups of mineral soils (BROWN SOILS, GLEY SOILS (GLEYSOLS), LITHOMORPHIC SOILS, MAN-MADE SOILS, PELOSOLS and PODZOLIC SOILS. The subgroups number 108, so only a few are covered in this dictionary. SOIL HORIZON, SOIL PROFILE. G.

soil creep the slow downward movement of SOIL under the force of gravity. MASS MOVEMENT, SOLIFLUCTION. G.

soil erosion the removal of soil by EROSION, the main types being GULLY EROSION, RILL EROSION, SHEET EROSION and wind erosion (DEFLATION), assisted in many cases by human activities and grazing animals, especially in the removal of vegetation acting as a soil protection. ACCELERATED EROSION. G.

soil horizon a distinctive SOIL layer with features produced by soil-forming processes within the surface layer of the earth's crust. If undisturbed, e.g. by ploughing or similar activities, or by EROSION, soils tend to develop a succession of layers, horizons, commonly identified from the surface downwards as the A HORIZON (often subdivided in soil studies), containing HUMUS, from which material is washed downwards by percolating water

(LEACHING) so that it is termed an eluvial horizon (ELUVIATION). Under the A horizon there may or may not be the B HORIZON, a horizon of deposition, of secondary enrichment, an illuvial horizon (ILLUVIATION) into which material (e.g. clay minerals, iron-aluminium oxides from the A horizon) is washed. Underneath is the C HORIZON, with the PARENT MATERIAL for the existing soil, the little altered, though weathered, BED ROCK. The above are the standard horizons in general use, but some soil scientists also distinguish a D horizon, with unweathered rock, underlying the C horizon; an F horizon or layer, a layer of forest soil consisting of partly decomposed plant residues; a G horizon, the layer where GLEY occurs; and H horizon or layer, an organic layer of forest soils with dark-coloured, structureless humus. Inferior (SUBSCRIPT) numbers are sometimes added to the principal capital letter to indicate small differences within the horizon, e.g. A_2, a leached A layer. Some soils scientists now add inferior lower case letters to the capital letter distinguishing the horizon in order to indicate some special feature, e.g. A_p, an A horizon which is ploughed; B_h, a B horizon with an accumulation of organic matter; B_{ca}, with an accumulation of calcium carbonate; B_s, enriched with translocated SESQUIOXIDES of iron an aluminium; and, in relation to GLEYING, (g) mottled; g, gleyed; G, intensely gleyed.

Another classification of soil horizons is based on organic (O HORIZON) and mineral (MINERAL HORIZONS) content. The O horizons (L, F, H, horizons) correspond approximately to the A_{oo} and A_o horizons. Below the O horizons lie the A_p (ploughed) and A_h (mineral and organic), approximately the A_1 horizon, grading downwards to E_b (brown horizon with CLAY removed) and E_a (lighter in colour, with sesquioxides as well as clay removed) to

A B, a transitional zone (E_b to A B layers approximate to A_2 and A_3). Then come B_h (high organic content), B_{fe} (iron pan, HARD PAN), B_t (with illuvial clay), B_s (with illuvial clay and sesquioxides), B C (transitional zone), the layers corresponding to B_1, B_2, B_3. Below these lies the C horizon (parent material), corresponding broadly to the traditional C horizon. A HORIZON, B HORIZON, C HORIZON, DIAGNOSTIC HORIZON, SOIL, SOIL CLASSIFICATION, SOIL PROFILE. G.

soil mechanics 1. the scientific study of the physical properties of soils, of the effects of forces on SOILS 2. engineers' term applied to the behaviour of both soils and surface rocks especially in relation to the foundations of buildings.

soil particles GRADED SEDIMENTS, PARTICLE.

soil phase in SOIL CLASSIFICATION, a subdivision (usually the lowest in the hierarchy) of any class of any category, based on some unusual feature of the soil (e.g. stoniness, salinity, depth of soil, slope), important in the use and management of it. SOIL TYPE.

soil profile a vertical section of soil showing the sequence of horizons (SOIL HORIZON) downwards from the surface to the PARENT MATERIAL. It may be cut and studied in the field, or, as a sample section, taken away for study. MONOLITH-2. G.

soil science the scientific study of soils, including PEDOLOGY and SOIL MECHANICS-1, and the study of soils in relation to plant growth (EDAPHOLOGY). Soil science has given rise to an extensive, specialist terminology, some common terms, e.g. structure, texture (SOIL STRUCTURE, SOIL TEXTURE) being given restricted meanings, many foreign words being introduced into international literature (e.g. PODZOL, REGUR), special terms being created (e.g. PEDALFER, LITHOSOL). Some authors have developed a personal terminology which may or may not be adopted eventually, and the terminology continues to grow. Only terms most commonly used at present and in the immediate past have been included in this dictionary.

soil series a group of soils the members of which, formed from similar PARENT MATERIAL, have horizons (SOIL HORIZON) similar in distinguishing characteristics and arrangement in the SOIL PROFILE, except for the texture of the surface layer and its state of erosion. The series is the group most commonly used as the basic unit in mapping soils (comparable with the SPECIES in biology), and is usually the lowest in the hierarchy of a system of SOIL CLASSIFICATION. G.

soil solum SOLUM.

soil structure in soil science, the character of a soil shown by the ability of its particles to come together and to hold together to form AGGREGATES-3 or CRUMBS (PEDS) and by the way they do so. Of various structures identified, CRUMB STRUCTURE consists of small, soft, porous, irregularly shaped aggregates; granular structure (GRANULE) of irregularly shaped, rounded aggregates up to 10 mm in diameter; angular and blocky (aggregates in the shape of blocks with sharp edges and faces generally at right angles to each other), subangular blocky (edges rounded, angles generally less than right angles); columnar (round-topped columns), prismatic (columns with sharp-edged vertical faces, resembling a prism); platey (aggregated in the shape of flat, thin flakes); or the soil may be classified as structureless (devoid of aggregation), e.g. a sand-dune.

soil texture 1. UK, the composition of soil in respect of particle size. The texture may be described as coarse-grained and gritty (e.g. a sandy soil with particles 0.02 to 2.00 mm in diameter), fine-grained and sticky (e.g. clayey, with particles lower than 0.002

in diameter), intermediate and silty (e.g. silt, with particles between 0.02 and 0.002 mm in diameter) or mixed and loamy (LOAM) 2. (American) a classification of soil based on the relative amounts of the various size groups of individual soil grains. Internationally recognized particle sizes are given under GRADED SEDIMENTS, but gravel should be distinguished (under stones) as being stones 2 to 20 mm in diameter. Particle size analysis is termed MECHANICAL ANALYSIS. SOIL, SOIL SCIENCE and its terminology. G.

soil type 1. in SOIL CLASSIFICATION, USA, a group of soils developed from a particular kind of PARENT MATERIAL, and having horizons similar in distinguishing characteristics and arrangement, the lowest unit in the American soil classification (although a SOIL PHASE may be distinguished). Compare SOIL SERIES in the UK classification 2. a subdivision of a SOIL SERIES based on variations in the SOIL TEXTURE of the A HORIZON. G.

soil water in soil science, water held in the soil and available to the roots of plants. CAPILLARITY, CAPILLARY MOISTURE, FIELD CAPACITY, FILM WATER, GRAVITATIONAL WATER, MOISTURE, WATER DEFICIT, WATER-LOGGING, WILTING POINT. G.

soke historically in Britain, a small district with various local rights of jurisdiction, surviving only in the place-name Soke of Peterborough. G.

solano (Spanish) a hot, oppressive easterly or southeasterly wind blowing from the Mediterranean to south-eastern Spain, often bringing rain in summer. LEVANTE.

solar *adj.* of or relating to the SUN.

solar battery, solar cell an apparatus that uses SOLAR RADIATION or its heating effect to produce an electrical current.

solar constant the rate per unit area at which RADIATION from the sun reaches the outer margin of the earth's atmosphere, averaging approximately 2 gramme-calories per sq cm per minute (139.6 mW per sq cm).

solar day   the time interval between two successive transits of the sun over the same meridian, varying slightly at different times of the year because the orbit of the EARTH round the sun is elliptical and inclined to the equator (EQUATION OF TIME). A mean solar day of 24 hours is the calculation commonly used, about 4 minutes longer than the SIDEREAL DAY. SOLAR MONTH, SOLAR YEAR.

solarimeter   an instrument for measuring the intensity of total radiation per unit area received on the ground. SOLAR RADIATION.

solar month   one-twelfth of a SOLAR YEAR.

solar radiation   ELECTROMAGNETIC WAVES emitted by the sun. The wavelengths range widely outside the earth's atmosphere, but absorption in the STRATOSPHERE ensures that the electromagnetic waves reaching the earth's surface are limited to certain bands. These include the wavelengths of VISIBLE LIGHT. ACTINOMETER, ELECTROMAGNETIC SPECTRUM, INSOLATION, SOLAR BATTERY, SOLARIMETER.

solar system   the SUN and the celestial bodies orbiting round it under the force of gravity, i.e. the nine PLANETS (in sequence measured in distance from the sun, Mercury, Venus, Earth, Mars, Jupiter, Saturn, Uranus, Neptune, Pluto), the natural SATELLITES which revolve round the planets (e.g. the moon around the earth), the ASTEROIDS, COMETS, METEORS, METEORITES. The orbits of the nine planets are elliptical and approximately in the same plane, the solar system as a whole moving through space at about 18.5 km (11.5 mi) per second. NEBULAR HYPOTHESIS.

solar wind   a flow of atomic particles from the sun.

solar year   the astronomical, equinoctial, natural or tropical year, the average time taken by the earth to complete one orbit round the sun with reference to the vernal EQUINOX as shown by the First Point of Aries (RIGHT ASCENSION), now 365.2422 SOLAR DAYS, i.e. 365 days 5 hours 48 minutes 45.51 seconds, decreasing by some 5 seconds in 1000 years.

sole   **1.** in geology, the lowest THRUST-PLANE in an area of intense compressional movements in the earth's crust, e.g. the Sole, the name applied to the lowest of the Caledonian thrust-planes in the northwest Highlands of Scotland **2.** in glaciology, the ice base of a glacier.

sole pasture, sole vesture   sole pasture is the exclusive right to graze cattle; sole vesture is the exclusive right to take the production of the soil. In neither case may the owner of the soil participate in the right.

solfatara   (Italian, derived from the name of a small volcano in Phlegraean Fields near Naples) a volcanic vent through which vapours and gases, especially sulphurous gases, gently issue, usually marking a late stage in volcanic activity. FUMAROLE, MOFETTE, SOFFIONE, SOLFATARIC STAGE.

solfataric stage   a dormant or decadent stage in volcanic activity, characterized by the emission of gases and vapours. FUMAROLE, MOFETTE, SOLFATARA. G.

solid   matter with a definite volume and shape, the structure being determined by the arrangement in space of its molecules, atoms or ions which, unable to move freely, vibrate about a fixed position. FLUID, GAS, LIQUID.

solid geology   the geology of the rocks underlying the layers of superficial deposits (DRIFT-1). The British Geological Survey 'solid edition' maps exclude DRIFT-1 but, where appropriate, extensive thick stretches of alluvium are included to give a true impression. GEOLOGY. G.

solifluction, solifluxion the slow movement of rock debris, saturated with water and not confined to definite channels, down a slope under the force of gravity. It occurs particularly when thawing releases such surface deposits while the underlying layers are still frozen. Formerly considered to be synonymous with SOIL CREEP, but now usually applied only to saturated deposits. DRY VALLEY, FREEZE-THAW, MASS MOVEMENT. G.

soligenous *adj.* applied to a saturated PEAT area where the water level is fed by water flowing in the underlying layer.

solitary *adj.* applied to an organism existing alone, separately from others of the same kind, because there is no community for it to live in. In animals the term is applied to the usual behaviour of the species. Solitary should not be confused with lone, *adj.* applied to an animal living by itself, but usually living as one of a herd. SOCIAL-2.

solod, soloth (Russian: soil science) a leached saline soil (degraded SOLONETZ), with a pale A_2 horizon and a degraded, fine-textured B horizon. The pl. form, solodi, is applied in Russian to degraded solonetz soils. SOIL HORIZON. G.

solonchak (Russian: soil science) a saline soil, without structure, occurring in arid and semi-arid regions. SOIL STRUCTURE. G.

solonetz (Russian: soil science) a formerly saline soil from which the salts have been leached. In Russian, solonets, pl. solontsy, is applied to a soil with surface rock salt. G.

solstice a term conveying the idea of the SUN-2 standing still, i.e. the point in the ECLIPTIC when the sun is farthest from the EQUATOR, either north or south (i.e. approximately 23°30'N, the Tropic of Cancer and 23°30'S, the Tropic of Capricorn). Thus in the northern hemisphere the summer solstice is 21-22 June, the longest day, when at noon the sun is shining vertically over latitude 23°30'N, the Tropic of Cancer; the winter solstice is 21-22 December (the shortest day) when it is shining vertically at noon over latitude 23°30'S, the Tropic of Capricorn. SOLSTITIAL POINT. G.

solstitial point the point reached by the sun in its ECLIPTIC at the time of the SOLSTICE.

solum the term applied by soil scientists to the part of the earth's crust influenced by climate and vegetation, i.e. the soil layers above and excluding the PARENT MATERIAL. SOIL, SOIL HORIZON. G.

solute the solid or gaseous substance which dissolves in the SOLVENT to form a SOLUTION.

solution 1. a homogeneous mixture of two or more substances, in which a SOLID, LIQUID or GAS forms a single phase with another solid, liquid or gas (usually a liquid) (SOLUTE, SOLVENT) which has constant physical and chemical properties throughout at any selected concentration up to its SATURATION point. A standard solution is a solution of known concentration. COLLOID, SUSPENSION 2. the act by which a substance is put into solution, or the state of being put into solution, e.g. in the chemical WEATHERING of rocks (CORROSION) the salts they contain are commonly dissolved by water to form a solution; rainwater charged with CARBON DIOXIDE dissolves (forms a solution with) CALCIUM CARBONATE, to remove it as CALCIUM BICARBONATE (CARBONATION, HYDROLYSIS); and rivers, in their work of transporting debris, carry a variety of substances in solution 3. the answer to a problem, or the act, method, process by which the answer is achieved.

solution collapse, solution subsidence in KARST, any subsidence of the surface due to CARBONATION of the subsurface rocks.

solution pipe a vertical cylindrical or cone-shaped hole, varying in size, sometimes not apparent at the surface, filled with debris, and occurring especially in chalk country, resulting from CARBONATION concentrated along fissures and joints. The debris comes from the overlying deposits, e.g. of sand, clay or Tertiary rocks (GEOLOGICAL TIMESCALE). In Belgium the term aard pijpen or orgel pijpen is applied to such a feature.

solvent the part of a SOLUTION which is present in greater bulk, i.e. usually the LIQUID in which the SOLUTE is dissolved. If the solvent is not water, this fact is usually noted, e.g. non-aqueous solvent.

soma all the cells of an organism apart from the germ-cells (the reproductive cells).

somatic *adj.* of or relating to the body or the body cells of an organism, as distinct from the germ-cells. SOMA.

somatic cell one of the cells which form TISSUE, the organs, etc. of a body. SOMA.

somatology physical anthropology, the comparative study of the structure, development, functions of the human body, concerned especially with body measurement and comparative descriptions of the distinguishing characteristics of various groups of people. G.

somma (Italian, Monte Somma) the old crater wall of Vesuvius, forming an arc around the present crater (named Monte Somma in Italy). The term has been extended and applied to similar volcanic crater rims elsewhere. It should not be applied to small, subsidiary craters inside a CALDERA, almost exactly the reverse of the proper application. G.

sonar Sound Navigation Ranging, echo sounding, a device for locating an underwater object by sending out high frequency sound waves which are reflected from the object and registered on the apparatus, the time delay and nature of the echo giving information about shoals of fish, underwater obstructions, and ocean depths (LIDAR, RADAR). Some animals, e.g. bats (in air), dolphins and whales (underwater) use high frequency sound waves in a similar way, to locate objects and to communicate with each other.

sonde (French) an apparatus designed to measure and record conditions in the atmosphere at certain altitudes, e.g. BALLOON-SONDE, RADIOSONDE, RAWINSONDE, variously equipped with a lifting device and sensing and recording instruments.

sonic *adj.* 1. of sound waves audible to the human ear 2. of or relating to the speed of sound in air (about 1188 kph, or 340 m per second at sea-level).

Sonnenseite (German) a sunny slope in a valley, synonymous with ADRET. SCHATTENSEITE, UBAC.

sonograph a graph showing ocean-floor patterns as revealed by SONAR.

sop a vertical deposit of iron ore occurring especially in Carboniferous Limestone where mineralized solutions from former overlying TRIASSIC sandstone have penetrated lines of weakness.

sorghum *Sorghum vulgare*, great millet, termed dura (northern Africa), kaffir corn (southern Africa), guinea corn (west Africa), a small-grained CEREAL with seeds larger than, and not storing so well as, those of most MILLETS, regarded by some botanists as a millet, but not by others. Possibly native to Africa, it is cultivated there and in other semi-arid tropical and subtropical regions. The white-grained variety is preferred for cooking,the red-grained for beer making. Sugar is obtained from a sweet variety grown in the USA. The grain of all varieties, some of which enters world trade, is fed to livestock.

soroche (Peru) mountain sickness, an

affliction suffered by some people unaccustomed to high altitudes. G.

sorting separation and putting into groups or classes according to some special quality or kind (shape, size, weight, age etc.), e.g. the sorting of sediments by the natural processes of flowing water and of wind. DEFLATION-1, GRADED SEDIMENTS.

sotch (French) term applied to a DOLINE in Causses, France (coup in Aquitaine). G.

souk, suq (Arabic; various other spellings) a market.

soum (Irish and Scottish) **1.** a unit of stock in common grazing **2.** the area of pasturage needed to support a cow or proportional number of other animals **3.** the number of cattle or other animals which can be supported by a certain area of pasture. Compare STINT. G.

sound (in Scandinavia: sund, a strait or inlet) **1.** a stretch of water connecting two larger bodies of water, e.g. a sea or large lake with another sea or the ocean, wider than a STRAIT **2.** the channel between an island and the mainland. CONCORDANT COAST **3.** an inlet of the sea **4.** a LAGOON fringing the southern and southeastern coast of USA. G.

sounding **1.** a method by which the depth of a sea or lake can be determined, formerly by a weighted line (a sounding line) dropped overboard, now usually by an ECHO-SOUNDER (SONAR) **2.** a measure of the depth of water determined by those means.

sounding balloon BALLOON-SONDE.

source of a river, the point at which a RIVER, identifiable as such, begins its flow. This may be at a SPRING-2, or from a lake, glacier, cave, marsh or swamp, or formed from the coalescence of trickles of water in RUNOFF on a hillside. G.

south **1.** one of the four cardinal points of the COMPASS, directly opposite the north, lying on the right side of a person facing due EAST **2.** towards or facing the south, the southern part, especially of a country, particularly of the southern states (The South) of the USA. BRANDT REPORT.

south *adj.* of, pertaining to, belonging to, situated towards, coming from, the south, e.g. of winds blowing from the south.

southeaster a strong wind or storm coming from southeast of the observer.

southeast trades the TRADE WINDS of the southern hemisphere.

southerly burster, southerly buster a strong, dry cold wind blowing from the south, most frequently in spring and summer, in Australia and New Zealand, in the wake of a trough of LOW pressure. BRICKFIELDER. G.

Southern Cross POLE STAR.

southern hemisphere the half of the earth south of the EQUATOR. HEMISPHERE.

southing **1.** nautical term, sailing southwards, or the distance so travelled since the last reckoning point **2.** in astronomy, the distance, measured in degrees, of any heavenly body south of the CELESTIAL EQUATOR.

south magnetic pole MAGNETIC POLE.

South Pole the geographical South Pole, the southern extremity of the earth's axis. MAGNETIC POLE, POLE, TRUE SOUTH.

southwester, sou'wester a strong wind or storm coming from southwest of the observer. G.

sovereign a person (e.g. a monarch or an emperor) or a body of persons holding supreme authority, supreme power, in a state.

sovereign *adj.* **1.** of or related to a SOVEREIGN **2.** having absolute, unlimited power, the right to make decisions, and accordingly to act, thus having complete freedom and power to govern or act. SOVEREIGNTY.

sovereign state a STATE-2 with the supreme authority, the supreme power, held within the state itself, a state which is therefore independent and fully self-governing. NATION, NATION STATE.

sovereignty **1.** the status or authority of a SOVEREIGN **2.** the state or quality of being SOVEREIGN-1,2, *adj*. **3.** the authority to act or govern, to make or amend laws laid down by the rules of a legal system.

soviet (Russian soviet, council) **1.** an elected governing council in the USSR, at local, provincial and national level, the latter (the Supreme Soviet) comprising delegates from all the Soviet Republics **2.** any of the associated republics of the USSR.

soviet *adj*. of, relating to, or pertaining to, the Union of Soviet Socialist Republics (USSR).

sovkhoz (Russian) a state farm (STATE FARMING) in the USSR, usually large-scale. KOLKHOZ. G.

sow **1.** an adult female PIG **2.** the adult female of some other animals, e.g. the badger.

soya bean, soybean, soja bean *Glycine max*, an ANNUAL leguminous herb (LEGUMINOSAE), susceptible to frost, probably native to southwestern Asia, a wide number of varieties being grown there and in the USA in areas with warm summers. The plant produces highly nutritious seeds (high in protein, low in carbohydrates) which yield oil (used in cooking and industrially for making paints, plastics, etc.), provide flour and 'milk', or are destined (fresh, fermented or dried) for human consumption, as are the young seedlings (bean sprouts). Soy sauce, a piquant sauce, is made from fermented soya beans soaked in brine. The mature plant and the residue remaining after the extraction of oil from the seed are fed to livestock.

spa a watering place, a resort with SPRINGS-2, the water of which contains minerals reputedly of medicinal value. The term is derived from Spa, a watering place near Liège, Belgium,

celebrated for the curative properties of its mineral spring water. G.

space **1.** that which objects occupy as a result of their volume, the amount of space occupied being the volume of the object **2.** a part of space, a volume, area or length that may be occupied by something, or may be empty, e.g. an extent or area of the earth's surface, or the distance between two points, or two objects, or two lines on a page **3.** a period of time, e.g. between two events **4.** that which is beyond the limit of the earth's ATMOSPHERE-1. SPATIAL.

space cost curve a CROSS SECTION-1 through a COST SURFACE, related either to total production costs, or to single items in such costs. COST CURVE.

Space Shuttle a NASA SATELLITE-1 launched in 1981 with a crew aboard. It was designed to make repeated return trips between the earth and SPACE-4, to act as a platform from which other orbiting satellites could be serviced, and to carry REMOTE SENSING equipment. LANDSAT, SKYLAB, SPOT.

space-time constraints the boxing-in of human activity produced by the simultaneous operation of the limitations of available time (which may include hours of daylight, or biological constraints such as the CIRCADIAN RHYTHM) and physical ACCESSIBILITY. For example, it is possible to travel only a certain distance from home to another place, to spend time there in some activity (working, shopping, playing, visiting friends, etc.) and to return home in daylight, or in waking hours.

spalling (American) EXFOLIATION.

Spanish moss *Tillandsia usneoides*, an EPIPHYTE, a stemless HERBACEOUS plant, forming long, loose, grey-green hanging tufts in warm humid conditions in southern USA (e.g. in the Everglades, EVERGLADE) and West Indies; or *Ramalina reticulata*, a

LICHEN forming lace-like nets on trees in the humid coastal regions of western USA.

spar any of the light-reflecting crystal-line minerals which cleave easily into flakes or chips.

sparselands semi-desert pastoral regions.

spate a sudden increase of volume of water in a river, a flood, resulting from heavy rainfall or the sudden melting of snow in the upstream area.

spatial *adj.* **1.** of, pertaining to, or relating to, SPACE **2.** consisting of, or having the character of, space **3.** extending in or occupying space **4.** subject to or controlled by the conditions of space (in contrast to TEMPORAL) **5.** existing in, occurring in, caused by, or involved by, space **6.** as applied to a faculty or a sense, perceiving space.

spatial analysis an approach to geography in which the LOCATIONAL variations of a phenomenon or a series of phenomena is studied and the factors influencing or controlling the patterns of distribution investigated. SPATIAL-1,3 patterns are broken down into simple elements so that measurements can be made of single patterns. This allows the comparison of two or more patterns, e.g. the pattern of a single phenomenon in different areas, or the pattern of different phenomena or VARIABLES in one area; and it allows tests to be developed to show whether a pattern differs significantly from a random pattern. ECOLOGICAL ANALYSIS, REGIONAL COMPLEX ANALYSIS. G.

spatial autocorrelation AUTOCORRELATION.

spatial diffusion the process of spreading out or scattering over a part of the earth's surface. DIFFUSION-1.

spatial equality a condition in which all consumers enjoy the same degree of proximity to the services which meet their needs. A state of spatial inequality exists if the degree of proximity varies.

spatial inequality SPATIAL EQUALITY.

spatial injustice, spatial discrimination the inferior or unfair treatment experienced by people living in a particular area (e.g. a GHETTO, or a rural area distant from health care and education facilities etc.), due to the location of that area.

spatial interaction 1. the movement, contact, relationship and linkage between points in space, e.g. the movement of people, goods, traffic, energy, information, capital etc. between one place and another **2.** applied by E. L. Ullman to the interdependence of geographic areas. INTERACTION-1, ULLMAN'S BASES FOR INTERACTION. G.

spatially *adverb* **1.** by means of space **2.** with reference to space.

spatial margin spatial margin to profitability, the limit where the total cost of production of a given volume of output equals the total revenue possible from the sale of that volume of output. It therefore indicates the area within which an activity should be profitable. On a COST SURFACE map the spatial margin of profitability appears as the contour which encloses the area within which production is economically feasible, but outside of which an operation will probably lose money. G.

spatial monopoly a MONOPOLY which owes its exclusive control of the market to its location rather than to its competitive success.

spatial organization the aggregate pattern of the use of space by a SOCIETY-2,3. G.

spatial patterns the patterns (e.g. of land use) related to SPATIAL PROCESSES, falling into three categories: regular, random, clustered.

spatial process any of the mechanisms, e.g. movement over the earth's surface, which produce the SPATIAL STRUCTURES of distributions. Four types of spatial process, or change, are commonly identified: uniform, covering changes affecting all parts of an area simultaneously; random, changes occurring anywhere in an area, regardless of previous changes; contagious, new changes likely to occur near to changes which have already occurred; and competitive, that which exists, which has specific needs, inhibiting the arrival of something new with the same specific needs, which must therefore seek a new area elsewhere. G.

spatial segregation in urban studies, the extent to which various parts of urban areas are associated with and lived in by different social and ethnic groups. SOCIAL SEGREGATION.

spatial sorting the separation of things which are unrelated or incompatible in some way, e.g. high status housing and a noxious chemical works. SORTING.

spatial structure the arrangement of phenomena on the earth's surface produced by SPATIAL PROCESSES, the way in which space is organized by and takes a part in such processes (physical or social). Thus the spatial structure of a distribution relates to the location of each element relative to each of the others, and relative to the aggregate of the others. G.

spatter cone, driblet cone a small VOLCANIC CONE, between some 3 and 6 m (10 and 20 ft) in height, formed when LAVA erupts violently from the side of a volcanic VENT, or in some cases through a fissure, and spatters (falls in drops) to the ground, where it rapidly congeals. G.

Spearman's rank correlation coefficient a measure of the strength of the relationship between variables used when only the relative size, or rank, of each variable is known. CORRELATION-3.

spearmint *Mentha spicata*, common mint, a PERENNIAL herb native to central Europe, widely grown in temperate lands, especially in the British Isles, for its flavouring properties in food preparation. The plant yields the ESSENTIAL OIL carvone, obtained by distillation, and used medicinally and for flavouring.

special district in USA, a single-purpose unit of local government, responsible for providing a particular service within its territory, e.g. a fire service. LOCAL GOVERNMENT IN USA.

speciation the evolutionary process involved in the formation of a new SPECIES.

specient an individual member of a SPECIES.

species 1. in biology, the smallest unit of clasification commonly used (BINOMIAL NOMENCLATURE, CLASSIFICATION OF ORGANISMS), the group with members having the greatest mutual resemblance, able to breed with each other but not with organisms of other groups. Local differences may occur through reproductive isolation, recognized in classification as a sub-species. The common names of familiar animals and plants often refer to species. ALLOPATRIC, INDEX SPECIES, INDICATOR PLANT, INDIFFERENT SPECIES, STRAIN-1, TYPE SPECIMEN, VARIETY-2 2. in chemistry, entities, such as atoms, molecules, ions, free radicals, and activated atoms or molecules, which are active in chemical reactions.

specific *adj.* 1. clearly distinguished, peculiar to or characteristic of, something 2. in biology, characteristic of a SPECIES 3. of disease, caused by a particular infection.

specific complementarity the COMPLEMENTARITY which exists when a surplus (i.e. when supply exceeds demand) for a particular product in one area is matched by a deficit of that product in another area.

specific gravity the ratio between the weight, at any chosen place, of a given volume of a substance and the weight of an equal volume of water at 4°C (39.2°F) at the same place. Specific gravity and density are numerically the same, but the former is a relative quantity, DENSITY is absolute.

specific heat the amount of heat needed to raise the temperature of 1 gram of a substance through 1°C.

specific humidity the ratio of the weight of WATER VAPOUR in a parcel of the ATMOSPHERE-1 to the total weight of air (i.e. including water vapour), measured in grams of water vapour per kilogram of air. The specific humidity of very cold dry air is low, that of very warm humid air is high. ABSOLUTE HUMIDITY, HUMIDITY, MIXING RATIO, RELATIVE HUMIDITY.

specific regional system a regional system in which types of places not only resemble one another in accordance with a combination of intrinsic attributes, but are also contiguous (as distinct from those in a GENERIC or GENERAL REGIONAL SYSTEM, which may be widely spaced). G.

specific variance in statistics, in FACTOR ANALYSIS, that part of the total VARIANCE which does not correlate with any other VARIABLE. COMMON VARIANCE, ERROR VARIANCE, RESIDUAL VARIANCE, TRUE VARIANCE.

spectrum 1. an arrangement of the components of a beam, waveband or sound, separated and exhibited in an order determined by a varying factor, e.g. energy, wavelength **2.** the colours displayed when white light (i.e. the main radiation received by the earth from the sun) is dispersed by a prism or by a diffraction grating (red, orange, yellow, green, blue, indigo, violet). ELECTROMAGNETIC SPECTRUM, FOGBOW, RAINBOW, SOLAR RADIATION.

speleology, spelaeology the scientific study of caves, including organisms living in caves. G.

speleotherm cave formation, a secondary mineral deposit formed by the accumulation, dropping or flowing of water in a cave, e.g. a STALACTITE, STALAGMITE. G.

spelter commercial ZINC, about 97 per cent pure, containing some LEAD.

spermaceti a white, waxy substance, a component of SPERM OIL, used in textile finishes, candles, cosmetics.

sperm oil a pale yellow liquid wax (not a true oil) in the head cavity of a sperm whale, used as a lubricant or dressing for leather after the SPERMACETI has been extracted. WHALE OIL.

sphagnicolous *adj.* living in SPHAGNUM, bog moss.

sphagniherbosa plant communities containing much SPHAGNUM and growing on peat.

sphagnum bog moss, a member of *Sphagnum*, a genus of soft MOSSES with erect stems, growing in swamps or in water in cold temperate zones. Minute holes in some of the cell walls of the leaves promote the absorption of water, even when the leaves are dead. The conditions of the environment prevent decay, so as the plant grows upwards the depth of the moss bed increases, becomes compacted, and forms PEAT. BOG.

sphalerite ZINC BLENDE, the mineral zinc sulphide, commonly occurring with GALENA in HYDROTHERMAL deposits, metasomatosed limestones (METASOMATISM) and in sedimentary STRATIFORM deposits. ZINC.

sphere 1. in geometry, a solid figure made when a circle rotates about a diameter, any point on the surface of this solid figure being equidistant from its centre (the centre of the sphere) **2.** in general, loosely, an object with a shape approximately like that of a sphere, i.e. not necessarily a perfect sphere. SPHEROID **3.** a realm, an area which is limited in extent but within which something is effective. SPHERE OF INFLUENCE **4.** a field of knowledge **5.** in CULTURAL GEOGRAPHY, the zone in which a CULTURE-1, having spread from its CORE AREA or CULTURAL

HEARTH and over its DOMAIN-2, is still effective and influential. OUTLIER.

sphere of influence 1. a territory or part of a territory in which a foreign power has political and economic interests, special rights and priviliges. SPHERE-3 2. in urban studies, an area which depends on an urban centre for various services, or with which it has special relations (the terms UMLAND, URBAN FIELD or URBAN HINTERLAND are now more commonly applied to such an area). CENTRAL PLACE THEORY, DAILY URBAN SYSTEM. G.

spherical, spheric *adj.* shaped like, or relating to, a SPHERE-1,2.

spherical triangle a three-sided enclosed figure bounded by the arcs of three GREAT CIRCLES and formed on the surface of a sphere.

spheroid a figure, a body, resembling a SPHERE-1 (especially an ELLIPSOID). GEOID. G.

spheroidal weathering onion weathering, a form of underground CHEMICAL WEATHERING occurring particularly in tropical regions in well-jointed rocks such as BASALTS and DOLERITES. Water penetrates the intersecting joints, attacking each separate block from all sides simultaneously, breaking off a succession of shells or skins, so that a succession of new surfaces is presented to the weathering solution, leaving a mass of unweathered rock in the centre which, on EXHUMATION, appears at the surface as a rounded mass. The process (HYDROLYSIS) is similar to that involved in EXFOLIATION. G.

spherulite a spherical concretion of radiating crystals, especially of QUARTZ and FELDSPAR, occurring in some rocks. RHYOLITE.

spices any of the many vegetable substances, mainly plant parts, from which ESSENTIAL OILS have not been extracted, nearly all derived from tropical or subtropical plants, used from earliest times to give their strong, aromatic flavours to food for human beings and,

to a less extent, to preserve it from decay. They were particularly valued in midlatitudes, especially in Europe, when the quality of meat was poor and preservation by refrigeration or canning unknown. In the MIDDLE AGES spices, purchased only by the wealthiest families, fetched enormous prices in Europe; so they were amongst the most valued of cargoes, tempting mariners to the most hazardous voyages, especially to the Far East. Some of the most popular entering international trade today are ALLSPICE, ANISE, CARDAMOM, CHILI, CINNAMON, CLOVE, CORIANDER, CUMIN, GINGER, MACE, MUSTARD, NUTMEG, PAPRIKA, PEPPER, TURMERIC.

Spiegeleisen (German Spiegel, mirror; eisen, iron) PIG IRON, containing manganese, used in making steel by the BESSEMER PROCESS.

spilite a BASALT, rich in SODIUM.

Spilitic suite a PETROGRAPHIC PROVINCE, marked by volcanic action and slow submergence. ATLANTIC SUITE, PACIFIC SUITE, PILLOW LAVA. G.

spill bank in engineering, a natural bank of coarse alluvium over which a river in flood flows.

spillway 1. the area below a dam or a natural obstruction over which excess water from a reservoir or lake above is allowed to drain away 2. the part of a dam over which water flows 3. a passage for the overflow of water from a reservoir etc. G.

spinach *Spinacia oleracea*, a leafy, herbaceous ANNUAL plant, native to southwest Asia, long cultivated there and in other temperate regions for its edible leaves, higher in protein and vitamin A than most other leaves, usually eaten cooked.

spinach beet, perpetual spinach *Beta vulgaris* Cicla, a variety of BEET, usually grown as an ANNUAL, for the sake of its edible leaves, which taste like those of SPINACH, and are similarly eaten cooked.

spinel any member of a group of minerals with a content of magnesium, iron, zinc, manganese or nickel combined with aluminium, iron or chromium, formed in IGNEOUS and METAMORPHIC rocks. Included are CHROMITE, MAGNETITE and the gem spinels, which are red, carmine, green, brown, blue or black in colour, some of the darker colours being opaque. The red variety (spinel ruby) sometimes misleadingly resembles a true oriental RUBY, but can be detected because it is less hard than the more valuable gemstone.

spinifex one of the genus *Tricupsis*, a coarse grass with sharp, pointed, spiny leaves growing in large tufts, separated by bare ground, over large areas of semi-desert in the heart of Australia. G.

spinney, spinny 1. a small wood with undergrowth, a clump of trees, a small grove **2.** a plot of land with thorny briars. G.

spiritualism in philosophy, the doctrine that spirit exists independently of matter, contrasting with the theory of MATERIALISM-1.

spit a narrow ridge of sand and shingle, resulting from LONGSHORE DRIFT, attached to the sea shore at one end, extending some distance seawards, and terminating in open water at the other. The outer end is often deflected landward to form a hook, or a recurved spit; and development of the hook may produce a compound recurved spit or compound hook. BARRIER ISLAND.

spitskop (Afrikaans) a hill with a pointed top, as distinct from a TAFELKOP. G.

splash zone SHORE ZONES.

splays FLOODPLAIN.

spodic horizon a SOIL HORIZON enriched with SESQUIOXIDES and/or organic matter, with or without CLAY, and occurring in humid climates.

spodosols in SOIL CLASIFICATION, USA, an order of soils associated with a cool and cool-humid climate and forest or heath vegetation, with a leached, acid, ash-grey A HORIZON, low in plant nutrients, overlying a B HORIZON rich in iron oxide and aluminium and enriched by organic material from the A horizon (ELUVIATION, ILLUVIATION), e.g. a PODZOL. G.

spoil waste material from mining or quarrying operations, piled up in spoil banks, spoil tips, spoil dumps or tipheaps. G.

sponge a member of the phylum Porifera (CLASSIFICATION OF ORGANISMS), an immobile marine animal, permanently attached to a rock, alone or in COLONIES-6, the body being supported by a skeleton of lime, silica or a special type of protein material (spongin), no nervous system, collecting food from the water which is drawn in through small, and flows out through large, pores. The sponges traded commercially are of the genera *Spongia* or *Hippospongia*, highly valued because their cleaned skeletons, consisting of the interlocking fibres of spongin, have the ability to absorb fluid and give it up readily under pressure.

spongy *adj.* having the attributes of a SPONGE, i.e. porous, absorbent, elastic, applied to many substances, including a metal in a finely porous, absorbent state.

SPOT Satellite Probatoire pour l'Observation de la Terre, a SATELLITE-1 orbiting the earth, launched by France, carrying REMOTE SENSING equipment similar to that aboard LANDSAT satellites in order to gather data on natural resources, environmental conditions, land use etc. needed for certain earth resource development projects. SKYLAB, SPACE SHUTTLE.

spot height a precise point on a map showing the height of the ground at that place measured from a given datum (e.g. in British Ordnance Survey maps, height above OD). It is not necessarily indicated physically on the ground, thus differing from BENCH MARK. G.

sprawl   URBAN SPRAWL.

spread and backwash effect   CIRCULAR AND CUMULATIVE GROWTH.

spring   1. one of the SEASONS of the year, occurring between winter and summer in midlatitudes, a period variously defined in the northern hemisphere as being March, April, May or (astronomically) from the spring EQUINOX, 22 March, to the summer SOLSTICE, 21 June 2. a continuous or INTERMITTENT natural flow of water issuing strongly or seeping gently from the earth's surface under its own pressure, the site being related to the nature and relationship of rocks (especially PERMEABLE and IMPERMEABLE layers), the level of the WATER TABLE, the surface relief. CRYSTOCRENE, DIKE-SPRING, FAULT SPRING, MINERAL SPRING, SCARP-FOOT SPRING, SPRING-LINE, VAUCLUSIAN SPRING.

spring alcove   ALCOVE.

spring-line   a line of SPRINGS-2 occurring roughly at the level where, by reason of deposition of the strata, the WATER TABLE reaches the surface, as at the foot of an ESCARPMENT. Where such springs are copious and constant they provided in the past a reliable water supply, one of the factors likely to influence the choice of a site for a village, hence the term spring-line village. G.

spring onion   ONION.

spring-sapping, spring-head sapping   the undermining of a hillside at the point of issue of a SPRING-2, caused by the erosive action of swiftly flowing water, leading to small slips and the formation of an ALCOVE or amphitheatre which cuts backwards into the slope and results in the retreat of the valley head. G.

spring tide   a tide with a range (TIDAL RANGE) greater than that of ordinary tides, i.e. the HIGH TIDE is higher, the LOW TIDE lower, occurring twice

monthly, when the moon, sun and earth are almost in the same straight line (SYZGY), either in CONJUNCTION (at the time of the new moon) or in OPPOSITION (the time of the full moon), so that the gravitational effects are reinforced. MOON, NEAP TIDE.

sprouting broccoli   *Brassica oleracea*, a variety of European brassica, a leafy plant grown in temperate regions as an ANNUAL, derived from the BIENNIAL or PERENNIAL wild CABBAGE, similar to cauliflower, but producing a loose terminal cluster of purple or white flower-heads, the leaves and flower-heads being cooked for eating. There is a more tender, Italian, variety of green sprouting broccoli (calabrese), liable to frost damage in Britain.

spruce beer   an alcoholic beverage made by fermenting a mixture of spruce twigs, leaves and sugar with yeast.

spruit   (Afrikaans) a small river or rivulet which may periodically dry up some time, but which is subject to sudden floods. The use of the term has spread from southern to eastern Africa. G.

spur   a marked projection of land from a mountain or a ridge. INTERLOCKING SPUR, NESS, TRUNCATED SPUR.

spurious correlation   a false CORRELATION in which there is no direct causal link (CAUSALITY) between two variables even though a relatively high correlation COEFFICIENT may be found. This kind of correlation occurs when one or more other variables underly both the variables falsely correlated. TEST FACTOR.

sputnik   (Russian) the first artificial satellite built in the USSR and sent into orbit by the Soviet Union in 1957.

squall   1. a sudden, violent gusty wind which lasts a minute or two and then subsides, usually accompanied by rain or hail 2. a storm characterized by a series of squalls, the gusts of wind being short-lived and blowing at speeds half as great again as the average wind speed. LINE SQUALL.

squash a member of the family Cucurbitaceae, a trailing or climbing herb, various varieties, some more tender than others, grown particularly in warm temperate regions of North America for its fruit, using in cooking, wine-making, or fed to livestock.

squatter 1. one who occupies (especially premises) without legal entitlement 2. in USA, one who occupies government land in order to gain legal title to it 3. in Australia, a large-scale sheep farmer, or one occupying a tract of grazing land as a crown tenant.

squatter settlement SHANTY-TOWN.

squattocracy (Australia) a class of wealthy landowners who obtained their land in the period before the Free Selection Acts, Australia. Comparable with the SQUIREARCHY in Britain.

squire historically in Britain, a landed proprietor, a country gentleman, especially one who was the main landowner in a district.

squirearchy historically in Britain 1. SQUIRES as a collective body 2. the class to which squires belonged, especially in relation to their social and political influence.

SSSI Site of Special Scientific Interest, in planning in Britain, an area of land judged by the Nature Conservancy to be of special interest on account of the fauna or flora it supports, or of its special geological or physiographical features.

stability the property of being STABLE *adj.*, of being in equilibrium.

stabilized dune a coastal DUNE fixed by vegetation and therefore usually not liable to DEFLATION-1. FORE-DUNE, MOBILE DUNE.

stable 1. a building in which horses (and, formerly, cattle) are housed 2. a group of horses kept in such a building 3. an establishment for training racehorses, or the proprietor(s) and staff of such an establishment 4. the racehorses which belong to a particular racing-stable.

stable *adj.* 1. staying or able to stay unchanged in form, structure, character etc. in conditions which would normally induce changes 2. not easily decomposing or changing 3. able to recover, to return to original condition after slight displacement 4. enduring, permanent. STABILITY.

stable equilibrium the state of the ATMOSPHERE-1 where the ENVIRONMENTAL LAPSE RATE of an air mass is less than the DRY ADIABATIC LAPSE RATE. If a pocket of air at the earth's surface is moved upwards it cools at the dry adiabatic lapse rate, becomes colder (thus denser) than the air around it, and sinks to its original level, conditions typical in an ANTICYCLONE. Contrast UNSTABLE EQUILIBRIUM.

stable population, stationary population a POPULATION-1 in which the proportion of members at each age is constant and the rate of annual growth or decline is fixed, due to the fact that there is no change in the chance of dying at a particular age or of a female giving birth at a particular age. If the annual rate of growth is zero, the population is termed stationary. REPRODUCTION RATE.

stabling 1. the act of putting horses in buildings where they can be tended and housed 3. the buildings so used.

stac (Gaelic) a mass of hard IGNEOUS ROCK, steep-sided, varying in height, especially such a mass forming an offshore island, e.g. St Kilda, northwest Scotland.

stack an isolated mass of rock near a coastline, detached from the main mass usually by marine erosion (especially by wave action), rising steeply from the surrounding sea. It represents the midway stage in the marine erosion cycle: CAVE-1, ARCH, STACK, STUMP, REEF-1.

stadial moraine a RECESSIONAL MORAINE, stadial in this case meaning, of or relating to a stage in development. INTERSTADIAL. G.

stage a point, level, period in progressive development or change **1.** the distance between two stopping places (staging posts) on a journey. STAGE-COACH **2.** in geology, the fourth rank in the division of stratified rocks, ranking below a SERIES, corresponding to the rank 'age' in chronological terms. GEOLOGICAL TIMESCALE **3.** in erosion cycle, the form reached by a geomorphic feature in its process of development, displaying characteristics assessed to be typical of a certain point **4.** in phytogeography, the point of development in a plant SUCCESSION-2, when there is a marked change in floristic composition or a noticeable extension of some species, the stages being distinguished as pioneer, transition, final. G.

stage-coach historically, a horse-drawn public vehicle carrying passengers and goods over a set route divided into STAGES, the stage being the distance which could be covered without stopping for a rest or change of horses.

stagnohumic gley soil GLEY SOILS.

stagnogley soils GLEY SOILS.

stagnopodzols PODZOLIC SOILS.

stainless steel steel alloyed with 12 to 20 per cent CHROMIUM, which makes it resistant to rust and unwanted marks.

staith, staithe (northern England) an elevated wharf with chutes for loading coal. G.

stalactite a cylindrical or conical deposit of mineral matter, hanging from an elevated point, formed by dripping water, i.e. a mass of CALCITE hanging from the roof of a limestone cave, formed from water containing dissolved CALCIUM BICARBONATE (CARBONATION) which has seeped through joints and crevices to drip very slowly from the cave roof. Some of the carbon dioxide in the water is released so that some of the dissolved calcium bicarbonate reverts to CALCIUM CARBONATE; and this process, combined with evaporation, leads to cumulative downward deposits of calcite with every drip of water. STALAGMITE.

stalagmite a mineral mass resemblilng a STALACTITE and formed by the same process, but markedly conical in shape and rising from the floor as a result of drops of water falling from the roof to the floor. It may be formed by water dripping from a STALACTITE in a limestone cave, leading to the eventual union of stalagmite and stalactite in a pillar connecting the roof to the floor.

stand 1. an aggregation of plants of roughly the same age, condition, species, composition, so as to make them distinguishable from neighbouring vegetation **2.** in forestry, a group of growing trees of a particular species, or the amount (commonly expressed as volume) of standing TIMBER-2 per unit area.

standard atmosphere the average condition of the ATMOSPHERE-1 in respect of its pressure and temperature at a selected altitude, used as a unit of measure in aviation in calibrating instruments, assessing aircraft performance etc. ATMOSPHERE-2, ATMOSPHERIC PRESSURE.

standard deviation in statistics, the square root of the VARIANCE, a widely used measure indicating the spread of values on each side of the MEAN in a FREQUENCY DISTRIBUTION. It is thus a measure of DISPERSION. If there is no variation between values in the distribution the standard deviation will be zero; it increases with the increase in variation. A relatively low standard deviation is associated with a close grouping around the MEAN; a relatively high standard deviation, a wide spread of values about the mean. COEFFICIENT OF VARIATION, MEAN DEVIATION, MEASUREMENT in statistics.

standard error in statistics, the STANDARD DEVIATION of a whole set of estimates. It owes its existence to the fact that if an infinite number of samples of the same size are drawn from a POPULATION-4 at random with replacement (SIMPLE RANDOM SAMPLE), not only the MEANS but also the standard deviations and the VARIANCES etc. of these SAMPLES-3 always vary. STANDARD ERROR OF THE MEAN.

standard error of the mean in statistics, the STANDARD DEVIATION of an infinite set of means of SAMPLES-3, all the samples being the same size and selected at random with replacement (SIMPLE RANDOM SAMPLE) from the same POPULATION-4.

standard hillslope a hillslope comprising four components (SLOPE ELEMENTS), stated by some authors to be the outcome of all types of slope process, independent of climatic influence. But a strong, massive bed rock is essential if all four components are to be supported; if it is weak only the WAXING and the WANING SLOPES may be formed.

standard man-day smd, a term applied in the agricultural statistics of the UK to eight hours' productive work by an adult male worker, under average conditions. MAN-HOUR, TYPE OF FARMING.

Standard Metropolitan Statistical Area SMSA, in USA, LOCAL GOVERNMENT IN USA, METROPOLITAN AREA.

standard of living 1. the conditions in which people live or would like to live 2. the conditions of living considered to be desirable as defined by national or international convention or agreement for a specific purpose (e.g. in order to establish a minimum wage, working hours etc.). SOCIAL WELL-BEING, QUALITY OF LIFE.

standard meridian STANDARD PARALLEL.

standard parallel a parallel of latitude selected for its appropriateness for making the necessary calculations and for drawing a particular map projection; or for acting as the horizontal axis of a grid-system. A meridian selected

for the same reason and purpose, but providing the vertical axis, is termed a standard meridian.

standard time the mean time of a meridian centrally located over a country or part (zone) of a country (TIME ZONE), and used for the whole of that area in order to avoid the inconvenience resulting from the use of LOCAL TIME. DAYLIGHT SAVING.

standing crop 1. in agriculture, a crop not yet cut 2. in ecology, the amount of living matter (BIOMASS-1) in a population of one or more species within a given area 3. the net amount of energy available from one TROPHIC LEVEL to the next higher level (FOOD CHAIN), e.g. plants provide the standing crop for HERBIVORES.

standing wave, stationary wave a wave form produced by the interaction of two wave motions (transverse or longitudinal) of identical amplitude, frequency and velocity, superposed and moving simultaneously through a medium in opposite directions, making a pattern of alternating points of no displacement (at nodes) and most displacement (at antinodes), the intermediate displacement between the two being a smooth curve. LEE WAVE, OSCILLATORY WAVE THEORY OF TIDES.

stannary 1. a tin mine, tin works 2. historically, in England, The Stannaries, the tin mining district of Devon and Cornwall with its smelting works, under the jurisdiction of special courts, Stannary Courts 3. the customs and privileges attached to the tin mines in The Stannaries.

stannic *adj.* of, relating to, containing, tin, especially of a compound including tin with a VALENCY of 4. STANNOUS.

stannite a natural sulphide of tin, copper and iron, colour grey to black, lustrous.

stannous *adj.* of, relating to, containing, tin, especially of a compound including tin with a VALENCY of 2. STANNIC.

staple **1.** the main commoditiy grown, produced, traded in, in a particular area **2.** a commodity in constant demand and regularly kept in stock **3.** a chief ingredient or constituent, e.g. the foodstuff forming the main constituent of the diet of the people of an area **4.** historically in Britain, fourteenth and fifteenth centuries, a town or place, especially a staple port, at home or abroad, designated a market by the monarch in which certain merchants (the Staplers) had the exclusive right of buying certain goods for export, duties being there collected. At various times these goods were wool, sheepskins with wool on, leather or hides, tin; at other times lead, cheese, butter, alum, tallow and, rarely, worsted were involved **5.** the merchant group having the exclusive purchase rights at those markets **6.** a measure of fineness, of length of a textile fibre, originally of wool, extended to linen, cotton etc.

star a hot, usually luminous celestial body, the energy probably derived from thermonuclear conversion of hydrogen to helium, thought to consist of varying amounts of most of the known elements in a much ionized state, countless numbers of stars being distributed in galaxies in the universe, e.g. the SUN in the GALAXY in which the EARTH lies. The magnitude, density and diameter of stars vary, but the mass of those so far measured lies between one-tenth and ten times the mass of the earth's sun.

star dune, star-dune a large, pyramidal, fairly permanent SAND DUNE, with a relatively high peak from which ridges radiate.

state **1.** condition with respect to growth, development, arrangement **2.** a self-governing group of people occupying a defined territory, or the territory thus occupied. BUFFER STATE, NATION STATE, SOVEREIGN STATE **3.**

the political organization forming the basis of civil government, the supreme civil power and government (sometimes with initial capital letter) **4.** in neo-Marxism, the political power centre of a society **5.** a unit of regional government with some independence in relation to internal affairs, forming part of a larger political unit, e.g. as in the USA, or the Commonwealth of Australia.

state capitalism an economic system (or the political doctrine advocating it) in which the STATE-3 owns or controls a major part of the economy, consistently supporting the interests of private capital, or, alternatively, regarding private capital as subordinate to the interests of the State and its own enterprises. In the latter case state capitalism is often related to CORPORATISM. State capitalism is usually equated with the MIXED ECONOMY now common in most western nations; but some authors apply the term to the economic systems now operating in the USSR and other eastern European countries. CAPITALISM, LIBERAL CAPITALISM, MARKET ECONOMY, STATE SOCIALISM.

state farming a form of agricultural organization in some countries with a centrally controlled economy, especially in the USSR (SOVKHOZ), in which the land is owned by the STATE-3 and the farm workers are employed, as wage-earners, by the state. They do not have a share in the produce of the farm (unlike the workers in COLLECTIVE FARMING). LAND TENURE.

state intervention activity by the STATE-3 in which it assumes responsibility for some of the processes usually associated with the market in a capitalist (CAPITALISM-1) or MARKET ECONOMY, e.g. in a MIXED ECONOMY.

state socialism a political doctrine or economic system based on the belief that social reform, improved social conditions, equalization of income and opportunity etc., accompanied by an economic role for the STATE-3 in controlling the market mechanism, can be accomplished peacefully through legislation by an existing state administration rather than through revolution. Thus economic and political power is centralized in the hands of the state, and the market is dominated by the state. STATE CAPITALISM.

static *adj.* **1.** at rest, not moving in relation to the earth **2.** unchanging **3.** in equilibrium. The opposite of DYNAMIC.

static lapse rate synonymous with ENVIRONMENTAL LAPSE RATE.

static rejuvenation REJUVENATION of a river caused by an increase in its eroding capability, not by a lowering in BASE-LEVEL, e.g. occurring when increased rainfall leads to increased RUNOFF; or the volume of a stream increases owing to RIVER CAPTURE; or there is a decrease in the load carried by a stream, all of which increase the eroding power of the drainage system. DYNAMIC REJUVENATION.

station **1.** in biology, a geographical location in which an organism or community occurs **2.** a place with its associated buildings on a route, where buses, trains etc. habitually stop to take on and discharge passengers, goods, mail **3.** a place with buildings and equipment for the transmission and reception of radio, television signals, etc. **4.** in Australia, a sheep or cattle run with associated buildings.

stationary population REPRODUCTIVE RATE, STABLE POPULATION.

stationary wave STANDING WAVE.

statistic a number which summarizes and describes a particular aspect of a set of observations.

statistical explanation an EXPLANATION in which the truth of the conclusion cannot be inferred logically (as in logical explanation), but which can be inferred as being more probable than false. Thus statistical explanation expresses, given two sets of observations, the degree to which one set may be predicted from the other. There may not be any logical connexion between the VARIABLES and it is usually impossible to infer the direction of causation.

statistical inference the use of appropriate statistical techniques to make estimates or predictions or to draw conclusions from a set of data. Statistical inference is particularly concerned with the estimation of population PARAMETERS from samples. INFERENTIAL STATISTICS.

statistical test an accepted method of deciding from the given data whether to retain the initial hypothesis (i.e. the NULL HYPOTHESIS) or to set it aside in favour of a specified ALTERNATIVE HYPOTHESIS, e.g. CHI-SQUARED TEST, MOOD'S MEDIAN TEST, STUDENT'S T-TEST, WILCOXON MATCHED PAIRS TEST, Z-TEST.

statistics **1.** the branch of mathematics concerned with the collection, analysis and interpretation of numerical data, deductions being made on the assumption that the relationships between a sufficient sample of numerical data are characteristic of those between all such data **2.** the numerical data relating to an AGGREGATE-1 of individuals **3.** the methods used in the collection, processing or interpretation of quantitative data. DESCRIPTIVE STATISTICS, INFERENTIAL STATISTICS, STATISTICAL EXPLANATION, STATISTICAL INFERENCE.

status SOCIAL STATUS.

statute **1.** the written or printed record of STATUTE LAW or part of it **2.** a decree or order of some chartered body or coporation etc.

Statute Law the body of law passed by a legislative body (e.g. a parliament) and recorded in a written or printed form. COMMON LAW.

Staublawine (German) an AVALANCHE of fine, powdery snow. SNOW-AVALANCHE. G.

steady *adj.* remaining at a certain position, posture, quality or value, but with an ability to vary.

steady state an OPEN SYSTEM in which external and internal relationships produce equilibrium, or internal balance. DYNAMIC EQUILIBRIUM. G.

steady time in geomorphology, a period of short duration during which small features such as hillslopes, river channels etc. form; one of the categories in the GEOMORPHOLOGICAL TIME-SCALE devised by Schumm and Lichty.

steam coal HARD COAL, carbon content between that of BITUMINOUS COAL and ANTHRACITE, used especially in steam boilers.

steamer 1. something driven by steam, usually applied to a steamship 2. a container in which things are steamed, e.g. a cooking utensil.

steam fog a FOG resulting from the passing of cold air over the surface of warmer freshwater, the moisture from the latter condensing into tiny visible droplets in the air so that its surface seems to steam. In very low temperatures the droplets are converted immediately to ice particles and form ICE FOG.

steam point the point at which pure water at sea-level boils at a standard pressure of ATMOSPHERE-2.

stearic acid a white, crystalline fatty acid obtained from animal and hard vegetables fats, used in making soap, candles etc.

steatite SOAPSTONE, a variety of TALC.

steel any of the many alloys of iron and 0.1 to 1.5 per cent carbon in the form of iron carbide, especially cementite, often alloyed with other metals if a special steel with a particular property is needed. In its solid state steel is hard, with great tensile strength, and it can be cast, rolled, drawn. It is used in construction work, all kinds of machinery, installations, equipment, vehicles, domestic goods etc. It is made by reducing the carbon in cast iron (PIG IRON) or by the diffusion of carbon into WROUGHT IRON. The methods formerly used in steel making (BASIC BESSEMER PROCESS, BESSEMER PROCESS, OPEN HEARTH PROCESS) have now been largely superseded by the use of the electric furnace in large integrated steelworks (ARC FURNACE, BLAST FURNACE, IRON, PIG IRON). Converters lined with dolomite bricks in the basic Bessemer and basic open hearth processes produce basic, or mild, steel; acid steel is produced from non-phosphoric ores processed in converters lined with silica bricks. Very full accounts of the history of the iron and steel industry, and the methods employed, appear in L. D.Stamp and S. H. Beaver. *The British Isles: A Geographic and Economic Survey*, 6 edn, Longman, 1971.

Some of the special steels and the metals used in alloys are chrome steel (about 2 per cent chromium, 0.8 per cent carbon) which, when cooled, is not only very hard but very elastic; high-speed steel (about 5 to 6 per cent chromium, about 18 per cent tungsten, under 1 per cent carbon), remains hard at temperatures even up to 400°C (760°F), used in the manufacture of turning lathes, a process in which great friction develops at high temperatures; stainless steel (13 per cent chromium, and other metals acording to type needed), non-toxic and resistant to corrosion; manganese steel (12 to 14 per cent manganese, 1.5 per cent carbon), great tenacity, extremely and irreducibly hard, needs special cutting tools, expensive; nickel steel (3 to 3.5 per cent nickel, about 0.25 per cent carbon), tough, strong, ductile. Vanadium combined with chromium in the alloy produces lightness, hardness, resistance to shock, useful in the manufacture of motor vehicles; steel with nickel and chromium is used in aircraft production.

steel-collar worker a robot. BLUE-COLLAR WORKER, WHITE-COLLAR WORKER, PROFESSIONAL.

steer a young castrated BULL. BULLOCK.

steering in meteorology, the role of an atmosperic phenomenon in causing the movement of another atmospheric phenomenon.

Steilwand (German) GRAVITY SLOPE, a term used by Penck. G.

stellar *adj.* of, relating to, having the nature of, a STAR.

stelliform *adj.* star-shaped.

stellular *adj.* having the shape of a small star.

stenohaline *adj.* applied to **1.** a marine organism tolerant of only a small range of salinity, a characteristic typical of those living in the PELAGIC ZONE **2.** a plant tolerant of only a small range in osmotic pressure (OSMOSIS) of soil water.

stenothermic, stenothermous *adj.* applied to an organism tolerant of only a small range of temperature. EURYTHERMIC, STENOHALINE. G.

stenotopic *adj.* applied to an organism restricted in its distribution. EURYTOPIC.

stentorg (Swedish sten, stone or boulder; torg, square or market place) a well-defined stone or boulder field, mainly on the crest of an ESKER, but sometimes extending down its flanks, sometimes exhibiting former SHORELINES and BEACH RIDGES. G.

step faults a series of parallel faults each with a THROW that projects in the same direction but to a greater distance than the one above, thus producing step-like changes of level of strata. Rift valleys are often bounded by such a series. G.

steppe (Russian step', pl. stepi) the treeless midlatitude grassland stretching from central Europe to southern Siberia in Asia. It is similar to midlatitude grassland elsewhere (e.g. PAMPAS, PRAIRIE) but if the term is applied to these it is best restricted to their drier parts. G.

Steppenheide (German, steppe-heath) a plant association defined by a German geographer, R. Gradman, at the end of the nineteenth century, as consisting of herbaceous plants commonly found in the STEPPES of southern Russia and some parts of the gravel wastelands of Bavaria (there termed Heiden), mixed with scrub and poorly developed trees (heather being absent), occurring on sunny sites in relatively dry conditions, on CALCAREOUS soils that have been neither worked nor manured. HEIDE. G.

steptoe KIPUKA, an island-like mass of rock in a sea of lava, occurring where a very fluid lava flow has encircled mountain spurs, thereby isolating the summits and converting them to islands in a lava sea. G.

stereoscope a binocular optical instrument designed to give photographs a three-dimensional character, the simplest comprising two lenses mounted horizontally in a frame with legs some 20 cm (8 in) in height, the distance between the eyepieces being approximately the same as the distance between human eyes. Two photographs of the same area but taken from slightly different angles are viewed through the lenses, with the result that each photograph is viewed by only one eye, giving the impression of a three-dimensional view. Stereoscopes, some very complex, are used especially in the interpretation of aerial photographs.

stereotaxis THIGMOTAXIS, TAXIS.

sterile *adj.* as applied to land, unproductive.

Stevenson screen a standardized METEOROLOGICAL SCREEN, designed by Thomas Stevenson, an engineer.

stigmatizing effect LABELLING.

stillstand, still-stand a period of time when the BASE-LEVEL does not vary, i.e. the level of the sea in relation to the land is stationary, almost unchanging. G.

stint (northern England gait or gate) the number and kinds of animals which a holder of COMMON RIGHT is entitled to put on a common. CATLE GATE, SOUM. G.

stitchmeal, stitch-meal historical, land in separated pieces (stitches) in a common arable field, wide balks of earth and stones separating the cultivated strips. SELION. G.

stochastic *adj.* pertaining to chance or conjecture; in mathematics, random. SYSTEMATIC-4.

stochastic model in statistical analysis, a MODEL-4 concerned with PROBABILITY, containing an element of randomness (as opposed to a DETERMINISTIC MODEL).

stock 1. an accumulating of goods held for future use or maintained as a constant source of supply 2. a RESOURCE 3. a group of plants or animals having the same line of descent 4. LIVESTOCK 5. the stem of a tree or bush used in grafting, the graft being inserted in a cut made in it; or a plant from which cuttings are taken 6. in geology, a small BATHOLITH, a discordant igneous intrusion, defined by some authors as having an upper surface area, when exposed by denudation, of only a few square km, i.e. under 100 sq km (38 sq mi). Termed a BOSS if in cross-section the exposed surface is more or less circular. G.

stock farming LIVESTOCK FARMING.

stock resources NATURAL RESOURCES.

stoma pl. stomata 1. a tiny pore in the epidermis (the outermost layer of cells) of a plant, present in large numbers, especially in leaves, through which gaseous exchange (WATER VAPOUR, OXYGEN, CARBON DIOXIDE) occurs. Each stoma has surrounding it two bean-shaped guard cells which change shape according to the absorption or loss of water. These alterations in shape control the size of the opening of the stoma. WILT.

stone 1. rock, hard mineral matter (other than metal) 2. a piece of rock of imprecise size, but larger than a particle of coarse sand, smaller than a boulder. GRADED SEDIMENTS 3. a small piece of rare or ornamental mineral cut and polished to display its properties to best advantage. GEM, GEMSTONE, PRECIOUS STONE, SEMI-PRECIOUS STONE 4. in botany, the hard case of the kernel in a DRUPE 3. a British unit of weight, equal to 14 lb (POUND) AVOIRDUPOIS or 6.350 kg (KILOGRAM). CONVERSION TABLES

Stone Age the period generally defined as extending from the beginning of the PLEISTOCENE (GEOLOGICAL TIMESCALE) to the beginning of the BRONZE AGE, when stone, bone or wooden (not metal) tools and weapons were used, divided into the culture periods EOLITHIC, PALAEOLITHIC, MESOLITHIC, NEOLITHIC. The term is sometimes applied loosely to the very few present-day groups of people who live by hunting and gathering.

stone fruit DRUPE.

stone polygon, stone stripe PATTERNED GROUND. G.

stone river a belt of boulders following the direction of flow of a shallow stream in arctic regions.

stoniness in SOIL CLASSIFICATION, the extent to which stones occur in a soil or on its surface.

stony iron a METEORITE of nickel-iron and silicate minerals.

stony rises (Australia: Victoria) ridges separated by valleys in an extensive stretch of lava, resulting from variations in the flow of the MAGMA before it cooled and became consolidated. G.

stop-and-go determinism a term introduced by T. Griffith Taylor, who maintained that in the progress of its economic development it is wise for a country to follow the directions indicated by natural environmental factors, but that the rate of use of the opportunities offered by the ENVIRONMENT may be controlled (i.e. speeded up, slowed down, or even stopped, according to circumstances) without ill-effect. DETERMINISM. G.

stope in mining, a cavern, the underground excavation formed when ore is mined by driving a horizontal tunnel along a steeply inclined lode (a method termed stoping), the ore being extracted from above and below the tunnel, resulting in the stope.

stoping 1. MAGMATIC STOPING 2. a mining method. STOPE.

storage capacity in soil science, FIELD CAPACITY, WILTING POINT.

store cattle cattle bought and kept for fattening and, as fat cattle, destined for the butcher. G.

storm 1. any violent disturbance of the ATMOSPHERE-1 and the effects associated with it, e.g. SANDSTORM, THUNDERSTORM 2. a gale-force wind. BEAUFORT SCALE.

storm-beach a deposit of coarse beach material, including COBBLES and BOULDERS, thrown high up on the shore (usually farther inland than the level reached by a high SPRING TIDE) by unusually strong waves in a storm.

storm-surge an unusual, rapid rise in tide level, above normal heights, caused by atmospheric factors, especially by the passage of an intense atmospheric DEPRESSION-3 producing gale-force (BEAUFORT SCALE) onshore winds.

storthe (England: Yorkshire dialect) a small wood in east Yorkshire, England.

stoss (German) the direction from which ice has come, thus the side facing the direction of flow of oncoming ice, as opposed to LEE, the side facing in the direction of the flow; hence stoss-and-lee relief. STOSS END.

stoss-and-lee relief a relief feature characteristic of a glaciated region, where small hills are markedly assymetric, their gentle, smooth slopes (STOSS, STOSS END) having been abraded by the oncoming ice, their steeper, rougher

slopes on the LEE side having been subjected to PLUCKING-1.

stoss end, stossend (German and English) the side of a prominent crag, hill or knob of rock facing the oncoming movement of an ice sheet or glacier, scratched (STRIAE) by the ice. CRAG AND TAIL, STOSS-AND-LEE RELIEF. G.

stow (Old English; obsolete) a place (many applications). The term was resuscitated by 1. J. F. Unstead, 1933, who applied it to the smallest unit of area, a region of the first or lowest order in a geographical study (stows can be combined to form TRACTS-3) 2. D. L. Linton, 1951, who applied the term to the second order (following the smallest, the SITE-3) in his hierarchy of a MORPHOLOGICAL REGION. G.

straate (Afrikaans, streets; Hottentot kivas) troughs between DUNES southwest of the Kalahari, in some cases floored with a clayey-sand. GRADED SEDIMENTS. G.

Strahler ordering STREAM ORDER.

strain 1. in biology, a sub-species not forming a VARIETY-2 because its members do not differ enough genetically from the rest of the SPECIES-1 (CLASSIFICATION OF ORGANISMS). Breeding occurs freely within the sub-species, and is possible (but does not occur so readily) between members of different sub-species of the same species; so reproductive isolation is incomplete and sub-species tend to grade into each other 2. in physics, the deformation of a body as a result of STRESS, termed homogeneous if the deformation is equal in all directions, heterogeneous if otherwise. Strain is measured by the ratio of the dimensional change produced to the original dimension (in linear measure, area or volume), e.g. the ratio of the change in area to the original area.

strait, straits a narrow passage of water connecting two larger bodies of water. CURRENT-4,5.

strand (English and German) shore or beach, a term used poetically, or only in certain parts of England. The German term is far more comprehensive, so some authors tend to use it in the German sense, in such combinations as strand-line (RAISED BEACH), strand-wall (SHINGLE beach, BAR-1).

strandflat (Norwegian) a wave:cut platform off the rocky coast of west and northeast Norway, exceptionally wide (locally up to 37 mi: 50 km), raised above the sea-level by isostatic uplift (ISOSTASY). ABRASION PLATFORM. G.

Strassendorf (German, street village) a village in which the houses line each side of a main street. G.

strata STRATUM.

strath (Scottish, from Gaelic; also Northern Ireland) a wide valley with a flat floor and a river meandering through an alluvial flat, where the valley is becoming aggraded (AGGRADATION). In Scotland it contrasts with GLEN, which is narrower or smaller and lacks the broad flat floor characteristic of the strath. G.

stratification (*verb* to stratify, to arrange or form in strata, or to become arranged in strata) formation into layers 1. in geology, the accumulation of SEDIMENTARY ROCKS and of some IGNEOUS ROCKS in layers or STRATA; the condition or manner of being stratified; the arrangement in strata (LAMINATION) 2. in meteorology, the formation of stable horizontal layers in the ATMOSPHERE-1, occurring when the LAPSE RATE is less than the ADIABATIC LAPSE RATE 3. in statistics, the division of a POPULATION-4 into layers or strata appropriate to the question at issue in order to be able to draw a representative RANDOM SAMPLE from each layer, thereby increasing the precision of the sample without increasing its size 4. in oceanography and hydrology, THERMAL STRATIFICATION.

stratified arranged or formed in

STRATA, in layers; the antonym is unstratified.

stratified society a SOCIETY-3 in which there are obvious social levels, identifiable by marked differences in attitude and inequality of access to scarce resources. Once defined as scarce by the society, these resources acquire a market value, and a MARKET ECONOMY develops which creates wealth in the society as a whole (but some would add depends on the maintenance of scarcity), and large-scale URBANIZATION becomes possible. CASTE-1, EGALITARIAN SOCIETY, PEASANT SOCIETY, RANK SOCIETY.

stratiform *adj.* having a stratified arrangement. STRATIFICATION 2. having the form of STRATUS, applied to all sheet CLOUDS, including ALTOSTRATUS, CIRROSTRATUS, NIMBOSTRATUS, STRATOCUMULS, STRATUS.

stratigraphical geology see STRATIGRAPHY. G.

stratigraphy a branch of GEOLOGY, sometimes termed historical geology, concerned with the study of the occurrence, lithology, composition, sequence, fossils and correlation of rock STRATA, and especially with the chronological order of succession of rock formations, by which historical changes in the geography of the earth can be traced. Studies concerned mainly with differentiation of strata on the basis of chronological order are termed chronostratigraphy; on the basis of fossil assemblages in the strata, biostratigraphy; and on the basis of lithology and petrography, lithostratigraphy. G.

stratocumulus a continuous extensive sheet of dark grey, heavy CLOUD, consisting of circular contiguous mounds, usually fairly low, but sometimes reaching an altitude of 2400 m (8000 ft). It usually occurs in the northern hemisphere in winter, and does not bring rain.

stratopause the layer in the earth's ATMOSPHERE-1 lying at a varying altitude of some 56 km (35 mi), separating the STRATOSPHERE (below) from the MESOSPHERE (above) and within which the rising temperature reaches some 80°C (176°F).

stratosphere the layer in the atmosphere between the TROPOPAUSE and the STRATOPAUSE, extending upwards from an altitude of some 11 km (about 7 mi), depending on latitude and season and condition of the weather in the TROPOSPHERE. It is a generally tranquil zone in which, at the lower level (sometimes termed the isothermal region), the temperature is relatively constant, moisture content is low, clouds do not form, and large convection currents are absent, ideal flying conditions for aircraft. At greater altitude, nearing the top of the stratosphere, where OZONE absorbs shortwave SOLAR RADIATION, temperatures resemble those on earth; but approaching the STRATOPAUSE the temperature begins to rise. G.

strato-volcanic cone (American) a COMPOSITE VOLCANIC CONE.

stratum (Latin, something spread or laid down) pl. strata, in geology, a generally distinct, roughly horizontal individual layer of homogeneous material, its surfaces parallel to layers of different material lying above and below. The term is usually restricted to a bed or layer of SEDIMENTARY ROCK or to a layer of pyroclastic material (PYROCLAST). GEOLOGICAL INVERSION, LAW OF SUPERPOSITION, STRATIFICATION, STRATIGRAPHY.

stratus a continuous extensive sheet of grey uniform CLOUD, often persistent, usually low-lying, but occurring at any altitude up to 2400 m (8000 ft), sometimes bringing fine drizzle, but never heavier rainfall. It is termed fractostratus (FRACTUS) if it is fragmented to form irregular patches. STRATIFORM.

straw the stem of certain CEREAL crops remaining after cutting and threshing.

strawberry a member of the genus *Fragaria*, a PERENNIAL herb, native to midlatitudes, having a leafy crown from which prostrate runners grow, bearing fruit, an enlarged, edible, sweet, red, fleshy receptacle in which the seeds are embedded. Many cultivated varieties enter commerce, the fruit being eaten fresh, or canned, frozen, made into jam.

stray historically, the right of allowing cattle to stray and graze COMMON LAND, an obsolete term, but retained in names of sections of common land in northern England, e.g. The Stray at Harrogate; or The Strays of York. G.

stream 1. a body of flowing water, permanent or intermittent, from the smallest to the largest, on land, underground, or in the ocean or sea, e.g. Gulf Stream. BED LOAD, DISSOLVED LOAD, INTERCISION, RILL, RIVER (and entries qualified by river), RIVULET, SUSPENDED LOAD 2. the flow or current of a FLUID, or the direction of that flow 3. a quantity of something fluid in motion 4. a linked succession of events. G.

streamlet a small STREAM-1.

streamline 1. the direction of smooth movement within a GAS or LIQUID past a solid body 2. the direction of movement of all parcels of air in an area, measured individually and simultaneously, providing an immediate general picture of air motion; compare TRAJECTORY 3. the shape given to a solid body to enable it to move with the least resistance through a fluid.

stream order a hierarchical classification of streams (suggested by A. N. Strahler, 1957, modified from R. E. Horton) based on the magnitude of their channels and position in a DRAINAGE AREA, the outermost tributaries being designated fingertip tributaries or first order streams, two first order streams uniting to form a second order stream, two second order streams joining to form a third order, and so on until the main river or trunk stream, opening to the mouth, is reached. At least two streams of any given order are needed to form a stream of the next higher order. A more recent classification, proposed by R. L. Shreve, is simpler, using only first order links as an index of magnitude. ALLOMETRIC GROWTH, BIFURCATION RATIO, HORTON'S LAW. G.

stream stage the height of the surface of a stream at any particular point in time. BANKFULL STAGE, FLOOD STAGE, OVERBANK STAGE.

stream tin CASSITERITE occurring as detrital particles in alluvial deposits. PLACER.

street 1. a road in a village or town, usually with a hard surface, with drainage and artificial lighting, lined on each side with buildings 2. that road and the associated buildings 3. in hot deserts, a gap of bare desert floor between chains of dunes. STRAATE. G.

street village STRASSENDORF. G.

strength theory of MIGRATION-1,2, a theory which suggests that people move when they are in a position of economic strength, not weakness, i.e. conditions are satisfactory in the place where they live, but conditions in another (e.g. higher wages, higher social status, more pleasing environment) seem more desirable. PUSH-PULL THEORY, WAGE DIFFERENTIAL THEORY.

stress 1. in physics, the force acting on an object, expressed as force per unit area, measured in newtons per metre squared (i.e. the force needed to give a mass of 1 kg an acceleration of 1 m per second per second, symbol N), calculated by dividing the total force by the area to which it is applied. When stress is applied to a body, e.g. a rock, it produces STRAIN, and the body can be distorted or deformed, according to its ELASTICITY. If two forces press towards each other COMPRESSION or THRUST results; if they pull apart there is TENSION; and if they act in parallel the result is shearing stress (SHEAR) **2.**

the state produced by that force **3.** the force exerted by environmental factors on the nervous system of an animal or human being, or the state produced by that force.

striae, striation scratches and narrow grooves on the surface of ice-worn rocks. They are cut as a glacier moves along, overriding boulders and bed rock in its path, by the rocks and the small fragments of rock that are frozen into its underside. Striae therefore indicate the direction of movement of the ice. Similar scratches and grooves on rocks may be the result of other movements, e.g. soil creep, so the distinction 'glacial striae' is sometimes made. G.

strike the direction of a horizontal line on an inclined rock stratum at right angles to the direction of the TRUE DIP of the rocks. Strike is applied as an *adj.* to features roughly parallel to the strike, e.g. strike fault, strike joint, strike valley. G.

strike fault a FAULT of which the STRIKE is parallel to the STRIKE of the strata affected.

strike-slip fault, strike-slip faulting tear fault, a SLIP-1 or FAULT characterized by movement transverse to the STRIKE of the folded strata. G.

strike valley a valley of which the direction is parallel to the regional STRIKE of the strata in an area, e.g. the valley of a subsequent stream in TRELLIS DRAINAGE.

strip 1. in general, a long narrow piece of generally uniform width 2. of ice, a long narrow stretch of PACK ICE, under 1 km (0.6 mi) wide, consisting of small fragments detached from ice masses and pushed together by wind, swell or current. G.

strip cultivation 1. the system of dividing a large field into STRIPS-1, each worked by a separate owner or occupant (STRIP FIELD), as in the FIELD SYSTEM in medieval England and other parts of western Europe 2. the growing of different crops in contiguous strips along the contours of a hillside, to counteract soil erosion.

striped farm   LADDER FARM. G.

striped ground   PATTERNED GROUND.

strip field   a field cultivated in long narrow strips, usually each by a separate owner or occupant. COMMON FIELD, FIELD SYSTEM, SELION. G.

strip map   a map showing only a narrow band of country in which the user is interested, e.g. on each side of a route.

strip mining   (American) OPENCAST MINING or QUARRYING. AREA STRIP MINING. G.

strip of ice   STRIP-2.

Strombolian eruption   a volcanic eruption characteristic of Stromboli, a VOLCANO in the Lipari Islands off the coast of Italy. Gases can readily escape from the molten lava in the crater, so pressure does not build up. Instead the volcano ejects incandescent dust (VOLCANIC DUST), SCORIA and bombs (VOLCANIC BOMB) with a little water vapour fairly frequently, the lava being somewhat less BASIC-1 than that typical of the HAWAIIAN VOLCANIC ERUPTION (BASIC LAVA). This is one of the four types of volcanic activity commonly distinguished, the others being Hawaiian, VULCANIAN (or VESUVIAN), and PELEAN. G.

strontium   a metallic element, the oxide of which is used in sugar refining. Its radioactive ISOTOPE, strontium 90, released by the explosion of a hydrogen bomb, is a health hazard.

structural geology   a subdivision of PHYSICAL GEOLOGY, concerned with the scientific study of the form, origins and spatial distribution of rock structures, as distinct from DYNAMIC GEOLOGY. G.

structuralism   a movement, an approach to knowledge concerned not so much with the apparent 'surface' STRUCTURES-1,2 of a subject but with the deep structures (e.g. the social

and ideological values, the rules, conventions and restraints etc.) which underlie and generate the phenomena being observed. STRUCTURALIST APPROACH.

structuralist approach   an analytical approach to knowledge which looks beneath the surface of a subject to try to find the logic which is presumed to bind the STRUCTURES-1,2 that are presumed to form it. STRUCTURALISM.

structural landscape   a landscape adjusted to tectonic STRUCTURE-3, a TECTONIC landscape. TECTOSEQUENT. G.

structuration   in HUMAN GEOGRAPHY, the interaction of SOCIAL STRUCTURES and HUMAN AGENCY, whereby the social structures are reproduced, reinforced and reasserted and continue through time as a result of the conscious or subconscious actions of people in conforming to the rules, constraints and conventions of the social system; or are modified deliberately (e.g. by government policy) or unintentionally (e.g. by an amalgam of unofficial actions). STRUCTURATIONIST APPROACH.

structurationist approach   an analytical approach in HUMAN GEOGRAPHY which uses the concept of STRUCTURATION as a basis for explanation, with the intention of setting HUMAN AGENCY and structural influences in an appropriate context.

structure   1. in general, something made of parts fitted or joined together 2. the way in which the constituent parts (of something) are fitted or joined together or arranged in a way that gives that thing its peculiar character. SOCIAL STRUCTURE, STRUCTURALISM 3. in geology, geological structure, most commonly (and best) applied to the arrangement and disposition of the rocks in the earth's crust, as a result of (or the absence of) earth movements; but also applied to the morphological features (MORPHOLOGY) of rocks, e.g.

columnar structure. As used by W. M. Davis in his explanatory descriptions of landforms (structure, process, stage), structure 'indicates the product of all construction agencies' ... including 'the nature of the material, its mode of aggradation, and even the form before the work of erosive agencies begins', i.e. it applies to 'that on which erosive agents are and have been at work' **3.** in ecology, the spatial and other arangement of species within an ECOSYSTEM.

struga (Slavic) a passage formed along a BEDDING PLANE in KARST. BOGAZ. G.

stubble the short stalk of CEREAL crops left standing in the ground after reaping.

stud 1. a group of horses kept for breeding, or the place where they are kept **2.** any male animal, e.g. a stud-horse, kept for breeding.

studbook an official register of pedigrees of pure-bred animals, e.g. cattle, horses, sheep.

Student's _t_-test (Student, nom de plume of W. S. Gosset) in statistics, an hypothesis test used to analyse one sample of data in order to compare a population mean with a particular value; or to analyse two paired or matched samples of data (PAIRING) in order to compare two population means; or to analyse two unrelated samples of data in order to compare two population means. The samples of data must be small, and the sample data on INTERVAL SCALE; the population distribution(s) is/are assumed to be normally distributed (NORMAL DISTRIBUTION). The two-sample _t_-test should be used only if it can be assumed that the two population STANDARD DEVIATIONS are approximately equal. For the matched pairs _t_-test the two samples must not only be paired: it should be possible to assume that the population distribution of the differences between the matched pairs is normal. The

STANDARD ERROR OF THE MEAN is used to find out if two samples are truly representative of the original POPULATION-4, if the means of two samples differ so much that the samples are unlikely to be drawn from the same population, or if the difference between the sample means is a chance or random occurrence; _t_ is given as the ratio of the difference between the sample means to the STANDARD ERROR of the difference between the sample means. The value of the calculation can then be checked in a table of _t_ values to assess the probability of its being a chance occurrence.

stud farm an establishment where horses are bred.

stump a worn-down STACK, the penultimate stage in the cycle of marine erosion of part of a coastline.

sty 1. PIGSTY **2.** in Cumbria, northwest England, the steep side of a hill.

sub- extensively used as a prefix in geographical terms, with or without a hyphen, mainly in the sense of underneath, below, inferior to, bordering on, near the base of, to a lesser degree (i.e. not completely, only partly), a smaller division, or less important part. Only a few of the geographical terms incorporating sub- appear below.

subaerial _adj._ occurring, forming or operating on the surface of the EARTH-1 (as distinct from SUBAQUEOUS, under the water; or submarine, under the sea; or subterranean, under the surface of the ground). It is applied particularly to atmospheric weathering, or to the deposits carried by the wind and laid down on the land, etc. G.

subalpine _adj._ of or relating to the lower Alpine slopes, or to the higher mountain slopes just below the TREE LINE.

subantarctic, Sub-Antarctic _adj._ of or relating to the region in latitudes near or just north of the ANTARCTIC CIRCLE and the phenomena associated with it.

subaquatic *adj.* **1.** partly AQUATIC **2.** of or relating to plants growing under water.

subaqueous *adj.* existing under the surface of the water or formed under the water **2.** equipment made to be used under water. SUBAERIAL.

subarctic, Sub-Arctic *adj.* **1.** the region in latitudes near or just south of the ARCTIC CIRCLE, and the phenomena associated with it **2.** a group of cold climatic types in KOPPEN'S CLIMATIC CLASSIFICATION, with low precipitation and evaporation, the mean temperature of the warmest month sometimes exceeding 10°C (50°F), of the coldest falling below −3°C (26.6°F).

Sub-Atlantic the climatic phase in which we live, dating from about 500 BC (preceded by the SUB-BOREAL), when summer temperatures became lower than those of the Sub-Boreal, but conditions generally became more mild and humid. In Britain the lime forests declined, but the alder, oak, elm, birch, hornbeam and beech flora (with the latter usually dominant in southern areas) spread. PRE-BOREAL.

subatomic *adj.* of or relating to particles smaller than an atom.

Sub-Boreal the climatic period dating from about 3000 BC to 500 BC, when conditions became cooler, reverting to the conditions prevailing before 5500 BC, but being generally drier, the dominant tree of the Atlantic stage, the oak, giving way slowly to ash, birch and pine. PRE-BOREAL.

subclimax the stage immediately preceding the CLIMAX in a PLANT SUCCESSION. The subclimax is the stage at which, under the influence of an arresting factor or factors, a plant community may become stabilized, proceeding to the climax only if the arresting influence is removed. DISCLIMAX, PLAGIOCLIMAX, PROCLIMAX.

subcloud layer in general, all the air between the CLOUD base and the earth's surface; or more specifically **1.**

the layer of air immediately below the cloud base **2.** shallow stable air immediately under a convection cloud.

sub-consequent stream a SECONDARY CONSEQUENT STREAM.

subcontinent 1. a large land mass forming part of a continent, and having a certain geographical entity, e.g. the Indian subcontinent **2.** a very large land mass smaller than one usually termed a continent, e.g. Greenland.

subculture the attitudes, beliefs, values and behavioural habits shared by a group of people within a SOCIETY-2,3 which differ from those of the accepted surrounding culture held to be characteristic of the society as a whole. FOLK CULTURE.

subdominant plant species which seem to be more dominant than the true DOMINANT in a CLIMAX at a particular time of the year.

subduction zone ANDESITE, DEEP, OCEANIC TRENCH, PLATE TECTONICS.

suberin a compound of fatty substances which constitutes the basic part of the cell walls of CORK, makes it waterproof, and yields suberic and other acids.

suberization in botany, the process in which, by the infiltration of SUBERIN, the cell walls of plants become cork-like tissue, impervious to water.

subglacial *adj.* of, relating to, formed in or by the underside of a GLACIER. ENGLACIAL, SUPERGLACIAL.

subglacial channel a MELT-WATER CHANNEL formed by melt-water flowing beneath an ice sheet or glacier, and now usually dry or carrying only a small stream.

subglacial moraine debris (MORAINE) frozen into and carried by the ice of a GLACIER at or near its base.

subhumid *adj.* applied to a climate in relation to its PRECIPITATION which (combined with other climatic and physical factors) is neither too little, resulting in arid conditions with the growth of XEROPHYTES, nor adequate in amount and evenness of seasonal fall for tree growth. Such a climate leads to the growth of tall grass. SAVANNA.

subjective *adj.* seen from the viewpoint of the thinking subject, and therefore conditioned by personal characteristics and feelings. HUMANISTIC GEOGRAPHY, OBJECTIVE.

sublimation a process whereby a SOLID changes directly to a VAPOUR without passing through a LIQUID state; or, similarly, a vapour is changed directly to a solid (e.g. the formation of ice-crystals directly from water vapour when condensation occurs at a temperature lower than that of FREEZING-POINT). LATENT HEAT.

sublittoral zone 1. the marine zone extending from low tide level to the edge of the CONTINENTAL SHELF, part of the NERITIC ZONE underneath the LITTORAL ZONE **2.** in a lake, the part in which the water is too deep for the growth of rooted plants.

submarine *adj.* existing, occurring, operating, under the sea. SUBAERIAL.

submarine canyon a steep-sided valley in the CONTINENTAL SHELF which in some cases may extend right across the shelf into the CONTINENTAL SLOPE. The origin of these canyons is obscure and has provoked much discussion; some are obviously continuations of valleys of the adjoining land. G.

submarine ridge OCEANIC RIDGE.

submerged coast, submerged shoreline a coastline formed when a rise of sea-level in relation to the land leads to the flooding of the former land surface. BODDEN, DALMATIAN COAST, FJARD, FJORD, FOHRDE, RIA, SUBMERGED FOREST.

submerged forest the remains of a former forest, now completely covered by the sea except occasionally at very low tide, resulting from the submergence of the COAST caused either by a rise of sea level relative to the land (EUSTATISM) or by a lowering of the land surface (ISOSTASY). SUBMERGED COAST. G.

submerged sill FJORD, SILL.

subsample in statistics, a sample of a SAMPLE-3, which may or may not be selected by the method used in selecting the original sample.

subscript a diacritic (i.e. a mark used to distinguish uses of the same letter for different sounds, e.g. an accent as in ç) placed below, or a character (commonly a number or lower case letter) placed after and slightly below, another character, e.g. A_1, A_p (SOIL HORIZON). In statistics subscript symbols are used to specify the location or relationship of values, e.g. in a series or a matrix. A superscript is a diacritic placed above, or a character placed after and slightly above, another character.

subsequent boundary, subsequent international boundary a political boundary drawn up after a region has been settled and has developed an identity or CULTURAL LANDSCAPE, the cultural and ethnic characteristics of the area to be divided being taken into account. ANTECEDENT BOUNDARY, RELICT BOUNDARY, SUPERIMPOSED BOUNDARY. G.

subsequent fall a waterfall resulting from the uncovering of hard rock behind soft rock as a down-cutting stream excavates its valley. G.

subsequent stream a stream indicating that its development was subsequent to a CONSEQUENT STREAM. A consequent stream flows down the DIP, but a subsequent stream excavates its valley along an outcrop of weak rocks such as clays or shales, or other lines of weakness, e.g. a FAULT. If these lines of weakness occur in the direction of the STRIKE, i.e. more or less at right angles to the valleys of the consequent streams, the subsequent streams will meet the consequents at right angles, resulting in TRELLIS DRAINAGE. ANNULAR DRAINAGE. G.

subsere 1. a secondary sere, a series of plant communities that make up the stages in a SECONDARY SUCCESSION, classified according to the conditions at the time of their initiation, i.e. hydrarch (damp), mesarch (intermediate between damp and dry), xerarch (dry) 2. any stage in a secondary succession.

sub-shrub a SHRUB with lower parts woody but with upper parts soft.

subsidence 1. in general, a sinking in level, settling downwards to a lower position, or returning to a normal state 2. in the earth's crust it may be readjustment on a large scale, as in a RIFT VALLEY, or on a smaller, more superficial scale, e.g. the collapse of a cave roof in KARST or the collapse of a land surface resulting from mining activity; or, in MASS MOVEMENT, a landslide in which there is downward displacement of relatively dry superficial earth material without a free surface (compare ROCK FALL) and without horizontal displacement 3. in meteorology, the slow descent of a large air mass, as in an ANTICYCLONE (SUBSIDENCE INVERSION).

subsidence inversion a condition of high level inversion in the ATMOSPHERE-1 caused by the slow descent of a large air mass which, as it approaches the ground, decelerates and spreads horizontally, resulting in ADIABATIC compression, the warming of the descending air, and a stable LAPSE RATE. SUBSIDENCE-3.

subsilic rock BASIC ROCK.

subsistence agriculture, subsistence farming farming in which the products are grown or raised primarily (but not necessarily solely) for the support of the farmer and the farmer's dependants, not primarily for sale or trading. The opposite is COMMERCIAL AGRICULTURE which is primarily concerned with the growing of crops or raising of livestock for sale. AGRIBUSINESS, FARMING. G.

subsistence crop a crop (commonly a FARINACEOUS food crop) grown as the basic item of diet to be eaten by the farmers and their dependants, not for sale or trading. CASH CROP, STAPLE-3.

subsistence economy an economic system in which there is little if any buying and selling, although there may be bartering. MARKET ECONOMY.

subsoil an imprecise term for the SOIL layer consisting of weathered parent material lying immediately below the soil proper (SOLUM, TOPSOIL) and above the BED ROCK, corresponding approximately to the C HORIZON. The term is seldom used by pedologists. PEDOLOGY, SOIL HORIZON, SOIL PROFILE. G.

subsolar point the point at which the sun's rays are perpendicular to the earth's surface.

subspecies, sub-species a subdivision of a SPECIES-1, a group of organisms (commonly occupying a well-marked geographical area) in which the individuals are like each other in some characteristics but which differ from other members of the species. While breeding may occur between members of different subspecies of the same species, it dos not happen so readily as within the subspecies itself; but because reproduction is not totally isolated, subspecies tend to grade into each other; and some subspecies are considered to be new species in the making. In botany a subspecies ranks between the species and the VARIETY. CLASSIFICATION OF ORGANISMS.

substrate a substratum 1. the surface to which an organism is attached or on which it moves 2. the medium on which a microorganism grows 3. in biochemistry, a substance acted upon, especially that which an ENZYME activates.

substratum pl. substrata, or substratums 1. in general, the layer lying underneath, especially a supporting layer for that which lies above it 2. in soil science, the soil layer below the

SOLUM, said to be conforming if this is the C HORIZON, unconforming if it is the D HORIZON. SOIL HORIZON.

sub-surface wash, subsurface wash the processes involved in the carrying by water of SOLUTES and particles within the REGOLITH, the flow of water being termed THROUGHFLOW. If the particles are carried down a slope, horizontally to the surface of the regolith, the process is termed lateral ELUVIATION; if the throughflow occurs along definite lines of seepage, the process is termed tunnelling.

subsystem a part of a larger SYSTEM which can itself constitute a system.

subtemperate *adj.* of or pertaining to the cooler parts of the TEMPERATE ZONE.

subterranean *adj.* situated, existing, functioning or operating underground.

subtopia (derived from suburb and utopia) an ironic term applied by Ian Nairn in the 1950s to the sprawling, low-density suburban development, fake rusticity, plethora of noticeboards, overhead wires, ill-designed car parks, wartime sites and installations abandoned by the armed services, and the similar general ugly untidiness that littered the landscape in Britain in the 1950s, despite comprehensive town and country planning legislation which, it had been hoped, would produce the utopia envisaged by planners. The term came to be used later, less precisely, to attack any low-density suburban development. SUBURB.

subtropical *adj.* applied **1.** imprecisely to climatic conditions that are tropical (TROPICAL CLIMATE) for part of the year, or to those that are 'nearly' tropical throughout the year, to the vegetation growing in, and to the lands with, those conditions, i.e. polewards beyond the TROPICS, merging into the warm TEMPERATE zone **2.** more precisely, to latitudinal zones between the TROPIC OF CANCER and about 40°N and between the TROPIC OF CAPRI-

CORN and about 40°S **3.** to belts of atmospheric high pressure occurring in those zones. SUBTROPICAL HIGH PRESSURE BELTS **4.** to climatic regions where the temperature in any month does not fall below about 6°C (43°F), e.g. MEDITERRANEAN CLIMATE. The term extra-tropical is sometimes preferred when the reference is to something occurring just outside the tropics, e.g. extra-tropical high pressure belt. G.

subtropical high pressure belts the belts of persistent HIGH atmospheric pressure (ANTICYCLONE) with an east to west trend, centred generally about latitudes 30°N and 30°S.

subtropical jet stream a JET STREAM forming at the TROPOPAUSE immediately over the HADLEY CELL.

subtropical zone SUBTROPICAL-2.

suburb, suburbs 1. the outer, socially homogeneous, mainly residential or dormitory part of a continuously built-up urban area, town, or city, distinguished from the inner area by a lower housing density, and characterized by a high level of commuting (COMMUTER) to central locations in the inner area. CONCENTRIC ZONE GROWTH THEORY, CENTRAL BUSINESS DISTRICT, INNER CITY, URBAN-RURAL CONTINUUM **2.** in USA, a suburb, a separate administrative unit of local government outside a central city, being independent of the central city in FISCAL affairs and land use planning, and having in some cases independent SCHOOL DISTRICTS. The suburb commonly houses higher income, higher status groups.

suburban *adj.* **1.** of or pertaining to the SUBURBS-1,2 **2.** having qualities considered to be characteristic of the suburbs or of the people who inhabit them.

suburbanization the process of creating SUBURBS.

suburbia a synonym for SUBURBS, sometimes used contemptuously.

subway 1. an underground passage for

pedestrian use under a street or streets, railway, large building etc. **2.** (American) an underground railway.

succession in general, the coming of one thing after another in order or time, a series of things in order, a sequence **1.** in geology, the order, in time, of beds of rock; the vertical sequence of rock in a certain locality **2.** in ecology, the progressive natural development of vegetation from the initial pioneer community (PIONEER STAGE) to the CLIMAX, one community being gradually replaced by another under the influence of physical factors (the living organisms responding to topographical features) and biotic factors (organisms reacting to one another). PRIMARY SUCCESSION, SECONDARY SUCCESSION, SERE.

succession and invasion in urban land use studies, the sequence of changes by which units of population or of land use replace those existing in another area, the invading population or land use type eventually achieving numerical superiority and controlling the area invaded, e.g. as in GENTRIFICATION. The term is used particularly to explain the outward growth in the CONCENTRIC ZONE GROWTH THEORY. G.

succession phenomenon a problem resulting from the control of pests and diseases, in that by removing one pest or disease another is given the opportunity to operate.

succulent a plant adapted to meet water loss. SUCCULENT-1 *adj.*

succulent *adj.* applied to **1.** a xerophytic (XEROPHYTE) plant with enlarged tissue for water storage in leaves and/or stems, which allows it to withstand drought **2.** the parts of a plant enlarged to hold water or sugar, e.g. leaves, fruit.

sudd (Sudan: Arabic) **1.** floating, compact masses of vegetation in the upper Bahr el-Jebel (White Nile) which obstructs navigation **2.** the marshes in Sudan resulting from this obstruction. G.

sugar one of a class of crystalline carbohydrates soluble in water, the term generally applied to sucrose, manufactured by all green plants and commonly stored in roots, bulbs, flowers, fruit. The plants from which sugar is refined for commercial use are mainly SUGARCANE (yielding cane sugar) and SUGAR BEET (yielding beet sugar); but some is also regularly obtained from the wild date palm, *Phoenix sylvestris*; the sugar palm, *Aranga saccharifera*; the PALMYRA PALM; and the SUGAR MAPLE. In addition to its use as a foodstuff, sweetener and preservative for other foods (e.g. fruit, as in jam making), sugar (especially that obtained from sugarcane) provides alcohol which can be used as a fuel or a fuel additive; and current experiments indicate that it may be used as a substitute for petrochemicals in various industrial processes, e.g. the making of detergents and plastics.

sugar beet *Beta vulgaris*, a BIENNIAL plant with a white, conical, swollen root yielding sucrose (SUGAR), native to midlatitude lands in continental Europe, grown there and in similar conditions elsewhere for a sugar supply. The leafy tops and the pulp remaining after the sugar has been extracted are used for cattle feed; molasses, another by-product of the processing of the roots, is also used as stock feed and for making industrial alcohol, the filter cake remaining after the juice has been purified by filtration being used as manure. Cultivation in continental Europe was encouraged by Napoleon I of France in the nineteenth century in order to render France independent of sugar supplies from the British colonies.

sugarcane *Saccharum officinarum*, a species of tall, coarse PERENNIAL grass, native to tropical areas of the OLD WORLD, still cultivated there but even more in similar conditions elsewhere, especially in the West Indies, the

USA, Central and South America, and in Australia, for its high yield of sucrose (SUGAR) obtained from the stems of the cane. The plant needs a fertile soil, heavy dressings of manure and fertilizer, a good rainfall, or irrigation. Sugar is extracted on the estates, molasses and rum being important by-products. Columbus introduced sugar-cane to America and the West Indies, and because it needs much labour it gave rise to the slave trade, becoming the basis of plantation industry in the region. Today sugarcane is one of the most scientifically produced crops, producing more human food per unit area than any other crop.

sugar limestone a soft, crumbly lime-stone resulting from THERMAL META-MORPHISM of CARBONIFEROUS LIME-STONE.

sugar loaf 1. a conical mass of refined sugar, the form in which sugar used to be retailed **2.** a conical hill resembling such an object, a steep-sided residual hill with smoothed faces, an INSEL-BERG, e.g. the most famous, the Sugar Loaf Mountain, Rio de Janeiro, Brazil. G.

sugar maple *Acer saccharum*, a large tree, native to northeastern North America, with sweet sap used for making syrup (maple syrup) and SUGAR. The black maple, *Acer nigrum*, closely allied to the sugar maple, is similarly used.

suitcase farming (American) an agricul-tural landholder whose holdings are scattered, and who moves from one holding to another at crucial times, e.g. seed-time, harvest-time, to make the best use of farm machinery. The crop is usually a CEREAL which needs little attention in the growing period. SIDE-WALK FARMER.

suk, souk SUQ. G.

sukhovey (Russian, pl. sukhovei) a hot, dry wind, blowing during the summer mainly from the southeast in the south-eastern part of European USSR and Kazakhstan. The air temperature may rise to 35°C to 40°C (95°F to 105°F) and the RELATIVE HUMIDITY may drop to 15 per cent or less, causing excessive evaporation and seriously in-juring crops and other vegetation.

sulfur, sulfuric (American) SULPHUR, SULPHURIC.

sulphate (American sulfate) a SALT or ESTER of SULPHURIC ACID.

sulphate (sulfate) wood pulp WOOD PULP made by boiling wood chips in a strongly ALKALINE solution made with caustic SODA and SODIUM sulphate.

sulphur (American sulfur) a nonmetal-lic element occurring in crystalline or amorphous form, abundant and wide-spread in nature in a free state or combined as sulphates and sulphides, an essential MACRONUTRIENT for plants. It is deposited near volcanic vents (e.g. FUMAROLES) and HOT SPRINGS, occurs in sedimentary rocks (Tertiary limestones and sandstones) and is associated with SALT DOMES. Sulphur is used in the manufacture of SULPHURIC ACID, dyes, fungicides, insecticides and many other industrial products; in medicines; and in vulcaniz-ing RUBBER. THIONIC.

sulphur (sulfur) dioxide SO_2, a stable oxide of SULPHUR, gaseous except at very unusual atmospheric pressure, but easily liquefied by pressure, produced by the burning of a wide range of fuels, but especially of coal and heavy fuel oil. It is released, with SULPHATES, in volcanic eruptions. Sulphur dioxide is used widely in manufacturing, e.g. in making SULPHURIC ACID, in reducing and bleaching in petroleum refining; and as a preservative, an insecticide, a disinfectant etc. ACID RAIN, SULPHUR TRIOXIDE.

sulphuric (sulfuric) acid H_2SO_4, a col-ourless, dense, rather oily liquid (oil of vitriol), dissolving and ionizing with ease in water, used extensively in petro-leum refining, in the chemical industry and in making detergents, explosives, dyes, rayon fabrics. It is formed in the atmosphere by the combination of SULPHUR TRIOXIDE with water, resulting in a stable mist of small acid droplets which fall to the ground and, sulphur being an important plant MACRONUTRIENT, contribute to soil fertility. Excessive deposition of sulphur from the atmosphere is, however, undesirable in an area with an acid soil, where it may lead to a high level of acidity in lakes and rivers, and a slowing down of growth, or even the death, of some plants; but it is not so disastrous in an area of ALKALINE SOIL where it may raise soil fertility. The term ACID RAIN is applied to precipitation charged with excessive sulphuric acid (and other acids) derived in part from SULPHUR DIOXIDE emitted in the burning of HYDROCARBON fuels and in other processes in towns and industrial plant.

sulphur (sulfur) trioxide an oxide of sulphur formed in the atmosphere by the oxidation of SULPHUR DIOXIDE.

sultana a small, dried seedless white GRAPE.

sulung YARDLAND.

sumatra a LINE SQUALL, accompanied by thunderstorms, occurring usually at night during the southwest monsoon in the Malacca Strait. G.

summer 1. in general,the warmest season of the year, as opposed to winter, the coldest 2. in the northern hemisphere, the months June, July, August; in the southern December, January, February 3. in astronomy, the period between the summer solstice (about 21 June) and the autumn equinox (about 22 September) in the northern hemisphere; and between the solstice about 22 December and the equinox about 21 March in the southern. EQUINOX, SEASON, SOLSTICE.

summer *verb* to summer, to provide pasture for sheep and/or cattle during the warmest season.

summer berm a BERM produced by

gentle waves during summer months in MIDLATITUDES. WINTER BERM.

summer solstice SEASON, SOLSTICE, SUMMER.

summit level the highest point or level of a mountain, road, railway or canal.

summit plane a plane passing through a series of accordant summit levels (ACCORDANCE OF SUMMIT LEVELS) and so inferred to be the level of a former peneplained surface. But it may also indicate a general balance in downwasting, the hill tops having been lowered simultaneously. GIPFELFLUR. G.

sum of squares in statistics, the sum of the squared differences between the individual values and the means of values. VARIANCE, VARIANCE ANALYSIS.

sump 1. a deep pool of subterranean water, i.e. under the level of the WATER TABLE, with an exit below the level of its own water, a common feature in KARST 2. a hole, usually subterranean, into which the drainage of an area can collect. G.

sun 1. in general, any heavenly body that is the centre of a solar system 2. in particular, the central body of 'our' SOLAR SYSTEM, lying at the main focus of the orbits of the earth and the other PLANETS, a dwarf yellow star in the spiral arm near the outer edge of the MILKY WAY. A nearly spherical gaseous body, diameter about 1 392 000 km (865 000 mi), it is thought to be composed of approximately 90 per cent hydrogen, 10 per cent helium mixed with small amounts of all other known elements, the temperature at the core being some 15 mn°C (59 mn°F), at the surface some 5700°C (10292°F). It rotates once in 24.5 days at its equator, and its mean distance from the earth is 150 mn km (92.9 mn mi) (APHELION, PERIHELION). Only a very small amount of the total SOLAR RADIATION reaches the earth, but it is responsible there, directly or indirectly, for most energy-requiring processes, including the growth of animal and plant life. DECLINATION-2, INSOLATION, STAR, SUNSHINE.

sundri, sundari (Indian subcontinent: Bengali, mangrove) the swamp forests (hence sundarbans) of the Ganga delta, in which the mangroves (*Heritiera fomes* and *H. littoralis*) are the dominant species. (The swamp forests of the Irrawaddy delta in Burma are also mainly of *Heritiera fomes*.) MANGROVE, MANGROVE SWAMP. G.

sunflower member of *Helianthus*, a tall ANNUAL or PERENNIAL herb with large flower heads, native to North America, Chile and Peru. *Helianthus annuus*, the common sunflower, is grown mainly in temperate lands, especially in eastern Europe, Argentina and Canada for its oil seeds, which give a high yield of edible oil, used in cooking, in making margarine and other foodstuffs, in varnishes and soaps, the residual oilcake being used as livestock (including poultry) feed. Sunflower seeds are also eaten raw, and fed to poultry and caged birds. The stem of the plant makes good fresh fodder or silage, can be used in paper-making, and yields fibre of textile quality.

sungei (Malay, usually abbreviated to S. in place-names) a river.

sunn hemp, kenaf *Crotalaria juncea*, a leguminous plant (LEGUMINOSAE) native to southern Asia, yielding a fibre of the same name, resemblilng JUTE, but lighter and stronger, used in making bags, ropes etc. It is not related to true HEMP, *Cannabis sativa*. C.

sunrise, sunset the times at which the SUN apparently rises in the morning and sets in the evening below the horizon, varying with latitude and with the sun's DECLINATION-2, due to the earth's rotation on its axis. Sunrise is defined in metereorology as the time at which the upper edge of the sun appears above the apparent horizon on a clear day; sunset, the time at which the upper edge of the sun appears to sink below the apparent horizon on a clear day.

sunshine the light received on earth directly from the SUN, the main components being visible radiation (i.e. light), ULTRAVIOLET radiation (a germicidal agent, producer of vitamin D), INFRARED radiation (heat-producing element, constituting half of all solar radiation reaching the earth's surface). The duration of sunshine, of vital importance to plant growth, is affected by latitude and daytime cloudiness. It is measured on a CAMPBELL-STOKES RECORDER, and data can be used to indicate duration in hours per day, per month, per year, or the percentage of the total possible amount. The lines of equal mean duration, plotted at various stations, are termed isohels (ISO-). INSOLATION.

sunspot a dark area on the visible surface of the SUN, consisting of a grey region surrounding a darker centre, causing an increase in the SOLAR RADIATION received on earth, and particularly affecting the earth's MAGNETOSPHERE and IONOSPHERE. Their number usually reaches a maximum every eleven years.

supercooling cooling to a temperature below the normal transition point for a change of state without the occurrence of that change, e.g. when water, if undisturbed, stays liquid at a temperature below 0°C (32°F). This can happen naturally if water droplets in clouds are not disturbed; but they become ice immediately on coming into contact with another body, e.g. an aircraft (which can be most dangerous), or tall trees; and supercooling can also create very large hailstones. SUPERHEATING.

superficial *adj.* of or relating to the surface, not penetrating below the surface.

superficial deposit unconsolidated PLEISTOCENE or HOLOCENE material lying on the surface of the earth's crust, not formed in situ from underlying BED ROCK, but carried and set down in position by the agencies of wind (AEOLIAN DEPOSIT, e.g. LOESS), water (e.g. ALLUVIUM), ice (e.g. GLACIAL DRIFT), or by gravity (e.g. COLLUVIUM). PEAT, formed in situ, is also included; but a SOIL formed in situ through the weathering of the underlying rock is not. CLAY-WITH-FLINTS, of uncertain origin, is usually referred to as a superficial deposit. SURFACE DEPOSIT.

supergene *adj.* applied to mineral deposits (e.g. ores and ore minerals) or to the processes involved in the downward percolation of water or an aqueous solution. HYPOGENE, SECONDARY ENRICHMENT.

superglacial, supraglacial *adj.* of or relating to the surface or to the environment at the surface, of a glacier. ENGLACIAL, SUBGLACIAL.

superglacial stream a rivulet of meltwater flowing in summer in a deep runnel on the surface of a glacier and descending into a crevasse. ENGLACIAL RIVER, MOULIN, SUBGLACIAL CHANNEL.

superglacial till an ABLATION MORAINE.

superheating heating to a temperature above the normal transition point for a change of state without the occurrence of that change, e.g. when water is heated above boiling point without boiling occurring. SUPERCOOLING.

superimposed boundary a political BOUNDARY superimposed on a CULTURAL LANDSCAPE, i.e. imposed after the development of settlements, language and culture in a region, the cultural and ethnic characteristics of the area to be divided being completely ignored. ANTECEDENT BOUNDARY, SUBSEQUENT BOUNDARY. G.

superimposed (superposed) drainage epigenetic drainage, a drainage pattern appearing to be independent of the structure of the underlying rocks be-

cause it was established on a former rock surface, since removed by denudation. G.

superimposed profile a diagram incorporating a series of PROFILES-3 constructed to show relief, using information taken from a physical map at regularly spaced intervals. These are placed one on top of another, complete in all detail (unlike those of a PROJECTED PROFILE or a COMPOSITE PROFILE). The result reveals marked coincidences, such as ACCORDANCE OF SUMMIT LEVELS and EROSION PLATFORMS etc. G.

supermarket a large, self-service store stocked with food and (usually) some small household goods displayed on open shelves. HYPERMARKET.

supernatural *adj.* of that which cannot be explained in terms of the known laws governing the material universe.

superparasite HYPERPARASITE.

superphosphate any of the various phosphate fertilizers containing soluble phosphorus pentoxide, P_2O_5, obtained by treating PHOSPHATE ROCK with SULPHURIC ACID. Triple superphosphate is superphosphate treated with phosphoric acid, which raises the P_2O_5 content. PHOSPHATE, PHOSPHORUS.

superposition LAW OF SUPERPOSITION.

supersaturation the state of a SOLUTION having a higher concentration of solute than at saturation (SATURATED), produced by heating and slow, steady, undisturbed cooling or by very fast cooling of a saturated solution. Supersaturation occurs in the ATMOSPHERE-1 when a cooling body of air with a RELATIVE HUMIDITY exceeding 100 per cent has enough water vapour to produce condensation, but condensation does not take place unless solid particles (e.g. dust) or negative IONS are available.

superstratum a STRATUM overlying another.

supply and demand the market forces that govern prices, operating freely in a

MARKET ECONOMY, apparent through the price mechanism as responses in quantities offered for sale (the supply) or the quantities that consumers are prepared to buy (demand) when the market price changes.

supply curve a plot on a graph showing the level of supply of a product in relation to the price obtainable, supply appearing on the horizontal axis, price on the vertical. DEMAND CURVE.

supraglacial *adj.* sometimes applied to the environment at the surface of a glacier. SUPERGLACIAL.

suq, suk, sook, souk, sôk (Arabic, other spellings, a market; pl. aswaq) a MARKET-PLACE or MARKET, applied to a town, village, or periodic market. The term is used widely in north Africa and Arabic-speaking areas generally; it also appears in place-names. G.

surazo (Brazil) friagem, a cold wind blowing in winter in the CAMPO of Brazil in the middle Amazon region, produced by an ANTICYCLONE. Temperatures may fall below 10°C (50°F), causing great discomfort. G.

surf the foaming water produced by a powerful wave as it breaks on rocks on the seashore. LITTORAL CURRENT.

surface **1.** the outside boundary of a solid, or the upper boundary of a liquid where the liquid is in contact with air or its vapours. Such a surface has length and breadth but no depth, i.e. it is two-dimensional **2.** in mathematics, the concept of an infinite set of points occupying a two-dimensional space (i.e. of no thickness) **3.** the boundary between two distinct materials or processes (as in BASAL SURFACE or a FRONT). INTERFACE.

surface deposit unconsolidated material overlying BED ROCK, weatheredd from the bed rock itself (RESIDUAL DEPOSIT), or weathered elsewhere and carried to the present position by wind (AEOLIAN DEPOSIT), by water (ALLUVIUM), by ice (GLACIAL DRIFT) or by gravity (COLLUVIUM). SUPERFICIAL DEPOSIT.

surface flow OVERLAND FLOW.

surface tension the surface force acting on the surface of a liquid with the effect of reducing the surface area to the minimum. This surface force is the result of forces within the liquid acting on the molecules of the surface in the absence of forces acting above the surface, and thereby drawing the surface together. In a small quantity of liquid a shape with the smallest area possible will be found, i.e. a sphere. Surface tension is usually reduced by a rise in temperature. CAPILLARY MOISTURE, POROSITY.

surface wash processes whereby soil and other unconsolidate material are carried by flowing water across the surface of the land. OVERLAND FLOW is the major process, but on sloping ground the action of large raindrops or a heavy downpour in dislodging soil particles, is also involved. SOIL EROSION.

surplus value **1.** in Marxism, the difference between the value of LABOUR POWER and the commodity produced by its expenditure. EXPLOITATION **2.** in non-Marxist usage, profit.

surrogate a substitute, a deputy for another, e.g. in statistics, a variable which represents or is a substitute for another, commonly used because it can be more easily investigated than the other one.

surroundings that which encircles, or almost encircles; the place around an organism, group or object. ENVIRONMENT.

survey **1.** a general inspection or viewing as a whole **2.** a careful consideration, inspection and examination as a whole and in detail, e.g. of a place, building, population, problem, state of affairs, condition etc. **3.** the presentation of the finding of such a survey in written, diagrammatic, cartographic, photographic form **4.** the process of gathering data relating to a chosen area, e.g. by REMOTE SENSING or by the measuring and recording of lines and angles of an area of land in order to make an accurate map of it **5.** an area that has been so measured and recorded etc. **6.** a group of people, or a department, engaged in the surveying indicated in **1,2,3,4 7.** in statistics, a method for estimating. SAMPLE SURVEY, SURVEY ANALYSIS.

survey analysis all the methods and procedures involved in estimating the characteristics of a POPULATION-1 from a SURVEY-2 of individual members of it, including the identification of the matter to be researched and of the data needed to test the HYPOTHESIS suggested, the selection of individuals to be approached (in some cases SAMPLING is used, in others the entire population is approached), the preferred techniques of investigation (questionnaire, sent by post or presented by interviewer; interview; observation). PILOT surveys usually precede the main survey; and the results of the latter are usually analysed by, and subjected to, statistical techniques, e.g. STATISTICAL INFERENCE.

suspended load the fine organic and inorganic materials (e.g. silt and clay) consisting of particles with diameters commonly less than 0.2 mm (sometimes termed wash load) carried by a STREAM-1 in SUSPENSION (without the aid of SALTATION), as distinct from the BED LOAD and the DISSOLVED LOAD. Measurements of suspended material are usually expressed in milligrams per litre or kilograms per cubic metre, and are used to calculate transport rates.

suspension a two-phase system in which a finely divided insoluble solid is uniformly dispersed in a liquid or gas. In moving water, e.g. in a STREAM-1, small particles are kept buoyant (in suspension) by turbulent upward eddies which prevent the particles sinking

under gravity; thus the finest particles may be carried long distances by a stream or river before they sink to the bottom or are carried out to sea. COLLOID, SALTATION, SOLUTION, SUSPENDED LOAD, TRACTION.

sustained yield forestry any practice or technique (e.g. PATCH CUTTING) used in FORESTRY which aims to conserve the forest and ensure that it remains a flow resource rather than a stock resource (NATURAL RESOURCES). CONSERVATION, RESOURCE CONSERVATION, RESOURCE MANAGEMENT.

suzerainty the position or power of a state which has political control over another.

swadeshi (Indian subcontinent: Bengali; Hindi, home-country things) in the subcontinent before partition in 1947, the national movement in favour of home-produced goods (especially cottons) and against imported foreign goods. G.

swag (England: Midlands) a shallow, water-filled hollow on the earth's surface resulting from subsidence caused by mining. FLASH-1, PITFALL. G.

swale 1. a low, long narrow depression approximately aligned with the coastline and lying between two ridges (FULLS) of shingle on a beach **2.** (American) a marshy or moist depression in level or rolling land, or a long narrow depression lying between the bars of a POINT BAR deposit on the floodplain of a river. SWELL AND SWALE TOPOGRAPHY. G.

swallow, swallow-hole, swallet 1. a deep vertical hole or opening in the earth, especially one produced by the solution of rocks in limestone country (KARST), down which a surface stream or rainfall disappears, in some cases as a waterfall (PONOR), synonymous with sink in American terminology **2.** a hole in a stream bed through which some of the stream water flows and disappears; or a point of no fixed location where a stream may dry up as its water percolates downwards. G.

swamp **1.** in general, wet spongy land saturated with water for much of the time, and its associated vegetation **2.** more precisely, the soil-vegetation type in which the normal summer water level is above that of the soil surface, and the characteristic vegetation is woody **3.** in ecological freshwater studies, the last phase of aquatic vegetation before it gives way to land vegetation (the sequence being, from the centre of a body of freshwater to the margins, aquatic, swamp, marsh). BOG, FEN, MANGROVE SWAMP, MARSH. G.

swash the body of water which rushes up a beach after an ocean WAVE-3 has broken. BACKWASH. G.

S-wave SHAKE WAVE.

swede *Brassica napus*, variety Napo-brassica, a BIENNIAL plant with a large, edible, strongly-flavoured 'root' (part of which is actually a swollen leaf stem), unknown in Europe until the seventeenth century, grown as fodder for animals and, when cooked, for human food. It can be left in the ground in winter.

sweet chestnut *Castanea sativa*, a large tree native to southern Europe, widely grown for its timber and edible KERNELS which may be eaten whole (roasted, boiled, preserved in sugar or syrup), or ground into flour, or fed to livestock, especially pigs.

sweet cicely *Myrrhis odorata*, an aromatic, erect PERENNIAL herb, native to Europe, formerly grown for medicinal use as well as for flavouring food (leaves and seeds being edible), now used mainly for the latter.

sweet corn MAIZE.

sweet orange ORANGE.

sweet pepper PEPPER.

sweet potato *Ipomoea batatas*, a trailing vine native to South America, now widely grown there and in wet tropical areas and warm temperate regions elsewhere for its sweet, edible, starchy tubers, eaten cooked, the plant itself providing livestock feed. YAM.

swell **1.** of a large body of water, especially in the ocean, the regular, undulating motion of the surface, the succession of waves which do not break **2.** on the deep sea floor, a RISE **3.** in geology, a very large domed area. G.

swell and swale (sag and swell) topography a type of undulating topography common on old glacial deposits. SWALE. G.

swidden farming a term applied by some authors to a type of SHIFTING CULTIVATION in which the existing vegetation is removed by cutting and burning (SLASH-AND-BURN), the cleared area being sown and cropped until the soil nutrients are exhausted and yields begin to fall. The area is then abandoned, the cultivators moving to a new patch which they similarly clear and sow (compare BUSH FALLOWING); but after some time they may return to the abandoned site where the natural vegetation has regenerated and soil fertility has built up naturally, to repeat the process. G.

swine sing. or pl., a member of the family Suidae, genus *Sus*, an omniverous, cloven-hoofed quadruped, a short-legged, thick-skinned MAMMAL. In general use the term has been superseded by PIG or HOG; but it is still used in dialect, and in poetic or zoological works.

swing bridge a bridge supported on a vertical axis which allows the bridge to be turned about, usually to allow unimpeded passage for vessels along a navigable waterway. TRANSPORTER BRIDGE.

swing of pressure belts (of wind systems) the seasonal shift of the atmospheric pressure belts, of the PLANETARY WINDS, northward in the northern summer, southward in the northern winter, following the apparent seasonal movement of the vertically overhead sun. THERMAL EQUATOR. G.

sword bean, horse bean *Canavalia gladiata*, a PERENNIAL leguminous plant

(LEGUMINOSAE) with red, pink or brown seeds in a curved pod, originating in Asia, cultivated in tropical Asia today, used in the same way as the JACK BEAN.

sword-dune, sword dune an alternative term for SEIF-DUNE. G.

syenite a group of coarse-grained, INTERMEDIATE IGNEOUS ROCKS, occurring in PLUTONIC form, characteristically composed mainly of alkali FELDSPAR. GNEISS, NEPHELINE.

sylviculture SILVICULTURE.

sylvite an EVAPORITE mineral, KCl, natural potassium chloride, occurring in colourless cubes or masses, one of the main sources of POTASSIUM, an important FERTILIZER.

symbiosis **1.** in biology, the very close association of dissimilar organisms to their mutual benefit. COMMENSALISM, LICHEN, MUTUALISM, MYCORRHIZA, PARASITISM, SYMBIOTICS **2.** in human ecology, the very close, mutually advantageous relations among dissimilar members of a human group or among dissimilar groups within a larger group, or among dissimilar institutions and activities. SYMBIOTIC RELATIONSHIP. G.

symbiotic *adj.* of or pertaining to or characterized by SYMBIOSIS.

symbiotic relationship the relationship of the dissimilar organisms in SYMBIOSIS-1, applied by analogy to other phenomena, e.g. the mutually beneficial relationship between a city or large town and its sourrounding area, or between different groups within a human community when the groups are unlike and the relations complementary.

symbiotics the study concerned with SYMBIOSIS-1,2.

symbol a sign, shape or object accepted as representing or typifying a thing, person, idea, quality or value. In cartography an explanation of the conventional signs used to represent specific objects commonly appears on the map itself; but if very many symbols are used, reference is usually made to a CHARACTERISTIC SHEET. G.

symbolic model a MODEL-4 in which selected aspects of reality are all expressed by mathematical expressions. ANALOGUE MODEL, ICONIC MODEL, MATHEMATICAL MODEL. G.

syminct varve VARVE.

symmetric, symmetrical *adj.* showing or possessing SYMMETRY.

symmetrical fold a FOLD in which the limbs dip away symmetrically (SYMMETRY) from the AXIS OF FOLD.

symmetry the quality of having a form so regular that one or more axes exist which divide the structure in exactly corresponding and equal parts.

sympatric *adj.* applied to different SPECIES-1 or sub-species of which the distribution areas coincide or overlap. ALLOPATRIC.

synchronic *adj.* LINGUISTICS, SYNCHRONOUS.

synchronic analysis the study of the pattern of linkages within a SYSTEM-1,2,3 at a certain point in time. DIACHRONIC ANALYSIS.

synchronous, synchronic, synchronal *adj.* **1.** occurring, existing at the same time **2.** exactly coinciding in rate, time etc.

synclastic *adj.* applied to a surface having the same curvature in every direction around any given point, e.g. a SPHERE-1.

synclinal valley a valley which follows a SYNCLINE in the underlying rocks, i.e. a valley formed by a downfold. G.

syncline a downfold in the STRATA of the earth's crust, the rocks dipping inwards to a central axis (AXIS OF FOLD), caused by COMPRESSION. The *adj.*, synclinal, was formerly used as a noun, but syncline is now the preferred noun. ANTICLINE, PITCH, SYNCLINORIUM. G.

synclinorium a broad downfold (SYNCLINE) of the rocks over a considerable tract of country on which numerous minor upfolds (ANTICLINE) and downfolds are superimposed. G.

syndynamics the study of changes in SUCCESSION-2 in plant communities.

synecology the scientific study of the relationships between communities of SPECIES-1 and their environment, as opposed to AUTECOLOGY. G.

synergism in biology, the combined activity of agencies (e.g. drugs, hormones) which individually and separately influence a certain process in the same direction, whereby the combined effect produced is greater than the sum of the effects produced by each agency acting by itself (i.e. the whole is greater than the sum of the parts).

synergy 1. the additional benefit accruing to a number of SYSTEMS-1,2 should they come together to form a larger system **2.** a synonym for SYNERGISM.

synform a general term applied to a downfold in the STRATA of the earth's crust when the stratigraphical relationships of the rocks is not known precisely.

syngameon a group of SPECIES-1 between which hybridization takes place. SUBSPECIES.

syngenetic *adj.* applied to ore deposits that are contemporaneous with the enlosing rocks, in contrast to epigenetic. EPIGENETIC MINERAL. G.

synodic (lunar) month MOON.

synoecism the union of towns and small settlements to form one administrative unit, e.g. as in the Greek city-state.

synopsis a brief condensed statement, a summary or outline.

synoptic *adj.* providing a general summary or a general view.

synoptic chart a chart giving a summary, a general view of the meteorological conditions (isobars, temperatures, winds etc.) over a large area at a given time, an essential tool in weather forecasting. WEATHER CHART.

synoptic climatology the comprehensive study of the condition of the ATMOSPHERE-1 over a very large area (e.g. one of the hemispheres) at a particular time, of prime importance in weather forecasting. DYNAMIC CLIMATOLOGY, METEOROLOGY, SYNOPTIC METEOROLOGY.

synoptic image a general view of a part of the earth's surface, usually provided by a SENSOR mounted on a high-altitude SATELLITE-1. REMOTE SENSING.

synoptic map a map displaying all the factors to be taken into account in solving a problem. The factors are shown either on one sheet or on a series of OVERLAYS placed over the base map, each overlay showing a separate factor or a selected combination of factors. SIEVE MAP.

synoptic meteorology the branch of METEOROLOGY concerned with the collection of meteorological data (over a smaller area than that involved in SYNOPTIC CLIMATOLOGY), the making of SYNOPTIC CHARTS from the information gathered and the interpretation of the pressure patterns etc revealed by the charts in order to anticipate changes likely to affect the weather in a particular area. MICRO-METEOROLOGY.

synphylogeny the study of the history, the evolutionary trends and changes taking place within plant communities. PHYLOGENY.

synpiontology the scientific study of ancient patterns of distribution of plant communities.

synthesis 1. the building up, the putting together of separate parts, elements, substances etc. to form a whole or a SYSTEM-1,2,3. ANALYSIS **2.** the product of such a process, also termed a SYNTHETIC **3.** the end phase of the dialectic process. In this process, the thesis, the first phase, presents the affirmation of a particular thought. But the thesis is seen to be unsatisfactory in some way. It gives rise to a second phase, the affirmation of its opposite, i.e. the antithesis. Further consideration shows the antithesis to be unsatisfactory too, so the third phase, the synthesis, the higher unity, comes into being, representing the more rational and acceptable parts of the now discarded thesis and antithesis. DIALECTIC-2, DIALECTICAL MATERIALISM, DIALECTICS-2.

synthetic a product made by SYNTHESIS-1.

synthetic *adj.* **1.** relating to, involved in, produced by SYNTHESIS-1 **2.** artificial.

synthetic fibre an ARTIFICIAL-1 fibre which can usually be spun, woven, felted etc., made to rival the natural FIBRES-1 of cotton, wool, flax, silk. The earliest production of any significance dates from the period following the First World War, when some machines used in that war were adapted to the manufacture of very fine fibres of rayon or artificial silk made by forcing a jelly-like substance derived from CELLULOSE through very tiny apertures or tubes of small bore. The two main types of synthetic fibre made today consist of such a regenerated cellulosic fibre, made by dissolving a natural product and extruding it (e.g. rayon), and a fibre made by polymerizing chemical compounds (e.g. nylon, made from BENZENE obtained from COAL TAR or PETROLEUM). In trade statistics the origin of the fibres (cellulosic, non-cellulosic) is usually distinguished.

synthetic rubber ARTIFICIAL-2 RUBBER made from HYDROCARBONS derived from PETROLEUM.

syntype in the CLASSIFICATION OF ORGANISMS, any specimen of a series chosen to designate a SPECIES-1 if neither the HOLOTYPE nor the PARATYPE are used.

synusia pl. synusiae, a group of plants of similar LIFE FORM, each filling the same ecologicial NICHE-1 and playing a similar role, and making a contribution to a BIOCOENOSIS-1.

system **1.** a set of related elements organized for a particular purpose, the whole being identifiable by the interconnexion of the elements **2.** a set of things or substances, associated, interdependent, governed by physical laws, and making a whole (as in the SOLAR SYSTEM) **3.** a set of things, structures, processes, activities (e.g. human activities) associated and interconnected, forming and functioning as a complex whole through a regular set of relations (as in the nervous system of a MAMMAL) and in many cases forming part of a larger system, the larger system itself forming part of an even larger system **4.** a set of principles linked to form a coherent doctrine **5.** a method of organization, administration, procedure (as in a legal system) **6.** a formal method of classification, nomenclature, notation, governed by well-defined rules (e.g. BINOMIAL NOMENCLATURE) **7.** in geology, a division of the succession of stratified rocks deposited during a geological period (GEOLOGICAL TIMESCALE). CASCADING SYSTEM, CATASTROPHE THEORY, CLOSED SYSTEM, CONTROL SYSTEM, ECOSYSTEM, ENTROPY, FEEDBACK, GENERAL SYSTEMS THEORY, GROUP-4, HOMEOSTASIS, MORPHOLOGICAL SYSTEM, MORPHOSTASIS, OPEN SYSTEM, PROCESS-RESPONSE SYSTEM, REGIONAL SYSTEMS, RELAXATION TIME, SUBSYSTEM, SYSTEMIC, SYSTEMS ANALYSIS.

systematic *adj.* **1.** working in an orderly, methodical way, in accordance with a SYSTEM-1,2,3,4,5,6 (SYSTEMATIC GEOGRAPHY) **2.** constituting a SYSTEM-2 **3.** in biology, relating to classification **3.** in statistics, the opposite of random or STOCHASTIC.

systematic control in social science research, the neutralizing of the effects of certain specified extraneous VARIABLES by ensuring that all groups are affected equally, e.g. if age were considered to be an important extraneous variable in a population survey, equal numbers of people of the same age group would be included in the experimental group and in the control group.

systematic error in SAMPLING-2, an error which is in some way biased, occurring predominantly in one direction throughout the sampling measurements, e.g. in a survey of people charged with exceeding the speed limit in driving, most of the accused will underestimate rather than overestimate their speed. A systematic error, unlike a RANDOM ERROR, does not balance out and will BIAS the results of a survey.

systematic geography a geographical approach which selects a particular aspect of the physical or human (social) environment in a defined geographical space, and studies and presents it in an orderly manner according to a system or plan. SYSTEMATIC-1,2. G.

systematics the scientific study of the CLASSIFICATION OF ORGANISMS, a term restricted in use by some authors as being synonymous with TAXONOMY; but usually applied more widely to include PHYLOGENY, and the identification and nomenclature of organisms as well as their classification.

systematic sample a SAMPLE-3 obtained by a SYSTEMATIC-4 or regular method, not by RANDOM SELECTION, e.g. a sample selected by randomly choosing an individual name on an electoral roll and then continuing at a fixed interval (the sampling interval, SAMPLING FRACTION) thereafter (say, every twentieth name); or sampling from an area by determining a pattern of points on a map, e.g. in land use survey, using points determined by the intersection of lines on a grid, or using a series of regularly spaced transects across an area to determine the relative proportion of each land use along each transect, or using sample areas equally spaced across the study area to measure the relative amount of each land use in

each. There is a disadvantage in using a systematic sample: it is possible for the regular pattern of the sample points to coincide with a regular distribution within the body of the study data, or within the study area. An alternative is the RANDOM SAMPLE, which has no such disadvantage. STRATIFIED SAMPLE.

system dynamics a type of model-building involving the construction of a large MODEL-4 in which little use is made of empirical evidence or of previous knowledge of the subject.

Système International d'Unité SI.

systemic *adj.* **1.** of or relating to a SYSTEM **2.** generally distributed throughout the body of a multicellular organism.

system of cities URBAN SYSTEM.

systems analysis, systems approach an approach which uses the concept of the SYSTEM-1,2,3 (i.e. broadly, a complex whole with interrelated elements) as an analytical tool. The system is identified by defining its boundaries, its purpose, and (if it is a subsystem), its position, its role, in a larger system. The structure and function of the system are investigated, and the level of abstraction at which it is to be treated defined (MODEL-4). As in GENERAL SYSTEMS THEORY, attempts are made to discover ISOMORPHISMS between systems so that ANALOGUE MODELS can be constructed. ANALOGUE THEORY, BLACK BOX APPROACH, CATASTROPHE THEORY, EQUIFINALITY, GREY BOX APPROACH, LINKAGES-2, MULTIFINALITY, WHITE BOX APPROACH. G.

syzygy in astronomy, the point at which two heavenly bodies are in CONJUNCTION or OPPOSITION, especially the MOON and the SUN; applied particularly to the point when the sun, moon and earth are in the same straight line, thus coinciding with the new moon and full moon as seen from the earth.

szik soil saline or alkaline soil in Hungary. G.

T

tabetisol TALIK. G.

tabki (Nigeria: Hausa) in the savanna area of northern Nigeria, a small semi-permanent pond in a saucer-shaped depression with an impervious clay bed which prevents the seepage of rain-water. G.

table cloth the popular name applied to the white cloud covering the flat top and hanging down the sides of Table Mountain in South Africa during a SOUTH-EASTER. The cloud is formed on the windward side, rolls over the top and descends the steep northern slopes like a waterfall, evaporating by adiabatic heating before reaching the lower slopes. The movement in the cloud is visible from the streets of Cape Town, but viewed from the north it appears to be stationary. It is not a LENTICULAR CLOUD. G.

tableland a flat or gently undulating area of high relief, a term generally super-seded by PLATEAU, but sometimes applied specifically to a plateau with abrupt cliff-like edges rising sharply from surrounding lowlands, e.g. a MESA.

table linen cloths, mats, napkins etc. laid on a table at mealtimes.

tablemount synonym for GUYOT.

tableware plates, dishes, glasses, cutlery etc. used in setting a table for a meal and for serving food and drinks.

taboleiro (Brazil: Portuguese) a flat-topped MESA-like coastal landform held up by a cover of fairly recent sedimentary strata. G.

tabular iceberg a very large, floating ice mass with a flat top, broken off from the Ross ICE SHELF and floating in the Southern Ocean. ICEBERG.

tacheometer, tachymeter a special THE-ODOLITE used in surveying designed to measure the direction, elevation, horizontal position and distance of distant points quickly, by instrumental observation.

tacit knowledge a form of knowing existing within a person, of which that individual is largely unaware.

taconite a low-grade IRON ore containing unleached haematite, magnetite, siderite and hydrous iron silicates, the iron content being less than 25 per cent. Vast deposits of it occur in the Lake Superior region. BANDED IRONSTONE.

tafelberg (Afrikaans) a MESA, a large TAFELKOP. G.

tafelkop (Afrikaans) a BUTTE, a KOP, an isolated hill with a flat top. SPIT-SKOP. G.

tafoni (Corsica: Italian) 1. hollowed-out granite blocks occurring in dry regions, as in northern Corsica. Dew at night and heating in the day cause mineral solutions to form a hard crust behind which the interior of the rock mass decomposes as the cementing material is removed. In extreme examples the interior disintegrates completely, and only the surface crust, an empty shell, remains 2. small and large recesses in rock faces in dry regions and in sea cliffs caused by differential SAPPING-2, a term adopted by Penck from the Corsican patois. G.

tafrogenesis, taphrogenesis the formation of RIFT phenomena in general, not restricted to those which can be attributed to TENSION alone. G.

tahsil, tehsil, taluk, taluq (Indian subcontinent: Urdu-Hindi) an administrative division in northern India, consisting of several villages. The states, if large enough, are divided into districts, each district being subdivided into tahsils (in southern India into taluks), normally three to eight to a district. In Bengal the unit smaller than the district is the THANA (police station area), much smaller than the average tahsil or taluk. TALLUQA. G.

taiga (Russian from Yakut, forest) cold woodland, the predominantly CONIFEROUS forest stretching in a broad zone in the northern hemisphere, adjacent to the TUNDRA. There does not seem to be any authority for the spelling taïga: etymologically tayga is more correct. G.

tail in statistics, the tapering end of a DISTRIBUTION-4 farthest away from the MEAN. If in a test only one end of the distribution is considered, the test is termed one tailed; if both end are considered it is termed two tailed.

tail dune a SAND DUNE of varying length, formed in the LEE of an obstacle and tapering away from it.

tailings the residual material remaining after mining or milling activities or after the extraction of metallic minerals from ores etc., considered at the time not to be worth further processing (in economic terms). TAIL-RACE-3.

tailo (Indian subcontinent) ADRET. SAYLO.

tail-race 1. in a WATER MILL, the part of a mill-race immediately beyond the mill-wheel 2. an artificial channel through which waste water is led back to a natural channel, having operated a water mill or passed through a generating plant 3. a channel in which the TAILINGS from treated ore are washed away. G.

takyr (Russian; Turkish, dry and hard) 1. an area of barren alkaline, unstructured clay soil, the surface forming a solid crust in summer, below which salts accumulate. In winter it may form a marshy bog 2. barren surface crusts of salt and clay in arid regions. G.

takyrisation (Russian) the formation of clay-salt crust in arid regions. TAKYR. G.

tala a minor basin with local interior drainage occurring in the great basin of the Gobi. GOBI. G.

talc hydrous magnesium silicate, occurring widely in nature, particularly in association with DOLOMITE, MARBLE or BASIC IGNEOUS ROCK rich in MAGNESIUM (e.g. PERIODOTITE), varieties being SOAPSTONE and FRENCH CHALK. It feels soapy, or greasy, and is used in solid form as a lubricant, or in powder form in the manufacture of paper, paint, cosmetics, ceramics, pharmaceutical products etc.

talik (Russian) 1. a layer of unfrozen ground between the seasonally frozen ground (ACTIVE LAYER) and the PERMAFROST 2. an unfrozen layer within the permafrost 3. unfrozen ground between permafrost. G.

tallow 1. a hard fat obtained by rendering the fat of sheep or cattle carcasses, used in candle and soap making 2. any of several fats resembling this hard fat but obtained from plants.

tālluqā, taluq, taluk (Indian subcontinent: Urdu, anglicized as taluk) in the Indian subcontinent, a political division of a district, larger than a TAHSIL. G.

talus 1. commonly applied to a SCREE a sloping heap of rock debris at the foot of a cliff or mountain slope 2. applied specifically by some authors to the landform produced by such rock debris, the term scree being restricted to the rock debris itself. DAMAN, TALUS CREEP. G.

talus creep the slow movement of TALUS-1 or SCREE down a slope under the influence of gravity. MASS MOVEMENT. G.

Talweg (German, valley way; older spelling Thalweg), the LONGITUDINAL-2 profile of a river. LONG PRO-

FILE OF A RIVER, RIVER PROFILE, VALLEY AXIS, VALLEY LINE. G.

Talwind (German, valley wind) a wind blowing up a valley, usually coinciding with an ANABATIC wind, in contrast to BERGWIND. G.

tangerine, mandarin, satsuma *Citrus reticulata*, a species of ORANGE, with many VARIETIES-2, hybridizing easily with many other CITRUS species, a small tropical or subtropical tree producing sweet fruit with a loose skin. It can withstand temperatures lower than those tolerated by the sweet orange, *Citrus sinensis*. The fruit is eaten raw, canned, or used in preserves.

tank in Indian subcontinent and Sri Lanka, a reservoir for irrigation, a small lake or pool made by damming the valley of a stream to retain the monsoon rain for later use. G.

tanker any large rail or road vehicle or ship especially designed with a large vessel (a tank) for the carrying of liquids in bulk.

tannery a place where hides and skins are tanned, i.e. treated with tanning agents (usually of vegetable origin for the heavyweight hides, a solution of chromium salts for the lightweight skins) to convert them to LEATHER. TANNIN.

tannia, yautia, new cocoyam *Xanthosoma sagittifolium*, a plant native to wet tropical areas, cultivated there on drier ground for its starchy tubers, used locally for food.

tannin, tannic acid any of the astringent, phenolic substances present in some plants, obtained from the bark and other tissues of selected species (e.g. WATTLE-3), used in tanning hides and skins, and in the manufacture of ink, dyes and pharmaceuticals. CULCH, LEATHER, TANNERY. C.

tansy *Tanacetum vulgare*, an aromatic PERENNIAL herb, native to the northern hemisphere midlatitude grassland, introduced to North America and New Zealand, formerly grown for

medicinal and culinary use, now used mainly as a flavouring in the latter.

tantalite an oxide of IRON, MANGANESE, TANTALUM and columbium.

tantalum an unreactive metallic element with a high melting-point, used in making special heat-resistant and corrosion-resistant alloys and in the manufacture of surgical instruments. Combined with carbon, it is used in less specialized tools, and as an abrasive. NIOBRIUM.

taphrogenesis TAFROGENESIS.

taphrogeosyncline a GRABEN or RIFT-VALLEY. G.

tapioca MANIOC.

tar a dark, thick, viscous substance obtained by the destructive DISTILLATION of coal, wood, or other organic material (COAL TAR). It is used in road surfacing (MACADAM), as a preservative for wood and iron, and in making dyes, antiseptics etc.

tarbet, tarburt (Scottish) 1. an ISTHMUS 2. a village situated on such a neck of land 3. portage between two navigable channels, or between two lochs.

tare the weight of a container or vehicle in which goods are packed and weighed, deducted from the gross weight (total weight) to deduce the net weight.

target area an area at which a particular effort or programme is aimed, e.g. a deprived neighbourhood in a city selected for a programme designed to upgrade the quality of the environment, social services etc.

tariff 1. a scale of duties imposed by a government on goods exported from or imported into the territory under its jurisdiction; or the classificatory instrument embodying such duties 2. the duty imposed 3. a scale of rates or charges, e.g. for hotel accommodation.

tarmac abbreviation for tarmacadam, MACADAM.

tarn a small lake among mountains, usually of glacial origin, fed by rainwater from the surrounding steep slopes rather than by a distinct feeder stream. Originally a local term in the north of England, it is now widely used by authors in the UK, some of whom apply it especially to a nearly circular lake occupying a corrie or CIRQUE; for that usage American authors prefer the term cirque lake. G.

taro (Pacific islands: Polynesian; West Indies eddo or dasheen; West Africa old cocoyam) *Colocasia antiquorum* **1.** a plant with an edible root, grown in wet or even slightly swampy parts of wet tropical areas for the sake of its easily digested, fine-grained starch, obtained from the thickened root (an underground corm) **2.** the thickened root of this plant.

tarragon *Artemisia dracunculus*, an aromatic, bushy PERENNIAL herb, native to southern Europe, grown preferably in warm, dry areas for seasoning savoury dishes and sauces. It should be used fresh because its ESSENTIAL OILS are dissipated in drying.

tar-sand OIL-SAND.

tartar a substance consisting mainly of acid potassium tartrate, deposited in wine barrels from grape juice. When mixed with sodium bicarbonate, it forms cream of tartar, a raising agent used in baking.

tartaric acid an acid occurring widely in plants, used in dyeing, textile printing, pharmaceuticals.

taung, daung (Burma) a mountain. G.

taungya, taung-ya, toungya (Burmese, mountain or hill field or plot) **1.** a temporary hillside clearing in SHIFTING CULTIVATION as practised by hill people in Burma **2.** a system of tropical forest management in which such clearings are sown with seeds of useful timber trees as well as with those of food crops in the last sowing before the plot is abandoned. G.

taxis pl. taxes, in biology, the movement or orientation of an organism or cell in response to the source of a stimulus. It may be towards (positive taxis) or away from (negative taxis) the stimulus. CHEMOTAXIS, GEOTAXIS, HYDROTAXIS, PHOTOTAXIS, RHEOTAXIS, THERMOTAXIS, THIGMOTAXIS; and KINESIS, TROPISM.

taxon any named group in TAXONOMY.

taxonomy the science of classification, usually restricted to the classification of living and extinct plants and animals and covering the principles and methods employed in classifying organisms in hierarchical groups. BIOSYSTEMATICS, CLADISTICS, CLASSIFICATION OF ORGANISMS.

tea *Camellia sinensis*, a product used in making the beverage of the same name, manufactured from the young leaves of a small EVERGREEN tree, usually pruned to form a shrub, native to hill lands of southeast Asia, now grown there and elsewhere in Asia, in east Africa and in areas bordering the Black Sea. It flourishes in high rainfall if the soil is well-drained, yields well on highly acid soils, withstands frost. Black tea is produced from fermented dried leaves, green tea from unfermented, oolong from the partially fermented leaves. The leaves contain CAFFEINE and THEOBROMINE. C.

teak *Tectona grandis*, a large DECIDUOUS tree, native to the relatively wet areas of tropical regions of southeast Asia, providing very hard, close-grained timber, used in shipbuilding and furniture-making.

tear fault a STRIKE-SLIP FAULT, sometimes termed a transcurrent fault.

technocracy **1.** organization and control by technical experts **2.** rule by technical experts. AUTOCRACY.

technological determinism in Marxism, the theory that the predominant type of TECHNOLOGY-3 of a society determines its other features, particularly its social organization.

technology **1.** the scientific study concerned with the practical and industrial arts **2.** the practical arts or practical science 3. the systematic application of scientific knowledge to industrial processes or to the problems arising from the interaction of people with their environment. LOW TECHNOLOGY, HIGH TECHNOLOGY 4. the technical terminology of a particular art or subject 5. as applied in the context of pre-historical or pre-industrial periods, the knowledge available for the making of tools and other artifacts, for the practice of manual skills and crafts, and for the extraction or collection of materials other than foodstuffs or those used in religious rituals etc.

technosphere the part of the physical environment created or modified by human action. PSYCHOSPHERE.

tectogenesis a collective term for all the processes involved in the deformation of the rocks of the earth's crust and the formation of structural features. OROGENESIS.

tectonic (Greek tektōn, builder) *adj.* of, relating to, or arising from, the processes which build up or form features of the earth's crust. PLATE TECTONICS, TECTOGENESIS.

tectonic creep very slow movement along a FAULT.

tectonics a branch of geology concerned with the study of the processes involved in the formation of structural features of the earth's crust. PLATE TECTONICS, TECTOGENESIS.

tectosequent *adj.* following the structure, applied to surface features, e.g. valleys, which reflect the underlying geological structure (STRUCTURAL LANDSCAPE) in contrast to MORPHOSEQUENT. G.

teff *Eragrostis abyssinica*, a tropical MILLET, the most widely grown food grain in Ethiopia (rarely cultivated elsewhere), yielding a fine white flour.

tegular *adj.* consisting of, being like, pertaining to, tiles.

tektite a small, greenish-black, glassy, rounded body, probably of meteoritic origin, found in groups in scattered

areas of the earth's surface, termed Australites in Australia. FULGURITE. G.

tele (Norwegian, older form taele; Swedish tjäle) frozen ground, a term often (many consider wrongly) applied to permanently frozen ground. TJALE. G.

teleologic, teleological *adj.* of, pertaining to, involving, TELEOLOGY, especially applied to the explanation of natural phenomena.

teleology consequentialism, the study of ends, goals or purposes, the doctrine of final causes. The belief that an EXPLANATION of anything (process, object, act, event etc.) can be achieved only if the ends to which it is directed are considered; that explanation restricted to terms of CAUSALITY is insufficient.

tell (Israel: Hebrew) **1.** a hill **2.** a hill site occupied many times or over successive periods. Tell appears as tel in place-names. G.

telluric *adj.* **1.** of or pertaining to the EARTH-1 **2.** of or pertaining to TELLURIUM, especially tellurium of high VALENCY.

tellurium a semi-metallic chemical element, similar to SULPHUR in its chemical behaviour, used as a contributary vulcanizing agent in the making of heavy-duty rubbers, and as an alloying element, especially in steels.

temperate *adj.* moderate, without extremes, of equable temperature, especially a climate without extremes of temperature. TEMPERATE ZONE. G.

temperate glacier WARM GLACIER.

temperate zone the term applied in classical times to one of the three latitudinal temperature ZONES-2, lying between the FRIGID and the TORRID, i.e. between the tropics (23°30′N or S) and the polar circles (66°30′N or S), the midlatitudes. But temperatures of great extremes occur in these midlatitude belts (especially in the interiors of the continents), so in making specific reference to them in terms of a climatic zone the term temperate is inappropriate: it is better to use the *adj.* MIDLATITUDE. Temperate zone does however appear in KOPPEN'S CLIMATIC CLASSIFICATION. SUBTEMPERATE ZONE. G.

temperature 1. the property of an object which indicates the direction in which heat energy will flow if the object is put in thermal contact with another, heat energy flowing from places of higher to places of lower temperature **2.** as a climatic element, the degree of sensible heat or cold in the ATMOSPHERE-1 (SENSIBLE TEMPERATURE), measured on various scales (TEMPERATURE SCALE). ABSOLUTE as applied to range, temperature, zero; and COMFORT ZONE, DIURNAL RANGE, INVERSION OF TEMPERATURE, POTENTIAL TEMPERATURE, TEMPERATURE ANOMALY, THAW, THERMOMETER. G.

temperature anomaly the difference between the mean temperature of a place and the mean temperature along its parallel of latitude (in both cases adjusted to sea-level). If the mean temperature of the place is higher than that of the mean along its parallel of latitude, the anomaly is qualified as positive; if lower, as negative. The British Isles is much warmer than the average of places on the same latitude, and thus has a positive temperature anomaly, especially in winter. ANOMALY, ISANOMAL. G.

temperature-humidity index an index drawn up according to various formulae to indicate the effects of weather conditions on human comfort. COMFORT ZONE, SENSIBLE TEMPERATURE.

temperature inversion INVERSION OF TEMPERATURE.

temperature scale a sequence of values representing TEMPERATURE, two fixed points with specific properties (e.g. ice point, steam point of water) being selected and subdivisions made between them, commonly shown on a

THERMOMETER. CELSIUS, CENTIGRADE, FAHRENHEIT, KELVIN, REAUMUR.

tempest a literary term applied to a violent wind, accompanied by heavy rain, hail or snow.

temporal *adj.* of, related to, pertaining to, time (as distinct from SPATIAL).

temporales (Spanish) a strong, monsoon-type wind blowing from the southwest in summer on the Pacific coasts of Central America. G.

temporal mode in explanation explanation in which the tracing of origins and the subsequent development of phenomena according to process laws is used to account for the current state of spatial distributions, i.e. the explanation of pattern through process, involving time as well as space. G.

temporary exclave EXCLAVE.

temporary grass ROTATION GRASS. G.

tenancy LAND TENURE.

tenant 1. in general, current use, someone who holds a piece of land, a building, part of a building etc. by lease for a set time **2.** in law, in England and Wales, someone who holds REAL ESTATE by any kind of right. LAND TENURE, PUBLIC HOUSING.

tenant capital the movable equipment contributed by the tenant in an agricultural tenancy system, i.e. stock, seed, fertilizers, machinery and cash. The responsibility for the provision of these is however sometimes varied by agreement with the landlord. LANDLORD CAPITAL, LAND TENURE, SHARECROPPING.

tenant farmer one who farms land over which another has superior rights of tenure, and to whom in return the farmer pays rent in money or produce. LAND TENURE, TENANT, TENANT CAPITAL, SHARECROPPING.

tension a pulling force, tending to stretch, to cause the extension of a body or to restore the shape of an extended elastic object. In the rocks of the earth's crust tension extends strata, resulting in

JOINTS and NORMAL FAULTS. Compare COMPRESSION, STRESS, THRUST.

tent-hill (Australia) a TEPEE BUTTE, a residual, conical, isolated hill in Australia, so-named on account of an apparent resemblance to a canvas tent, although the top of the tent-hill is often a flat remnant of a former plateau surface. G.

tenure 1. the act, manner or right of holding office or property, especially landed PROPERTY-2 2. the period of holding this. LAND TENURE.

tepee, teepee a conical tent, especially of skins stretched over a framework of poles, formerly used by North American Indians. WIGWAM.

tepee butte a residual conical hill, especially in the Painted Desert, Arizona. The soft horizontal rocks of the plateau edge have been easily and rapidly eroded, but in places a capping of harder sandstone protects the underlying strata so that an isolated conical hill, similar to a TENT-HILL in Ausralia, may form.

tephigram (derived from t, temperature; phi, entropy) a thermodynamic diagram showing the condition of the ATMOSPHERE-1 at different levels in terms of its temperature and entropy. G.

tephra a collective term for all the solid material ejected into the air from a volcanic vent during an eruption, i.e. PYROCLAST, e.g. ASH, BOMBS, CINDERS, LAPILLI, PUMICE, SCORIA, VOLCANIC DUST. TUFF. G.

tephrochronology the science of dating the layers of volcanic ash deposited and buried around volcanoes, successfully used as a basis for the drawing-up of a GEOLOGICAL TIMESCALE in a land where volcanic eruptions are frequent, e.g. Iceland. G.

tequila a strong, intoxicating beverage made by distillation from various species of agave, especially *Agave tequilana* in Mexico.

terai (Indian subcontinent: Urdu-Hindi; duār in Bengal) originally a swampy, unhealthy area in the north of the subcontinent, some 80 to 95 km (50 to 60 mi) wide between the Himalayas and the plains, now largely drained and cultivated. G.

teras (Sudan: colloquial Arabic, pl. turus) an artificial ridge built with hand tools to contain runoff rainwater on an area selected for cultivation in the drier parts of northern Sudan. G.

teratogenesis the production of monsters, of malformed or otherwise defective organisms. TERATOGENETIC.

teratogenetic, teratogenic *adj.* of or pertaining to TERATOGENESIS.

teratogenic pollution the kind of POLLUTION which gives rise to the birth of malformed or otherwise defective organisms.

terlough in western Ireland, a shallow depression with a hole through which water drains, but which holds water when the WATER TABLE rises sufficiently.

terminal TERMINUS.

terminal *adj.* 1. of or forming the end 2. growing at the end of a stem or branch 3. of disease, likely to be fatal 4. occurring each term 5. of or relating to a TERMINUS.

terminal moraine, end moraine a crescent-shaped MORAINE forming a ridge beyond the SNOUT of a GLACIER or at the end of an ice sheet, marking the furthest extent of the ice. If it is large it indicates a long pause in the retreat of the ice. G.

terminus, terminal 1. an end or extremity, e.g. either end of a rail, air or bus route 2. the station buildings, town, etc. at either of the ends of a transport route.

termite a member of Isoptera, an order of very small tropical social insects which build an elaborate nest (termitarium) consisting of a chambered earthmound, frequently 3 to 3.5 m (10 to 12 ft) in height, some reaching 8 m (26 ft) or more. The termites, of which there are very many species, cause great damage to wood.

terms of trade the quantity of purchases (imports) for which a given quantity of sales (exports) will exchange. The concept is usually applied to the imports and exports of a country, although it can be applied to dealings within the country, e.g. between regions, or between one product and another, or between one particular industry and the rest of the economy. Changes in terms of trade are usually shown by an index number which, as it rises, indicates improvement. It is calculated by dividing an index number of prices of sales (exports) by an index number of prices of purchases (imports). BALANCE OF PAYMENTS, BALANCE OF TRADE.

terrace 1. in agriculture, one of a series of horizontal steps cut into a hillside to provide cultivable land in an area of steep relief, and to reduce soil erosion. TERRACE CULTIVATION 2. in architecture, a row of houses joined together 3. a row of houses built on raised ground or on a hillside 4. a level, raised area, usually paved, adjoining a building, or made on sloping ground, or on a river bank, where pedestrians may walk or sit 5. in geology, a level or nearly level and horizontal or nearly horizontal strip of land, usually narrow and bordering the sea, a lake, or a river (RIVER TERRACE), lying between a rising slope on one side and a downward, often abrupt, slope on the other. ALLUVIAL TERRACE, CONTINENTAL TERRACE, KAME TERRACE, MEANDER TERRACE; and REJUVENATION, TERRACE GRAVEL, TERRACETTE. G.

terrace cultivation a system of cultivation which reduces SOIL EROSION, practised in areas of steep relief and scarcity of level land. A series of artificial horizontal steps is cut into the hillside, the soil being retained by stone walls or earth banks, behind which crops are sown. If irrigation is necess-

ary (e.g. in rice cultivation) the water is allowed to move by gravity from the upper to the lower terraces. G.

terrace gravel, terrace-gravel a gravel deposit remaining on a RIVER TERRACE after the erosion of the finer alluvium with which it was combined in the original deposition.

terracette one of a series of narrow horizontal steps from a few centimetres to 60 cm (up to 2 ft) in height, making a ribbed pattern on a steep slope on which SOIL CREEP occurs. The alternative term, sheep track, is considered by some authors to be misleading. They maintain that, once formed, terracettes may be used by sheep and other animals; but that there is no positive proof that they have been formed by animal treading. LYNCHET. G.

terracing the work of making terraces for TERRACE CULTIVATION.

terra firma (Latin, firm earth) land as opposed to the water.

terra firme (Brazil: Portugese) the higher well-drained uplands, free from flood water, in the Amazon basin. VARZEA. G.

terra fusca a neutral or slightly alkaline brown clay LOAM developed on limestone in areas with a warm temperate, seasonally dry climate, e.g. in parts of the Mediterranean area, where it supports MAQUIS.

terrain (American terrane) an area of land in respect of its physical characteristics or condition, especially if considered for its fitness or use for a special purpose, e.g. for laying a railway track, or for a military operation. TERRANE. G.

terrain analogue a type of geomorphic spatial analogue, in which a region is divided into component landscapes, each defined in terms of four measurable terrain factors (characteristic slope, relief, plan profile, occurrence of steep slopes) to form terrain types with which other regions can be compared and classified. G.

terral (South America) the land breeze

blowing along the coasts of Peru and Chile. VIRAZON. G.

terrane in geology, a term now seldom used, formerly applied to a group of strata, a zone, or a series of rocks. G.

terra rossa (Italian) a red-coloured thin clay LOAM soil, rich in iron sesquioxides, developed in limestone areas with a warm temperate, seasonally dry climate, occurring especially in KARST in Yugoslavia, and elsewhere in the Mediterranean region. It used to be considered to be a residual soil derived from the weathering and partial solution of the limestone; but its lack of humus and the fact that it supports GARIGUE suggests that deforestation may have contributed to its existence. G.

terra roxa (Brazil: Portuguese, purple soil) a deep, dark red-purple, porous soil, rich in humus, a type of LATOSOL, formed on the Parana plateau in Brazil, especially suitable for coffee cultivation.

terrestrial *adj.* **1.** of or pertaining to land, or to the EARTH-1 (as opposed to CELESTIAL) **2.** consisting of land (as opposed to water) **3.** in botany, growing on land **4.** in zoology, living on land. AQUATIC.

terrestrial deposits in geology, deposits laid down on land, as opposed to MARINE-1 deposits. G.

terrestrial magnetism the MAGNETIC FIELD of the earth as a whole, weak and varying in intensity and direction (MAGNETIC DECLINATION, PALAEO-MAGNETISM) which causes the needle of a magnetic COMPASS, swinging freely in a horizontal plane, to come to rest, indicating the north and south MAGNETIC POLES. The origin of terrestrial magnetism has not yet been satisfactorily accounted for. MAGNETISM, MAGNETOSPHERE.

terrestrial radiation long-wave RADIATION given out by the earth. G.

terrier, terrar (other spellings) a register of landed property, especially a book complete with maps, plans, details of tenancies etc., usually kept for a large estate, some dating from the Middle Ages in England, of immense value to the historical geographer. G.

terrigenous *adj.* derived from the land.

terrigenous deposits inorganic deposits (sand, gravel, pebbles etc.) derived from the denudation of the land and laid down in the LITTORAL ZONE-1 of the sea floor, as distinct from PELAGIC DEPOSITS. G.

terriherbosa herbaceous vegetation growing on dry land.

territorial *adj.* of, belonging to or relating to TERRITORY-1,2,3,4,5 or a territory **2.** of or limited to a specific territory.

territorial sea TERRITORIAL WATERS.

territorial social indicator in WELFARE GEOGRAPHY, a SOCIAL INDICATOR that indicates areal differentiation at various spatial scales. It is used particularly in measuring and comparing the distribution of social goods and bads in different areas.

territorial waters the coastal waters with the sea bed below and all that lies or lives therein (and the air space above) over which a coastal state has sovereignty. The seaward limit recognized in the eighteenth century was usually 5.5 km (3 international nautical mi) from the shore (more in estuaries), then the range of shore-based cannon; but this limit was not adopted by all coastal states (later variations are listed in successive editions of the *Statesman's Yearbook*, Macmillan Press). After the Second World War, the United Nations organized a series of Law of the Sea Conventions in an attempt to introduce uniformity.

The Territorial Sea Convention 1958 established the use of the term 'Territorial Sea', applied to the zone over which a coastal state has sovereignty, and provided a means for defining its landward baseline, a difficult problem for countries with a deeply indented coastline. A line drawn to link major promontories is deemed to be the baseline; for the treatment of bays, see BAY-2. But the 1958 Convention did not prescribe a maximum outer (i.e. seaward) limit. This eventually appeared in the Law of the Sea Convention opened for signature and ratification in 1983: it is 22.2 km (12 international nautical mi) from the baseline.

The 1958 Convention did however introduce the idea of a Contiguous Zone, a buffer zone stretching 44.45 km (24 international nautical mi) from the baseline, over which a coastal state has extended powers of policing (to prevent infringement of its customs, fiscal, immigration and sanitary regulations etc.).

And another Law of the Sea Convention introduced the idea of the so-called Continental Shelf, an arbitrarily defined area over which a coastal state has sovereign rights over the sea bed, to explore and exploit its natural resources. The Law of the Sea Convention ratified in 1983 has provided this Continental Shelf with a seaward limit of 370.4 km (200 international nautical mi) or such a limit coinciding with the outer edge of the CONTINENTAL MARGIN, providing this is not beyond 648.2 km (350 international nautical mi) from the baseline. This Law of the Sea Convention also introduced another area of coastal jurisdiction, the Exclusive Economic Zone, in order to extend a coastal state's rights of sovereignty over the natural resources of the waters as well as those over the natural resources of the sea bed. The Exclusive Economic Zone also extends from the baseline to a seaward limit of 370.4 km (200 international nautical miles). Where the distance between coastal states is less than 740.8 km (400 international nautical mi) a median line is usually drawn between the closest points of the baselines of the states concerned.

The traditional freedom of the High Seas retains for all nations the right of navigation, overflight and laying of submarine cables in the areas of the so-called Continental shelf and the Exclusive Economic Zone. Legal problems relating to rights of ownership of the deep sea bed and of rights of exploitation of its resources have yet to be resolved.

territory 1. the area of land and adjacent seas, and the air space over both, ruled by a sovereign authority 2. an area dependent on a sovereign state, but having some autonomy, e.g. an area supervised by a sovereign state on behalf of the United Nations 3. historically, in Australia, Canada, USA, an area not admitted to full rights as a state or province, having a separate legislature under an administrative authority appointed by the central government 4. any large tract of land, a region, a district, with undefined boundaries 5. in zoology, the area of the habitat occupied by an individual animal or group of animals which will be defended by them, attacks being made especially against a trespasser or trespassers belonging to the same SPECIES-1 as the occupant(s). G.

tertiary third in order or rank.

Tertiary *adj.* a much discussed term applied to a division of geological time. It is now generally accepted that it applies to the third period of geological time with its associated system of rocks (following the CRETACEOUS and preceding the QUATERNARY) consisting of the PALAEOCENE, EOCENE, OLIGOCENE, MIOCENE, PLIOCENE epochs (of time) and series (of rocks) when MAMMALS became dominant and mountain chains such as the Alps and Himalayas were formed, i.e. the earlier of the two periods of the CAINOZOC era (GEOLOGICAL TIMESCALE). But some geologists regard Tertiary as one of the four eras, with its asociated rocks and with its four periods, Eocene, Oligocene, Miocene, Pliocene; others regard it as an era, with its associated rocks, but divide it into two periods (systems) entitled PALAEOGENE and NEOGENE, the latter including the HOLOCENE (Recent), the term QUATERNARY being avoided.

tertiary industry, tertiary activity, tertiary sector one of the main categories of INDUSTRY, the activity concerned with service to the PRIMARY and SECONDARY INDUSTRIES, to the community and to the individual, e.g. financial, commercial and educational institutions, distributive trades, professions, transport and communications, construction, repairs, maintenance, defence, personal services. QUATERNARY INDUSTRY.

test factor in statistics, a third variable brought into an analysis with the purpose of discovering whether an apparent correlation between the other two variables is genuine or not. SPURIOUS CORRELATION.

Tethys in PALAEOGEOGRAPHY, the name applied to the ocean and the GEOSYNCLINE which it occupied, separating LAURASIA and GONDWANALAND. CONTINENTAL DRIFT, PANGAEA, PLATE TECTONICS. G.

tetraethyl lead a poisonous liquid, a compound of lead, added to petrol to increase its octane number and thus prevent 'knocking' in the internal combustion engine.

tetrahedral theory the hypothesis formulated by Lothian Green, 1875, to account for the apparent symmetry of the great land masses. He maintained that the crust of the earth, in cooling, was warping towards the shape of a TETRAHEDRON, with its apex at the South Pole, its base at the North Pole; a theory now discarded. TETRAHEDRON. G.

tetrahedrite a grey sulphide mineral, usually consisting of tetrahedral (TETRAHEDRON) crystals, occurring in

HYDROTHERMAL veins, an ore of copper and of antimony.

tetrahedron in geometry, a solid figure contained by four plane triangular faces. TETRAHEDRAL THEORY.

tetrapod a four-footed animal, especially one of the higher vertebrates. TETRAPODA.

tetrapod *adj.* having four feet or four limbs.

tetrapoda a group of land-living VERTEBRATES, characterized by having two pairs of limbs (including amphibia, reptiles, birds and mammals), sometimes used in classification.

textile a woven fabric.

texture 1. in geology, petrology, the size, shape, arrangement and distribution of particles constituting a rock or a surface deposit, as opposed to the chemical character of such particles 2. the arrangement and relationship of particles of a soil (SOIL TEXTURE) as opposed to their chemical character 3. occasionally also applied to topography, i.e. coarse-textured topography (relief in regions of massive and resistant rocks, where the plateaus are large and bold and the streams far apart), or the opposite, fine-textured topography (relief with streams closely spaced) which, if extremely fine, becomes BADLAND topography. G.

thal (Indian subcontinent: Panjabi) a sandy waste or desert, especially in the central part of the Sind-Sagar DOAB between the Jhelum-Chenab and the Indus. THAR. G.

thalassic *adj.* of or relating to the sea or ocean, sometimes distinguished from oceanic in being of or relating to the bays, gulfs, small bodies of salt water or of inland seas. THALASSOGRAPHY.

thalassography OCEANOGRAPHY. G.

thalassostatic *adj.* 1. related to a period of static sea-level 2. applied to a RIVER TERRACE resulting from fluctuations in sea level. EUSTATISM. G.

thallium a malleable, metallic element,

very poisonous in compounds, used in pesticides.

Thallophyta a large group or division of primitive plants (algae, bacteria, fungi, lichen, slime fungi) characterized by a simple plant body (THALLUS), probably the main plants living in Precambrian - early Palaeozoic times. GEOLOGICAL TIMESCALE.

thallus a simple, vegetative plant-body, not differentiated into root, stem, leaf. It may be unicellular, or multicellular with branched or unbranched filaments, flattened and ribbon-like (as in brown SEAWEED). THALLOPHYTA.

Thalweg TALWEG.

thānā (Indian subcontinent: Urdu) a political division of a district in Bengal, which is under the jurisdiction of a single police station; it may thus be defined as a police station area. MOUZA, TAHSIL. G.

thanatocoenosis a group of fossils consisting of the remains of organisms assembled after death, in contrast with BIOCOENOSIS-3.

thane, thegn historical 1. in England, Anglo-Saxon period, a freeman holding land in return for service to a nobleman 2. in Scotland, someone holding lands of the king.

thar (Indian subcontinent: Sindhi) a desert, a sandy waste, especially The Thar, the great desert in the northwest of the Indian subcontinent. THAL. G.

thatch 1. a covering, especially of a roof, made with reeds, straw, rushes etc. 2. the material used for that purpose.

thaw 1. the physical change as anything frozen (e.g. ice, snow) becomes soft and liquid due to the rising of the TEMPERATURE above FREEZING POINT 2. the period of time when this occurs. THERMOKARST.

thaw depression a depression in the ground resulting from subsidence following the thawing (THAW) of perennially frozen ground. THERMOKARST.

thaw lake CAVE-IN LAKE, a lake occupying a THAW DEPRESSION. G.

thaw sink a closed depression with subterranean drainage, believed to have originated as a THAW LAKE. G.

thematic *adj.* of, pertaining to or constituting the topic, the main subject with which a study, discussion, piece of writing etc. is concerned. G.

thematic map a map on any scale representing a specific spatial distribution, theme, topic or aspect under discussion. G.

theobromine a bitter, crystalline alkaloid similar to CAFFEINE contained in CACAO beans, KOLA nuts and the leaves of TEA, used medicinally as a heart stimulant and diuretic.

theocracy 1. government by those claiming to know the will of God 2. a state governed by such people. AUTOCRACY.

theodolite an optical surveying instrument used to measure vertical and horizontal angles by means of a small telescope, a spirit level, and graduated arcs, mounted on a tripod. TACHEOMETER, TRIANGULATION.

theory 1. in general, an organized body of ideas, an integrated system of HYPOTHESES, put forward as the truth of something, supported by a number of facts relating to it, but sometimes resulting from speculation. FORMAL THEORY, GROUNDED THEORY 2. scientific, a structure resting on a series of steps of observations and assumptions, each supported by the preceding step, put forward to explain a particular class of phenomena 3. a process of investigation based on logical or mathematical reasoning rather than on experiment. HYPOTHESIS, LAW.

thermal a current or updraught of air rising vertically in the ATMOSPHERE-1 (of great advantage to birds and glider-pilots), the result of differential heating by the sun's rays on a sunny day of small parts of the earth's surface. It is the parts that warm up more rapidly which give rise to conductional heating and ABSOLUTE INSTABILITY and thus to rapidly ascending vertical currents. A thermal may rise high enough for condensation to occur, leading to the formation of CUMULUS cloud, in some cases associated with heavy CONVECTION RAIN and a THUNDERSTORM.

thermal *adj.* pertaining to heat.

thermal conduction the process of transfer of heat through a body where there is a temperature gradient, the heat energy diffusing through the body by the action of particles of high KINETIC ENERGY on particles of lower kinetic energy (from a high to a low temperature point). There is no visible movement of any part of the body. CONVECTION, RADIATION.

thermal depression a LOW pressure system, usually intense, varying in size, caused by localized heating of the earth's surface, leading to convectional rising of air, resulting in heavy rainfall and thunderstorms if the warmed air rises high enough for condensation to occur (THERMAL). Small scale thermal depressions lead to DUST DEVILS in hot deserts and the SIMOOM in the Sahara; on a larger scale they are associated with MONSOON conditions.

thermal efficiency index THORNTHWAITE'S CLIMATIC CLASSIFICATION.

thermal electricity electricity produced by a process in which the mechanical energy (converted to electricity by a dynamo) is produced by steam-turbines, fuelled by PEAT, BROWN COAL, LIGNITE, COAL, GAS, OIL or NUCLEAR ENERGY. CARBOELECTRICITY, HYDROELECTRICITY.

thermal equator, heat equator an imaginary line drawn round the earth joining the places on each meridian with the highest mean temperature for any particular period. The line moves north and south with the seasons, following the apparent movement of the sun; but the greater land masses in the northern hemisphere result in greater heating, so

on the whole the thermal equator lies to the north of the 'true' EQUATOR. SWING OF PRESSURE BELTS. G.

thermal erosion a type of EROSION occurring in areas of PERMAFROST where, if the organic layer protecting the land surface is removed, the GROUND ICE-2 melts and the land surface breaks up.

thermal fracture the cracking or fissuring of rocks caused by a sudden change of temperature, occurring particularly in rocks containing minerals of differing rates of expansion and contraction; or on a steep rock face exposed in such a way to the sun's rays that it quickly passes from being in the sunlight to being in the shade; or where there is a very rapid and considerable drop in temperature at sundown, all occurring especially in the presence of some moisture.

thermal (infra-red) sensing REMOTE SENSING in which a thermal scanner, penetrating darkness and cloud, is used rather than a camera to sense the natural RADIATION emitted by features on the earth's surface and, by revealing relative temperature differences, to detect pollution in water, the source of forest fires etc. INFRA-RED RADIATION.

thermal metamorphism contact metamorphism, the alteration of pre-existing rock to form a new, well-defined type of rock, caused by a rise in temperature, usually brought about by an INTRUSION into the pre-existing rock of very hot, molten IGNEOUS ROCK. METAMORPHIC AUREOLE, METAMORPHISM, REGIONAL METAMORPHISM. G.

thermal pollution the heating of part of the environment by the discharge of substances with TEMPERATURES higher than that of the AMBIENT. The effect is particularly detrimental in freshwater because heating tends to lower its content of free dissolved oxygen, needed by most of the organisms living in it. POLLUTION.

thermal spring hot spring, a continuous flow of hot water from the ground, usually (but not always) associated with present or former volcanic activity. It contrasts with a GEYSER, with its violent, intermittent emission. GEOTHERMAL ENERGY.

thermal strain diagram a climogram or CLIMOGRAPH showing the effects of climate, especially of heat and moisture, on human beings. G.

thermal stratification the succession of well-defined layers of water of different temperatures lying at various depths in the ocean or a deep lake, the top layer being the warmest, except under ice, where there may be an inversion or, in very cold areas, just one unstratified layer of very cold water. EPILIMNION, HYPOLIMNION, THERMOCLINE.

thermoclasty the disintegration of rocks caused (theoretically) by their alternating expansion and contraction in accordance with the daily range of temperature. The theory has not been substantiated by laboratory experiments. HALOCLASTY, WEATHERING.

thermocline the layer of water in an ocean or deep lake, lying between the non-circulating HYPOLIMNION and the warmer EPILIMNION, through which the temperature falls swiftly with increasing depth, commonly exceeding 1°C per metre (about 11°F per foot) of descent. THERMAL STRATIFICATION.

thermodynamics the branch of physics concerned with the relationship between heat and other forms of energy.

thermoelectricity electric current produced by the direct transformation of heat energy into electrical energy.

thermogenesis the generation or production of heat by physical or chemical changes, especially within the body of an animal, e.g. by the oxidation of foodstuffs.

thermogenetic *adj.* of THERMOGENESIS.

thermogenic *adj.* of or relating to THERMOGENESIS.

thermograph a self-recording thermometer. A continuous trace of the air temperature is recorded on a chart fixed to a drum rotated by clockwork.

thermohaline circulation the circulation of the water of the ocean caused by differences in salinity and temperature, a process fundamental to the flow of ocean CURRENTS-1.

thermokarst, cryokarst a KARST-like landform with irregular depressions formed in periglacial or former periglacial superficial deposits as a result of the melting of GROUND ICE-2 and the subsequent settling or caving of the ground. The term is not widely used by British geomorphologists, but is favoured by authors of the USSR. THAW DEPRESSION. G.

thermokarst pit a steep-walled depression formed by the thermokarstic process. THERMOKARST. G.

thermomeion MEION, an area with a high negative TEMPERATURE ANOMALY. ANTIPLEION.

thermolabile *adj.* unstable, losing characteristic properties when heated, the opposite of THERMOSTABLE.

thermometer an instrument used to measure TEMPERATURE on any TEMPERATURE SCALE, commonly consisting of a graduated glass tube with a bulb at one end containing mercury or alcohol which is heat-sensitive, expanding with increase in heat, and therefore rising in the tube; contracting with decrease, and falling in the tube (DRY-BULB THERMOMETER, MAXIMUM-MINIMUM THERMOMETER, WET-BULB THERMOMETER). Less common types incorporate heat-sensitive metals which expand or contract with temperature change, at a known rate; or which, with temperature change, have varying resistance to the passage of electricity. There is also a gas thermometer which measures pressure variations in a gas maintained at constant volume.

thermonasty NASTIC MOVEMENT.

thermonuclear *adj.* relating to NUCLEAR FUSION.

thermophile an organism which grows best at temperatures exceeding 45°C (113°F), e.g. BACTERIA in dung heaps. MESOPHILE, PSYCHROPHILE.

thermophilic *adj.* of or relating to a THERMOPHILE.

thermoplastic a synthetic substance, such as polystyrene, which becomes soft or plastic when heated and rigid again when cooled. PLASTIC MATERIAL.

thermoplastic *adj.* becoming soft or plastic when heated and rigid again when cool.

thermopleion an area with a high positive TEMPERATURE ANOMALY. ANTIPLEION, ISANOMAL, PLEION.

thermosetting *adj.* of RESINS-2 and PLASTIC MATERIALS which, having been once heated and compressed, are resistant to further heat treatment.

thermosphere the layer of the upper atmosphere (above the MESOPAUSE) in which temperature increases with increasing height.

thermostable *adj.* not losing characteristic properties when heated, the opposite of THERMOLABILE.

thermotaxis TAXIS in which heat is the source of stimulus.

thermotropism TROPISM in response to the stimulus of variation in temperature.

therophytes a class of RAUNKIAER'S LIFE FORMS, plants which complete their life cycle within a single favourable season and survive the unfavourable season by remaining dormant in the form of seeds, i.e. ANNUALS. G.

thicket in general, a WOOD-2, usually small, with closely set trees and dense undergrowth, a term which has no precise ecological application. G.

Thiessen polygon DIRICHLET POLYGON.

thigmotaxis stereotaxis, TAXIS in which touch is the source of stimulus.

thigmotropism haptotropism, TROPISM in response to the stimulus of the touch of a solid body in localized contact.

thionic *adj.* relating to or containing SULPHUR.

Third World (from French tiers monde) a term of variously defined origin. It was applied (as Tiers Monde) by Alfred Sauvy in the 'cold war' period of the 1950s to countries committed to neither of the two major power blocs, i.e. neither to the generally free market economy Western 'capitalist' bloc (the FIRST WORLD) nor to the centrally controlled economy Eastern 'communist' bloc (the SECOND WORLD). The term Third World is now generally applied to countries considered to be not yet fully DEVELOPED (UNDERDEVELOPED) primarily in economic but also in cultural and social terms, though it is still perhaps seen as consisting of unaligned countries, outside both the 'capitalist' and 'communist' world. Conventionally the Third World includes most countries in Asia, Africa, Latin America; but opinions differ as to whether China, Vietnam, Cuba, South Africa, Israel and the oil-rich nation states of the Middle East should be included; and the United Nations has singled out from the Third World some 25 countries designated as least developed. This group is sometimes termed the Fourth World. BRANDT REPORT.

thixotropy the property of certain gels, of certain solid/liquid materials (e.g. some CLAY minerals, some types of paint) to become liquefied or mobile on being disturbed (e.g. by being shaken, stirred, subjected to shearing stress), settling to a solid/liquid again when the disturbance ceases. G.

tholoid a dome-shaped volcanic plug. VOLCANIC NECK. G.

thorite a naturally occurring silicate of THORIUM.

thorium a dark grey, heavy, radioactive metallic element compound, occurring in MONAZITE and THORITE and in small quantity in PITCHBLENDE, used in alloys and in nuclear reactors. RADIOGENIC HEAT.

thorn forest, thorn woodland a general term applied to tropical or subtropical forest or woodland of small, XEROPHILOUS thorny trees, the thorns giving partial protection from browsing animals. G.

Thornthwaite's climatic classification a classification of climates developed by C. W. Thornthwaite, American climatologist, between 1931 and 1948, based on the effectiveness of climate in the development of plant communities. Identifying rainfall and temperature as the dominant, variable influences, he drew up formulae to assess them. His index for measuring PRECIPITATION in terms of its usefulness to plant growth is:

$$P/E = 11 \cdot 5 \left(\frac{p}{T-10} \right) \frac{10}{9}$$

P/E being precipitation efficiency, p the monthly mean precipitation in inches, T the monthly mean temperature. Five major regions were distinguished on that basis. He subdivided these on the basis of thermal efficiency (the heat received by a particular area in relation to the production of POTENTIAL EVAPOTRANSPIRATION). His index for thermal efficiency being:

$$TE = \frac{T-32}{4}$$

where TE = thermal efficiency, T = mean monthly temperature in °F. Finally he drew up a moisture index, to show whether an area has a positive or negative water balance, expressed as:

$$MI = \frac{100(P-PE)}{PE}$$

where MI = moisture index, P = precipitation, PE = potential evapotranspiration. Using those indices he distinguished the climatic zones: per-

humid (A), MI exceeding 100; humid (B, with four subdivisions), MI 20 to 100; moist subhumid (C$_2$), MI 0 to 20; dry subhumid (C$_1$), MI -20 to 0; semi-arid (D), MI -40 to -20; arid (E) with MI below -40. KOPPEN'S CLIMATIC CLASSIFICATION.

thoroughfare a road or passage open at each end and available to pedestrian and/or wheeled traffic.

three-field system a system of cultivation common in medieval England whereby the arable land was divided in three large fields (open fields), worked in common by the village, one being fallow or resting each year while the other two were cropped with wheat or rye and barley or oats. The system disappeared with enclosure, the discovery that CLOVERS enrich the soil and that fallow is largely unnecessary. G.

threshold **1.** a plank or stone at the bottom of a doorway, the sill of a doorway (FJORD) **2.** the point of beginning **3.** the point at which a stimulus of increasing strength is first perceived or provokes a specific response, the point of transition from one state to another **4.** in CENTRAL PLACE THEORY, the threshold of success, the lowest demand necessary to ensure that any good, service or function will be offered at a CENTRAL PLACE. Demand may be measured in terms of population (THRESHOLD POPULATION) or income per caput.

threshold population the minimum number of people needed in an area to support a function, service or provision of goods. CENTRAL GOOD, CENTRAL PLACE THEORY, THRESHOLD-4.

throughflow the flow of water down a slope through the REGOLITH, as distinct from OVERLAND FLOW. It occurs when the quanity of water falling on the ground surface, or the rate at which it falls, is too great for it all to percolate sufficiently swiftly downwards through the upper SOIL HORIZON(s). Lateral ELUVIATION results from the carrying of soil particles by the throughflow. PIPE-4, RUNOFF, SUB-SURFACE WASH.

throw of a fault the vertical displacement of strata or rocks in a FAULT, varying from a few millimetres to hundreds of metres in extent. The rocks on one side of the fault-line are termed upthrow, on the other side downthrow, indicating the displacement of each in relation to the other. HEAVE OF FAULT.

thrust 1. a force tending to compress, similar to TENSION but acting in the opposite direction. STRESS 2. in geology, a compressional force affecting strata in an almost horizontal plane, leading to a very marked RECUMBENT FOLD or a REVERSE FAULT of very low angle. OVERTHRUST, THRUST FAULT, THRUST PLANE.

thrust fault a low-angled REVERSE FAULT, with the beds of the upper limb pushed far forward over the beds of the lower. FAULT, NAPPE, THRUST PLANE. G.

thrust plane the surface, usually inclined at a low angle and not strictly a plane, over which the upper strata of a REVERSE FAULT are pushed. FAULT, MYLONITE. G.

thunder the sound produced by the explosive expansion of suddenly heated gases in the ATMOSPHERE-1 resulting from the expending of electrical energy in LIGHTNING.

thunderhead the top part of a CUMULONIMBUS cloud, apparently with shining white edges, associated with the onset of a THUNDERSTORM.

thunderstorm a STORM of heavy rain and/or hail and wind, with LIGHTNING and THUNDER, occurring when intense heating of the ground surface leads to strong upward air currents, great ATMOSPHERIC INSTABILITY and the formation of CUMULONIMBUS clouds. Thunderstorms are frequent in tropical and equatorial regions, where air masses are warm and moist; and they are also associated with the passage of

a COLD FRONT (termed a frontal thunderstorm).

thyme a member of *Thymus*, family Labiateae, a genus of low, aromatic, small sub-shrubs with small greyish or green aromatic leaves, native to the Mediterranean region, used (especially *Thymus vulgaris*) for seasoning food. The leaves yield an ESSENTIAL OIL used in THYMOL.

thymol an aromatic phenol derived from the oil of THYME leaves, used as an antiseptic.

tibba (Indian subcontinent: Panjabi) 1. sand or sand hills 2. desert. G.

tidal *adj*. or of pertaining to, due to, affected by, the TIDE.

tidal barrage a barrier built on the seaward side of the reservoir in a TIDAL POWER STATION.

tidal basin a BASIN-10 filled with water at high tide that can be held and released at low tide, the force of the outflowing water scouring the neighbouring HARBOUR.

tidal current a powerful horizontal movement of sea water in areas affected by the TIDE. It is sometimes regarded as synonymous with TIDAL STREAM; but more specifically it is applied to conditions in a STRAIT where differing tidal regimes result in water-levels which differ at each end of the strait, the tidal current effectively equalizing the water level. CURRENT.

tidal dock a DOCK in which the level of water rises and falls with the tide.

tidal flat an area of sand or mud uncovered at low tide.

tidal glacier, tide-water glacier a VALLEY GLACIER which reaches the sea, where part of it may become detached, forming ICEBERGS or ICE FLOES.

tidal power station a coastal power installation which uses the natural POWER-3 of the ebb and flow of tides as they rush out of and into an enclosed reservoir (TIDAL BARRAGE) to drive

HYDRAULIC turbines which generate ELECTRICITY.

tidal prism the total volume of sea water (i.e. fresh water excluded) which the FLOOD TIDE brings into or the EBB TIDE removes from an estuary or bay, measured in cubic metres.

tidal range the difference in the height of the water at high and low tide at a place, varying from day to day. The fortnightly NEAP TIDES have a small range, the SPRING TIDES have a greater range. G.

tidal stream the normal movement of sea water in a coastal inlet, the inward flow with the FLOOD TIDE, the outward with the EBB TIDE, usually resulting in SCOUR. SAND RIBBON, TIDAL CURRENT.

tidal wave 1. an unusually large wave at high water resulting from tidal movement 2. popularly, but inaccurately, applied to a giant destructive wave caused by an earthquake, the correct name for which is TSUNAMI. G.

tidal zone LITTORAL ZONE.

tide one of the original applications was to a space of time or a period (e.g. Christmastide), and particularly to the space of time between two successive high water periods. Now tide is usually applied to the regular periodic alternating rise and fall of the level of the water in the oceans (often accentuated in adjoining seas, bays, gulfs). The rising of the water is termed the FLOOD TIDE, the falling is the EBB TIDE. The tide flows and ebbs twice in a LUNAR DAY of 24 hours 51 minutes and is caused by the gravitational pull of the SUN and the MOON, the latter being the more powerful. When the sun and moon act together (CONJUNCTION) a higher tide results (SPRING TIDE); when they do not reinforce each other (OPPOSITION) there is a smaller tide (NEAP TIDE). See entries qualified by tidal and tide, as well as AMPHIDROMIC SYSTEM, APO-GEAN TIDE, DIURNAL TIDE, DOUBLE TIDE, HIGH WATER, LOW WATER,

OSCILLATORY WAVE THEORY OF TIDES, PERIGEAN TIDE, PROGRESS-IVE WAVE THEORY OF TIDES. G.

tide crack of ice, the fissure at the junction of an immovable ICE FOOT or ICE WALL and FAST ICE, the latter reacting to tidal rise and fall. G.

tide mill SEA MILL.

tide rip rough, choppy water occuring at the meeting of opposing currents or TIDES.

tidewater seawater flowing over a land surface at FLOOD TIDE.

tied cottage a dwelling house in which a worker (especially a farm worker) can live while employed by the owner.

tierra caliente (Spanish, hot land) one of the four altitudinal zones of low latitudes in the northern Andes, central America and Mexico (the others being TIERRA FRIA, TIERRA TEMPLADA, TIERRA HELADA). Tierra caliente is the lowest zone, the zone of tropical products, the hot tropical coastland from sea level to about 1000 m (3000 ft), where the climate is humid, the temperature varying little between 24°C and 27°C (74°F and 80°F) throughout the year, and with the difference between the coldest and warmest months not more than three or four degrees. The natural vegetation is luxuriant and tropical, with dense forests in wetter parts. The crops are bananas, sugar, cocoa; with maize, tobacco and coffee on the mountain slopes. G.

tierra fria (Spanish, cold land) one of the four altitudinal zones of low latitudes in the northern Andes, central America and Mexico (TIERRA CALI-ENTE), the tierra fria, the zone of grains, lying higher than the TIERRA TEM-PLADA, lower than the TIERRA HELADA, i.e. from about 1800 m to 3000 m (6000 to 10 000 ft). The average annual temperature lies between 12.5°C and 18°C (55°F and 65°F), there is little difference in temperature from one month to another, and the natural vegetation is coniferous forest giving way to scrub and grassland with increasing height and decreasing temperature. The crops are wheat and vegetables; the fruits are those common at sea level in higher latitudes in the northern hemisphere; and there is much pasture. G.

tierra helada the highest of the altitudinal zones (TIERRA CALIENTE) in the northern Andes, central America and Mexico, the permanently snow-covered region of mountain summits, lying higher than the TIERRA FRIA.

tierra templada (Spanish, temperate land) in tropical central America, Mexico and the northern Andes, the altitudinal zone known as the zone of coffee (the zone of perpetual spring in the northern Andes), lying between the TIERRA CALIENTE and the TIERRA FRIA, from about 1000 m to 1800 m (3000 to 6000 ft), average annual temperature varying between 18.3°C and 24°C (65°F and 75°F), the range between the coldest and the warmest month rather less than that in the tierra caliente. The natural vegetation is savanna with open forest; the crops are maize, coffee, tobacco; but rainfall is too low for pasture. G.

tilā (Indian subcontinent: Bengali) a low isolated hill at the foot of an escarpment. G.

till unstratified, unconsolidated DRIFT-1 consisting of a heterogeneous mixture of angular and/or subangular clay, sand, gravel and boulders carried by ice and deposited, with little or no subsequent sorting or transportation by water. The term BOULDER CLAY, long in use, has now been replaced by till; and recently refinements in the nomenclature of till have been proposed, taking into account the processes which form the sediments and the position of their deposition. The term ABLATION TILL would be dropped, but LODGEMENT TILL retained (to be termed deformation till or deformed lodgement till if glacially re-worked after deposition). Super-

glacial till would be termed sublimation till; melt-out till would apply to material formerly buried in ice and exposed when the top surface of the ice melts, the structure of the material as it was when buried in the parent ice still being apparent. BASAL TILL, FLOW TILL, FLUVIOGLACIAL DEPOSITION (stratified drift), TILLITE. G.

tillage 1. the process of cultivating land so as to make it fit for raising crops 2. land ploughed or hoed in the current year 3. arable land, excluding ROTATION GRASS and CLOVER 4. in law, with reference to TITHES-2, agricultural land excluding gardens and orchards. G.

tillite consolidated TILL (or, formerly, BOULDER CLAY), compacted and lithified to form a SEDIMENTARY ROCK, especially such a deposit of a pre-QUATERNARY glacial period. G.

tilt block a block of rock standing between prominent fault lines and slanted at an angle in such a way that its slopes contrast with those that border it. BASIN AND RANGE, BLOCK FAULTING, TILTING.

tilth 1. cultivated or tilled land, as distinguished from PASTURE 2. the depth of soil prepared for sowing or planting 3. in soil science, the physical condition of the soil in respect of its suitability for cultivation, a soil in good tilth being FRIABLE and POROUS, with a GRANULAR STRUCTURE. G.

tilting the condition of slanting or being slanted from the horizontal or vertical.

timber 1. a wood suitable for or processed for use in construction 2. a tree yielding such wood. HARDWOOD, LUMBER, SOFTWOOD.

timber-line, tree-line 1. the upper limit of tree growth on a mountain, sometimes termed the cold timber-line to distinguish it from the dry timber-line of arid regions 2. the lower limit of tree growth on mountains in ARID regions where precipitation decreases with descent down the slope **3.** the latitudinal limit at which tree growth ceases **4.** the upper limit on a mountain, or the altitudinal limit, at which trees sufficiently large to be of use for TIMBER cease to grow. TREE-LINE. G.

time AM, PM, ANALEMMA, APPARENT TIME, DAYLIGHT SAVING, EQUATION OF TIME, GEOCHRONOLOGY, GEOLOGICAL TIME, GEOLOGY, GMT, INTERNATIONAL DATE LINE, LOCAL TIME, MEAN SOLAR TIME, SIDEREAL DAY, SIDEREAL MONTH (MOON), SIDEREAL YEAR, STANDARD TIME; and entries qualified by time.

time distance the time taken in travelling a certain distance. COST SPACE.

time-divisible services inherently indivisible but mobile services (INDIVISIBLE SERVICES) which visit different isolated locations on different, usually regular, days, e.g. mobile libraries.

time signal a radio signal indicating the exact time, transmitted frequently and regularly so that CHRONOMETERS can be synchronized and LONGITUDE calculated.

time space, time distance COST SPACE, COST DISTANCE. G.

time-space constraints SPACE-TIME CONSTRAINTS.

time utility the addition of the dimension of time to a good or service by storage and/or seasonal marketing which adds to the VALUE-1 of the good or service.

time zone the division represented by 15° longitude (less in small countries) within which the mean time of the central (or near central) meridian is selected to represent the whole division. APPENDIX 3, STANDARD TIME.

tin a silvery-white, soft, MALLEABLE, DUCTILE, stable metallic ELEMENT-6, resistant at ordinary temperatures to the chemical action of air and water. It occurs mainly as the oxide CASSITERITE, SnO_2, either in PLACER deposits weathered from veins in granite, or directly from LODES-1. It is used in the making of TIN PLATE, and as an ALLOY in solder, TINFOIL, BRONZE etc.

tinaja (Spanish) **1.** in general, a temporary pool **2.** more precisely, a natural bowl or bowl-shaped cavity, specifically the cavity below a waterfall, especially if partly filled with water. G.

tinajita (Spanish) solution pan, a shallow basin formed by solution on bare limestone, usually with a flat bottom and overhanging sides. G.

tind (Norwegian) a HORN in a mountain range. G.

tinfoil an ALLOY of LEAD and TIN or ALUMINIUM beaten into very thin sheets for wrapping, cooking purposes, etc.

tinning the process of applying a very thin coating of tin to the surface of iron. TINPLATE.

tinplate a thin sheet of steel or iron coated with a very thin layer of TIN to prevent rusting.

tin pyrites STANNITE.

tinstone CASSITERITE. TIN.

tinware articles made from TINPLATE.

tir comin (Welsh) true common. G.

tir cyd (Welsh) areas of moorland grazed in common. COMMON. G.

tirr the loose, slightly decomposed top layer of a raised MOSS-2. G.

tirs a black clay soil of north Africa, resembling REGUR. Soils resembling tirs, or in the process of becoming tirs, are termed tersified or tersoid. GRUMUSOL. G.

tissue in biology, a group of cells mainly of similar structure and performing a similar function, associated in large numbers and bound together by cell walls in plants or by intercellular material in animals.

tissue culture the process or technique of keeping cells alive in an artificial medium after their removal from an organism.

titanium a silver-grey, heat-resistant, metallic ELEMENT-6 of the carbon, silicon group, resembling IRON, the chief ores being ILMENITE and RUTILE. It is used in high grade steel alloys for the aircraft and aerospace industries; and as a pigment in paints and dyestuffs. TITANIUM DIOXIDE. C.

titanium dioxide TiO_2, a compound occurring naturally in various crystalline forms, e.g. RUTILE, used as a pigment to improve the light reflecting qualities of paints, plastics, rubber. Some crystalline forms are used as GEMSTONES.

tithable *adj*. subject to TITHE.

tithe 1. from Old English, a tenth or tenth part, an application now obsolete 2. historically, in England, a tenth part of annual agricultural produce etc. given as an offering or paid as a tax, especially the tax levied by the church for its support.

tithe barn a large barn where the agricultural produce representing TITHES-2 was delivered and stored.

tjäle (Swedish; Norwegian TAELE or TELE) anglicized as tjaele and taele. 1. in common use in Sweden, the frozen soil or ground frost showing a peculiar structure, with layers of lenses of pure ice within the ACTIVE LAYER. Tjäle thus designates not only the frozen soil but also the condition of the ground 2. tjäle or tjaele, terms used by some authors to designate perennially frozen ground. PERMAFROST. Long discussion in G.

toadstone a layer of impermeable, dark, basaltic rock in the CARBONIFEROUS LIMESTONE of Derbyshire, England, sometimes making a PERCHED WATER TABLE, a favourable factor in the siting of villages. The derivation is uncertain: its rough dark surface is similar to a toad's skin; but the term may be derived from German Todstein, worthless stone, the toadstone lacking the metallic ores present in the mineral veins of the neighbouring rocks.

tobacco 1. a member of *Nicotiana*, family Solanaceae, a genus of frost-sensitive, ANNUAL or PERENNIAL herbs with large leaves, native to tropical America, now cultivated there and elsewhere where the summer is hot and dry. The dried and cured leaves of *Nicotiana tabacum* are further processed for smoking, chewing, use as snuff, varieties having been developed to suit the varying conditions of climate, soils and methods of curing. 2. the manufactured products made from the leaves of *N. tabacum*. ALKALOID, NICOTINE.

toddy the sap of several palms native to southeast Asia, e.g. PALMYRA.

tokay a sweet Hungarian dessert WINE.

toich (Sudan: Dinka) annually flooded marshland close to a watercourse, providing valuable pasturage for cattle during the dry season. G.

tolerance in biology, the ability of an organism to withstand processes or acts harmful to it, or to survive in conditions which are difficult or even hostile to it, e.g. drought or excessive water; too much or too little heat or cold.

toll a tax or charge levied by an individual or authority for the protection of rights, especially a payment for permission to pass over a bridge (toll bridge) or along a road, where payment is made at a toll gate. G.

tomato *Lycopersicon esculentum*, a trailing, herbaceous plant, native to the lower Andes in South America, PERENNIAL, but normally now cultivated there and widely elsewhere as an ANNUAL, in the open or under glass, for its fleshy, juicy fruit (a BERRY), varying in colour when ripe from yellow to deep red, with a high sugar content, rich in vitamins A and C. The fruit is eaten raw or cooked.

tombolo (Italian) a bar of sand or shingle linking an island to the mainland, or one island to another. G.

ton, tonne a measure of weight. In avoirdupois 1 long ton (lgt) is equal to 2240 lb (20 cwt of 112 lb) or 1.016

metric ton (tonne). A metric ton (tonne) is equal to 1000 kilograms or 2204.62 lb or 0.984 long ton, or 1.1 short ton. A short ton (American ton) equals 1000 lb (i.e. 20 cwt of 100 lb) or 0.907 tonne, or 0.892 long ton. CONVERSION TABLES.

tongue of ice a long projection of ICE EDGE caused by wind and current. G.

ton-kilometre the product of a ton load multiplied by the distance travelled in kilometres along a railway, road, waterway, indicating work done.

tonnage **1.** the carrying capacity of a vessel or the total carrying capacity of a fleet measured in tons **2.** the duty based on the cargo capacity of a vessel **3.** the charge per ton of cargo carried on canals or in some ports. SHIPPING TONNAGE.

topaz a SEMI-PRECIOUS transparent GEMSTONE, a compound of alumina, silicon, oxgyen, fluorine etc., varying in colour from canary yellow to a deep orange, some white examples. 'True' topaz has shades of light blue, light green, pink and straw yellow; 'oriental' topaz, or yellow sapphire, is a variety of CORUNDUM; 'false' topaz is a variety of QUARTZ.

topographic *adj.* of or relating to TOPOGRAPHY.

topographic adolescence a term applied by some authors to the condition of an area in the early stages of erosion, when the main branches of streams have developed their narrow valleys but the land between the streams remains relatively unaffected. TOPOGRAPHIC INFANCY, TOPOGRAPHIC MATURITY, TOPOGRAPHIC OLD AGE. G.

topographical *adj.* **1.** of or relating to TOPOGRAPHY **2.** relating to the representation of or reference to a place in a painting, literature, etc.

topographic desert a local hot desert caused by relief and descending air currents, as distinct from a PLANETARY DESERT occurring in arid zones. G.

topographic infancy the condition of an area newly-exposed to erosion by surface waters, the original hollows being still water-filled, the plains only partly cut by narrow valleys of streams. TOPOGRAPHIC ADOLESCENCE, TOPOGRAPHIC MATURITY, TOPOGRAPHIC OLD AGE. G.

topographic map a map, usually on a fairly large scale (e.g.1 : 50 000), representing surface features, e.g. landforms and other natural phenomena as well as features produced by human activities. The term should not be applied to a map showing only relief features. Some American authors apply the term topographic map to a large scale map of a small area (as distinct from a chorographic map on a scale of, say, between 1 : 500 000 and 1 : 5 mn, and a global map on an even smaller scale). CHOROGRAPHY, TOPOGRAPHY.

topographic maturity the condition of an area which has been eroded for some time, so that the original upland is completely dissected, many individual river valleys are mature (although some headwaters of tributaries may be youthful), the land is reduced to slopes, but a new erosion plain has yet to appear. TOPOGRAPHIC ADOLESCENCE, TOPOGRAPHIC INFANCY, TOPOGRAPHIC OLD AGE. G.

topographic old age the condition of an area reduced by erosion almost to base level. TOPOGRAPHIC ADOLESCENCE, TOPOGRAPHIC INFANCY, TOPOGRAPHIC MATURITY. G.

topography a term which has given rise to so much confusion that geographers today tend to avoid it. Those who do use it apply it in general to the description or representation on a map of all the surface features of an area, natural and ARTIFICIAL-1 (TOPOGRAPHIC MAP). A few restrict it to the relief features, but this use is not generally accepted. CHOROGRAPHY. G.

topological *adj.* applied in geographical studies as relating to, pertaining to,

concerned with enclosure, order, CON-
NECTIVITY, CONTIGUITY and relative
position rather than with actual distance and orientation, topological relationships being frequently expressed in
terms of NETWORKS. TOPOLOGICAL
MAP, TOPOLOGY.

topological diagram or map a diagrammatic map which shows the actual
relationship of certain features (e.g.
positions of towns) but on which true
scale is deformed to accommodate
some other consideration(s), e.g. the
best way to show a communications
system. The diagrammatic maps of
underground railway systems are good
examples (e.g. of the London
Underground, or of the Paris Métro):
they show connectivity, i.e. the way in
which lines connect the stations, but are
not concerned with the correct orientation of the stations, and they are not
drawn to scale. GRAPH-2, HOMALO-
GRAPHIC, TOPOLOGY.

topology a branch of mathematics concerned with preserving certain relationships, e.g. closeness. The definition
which restricts topology to the study of
the properties of a geometrical figure
that are unaffected when the figure is
continuously transformed or deformed
is too narrow in its concept. TOPO-
LOGICAL, TOPOLOGICAL DIAGRAM.

toponym a place-name. G.

toponymy the study of place-names.
ONOMASTICS. G.

topophilia the love of place, the coup-
ling of sentiment with place. MENTAL
MAP.

topotype in ecology, a population with
characteristics associated with a particular region, which are distinct from
the characteristics of the population in
another region.

topset beds, top-set beds the fine-
grained sedimentary layer overlying the
foreset beds of a delta. DELTA, DELTA
STRUCTURE. G.

topsoil an imprecise term applied by
agriculturalists rather than by soil

scientists to the cultivated layer of the
mature SOIL, whatever SOIL HORI-
ZONS were originally involved, or to the
surface soil as distinct from the SUB-
SOIL. G.

tor a prominent, isolated mass of
jointed, weathered rock, usually
granite, especially one rising from the
moorland of Dartmoor, southwest
England. G.

tornado 1. African tornado, a violent
storm over the lands of west Africa,
consisting of a SQUALL, usually with
torrential rain, sometimes of short
duration, but extending over a long
front (up to 320 km: 200 mi) associated
with a THUNDERSTORM. It occurs
most frequently in daytime between the
wet and dry seasons when humid
monsoon air from the southwest meets
the dry northeasterly HARMATTAN
from the Sahara 2. a violent, counterclockwise, very destructive, short-lived
revolving storm (sometimes termed a
twister in USA), usually accompanied
by rain and thunder, associated with an
intensely LOW pressure system, with
wind velocities estimated to exceed 320
kph (200 mph), in some examples
travelling a nearly straight track at
between 16 and 80 kph (10 and 50
mph), and with a dark, funnel-shaped
cloud (FUNNEL CLOUD), small in
diameter, appearing to grow downwards from dark CUMULONIMBUS
cloud. Such a tornado is particularly
common in the Mississippi basin in the
afternoons in spring and early summer
where warm humid air from the Gulf of
Mexico meets cool, dry air from the
north, and when the heating of the land
surface is at its greatest. It may travel
only a short distance (some 30 km: 18
mi) and last uder two hours, but in that
time it mows down anything in its path.
TROPICAL REVOLVING STORM.

torrent a stream flowing swiftly and
violently.

torrid *adj.* very hot, exposed to great
heat.

torrid zone a term applied in classical times to the warmest of the three latitudinal temperature zones, i.e. the hot, burning zone lying between the TROPICS of CANCER and CAPRICORN, the other two being the FRIGID ZONE and the TEMPERATE ZONE. ZONE-2. G.

total eclipse ECLIPSE.

totalitarian *adj.* applied to a form of government or a state in which all the social and economic activities of every individual and of every enterprise, all public argument, and the flow of information are controlled by a dictator or a dictatorial caucus, any dissidence being forcibly suppressed.

toun, tounship Scottish spelling for TOWN-2.

tourelle (French, little tower) a small, flat-topped limestone BUTTE in KARST. COUPOLE, KEGELKARST, PITON, TOWER KARST. G.

tour a journey made for pleasure or for reasons of business, inspection, education etc. which ends at the place of origin and takes in several places or points of interest on the way. TOUR-ISM.

tourism 1. the practice of making TOURS for pleasure 2. synonym for tourist industry, the whole business of providing hotel and other accommodation, facilities and amenities for those travelling, or visiting, or staying in a place for a relatively limited period of time primarily for pleasure.

tow coarse, broken fibre removed in the processing of FLAX, HEMP, JUTE and made into twine (a strong cord or string, consisting of two or more threads or strings twisted together).

towan a coastal SAND DUNE, especially in Cornwall, southwest England. TWYN. G.

tower a tall, vertical structure, with a height great in proportion to its lateral dimensions.

tower karst, turmkarst KARST formed in tropical conditions (TROPICAL KARST), with isolated limestone hills, generally flat-topped with steep, forest-covered sides (TOURELLE), interspersed with stretches of alluvium or other detrital sand. KEGELKARST.

town 1. in general, a place larger than a VILLAGE consisting of a compact agglomeration of dwellings, shops, offices, public buildings etc., usually with paved roads, street lighting, public services, an organized local government, and a community pursuing a distinctive, URBAN way of life. Specialized functions are commonly defined by qualification (e.g. market town, mining town, railway town) as are locational features (e.g. gap town, seaside town), or special characteristics (e.g. GHOST TOWN, SHANTY TOWN) 2. in Scotland, a collection of CROFTS. Extensive discussion in G.

town house in England, the London house of one who also owns a COUN-TRY HOUSE.

townland 1. in Scotland, the enclosed infield of a farm. INFIELD-OUTFIELD 2. in Ireland, an administrative land unit. G.

town planning REGIONAL PLANNING.

townscape the physical forms and arrangements of the buildings and spaces that make the urban landscape. FARMSCAPE, WATERSCAPE, WILD-SCAPE. G.

township 1. historically, one of the local divisions of, or districts comprised in, a large, original PARISH, each having within it a small town or village, usually with its own church. PRECINCT-4 2. in Scotland, a CROFTING township, a district in the CROFTING COUNTIES, comprising individually held croft land and the common grazing, the croft land being separated from the common pasture by the township dyke (a stone or turf wall), the number of crofts varying widely (from six to fifty or more) 3. (American) a term applied in two ways in PUBLIC LAND SURVEY, USA, the first being to the congressional township of 6 mi square, whether it is settled or not, the second to the northern component drawn up for locating and identifying the townships **4.** in Australia, a tract of land laid out with streets and subdivided into lots for future urban development, or a temporary settlement on such a site. G.

towpath a path alongside a navigable river or canal, used originally by draught animals or people towing boats.

toxic *adj.* of, pertaining to or caused by a substance seriously injurious (poisonous) to the health of a living orgaism, e.g. toxic waste material produced by an industrial process. OBNOXIOUS, POLLUTION.

trace element one of the chemical elements present in relatively small quantities in the earth's crust, and essential in very small amount for the normal health of plants and animals, e.g. BORON, COPPER, MANGANESE, MOLYBDENUM, ZINC for higher plants, COBALT for cattle and sheep, IODINE for human beings. An insufficiency or lack of intake of trace elements leads to deficiency diseases. An excess is in some cases harmful. MACRONUTRIENT, MICRONUTRIENT.

trace fossil the trace of plant or animal activities (e.g. tracks or burrows of animals, root passages of plants) preserved in the rocks of the earth's crust. FOSSIL.

trachyte a fine-grained, alkali INTERMEDIATE EXTRUSIVE IGNEOUS ROCK, composed mainly of alkali FELDSPAR, characterized by the distinctive arrangement of the feldspar crystals in long, narrow, parallel strips. PHONOLITE is a variety of trachyte.

track **1.** a path made by the passage of animals, including people **2.** a rough, uneven, narrow, unmetalled road.

tract **1.** in general, an expanse of land without defined boundaries **2.** the third order of unit in D. L. Linton's HIERARCHY OF MORPHOLOGICAL REGIONS **3.** the second smallest order of magnitude in J. F. Unstead's hierarchy of regional division. STOW **4.** a unit of area in the Canadian census and the USA census. G.

traction a process in the transportation of debris by a river, in which the debris rolls and slides along parallel with and close to the bed, other debris being carried in SOLUTION-1 and SUSPENSION. SALTATION. G.

traction load BED LOAD. TRACTION.

trade the business of distributing, selling and exchanging commodities.

trade cycle the recurring succession of alternating booms and slumps in trading activities. TRADE.

trade gap in the trade of a country, the amount by which, over a period of time, the value of imports exceeds those of exports.

trade-off theory of land use, a theory which maintains that consumers are prepared to trade-off (balance out) rents and transport costs, i.e. to trade-off the accessibility offered by city centre sites (where the plots of land are small and the cost of land is high, but transport costs involved in living there are low) and the lower land costs but higher transport costs incurred in living in suburban sites. Trade-off theory thus relies on the ideas of indifference theory (a theory which states that, within certain limits, consumers are indifferent to varying combinations of goods, e.g. of high rents and low transport costs, or low rents and high transport costs), and is used particularly in studies of residential land use. It is similar to bid rent theory (BID PRICE CURVE) in that it explains residential location decision solely in terms of minimizing travel costs (accessibility) and housing costs (location rent). If a graph is drawn to show transport costs, land costs and overall cost in relation to distance from the city centre, there will be various points where the mix of land and transport costs produces the same overall cost. These are the points of access/ space trade-off which, in those terms, present the consumer with a choice of best location, the consumer in this case being indifferent to the alternative combinations of factors which make up the total costs.

trade wind (from nautical phrase to blow trade, to blow a regular course) a constant wind which blows (more strongly over the ocean than over the continents) from the tropical HIGH pressure belts towards the equatorial LOW in the northern and southern hemispheres, i.e. from the northeast in the northern, from the southeast in the southern, the typical PLANETARY WINDS. ANTI-TRADE WIND, WESTER-LIES. G.

traffic principle, transportation principle one of the principles used by W. Christaller to account for the varying levels and distribution of CENTRAL PLACES in a CENTRAL PLACE SYSTEM. Assuming ease and efficiency of transport between central places to be the dominant consideration, as many lower order centres as possible will lie on the traffic routes between the higher order centres (shown by a K-4 hierarchy, where a higher order place serves three adjacent lower order places). There will therefore be a greater number of higher order centres than is accounted for by the MARKETING PRINCIPLE. ADMINISTRATIVE PRINCIPLE, K-VALUE.

train oil oil obtained from the blubber of WHALES.

trajectory 1. the path of a body in space 2. the path of an individual parcel of air over some duration of time, in contrast to STREAMLINE-2.

tramontana, tramontane, tramontanto (Spanish; Italian, north wind) a cold dry north or northeast wind descending towards the sea from cold, dry plateaus in the western Mediterranean area. G.

transaction flow analysis a statistical technique that uses the flows of com-munication and information to measure the extent to which spatially separate points are unified.

transalpine (Latin transalpinus, across the Alps, i.e. from Italy) *adj.* on, or relating to, the north side of the Alps. CISALPINE.

transcontinental *adj.* extending over or going across a continent.

transcurrent fault TEAR FAULT.

transect a section taken across a tract of country for the purpose of studying the vegetation in relation to soil and relief.

transferability the handling characteristics and value of PRODUCTS, a term used in ULLMAN'S BASES FOR INTER-ACTION, to cover the cost of moving different products (a cost which reveals the nature of the product as well as that of the transport system) and the ability of the product to carry transport costs. Goods are said to have a high transfer-ability if they are easy to handle and transport costs are low in proportion to the value of the goods; and to have a low transferability if they are difficult to handle and transport costs are high in proportion to the value of the goods. COMPLEMENTARITY. INTERVENING OPPORTUNITY.

transfer costs in international trade, TRANSPORT COSTS combined with the costs of overcoming other obstacles to commodity movements, e.g. tariff walls. G.

transformation in statistics, the systematic manipulation of all the observations in a set of data. If the original data may be restored by reversing the operation, no information is lost, and the process is termed reversible transformation. But if there is no logical way of returning to the original, information having been lost in the manipulation, the process is termed irreversible transformation.

transform fault in theory of PLATE TECTONICS, a massive TEAR FAULT that marks the divide where two plates slide past each other, the edges moving jerkily and jostling each other but staying close together. The plates do not dive into the mantle, no material is added to or subtracted from the earth's crust, but the friction arising from the movement of the plates usually causes severe EARTHQUAKES and earth tremors, e.g. the San Francisco earthquake, 1906, caused by the movement along the San Andreas transform fault in California, USA.

transgression in geology 1. the invasion of a land surface by the sea, the result of positive eustasy or of negative isostasy, and the strata associated with such a movement. EUSTASISM, ISO-STATISM, TRANSGRESSIVE 2. of igneous rocks, an INTRUSION (transgressive intrusion) cutting across the bedding planes of sedimentary rocks from one horizon to another. DISCORDANT INTRUSION. G.

transgressive *adj.* having the quality of overlapping, applied 1. in biology, to a SPECIES-1 which spills over from one community to another 2. in geology, to the water or sediments deposited by it as it invades the land surface. TRANS-GRESSION-1.

transhumance a periodic or seasonal movement of pastoral farmers and livestock seeking fresh pasture between two areas of different climatic condition, e.g. in mountainous areas, the movement from valley floor, the winter location, to mountain pasture for the summer, and the return to the valley in autumn; or the movement from drought-stricken lowlands in summer to cooler higher land, as in Spain. Some authors also use the term as equivalent to NOMADISM, the migration of nomadic pastoralists in search of fresh pasture in a regular, seasonal pattern according to the rainfall regime. MAYEN, SAETER, SHIELING. G.

transit 1. in general, the act of passing or being passed over, across or through, or from one place to another

2. in astronomy, the passage of a heavenly body across the disc of another one, or its apparent passage across a MERIDIAN.

transit compass a THEODOLITE used in measuring horizontal angles.

transition zone in CONCENTRIC-ZONE GROWTH THEORY, the zone surrounding the CENTRAL BUSINESS DISTRICT, with residential areas invaded by business and light manufacturing, mainly from the CORE-2, but also from elsewhere. SEED BED. G.

transliteration of geographical names the replacing of letters or characters of one language or alphabet by those of another with the same phonetic sound. There is now general international agreement that the place-names used in the official language of the country concerned should be used internationally. But some conventional English forms are so firmly established in English-speaking countries that their use is not likely wholly to disappear for some time, e.g. Moscow (Moskva), Warsaw (Warszawa), Vienna (Wien), Rome (Roma) etc.; and even when local names are used there is a tendency to drop accents. Languages in which the Roman alphabet is not used present special difficulties, e.g. Chinese, in which the long-established Wade-Giles system of romanization was officially replaced by the Pinyin system by China on 1 January 1979. Pinyin is considered to be phonetically more accurate than Wade-Giles, and the Chinese government decreed that Pinyin should be used on all 'official documents and other publications used overseas'.

translocation 1. in botany, the transfer of dissolved food materials, e.g. mineral salts, from one part of a plant to another part 2. in soil science, the transfer of substances in SOLUTION-1 or SUSPENSION from one SOIL HORI-ZON to another.

transmigrant one passing through a

country on the way to settling in another country.

transmission capacity of soil, the ability of a soil to permit the passage of water through it. If this capacity is low the upper SOIL HORIZONS will become saturated. OVERLAND FLOW, SATURATED-3.

transpiration the loss of WATER VAPOUR from a plant mainly through the stomata (STOMA) of the leaves, resulting in a stream of water with dissolved mineral salts surging up through the plant. Transpiration is unlike EVAPORATION in that it takes place through living tissue under the influence of the physiology of the plant. EVAPOTRANSPIRATION.

transport the act of carrying material or a person from one place to another. In British usage the terms transport and TRANSPORTATION are commonly interchangeable, transport being preferred except in geomorphology in the cases cited below. American usage generally favours transportation. TRANSPORT GEOGRAPHY. G.

transportation 1. (American) the carrying or the conveying of material or a person from one place to another 2. a phase in the process of DENUDATION concerned with the conveying of loose material of the earth's crust by a natural agent (other than MASS MOVEMENT by GRAVITY) to the site of DEPOSITION-1, the agents being running water, ice (glaciers and ice sheets), wind, the ocean (waves, tides, currents). The material carried, termed the load (LOAD OF A RIVER), may itself act as an eroding agent (ABRASION, CORRASION), suffering ATTRITION as the particles rub against each other and the surface over which they are being carried. G.

transportation slope a SLOPE-2 on which at each point the amount of material received from points upslope is balanced by the loss of material passing downslope.

transport costs all the costs involved in the moving of goods from one place to another, e.g. inventory, paperwork, handling, packaging, insurance, freight rates, temporary warehousing en route etc. TRANSFER COSTS.

transported soil SECONDARY SOIL. G.

transporter bridge a high bridge from which hangs a movable platform on which loads can be conveyed across a navigable waterway without impeding shipping. SWING BRIDGE.

transport geography an approach in HUMAN GEOGRAPHY which concentrates on the part played by the means by which people and/or goods are carried from one place to another over short or long distances, the patterns made by such movements, the volume of numbers/quantities of the people and/or goods carried, the costs involved; on the part played by transport facilities in economic, political and social development; and on the relationship between transport and other environmental factors.

transverse *adj.* crosswise, lying across, crossing from one side to another.

transverse coast an ATLANTIC TYPE COASTLINE, or DISCORDANT COAST. G.

transverse crevasse a crack or fissure cutting across a glacier, occurring where the slope increases and the glacier falls over a step. CREVASSE, ICE FALL, LONGITUDINAL CREVASSE.

transverse dune a DUNE with its crest running at right angles to the direction of the prevailing wind. LONGITUDINAL DUNE, SAND SEA.

transverse valley a VALLEY which cuts across a ridge, at right angles to the ridge. LONGITUDINAL VALLEY. G.

transverse wave SHAKE WAVE.

transverse wave motion 1. a form of transference of energy in a material medium, the disturbance of the particles displacing the particles in a direction at right angles to the direction of propagation 2. a form of transmission of electromagnetic waves in a material medium or in space, in each case the disturbance in the medium producing variations in the electric and magnetic properties in the medium which are at right angles to each other and at right angles to the direction of propagation. LONGITUDINAL WAVE MOTION, SHAKE WAVE, WAVE.

trap 1. in geology, a term formerly applied to any of the various dark-coloured, fine-grained IGNEOUS ROCKS, e.g. BASALT, when its character was unknown or not understood. The term was derived from the Swedish trappa, a stair, because hills formed from the denudation of such rocks were often terraced, or had a stair-like profile. The term has been dropped in favour of one which is more precise, e.g. the Deccan traps are now termed Deccan lavas or Deccan basalts 2. a structural formation in rocks in which PETROLEUM or NATURAL GAS may build-up and be prevented from escaping upwards by IMPERMEABLE layers 3. a formation in a cave system, where the roof of a chamber or the passage of a cave dips under the WATER TABLE but rises above it again farther on.

trass a volcanic rock with a high PUMICE content, used in the making of HYDRAULIC CEMENT, POZZOLANA.

traumatropism the curving of a plant during growth caused by a wound. TROPISM.

traverse a surveyed line consisting of a series of observations (legs), measured in distance and direction from a known starting-point, the end of one leg being the beginning of the next, thus incorporating cumulative error. The term closed traverse is applied if the legs are joined to link a known starting-point to a known finishing-point. If the position of only the starting-point or the finishing-point is accurately known, and the legs are joined, the traverse is termed an open traverse. TRIANGULATION.

travertine a crusty, generally buff-coloured deposit of calcium carbonate

from HOT SPRINGS (GEYSERITE, HYDROTHERMAL, SINTER, TUFA), sometimes used as an oranmental building stone in warm lands, e.g. in Italy. G.

trawl, trawler a large bag-shaped fishing net with a wide mouth which is dragged along the bottom of the sea by a boat; hence trawler, a fishing boat dragging a trawl for catching fish.

treaty port a sea or river port (later also an inland town) opened by treaty to foreign trade, applied especially in the nineteenth century to certain ports in China, Japan, Korea. In 1842 the first in China were opened with extraterritorial rights to foreigners. The system ended in 1943. EXTRATERRITORIALITY. G.

tree a woody PERENNIAL rising from the ground with a strong, distinct trunk. HERB, SHRUB.

tree-line, tree-limit the line or zone beyond which trees do not grow, latitudinally from the equator to the poles, altitudinally in the ascent of mountains (SUBALPINE); synonymous with TIMBER-LINE in American usage.

tref (Welsh) a homestead, hamlet, town. HENDREF, MAERDREF, PENTREF. G.

trek (Afrikaans, a journey) 1. specifically, a journey made by means of ox-wagon, especially the Great Trek of the Boers from the Cape to found the Transvaal and Orange Free State 2. the distance so covered 3. the company of people so travelling. G.

trekker (Indonesia: Dutch) a foreigner, especially European or Chinese, working in Indonesia, but not intending to settle there on retirement, as distinct from a blijver, who does so intend. G.

trellis drainage, trellised drainage DRAINAGE-2 with a rectilinear pattern, occurring particularly in areas of folded sedimentary rocks, e.g. in scarplands, where CONSEQUENT, SUBSEQUENT, OBSEQUENT and SECONDARY CONSEQUENT STREAMS cut channels through the less resistant rocks at right angles to the initial slope, and thus meet the main streams at right angles. ANGULATE DRAINAGE. G.

trench 1. a deep elongated submarine trough. DEEP, OCEANIC TRENCH **2.** a long narrow valley between two mountain ranges, especially a RIFT VALLEY or a U-SHAPED VALLEY. G.

trend 1. a tendency, a general direction **2.** a dominant, smooth, movement shown by a statistical process.

trend line the GRAIN, the pattern of the main structural lines, e.g. of folding (FOLD-2) and faulting (FAULT) in a region. G.

trend surface analysis a statistical technique particularly useful to the geographer in that it extends REGRESSION ANALYSIS to three dimensions. It entails the fitting of a statistical surface to values which are distributed in space, using a mathematical power function, usually with the aid of a computer. The observations of the DEPENDENT VARIABLE represent a series of sample points on a map, and the INDEPENDENT VARIABLES are the COORDINATES (latitudinal and longitudinal) of those points; but in order to describe the surface accurately some transformation of the independent variables may be necessary.

trend surface map a map showing a geographical pattern by isolating, measuring and plotting its quantitative components and using either a filter mapping or NESTED SAMPLING method. In the former, appropriate if complete data are available, the area is covered by a squared GRID-2; for each square of the grid the pattern is expressed as a value or ratio, and ISOPLETHS are drawn. This produces a response surface, a surface derived from both local and regional factors. It can be used to show regional trends (regular patterns extending over the whole area of study) or filtered to show local anomalies (irregular or spot variations from the general trend, without a regular pattern). The nested sampling method is used if data is incomplete, e.g. in an exploratory survey. In this case a few areas of equal size are selected at random and broken down, at random, into smaller units. Values for each level are then determined by the use of an appropriate type of VARIANCE ANALYSIS. From either method the variability of areal patterns can be broken down and sampled. DIFFUSION WAVE.

triangulation in surveying, the series or network of triangles into which a land area of any size may be divided in a TRIGONOMETRICAL SURVEY in order to provide a geodetic framework for a topographical survey. From a predetermined BASE LINE, which serves as one side of the primary triangle, triangles are constructed, their angles being measured with a THEODOLITE, the length of their sides being calculated by TRIGONOMETRY as the equipment is moved from one point that is to be determined to another. Triangulation may be of primary, secondary or tertiary order, in accord with the area of the triangles and the standard of accuracy needed. TRAVERSE, TRIGONOMETRIC, TRIGONOMETRIC POINT, TRIGONOMETRIC SURVEY, TRIGONOMETRY. G.

Triassic *adj.* of or relating to the earliest period (of time) or system (of rocks) of the MESOZOIC era, when reptiles were dominant, gymnosperm plants (plants producing seeds not enclosed in an ovary) appeared, and sandstone and pebble beds, shelly limestone, red sandstones and marls, with layers of rock salt and gypsum, were laid down. GEOLOGICAL TIMESCALE, RHAETIC.

tributary a stream or river flowing into a larger one. G.

tributary *adj.* auxiliary, contributory, subsidiary.

trigger action in meteorology, any process causing CONDITIONAL INSTABILITY in an AIR MASS.

trigonometric(al) *adj.* of, relating to, or accomplished by TRIGONOMETRY.

trigonometrical point a fixed point determined (astronomically) with great accuracy in triangulation, the vertex of a triangle. G.

trigonometric(al) survey a survey carried out by TRIANGULATION and trigonometrical calculation.

trigonometry the branch of mathematics that deals with the relationship between the sides and angles of triangles (plane figures bounded by three straight lines).

trip a short journey, especially one of a series of journeys over a particular route.

trip generation analysis the identification and QUANTIFICATION of journeys which begin and end in a study area, an analysis carried out in order to establish the functional relationship between the volume of the TRIPS at the places of origin and destination, and the land use and socio-economic character at those ends of journeys. DESIRE LINE. G.

tripton abioseston, SESTON.

troglo- a prefix derived from Greek, a hole, appearing in the terms that follow.

troglobite an animal living permanently in a dark underground cave, leaving it only by accident.

troglodyte an inhabitant of a cave or rock shelter, sometimes restricted to a person, but also applied to any other animal. G.

troglophile an animal inhabiting the dark zone of a cave but having to search for food outside, e.g. some species of bat.

troglophobe a human being or another animal unable for psychological or physical reasons to enter the dark zone of underground areas.

trogloxene an animal which does not permanently inhabit a cave but which enters it for various reasons.

trophic *adj.* of or related to a food supply, or the obtaining of nutrition.

trophic level 1. one of the levels in a FOOD CHAIN or FOOD WEB 2. the nutrient level of a body of water, especially in relation to nitrate and phosphate content. DYSTROPHIC, EUTROPHIC, MESOTROPHIC, OLIGO-TROPHIC.

trophobiosis a type of SYMBIOSIS in which two organisms of different species feed one another. The organisms concerned are termed trophobionts.

trophogenic region the region of a water body where organic material is produced by PHOTOSYNTHESIS. PHOTIC REGION.

trophotropism TROPISM in response to the location of a food supply, a form of CHEMOTROPISM.

tropic one of the two parallels of latitude of approximately 23°30′N (TROPIC OF CANCER) and 23°30′S (TROPIC OF CAPRICORN). G.

tropical *adj.* of or pertaining to the TROPICS, relating either to the specific parallels of latitude, 23°30′N or 23°30′S; or to the zone lying between those two parallels; or to that zone with the adjacent areas, since major climatic and other changes take place more nearly at 30°N or S than at 23°30′N or S. But some authors exclude as distinct the DOLDRUMS or belt of calms (ITCZ) or the EQUATORIAL BELT. G.

tropical air mass an air mass, symbol T, originating within the SUBTROPICAL HIGH PRESSURE BELTS, either the warm, moist maritime tropical (mT), originating in the TRADE WIND belt and subtropical waters of the ocean, or the hot, very dry, unstable continental tropical (cT) originating in low latitude deserts (especially the Sahara and Australian deserts). POLAR FRONT. G.

tropical climate any of several types of climate occurring in the tropics (in this case the term tropics is understood to exclude the EQUATORIAL BELT with its EQUATORIAL CLIMATE), i.e. one of the belts which for part of the year comes under the influence of TRADE WINDS but for the rest of the year is subject to convectional rain. There is no cold (winter) season, but in general there are three others, i.e. hot rainy, cool dry, hot dry, with average monthly temperatures exceeding 18°C (64.4°F) and with considerable rainfall, mainly convectional, the maximum falling in the hot rainy period. The two main types are marine, which lacks a pronounced dry season (tropical marine, tropical marine monsoon) and continental, with a pronounced dry season (tropical continental, tropical continental monsoon). HUMID TROPICALITY. G.

tropical cyclone TROPICAL REVOLVING STORM.

tropical easterly jet stream a JET STREAM moving from east to west at very high altitudes over southeast Asia.

tropical forest the natural vegetation covering the wooded parts of the TROPICS which have a dry season. TROPICAL CLIMATE, RAIN FOREST.

tropical grasslands CAMPO, LLANO, SAVANNA. G.

tropicality HUMID TROPICALITY. G.

tropical karst KARST formed in tropical conditions, i.e. high temperature and high rainfall, the two main forms being KEGELKARST and TOWER KARST.

tropical revolving storm a small, localized, very deep LOW pressure area occurring most commonly in late summer or early autumn in tropical latitudes over the western margins of the great oceans, usually moving very slowly along fairly well-defined tracks and causing much destruction. An intense cyclonic circulation is set up about the centre, with violent winds sometimes exceeding 160 kph (100 mph), accompanied by dense dark clouds, heavy rain, sometimes thunder and lightning; but near the centre, the eye, where pressure is at its lowest, there is an area of calm with a clear sky. In the areas affected the storms are given special, local names. CYCLONE, HURRICANE, TYPHOON, WILLY-WILLY. G.

tropical year SOLAR YEAR.

Tropic of Cancer an imaginary line encircling the earth at approximately 23°30′ north of the equator, where the sun's midday rays are vertical about 21 June, the most northerly point of the ECLIPTIC, and the northern limit of the TROPICS. G.

Tropic of Capricorn an imaginary line encircling the earth at approximately 23°30′ south of the equator, where the sun's midday rays are vertical about 21 December, the most southerly point of the ECLIPTIC, and the southern limit of the TROPICS. G.

Tropics, tropics the zone of the earth's surface lying between the TROPIC OF CANCER and the TROPIC OF CAPRICORN (i.e. between 23°30′N and 23°30′S), where the sun's rays strike vertically at noon on at least two days in the year, termed the TORRID ZONE in classical times. HUMID TROPICALITY (humid tropics), TROPIC, TROPICAL, TROPICAL AIR MASS, TROPICAL CLIMATE and other entries qualified by tropical; and TROPICALITY. G.

tropism 1. in biology, a growth response by plants or sedentary animals towards (positive tropism) or away from (negative tropism) a source of stimulus, shown by the curvature in the growth, the curve being determined by the direction from which the stimulus originates. AEROTROPISM, APHELIOTROPISM, CHEMOTROPISM, DIAGEOTROPISM, GEOTROPISM, HYDROTROPISM, PHOTOTROPISM, PLAGIOTROPISM, RHEOTROPISM, THERMOTROPISM, THIGMOTROPISM, TRAUMATATROPISM, TROPHOTOTROPISM, 2. in zoology, formerly a synonym for TAXIS, now seldom so used.

tropoparasite a PARASITE which is an OBLIGATE PARASITE for part of its life cycle, but lives non-parasitically for the remainder.

tropopause a zone of the ATMO-

SPHERE-1 consisting of several, overlapping levels, separating the TROPOSPHERE from the STRATOSPHERE. In the tropopause temperatures cease to decrease with increasing height (as in the troposphere): there is a pause for some kilometres in height before they begin to increase with increasing height (as in the stratosphere). The level of the tropopause varies daily, seasonally and latitudinally, but in general it is some 18 km (11 mi) above the earth's surface at the EQUATOR, some 6 km (4 mi) at the poles. G.

tropophyte a plant adapted to living in moist or dry conditions according to seasonal variation, e.g. DECIDUOUS trees, which shed their leaves when adequate water is unavailable. HYDROPHYTE, HYGROPHYTE, MESOPHYTE, XEROPHYTE. G.

troposphere a layer of the ATMOSPHERE-1 in which temperature decreases with height at a mean rate of some 6.5°C per km (3.6°F per 1000 ft), the layer nearest to the earth's surface, i.e. below the TROPOPAUSE. The thickness of the troposphere varies from 7 to 8 km (11 to 13 mi) near the poles to some 16 km (26 mi) over the equator. It contains nearly all the dust and liquid particles, some 90 per cent of the water vapour, and 75 per cent of the total gases in the atmosphere, so that most of the WEATHER activity affecting human life on earth takes place in its lower layers. G.

trough 1. a DEEP or TRENCH in the ocean floor **2.** an elongated U-shaped valley **3.** a SYNCLINE **4.** of a WAVE, the position of displacement or disturbance opposite to the position of a CREST, the depression between any two crests of a regular wave motion **5.** in meteorology, a narrow, elongated region of low barometric pressure between two areas of higher pressure. G.

trough end a steep wall of rock at the head of a glaciated valley. TROUGH-3.

trough fault a structure resulting from

two parallel NORMAL FAULTS between which a block of country has been let down, a GRABEN. G.

trough lake a lake occupying part of the floor of a trench made by an ALPINE GLACIER. GLACIAL TROUGH.

troy weight (initially used in Troyes, France) a system of units of weight used particularly in measuring PRECIOUS METALS (gold, platinum, silver). 1.215 lb troy is 1 lb AVOIRDUPOIS.

truck farming (American) the cultivation of vegetables and fruit for market. Truck farming differs from British market gardening mainly in that it is on a larger scale, there is a concentration on one or more crops, and the distance of the enterprise from the market is greater. G.

true bearing BEARING.

true dip in geology, DIP-2, i.e. the maximum inclination of a stratum, as distinct from APPARENT DIP. STRIKE.

true north, true south the direction determined by the geographical NORTH POLE or SOUTH POLE of the earth, i.e. the direction of the geographical North or South Pole from the observer, i.e. along the MERIDIAN passing through the observer, as distinct from MAGNETIC NORTH and GRID NORTH.

true origin the point on which the GRID-1 system on a map is based, at the intersection of the projection axes (the central meridian and a line drawn at right angles to it). FALSE ORIGIN.

true south TRUE NORTH.

true variance in statistics, all that part of the VARIANCE in a test which cannot be attributed to ERROR VARIANCE, i.e. variance which may be attributed to common and to specific factors. COMMON VARIANCE, SPECIFIC VARIANCE.

truncated soil a SOIL which has lost all or part of its upper horizons by erosion. G.

truncated spur a SPUR which projected into the side of a pre-glacial valley until the valley became glaciated, when it was sharply cut and shortened by the glacier as it moved down the valley. U-SHAPED VALLEY.

trunk valley the chief valley down which the main stream flows in a RIVER BASIN.

trust territory, trusteeship a territory (which is not self-governing) under the authority of the United Nations, or an authority deputed by the United Nations Trusteeship Council. It may be a former MANDATED TERRITORY not yet independent, a territory taken away from a state after the Second World War, a territory placed under the authority of the United Nations by the state governing it. The Charter of the Trusteeship Council of the United Nations provides a system to safeguard the interests of the inhabitants of the territories not yet fully self-governing.

Trypanosomatidae a family of parasitic flagellate PROTOZOA, some of which cause diseases in domestic animals and people, e.g. *Trypanosoma gambiensis* and *T. rhodesiensis* which cause sleeping sickness, transmitted to humans by the bite of the TSETSE FLY; and those species together with *T. Brucei* which cause African cattle sickness, but to which ANTELOPE-1 are immune.

trypanosomiasis any disease caused by a trypanosome, a member of TRYPANOSOMATIDAE, e.g. sleeping sicknes.

tsetse fly a member of *Glossina*, family Glossinidae, a bloodsucking fly of central and south Africa, carrying nagana, a disease fatal to domestic animals, notably cattle, and transmitting sleeping sickness and other diseases to human beings. The presence of tsetse fly restricts the areas where cattle rearing is possible in Africa.

tsunami (Japanese, harbour wave) a large-scale seismic ocean wave (incorrectly termed a TIDAL WAVE) caused by a submarine earthquake or a volcanic eruption. It travels at great speed in the open ocean (between some 600 and 1000 kph: 370 and 1600 mph), with enough energy in some cases to travel

halfway round the world. The wave height is low in the open ocean, but on entering shallow water the energy is concentrated, resulting in a wave of great height (up to 15 m: 50 ft), inundating low-lying areas on the shore. G.

t-test STUDENT'S T-TEST.

tuba cloud a CLOUD in the shape of a cone or column emerging from a CLOUD BASE, e.g. a FUNNEL CLOUD.

tuber the swollen food-storing part of an underground stem (e.g. POTATO) or root of a plant.

tube well a WELL dug or drilled to reach a deep-seated supply of water, lined with a pipe (a tube). G.

tufa (Italian) calc tufa, a soft porous SEDIMENTARY ROCK composed of CALCIUM CARBONATE or SILICA-1, deposited by evaporation of circulating GROUND WATER, or water in lakes, or water near the point of issue of a spring. The tufa associated with the issue of a HOT SPRING is a spongy, cellular type of TRAVERTINE. G.

tuff a rock formed from compacted or cemented PYROCLASTIC material (fine volcanic dust, ash etc. thrown out of a VOLCANO in eruption), with particles smaller than 4 mm in diameter. TEPHRA, WELDED TUFF (ignimbrite). G.

tuffisite fragmentary pyroclastic material (PYROCLAST) which is deposited and lithified in the pipe of a VOLCANO, unlike TUFF which forms on the ground surface. LITHIFICATION.

tumulus 1. an ancient burial mound. BARROW **2.** a small mound or hummock of solid lava on a lava flow, in some cases up to 9 m (30 ft) in height and 18 m (60 ft) across at the base, produced by the resistance of the lava surface to the spreading of a more fluid lava below, thus resembling a LACCOLITH in origin. G.

tundra (Russian, from Finnish tunturi, or Lap tundra) a treeless region and its associated vegetation north of the northern latitudinal TREE-LINE, characterized by long, very cold winters and PERMAFROST, and supporting a vegetation of MOSSES, LICHENS, HERBACEOUS plants and dwarf SHRUBS, infested with insects (black flies, midges, mosquitoes) in the short summer. The mean monthly summer temperature lies below 10°C (50°F), warm enough to thaw the snow and the surface of the permafrost, providing ideal marshy conditions in the ill-drained soils for insect-breeding. Köppen included tundra in his polar zone. CARIBOU, KOPPEN'S CLIMATIC CLASSIFICATION, REINDEER, TUNDRA SOIL. G.

tundra soil a dark-coloured shallow soil with a highly organic surface layer and usually (but not necessarily) with frozen subsoil. Comparatively little moisture is available but where it is the typical TUNDRA plants in decay give rise to a surface layer of organic material which accumulates and forms a peaty layer over an ANAEROBIC layer. In general soil profiles are not well developed, and the soils range from polar desert soil in the arid north to Arctic brown earths in the more humid upland regions with better drainage. G.

t'ung (Chinese) wood-oil tree, *Aleurites cordata*, TUNG OIL. G.

tung oil a poisonous, pungent, fast-drying oil expressed from the seeds of any of the Chinese trees of the genus *Aleurites*, especially from *A. cordata* and *A. fordii*, used mainly in the manufacture of hard-drying paints, varnishes etc. and as a waterproofing agent. G.

tungsten a grey, hard, MALLEABLE, DUCTILE metallic element with the highest melting point of all metals, and with great tensile strength, obtained mainly from the ore WOLFRAM, sometimes termed wolframite (a name also used for the metal), most commonly occurring in veins in metamorphosed sedimentary rocks. Tungsten is used in electric light filaments and in the making of special, hard, tough (high speed) STEELS. SCHEELITE.

tunnelling of THROUGHFLOW, SUB-SURFACE WASH.

turbary 1. a place where peat or turf is dug 2. rights of turbary, in British law, the right to cut peat for fuel. COMMON RIGHTS. G.

turbidite the sediment deposited by a TURBIDITY CURRENT, usually in a GEOSYNCLINE. GRAYWACKE.

turbidity cloudiness in a FLUID caused by disturbance which results in the holding in SUSPENSION of finely-divided particles.

turbidity current a current occurring in the ocean when sediment is locally churned up (e.g. by an earthquake, or by material sliding down the CONTINENTAL SLOPE), raising the density of the water to higher values than that of the surrounding clear water. The heavier water flows very swiftly, under the influence of gravity, down any available slope, spreading out on a horizontal floor. TURBULENCE due to the flow tends to keep the sediment in SUSPENSION until the flow itself ceases; the sediment is then deposited, forming TURBIDITE. Turbidity currents near the ocean floor have an erosional effect and are thought to have cut deep valleys in the CONTINENTAL SLOPE. SUBMARINE CANYON. G.

turbulence movement of a FLUID in which the flow is not in smooth, parallel layers, but in eddies (EDDY), so that mixing occurs, in contrast to a smooth LAMINAR FLOW. The term is used especially in meteorology in connexion with the flow and mixing of air (DIFFUSION-2). Turbulence in river flow helps to carry material in SUSPENSION; and it is a characteristic feature of ocean DRIFT-2. HYDRAULIC FORCE.

turgid TURGOR PRESSURE.

turgidity in plants, the state of being turgid. TURGOR PRESSURE.

turgor pressure in plants, the HYDRO-STATIC PRESSURE built up in a plant cell by the intake of water by OSMOSIS. The pressure makes the cell walls rigid, and in this state the cell is described as turgid. If the loss of water through TRANSPIRATION exceeds the water intake the pressure falls and the plant wilts. WILT, WILTING POINT.

turkey *Maleagris gallopavo*, a large North American BIRD, domesticated in many parts of the world, reared for its meat.

turlough (Ireland; from the Gaelic turloch) a hollow periodically flooded, depending on the fluctuations of the WATER TABLE. G.

turmeric *Curcuma longa*, family Zingiberaceae (of which GINGER is also a member), a broad-leaved, low-growing plant native to tropical Asia, cultivated for its swollen stem, a starchy yellow rhizome (rich in pungent oil), which is washed, peeled, dried and used as spicy flavouring, especially in CURRY, to which it imparts its characteristic yellow colour. It is also used as colouring matter in foodstuffs, as a chemical indicator of ALKALIS (which react by turning a red-brown), and as the source of a red-brown dye.

Turmkarst (German) TOWER KARST.

turnip *Brassica rapa*, a BIENNIAL plant with a swollen starchy 'root' (which is part of the axis lying between the true root and the seedling leaves, i.e. the hypocotyl), known to have existed in Europe in prehistoric times, now cultivated there and in many other parts of the world for its 'root' and leaves, which provide food for animals and for humans (the root being eaten cooked, as are the leaves, which provide 'spring greens').

turnpike historically, a spiked barrier fixed across a passage for defence purposes; later a gate placed across a road (hence turnpike road) where a toll was paid by anyone wishing to use the road. In Britain in the eighteenth century the roads were privately owned (especially by 'turnpike trusts') and maintained by these tolls. Later the term turnpike was applied to the road itself; and later still revived, especially in the USA, and applied to motorways subject to a charge or toll. G.

turpentine an ESSENTIAL OIL, mainly pinene, obtained by distilling the RESIN (oleoresin) of certain CONIFEROUS trees, notably of the pines of the southeastern USA, e.g. the terebinth.

tussock grass BUNCH GRASS.

tweed (Scottish tweel, twill) a strong, fairly loosely woven twilled (TWILL) woollen cloth, originally woven with yarns of several colours in southern Scotland.

twilight 1. the faint light of the sun reflected from the upper ATMOSPHERE-1 on to the earth before the sun itself rises above the horizon in the morning (dawn or sunrise) and after it has sunk below the horizon (sunset) in the evening, its duration depending on the date and latitude, i.e. brief in tropical regions, longer in higher latitudes. SUNRISE, SUNSET 2. astronomical twilight begins in the morning when the centre of the sun is 18° below the horizon, lasting until dawn; in the evening from sunset until the centre of the sun is 18° below the horizon. About the time of the SOLSTICE in high latitudes the sun's centre never sinks to 18° below the horizon, so twilight is continuous from sunset to sunrise for some nights, the number of such nights increasing towards the poles 3. civil twilight is classified as commencing or ending when the sun's centre is 6° below the horizon, when the light is judged to be sufficient for outdoor work 4. nautical twilight commences or ends when the sun's centre is 12° below the horizon, when the light should be sufficient to allow vague shapes to be seen. CREPUSCULAR.

twilight area the part of a town where old buildings in need of repair and inadequate facilities lead to poor living conditions.

twill a weaving pattern in which the weft threads pass over one and under two (or possibly more) warp threads, giving a diagonal line in the textile fabric. TWEED.

twine a strong cord made from TOW, especially from Manila HEMP fibres.

twister a TORNADO-2 in USA, a WATERSPOUT.

two-field system a simple system of cultivation practised in parts of MEDIEVAL England, in which half the land was cultivated, half left fallow each season. THREE-FIELD SYSTEM. G.

twyn (Welsh) a hillock or knoll, usually earthy. Twyn is derived from the same root as TYWYN or TOWYN, but the meanings have diverged. G.

tyddyn, ty'n (Welsh) 1. a small farm, a holding 2. the dwelling of a GWELY. G.

type of farming a classification of agricultural activity based broadly on farm organization and practice. Various criteria have been used to determine the types, e.g. the proportion of land under the plough or under permanent grass, and the dominant enterprise; a qualitative analysis of intensity of land use, association of crops and livestock, degree of mechanization; the propor-

tion of total cash output produced by each activity on an individual farm; a calculation of the STANDARD MANDAYS spent on each farm operation or each enterprise.

type species in biology, the SPECIES-1 judged to be the most typical of a GENUS, the name of which is accordingly added to the genus in BINOMIAL NOMENCLATURE. CLASSIFICATION OF ORGANISMS, TYPE SPECIMEN.

type specimen, holotype the original plant or animal specimen(s) from which the description of a new SPECIES-1 or a smaller group is made. ISOTYPE, LECTOTYPE, NEOTYPE, PARATYPE, TYPE SPECIES; and BINOMIAL NOMENCLATURE, CLASSIFICATION OF ORGANISMS.

typhoon (Chinese dialect tai fung, big wind) a violent TROPICAL REVOLVING STORM in the China Sea and adjacent regions, particularly around the Philippines, commonly occurring in the period from July to October. G.

typology 1. a study based on the classification or comparison of types (e.g. of social groups, of archaeological remains). URBAN TYPOLOGY 2. the doctrine or study of symbols or types, especially those of the Scripture.

tywyn, towyn (Welsh) a coastal hillock, usually sandy. TOWAN, TWYN. G.

U

ubac (French dialect; Italian opaco; German Schattenseite) the shady side of a valley facing away from the sun, in contrast to ADRET. G.

ubehebe (USA: from the Ubehebe craters, Death Valley, California) a crater formed by the expulsion of VOLCANIC ASH, LAPILLI etc. round a volcanic VENT, as in Death Valley, California. G.

ubiquitous materials the raw materials used in MANUFACTURING INDUSTRY which are available anywhere, not localized, and therefore do not influence the selection of location of the industry concerned. RESOURCE ORIENTATION.

ugli a CITRUS fruit, a grapefruit-tangerine hybrid, resembling the GRAPEFRUIT in appearance and use.

Uinta structure (USA: from the Uinta mountains, Utah) a broad, flattened ANTICLINAL FLEXURE from which STRATA descend sharply on each flank before resuming their horizontal state. In the Uinta mountains there is a classic example of subsidence and uplift on a giant scale, a flattened anticlinal flexure being there raised up in Cretaceous times, later extensively denuded to expose Precambrian rocks, uplifted again at the end of the Eocene, with large faults on its north and south flanks, and again uplifted in the late Pliocene-Pleistocene epochs. GEOLOGICAL TIMESCALE. G.

uitlander (Afrikaans, outlander, foreigner) historically, a Boer term applied to an alien or foreigner (especially if British) in the Boer Republic (Orange Free State, Transvaal) before the Anglo-Boer war in South Africa, 1899-1902. G.

Ullman's bases for interaction the concepts of COMPLEMENTARITY, INTERVENING OPPORTUNITY, TRANSFERABILITY originally suggested by E. G. Ullman in 1951 in his theory of commodity flow, but applicable to the flow of ideas, the movement of people. SPATIAL INTERACTION.

ulotrichous (Greek oulos, crisp; thrix (trichos), hair) *adj.* having crisp, very curly hair. CYMOTRICHOUS, LEIOTRICHOUS.

ultisols in SOIL CLASSIFICATION, USA, an order of soils which are deeply weathered and relatively infertile, lacking base minerals, resembling LATOSOLS, having red and yellow CLAY constituents, associated with humid temperate to tropical climates. Characteristically the A HORIZON is marked by residual iron oxides and the B HORIZON has accumulations of clay.

ultrabasic rock an IGNEOUS ROCK, generally PLUTONIC and containing very little QUARTZ or FELDSPAR, with a SILICA content less than 45 per cent and a BASIC-1 oxide content more than 55 per cent, largely FERROMAGNESIAN MINERALS, metallic oxides, sulphides, e.g. PERIDOTITE (mainly OLIVINE), perknites (consisting of ferromagnesian minerals other than olivine), and picrites (over 90 per cent ferromagnesian minerals, up to 10 per cent feldspar) which grade into such BASIC ROCK-1 as GABBRO. Ultrabasic rocks usually occur in association with other basic rocks in layered igneous INTRU-SIONS which often contain CHROMITE, ILMENITE, MAGNETITE as well as the PLATINUM group of metals. LAVA. G.

ultramafic igneous rock an IGNEOUS ROCK composed mainly of MAFIC minerals.

ultramarine (Latin ultra, beyond; mare, the sea; beyond the sea, from whence came lapis lazuli) a vivid blue pigment originally obtained by crushing LAPIS LAZULI to a powder, now by heating a mixture including soda ash, sulphur, charcoal; used in paints, printing inks, dyes etc.

ultraviolet radiation very short ELEC-TROMAGNETIC WAVES emanating from the SUN, with wavelengths between those of X RAYS and those of the violet end of VISIBLE LIGHT. Most ultraviolet radiation is absorbed by OZONE molecules in the upper atmosphere, but some reaches the earth's surface in sunlight, especially in high mountainous areas. It plays an important part as a PHOTOCHEMICAL agent in some life processes, e.g. it acts on the skin of some animals, including people, as a factor in helping to produce vitamin D; but an excess of ultraviolet radiation can be lethal to organic life. ACTINIC RAYS, ELECTROMAGNETIC SPECTRUM.

umber a dark brown earth with a high content of FERRIC oxides and MANGANESE, used as a pigment in paints etc.

umbra 1. the central, dark, complete shadow produced when an opaqe object stands in the path of a beam of light from an extended source (the outer, less dark area surrounding the umbra is the penumbra), e.g. as in an ECLIPSE **2.** the central, darker part of a SUNSPOT.

umlak (Eskimo) the traditional Eskimo open boat made from skins stretched tightly over a wooden frame, paddled especially by women and children.

umland (German, around land) for-merly applied in a general way to surroundings, and included in HINTER-LAND; now more precisely applied to an area which is culturally, economically and politically related to a particular town or city. AUSLAND, URBAN FIELD. G.

unaka (USA: from Unaka mountains, southern Appalachians) a MONAD-NOCK. G.

unavailable water water in the soil that is useless to plants because its molecules adhere to the surfaces of the soil particles. ADSORPTION.

uncertainty 1. the possibility that several outcomes will occur as a consequence of a decision or action, the form but not the probability of each being known. Thus uncertainty is incalculable, while RISK is calculable **2.** in statistics, the degree to which a sample statistic (e.g. the MEAN or MEDIAN) does not agree with (is in error from) the 'true' value which would have been found if the whole POPULATION-4 had been measured. The degree of uncertainty is sometimes termed the limits of error.

unconformable *adj.* applied to an overlying rock STRATUM which does not conform in dip and strike to the underlying strata. CONFORMABLE, UNCONFORMITY. G.

unconformity in geology, a break or gap in the continuity of a stratigraphical sequence, between two beds that are in contact, where the overlying younger rocks have been laid down on a surface resulting from a very long period of denudation, the older, lower set of beds having been laid down then uplifted, tilted, warped or folded and denuded to a greater or less degree before the deposition of the upper, younger series. The plane or the division between two such sets is the unconformity, and it implies a break (of any duration) in a geological record. The various types of unconformity are ANGULAR UNCON-

FORMITY, DISCONFORMITY, NONCON-
FORMITY. G.

UNCTAD United Nations Conference
on Trade and Development, a forum
set up by the United Nations where
matters relating to trade and development are discussed by developed and
developing countries.

undation theory of the development of
the earth's crust, the theory that differentiation of magma in the deeper parts
of GEOSYNCLINES creates waves in the
earth's crust. G.

underclay SEAT-EARTH.

undercliff a large mass of unstable rock
debris lying below a cliff, consisting of
material which has slipped as a result of
weathering, occurring particularly if
chalk overlies clay. G.

undercut slope the steeper slope on the
outside curve of a meandering stream,
the opposite to SLIP-OFF SLOPE.
UNDERCUTTING.

undercutting the carving away, eroding,
of material from the undersurface as by
1. the current of a meandering stream,
cutting into its bank on the outside of a
bend. MEANDER **2.** a sand-laden wind in
deserts, eroding the base of exposed
rocks **3.** wave action against cliffs along
sea coasts. G.

**underdeveloped, underdeveloped lands,
underdevelopment** terms that came
into prominence in 1949 as a result of
President Truman's inaugural address
in the USA, applied to those countries
to which the adjective 'backward' had
previously been applied but to which
exception was naturally taken by their
people. Underdeveloped (of a country
or region) came to be applied to one
not achieving the level that could be
reached, given its natural and economic
resources, if the necessary capital,
skills, machinery etc. were available, i.e.
underdevelopment was assessed in economic and technical terms. From an
economist's viewpoint, underdeveloped
can be applied to a country which could
use more capital, labour or more

available resources (or all of these) to
support its present population on a
higher living standard or, if its per
caput income is already fairly high,
could support a larger population on a
living standard which would not be
lower. In time the narrow economic
view of underdevelopment came to be
regarded as unsatisfactory: it did not
take into account cultural or social conditions in the country or region concerned. It is thus now usual to include a
consideration of social and cultural (as
well as economic and technical) elements in references to development or
underdevelopment. Over time other
terms have been introduced relating to
underdevelopment, e.g. in relation to
poverty or to the degree of realized
potential, less developed country
(LDC), moderately developed country
(MDC) or highly developed country
(HDC). Other terms refer to the stage
reached in INDUSTRIALIZATION, e.g.
advanced industrial country (AIC) and
newly industrializing country (NIC).
BRANDT REPORT, DEVELOPED, DE-
VELOPING, DEVELOPMENT, INTER-
NATIONAL DIVISION OF LABOUR,
THIRD WORLD. G.

underfit river MISFIT RIVER. G.

undergrowth the low plants (herbaceous, shrubs and saplings) under the
trees in a forest.

underpopulation too few people in relation to resources, a term applied to the
population in an area where the available RESOURCES-1 are not used so
fully as they might be because very few
people live there.

undertow a strong current flowing near
the bottom of the sea close to the shore,
pulling away from or aligned with the
coastline. It is caused by the flowing
seawards of water thrown up on the
beach by a wave.

undissected *adj.* not DISSECTED, applied especially to land surfaces not cut
up by erosion. DISSECTION.

undiscovered reserves RESERVES which

are thought, on the basis of current scientific knowledge, to exist; or are known to exist but of which the extent and quantity are uncertain. RESOURCES.

unearned increment BLIGHT.

unenclosed not fenced in. INCLOSURE.

unemployment 1. lack of employment, the state of being unable to secure paid employment **2.** the inability to find a paid job at the current wage rate, commonly classified as demand deficient (no vacant jobs of any type), structural (the skills of the unemployed do not fit the jobs available; the unemployed in this category can usually get jobs only by retraining or by accepting work needing less skill than they possess, DE-SKILLING), or frictional (vacant jobs available and the unemployed have the appropriate skills but for some reason, e.g. immobility or lack of information, the unemployed do not fill them).

UNEP United National Environment Programme, an agency of the United Nations formed in 1972 to coordinate intergovernmental measures for the monitoring and protection of the environment.

unequal slopes, law of LAW OF UNEQUAL SLOPES.

UNESCO United Nations Educational, Scientific and Cultural Organization, an agency of the United Nations established in 1946 to promote international collaboration in education, science and culture in the cause of peace and security.

unguiculate a MAMMAL with claws or nails, as distinct from one with hooves.

unguligrade in zoology, a MAMMAL which walks on the tips of the digits which have developed hooves, e.g. a horse. DIGITIGRADE, ORTHOGRADE, PLANTIGRADE.

uniclinal *adj.* applied by some authors to strata dipping steadily and uniformly in one direction, hence uniclinal structure. The term monoclinal, applied by some authors in the same way, is apt to be confusing and should be avoided, a MONOCLINE being a fold with only one limb. HOMOCLINE. G.

uniclinal shifting the process of asymmetrical development of a valley due to a stream which, following the line of the geological STRIKE, tends to cut sideways in the direction of the DIP, i.e. the valley is gradually eroded and moved laterally down the dip. Some authors apply the term monoclinal shifting to this process, but others consider this confusing (UNICLINAL), preferring uniclinal because it implies a uniform dip in one direction.

unified system of settlements in the COMECON countries of eastern Europe, a network of interdependent settlements of small and medium size, organized so that industrial development and public services are equally spread to the benefit of society, and urbanization of the population is increased.

uniformitarianism the principle that the processes and natural laws existing in the past, which steadily and slowly brought about changes in and on the surface of the earth, can still be seen at work today. The opposite view is expressed in CATASTROPHISM-2. NEO-CATASTROPHISM, NEPTUNISM, PLUTONIC THEORY. G.

uniform region a REGION-1 defined by the presence or absence of a disinguishing feature and by the observation that the variation of specified criteria within the area of the region is much less than that between the area and other areas.

unimodal MODE-3.

unimproved *adj.* not used well, not made better, not enhanced in value, e.g. as applied to land which has not been cultivated.

unilineal *adj.* in a social system, applied to a single line of descent, either through females only (MATRILINEAL), or males only (PATRILINEAL). If both matrilineal and patrilineal descent is accepted, the social system is described as double-unilineal; if the line of descent is a matter of choice, either matrilineal or patrilineal, the *adj.* ambilineal is applied.

uninverted relief a landscape where the surface relief reflects the underlying geological structure, where hill ridges coincide with ANTICLINES and valleys with SYNCLINES, the opposite of INVERTED RELIEF. G.

unitary hierarchy a single hierarchical chain of command in administration, e.g. in a small scale enterprise concentrating on a limited number of products or services, in which authority is centralized. POLYCENTRIC HIERARCHY.

unitary system of government, a state system in which authoritative control is vested in a central or national government which delegates power to local government authorities. FEDERAL SYSTEM.

United Nations an association of countries which by signing the Charter pledge themselves to maintain international peace and so to help the political, social and economic progress of the world. The organization is not authorized to intervene in the domestic affairs of any state. It originated during the Second World War, statesmen from the UK, USA and USSR (including Winston Churchill, Franklin Roosevelt and Joseph Stalin) agreeing that such a conflict must never happen again. They met at Washington D.C. in August-September 1944, talks with China came later, and the United Nations came formally into existence on 24 October 1945, headquarters New York, the working languages being English, French and (in the General Assembly) Arabic, Chinese, Spanish and Russian. The official languages are English, French, Russian, Spanish and Chinese. The principal organs are the General Assembly, the Security Council, the Economic and Social Council, the Trusteeship Council, the International Court of Justice, the Secretariat. A very

full account of the United Nations, its operations and agencies, appears in *The Statesman's Year-Book*, The Macmillan Press Ltd.

United Nations Conference on Trade and Development UNCTAD.

units of account in EEC, the units used in operating the Community Budget and in fixing support prices, e.g. under the Community Agricultural Policy for fixing agricultural support prices. CAP, GREEN POUND.

univariate analysis REGRESSION ANALYSIS.

univariate frequency distribution FREQUENCY DISTRIBUTION.

unroofed anticline a BREACHED ANTICLINE.

unstable air mass an AIR MASS with a high WATER VAPOUR content, liable to spontaneous convectional activity (CONVECTION) which results in heavy showers and THUNDERSTORMS.

unstable equilibrium the state of the ATMOSPHERE-1 where the ENVIRONMENTAL LAPSE RATE of an AIR MASS exceeds the DRY ADIABATIC LAPSE RATE. Such an air mass, being warmer and lighter than the air around it, will rise and go on rising. If it is very moist it will cool very slowly, at the SATURATED LAPSE RATE, and be even more unstable, leading to the formation of large CUMULUS clouds which may be associated with heavy rainfall, HAIL, THUNDERSTORMS. It will stop rising only when its temperature equals that of the surrounding air. STABLE EQUILIBRIUM.

up-country the interior of a country or region.

up-country *adj.* of, pertaining to, or situated in, the interior of a country or region.

updraught a very strong rising current of air. THERMAL.

upland, uplands a general, unspecific term applied to higher ground, in contrast to lowland or lowlands. It usually implies an area of relatively subdued relief inland, away from the coast. G.

upland, uplands *adj.* situated or living on, growing in, UPLAND, UPLANDS. G.

upland plain a comparatively high, level tract of land.

upper atmosphere EXOSPHERE.

upstream **1.** *adj.* at a location relatively nearer the source of a stream from a given point, or having direction towards the source of a stream **2.** *adv.* (moving) in the direction of the source of a stream, i.e. against the flow. DOWNSTREAM.

upthrow of fault THROW OF A FAULT.

upward transition region CORE-PERIPHERY MODEL.

upwelling in a deep body of water, the upward movement of colder water from lower layers to the warmer zone above, caused by a CURRENT-1, of economic importance in the ocean because the colder water, rich in PLANKTON and other nutrients, attracts fish. Many of the world's most important fisheries are associated with such upwelling, e.g. the cold Benguela current off southwest Africa.

uraninite a mineral, UO_2, consisting mainly of URANIUM oxide, with some THORIUM, RARE EARTH minerals and LEAD, occurring in GRANITE and PEGMATITE, in HYDROTHERMAL veins, in certain SEDIMENTARY ROCKS. It is the main ore of uranium.

uranium a radioactive, hard, white metallic element in the CHROMIUM group, the heaviest of the elements occurring in nature, found in PITCHBLENDE and some other minerals. Natural uranium consists of isotopes U 238, which converts into PLUTONIUM, U 235 and some U 234. Plutonium and U 235 are used as a source of NUCLEAR ENERGY. RADIOGENIC HEAT.

uranous *adj.* relating to or containing URANIUM.

urban (Latin urbs, town, city) *adj.* **1.** relating to, belonging to, characteristic of, constituting, forming part of, a town or city, the opposite of RURAL, i.e. applied to any settlement in which most of the working inhabitants (some authorities specify over 60 per cent) are engaged in non-agricultural occupations (retail and wholesale trades, handicrafts, manufacturing industries and commerce, with associated service occupations, etc.) **2.** relating to, belonging to, people living in such a place. URBAN SETTLEMENT. G.

urban climate the LOCAL CLIMATE of a built-up area, where the buildings affect temperature (by their heated interiors) and the pattern and speed of winds (by their layout), and they together with the paved surfaces create an impermeable layer which increases and speeds-up RUNOFF; and where the emissions from the burning of HYDROCARBON fuels in motor vehicles combined with those and other emissions from industrial plant may pollute the atmosphere, affecting cloud formation and precipitation. HEAT ISLAND, POLLUTION, SMOG, VENTURI EFFECT.

urban district LOCAL GOVERNMENT IN BRITAIN.

urban ecology the interrelationship of people and the identified urban environment in which they live, particularly the environment of a city. A concept developed from ECOLOGY, urban ecology is equated by some authors with HUMAN ECOLOGY-1.

urban economic base a two-fold classification of urban economic function, based on space relationships, providing a means of identifying economic ties between a large URBAN SETTLEMENT and other areas, of classifying and drawing up a comparative analysis of settlements, and of classifying individual economic activities within the urban area itself. The two categories are: basic activities (functions), which serve areas outside the TOWN or metropolitan area (METROPOLIS-2), and thus link the activities of the town/

metropolitan area with other parts of the earth's surface; and non-basic activities (functions), which serve the inhabitants of the town or metropolitan area and form links within the settlement itself. The extent to which a given function is 'basic' is measured by calculating the ratio of people engaged in that activity to the total population of the town/metropolitan area, and comparing this with a similar ratio for the country as a whole. BASIC ACTIVITY, NON-BASIC ACTIVITY. G.

urban fence a boundary line drawn around a town or larger urban settlement on planning maps in Britain in the 1940s to define land where urban influences were dominant, i.e. where land was built over with industrial works, public buildings, offices, shops, housing (including houses with gardens) and where such open land as existed was devoted primarily to urban use, e.g. playing fields, parks, cemeteries, or was so built round as to have little value for farming. It was considered that development of land within the urban fence was the concern of the town planner because agricultural and rural interests would be little, if at all, affected by such development. G.

urban field, urban region the sphere of influence of a town, the territory around a town with which it is functionally linked. UMLAND. G.

urban form the juxtaposition of land use zones in an urban area, regarded as the response to variation in accessibility.

urban fringe the area of social change around a town or metropolitan area (METROPOLIS-2), where urban development impinges on agricultural land, population density increases and land values rise.

urban geography the branch of geography concerned with the site, evolution, morphology (URBAN MORPHOLOGY) and classification of villages, towns and cities, their location in relation to a region or the country, the general processes (economic, political, social) at work within them, and the pattern of their relationship to other urban areas. G.

urban growth phases, USA the four stages recognized by B. J. L. Berry, American geographer, in his theory concerning the emergence of cities in the USA, the first phase being the mercantile (growth of towns on the Atlantic seaboard), the second the industrial (1840-50, the development of towns linked to the availability of coal and iron, i.e. the towns in the heavy industry area, thrusting westward towards the USA HEARTLAND-2), the third the heartland-periphery (1850-1950, contrasts between the CORE AREA, the heartland, and the peripheral centres, the core becoming richer), and fourth, the decentralized phase (post-1950, the location of amenity resources, e.g. sunny climate, becoming important, leading to rapid urban growth in Arizona and the southwest).

urban hearth the place of origin of URBAN development, of urban culture, e.g. the land between the rivers Tigris and Euphrates, Mesopotamia; the Nile valley; the Indus valley in the Indian subcontinent; the valley of the Hwang Ho, northern China. HYDRAULIC HYPOTHESIS.

urban hierarchy a ranked classification (HIERARCHY) of towns according to size (e.g. in England and Wales, major cities, cities, minor cities or major towns, towns, sub-towns), or to a stage of development (e.g. sub-infantile, infantile, juvenile, adolescent, mature, late mature and senile), or to function (CENTRAL PLACE HIERARCHY). Some authors prefer to use the term urban typology rather than urban hierarchy, on the grounds that hierarchy implies some degree of dependence or interdependence, whereas typology, the study of types, is non-committal. G.

urban hinterland the hexagonal trade

areas in CENTRAL PLACE THEORY. ISOTROPIC SURFACE, SPHERE OF INFLUENCE-2.

urbanism 1. town character, the typical condition of a town, or way of life characteristic of a town 2. used by some authors as an alternative term for URBANIZATION 3. sometimes applied as a synonym for town planning, e.g. in USA. ANTI-URBANISM. G.

urbanistics the technique of town (city) planning. G.

urbanization 1. the continuous process of transformation from being of RURAL to being of URBAN character, and the continuous change within the urban area itself as it grows by natural increase and by MIGRATION-1,2 from other (usually rural) areas. The result is that an increasing proportion of the population of an extensive area is concentrated in defined urban places, with resulting changes in land use, landscape, way of life, economic activities etc. In the process most urban places grow but the population tends to concentrate most quickly and in the greatest numbers in the largest places 2. the state reached in the process. ANTI-URBANISM, COUNTER-URBANIZATION, ECONOMIES OF URBANIZATION, OVER-URBANIZATION, URBANISM. G.

urbanization economies ECONOMIES OF URBANIZATION.

urban land the land on which an URBAN SETTLEMENT is built.

urban land use the concentrated use of land for urban purposes over a relatively wide area, i.e. for an URBAN SETTLEMENT (under this definition the main shopping street of a small village would not be classified as urban). A distinction has been made between urban land use and rural land use, the former being use *on* the land, the latter *of* the land.

urban mesh the geometrical pattern of the relations of urban centres (URBAN HIERARCHY). The term is derived from W. Christaller's theoretical pattern for

CENTRAL PLACES, i.e. his hexagonal mesh. CENTRAL PLACE THEORY, ISOTROPIC SURFACE. G.

urban morphology the systematic study of the origin, growth, form, plan, structure, functions and development of a TOWN, of the urban habitat.

urban renewal a process in which the obsolete fabric of an urban area is restored, renovated and improved in order to meet contemporary needs or standards. In most cases an attempt is made to retain its original external character, but if this is impossible, a certain amount of redevelopment may be included. Redevelopment involves the total destruction of all or part of the obsolete fabric before the work of new building, creation of open spaces, possibly new roads, etc. begins.

urban-rural continuum rural-urban continuum, the merging of town and country, a term used in recognition of the fact that in general there is rarely, either physically or socially, a sharp division, a clearly marked boundary, between the two, with one part of the population wholly urban, the other wholly rural. RURAL, RURBAN, RURBAN FRINGE.

urban settlement a term loosely applied to a relatively densely built-up area with its associated open spaces where the majority of the economically occupied inhabitants are engaged in activities mainly concerned with SECONDARY, TERTIARY, QUATERNARY INDUSTRIES, i.e. the definition is based on the function, not on the number, of the inhabitants. URBAN, URBAN LAND, URBAN LAND USE.

urban size ratchet the point at which an urban centre, having grown to a sufficient size, reached a certain THRESHOLD-4, will not decline but will survive no matter what troubles beset it.

urban sprawl an irregular, unplanned, untidy spread of buildings around a town, sometimes linking-up with similar development around a neighbouring town, and usually consisting of residential areas, small shopping centres, small industrial enterprises.

urban system 1. synonymous with system of cities, a set of interdependent, integrated urban places, i.e. towns and cities with their HINTERLANDS. The functions of the component parts are distributed in accordance with economic, social or political needs, some being concentrated in particular centres, others occurring widespread throughout the system (CENTRAL PLACE THEORY). The system is served by a network of routes along which flow goods, services, labour, capital, ideas, so that economic and social change in one component has major repercussions in the others. LOSCH'S THEORY, VANCE'S MERCANTILE MODEL **2.** a set of interdependent parts within an individual urban area.

urban typology URBAN HIERARCHY.

urban village VILLAGE-2.

Urlandschaft (German) a landscape of the past.

Urstromtal (German, ancient river valley; Polish PRADOLINY) a wide, shallow valley excavated by a meltwater stream flowing in front of a continental ice sheet in the North European Plain, corresponding to static periods in the northwards retreat of the edge of the Scandinavian ice sheet. G.

usar (Indian subcontinent: Hindi) saline land. G.

useful energy the amount of ENERGY needed for satisfactory performance.

use value the usefulness of a COMMODITY to the one who possesses it. UTILITY, VALUE-1,3.

U-shaped valley a glaciated valley, a valley which in cross section has the shape of a U, the floor being generally flat, the sides usually steep, due to the work of a VALLEY GLACIER moving down the V-SHAPED VALLEY of a preglacial river. The glacier gouges out the floor and erodes the valley sides up to the level of the surface of the ice. If the ice does not fill the valley a prominent SHOULDER-2 commonly occurs where the steepened sides meet the more gentle slopes of the pre-glacial valley. The glacier straightens the valley, shortens projecting spurs (TRUNCATED SPUR) and creates HANGING VALLEYS. The head of the valley may end in a steep wall (TROUGH END); and PATERNOSTER LAKES may be formed. G.

utilitarianism in ETHICS, a doctrine (expounded broadly by David Hume, 1711-76, but fully developed by Jeremy Bentham, 1748-1832) that the greatest good is the greatest happiness of the greatest number; and that therefore the moral and political rightness of an action is determined by its UTILITY, i.e. its contribution to the greatest happiness.

utility in economic theory, VALUE-1, the capacity (especially the capacity of goods or services) to satisfy human wants. The worth to the consumer is determined by the extent to which these satisfy the wants; and it is reflected in the price which the consumer is prepared to pay. MARGINAL UTILITY, NEOCLASSICAL ECONOMIC THEORY, PLACE UTILITY, USE VALUE.

uvala, ouvala, vala (Slavic) a large, closed depression in KARST, smaller than a POLJE but usually more than a kilometre in diameter, and having a broad but rather irregular bottom, formed by the coalescence of several neighbouring DOLINES. G.

V

vacoua VICUA.

vadose water water wandering in the ground above the permanent WATER TABLE, varying in amount and position as it moves through PERMEABLE rock. PHREATIC WATER. G.

val (French, valley) a valley, specially a synclinal valley (LONGITUDINAL VALLEY) in a range of fold mountains, originally in the Jura mountains. G.

vala Montenegrin term for UVALA (Bosnian term). G.

vale imprecise term applied to a broad, flat, extensive valley (e.g. the Vale of St Albans), or simply a gently undulating lowland (e.g. the Vale of Glamorgan). It is also used poetically, and is best avoided as a geographical term. G.

valency, valence the measure of the capacity of one ATOM-1 of an element or of a RADICAL to combine with or displace other atoms in a molecule, expressed as a number indicating the number of BONDS that the atom of the element or radical makes with other atoms in a molecule, e.g. in methane, CH_4, CARBON forms four bonds with hydrogen and is termed quadrivalent (four valent), while the HYDROGEN is univalent. The valency of an element varies in different compounds. Latin prefixes (uni-, bi-, ter-, quadri- etc.) express the number of bonds (valencies) exhibited; but Greek prefixes (mono-, dri-, tri-, tetra- etc.) express the specific valency of an atom or radical. It is now usual to use the terms electrovalency and covalency to indicate the means by which the bonds are formed. Electrovalency is the number of electrons

which an atom of an element has available for transfer to another element in the case of metallic elements, and the number of unpaired electrons available for pairing by electron transfer from an atom of a metallic element. The bond formed is an IONIC BOND. Covalency is the number of unpaired electrons which an atom of an element has available for sharing with another atom to form a covalent bond.

valley an elongated depression, usually with an outlet, sloping down to an area of inland drainage, a lake, or to the sea, sometimes (but not always) occupied by a river. See valley qualified by BEADED, DRY, HANGING, LONGITUDINAL, TRANSVERSE, U-SHAPED, V-SHAPED.

valley axis a term applied by some authors to the surface profile along the centre line of a valley, in preference to the term TALWEG. G.

valley bog BOG vegetation, usually dominated by SPHAGNUM at the head and margins, *Molinia* and sometimes *Schoenus nigricans* in the lower parts, occupying a waterlogged, flattened depression of a flat valley bottom, where drainage from acidic rocks is obstructed by the underlying IMPERMEABLE rock, so that the sluggish stream overflows its course, and heavy rain, falling on the adjacent saturated ground, cannot run easily into the stream.

valley glacier alpine glacier, mountain glacier, outlet glacier, a GLACIER which occupies an existing valley, i.e. a preglacial valley, termed an alpine glacier if it is formed by the merging of several CIRQUE GLACIERS; an outlet glacier if it originates from the margin of an ice cap or ice sheet. TIDAL GLACIER. G.

valley line, valley-line a term sometimes used in translation of TALWEG. G.

valley train generally stratified FLUVIO-GLACIAL material distributed by meltwater from the SNOUT of a glacier, also termed OUTWASH APRON or frontal apron. G.

valley wind, valley breeze a general term applied to cold air draining down a valley especially by night (KATABATIC WIND) or a wind blowing up a valley (ANABATIC WIND) by day, the result of the differential heating of the mountains above and the low land below. TALWIND. G.

vallon de gélivation (French) DRY VALLEY.

value 1. the measure of how much something is wanted for its special quality (e.g. beauty, rarity, UTILITY) expressed in terms of the money, effort etc. someone is prepared to expend in order to acquire, hold in possession, preserve it **2.** a quality, principle etc. that excites such a desire (e.g. moral values) **3.** in economics, the monetary equivalent of a product or a factor of production which satisfies three criteria, i.e. it is capable of being owned, it has UTILITY (satisfies needs or desires), and is in limited supply. EXCHANGE VALUE, SURPLUS VALUE, USE VALUE **4.** in mathematics, the amount represented by a symbol or expression, or **5.** the category of a VARIABLE.

value added the difference between the revenue of a firm obtained from a given volume of output and the cost of the inputs (the materials, components, services) used in producing that output.

value judgement a judgement that attributes worth or goodness, evil, beauty or some other VALUE-1,2, to something; or which asserts that some action ought or ought not to occur.

vanadium a MALLEABLE, white metallic element occurring in a few rare minerals in igneous and sedimentary rocks, e.g. in petronite, used in alloying in making very hard steel and, as vanadic acid, as a catalyst in oxidizing ANILINE.

Van Allen radiation belt either of the two layers of intense ionizing radiation, with high energy particles, which envelop the earth in its outer atmosphere, the inner occurring at some 3000 km (1865 mi) above the earth's surface, the outer at some 13-19 000 km (8080 to 11 800 mi). The particles probably come from cosmic rays and from the sun. Their movement is influenced by the MAGNETISM of the earth rather than by GRAVITATION.

Vance's mercantile model MERCANTILE MODEL.

vanilla *Vanilla planifolia*, a twining, epiphytic (EPIPHYTE) orchid, native to Mexico and Central America, now cultivated there and elsewhere in tropical regions where rainfall is adequate for its growth, for the sake of its pods, from which vanillin (among other aromatic substances) is extracted, for use in flavouring foodstuffs. Synthetic vanillin is available. CLOVE OIL, ESSENCE.

vanity effect a bias introduced in the results of surveys of human populations through the response of the individuals who, rather than accurate replies, give answers pleasing to their personal vanity.

vapour in chemistry and physics, a substance in the gaseous state (GAS-2) which separates into two phases when compressed. CONDENSATION, CRITICAL TEMPERATURE.

vapour pressure the pressure exerted by the vapour of a substance, e.g. the pressure exerted by water vapour in the atmosphere. If the air is SATURATED the term saturated vapour pressure is applied. ATMOSPHERIC PRESSURE.

vapour trail CONDENSATION TRAIL.

vardarac (from river Vardar) a cold

wind, similar to the MISTRAL, which blows down the valleys of Macedonia, including that of the river Vardar, to the Aegean sea in winter. G.

variability 1. the state or quality of being, or the tendency to be, VARIABLE-1 *adj.* 2. in statistics, the extent to which a set of SCORES spreads about the mean value, termed low variability if they are closely bunched around the average, high variability if they are widely spread.

variable 1. something VARIABLE-1 *adj.* 2. in mathematics, a quantity which may take two or more values, may take any one of a specified set of VALUES-4 (also applied to denote non-measurable characteristics, e.g. sex is a variable in that any human individual may take one of two 'values', i.e. male or female), or a symbol for such a quantity. ATTRIBUTE, VARIATE; and variable qualified by CONFOUNDED, CONSTANT, CONTINUOUS, DEPENDENT, DICHOTOMOUS, DISCRETE, INDEPENDENT, INTERVAL, INTERVENING, NOMINAL, ORDINAL, RANDOM, RATIO, QUALITATIVE.

variable *adj.* 1. having the quality of being able to change, able to be changed 2. in biology, not true to type. CLASSIFICATION OF ORGANISMS 3. in mathematics, characteristic of a quantity which may have different values. VARIABLE-2.

variable capital in Marxism, the living labour expended in the production process, as distinct from CONSTANT CAPITAL which represents 'dead' labour. It is termed variable because, being produced by living labour, the only creator of new value, its value increases during the production process. LABOUR POWER, LABOUR THEORY OF VALUE, SOCIAL CAPITAL, SOCIAL CONSUMPTION.

variable cost analysis an approach to the explanation of industrial (or other) location based on spatial variation in production costs. COMPARATIVE COST ANALYSIS, COST BENEFIT ANALYSIS, VARIABLE REVENUE ANALYSIS.

variable costs 1. costs that vary with volume of output, unlike FIXED COSTS 2. in spatial economic analysis, costs that are subject to spatial variation. COST SURFACE, VARIABLE COST ANALYSIS.

variable revenue analysis an approach to the explanation of industrial location based on spatial variation in revenue. VARIABLE COST ANALYSIS.

variance in statistics, the square of the STANDARD DEVIATION, a statistic which measures the VARIABILITY-2 in a set of SCORES-1 or observations. COMMON VARIANCE, COVARIANCE, ERROR VARIANCE, RESIDUAL VARIANCE, SPECIFIC VARIANCE, TRUE VARIANCE.

variance analysis in statistics, an analysis in which the total VARIATION-4 displayed by a set of observations (as measured by the sums of squares deviation from the mean, VARIANCE) may in some circumstances be separated into components, the components expressing well-defined, isolated sources of variation used as criteria of classification for the observations, i.e. the VARIABILITY-2 found in a POPULATION-4 is allocated to different, isolated sources. This allows a comparison to be made not only of the total variability in a population with the variability contributed by each of the sources of variance studied, but also of the sources of variance themselves, one with another, with the objective there of discovering whether any source or sources differ(s) significantly from another or others. The SIGNIFICANCE TEST associated with variance analysis indicates the probability that this same difference will occur in the parent population.

variate in statistics, an individual observation, one member of a set of values for one variable. A variate is a quantity that may take any of the range of values of a specified set with a specified relative FREQUENCY-3 or PROBABILITY-2, and is therefore sometimes termed a random variable. It is particularly associated with frequency (probability) function and expresses how often those values appear in the situation being observed. DATA IN STATISTICS.

variation 1. departure from the norm or from a standard 2. an example of that departure 3. in biology, the process that leads to structural or functional differences between individuals within a SPECIES-1 at corresponding stages in their life cycles. Such variations may be heritable, the consequence of genetic differences, propagated to the next generation, small and continuing with a tendency towards the average, or great and abrupt, e.g. a MUTATION; or they may be phenotypic (PHENOTYPE), the consequence of differences of ENVIRONMENT or upbringing and not heritable 4. in statistics, a synonym for DISPERSION; but in some cases variation implies a spread of VALUES-4 due to some definite, non-random cause.

varietal *adj.* of or pertaining to, characteristic of, a VARIETY-1 or constituting a VARIETY-2.

variety 1. in biology, loosely applied to any kind of VARIATION-3 occurring in a plant or animal species 2. more specificially, in CLASSIFICATION OF ORGANISMS, a taxonomic group below a subspecies, i.e. a group of organisms with certain qualities in common which distinguishes the group from others in the subspecies. The qualities may be, but are not necessarily, inherited. STRAIN-1.

Variscan orogeny in geology, a phase of the ARMORICAN orogeny, of late Carboniferous, early Permian times (GEOLOGICAL TIMESCALE). The term is used by some authors as a synonym for HERCYNIAN, but by others is applied only to the eastern arc of the Hercynian orogeny. G.

varve (Swedish varv, a layer) a distinct two-layered sediment deposited annually in a lake or other body of still water, especially in lakes near the margins of retreating ice sheets, where, of the laminated sediment, the lower, thicker layer, light in colour, consists of coarser material (deposited by meltwater from the rapidly thawing ice in summer), the upper, thinner layer, darker in colour, of very fine-grained material (settling during the slow melting of ice in winter). Each varve thus represents a year, so by counting the varves the time involved in the formation of the sediment can be estimated; and by correlations over a fairly extensive area, a glacial chronology be established (as by de Geer in Sweden, 1910). The sediments are termed varve clays, varved clays, varved sediments. A varve with bedding of graded particles is termed a diatectic varve, the term syminct varve being applied to a varve with bedding of flocculated, not graded, material. FLOCCULATION-2. G.

várzea (Brazil: Portuguese) one of the series of alluvial floodplains in the Amazon basin, regularly inundated by sediment-rich river water, poorly drained and covered with forest more dense and varied than that occurring on the higher intervening areas of the well-drained TERRA FIRME-2. G.

vauclusian spring (from Fontaine de Vaucluse, southern France) a gushing spring, a large spring, the resurgence of an actively eroding underground stream, commonly occurring in limestone country and varying greatly in output. RESURGENCE. G.

veal CATTLE.

vector 1. in biology, an agent carrying a disease-producing agent (e.g. a VIRUS or BACTERIUM) from one organism to another 2. in mathematics, a quantity which is specified by direction in addition to magnitude.

veering of wind a change of direction of

wind, in a clockwise direction, i.e. from north through east to south in the northern hemisphere, equivalent to a BACKING WIND in the southern hemisphere. G.

vega (Spanish) **1.** irrigated lowland of Granada, Spain, usually applied to an area yielding only one crop a year, as distinct from HUERTA **2.** in other parts of the world, an extensive fertile, grassy plain or, in Cuba, a tobacco field. G.

vegetable ivory the seeds of *Phytelephas macrocarpa* (a tree allied to palm, native to Ecuador, Peru, Colombia), the seeds being known as corozo nuts, and of *Hyphaene thebaica* (the dum palm of Sudan and other parts of north Africa) used, when hardened, as an ivory substitute in the manufacture of buttons, toys etc. until replaced by plastic materials. C.

vegetable oils the various oils obtained from plants, used mainly in food products, cosmetics, soap-making and other industrial processes, the residue left after the extraction in many cases being fed to livestock (OIL CAKE). Vegetable oils can be classified as DRYING OILS, oils which in drying form a thin elastic film (e.g. LINSEED, SOYA BEAN oil); semi-drying, those which form a soft film only after long exposure (e.g. COTTONSEED oil); non-drying, those which remain liquid at ordinary temperatures (e.g. OLIVE oil); and vegetable fats or tallow which are more or less solid at ordinary temperatures (e.g. COCONUT oil). Allied are the waxes, harder than oil, occurring on leaf surface (e.g. CARNAUBA WAX); and lather-forming products of leaves and stems used as soap-substitutes (e.g. cultivated soapwort, SAPONARIA OFFICINALIS). The most important commercial oils are COCONUT, COTTONSEED, ESSENTIAL OILS, GROUNDNUT, LINSEED, OLIVE, PALM KERNEL, PALM, RAPE, SESAME, SOYA, SUNFLOWER.

vegetation a general term for the total plant cover in an area or on the surface of the earth as a whole. CLIMATIC FORMATIONS, EDAPHIC FORMATIONS, NATURAL VEGETATION.

vein LODE-1, a general term for a crack or fissure in the earth's crust in which highly heated waters from below have deposited CRYSTALLINE minerals (especially vein quartz) from solution and, under special circumstances, metallic minerals of economic importance. G.

veld (Afrikaans; historical spelling veldt) any unenclosed country in South Africa with vegetation suitable for pasture. Many different types are distinguished: high, middle, low, or mountain veld according to elevation; sour or sweet veld according to lime content, sour veld being deficient; bush, grass, karoo veld according to vegetation; sand veld, hardeveld, according to soil condition. G.

vellum very fine treated SKIN-2 of a young animal, usually of a calf, kid or lamb, prepared for writing on or for special book binding.

velocity the speed of movement in a certain direction, absolute or relative swiftness, high speed of action or operation.

vendavales (Spanish) **1.** strong southwesterly gale-force winds blowing in the strait of Gibraltar and along the east coast of Spain in winter, accompanied by heavy rain and high seas **2.** a thunder squall on the Mexican coast. G.

venison the MEAT of DEER. LIVESTOCK.

vent an orifice in the earth's surface, especially that of a volcano, through which molten material erupts during volcanic activity. In a volcano it may become choked as the lava solidifies to form a PLUG or VOLCANIC NECK. G.

vent d'Autan Autan, a strong, hot dry wind blowing from southern France towards the centres of low pressure which come from the ocean into the Bay of Biscay. G.

ventifact glyptolith or rillstone, a PEBBLE with several flat facets which meet at fairly sharp angles, worn and polished by wind-blown sand, usually in desert conditions. If only three facets are present the term DREIKANTER is preferable; and some authors use the term EINKANTER if there is only one facet, ZWEIKANTER if there are two. G.

Venturi effect the effect produced by the narrowing of the channel or opening along or through which a gas or a liquid is passing, the flow of either being speeded up, e.g. as indicated by the gusts of wind in narrow streets and passages between tall buildings in towns. URBAN CLIMATE.

veranillo (tropical South America: Spanish, little summer) the short dry season, a period of drier, brighter weather which breaks the wet season, e.g. about the end of December in Ecuador. VERANO. G.

verano (tropical South America: Spanish, summer) the long dry season. INVIERNO, VERANILLO. G.

verglas a thin, clear, hard, smooth film of ice formed on exposed rock surfaces when a hard frost follows rain or the thawing of snow. GLAZE.

vermiculite a flaky, silicate mineral of the MICA group which, when subjected to great heat, yields a lightweight material used for thermal and acoustic insulation, in horticulture in seedbeds etc., and in packaging.

vermilion, vermillion mercuric sulphide, used as a PIGMENT to produce a brilliant red colour.

vernacular the indigenous language or dialect of a particular locality, region, country.

vernacular *adj.* **1.** of, relating to, speaking, using, the local indigenous speech of a region or of an ethnic group (as distinct fron an introduced speech or a dead classical language) **2.** applied to the indigenous arts, particularly in architecture to an indigenous style of building.

vernal *adj.* relating to, occurring in, the SPRING, e.g. vernal EQUINOX.

vernalization the act or process of inducing a plant to flower and fruit prematurely by special treatment of the seed or bulb, e.g. by subjecting it at a specific time for a specified duration to exposure to a low temperature, to darkness and to an amount of moisture abnormal to its natural environment. By these means the plant embryo completes part of its development independently of its rate of growth. The vernalization of the seed of winter WHEAT allows the crop to be sown in spring and harvested in the summer of the same year.

vernier (Pierre Vernier, 1580-1637, a French mathematician) a small auxiliary scale which slides along the main scale of an instrument such as a BAROMETER or a THEODOLITE and makes it possible to measure intervals smaller than those shown on the main scale.

versant (French) the slope, side or descent of a mountain or mountain chain, often applied to a large area with such a general inclination, e.g. the Pacific versant of the USA. G.

vertebrate in zoology, a member of Vertebrata, a subphylum of Chordata, the animals with a skull and with a segmented spinal column of bone or cartilage, imperfectly developed in some species. Vertebrates include Agnatha (primitive, jawless, fish-like vertebrates, e.g. the lamprey), true fishes, amphibians, reptiles, birds, mammals. INVERTEBRATE.

vertebrate *adj.* having a segmented spinal column. INVERTEBRATE.

vertex **1.** the top, the highest point **2.** in astronomy, the ZENITH **3.** in geometry, the point opposite a base, the meeting point of two lines opposite to the base of a plane or solid figure, e.g. the topmost angular point of a triangle or cone, i.e. the apex of a triangle or cone.

vertical *adj.* PERPENDICULAR, having a direction pointing towards the centre of the earth (HORIZONTAL), e.g. aerial photographs may be taken from an aircraft with the lens of the camera pointing down at a vertical angle.

vertical circle AZIMUTH-2.

vertical exaggeration the increase necessary in the vertical scale in comparison with the horizontal scale in order fairly to represent the height of an ELEVATION-1 in a MODEL-3 or a SECTION-2.

vertical interval the difference in vertical height between two points. HORIZONTAL EQUIVALENT.

vertical temperature gradient LAPSE RATE.

vertical theme a concept characteristic of DIACHRONIC ANALYSIS and used in HISTORICAL GEOGRAPHY, which traces the changing form of landscape and the human activities which effect such change over time. H. C.Darby identifies his vertical themes of landscape change in England as clearing the woodland, draining the marsh, reclaiming the heath, the changing arable, the landscape garden, towns and seats of industry. CROSS-SECTION APPROACH.

vertical zone altitudinal zone, in South and Central America, TIERRA CALIENTE, TIERRA FRIA, TIERRA HELADA, TIERRA TEMPLADA. G.

vertisols in SOIL CLASSIFICATION, USA, an order of clay-rich soils which swell and crack in seasonally alternating wet and dry conditions, thereby mixing or inverting their horizons, e.g. GRUMUSOL.

vesicle 1. in geology, a small, generally rounded or oval cavity in a mineral or rock, particularly in glassy volcanic rock, formed by the trapping of steam or gas bubbles in the molten material as it solidified. AMYGDALE 2. in biology, a small globular or bladder-like air space

in tissue; or a small sac or cavity commonly containing fluid.

vesicular *adj.* covered with vesicles, or resembling a VESICLE in form or structure, applied especially in geology to the TEXTURE-1 of rock.

vessel 1. a container for liquids 2. a ship or boat, other than one moved by oars or poles, or a small sailing boat 3. in biology, a tube-like structure through which fluids pass.

vestigial *adj.* applied to the now barely visible or discernible evidence of a structure, function, or behaviour pattern which formerly existed or was present.

Vesuvian eruption VULCANIAN ERUPTION. G.

viaduct a high bridge supported by a series of pillars built to carry road or rail raffic over a dry valley, river, marshy ground or road. AQUEDUCT.

Vico, Giambattista 1668-1744, Italian philospher and historian who spent most of his life in Naples and who maintained that the social world (having been created by human beings) was more readily intelligible than the natural world (created by God). He allowed that mathematics provided certain knowledge, but only because it was a human creation. He is deemed to be the first historian to interpret history in terms of the rise and fall of human societies; and to look upon and use legends, poetry and the study of linguistics as historical evidence.

vicua, vacoua coarse sacking material made from the fibres of the leaves of the screw pine, *Pandanus odoratissimus*, native to southern Asia, Madagascar, the Pacific islands. PIASSAVA. C.

vicuña (Spanish from Quechuan) *Lama vicugna*, a mammal allied to the LLAMA and ALPACA, native to the high Andes, not domesticated, hunted for the small quantity of the very fine wool yielded by its coat, which makes soft warm fabric. The numbers have decreased greatly in recent years, and the vicuña is now a protected animal. C.

vidda (Norwegian) **1.** in general, a high, treeless undulating area **2.** more precisely, the old land surface predominant in eastern Norway. G.

vill historically, a territorial division in the FEUDAL SYSTEM in England, akin to the later civil PARISH, consisting of lands and buildings within a common social organization; sometimes used as a synonym for VILLAGE.

villa (Latin) **1.** historical, Roman times, a large residence, usually with extensive grounds and an agricultural estate attached to it **2.** in nineteenth century Britain, any residence in the suburbs or in a residential district, such as one occupied by middle class people **3.** more recently, a small suburban house standing in a garden, or a holiday house near the sea or in a mountainous region, sometimes a substantial country house.

village 1. a collection of dwelling houses and other buildings, especially in rural surroundings; a nucleated settlement as contrasted with dispersed habitations **2.** a close-knit, small community forming an 'island' in an URBAN environment and sometimes termed an urban village, often situated in the INNER CITY or the TRANSITION ZONE. People of similar ethnic or cultural characteristics may cluster in some urban villages. G.

village green, town green a piece of open land in a village or town on which the inhabitants of the village or town have a customary right of playing lawful games and enjoying it for recreation.

ville (French) a town, used as a suffix in English place-names which are derived from a proper name or product, e.g. Sharpville, Waterlooville, Coalville. G.

vine GRAPE.

vinery a greeenhouse in which GRAPE vines are cultivated for dessert grapes.

vineyard a plot of land devoted to the cultivation of GRAPE vines.

virazon (South America) the sea breeze blowing regularly on the coasts of Chile and Peru, especially on summer afternoons. TERRAL. G.

virga fragments of cloud trailing beneath the dark cloud of a passing FRONT.

virgate YARDLAND.

virgation (Latin virga, a twig) of mountain ranges, a bunching together and divergence of ranges of fold mountains from a central knot, as in the Pamir Knot in Asia. G.

virgin adj. applied to soil, earth, forest etc. not yet cultivated, exploited or brought into use. G.

virus a member of a group of submicroscopic entities that invade the cells of plants and animals (some being able to attack BACTERIA and the cultivated mushroom). Viruses cannot exist outside the host tissue. They are not self-reproducing, but they are able to use the machinery of the host cell in order to produce copies of themselves, at the expense of the host cell. Most show their presence by causing disease, e.g. influenza, poliomyelitis, hepatitis in people; foot and mouth disease, rabies in animals; MOSAIC DISEASES of cultivated plants. Viruses which invade animals are spread by contact, droplet infection or by insect vectors (e.g. the mosquito *Aedes aegypti*, the carrier of yellow fever); those which affect plants are usually spread by insects such as aphids, but some are transmitted by eelworms.

viscosity 1. the property or quality of a FLUID which makes it resistant to flow **2.** the degree to which a LIQUID is resistant to flow.

viscous adj. having VISCOSITY, being sticky, slow-flowing, e.g. the lava forming a volcanic CONE.

visibility 1. the state or fact of being visible **2.** the range of vision of an observer, which depends on the time of day, the quality of light, the clarity of the atmosphere (presence or absence of dust, fog, mist), the height above sea-level at which the observer stands. DEAD GROUND, HORIZON.

visible light the part of the ELECTRO-MAGNETIC SPECTRUM which may be perceived by the human eye, the wave-lengths ranging between the limits of INFRA-RED and ULTRAVIOLET radiation. ACTINIC RAYS, ELECTROMAG-NETIC RADIATION.

vital rates in demography, the measures relating to changes in size and structure of a population.

vitamin any of a range of organic food substances essential for metabolism in HETEROTROPHIC organisms, only a small, regular intake being necessary. MACRONUTRIENT, MICRONUTRIENT, TRACE ELEMENTS.

viticulture the cultivation of the GRAPE vine for the production of grapes and WINE. G.

vlei (Afrikaans, historically vley) in southern Africa, a shallow lake or swamp, a depression with poor drainage which becomes marshy after rain. It differs from a PAN-4 in that it is in many cases connected with a river system, has a more permanent water supply and is more irregular in shape. G.

vloer (Afrikaans, floor) in southern Africa, a very large shallow hollow occurring on an arid plateau surface with so small a slope that well defined streams are non-existent. The rain water therefore disappears by evaporation rather than by runoff. G.

vodka a spirit with a high alcoholic content, distilled from POTATOES, RYE, WHEAT.

voe (Scottish: Shetland and Orkney dialect) 1. a narrow gully cut in a cliff, in many cases ending in a cave or tunnel with a BLOWHOLE 2. a bay, creek, inlet, specifically in the Shetland and Orkney islands, but the term is applied elsewhere. G.

volatile 1. a substance that is quickly and easily evaporated 2. an ELEMENT or COMPOUND dissolved in MAGMA which would be in a gaseous state (GAS-2) if it were at the temperature of the magma under ATMOSPHERIC PRESSURE.

volatile adj. evaporating quickly and easily.

volcanic adj. of or pertaining to, like a VOLCANO covering all types of extrusive IGNEOUS activity, as distinct from intrusive PLUTONIC activity.

volcanic ash the unconsolidated PYRO-CLASTIC material consisting of finely comminuted fragments of rock and lava which have been ejected explosively from a volcano. The term ASH, which dates from the time when a volcano was thought to be a 'burning mountain', is a misnomer. POZZOLANA. G.

volcanic bomb a lump of LAVA ejected into the air from a volcano, whirled around and partly solidifed by cooling before it falls to the earth in a generally round shape. BREADCRUST BOMB. G.

volcanic breccia a PYROCLASTIC rock, like VOLCANIC ASH but consisting of larger, angular fragments, and more or less consolidated. G.

volcanic cinders LAPILLI. G.

volcanic cone CONE-1.

volcanic dome DOME VOLCANO.

volcanic dust the finest particles thrown out of a volcano in eruption. It may be shot high in the air by the explosive force and carried great distances by wind, e.g. dust from the 1883 eruption of Krakatoa in Indonesia is said to have been carried round the earth three time before settling. OOZE. G.

volcanic eruption ERUPTION. The various types are: HAWAIIAN, PELEAN, STROMBOLIAN, VULCANIAN.

volcanicity, volcanism VULCANICITY, VULCANISM.

volcanic mud and sand 1. volcanic deposits laid down around volcanic islands and coastlines 2. VOLCANIC ASH washed down by the heavy rains which often accompany or follow an eruption. G.

volcanic neck or plug strictly the orifice of a volcano through which lava reaches the surface, and in which the lava eventually solidifies as a plug, the plug itself also being termed a neck. This may later stand in isolation if the material of the surrounding cone is denuded. DENUDATION, PIPE, THOLOID, VENT. G.

volcanic rock an IGNEOUS ROCK formed by volcanic action at the earth's surface, consisting of solidified material which has issued in the molten state from the depths of the earth, i.e. EXTRUSIVE ROCK, in contrast to HYPABYSSAL and PLUTONIC rocks. That is the most common application, but some authors include rock formed in association with intrusive activity (INTRUSION), and therefore include some hypabyssal rocks; and others also include plutonic rocks. G.

volcanic sink a large depression, commonly several hundred metres deep and more than 1.5 km (0.9 mi) in diameter, in a large SHIELD VOLCANO or lava dome, resulting from subsidence as the stream of molten lava is reduced in volume and may finally cease to flow.

volcano a rift or vent in the earth's crust through which molten material is erupted and solidifies on the surface as LAVA or through which the molten rocks, charged with gases and vapours, are ejected with explosive force and fall back as VOLCANIC ASH and VOLCANIC DUST etc. (PYROCLAST). A volcano may be a central type (the eruption taking place through a more or less cylindrical pipe) or a fissure type (the lava issuing through a line of weakness in the earth's crust, FISSURE ERUPTION). It is described as active whilst in eruption or liable to eruption; dormant during a long period of inactivity; extinct after all eruptions are presumed to have ceased. Four types of eruption are identified: HAWAIIAN, PELEAN, STROMBOLIAN, VULCANIAN. See entries qualified by volcanic, as well as ADVENTIVE CONE, AGGLOMERATE, ASH CONE, CALDERA, CINDER CONE, COMPOSITE VOLCANIC CONE, CONE, CRATER, DOME VOLCANO, DRIBLET CONE, MUD-VOLCANO, NUEE ARDENTE, OCEANIC TRENCH, PAROXYSMAL VOLCANIC ERUPTION, PLANEZE, PUMICE, PUY, PYROCLAST, SHIELD VOLCANO, SPATTER CONE, TEPHRA, THOLOID, TUFF. G.

volcanology VULCANOLOGY.

volume the measure of the space occupied by a solid or liquid, or the quantity of a substance or material held in a container. CONVERSION TABLES.

Von Thünen model a model devised by J. H. Von Thünen, published 1826 in his *Der Isolierte Staat*, to explain the principles which govern the prices of agricultural products and the way in which these variable prices control the pattern of agricultural land use. Using the ISOLATED STATE as his basic assumption and applying his theory of ECONOMIC RENT, he further assumed that all farmers would produce the crop giving them individually the maximum net profit (LAND RENT). To accomplish this each farmer would produce a crop or adopt an agricultural system for which the location of the farm land in relation to the market was most advantageous. Von Thünen also assumed that the value of a unit of produce to the farmer would be equal to its market price less the cost of transport to the market. The cost of transport of agricultural produce from farm to market was the only variable in the model. Land rent for any product declined with increasing distance from the market point, but the rate of decline varied for each product according to its particular transport cost. The market price for each product determined the highest land rent possible. Von Thünen also considered the technology available for production and transport, and the kinds and quantities of produce needed by the central, large town. From all this he postulated a model with concentric rings or zones of agricultural land use centred on the large town, the central market (ISOLATED STATE). In this simple, original model (reflecting the needs, conditions, equipment and technology of 1826), the zone nearest to the town was devoted to market gardening and milk production, the zone beyond this to forestry (for fuel, building timber, wood products), the next one to intensive crop rotation (without FALLOW), the next to crop farming, fallow and pasture, the next to a three-field system, the next to LIVESTOCK farming, which gave way to waste. This original model thus concentrated on the distance from MARKET-3 as the governing, independent variable. Later Von Thünen considered the effects of varying soil fertility on production costs, and modifed his assumptions of transport costs' being uniform in all directions. ALONSO MODEL. G.

voortrekker (Afrikaans) in South Africa, one who travels in front, especially one of the pioneers of the Great Trek (TREK) of the Boers, 1835-7. G.

Voralp (German) particularly in Switzerland, the lower pastures of an alpine valley, i.e. those above the valley floor but lower than the ALP-1 proper. MAYEN. G.

voyageur (French, a traveller) in North American literature, used with special reference to the French explorers of the continent; but originally applied to any traveller, including those employed by trading companies. G.

V-shaped depression a term formerly applied to a trough of low pressure (LOW, ATMOSPHERIC) in which the isobars form the shape of a V.

V-shaped valley a valley eroded by a RIVER, V-shaped in cross-section in contrast to the U-shape common to a glacially modified valley. In W. M. Davis's CYCLE OF EROSION the V-shape is cited as evidence of youth in the stage of river erosion; but nearly all river valleys, if not subjected to glaciation, are V-shaped in the upper course of the river. Among other factors the valley shape will be influenced by the type of rocks through which the river is flowing, their resistance to weathering and erosion, and climate. U-SHAPED VALLEY. G.

vug, vugh (England, Cornwall: miner's term) a cavity in a metalliferous vein, in some cases quite large and sometimes lined with well-formed crystals. LOCH, LOCH-HOLE. G.

Vulcanian eruption, Vesuvian eruption a VOLCANIC ERUPTION characterized by Vesuvius, the active volcano southeast of Naples, the first known eruption of which destroyed the towns of Pompeii and Herculaneum in 79 AD. The eruptions are less frequent, the magma more viscous, the ejected lava less basic than the phenomena associated with a STROMBOLIAN ERUPTION; but the ejected material is not so sticky and acid as that of a PELEAN ERUPTION. Some authors draw a distinction between Vulcanian and Vesuvian, associating the latter with the more explosive eruptions occurring when the plug of viscous lava in the vent of the VOLCANO is blown out. ACID LAVA, BASIC LAVA. G.

vulcanicity, vulcanism all the processes in which molten rock (MAGMA), liquids or gaseous materials move towards and erupt at the earth's surface, or are forced into the earth's crust, covering all igneous activity, not only the phenomena associated with a VOLCANO. G.

vulcanology the scientific study of volcanic phenomena. G.

W

Wächte (German) a snowdrift formed along a ridge, resembling in form the crest of a BREAKER, curling towards the steeper slope as strong winds sweep snowflakes over the ridge, the canopy so formed gradually curling downwards under its own weight. G.

wadden (Dutch) a tidal flat, synonymous with WATTE. G.

wadi (Arabic) a stream course or valley in hot desert or semi-arid areas, especially in north Africa, usually dry but occasionally carrying a stream following heavy rain. In those parts of Spain once under Moorish domination 'wadi' has become 'guadi' in river names. Guadiana thus translates as River Ana (tautologous, 'ana' signifying a river). On some maps the river has been marked River Guadiana, or even River Rio Guadiana, which translates as river-river-river-river, from four languages: English, Spanish, Arabic, Latin. ARROYO, NALA. G.

wage differential theory of MIGRATION-1,2, a theory which asserts that differences in wages in different places are the main cause of migration. PUSH-PULL THEORY, STRENGTH THEORY.

wake 1. the disturbance of water behind a moving ship 2. the disturbance of air behind a moving object or to leeward of a fixed object in an air stream. WAKE DUNE.

wake dune a SAND DUNE formed in the LEE of a larger dune, trailing away DOWNWIND.

Waldhufendorf (German, forest village) in Germany, applied specifically to a village, without common land, which originated during the clearing period (ninth to fourteenth century) and in which a single or double row of farmhouses border a stream at a valley bottom, the consolidated farmland stretching behind each farmhouse to the parish boundary. G.

Wallace's line a 'line' drawn by the English naturalist and geographer, Alfred Russell Wallace, 1823-1913, separating the distinct flora and fauna of south-east Asia and Australasia (Oriental and Australian ZOOGEOGRAPHICAL REGIONS), running southeast of the Philippines through the Macassar strait and following the deep water channel between the islands of Bali and Lombok in Indonesia. In the Oriental region all the characteristic MAMMALS are placental (as are the great majority of all living mammals, PLACENTA-2); in the Australian region the characteristic mammals are marsupial (e.g. kangaroo) and monotreme (e.g. duck-billed PLATYPUS). Wallace developed an hypothesis of evolution by natural selection independently of DARWIN, the two men publishing their work simultaneously in 1858. G.

wall-sided glacier a VALLEY GLACIER projecting from the valley it is occupying on to an adjoining plain so that its steep sides are visible, unrestricted by the valley sides.

walnut *Juglans regia*, a tall DECIDUOUS tree (up to 30 m: 100 ft), native to southeastern Europe, west and central Asia and China, cultivated in the British Isles probably since the fifteenth century. A valuable timber tree, the wood is used in cabinet-making. The fruit is a green DRUPE with a wrinkled stone, the edible wrinkled kernel (the nut) of which is eaten fresh or used in baking etc.; the unripe green drupe is pickled in vinegar; and an edible oil is extracted from the nuts. BLACK WALNUT.

wane 1. the act or process of waning, i.e. of diminishing in size or brilliance, declining or decreasing in strength or power 2. the period from full MOON to new moon.

waning slope, waxing slope in the study of hillslopes (SLOPE) four elements are commonly identified: the waxing slope near the top which becomes steeper as the hillside is worn back; the FREE FACE; the CONSTANT SLOPE; and the waning slope (sometimes termed the wash slope, or wash controlled slope) at the foot which becomes less steep as it develops by accumulating debris brought down by SURFACE WASH from above, merging into the widening valley floor. STANDARD HILLSLOPE. G.

wapentake an old subdivision of certain English counties where the Danish influence was strong, in Yorkshire and the Midlands. Corresponding divisions elsewhere are the HUNDRED and the SOKE. YARDLAND. G.

ward 1. an small administrative division in a district council etc. in Britain which elects its own representative to serve on the local authority 2. the people, collectively, of such an administrative division 3. the division of a forest. G.

warehouse 1. a building in which goods are stored 2. such a building under the control of customs officials where goods are stored until the duty is paid on them.

warm-blooded *adj.* applied to an animal (e.g. a mammal) which maintains a constant body temperature regardless of variations in external temperature. The state of being warm-blooded, i.e. warm-bloodedness, is termed homo-

iothermy (homoethermy). POIKILO-THERMIC.

warm front the boundary zone between the advancing mass of warm air forming part of a DEPRESSION-3 and the colder air which it is overriding and overtaking. In advance of the passing of a warm front over a fixed point, pressure falls, the cloud base lowers, the wind backs (BACKING OF WIND). A characteristic series of cloud forms accompanies its approach and passing (CIRRUS, CIRROSTRATUS, ALTO-STRATUS, STRATUS, NIMBOSTRATUS), intermittent DRIZZLE and then steady, heavy precipitation accompanying the nimbostratus. The wind veers (VEER-ING OF WIND) as the front passes over, and the precipitation dies away. COLD FRONT, FRONT, OCCLUSION, WARM SECTOR. G.

warm glacier temperate glacier, a moving ice mass, the surface of which melts through THERMAL CONDUCTION, the resultant surface water percolating through the ice mass, releasing LATENT HEAT in refreezing, thus raising the temperature of the ice mass itself (the summer temperature of the whole mass is about 0°C: 32°F but in winter the surface is colder). COLD GLACIER. G.

warm occlusion an OCCLUSION where the cold, overtaking air is not so cold as the air mass ahead of it.

warm sector a region, a bulge, of warm air in a DEPRESSION-3, where the air temperature and RELATIVE HUMIDITY rise. It is preceded by the WARM FRONT and followed by a COLD FRONT. OCCLUSION.

warp, warp-clay, warpland 1. the alluvial silt, mud or clay deposited in a tidal estuary 2. the land so covered. G.

warping 1. the process (natural or ARTIFICIAL-1) whereby the low-lying land of a tidal estuary is flooded, leading to deposition of silt, mud or clay. WARP 2. a gentle, slow deformation of the earth's crust over a wide area, resulting in the raising or lowering of the surface. G.

warren, warrener originally a piece of land enclosed and preserved for the breeding of game; later a piece of land appropriated for the breeding of rabbits (rabbit-warren) and at times hares. In MEDIEVAL Britain there were certain legal rights of warren, there were legal definitions of beasts and fowls of warren, and keepers known as warreners. With changing social conditions and the spread of wild rabbits, controlled warrens with their warreners disappeared, and the term warren or rabbit-warren came to be applied to a piece of uncultivated land in which wild rabbits burrowed, having taken it over as breeding territory. G.

Warsaw Pact a treaty of defence and mutual assistance signed in 1955 by Albania, Bulgaria, Czechoslovakia, East Germany, Hungary, Poland, Romania and the USSR in response to the formation of NATO. Albania withdrew in 1968 when the USSR invaded Czechoslovakia.

wash 1. the surging movement of the sea or other large body of water; the surge of water up a beach following the breaking of a wave, in contrast to the BACKWASH **2.** an area of sand and mud washed by the tide **3.** fine material moved down a slope. SUB-SURFACE WASH, SURFACE WASH, WASH SLOPE. G.

washboard moraine a series of nearly parallel, relatively low, sandy morainic ridges, so closely spaced as to resemble the ribbed surface of a wooden washboard, formerly used for scrubbing clothes, and occurring in an area that has been subjected to the advance, stagnation and melting of an ice sheet or glacier.

washland low land bordering a river, usually part of the natural FLOOD-PLAIN over which flood water is permitted to flow to prevent damage elsewhere, being controlled by pumping etc. Where such land lies between the normal river bank and a main LEVEE or DIKE the term foreland has sometimes been applied, but is not recommended in view of the other applications of the term. FORELAND.

wash load SUSPENDED LOAD.

washout 1. an erosion channel cut in a sedimentary deposit and filled by material from a later deposit **2.** a channel of sandstone in a coal seam (indicating stream erosion in the formation of the coal), formed when a stream, similar to a distributary in a delta of today, flowed through the forest swamps, its channel subsequently becoming filled with sand **3.** the washing away of soil, scouring, undermining of river banks etc. as a result of torrential rain, flooding **4.** the place where soil has been washed away. G.

wash-over sometimes termed wave-delta, the material deposited in a lagoon by a storm wave which breaks over the low part of the BAR sheltering the lagoon. Nearly every bar has on its lagoon side a row of deltas formed in this way.

wash slope, wash-controlled slope GRAVITY SLOPE, HALDENHANG, WANING SLOPE, WAXING SLOPE.

waste, waste land 1. commonly, any wild, uncultivated, uninhabited land **2.** formerly, the little-used common land, usually on light soil, which failed to yield a return to the medieval and later cultivator. Now rarely applied to such land because in so many cases it is valued as an open space **3.** now applied to land previously used but abandoned, for which further use has yet to be found **4.** in law (the plain and common acceptation of the word), desolate or uncultivated ground, land unoccupied, or that lies in commons. G.

waste of the manor, manorial waste part of the demesne of a MANOR, uncultivated and unenclosed, over which the freehold and customary tenant might have rights of common; but not all manorial waste was subject to COM-MON RIGHTS. G.

waste-mantle REGOLITH, the weathered material, broken and rotted rock etc. accumulated on the earth's surface. G.

wat (Thai) a Buddhist shrine (wrongly translated as a temple) in Thailand. G.

water the transparent, colourless liquid, H_2O, forming as ocean, lakes, rivers etc. some three quarters of the earth's surface. AGGRESSIVE WATER, HYDROLOGICAL CYCLE, HYDROSPHERE, ICE, OCEAN, PRECIPITATION-1, SPRING-2, TIDE, WATER HEMISPHERE.

water-bearing strata AQUIFER. G.

water buffalo *Bubalus bubalis* or *Bos bubalis*, the common domesticated buffalo, widely used as a draught and dairy animal in the warm parts of Asia.

water chestnut caltrops, *Trapa natans*, an aquatic plant native to the warmer parts of Europe, Asia, Africa, cultivated for its edible seed, rich in starch, eaten raw, roasted, boiled or made into flour. Related species with edible seeds similarly used are LING and singhara nut, *Trapa bispinosa*, native to tropical Asia, grown especially in Kashmir, the seed being eaten raw or cooked. CHINESE WATER CHESTNUT.

watercourse 1. the bed of a stream which flows continuously or intermittently 2. a stream of water, e.g. a river 3. a natural or artificial channel for carrying water. G.

watercress variously classified as *Rorippa Nasturtium-aquaticum* or *Nasturtium officinale*, a PERENNIAL aquatic herb native to the temperate and warmer regions of the northern hemisphere, cultivated (in running shallow water, preferably with a clear bed free of mud) for its pungent leaves, a valuable source of vitamins, containing some proteins, eaten raw or cooked.

water cycle HYDROLOGICAL CYCLE.

water deficit in soil science, the difference between SOIL WATER present in the soil and the storage capacity (FIELD CAPACITY) of the soil.

waterfall a sudden, steep or perpendicular descent of water in the bed of a river, occurring where the flow of the river is broken by a nearly horizontal bed of hard rock overlying easily eroded soft rock; or by the sharp edge of a plateau; or the abrupt end of a HANGING VALLEY, high on the slopes of a U-SHAPED VALLEY; or a FAULT-LINE SCARP; or the edge of a coastal cliff. The term waterfall should not be confused with RAPIDS, which are associated with an unbroken, continuous slope in the river bed. G.

water-gap a cutting or gap, a low-lying valley made by a river as it flows across a ridge. Water-gaps tend to be associated with ANTECEDENT DRAINAGE and may be cut by either CONSEQUENT or OBSEQUENT STREAMS. Unlike a WIND-GAP, a water-gap is occupied by a stream which continues to form part of the drainage system. G.

water gas a mixture of carbon monoxide and hydrogen made by passing steam over or through very hot COKE, used in the manufacture of liquid fuels, and as a source of HYDROGEN. COAL GAS, GAS.

water glass a jelly-like solution of sodium or potassium silicate, formerly widely used in preserving eggs, now mainly used in the making of soap and detergents, or as a protective coating in industrial processes.

water hemisphere the half of the earth opposite to the LAND HEMISPHERE. Its centre lies near New Zealand and water covers six-sevenths of its surface. G.

water hole, waterhole 1. a hollow, natural or ARTIFICIAL-1, where water gathers, especially in savanna and hot desert lands, in some cases fed by a spring 2. a depression in the bed of an intermittent stream 3. a hole on the surface of ice. G.

watering place a term formerly applied to 1. a place to which animals were taken for drinking **2.** a place at which ships called to lay in a supply of water **3.** a spa or locality with a mineral spring to which people went to drink the waters or to bathe. G.

water level the surface level of a body of water.

waterline any of the levels shown on the hull of a ship up to which it may legally be submerged in water when loaded, e.g. PLIMSOLL LINE.

waterlogged *adj.* applied in soil science to the state of a soil when it is SATURATED-2 with water. G.

water meadow (American washland) a low-lying meadow by the side of a stream, artificially irrigated by flooding in the early part of the year to encourage an early growth of grass, a practice especially common in the chalk valleys of southern England where the lime in the water was beneficial. Most of these English water meadows have now fallen into disuse on account of the high cost (especially the labour cost) of maintaining the elaborate series of miniature canals and drains. G.

water mill a MILL-1 powered by water. HEAD-RACE, TAIL RACE-2.

water-parting WATERSHED. G.

water power, waterpower the ENERGY of moving water converted into mechanical energy, formerly used directly to drive water mills etc. but now nearly always used to generate ELECTRICITY. HYDRAULIC, HYDROELECTRIC POWER, POWER-3. G.

waterscape as a land use category, a large body of water which dominates the landscape, that is one larger than a farm pond, ornamental lake in a town park, or a small natural lake, all of which are subordinate to another land use, e.g. to FARMSCAPE, TOWNSCAPE, WILDSCAPE. G.

watershed **1.** (British) a water parting, the elevated line which may or may not be sharply defined, separating two contiguous drainage areas from which the headsteams flow in different direc-

tions, into different river systems or basins **2.** (American) a water parting, as in British usage; but also the whole catchment area or DRAINAGE BASIN of a single river system. ANOMALOUS WATERSHED, NORMAL WATERSHED. G.

waterspout the product of an intense, localized small scale cyclonic storm (CYCLONE) occurring over the ocean or a lake, usually in tropical and subtropical regions. From the underside of a CUMULONIMBUS cloud a spinning FUNNEL CLOUD (carrying water droplets formed by condensation) descends to meet spray thrown up from the water surface by whirling winds, the combination forming a rotating column of mist, water and spray which is sometimes vertical and straight, sometimes bent (when the top part moves faster than the base), moving swiftly over the surface of the water. TORNADO. G.

waterstones a term sometimes applied to the upper part of the Keuper sandstone of the English Midlands from which, in some places, SPRINGS-2 flow. The term should not be used as a synonym for AQUIFER. G.

water table, water-table the surface below which PERMEABLE rocks are SATURATED-2 with water. In areas with PERVIOUS soil and pervious subsoil-rocks it tends to follow generally, but not in detail, the form of the land surface. Where the water table lies below the land surface its height corresponds to the level of water in wells (GROUND WATER, PHREATIC WATER), and similarly fluctuates seasonally. Where the water table reaches the land surface a SPRING-2 results; fluctuations in the water table account for the intermittent flow of BOURNES. A permanent marsh or lake results when the theoretical water table is above the land surface. In some circumstances there is no regular water table, e.g. where underlying rocks are irregularly fissured, as in the ancient METAMOR-PHIC plateau of Africa. In other cases there is a PERCHED WATER TABLE. ARTESIAN WELL, CAPILLARY FRINGE, HYDRAULIC GRADIENT, PIEZOMETRIC LEVEL, VADOSE WATER, WELL. G.

water vapour in the earth's ATMO-SPHERE-1, water in the VAPOUR state and below the CRITICAL TEMPER-ATURE for water. HUMIDITY.

waterway 1. a navigable stretch of inland water, i.e. of a lake, river, canal, which is or can be used for transport **2.** the route followed by inland water traffic. G.

waterwheel a wheel turned by moving water, formerly used to work machinery, e.g.in a MILL-2.

Watte (German) a tidal flat between the mainland coast of Denmark, West Germany and the Netherlands and the dune-covered offshore islands; termed WADDEN in the Netherlands. G.

wattle 1. vertical wooden stakes interwoven with their branches and thick twigs, e.g. of hazel, forming a framework used for making fencing, roofs, walls, etc., e.g. wattle and daub, wattle plastered with mud, formerly used for walls of buildings **2.** a HURDLE **3.** popular name for any of the trees and shrubs of the genus *Acacia*, especially *Acacia pycnantha*, the black wattle, native to Australia, the bark of which yields TANNIN. C.

wave 1. in physics, a self-propagating disturbance which passes on energy through a material medium by means of the elastic and inertial characteristics of the medium, the particles displaced by the disturbance returning to their position of rest when the disturbance subsides **2.** in physics, a disturbance which passes on energy through empty space by variations in the electric and magnetic properties of space, i.e. by an ELECTROMAGNETIC WAVE. ELEC-TROMAGNETIC SPECTRUM, LONGI-TUDINAL WAVE, SEISMIC WAVE,

SHAKE WAVE **3.** specifically, in a body of water, particularly in the ocean, the rise and fall in the forward movement in the surface area of the water, due to the oscillation of water particles, usually caused by friction of wind on the water surface. The motion of the water particles is perpendicular to the direction of the movement of the water, each particle moving up to the CREST and falling back almost to its original position in the TROUGH-4. The size of wave depends on the cause, e.g. speed and direction of wind (FETCH); the height constitutes the distance between the trough and the crest; length, the distance between two successive crests; steepness, the ratio of height to length (BREAKER); velocity, the speed of movement of an individual crest. BACKWASH, CONSTRUCTIVE WAVE, DOMINANT WAVE, HYDRAULIC FORCE, LONGSHORE DRIFT, OSCIL-LATORY WAVE THEORY OF TIDES, SWASH, TIDAL WAVE, TSUNAMI, WAVE BASE; and the entries, qualified by wave, which follow.

wave-base the greatest depth at which sea floor sediment can be just slightly moved by oscillating water. G.

wave-built terrace a TERRACE-1 formed by marine deposition seawards from a WAVE-CUT BENCH.

wave-cut bench, wave-cut beach bench a marine erosion plane formed at the base of a sea cliff, sloping down towards the sea, in some cases merging imperceptibly into a WAVE-BUILT TERRACE. ABRASION PLATFORM, BENCH. G.

wave delta, wave-delta WASH-OVER. G.
wave ogive OGIVE.

wax any of the various natural or synthetic substances made to resemble natural wax in physical or chemical properties or both, consisting of mixtures of higher fatty acids, being harder and less greasy than fats. It may be of animal, vegetable or mineral origin (e.g. BEESWAX, CARNAUBA WAX,

PARAFFIN WAX); used mainly in making polishes etc.

waxing slope WANING SLOPE. G.

wax palm *Copernicia cerifera,* carnauba palm, native to South America, grown particularly in Brazil, the leaves of which yield a fine, valuable resinous WAX, carnauba wax, used in making polishes, floor waxes and carbon paper. C.

Weald (Old English, forest; German Wald) in England, a tract of country covered in Saxon times by the Forest of Andred, lying between the scarps of the chalk of the North and South Downs in the English counties of Kent, Surrey and Sussex, restricted by some authors to the area between Greensand ridges. It includes the clay vales of Kent and Sussex and the sandstone ridges of Ashdown Forest. Formerly applied, especially poetically, to similar areas elsewhere, the term is now used only as the specific, regional name. G.

wealth consuming sector of an economy, the services sector.

wealth creating sector of an economy, the manufacturing sector.

weather a general term for the conditions prevailing in the ATMOSPHERE-1, especially in the layer near the ground (TROPOSPHERE), over a short period of time (in contrast to CLIMATE) or at a specific time, at any one place, and as affecting human beings. Temperature, sunshine, pressure and wind, humidity, amount of cloud, precipitation (rain, sleet, hail, snow), the presence of fog or mist are all taken into account. ANALOGUE-1, LONG-RANGE WEATHER FORECAST, METEOROLOGY, SYNOPTIC-CHART, WEATHER CHART. G.

weather chart, weather map a chart or map showing weather details for a selected area at a specific time. Data collected at observation posts and transmitted to meteorological stations provides the basis for a series of charts relating to pressure conditions, etc. affecting the selected area at set, regular intervals; and from this series the final maps, giving a summary of isobars, temperature, winds etc. at a selected time is drawn. Some authors prefer to use the term SYNOPTIC CHART, as indicating the summary character of the final chart.

weathered ice hummocked POLAR ICE and PRESSURE RIDGES worn by WEATHERING-1 to a smooth, rounded form. Continued weathering may result in a generally even surface. G.

weathering 1. the action of the WEATHER on objects exposed to it **2.** in geology, the mechanical or physical, chemical and biological PROCESSES (CHEMICAL WEATHERING, MECHANICAL WEATHERING, ORGANIC WEATHERING) by which rocks are decomposed or disintegrated by exposure at or near the earth's surface to water, the atmosphere, organic matter (DENUDATION), a mantle of rock debris being produced in situ. Transport (except by gravity) is not involved (EROSION). The main mechanical or physical agents are SHATTERING, frost action and temperature change (THERMOCLASTY), assisted by the biological processes, the organic agents being plant roots, mosses, lichens, the burrowing of animals. The chemical processes include CARBONATION, HALOCLASTY, HYDRATION, HYDROLYSIS, OXIDATION, SOLUTION. CORROSION. G.

weathering front the boundary between weathered and unweathered rock, an alternative term for BASAL SURFACE. WEATHERING. G.

weather ship a vessel at a fixed station at sea, equipped with meteorological instruments used for routine observations and, in some cases, for research.

weather vane wind vane, an apparatus for indicating wind direction. A broad, thin strip, usually of metal, is fixed to a pivoted, freely rotating support, so that it may swing round easily in an air currrent.

Weberian analysis a theory of the optimum location of firms (manufacturing enterprises) formulated by Alfred Weber, 1909, German economist, who maintained that transport costs were the major factor determining location; that optimum location was primarily the point where the costs of the transport of raw materials to the factory and of supplying goods to the necessary market were at their lowest; but that if variations in other costs (e.g. of labour) were high enough, location determined solely by transport costs might not be the optimum one. LEAST COST LOCATION, LOCATION, LOCATIONAL TRIANGLE, LOCATION THEORY, MINIMAX LOCATION, OPTIMUM LOCATION. G.

wedge of high pressure a region of HIGH atmospheric pressure, indicated by a V-shaped pattern of isobars, narrower than a RIDGE OF HIGH PRESSURE, occurring between two DEPRESSIONS-3, bringing a brief period of fine weather in a generally rainy period.

weed any plant flourishing in an area where it is not wanted, e.g. an invasive wild plant on cultivated land.

weir an obstruction built across a river to impound or raise the level of the water for fishing purposes, for creating a head for a WATER MILL, for the control of the current and maintenance of the water depth to aid navigation, for irrigation, or to divert the flow. The term is limited to small, low constructions over which the water may flow, the larger being termed DAMS and BARRAGES. G.

welded tuff, ignimbrite a fused mass of hot VOLCANIC ASH forming a rock such as RHYOLITE. TUFF.

welfare 1. the state or condition of being well, thriving, happy, prospering 2. work organized to bring about this state in needy members of a community. DEVELOPMENT, PARETO OPTIMUM, SOCIAL WELL-BEING, WELFARE GEOGRAPHY, WELFARE MAXIMIZATION.

welfare geography an approach in HUMAN GEOGRAPHY concerned with social inequality, which considers the areal differentiation and spatial organization of human activity from the perspective of the WELFARE-1 of the people involved. It touches on everything, positive or negative, contributing to the quality of human life, covering everything differentiating one state of society from another, the 'good' and the 'bad' things consumed in society, what these are, to whom and where they are distributed; and how the observed differences arise (i.e. who gets what, where and how). PARETO OPTIMUM, QUALITY OF LIFE, SOCIAL GEOGRAPHY, WELFARE, WELFARE MAXIMIZATION. G.

welfare maximization the optimum benefit to be achieved by the interrelationship of production techniques, the combination of goods and services produced, and distribution among the population. It depends not only on the technical relationships of how to obtain the greatest output possible from available resources, but also on the social relationships behind the preferences that are actually implemented. PARETO OPTIMUM, WELFARE. G.

welfare state a state with a political system based on the principle that the protection, social security, SOCIAL WELL-BEING and WELFARE of the individual is the concern, the responsibility, of the community as represented in the state. The state therefore provides the facilities and services necessary to bring this about (e.g. by providing medical care, education, public housing, pay for the unemployed and the aged etc. on the cradle-to-grave basis), financed by taxation and compulsory contributions from the population. LAISSEZ-FAIRE.

well originally a natural SPRING-2 or pool fed from a spring. The term is now restricted to a deep hole, usually cylindiracal, or a shaft, dug in the ground to obtain water, oil or gas. A well sunk for water is usually lined with brick or masonry, but may be unlined (e.g. if sunk through hard rock) and normally fills with water up to the level of the WATER TABLE (PHREATIC WATER), the surface of water fluctuating seasonally with the height of the water table. A well sunk into an ARTESIAN BASIN taps water held under considerable pressure. ARTESIAN WELL, OIL WELL, TUBE WELL. G.

Wentworth scale a scale devised by C. K. Wentworth, 1922, to measure the size of PARTICLES in sediments, a geometric scale of factor 2. This scale ranges from clay particles of 0.004 mm diameter, through silt, sand, granule, pebble, cobble to boulder, exceeding 256 mm diameter. GRADED SEDIMENTS.

WE-ocratic a term applied by T. Griffith Taylor, 1936, to human control as opposed to environmental control. G.

west 1. one of the four cardinal points of the COMPASS, directly opposite the EAST on the side of someone facing due NORTH, i.e. the direction of the setting sun at the EQUINOX 2. towards or facing the west 3. the western part, especially of a country, particularly the states west of the Mississippi, USA 4. the West, the countries of Western Europe and North America as distinct from the centrally-controlled countries of Europe and Asia.

west *adj.* of, pertaining to, belonging to, situated towards, coming from, the west, e.g. of winds blowing from the west.

Westerlies winds which blow frequently from the subtropical high pressure area to the temperate low pressure area, between 35° N and 65°N and 35°S and 65°S, blowing predominantly from the southwest in the northern hemisphere, predominantly from the northwest in the southern hemisphere. In winter in the northern hemisphere their presence makes the North Atlantic ocean one of the stormiest regions in the world; and in winter too they move southwards, carrying winter rain to the Mediterranean region. In the northern hemisphere their force and direction vary, and they are associated with the succession of DEPRESSIONS-3 and ANTICYLONES characteristic of the weather of the area in which they blow. But in the southern hemisphere they blow strongly and with greater regularity throughout the year over the great expanse of ocean, giving the region the name ROARING FORTIES. Westerlies gain strength with height, evolving into JET STREAMS. The old term applied to the Westerlies, the ANTI-TRADE WINDS, is misleading and no longer used. G.

wet adiabatic lapse rate SATURATED ADIABATIC LAPSE RATE.

wet-bulb thermometer a THERMOMETER with a bulb covered by wet muslin, thereby being cooled by evaporation. The temperature recorded is accordingly lower than that shown by a DRY-BULB THERMOMETER; and the two different readings, combined with reference to a set of statistical tables, enable DEW-POINT, RELATIVE HUMIDITY and the VAPOUR PRESSURE of the air to be calculated. G.

wet-day in UK, officially, a day of 24 hours beginning at 0900 hours during which at least 1 mm (0.04 in) of rain falls. RAIN-DAY, PRECIPITATION-DAY. G.

wet dock a large BASIN-9 in which the water level is maintained at the level of HIGH TIDE so that the vessel in it stays afloat. DOCK.

wether 1. a male SHEEP, a RAM, but especially a castrated ram 2. the FLEECE from the second or subsequent shearing of a sheep.

wetland, wetlands a general term applied to an ECOSYSTEM intermediate between the TERRESTRIAL-2 and the AQUATIC, a natural or artificial landscape in which fresh or salt water plays a key role, i.e. where the soil is waterlogged, the WATER TABLE is at or near the surface, or the land is covered occasionally, periodically or permanently, by shallow fresh or salt water (e.g. BOG, CARR, FEN, MARSH, SWAMP, flooded pasture land, intertidal mud flats).

wet-point settlement a settlement the site of which was related to the availability of a water supply, especially to a constant SPRING-2, in contrast to a DRY-POINT SETTLEMENT in lands liable to flood. G.

wet spell in UK, officially, an unbroken succession of 15 or more consecutive WET-DAYS, a definition not accepted internationally. DRY SPELL. G.

whale a member of any of the species of very large marine mammals of the order Cetacea, some of which have been hunted nearly to extinction for the sake of their flesh, for WHALEBONE, AMBERGRIS, SPERM OIL, SPERMACETI and for the oil (TRAIN OIL) from their blubber. MICROPHAGOUS.

whaleback 1. a rounded elongated mass of coarse-grained sand in a hot desert, a very large LONGITUDINAL DUNE, possibly formed by the merging of a succession of SEIF DUNES **2.** a rounded, elongated mass of rock, usually granite. RUWARE. G.

whalebone a horny substance forming toothed plates in the palate of certain WHALES, used as a flexible stiffening, e.g. formerly for corsets.

whale oil TRAIN OIL, a true fat obtained from WHALES, used in making soap, margarine and other edible fats.

whare (New Zealand) in New Zealand **1.** a Maori sleeping house; a term now applied to any small, usually old, house **2.** the accommodation provided for a shearer or general hand on a farm. G.

wharf a landing stage to which barges and ships may be moored while loading and unloading.

wheat any of the grasses of the genus *Triticum*, probably native to southwestern Asia and eastern Mediterranean regions, an important grain crop, probably the first crop cultivated by Stone Age people, now grown especially in the former grasslands of the midlatitudes, providing the STAPLE-3 food in temperate climates, the grain being ground into FLOUR used in the making of bread, biscuits, cakes etc. If the whole grain is ground 'wholemeal' flour is produced; if the husk of the grain is removed first, a fine white flour, which usually has some lime added, is produced. *Triticum aestivum* is the most widely grown for this purpose. Hard wheat, durum wheat, *Triticum durum*, with small hard grains, grown in dry regions in the Mediterranean region, the USSR, Asia, North and South America, is the best for making pasta, macaroni etc.

Most wheats need about 100 days to grow and ripen between the last killing frost of spring and the first killing frost of autumn, and this sets the northern limit in Canada and USSR where, however, wheat is being bred to ripen in 90 days. Wheat needs a good firm moisture-holding soil, such as a heavy loam. The black earths or CHERNOZEMS of the USSR and North America are particularly favourable. Climatically the best conditions are a cool moist spring, which causes the grain to 'tiller' (produce a number of stalks capable of bearing a head of grain), followed by warm sunny weather when the heads have formed, and some rain or moisture just before harvest to swell the grain. A total rainfall of between 375 and 875 mm (15 and 35 in) is about right. With a lower rainfall harvests are poor and irregular (as in Spain and the drier parts of Australia). In countries with a very cold winter (e.g. Canada) wheat is sown in spring; in countries with a mild winter (e.g. Britain) it can be sown in the autumn or fall, the seeds remaining in the ground during the winter to sprout as soon the temperature rises in spring. Hence the distinction between 'spring' wheat and 'winter' or 'fall' wheat (VERNALIZA-TION). In tropical regions (e.g. in Egypt or in parts of the Indian subcontinent) wheat is grown as a winter crop to be reaped before the heat of summer.

whey the watery substance in milk remaining after CURDS have formed and separated.

whinstone 1. in northern England, quarryman's term for DOLERITE, hence the name Great Whin Sill **2.** any of various dark, hard rocks, especially a basaltic rock. G.

whirlpool a circular eddy or current in a river or the sea produced by the configuration of the channel, by the effect of winds on tides, by the meeting of currents, or by similar phenomena. G.

whirlwind a rapidly rotating column of air revolving around a local centre of low pressure, caused by local surface heating and exceptionally strong CON-VECTION. It is limited in extent, formed round a vertical or slightly inclined axis, the inward and upward spiral movement of the lower air spreading to an outward and upward spiral, the whole moving progressively over land or water. CYCLONE, DUST DEVIL, TORNADO, WATERSPOUT. G.

white box approach an approach in SYSTEMS ANALYSIS which tries to identify in as much detail as possible the sub-systems, components, processes etc. present within the system in order to build up as complete an under-standing as possible of the system's internal structure and functioning. BLACK BOX APPROACH, GREY BOX APPROACH.

white-collar worker a person who works in an office, and is employed in non-manual work (usually excluding those engaged in the professions).

BLUE-COLLAR WORKER, PROFES-SIONAL-1, STEEL-COLLAR WORKER.

white currant a shrub, a seedling from the red currant, *Ribes sativum*. The berries lack the red pigment and are not so acid as those of the red currant, but have the same use.

white fish, whitefish any of the various marine fish with white flesh used for food, especially flat fish (e.g. sole) and species of the cod family.

white lead basic lead carbonate, used as a pigment, valued for its good covering power.

white man's grave an outmoded term applied to the hot, humid coastlands of west Africa, especially of Sierre Leone, once considered to be unhealthy for Europeans. With the introduction of inoculations against such former killing diseases as yellow fever, of prophylatic drugs (i.e. drugs that guard against disease), of pest control, and improved sanitation, the reputation became un-justified.

white metal any of the alloys based on tin, used for bearings, castings etc.

white-out a condition in a BLIZZARD when the snow cover is extensive and the falling snow so great that visibility is reduced to the minimum and finding direction almost impossible.

white pepper PEPPER-2.

white pine *Pinus strobus*, a tall pine native to eastern North America, valued for its soft, pale-coloured timber. SOFTWOOD.

WHO World Health Organization, an international body established in April 1948, headquarters Geneva, with an executive board consisting of techni-cally qualified health experts, to foster the highest possible level of health in the world. Its work includes dealing with matters of international health, helping governments to strengthen health services (especially so that pri-mary health care reaches the maximum number of people, and that diseases endemic in under-developed areas are combated); promoting maternal and child care; stimulating work and research in mental health, medical research, the prevention of accidents and the eradication of disease; and encouraging the improvement of standards in teaching and training in the health professions etc.

Wiesenboden (German, wet meadow soil) in soil science, a poorly drained soil with an A_1 horizon rich in HUMUS, grading into grey gleyed mineral soil. A HORIZON, GLEY. G.

wigwam a tent-like construction consisting of hides, bark etc. spread over a semicircular framework of poles, traditionally used as a dwelling by North American Indians in the area eastward of and around the Great Lakes. TEPEE.

Wilcoxon matched pairs test in statistics, an hypothesis test used to analyse two samples of paired data (PAIRING) in order to compare population MEDIANS-3. The sample data must be on INTERVAL SCALE, the sample sizes small, the data must be paired. The results may not be totally reliable if the population distribution of differences between the data values is not symmetric.

wildcat a test well for petroleum or natural gas bored as a speculation without any detailed geological evidence of the existence of either of them.

wilderness 1. an uncultivated, uninhabited region **2.** in nature conservation 'wilderness areas' are those left in a wild state as natural habitats, in contrast to those nature reserves which may need careful management to maintain small communities of plants and animals. G.

wild rice AMERICAN WILD RICE.

wildscape an area of land dominated by natural or semi-natural vegetation, rock outcrops, swamps, water, glaciers etc., as distinct from areas where the results of human activities are dominant. FARMSCAPE, TOWNSCAPE, WATERSCAPE. G.

Wildschnee (German, wild snow) very powdery snow commonly falling in calm weather in high mountains in Switzerland.

williwaw, willywaw sailor's term for a sudden violent SQUALL, especially in the ROARING FORTIES, originally in the Strait of Magellan. G.

willow a member of *Salix*, family Salicaceae, a genus of trees or shrubs with many species, most being native to northern temperate and arctic regions, growing near surface water, in many cases POLLARDED to provide a supply of pliable shoots used in basketry etc. The wood of some species is used in making cricket bats. OSIER.

willy-willy, willi-willi (Austalian) a type of TROPICAL REVOLVING STORM originating off the coast of western Australia, in some cases moving on to the land. G.

wilt a condition of plants in which the cells lose their turgidity (TURGOR PRESSURE), so that the leaves, young stems and tops of older stems become limp. It is usually caused by an excess of water loss through TRANSPIRATION in relation to water absorption (WILTING POINT); or it may be due to functinal disorder or the action of fungus parasites.

wilting point in soil science, the point below which the amount of water stored in the soil cannot be absorbed by plants quickly enough to meet their needs, causing WILT in any plant not adapted to drought. Wilting point is used as a measure of storage capacity (FIELD CAPACITY) of a soil.

wind air in motion, usually restricted to natural horizontal movement, varying in strength from light to hurricane (BEAUFORT SCALE). The term CURRENT is usually applied to the vertical movement of air (THERMAL). There are very many names for local winds (e.g. FOHN, MISTRAL, NORTHER). ANABATIC, DEFLATION, DOMINANT WIND, KATABATIC, PLANETARY

WINDS, PREVAILING WIND, TRADE WIND. G.

windbreak, wind-break something, but especially a line of trees or a thick hedge or a HURDLE, designed to break the force of the wind and provide shelter for animals or, more often, for growing plants. A windbreak is particularly important where a cold wind, e.g. the MISTRAL, would damage unprotected crops. DUST BOWL. G.

wind chill the cooling power of wind and temperature on shaded dry human skin. It was originally measured as the product of wind speed in metres per second and air temperature in degrees Centigrade below zero; but a later formula measures the cooling power of wind and temperature in complete shade regardless of evaporation. G.

wind erosion DEFLATION, EROSION.

wind-gap, air-gap a dry gap, a notch or gap in the crest of a hill range, or a pass, originally cut by a stream, from which the water has disappeared, e.g. a dry COL in an escarpment through which a CONSEQUENT STREAM may have flowed before RIVER CAPTURE. In many cases a wind-gap lies at a higher level than that of a neighbouring WATER-GAP. G.

wind gauge ANEMOMETER.

windmill a MILL-1 operated by rotating sails which are turned by the wind. WIND POWER.

window 1. FENETRE 2. one of the bands in the ELECTROMAGNETIC SPECTRUM within which energy, radiated through the atmosphere, escapes into outer space. ATMOSPHERIC WINDOW.

wind power mechanical or electrical POWER-3 generated by a WINDMILL.

wind pump a pump activated by the wind's POWER-3 in rotating a propeller wheel composed of vanes (blade-like, thin, flat strips, often curved).

wind rose a diagram with radiating arms constructed to show the frequency (and usually the speed as well) of winds blowing from the eight chief points (but sometimes from twelve points) of the COMPASS. The length of each arm shows the frequency recorded over a specific period of time, and gradations on the arms show the frequency of wind speeds. G.

wind shadow an area of seemingly completely calm (but actually eddying) air, immediately to LEEWARD of an obstracle in the wind's path, e.g. of a SAND DUNE.

wind-slab a sheet of snow packed hard by the wind.

wind sock a cone-shaped fabric bag, open at each end, mounted on a pivot on a high pole so that it becomes inflated when the wind blows, swinging round and indicating the direction from which the wind is blowing, erected especially on an airfield. WEATHER VANE.

windward the direction from which the wind is blowing, facing into the wind, as opposed to LEEWARD. G.

wine an alcoholic drink made from the fermented juice of the grape (but the term is also applied to other juices fermented and containing alcohol). Wine is produced in nearly all vine-growing countries (GRAPE), where most of it is consumed, although much (especially the better quality) enters international trade. Even in a small area variations in soil and microclimate, as well as in the weather and manufacturing processes, produce great differences in yield and the quality of the wine. The colour of red wine comes from the dark-coloured grape skins which are left in the fermenting juice for some time before the fruit is pressed. White wine comes from the juice of paler-skinned grapes, the fruit being pressed as soon as it is picked. White wines with a high sugar content are 'sweet'; where the process of fermentation has been allowed to continue so that more of the sugar is converted to alcohol, the wines are 'dry'. The alcholic content depends on the type of grape and condition of ripening. Fermentation is usually completed in casks or vats, but for some white wines it is allowed to continue in the bottle, producing a 'sparkling' wine which includes a proportion of natural gas, e.g. the champagne from the dry limestone chalk hills east of Paris.

Many districts are famous for special types: hock from the terraced sunny hillsides of the Rhine valley, moselle from similar lands along the Moselle valley; claret from southwest France, burgundy from the old French province of that name near Dijon. Port is a name protected by international agreement and may be used only for certain wines from the Douro valley of Portugal. True sherry, a fortified wine, comes from southern Spain, the name being a corruption of the place-name Jerez. Wines are produced in the 'newer' Mediterranean lands of California, Australia, South Africa and South America and are often named after the old European types they most closely resemble. Some wines are distilled to produce BRANDY, the alcoholic content of which is much higher than that of wine. C.

winnowing the act of blowing chaff (the outer husk) free of GRAIN-5. DEFLATION-1.

winter 1. the colder part of the year, in contrast to summer, the hotter **2.** loosely, the cold season; in tropical regions the term winter is usually dropped, the term cool season being preferred **3.** one of the seasons in mid- and high-latitudes, popularly December, January, February in the northern hemisphere (the other seasons being SPRING, SUMMER, AUTUMN), or June, July, August in the southern **4.** astronomically, from the winter SOLSTICE to the spring EQUINOX, i.e. from about 22 December (also paradoxically termed midwinter day) to 20 March in the northern hemisphere, 22 June to 21

September in the southern hemisphere. G.

winter ice generally unbroken, level sea ice, 15 cm to 2 m (6 in to 6.5 ft) thick, originating from YOUNG ICE and of no more than one winter's growth. G.

winter solstice 21-22 December in the northern hemisphere, 21-22 June in the southern. SEASON, SOLSTICE, WINTER.

wirescape a landscape with a proliferation of electricity pylons and poles supporting wires for various purposes. G.

woad *Isatis tinctoria*, a BIENNIAL plant native to Europe, the source of a dark blue dye (also termed woad), notable for having been used by ancient Britons as a body colouring.

woina-dega, voina-dega, voina dega (Ethiopia) DEGA.

wold (Old English weald, forest; German Wald) originally, forest or wooded upland, later open land, then restricted to an elevated tract of open country. The term is applied specifically to open rolling chalk upland as part of a place-name, e.g. Yorkshire Wolds, Lincolnshire Wolds on CHALK; but also in some cases to uplands of other rocks, e.g. the Cotswolds on JURASSIC rocks. G.

wolfram, wolframite the major ore of TUNGSTEN, ferrous tungstate. It occurs in VEINS associated with granitic rocks and in PLACERS.

wood 1. the hard, fibrous vascular tissue of mature plants, forming stems, roots and the trunks of trees, providing mechanical support, and through which water containing dissolved mineral salts passes **2.** with indefinite article, i.e. a wood, or pl. woods, imprecise terms applied to a piece of ground (small in relation to that supporting a FOREST-1) covered with relatively widely-spaced trees growing naturally (as distinct from a PLANTATION-1), with or without undergrowth. COVERT, HANGER, WOODLAND. G.

woodland, woodlands land covered with

trees, sometimes defined as an open stand of widely-spaced trees without a continuous canopy of overhead foliage (sometimes specifically as a canopy coverage between 25 and 60 per cent). In UK statistics, the total woodland area is commonly divided into high forest, COPPICE and coppice-with-standards, scrub, felled or devastated and uneconomic, the term forest being reserved for the more closely wooded areas, or those with larger trees. In FAO statistics the term woodland is not used, all wooded land being classed as forested land (French terrains boisés). FOREST, FORESTRY, WOOD-2.

wood pulp the fibre of wood processed by mechanical means and chemicals to form a mixture of water and cellulose fibres, used as raw material in making PAPER or RAYON. SULPHATE WOOD PULP.

wool the fibrous growth on the skin of some animal species, especially of the SHEEP. The fibres are covered with overlapping scales which hook into each other when the fibre is spun into yarn, entrapping air and making any fabrics or articles made from it warm in wear. FELT, FLEECE, GOAT, WOOL GREASE, WOOLLEN, WORSTED.

wool grease a fatty, waxy coating on the fibres of sheep's WOOL, used in dressing leather and furs, and the source of lanoline, a waxy substance easily absorbed by the human skin, used in cosmetics and ointments.

woollen *adj.* applied to a fabric made from WOOL.

working class 1. in general and imprecisely, lower income manual workers 2. in Marxism, the proletariat, those who sell labour power to the capitalist class. SOCIAL CLASS.

workplace-residence ratio a ratio of the number of persons at place of residence to the number of persons at place of work, values greater than unity indicating a place specializing in the provision of jobs.

World Health Orgnization WHO.

world-island a term applied by H. J. Mackinder to the world's largest landmass, the combined continents of Europe, Asia and Africa. Being surrounded by water this vast landmass is, by conventional definition, an island. In the same way the two Americas are an island, as are the continents of Australia and Antarctica. HEARTLAND. G.

wormwood *Artemisia absinthium,* a PERENNIAL herb native to temperate regions of Europe and the northern part of western Asia, used medicinally and in making the liqueur, absinthe.

worsted (England, from Worsted, a village in Norfolk) 1. a woollen yarn spun from long-stapled wool, combed so that the fibres lie as nearly as possible parallel to each other. STAPLE-5 2. woollen fabric woven from such yarn.

wrack a SEAWEED.

wrench-fault, wrench-faulting a nearly vertical STRIKE-SLIP FAULT. G.

wrought iron cast iron, PIG IRON, STEEL.

X

xeno- (Greek) a stranger, a foreigner, a prefix used in that sense in many scientific terms.

xenocryst a crystal which is foreign to the IGNEOUS ROCK in which it occurs.

xenogamy in botany, cross fertilization.

xenolith a piece of pre-existing rock picked up and in some cases partly dissolved by molten MAGMA and embedded in an IGNEOUS body, commonly occurring near the edge of a BATHOLITH. A xenolith may be a piece of the intruded COUNTRY ROCK, or a piece of rock from the METAMORPHIC AUREOLE, or just a stray piece of the IGNEOUS ROCK itself which, partly formed, had failed to be absorbed in the main igneous body at the time of its solidification. G.

xenoparasite a PARASITE which lives on an organism which is not its usual host; or which can live only by invading an injured organism.

xenophobia a fear or dislike of individuals or groups thought of as strangers or foreigners.

xerarch SUBSERE.

xeric adj. 1. having a low or inadequate supply of moisture to sustain plant life 2. adjusted to arid conditions, applied particularly to a plant or animal having such a quality. HYDRIC, MESIC.

xero- (Greek) dry, a prefix used in that sense in many scientific terms.

xerophyte a plant adapted to a dry habitat (in desert conditions, or in an alkaline, acid, salt or dry soil) and able to withstand prolonged drought (XEROPHYTIC CONIFEROUS FOREST). HYDROPHYTE, HYGROPHYTE, MESOPHYTE, TROPOPHYTE. G.

xerophytic, xerophilous adj. of, pertaining to, characteristic of a XEROPHYTE. G.

xerophytic coniferous forest forest occurring at high elevations in semi-arid zones, e.g. in southwestern USA, the species including juniper. HYDROPHYTE, HYGROPHYTE, MESOPHYTE, TROPOTHYTE. G.

xerorendsina in soil science, a soil on solid parent rock (PARENT MATERIAL) with strongly varying structure and composition, occurring in mountainous places. G.

xerosere a PRISERE in which the pioneer community develops on dry materials, e.g. a LITHOSERE, which develops on bare rock, or a PSAMMOSERE, developing on dry sand. HALOSERE, HYDROSERE, MESOSERE, XERARCH.

xerothermic adj. related to both dryness and heat.

xerothermic index an index which enables the measurement of drought as it affects plant growth to be made. OMBROTHERMIC. G.

X rays extremely short wavelength (high frequency), high energy radiation in the ELECTROMAGNETIC SPECTRUM.

xylophagous adj. wood eating. ANIMAL-1, MONOPHAGOUS, POLYPHAGOUS, PHYTOPHAGOUS, ZOOPHAGOUS.

Y

ya (Burmese) a field or plot of land. G.

yak *Poephagus grunniens* or *Bos grunniens*, a large, domesticated RUMINANT, with long, shaggy hair and outward pointing horns, similar to an OX, native to the mountains of central Asia, used as a transport animal, particularly in Tibet and the Himalayan foothills.

yam 1. a cultivated climbing plant, species of the genus *Dioscorea*, most being native to tropical regions and grown for their edible, large starchy tubers, which store well and, cooked, provide a STAPLE-3 food in the wetter parts of tropical lands **2.** in general, any tropical root crop **3.** in USA, SWEET POTATO.

yard 1. a small plot of open land attached to a building, partly or completely enclosed, in some cases by the building itself **2.** an enclosed plot of open land set aside for a particular activity or business (e.g. a brickyard, used for brick making, RAILWAY YARD, SHIPYARD, VINEYARD) **3.** a clearing in a forest (especially in North America) where deer etc. gather for protection and feeding in winter months **4.** the standard unit of length in the British system, defined as the distance between the centres of two gold plugs in a bronze rod at a temperature of 17°C kept at the Standards Office of the Board of Trade, London. It is equal to 3 feet, each foot being 12 inches; 1760 yards (5280 feet) equal one mile. One yard is equal to 0.914399 metres or 91.44 centimetres (0.91440183 metres in American measures). CONVERSION TABLES.

yardang a narrow, steep-sided crest in desert, particularly in central Asia, separated from others lying parallel to it by grooves or corridors cut in the desert floor by wind carrying sand. A yardang may reach 6 m (20 ft) high and from 9 to 36 m (30 to 120 ft) in width. CORRASION.

yardland in Britain, in agricultural history, virgate, an imprecise unit of land measurement, representing a tenement varying in size measured in customary acres (ACRE), including arable with adjoining meadow and pasture. In some cases it might be divided into two ox-gangs (an ox-gang being half a virgate) or four ferlings (quarter-virgate), again of inexact area. In areas under Danish influence the equivalent to ox-gang was bovate (the HIDE being termed carucate, the HUNDRED being termed WAPENTAKE). In south-eastern England the equivalent of hide was sulung (subdivided into four yokes). G.

yarn any spun thread (of cotton, flax, silk, wool, synthetic fibre, etc.) produced for weaving, knitting, rope-making etc.

yazoo a DEFERRED JUNCTION of a tributary, the name derived from the Yazoo river, the type example of a tributary which flows for some distance parallel to the main river (the lower Mississippi, in the case of the Yazoo) before merging with it. G.

year the period of time taken by the earth to complete one orbit round the SUN, but see: ANOMALISTIC YEAR (365.25964 days), GREGORIAN

CALENDAR YEAR or CIVIL YEAR, LUNAR YEAR, SIDEREAL YEAR (365 days 6 hours 9 minutes 9.54 seconds), SOLAR YEAR (365 days 5 hours 48 minutes 45.51 seconds).

years' purchase YP, the number of years' income a buyer is prepared to pay for an investment, e.g. if the rent of a PROPERTY-2 were 50 000x per annum and the buyer were willing to pay 500 000x for the property, the years' purchase would be ten.

yeast any of the various unicellular FUNGI belonging mainly to *Ascomycetes*, family Saccharomycetaceae, which secrete ENZYMES that cause the fermentation of sugars, with the production of ALCOHOL and CARBON DIOXIDE, and multiply by budding. They are used in brewing, as a raising agent in baking, and as a source of protein and vitamins.

yellow earth LOESS of northern China. G.

yellow ground (South Africa) in South Africa, the diamond-bearing oxidized zone of soft KIMBERLITE, overlying the BLUE GROUND. G.

yeoman in Britain, historical, a man (not a member of the nobility or LANDED GENTRY) holding a small FREEHOLD estate, hence a commoner or countryman of standing, especially one who cultivates his own land. G.

yerba maté MATE.

yield 1. in agriculture, output, product, amount of produce, result, e.g. output or production expressed in relation to units of land or livestock or to units of capital or labour applied. PRODUCTIVITY 2. from investment, the rate of return from the investment of CAPITAL-4, e.g. with a 5 per cent yield the capital invested should be recouped in 20 years (5 x 20 = 100); with a 20 per cent yield in 5 years (20 x 5 = 100).

yoke historical, one quarter of a sulung. YARDLAND. G.

yoma (Burmese; older form yomah) a mountain range. G.

young *adj.* in the early stages of development, not far advanced. MATURE, OLD AGE, SENILE, YOUTH, YOUTHFUL.

Younger Drift Newer Drift, in Geological Survey of Britain, the TILL laid down after the last interglacial period (in the PLEISTOCENE) and overlying parts of the tills of the OLDER DRIFT.

young ice new LEVEL ICE, 5 to 15 cm (2 to 6 in) thick, in transition from the stage of ICE RIND or PANCAKE ICE to WINTER ICE. G.

young fold mountains FOLD MOUNTAINS of the Alpine orogeny, in contrast to Armorican, Caledonian and other earlier orogenies. OROGENESIS. G.

young mountains mountains so recently formed that their surface configuration of jagged peaks etc. has not yet been smoothed by the agents of erosion. G.

youth the first stage of development, applied specifically (as are YOUNG and YOUTHFUL) in the CYCLE OF EROSION to the first stage in the development of landforms, i.e. when the original upland surfaces are undissected, not yet attacked by CORRASION or EROSION, when RIVERS flow swiftly, slopes are steep and gradients irregular. MATURITY, OLD AGE, SENILITY. G.

youthful *adj.* of or pertaining to youth, e.g. applied to a landform which has suffered little erosion. MATURE, OLD AGE, SENILE, YOUNG, YOUTH.

yungas 1. a zone in Bolivia characterized by deep valleys and rainy and heavily forested ridges, lying on the eastern side of the Cordillera 2. in Bolivia, in mining, a region of low plains; an alluvial basin, in some cases containing valuable PLACERS. G.

ywa (Burmese) a village. G.

Z

zaaidam SAAIDAM.

zāid-rabi (Indian subcontinent: Urdu) an additional RABI-2 crop, usually melons or cucumber, sown some time in April or May and harvested in June or July. G.

zambo in South America, a person who is of mixed African and American Indian descent. MESTIZO, MULATTO. G.

zariba, zareba, zareeba (Sudan: Arabic) in northern Sudan, a stockade or enclosure for domestic animals, especially one made of thorn bushes. G.

zastrugi SASTRUGI. G.

zawn (England, Land's End, Cornwall) specifically in Cornwall, a small narrow inlet of the sea in a coast with cliffs. G.

ZEG zero economic growth.

zemindār, zamindār (Indian subcontinent: Urdu-Hindi) various other spellings, formerly a collector of revenue, now 1. a landlord, usually a large landowner, from whom peasant farmers hold their lands and who is responsible for the payment of taxes on the whole estate; hence zemindary, the system of LAND TENURE involved, in contrast to ryotwary. RYOT 2. a peasant proprietor or owner-occupier of a small piece of land. G.

zenith the point where the line joining the earth's centre to the observer cuts the CELESTIAL SPHERE, the opposite of NADIR. G.

zenith distance in astronomy, the angular distance of a heavenly body from the ZENITH.

zeolite one of a large group of hydrated alluminium SILICATES, having an open structure with channels which act as a molecular sieve, occurring in cavities in some basic volcanic rocks. BASIC ROCK.

zero annual population growth STABLE POPULATION.

Zeuge (German) in hot deserts, a tabular mass of resistant, harder stratum capping and supported by softer rock (SHALE, MUDSTONE etc.) which has been undercut by CORRASION of wind-blown sand, i.e. the result of differential erosion, varying in height from 1.5 to 45 m (5 to 150 ft). G.

Zeugenberg (German) a hill or mountain of similar form and origin to a ZEUGE but on a larger scale. BUTTE, BUTTE TEMOIN, KOP. G.

zeyat (Burmese) a shelter or rest house for travellers erected by the wayside by pious Buddhists as a work of merit. G.

zinc a hard, blue-white, corrosion-resistant metallic element, a TRACE ELEMENT, an essential MICRO-NUTRIENT, often occurring in association with lead and silver, obtained mainly from CALAMINE and ZINC BLENDE (CALCOPHILE). It is used especially in coating sheet iron to prevent rust (galvanized iron), in alloys (with copper to make brass), in electric cells; in its oxide form, zinc oxide, as a white pigment (zinc white); and as a filler in ointments. G.

zinc blende naturally occurring zinc sulphide, ZnS, SPHALERITE, a major ore of ZINC.

zinc carbonate a crystalline salt occurring naturally as smithsonsite, known formerly in Britain as CALAMINE.

zinc oxide ZINC.

zircon ZIRCONIUM silicate, a transparent mineral, found (sometimes in PLACERS) as tetragonal crystals, the colour varying from red or brown, pale yellow, and smoky blue, some colourless, some varieties being cut for use as GEMSTONES. Tetragonal crystals have three axes at right angles, the two lateral axes being equal.

zirconium a resistant mineral, occurring in IGNEOUS and SEDIMENTARY ROCKS. Zirconium compounds are used in the making of ceramics and refractory materials, the mineral itself as construction material for containers in nuclear reactors. Being so resistant it is also used to date sedimentary or igneous rocks.

zodiac an imaginary belt on the CELESTIAL SPHERE, 16° in width, the boundaries being two circles each lying 8° away from and parallel to the ECLIPTIC. Within the belt lie the apparent paths of the sun, moon and principal planets. It is divided into 12 equal parts (signs) each 30° and taking the name of the constellation which some 2000 years ago occupied that part of the sky, but which now (owing to the PRECESSION OF THE EQUINOXES) lie some 30° to the west. Starting from the First Point of Aries (RIGHT ASCENSION) the signs are Aries (the ram), Taurus (the bull), Gemini (heavenly twins), Cancer (the crab), Leo (the lion), Virgo (the virgin), Libra (the scales), Scorpio (the Scorpion), Sagittarius (the archer), Capricornus (the goat), Aquarius (the waterer), Pisces (the fish); but because today each sign lies one constellation to the west, the First Point of Aries is no longer in Aries, but in Pisces. G.

zonal *adj.* of, resembling, forming, characteristic of, relating to, a ZONE or zones.

zonal flow atmospheric circulation in which the dominant airflow follows the lines of latitude, e.g. TRADE WINDS WESTERLIES, in contrast to MERIDIONAL FLOW.

zonal inselberg an INSELBERG formed not as a result of the exposure of the BASAL SURFACE OF WEATHERING, as is a SHIELD INSELBERG, but by the PARALLEL RETREAT OF SLOPE and allied PEDIMENTATION.

zonal model CONCENTRIC ZONE GROWTH THEORY.

zonal soil a soil with a profile showing a dominant influence of climate and vegetation in its development, as contrasted with an AZONAL SOIL. INTRAZONAL SOIL, PEDALFER, PEDOCAL, SOIL, SOIL CLASSIFICATION. G.

zonda 1. in Argentina and Uruguay, a hot, humid wind bringing tropical air from the north in front of a low pressure system, and having an enervating effect on people 2. in Argentina, a strong, hot, dry and dusty FOHN-like wind, blowing from the west down the eastern slopes of the Andes, most frequently in spring, often preceding the PAMPERO. G.

zone 1. frequently applied more or less loosely to a region, belt, tract or area of the earth (i.e. of the atmosphere, lithosphere, hydrosphere, or of any place or space), with or without defined limits, with some characteristic or characteristics or activity particular to it (e.g. of climate, rocks, soil, plant and animal life, condition), indicated by a qualifying word or phrase which differentiates it from other regions, belts, tract or areas etc. 2. in classical times, one of the latitudinal climatic belts into which the earth's surface (and in ancient cosmography the CELESTIAL SPHERE) was divided. FRIGID, TEMPERATE, TORRID ZONES 3. any region within defined limits encircling the surface of the earth, of a planet, or of the sun; or a similar region in the celestial sphere 4. in geology, a group of strata of limited but variable thickness, characterized by a definite assemblage of FOSSILS which distinguishes it from all other deposits, the zone being named after one of the characteristic species. HEMERA 5. the area of contact around an igneous mass (e.g. granite) in which an alteration of COUNTRY ROCK has taken place. METAMORPHIC AUREOLE 6. a layer or part of the earth's crust (e.g. zone of weathering or, deeper in the crust, zone of fracture; and, deeper still, zone of flow) 7. in land use planning, an area designated (zoned) for a specific purpose. G.

zone of assimilation, zone of discard related concepts indicating the movement of city centre growth and decline. City centres tend to grow outwards systematically; thus the zone of new growth becomes the zone of assimilation, and the zone relatively in decline is termed the zone of discard.

zone of indifference in CENTRAL PLACE THEORY, the area between the hinterlands of competing centres within which no one centre exerts a dominant influence. THRESHOLD-4.

zoning in land use planning, the designation of specific sites for specific uses, e.g. for residential use, for industrial use etc. ZONE-7.

zoobenthos the animals that live on the floor of the ocean or lake. BENTHOS, PHYTOBENTHOS.

zoobiotic *adj.* applied to an organism living as a PARASITE on an animal.

zoocenose animal community.

zoogeographical region a region distinguished by the natural distribution of a distinctive fauna, i.e. Arctogea, comprising the Palaearctic Region (Europe, North Africa, Asia north of the Himalayas) and the Nearctic Region (Greenland, North America north of Central Mexico), the Ethiopian Region (Africa south of the Sahara) and the Oriental region (the Indian subcontinent, the Indochinese peninsula, Malaysia, the Philippines and the Indonesian islands west of WALLACE'S LINE); and Neogea, the

Neotropical Region, comprising South America, Central America, the Mexican lowlands (southern Mexico) and the West Indies; and Notogea, the Australian Region, comprising Australia, New Zealand, most of the Pacific Islands, and the Indonesian islands east of Wallace's Line.

zoogeography the scientific study of the natural distribution of animals. G.

zooid an individual, mainly independent, member of a colony of animals, e.g. a polyp in a colony of coelenterate polyps. CORAL.

zoology 1. the scientific study of the classification, structure and functions of all forms of animal life, including by convention those PROTOZOA (i.e. the sub-kingdom and phylum of unicellular or non-cellular animals) which contain chlorophyll and which, for that reason, could equally well be classified as plants.

zoonosis a disease of animals naturally transmitted between VERTEBRATE animals and human beings. G.

zoophagous *adj.* feeding on animals. PHYTOPHAGOUS.

zoophyte an animal resembling a plant in appearance or growth, e.g. CORAL, SPONGE. G.

zooplankton minute animals, many microscopic and including the larvae of molluscs and other INVETEBRATES, which float or very feebly swim in bodies of fresh or salt water. PHYTO-PLANKTON, PLANKTON. G.

z-test in statistics, an hypothesis test used either to analyse one sample of data in order to compare a population mean with a particular value, or to analyse two unrelated samples of data in order to compare two population means. The sample data must be interval (INTERVAL SCALE) and the sample sizes must be large, the assumption being that the sampling distribution of the mean is approximately normal (NORMAL DISTRIBUTION); and, for the two sample test, the data must be totally unrelated.

zucchini COURGETTE.

Zweikanter (German) a VENTIFACT with two facets, as distinct from DREIKANTER with three facets, or EINKANTER with one.

zymogenous *adj.* applied to soil organisms whose metabolic and reproductive rates increase if organic material is added to the soil. AUTOCHTHONOUS-1.

Appendix 1
Greek and Latin roots

Greek and Latin roots commonly used in construction of terms

This Appendix gives the more common prefixes, suffixes and syllables derived from classical Greek and Latin which have been used in the construction of geographical terms. It should enable a number of terms not included in the Dictionary to be interpreted and understood. Some root words are common to Greek and Latin; it is generally agreed that a mixture of Greek and Latin derivatives in a single word is undesirable though such a mixture is by no means uncommon.

Greek

a- (used before a consonant), **an-** (before a vowel), from ἀ, ἀν (a, an) without, not, -less. Equivalent to in- (Latin), un- or non-
    abyssal, aclimatic, azonal, anaerobic, axeric

aer- from ἀηρ (aer) the air (also Latin)
    aerology

agri-, agro- from ἀγρός (agros) a field; also Latin, *ager*
    agronomy

agrost- from ἀγρωοτις (agrostis) a certain wild grass
    agrostology

allo- from ἀλλος (allos) another, strange
    allogenic

ana- from ἀνά- (ana-) up, in place or time, back, again, anew
    anabranch

anemo- from ἄνεμος (anemos) wind
    anemometer

anthropo- from ἄνθρωπος (anthropos) man
    anthropogeography, anthropoid

anti-, ant-, anth- from ἀντί, ἀντ-, ἄνθ (anti-, ant-, anth-) opposite, against
    anticline

apo-, ap- from ἀπό- (apo-) off, from, away, detached
    apogee, aphelion

archaeo-, archeo- from ἀρχαῖος (archaios) ancient, primitive
    archaeology, archaean

arch- (1) from ἀρχί-, ἀρχός (archi; archos) chief
    archipelago, lit. the chief sea

-arch (2) from ἄρχω (archo) to command, to rule
    autarchy, monarch

argill- from ἄργιλλος (argillos) clay; also Latin
    argillaceous

aster, astro- from ἀστήρ (aster) a star, pertaining to stars; also Latin
    astronomy
astheno- from ἀσθένεια (astheneia), from α and σθενος (asthenos) lack of strength
    asthenosphere
aut-, auto-, auta- from αὐτός (autos) self; by oneself, independently
    autarchy, autochthon
    Note: automobile, a vehicle mobile by itself, *i.e.* not drawn by animals;
    shortened to auto and then used in such combinations as autobahn.

bar, baro- from βάρος (baros) weight
    barometer, isobar, millibar
batho-, bathy- from βάθος (bathos) depth
    batholith, bathymetric
benthos, benthic from βένθος (benthos) poetical for βάθος, the depth of the sea
    benthos
bio- from βίος (bios) life, course, way of living
    biosphere, biogeography
boreal from βορέας (boreas) the north wind, hence the north and Latin borealis,
    pertaining to the north
    boreal forests
brachy- from βραχύς (brachus) short
    brachycephalic, brakeph
brady- from βραδύς (bradus) slow
    bradyseism
bysma-, bysm, -byssal from βυσσός (bussos) or βυθός (buthos) the depth, the sea, the
    bottom
    bysmalith, abysm, hypabyssal

caino-, caeno-, ceno-, kaino- from καινός (kainos) recent
    cainozoic
cata-, kata-, cat-, cath- from κατα-, κατ-, καθ- (kata-, kat-, kath-) down, away, entirely
    catabatic (katabatic), katothermal
ceph, cephal, keph- from κεφαλή (kephale) the head
    cephalic index, brachycephalic, brakeph
chalyb- from χαλυβηίς (chalubeis), the Chalybes, an ancient nation in Asia Minor
    famed for their work in iron and steel hence chalybeate, impregnated or flavoured
    with iron, applied to mineral springs.
choro- from χώρα (chōra), a place, a district
    chorography, choropleth
chrom-, chromo- from χρῶμα (chrōma) colour
chrono- from χρόνος (chronos) time
    chronology, isochrone, isophytochrone, tautochrone
clima from κλίμα (clima) slope
    climate
-cline, clino-, clinal from root κλιν- (klin-) sloping
    anticline, syncline, isoclinal, clinometer
coeno- from κοινος (koinos) common
    coenosis, biocoenosis
-cole, -colous from κόλον (kolon) fruit, juice; τὸ κόλον, fodder; but see also Latin
    colere, to inhabit
    calcicole (growth on limestone), (inhabiting limestone)

copro- from κόπρος (kopros) dung
    coprolite
cosmo-, cosmic from κόσμος (kosmos) the world considered as an organized entity
    cosmography, cosmopolitan
crat-, crato-, -crat from κράτος (kratos) authority, rule, sovereignty, power
    craton, orocratic, autocrat
cryo-, from κρύος (kruos) frost
    cryology, etc.
crypto- from κρυπτός (kruptos) hidden, secret
    cryptocrystalline
cryst-, crystal- from κρύσταλλος (krustallos) clear ice, rock crystal
    crystalline
cyclo- from κύκλος (kuklos), circle
    cycle, cyclone

dasy- from δασύς (dasus), hairy, hence thick, dense
    dasymetric
demo- from δῆμος (dēmos) the people
    demography, demopleth
dendro- from δένδρον (dendron) a tree
    dendritic (shaped like a tree with branches)
deutero- from δεύτερος (deuteros) second, hence secondary
    deuterozoic
di- from δι- (di-) for δις (dis), twice, two (see also Latin dis-)
    dimorphous, dicotyledon, diarchy
dia- from δι- (di-) for διά- (dia-), through, during, across
    diachronism, diaclinal
dolicho- from δολιχός (dolichos) long
    dolichocephalic, dokeph
dynamo- from δύναμις (dunamis) power
    dynamo-metamorphism
dys- from δυς- (dus-) ill, bad; used as a negative with a sense of pain or hardness in fulfilment
    dysgeogenous, dysgenic, dystrophy

eco-, oeco-, ek- from οῖκος (oikos) a house; οικονομία (oikonomia), house management (also Latin).
    ecology, economic, ekistics
ecto- from ἐκτο- (ekto) outside; but exo- is more often used
    ectogenic
edaph- from ἔδαφος (edaphos) basis, floor
    edaphic
endo- from ἔνδον (endon) within
    endogenic
eo- from ἠῶς (ēos) the dawn
    eozoic, eocene
ep-, epi- from ἐπί (epi-) on, upon, over, in addition, near; to, towards
    epigenic, epicycle, epicontinental
epeiro- from ἐπειρύω (epeiruō) to pull to, to collect
    epeirogenetic
erem- from ἐρήμος (erēmos) desolate, lonely; ἐρημία (erēmia) a solitude, a desert,

hence ἐρημίτες (erēmites) belonging to the desert. Usually applied to hermits as dwellers in solitude but applied by Gaussen to deserts
ethni-, ethno- from ἔθνος (ethnos) a nation
    ethnology
eu- from εὖ (eu) well, good, easy
    eustatic, entrophy
ex- from ἐξ (ex-) out (also Latin)
    exogenous

-gam from γάμος (gamos) marriage
ge-, geo- from γῆ (ge), the earth
    geography, geology
-gen, -genic, -genous, -geny from γεννάω (gennao) to produce; γενεά (genea) race, stock, family; γένεσις (genesis) origin, source. -gen: that which produces, or is produced. Also γένος (genos) kind
    endogenous, cratogen
-glot from γλῶττα (glotta) tongue
    monoglot, polyglot
-glyph, -glyphic from ἡ γλυφή (gluphē) a carving, incision and γλύφω (gluphō) to carve, write
    petroglyph, hieroglyph
-gon from γωνία (gōnia) an angle
    polygon, agonic
-gram from γράμμα (gramma) something written; also τὸ γράφος (to graphos)
    cartogram, diagram
-graph, -graphy from ἡ γραφή (graphe) written; γράφω (graphō) to write, to draw
    geography, topography, barograph

hal-, halo- from ἅλς (hals) salt; αλο- (halo-)
    halophyte, isohaline
helio-, -hel from ἥλιος (helios) the sun
    isohel, aphelion
hemi- from ημι (hēmi) half (semi- in Latin)
    hemisphere
hetero- from ἕτερος (heteros) another, the other of two. Prefix denoting difference as opposed to resemblance
    heterogeneous
hiero- from ἵερος (hieros) sacred
    hieroglyph
hol-, holo- from ὅλος (holos) whole, entire
    holism, holokarst
homo- from ὁμός (homos) same: opposite of hetero-
    homocline, homologue
hydra-, hydro- from ὕδωρ (hudōr) water
    hydraulics, hydrology, hydrography
hyet-, -hyet from ὑετός (huetos) rain
    isohyet
hygro- from ὑγρός (hugros), wet
    hygrophyte, hygrophilous
hypa-, hypo- from ὑπό (hupo) under, within
    hypogene

hyper-, hypa- from ὑπέρ (huper) above, beyond (Latin super)
    hypabyssal (better hyperbyssal)
hypso- from ὕψι (hupsi) high
    hypsometry

-id from -ις (is), -ιδα (-ida); also -ιδης (idēs) son of; a member of a group
    altaid, altaides, caledonids, caledonides
iso-, is- from ἴσος (isos) equal
    isopleth, isohyet, etc.

kata-, kaino-, kephalo-, kosmo-, etc.; see cata-, caino-, cephalo-, cosmo-, etc.

lacco- from λάκκος (lakkos) a hole, a pit, a reservoir
    laccolith
limn- from λίμνη (limnē) a marsh
    limnology, monimolimnion
litho-, -lith, -lite from λίθος (lithos) a stone
    lithology, megalith, coprolite, laccolite
-logy, -ology from λόγος (logos) that which is spoken or said, story of, a treatise about
    geology, zoology, phytology
-lysis from λύσίς (lusis) a setting free, loosing

macro- from μακρός (makros) long, large as opposed to μικρός (micros) small
    macrogeography
mega- from μέγας (megas) great, large; the combining form is mega-, megal-, megalo-
    megalith, megalopolis
-mene from μην, μηνος a month
    isohyetomene
mero- from μέρος (meros) part, fraction
    merokarst
meso-, mes- from μέσος (mesos) middle
    mesozoic, mesoclimate
meta- from μετά- (meta-) a prefix with various meanings but especially change (of place, order, condition or nature); *cf.* Latin trans.
    metamorphism
meteor from μετέωρος (meteōros) raised above the earth, soaring in the air, and μετεωρὸν (meteōron), a meteor
    meteorology
metro-, -meter from μέτρον (metron) a yardstick, that by which something is measured
    dasymeter
metro- from μήτρος (mētros) genitive singular of μήτηρ (mēter) a mother
    metropolis
micro- from μικρός (mikros) small
    microclimate
mio- from μεῖον (meion) less
    miocene
mono-, mon- from μόνος (monos) alone, single
    monocline, monolith, monoculture
morphe-, morph-, -morph from μορφή (morphē) form, shape
    morphology, pseudomorph, geomorphology

nem- from νῆμα (nema) a thread
    nematoid
neo- from νέος (neos) new
    neogene, neomalthusianism
-nomy (1) from νόμος (nomos) established custom, usage of law or (2) νέμω(genitive) (nemō) administration
    agronomy
noso- from νόσος (nosos) disease
    nosopleth
-nym from ὄνομα (onoma) a name
    exonym, toponymy

oec-, see **eco-**
    oecology now ecology
-oid from εἶδος (eidos) form, appearance, shape. A suffix used to denote resemblance
    caledonoid
oligo- from ὀλίγος (oligos) few, little
    oligocene, oligomict, oligotrophy
-ology, see **-logy**
-ombro from ὄμβρος (ombros) rain
    ombrothermic, isothermombrose
oro-, **oreo-** from ὄρος (oros) a mountain
    orography
ortho- from ὀρθός (orthos) straight, upright, regular
    orthogneiss

pachy-, **-pach** from παχύς (pachus) thick
    isopach, isopachyte
palaeo-, **paleo-** from παλαίος (palaios) ancient
    palaeogeography, palaeomagnetism
pan- from πᾶν (pan), neuter of πᾶς (pas) all. As prefix denoting all, everything, altogether, everyway
    panplanation, panfan, pan-African
para- from παρά (para) from the side of, alongside; *cf.* parallel
    paragneiss
ped- from πίδον, the ground hence related to soil. See also Latin pes, pedis, a foot (see pod-). Not to be confused with Greek παῖς, παιδός, a boy (pais, paidos) used in pedagogue
    ped, pedology
pelag- from πέλαγος (pelagos) the sea, the ocean (also pelagus in Latin)
    pelagic
peri- from περί- (peri-) around, about, near (*cf.* Latin circum)
    periglacial
petro-, **petra-**, **petri-** from πέτρα (petra) a stone, rock (also Latin)
    petrology
phaco- from φακός (phakos) a lentil bean, anything shaped like a bean, lenticular
    phacolith
phanero- from φανερός (phaneros) visible, evident (opposite of crypto-, hidden, secret)
    phanerophyte, phanerozoic

pheno- from φαίνειν (phainein) to show
    phenocryst
-phil, philo- from φίλος (philos) loving, fond of, cultivating
    hygrophilous, anglophil
-phobe-, -phobus from φόβος (phobos) fear
    calciphobe
photo-, phot- from φῶς (phōs) light; genitive singular φωτός (phōtos)
    aphotic, photic
-phyll from φύλλον (phullon) a leaf
    schlerophyll
phylo- from φύλον (phulon) a tribe or race
    phylogenetic, phylum
physic-, physio- from φυσικός (phusikos) pertaining to nature, natural; φύσις (phusis) nature
    physiography, physical geography
phyto-, -phyte from φυτόν (phuton) that which has grown, a plant, also a creature
    halophyte, mesophyte, xerophyte, phytogeography
plagio-, plagi- from πλάγος (plagios) oblique, slanting
    plagioclase, plagiotropism
plat- from πλατύς (platus) broad, flat
    platform
-pleth from πλέθρον (plethron) a measure
    isopleth
plio- from πλειών (pheiōn) more
    pliocene
pleisto- from πλεῖστος (pleistos) most
    pleistocene
pluto- from πλούτων (Ploutōn) Plouton or Pluto, God of the Underworld
    plutonic rocks, plutonism
pneumat- from πνευματικός (pneumatikos) belonging to the air or wind or gases
    pneumatolysis
pod-, podo- from πούς, πόδος (pous, podos) a foot
-polis, -opolis, -politan from πόλις (polis) a city, city-state; πολίτης (politēs) a citizen
    metropolis, cosmopolitan
potamo- from ποταμός (potamos) a river, marsh
    potamology
poly- from πολύς (polus) many, much
    polycyclic, polygon
pro- from πρό- (pro-) before (also Latin)
    proglacial, progradation
protero- from πρότερος (proteros) before
    proterozoic
proto-, prot- from πρῶτος (prōtos) first
    prototype, proto-Thames, etc.
psamm- from ψαμμός (psammos) sand
    psammitic rocks
pseudo-, pseud- from ψευδής (pseudēs) false, having a deceptive resemblance
    pseudomorph
psycho- from ψυχή (psuchē) breath, the soul
    psychosphere

pyro-, pur- from πῦρ (pur) fire
    pyrometamorphism, pyroclastic
rheo-, -reic from ῥέω (rheo) to flow
    rheology, areic, endoreic
rhiza-, rhizo-, rhiz- from ρίζα (riza) a root
    rhizosphere

sapro- from σαπρός (sapros) putrid, rotten
    saprophyte, saprolite
seismo-, -seism from σεισμός (seismos) an earthquake
    seismology, isoseismic
spher-, sphaer-, -sphere from σφαῖρα (sphaira) a ball or sphere
    spheroidal weathering, lithosphere
stadia from στάδιον (stadion) a measure of length
    stadial moraine
stat- from στατικός (statikos) at a standstill; from στα- (sta) the root of to stand
    isostasy, isostatic
steno-, sten- from στενός (stenos) narrow, within narrow limits
    stenohaline
-strophe from στροφή (strophē) a turning; στρόφος (strophos) a twisted band
    catastrophism
syn- from σύν (sun) with, together, similarly, alike; also in form σύμ (sum)
    syncline, synecology, symbiosis

tauto- from ταῦτό (tauto) (for τὸ αὐτό- to auto) the same
    tautochrone
taxi-, -taxis from τάξις (taxis) order, arrangement, line of battle
    taxonomy
tecton- from τέκτων (tektōn) a carpenter, hence builder; and τεχτονικός (techtonikos) relating to construction
    tectonic
thalass- from θάλασσα (thalassa) the sea, ocean
    thalassography
-them from θέμα (thema) that which is laid down, a theme
    cyclothem
therm-, -therm from θέρμη (therme) heat; θερμός (thermos) hot
    isotherm
-tone from τόνος (tonos) strain
    ecotone
topo-, -tope from τόπος (topos) a place
    topography, ecotope
trach- from τραχύς (trachus) rough, hairy
    trachyte
-trope from τρέπω (trepo) to turn; τρόπέ (tropē) a change, variation, turn; see also τρεφειν, to nourish
    entropic
tropo- from τρόπος (tropos) turning
    tropophyte, troposphere
-trophe from τροφή (trophē) nourishment, rearing; τρεφειν, to nourish
    oligotrophy, eutrophy

xeno- from ξήνος (xenos) a stranger
    xenolith
xero- from ξερός (xeros) dry
    xerophyte
zoo- from ζωον (zōon) an animal, life
    zoology, azoic, eozoic

Latin

ab-: off, away, from
    ablation, abrade, abrasion
-acy, -cy from -acia, atia. Used as suffix to change an adjective of quality state, or condition into a noun
    pirate, piracy; potent, potency
ad-: (becomes ac- before c, k and qu; af- before f; ag- before g; al- before l; ar- before r; as- before s; at- before t) to, with sense of motion to, change into, addition, or intensification
    acclimatize, accumulate, advection, afforestation, agglomerate, association, attrition
aer-, air from aër, the air (also Greek)
    aeration
agri-, agro- from ager, a field, agris of or pertaining to a field (also Greek)
    agriculture
-al from -alis, of the kind of, pertaining to. Used as suffix to change a noun to an adjective
    fluvial, spherical
alti-, alto- from altus, high
    altitude
ambi- from ambo, both
-an, -ian from -anus, -ana, -anum (also -ianus) of or belonging to. Suffix to change a noun (especially a country or place) to an adjective. Occasionally -ane; also -ian
    America, American; Paris, Parisian
annum: year, hence per annum, p.a., also annual; perennial, through the years
annular from annulus or anulus, a ring
    annular drainage
ante: before (in time or place)
    antecedent
aqua: water
    aqueous, aquifer or aquafer
ara: a plough; arare to plough; arabilis, able to be ploughed
    arable
arena: sand
    arenaceous
argilla: clay (also Greek)
    argillaceous
arti-, arte- from ars, artis art, human skill as opposed to nature
    artefact
aster, astro-: a star, pertaining to stars (also Greek)
    astronomy

auri-, auro- from aurum, gold
    auriferous
auster, austral-: the south wind, hence the south; australis, pertaining to the south (wind)
    austral, australia(n)

balnea-, balneo- from balneum, a bath
    balneology
bi-: twice, doubly, having two, two- (see also di-, and Greek)
    bipolar
boreal: borealis, pertaining to the north (wind) (also Greek)
    boreal forests

calc- from calx, calcis, lime; calcarius, of or pertaining to lime
    calcareous, calcicole, calcifuge, pedocal, calcrete
capilla: a hair, capillarius, hair-like
    capillary fringe, capillarity
carbo-, carboni- from carbo, carbonis, coal
    carboniferous
carta-, carto-, chart from charta, carta, a map (also Greek)
    cartography, chart
catena: a chain, a connected series
    catena, catenary
centrum: centre; hence centri-, centro- (also Greek)
    centrosphere
centum: a hundred
    per centum, usually shortened to per
circum-: around, round about
    circumdenudatio
-cide, -cision from caedere, to cut; decidere, to cut down
    incised, incision, deciduous
cis-: on this side of; as opposed to trans- or ultra- on the other side of
    cis-Alpine
co-: col- (before 1), com- (before b, p, m, etc.), con-, cor- (before r) from Latin cum-,
    together, together with, in combination or union
    confluent, conformable, congelifraction, conglomerate
contra-: against, in opposition to, opposite
    contraposed shoreline
creta: chalk
    cretaceous
cult from cultus, worship
    cultural geography
-culture from cultura, culture
    agriculture, silviculture
cultivate from cultivare, to till, plough
    cultivated land
cumulus: a heap, a pile
    accumulation mountains

de-: down, down from; as a prefix to undo the action
    degrade; deglaciation
demi-: half from Latin dimidius, through French demi

dexter, dextra from dexter, dextr- right; dextra the right hand

di-, dis-: used with a variety of meanings but in general to express negation; away from, denoting the opposite or lack of the characteristic in question
discordant, divagation

digit- from digitus, finger
digitate

diluvi- from diluvium, a flood
diluvium, diluvial

dis-: prefix expressing the opposite of the thing or characteristic in question
disconformity

dom- from domus, a house, home
domestic

duro-, dura- from durus, dura, durum hard; duro to last
duricrust

en-, -in- from French en-, Latin in-, Greek ἐν (en), in
enclosure, inclosure

equi- from aequus, equal
equinox

erode, erosion from erodo, erodere, to gnaw away

erro, errare: to wander; erratum, wandered
erratic block

-escens, -escent: a Latin suffix which conveys the sense of 'becoming'; crescent
obsolescent = becoming obsolete; senescent = becoming senile, etc.

ex-: ex- out (also Greek)

extra-: beyond, farther than, except
extraterritorial

-etum: a suffix of Latin form used by ecologists to designate a plant association dominated by a single genus, *e.g.* calluna, callunetum; sphagnum, sphagnetum

-fact, -faction, -fication from factum a thing done (neuter singular of factus from verb facio, facere; French faire, to do). Also factio, a doing, and -ficare
Conveys the idea of something done or made
artifact, ventifact
Conveys the idea of doing or making
petrifaction, petrification

-fer, -ferous from fero, ferre, to bear; bearing, carrying, producing
aquifer, conifer, coniferous (cone-bearing)

ferrum, ferrous from ferrum, iron

fluvi-, fluo-, -fluction from fluvius, a river; fluo, fluere to flow. Compare French fleuve.
fluvial, fluviatile, interfluve, solifluction, affluent

for-, fore- from foris, outside
forest
More often the prefix fore- is from the Anglo-Saxon fore meaning before, as in fore-deep, fore-set beds

fossa from fodio, fodere, to dig; fossus, dug, hence fossa terra, land dug up and fossa, a ditch; fossilis, dug up
fossa, fossil

-fract, -fraction from frango, frangere to break, fractus broken, something broken, the act or result of breaking
congelifraction

713

-fuge from fugio, fugere, to flee; fleeing from
    calcifuge

gel- from gelo, gelare, to congeal; gelu, frost
    congeliturbation, regelation
glac- from glacies, ice
    glacial, glaciation, deglaciation
glob- from globo, globere, to make into a ball; globus, a globe
    global
glomer- from glomero, glomerare, to collect into a ball, glomus, a ball
    agglomerate, conglomerate
grad- from gradus, a step
    aggrade, degrade, grade
gran- from granum, a grain
    granite, granular
grav- from gravis, heavy
    graviplanation

haema-, haemat-, haemato-, hema-: Latin from Greek $\alpha\hat{\iota}\mu\alpha$ blood
    haematite
horti- from hortus, a garden
    horticulture
humi- from humus, the ground; humi, on the ground; also
    humidus, umidus, moist

igne-, igni- from ignis, fire
    igneous rocks
in-, il-, im-: the Latin negative in-, which becomes il- before l; im- before m, etc. With
    many words the Anglo-Saxon negative un- is commonly preferred
    immature
infra-: within, below
insula: island
    peninsula, insularity
inter-: between
    interfluve
intra-: on the inside, within
    intratelluric
inver-: inversus, inverted, from in- and verto, to turn

lac-: lacus, a lake
    lacustrine
lam(m)ina, lamella: a thin plate
    lamination
later (1) lateral from latus, a side, lateralis
    unilateral
later (2) a brick
    laterite
lav-, luv-, -luv from lavo, lavare, to wash; also luere
    eluvial, illuvial
litor-, littor- from lit(t)us, the shore; littoralis, pertaining to the shore
    littoral deposits

loc-, loco- from locus, a place
   location, localization

mal- from malus, bad, ill
man-, manu- from manus, the hand
   manufacture
mar-, marine from mare, the sea; marinus of the sea
   marine, maritime
medi- from medius, middle
   Mediterranean
mil-, mill-, milli-, mille- from mille, a thousand
   mille map, mile
minut- from minutus, small
   minute
mort- from mors, mortis, death
   mortlake
mult-, multi- from multus, many
   multi-cycle landscape

navi- from navis, ship; navigo, navigare, to navigate
   navigation
niv-, nif from nix, nivis, snow
   nivation, isonif, niveo-eolian
non-: negative prefix
   non-ferrous
nud-, nudo- from nudus, nude, naked
   denudation

ob-: a common prefix with many meanings, sometimes intensive, sometimes to denote
   inversion or on the back of. Becomes oc-, of-, op-, etc., before c, f, p
oper- from opus, operis (*pl.* opera), work
optimum: optimus, best
   optimum population
ordin- from ordo, ordinis, order arrangement
-origine from origo, originis, origin, a beginning
   aborigine
oro-: usually from the Greek oros, a mountain, but also Latin os, oris
   the mouth

ped- from pes, pedis, a foot
   pedology, pediment
pelag- from pelagus, the sea, ocean (also Greek)
   pelagic
pen-, pene- from pene, almost
   peninsula, peneplane
per: by, through
   per annum, per cent(um), perennial
petra: a rock, stone (also Greek)
   petrifaction
pinna: a feather, a fin
   pinnate drainage

plan-, plano-, planus: level, flat
    plan, planation, peneplane
plen: plenus, full; plenarius, entire
    plenary
pluvi: pluvia, rain
    pluvial period
post-: after, behind, since
    post-glacial
pre-, prae: before (in time, place, etc.)
    pre-glacial
prima-, primo-: primus, first
    primate city
pro-: before (also Greek)
    proglacial

re- (i) ablative of Latin res, thing. Referring to, in the matter of (also 'in re')
    (ii) prefix denoting repetition of an action
    resequent, rejuvenation
retro-: back, backward
    retrogradation
ripa: a bank of a river
    riparian
rur- from rus, ruris, country (as opposed to urbs, urbis, town)
    rural

sal: salt
    saline
salto, saltare, saltatio: to leap, leaping, jumping
    saltation
sect-, secto-: seco, secere to cut; sectus, cut
    transect diagram
semi-, sem-: half
senile: senex, senilis, old; senescens, becoming old
    senile topography
sequent, sequence from sequens, following; sequentia, a following
    resequent, consequent, obsequent
silva-, silvi- from silva, a wood
    silviculture
socio-, social from socius, a companion; socio, to accompany
    sociology
sol: the sun
    insolation
sol from solea, the sole of the foot: hence the ground, or soil it touches
    : solifluxion, latosol
spelaeum: a cave (also Greek σπήλαιον, spēlaion)
    spelaeology, usually speleology
stratum, strata (*pl.*): layer, layers (*lit.* that which is laid flat)
    stratigraphy, stratosphere
sub-: under; also, to some degree
    submarine, subsoil, subnival, suburb

super-: above
  superposition
syn-: a Latinized form of Greek σύν (sun) with, together
  synecology

tellus: the earth
  telluric
tempor-: tempus, temporis, time
terra: land, the earth
  Mediterranean, terrigenous
trans-: across, beyond, through, with idea of change
  transgression, trans-Alpine, transhumance

ultra-: beyond
  ultrabasic rocks
un-: an English prefix denoting negation frequently used with words of Latin origin,
  though the Latin is in-
  unconformity
unda: a wave; undula, a small wave
  undation theory
uni-: unus, una, unum, one
  unilateral shifting
urb-: urbs, urbis, a town
  urbanism

vado-: vado, vadare, to wander
  vadose water
vect- from veho, vehere, to carry; vectus, carried; hence advect-, convect-
  advection, convection
vent- (1) from venio, venire, to come; ventus, come; ad-venio, to come to
  adventitious
vent-, **venti-** (2) ventus, wind
  ventifact
vitri, from vitrum, glass
  vitrifaction
vulcan: Vulcanus, the god who presided over the smelting of metals
  vulcanism

Appendix 2
Conversion Tables

Temperature

°F	°C	°C	°F
0	− 17.8	0	32.0
32	0.0	5	41.0
40	4.4	10	50.0
50	10.0	15	59.0
60	15.6	20	68.0
70	21.1	25	77.0
75	23.9	30	86.0
80	26.7	35	95.0
85	29.4	40	104.0

To convert degrees Fahrenheit to degrees Centigrade (Celsius):

$$(°F - 32°) \times \tfrac{5}{9} = °C$$

To convert degrees Centigrade (Celsius) to degrees Fahrenheit:

$$(°C \times \tfrac{9}{5}) + 32° = °F$$

To convert degrees Centigrade (Celsius) to Kelvin (K): add 273°C to the Centigrade value.

Length

$$1 \text{ inch} = \begin{cases} 2.54 \text{ centimetres} \\ 25.4 \text{ millimetres} \end{cases}$$

1 foot = 0.3048 metre
1 yard = 0.914 metre
1 mile = 1.609 kilometres

1 millimetre = 0.0394 inches
1 centimetre = 0.394 inches

$$1 \text{ metre} = \begin{cases} 39.37 \text{ inches} \\ 3.28 \text{ feet} \\ 1.09 \text{ yards} \end{cases}$$

1 kilometre = 0.62 mile

Approximations

10 inches	=	250 millimetres
40 inches	=	1 000 millimetres
10 feet	=	3 metres
500 feet	=	150 metres
100 yards	=	90 metres
5 miles	=	8 kilometres
100 miles	=	160 kilometres

25 millimetres	=	1 inch
1 500 millimetres	=	60 inches
10 centimetres	=	4 inches
100 metres	=	110 yards
8 kilometres	=	5 miles
250 kilometres	=	155 miles

10 fathoms = 18.3 metres
10 nautical miles (international)
    = 11.51 statute miles
10 nautical miles (international)
    = 18.52 kilometres

10 metres   = 5.5 fathoms
10 statute miles = 8.69 nautical
    miles (international)
10 kilometres = 5.40 nautical
    miles (international)

Area

1 sq inch = 6.45 sq centimetres
1 sq foot = 0.093 sq metres
1 sq yard = 0.84 sq metres

1 sq centimetre = 0.155 sq inches
1 sq metre $\begin{cases} = 10.8 \text{ sq feet} \\ = 1.20 \text{ sq yards} \end{cases}$

1 acre   =   0.4 hectare
100 acres   = 40.47 hectares
1 sq mile $\begin{cases} = & 2.59 \text{ sq kilometres} \\ = 258.99 \text{ hectares} \end{cases}$
 = 640 acres
150 sq miles = 388.5 sq kilometres

1 hectare   =   2.471 acres
100 hectares = 247.11 acres
1 sq kilometre $\left.\begin{array}{l} = & 0.386 \text{ sq miles} \\ = 247.1 \text{ acres} \end{array}\right.$
 = 100 hectares
150 sq kilometres = 57.91 sq miles

Capacity and volume

British liquid and dry measure

1 pint $\left.\begin{array}{l} = & 0.568 \text{ litre} \\ = 34.68 \text{ cubic} & = 568.26 \text{ cubic} \\ \text{inches} & \text{centimetres} \end{array}\right\}$

1 litre $\left.\begin{array}{l} = & 1.76 \text{ pints} \\ = 1000 \text{ cubic} & = 61.025 \text{ cubic} \\ \text{centimetres} & \text{inches} \\ & = 0.22 \text{ gallon} \end{array}\right\}$

1 gallon
 = 4 quarts
 = 277.42 cubic $\left.\begin{array}{l} \\ \\ \end{array}\right\} = \ 4.545 \text{ litre}$
  inches

1 hectolitre $\begin{cases} = 22 \text{ gallons} \\ = 2.7 \text{ bushels} \end{cases}$

8 gallons
 = 1 bushel $\left.\begin{array}{l} = & 36 \text{ litres} \\ = & 0.036 \text{ cubic} \\ = 2219.3 \text{ cubic} & \text{metres} \\ \text{inches} \end{array}\right\}$

1 quintal $\left.\begin{array}{l} \\ = 100 \text{ kilograms} \end{array}\right\} = \ 3.7 \text{ bushels}$

US liquid measure

1 pint $\left.\begin{array}{l} = & 0.473 \text{ litre} \\ = 28.875 \text{ cubic} & = 473 \text{ cubic} \\ \text{inches} & \text{centimetres} \end{array}\right\}$

1 litre
 = 1000 cubic $\left.\begin{array}{l} \\ \\ \end{array}\right\} = 0.264 \text{ gallons}$
  centimetres

1 gallon
 = 4 quarts
 = 231 cubic $\left.\begin{array}{l} \\ \\ \end{array}\right\} = \ 3.785 \text{ litres}$
  inches

1 hectolitre   = 26.4 gallons

(1 British gallon = 1.20094 US gallon)
(1 US gallon   = 0.83268 British gallon)

US dry measure

1 pint	$\Big\}$	= 0.550 litre	1 litre	$\Big\}$	= 1.8 pints
= 33.6 cubic inches		= 550 cubic centimetres	= 1000 cubic centimetres		= 60.48 cubic inches

1 bushel
= 2150.42 cubic inches $\left.\begin{array}{l}\\\\\\\end{array}\right\}$
= 0.35 hectolitre
= 35.328 litres
= 35328 cubic centimetres

1 hectolitre = 2.83 bushels

(1 cubic foot = 0.028 cubic metres)
(1 cubic metre = 35.315 cubic feet)

Petroleum

1 barrel $\left\{\begin{array}{l}\\\\\\\end{array}\right.$
= 42 gallons (US)
= 35 gallons (UK)
= 350 lb
= 1.59 hectolitres

1 tonne $\left\{\begin{array}{l}\\\\\\\end{array}\right.$
= 8–6.6 barrels crude
= 9–8.1 b motor spirit
= 9.1–8.2 b aviation spirit
= 6.9–6.5 b fuel oil

7 barrels = 1 metric ton (tonne) approx., depending on density
1 million barrels per day (bpd) = 50 million tons per year

Weight (avoirdupois)

1 lb $\left\{\begin{array}{l}\\\\\end{array}\right.$ = 453.592 grams
= 0.45 kilogram

100 lb = 45.36 kilograms
1 long ton = 2 240 lb
1 metric ton (tonne) $\Big\}$ = 2 204.6 lb
1 short ton = 2 000 lb

1 kilogram $\left\{\begin{array}{l}\\\end{array}\right.$ = 35.274 oz
= 2.20 lb

100 kilograms = 1 quintal $\Big\}$ = 220.46 lb

1 tonne = 1000 kilograms $\left.\begin{array}{l}\\\\\end{array}\right\}$ = $\left\{\begin{array}{l}\\\\\\\end{array}\right.$ 0.98 UK (long) ton
1.1 US (short) tons

Metals

1 troy oz $\left\{\begin{array}{l}\\\\\end{array}\right.$
= 480 grains
= 31.1 grams
= 0.03 kilogram

32 troy oz = 1 kilogram
35.3 avoirdupois oz = 1 kilogram

Yield

1 lb per acre   =   1.12 kilograms per hectare
100 lb per acre   = 112 kilograms *or* 1.12 quintals per hectare
1 cwt per acre   = 125.54 kilograms *or* 1.3 quintals per hectare
10 bushels (UK and USA) per acre = 9 hectolitres per hectare

1 000 kilograms per hectare = 892 lb per acre
10 quintals per hectare   =   8 cwt per acre
1 quintal per hectare   =   1.49 bushels (UK and USA) per acre
10 hectolitres per hectare   = 11.1 bushels (UK and USA) per acre

(1 quintal = 100 kilograms)

The bushel and the hectolitre are now in general use as standard units of weight, but it should be remembered that they vary for different commodities and for different countries:

Rice (@45 lb per bushel)
10 bushels per acre =   5 quintals per hectare
50 bushels per acre = 25.1 quintals = 2 500 kilograms
                    =   2.5 tonnes per hectare

Wheat (@60 lb per bushel)
10 bushels per acre =   6.7 quintals per hectare
50 bushels per acre = 33.6 quintals per hectare

Population density

10 persons per sq kilometre =   26 per sq mile
39 persons per sq kilometre = 100 per sq mile (approx.)
90 persons per sq kilometre = 250 per sq mile (approx.)
100 persons per sq kilometre = 259 per sq mile
193 persons per sq kilometre = 500 per sq mile

Density per sq kilometre $\times$ 2.59 = density per sq mile
Density per sq mile $\times$ 0.386   = density per sq kilometre

Comprehensive conversion tables will be found in *Geographical Conversion Tables*, compiled and edited by D. H. K. Amiran and A. P. Schick, The Hebrew University of Jerusalem, an International Geographical Union publication, obtainable from Kümmerly and Frey, Hallerstrasse 6–8, 3 000 Berne, Switzerland.

Appendix 3
Map of time zones

Time zones

The surface of the earth is divided into 24 time zones, each representing 15° longitude (one hour of time). The map overleaf shows the time of the initial (zero) zone, based on the central meridian of Greenwich, England (0°), its boundaries fixed at $7\frac{1}{2}$° to the west and $7\frac{1}{2}$° to the east of that central meridian. From the initial zone the other zones are identified by a number which shows the hours and/or minutes (+ or −) by which the standard time of the zone differs from that of Greenwich mean time (GMT). From the base of GMT time to the west is 'slow' (behind), time to the east is 'fast' (in advance). For example, disregarding any daylight saving measures that may be in force, the standard time of the zone in which Accra, West Africa, lies coincides with GMT. But when it is 12 noon by the standard time of the zone in which New York (N.Y. on the map) lies it is 5 p.m. (1700 hours) GMT; when it is 12 noon GMT it is 7 a.m. (0700 hours) standard time in the New York zone. When it is 12 noon by the standard time of the zone in which Hong Kong (H.K.) lies it is 4 a.m. (0400 hours) GMT; and when it is 12 noon GMT it is 8 p.m. (2000 hours) standard time in the Hong Kong zone.

Local variations in standard time arise from social, political or economic considerations. The untinted land areas on the map indicate intermediate time zones or areas lacking a standard time.

The map is based on Miller's cylindrical projection, in which the lines of latitude and longitude appear as straight lines intersecting at right angles.

See DAYLIGHT SAVING, GMT, INTERNATIONAL DATE LINE, LATITUDE, LOCAL TIME, LONGITUDE, MAP PROJECTIONS, SOLAR DAY, STANDARD TIME.

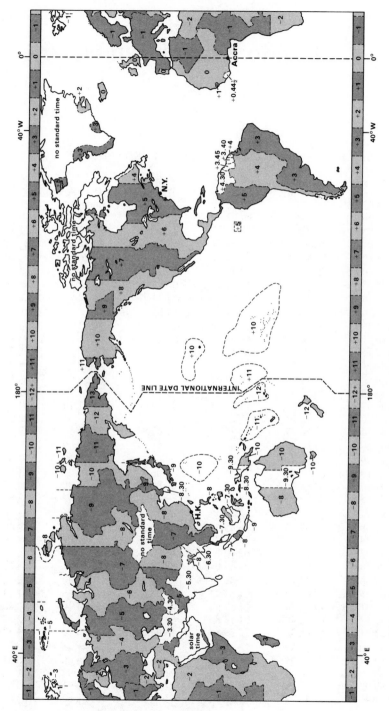

Appendix 4
Random number table

Random digits produced by throws of a ten-sided dice marked 0, 1 ... 9.

39 56 14	02 45 65	16 86 78	90 46 39	58 62 66	96 12 56
32 53 16	30 76 36	80 52 65	02 10 07	81 40 80	33 18 70
98 43 67	05 82 06	19 24 86	24 30 44	06 15 54	29 00 60
53 08 00	94 46 80	60 94 01	83 94 45	42 43 55	52 27 23
28 21 05	43 60 40	73 70 75	33 10 74	91 83 95	25 43 89
89 79 63	50 98 53	56 42 12	76 48 56	34 46 82	02 58 68
61 48 17	25 59 95	19 14 31	68 94 23	83 40 83	53 36 90
41 98 20	72 70 69	39 46 17	37 70 37	81 75 23	82 31 79
51 08 35	35 16 20	92 94 25	05 04 01	65 33 82	87 28 54
73 97 76	94 92 07	24 89 41	98 35 91	96 52 82	62 63 42
43 74 49	01 59 38	60 29 94	61 02 11	61 86 36	95 57 95
94 94 39	87 49 44	54 02 52	56 28 49	34 49 25	35 65 55
52 10 65	11 34 68	68 65 58	90 17 33	98 36 82	93 87 17
54 42 73	62 51 54	80 63 36	65 12 44	52 16 12	64 41 70
73 27 51	94 71 14	37 55 00	05 32 36	59 89 86	79 08 65
77 69 59	62 33 99	26 67 95	72 77 16	02 28 96	75 17 45
08 19 98	26 68 06	02 05 57	21 73 55	35 07 79	91 04 44
50 83 92	60 44 28	52 83 25	39 83 60	92 71 10	34 33 73
16 89 30	82 48 70	63 82 71	48 72 82	77 37 56	22 90 95
21 41 74	65 08 73	82 94 72	22 67 92	34 74 33	69 86 14
99 08 47	77 43 94	17 07 76	57 93 68	61 15 97	78 76 99
20 02 69	70 87 44	57 23 35	99 94 16	63 40 99	72 64 82
93 95 15	81 21 75	71 39 23	31 06 43	87 44 21	81 55 34
10 91 65	40 88 43	50 57 83	50 82 34	12 78 80	00 34 07
91 72 35	36 80 19	49 49 37	17 40 98	02 53 59	18 91 30
23 82 82	20 56 34	76 49 27	40 78 29	99 07 22	01 40 97
21 02 08	25 07 15	36 45 19	21 30 48	30 76 99	24 46 39
82 45 49	85 02 33	58 84 03	74 63 52	15 47 04	09 50 45
44 33 94	98 75 51	62 00 17	59 00 42	09 39 66	86 57 76
96 00 26	82 60 22	02 60 69	99 09 67	01 12 01	88 58 15
20 67 56	12 77 16	78 04 36	38 95 35	71 26 49	34 20 46
64 60 21	12 41 60	04 63 93	45 25 52	75 50 35	51 13 61
64 76 41	17 07 54	01 29 86	41 93 16	55 54 40	32 80 30
93 46 82	67 64 48	91 74 85	94 40 51	30 93 08	42 35 24
82 64 44	58 45 94	30 39 86	19 64 84	35 30 19	04 77 69
61 46 40	89 21 47	20 85 91	90 56 67	40 31 46	30 97 14
92 80 33	89 23 96	24 33 16	80 45 20	35 36 00	76 31 13
45 65 20	02 56 40	21 35 17	71 33 07	36 71 90	71 56 81
40 99 02	66 37 59	24 79 35	21 61 29	96 50 01	27 51 87
50 31 47	84 44 30	70 33 12	63 54 86	63 08 62	63 07 30